process dynamics, modeling, and control

TOPICS IN CHEMICAL ENGINEERING
A Series of Textbooks and Monographs

SERIES TITLES

Receptors: Models for Binding, Trafficking, and Signalling
  D. Lauffenburger and J. Linderman

Process Dynamics, Modeling, and Control   *B. Ogunnaike and W. H. Ray*

process dynamics, modeling, and control

BABATUNDE A. OGUNNAIKE

E. I. DuPont de Nemours,
Experimental Station,
and
Adjunct Professor,
Department of Chemical Engineering
University of Delaware

W. HARMON RAY

Department of Chemical Engineering
University of Wisconsin

New York   Oxford
OXFORD UNIVERSITY PRESS
1994

Oxford University Press

Oxford   New York
Athens   Auckland   Bangkok   Bombay
Calcutta   Cape Town   Dar es Salaam   Delhi
Florence   Hong Kong   Istanbul   Karachi
Kuala Lumpur   Madras   Madrid   Melbourne
Mexico City   Nairobi   Paris   Singapore
Taipei   Tokyo   Toronto

and associated companies in
Berlin   Ibadan

Published by Oxford University Press, Inc.,
200 Madison Avenue, New York, New York 10016

Oxford is a registered trademark of Oxford University Press

Library of Congress Cataloging-in-Publication Data
Ogunnaike, Babatunde A. (Babatunde Ayodeji)
Process dynamics, modeling, and control /
Babatunde A Ogunnaike, W. Harmon Ray.
p.   cm. —(Topics in chemical engineering)
Includes indexes.
ISBN 0-19-509119-1
1. Chemical process control.
I. Ray, W. Harmon (Willis Harmon), 1940–   .
II. Title.   III. Series: Topics in chemical engineering
(Oxford University Press)
TP155.75.036   1994   660'.2815—dc20   94-28307

1 3 5 7 9 8 6 4 2

Printed in the United States of America
on acid-free paper

To Anna and "the boys" (Damini and Deji),
... *Agbajo owo ni n'gberu d'ori;*

and

To decades of superb graduate student teachers,
... the heart of process control at Wisconsin.

CONTENTS

---

part I

INTRODUCTION

---

part II

PROCESS DYNAMICS

part III

PROCESS MODELING AND IDENTIFICATION

part IV

PROCESS CONTROL

PART IVA: SINGLE-LOOP CONTROL

PART IVB: MULTIVARIABLE PROCESS CONTROL

PART IVC: COMPUTER PROCESS CONTROL

part V

SPECIAL CONTROL TOPICS

part VI

APPENDICES

preface

Over the last two decades there has been a dramatic change in the chemical process industries. Industrial processes are now highly integrated with respect to energy and material flows, constrained ever more tightly by high quality product specifications, and subject to increasingly strict safety and environmental emission regulations. These more stringent operating conditions often place new constraints on the operating flexibility of the process. All of these factors produce large economic incentives for reliable, high performance control systems in modern industrial plants.

Fortunately, these more challenging process control problems arise just at the time when inexpensive real time digital computers are available for implementing more sophisticated control strategies. Most new plants in the chemical, petroleum, paper, steel, and related industries are designed and built with a network of mini- and microcomputers in place to carry out data acquisition and process control. These usually take the form of commercially available distributed control systems. Thus digital computer data acquisition, process monitoring, and process control are the rule in process control practice in industry today.

Because of these significant changes in the nature of process control technology, the undergraduate chemical engineer requires an up-to-date textbook which provides a modern view of process control engineering in the context of this current technology. This book is directed toward this need and is designed to be used in the first undergraduate courses in process dynamics and control. Although the most important material can be covered in one semester, the scope of material is appropriate for a two-semester course sequence as well. Most of the examples are taken from the chemical process industry; however, the text would also be suitable for such courses taught in mechanical, nuclear, industrial, and metallurgical engineering departments. Bearing in mind the limited mathematical background of many undergraduate engineers, all of the necessary mathematical tools are reviewed in the text itself. Furthermore, the material is organized so that modern concepts are presented to the student but the details of the most advanced material are left to later chapters. In this way, those preferring a lighter treatment of the subject may easily select coherent, self-consistent material, while those wishing to present a deeper, more comprehensive coverage, may go further into each topic. By providing this structure, we hope to provide a text which is easy to use by the occasional teacher of process control courses as well as a book which is considered respectable by the professor whose research specialty is process control.

The text material has been developed, refined, and classroom tested by the authors over many years at the University of Wisconsin and more recently at the University of Delaware. As part of the course at Wisconsin, a laboratory has been developed to allow the students hands-on experience with measurement instruments, real-time computers, and experimental process dynamics and control problems. The text is designed to provide the theoretical background for courses having such a laboratory. Most of the experiments in the Wisconsin laboratory appear as examples somewhere in the book. Review questions and extensive problems (drawn from many areas of application) are provided throughout the book so that students may test their comprehension of the material.

The book is organized into six parts. In Part I (Chapters 1–2), introductory material giving perspective and motivation is provided. It begins with a discussion of the importance of process control in the process industries, with simple examples to illustrate the basic concepts. The principal elements of a modern process control scheme are discussed and illustrated with practical process examples. Next, a rudimentary description of control system hardware is provided so that the reader can visualize how control schemes are implemented. This begins with a discussion of basic measurement and computer data acquisition methodology. Then the fundamentals of digital computers and interfacing technology are presented in order to introduce the basic concepts to the reader. Finally, control actuators such as pumps, valves, heaters, etc. are discussed. The purpose of the chapter is to provide some practical perspective, before beginning the more theoretical material which follows.

Part II (Chapters 3–11) analyzes and characterizes the various types of dynamic behavior expected from a process and begins by providing an introduction to the basic mathematical and analysis tools necessary for the engineering material to be studied. This is followed by a discussion of various representations and approaches in the formulation of dynamic models. The emphasis is on learning how to select the model formulation most appropriate for the problem at hand. The essential features of state-space, transform-domain, frequency-response, and impulse-response models are presented and compared. Then comes a discussion of the fundamental dynamic response of various model types. Processes with time delays, inverse response, and nonlinearities are among the classes considered in some detail. The fundamentals of process stability analysis are then introduced and applied to the models under discussion.

Methods for constructing process models and determining parameters for the model from experimental data are discussed in Part III (Chapters 12–13). Both theoretical and empirical models are discussed and contrasted. Complementing the usual material on step, pulse, or frequency response identification methods, is a treatment of parameter estimation for models represented by difference and differential equations. Sufficient examples are provided to allow the student to see how each method works in practice.

In Part IV we begin the treatment of control system design. Part IVA (Chapters 14–19) deals with single loop control systems and introduces the basic principles of controller structure (e.g. feedback, feedforward, cascade, ratio, etc.) and controller tuning methodology. The choice of controller type is discussed for processes having the various types of process dynamics described in Part II. Physical examples are used to illustrate the control system design in practical engineering terms.

Control system design for multivariable processes having interactions is introduced in Part IVB (Chapters 20–22). Methods of characterizing loop interactions, choosing loop pairing, and designing various types of multivariable controllers are presented and illustrated through physical process examples. While not bringing the reader to the frontiers of research, this section of the book acquaints one with the most important issues in multivariable control and provides approaches to control system design which will work adequately for the overwhelming majority of practical multivariable control problems encountered in practice.

Part IVC (Chapters 23–26) introduces the principles of sampled-data process control. This begins with the modeling and analysis of discrete-time systems, develops stability analysis tools, and finally provides control system design methods for these dynamic systems.

In Part V (Chapters 27–30), we provide the reader an overview of important special topics which are too advanced to be covered in great depth in this introductory book. Among the subjects included are model predictive control, statistical process control, state estimation, robust control system design, control of spatial profiles (distributed parameter systems), on-line intelligence, and computer-aided-design of control systems. The practicing control engineer will find these approaches already in place in some industrial control rooms and thus needs to be aware of the basic concepts and jargon provided here. The last chapter in Part V consists of a series of case studies where the reader is led through the steps in control system synthesis for some representative chemical processes and then shown the performance of the process after employing the controller. Through these more involved example applications, which draw upon a variety of material from earlier chapters, the reader will have a glimpse of how modern process control is carried out by the practicing engineer.

An important aspect of the book is the substantial material provided as Appendices in Part VI. Appendix A is devoted to a summary of modern instrumentation capabilities and P & I diagram notation. Appendix B provides a basic review of complex variables and solution methods for ordinary differential and difference equations. Appendix C provides a summary of important relations and transform tables for Laplace transforms and z-transforms. Matrix methods are reviewed in Appendix D. Finally in Appendix E, existing computer packages for computer-aided control system design are surveyed.

There are many who contributed their efforts to this book. Undergraduate students at Wisconsin, Delaware, Colorado, and I.I.T. Kanpur provided extensive feedback on the material and helped find errors in the manuscript. Many graduate students, (especially the teaching assistants) tested the homework problems, caught many of the manuscript errors and contributed in other ways to the project. Special thanks go to Jon Debling, Mike Kaspar, Nolan Read, and Raymond Isaac for their help. We are indebted to Derin

Adebekun, Doug Cameron, Yuris Fuentes, Mike Graham, Santosh Gupta, and Jon Olson who taught from the manuscript and provided many helpful suggestions. Also we are grateful to Dave Smith who read the manuscript and provided detailed comments, to Rafi Sela who provided help with examples, and to I-Lung Chien who contributed to Chapter 27. We are indebted to Joe Miller, Jim Trainham, and Dave Smith of Dupont Central Science and Engineering for their support of this project. Our thanks to Sally Ross and others at the U.W. Center for Mathematical Sciences for their hospitality during several years of the writing. The preparation of the camera-ready copy for such a large book required an enormous amount of work. We are grateful to Andrea Baske, Heather Flemming, Jerry Holbus, Judy Lewison, Bill Paplham, Stephanie Schneider, Jane Smith, and Aimee Vandehey for their contributions along the way. Special thanks to Hana Holbus, who gave her artistic talents to the figures and provided the diligence and skill required to put the manuscript in final form. Thanks also to David Anderson who did the copy editing. The book could not have been completed without the patience, support, and forbearance of our wives, Anna and Nell; to them we promise more time. Finally, the authors credit the atmosphere created by their colleagues at the University of Wisconsin and at the Dupont Company for making this book possible.

May 15, 1994

Babatunde A. Ogunnaike
W. Harmon Ray

process
dynamics,
modeling,
and control

part I

INTRODUCTION

*"If we could first know where we are
and whither we are tending,
we could better judge what to do
and how to do it."*

Abraham Lincoln (1809–1865)

part I

INTRODUCTION

In embarking upon a study of any subject for the first time, the newcomer is quite likely to find the unique language, idioms, and peculiar "tools of the trade" associated with the subject matter to be very much like the terrain of an unfamiliar territory. This is certainly true of *Process Dynamics, Modeling, and Control* — perhaps more so than of any other subject matter within the broader discipline of Chemical Engineering. It is therefore frequently advantageous to begin a systematic study of such a subject with a panoramic survey and a general introduction. The panoramic survey provides *perspective*, indicating broadly the scope, extent, and constituent elements of the terrain; an initial introduction to these constituent elements in turn provides *motivation* for the subsequent more detailed study. Part I, consisting of Chapters 1 and 2, provides just such an orientation tour of the *Process Dynamics, Modeling, and Control* terrain before the detailed exploration begins in the remaining parts of the book.

1

INTRODUCTORY CONCEPTS
OF PROCESS CONTROL

A formal introduction to the role of process control in the *chemical process industry* is important for providing motivation and laying the foundation for the more detailed study of *Process Dynamics, Modeling, and Control* contained in the upcoming chapters. Thus this chapter is an introductory overview of process control and how it is practiced in the chemical process industry.

1.1   THE CHEMICAL PROCESS

In the chemical process industry, the primary objective is to combine chemical processing units, such as *chemical reactors, distillation columns, extractors, evaporators, heat exchangers, etc.*, integrated in a rational fashion into a *chemical process* in order to transform raw materials and input energy into finished products. This concept is illustrated in Figure 1.1 and leads to the definition:

> *Any single processing unit, or combinations of processing units, used for the conversion of raw materials (through any combination of chemical, physical, mechanical, or thermal changes) into finished products, is a* **chemical process.**

A concrete example of these somewhat abstract ideas is the crude fractionation section of a typical oil refinery illustrated in Figure 1.2. Here, the *raw material* (in this case, crude oil) is pumped from the "tank farms," through the gas-fired preheater furnace, into the fractionator, where separation into such useful *products* as napthas, light gas oil, heavy gas oil, and high boiling residue takes place.

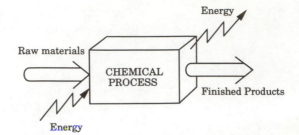

Figure 1.1.   The chemical process.

Let us now compare this actual *process* with the abstract representation in Figure 1.1:

- The *processing units* are the storage tanks, the furnace, and the fractionator, along with their respective auxiliary equipment.
- The *raw material* is basically the crude oil; the air and fuel gas fed into the furnace provide the *energy input* realized via firing in the furnace. In addition there are often other sources of energy input to the fractionator.
- The condensation of lighter material at the top of the fractionator, effected by the cooling unit, constitutes *energy output*.
- The (*finished*) *products* are the naptha and residue streams from the top and bottom, respectively, and the gas oil streams from the mid-sections.

The basic principles guiding the operation of the processing units of a chemical process are based on the following broad objectives:

1. *It is desirable to operate the processing units* **safely.**
   This means that no unit should be operated at, or near, conditions considered to be potentially dangerous either to the health of the operators *or* to the life of the equipment. The safety of the immediate, as well as the remote, environment also comes into consideration here. Process operating conditions that may lead to the violation of environmental regulations must be avoided.

2. *Specified* **production rates** *must be maintained.*
   The *amount* of product output required of a plant at any point in time is usually dictated by market requirements. Thus, production rate specifications must be met and maintained, as much as possible.

3. **Product quality** *specifications must be maintained.*
   Products not meeting the required quality specifications must either be discarded as waste, or, where possible, reprocessed at extra cost. The need for economic utilization of resources therefore provides the motivation for striving to satisfy product quality specifications.

For the process shown in Figure 1.2, some operating constraints mandated by *safety* would be that the furnace tubes should not exceed their metallurgical temperature limit and the fractionation unit should not exceed its pressure rating.

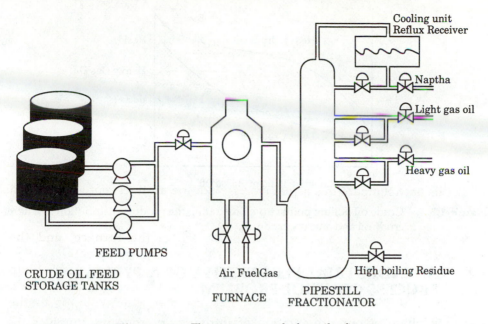

Figure 1.2. The upstream end of an oil refinery.

The issues of maintaining *production rates* and *product quality* are linked for this process. The products available from crude oil are determined by their boiling points, as shown in Figure 1.3. Thus a lighter crude oil feed could produce more naptha and light gas oil, while a heavier crude oil would produce more heavy gas oil and high boiling residue. Hence the production rate possible for each of the products depends on the particular crude oil being fractionated and the quality specifications (usually a maximum boiling point for each fraction above the bottom). Thus by shifting the maximum boiling point upwards for a product such as naptha or gas oil, one could produce more of it, but it would have a lower quality (i.e., more high-boiling materials).

Now, chemical processes are, by nature, *dynamic,* by which we mean that their variables are always changing *with time.* It is clear, therefore, that to achieve the above noted objectives, there is the need to *monitor,* and be able to induce change in, those key process variables that are related to *safety, production rates*, and *product quality.*

This dual task of:

1. Monitoring certain process condition indicator variables, and,
2. Inducing changes in the appropriate process variables in order to improve process conditions

is the job of the *control system.* To achieve good designs for these control systems one must embark on the study of a new field, defined as follows:

Process Dynamics and Control is that aspect of chemical engineering concerned with the analysis, design, and implementation of control systems that facilitate the achievement of specified objectives of process safety, production rates, and product quality.

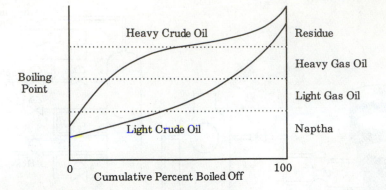

Figure 1.3.   Crude oil boiling point curve illustrating the product distribution of a light crude oil and a heavy crude oil.

1.2   AN INDUSTRIAL PERSPECTIVE OF A TYPICAL PROCESS CONTROL PROBLEM

The next phase of our presentation of introductory concepts involves the definition of certain terms that are routinely used in connection with various components of a chemical process, and an introduction to the concept of a process control system. This will be done in Sections 1.3 and 1.4. To motivate the discussion, however, let us first examine, in this section, a typical industrial process control problem, and present what may well be a typical attempt to solve such problems, by following a *simulated*, but plausible, discussion between a plant engineer and a control engineer.

As industrial systems go, this particular example is deliberately chosen to be simple, yet possessing enough important problematic features to capture the essence of control applications in the process industry. This allows us to focus on the essentials and avoid getting bogged down with complex details that may only be distracting at this point.

Close attention should be paid to the *jargon* employed in communicating ideas back and forth during the dialogue. These terms will be defined and explained in the next portion of this chapter, and their importance will become obvious in subsequent chapters.

1.2.1   The Problem

The process unit under consideration is the furnace in Figure 1.2 used to preheat the crude oil feed material to the fractionator. A more detailed schematic diagram is shown in Figure 1.4. Such units are typically found in refineries and petrochemical plants.

The crude oil flowrate F and temperature T_i at the inlet of the furnace tend to fluctuate substantially. The flowrate and temperature of the crude oil at the outlet of the furnace are, respectively, F_o and T.

It is desired to deliver the crude oil feed to the fractionator at a *constant* temperature T^*, regardless of the conditions at the furnace inlet. For plant safety reasons, and because of metallurgical limits, it is mandatory that the furnace tube temperature not exceed the value T_m.

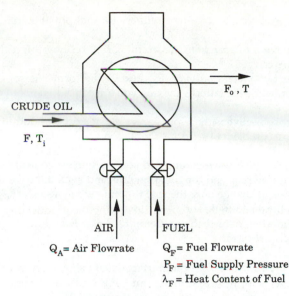

CRUDE OIL

F_o , T

F, T_i

AIR | FUEL

Q_A = Air Flowrate Q_F = Fuel Flowrate

P_F = Fuel Supply Pressure

λ_F = Heat Content of Fuel

Figure 1.4. Crude oil preheater furnace.

The heat content of the heating fuel, as well as the fuel supply pressure, are also known to vary because of disturbances in the fuel gas coming from a different processing unit in the refinery complex.

The furnace control problem may be summarized as follows:

Deliver crude oil feed to the fractionator at a **constant** *temperature T^*, and flowrate F_o, regardless of all the factors potentially capable of causing the furnace outlet temperature T to deviate from this desired value, making sure that the temperature of the tube surfaces within the furnace does not, at any time, exceed the value T_m.*

Observe the presence of the three objectives related to *safety, product quality, and production rate*, namely: furnace temperature limit T_m, the required target temperature T^*, for the furnace "product", and the crude oil throughput F_o, respectively.

1.2.2 Evolving Effective Solutions

The various phases in the evolution of an acceptable solution to typical industrial control problems are illustrated by the following dialogue between a *plant engineer* (PE), charged with the responsibility of smooth operation of the plant (in this case, the furnace), and the *control engineer* (CE), who is responsible for assisting in providing solutions to *control-related* process operation problems.

Phase 1

    CE: What are your operating objectives?

    PE: We would like to deliver the crude oil to the fractionation unit downstream at a consistent target temperature T^* . The value of this *set-point* is usually determined by the crude oil type, and desired refinery throughput; it therefore changes every 2–3 days.

Also, we have an *upper limit constraint* (T_m) on how high the furnace tube temperatures can get.

CE: So, of your two *process outputs*, F_0, and T, the former is set externally by the fractionator, while the latter is the one you are concerned about controlling?

PE: Yes.

CE: Your control objective is therefore to *regulate* the process output T as well as deal with the *servo* problem of *set-point* changes every 2–3 days?

PE: Yes.

CE: Of your *input variables* which ones do you really have control over?

PE: Only the air flowrate, and the fuel flowrate; and even then, we usually preset the air flowrate and change only the fuel flowrate when necessary. Our main *control variable* is the air-to-fuel ratio.

CE: The other input variables, the crude oil feed rate F, and inlet temperature T_i, are therefore *disturbances*?

PE: Yes.

CE: Any other process variables of importance that I should know of?

PE: Yes, the fuel supply pressure P_F, and the fuel's heat content λ_F; they vary significantly, and we don't have any control over these variations. They are also *disturbances*.

CE: What sort of *instrumentation* do you have for *data acquisition* and *control action implementation*?

PE: We have thermocouples for measuring the temperatures T and T_i; flow meters for measuring F, Q_F; and a control valve on the fuel line. We have an *optical pyrometer* installed for monitoring the furnace tube temperature. An *alarm* is tripped if the temperature gets within a few degrees of the upper limit constraint.

Phase 2

CE: Do you have a *process model* available for this furnace?

PE: No; but there's an *operator* who understands the process behavior quite well. We have tried running the process on *manual* (control) using this operator, but the results weren't acceptable. The record shown below, taken off the outlet temperature *strip-chart recorder*, is fairly representative. This is the response to a *step increase* in the inlet feedrate F. (See Figure 1.5).

CE: Do you have an idea of what might be responsible?

PE: Yes. We think it has to do with basic human limitations; his anticipation of the effect of the feed disturbance is ingenious, but imperfect, and he just couldn't react

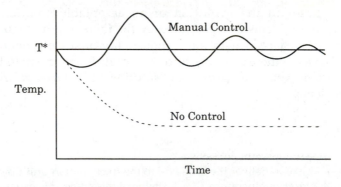

Figure 1.5. System performance under manual control.

fast enough, or accurately enough, to the influence of the additional disturbance effects of variations in fuel supply pressure and heat content.

CE: Let's start with a simple *feedback* system then. Let's install a temperature *controller* that uses measurements of the furnace outlet temperature T to adjust the fuel flowrate Q_F accordingly [Figure 1.6(a)]. We will use a *PID controller* with these *controller parameter values* to start with (*proportional band* = 70%, *reset rate* = 2 repeats/min, *derivative time* = 0). Feel free to *retune* the controller if necessary. Let's discuss the results as soon as you are ready.

Phase 3

PE: The performance of the feedback system [see Figure 1.6(b)], even though better than with manual control, is still not acceptable; too much low-temperature feed is sent to the fractionator during the first few hours following each throughput increase.

CE: (After a little thought) What is needed is a means by which we can change fuel flow the instant we detect a change in the feed flowrate. Try this *feedforward* control strategy by itself first (Figure 1.7); augment this with *feedback* only if you find it necessary (Figure 1.8(a)).

Phase 4

PE: With the *feedforward* strategy by itself, there was the definite advantage of quickly compensating for the effect of the disturbance, at least initially. The main problem was the nonavailability of the furnace outlet temperature measurement to the controller, with the result that we had *offsets*. Since we can't afford the persistent offset, we had to activate the feedback system. As expected the addition of feedback rectified this problem (Figure 1.8(b)).

PE: We have one major problem left: the furnace outlet temperature still fluctuates, sometimes rather unacceptably, whenever we observe variations in the fuel delivery pressure. In addition, we are pretty sure that the variations in the fuel's heat content contributes to these fluctuations, but we have no easy way of *quantitatively* monitoring these heat content variations. At this point, however, they don't seem to be as significant as supply pressure variations.

CE: Let's focus on the problem caused by the variations in fuel supply pressure. It is easy to see why this should be a problem. The controller can only adjust the valve on the fuel line; and even though we expect that specific valve positions should correspond to specific fuel flowrates, this will be so only if the delivery pressure is constant. Any fluctuations in delivery pressure means that the controller will not get the *fuel flowrate* it asks for.

We must install an additional *loop* to ensure that the temperature controller gets the actual flowrate change it demands; a mere change in valve position will not ensure this.

We will install a *flow controller* in between the temperature controller and the control valve on the fuel line. The task of this *inner loop controller* will be to ensure that the fuel flowrate demanded by the temperature controller is actually delivered to the furnace regardless of supply pressure variations. The addition of this *cascade control system* should work well. (See Figure 1.9 for the final control system and its performance.)

Having overheard the successful design and installation of a control system, let us now continue with our introduction to the basic concepts and terminology of process control.

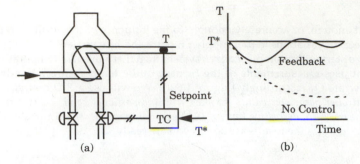

Figure 1.6. The feedback control system.

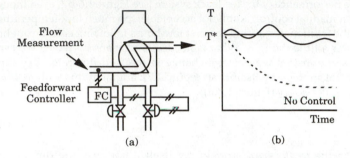

Figure 1.7. The feedforward control system.

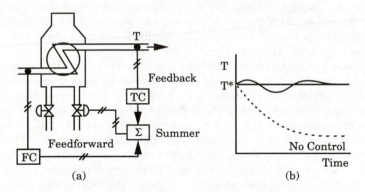

Figure 1.8. The feedforward/feedback control system.

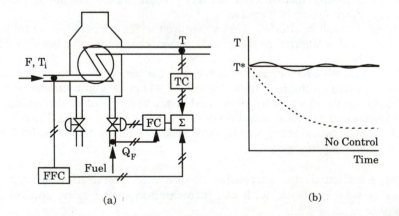

Figure 1.9. The final control system (feedforward/feedback-plus-cascade).

1.3   VARIABLES OF A PROCESS

The state of affairs within, or in the immediate environment of, a typical processing unit is usually indicated by such quantities as *temperature, flowrates in and out of containing vessels, pressure, composition, etc.* These are referred to as the *variables* of the process, or *process variables*. Recall that in our discussion of the furnace control problem we frequently referred to such variables as these.

It is customary to classify these variables according to whether they simply *provide information about process conditions,* or whether they are *capable of influencing process conditions.* On the first level, therefore, there are two categories of process variables: *input* and *output* variables.

Input variables *are those that independently stimulate the system and can thereby induce change in the internal conditions of the process.*

Output variables *are those by which one obtains information about the internal state of the process.*

It is appropriate, at this point, to introduce what is called a *state* variable and distinguish it from an *output* variable. *State variables* are generally recognized as:

That minimum set of variables essential for completely describing the internal state (or condition) of a process.

The state variables are therefore the *true* indicators of the internal state of the process system.   The actual *manifestation* of these *internal state*s by measurement is what yields an *output*. Thus the *output* variable is, in actual fact, some *measurement* either of a single state variable or a combination of state variables.

On a second level, it is possible to further classify *input* variables as follows:

1.   Those input variables that are at our disposal to manipulate freely as we choose are called *manipulated (or control) variables*.

2.   Those over which we have no control (i.e., those whose values we are in no position to decide at will) are called *disturbance variables*.

Finally, we must note that some process variables (*output* as well as *input* variables) are directly available for measurement while some are not. Those process variables whose values are made available by direct on-line measurement are classified as *measured variables;* the others are called *unmeasured variables* (see Figure 1.10.)

Although output variables are defined as measurements, it is possible that some outputs are not measured on-line (no instrument is installed on the process) but require infrequent samples to be taken to the laboratory for analysis. Thus for control system design these are usually considered unmeasured outputs in the sense that the measurements are not available frequently enough for control purposes.

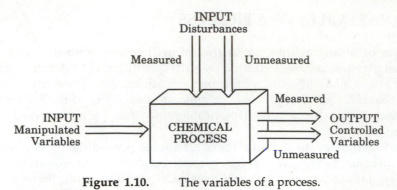

Figure 1.10.   The variables of a process.

Let us now illustrate these ideas with the following examples.

Example 1.1   THE VARIABLES OF A STIRRED HEATING TANK PROCESS.

Consider the stirred heating tank process shown in Figure 1.11 below, in which it is required to regulate the temperature of the liquid in the tank (as measured by a thermocouple) in the face of fluctuations in inlet temperature T_i. The flowrates in and out are constant and equal.

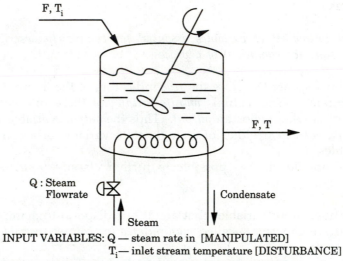

INPUT VARIABLES: Q — steam rate in [MANIPULATED]
T_i— inlet stream temperature [DISTURBANCE]
OUTPUT VARIABLE: T — tank temperature [MEASURABLE]

Figure 1.11.   The stirred heating tank process.

In this case, clearly our main concern is with the temperature of the liquid in the tank; thus T is the *output variable*. It is, in fact, a *measured* output variable, since it is measured by a thermocouple.

Observe now that the value of this variable T is affected by changes in the values of both T_i and Q. These are therefore the *input* variables. However, only Q can be manipulated at will. Thus, T_i is a *disturbance variable*, while Q is the *manipulated variable*.

Let us now formally consider the variables of the industrial furnace discussed in Section 1.2.

Example 1.2   THE VARIABLES OF AN INDUSTRIAL FURNACE

Referring back to the description of the process given earlier in Section 1.2, it is clear that T, the outlet temperature, is our *output variable*. Next, we note that the value of this variable is affected by a host of other variables that must be carefully considered in order to classify them properly.

Of all the variables that can affect the value of T, only Q_A, the air flowrate, and Q_F, the fuel flowrate, can be manipulated at will; they are therefore the *manipulated (or control) variables*.

The other variables, F (the inlet feedrate), T_i (the inlet temperature), P_F (the fuel supply pressure), and λ_F (the fuel's heat content), all vary in a manner that we cannot control; hence they are all *disturbance variables*.

This process, therefore, has *one output variable, two manipulated input (i.e., control) variables, and four disturbance variables.*

One final point of interest. As we shall see later on in Chapter 4, when we take up the issue of mathematical descriptions of process systems, it is fairly common to represent the process variables as follows:

> y — the output variable
> u — the input (control) variable
> d — the disturbance variable, and
> x — the state variable (whenever needed)

The appropriate corresponding vector quantities, **y, u, d**, and **x**, are used whenever the variables involved in each category number more than one. We shall adopt this notation in our subsequent discussion.

1.4   THE CONCEPT OF A PROCESS CONTROL SYSTEM

As earlier noted, the dynamic (*i.e.,* ever changing) nature of chemical processes makes it imperative that we have some means of effectively monitoring, and inducing change in, the process variables of interest.

In a typical chemical process (recall, for example, the furnace of Section 1.2) the *process control system* is the entity that is charged with the responsibility for *monitoring outputs, making decisions about how best to manipulate inputs so as to obtain desired output behavior, and effectively implementing such decisions on the process.*

It is therefore convenient to break down the responsibility of the control system into the following three major tasks:

1.   Monitoring process output variables by *measurement*
2.   Making rational *decisions* regarding what corrective action is needed on the basis of the information about the current and desired state of the process
3.   Effectively *implementing* these decisions on the process

When these tasks are carried out *manually* by a human operator we have a *manual control system*; on the other hand, a control system in which these tasks are carried out in an *automatic* fashion by a machine is known as an *automatic control system*; in particular, when the machine involved is a computer, we have a *computer control system*.

With the possible exception of the manual control system, all other control systems require certain hardware elements for carrying out each of the above itemized tasks. Let us now introduce these hardware elements, reserving a more detailed discussion of the principles and practice of control system implementation to Chapter 2.

1.4.1   Control System Hardware Elements

The hardware elements required for the realization of the control system's tasks of *measurement, decision making, and corrective action implementation* typically fall into the following categories: *sensors, controllers, transmitters, and final control elements*.

Sensors

The first task, that of acquiring information about the status of the process output variables, is carried out by *sensors* (also called *measuring devices* or *primary elements*). In most process control applications, the sensors are usually needed for pressure, temperature, liquid level, flow, and composition measurements. Typical examples are: *thermocouples* (for temperature measurements), *differential pressure cells* (for liquid level measurements), *gas/liquid chromatographs* (for composition measurements), *etc.*

Controllers

The decision maker, and hence the "heart" of the control system, is the *controller*; it is the hardware element with "built-in" capacity for performing the only task requiring some form of "intelligence."

The controller hardware may be *pneumatic* in nature (in which case it operates on air signals), or it may be *electronic* (in which case, it operates on electrical signals). Electronic controllers are more common in more modern industrial process control applications.

The pneumatic and electronic controllers are limited to fairly simple operations which we shall have cause to discuss more fully later. When more complex control operations are required, the *digital computer* is usually used as a controller.

Transmitters

How process information acquired by the sensor gets to the controller, and the controller decision gets back to the process, is the responsibility of devices known as *transmitters*. Measurement and control signals may be transmitted as *air pressure signals*, or as *electrical signals*. Pneumatic transmitters are required for the former, and electrical ones for the latter.

Final Control Elements

Final control elements have the task of actually implementing, on the process, the control command issued by the controller. Most final control elements are control valves (usually pneumatic, *i.e.*, they are air-driven), and they occur in various shapes, sizes, and have several modes of specific operation. Some other examples of final control elements include: *variable speed fans, pumps, and compressors; conveyors; and relay switches.*

Other Hardware Elements

In transmitting information back and forth between the process and the controller, the need to *convert* one type of signal to another type is often unavoidable. For example, it will be necessary to convert the electrical signal from an electronic controller to a pneumatic signal needed to operate a control valve. The devices used for such signal transformations are called *transducers*, and as will be further discussed in Chapter 2, various types are available for various signal transformations.

Also, for computer control applications, it is necessary to have devices known as *analog-to-digital* (A/D) and *digital-to-analog* (D/A) *converters*. This is because, as will be elaborated further in Chapter 2, while the rest of the control system operates on *analog* signals (electric voltage or pneumatic pressure), the computer operates *digitally*, giving out, and receiving, only binary numbers. A/D converters make the process information available in recognizable form to the computer, while the D/A converters make the computer commands accessible to the process.

1.4.2 Control System Configuration

Depending primarily upon the structure of the decision-making process in relation to the information-gathering and decision-implementation ends, a process control system can be configured in several different ways. Let us introduce some of the most common configurations.

Feedback Control

The control system illustrated in Figure 1.12 operates by *feeding* process output information *back* to the controller. Decisions based on such "fed back" information is then implemented on the process. This is known as a *feedback control structure*, and it is one of the simplest, and by far the most common, control structures employed in chemical process control. It was introduced for the furnace example in Figure 1.6(a).

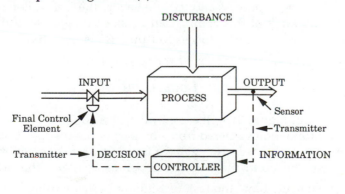

Figure 1.12. The feedback control configuration.

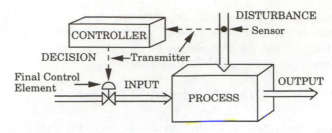

Figure 1.13.   The feedforward control configuration.

It is important to point out the intuitively appealing nature of this control structure. Observe that it makes use of *current* information about the output of the process to determine what action to take in regulating process behavior. We must note, however, that with such a structure, the effect of any disturbance entering the process must first be registered by the process as an upset in its output before corrective control action can be taken; *i.e.,* controller decisions are taken "after the fact."

Feedforward Control

In Figure 1.13 we have a situation in which it is information about an incoming disturbance that gets directly communicated to the controller instead of actual system output information. With this configuration, the controller decision is taken *before* the process is affected by the incoming disturbance. This is the *feedforward control* structure (compare with Figure 1.12) since the controller decision is based on information that is being "fed forward." As we shall see later, feedforward control has proved indispensable in dealing with certain process control problems.

The main feature of the feedforward configuration is the choice of measuring the *disturbance* variable rather than the output variable that we desire to regulate. The potential advantage of this strategy has already been noted. Further reflection on this strategy will, however, also reveal a potential drawback: the controller has *no information* about the conditions existing at the process output, the actual process variable we are concerned about regulating.

Thus the controller detects the entrance of disturbances and before the process is upset attempts to compensate for their effects somehow (typically based on an imperfect process model); however, the controller is unable to determine the accuracy of this compensation, since this strategy does not call for a measurement of the process output. This is often a significant disadvantage as was noted in Section 1.2.

Open-Loop Control

When, as shown in Figure 1.14, the controller decision is *not* based upon any measurement information gathered from any part of the process, but upon some sort of internally generated strategy, we have an *open-loop control structure.* This is because the controller makes decisions *without* the advantage of

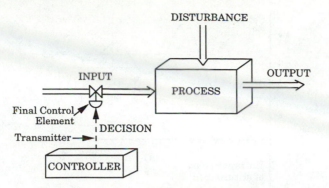

Figure 1.14. The open loop control configuration.

information that "closes the loop" between the output and input variables of the process, as is the case with the feedback control configuration (see Figure 1.12.) This otherwise vital loop is "open." However, this does not necessarily constitute a handicap.

Perhaps the most common example of an open-loop control system can be found in the simple timing device used for some traffic lights. Regardless of the volume of traffic, the timer is set such that the period of time for which the light remains green, yellow, or red is predetermined.

We shall study these and other control system structures in greater detail later.

1.4.3 Some Additional Control System Terminology

Important process variables that have been selected to receive the attention of the control system typically have target values at which they are required to be maintained. These target values are called *set-points*. Maintaining these process variables at their prescribed set-points is, of course, the main objective of the process control system, be it manual or automatic. However, output variables deviate from their set-points:

1. Either as a result of the effect of disturbances, or
2. Because the set-point itself has changed.

We have *regulatory control* when the control system's task is solely that of counteracting the effect of disturbances in order to maintain the output at its set-point (as was the case in the furnace example of Section 1.2). When the objective is to cause the output to track the changing set-point, we have *servo control* (see Figure 1.15.)

1.5 OVERVIEW OF CONTROL SYSTEM DESIGN

The design of effective control systems is the main objective of the process control engineer. The following is an overview of the steps involved in successfully carrying out the task of control system design, followed by an illustrative example.

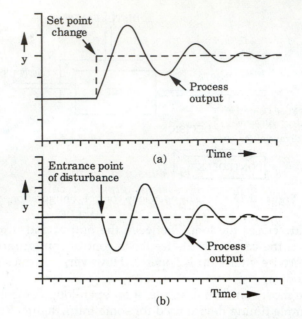

Figure 1.15.   Possible process responses under (a) servo; (b) regulatory control.

1.5.1   General Principles

***Step 1.** Assess the process and define control objectives.*

The issues to be resolved in this step include the following:

1.   Why is there a need for control?

2.   Can the problem be solved only by control, or is there another alternative (such as redesigning part of the process)?

3.   What do we expect the control system to achieve?

***Step 2.** Select the process variables to be used in achieving the   control objectives articulated in Step 1.*

Here we must answer the following questions:

1.   Which output variables are crucial and therefore must be measured in order to facilitate efficient monitoring of process conditions?

2.   Which disturbances are most serious? Which ones can be measured?

3.   Which input variables can be manipulated for effective regulation of the process?

***Step 3.** Select control structure.*

What control configuration is chosen depends on the nature of the control problem posed by the process system. The usual alternatives are: *Feedback, Feedforward, Open Loop,* and others which we shall discuss later.

Step 4. *Design controller.*

This step can be carried out to varying degrees of sophistication, but it essentially involves the following:

Obtain a **control law** *by which, given information about the process (current and past outputs, past inputs and disturbances, and sometimes even future predictions of the system output), a* **control decision** *is determined which the controller implements by adjusting the appropriate manipulated variables accordingly.*

The process control engineer requires a thorough understanding of the process itself as well as a proper understanding of the principles of *Process Dynamics and Control* in order to accomplish these steps to a successful control system design.

The task of designing effective controllers is central to the successful operation of chemical plants. This task is made easier if we understand both the *dynamic* as well as the *steady state* behavior of the process. Observe that to make rational control decisions, it helps to know how process outputs respond to process inputs. This is the key issue in *Process Dynamics*. This is also why *Process Dynamics* is the usual precursor to a study of *Process Control*.

To assist in this study of process dynamics, it is important to have some sort of *quantitative* assessment of process behavior; this is usually in the form of a *mathematical model*. The use of process models in analyzing the dynamic behavior of process systems will be discussed in Part II of this book, while the issue of *how* process models are developed will be taken up in Part III. The design of controllers of various types is the main issue taken up in Part IV.

1.5.2 An Illustrative Example

Thus far, our introduction to the fundamental principles and philosophies of process control has been somewhat *qualitative* in nature. We now illustrate the *quantitative* aspects with the following example.

The Process

The process under consideration is the simple cylindrical liquid receiver shown in Figure 1.16. Liquid flows in at the rate F_i and is permitted to flow out at a possibly different flowrate F. The tank's cross-sectional area is considered uniform, with a constant value A_c. Control valves are available at both the inlet and outlet pipes; a differential pressure level measuring device is available for providing liquid level measurements; flow measurement devices are available for monitoring both inflow and outflow.

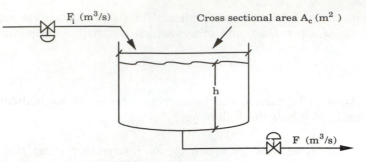

Figure 1.16.   The liquid receiver.

The Problem

It is desired to design a control system that will maintain a *constant* liquid level in the tank. The inlet and outlet flow valves are allowed to vary around their nominal values. An electronic controller is available.

Remarks

1. Note that even for such a simple process, with a very straightforward control problem, it is possible to *configure* the system in a number of different ways.

2. The different configurations will obviously give rise to different control systems whose performances we shall now proceed to demonstrate.

Configuration 1

The first configuration we will consider is as shown in Figure 1.17 where the outflow is used to regulate the liquid level, using a *feedback* strategy. The main features of this configuration are as follows:

1. Process Variables

Input variables: F_i (a *disturbance variable*)
                      F (the *control variable*)
Output variable: h, (the liquid level *measurement*)

2. Mathematical Model

It can be shown (see Part III) that an appropriate mathematical model for the process configured this way, obtained by carrying out a material balance around the system, is:

$$A_c \frac{dh}{dt} = F_i - F \tag{1.1}$$

At steady state, the time derivative vanishes, and if the inflow and outflow have respective *steady-state* values of F_{is} and F_s, then Eq. (1.1) becomes:

$$0 = F_{is} - F_s \tag{1.2}$$

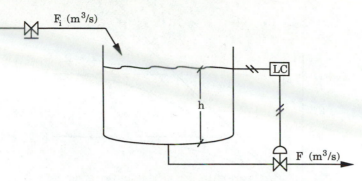

Figure 1.17. Configuration 1: feedback strategy.

Subtracting Eq. (1.2) from Eq. (1.1) now gives:

$$A_c \frac{dh}{dt} = (F_i - F_{is}) - (F - F_s) \qquad (1.3)$$

For reasons that will become clearer later, we now find it convenient to define the following *deviation variables:*

$$y = (h - h_s)$$
$$d = (F_i - F_{is})$$
$$u = (F - F_s)$$

so that Eq. (1.3) becomes, in these new variables:

$$\frac{dy}{dt} = \frac{1}{A_c} d - \frac{1}{A_c} u \qquad (1.4)$$

3. Process Behavior

Let us consider how this system will respond to the situation in which the inflowrate F_i *suddenly* increases from its steady-state value of F_{is} to $(F_{is} + \xi)$. Keep in mind that with this configuration, the variable F_i is now a disturbance.

The simplest feedback control law, and perhaps the most intuitive, is that which requires the automatic level controller to change the outflow proportionately to the observed deviation of h from h_s. Mathematically, this translates to:

$$F = F_s + K(h - h_s)$$

or, in terms of the deviation variables introduced above:

$$u = Ky \qquad (1.5)$$

This is known as *proportional* control — for obvious reasons — and the parameter K is known as the *proportional controller gain.* (Other types of control laws will be encountered later on.)

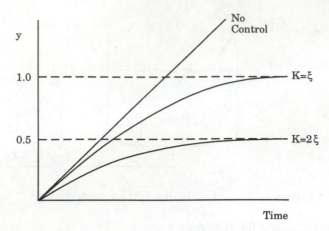

Figure 1.18.   Tank level response to inlet flowrate (proportional control).

Let us note here that with the above control law, an increase in the value of h over its steady-state value h_s provokes an increase in outflow F, an action that brings down the liquid level. Observe, therefore, that this controller appears to make sensible control decisions.

We shall now investigate the actual overall system response to the disturbance $d = \xi$, for various values of the controller parameter K, by substituting Eq. (1.5) for u in Eq. (1.4), and solving the resulting equation:

$$\frac{dy}{dt} = \frac{\xi}{A_c} - \frac{Ky}{A_c} \tag{1.6}$$

This is a linear, first-order, ordinary differential equation whose solution for constant ξ is easily written down (see Appendix B) as:

$$y(t) = \frac{\xi}{K}\left[1 - \exp\left(-\frac{K}{A_c}t\right)\right] \tag{1.7}$$

A sketch of this response for various values of the parameter K is shown in Figure 1.18.

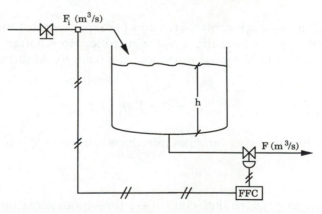

Figure 1.19.   Configuration 2: feedforward strategy.

The important points to note here are:

1. Without control action, (i.e., $u = 0$ in Eq. (1.4), or, equivalently, $K = 0$ in Eq. (1.6)), the system response is obtained by solving the equation:

$$\frac{dy}{dt} = \frac{\xi}{A_c} \tag{1.8}$$

    In other words, the system response is given by:

$$y = \frac{\xi}{A_c} t \tag{1.9}$$

    implying a perpetual increase in the liquid level. (Of course, there is a physical limit imposed by the finite capacity of the tank.) A little reflection upon the physics of this process confirms that this should indeed be the case.

2. With proportional feedback control, as indicated in Figure 1.18, there is substantial improvement in the system response; the otherwise indefinite increase in liquid level when no controller is employed is now contained, for all finite values of K. The system now settles to a new steady-state height, even though different from the desired value. Recall that the desired situation is for h to remain at h_s, or, equivalently, $y = 0$; observe that this is not achievable for any finite values of K. For $y = 0$ in Eq. (1.7), we require $K = \infty$.

3. When a process output settles down to a steady-state value different from the specified set-point, there is said to be an *offset*. Note that the offset is reduced as K increases and is actually eliminated as $K \rightarrow \infty$.

Configuration 2

Next, consider the configuration shown in Figure 1.19, where, as in Configuration 1, the outflow is still the control variable used to regulate the liquid level, but the feedforward strategy is now employed. The input, output, and disturbance variables for this configuration remain exactly as in Configuration 1; only the control strategy is different.

With this new configuration, any change observed in the inlet flowrate is communicated directly to the feedforward controller which adjusts the outflow accordingly.

How do we determine a control law for this feedforward controller? It turns out that we may easily take advantage of our knowledge of the characteristics of this simple process and employ the following purely commonsensical arguments in arriving at such a control law:

1. Because of the nature of this process, if inflow equals outflow, then whatever quantity of liquid comes in also leaves, so that whatever material was initially accumulated within the system is neither

increased nor depleted. Under these circumstances, the liquid level must remain constant.

2. The required control law is therefore the one that requires outflow F to be set equal to the inflow F_i *at all times.*

Mathematically, this translates to:

$$F = F_i \tag{1.10}$$

Substituting this into the mathematical model for the process, Eq. (1.1) gives the interesting result:

$$A_c \frac{dh}{dt} = 0 \tag{1.11}$$

which requires that:

$$h = \text{constant} \tag{1.12}$$

The implication of this result is that, so long as the inflow can be measured *with sufficient accuracy,* and the outflow can be regulated *also with sufficient accuracy,* this strategy will keep the liquid level constant at all times; i.e., we have perfect regulatory control. However, this particular scheme is incapable of changing the setpoint for liquid level (i.e., of servo control). Practical issues such as how accurately inflow can be measured and outflow regulated constitute important limitations here and would probably make this scheme unsuitable.

Configuration 3

A third possibility for configuring this system is shown in Figure 1.20, where the liquid is allowed to flow out by gravity, at a rate determined jointly by the hydrostatic pressure and the valve resistance. The outlet valve stem position is fixed and does not vary with time. The *inflow* — not the outflow this time — is used to regulate the level, employing a feedback strategy.

Observe now that where Configuration 2 is a mere variation on the theme of Configuration 1, this new configuration is totally different in philosophy. Let us confirm this by examining its main features, and comparing with those of Configuration 1.

1. Process Variables

Input variable: F_i (now the *control variable* and the only input)

Output variable: h (the liquid level measurement)

The only other variable, F, the outflow, which played a major role in Configuration 1 (and its variation, Configuration 2), is now *solely* dependent on the hydrostatic pressure and the valve resistance; it has lost its status as a manipulated variable.

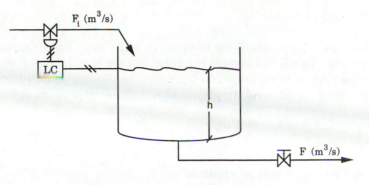

Figure 1.20. Configuration 3.

Note also that we now no longer have a *disturbance variable* to contend with. By virtue of the installation of a controller to manipulate F_i for the purpose of regulating the liquid level, this variable has given up its status as a *disturbance variable* to become a *control variable*, one whose value we can determine at will.

2. *Mathematical Model*

If we assume, for now, that the outflow from the tank is *directly* proportional to the liquid level (strictly speaking, this is not always true), and that the constant of proportionality is the valve resistance c, then one can easily show, once more, that a material balance around the system yields the following mathematical model:

$$A_c \frac{dh}{dt} = F_i - ch \tag{1.13}$$

or, in terms of the deviation variables:

$$y = h - h_s$$
$$u = F_i - F_{is}$$

Eq. (1.13) becomes

$$\frac{dy}{dt} = -\frac{c}{A_c} y + \frac{1}{A_c} u \tag{1.14}$$

A comparison of this equation with Eq. (1.4) shows that we are now dealing with an *entirely* different system configuration.

Some Concluding Remarks

In order not to encourage an unduly false, and simplistic, view of chemical process control problems on the basis of the above illustrative example, the following is just a sample of some typical complications one would normally expect to encounter in practice.

1. Nonlinearities

The process model equations we have dealt with have been linear, and thus easy to analyze. This is not always the case. A case in point is the model in Eq. (1.13) for the system in Figure 1.20. In reality, outflow is often proportional to *the square root* of liquid level, implying that a more realistic model will be:

$$A_c \frac{dh}{dt} = F_i - c \sqrt{h} \qquad\qquad (1.15)$$

which is not nearly as easy to analyze.

2. Modeling Errors

With the exception of the most trivial process, it is *impossible* for a mathematical model to represent *exactly* all aspects of process behavior. This fact notwithstanding, however, the usefulness of the mathematical model should not be underestimated; we just need to keep its limitations in proper perspective. The effectiveness of any control system designed on the basis of a process model will, of course, depend on the *integrity* of such a model in representing the process.

3. Other Implementation Problems

The illustrative example is strictly a flow process involving a reasonably incompressible fluid. This fact frees us from worrying about problems created by fluid transportation through connecting pipes.

Observe that were we dealing with a thermal system, in which liquid streams at different temperatures are moved around in the pipes, or if our system were to involve mixing streams of different liquid compositions, then the situation would be different. To effect temperature, or composition, changes by moving such liquids around, we must now consider the fact that the time it takes to flow from one point to the other within a pipe can quite often be so significant as to introduce a *delay* in the system's response to the effect of control action. As we will see later, the influence of such delays can become a most serious consideration in the design of a control system.

Even when a control system is impeccably designed, perfect implementation may be limited by, among other things, such factors as imperfect measurements, inaccurate transmission, or control valve inertia (leading to inaccurate valve actuation), factors that by and large are unavoidable in practice.

4. Complicated Process Structure

It is the rare chemical process that is as simple as the example just dealt with; the variables involved are usually more numerous, and their interrelationships more complicated. It is in fact not unusual to have to deal with an integration of several such processes. Nevertheless, the knowledge gained from investigating such simple processes can be gainfully applied to the more complicated versions, sometimes with only minimal, quite often obvious, additional considerations, and sometimes with considerable modifications that may not be immediately obvious.

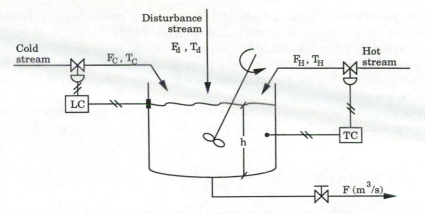

Figure 1.21. Dual liquid level and temperature control system.

To put this important point in proper perspective, consider the situation in which our previous example process, as configured in Figure 1.20, now has an additional problem: the temperature of the liquid leaving the tank is now to be regulated because of the introduction of a new *disturbance* stream bringing in liquid at an erratically varying temperature T_d, at a flowrate F_d. To facilitate this dual level and temperature control task, a hot liquid line is provided, from which liquid at temperature T_H and flowrate F_H is made available. The original inlet stream, the only one available in Configuration 3 (Figure 1.20), is now treated as the cold stream. The new situation arising from these additions is shown in Figure 1.21.

The transformation from the system of Figure 1.20 to the one in Figure 1.21, to be sure, is not trivial. We have tacitly assumed that the *cold* inlet stream will retain its assignment as the control variable for liquid level regulation, leaving the task of temperature control to the hot stream, by *default*.

The first additional consideration raised here, therefore, is this:

How are we sure that the hot stream will not be more effective as a control variable for regulating the level?

Or, put more fundamentally:

Which control variable should be used in regulating which output variable to achieve maximum effectiveness?

Even after we have resolved the nontrivial issue of which control variable to *pair* with which output variable, we are faced with a second important consideration:

Assuming the input/output pairing implied in the configuration of Figure 1.21, can the cold stream regulate the liquid level without upsetting the hot stream's task of regulating the liquid temperature?

The answer, of course, is in the negative; the level controller will always interfere with the liquid temperature, and hence with the temperature controller. The converse is also true; the temperature controller will affect the liquid level and hence interfere with the level controller. This situation of mutual interference from otherwise *independent* controllers is known as *interaction* and is a key element in the analysis of the behavior of processes having *multiple input* and *multiple output* variables.

To transfer the knowledge gained from investigating the system in Figure 1.20 to the one in Figure 1.21 requires due consideration of, at least, the two issues raised above.

This last example should motivate our study of the more advanced control strategies discussed in Part IVB and Part IVC.

1.6   SUMMARY

The objective in this introductory chapter has been to provide no more than a panoramic overview of the basic concepts and philosophies of automatic control as they apply in the chemical process industry.

We have attempted to define that usually abstract entity known as the *chemical process,* and, in identifying the main objectives of process operations, i.e., meeting *safety, production rate,* and *product quality* specifications, it has become clear that these objectives cannot be realized without the aid of a control system. This is because of the fundamentally dynamic character of these processes.

Some of the most frequently used process control concepts and terminologies were first introduced, in their *natural settings,* via a discussion arising between a plant engineer and a control engineer in their search for a solution to a typical industrial control problem involving a preheater furnace. This motivated the somewhat more formal definitions of these terms which followed.

The role of the control system, i.e., that of process *monitoring, decision making* about how best to maintain control, and *control action implementation,* have been highlighted; the hardware elements that make it possible for this role to be carried out (sensors, transmitters, controllers, final control elements) have been identified.

We have seen that a typical control system can be configured in several different ways, depending on what process system information it uses to make control decisions, and how these decisions are actually implemented. A few of the most important control system configurations, namely, the ubiquitous feedback, the industrially indispensable feedforward and cascade, and the (infrequently used) open loop-configurations, were introduced and illustrated before the central issue of what control systems design entails was reviewed.

The illustrative example introduced in the last section was for the purpose of providing a quantitative illustration of the most vital ideas which, up until then, had only been discussed at a qualitative level. The simplicity of the example was for pedagogic convenience only; to avoid the pitfall of erroneously thinking that all process control problems have the same facile bearings, the additional complications that usually accompany more realistic systems were quickly noted before rounding up the discussion.

Before proceeding to a detailed study of *Process Dynamics, Modeling,* and *Process Control,* one more task remains: that of providing a more detailed introduction to the principles and practice of data acquisition and control action implementation. The next chapter is devoted to this.

SUGGESTED FURTHER READING

A historical perspective of what the theory and practice of Process Dynamics and Control used to be in the past may be found in several *classic* texts. A few of these are:

1. Eckman, D. P., *Principles of Industrial Process Control*, J.Wiley, New York (1948)
2. Truxel, J. G., *Automatic Feedback Control System Synthesis*, McGraw-Hill, New York (1955)
3. Ceaglske, N. H., *Automatic Process Control for Chemical Engineers*, J. Wiley, New York (1956)
4. Eckman, D. P., *Automatic Process Control*, J. Wiley, New York (1958)
5. Del Toro, V. and S.R. Parker, *Principles of Control Systems Engineering*, McGraw-Hill, New York (1960)
6. Williams, T. J., *Systems Engineering for the Process Industries*, McGraw-Hill, New York (1961)

The industrial perspective of process control is presented in greater detail, complete with practical considerations based on actual experience in the following references:

7. Buckley, P. S., *Techniques of Process Control*, J. Wiley, New York (1964)
8. Shinskey, F. G., *Process Control Systems* (2nd ed.), McGraw-Hill, New York (1979)
9. Lee, W. and V. W. Weekman, Jr., "Advanced Control Practice in the Chemical Process Industry: A View from Industry," *AIChE J*, **22**, 27 (1976)

The more theoretical foundations of modern control techniques may be found in several textbooks and research monographs. Two of the more easily accessible ones are:

10. Brogan, W. L., *Modern Control Theory*, Quantum Publishers, New York (1974)
11. Ray, W. H., *Advanced Process Control*, Butterworths, Boston (1989) (paperback reprint of original 1981 book)

In each first chapter of the following books, one will find introductory examples that may be of interest in the way they are used to provide a preview of the design, analysis, and implementation of control systems. Reference 14, in particular, contains an interesting catalog of examples of control systems that occur in nature.

12. Coughanowr, D. R. and L. B. Koppel, *Process Systems Analysis and Control*, McGraw-Hill, New York (1965) (2nd ed., D. R. Coughanowr, McGraw-Hill, New York (1991))
13. Luyben, W. L., *Process Modeling, Simulation, and Control for Chemical Engineers* (2nd ed.), McGraw-Hill, New York (1991)
14. Weber, T. W., *An Introduction to Process Dynamics and Control*, Wiley-Interscience, New York (1973)

REVIEW QUESTIONS

1. What are the three broad objectives on which the basic guiding principles of process operation are based?

2. Based on the guiding principles in operating a chemical process (Section 1.1), can you guess why pneumatic controllers, actuators, and transmitters can be found in many plants?

3. What are the main concerns of *Process Dynamics and Control* as a subject matter within the chemical engineering discipline?

4. What is the difference between the input and the output variables of a chemical process?

5. What is a state variable? How are they related to output variables?

6. How can you distinguish a manipulated (control) variable from a disturbance variable?

7. What are the three main responsibilities of the control system? Assign to each of these responsibilities the hardware elements required for carrying out the indicated tasks.

8. Differentiate between a manual and an automatic control system.

9. What makes an automatic control system a computer control system?

10. What are transducers used for?

11. What differentiates a feedback control system configuration from the feedforward configuration?

12. What is unique about the open-loop control system configuration?

13. Differentiate between a servo control problem and a regulatory control problem. Can you guess which will be more common in a plant in which the processes operate predominantly in the neighborhood of steady-state conditions for long periods of time?

PROBLEMS

1.1   The process shown in Figure P1.1 is a distillation column used to separate a binary mixture of methanol and water. It is desired to regulate the composition of methanol in both overhead and bottoms product streams: x_D and x_B respectively.
Let us assume that the reflux flowrate L can be varied, but both overhead and bottoms product flowrates, respectively, D and B, are determined by the steady-state

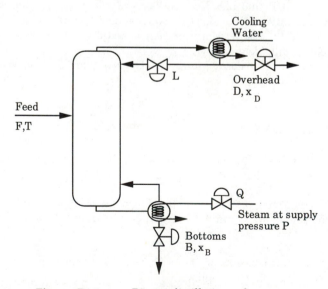

Figure P1.1.   Binary distillation column.

material balance $F = D + B$. The feedrate to the column F, as well as the feed temperature T, are known to vary. The steam supply pressure to the reboiler P is also known to vary; Q is the steam flowrate through the thermal siphon reboiler.
(a) Identify the output, manipulated (i.e., control), and disturbance variables of this process.
(b) Assume that feed temperature measurements are available, and that this information is used to adjust the steam flow into the reboiler in order to control the bottoms product composition. What type of control system configuration is this? Will this be an effective way of regulating the value of x_B? Explain what control problems one might expect this control system configuration to suffer from.
(c) If the steam supply pressure variations become too frequent, what control system configuration would you suggest to minimize the effect which such variations will have on the regulation of x_B? Sketch this configuration.

1.2 Consider the crude oil preheater furnace introduced in Section 1.2. The following simplified mathematical model for the process may be obtained by applying a *steady-state* energy balance which requires that:
Heat input through fuel = Heat required to heat feed from T_i to T, i.e.:

$$\lambda_F\, Q_F \;=\; F\, C_p\!\left(T - T_i \right) \qquad\qquad \text{(P1.1)}$$

Here, C_p is the specific heat capacity of the crude oil feed.
(a) Use this information to derive a control law that, given measurements of the disturbance variable T_i , and assuming that the physical properties of the fuel and the crude oil are available, prescribes how to adjust Q_F so that T attains the desired value T^*.
(b) Identify the potential problems a controller based on such a control law is likely to suffer from.

1.3 A liquid receiver of the type used in the example of Section 1.5 has a cross-sectional area of 1.5 m². Under steady state conditions, the inlet and outlet flows are steady at 0.05 m³/s with accumulated liquid at a level of 3 m.
Under proportional-only feedback control with $K = 0.05$, derive the system response to a sudden increase in inlet flowrate from 0.05 to 0.075 m³/s and *draw* this response as a function of time.
To what new level does the liquid in the tank settle down after steady state is achieved? What is the *offset?*

1.4 For the Problem 1.3 situation, consider now that a new feedback control law:

$$u \;=\; K\!\left[y \;+\; \int_0^t y(\tau)\, d\tau \right] \qquad\qquad \text{(P1.2)}$$

is to be applied. Rederive the new system response and establish that, in this case, there will be *no* offset.

1.5 To illustrate the effect of inaccurate measurements on feedforward control systems, consider the situation where, for the system of Configuration 2 in Section 1.5, there is a consistent error δ in the measurement of the inflow F_i, so that the feedforward control law employed is no longer as in Eq. (1.10) but as shown below:

$$F \;=\; F_i + \delta \qquad\qquad \text{(P1.3)}$$

Show that in this case the liquid level in the tank does not remain constant but in fact decreases linearly with time for $\delta > 0$.

2

INTRODUCTION TO CONTROL SYSTEM IMPLEMENTATION

In the rest of this book, we will be occupied primarily with just *two* issues: *how to characterize the dynamic behavior of chemical processes quantitatively* (in Parts II and III), and *how to use such knowledge to design effective control systems* (in Parts IV and V). But before beginning this detailed study of chemical process characterization and controller design, it will be advantageous to become acquainted with the essential aspects of control system implementation: i.e., *how* process information actually flows from the *process* to the controller and back, and *how* the computations required of the controller are actually carried out. Since entire books have been written on these control system implementation issues, the intention is to provide only a brief introductory overview in this chapter.[†] While the material covered here is important in providing a well-rounded perspective for the reader, certain aspects are not crucial to the discussion that follows in the rest of the book; thus it is possible, if desired, to bypass such material for the moment and return to it later.

2.1   INTRODUCTION

In Chapter 1 we identified the process control system as that part of the overall chemical process that is charged with the responsibility of monitoring the outputs, making decisions about how best to manipulate the inputs, and implementing these decisions on the process. Figure 2.1 shows a simplified structure of a process incorporating such a control system. The process variables are measured and "conditioned," and the data transmitted to the controller; this measurement information is then incorporated into the controller's algorithm and desired control actions are computed; the computed control actions are then transmitted to the process via local actuators (such as valves and heaters) which cause changes in the manipulated variables of the process.

[†] Part of the discussion in this chapter has been adapted and updated from Ray [4].

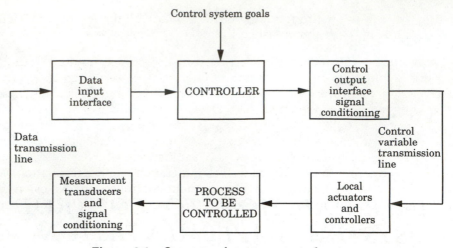

Figure 2.1.   Structure of a process control system.

In interfacing the controller with the process, control system implementation is concerned broadly with:

- Information acquisition, conditioning, and transmission
- Calculation of control actions to be taken

The technology by which these tasks are accomplished must be generally understood by the control engineer so that the designs are realistic and effective.   Thus this chapter will be devoted to providing the reader an overview of the considerations necessary in actually implementing a process control scheme.

We shall begin by giving a brief historical overview of the evolution of control system implementation so as to put in perspective the broad appeal of digital implementation, and the role of the digital computer. This is followed by a description of basic digital computer architecture, capabilities, and peripherals. This should give one an idea of the computer resources that might be typically present in an industrial computer control installation. We shall also discuss the types of interfaces normally used for data acquisition and control signal generation. Then we consider commonly used sensing devices as well as techniques for transducing, multiplexing, and conditioning the data and control signals. This last material should provide the control system designer with a grasp of the quality of the signals that must be dealt with in design, and can guide the choice of signal mode (e.g., analog voltage, analog current, or digital signal). Finally we present some typical process control computer configurations in industrial use today to indicate how all of these component parts can be synthesized into effective process computer control systems.

2.2   HISTORICAL OVERVIEW

A historical perspective on the evolution of control system implementation philosophy and hardware elements is essential for a proper appreciation of the current state of affairs in industrial practice. At the heart of this evolution is how transmission (information flow) and calculation of control actions (decision making) are accomplished.

1. **Pneumatic Implementation:** In most of the earliest implementations of automatic control systems, information flow was accomplished by *pneumatic* transmission, and computation was done by mechanical devices using bellows, springs, etc. Since the medium of transmission was air, there was a fundamental limitation to the speed of information flow. Also, the mechanical nature of the computational devices meant that the actual calculations that could be carried out could not be too sophisticated. Nevertheless, basic summation and multiplication, as well as integration and differentiation, could be carried out by these devices. Also, pneumatic controllers are explosion proof, so that safety is not an issue. For this reason, some parts of the local control system are still pneumatic even in modern installations.

Thus the primary problems associated with pneumatic implementation are:

- *Transmission:* pneumatically transmitted signals are slow responding and susceptible to interference,
- *Calculation:* mechanical computation devices must be relatively simple and tend to wear out quickly.

2. **Electronic Analog Implementation:** The medium of transmission with this implementation mode is electrons, and therefore signals could potentially travel at a rate close to the speed of light. Also, electronic analog computational devices are not subject to the same degree of wear as their mechanical counterparts.

Despite the advantages of electronic analog implementation, many final control elements are still pneumatic for the reasons noted above. This co-existence of pneumatic and analog elements is made possible by devices that can convert electrical signals to pressure signals (E/P transducers) and vice versa (P/E transducers).

The primary problems associated with electronic analog implementation are:

- *Transmission:* analog signals are susceptible to contamination from stray fields, and signal quality tends to degrade over long transmission lines.
- *Calculation:* the computations possible with electronic analog devices, while less restrictive than for pneumatic devices, are still limited.

3. **Digital Implementation:** The need to preserve signal integrity in the face of interference, in conjunction with the need for a more versatile computational device, led to the development of modern digital implementation strategies. The transmission medium is still electrical, but the signals are transmitted as binary numbers. Such digital signals are far less sensitive to noise. Also, the computation device is the digital computer, more flexible because it is programmable (it carries out computations on the basis of a sequence of instructions that can be easily altered as needed), and more versatile because there is virtually no limitation to the complexity of the computations it can carry out. In addition, with digital implementation, it is possible to carry out computation with a single computing device, or with a network of such devices.

Because many sensors naturally produce voltage or current signals, the digital revolution in control system hardware has not eliminated all electronic analog elements. Transducers that convert analog signals to digital signals (A/D) and vice versa (D/A) provide the interface between analog and digital elements of the control system.

In conclusion let us note that this evolution of strategies for control system implementation — from pneumatic through electronic analog and finally to digital — has not meant a total supplanting of one mode by the other; rather it has resulted in the "optimal" utilization of each of the modes where they are the most logical. This is why, for example, control valves are still pneumatic, a good number of sensors are still electronic analog, and most controllers are now digital.

It is also interesting to note that at one time the bandwidth of process operation (i.e., how fast the control system could perform its tasks) was primarily limited by the transmission medium, and the translation of control theory to practice was primarily limited by the computational device. However, today with digital transmission and the digital computer, it is possible to implement many sophisticated control strategies on a very fast timescale.

2.3   BASIC DIGITAL COMPUTER ARCHITECTURE

There are many different manufacturers of computers designed for process control applications and many different types of architecture. However, the functions of the computer itself are much the same, and peripherals have been standardized and made modular in order to be compatible with a wide variety of computer types. Thus it is possible to describe generally a basic digital computer architecture. The elements are shown in schematic in Figure 2.2.

The basic components of a computer are the same for the lowliest personal computer (PC) and for the largest *distributed control system* (DCS). The differences are in the design and speed of each component and in the fine details of the architecture. As illustrated in Figure 2.2, the computer is composed of a *power supply* which converts the available 110 V AC (or other power) to the voltage and currents used by the computer circuitry (e.g., ± 5 V DC). The power supply provides power to the *backplane* which is simply a box that has slots for *integrated circuit boards*. For a given backplane, there are certain boards that are essential (e.g., CPU board with minimal memory) and others that are optional (e.g., floating-point coprocessor or expansion memory boards or controller boards for peripherals). One can see this modular design in essentially every modern computer.

Let us now discuss the function of each of the elements of the computer:

1. *Central processing unit (CPU)* — This is the heart of the computer and has overall control of the system. The computer uses the *binary digits* (0 or 1) as the basic computational and communication code. This is accomplished electronically by manipulating and transmitting square pulses that need be interpreted only as a 0 or a 1. The resolution of a machine is dependent on the number of binary digits (abbreviated *bits*)

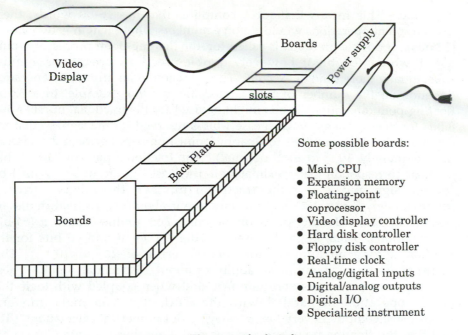

Some possible boards:

- Main CPU
- Expansion memory
- Floating-point
  coprocessor
- Video display controller
- Hard disk controller
- Floppy disk controller
- Real-time clock
- Analog/digital inputs
- Digital/analog outputs
- Digital I/O
- Specialized instrument

Figure 2.2.   Elements of a digital computer.

in the basic unit of information, called the *word*. Computers today typically have 8-, 16-,32-, or 64-bit words. Usually a smaller unit, a *byte* consisting of 8 bits, is used as well, because a byte suffices to specify characters (alphabet letters, numbers, control characters, etc.) according to industry standards (e.g., ASCII). Frequently, 7 bits of the byte are used to specify the character, and one is used for error checking (parity). Thus $2^7 = 128$ different characters may be defined by a byte. Often the bits in a word will be arranged in *octal* form by grouping them in 3s, for example, a 16-bit (2 byte) word might have the binary code

$$110 \qquad 010 \qquad 100 \qquad 001 \qquad 101$$

Since 3 bits allow one to obtain the numbers 0 to 7, as shown in Table 2.1, octal integer notation is a convenient shorthand for addresses, machine instructions, etc. Thus the 15 bits shown above represent the octal number 62415, which is a much more efficient notation than the original binary form. Alternatively, one may find some machines use *hexadecimal* notation (base 16) in machine language. The conversions between binary, octal, hexadecimal, and decimal numbers are shown in Table 2.1.
A 16-bit machine, therefore, can use 1 bit for sign (+ or −) and 15 bits to represent integer numbers. Since $2^{15} = 32,768$, the maximum integer range for a computer with a 16-bit word is from − 77777 to + 77777 octal, which corresponds to − 32,768 to + 32,767 decimal. By contrast, a computer using a 32-bit word can handle integers from − 2,147,483,648 to + 2,147,483,647. Thus the larger the word used by the computer, the wider range of integers that can be handled. Associated with the CPU, and working in parallel, many computers have a *hardware floating-point processor* which performs all floating-point arithmetic operations at high speed and with multiword

precision. This means that for a computer that has 16-bit words, these floating-point operations would involve numbers whose length is two words (32 bits). If double precision is called for, the floating-point processor would use 4 words (64 bits); however, double-precision operations require substantially more time to execute. The precision of an arithmetic operation depends on the number of bits carried along. For example, in a 32-bit floating-point operation, 24 bits might be used for the fractional number and 8 bits for the exponent. In each case 1 bit is reserved for the sign, so that the exponent would be limited to floating-point numbers between 2^{-2^7} and 2^{2^7} (approximately 10^{-38} to 10^{+38} decimal). The fractional part, of 24 bits, has an error corresponding approximately to the least significant bit. With 1 bit for the sign and 23 for the fractional number, the relative error is approximately $2^{-23} \cong 10^{-7}$, so that only about about six decimal digits are reliable in 32-bit floating-point operations. For double-precision (64-bit) computations, 8 bits might be used for the exponent and 56 bits for the fraction, giving an approximate relative error of $2^{-55} \cong 10^{-17}$. This corresponds to a maximum reliability of about 16 decimal digits. These floating-point processors are quite fast, and when coupled with logic that allows operation in parallel with the CPU, they can make modern, inexpensive computers very fast in carrying out numerical calculations. The fact that the floating-point processor is really another computer processor running in parallel with the CPU introduces us to the concept of *parallel computing*, in which several processors work on a computing task simultaneously. More will be said about parallel computing later.

2. *Memory* — The memory devices associated with computers today are very fast and very cheap when compared to those even 10 years ago; thus even the most inexpensive PC comes equipped with 4 million bytes (megabytes) or more of 50 ns cycle time memory at a cost of $50 per megabyte. A decade ago, 10 times less memory with 10 times less speed would have had a factor 10 higher price. Memory is of two basic types; the first, *dynamic random access memory (DRAM)*, is used for working storage during computations and loses stored information in a power failure; thus battery power backup is often available. The second type is a special purpose memory, *read-only memory (ROM)*, used to store permanently basic instructions for starting up the machine, input/output commands, etc., or for crucial, frequently used data files. ROM devices require special programming hardware in order to make any changes in the stored information and often are WORM (write once, read many) devices.

3. *Mass storage devices* — There are a number of different types of mass storage devices available for computers (see Table 2.2). These are used for inexpensive high-capacity storage with access times and transfer rates much slower than memory. *Magnetic tape* is the cheapest bulk storage device, but has the disadvantage that it has slow access time (due to the time required to move back and forth on the tape drive to find the desired files). For sequentially stored files, magnetic tape has reasonably fast transfer rates and is thus suitable for backup storage of large programs, data banks, etc. A much faster mass storage device is the *rotating disk*, which comes in several types: floppy disks, magnetic hard disks, optical disks, and compact read-

Table 2.1
Binary, Octal, Hexadecimal, and Decimal Conversion

Binary	Octal	Hexadecimal	Decimal
000 000	0	0	0
000 001	1	1	1
000 010	2	2	2
000 011	3	3	3
000 000	4	4	4
000 101	5	5	5
000 110	6	6	6
000 111	7	7	7
001 000	10	8	8
001 001	11	9	9
001 010	12	A	10
001 011	13	B	11
001 100	14	C	12
001 101	15	D	13
001 110	16	E	14
001 111	17	F	15
010 000	20	10	16
011 000	30	18	24
100 000	40	20	32
101 000	50	28	40
110 000	60	30	48
111 000	70	38	56

Octal – binary conversion: For 3 bits octal $= b_0 + 2b_1 + 4b_2$, where b_0, b_1, b_2 are binary digits.

Decimal – octal conversion: Decimal $= a_0 + 8a_1 + 64a_2 + ... + 8^n a_n$, where $a_0, a_1, ..., a_n$ are the octal digits.

Hexadecimal – binary conversion: For 4 bits hex $= b_0 + 2b_1 + 4b_2 + 8b_3$, where b_0, b_1, b_2, b_3 are binary digits.

Decimal-hexadecimal conversion: Decimal $= a_0 + 16a_1 + 256a_2 + .. + 16^n a_n$ where $a_0, a_1, ... a_n$ are hexadecimal digits

only disks (sometimes called CD-ROM). *Floppy disks* are low-capital-cost, small-capacity, low-performance removable storage devices in common use with workstations and PC's. *Magnetic hard disks* are high-capacity, high-speed mass storage devices that rotate at ~1800 RPM and transfer information to a magnetic sensing head that moves in and out on the disk. The access time and transfer rates are directly tied to the speed of rotation of the disk and the speed of head movement. *Optical disks* are removable media much like magnetic floppies in function, but store 100 times the data and have 10 times the speed of a magnetic floppy. The *compact disk read-only memory* (CD-ROM) is an inexpensive, removable, high-capacity mass storage device for loading permanent databases, system software, etc., when needed. It is basically a home CD player adapted for computers.

Table 2.2
Capabilities of Storage Devices

Device	Typical capacity/ megabytes	Typical access times	Typical transfer rates (megabytes/s)	Approx. cost $/megabyte
Memory:			(cycle time)	
DRAM	4 – 96	80 ns	50 ns	50
Mass storage:				
Magnetic hard disk	100 – 3000	1 – 20 ms	1 – 5	1
Optical disk	200	30 ms	0.6	10
Floppy disk	1 – 5	100 ms	0.05	100
Magnetic tape (DAT)	2000		0.1 – 1	1
CD-ROM	1000	500 ms	0.1 – 1	1

4. *Real-time clock* — The scheduler and timekeeper of the computer system is the *real-time clock*. This allows the computer to schedule tasks to be performed on a regular basis, such as taking data or changing control variables at specified time intervals, and timesharing the CPU's attention between various terminals and other peripherals. When the computer is installed, the correct date and time are entered, and the real-time clock keeps track of chronological time after this initialization. A different means of sharing the CPU's attention is through *interrupts*. Each task running on the computer has a certain *priority* level; when it needs the attention of the CPU (for example, to record a data point), the task generates an *interrupt*, and if this task has a higher priority than the task currently being serviced by the CPU, the new task is immediately serviced. Thus one may establish a hierarchy of priorities and interrupts in the computer and have the most urgent tasks (such as data taking) serviced immediately while those less pressing (such as printing or responding to a request for a new video display) must wait. Typical waiting times are a fraction of a second, so that the user is usually unaware of waiting in a queue.

5. *Input/output interfaces for data* — These interfacing devices deal with either *digital* or *analog* signals. *Digital I/O* interfaces are usually capable of either (*a*) *parallel* two-way simultaneous transmission (for example, 16 bits input and 16 bits output for a 16-bit machine) at speeds comparable to those within the computer system, or (*b*) one-way *serial* transmission (16 bits in only one direction at once for a 16-bit machine) at specified transmission rates.
Parallel-transmission interfaces are typically used for high-speed communication between computers or to high-speed digital logic devices.

Table 2.3
Some Communication Peripherals

Peripheral	Communication rates characters/s	Typical capital cost, $
Printers	10 – 100	100 – 10,000
Video displays and terminals	240 – 1440	200 – 5000
Modem	240 – 1440	100 – 500
Telefax	240 – 1440	200 – 1000

Serial-transmission interfaces are used for communication with slower digital devices, such as terminals or graphical display units. Serial-transmission rates are indicated as *baud rates*, where baud rate = $10 \times$ (# characters/s). Typical rates range from 240 to 1440 characters/s, depending on the device being interfaced (see Table 2.3). Both types of digital interface devices can be put under program control and the information analyzed as individual bits, bytes, or words, depending on the application. Table 2.4 lists some typical applications of digital I/O interfaces.

Analog I/O interfaces involve *analog-to-digital (A/D)* and *digital-to-analog (D/A) conversion*. A/D converters convert an analog voltage signal lying within a specific range (such as $\pm$ 10 V, 0 to 5 V, etc.) to an integer number. The maximum size of the integer number, and hence the resolution of the conversion, depends on the number of bits handled by the converter. The range of the integer I is given by $I_{range} = 2^N$, where N is the number of bits. A 12-bit converter with an input voltage range of $- 10\ V \le V \le + 10\ V$ *differential* is shown in Figure 2.3. The converter handles a *differential* signal because the reference voltage is also measured. A *single-ended input* would have only one voltage converted and would be referenced to the computer ground voltage. The maximum integer range for the 12-bit converter is $2^{12} = 4096$. This number of integers is chosen to run between $- 2048 \le I \le + 2047$, where $I = 0$ falls between the negative and positive integers. For the positive range (2047), the *resolution* of the converter is the voltage range divided by the intervals between integers, or:

$$\text{Resolution} = \frac{10\ V}{2047} = 0.00489\ V$$

This gives an *expected error* of $\pm$ 1/2 the resolution, or $\pm$ 0.00244 V. In this case the *relative error* in conversion is approximately given by:

$$\text{Relative error} \cong \frac{0.00244\ V}{|\text{ measured voltage }|}$$

so that more precision is obtained if the measured voltages are close to the full-scale voltage (10 V in this example). This can be accomplished through *signal conditioning*, which will be discussed in the next section.

Table 2.4
Some Typical Applications of Data I/O Interfaces

Type of interface	Typical connections and data
Digital inputs	From other computers
	Digital instrumentation
	Communication terminals
	Telefax
	Modem
	Switch settings
	Relay positions
	Multiplexer status
	Alarms
	Counters
	Logic devices
Digital outputs	To other computers
	Commands to relays and solenoids
	Commands to switches
	Alarm messages
	Stepping motors
	Sequencing of multiplexers
	Logic devices
	Commands to digital instrumentation
	Communication terminals
	Telefax
	Modem
	Hard copy and graphical output devices
	Various control signals
Analog inputs	Analog instrumentation:
	Temperatures
	Pressures
	Flows
	Composition
Analog outputs	Graphical plotters
	Oscilloscopes
	Control signals:
	Heater controls
	Set-point changes
	Valve drivers

The throughput speed of A/D converters is routinely 50,000 – 100,000 conversions/s, and high-performance models can achieve even higher rates.

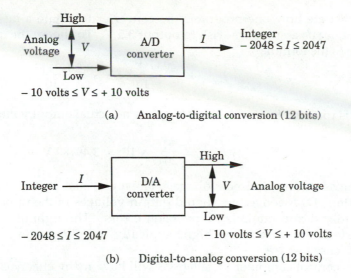

(a)   Analog-to-digital conversion (12 bits)

(b)   Digital-to-analog conversion (12 bits)

Figure 2.3. 12-bit A/D and D/A conversion.

As an illustration of A/D conversion, suppose the 12-bit converter shown in Figure 2.3 measures a voltage and reports the integer 1261 as a result of the conversion. To determine the actual measured voltage, one uses the formula:

$$V_{measured} = \frac{10\,I}{2047} = \frac{12610}{2047} = 6.16023 \text{ V}$$

where for this input voltage, the relative error is:

$$\text{Relative error} = \frac{\pm\,0.00244}{6.16023} = \pm\,0.000397 = \pm\,0.0397\%$$

a rather small value.

However, if the same 12-bit converter reported the number 21, the actual voltage measured would be:

$$V_{measured} = \frac{210}{2047} = 0.102 \text{ V}$$

and now the relative error is:

$$\frac{\pm\,0.00244}{0.102} = \pm\,2.4\%$$

which is a significant error. Thus for this converter one should convert voltages larger than 1 V to ensure negligible A/D conversion error.

D/A converters operate as just the reverse of A/D conversion. An integer with range 2^N is placed in a register and converted to an analog voltage output. The 12-bit D/A converter shown in Figure 2.3 has a voltage output range of $-10 \text{ V} \le V \le +10 \text{ V}$, corresponding to input integers ranging from $-2048 \le I \le +2047$. Thus the resolution and expected error of this D/A converter are the same as those of the A/D converter discussed above. To demonstrate how one programs the converter to obtain a desired voltage output, let us assume we wish to output 3.5 V. To choose the input integer, one uses the formula:

$$I_{input} = \frac{2047 \ V_{output}}{10} = \frac{(2047) \ (3.5)}{10} = 716.45$$

which is rounded to I_{input} = 716. This gives an actual output voltage of:

$$V_{output} = \frac{716}{2047} \times 10 = 3.49780 \ V$$

very close to the desired voltage. However, just as the case for A/D converters, D/A converters should output voltages in the upper 90% of the range in order to minimize conversion errors. Throughput rates for D/A converters are also very high, typically on the order of a million conversions/s.

Process control computer installations will have many channels of A/D and D/A converters in order to allow many loops to interface with the computer at one time.

6. *Communication peripherals* — There are many different types of devices for communicating between the user and the computer system (see Table 2.3). For printed text and graphical output, there are laser printers, ink-jet printers, or even color printers. For video I/O one can use local video display units or remote video display terminals (which have their own CPU) or small PC's which can be terminal emulators. There are also modem connections which allow telephone data transfer to and from worldwide remote locations.

2.4   DATA ACQUISITION AND CONTROL

In order to use a real-time computer to implement a process control algorithm, one must be concerned with the hardware realization of data acquisition and control. There are many different computer configurations that will accomplish these goals. One may use a large real-time computer system linked to many laboratories or processes, or have a digital computer network involving many smaller microcomputers, each dedicated to a specific process or laboratory. Later in this section we shall discuss some possible computer structures; however, first we will develop some essential elements of data acquisition and control.

2.4.1   Transducers for Data Acquisition and Actuators for Control Output

It is useful to consider some of the transducers and actuators most frequently encountered in the process industries. A representative sample of data acquisition sensors for some of the more common process variables is listed in Table 2.5. Depending on the sensor chosen, the primary output can be either digital, pneumatic (air pressure), current, or voltage. Because the digital

Table 2.5
Some Common Data Acquisition Sensors

Process variable	Device	Output	Comments
1. Temperature	Thermocouple	mV	Cheap, rugged
	Resistance sensor (e.g., thermistors)	Resistivity	High sensitivity
	Oscillating quartz crystal	Oscillating frequency	High sensitivity
	Radiation pyrometer	mV	For high-temperature applications
2. Pressure	Diaphragm bellows	Capacitance	Low pressure
	Bourdon tube	Inductance	Both low and high pressure
	Strain gauge	Resistance	
	Piezoelectric	Reluctance	
	Ionization gauge	Emf	From low to very high pressures
3. Flow	Pitot tube	Differential pressure	Ultra-low pressure
	Orifice	Differential pressure	
	Turbine	Rotational speed	
	Electromagnetic field	Generated emf	For difficult flow measurements
	Neutron bombardment	Nuclear radiation	For difficult flow measurements
	Ultrasound	Doppler shift	
	Hot-wire anemometry	Resistivity	High precision
	Angular momentum	Force	Rugged
4. Density	Vibrating tube	Frequency shift	
	X-ray	Nuclear radiation	
	Bubbles	Differential press.	
	Differential head pressure	Differential press.	

Table 2.5
(continued)

Process variable	Device	Output	Comments
5. Liquid level	Float	Mechanical deflection	
	Sonic resonance	Resonant frequency	
	Conductivity	Resistance	
	Dielectric variability	Capacitance	
	Thermoelectric probe	Resistivity	
	Nuclear radiation	Radiation intensity	
	Photoelectric cell	Voltage	
	Head pressure	Differential press.	
6. Viscosity	Capillary	Differential pressure	
	Vibrating ball	Frequency shift	
	Rheometry	Frequency response	For polymer properties
7. Composition	Potentiometry	mV	
	Moisture content:		
	Hygrometry	Resistivity	
	Psychrometry	Wet- and dry-bulb temperatures	
	Chromatography	mV	Long analysis times
	Refractive index	mV	
	Ultrasound	Sound velocity, freq.	
	Spectroscopy: UV, visible, IR, Mössbauer, Raman, atomic-emission, X-ray, electron, ion, nuclear magnetic resonance	Intensity spectra	
	Polarography	mV	
	Conductimetry	Resistance	
	Mass spectrometry	mV	
	Differential thermal analysis	Heat transferred	
	Thermogravi-metric analysis	Mass changes	

Table 2.6
Some Typical Control System Actuators

Control action	Device	Input
Electrical heating	Electric resistance or microwave heater	Voltage or current
Flow adjustment	Pneumatic valve	Air pressure
	Motor-driven valve	Voltage, current, or digital pulse
	Variable-speed pump	Voltage or current
	Fluidic control valve	Pressure
Alarms	Lights, bells	Digital pulse
On-off signals	Relays, switches	Digital pulse

computer can receive only voltage or digital signals, the pneumatic and current signals must be converted to one of these acceptable inputs. In practice this involves special transducers for pressure/voltage conversion (P/E transducer) and current/voltage conversion (I/E transducer, which is usually just a high-precision resistor).

In a similar way control system actuators come in several different forms. Table 2.6 includes some of the most common types. Notice that in the process industries, the most frequently seen actuators are for flow control. This is because heating and cooling processes often involve such factors as coolant flow-rate adjustment, steam injection rates, or furnace fuel flowrates. Similarly, pressure adjustments usually require controlling inflow or outflow from a vessel, as does liquid level control. Although the digital computer outputs are limited to digital or voltage signals, some of the control actuators require current or pneumatic inputs. Thus special transducers are needed to provide for voltage/pressure (E/P) and voltage/current (E/I) conversion.

2.4.2 Signal Conditioning

An additional consideration in data acquisition and control is conditioning the sensor or actuator signal so that it will have high accuracy, have low noise levels, and make efficient use of the interface equipment. Usually the steps involved in signal conditioning are multiplexing and amplification of the signals, suppression of noise on the signal, and selection of a transmitting mode for the signal. We shall discuss each of these considerations in turn.

In some applications, there will be many similar measurements at a remote location, and it is usually most efficient *to multiplex* these at the source and transmit them sequentially over only a few lines. As an example, consider 10 thermocouple measurements from a process, each with a voltage signal of ~ 10 mV. First of all, one may not wish to run 10 lines and use 10 channels of our A/D converters just for these signals, so multiplexing is necessary. In addition, if such low-voltage signals are to be transmitted some distance, they will be heavily corrupted by noise in transmission and will have very low resolution

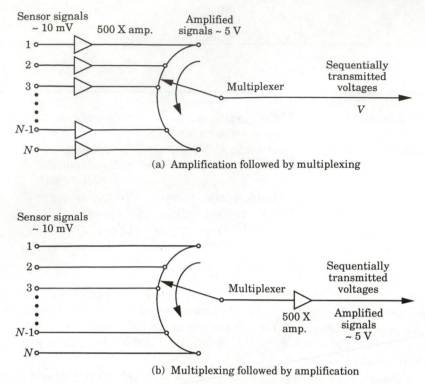

(a) Amplification followed by multiplexing

(b) Multiplexing followed by amplification

Figure 2.4. Two different strategies for multiplexing and amplifying low-voltage signals.

when they reach the A/D converter; thus *amplification* as well as *multiplexing* would be needed. Figure 2.4 illustrates two possible approaches to solving this problem. Scheme (*a*) involves amplification of each low-level signal (requiring *N* high-gain amplifiers) up to ~ 5 V, then multiplexing with solid-state relays. By multiplexing high-level signals, one minimizes the effects of noise introduced by the relays; however, this is at the cost of an expensive amplifier for each signal. Scheme (*b*), on the other hand, multiplexes first and then requires only a single high-gain amplifier. This is more efficient, but requires care in selecting a low-noise-level multiplexer. An alternative procedure would be to use a local A/D converter with multiplexing capability so that the signals are transmitted digitally, with no noise problems. Both types of multiplexing schemes are known to be successful.

Noise suppression, either before or after signal transmission to the computer, is also an important part of signal conditioning. There are times when relatively high-frequency noise, such as 60 Hz coming from power lines or motors, appears on the signal. Fortunately, in the process industries, such high frequencies are very rarely contained in the desired signal; thus *filtering* to remove these frequencies can be a very successful means of noise suppression. A simple *low-pass filter* with frequency-response characteristics like those shown in Figure 2.5(a) will often suffice. Other common types of filters include *high-pass filters*, which reject low frequencies, and *notch filters*, which pass only signals within a specified frequency band. Figure 2.6 shows an example of the performance of a low-pass filter applied to a thermocouple signal afflicted with motor noise.

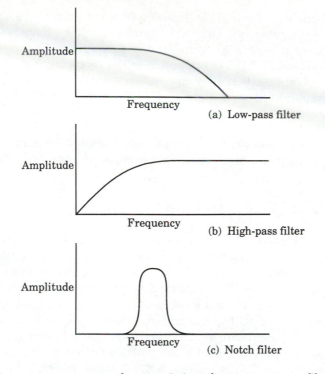

Figure 2.5. Frequency response characteristics of some common filters.

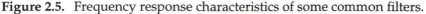
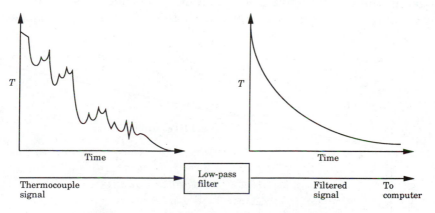

Figure 2.6.   Performance of a low-pass filter.

The last part of signal conditioning involves a decision on how the signal is to be transmitted to the computer. There are several alternatives available for signal transmission:

1. **Voltage signals** — These work well over short distance cables (up to ~ 300 ft) where voltage losses and cable capacitance do not cause signal deterioration.
2. **Current signals** — This method works well with hardwire connections over longer distances but requires special transmitters and I/E conversion at the receiving end.

3. **Digital signals** — These signals can be transmitted most easily and are the natural choice when the original signal is digital. Even voltage signals may be transmitted in a digital mode through A/D conversion at the sending location and D/A conversion at the receiving end. Digital signals have the advantage that they are less susceptible to noise problems and can be transmitted over extremely long distances through telephone connections.

Of course, the selection of transmission mode should be made based on the particular application, noise levels to be encountered, and distances to be covered.

2.4.3 Modes of Computer Control

Computer control is usually carried out in one of two modes: *supervisory control* or *direct digital control* (DDC). Supervisory control, illustrated in Figure 2.7(a), involves resetting the set-point of the local controller according to some computer algorithm. Thus the computer control scheme need only supervise and coordinate the actions of the local controllers. *Direct digital control*, by contrast, requires that all the controller action be carried out by the digital computer. Measurements are sent to the computer and compared with the set-point; then the computed control action is transmitted to the actuator. This is illustrated in Figure 2.7(b). For DDC the computer samples the flow measurement at discrete instants of time and sends as control signals step changes in the control valve stem position. The time interval Δt between samples and controller changes must be chosen with care. If Δt is taken too

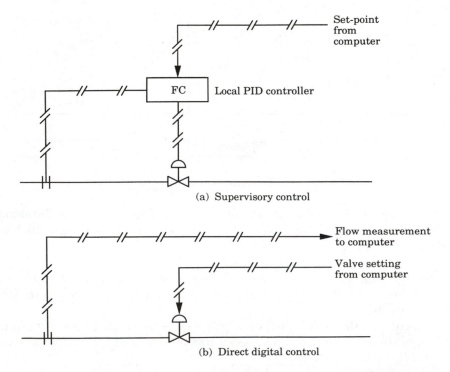

Figure 2.7. Supervisory and direct digital control of a fluid flowrate.

large, the controller performance will deteriorate, while having Δt too small will put an unnecessarily high computational load on the computer. Both supervisory control and DDC are in wide use in industrial applications, and both may be readily used to implement modern control strategies. In most processes, the local controller dynamics (whether analog or digital) are short compared with the time constants of the process under control; thus most sophisticated control strategies are implemented in a cascade fashion — involving programmed set-point changes in the local controllers. Hence, it matters little whether these local controllers are analog or digital. As an example, suppose the flowrate control loop in Figure 2.7 is a steam flow into a heat exchanger. Suppose further that an optimization algorithm calculates the optimal steam flowrate program $F(t)$.

Clearly this program could be implemented by either supervisory control or DDC when the steam valve dynamics are fast compared with the heat exchanger time constants. In either case the flow loop set-point would be programmed to follow the function $F(t)$.

2.4.4 Computer Data Acquisition and Control Networks

Depending on the specific application, a wide variety of tasks may be required from a data acquisition and control network: control of plant processes or pilot plants, servicing of video display units in various laboratories and control rooms, logging of data from analytical and/or research laboratories, etc. The computer network could be as simple as an array of inexpensive PC's (some smaller plants are actually controlled this way) or it could be a large commercial *distributed control system* (DCS). Let us discuss each case in some detail.

Small Computer Networks

For small processes such as laboratory prototypes or pilot plants, having a relatively small number of control loops, one can configure a network of personal computers to provide a satisfactory and inexpensive data acquisition and control system. Figure 2.8 provides an example configuration of a PC network. The process has a two-way direct link to the main control computer (#1) which sometimes may take some measurements and implement direct digital control on some loops itself or it may only coordinate and direct the activities of the other computers in the network. Computer #2 in the network is dedicated to some on-line analytical instrument (e.g., FTIR, Mass Spectrometer, etc.) and has the task of collecting the spectra, translating the spectra into composition measurements, and sending these to the main control computer for action. Computer #3 has a task similar to that of computer #2, but its measurement results are sent to computer #4 as well as to the main computer. Computer #4 has responsibility for controlling some of the loops in the process based on its own measurements as well as the measurements supplied by computer #3. Computer #4 reports its measurements and control actions to the main computer and receives instructions back from the main computer as to set-point, controller tuning parameters, etc. Computer #5 is similar to computer #4 and provides only local control action based on instruction from the main computer and its local measurements. All of the computers operate with a multitasking operating

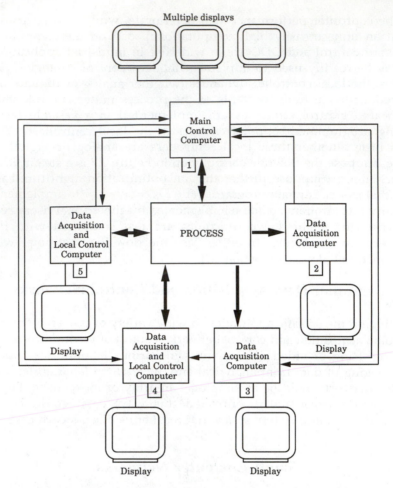

Multiple displays

Figure 2.8.   A small data acquisition and computer control network based on personal
                   computers.

system and have local memory. They would normally be configured with local
disk storage and often have shared disk storage with a server. Each computer
in the network could have its own video display, and the main control computer
could have multiple displays.

   Note that this relatively small network has the essential features of
network computing: distributed CPU's with assigned tasks, assignment of high-
volume data transfer (e.g., entire spectra, chromatograph traces, etc.) to the
local computers with only low-volume data transfer (final composition result,
etc.) between computers in the network, ease of expansion to additional loops,
etc.

Commercial Distributed Control Systems

For larger processes such as more complex pilot plants and full scale-plants,
where there can be hundreds of control loops, commercial distributed control
systems are often the network of choice. There are many vendors who provide
these DCS networks — the largest are Bailey, Foxboro, and Honeywell — with

a variety of proprietary features. However, in our discussion here we will provide only an overview of the generic features of DCS networks and refer the reader to the vendors themselves for a discussion of their distinguishing attributes.

Commercial DCS networks are similar in philosophy to the simple network described above. The difference is that they are designed, in both hardware and software, to be flexible (i.e., easy to configure and modify) and to be able to handle a large number of loops. Furthermore, the latest commercial networks come with optimization and high-performance model building and control software as options. This means that engineers who understand the process dynamics, the process control system goals, and the theory behind modern control systems can quickly configure the DCS network to implement high-performance controllers. Chapter 27 provides an overview of some of the control strategies readily available in commercial DCS networks.

The elements of a commercial DCS network are illustrated in Figure 2.9. Connected to the process are a large number of local data acquisition and control computers which play the same role as computers 2–5 in Figure 2.8. Each local computer is responsible for certain process measurements and part of the local control action. They report their measurement results and the control actions taken to other computers via the data highway. Although Figure 2.9 shows only one data highway, in practice there can be several levels of data highways. Those closest to the process handle high raw data traffic to local computers while those farther away from the process deal only with processed data but for a wider audience. For each data highway, any computer on that highway, in principle, could access any of the data on the highway. The data highway then provides information to the multidisplays on various operator control panels, sends new data and retrieves historical data from archival storage, and serves as a data link between the main control computer and other parts of the network. Sitting above the lower part of the DCS network is a supervisory (or host) computer that is responsible for performing many higher level functions. These could include optimization of the process operation over varying time horizons (days, weeks, or months), carrying out special control procedures such as plant startup or product grade transitions, and providing feedback on economic performance. The data link between the host and the main control computer is usually much smaller than those on the data highway because the data transfers are much less frequent and are much less voluminous.

A DCS network such as shown in Figure 2.9 is very powerful because the engineer or operator has immediate access to a large amount of current information from the data highway and can see displays of past process conditions by calling on archival data storage. The control engineer can readily install new on-line measurements together with local computers for data acquisition and then use the new data immediately in controlling all the loops of the process. It is possible to change quickly among standard control strategies and readjust controller parameters in software. Finally, the imaginative control engineer can use the flexible framework to implement quickly his latest controller design ideas on the host computer or even on the main control computer. These are the attributes that will lead to every process of any size having a DCS network in the future.

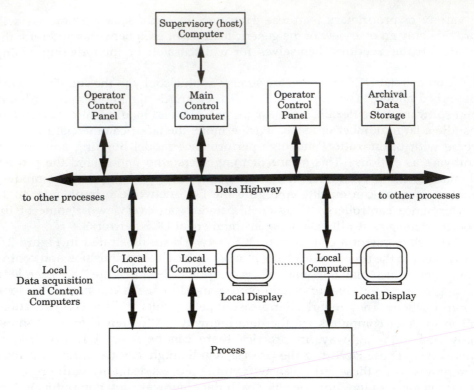

Figure 2.9.   The elements of a commercial distributed control system network.

2.5   SOME EXAMPLES

To illustrate some of the principles discussed in the earlier sections, several example case studies shall be presented. It is hoped that these will give the reader a practical grasp of what is involved in interfacing processes to real-time computers.

2.5.1   Computer Control of the Pressure in a Gas Storage Tank

A simple interfacing problem arises for a gas storage tank system which has been used in the teaching laboratory at the University of Wisconsin.

A drawing of the interfacing scheme is shown in Figure 2.10. A bourdon tube transducer is used to measure the pressure in the tank. This transducer is linear and transforms a pressure range of 0 to 60 psig to a voltage range of 0 to 50 mV. This low-voltage signal is then amplified to a voltage range of 0 to 10 V before being sent to the 12-bit A/D converter.

The digital computer must be programmed to receive the pressure measurements and calculate control signals based on some controller algorithm. For example, a simple PID controller in discrete form would be (see Chapter 26)

$$u(t_k) = u_0 + K_c \left\{ \left[y_d - y(t_k) \right] \right.$$

$$\left. + \frac{\Delta t}{\tau_I} \sum_{n=1}^{k} \left[y_d - y(t_n) \right] + \frac{\tau_D}{\Delta t} \left[y(t_{k-1}) - y(t_k) \right] \right\}$$

where $u(t_k)$ is the control signal at time t_k, u_0 is the zero-error control signal, K_c is the proportional controller gain, τ_I is the integral time, and τ_D is the derivative time. The measured output (pressure in this case) at time t_k is denoted by $y(t_k)$, and y_d is the controller set-point. The quantity Δt is the sampling/control time interval and is usually selected to be short compared with the time constants of the process.

The control signal, which is transmitted as an analog voltage, must be converted to a 3–15 psig air-pressure signal in order to drive the pneumatic valve. This could be done in one step with an E/P transducer; however, to make use of existing equipment, this conversion is done in two steps, as shown in Figure 2.10.

After installation, the transducers must be calibrated. For example, 0 to 60 psig corresponds to 0 to 10 V analog voltage for our system, while 0 to 10 V analog control signal corresponds to 3 (full shut) to 15 psig (full open) for the control valve. Because the analog signals were sent over shielded cables for relatively short distances and were not in the vicinity of power lines or motors, no noise suppression devices were needed in this application. Figure 2.11 illustrates the dynamic behavior of the implemented control scheme, showing both the measured tank pressure and the computer-specified control valve position. The control time interval Δt was taken to be only a few seconds, so that the DDC step changes in valve position appear as a continuous curve in the plot. The actual valve stem position does not exactly follow the specified control signal due to the dynamics of the valve itself. Nevertheless, the control system performs well, as might be expected.

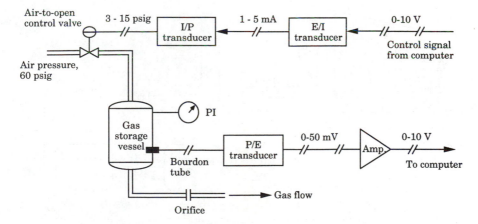

Figure 2.10.   Interfacing for the gas storage vessel.

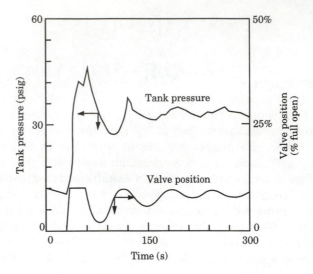

Figure 2.11.   Dynamic behavior of the gas storage vessel under DDC control.

2.5.2   A Commercial DCS Network Example

To provide an example of an actual commercial DCS installation, consider the process for manufacturing an engineering plastic shown in Figure 2.12. The process begins with a feed solution in a hold tank which must be concentrated by two-stage evaporation. Then it is passed to reactor #1 where polymerization begins.

From reactor #1 the material goes through gear pumps into a flashing section where volatile byproducts are removed. Then it is sent to reactor #2 with its flashing unit and finally to a finishing stage where there is an on-line rheometer for monitoring polymer product quality. At the final stage, the material is pelletized and packaged for shipment. Note that there is a great deal of redundancy in the process with duplicate pumps, reactors, and

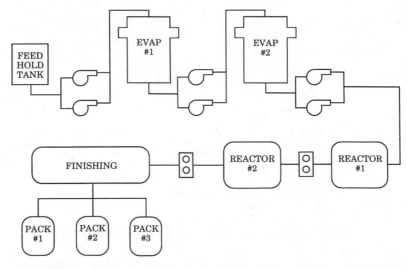

Figure 2.12.   DCS screen showing process overview.

pelletizing lines. This allows one of the redundant units to be shut down for cleaning while the other carries the processing load.

The task of the DCS for this process is to handle approximately 100 control loops (some of them cascade) involving temperature, pressure, flow, chemical composition, and product quality. The DCS has many other detailed graphical displays of the process flowsheet showing all of the loops in whatever level of detail desired. For example, the upstream evaporation section of the plant can be seen in more detail in Figure 2.13. Note that each element is denoted by a letter A–Z and the measuring devices, transmitters, controllers, etc., are denoted by standard symbols (see Appendix A). This graphic is sometimes called a *process and instrumentation* (P&I) diagram. For most measurements, the current values are displayed on the screen. For example, the feed solution level in the hold tank (top left) is controlled by a level measurement and transmitter, LT, which sends a signal to the level controller, LC, which adjusts the valve controlling the flow of the feed solution to the hold tank. The graphic hides the fact that a multitasking DCS computer actually receives the level measurement and transmits the command to the feed solution valve.

The architecture of the relatively small commercial DCS network that is controlling this plant is shown in Figure 2.14. The process has analog and digital signals which are sent via transmission lines to the termination units, TU, to local A/D, D/A, and digital I/O *slave units*; there is also a *field bus* interface that takes processed information from sensors and actuators having their own CPU (e.g., mass flowmeters, FTIR sensors, digital valve positioners, etc.). This information goes to the lowest level data highway, the *slave bus*.

EVAPORATOR DETAIL

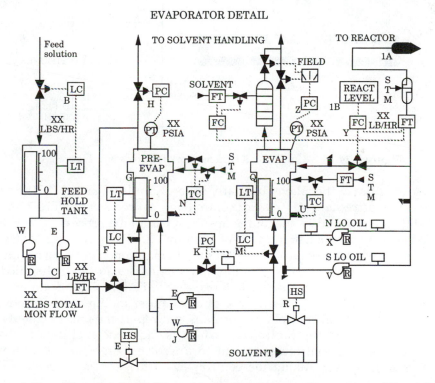

Figure 2.13.   More detailed DCS graphic display of evaporation section of Figure 2.12.

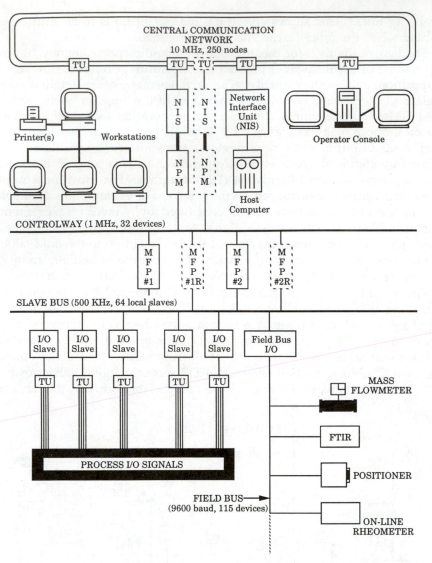

Figure 2.14.   Commercial DCS network for process in Figure 2.12.

Sitting above the slave bus are several multitasking CPU's called *multifunction processors* (MFP's). For controlling the process in Figure 2.12, only two MFP's are required — one for the upstream part of the process (up to reactor #2) and one for the remaining part of the process. Each of these MFP's has limited memory and uses the same CPU as found in common personal computers. However, by using multiple MFP's the computational load can be handled. In order to ensure against computer failures, each MFP has a duplicate which is sitting idle ready to take over in case of a computer malfunction. The MFP's handle all of the local controller algorithms and report the measurements and controller actions to the *controlway*, which is the next level data highway. The data from the controlway is sent to a *Network Processing Unit* (NPU) which is a computer linking the controlway to the *Central Communication*

Network. On this central network hang the operator's console and displays, engineering workstations, and the host (supervisory) computer. The *Central Communication Network* can be accessed by computers throughout the plant, and by remote connections anywhere in the world. Thus the engineer at the company headquarters on the east coast can see on his computer screen the same displays the operators see at the plant in Texas.

To illustrate how this DCS works, consider the level control of the feed hold tank discussed above. The level sensor $\boxed{\text{LT}}$ sends the level signal to an A/D slave unit; this measurement goes on the slave bus and is picked up by MFP #1 (which is responsible for the upstream part of the plant). MFP #1 calculates the change in the feed solution valve position to be sent to the valve. This value is put on the slave bus and picked up by a D/A slave I/O device which transmits the command to the valve. Simultaneously, the MFP sends a report of measured level and control action taken to the *controlway*. This report is picked up by the NPM and transmitted to the *Central Communication Network*. The data appear on the operator console graphic display unit as a new point on a trend chart, and appear in numerical form next to the valve in the display shown in Figure 2.13. The data are also logged by the host computer and filed in archival data storage.

To minimize the data traffic on each of the data highway levels, there are usually several timescales at which sensors report data and control action taken. Thus for variables that change slowly (such as temperatures in large processes), measurements and control action can be taken infrequently (e.g., every minute); for variables that change rapidly (e.g., flowrates in a transfer line), it is necessary to use a fast sampling and control action timescale (e.g., 1 s). A second way to minimize data traffic is to only report *changes* in variables within some band of values. Thus if the last reported value of temperature was 41.5°C and the report bounds are ± 0.5°C, then no changes would be reported until the temperature fell below 41.0°C or above 42.0°C.

2.6   SUMMARY

In this chapter we have presented a brief introduction to the hardware and software elements required for control system implementation, with particular emphasis on on-line data acquisition and computer control. It is hoped that this brief overview will provide the reader with some perspective about how control system designs are currently brought to the plant and implemented. Since any control system design that is not implemented and used by the plant is economically useless, having a good grasp of these practical realities is essential to a successful control system design project. Fortunately, if the necessary sensors and actuators are available, almost any control algorithm that can be programmed into a digital computer can be implemented in hardware. This situation provides a challenge to the imagination of the control engineer as well as an incentive to the analytical chemist to develop a wider range of reliable on-line instrumentation.

SUGGESTED FURTHER READING

1. Albert, C. L. and D. A. Coggan (Ed.), *Fundamentals of Industrial Control*, ISA Publ., Research Triangle Park, North Carolina (1992)
2. Harrison, T. J. (Ed.), *Minicomputers in Industrial Control*, ISA Publ., Pittsburgh (1978)
3. Mellichamp, D. A. (Ed.), *Real-Time Computing with Applications to Data Acquisition and Control*, Van Nostrand Reinhold, New York (1983)
4. Ray, W. H., *Advanced Process Control*, Butterworths, Boston (1989) (paperback reprint of original 1981 book)

REVIEW QUESTIONS

1. What is the general structure of a digital computer data acquisition and control system?

2. What are the basic elements of a digital computer? What are their characteristics and cost?

3. What is the formula for converting from binary to octal to decimal numbers?

4. How do A/D and D/A converters work? How can one compute the resolution of a 20-bit A/D converter?

5. List some analog inputs/outputs to a control computer.

6. List some digital inputs/outputs to a control computer.

7. If one wished to measure temperature, pressure, and chemical composition on-line in a chemical reactor, what would be a good choice of sensors?

8. Suppose there is an electric motor nearby adding electrical noise to the level measurement signal from a tank. How would one condition this signal before sending it to the computer?

9. If one wished to control the temperature in a tank by adjusting external heating, what are the choices of actuators usually available?

10. What are the two modes of computer control? How do they differ in hardware requirements?

11. What are the essential elements of a distributed control system (DCS)?

12. How does one ensure safe process operation in case one of the computers in a DCS fails?

13. What is the likelihood of encountering a DCS in each new plant you see in the next decade?

PROBLEMS

2.1   Convert the following binary numbers to octal numbers:

(a) 100	111	000	111	101
(b) 101	010	101	110	011
(c) 111	001	110	100	011

2.2 Convert the following octal numbers to decimal numbers:

(a) 747   (c) 100
(b) 440   (d) 556

2.3 A Fortran program consists of 100 lines with approximately 25 characters per line. Estimate the number of words of memory in a 16-bit machine required to store this Fortran code.

2.4 In responding to interrupts from various peripheral devices, a computer manufacturer recommends the following priority levels:

Terminal – 4   Line printer – 4
Real-time clock – 7   A/D converter – 6
Disk drive – 7
Magnetic tape unit – 7

Discuss why some of these peripherals are assigned higher priorities than others.

2.5 A computer has a 12-bit A/D converter with – 10 V to + 10 V span. Determine the analog input voltages and relative error for the following integers coming from the converter:

(a) 570   (c) – 960
(b) 2000   (d) 25

2.6 Provide the answers to Problem 2.5 for a 10-bit A/D converter with a – 5 V to + 5 V span and for the following integers:

(a) 450   (b) –200   (c) 100   (d) 25

2.7 It is desired to output the following voltages through a 12-bit D/A converter with a – 10 V to + 10 V span. Determine the integer that should be loaded into the D/A in each case.

(a) –1 V
(b) + 2.7 V
(c) + 8.7 V
(d) –6 V

2.8 Two computers are linked over a 2400-baud telephone line. The operators of one computer would like to transfer a Fortran program to the other computer over this telephone link. The Fortran program in question has about 500 lines of code with approximately 30 characters per line. Estimate how long it will take to transfer this program. An engineer suggests connecting the two computers via a 9600-baud cable link. How long would the program transfer take in this case?

part II

PROCESS DYNAMICS

The main justification for a separate study of *Process Dynamics* as a precursor to the study of *Process Control* lies in the simple fact that to succeed in *forcing* a system to behave in a desired fashion, we must first understand the inherent, dynamic *behavior* of the system *by itself*, without assistance or interference from the controller. The task of designing effective control systems can best be carried out if it is based on a sound understanding of inherent process dynamics.

Thus Part II, consisting of Chapters 3–11, will be devoted to analyzing and characterizing the various types of dynamic behavior expected from a chemical process. The concept of employing a *process model* as a "surrogate" for the actual process is central to these studies. It allows us to carry out systematic analysis of process behavior most efficiently; it also provides the criteria for characterizing the behavior of an otherwise limitless number of actual processes into a relatively small number of well-defined categories.

part II

PROCESS DYNAMICS

"... always form a definitio
... of whatever presents itself to the mind
strip it naked and look at its essential natur
contemplating the whole through its separate part
and these parts in their entirety.

Marcus Aurelius, *Meditation*

3

BASIC ELEMENTS
OF DYNAMIC ANALYSIS

There are important fundamental concepts, terminology, and mathematical tools employed in the analysis of process dynamic behavior. Familiarity with these is essential for the discussion in Part II. Thus, the objective of this chapter is to introduce these *elements of dynamic analysis*.

3.1   INTRODUCTION

As defined in Chapter 1, the input variables of a process are those variables capable of independently stimulating the process, and inducing changes in its internal conditions. Such changes in the internal state of the process are usually apparent from observing changes in the output variables of the process.

By *how much*, and *in what manner*, the process responds to a change in the input variable clearly depends on the *nature* of this input change; it also depends on the intrinsic nature of the process itself. For example, the insertion of a hot piece of iron into a beaker of water will elicit a water temperature response different in *magnitude* and *character* from the response obtained by inserting the same piece of iron into a larger body of water such as a lake.

For any given input change, the observed process response provides information about the intrinsic nature of the process in question. By the same token, if the intrinsic nature of the process is known, and can be properly characterized, then the response of the process to any input change can be predicted.

Process Dynamics is concerned with analyzing the dynamic (i.e., time-dependent) behavior of a process in response to various types of inputs. It is by such studies that we are able to characterize a wide variety and limitless number of actual processes into a relatively small number of well-defined categories. The advantages that such process characterization provides cannot be overstated, as will become clearer when we take on the issue of control system design in Part IV.

To illustrate this point, consider the possible responses to a step change in

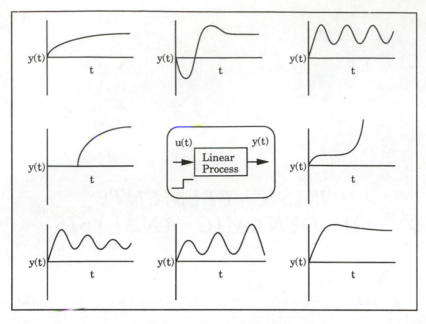

Figure 3.1.   Possible responses of a linear process.

the input $u(t)$ for a linear dynamic process. Depending on the process, the dynamic response could take one of the forms shown in Figure 3.1. It is important for the control system designer to understand the type of process model which could produce such dynamic behavior. This will be one of the goals of our study of process dynamics.

The tools required for carrying out the analysis of dynamic process behavior will be reviewed in the next section.

3.2   TOOLS OF DYNAMIC ANALYSIS

3.2.1   The Process Model and Ideal Forcing Functions

The main objective of dynamic analysis is to investigate and characterize system behavior when a process is subjected to various types of input changes. To be sure, such investigations can be carried out on the actual physical processing units themselves: changing the inputs, and recording the corresponding responses in the output variables for subsequent analysis. However, this approach suffers from two major problems that are immediately obvious:

1. *The physical process may not even exist yet.*
    This situation arises when the issues of control systems design and analysis are being confronted before the actual construction of the plant.

2. *It is time consuming and expensive.*
    Most real chemical processes have such large capacities that once disturbed, a considerable amount of time must elapse before they

finally "settle down" again. Carrying out several such investigative experiments must therefore be time consuming. Furthermore, regular process operation must often be suspended while carrying out such investigations; the financial implications of tying up profit-making processing units for extended periods of time are quite obvious. This is in addition to the cost of acquiring and analyzing data, which, in some cases, may not be trivial.

There is an additional, and perhaps not so obvious, setback: certain critical *theoretical* information, useful in characterizing the behavior of entire *classes* of processes, are not easily available from the results of such strictly experimental investigations carried out on *single*, specific processes.

Thus the dynamic analysis of process behavior is carried out with the aid of some form of a *mathematical representation* of the actual process. Such theoretical analyses, of course, only provide knowledge about *idealized* process behavior, but this has proven very useful in understanding, and in characterizing, real behavior.

The essence of dynamic analysis may therefore be stated:

Given some form of mathematical representation of the process, investigate the process response to various input changes. That is, given a **process model***, find y(t) in response to inputs u(t), d(t).*

Observe then that any successful dynamic analysis requires first and foremost:

1. A process model, and
2. Well characterized input functions (also referred to as "forcing functions" because they are used to *force* the observed output response).

In addition to these requirements, however, we note that the actual task of obtaining the process response involves *solving* the mathematical equations that arise when the specified forcing function is incorporated into the process model. This is the central task in process dynamic analysis. As is to be expected, the successful performance of this task requires the employment of certain tools of mathematical analysis. The most commonly utilized of these tools will now be summarized.

3.2.2   Mathematical Tools

As will be seen later, process models are usually in the form of *differential equations*: linear and nonlinear, ordinary and partial. A familiarization with such equations, especially how they are solved, will therefore be important in carrying out dynamic analysis.

In this regard, one of the most important tools we will use is the *Laplace transform*, a technique that, among other things, converts linear, ordinary differential equations into the relatively easier-to-handle algebraic equations. It also converts partial differential equations into the ordinary type. Beyond this, however, by far the most significant contribution of the theory of Laplace transforms to dynamic analysis lies in the fact that it is responsible for the development of the *transfer function* model form, which is still the most widely used model form in process control studies.

When process models are nonlinear, accurate analysis requires the use of *numerical methods* in carrying out a computer simulation of process behavior. If the emphasis is on *simplicity* rather than accuracy, the nonlinear equations may be *linearized* and approximate process behavior deduced from such linear approximations of the process model.

Process models used for computer control take the form of *difference equations*, which, like their differential equation counterparts, can be solved analytically when linear, or reduced to equivalent transfer function forms by the use of *z-transforms* — the *discrete-time* equivalent of the Laplace transform.

Processes that have multiple input and output variables, i.e., multivariable systems, are modeled, as we would expect, by several differential equations. The employment of *matrices* in dealing with such equations sometimes dramatically reduces an otherwise tedious or intractable problem to a manageable one.

We have thus identified the mathematical tools usually considered most important to the study of dynamic analysis, viz., *differential (and difference) equations, numerical methods, Laplace and z–transforms, and matrices.* Thus a discussion of these tools, specifically designed to provide the required working understanding, appears desirable. We observe, however, that an insertion of a discussion of each of these topics at this point either must constitute a prolonged distraction from the main task at hand, or must be too sketchy to be useful. Such discussions can be found in Appendices B–D, and the reader is encouraged to refer to these as often as needed. Thus we will introduce only Laplace transforms here — and z–transforms in Chapter 24 — leaving a fuller discussion of each of these topics for Appendix C.

3.2.3 The Digital Computer

On a strictly conceptual level, *Process Dynamics* has been presented simply as the analysis of the dynamic response of processes to various types of inputs, with the understanding that such analyses are typically carried out using mathematical process models and applying the mathematical tools listed above. Beyond the conceptual, however, there is the actual, mechanical aspect of dynamic analysis: the solution of the describing equations and the presentation of the results, usually in graphical form.

Although there are classes of processes represented by mathematical models that allow analytical solution of the describing equations, a substantial number of processes of interest are modeled by equations that are not easily amenable to analytical solutions. Whatever the case, the graphical presentation of the dynamic response (whether such results are obtained analytically or otherwise) is often a tedious exercise. The use of the digital computer in carrying out dynamic analysis makes it possible to obtain numerical solutions to modeling equations when analytical methods are inapplicable, and relieves us of the sometimes tedious work involved even when analytical solutions are possible. Of no less importance are the graphics capabilities of the computer, facilitating the presentation of the dynamic responses directly in graphical form. It is therefore more customary to carry out dynamic analysis studies with the aid of the digital computer and to employ one of the several currently available computer software packages specifically designed for *simulating* the process response to various inputs. In general, these programs are designed to accept a process model, and a definition of the desired process input to be employed in "forcing" the response. These will then be used to produce graphical displays of the simulated process response.

Appendix E has a description of computer-aided control system analysis and design software suitable for this task.

3.3   THE LAPLACE TRANSFORM

3.3.1 Definition

The Laplace transform of a function $f(t)$ is defined as:

$$\bar{f}(s) = \int_0^\infty e^{-st} f(t) \, dt \qquad (3.1)$$

The *operation* of (1) multiplying the function $f(t)$ by e^{-st}, followed by (2) integrating the result with respect to t from 0 to ∞ is often compactly represented by $L\{f(t)\}$, so that Eq. (3.1) may be written as:

$$\bar{f}(s) = L\{f(t)\} \qquad (3.2)$$

Remarks

1. The function of the variable t (which, in Process Control applications is almost always time) is transformed to a function of the Laplace variable s, normally understood to be a *complex* variable. The Laplace transform operation is thus often thought of in terms of a *mapping* from the *t-domain* (or time domain) to the *s-domain* (or Laplace domain).
2. To be precise it is necessary (and instructive) to differentiate between a function and its transform by employing different nomenclature. Thus here we have represented the transform of $f(t)$ by $\bar{f}(s)$; but we shall do this for this chapter only. In later chapters, for simplicity, we shall drop the overbar so that the transform of $f(t)$ will be represented as $f(s)$ where it will be enough to retain the corresponding arguments s and t to differentiate the transform from the original function.

The following examples illustrate how to obtain Laplace transforms using the definition in Eq. (3.1).

Example 3.1   LAPLACE TRANSFORM OF SIMPLE FUNCTIONS.

Find the Laplace transform of the simple function $f(t)=1$.

Solution:

By definition:

$$\bar{f}(s) = L\{1\} = \int_0^\infty e^{-st} 1 \, dt \qquad (3.3)$$

which, upon performing the indicated integration, yields:

$$\bar{f}(s) = -\left.\frac{e^{-st}}{s}\right|_{t=0}^{t=\infty} \tag{3.4}$$

to give:

$$\bar{f}(s) = L\{1\} = \frac{1}{s} \tag{3.5}$$

Example 3.2   LAPLACE TRANSFORM OF SIMPLE FUNCTIONS.

Find the Laplace transform of the function $f(t) = e^{-at}$.

Solution:

In similar fashion we obtain, for the exponential function, that:

$$\bar{f}(s) = L\{e^{-at}\} = \int_0^\infty e^{-st}\, e^{-at}\, dt \tag{3.6}$$

which yields:

$$\bar{f}(s) = -\left.\frac{e^{-(s+a)t}}{s+a}\right|_{t=0}^{t=\infty} \tag{3.7}$$

and finally :

$$\bar{f}(s) = L\{e^{-at}\} = \frac{1}{s+a} \tag{3.8}$$

Similar results for any Laplace-transformable function can, in principle, be obtained in the same manner as in the foregoing examples. However, this is hardly ever done in practice. Firstly, note that because of the integration involved, the exercise can be quite tedious for all but the very simple functions. However, more important is the fact that extensive tables of various functions and their corresponding transforms have been compiled and are commonly available (see Appendix C). It is therefore more customary to make use of such tables.

Some Properties of the Laplace Transform

The following are certain properties of the Laplace transform that are of importance in Process Control applications.

1.   Not all functions $f(t)$ possess Laplace transforms. This is because the Laplace transform involves an indefinite integral whose value must be guaranteed to be finite if the Laplace transform is to exist. It is, in general, not possible to guarantee that the integral in Eq. (3.1) will be finite for all possible functions $f(t)$.

2. We note, however, that all the functions of interest in our study of Process Dynamics and Control are such that they possess Laplace transforms.

3. The Laplace transform $\bar{f}(s)$ contains no information about the behavior of $f(t)$ for $t < 0$ since the integration involved is from $t = 0$. This is, however, not a problem for our application since t usually represents time and we are normally interested in what happens at time $t > t_0$ where t_0 is some initial time that can be arbitrarily set to zero. (As will be seen later, defining our process variables such that $f(t) = 0$ for $t < 0$ is a common feature in Process Dynamics and Control.)

4. The function $f(t)$ and its corresponding transform $\bar{f}(s)$ are said to form a *transform pair*. A most important property of Laplace transforms is that *transform pairs are unique*. This implies that no two distinct functions $f(t)$ and $g(t)$ have the same Laplace transform. The usefulness of this property will become clearer later.

5. The Laplace transform operation is a linear one, by which we mean that for two constants c_1 and c_2, and two (Laplace-transformable) functions $f_1(t)$ and $f_2(t)$, the following is true:

$$L\{c_1 f_1(t) + c_2 f_2(t)\} = c_1 L\{f_1(t)\} + c_2 L\{f_2(t)\} \tag{3.9}$$

The validity of this assertion is easily proven (see Problem 3.2.)

3.3.2 Some Useful Results

The utility of the Laplace transform rests on the following collection of results. (Proofs and detailed discussions are taken up in Appendix C.)

The Inverse Transform

It should not be difficult to see that for the Laplace transform to be useful, we must have a means by which the original function $f(t)$ may be recovered if we are given the transform $\bar{f}(s)$.

This is accomplished, *in principle*, by the following expression known as the Laplace inversion formula:

$$f(t) = L^{-1}\{\bar{f}(s)\}$$

$$= \frac{1}{2\pi j} \int_C e^{st} \bar{f}(s) ds \tag{3.10}$$

a complex contour integral over the path represented by C, called the Bromwich path (see Appendix C).

However, just as was the case with the transform formula itself, this inversion formula is hardly ever used in practice. Owing to the very important fact that transform pairs are unique, tables of such transform pairs can be used not only to obtain the transform $\bar{f}(s)$ given $f(t)$, but also to obtain the inverse transform, $f(t)$, from a given $\bar{f}(s)$. We shall demonstrate precisely how this is done shortly.

The Transform of Derivatives

Under conditions that are always satisfied for most of the Process Control systems of practical interest, the Laplace transform of the derivative of a given function $f(t)$ (whose transform $\bar{f}(s)$ is known) is given by the following expression:

$$L\left\{\frac{df(t)}{dt}\right\} = s\ \bar{f}(s) - f(t)\Big|_{t=0} \tag{3.11}$$

Thus if we agree to designate the initial condition of the *time* function $f(t)$ as $f(0)$, it is more convenient to recast Eq. (3.11) as:

$$L\left\{\frac{df(t)}{dt}\right\} = s\ \bar{f}(s) - f(0) \tag{3.12}$$

The transforms of higher derivatives follow directly from here:

$$L\left\{\frac{d^2f}{dt^2}\right\} = s^2\ \bar{f}(s) - sf(t)\Big|_{t=0} - \frac{df(t)}{dt}\Big|_{t=0} \tag{3.13}$$

If we now designate the initial condition on the first derivative as:

$$\frac{df(t)}{dt}\Big|_{t=0} = f'(0) \tag{3.14}$$

we have that:

$$L\left\{\frac{d^2f}{dt^2}\right\} = s^2\ \bar{f}(s) - sf(0) - f'(0) \tag{3.15}$$

Similarly:

$$L\left\{\frac{d^3f}{dt^3}\right\} = s^3\ \bar{f}(s) - s^2f(0) - sf'(0) - f''(0) \tag{3.16}$$

In general:

$$L\left\{\frac{d^nf}{dt^n}\right\} = s^n\ \bar{f}(s) - s^{n-1}f(0) - s^{n-2}f'(0) - \ldots - f^{(n-1)}(0) \tag{3.17}$$

or

$$L\left\{\frac{d^n f}{dt^n}\right\} = s^n \bar{f}(s) - \sum_{k=0}^{n-1} s^k f^{(n-1-k)}(0) \tag{3.18}$$

where $f^{(i)}(0)$ denotes the value of the ith derivative of $f(t)$ evaluated at $t = 0$.

Let us note here that, if the initial conditions on $f(t)$ and all its derivatives are zero, then Eq. (3.18) reduces to:

$$L\left\{\frac{d^n f}{dt^n}\right\} = s^n \bar{f}(s) \tag{3.19}$$

a very important result indeed.

The Transform of Integrals

If the Laplace transform of a function $f(t)$ is known to be $\bar{f}(s)$, then the Laplace transform of the integral of this function is given by:

$$L\left\{\int_0^t f(t')\, dt'\right\} = \frac{1}{s}\,\bar{f}(s) \tag{3.20}$$

Shift Properties of the Laplace Transform

1. Shift in s

If $L\{f(t)\} = \bar{f}(s)$, then:

$$L\{e^{at}f(t)\} = \bar{f}(s-a) \tag{3.21}$$

implying that the effect of multiplying a function by e^{at} is that the s variable in its transform is shifted by a units, as indicated in Eq. (3.21).

For example, recall from Eq. (3.5) that we obtained the Laplace transform of $f(t)=1$ as $1/s$. We could now use this fact and the shift property indicated above to obtain $L\{e^{-at}\}$ directly as being $1/(s + a)$ without having to carry out the integration as was done in Eqs. (3.6)–(3.8).

The main utility of this particular shift property, however, lies in carrying out inversions of transforms. Thus a more useful version of Eq. (3.21) is:

$$L^{-1}\{\bar{f}(s-a)\} = e^{at}f(t) \tag{3.22}$$

The implication is that if the transform pair $[f(t), \bar{f}(s)]$ is known, then to find the inverse of a transform involving a shift in s by a, we merely multiply $f(t)$ by e^{at}.

An example will serve to illustrate the application of this property.

Example 3.3

Given that the Laplace transform of $f(t) = 1$ is:

$$\bar{f}(s) = L\{1\} = \frac{1}{s}$$

find

$$L^{-1}\left\{\frac{1}{s+a}\right\}$$

Solution:

From Eq. (3.22) we obtain immediately that:

$$L^{-1}\left\{\frac{1}{s+a}\right\} = e^{-at}\cdot 1$$

$$= e^{-at} \tag{3.23}$$

2.   Shift in Time

If $L\{f(t)\} = \bar{f}(s)$, then it is possible to find the Laplace transform of $f(t-a)$, a time-shifted version of $f(t)$ (see Figure 3.2), in terms of $\bar{f}(s)$ as follows:

$$L\{f(t-a)\} = e^{-as}\,\bar{f}(s) \tag{3.24}$$

Thus the presence of an exponential function of the Laplace variable s in a transform indicates that the inverse transform will contain a "shifted" time argument. We shall encounter such transforms and the physical systems with which they are associated in Chapter 8.

Situations do arise in practice in which one needs to find, from $\bar{f}(s)$, only the initial, or the final value of $f(t)$, and not the entire function itself. It is possible to do this by making use of the following results.

The Initial-Value Theorem

$$\lim_{t \to 0} [f(t)] = \lim_{s \to \infty} [s\,\bar{f}(s)] \tag{3.25}$$

The Final-Value Theorem

Provided $s\,\bar{f}(s)$ does not become infinite for any value of s in the right half of the complex s-plane (see Appendix C), then the final value of $f(t)$ is finite and may be obtained from $f(s)$ as follows:

$$\lim_{t \to \infty} [f(t)] = \lim_{s \to 0} [s\,\bar{f}(s)] \tag{3.26}$$

If the above stated conditions are *not* satisfied, the implication is that $f(t)$ has no limit as $t \to \infty$.

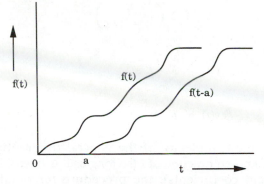

Figure 3.2.   The time-shifted function.

3.3.3   Application to the Solution of Differential Equations

Consider the problem of finding the solution to the simple differential equation:

$$\tau \frac{dy}{dt} + y = K u(t) \tag{3.27}$$

for the situation where $u(t) = 1$, and $y(0) = 0$. Now, because of the uniqueness of transform pairs, equality of two functions implies equality of their transforms. Hence we can take Laplace transforms of both sides of Eq. (3.27) to obtain an *equivalent* equation in the Laplace variable s, i.e.:

$$\tau s \, \bar{y}(s) + \bar{y}(s) = \frac{K}{s} \tag{3.28}$$

since $u(t) = 1$, and therefore $\bar{u}(s) = 1/s$ (recall Eq. (3.5)).

   Observe that Eq. (3.28) is now an algebraic equation that can be solved easily for $\bar{y}(s)$. The result is:

$$\bar{y}(s) = \frac{K}{s(\tau s + 1)} \tag{3.29}$$

If we now know of a function $y(t)$ whose transform is as given in Eq. (3.29), the problem will be solved. The next step is therefore that of obtaining $y(t)$, the inverse transform of the $\bar{y}(s)$ function given in Eq. (3.29).

   We note, firstly, that Eq. (3.29) may be rewritten in the entirely equivalent form:

$$\bar{y}(s) = \frac{K}{s} - \frac{K\tau}{(\tau s + 1)} \tag{3.30}$$

that is, the right-hand side of Eq. (3.29) has been broken into two *partial fractions*. By applying the linearity property of Laplace transforms we know that Eq. (3.30) is the Laplace transform of the sum of two functions, the first being K, a constant. By recalling the shift property, and noting that the second function can be expressed as:

$$\left(\frac{K}{s + 1/\tau}\right)$$

we obtain that:

$$y(t) = K - K e^{-t/\tau}$$

or

$$y(t) = K (1 - e^{-t/\tau}) \tag{3.31}$$

Regardless of the complexity of the general differential equation in the variable y as an unknown function of t (so long as it is linear, and of the ordinary type, with constant coefficients), the procedure for obtaining the solution illustrated by the preceding simple example may be generalized and summarized as follows:

1. Take the Laplace transform of both sides of the equation. (Note that by the definition of the Laplace transform of a derivative of any order, the initial conditions are automatically included.)
2. Solve the resulting algebraic equation for the Laplace transform of the unknown function, i.e., $\overline{y}(s)$.
3. Obtain $y(t)$ by inversion of the solution obtained for $\overline{y}(s)$ in Step 2. This step usually involves breaking up the $\overline{y}(s)$ function into simpler functions (whose inverse transforms are more easily recognizable) by a procedure known as partial fraction expansion. (See Appendix C for a detailed discussion of how partial fraction expansions are carried out.)

Let us illustrate the above procedure with the following example.

Example 3.4   SOLUTION OF LINEAR ODE'S BY LAPLACE TRANSFORMS.

Solve the following differential equation:

$$y''(t) + 5y'(t) + 6y(t) = f(t) \tag{3.32}$$

by the method of Laplace transforms given the following conditions:

$$f(t) = 1; \; y(0) = 1, \; y'(0) = 0$$

Solution:

Taking Laplace transforms of both sides of the equation gives:

$$\left[s^2 \, \overline{y}(s) - sy(0) - y'(0)\right] + 5 \left[s \, \overline{y}(s) - y(0)\right] + 6 \, \overline{y}(s) = \frac{1}{s}$$

and upon introducing the initial conditions and rearranging, we have:

$$(s^2 + 5s + 6) \, \overline{y}(s) = \frac{1}{s} + (s + 5)$$

Solving this equation for $y(s)$ gives:

$$\bar{y}(s) = \frac{1 + s(s + 5)}{s(s^2 + 5s + 6)} \tag{3.33}$$

By partial fraction expansion of the right-hand side, expression (3.33) becomes:

$$\bar{y}(s) = \frac{1/6}{s} + \frac{5/2}{s + 2} - \frac{5/3}{s + 3} \tag{3.34}$$

Inverting Eq. (3.34) finally leads to the required solution:

$$y(t) = \frac{1}{6} + \frac{5}{2} e^{-2t} - \frac{5}{3} e^{-3t} \tag{3.35}$$

3.3.4 Main Process Control Application

The main application of the Laplace transform in Process Control is intimately related to the application discussed in the last subsection — solving differential equations. This is illustrated in Figure 3.3.

The original problem, formulated in the time domain in terms of the time variable t, is usually easier to solve after it has been transformed to the s-domain, in terms of the Laplace variable s. Since the solution of the transformed problem is related to the solution of the original time-domain problem by Laplace transformation, the final step involves recovering the desired time-domain solution from the Laplace domain by the inverse Laplace transform operation.

A significant proportion of the dynamic analysis to be carried out in Part II will be carried out, as indicated in Figure 3.3, in the Laplace domain making use of a particular process model form (the transfer function model) which will be discussed in detail in the next chapter.

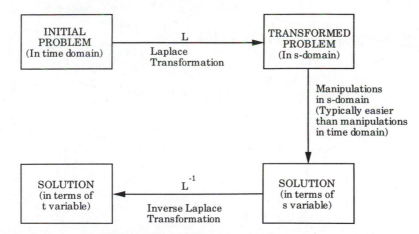

Figure 3.3. The application of the Laplace transform to differential equation

3.4   CHARACTERISTICS OF IDEAL FORCING FUNCTIONS

As noted earlier, the analysis of dynamic process behavior is carried out by investigating the process response forced by the application of certain well-characterized input functions. The characteristics of the various types of "forcing functions" typically used in dynamic analysis will be discussed in this section.

In the upcoming chapters, we will find that the dynamic analysis of various types of processes is carried out almost entirely in the Laplace domain. In anticipation of this, therefore, the Laplace transforms of these forcing functions will also be given.

The functions we shall be concerned about are:

1. The step function
2. The rectangular pulse function
3. The impulse function
4. The ramp function
5. The sinusoidal function

3.4.1   The Ideal Step Function

The ideal step function of magnitude A, as illustrated in Figure 3.4, is a function whose value remains at zero until it takes on the value of A instantaneously at the "starting time" set arbitrarily to $t = 0$.

Such a function is represented mathematically as:

$$u(t) = \begin{cases} 0; & t < 0 \\ A; & t > 0 \end{cases} \qquad (3.36)$$

or in terms of the unit step function, $H(t)$ (also called the Heaviside function):

$$u(t) = AH(t), \qquad (3.37a)$$

where the Heaviside function is defined as:

$$H(t) = \begin{cases} 0; & t < 0 \\ 1; & t > 0 \end{cases} \qquad (3.37b)$$

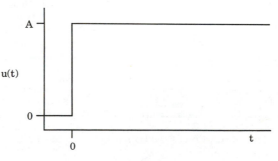

Figure 3.4.   The ideal step function.

The Laplace transform of the step function is given by:

$$\bar{u}(s) = L\{u(t)\} = \frac{A}{s} \tag{3.38}$$

A step function may be realized in practice by implementing a sudden change in the position of the actuator (e.g., control valve, etc.) which will correspond to a change of magnitude A in the input variable. For example, such a step change may be realized in a liquid flow process by making a sudden change in the valve position which will yield a change in liquid flowrate of A units.

It should be clear that the step function, as represented here, is idealized and cannot, in general, be realized *exactly* in practice.

3.4.2 The Ideal Rectangular Pulse Function

The ideal rectangular pulse function of magnitude A, and duration b time units is illustrated in Figure 3.5. It is a function whose value remains at zero until it takes on the value of A instantaneously at the "starting time" $t = 0$, maintains this new value for precisely b time units, and returns to its initial value thereafter.

Mathematically, this function is represented as:

$$u(t) = \begin{cases} 0; & t < 0 \\ A; & 0 < t < b \\ 0; & t > b \end{cases} \tag{3.39}$$

In finding the Laplace transform of this function, however, it is more instructive to consider $u(t)$ as a combination of two step functions: one starting at $t = 0$, and the other starting at $t = b$ (i.e., shifted by b time units). Recalling the definition of the Heaviside function, $H(t)$, it is easy to see that the rectangular pulse function is given by:

$$u(t) = A[H(t) - H(t - b)] \tag{3.40}$$

where, of course, $H(t - b)$ is understood to be the unit step function delayed by b time units so that its value remains at zero until $t = b$, i.e.:

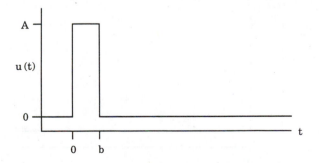

Figure 3.5.   The rectangular pulse function.

$$H(t - b) = \begin{cases} 0; & t < b \\ 1; & t > b \end{cases} \tag{3.41}$$

Taking Laplace transforms of Eq. (3.40), and recalling the shift properties of Laplace transforms discussed in the preceding section, we immediately obtain:

$$\bar{u}(s) = \frac{A}{s}(1 - e^{-bs}) \tag{3.42}$$

as the transform of the ideal rectangular pulse function illustrated in Figure 3.5.

Just as with the step function, the ideal rectangular pulse function is difficult to realize perfectly in actual practice. The procedure for its realization, however, involves making an instantaneous change in actuator position, maintaining this change for a prespecified time period (b time units), and then returning to the initial state.

3.4.3   The Ideal Impulse Function

The ideal impulse function of magnitude A is a function represented mathematically as:

$$u(t) = A\,\delta(t) \tag{3.43}$$

where $\delta(t)$ is the "Dirac delta function" illustrated in Figure 3.6, an unusual function that is somewhat easier to define in terms of what it *does* rather than what it *is*.

However, for our purposes, it is *convenient* (if not entirely rigorous) to define it as a "function" whose value is zero elsewhere and *infinite* at the point $t = 0$, and yet has the property that the total area under the "curve" is 1, i.e.:

$$\delta(t) = \begin{cases} \infty; & t = 0 \\ 0; & elsewhere \end{cases} \tag{3.44}$$

and

$$\int_{-\infty}^{+\infty} \delta(t)\,dt = \int_{-\varepsilon}^{+\varepsilon} \delta(t)\,dt = 1 \tag{3.45}$$

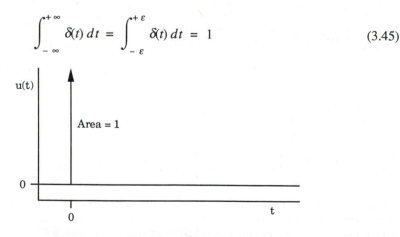

Figure 3.6.   The Dirac delta function.

The impulse function of area A (given by Eq. (3.43)) has the Laplace transform:

$$\bar{u}(s) = L\{u(t)\} = A \tag{3.46}$$

The following are some properties of the Dirac delta function:

1. It can be obtained as the limit of the rectangular pulse function of the previous subsection. Let $A = 1/b$ so that the area of the rectangle, i.e., $A \cdot b$, is 1. In the limit as $b \to 0$ (and therefore $A \to \infty$) while maintaining the constant "rectangular" area of unity the result is the Dirac delta function; i.e., the Dirac delta function is a rectangular pulse function of zero width and unit area. We will make use of this property in subsequent chapters.

2. For any function $f(t)$:

$$\int_{t_0 - \varepsilon}^{t_0 + \varepsilon} \delta(t - t_0)\, f(t)\, dt = f(t_0) \tag{3.47}$$

where $\delta(t - t_0)$ is the general, "shifted" delta function whose value is zero everywhere except at the point $t = t_0$. (Note that the function defined in Eq. (3.44) is obtained by setting $t_0 = 0$.) This property may be used, for example, to establish Eq. (3.46).

3. If $H(t)$ represents the unit step function, then:

$$\frac{d}{dt}[H(t)] = \delta(t) \tag{3.48}$$

i.e., the Dirac delta function is the derivative of the unit step (or Heaviside) function.

In terms of realization on a physical process, it is of course obvious that it is impossible to implement the impulse function exactly. However, a pulse function implemented over as short a time interval as possible will constitute a good approximation, especially for processes with slow dynamics.

3.4.4   The Ideal Ramp Function

The ideal ramp function, as illustrated in Figure 3.7, is a function whose value starts at 0 (at $t=0$) and increases linearly with a constant slope which has been designated as A.

Mathematically, this is represented as:

$$u(t) = \begin{cases} 0; & t < 0 \\ At; & t > 0 \end{cases} \tag{3.49}$$

Observe that this function can be obtained by integrating the step function of magnitude A. Alternatively a first derivative of the ramp function in Eq. (3.49) gives a step function of magnitude A.

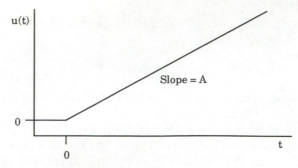

Figure 3.7.   The ramp function.

The Laplace transform of this ramp function is given by:

$$\bar{u}(s) \; = \; L\{u(t)\} \; = \frac{A}{s^2} \tag{3.50}$$

To implement a ramp function as an input to a physical system requires a systematic, graduated change in actuator position in such a way that the rate of change (increase or decrease) is constant. We must note, however, that even though the mathematical representation of this type of input indicates a function with no limiting value as $t \rightarrow \infty$, in real processes, physical limits are naturally imposed. For example, in implementing a ramp input in the inlet flowrate of a process flow system, we can only increase the flowrate up to the value obtained when the valve is fully open; beyond this no flowrate increase is physically possible.

3.4.5   The Ideal Sinusoidal Function

The sinusoidal input function of amplitude A and frequency ω is illustrated in Figure 3.8.

It is represented mathematically as:

$$u(t) \; = \; \begin{cases} 0; & t < 0 \\ \\ A \sin \omega t; & t > 0 \end{cases} \tag{3.51}$$

The Laplace transform is given by:

$$\bar{u}(s) \; = \; L\{u(t)\} \; = \; \frac{A\omega}{s^2 + \omega^2} \tag{3.52}$$

As might be expected, it is quite difficult to implement a perfect sinusoidal input function in practice. However, the theoretical response of a process system to a sinusoidal input provides valuable information that is useful not just for process dynamic studies, but also for designing effective controllers.

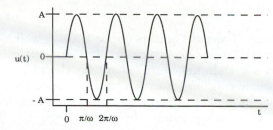

Figure 3.8. The sinusoidal function.

3.4.6   Realization of Ideal Forcing Functions: An Example

We now illustrate how these ideal forcing functions might be implemented on real processes by considering the following example system, the stirred heating tank first introduced in Example 1.1 of Chapter 1 (see Figure 3.9).

Here, the input variable of interest is the steam flowrate. How the various forcing functions we have discussed in this section will be implemented on this particular process will now be summarized.

1. **STEP:** Open the steam valve a given percentage at $t = 0$ such that Q changes by A units.

2. **PULSE:** Open the steam valve at $t = 0$, hold at the new value for a duration of b time units, and return to the old value.

3. **IMPULSE:** (Impossible to realize perfectly.) Open the steam valve (wide open) at $t = 0$ and instantaneously (or as soon as physically possible thereafter) return to the initial position.

4. **RAMP:** Gradually open the steam valve such that Q increases linearly. Ramp ends when the steam valve is fully open.

5. **SINUSOID:** The only practical way to realize this is to connect a sine wave generator to the steam valve. Realizing higher frequency sinusoidal inputs may be limited by the valve dynamics.

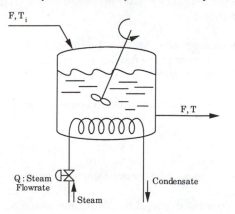

Figure 3.9.   The stirred heating tank.

3.5   SUMMARY

A description of *Process Dynamics* and its importance, as well as an introduction to the basic elements that are necessary for dynamic analysis, are the issues that have occupied our attention in this chapter. The idea of using a process model and ideal forcing functions for such studies was introduced. The role of the digital computer in the actual carrying out of the numerical and graphical aspects of dynamic analysis was briefly discussed.

The ideal forcing functions that are typically employed in dynamic analysis studies and their unique characteristics have been introduced. In later chapters we shall see how they are employed and the sort of information that can be extracted from the responses forced by each of these input functions.

The importance of several indispensable mathematical tools for dynamic analysis has been stressed. Even though a detailed treatment of most of these tools was deferred until later in the book, the Laplace transform and its usefulness in analyzing dynamic models were discussed.

REVIEW QUESTIONS

1. What is the essence of dynamic analysis?

2. Why is a process model a useful tool of dynamic analysis?

3. In addition to a process model, what else is needed in order to carry out successful dynamic analysis?

4. The following mathematical tools:

    - Differential equations
    - Laplace transforms
    - Numerical methods
    - z-transforms
    - Matrices

    are each useful for what aspects of process dynamic analysis?

5. In what sense was the digital computer presented in this chapter as an indispensable tool of dynamic analysis?

6. What do you think is responsible for some functions not being Laplace transformable; and can you guess why most of the functions of practical interest in process dynamics and control problems are in fact Laplace transformable?

7. Why is it not possible to use the final-value theorem in determining the "final value" of the function $f(t)$ given that:

$$f(s) = \frac{K}{s - a}$$

8. What is the most common method for obtaining inverse Laplace transforms?

9. Why is the Laplace transform useful in solving linear ODE's?

10. Outline the main application of the Laplace transform in process control.

11. Of all the ideal forcing functions discussed in Section 3.4, which do you think is most easily implementable in practice?

12. If your objective is to cause the least possible upset in your process, which of the ideal forcing functions should you employ in an experimental study?

PROBLEMS

3.1   By carrying out the indicated integration explicitly, use the definition of the Laplace transform given in Eq. (3.1) to establish that:

(i) $$L\{\cos \omega t\} = \frac{s}{s^2 + \omega^2}$$

(ii) $$L\{\sin \omega t\} = \frac{\omega}{s^2 + \omega^2}$$

3.2   Use the definition given in Eq. (3.1) to establish the linearity property given in Eq. (3.9).

3.3   By expressing $\cos \omega t$ and $\sin \omega t$ respectively as:

(1) $$\cos \omega t = \frac{1}{2}\left(e^{j\omega t} + e^{-j\omega t}\right)$$

(2) $$\sin \omega t = \frac{1}{2j}\left(e^{j\omega t} - e^{-j\omega t}\right)$$

use the linearity property of Laplace transforms given in Eq. (3.9) along with the fact that:

$$L\{e^{-at}\} = \frac{1}{s + a} \tag{P3.1}$$

to reestablish the results of Problem 3.1.

3.4   By differentiating the function:

$$f(t) = t^n \tag{P3.2}$$

n times with respect to t, and noting that:

$$\Gamma(n + 1) = n(n - 1)(n - 2) \dots 1 \tag{P3.3}$$

(when n is an integer) use the results on the transform of derivatives in Section 3.3.2 to establish that:

$$L\{t^n\} = \frac{\Gamma(n + 1)}{s^{n + 1}} \tag{P3.4}$$

3.5 Given that the Laplace transform of a certain process response to a unit step input is given by:

$$y(s) = \frac{K(\xi s + 1)}{s(\tau s + 1)} \tag{P3.5}$$

by using the initial- and final-value theorems, or otherwise, find:

(1) $\lim\limits_{t \to 0} y(t)$, the initial value of the response, and

(2) $\lim\limits_{t \to \infty} y(t)$, the ultimate, steady-state value of the response.

Under what condition will the initial value of the response be greater than the final value? (The process in question will be studied more fully in Chapter 5.)

3.6 Find $\lim\limits_{t \to \infty} y(t)$, for the function whose Laplace transform is given as:

$$y(s) = \frac{3(s + 2)(s - 2)}{5s^4 + 6s^3 + 2s^2 + 3s} \tag{P3.6}$$

3.7 The input function $u(t)$ shown in Figure P3.1

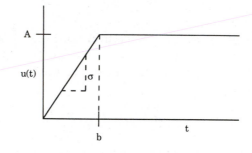

Figure P3.1.

may be represented mathematically as:

$$u(t) = \begin{cases} 0 & t < 0 \\ \sigma t & 0 \leq t < b \\ A & t \geq b \end{cases} \tag{P3.7}$$

Find $u(s)$, its Laplace transform.

CHAPTER
4

THE PROCESS MODEL

The idea of utilizing a collection of mathematical equations as a "surrogate" for a physical process is at once ingenious as well as expedient. More importantly, however, this approach has now become indispensable to proper analysis and design of process control systems.

The *concept* of representing process behavior with purely mathematical expressions, its benefits and shortcomings, the basic principles of how such mathematical models are obtained, the various forms they can assume, as well as the interrelationships between these various model forms, are some of the issues to be discussed in this chapter.

The most important model form in Process Dynamics and Control studies still remains the *transfer function* form. We will therefore examine transfer functions more closely towards the end of the chapter, and discuss some fundamental principles of how transfer functions are used for dynamic analysis.

4.1 THE MATHEMATICAL DESCRIPTION OF CHEMICAL PROCESSES

As stated earlier on in Chapter 3, the design of effective controllers is significantly facilitated by a proper understanding of process dynamics. This understanding in turn has been shown to be dependent on the availability of a *process model*; and process models are most effectively couched in the language of mathematics. It is thus fairly accurate to say that the first step in the analysis and/or design of a control system is the development of an appropriate mathematical model.

The *mathematical model* is, in principle, a collection of mathematical relationships between process variables which purports to describe the behavior of a physical system. The main use of the mathematical model is therefore as a convenient "surrogate" for the physical system, making it possible to investigate system response under various input conditions both rapidly, and inexpensively, without necessarily tampering with the actual physical entity.

We must, however, exercise caution in distinguishing between the usually quite complex physical entity that exists in the "real world" — that we have referred to as a *process* — and its mathematical description, called a *process model*. The latter at best captures only the salient features of the former, and should never be thought of as being its *exact equivalent*.

This fact notwithstanding, mathematical descriptions in the form of models have provided very valuable insight into the behavior of processes in general, making it possible to characterize efficiently the behavior of a wide variety of actual processes.

The extent of the usefulness of these mathematical descriptions is underscored by the fact that it is now customary to classify chemical processes according to the *nature* of the models used for their mathematical description. Thus a process described by linear equations is classified as *linear*, while the *nonlinear* process is the one described by nonlinear equations.

Furthermore, a *lumped parameter process* (which may be linear or not) is one in which, physically, its process variables change only with time but not with spatial position. They are described by ordinary differential equations. The process variables of a *distributed parameter process* on the other hand change with spatial position as well as with time.

The mathematical model representing a process is comprised of a system of equations, and thus, strictly speaking, the term "system" refers to the process model rather than the process itself. However, in the jargon of engineering practice, the term "system" is used to denote the process itself as well as its model. Thus in the discussion to follow, the term "system" will refer to both the process and its model.

Let us examine all these issues a little more closely.

4.1.1   Process Characteristics and Process Models

The terminology used in characterizing processes on the basis of their inherent characteristics, and the types of mathematical representation that best describe them, will now be discussed. It is important to note that in what follows, whenever the terms *dependent* and *independent* variables are used, they should be understood in the following context: All the *process variables* we identified in Chapter 1, be they state, output, input, or disturbance variables, *depend* on time and/or spatial position; they are therefore *all* considered as *dependent variables*. Time, and the spatial coordinate variables, are the *independent* variables.

Linear and Nonlinear Systems

A system described by linear equations (i.e., equations containing only linear functions) is said to be linear. The system which is *not* linear is said to be nonlinear. The issues involved with this classification will be further elaborated on in Chapter 10.

For now, we just note that most real chemical processes exhibit nonlinear *behavior* to a greater or lesser extent. It turns out, however, that some of the observed nonlinear behavior can indeed be effectively approximated by linear equations. Such systems are then usually classified as *linear*, but only for convenience; such classifications should, of course, not be taken too rigidly.

Lumped Parameter Systems

There are processes (linear or not) in which the dependent variables (i.e., all the process variables) may be considered, for all intents and purposes, as being *uniform* throughout the entire system, varying only with time. As will soon be shown, the theoretical models for such systems naturally occur as *ordinary differential equations*, with time as the only independent variable.

These processes are referred to as *lumped (parameter) systems*, since, in a sense, the dependency of all the observed variations have been *lumped* into one single dependent variable, i.e., time.

The *order* of the differential equation that describes the dynamic behavior of a lumped parameter system is used for further classification. A process described by a *first-order* differential equation is called a *first-order* system, and, in general, the nth-order system is a process described by an nth-order differential equation or by a system of n first-order differential equations.

Distributed Parameter Systems

When the variables of a process vary from point to point within the system in addition to varying with time, the fundamental mathematical description must now take the form of *partial differential equations* in order that the additional variation with position be accounted for properly.

Such processes are called *distributed (parameter) systems*. There are more independent variables, and the observed process variations are "distributed" among them.

Discrete-Time Systems

Even though the tacit assumption is that the variables of a process do not ordinarily change in "jumps," that they normally behave as smooth *continuous* functions of time (and position), there are situations in which output variables are deliberately *sampled* — and control action implemented — only at discrete points in time. The process variables now appear to change in a piecewise constant fashion with respect to the now discretized time. Such processes are modeled by *difference equations* and are referred to as *discrete-time systems*.

4.1.2 Various Forms Of Process Models

As noted earlier, the process model is a collection of mathematical relationships between the process variables that were classified in Chapter 1 as state, input, and output variables. Those mathematical models in which the state variables occur *explicitly* along with the input and output variables are called *state-space models*. As we will soon demonstrate, when the process model is formulated from first principles, it often naturally occurs in the state-space form in the time domain.

The mathematical models that, on the other hand, strictly relate only the input and output variables — entirely excluding the state variables — are called *input/output models*. As opposed to the state-space models which usually occur only in the time domain, input/output models can occur in the Laplace or z-transform domain, or in the frequency domain, as well as in the time domain. In the Laplace or z-transform domain, these input/output models usually occur in what is known as the "transform-domain transfer function" form; in the frequency domain, they occur in the "frequency-response" (or complex variable) form; and in the time domain, they occur in the "impulse-response" (or convolution) form.

In general, input/output models occur as a result of appropriate transformations of the state-space form, but they can also be obtained directly from input/output data correlation.

Thus, it is usual to cast the mathematical model for any particular process in one of four ways:

1. The state-space (differential or difference equation) form
2. The transform-domain (Laplace or z-transforms) form
3. The frequency-response (or complex variable) form
4. The impulse-response (or convolution) form

Each of these model forms and their applications will be discussed in more detail in the sections to follow.

Because these model types are obviously interrelated, it is possible to convert from one form to another. For Process Dynamics and Control applications, this is quite advantageous. Later on in Sections 4.3 through 4.6 when we examine the salient features of each model form, the importance of being able to switch from one model to the other will be better appreciated.

Before examining each model form critically and discussing the interrelationships between them, let us first consider the important issue of *how* process models are formulated.

4.2   FORMULATING PROCESS MODELS

The theoretical principles of how process models used in Process Dynamics and Control studies are formulated will now be introduced.

4.2.1   The General Conservation Principle

The basis for virtually all theoretical process models is the general conservation principle which, in essence, states that:

> *What remains accumulated within the boundaries of a system is the difference between what was added to the system and what was taken out, plus what was generated by internal production.*

While this may appear so straightforward, almost to the point of being simplistic, it is nevertheless the basis for a countless number of mathematical models.

The conservation principle written as an equation is:

$$\text{Accumulation} = \text{Input} - \text{Output} + \text{Internal Production} \tag{4.1}$$

It is often more appropriate to present the conservation principle in terms of *rates*, in which case, we have that:

The rate of accumulation of a conserved quantity q within the boundaries of a system is the difference between the rate at which this quantity is being added to the system and the rate at which it is being taken out plus the rate of internal production.

That is:

$$
\begin{array}{cccccc}
\text{Rate} & & \text{Rate} & \text{Rate} & \text{Rate} \\
\text{of} & = & \text{of} & - & \text{of} & + & \text{of} \\
\text{Accumulation} & & \text{Input} & \text{Output} & \text{Production} \\
\text{of q} & & \text{of q} & \text{of q} & \text{of q}
\end{array}
\qquad (4.2)
$$

4.2.2 Conservation of Mass, Momentum, and Energy

Despite the fact that there is a wide variety of chemical process systems, the fundamental quantities that are being conserved in all cases are either *mass*, or *momentum*, or *energy*, or combinations thereof. Thus, the application of the conservation principle specifically to the conservation of mass, momentum, and energy provides the basic blueprint for building the mathematical models of interest. It is customary to refer to the mathematical equations obtained using Eq. (4.2) as mass, momentum, or energy balances, depending on which of these quantities *q* represents. Such *balances* could be made over the entire system, to give "overall" or *macroscopic* balances, or they could be applied to portions of the system of differential size, giving "differential" or *microscopic* balances.

We shall now give the equations to be used in carrying out mass, momentum, and energy balances over a system's "volume element," be it microscopic or macroscopic.

Mass Balance

The mass balance equation may be with respect to the mass of individual components in a mixture (a component mass balance), or with respect to total mass. The general form of the equation is no different from Eq. (4.2), i.e.:

$$
\begin{array}{ccccc}
\text{Rate of} & & \text{Rate of} & & \text{Rate of} \\
\text{Accumulation} & = & \text{Input} & + & \text{Generation of} \\
\text{of } \mathbf{mass} & & \text{of } \mathbf{mass} & & \mathbf{mass}
\end{array}
$$

$$
\begin{array}{ccccc}
 & & \text{Rate of} & & \text{Rate of} \\
 & - & \text{Output} & - & \text{Depletion of} \\
 & & \text{of } \mathbf{mass} & & \mathbf{mass}
\end{array}
\qquad (4.3)
$$

If there are *n* components in a mixture, then *n* component mass balances of this form will give rise to *n* equations, one for each component. If one also formulates a total mass balance, it is important to remember that only *n* of the resulting *n* + 1 equations are independent.

It should be stressed that the *total* mass balance equation will not have any generation or depletion terms; these will always be zero. The reason is that even though the mass of a component within the system may change (primarily because of chemical reaction) the total mass within the system will be constant; whatever disappears in one component appears as another component, since matter can never be totally created or totally destroyed in chemical process systems that operate under nonrelativistic conditions.

Momentum Balance

The general form of a momentum balance equation is as follows:

$$
\begin{array}{ccccc}
\text{Rate of} & & \text{Rate of} & & \text{Rate of} \\
\text{Accumulation} & = & \text{Input} & + & \text{forces acting} \\
\text{of } \mathbf{momentum} & & \text{of } \mathbf{momentum} & & \text{on volume element}
\end{array}
$$
$$\tag{4.4}$$
$$
\begin{array}{ccc}
 & \text{Rate of} \\
- & \text{Output} \\
 & \text{of } \mathbf{momentum}
\end{array}
$$

where now we have paid due respect to the fact that any "generation" of momentum must be due to forces acting on the volume element, and that this is done at a rate equal to the total sum of these forces.

It is important to note that this equation is in fact just a generalization of Newton's law of motion which states that force is the product of mass and acceleration. For systems in which mass could be variable, it is more accurate to express force as the rate of change of momentum. Thus, Eq. (4.4) could also be considered as an equation of force balance. Since force is a vector quantity, possessing both magnitude as well as direction, as discussed further in Bird *et al.* [1], a complete force (or momentum) balance will consist of three such equations written for each of the three directions in a three-dimensional space. For our current purposes, we note that force (or momentum) balances are very rarely the bases for modeling process control systems; therefore we will not pursue this aspect further.

Energy Balance

Modifying Eq. (4.2) specifically for energy conservation gives rise to the following general energy balance equation:

$$
\begin{array}{ccccc}
\text{Rate of} & & \text{Rate of} & & \text{Rate of} \\
\text{Accumulation} & = & \text{Input} & + & \text{Generation of} \\
\text{of } \mathbf{energy} & & \text{of } \mathbf{energy} & & \mathbf{energy}
\end{array}
$$
$$\tag{4.5}$$
$$
\begin{array}{ccccc}
 & & \text{Rate of} & & \text{Rate of expenditure} \\
 & - & \text{Output} & - & \text{of } \mathbf{energy} \text{ via work} \\
 & & \text{of } \mathbf{energy} & & \text{done on surrounding}
\end{array}
$$

The important point to note about this equation is that it is easily recognizable as an equivalent statement of the first law of thermodynamics.

4.2.3 Constitutive Equations

The next step in formulating the mathematical description of a process system involves the introduction of *explicit* expressions for the rates that appear in these balance equations. More often than not, these expressions are based on basic physical and chemical laws, and the underlying mechanisms by which changes within the process are presumed to occur. Fortunately, these so-called constitutive laws are fairly well known in quite a wide variety of cases and are typically discussed in undergraduate courses in Transport Phenomena, Kinetics and Reactor Design, Thermodynamics, etc.

The most widely used of these constitutive equations include:

1. Equations of Properties of Matter

Basic definitions of mass, momentum, and energy in terms of physical properties such as density, specific heat capacities, temperatures, etc.

2. Transport Rate Equations:

- Newton's law of viscosity (for momentum transfer)
- Fourier's heat conduction law (for heat transfer)
- Fick's law of diffusion (for mass transfer)

3. Chemical Kinetic Rate Expressions

Based on:
- The law of mass action
- Arrhenius expression of temperature dependence in reaction rate constants

4. Thermodynamic Relations

- Equations of state
  (e.g., ideal gas law, van der Waal's equation)
- Equations of chemical and phase equilibria

Perhaps the best way to illustrate the foregoing discussion is by presenting some examples.

4.2.4   Some Examples of Theoretical Modeling

The Stirred Heating Tank

The process shown in Figure 4.1 is the stirred heating tank we encountered earlier on in Chapters 1 and 3. The purpose here is to develop a mathematical model for this process.

The Process

Liquid at temperature T_i flowing into the tank at a *volumetric flowrate F*, is heated by steam flowing through the steam coil arrangement at a rate Q (mass/time). The heated fluid, now at temperature T, is withdrawn at the same volumetric rate as in the inlet stream. The tank volume is V; the liquid density and specific heat capacity are, respectively, ρ, and C_p. The latent heat of vaporization of steam is λ.

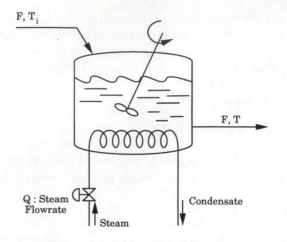

Figure 4.1. The stirred heating tank.

Assumptions

1. We will assume that the content of the tank is well mixed so that the liquid temperature in the tank and the temperature of liquid in the outlet stream are the same.

2. The physical properties ρ, C_p, and λ do not vary significantly with temperature.

3. All the heat given up by the steam (through condensation) is received by the liquid content of the tank; i.e., no heat from the steam is accumulated in the coils.

4. Heat losses to the atmosphere are negligible.

We are now in a position to obtain a mathematical model for this process. As was discussed in the preceding subsection, we do this by carrying out mass and energy balances, and introducing appropriate constitutive equations.

Overall Mass Balance

Rate of mass accumulation within the tank: $\dfrac{d}{dt}(\rho V)$
Rate of mass input to the tank : ρF

Rate of mass output from the tank: ρF

There is neither generation nor consumption of material in this process. Thus from Eq. (4.3) the overall mass balance equation is:

$$\frac{d}{dt}(\rho V) \;=\; \rho F - \rho F \tag{4.6}$$

and on the basis of the assumption of constant density ρ, we obtain:

$$\frac{dV}{dt} = 0 \tag{4.7}$$

implying that:

$$V = \text{constant} \tag{4.8}$$

Thus the material balance equation merely tells us that we have a *constant volume* process. This should not come as a surprise; it is in perfect keeping with common sense that if inlet flowrate equals outlet flowrate then the volume within the system must remain unchanged.

Overall Energy Balance

For this system let us note that even though the total energy is a sum of the internal, potential, and kinetic energies, the rate of accumulation of energy will involve only the rate of change of internal energy. This is for the simple reason that the tank is not in motion, thus eliminating the involvement of any kinetic energy terms; and since there is also no change in the position of the tank, the rate of change of potential energy will be zero. One final point: for liquid systems, the rate of change of total enthalpy is a good and convenient approximation for the rate of change of internal energy.

Let us now proceed with the energy balance. (We shall use the variable T^* to represent a reference temperature at which the specific enthalpy of the liquid is taken to be zero. As we shall see, this is only for exactness; the variable cancels out in the long run.)

Rate of accumulation of energy : $\rho C_p \dfrac{d}{dt}[V\,(T - T^*)]$
(within the tank)

Rate of heat input: $\rho F C_p\,(T_i - T^*)$
(from inlet stream)

Rate of heat input: λQ
(through steam heating)

Rate of heat output: $\rho F C_p\,(T - T^*)$
(through outlet stream)

Thus, from Eq. (4.5) the energy balance equation is:

$$\rho C_p \frac{d}{dt}[V\,(T - T^*)] = \rho F C_p\,(T_i - T^*) + \lambda Q - \rho F C_p\,(T - T^*) \tag{4.9}$$

which simplifies to:

$$\rho V C_p \frac{dT}{dt} = \rho F C_p\,(T_i - T) + \lambda Q \tag{4.10}$$

since, by the mass balance, we know that V is constant. The equation may be further simplified to:

$$\frac{dT}{dt} = -\frac{1}{\theta}T + \frac{\lambda}{\rho VC_p}Q + \frac{1}{\theta}T_i \qquad (4.11)$$

where $\theta = V/F$ is the residence time.

Equation (4.11) is a *differential equation* model that represents the time variation of the temperature within the stirred heating tank as a function of inlet temperature, rate of steam heating, tank residence time, and other physical parameters.

The Model in Terms of Deviation Variables

Under steady-state conditions, nothing changes with time within the system any longer. As such, time derivatives in the process models vanish, and Eq. (4.11) becomes:

$$0 = -\frac{1}{\theta}T_s + \frac{\lambda}{\rho VC_p}Q_s + \frac{1}{\theta}T_{is} \qquad (4.12)$$

where the subscript s has been introduced to designate the steady-state values of the variables in question.

   If we now define the following variables, in terms of deviations from the steady-state values indicated above:

$$x = T - T_s; \qquad u = Q - Q_s; \qquad d = T_i - T_{is}$$

and assuming that the tank temperature is available for measurement directly, then subtracting Eq. (4.12) from Eq. (4.11) gives:

$$\frac{dx}{dt} = -\frac{1}{\theta}x + \beta u + \frac{1}{\theta}d \qquad (4.13a)$$

with

$$y = x \qquad (4.13b)$$

where $\beta = \lambda/\rho VC_p$ and $x(0) = 0$ is the proper initial condition if $T = T_s$ at $t = 0$. Note that Eqs. (4.13a,b) are a state-space model representation of the heating tank, where in this case the measured output y is equal to the state x from Eq. (4.13b).

The Transform-Domain Model

One of the main advantages of rewriting the mathematical model in Eq. (4.11) in terms of deviation variables is that if we now consider the process to be initially at steady state, then the initial conditions on the deviation variables used in Eq. (4.13) will all be zero; and Laplace transformation can be done without carrying along extraneous terms that have to do with initial conditions.

Taking Laplace transforms of Eq. (4.13) then gives:[†]

$$sy(s) = -\frac{1}{\theta}y(s) + \beta u(s) + \frac{1}{\theta}d(s)$$

This is now a *transform-domain model* for the dynamic response of the stirred mixing tank, an algebraic equation that may now be rearranged to give an expression for $y(s)$ in terms of its dependence on $u(s)$ and $d(s)$, i.e.:

$$y(s) = \left(\frac{\beta\theta}{\theta s + 1}\right) u(s) + \left(\frac{1}{\theta s + 1}\right) d(s) \qquad (4.14)$$

If we now introduce the following definitions:

$$g(s) = \frac{\beta\theta}{\theta s + 1} \qquad (4.15a)$$

$$g_d(s) = \frac{1}{\theta s + 1} \qquad (4.15b)$$

then Eq. (4.14) becomes:

$$y(s) = g(s)u(s) + g_d(s)\, d(s) \qquad (4.16)$$

The expressions in Eqs. (4.15a,b) are called *transfer functions* (see Section 4.8 for a detailed discussion) and Eq. (4.14) is a *transfer function model* for the stirred heating tank. Equation (4.16) is the general expression for transfer function models in the transform domain.

The Frequency-Response Model

If we make a variable change $s = j\omega$ (where ω is a frequency and $j = \sqrt{-1}$) in Eqs. (4.15) and (4.16), the Laplace transform, $g(s)$, becomes a Fourier transform, $g(j\omega)$, and we can convert the transform-domain model to a frequency-response model. In this case the model becomes:

$$g(j\omega) = \frac{\beta\theta}{j\theta\omega + 1} = \frac{\beta\theta(1 - j\omega\theta)}{(\theta\omega)^2 + 1} \qquad (4.17a)$$

$$g_d(j\omega) = \frac{1}{j\theta\omega + 1} = \frac{1 - j\omega\theta}{(\theta\omega)^2 + 1} \qquad (4.17b)$$

$$y(j\omega) = g(j\omega)\, u(j\omega) + g_d(j\omega)\, d(j\omega) \qquad (4.18)$$

Thus $g(j\omega)$, $g_d(j\omega)$ are functions of a complex variable having both real and imaginary parts; for example, the real and imaginary parts of $g(j\omega)$ are:

$$\text{Re}_g(\omega) = \frac{\beta\theta}{(\theta\omega)^2 + 1}, \quad \text{Im}_g(\omega) = \frac{-\beta\omega\theta^2}{(\theta\omega)^2 + 1}$$

[†] Note that here and from now on in this book we have dropped the overbar on the Laplace transform for notational convenience.

Equation (4.17) can be interpreted as the frequency response of the process. If one stored as data the real and imaginary parts of g over the frequency range of relevance, these data files, $g(j\omega)$, would constitute the frequency-response model for the process. Equation (4.18) is also a *transfer function* model, but this time in the frequency domain. An expanded discussion of this type of model can be found in Chapter 9.

The Impulse-Response Model

The process model as presented in Eq. (4.13) is a linear, first-order differential equation which is easily solved analytically for $y(t)$, for any arbitrary value of the input function $u(t)$, and for the disturbance function $d(t)$. Applying the methods described in Appendix B, the solution is easily shown to be:

$$x(t) = e^{-t/\theta}x(0) + \int_0^t e^{-(t-\sigma)/\theta}\beta u(\sigma)\,d\sigma + \int_0^t e^{-(t-\sigma)/\theta}\frac{1}{\theta}d(\sigma)\,d\sigma$$

which becomes, upon introducing Eq. (4.13b), and as a result of the zero initial condition on $x(t)$:

$$y(t) = \int_0^t e^{-(t-\sigma)/\theta}\beta u(\sigma)\,d\sigma + \int_0^t e^{-(t-\sigma)/\theta}\frac{1}{\theta}d(\sigma)\,d\sigma \qquad (4.19)$$

This equation is an alternative mathematical model for the stirred heating tank, equivalent in every respect to the original differential equation model (4.13) from which it was derived.

It is now important to make the following observations:

1.   Let us obtain the inverse Laplace transforms of the transform-domain transfer functions given in Eqs. (4.15a,b). The result is:

$$g(t) = \beta e^{-t/\theta} \qquad (4.20a)$$

$$g_d(t) = \frac{1}{\theta}e^{-t/\theta} \qquad (4.20b)$$

2.   This being the case, Eq. (4.19) becomes:

$$y(t) = \int_0^t g(t-\sigma)u(\sigma)\,d\sigma + \int_0^t g_d(t-\sigma)d(\sigma)\,d\sigma \qquad (4.21)$$

3.   Because the Laplace transform of the unit impulse function $\delta(t)$ is 1 (see Chapter 3), then from the transform-domain model form in Eq. (4.16), the response of this process to the input $u(t) = \delta(t)$, (with $d(t) = 0$) will simply be:

$$y(t) = g(t)$$

and, similarly, the response to $d(t) = \delta(t)$ (with $u(t) = 0$) will be:

$$y(t) = g_d(t)$$

Thus, $g(t)$, $g_d(t)$ in Eq. (4.20) represent the response of this process to an impulse in the manipulated variable and disturbance respectively. It is for the foregoing reasons, therefore, that the model form in Eq. (4.20) is referred to as the *impulse-response* model form. As will be explained in Section 4.6.1, Eq. (4.21) is an *impulse-response* transfer function model.

4. One final point to note is that it is possible to obtain the impulse-response transfer function model in Eq. (4.21) directly from the transform-domain transfer function in Eq. (4.16) by taking inverse Laplace transforms directly. The important result we need to enable us carry out this task is what is known as the Convolution Theorem of Laplace transforms (see Appendix C) which states that:

For any two functions f(t) and g(t), if L{f(t)} = f(s) and L{g(t)} = g(s) then the inverse Laplace transform of the product {f(s)g(s)} is given by the following integral:

$$L^{-1}\{f(s)\,g(s)\} = \int_0^t g(t-\sigma)f(\sigma)\,d\sigma \qquad (4.22)$$

known as the **convolution integral**.

If we now apply this convolution theorem directly to Eq. (4.16), we obtain Eq. (4.21) immediately. This is why the impulse-response model form is sometimes referred to as the convolution model form.

The Nonisothermal CSTR

Our next example is a nonisothermal continuous stirred tank reactor (CSTR) shown in Figure 4.2, in which a first-order, irreversible chemical reaction A $\rightarrow$ B (with a reaction rate constant k) is taking place. Feed material containing c_{Af} moles/volume of A enters the reactor at temperature T_f, and constant volumetric flowrate F. Product is withdrawn from the reactor at the same volumetric flowrate F. The mixing is assumed to be efficient enough to guarantee homogeneity of the liquid content within the reactor. This means that composition and temperature are uniform throughout the reactor and that the exit temperature, T, and composition c_A, will be the same as within the reactor.

As in the previous example, the mathematical model for this process may be formulated by carrying out mass and energy balances, and introducing appropriate constitutive equations. In this particular case we shall, in principle, need component mass balances for each of the two components A and B, in addition. However, since matter can be neither created nor destroyed, the amount of B present in the reactor can always be deduced from the total amount of A in the feed and the unreacted amount in the reactor. Thus a separate component mass balance for B is superfluous.

The required mass and energy balances are now carried out in the same manner as in the last example.

Overall Mass Balance

Just as with the stirred heating tank of the previous example, all the overall mass balance tells us is that we have a constant volume system. (Note that once again, rate of material inflow exactly balances rate of outflow.)

Component Mass Balance on A

The mass balance equation applied to component A requires the following information:

Rate of accumulation of A within the reactor: $\dfrac{d}{dt}(Vc_A)$

Rate of A Input into the reactor: Fc_{Af}

Rate of A Output from the reactor: Fc_A

Rate of consumption of A
by first-order chemical reaction: kVc_A

Thus the required component A material balance equation is:

$$\frac{d}{dt}(Vc_A) = Fc_{Af} - Fc_A - kVc_A$$

By virtue of constant reactor volume, this may be simplified to:

$$\frac{dc_A}{dt} = -(k + 1/\theta)c_A + \frac{1}{\theta}c_{Af} \qquad\qquad (4.23)$$

where θ is the reactor residence time (V/F).

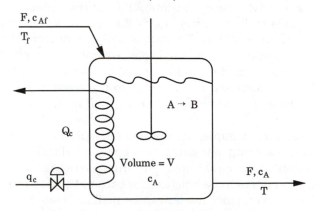

Figure 4.2. The nonisothermal CSTR.

Energy Balance

The following quantities are required for the energy balance:

Rate of energy accumulation in the reactor: $\rho C_p \dfrac{d}{dt}[V\,(T-T^*)]$

Rate of energy input into the reactor : $\rho F C_p\,(T_f-T^*)$

Rate of energy output from the reactor: $\rho F C_p(T-T^*)$

Rate of heat transfer to the cooling coil: Q_c

Rate of heat generation by chemical reaction: $(-\Delta H)kVc_A$

Here, the density and specific heat capacity of the reacting material, respectively represented by ρ and C_p, are assumed constant. T^* is the datum temperature introduced in the previous example; once again it does not show up in the final expression. The quantity $(-\Delta H)$ is the heat of reaction, with the convention that it is positive for exothermic reactions and negative for endothermic reactions.

Thus the energy balance equation is:

$$\rho C_p \frac{d}{dt}[V\,(T-T^*)] = \rho F C_p\,(T_f-T^*)-\rho F C_p\,(T-T^*)+(-\Delta H)kVc_A-Q_c \quad (4.24)$$

which may be simplified to:

$$\frac{dT}{dt} = -\frac{1}{\theta}T + \beta k c_A + \frac{1}{\theta}T_f - \chi \quad (4.25)$$

where $\beta = (-\Delta H)/\rho C_p$, and $\chi = Q_c/\rho C_p V$.

If we now introduce the Arrhenius expression for the temperature dependence of the reaction rate constant, the two equations that constitute the mathematical model for this process are now given from Eqs. (4.23) and (4.25) as:

$$\frac{dc_A}{dt} = -\frac{1}{\theta}c_A - k_0\,e^{-(E/RT)}c_A + \frac{1}{\theta}c_{Af} \quad (4.26)$$

$$\frac{dT}{dt} = -\frac{1}{\theta}T + \beta k_0\,e^{-(E/RT)}c_A + \frac{1}{\theta}T_f - \chi \quad (4.27)$$

It is important to note the following about these modeling equations:

1.   Both of these equations contain nonlinear functions of T and c_A.

2.   They are coupled, in the sense that it is not possible to solve one equation independently of the other.

3. As a result of the nonlinearities, it is not possible to take Laplace transforms of these equations. It is also impossible to obtain a closed form, analytical solution to these equations.

In closing, let us make the following observations:

1. Using the conservation principle along with constitutive equations has given rise to differential equation model forms. This will always be the case.

2. The equivalent transform-domain, frequency-response, or impulse-response versions of theoretical models are derived from the original differential equation representations: the first by Laplace transforms, the second by variable substitution, the last by solving analytically.

3. Only when the differential equation model is *linear* with constant coefficients can the equivalent transform-domain form be obtained by taking Laplace transforms; and only when analytical solutions are possible can the impulse-response form be derived from the differential equation model form.

The objective in this section has been merely to use some rather straightforward examples to do no more than illustrate the basic principles of *formulating* mathematical models. A broader spectrum of examples are given in Chapter 12, in Part III.

It should be noted that there are many more issues involved in the actual development of a working model for any particular process than have been discussed here. The two chapters (12 and 13) that make up Part III are devoted to a fuller treatment of these and other issues having to do with process modeling and process identification.

We now turn our attention to some general discussions on each of the various process model forms, their individual salient features, and what each model form is best suited for.

4.3   STATE-SPACE MODELS

4.3.1   Continuous Time

For *continuous-time* processes, i.e., those whose variables are *not* sampled at discrete points in time, the *differential equation models* that relate the *state* variable to the input and output variables are referred to as *state-space models*. The origin of the term "state-space" dates back to the early beginnings of modern control theory where the ideas of a state variable were first elucidated. According to the original concept, the *state* of the process at any time is considered to belong to an *abstract*, n-dimensional space called the *state space*, hence the term.

Typical state-space model forms for different classes of systems are given below.

Linear, Lumped Parameter, Single-Input, Single-Output Process

A typical linear, lumped parameter system having one input variable $u(t)$, one output variable $y(t)$, and *one* state variable $x(t)$ is modeled by the following differential equation:

$$\frac{dx(t)}{dt} = ax(t) + bu(t) + \gamma d(t) \qquad (4.28)$$

$$y(t) = cx(t)$$

This is the model for a *first-order* system because the describing equation is a *first-order* differential equation. A good example is the stirred heating tank of the last section whose model was obtained in Eq. (4.13). Note that in this case, $a = (-1/\theta)$, $b = \beta$, $\gamma = (1/\theta)$, and $c = 1$. Other examples of models of this type will be discussed in Chapters 5 and 6.

It is quite possible for a system to have only one input, and one output but have multiple state variables. In this case, the typical model takes the following form:

$$\frac{dx(t)}{dt} = \mathbf{A}x(t) + \mathbf{b}u(t) + \gamma d(t) \qquad (4.29)$$

$$y(t) = \mathbf{c}^T\mathbf{x}$$

where vectors such as $\mathbf{x}$ have the form:

$$\mathbf{x} = \begin{bmatrix} x_1 \\ x_2 \\ . \\ . \\ . \\ x_n \end{bmatrix}$$

and the matrix $\mathbf{A}$ has the form:

$$\mathbf{A} = \begin{bmatrix} a_{11} & a_{12} & \cdots & a_{1n} \\ a_{21} & a_{22} & \cdots & a_{2n} \\ \cdots & \cdots & \cdots & \cdots \\ a_{n1} & a_{n2} & \cdots & a_{nn} \end{bmatrix}$$

Here the fact of the multiplicity of state variables is reflected in the presence of a state *vector* $\mathbf{x}(t)$, (not just a single state variable $x(t)$) accompanied by the matrix $\mathbf{A}$ and vectors $\mathbf{b}$, $\mathbf{c}$, and γ. This is the state-space representation for higher order single-input, single-output systems (see Chapter 6).

Linear, Lumped Parameter, Multivariable Process

The typical state-space model for the linear, lumped parameter system having multiple input and output variables is:

$$\frac{d\mathbf{x}(t)}{dt} = \mathbf{A}\mathbf{x}(t) + \mathbf{B}\mathbf{u}(t) + \mathbf{\Gamma}\mathbf{d}(t) \qquad (4.30)$$
$$\mathbf{y}(t) = \mathbf{C}\mathbf{x}(t)$$

where $\mathbf{A}$, $\mathbf{B}$, $\mathbf{C}$, and $\mathbf{\Gamma}$ are appropriately dimensioned system matrices. The brevity indicated by this model form (and the one in Eq. (4.29)) was achieved by employing the compact vector-matrix notation. (See Appendix D and Chapter 20.) It is fairly standard practice to use this vector-matrix notation in representing linear multivariable process models.

Nonlinear, Lumped Parameter Systems

Nonlinear lumped parameter systems having multiple-input and -output variables are represented in the following general compact form:

$$\frac{d\mathbf{x}(t)}{dt} = \mathbf{f}(\mathbf{x},\mathbf{u},\mathbf{d}) \qquad (4.31)$$
$$\mathbf{y}(t) = \mathbf{h}(\mathbf{x}(t))$$

where $\mathbf{f}(\cdot)$ and $\mathbf{h}(\cdot)$ are usually vectors of nonlinear functions. For single-input, single-output nonlinear systems, the vectors are simply replaced by single functions. In the example of the nonisothermal reactor of the preceding section, the state variables are c_A and T and the explicit forms of the nonlinear functions are shown in Eqs. (4.26) and (4.27). In this case, the model form in Eq. (4.31) is obtained with:

$$\mathbf{x} = \begin{bmatrix} c_A \\ T \end{bmatrix} = \begin{bmatrix} x_1 \\ x_2 \end{bmatrix}; \qquad u = \chi; \qquad \mathbf{d} = \begin{bmatrix} c_{Af} \\ T_f \end{bmatrix}$$

$$f = \begin{bmatrix} -\left[\dfrac{1}{\theta} + k_0 \exp\left(\dfrac{-E}{Rx_2} \right) \right] x_1 + \dfrac{1}{\theta} d_1 \\[4mm] \beta k_0 \exp\left(\dfrac{-E}{Rx_2} \right) x_1 - \dfrac{1}{\theta} x_2 + \dfrac{1}{\theta} d_2 - u \end{bmatrix}$$

Other examples of this class of models will be encountered in Chapter 10.

We will introduce time delays in Chapter 8, and there we will indicate how these state-space models can be augmented to include delays. A hint is given in Example 4.3 of this chapter.

Distributed Parameter Systems

A simple example of a distributed parameter system is the steam-heated shell-and-tube heat exchanger shown in Figure 4.3. As will be discussed further in Chapter 8, this particular system is modeled by the following equation:

$$\frac{\partial T}{\partial t} + v \frac{\partial T}{\partial z} = \frac{UA_s}{\rho A C_p} (T_s - T) \qquad (4.32)$$

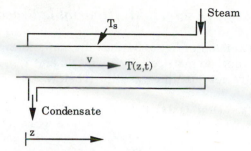

Figure 4.3. Steam-heated shell-and-tube heat exchanger.

where the fluid being heated is flowing inside the tube (cross-sectional area A) with constant velocity v along the length of the tube (the z-direction). The fluid density ρ and specific heat capacity C_p are assumed constant. The heat exchanger arrangement is assumed to have a heat transfer area A_s per unit length of tube, and U is the overall heat transfer coefficient. $T(z,t)$ is the fluid temperature profile as a function of time and position along the length of the tube; T_s is the steam temperature on the shell side.

A more general model form for distributed parameter systems is:

$$\frac{\partial \mathbf{x}}{\partial t} = \mathbf{f}(\mathbf{x}, \mathbf{u}, \nabla \bullet \mathbf{x}, \nabla^2 \mathbf{x}, \ldots) \tag{4.33}$$

Here, ∇ is recognized as the Laplacian operator, representing the vector of partial derivative operators in the spatial coordinate variables. The right-hand side represents a vector of general (possibly nonlinear) functions in the indicated arguments.

A full-scale discussion of the dynamic analysis and control of distributed parameter systems (DPS) is excluded from the intended scope of this book. Nevertheless, some example systems that fall under this category are treated later in Chapters 8 and 27.

4.3.2 Discrete Time

When process variables are available at discrete points in time, the state-space models take the form of difference equations. The following are typical discrete-time models.

Linear, Discrete, Single-Input, Single-Output Processes

In this case, $x(k)$, $u(k)$, $d(k)$, and $y(k)$ respectively represent the state, control, disturbance, and output variables at the discrete-time instant $t_k = k\Delta t$, where Δt is the sampling interval. Thus the model is

$$x(k + 1) = qx(k) + ru(k) + \psi d(k) \tag{4.34}$$

$$y(k) = cx(k)$$

Linear, Discrete, Multivariable Processes

The multivariable version of Eq. (4.34) has the same form, but with q, r, and ψ replaced with appropriate matrices, $\mathbf{Q}$, $\mathbf{R}$, and ψ, and x, u, d, and, y replaced by the corresponding vector quantities, as shown below:

$$\mathbf{x}(k+1) = \mathbf{Q}\mathbf{x}(k) + \mathbf{R}\mathbf{u}(k) + \psi\mathbf{d}(k) \qquad (4.35)$$
$$\mathbf{y}(k) = \mathbf{C}\mathbf{x}(k)$$

Nonlinear, Discrete Processes

The most general representation for the nonlinear discrete process takes the following form:

$$\mathbf{x}(k+1) = \mathbf{f}(\mathbf{x}(k), \mathbf{u}(k), \mathbf{d}(k)) \qquad (4.36)$$
$$\mathbf{y}(k) = \mathbf{h}(\mathbf{x}(k))$$

Again, $\mathbf{f}(\cdot)$ and $\mathbf{h}(\cdot)$ are usually vectors of nonlinear functions. A fuller discussion of discrete-time systems may be found in Chapters 23–26.

4.3.3   Uses of State-Space Models in Dynamic Analysis

Since the modeling equations are formulated with time as the independent variable, state-space models are most useful for obtaining real-time behavior of process systems. The discrete-time formulations are especially well suited to computer simulation of process behavior.

State-space models are also used, almost exclusively, for the analysis of nonlinear system behavior. This is because most of the other model forms can represent only linear dynamic behavior.

Another important use of state-space models is in representing processes in which some states are *not* measured. As we will soon see, all the other model forms are based on input/output relationships so that any aspect of the process not manifested as either input or measured output is not likely to be represented.

4.4   TRANSFORM-DOMAIN MODELS

4.4.1   Continuous Time

These models relate the process inputs (i.e., control and disturbance variables) to the process outputs according to the following algebraic equation which we introduced earlier in Eq. (4.16):

$$y(s) = g(s)\, u(s) + g_d(s)d(s) \qquad (4.37)$$

where $g(s)$ and $g_d(s)$ are referred to, respectively, as the *process* and *disturbance transfer functions* in the transform domain. Thus they are input/output models in the transform domain. The process transfer function for the stirred heating tank of Section 4.2, for example, is given in Eq. (4.15a), and the disturbance transfer function is given in Eq. (4.15b).

When the problem has multiple inputs and multiple outputs it is possible to define transform-domain matrices $\mathbf{G}(s)$, $\mathbf{G}_d(s)$ relating inputs and outputs as:

$$\mathbf{G}(s) = \begin{bmatrix} g_{11}(s) & g_{12}(s) & \cdots & g_{1m}(s) \\ g_{21}(s) & g_{22}(s) & \cdots & g_{2m}(s) \\ \cdots & \cdots & \cdots & \cdots \\ g_{n1}(s) & g_{n2}(s) & \cdots & g_{nm}(s) \end{bmatrix} ;$$

$$\mathbf{G}_d(s) = \begin{bmatrix} g_{d_{11}}(s) & g_{d_{12}}(s) & \cdots & g_{d_{1l}}(s) \\ g_{d_{21}}(s) & g_{d_{22}}(s) & \cdots & g_{d_{2l}}(s) \\ \cdots & \cdots & \cdots & \cdots \\ g_{d_{n1}}(s) & g_{d_{n2}}(s) & \cdots & g_{d_{nl}}(s) \end{bmatrix}$$

Thus, the multivariable transform-domain transfer function model relates the *vector* of input variables to the *vector* of output variables as follows:

$$\mathbf{y}(s) = \mathbf{G}(s)\,\mathbf{u}(s) + \mathbf{G}_d(s)\,\mathbf{d}(s) \tag{4.38}$$

where the *scalar* transfer functions of Eq. (4.37) have been replaced by *matrix transfer functions*.

As we briefly showed with the example of the stirred mixing tank, the transform-domain model representation arises, in theory, from Laplace transformation of the state-space model. Because Laplace transformation is a linear operation (see Appendix C), these models exist therefore only for linear systems. There is thus no easy way of representing nonlinear process systems by such transform-domain models.

4.4.2 Discrete Time

By direct analogy to the continuous-time models discussed above, the transform-domain model for discrete-time systems relates the *z-transform* of the *sampled* input signal to that of the output signal, as shown below:

$$\hat{y}(z) = \hat{g}(z)\,\hat{u}(z) + \hat{g}_d(z)\,\hat{d}(z) \tag{4.39}$$

Here $\hat{g}(z)$, $\hat{g}_d(z)$ are the z-transform transfer functions of the discrete-time system (see Chapters 24–25 for more detailed discussion). The multivariable version follows from Eq. (4.38) by direct analogy:

$$\hat{\mathbf{y}}(z) = \hat{\mathbf{G}}(z)\,\hat{\mathbf{u}}(z) + \hat{\mathbf{G}}_d(z)\,\hat{\mathbf{d}}(z) \tag{4.40}$$

where $\hat{\mathbf{G}}(z)$ and $\hat{\mathbf{G}}_d(z)$ are z-transform transfer function matrices.

4.4.3   Uses of Transform-Domain Models

The transform-domain transfer function model is used extensively for dynamic analysis and control systems design. In fact, the very concept of a transfer function (by no means unique to this model form alone) plays a very central role in process dynamics and control. As such we shall devote Section 4.8 to examining *transfer functions* more closely.

4.5   FREQUENCY-RESPONSE MODELS

4.5.1   Continuous Time

Recall from Section 4.2.4 that if we let $s = j\omega$, in the transform-domain models we convert them to frequency-response models of the form:

$$y(j\omega) = g(j\omega)\, u(j\omega) + g_d(j\omega)d(j\omega) \tag{4.18}$$

where $g(j\omega)$, and $g_d(j\omega)$ are frequency-response transfer functions. These are therefore input/output models in the frequency domain. In general:

$$g(j\omega) = \mathrm{Re}(\omega) + j\,\mathrm{Im}(\omega) \tag{4.41}$$

i.e., such functions will have a real part, $\mathrm{Re}(\omega)$, and an imaginary part, $\mathrm{Im}(\omega)$. Even if these are stored as numbers in a data record as a function of frequency, they still constitute the frequency-response transfer function although in numerical form.

   Just as in the transform domain, frequency-response models for multivariable systems may be formulated in terms of vectors and matrices. Such a multivariable model would take the form:

$$\mathbf{y}(j\omega) = \mathbf{G}(j\omega)\, \mathbf{u}(j\omega) + \mathbf{G}_d(j\omega)\, \mathbf{d}(j\omega) \tag{4.42}$$

More details of frequency-response models are discussed in Chapter 9.

4.5.2   Discrete Time

Again as with the transform-domain models, there is a discrete-time form of frequency-response models. If we let $z = e^{j\omega\Delta t}$ in the z-transform transfer functions (4.39) and (4.40) we obtain the frequency response of the discrete-time system:

$$\hat{y}(e^{j\omega\Delta t}) = \hat{g}(e^{j\omega\Delta t})\, \hat{u}(e^{j\omega\Delta t}) + \hat{g}_d(e^{j\omega\Delta t})\, \hat{d}(e^{j\omega\Delta t}) \tag{4.43}$$

As in the continuous case, any frequency-response transfer function $\hat{g}(e^{j\omega\Delta t})$ will have a real and imaginary part, i.e.:

$$\hat{g}(e^{j\omega\Delta t}) = \mathrm{Re}(\omega) + j\,\mathrm{Im}(\omega) \tag{4.44}$$

A data record of these as a function of frequency would constitute the numerical frequency-response transfer function of the sampled system.

4.5.3 Uses of the Frequency-Response Model

Frequency-response models are used primarily as nonparametric models for arbitrary dynamic processes. By taking experimental response data from pulse, sine wave or other inputs, the frequency response of the process is obtained directly as:

$$g(j\omega) \;=\; \mathrm{Re}(\omega) + j\,\mathrm{Im}(\omega) \qquad\qquad (4.41)$$

where $\mathrm{Re}(\omega)$, $\mathrm{Im}(\omega)$ are data records over a range of frequencies. The details and some examples of this procedure can be found in Chapter 13.

Frequency-response models are also used in control system analysis and design because, as we shall see later, a number of control system design procedures are based on the frequency domain.

4.6   IMPULSE-RESPONSE MODELS

4.6.1   Continuous Time

Consider the situation in which a process has its input in the form of a unit *impulse* function (one of the idealized, input functions used in dynamic analysis, which was discussed in Chapter 3). The response to this input is referred to as the *impulse response*, $g(t)$.

Once the impulse response of a process is known, it can be shown that the response of this process to *any* arbitrary input $u(t)$ is given by the convolution integral:

$$y(t) \;=\; \int_0^t g(t - \sigma)\, u(\sigma)\, d\sigma \qquad\qquad (4.45)$$

where σ is a dummy time argument which gets integrated out. This is called an *impulse-response model*. The general impulse-response transfer function model takes the form:

$$y(t) \;=\; \int_0^t g(t - \sigma)\, u(\sigma)\, d\sigma + \int_0^t g_d(t - \sigma)\, d(\sigma)\, d\sigma \qquad\qquad (4.21)$$

when $u(t)$, $d(t)$ are scalar quantities. These are input/output models in the time domain since they relate the inputs $u(t)$ and $d(t)$ to the output $y(t)$. We saw an example of this model form for the stirred heating tank in Section 4.2.

For multivariable systems, where $\mathbf{y}$, $\mathbf{u}$, $\mathbf{d}$ are vectors, if we define impulse-response matrices of the form:

$$
\mathbf{G}(t) = \begin{bmatrix} g_{11}(t) & g_{12}(t) & \cdots & g_{1m}(t) \\ g_{21}(t) & g_{22}(t) & \cdots & g_{2m}(t) \\ \cdots & \cdots & \cdots & \cdots \\ g_{n1}(t) & g_{n2}(t) & \cdots & g_{nm}(t) \end{bmatrix} ;
$$

$$
\mathbf{G}_d(t) = \begin{bmatrix} g_{d_{11}}(t) & g_{d_{12}}(t) & \cdots & g_{d_{1l}}(t) \\ g_{d_{21}}(t) & g_{d_{22}}(t) & \cdots & g_{d_{2l}}(t) \\ \cdots & \cdots & \cdots & \cdots \\ g_{d_{n1}}(t) & g_{d_{n2}}(t) & \cdots & g_{d_{nl}}(t) \end{bmatrix}
$$

then the multivariable impulse-response model is:

$$
\mathbf{y} = \int_0^t \mathbf{G}(t - \sigma)\,\mathbf{u}(\sigma)\,d\sigma + \int_0^t \mathbf{G_d}(t - \sigma)\,\mathbf{d}(\sigma)\,d\sigma \tag{4.46}
$$

These models are also called convolution models because they involve the convolution integral.

An important property of the convolution integral Eq. (4.45) concerns the interchangeability of the arguments associated with each function under the integral (see Appendix C), i.e.:

$$
\int_0^t g(\sigma)\,u(t - \sigma)\,d\sigma = \int_0^t g(t - \sigma)\,u(\sigma)\,d\sigma \tag{4.47}
$$

It is very useful in dealing with these models. By virtue of this relationship, the model in Eq. (4.21) may also be written as:

$$
y(t) = \int_0^t g(\sigma)u(t - \sigma)\,d\sigma + \int_0^t g_d(\sigma)d(t - \sigma)\,d\sigma \tag{4.48}
$$

An *entirely equivalent* representation — the *step-response* model form — uses the step response function, $\beta(t)$ (the response of the system to a *unit* step function) in place of the impulse-response function $g(t)$. It can be shown that these two functions are related by:

$$
\frac{d}{dt}[\beta(t)] = g(t) \tag{4.49a}
$$

or, conversely:

$$
\beta(t) = \int_0^t g(\sigma)\,d\sigma \tag{4.49b}
$$

The actual *step-response model* takes the form:

$$y(t) = \int_0^t \frac{d\beta(\sigma)}{d\sigma} u(t - \sigma)\, d\sigma \tag{4.50a}$$

$$= \int_0^t \beta(\sigma) \frac{du(t - \sigma)\, d\sigma}{d\sigma} \tag{4.50b}$$

4.6.2 Discrete Time

The discrete-time impulse-response model relates the sampled input $u(k)$ to the sampled output $y(k)$, using the discrete impulse-response function $g(k)$. By analogy with Eq. (4.45) and using Eq. (4.47) we obtain:

$$y(k) = \sum_{i=1}^k g(k - i)\, u(i) \tag{4.51a}$$

$$= \sum_{i=1}^k g(i)\, u(k - i) \tag{4.51b}$$

where the second form is used for computational convenience.

Thus the discrete impulse-response model takes the form:

$$y(k) = \sum_{i=1}^k g(i)\, u(k - i) + \sum_{i=1}^k g_d(i)\, d(k - i) \tag{4.52}$$

when y, u, d are scalars. By analogy with the continuous case, the discrete-time, multivariable impulse-response model is:

$$\mathbf{y}(k) = \sum_{i=1}^k \mathbf{G}(i)\mathbf{u}(k - i) + \sum_{i=1}^k \mathbf{G_d}(i)\mathbf{d}(k - i) \tag{4.53}$$

where

$$\mathbf{G}(i) = \begin{bmatrix} g_{11}(i) & g_{12}(i) & \cdots & g_{1m}(i) \\ g_{21}(i) & g_{22}(i) & \cdots & g_{2m}(i) \\ \cdots & \cdots & \cdots & \cdots \\ g_{n1}(i) & g_{n2}(i) & \cdots & g_{nm}(i) \end{bmatrix} \quad ;$$

$$\mathbf{G_d}(i) = \begin{bmatrix} g_{d_{11}}(i) & g_{d_{12}}(i) & \cdots & g_{d_{1l}}(i) \\ g_{d_{21}}(i) & g_{d_{22}}(i) & \cdots & g_{d_{2l}}(i) \\ \cdots & \cdots & \cdots & \cdots \\ g_{d_{n1}}(i) & g_{d_{n2}}(i) & \cdots & g_{d_{nl}}(i) \end{bmatrix}$$

Also the continuous forms Eqs. (4.49) and (4.50) become in the discrete case:

$$\beta(k) = \sum_{i=1}^{k} g(i) \tag{4.54}$$

and

$$g(k) = \beta(k) - \beta(k-1) \tag{4.55}$$

so that the discrete-time version of the step-response model is:

$$y(k) = \sum_{i=1}^{k} [\beta(i) - \beta(i-1)]u(k-i) \tag{4.56a}$$

$$= \sum_{i=1}^{k} \beta(i)\Delta u(k-i) \tag{4.56b}$$

where

$$\Delta u(k) = u(k) - u(k-1) \tag{4.57}$$

Note that it becomes merely a question of preference in deciding whether to use the impulse-response model form, or the step-response alternative since they are both entirely equivalent.

4.6.3 Uses of Impulse-Response Models

The impulse-response model finds its main application in dynamic analysis problems involving arbitrary input functions $u(t)$. The discrete impulse-response model is particularly useful for such problems because it requires only that the values of the input be available at discrete points in time; the particular functional form it takes is immaterial.

In building process models from experimental data sampled at an interval Δt, discrete impulse-response models are most useful because the model requires only a data record $g(k)$ from well-designed experiments. Observe that a very simple experiment of sending an impulse (or a unit step) function as input to the physical system will give this required impulse (or step) response data record. This may then be used for dynamic analysis through the convolution sum in Eq. (4.51).

At this point we should mention that the step function is easier to implement on a physical system than an impulse. However, Eq. (4.55) can be used to deduce the corresponding impulse-response function.

It is perhaps of interest to note one application of impulse-response models outside of process dynamics and control. The reader who is familiar with chemical reactor design will recognize the fact that models of this type are employed extensively in studying residence time distributions in chemical reactors.

4.7 INTERRELATIONSHIPS BETWEEN PROCESS MODEL FORMS

From the discussions in Sections 4.3 through 4.6, we may conclude the following about the principal uses of each model type in process dynamics and control.

1. **State Space**: real-time simulation of process behavior; nonlinear dynamic analysis.
2. **Transform domain**: (linear) dynamic analysis involving well-characterized input functions; control system design.
3. **Frequency response**: (linear) nonparametric models for processes of arbitrary mathematical structure; control system design.
4. **Impulse Response**: (linear) dynamic analysis involving arbitrary input functions; dynamic analysis and controller design for processes with unusual model structures.

Now, it is not unusual to find that a specific task will be easier to perform using a model form different from the one that is readily available. In such a case, it will be necessary to convert the available process model from its current form into the more useful form. Thus the relationships that exist between the various process model forms and the important issue of how to convert one model form to another will now be discussed. The reader may find Figure 4.4 a helpful roadmap of the discussion.

4.7.1 Continuous- and Discrete-Time Models

The basic difference between a continuous-time model and the discrete-time counterpart is that in the former representation, each variable is available *continuously* as a function of time, as opposed to being available only at discrete points in time in the latter representation. Continuous-time and discrete-time models are thus related by the process of discretization.

The conversion of continuous-time models to discrete-time form is achieved by discretizing the continuous process model. The reverse situation of converting from discrete to continuous time arises less frequently primarily because process models rarely originate in discrete form.

The mechanics of how to obtain discrete models by discretizing the continuous form and how to convert between various types of discrete models will be fully discussed and illustrated at a more appropriate stage in Chapters 23 and 24. For now, we shall focus on the interrelationships between continuous-time process model forms.

4.7.2 State-Space and Transform-Domain Models

Of all the interrelationships between process model forms, this is perhaps the most important. Most process models formulated from first principles occur in the state-space form, but the preferred form for dynamic analysis and controller design is the transform-domain transfer function form. However, after dynamic analysis and controller design, the task of control system simulation and real-time controller implementation are carried out in time, requiring the state-space form.

Observe, therefore, that a substantial amount of important process dynamics and control activities utilize the state-space and transform-domain transfer function model forms almost exclusively.

Now, whenever the state-space model is linear, the corresponding transform-domain form is obtained by Laplace transformation followed by simple algebraic manipulations to obtain the transfer functions. The reverse process of obtaining the state-space equivalent of a transform-domain model is called *realization*. Thus, state-space and transform-domain models are related by Laplace transformation and the reverse process of realization.

Let us illustrate the procedures for converting from one form to the other.

State Space to Transform Domain

The typical linear SISO system is modeled in the state space by Eq. (4.28), i.e.:

$$\frac{dx(t)}{dt} = ax(t) + bu(t) + \gamma d(t) \tag{4.28}$$

$$y(t) = cx(t)$$

Assuming the model has been presented in terms of deviation variables so that we have zero initial conditions, Laplace transformation gives:

$$sx(s) = ax(s) + bu(s) + \gamma d(s)$$

$$y(s) = cx(s)$$

which may be quite easily rearranged and solved for $y(s)$ in terms of $u(s)$ and $d(s)$ to give:

$$y(s) = \frac{bc}{s - a} u(s) + \frac{\gamma c}{s - a} d(s) \tag{4.58}$$

We may now compare this with the general transform-domain transfer function given in Eq. (4.37) to establish the following relationship between the variables $\{a,b,c,\gamma\}$ of the state-space model, and the transfer functions $g(s)$ and $g_d(s)$:

$$g(s) = \frac{bc}{s - a} \tag{4.59a}$$

$$g_d(s) = \frac{\gamma c}{s - a} \tag{4.59b}$$

As will be further illustrated in Chapter 20, the same procedure is used for multivariable systems, with scalars replaced by matrices. We note that the transformation from Eq. (4.28) above to Eqs. (4.58) and (4.59) is unique in the sense that for any given set of values $\{a,b,c,\gamma\}$, we can obtain only one set of transfer functions $g(s)$ and $g_d(s)$.

When the state-space model is *nonlinear*, only approximate transform-domain transfer function models obtained from linear approximations of the nonlinear models are possible. This is discussed in Chapter 10.

It should be reemphasized here that while the state variable $x(t)$ appears explicitly in the state-space model, it is entirely missing from the transform-domain transfer function representation. Thus the state-space model inherently contains more information than the equivalent transform-domain transfer function form.

Transform Domain to State Space

In converting from the transform-domain transfer function model to the state-space form, the issue is best thought of as that of recovering the "original" differential equation that "gave rise" to the given transform-domain transfer function (even though the transfer function may have been obtained directly from input/output plant data, independently of any such differential equation model).

There are several ways of obtaining such "realizations" of the transform-domain models. The general procedure is best illustrated using one of the most straightforward realization techniques on several examples.

Example 4.1 **REALIZATION OF A FIRST-ORDER TRANSFORM-DOMAIN TRANSFER FUNCTION.**

Given the following transfer function model:

$$y(s) \ = \ g(s)\, u(s) \tag{4.60a}$$

with

$$g(s) \ = \ \frac{K}{\tau s \ + \ 1} \tag{4.60b}$$

find the equivalent state-space representation.

Solution:

Introducing Eq. (4.60b) into Eq. (4.60a) and rearranging, the transform-domain model may be rewritten as:

$$(\tau s + 1)y(s) \ = \ Ku(s) \tag{4.61a}$$

From here it is now clear that the most obvious differential equation that would give rise to Eq. (4.61a) upon Laplace transform is:

$$\tau\frac{dy}{dt} + y(t) \ = \ Ku(t) \tag{4.61b}$$

with the initial condition, $y(0) = 0$. This may be rewritten as:

$$\tau\frac{dx}{dt} + x(t) \ = \ Ku(t) \tag{4.62a}$$
$$y(t) \ = \ x(t) \tag{4.62b}$$

This is therefore a state-space equivalent of the model given in Eq. (4.60). Observe, however, that for any arbitrary constant c, we may define:

$$y(t) = cx(t)$$

and by this device, rewrite Eq. (4.61b) as:

$$\tau \frac{dx}{dt} + x(t) = \frac{K}{c} u(t)$$

Thus, another acceptable state space equivalent of the model in Eq. (4.61a) will be:

$$\tau \frac{dx}{dt} + x(t) = \frac{K}{c} u(t) \tag{4.63a}$$

$$y(t) = cx(t) \tag{4.63b}$$

Thus for various values of the constant c, we are able to generate a corresponding number of equally acceptable state-space equivalents of Eq. (4.61a). Each of these equivalent forms will have the same input/output relation between $y(t)$ and $u(t)$, but will have different definitions of the state $x(t)$.

Some comments about this last example are in order here. Regarding the concept of realization, we note that:

1. Several, equally acceptable, state-space realizations are possible for the transfer function model in Eq. (4.60). Thus while the transformation from the state-space to the transform domain is unique, the reverse transformation back to the state space is not.

2. The nonuniqueness of the state-space realizations from the transform-domain model arises because the transform-domain representation inherently contains *less* information. The information about the state variables, missing from the transform-domain form but required in the state-space form, can be incorporated in an arbitrary number of equally acceptable ways as shown in Eq. (4.63).

3. In all realizations, the relationship between the input $u(t)$ and output $y(t)$ *is unique* even if the definition of the state variables is not.

Regarding the actual "mechanics" of the procedure we note:

1. Even though we utilized the inverse of the process of Laplace transformation in arriving at Eq. (4.62) (and Eq. (4.63)) from Eq. (4.60), the utility was only partial. A complete Laplace inversion would have provided the *solution* to the differential equation, not merely recover the differential equation in its unsolved form.

2. We may therefore consider the process of *realization* as a "reverse" Laplace transform, which produces differential equations from transform-domain transfer functions, not to be confused with the process of Laplace inversion (or inverse Laplace transformation) which provides the *solution* to the differential equation.

3. In realization, the desired end result is the (unsolved) differential equation itself; in Laplace inversion, it is the solution of the differential equation that is desired.

When the process model includes, in addition, a disturbance transfer function, the same basic principles of obtaining realizations apply. However, some additional considerations are called for. Let us illustrate with the following example.

Example 4.2   REALIZATION OF PROCESS AND DISTURBANCE TRANSFORM-DOMAIN TRANSFER FUNCTIONS.

Given the following transfer function model:

$$y(s) = g(s) u(s) + g_d(s) d(s) \qquad (4.64a)$$

with

$$g(s) = \frac{K}{\tau s + 1} \qquad (4.64b)$$

$$g_d(s) = \frac{K_d}{\tau_d s + 1} \qquad (4.64c)$$

find an equivalent state-space representation.

Solution:

Introducing Eqs. (4.64b,c) into Eq. (4.64a) the transfer function model may be rewritten as:

$$y(s) = x_1(s) + x_2(s) \qquad (4.65a)$$

where

$$x_1(s) = \frac{K}{\tau s + 1} u(s) \qquad (4.65b)$$

$$x_2(s) = \frac{K_d}{\tau_d s + 1} d(s) \qquad (4.65c)$$

If we now rearrange Eqs. (4.65b,c) as was done in Example (4.1) we will have:

$$(\tau s + 1)x_1(s) = Ku(s) \qquad (4.66a)$$

$$(\tau_d s + 1)x_2(s) = K_d d(s) \qquad (4.66b)$$

from where we may now deduce one possible set of differential equations which upon Laplace transformation would give rise to Eq. (4.66a,b) and which are, respectively:

$$\tau \frac{dx_1}{dt} + x_1(t) = Ku(t) \qquad (4.67a)$$

for the process, and

$$\tau_d \frac{dx_2}{dt} + x_2(t) = K_d d(t) \tag{4.67b}$$

for the disturbance. The process output is obtained as:

$$y(t) = x_1(t) + x_2(t) \tag{4.68}$$

Along with the initial conditions with $x_1(0) = 0$, and $x_2(0) = 0$, Eqs. (4.67a,b) and (4.68) constitute one possible state-space equivalent of the model given in Eq. (4.64.)

Remarks

In the special case for which $\tau = \tau_d$, we may add Eq. (4.66a) to Eq. (4.66b) to obtain (by means of Eq. (4.65a)):

$$(\tau s + 1)y(s) = Ku(s) + K_d d(s)$$

and the differential equation extracted from here will simply be:

$$\tau \frac{dy}{dt} + y(t) = Ku(t) + K_d d(t)$$

We may then set $y(t) = x(t)$ and obtain an equivalent state-space representation:

$$\tau \frac{dx}{dt} + x(t) = Ku(t) + K_d d(t)$$
$$y(t) = x(t)$$

In Chapter 8 we will discuss the dynamic behavior of a very important class of process systems called *time-delay* systems. The transform-domain transfer function for such a system has the term $e^{-\alpha s}$ in the numerator. The following example illustrates how to obtain a realization for such transfer functions involving time delays.

Example 4.3   REALIZATION OF A FIRST-ORDER-PLUS-TIME-DELAY TRANSFORM-DOMAIN TRANSFER FUNCTION.

Given the following transfer function model:

$$y(s) = g(s) u(s) \tag{4.69a}$$

with

$$g(s) = \frac{Ke^{-\alpha s}}{\tau s + 1} \tag{4.69b}$$

find an equivalent state-space representation.

Solution:

Again, introducing Eq. (4.69b) into Eq. (4.69a), the transfer function model may be rewritten as:

$$(\tau s + 1)y(s) \; = \; Ke^{-\alpha s}u(s) \tag{4.70}$$

To obtain a differential equation which, upon Laplace transform, would give rise to Eq. (4.69), we now only need to recall the time-shift property of Laplace transform given in Eq. (3.24) in Chapter 3. This, along with what we have seen earlier in Example 4.1, helps us to establish that one such differential equation is:

$$\tau\frac{dy}{dt} + y(t) \; = \; Ku(t - \alpha)$$

We may then set $y(t) = x(t)$ and rewrite this as:

$$\tau\frac{dx}{dt} + x(t) \; = \; Ku(t - \alpha) \tag{4.71a}$$
$$y(t) \; = \; x(t) \tag{4.71b}$$

and this will then provide a state-space equivalent of the model given in Eq. (4.69).

The issues involved in obtaining realizations for multivariable transform-domain transfer function matrix models will be investigated in Chapter 20.

4.7.3 State-Space and Frequency-Response Models

There is no strong analytical relationship between state-space and frequency-response models. To generate the frequency-response model, one must either go through the Laplace domain (as described below) or generate "data" from simulation of the state-space model with sine wave inputs.

State Space to Frequency Response

The easiest way to convert a linear state-space model to the frequency-response form is to use the following two-step procedure:

1. Convert to the transform-domain transfer functions $g(s)$, $g_d(s)$
2. Let $s = j\omega$ to produce $g(j\omega)$, $g_d(j\omega)$

This will produce an analytical functional form for the frequency-response model.

A second approach which is precise for linear systems and can be used in an approximate way even with nonlinear state-space models is to carry out pulse or sine wave response tests of the model and then use these "data" records to produce $g(j\omega)$, $g_d(j\omega)$. This approach requires no analytical solution and can be done numerically. The details of this procedure may be found in Chapter 13.

Frequency Response to State Space

There is no direct route to convert a frequency-response model to the state-space form. If the frequency-response model is in tractable analytical form, it is possible to convert to the transform domain and then perform a realization from the transform domain.

4.7.4   State-Space and Impulse-Response Models

The analytical, closed form solution of the differential equation in the state-space model (for arbitrary inputs $u(t)$ and $d(t)$) is what gives rise to the equivalent impulse-response form. Thus, the transformation from the state-space to the impulse-response form is only possible whenever analytical solutions can be obtained.

If the process of obtaining the analytical solution of a differential equation is conceived of in its most general sense as essentially involving the mathematical operation of "integration," then we would be led to the correct conclusion that the reverse process of generating a state-space model from the impulse-response form (whenever possible) involves differentiation.

Once again, let us illustrate the procedures for converting from the state-space to the impulse-response form, and vice versa.

State Space to Impulse Response

Recall that the typical linear SISO system is modeled in the state space by Eq. (4.28):

$$\frac{dx(t)}{dt} = ax(t) + bu(t) + \gamma d(t) \tag{4.28}$$
$$y(t) = cx(t)$$

The analytical solution to the linear, first-order differential equation in $x(t)$ is easily obtained as:

$$x(t) = \int_0^t e^{a(t-\sigma)}bu(\sigma)\,d\sigma + \int_0^t e^{a(t-\sigma)}\gamma\,d(\sigma)\,d\sigma \tag{4.72}$$

so that, multiplying by c, we obtain the expression for $y(t)$ in the required impulse-response transfer function form given in Eq. (4.21). The relationships between the variables $\{a,b,c,\gamma\}$ of the state-space model and the impulse-response functions $g(t)$ and $g_d(t)$ are easily seen to be:

$$g(t) = b\,c\,e^{at} \tag{4.73a}$$
$$g_d(t) = \gamma\,c\,e^{at} \tag{4.73b}$$

Impulse Response to State Space

The reverse process of starting from the impulse-response transfer function (Eq. (4.21))

$$y(t) = \int_0^t g(t-\sigma)u(\sigma)\,d\sigma + \int_0^t g_d(t-\sigma)d(\sigma)\,d\sigma \tag{4.21}$$

and deriving the corresponding state-space model is possible only if the impulse-response functions $g(t)$ and $g_d(t)$ are available in well-characterized and *differentiable* functional form.

Whenever this is the case, Eq. (4.21) may be differentiated to obtain the equivalent differential equation. This venture is usually carried out with the aid of Leibnitz's rule for differentiating under the integral sign:

$$\frac{d}{dx}\left(\int_{A(x)}^{B(x)} f(x, r)dr\right) = \int_{A}^{B} \frac{\partial f(x, r)}{\partial r} dr + f(x, B)\frac{dB}{dx} - f(x,A)\frac{dA}{dx} \qquad (4.74)$$

In most realistic cases, however, these impulse-response functions are available in numerical form only; as such, conversion to the state-space form is usually not possible.

Nevertheless, for the purpose of illustration, we will consider an example in which the impulse-response function is available as a well-characterized, differentiable function.

Example 4.4   OBTAINING A STATE-SPACE MODEL FROM AN IMPULSE-RESPONSE MODEL.

Find an equivalent state-space representation of the following impulse-response model:

$$y(t) = \frac{K}{\tau}\int_{0}^{t} e^{-(t-\sigma)/\tau} u(\sigma)\, d\sigma \qquad (4.75)$$

Solution:

Clearly, in this case, the impulse-response function $g(t)$ is given by:

$$g(t) = \frac{K}{\tau} e^{-t/\tau} \qquad (4.76)$$

which is differentiable. If we now differentiate Eq. (4.75) with respect to t (with the aid of Eq. (4.74)), we obtain:

$$\frac{dy}{dt} = \frac{K}{\tau}\int_{0}^{t} -\frac{1}{\tau} e^{-(t-\sigma)/\tau} u(\sigma)\, d\sigma + \frac{K}{\tau} u(t) - 0$$

which simplifies further (using Eq. (4.75) to eliminate the integral) to give:

$$\frac{dy}{dt} = -\frac{1}{\tau} y(t) + \frac{K}{\tau} u(t)$$

or

$$\tau\frac{dy}{dt} + y(t) = K u(t)$$

Again, we may now specify $y = x$, to obtain:

$$\tau\frac{dx}{dt} + x(t) = K u(t) \qquad (4.62a)$$

$$y(t) = x(t) \qquad (4.62b)$$

as required state-space representation.

It is important to recognize that the impulse-response model is a time-domain input/output model form that also does not explicitly contain information about the state variable $x(t)$; therefore as with the transform-domain model, conversions to the state space will not be unique.

4.7.5   Transform-Domain and Impulse-Response Models

Transform-domain and impulse-response models are *theoretically* quite closely related, because they are just representations of the *same* input/output model in different domains. The transform-domain model is the Laplace transform version of the impulse-response model; and conversely, the impulse-response model is the time-domain input/output version of the transform-domain model. The impulse-response function $g(t)$ is the inverse Laplace transform of the transform-domain transfer function $g(s)$; and $g(s)$ is the Laplace transform of the impulse-response function $g(t)$.

Transform Domain to Impulse Response

The conversion from the transform-domain to the impulse-response form involves Laplace inversion using the convolution integral. Observe that this merely requires replacing the product of two Laplace transforms $g(s)$ and $u(s)$ with the equivalent convolution integral in time.

Impulse Response to Transform Domain

In principle, the reverse process of converting the impulse-response form to the transform-domain transfer function form should require no more than replacing the convolution integral with the corresponding product of the Laplace transforms of the two functions contained in the integral. Since the starting point in this case is the impulse-response form, it should be clear that this conversion is possible only if the impulse-response function is Laplace transformable.

The following examples will illustrate these procedures.

> **Example 4.5**   **IMPULSE-RESPONSE FORM OF A FIRST-ORDER TRANSFORM-DOMAIN TRANSFER FUNCTION.**

Find the equivalent impulse-response representation of the following transform-domain transfer function model first given in Example 4.1:

$$y(s) \; = \; g(s) \, u(s) \qquad\qquad (4.60a)$$

where

$$g(s) \; = \; \frac{K}{\tau s \, + \, 1} \qquad\qquad (4.60b)$$

Solution:

In this case, by Laplace inversion of $g(s)$ we obtain the impulse-response function as:

$$g(t) = \frac{K}{\tau} e^{-t/\tau} \tag{4.76}$$

so that the required impulse-response model is given by the convolution integral:

$$y(t) = \frac{K}{\tau} \int_0^t e^{-(t-\sigma)/\tau} u(\sigma) \, d\sigma \tag{4.75}$$

obtained by taking the inverse Laplace transform of the transfer function model given in Eq. (4.60) and using the convolution theorem of Laplace transforms stated earlier on in Eq. (4.22).

We will now use a situation in which the impulse-response function is Laplace transformable to illustrate the reverse procedure.

Example 4.6   OBTAINING A TRANSFER-DOMAIN FUNCTION MODEL FROM AN IMPULSE-RESPONSE MODEL.

Find the equivalent transform-domain transfer function representation of the following impulse-response model:

$$y(t) = \frac{Kt}{\tau^2} \int_0^t e^{-(t-\sigma)/\tau} u(\sigma) \, d\sigma \tag{4.77}$$

Solution:

Observe that the impulse-response function in this case is:

$$g(t) = \frac{Kt}{\tau^2} e^{-t/\tau} \tag{4.78}$$

the Laplace transform of which is easily obtained as:

$$g(s) = \frac{K}{\left(\tau s + 1\right)^2} \tag{4.79}$$

The required transfer function model equivalent of Eq. (4.77) is therefore:

$$y(s) = \frac{K}{\left(\tau s + 1\right)^2} u(s) \tag{4.80}$$

In conclusion, we note that since we are concerned here with two fundamentally similar input/output models having the same information content, all the transformations involved are unique.

4.7.6   Transform-Domain and Frequency-Response Models

As already indicated several times, the conversion from the transform-domain to the frequency-response form merely requires the conversion from a Laplace transform to a Fourier transform by letting $s = j\omega$ in $g(s)$, $g_d(s)$. However, the reverse conversion, from the frequency-response form to the Laplace transform, can only be done when the transfer function, $g(j\omega)$, is analytically tractable so that reversing the substitution $j\omega = s$ can lead to a meaningful Laplace transform. This is impossible when the original model arises in nonparametric form from experimental data.

To illustrate, consider the heating tank of Section 4.2. The Laplace domain transfer functions relating the tank temperature to the tank inputs were determined to be:

$$g(s) = \frac{\beta\theta}{\theta s + 1} \tag{4.15a}$$

$$g_d(s) = \frac{1}{\theta s + 1} \tag{4.15b}$$

By making the substitution $s = j\omega$, we are able to convert to the frequency-response domain with transfer functions:

$$g(j\omega) = \frac{\beta\theta}{j\theta\omega + 1} = \frac{\beta\theta(1 - j\omega\theta)}{(\theta\omega)^2 + 1} \tag{4.17a}$$

$$g_d(j\omega) = \frac{1}{j\theta\omega + 1} = \frac{(1 - j\omega\theta)}{(\theta\omega)^2 + 1} \tag{4.17b}$$

Note that to begin with a model of the form (4.17), it is not enough to simply let $j\omega = s$ in order to recover the Laplace transform. It is necessary to recognize that everywhere in Eq. (4.17) where even powers of ω^{2n} appear, there is also a hidden factor $(-1)^n j^{2n}$. With simple rules such as this the reverse transformation is possible when the frequency-response model $g(j\omega)$ remains in analytical form. If the original model, $g(j\omega)$ arose from experimental data so that:

$$g(j\omega) = \text{Re}(\omega) + j\,\text{Im}(\omega) \tag{4.41}$$

was simply a nonparametric record of $\text{Re}(\omega)$, $\text{Im}(\omega)$ stored at a large number of frequencies, the transformation to $g(s)$ in analytical form is obviously impossible because the analytical dependence on ω is unknown.

4.7.7   Frequency Response and Impulse Response

As indicated earlier, it is impossible to convert a frequency-response model to the state-space form without using the reverse of the two-step procedure through the transform domain in the special case when that is possible. However, it is sometimes necessary to have a time-domain representation of frequency-response data; this is most conveniently done by transforming to an impulse-response model. The frequency-response model may be converted to an impulse-response model by using a computational form of the inverse Fourier Transform known as the Fast Fourier Transform (FFT) [2]. The impulse-response function, $g(t)$, may then be recovered numerically from such expressions as

$$g(t) = \frac{1}{\pi} \Delta\omega \sum_{n=1}^{N} g(j\omega_n) \cos\omega_n t \qquad (4.81))$$

where $\Delta\omega$ is the mesh size in frequency at which the data is stored. This $g(t)$ can then be used with the convolution integral to produce an impulse-response model of the form of Eq. (4.21).

In order to accomplish the reverse transformation from the impulse response to the frequency domain, there are two possibilities. When $g(t)$ is in a convenient analytical form, or when the numerical inversion methods of Chapter 13 apply, then the two-step procedure of obtaining the Laplace transform $g(s)$ and then making the substitution $s = j\omega$ will work. However, very often $g(t)$ is obtained directly from experiment as a data record stored at a large number of times. In this case, it is usually better to take the Fourier transform numerically, to generate $g(j\omega)$. A form of the expression often used in practice (see Chapter 13) is:

$$g(j\omega) = \int_0^{t_f} g(t) \cos\omega t \, dt - j \int_0^{t_f} g(t) \sin\omega t \, dt \qquad (4.82)$$

Thus there are reliable practical means of relating the frequency response and impulse response of a process even when the models consist of nonparametric data records.

4.7.8   Summary of Model Relationships

To summarize, the interrelationships between the model forms are now presented in the schematic diagram shown in Figure 4.4. The hatched arrows are used to indicate relationships that are of theoretical interest only, being of doubtful utility in practical situations.

Note that two of the model forms (Impulse Response and Frequency Response) allow the creation of nonparametric dynamic models — independent of any mathematical form — directly from experimental data records. Both the state-space and transform-domain models require a parametric functional form in order to fit experimental data.

It is important to keep Figure 4.4 in mind when thinking about possible alternative model forms, so that one can choose the most appropriate model form for the problem at hand. It is also important to reemphasize that the state variables appear explicitly only in the state-space model form; all the others are strictly input/output models — in different domains — and do not contain information about the system's state variables *explicitly*.

Finally, it is comforting to know that computer-aided-design programs are available for performing all the model conversions indicated by the dark arrows in Figure 4.4. Some of these programs are listed in Appendix E.

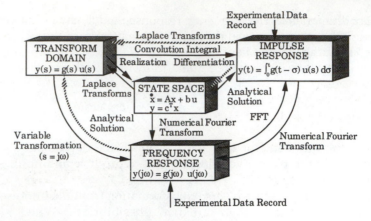

Figure 4.4. Interrelationships between process model forms.

4.8   THE CONCEPT OF A TRANSFER FUNCTION

Thus far in our discussion, we have only alluded to — without formally examining — the concept of the transfer function in the mathematical description of process behavior. However, as will become abundantly clear in our subsequent study throughout the rest of the book, this *transfer function* concept is very central to process dynamics and control. Thus we shall now take a closer look at this important concept.

4.8.1   A Generalized View of the Process Model

To reiterate, as far as process dynamics and control studies are concerned, a process model is any collection of mathematical equations by which the system output response to a given input can be predicted. The state-space model explicitly calculates the *state* variable and uses this to generate the output; the input/output model forms generate the output directly. However, even though these various model forms may differ in the form of input function they accept, the form of output response they generate, and in the way the output is generated, they are fundamentally similar.

We recall that:

1.   For the *state-space model*, the input, $u(t)$, is a time-domain function, and the output, the time-domain function $y(t)$, is generated by the process of solving differential or difference model equations.

2.   For the continuous *transform-domain model*, the input is $u(s)$, a function of the Laplace transform variable s, and the output, $y(s)$ (also a function of s), is generated by the process of multiplying $u(s)$ by the function $g(s)$.

3.   For the *frequency-response model*, the input, $u(j\omega)$ is a function of the indicated complex frequency variable, and the output $y(j\omega)$ (also a function of the same variable), is generated by the process of multiplying $u(j\omega)$ by the function $g(j\omega)$.

4.   For the *impulse-response model*, the input, $u(t)$ is a function of time, and the output, $y(t)$ (also a function of time), is generated by a convolution operation of $g(t)$ on $u(t)$.

We may now observe that in each case, the output is produced by the process of "operating" on the given input by an appropriate "function." Thus, in essence, each model form is seen to consist of two parts:

1. A function, and
2. An operation

The underlying principle in each case may now be stated as follows:

The function, by way of the indicated operation "transfers" (or transforms) the input, $u(\cdot)$, to the output, $y(\cdot)$, as depicted in the equation:

$$y(\cdot) \ = \ g(\cdot) \bullet u(\cdot) \qquad\qquad (4.83)$$

where $\bullet$ represents the particular operation demanded by the specific model form and $g(\cdot)$ represents the function employed for the "transfer" of the input $u(\cdot)$ to the output $y(\cdot)$ via the indicated operation. For any model form the function $g(\cdot)$ that performs the noted task is referred to as the "transfer function" of such a model form.

Observe now that the "transfer function" form of the mathematical model as represented in Eq. (4.83) is an input/output representation. It may be extended to include input disturbances to read:

$$y(\cdot) \ = \ g(\cdot) \bullet u(\cdot) + g_d(\cdot) \bullet d(\cdot) \qquad\qquad (4.84)$$

In principle, all the various model forms we have discussed in this chapter can be considered in this form, where we recognize that the argument $(\cdot)$ and the operator options $g \bullet u, g_d \bullet d$ will be different in each model form. The detailed mathematical expressions for the "transfer functions" of each model form are shown in Table 4.1. Let us now observe that in this context:

1.   The state-space model exists in the time domain, and its "transfer function" representation involves an operation consisting of the solution of differential equations and a "transfer function" in which the state variable $x(t)$ appears *explicitly*. Thus time, the fundamental variable of the domain of existence, shows up *implicitly* in this "transfer function" through the state variable $x(t)$.

2.   The Laplace transform-domain model, on the other hand, involves a simple multiplication operation, and a "transfer function" that is an *explicit* function of s, the fundamental variable of its domain of existence.

3. Similarly, the frequency-response model also involves a simple multiplication operation, and a "transfer function" that is an *explicit* function of $j\omega$, the fundamental variable of its domain of existence.

4. As for the impulse-response model, it exists in the time domain, involves the convolution operation, and is a "transfer function" which is an *explicit* function of t, again the fundamental variable of its domain of existence. The reader familiar with linear operator theory will recognize the convolution operation as the "continuous" version of the familiar dot product of two vectors, itself an extension of the even more familiar scalar operation of multiplication. Thus, *convolution* is in fact no more than *multiplication* in the functional analysis sense.

It is thus clear that this "transfer function" context is not as *natural* for the state-space model as it is for the others; the transform-domain, frequency-response, and impulse-response models, introduced at the very beginning of this chapter as input/output models, are the ones that more naturally conform to the transfer function representation in Eq. (4.84). Thus, although the linear *state-space model* can be considered conceptually in this input/output "transfer function" context, this is somewhat unnatural; it is therefore not customary to refer to the state-space representation in these terms. Thus subsequently we shall refer only to the other forms in Table 4.1 as transfer function representations.

It has proved very convenient to express the transfer function relation of Eq. (4.84) in terms of a *block diagrams* as shown in Figure 4.5. It shows in general, how the transfer function relates the manipulated variable u and disturbance d to the process output y.

The block diagram promotes a conceptual understanding of what happens to a process when changes are made in the value of its input variables. The indication that the transfer function (residing within the "block") "operates" on the input $u(\cdot)$ to produce the observed change $y(\cdot)$ in the output variable is not only intuitively appealing, it is in fact also mathematically exact. We shall have cause to use block diagrams extensively in the remaining portions of this book.

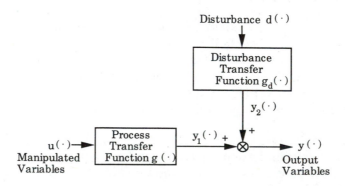

Figure 4.5.   The input/output form of a process model — the transfer function.

Table 4.1

CONTINUOUS LINEAR
INPUT/OUTPUT
TRANSFER FUNCTION MODELS

State-Space Model:

$$g(\cdot) \bullet u(\cdot) \quad = \quad \begin{cases} \dfrac{d\mathbf{x}(t)}{dt} = \mathbf{A}\mathbf{x}(t) + \mathbf{b}u(t); \ \mathbf{x}(0) = \mathbf{0} \\[2mm] y_1(t) = \mathbf{c}^T\mathbf{x}(t) \end{cases}$$

$$g_d(\cdot) \bullet d(\cdot) \quad = \quad \begin{cases} \dfrac{d\mathbf{x}(t)}{dt} = \mathbf{A}\mathbf{x}(t) + \gamma d(t); \ \mathbf{x}(0) = \mathbf{0} \\[2mm] y_2(t) = \mathbf{c}^T\mathbf{x}(t) \end{cases}$$

$$y(t) \quad = \quad y_1(t) + y_2(t)$$

Note: *Although the state-space model can be considered in this input/ output form, it is not usually referred to as a transfer function model.*

Transform-Domain Model:

$$\begin{aligned} g(\cdot) \bullet u(\cdot) &= g(s)u(s) \\ g_d(\cdot) \bullet d(\cdot) &= g_d(s)d(s) \\ y(s) &= g(s)u(s) + g_d(s)d(s) \end{aligned}$$

Frequency-Response Model:

$$\begin{aligned} g(\cdot) \bullet u(\cdot) &= g(j\omega)u(j\omega) \\ g_d(\cdot) \bullet d(\cdot) &= g_d(j\omega)d(j\omega) \\ y(j\omega) &= g(j\omega)u(j\omega) + g_d(j\omega)d(j\omega) \end{aligned}$$

Impulse-Response Model:

$$g(\cdot) \bullet u(\cdot) \quad = \quad \int_0^t g(t - \sigma)u(\sigma)d\sigma$$

$$g_d(\cdot) \bullet d(\cdot) \quad = \quad \int_0^t g_d(t - \sigma)d(\sigma)d\sigma$$

$$y(t) \quad = \quad \int_0^t g(t - \sigma)u(\sigma)d\sigma + \int_0^t g_d(t - \sigma)d(\sigma)d\sigma$$

4.8.2   The "Traditional" Transfer Function

Let us reiterate that while the "transfer function" as well as the required operations that correspond to the transform-domain, frequency-response, and impulse-response models are very obvious and therefore easily identifiable, the "transfer function" and the "operation" for the state-space model are neither obvious nor easy to identify. It is for this reason that, traditionally, only the transform-domain, frequency-response, and impulse-response models are exclusively referred to as transfer function models.

Another important fact to note is that of these three *natural* input/output model forms, the transform-domain transfer function $g(s)$ has the simplest and most compact form and has traditionally been the most widely used for block diagram analyses. When $g(j\omega)$ and $g(t)$ are obtained *independently* of $g(s)$ (by experimental means) they will occur in numerical form and will clearly not be able to match $g(s)$ in simplicity and mathematical convenience. It is only when $g(j\omega)$ and $g(t)$ are derived directly from $g(s)$ (the former by $s = j\omega$ variable substitution, the latter by inverse Laplace transform) that they retain many of the desirable properties of $g(s)$.

Thus we find that as a result of the mathematical convenience of form enjoyed by the transform-domain transfer function $g(s)$, it is by far the preferred model form for dynamic process analysis. For this reason, it has become customary that, unless specifically stated to the contrary, the term "transfer function" is tacitly assumed to be synonymous with the transform-domain transfer function, $g(s)$.

From now on, therefore, we shall also adopt this notation.

4.8.3   Some Properties of the Transfer Function

Linearity

The "operation" indicated by the transfer function relation in Eq. (4.84) is *linear*. This implies that the principle of superposition applies to all forms of the transfer function shown in Table 4.1. Thus if $y_1(\cdot)$ is the response to $u_1(\cdot)$, and $y_2(\cdot)$ is the response to $u_2(\cdot)$, then the response to $u(\cdot) = c_1 u_1(\cdot) + c_2 u_2(\cdot)$ will be $c_1 y_1(\cdot) + c_2 y_2(\cdot)$, i.e.:

$$
\begin{aligned}
y(\cdot) &= g(\cdot) \bullet [c_1 u_1(\cdot) + c_2 u_2(\cdot)] \\
&= c_1 g(\cdot) \bullet u_1(\cdot) + c_2 g(\cdot) \bullet u_2(\cdot) \\
&= c_1 y_1(\cdot) + c_2 y_2(\cdot)
\end{aligned}
\tag{4.85}
$$

Poles and Zeros of a Transfer Function

The poles and zeros of a transfer function are very useful in characterizing the dynamic character of the process. Although these can be defined for all model forms, we are usually concerned about poles and zeros primarily in the transform or the frequency domain. In what follows, therefore, we shall use the transform-domain transfer function $g(s)$.

With the exception of time-delay systems (to be dealt with in Chapter 8) $g(s)$ is, in general, a ratio of two polynomials in s, i.e.:

$$g(s) = \frac{N(s)}{D(s)} \tag{4.86}$$

where the numerator polynomial $N(s)$ is of order r, and the denominator polynomial, $D(s)$, is of order n. For real processes, we note that r is usually *strictly* less than n.

If these polynomials are factorized as follows:

$$N(s) = K_1\left(s - z_1\right)\left(s - z_2\right)\cdots\left(s - z_r\right) \tag{4.87}$$

$$D(s) = K_2\left(s - p_1\right)\left(s - p_2\right)\cdots\left(s - p_n\right) \tag{4.88}$$

then Eq. (4.86) becomes:

$$g(s) = \frac{K\left(s - z_1\right)\left(s - z_2\right)\cdots\left(s - z_r\right)}{\left(s - p_1\right)\left(s - p_2\right)\cdots\left(s - p_n\right)} \tag{4.89}$$

where $K = K_1 / K_2$.
Note that:

- For $s = z_1$, or z_2, ..., or z_r, $g(s) = 0$.
- For $s = p_1$, or p_2, ..., or p_n, $g(s) = \infty$.

The roots of the *numerator* polynomial, $N(s)$, i.e., $z_1, z_2, ..., z_r$, are thus called the *zeros* of the transfer function, while $p_1, p_2, ..., p_n$, the roots of the *denominator* polynomial, $D(s)$, are called the transfer function *poles*.

4.8.4 General Principles of Using Transfer Functions for Dynamic Analysis

The basic problem in dynamic analysis is that of using a process model to obtain $y(t)$, the system response to a given change $u(t)$ in the input. In using the transfer function model:

$$y(s) = g(s)u(s)$$

the dynamic analysis problem becomes that of finding $y(t)$, given $g(s)$ and $u(t)$ or its Laplace transform $u(s)$.

From, Eq. (4.86) it is clear that the product $g(s)u(s)$ will also be a ratio of polynomials, i.e.:

$$y(s) = g(s)u(s) = \frac{A(s)}{B(s)} \tag{4.90}$$

If r_i, $i = 1, 2, ..., n$, are the roots of the denominator polynomial $B(s)$, then, by factorization, Eq. (4.90) becomes:

$$y(s) = \frac{A(s)}{K_2\left(s - r_1\right)\left(s - r_2\right)\cdots\left(s - r_n\right)} \tag{4.91}$$

The right-hand-side expression may now be broken down by partial fraction expansions to give:

$$y(s) = \frac{A_1}{s - r_1} + \frac{A_2}{s - r_2} + \ldots + \frac{A_n}{s - r_n} \tag{4.92}$$

It is now easy to obtain the inverse Laplace transform of Eq. (4.92); the result is the general expression:

$$y(t) = A_1 e^{r_1 t} + A_2 e^{r_2 t} + \ldots + A_n e^{r_n t} \tag{4.93}$$

In subsequent chapters of Part II, we shall utilize the above outlined general strategy in carrying out dynamic analysis for various process systems.

4.9   SUMMARY

The discussion of the fundamental elements of dynamic analysis that began in Chapter 3 has been completed in this chapter with the detailed treatment of the concept of process models.

The objective in this chapter has been to treat, in some detail, the most pertinent aspects of the process model as a tool of dynamic analysis. An introductory discussion of the important concept of describing process behavior mathematically, the implication of such a strategy on how process systems are characterized, and a quick summary of the most important process model forms were used as a backdrop for the rest of the chapter.

The essential points of how process models are formulated using mass, momentum, and energy balances have been discussed and illustrated with some simple examples. A broader spectrum of examples and further discussions on the practice of process model development and usage have been reserved until Part III.

The most important process model forms have been identified as the state-space, the transform-domain, the frequency-response, and the impulse-response forms; the salient features of each one of them were examined both in the continuous time as well as in the discrete-time representations. We have shown how one process model form is related to the others and stressed the importance of being able to switch back and forth from one process model form to the other whenever this is called for. Several examples were used to illustrate the procedure for converting a process model from one form to another equivalent form.

In anticipation of its usefulness in the discussions to follow in the remaining portions of this book, an entire section was devoted to a discussion of the transfer function, its linearity properties, its poles and zeros, and how it is used, in general, for dynamic analysis.

With this background, we are now in a position to embark on the main task at hand: that of analyzing the dynamic behavior of the various classes of processes. The next seven chapters (from Chapter 5 through Chapter 11) will be concerned with just this task.

REFERENCES AND SUGGESTED FURTHER READING

1. Bird, R. B., W. E. Stewart, and E. N. Lightfoot, *Transport Phenomena*, J. Wiley, New York (1960)
2. Bracewell, R. N., *The Fourier Transform and its Applications* (2nd ed.), McGraw-Hill, New York (1986)

REVIEW QUESTIONS

1. What is a mathematical model and what is its main utility in process dynamics?

2. Can a process model be considered as the equivalent of the physical process it purports to model? Why, or why not?

3. What is:
    (a) a linear system
    (b) a nonlinear system
    (c) a lumped parameter system
    (d) a distributed parameter system
    (e) a discrete-time system?

4. What are the various ways in which a process model can be cast?

5. What is the fundamental basis for developing virtually all theoretical process models?

6. Give a statement of the general conservation principle.

7. What are the principal uses of each of the four model types?

8. What procedure is required in converting a state-space model to a transform-domain model? What is the reverse procedure for converting the transform-domain model to a state-space model called?

9. What conditions must apply for a state-space model to be convertible to the impulse-response form?

10. What variable transformation is required to convert a transform-domain model to a frequency-response model?

11. Define the concept of a "transfer function" in process modeling?

12. What is a block diagram, and why is it an intuitively appealing device for representing particularly a transfer function model?

13. Traditionally, with which model form is the term "transfer function" tacitly assumed to be synonymous?

14. What are the poles and zeros of a "traditional" transfer function?

PROBLEMS

4.1   This diagram is from a Department of Energy underground storage tank. Such tanks are used throughout the United States for storing gasoline for sale to the public.

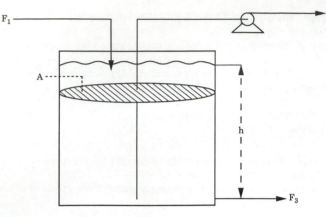

Figure P4.1.

Recently, engineers believe that a leak has developed, which threatens the environment. Your supervisor has assigned you the task of modeling the height in the tank as a function of the supply flow F_1, the sales flow F_2, and the unmeasured leakage F_3. It is assumed that this leakage is proportional to the height of liquid in the tank (i.e., $F_3 = \beta h$). The tank is a cylinder with constant cross-sectional area A. The specific gravity of the gasoline is 0.78 and is assumed to be constant. Find the transfer function model for the level in the tank as a function of the supply flow and the sales flow. Which of these input variables would you describe as a control variable and which would you describe as a disturbance variable?

4.2 (a) Figure P4.2 shows a system for heating a continuous flow water kettle using a hot plate. Assuming that the hot plate temperature can be changed instantaneously by adjusting the hot plate rate of heat input and assuming uniform water kettle temperature, show that the following is a reasonable model for the process:

$$\rho V C_p \frac{dT_1}{dt} = c(T_2 - T_1) + F\rho C_p T_0 - F\rho C_p T_1 \qquad (P4.1)$$

where ρ, C_p are, respectively, water density and heat capacity, and

c = a constant
F = volumetric flowrate in and out of the kettle.

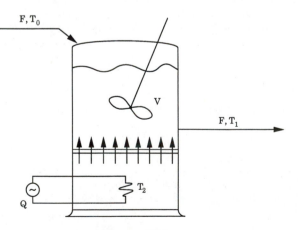

Figure P4.2.

(b) If the heat capacity of the hot plate material is assumed uniform and given as C_{p2}, and its mass and effective lateral area for heat transfer to the atmosphere are respectively given as m and A_c, obtain a second modeling equation that adequately describes the dynamics of T_2, the hot plate temperature in response to changes in Q, the rate of heat input. It may be assumed that the atmospheric temperature is T_a, a constant; the heat transfer coefficient may also be taken as h, another constant.

4.3 Assuming that the simpler conditions of Problems 4.2(a) apply and that the objective is to control the water temperature T_1 by adjusting T_2.
(a) Which of the variables indicated in Eq. (P4.1) is the input variable, and which is the disturbance variable?
(b) By defining appropriate deviation variables (in terms of T_1^*, T_2^*, and T_0^*, the steady-state values of T_1, T_2, T_0 respectively) rewrite Eq. (P4.1) in the form:

$$\frac{dy}{dt} = ay + bu + \gamma d \qquad (P4.2)$$

Explicitly state a, b, and γ in terms of process parameters ρ, C_p, c, F, and V.
(c) Express this differential equation model in the transform-domain transfer function form. What are $g(s)$ and $g_d(s)$ in this case?
(d) Express the differential equation model in the impulse-response form *either* directly, *or* using $g(s)$ and $g_d(s)$ obtained in (c).

4.4 The process of drug ingestion, distribution, and subsequent metabolism in an individual may be represented by the simplified diagram in Figure P4.3. A simplified mathematical model for the process is:

$$\frac{dx_1}{dt} = -k_1 x_1 + u \qquad (P4.3)$$

$$\frac{dx_2}{dt} = k_1 x_1 - k_2 x_2 \qquad (P4.4)$$

$$y = x_2 \qquad (P4.5)$$

where x_1, x_2, and u respectively represent (as *deviations* from initial steady-state values) the drug mass in the Gastrointestinal tract (GIT), the drug mass in the Bloodstream (BS), and the drug ingestion rate. (The immediate consequence is that the initial conditions for these *deviation variables*, x_1, x_2, and u, are all zero.)
The *constants* k_1 and k_2 are physiological properties of the individual in question and in general will *not* be equal.

Figure P4.3.

(a) By taking Laplace transforms in Eqs. (P4.3) and (P4.4) and rearranging appropriately, obtain a *transfer function* relationship of the form:

$$y(s) = g(s)u(s) \qquad (P4.6)$$

relating the measurement of drug mass in the BS, x_2, to u, the drug ingestion rate.
(b) To rewrite the model for this process in the vector matrix form:

$$\dot{x} = Ax + Bu$$

$$y = c^T x$$

what will A, B, and c^T be?

4.5 (a) The top portion of a distillation column has been modeled by the following transform-domain transfer function model:

$$y(s) = \left(\frac{12.8}{16.7s + 1}\right) u(s) + \left(\frac{3.8}{14.9s + 1}\right) d(s) \qquad \text{(P4.7)}$$

Obtain the corresponding set of differential equations to be solved in order to generate a time-domain response.
(b) Choose 10 frequency values in the range of $10^{-3} < \omega < 100$ and generate values for the $g(jw)$ and $g_d(jw)$ corresponding to the transfer functions given in Eq. (P4.7).

4.6 An approximate model for a chemical reactor has the form:

$$y(k) = ay(k-1) + bu(k-4) \qquad \text{(P4.8)}$$

Given that $u(k) = 0$ for $k \leq 0$; $y(0) = 0$, express this model in the impulse-response form:

$$y(k) = \sum_{i=1}^{k} g(i)u(k-i) \qquad \text{(P4.9)}$$

What is $g(i)$ in terms of a and b?

4.7 A process has a transform-domain transfer function model:

$$y(s) = g(s)u(s)$$

with

$$g(s) = \frac{K}{(\tau_1 s + 1)(\tau_2 s + 1)} \qquad \text{(P4.10)}$$

Obtain the equivalent impulse-response model. [Hint: Expand $g(s)$ by partial fractions first and then obtain $g(t)$.]

4.8 For each of the following transform-domain transfer functions, state the number *and* location of the poles and zeros:

(a) $g(s) = \dfrac{K}{\tau s + 1}$

(b) $g(s) = \dfrac{K(\xi s + 1)}{(\tau s + 1)}$

(c) $g(s) = \dfrac{K}{(\tau_1 s + 1)(\tau_2 s + 1)}$

(d) $g(s) = \dfrac{5}{-12s^2 + s + 1}$

(e) $g(s) = \dfrac{5}{12s^2 + 7s + 1}$

(f) $g(s) = \dfrac{5}{s^2 + s + 1}$

(g) $g(s) = \dfrac{3(s + 4)}{(5s + 1)(-3s + 1)}$

(h) $g(s) = \dfrac{2(3s + 1)(s + 1)(-5s + 1)}{(4s + 1)(2s + 1)(2.5s + 1)(1.5s + 1)}$

5

DYNAMIC BEHAVIOR OF LINEAR LOW-ORDER SYSTEMS

We begin our study of process dynamics in this chapter by looking first into linear systems with low orders. Systems that come into consideration here are primarily *first-order* systems, along with *pure gain* and *pure capacity* systems which, in actual fact, are variations of the first-order system.

These systems are characterized by transfer functions with denominator polynomials of order 1 or less. The lead/lag system is of importance in control system design, and since its characteristics fit in with our low-order classification, its dynamic behavior will also be investigated in this chapter. Systems whose transfer functions have denominator polynomials of order higher than 1 are considered in Chapter 6.

5.1   FIRST-ORDER SYSTEMS

Systems whose dynamic behavior is modeled by first-order differential equations of the type:

$$a_1 \frac{dy}{dt} + a_0 y = bu(t) \tag{5.1}$$

(where $y(t)$ represents the output and $u(t)$ represents the input) are referred to as first-order systems. It is customary to rearrange Eq. (5.1) to read:

$$\tau \frac{dy}{dt} + y = K u(t) \tag{5.2}$$

where, for $a_0 \neq 0$, the new parameters τ and K are given by:

$$\tau = a_1 / a_0 \tag{5.3a}$$
$$K = b / a_0 \tag{5.3b}$$

If the model in Eq. (5.1) had been presented in terms of *deviation variables* (see Chapter 4), then taking Laplace transforms in Eq. (5.2) and rearranging gives:

$$y(s) = \left(\frac{K}{\tau s + 1}\right) u(s) \tag{5.4}$$

It is therefore clear that the general transfer function[†] for a first-order system is given by:

$$g(s) = \frac{K}{(\tau s + 1)} \tag{5.5}$$

Characteristic Parameters

The characteristic parameters of a first-order system may now be identified as K and τ, respectively called the *steady-state gain* and the *time constant* for reasons which shall soon be made clear. The transfer function has a single pole at $s = -1/\tau$, and no zeros.

5.1.1   Physical Examples of First-Order Systems

1. Liquid Level Systems

Recall the mixing tank we encountered in Chapter 1, whose schematic diagram is shown in Figure 5.1.

Recall also that the mathematical model for this process was:

$$A_c \frac{dh}{dt} = F_i - ch \tag{1.13}$$

or in deviation variables:

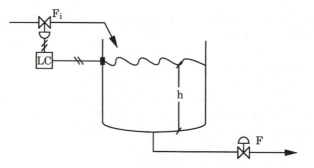

Figure 5.1.   Mixing tank with liquid level dynamics.

[†] As indicated in the previous chapter, the term "transfer function" can actually refer to any of the model forms; however, it will be quite clear from the functional form of $g(\cdot)$ which specific form is under consideration.

$$\frac{dy}{dt} = -\frac{c}{A_c} y + \frac{1}{A_c} u \tag{5.6}$$

Observe that this is a first-order differential equation of the type given in Eq. (5.1). Upon taking the usual Laplace transform and rearranging, we easily obtain the transfer function representation:

$$y(s) = g(s) u(s)$$

with the transfer function given by:

$$g(s) = \frac{1/c}{(A_c/c)s + 1} \tag{5.7}$$

Observe now that for this system, the time constant is A_c/c, while the steady-state gain is $1/c$.

2. Isothermal Continuous Stirred Tank Reactor

The process shown in Figure 5.2 below, a first-order, irreversible reaction $A \rightarrow B$, is taking place in an isothermal CSTR. Making the usual assumption of constant volume, and perfect mixing, it is easy to show that an appropriate mathematical model for this process, obtained from a component mass balance on A, is:

$$\frac{dy}{dt} + (k + 1/\theta)y = \frac{1}{\theta} u \tag{5.8}$$

Here, y is the concentration of A in the reactor, c_A, in terms of deviations from its initial steady-state value c_{As}; u is the inlet concentration of A, c_{Af} (also in deviation variables); k is the reaction rate constant; and θ is the reactor residence time (V/F).

Note that because this reactor is isothermal, there are no heat effects to warrant an energy balance as was the case with the nonisothermal example discussed earlier in Chapter 4.

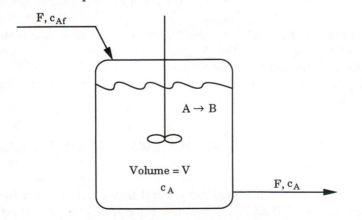

Figure 5.2. The isothermal CSTR.

Laplace transformation and subsequent rearrangement of the first-order differential equation in Eq. (5.8) gives the transfer function for this process as:

$$g(s) = \frac{\left(\dfrac{1}{\theta k + 1}\right)}{\left(\dfrac{\theta}{\theta k + 1}\right) s + 1} \tag{5.9}$$

so that in this case, the time constant and steady-state gain are respectively given by:

$$\tau = \frac{\theta}{\theta k + 1}$$

$$K = \frac{1}{\theta k + 1}$$

3. The Stirred Heating Tank

The dynamic behavior of the stirred heating tank process used as an example in Chapter 4 (see Figure 4.1) was shown to be modeled by the first-order differential equation (4.13); subsequent Laplace transformation gave rise to the transfer function model (4.14):

$$y(s) = \left(\frac{\beta\theta}{\theta s + 1}\right) u(s) + \left(\frac{1}{\theta s + 1}\right) d(s) \tag{4.14}$$

This, of course, is of the form:

$$y(s) = g(s) u(s) + g_d(s) d(s)$$

with the process transfer function, $g(s)$, and the disturbance transfer function, $g_d(s)$, given by :

$$g(s) = \left(\frac{\beta\theta}{\theta s + 1}\right) \tag{5.10a}$$

$$g_d(s) = \left(\frac{1}{\theta s + 1}\right) \tag{5.10b}$$

Observe therefore that this is another example of a first-order system, both in the effect of the input u on the system output, as well as in the effect of the disturbance d on the output. It should be easy to identify the time constants and the steady-state gains in each of the first-order transfer functions shown above in Eq. (5.10).

5.2   RESPONSE OF FIRST-ORDER SYSTEMS TO VARIOUS INPUTS

In what follows, we shall utilize the general first-order transfer function form introduced in Eqs. (5.4) and (5.5) to obtain the response of first-order systems to such input functions as were discussed in Section 3.4 of Chapter 3, i.e., step, rectangular pulse, impulse, ramp, and sinusoidal functions.

5.2.1 Step Response

When a step function of magnitude A is applied as an input to a first-order system, the dynamic response of the system output is obtainable from:

$$y(s) = \frac{K}{\tau s + 1} \frac{A}{s} \tag{5.11}$$

since, as we may recall from Chapter 3, $u(s)$, the Laplace transform for this step input function, is A/s.

By partial fraction expansion, Eq. (5.11) becomes:

$$y(s) = AK \left(\frac{1}{s} - \frac{\tau}{\tau s + 1} \right) \tag{5.12}$$

which, upon inverting back to the time domain, gives the required step response:

$$y(t) = AK \left(1 - e^{-t/\tau} \right) \tag{5.13}$$

A sketch of this response and the input (forcing) function is shown in Figure 5.3.

Characteristics of the Step Response

There are a number of important points to note about the response of first-order systems to step inputs:

1.   The output, $y(t)$, attains the value AK at steady state. The value of the input function, $u(t)$, responsible for this response is A. Observe, therefore, that the ultimate change in the output as $t \to \infty$ is the magnitude of the input step function A multiplied by the constant K. (This obviously also implies that the ratio of the steady-state value of the output to that of the input (AK/A) is K.) This is the reason for calling K the *steady-state gain*.

2.   When $t = \tau$, $y(t) = 0.632AK$; i.e., $y(t)$ attains 63.2% of its ultimate value in τ time units. It is easy to establish that by the time $t = 3.9\tau$, $y(t)$ has reached 98% of its ultimate value, reaching the 99% mark when $t = 4.6\tau$.

3.   The slope of the step response at the origin, obtained by differentiating Eq. (5.13) with respect to t, and setting $t = 0$, is given by:

$$\left. \left(\frac{dy}{dt} \right) \right|_{t=0} = \frac{AK}{\tau} \tag{5.14}$$

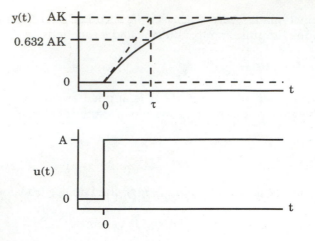

Figure 5.3.   Response of a first-order system to the step function.

This has the following implication: if the initial rate of change of $y(t)$ were maintained, the ultimate value of AK would be attained in exactly τ time units, providing an interesting interpretation of the first-order system *time constant* τ.

Thus, while K is a measure of "by how much" the process output will ultimately change in response to a step change in the input, τ is a measure of "how fast" the process responds to the input; i.e., the magnitude of the step response is indicated by K, while τ indicates the speed of response.

Observe that the larger the value of K, the larger the ultimate value of the process output in response to a unit step change; the converse is also true. On the other hand, observe that the larger the value of τ, the more *sluggish* the process response; and conversely, the *smaller* the value of τ the faster the process response. In the extreme case with $\tau = 0$ the process responds in the fastest possible manner, i.e., instantaneously, and $y = AK$ for all times.

4.   The difference between the fastest possible response (when $\tau = 0$; i.e., $y^* = AK$) and the actual step response for the first-order process (i.e., Eq. (5.13) with $\tau > 0$) when integrated over time gives a measure of the "reluctance" inherent in the first-order process, i.e.:

$$J = \int_0^\infty [y^* - y(t)]\, dt = AK \int_0^\infty [1 - (1 - e^{-t/\tau})]\, dt$$

$$J = AK \int_0^\infty e^{-t/\tau}\, dt = AK\tau \tag{5.15}$$

Thus, for a *unit* step input, the area under the curve bounded by the "instantaneous" response, and the actual first-order response is $K\tau$, as illustrated in Figure 5.4.

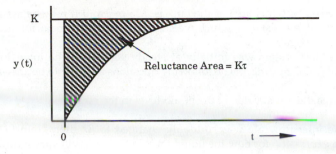

Figure 5.4. Response of a first-order process to a unit step input.

Process Information Obtainable from the Step Response

We are now in a position to summarize the information that can be gathered from a graphical representation of a first-order process step response.

Given $y(t)$, the response of a first-order system to a step input of magnitude A, observe that the two characteristic parameters of the system may be extracted from this response as follows:

1. Steady-state gain, K

$$K = \frac{y(\infty)}{A}$$

i.e., the final, steady-state value of the response divided by the input step size (see Characteristic 1 above).

2. Time constant, τ

Let σ be the slope of the $y(t)$ response at the origin. Then from Characteristic 3 above:

$$\sigma = \frac{AK}{\tau} = \frac{y(\infty)}{\tau}$$

so that:

$$\tau = \frac{AK}{\sigma} = \frac{y(\infty)}{\sigma}$$

We may also obtain τ by utilizing the fact that at $t = \tau$ the value of the output is $y(\tau) = 0.632\, y(\infty)$.

Let us illustrate some of the issues involved in analyzing the dynamic behavior of first-order systems with the following examples.

Example 5.1   DYNAMIC BEHAVIOR OF THE LIQUID LEVEL IN A STORAGE TANK.

A 250 liter tank used for liquid storage is configured as in Figure 5.1 so that its mathematical model and its dynamic behavior are as discussed in Section 5.1.1.

The cross-sectional area of the tank is 0.25 m², and it may be assumed uniform. The outlet valve resistance, which we shall assume to be linear for simplicity, has a value $c = 0.1$ m²/min.

The entire system was initially at steady state with the inlet flowrate F_i at 37 liters/min (0.037 m³/min). The following questions are now to be answered:

1.   What is the initial value of the liquid level in the tank?

2.   If the inlet flowrate was suddenly changed to 87 liters/min (0.087 m³/min), obtain an expression for how the liquid level in the tank will vary with time.

3.   When will the liquid level in the tank be at the 0.86 m mark? And to what final value will the liquid level ultimately settle ?

Solution:

1.   From Eq. (1.13), we observe that at steady state we have:

$$F_{is} = ch_s$$

If we now introduce the given values for the flowrate and the resistance c, we have the required value for the initial steady-state liquid level:

$$h_s = 0.37 \text{ m}$$

2.   Recalling the transfer function for this process given in Eq. (5.7), and introducing the given numerical values, we find that the steady-state gain and time constant are given by:

$$K = 1/0.1 = 10 \text{ (min /m}^2)$$
$$\tau = 0.25/0.1 = 2.5 \text{ (min)}$$

The magnitude of the step input in the flowrate is $(0.087 - 0.037) = 0.05$ m³/min. Thus from Eq. (5.13) the expression representing the *change* in liquid level in response to this input change is:

$$y(t) = 0.5 \left(1 - e^{-0.4t}\right)$$

Note that this is in terms of y, the deviation of the liquid level from its initial steady-state value. The actual liquid level time behavior is represented by:

$$h(t) = 0.37 + 0.5 \left(1 - e^{-0.4t}\right)$$

3.   Introducing $h = 0.86$ into the expression shown above, it is easily solved for t to give the time required to attain to the 0.86 m mark as 9.78 min. The final value to which the liquid level eventually settles is easily obtained as $0.37 + 0.5 = 0.87$ m.

Example 5.2   DYNAMIC BEHAVIOR OF LIQUID LEVEL, CONTINUED.

Suppose now that for the system of Example 5.1 the flowrate had been suddenly changed instead to 137 liters/min (0.137 m³/min) from its initial value of 37 liters/min (0.037 m³ /min). An experienced plant operator says that such a change in the input flowrate will result in liquid spillage from the tank. Is this true? If so, when will the liquid begin to overflow?

Solution:

To answer this question, we obviously need to know the height of the tank. From the given capacity (250 liters = 0.25 m³) assuming that the cross-sectional area is uniform, since:

$$\text{Volume} = \text{Area} \times \text{Height}$$

it is therefore easy to deduce that the tank is 1 m high.

The magnitude of the step change in this instance is 0.1 m³ /min, and since the steady-state gain for the process has been previously found to be 10 (min/m²), the ultimate value of the *change* in the liquid level will then be 0.1 x 10 = 1 m.

Thus as a result of this *change* in the inlet flowrate, the liquid level is scheduled to change from its initial value of 0.37 m by a total of 1 m, attaining to a final value of 1.37 m. Observe that this value is greater than the total height of the tank. The conclusion therefore is that the operator is correct: there *will* be spillage of liquid.

Liquid will begin to overflow the instant $h > 1$ or, equivalently, in terms of deviation from the initial steady-state value of 0.37 m, the instant $y > 0.63$.

The expression for the time variation of the liquid level is easily seen in this case to be given by:

$$h(t) = 0.37 + 1.0 \left(1 - e^{-0.4t}\right)$$

or

$$y(t) = 1.0 \left(1 - e^{-0.4t}\right)$$

We may now use either expression to find the time when $h = 1$, or $y = 0.63$; the result is $t = 2.5$ min.

We therefore conclude that for this magnitude of inlet flowrate change to the liquid level system, liquid spillage will commence at about 2.5 min after implementing the flow rate change.

5.2.2 Rectangular Pulse Response

The input function in this case is as described in Section 3.4.2 in Chapter 3:

$$u(t) = \begin{cases} 0; & t < 0 \\ A; & 0 < t < b \\ 0; & t > b \end{cases} \tag{3.39}$$

Since the Laplace transform of this forcing function is given by:

$$u(s) = \frac{A}{s}\left(1 - e^{-bs}\right) \tag{3.42}$$

the rectangular pulse response of the first-order process is obtained from:

$$y(s) = \frac{K}{\tau s + 1}\frac{A}{s}\left(1 - e^{-bs}\right)$$

$$= \frac{AK}{s(\tau s + 1)}\left(1 - e^{-bs}\right) \tag{5.16}$$

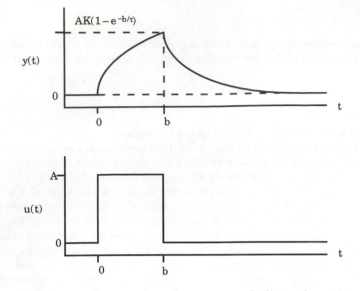

Figure 5.5. Rectangular pulse response of a first-order system.

The inversion of this apparently complicated function is made easier by splitting the terms into two parts as follows:

$$y(s) = \frac{AK}{s(\tau s + 1)} - \frac{AKe^{-bs}}{s(\tau s + 1)} \tag{5.17}$$

We may now observe that the two terms on the right-hand side of Eq. (5.17) are identical except for the presence of the e^{-bs} function which, as we recall, merely indicates a translation in time. If we now call upon our knowledge of the effect of the translation function in Laplace transforms, Eq. (5.17) is easily inverted into the time domain to give the following function which takes on different values over different time intervals:

$$y(t) = \begin{cases} AK\,(1 - e^{-t/\tau}) & ;\ t < b \\ AK[(1 - e^{-t/\tau}) - (1 - e^{-(t-b)/\tau})]; & t > b \end{cases} \tag{5.18}$$

This response is shown graphically in Figure 5.5. Note the change in direction experienced at $t = b$ in response to the change in the input function at this same instant. It is easy to establish that the value of the peak response at this point is $y_{max} = y(b) = AK(1 - e^{-b/\tau})$.

**Example 5.3   RECTANGULAR PULSE RESPONSE OF THE LIQUID
LEVEL SYSTEM.**

Let us return once again to the system of Example 5.1. The flowrate is again to be changed suddenly to 87 liters/min (0.087 m^2/min), but this time we hold it at this value only for 2.5 min; returning it to its initial value of 37 liters/min (0.037 m^3/min).

What is the maximum value the liquid level will attain during the course of this experiment?

Solution:

The expression for the peak value attained by a first-order system in response to a rectangular pulse input of magnitude A and duration b is given by:

$$y_{max} = y(b) = AK(1 - e^{-b/\tau})$$

(Note that this expression is in terms of deviations from the initial steady-state value for the liquid level.)

Since in this example, $A = 0.05$, $b = 2.5$, and from Example 5.1 we recall that $K = 10$, $\tau = 2.5$, with an initial liquid level of 0.37 m, the peak value for the liquid level is obtained from:

$$h_{max} = 0.37 + 0.5(1 - e^{-1})$$

or

$$h_{max} = 0.686 \text{ m}$$

5.2.3 Impulse Response

In this case, recall from Eqs. (3.43) and (3.46) that the input is a Dirac delta function with area A, $u(t) = A\delta(t)$ and $u(s) = A$; thus the impulse response is obtained from:

$$y(s) = \frac{K}{\tau s + 1} A \tag{5.19}$$

which is easily inverted in time to give:

$$y(t) = \frac{AK}{\tau} e^{-t/\tau} \tag{5.20}$$

The impulse response is shown in Figure 5.6.

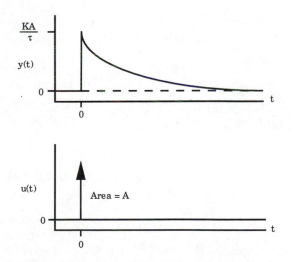

Figure 5.6.   Impulse response of a first-order system.

Note:

1. The impulse response indicates an immediate "jump" to a value AK/τ at $t = 0$ followed by an exponential decay.

2. If y_{step} represents the step response, and y_{impulse} represents the impulse response, then:

$$\frac{d}{dt}\left(y_{\text{step}}\right) = y_{\text{impulse}} \tag{5.21}$$

5.2.4 Ramp Response

Recalling that for the ramp input function:

$$u(t) = \begin{cases} 0; & t < 0 \\ At; & t > 0 \end{cases} \tag{3.49}$$

$$u(s) = \frac{A}{s^2} \tag{3.50}$$

the ramp response is therefore obtained from:

$$y(s) = \left(\frac{K}{\tau s + 1}\right)\frac{A}{s^2} \tag{5.22}$$

It is left as an exercise to the reader to show that upon Laplace inversion, Eq. (5.22) becomes:

$$y(t) = AK\tau\left(e^{-t/\tau} + t/\tau - 1\right) \tag{5.23}$$

Note:

1. As $t \to \infty$, $y(t) \to AK(t - \tau)$. Thus the response is asymptotic to a ramp function with slope AK, displaced by τ time units from the origin at $y = 0$.

2. If the step response is represented by y_{step} and the ramp response by y_{ramp} then it is easy to establish that:

$$\frac{d}{dt}\left(y_{\text{ramp}}\right) = y_{\text{step}} \tag{5.24}$$

The ramp response is shown in Figure 5.7.

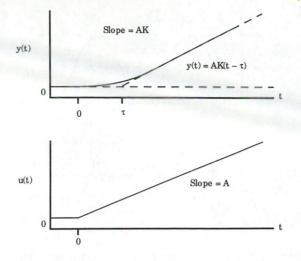

Figure 5.7.   Ramp response of a first-order system.

5.2.5   Sinusoidal Response

Recalling Eq. (3.52) for the Laplace transform of a sine wave input of amplitude A and frequency ω, the response of a first-order system to this sinusoidal input function is obtained from:

$$y(s) = \left(\frac{K}{\tau s + 1}\right) \frac{A\omega}{s^2 + \omega^2} \tag{5.25}$$

The inversion of this to the time domain (see tables in Appendix C) yields:

$$y(t) = AK\left[\frac{\omega\tau}{(\omega\tau)^2 + 1} e^{-t/\tau} + \frac{1}{\sqrt{(\omega\tau)^2 + 1}} \sin(\omega t + \phi)\right] \tag{5.26}$$

where

$$\phi = \tan^{-1}(-\omega\tau) \tag{5.27}$$

This response is shown in Figure 5.8.

There are a few important points to note in this sinusoidal response:

1. The first term in Eq. (5.26) is a transient term that decays with a time constant τ, and vanishes as $t \to \infty$; but the second term persists.

2. After the transients die away (after about four or five time constants in practice) the ultimate response will be a pure sine wave. Thus, we have, from Eq. (5.26):

$$y(t)\,\Big|_{t\to\infty} = \frac{AK}{\sqrt{(\omega\tau)^2 + 1}} \sin(\omega t + \phi) \tag{5.28}$$

which we shall refer to as the *Ultimate Periodic Response* (UPR) because it is to this — a function which is also periodic in its own right — that the sinusoidal response ultimately settles.

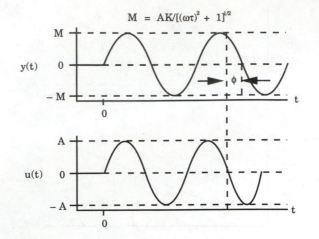

Figure 5.8. Sinusoidal response of a first-order system.

Characteristics of the Ultimate Periodic Response

The following are some important, noteworthy characteristics of the ultimate periodic response (UPR) given in Eq. (5.28):

1.   The input function ($u(t) = A\sin\omega t$) and the UPR are both sine waves having the same frequency ω.

2.   The UPR sine wave lags behind the input signal by an angle ϕ.

3.   The ratio of the UPR's amplitude to that of the input sine wave is $K/\sqrt{(\omega\tau)^2 + 1}$, a quantity termed the "amplitude ratio" (AR) for precisely this reason.

Thus we conclude by noting that the ultimate response of a first-order system to a sine wave, the UPR, can be characterized by the following two parameters:

$$AR = \frac{K}{\sqrt{(\omega\tau)^2 + 1}} \tag{5.29a}$$

and

$$\phi = \tan^{-1}(-\omega\tau) \tag{5.29b}$$

respectively known as the amplitude ratio and the phase angle. Note that both quantities are functions of ω, the frequency of the forcing sinusoidal function. Studying the behavior of AR and ϕ functions as they vary with ω is the main objective in *frequency response analysis*.

Frequency Response

The *frequency response* of a dynamic system is, in general, a summarization of its (ultimate) responses to pure sine wave inputs over a spectrum of frequencies ω. As indicated above, the primary variables are AR, the amplitude ratio, and ϕ, the phase angle. It is customary to refer to the phase angle as a *phase lag* if the angle is negative and a *phase lead* if the angle is positive. The most common method for summarizing frequency response information is the graphical representation suggested by Hendrik Bode [1]. Known as the Bode diagram (or Bode plot), it consists of two graphs:

1. log (AR) vs. log ω, in conjunction with
2. ϕ vs. log ω

It is often the case that the process steady-state gain K and time constant τ, are used as scaling factors in presenting these graphs. In such a case, the Bode diagram will consist of a log-log plot of the so-called *Magnitude Ratio*, MR = AR/K, in conjunction with a semilog plot of the phase angle ϕ, with $\log(\omega\tau)$ as the abscissa in each plot. The main advantage lies in the generalization that such scaling allows; the qualitative nature of the actual frequency response characteristics are not affected in any way.

The Bode diagram for the first-order system (consisting of MR and phase angle plots against $\log(\omega\tau)$) is shown in Figure 5.9. Detailed discussions about some important characteristics of this diagram will be taken up later in Chapter 9, which is devoted entirely to the study of frequency response analysis.

5.3   PURE GAIN SYSTEMS

Consider the first-order system with $\tau = 0$. (This corresponds to a physical system that, theoretically, is infinitely fast in responding to inputs. More realistically, one might imagine a situation in which the first-order system is so fast in responding that τ is so small as to be negligible.) In this case, Eq. (5.2) becomes:

$$y(t) \; = \; Ku(t) \tag{5.30}$$

and, either by taking Laplace transforms of Eq. (5.30), or by setting $\tau = 0$ in Eq. (5.4), we obtain:

$$y(s) \; = \; Ku(s) \tag{5.31}$$

so that the transfer function for such a process is identified as:

$$g(s) \; = \; K \tag{5.32}$$

A process having such characteristics is referred to as a *pure gain* process by virtue of the fact that its transfer function involves only one characteristic parameter: K, the process gain term. Such systems are always at steady state, moving instantly from one steady state to another with no transient behavior in between steady states.

The transfer function for the pure gain system has *no* poles and *no* zeros.

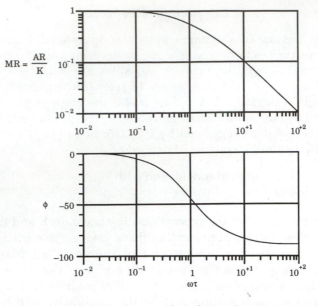

Figure 5.9.   Bode diagram for a first-order system.

5.3.1   Physical Examples of Pure Gain System

There are a few physical processes that truly exhibit pure gain characteristics. One example is the capillary system shown in Figure 5.10. There is a flow constriction that results in a pressure drop in the incompressible fluid flowing through the capillary. The upstream and downstream pressures are measured by liquid levels in manometers.

The value of the head h is observed to change whenever the liquid flowrate F changes. Thus in this case, the input variable is F, while the output is h. Owing to the fact that a capillary constitutes a laminar resistance, the head-flow relationship is given by the equation:

$$h = RF \tag{5.33}$$

where R is the resistance. In terms of deviation from an initial steady-state h_0, and F_0, Eq. (5.33) becomes:

$$y = Ru$$

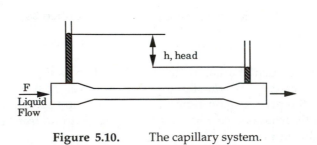

Figure 5.10.   The capillary system.

with

$$y = h - h_0; \ u = F - F_0$$

The important point to note here is that any change in F is instantaneously transmitted as a change in the head, h. Other physical systems that exhibit a pure gain response include electrical resistors and the mechanical spring.

First-order or higher order systems whose dynamics are extremely fast may be conveniently approximated as pure gain processes. (Observe that when $\tau \approx 0$, Eq. (5.4) becomes approximately the same as Eq. (5.32).) For example, a small pneumatic control valve with very rapid response may be approximated as a pure gain process if the other parts of the process have much larger time constants.

Perhaps the pure gain system with the most important application in process control is the *proportional controller*. As we shall discuss fully later, the output of this controller, the command signal c, is dependent on the input signal ε according to:

$$c(t) = K_c \ \varepsilon(t) \tag{5.34}$$

clearly indicating pure gain characteristics.

5.3.2 Response of Pure Gain System to Various Inputs

From the mathematical representation of the pure gain process in Eq. (5.30) or (5.31), it is easy to see that:

> The output of a pure gain process is **directly proportional** to the input, the constant of proportionality being the process gain.

Thus the responses are identical in form to the input (or forcing) functions, differing only in magnitude. The input function is amplified if $K > 1$, attenuated if $K < 1$, and left unchanged if $K = 1$.

The following is a catalog of the responses of a pure gain process to various input functions.

Step Response

Input

$$u(t) = \begin{cases} 0; & t < 0 \\ A; & t > 0 \end{cases} \tag{3.36}$$

Output

$$y(t) = \begin{cases} 0; & t < 0 \\ AK; & t > 0 \end{cases} \tag{5.35}$$

Rectangular Pulse Response

Input

$$u(t) = \begin{cases} 0; & t < 0 \\ A; & 0 < t < b \\ 0; & t > b \end{cases} \tag{3.39}$$

Output

$$y(t) = \begin{cases} 0; & t < 0 \\ AK; & 0 < t < b \\ 0; & t > b \end{cases} \tag{5.36}$$

Impulse Response

Input

$$u(t) = A\,\delta(t) \tag{3.43}$$

Output

$$y(t) = AK\,\delta(t) \tag{5.37}$$

Ramp Response

Input

$$u(t) = \begin{cases} 0; & t < 0 \\ At; & t > 0 \end{cases} \tag{3.49}$$

Output

$$y(t) = \begin{cases} 0; & t < 0 \\ AKt; & t > 0 \end{cases} \tag{5.38}$$

Sinusoidal Response

Input

$$u(t) = \begin{cases} 0; & t < 0 \\ A\sin\omega t; & t > 0 \end{cases} \tag{3.51}$$

Output

$$y(t) = \begin{cases} 0; & t < 0 \\ AK\sin\omega t; & t > 0 \end{cases} \tag{5.39}$$

We note from here that the *frequency response* of the pure gain process indicates an amplitude ratio AR = K, and a phase angle $\phi = 0$. In other words, the UPR of the pure gain process is a sine wave whose amplitude is K times the amplitude of the input sine wave, and perfectly in phase with the input. This, of course, is in keeping with the characteristic of the pure gain process.

The advantage in being able to approximate the dynamic behavior of a fast dynamic process with that of a pure gain process lies in the fact that the derivation of the dynamic responses (an exercise which normally involves Laplace inversion) is now considerably simplified to an exercise involving mere multiplication by a constant.

5.4   PURE CAPACITY SYSTEMS

Let us return to the general, first-order differential equation used to model the dynamic behavior of first-order systems:

$$a_1 \frac{dy}{dt} + a_0 y = b\, u(t) \tag{5.1}$$

Recall that the characteristic first-order system parameters, the steady-state gain, and time constant, are obtainable as indicated in Eq. (5.3), provided $a_0 \neq 0$.

It is now of interest to investigate what happens when $a_0 = 0$ in Eq. (5.1). Observe that in this case, the original equation becomes:

$$a_1 \frac{dy}{dt} = bu(t)$$

or

$$\frac{dy}{dt} = K^* u(t) \tag{5.40}$$

where

$$K^* = \frac{b}{a_1} \tag{5.41}$$

A process modeled by Eq. (5.40) is known as a *pure capacity* process. Observe that regardless of the specific nature of the input function $u(t)$, the solution to Eq. (5.40) is easily obtained by direct integration; the result is, assuming zero initial conditions:

$$y(t) = K^* \int_0^t u(\sigma)\, d\sigma \tag{5.42}$$

Because the output of this process involves an integration of the input function, pure capacity processes are sometimes called *pure integrators*.

Taking Laplace transforms in Eq. (5.40) gives:

$$y(s) = \frac{K^*}{s} u(s) \tag{5.43}$$

from which we obtain the transfer function for the pure capacity system as:

$$g(s) = \frac{K^*}{s} \tag{5.44}$$

The pure capacity system is therefore characterized by the presence of the integrator (or capacitance) element $1/s$, and the parameter K^*, which may be regarded as an integrator gain (see Eq. (5.42).)

From the expressions in Eqs. (5.3a,b) and Eq. (5.44), we may now note that the behavior of a first-order system approaches that of a pure capacity system in the limit as $\tau \to \infty$ and $K \to \infty$ while their ratio, K/τ, remains fixed at the value K^*. We may thus imagine a pure capacity system to be a first-order system whose time constant and steady-state gain are both extremely large but for which the ratio of these parameters is a fixed, finite constant.

The transfer function of a pure capacity system has one pole at the origin ($s = 0$) and no zeros.

5.4.1 Physical Example of a Pure Capacity System

The most common example of a pure capacity system is a storage (or surge) tank with an outlet pump such as shown in Figure 5.11.

Such a tank is typically used in the process industries for intermediate storage between two processes. The outflow, F (usually fixed), is set by the pump; its value is therefore independent of the liquid level, h, in the tank. The inflow F_i is usually the input variable, and its value can vary.

If A_c is the cross-sectional area of the storage tank (assumed uniform), then a material balance on the tank yields:

$$A_c \frac{dh}{dt} = F_i - F \tag{5.45}$$

At steady state, for a fixed value of F, Eq. (5.45) becomes:

$$0 = F_{is} - F \tag{5.46}$$

Subtracting Eq. (5.46) from Eq. (5.45) and defining the deviation variables $y = h - h_s$, $u = F_i - F_{is}$, we obtain:

$$\frac{dy}{dt} = \frac{1}{A_c} u \tag{5.47}$$

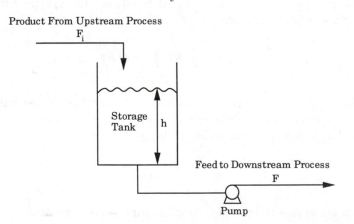

Figure 5.11.   The storage tank with an outlet pump.

from where Laplace transformation gives:

$$y(s) = \frac{1/A_c}{s} u(s) \qquad (5.48)$$

which is of the same form as Eq. (5.43), with $K^* = 1/A_c$.

Other examples of pure capacity processes include the heating of well-insulated batch systems, the filling of tanks with no outlet, the batch preparation of solutions by addition of chemicals to solvent, etc.

5.4.2   Response of Pure Capacity System to Various Inputs

Once again, by combining the transfer function of the pure capacity system (5.44) with the Laplace transform of the input function in question, we may use the transfer function model in Eq. (5.43) to derive this process system's response to various input functions.

Step Response

We have, in this case:

$$y(s) = \frac{K^* A}{s \ s} = \frac{AK^*}{s^2}$$

which is easily inverted back to time, giving:

$$y(t) = AK^*t \qquad (5.49)$$

the equation of a ramp function with slope AK^*.

A little reflection bears out the fact that this mathematically derived step response does in fact make physical sense. Observe that a sudden increase in the input flowrate to the example storage tank of Figure 5.11 results in a continuous increase in the liquid level in the tank — a situation which, in theory, can continue indefinitely but is limited in actual practice by the finite volume of the tank. (In actual practice, the liquid level in the tank increases until the tank is full and material begins to overflow.)

The following is a catalog of the pure capacity process response to the other input functions. It is left as an exercise to the reader to verify these results mathematically, and to confirm that they make physical sense.

Rectangular Pulse Response

$$y(t) = \begin{cases} 0; & t < 0 \\ AK^*t; & 0 < t < b \\ AK^*b; & t > b \end{cases} \qquad (5.50)$$

a response that starts out as a ramp function, and subsequently settles down to a new steady-state value AK^*b.

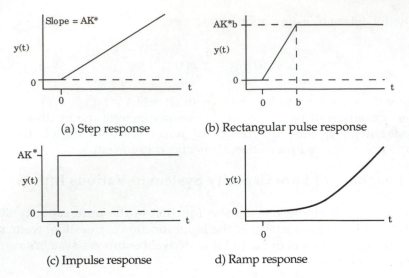

(a) Step response   (b) Rectangular pulse response

(c) Impulse response   d) Ramp response

Figure 5.12.   Pure capacity system responses.

Impulse Response

$$y(t) = AK*$$ (5.51)

a step function.

Ramp Response

$$y(t) = \frac{AK*t^2}{2}$$ (5.52)

Sinusoidal Response

$$y(t) = \frac{AK*}{\omega}(1 - \cos\omega t)$$

or

$$y(t) = \frac{AK*}{\omega}[1 + \sin(\omega t - 90°)]$$ (5.53)

implying that the *frequency response* of a pure capacity process has an amplitude ratio and a phase angle given by:

$$AR = \frac{K*}{\omega}$$ (5.54a)

$$\phi = -90°$$ (5.54b)

Diagrammatic representations of the step, pulse, impulse, and ramp responses are shown in Figure 5.12; the sinusoidal response is shown separately in Figure 5.13.

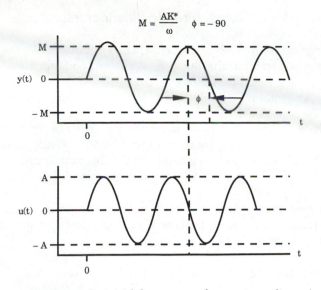

Figure 5.13. Sinusoidal response of a pure capacity system.

5.5 THE LEAD/LAG SYSTEM

The dynamic system whose transfer function is given by:

$$g(s) = K \frac{(\xi s + 1)}{(\tau s + 1)} \tag{5.55}$$

is known as a *lead/lag system* for reasons we shall soon explain.

It is important to note that the one — and very critical — difference between this transfer function and that for the first-order system is the presence of the additional first-order numerator term. Both transfer functions otherwise have denominator polynomials of identical order.

Let us now recall that when considering the frequency response of a first-order system, we discovered that the UPR sine wave *lags* behind the input sine wave by ϕ degrees. For this reason, first-order terms appearing in the denominator of a transfer function are sometimes referred to as *first-order lags*.

As will be shown in Chapter 9, for a system whose transfer function is of the form:

$$g(s) = K (\xi s + 1) \tag{5.56}$$

with no denominator polynomial, frequency response analysis results indicate that the UPR sine wave *leads* the input sine wave by:

$$\phi = \tan^{-1}(\omega \xi) \tag{5.57}$$

For this reason, first-order numerator terms in a transfer function are referred to as first-order *leads*.

Because the system currently under consideration has *both* a lead and a lag term, it is therefore naturally referred to as a lead/lag system.

Characteristic Parameters

The characteristic parameters of the lead/lag system are: K, the system gain; ξ, the lead time constant, τ, the lag time constant; and the "lead-to-lag" ratio, $\rho = \xi/\tau$.

The transfer function for the lead/lag system has one pole located at $s = -1/\tau$ and one zero located at $s = -1/\xi$.

Although it is rare to find a chemical process whose behavior is characterized by the lead/lag transfer function of Eq. (5.55), it is important to understand them because these systems find important applications in control system design such as that in the implementation of feedforward control systems.

Lead/lag systems are typically implemented in control systems as analog units, or electronic (microprocessor-based) systems that operate according to the transfer function form given in Eq. (5.55).

5.5.1   Obtaining the Dynamic Response

In investigating the response of the lead/lag system to various inputs, we find it more informative first to recast the somewhat strange transfer function in Eq. (5.55) into a more familiar form.   Observe that by partial fraction expansion, we may break up the transfer function into two terms as follows:

$$g(s) = K\frac{(\xi s + 1)}{(\tau s + 1)} = K\left[A_0 + \frac{A_1}{(\tau s + 1)}\right] \qquad (5.58)$$

where the constants A_0, and A_1 are easily found to be given by:

$$A_0 = \frac{\xi}{\tau} = \rho \qquad (5.59a)$$

$$A_1 = 1 - \frac{\xi}{\tau} = (1 - \rho) \qquad (5.59b)$$

Thus the dynamic behavior of the lead/lag system may be determined using the following transfer function representation:

$$y(s) = g(s)\, u(s) = \left[\rho K + (1 - \rho)\left(\frac{K}{\tau s + 1}\right)\right]u(s) \qquad (5.60)$$

A careful observation of Eq. (5.60) now reveals a very important fact about lead/lag systems:

> *The dynamic behavior of the lead/lag system is a **weighted average** of the dynamic behavior of a pure gain system and that of a first-order system, the weighting factor being ρ, the "lead-to-lag" ratio.*

This fact significantly simplifies the task of obtaining the dynamic response of lead/lag systems to various input functions. For example, observe that if $y_g(t)$ is the response of a pure gain system (gain = K) to a certain input function, and $y_1(t)$ is the response of the first-order system (gain K, time constant τ) to the same input function, then the response of the lead/lag system (gain K, lead ξ, lag τ, and lead-to-lag ratio ρ) is given by:

$$y(t) = \rho y_g(t) + (1 - \rho)y_1(t) \tag{5.61}$$

Since we are already very familiar with the dynamic responses of the pure gain, and the first-order systems, obtaining the dynamic responses of the lead/lag system is now very straightforward.

Process Example of a Lead/Lag System

As an example of a lead/lag process, consider the mixing tank shown in Figure 5.14. A large flowrate F_A of material A is fed into a tank of volume V, where a tiny quantity F_B of material B (e.g., a dye, catalyst, etc.) is mixed with it. However, it has been found that in order to achieve a fast changeover from one mixture to another, it is desirable also to feed a fraction ρ of the B feed stream directly into the tank outlet. In this case the outlet concentration of the B, c_B, would be given by:

$$V\frac{dc_{B_{tank}}}{dt} = (1 - \rho)F_B\, c_{Bf} - \left[F_A + (1 - \rho)F_B\right]c_{B_{tank}} \tag{5.62a}$$

$$c_{B_{outlet}} = \frac{\left[F_A + (1 - \rho)F_B\right] c_{B_{tank}} + \rho F_B c_{Bf}}{F_A + F_B} \tag{5.62b}$$

If we note that $F_A \gg F_B$, and define:

$$\tau = \frac{V}{F_A}$$

then it is reasonable to make the following approximations:

$$F_A + F_B \approx F_A + (1 - \rho)F_B \approx F_A$$

and the process model simplifies to:

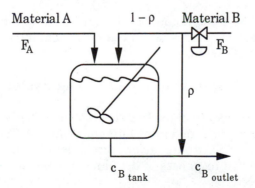

Figure 5.14.   Example of a lead/lag process.

$$\tau \frac{dc_{B_{tank}}}{dt} = \left[\frac{(1 - \rho)F_B}{F_A} \right] c_{Bf} - c_{B_{tank}} \tag{5.63a}$$

$$c_{B_{outlet}} = c_{B_{tank}} + \left(\frac{\rho F_B}{F_A} \right) c_{Bf} \tag{5.63b}$$

If we now define the following variables:

$$u = \frac{F_B}{F_A} c_{Bf} - \left(\frac{F_B}{F_A} c_{Bf} \right)_s$$

$$y = c_{B_{outlet}} - \left(c_{B_{outlet}} \right)_s$$

$$x = c_{B_{tank}} - \left(c_{B_{tank}} \right)_s$$

where the notation $(\cdot)_s$ is used for the steady-state value of the indicated variable, then Eq. (5.63) becomes:

$$\tau \frac{dx(t)}{dt} = (1 - \rho)\, u(t) - x(t) \tag{5.64a}$$

$$y(t) = x(t) + \rho u(t) \tag{5.64b}$$

and by taking the Laplace transform we obtain:

$$x(s) = \frac{(1 - \rho)}{\tau s + 1}\, u(s)$$

$$y(s) = x(s) + \rho u(s) = \left[\rho + \frac{(1 - \rho)}{\tau s + 1} \right] u(s)$$

or if we let $\rho = \xi / \tau$.

$$y(s) = \frac{(\xi s + 1)}{(\tau s + 1)}\, u(s)$$

We now note that the transfer function for a process will involve a lead term whenever the state-space model has an output depending directly on an input, as in this case (see Eq. (5.64b).

5.5.2   Unit Step Response of the Lead/Lag System

The most important response of the lead/lag system is the unit step response, its response to a step input function of *unit* magnitude. Taking advantage of Eqs. (5.60) and (5.61), since from previous discussions in this chapter we already know that the unit step responses of the pure gain and first-order components are respectively given by:

$$y_g(t) = K \; ; y_1(t) \; = \; K\,(1 - e^{-t/\tau})$$

the required unit step response of the lead/lag system is easily seen to be given by:

$$y(t) \; = \; K[\rho + (1 - \rho)(1 - e^{-t/\tau})] \qquad\qquad (5.65)$$

This may now be rearranged to give:

$$y(t) \; = \; K\,[1 - (1 - \rho)\,e^{-t/\tau}] \qquad\qquad (5.66)$$

The unit step response of the lead/lag system is shown in Figure 5.15 drawn for the situation with $\tau = 1$ minute.

The important points to note about this response are:

1. At $t = 0^+$, $y(t) = K\rho$; i.e., the output experiences an immediate "jump" to the value $K\rho$. The pure gain component is clearly responsible for this aspect of the response.

2. As indicated by Eqs. (5.65) or (5.66), when $t \rightarrow \infty$, $y(t)$ attains to its final value K, in exponential fashion, at a speed determined solely by τ, the lag time constant.

3. Observe that the initial value of $y(t)$ is ρK, while the final ultimate value is K; thus whether the initial value is higher or lower than the final value will be determined by whether ρ is greater or less than 1.

As we might therefore expect from note 3 above, three distinct types of responses are represented in this figure, and they depend on the parameter ρ.

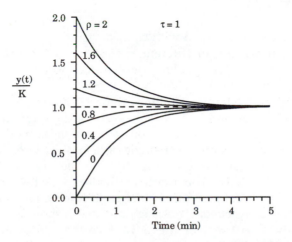

Figure 5.15.   The unit step response of the lead/lag system.

CASE 1. $\rho < 1$ (i.e., $\xi < \tau$)

This case is characterized by an initial jump to a value lower than the ultimate value followed by an exponential approach to the final, higher steady-state value.

To obtain this type of response requires $\xi < \tau$; i.e., a weaker lead (or a stronger lag). Such systems give the impression of being more reluctant to "move" by virtue of the dominant influence of the lag term. In the limit as $\xi \to 0$, the lead/lag system becomes a pure first -order system.

CASE 2. $\rho = 1$ (i.e., $\xi = \tau$)

Under conditions of identical lead and lag time constants, the lead/lag system becomes a pure gain system (by virtue of the cancellation of the identical numerator and denominator polynomials), attaining the ultimate value instantaneously; i.e., $y(t) = K$ for all times.

CASE 3. $\rho > 1$ (i.e., $\xi > \tau$)

The response in this case is characterized by an initial jump that overshoots the final value, followed by an exponential decay to the steady-state value.

This type of response is obtained when $\xi > \tau$; i.e., when we have a stronger lead (or a weaker lag). The swiftness of this type of response — indicative of the dominant influence of the lead term — tends to give the impression of an impulse response of a first-order system.

It is worthwhile to remark once again that the approach to the final steady state is governed solely by the lag time constant; but the nature and value of the initial response is determined by the lead-to-lag ratio.

The task of deriving the lead/lag system response to the remaining ideal forcing functions is quite straightforward and is left as an exercise to the reader.

5.6   SUMMARY

This chapter has been concerned with characterizing the dynamics of processes modeled by low-order, linear differential equations. We have found that the transfer functions for these systems involve denominators having at most a first-order polynomial.

The first-order system was identified as having two characteristic parameters: the steady-state gain K, a measure of the *magnitude* of the steady-state system response, and the time constant τ, a measure of *speed* of system response; its transfer function has a single pole and no zeros. Several physical examples of relatively simple processes that can be categorized as first-order systems were presented. The mathematical expressions representing the response of the first-order system to other ideal forcing functions were derived and the important characteristics of each of these responses highlighted.

When a_1, the coefficient of the only derivative term in the differential equation model of a first-order system vanishes, the resulting system responds instantaneously to input changes. Such a system was identified as a pure gain process. A *pure gain process* is therefore no more than a first-order system with zero time constant — one that is infinitely fast. The transfer function for this process was found to have no pole, and no zero: only a constant gain term. As such, the dynamic response of pure gain processes are very easy to compute since they are mere multiples of the forcing functions. Very few such systems occur in actual practice, but first-order systems whose time constants are quite small (systems whose dynamics are very fast, responding with hardly noticeable transients) behave approximately like pure gain processes.

The disappearance of a_0, the coefficient of $y(t)$, from the first-order differential equation model results in a unique situation in which the first derivative of the process output is directly proportional to the input function (immediately implying that the output is directly proportional to the integral of the input function). Systems of this type were identified as *pure capacity systems* or *pure integrators*. The transfer function for these processes has a single pole located at the origin ($s = 0$), and no zero. We have noted that in the limit as the time constant *and* steady-state gain of a first-order system both approach infinity, doing so in such a way that the ratio K/τ remains fixed, the behavior of the pure capacity system is obtained. The storage tank in which the outflow is independent of the liquid level was presented as the typical example of the pure capacity system. The dynamic responses of this class of systems were also derived and discussed.

The final class of systems to be discussed was the lead/lag system. The characteristic parameters of the lead/lag system were identified as the system gain, a lead time constant, a lag time constant, and the lead-to-lag ratio. The transfer function has one pole *and* one zero; and it is the presence of this zero that really distinguishes the lead/lag system from all the other low-order systems. We showed that the dynamic behavior of the lead/lag system is in fact a weighted average of the dynamic behaviors of the pure gain and the first-order systems. This fact was used to derive the unit step response of the lead/lag system, and its somewhat unusual characteristics were then discussed. What we have learned from this particular response will become useful in Chapter 16.

Our study of the dynamic behavior of process systems continues in the next chapter when we investigate the behavior of systems whose dynamics are described by models of higher order than the ones we have encountered in this chapter.

REFERENCES AND SUGGESTED FURTHER READING

1. Bode, H. W., *Network Analysis and Feedback Amplifier Design*, Van Nostrand, New York (1945)

REVIEW QUESTIONS

1. What is a first-order system and what are its characteristic parameters? Which of these characteristic parameters measures "how much" the process will respond to input changes, and which measures "how fast" the process will respond?

2. About how long in units of "time constants" will it take for the step response of a first-order system to attain *more* than 99% of its ultimate value?

3. What does the "inherent reluctance factor" measure?

4. How can the characteristic parameters of a first-order system be determined from its theoretical step response?

5. What is the relationship between a process *step* response and *impulse* response?

6. What is the relationship between a process *ramp* response and *step* response?

7. What are the main characteristics of the sinusoidal response of a first-order system?

8. What is frequency response in general?

9. What is a pure gain system and what are its main characteristics?

10. What limiting type of first-order system is a pure gain system?

11. What is the main characteristic of a pure capacity system?

12. What limiting type of first-order system is a pure capacity system?

13. A lead/lag system has how many poles and how many zeros?

14. In what way does the value of the lead-to-lag ratio affect the step response of a lead/lag system?

15. What relationship exists between the dynamic behavior of the lead/lag system and that of the pure gain system and the first-order system?

PROBLEMS

5.1 The process shown in Figure P5.1 is a *constant volume* electric water heater. On a particular day, as the tank water temperature reached the 80°C mark, the heater broke down and stopped supplying heat. At this time, water was withdrawn from this 100 liter capacity tank at the rate of 10 liters/min; cold water at 30°C automatically flowed into the tank at precisely the same rate. This exercise went on for exactly 5 min, and the water withdrawal was stopped. (Because of the heater design the cold water also stopped flowing in.)

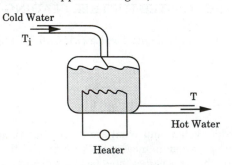

Figure P5.1.   The water heater.

By developing an appropriate mathematical model for this process, show that it is reasonable to consider this a first-order process; and by solving the resulting differential equation model, find the final tank water temperature at the end of this 5-minute period. *State all your assumptions.*

5.2 The temperature T on a critical tray in a pyrolysis fractionator is to be controlled as indicated in the following schematic diagram, employing a *feedforward controller* that manipulates the underflow reflux flowrate R on the basis of feed flowrate measurements F. By experimental testing, the following approximate transfer function models were obtained for the response of the critical tray temperature (1) to feed rate changes, and (2) to reflux flowrate changes:

$$(1) \qquad\qquad (T - T^*) = \frac{20}{30s + 1} (F - F^*) \qquad\qquad (P5.1)$$

and:

$$(2) \qquad\qquad (T - T^*) = \frac{-5}{6s + 1} (R - R^*) \qquad\qquad (P5.2)$$

the timescale is in minutes, the flowrates and temperature are, respectively, in Mlb/hr (thousands of pounds/hour) and °F, and the (*) indicates steady-state conditions.

(a) Rewrite the model in the form:

$$y(s) = g(s)u(s) + g_d(s)d(s)$$

stating explicitly what the variables y, u, and d represent in terms of the original process variables, and clearly identifying $g(s)$ and $g_d(s)$.

(b) Obtain an expression for the dynamic response of the tray temperature to a unit step change in the underflow reflux flowrate, given that the initial temperature on the tray is 250°F and that the feedrate remained constant. To what ultimate value will the tray temperature settle after this unit change in R?

(c) Obtain an expression for the dynamic response of the tray temperature, this time to a unit step change in the feed flowrate, once more given that the initial temperature on the tray is 250°F, and that this time the reflux rate has remained constant. Deduce the final steady-state tray temperature in response to this change in feedrate.

PYROLYSIS FRACTIONATOR

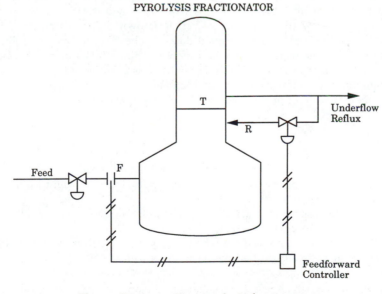

Figure P5.2. The pyrolysis fractionator.

5.3   (a) Using your favorite computer simulation software, or otherwise, plot the dynamic responses obtained in Problem 5.2(b) and (c) above on the same graph, and on the same scale. Which input variable affects the tray temperature *the most* (in terms of *absolute* change in the tray temperature), and which affects it *the quickest* (i.e., which response gets to steady state the quickest)? (Keep in mind that these are both *unit* step responses.)

(b) By summing up the individual expressions for the two responses (or otherwise) obtain an expression for the tray temperature response to a *simultaneous* implementation of unit step changes in both the feedrate and the reflux rate, and given the same initial temperature of 250°F. Plot this combined response and comment on its shape.

5.4   (a) Assuming that the model supplied for the pyrolysis fractionator in Problem 5.2 is accurate, given that an inadvertent step *change* of 2 Mlb/hr has occurred in the feedrate, what *change* would you recommend to be made in the reflux flowrate in order that *at steady state* no net change will be evident in the tray temperature?

(b) It has been recommended by an experienced control engineer that if the specified process model is accurate, "perfect" feed disturbance rejection can be obtained if the reflux flowrate is made to change with observed variations in the feedrate according to:

$$(R - R^*) \;=\; g_{FF}(F - F^*) \tag{P5.3}$$

where the indicated transfer function, g_{FF}, is given by:

$$g_{FF} \;=\; \frac{4(6s + 1)}{(30s + 1)} \tag{P5.4}$$

Using Eqs. (P5.3) and (P5.4), for the situation when the process is operating under the indicated strategy, obtain an expression for the reflux rate response to the step change of 2 Mlb/hr introduced in part (a) above. Plot the response and compare the *ultimate change* in the reflux flowrate recommended by this control scheme with your recommendation in part (a).

5.5   In *Process Identification* (see Chapter 13) the objective is to obtain an approximate transfer function model for the process in question by correlating input/output data obtained directly from the process. It is known that the frequency-response information derived from such plant data is richest in information content if the process input is an ideal impulse function. Since this function is, of course, impossible to implement perfectly in reality, it is customary to utilize instead the rectangular pulse input as a close approximation.

Given a first-order process with the following transfer function:

$$g(s) \;=\; \frac{4}{5s + 1} \tag{P5.5}$$

(a) Obtain the theoretical *unit* impulse response; let this be $y^*(t)$.

(b) Obtain the response to $u_1(t)$, a rectangular pulse input of height 1 unit and width 1 unit; let this be $y_1(t)$.

(c) Obtain the response to $u_2(t)$, a rectangular pulse input of height 2 units and width 0.5 unit; let this be $y_2(t)$.

(d) Compare these three responses by plotting $y^*(t)$, $y_1(t)$, and $y_2(t)$ together on the same graph and confirm that the rectangular pulse input with a narrower *width* provides a better approximation to the ideal impulse function.

5.6   The procedure for starting up some industrial reactors calls for the gradual introduction of reactants in several steps rather than all at once. Given that the dynamic behavior of one such reactor is reasonably well represented by the simple transfer function model:

$$y(s) = \frac{0.32}{15.5s + 1} u(s) \qquad\qquad (P5.6)$$

where y represents, in deviation variables, the reactor temperature (with the time scale in minutes), it is required to find the response of the process to a program of reactant feed represented by the staircase function described mathematically as:

$$u(t) = \begin{cases} 0; & t \le 0 \\ 1; & 0 < t \le 2 \\ 2; & 2 < t \le 4 \\ 3; & 4 < t \le 6 \\ 4; & t > 6 \end{cases} \qquad\qquad (P5.7)$$

and illustrated in Figure P5.3.
While it is possible to obtain first the Laplace transform of the input function $u(t)$ and then use Eq. (P5.6) directly, this may not be very convenient. Find the process impulse-response function $g(t)$, and recast the given process model in the impulse-response form:

$$y(t) = \int_0^t g(\sigma)u(t - \sigma)\, d\sigma \qquad\qquad (P5.8)$$

Now use this directly with Eq. (P5.7) to obtain the required response.

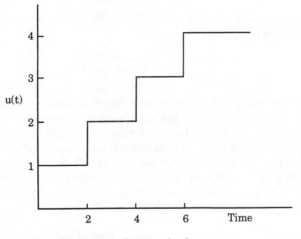

Figure P5.3. Reactor feed program.

5.7   A storage tank of the type shown in Figure 5.11 in the main text is subject to periodic loading from the upstream process. Because a steady rate of product withdrawal is maintained by the constant speed pump at the tank outlet, this is a true "pure capacity" process. The tank's cross-sectional area is given as 2.5 ft², and that the incoming feedstream fluctuates around its nominal flowrate in essentially sinusoidal fashion with *maximum* deviation of 10 ft³/hr; the frequency of the sinusoidal fluctuation is different from shift to shift, but it may be considered constant on any particular shift.

(a) *Derive* from first principles the expression for the response of the tank level (as a deviation from nominal operating conditions) assuming that the input is a perfect sinusoid with the given amplitude but a yet unspecified frequency ω. For a given frequency of 0.2 radians/hr, plot on the same graph, both the periodic input loading function and the tank level response. Compare these two functions.

(b) In terms of the frequency of the periodic loading, what is the maximum deviation from nominal tank level that will be observed?

(c) Given now that the tank is 10 ft high and that the nominal operating level is 5 ft, what condition must the loading frequency satisfy to guarantee that during any particular shift, the tank neither overflows nor runs dry?

5.8   Consider a household pressing iron along with its thermostat as a dynamic system of constant mass m and heat capacity C_p.

(a) Develop a mathematical model for this system assuming that the *rate* of heat input through the heating element is Q (appropriate energy units/min) and that the iron loses heat to the surrounding atmosphere according to Newton's "law" of cooling; i.e., the heat loss is given by $hA(T - T_a)$, where T_a is the (constant) temperature of the surrounding atmosphere, h is the heat transfer coefficient, and A is the effective area of heat transfer.

(b) Given T_s and Q_s as appropriate steady-state values for T and Q respectively, express your model in the transfer function form $y(s) = g(s) u(s)$ where $y = T - T_s$ and $u = Q - Q_s$; show that $g(s)$ has the first-order form. What are K and τ in this case?

(c) A specific pressing iron, with $K = 5$, has a thermostat set to cut off the standard energy supply ($u^* = 10$) when T attains 55°C. At $t = 0$, the iron was at an initial steady-state temperature of 24°C and the standard energy supply commenced; precisely 6 minutes after the commencement of this operation, the energy supply was automatically cut off. Find the effective time constant τ for this particular iron.

5.9   A certain process is composed of two first-order processes in parallel such that its transfer function is given by:

$$g(s) = \frac{K_1}{\tau_1 s + 1} + \frac{K_2}{\tau_2 s + 1} \qquad\qquad (P5.9)$$

(a) Consolidate this transfer function and identify its steady-state gain, its poles and its zeros, and thereby establish that the steady-state gain is the *sum* of the steady-state gains of the contributing transfer functions; the poles are identical to the poles of the contributing transfer functions; the *lead* time constant (from which the zero is obtained) is a weighted average of the two contributing *lag* time constants. (The dynamic behavior of such systems are investigated in Chapters 6 and 7.)

(b) What conditions must the parameters of the contributing transfer functions satisfy if the composite system zero is to be positive?

5.10   Consider creating a model for the bottom part of the distillation column shown below in Figure P5.4. Here, in terms of deviation variables:

V = vapor rate leaving the bottom; W_R = mass in reboiler
L = liquid rate entering the bottom Q = heat input rate
W_B = mass in column base B = bottom product
                                                                             withdrawal rate

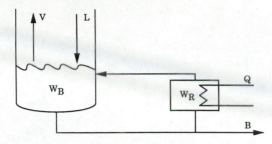

Figure P5.4. The bottom part of a distillation column.

Transfer function relations are given as:

$$\frac{W_R(s)}{Q(s)} = \frac{-8}{10s + 1} \quad ; \quad \frac{V(s)}{Q(s)} = \frac{0.01}{10s + 1} \quad ; \quad \frac{L(s)}{V(s)} = \frac{1}{1000s + 1}$$

in addition to the differential equation *in time*:

$$\frac{d\,[W_R(t) + W_B(t)]}{dt} = L - V - B$$

(a) Derive the transfer function $W_B(s)/Q(s)$ assuming that $B = 0$.

(b) Obtain the mathematical expression for $W_B(t)$ given that initially $W_B(0) = 0$ and a step change of magnitude Q^* is made in Q at $t = 0$.

DYNAMIC BEHAVIOR OF LINEAR HIGHER ORDER SYSTEMS

Having discussed the behavior of low-order systems in the last chapter, the next level of complexity involves those systems modeled by linear differential equations of higher (but still *finite*) order. Such systems are typically characterized by transfer functions in which the denominator polynomials are of order higher than 1. The objectives in this chapter are similar to those of the preceding chapter: to investigate and characterize the dynamic responses of this class of systems.

We will begin our investigation of such systems in the most logical place: with the composite system made up of two first-order systems connected in series. We will find this to be a natural extension of our earlier discussion on first-order systems. This opening discussion of two first-order systems in series will also serve two other purposes:

1.   As a precursor to the discussion of general, second-order systems; for, as we shall see, the former is, in fact, a special case of the latter, and

2.   As a natural launching pad for subsequently generalizing to Nth order systems (with $N>2$).

Our discussion in this chapter will end with a treatment of the dynamic behavior of one final class of higher order systems: those whose transfer functions have *zeros* in addition to the higher order denominator polynomials.

There is yet another class of systems — *time delay systems* — qualified to be classified as "higher order." However, these systems are, in principle, of *infinite* order and will be discussed at a later stage in Chapter 8.

6.1   TWO FIRST-ORDER SYSTEMS IN SERIES

Consider the situation in which two liquid holding tanks are connected in series, each one being of the type identified as a first-order system in Chapter 5

(Section 5.1.1). Such systems can be configured in one of two ways. Let us consider the first of such configurations shown in Figure 6.1 below. This indicates a situation in which the flow out of tank 1 discharges freely into the atmosphere before entering tank 2. The flowrate of this stream is therefore solely determined by the liquid level in tank 1 and is totally independent of the conditions in tank 2.

Thus while the conditions in tank 1 influence the conditions in tank 2, observe carefully that the conditions in tank 2 in no way influence tank 1; i.e., tank 2 does not *interact* with tank 1. This is the *noninteracting* configuration.

Contrast the arrangement in Figure 6.1 with the one shown in Figure 6.2. In this case, the flow out of tank 1 now depends on the *difference* between the levels in the two tanks, with the result that variations in the conditions in tank 2 *will* influence the conditions in tank 1. Such an arrangement gives rise to a situation in which tank 2 does in fact interact with tank 1; it is therefore called the *interacting* configuration.

As we might expect, the dynamic behavior of the interacting ensemble will differ from the dynamic behavior of the alternative noninteracting arrangement. Let us look at these in turn, beginning with the noninteracting system of Figure 6.1.

6.1.1 Two Noninteracting Systems in Series

Let us, for simplicity, assume a linear relation between liquid level and flowrate through the outlet valves in the system shown in Figure 6.1, so that outflows will be directly proportional to the liquid levels. If A_1 and A_2 represent the (uniform) cross-sectional areas of tanks 1 and 2 respectively, then the following mathematical models are obtained by carrying out material balances for each tank:

- For Tank 1:

$$A_1 \frac{dh_1}{dt} = F_0 - c_1 h_1 \tag{6.1}$$

- For Tank 2 the inflow is given by $F_1 = c_1 h_1$; thus we have:

$$A_2 \frac{dh_2}{dt} = c_1 h_1 - c_2 h_2 \tag{6.2}$$

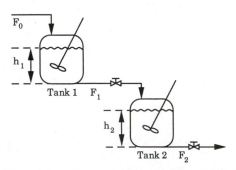

Figure 6.1. Two noninteracting tanks in series.

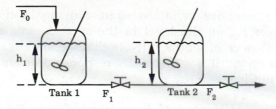

Figure 6.2. Two interacting tanks in series.

These two equations may be rearranged and presented in terms of deviation variables in the usual manner; the results are:

$$\tau_1 \frac{dy_1}{dt} = -y_1 + K_1 u \tag{6.3}$$

$$\tau_2 \frac{dy_2}{dt} = -y_2 + K_2 y_1 \tag{6.4}$$

where the deviation variables are given by $y_1 = h_1 - h_{1s}$; $y_2 = h_2 - h_{2s}$, and $u = F_0 - F_{0s}$; and the system parameters are given by:

$$\tau_1 = \frac{A_1}{c_1}; \ \tau_2 = \frac{A_2}{c_2}$$

and

$$K_1 = \frac{1}{c_1}; \ K_2 = \frac{c_1}{c_2}$$

It is important to note the following facts about this system:

1. Each component tank is modeled by a first-order differential equation; therefore the entire process is an ensemble of two first-order systems in series.

2. The individual transfer function representation for each of the component tanks may be obtained by taking Laplace transforms in Eqs. (6.3) and (6.4). The results are:

$$y_1(s) = \frac{K_1}{\tau_1 s + 1} u(s) = g_1 u(s) \tag{6.5}$$

$$y_2(s) = \frac{K_2}{\tau_2 s + 1} y_1(s) = g_2 y_1(s) \tag{6.6}$$

clearly identifying each tank as a first-order system in its own right.

Now, for this system, we are interested in studying the effect of changes in the input function F_0, on the level in the second tank; i.e., we wish to investigate the influence of u on y_2. (Note that we are already familiar with the influence of u on y_1; it is a first-order response problem similar to those studied in detail in Chapter 5.)

Differential Equation Model

The differential equation we seek for this system is the one that represents the behavior of y_2 and its *direct* dependence on u, not its *indirect* dependence through the unimportant intermediate variable y_1.

Such a model can be obtained by eliminating y_1 from Eqs. (6.3) and (6.4). This may be done as follows.

Differentiating Eq. (6.4) with respect to time and substituting Eq. (6.3) for the resulting first derivative of y_1 gives:

$$\tau_2 \frac{d^2 y_2}{dt^2} = -\frac{dy_2}{dt} + \frac{K_2}{\tau_1}\left(-y_1 + K_1 u\right) \tag{6.7}$$

By solving Eq. (6.4) for y_1 and introducing the result into Eq. (6.7), we may now eliminate the last surviving y_1 term. The final expression appropriately rearranged is:

$$\tau_1 \tau_2 \frac{d^2 y_2}{dt^2} + (\tau_1 + \tau_2)\frac{dy_2}{dt} + y_2 = K_1 K_2 u \tag{6.8}$$

a second-order differential equation in y_2, with u as the forcing function.

Transfer Function Model

The transfer function representation for the composite system relating $y_2(s)$ directly to $u(s)$ is shown in Figure 6.3 and can be obtained by substituting Eq. (6.5) for y_1 in Eq. (6.6); the result is:

$$y_2(s) = \frac{K_2}{\tau_2 s + 1}\frac{K_1}{\tau_1 s + 1}u(s) \tag{6.9}$$

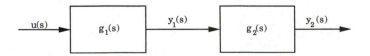

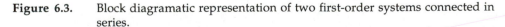

Figure 6.3.   Block diagramatic representation of two first-order systems connected in series.

The overall transfer function for this process is therefore now given by:

$$g(s) = \frac{K_1 K_2}{(\tau_1 s + 1)(\tau_2 s + 1)} \tag{6.10}$$

with the characteristic parameters:

- $K = K_1 K_2$, the "combined" steady-state gain,
- Two time constants, τ_1, and τ_2, the individual time constants of the contributing first-order elements.

The transfer function for this process therefore has two poles (at $s = -1/\tau_1$ and $s = -1/\tau_2$) and no zeros.

Alternatively, we may equally well obtain the transfer function model for this process by taking Laplace transforms of the differential equation model of Eq. (6.8), assuming zero initial conditions. It is an easy task — left as an exercise for the reader — to show that the results are identical.

A very important point to note here is that the overall transfer function for this process is in fact just a product of the two contributing transfer functions indicated in Eqs. (6.5) and (6.6). A block diagramatic representation of this system shown in Figure 6.3 emphasizes this point.

6.1.2 Unit Step Response of the Noninteracting Systems in Series

We may now use the transfer function of Eq. (6.9) to obtain expressions for how this system responds to various inputs. When the input function is the unit step function, for example, the system response may be obtained from:

$$y_2(s) = \frac{K}{(\tau_1 s + 1)(\tau_2 s + 1)} \frac{1}{s} \tag{6.11}$$

since, in this case $u(s) = 1/s$. Upon partial fraction expansion Eq. (6.11) is easily inverted (provided $\tau_1 \neq \tau_2$) to give:

$$y_2(t) = K\left[1 - \left(\frac{\tau_1}{\tau_1 - \tau_2}\right)e^{-t/\tau_1} - \left(\frac{\tau_2}{\tau_2 - \tau_1}\right)e^{-t/\tau_2}\right] \tag{6.12}$$

In the special case when $\tau_1 = \tau_2 = \tau$, Eq. (6.11) becomes:

$$y_2(s) = \frac{K}{(\tau s + 1)^2} \frac{1}{s} \tag{6.13}$$

and Laplace inversion yields:

$$y_2(t) = K\left(1 - e^{-t/\tau} - \frac{t}{\tau}e^{-t/\tau}\right) \tag{6.14}$$

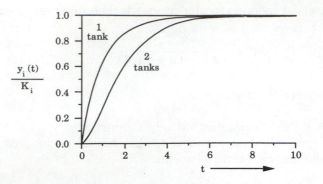

Figure 6.4. Unit step responses of the liquid level in the two-tank process.

A sketch of the two-tank system response is shown in Figure 6.4 and compared with the first-order response behavior of a single tank.

The important points to note about these responses are:

1. While the single first-order system response is instantaneous, the response of the two-tank system shows a sigmoidal behavior characterized by an initial sluggishness (at $t = 0$) followed by a speeding up prior to the final approach to steady state. Note the presence of an inflection point in the response curve.

2. It is left as an exercise to the reader to show that the slope of the two-tank response at the origin is zero, i.e.:

$$\left.\frac{dy_2}{dt}\right|_{t=0} = 0$$

    which is to be compared with the nonzero initial slope of the single, first-order response given in Eq. (5.14).

3. As was obtained for the single first-order system in Eq. (5.15), the same measure of "inherent reluctance" for the two (noninteracting) first-order systems in series may be obtained by integrating, over time, the difference between instantaneous response, $y^* = K$, and the actual unit step response given in Eq. (6.12), i.e.:

$$J_2 = \int_0^\infty [y^* - y(t)]\, dt$$

$$= K \int_0^\infty \left\{1 - \left[1 - \left(\frac{\tau_1}{\tau_1 - \tau_2}\right)e^{-t/\tau_1} - \left(\frac{\tau_2}{\tau_2 - \tau_1}\right)e^{-t/\tau_2}\right]\right\} dt$$

    which, when carefully evaluated, gives the interesting result:

$$J_2 = K\left(\tau_1 + \tau_2\right) \tag{6.15}$$

which is to be compared with the corresponding single, first-order system result: $J = K\tau$ when $A = 1$ in Eq. (5.15). The implication of this result is that the addition of another first-order system with a time constant of τ_2 increases the reluctance inherent within the system by an amount directly related to this time constant.

Thus two first-order systems in series are *more reluctant* to respond to changes than a single, first-order system by itself. Upon careful consideration of the physical system in Figure 6.1, a little reflection will confirm that this statement is in perfect keeping with what we would intuitively expect.

6.1.3   The Interacting System

Let us now consider the interacting system arrangement of Figure 6.2. The material balances for this arrangement are as shown below:

For Tank 1:

$$A_1 \frac{dh_1}{dt} = F_0 - F_1 \tag{6.16}$$

For Tank 2:

$$A_2 \frac{dh_2}{dt} = F_1 - F_2 \tag{6.17}$$

Assuming linear resistances, the flowrate out of tank 2, as before, is still given by $F_2 = c_2 h_2$; but now F_1 is given by:

$$F_1 = c_1(h_1 - h_2)$$

The mathematical models Eqs. (6.16) and (6.17) therefore become:

$$\frac{A_1}{c_1} \frac{dh_1}{dt} = -h_1 + h_2 + \frac{F_0}{c_1} \tag{6.18}$$

$$\frac{A_2}{c_2} \frac{dh_2}{dt} = \frac{c_1}{c_2} h_1 - \left(1 + \frac{c_1}{c_2}\right) h_2 \tag{6.19}$$

As was done in the noninteracting case (in producing Eqs. (6.3) and (6.4)), if we now introduce the same constants, and express these equations in terms of the usual deviation variables, taking Laplace transforms and rearranging the resulting expressions will give rise to the following transfer function model:

$$y_2(s) = \frac{K}{\tau_1 \tau_2 s^2 + (\tau_1 + \tau_2 + K_2 \tau_1)s + 1} u(s) \tag{6.20}$$

where, once again, $K = K_1 K_2$.

Observe now that the difference between the transfer function for the noninteracting case (given in Eq. (6.9)) and the one indicated in Eq. (6.20) for the interacting case lies in the presence of the term $K_2\tau_1$ in the coefficient of s in the denominator polynomial in Eq. (6.20). This is appropriately indicative of the interaction of tank 2 (via its steady-state gain K_2) with tank 1 (via its time constant τ_1).

The following example will assist us in appreciating the effect of interaction on the dynamic behavior of interacting first-order systems in series.

Example 6.1   COMPARISON OF THE DYNAMIC BEHAVIOR OF LIQUID LEVEL FOR INTERACTING AND NONINTERACTING CONFIGURATIONS.

Consider the situation in which the system of Figure 6.1 consists of two identical tanks with identical time constants $\tau_1 = \tau_2 = 1$ minute. The steady-state gains are also identical, and equal to 1; i.e., $K_1 = K_2 = 1$.

1.   Obtain an expression for the response of y_2 (the level in the second tank as a deviation from its initial steady-state value) to a unit step change in the inlet flowrate to tank 1.

2.   If these same tanks are now arranged as in Figure 6.2, obtain the unit step response for this interacting configuration and compare with the response obtained in (1) for the noninteracting case.

Solution:

1.   In this case since $\tau_1 = \tau_2$, the required response may be obtained by directly appealing to our earlier result in Eq. (6.14) for $\tau = 1$ and $K = 1$; what we have is the following result:

$$y_2(t) = \left(1 - e^{-t} - t\, e^{-t}\right) \qquad (6.21)$$

2.   In the interacting situation, introducing the given parameter values into Eq. (6.20), we find that the transfer function model is given by:

$$y_2(s) = \frac{1}{s^2 + 3s + 1} u(s)$$

which, for a unit step input, becomes (after factorization of the denominator quadratic):

$$y_2(s) = \frac{1}{(2.618s + 1)(0.382s + 1)} \frac{1}{s} \qquad (6.22)$$

Before carrying out the final step of Laplace inversion in Eq. (6.22), note first of all that this equation, when compared with Eq. (6.11), indicates that the *interacting* configuration of two first-order systems in series, each having a time constant of 1 minute, is equivalent to a *noninteracting* configuration of two first-order systems having unequal time constants 2.618 min and 0.382 min.

Thus the first effect of interaction is to alter the effective time constants of the individual contributing first-order systems.

Laplace inversion of Eq. (6.22) immediately yields:

$$y_2(t) = \left(1 - 1.171e^{-0.382t} + 0.171e^{-2.618t}\right) \qquad (6.23)$$

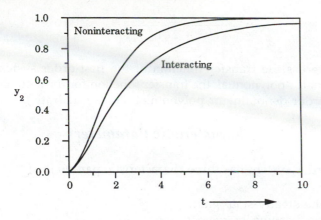

Figure 6.5.   Step response of interacting and noninteracting tanks in series.

A plot of the two responses (6.21) and (6.23) are shown in Figure 6.5. It is now clear that the main effect of interaction is to make the response more sluggish. It should be easy to see, from purely physical grounds, that this is indeed to be expected once we take cognizance of the fact that with the interacting arrangement, any increase in tank 2 level will reduce the flow from tank 1, further slowing down the rate at which this level will increase. This is not the case in the noninteracting arrangement.

6.2   SECOND-ORDER SYSTEMS

A second-order system is one whose dynamic behavior is represented by the second-order differential equation of the type:

$$a_2\frac{d^2y}{dt^2} + a_1\frac{dy}{dt} + a_0 y = bu(t) \tag{6.24}$$

where $y(t)$ and $u(t)$ are respectively the system's output and input variables.

Once again, it is customary to rearrange such an equation to a "standard form" in which the characteristic system parameters will be more obvious. In this case, the "standard form" is:

$$\tau^2\frac{d^2y}{dt^2} + 2\zeta\tau\frac{dy}{dt} + y = Ku(t) \tag{6.25}$$

and the newly introduced parameters are given (for $a_0 \neq 0$) by:

$$\tau = \sqrt{\frac{a_2}{a_0}} \quad ; \qquad 2\zeta\tau = \frac{a_1}{a_0} \quad ; \qquad K = \frac{b}{a_0}$$

Assuming, as usual, that the model in Eq. (6.25) is in terms of deviation variables, Laplace transformation and subsequent rearrangement gives the transfer function model:

$$y(s) = \frac{K}{\tau^2 s^2 + 2\zeta\tau s + 1} u(s)$$

so that the general transfer function for the second-order system is given by:

$$g(s) = \frac{K}{\tau^2 s^2 + 2\zeta\tau s + 1} \tag{6.26}$$

Note that whereas the transfer function of the first-order system has a *first-order* denominator polynomial, the transfer function for the second-order system has a *second*-order denominator polynomial (i.e., a quadratic).

Characteristic Parameters

The second-order system has three characteristic parameters:

1. K, the steady-state gain,
2. ζ, the *damping coefficient*, and
3. τ, the *natural period* (or the inverse natural frequency)

The reason for the terminology adopted for these parameters will soon become clear.

The second-order system's transfer function has no zero, but has two poles located at $s = r_1$ and $s = r_2$ where r_1 and r_2 are the two roots of the denominator quadratic.

6.2.1   Physical Examples of Second-Order Systems

1. Two First-Order Systems in Series

Let us now recall the opening discussion of this chapter involving the dynamic behavior of two first-order systems in series. We obtained the differential equation model for the noninteracting arrangement as the second-order differential equation in Eq. (6.8).

Comparing this now with the standard form in Eq. (6.25), we see that:

The system consisting of a noninteracting, series arrangement of two first-order systems (with time constants τ_1 and τ_2 and steady-state gains K_1 and K_2) is a second-order system with the following parameters:

$$\tau^2 = \tau_1\tau_2 \quad \Rightarrow \quad \tau = \sqrt{\tau_1\tau_2}$$

$$K = K_1 K_2$$

$$\text{and} \quad 2\zeta\tau = (\tau_1 + \tau_2) \quad \Rightarrow \quad \zeta = \frac{1}{2}\frac{(\tau_1 + \tau_2)}{\sqrt{\tau_1\tau_2}}$$

A comparison of the transfer function representation in Eq. (6.20) with Eq. (6.26) shows that the interacting configuration gives rise to a second-order system in which τ and K are as given above for the noninteracting system, but with ζ given by:

$$\zeta = \frac{1}{2}\frac{(\tau_1 + \tau_2 + K_2\tau_1)}{\sqrt{\tau_1\tau_2}}$$

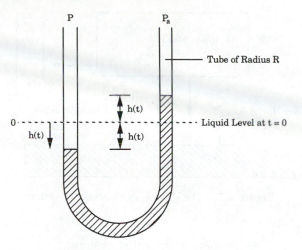

Figure 6.6. The U-tube manometer.

2. The U-Tube Manometer

The U-tube manometer shown in Figure 6.6 is a device used for measuring pressure.

The dynamic behavior of the liquid level in each leg of the manometer tube in response to pressure changes can be obtained by carrying out a force balance on this system. As shown in Refs. [1, 2], the resulting equation is:

$$\frac{d^2h}{dt^2} + \frac{6\mu}{R^2\rho}\frac{dh}{dt} + \frac{3}{2}\frac{g}{L}h = \frac{3}{4\rho L}\Delta P \tag{6.27}$$

where h is the displacement of the liquid level from rest position, L is the total length of liquid in the manometer, R is the radius of the manometer tube; ρ and μ are respectively the density and viscosity of the manometer liquid; ΔP is the pressure difference across the tops of the two manometer legs; and g is the acceleration due to gravity.

When Eq. (6.27) is arranged in the standard form, we observe that the dynamic behavior of this system is second order with:

$$K = \frac{1}{2\rho g}$$

$$\tau^2 = \frac{2}{3}\frac{L}{g} \Rightarrow \tau = \sqrt{\frac{2}{3}\frac{L}{g}}$$

and
$$2\zeta\tau = \frac{4\mu L}{\rho g R^2} \Rightarrow \zeta = \frac{\mu}{\rho R^2}\sqrt{\frac{6L}{g}}$$

3. The Spring-Shock Absorber System

Our final example is the spring-shock absorber system shown in Figure 6.7 (also known as the damped vibrator in mechanics).

Performing a force balance on this system gives the following differential equation model:

$$m\frac{d^2x}{dt^2} = -kx - c\frac{dx}{dt} + f(t) + g \tag{6.28}$$

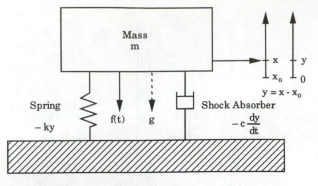

Figure 6.7.   The spring-shock absorber system.

where

m	is the mass of the entire system
x	is the displacement of the spring from the fully extended position
g	is the force of gravity
$f(t)$	is some forcing function imposed downwards
$-kx$	is the force exerted by the spring (k is the spring constant)
$-c\dfrac{dx}{dt}$	is the force due to the shock absorber, with c as the coefficient of viscous damping

If the initial rest position for the system is $x(0) = x_0$, with $f(0) = 0$ observe that the force due to the compressed spring will exactly equal the force of gravity, i.e.:

$$kx_0 = g$$

We may now rewrite Eq. (6.28) in terms of the deviation variable $y = x - x_0$; the result is:

$$m\frac{d^2y}{dt^2} = -ky - c\frac{dy}{dt} + f(t)$$

indicating a second-order system with:

$$u(t) = f(t); \quad \tau = \sqrt{\frac{m}{k}} \ ; \quad K = \frac{1}{k} \ ; \quad \zeta = \sqrt{\frac{c^2}{4mk}}$$

6.2.2   State-Space Representation of Second-Order Systems

We have just established that two first-order systems in series constitute an example of a second-order system. Let us now return to the original differential equation couple in Eqs. (6.3) and (6.4) from which the second-order Eq. (6.8) was obtained.

Further, for consistency in notation, let us redefine the deviation variables in Eqs. (6.3) and (6.4) as $x_1 = h_1 - h_{1s}$, $x_2 = h_2 - h_{2s}$; and since the actual process output is now x_2, we let $y = x_2$. Eqs. (6.3) and (6.4) now become:

$$\frac{dx_1}{dt} = -\frac{1}{\tau_1} x_1 + \frac{K_1}{\tau_1} u$$

and

$$\frac{dx_2}{dt} = -\frac{1}{\tau_2} x_2 + \frac{K_2}{\tau_2} x_1$$

with

$$y = x_2$$

These state-space modeling equations are represented more compactly in vector-matrix notation as:

$$\begin{bmatrix} \dot{x}_1 \\ \dot{x}_2 \end{bmatrix} = \begin{bmatrix} -\frac{1}{\tau_1} & 0 \\ \frac{K_2}{\tau_2} & -\frac{1}{\tau_2} \end{bmatrix} \begin{bmatrix} x_1 \\ x_2 \end{bmatrix} + \begin{bmatrix} \frac{K_1}{\tau_1} \\ 0 \end{bmatrix} u(t) \qquad (6.29a)$$

$$y = \begin{bmatrix} 0 & 1 \end{bmatrix} \begin{bmatrix} x_1 \\ x_2 \end{bmatrix} \qquad (6.29b)$$

which is of the form:

$$\frac{d\mathbf{x}(t)}{dt} = \mathbf{A}\mathbf{x}(t) + \mathbf{b}u(t)$$
$$y(t) = \mathbf{c}^T \mathbf{x}(t)$$

(see Eq. (4.29) in Chapter 4).

The general second-order system whose model is given by the standard, second-order differential equation form in Eq. (6.25) can also be represented in this state-space form. The following procedure is usually employed for this purpose.

Define two new variables x_1 and x_2 as follows:

$$x_1 = y \qquad (6.30a)$$

$$x_2 = \frac{dy}{dt} \qquad (6.30b)$$

From here, we take derivatives to obtain:

$$\frac{dx_1}{dt} = \frac{dy}{dt} \qquad (6.31a)$$

$$\frac{dx_2}{dt} = \frac{d^2y}{dt^2} \qquad (6.31b)$$

Now we introduce Eq. (6.30b) into Eq. (6.31a) for dy/dt; solve Eq. (6.25) for d^2y/dt^2 and introduce the result into the RHS of Eq. (6.31b); the result is:

$$\frac{dx_1}{dt} = x_2 \tag{6.32a}$$

$$\frac{dx_2}{dt} = -\frac{1}{\tau^2}x_1 - \frac{2\zeta}{\tau}x_2 + \frac{K}{\tau^2}u(t) \tag{6.32b}$$

and from Eq. (6.30a) we have the equation for the system output, i.e.:

$$y = x_1 \tag{6.32c}$$

These three equations may now be written in the vector-matrix notation of Eq. (6.29). It is easy to see that the vectors and matrices concerned are now given by:

$$\mathbf{x} = \begin{bmatrix} x_1 \\ x_2 \end{bmatrix} ; \quad \mathbf{A} = \begin{bmatrix} 0 & 1 \\ -\dfrac{1}{\tau^2} & -\dfrac{2\zeta}{\tau} \end{bmatrix} ; \quad \mathbf{b} = \begin{bmatrix} 0 \\ \dfrac{K}{\tau^2} \end{bmatrix} ; \quad \mathbf{c}^T = [\,1 \quad 0\,]$$

Thus any second-order system may be represented by the state-space differential equation form in Eq. (6.29) as an alternative to the standard, second-order differential equation form given in Eq. (6.25). Such state-space model forms are useful for obtaining real-time responses of second-order systems. A significant number of computer simulation packages utilize this model form.

As an exercise, the reader should introduce the deviation variables x_1 and x_2 in Eqs. (6.18) and (6.19) and obtain the state-space form for the interacting system's differential equation model.

6.3   RESPONSE OF SECOND-ORDER SYSTEMS TO VARIOUS INPUTS

6.3.1   Step Response

The response of the second-order system to a step function of magnitude A may be obtained using Eq. (6.26). Since $u(s) = A/s$ we have:

$$y(s) = \frac{K}{\tau^2 s^2 + 2\zeta\tau s + 1}\frac{A}{s} \tag{6.33}$$

We now choose to rearrange this expression to read:

$$y(s) = \frac{AK/\tau^2}{s\,(s - r_1)\,(s - r_2)} \tag{6.34}$$

where r_1 and r_2 are the roots of the denominator quadratic (the transfer function poles).

Ordinarily, one would invert Eq. (6.34), after partial fraction expansion, and obtain the general result:

$$y(t) = A_0 + A_1 e^{r_1 t} + A_2 e^{r_2 t} \qquad (6.35)$$

where, as we now know, A_0, A_1, and A_2 are the usual constants obtained during partial fraction expansion. However, the values of the roots r_1 and r_2, obtained using the quadratic formula, are:

$$r_1, r_2 = -\frac{\zeta}{\tau} \pm \frac{\sqrt{\zeta^2 - 1}}{\tau} \qquad (6.36)$$

By examining the quantity under the radical sign we may now observe that these roots can be real or complex, depending on the value of the parameter ζ.

Observe further that the type of response obtained in Eq. (6.35) depends on the nature of these roots. In particular when $0 < \zeta < 1$, Eq. (6.36) indicates complex conjugate roots, and we will expect the response under these circumstances to be different from that obtained when $\zeta > 1$, and the roots are real and distinct. When $\zeta = 1$, we have a pair of repeated roots, giving rise to yet another type of response which is expected to be different from the other two. The case $\zeta < 0$ can occur only in special circumstances and signals process instability. This will be discussed in detail in Chapter 11.

There are thus three different possibilities for the response in Eq. (6.35), depending on the nature of the roots r_1, r_2, that are dependent on the value of the parameter ζ. Let us now consider each case in turn.

CASE 1: $0 < \zeta < 1$ (r_1 and r_2 Complex conjugates)

It can be shown that in this case, Eq. (6.35) becomes, after some simplification:

$$y(t) = AK\left[1 - \frac{1}{\beta}e^{-\zeta t/\tau} \sin\left(\frac{\beta}{\tau}t + \phi\right)\right] \qquad (6.37)$$

where:

$$\beta = |\zeta^2 - 1|^{1/2} \qquad (6.38a)$$

$$\phi = \tan^{-1}(\beta/\zeta) \qquad (6.38b)$$

Note:

1. The time behavior of this response is that of a damped sinusoid with β/τ as the frequency of oscillation. The damping is provided by the exponential term $e^{-\zeta t/\tau}$ which gets smaller in magnitude with time and eventually goes to zero as $t \rightarrow \infty$.

2. The response ultimately settles, as $t \rightarrow \infty$, to the value AK.

CASE 2: $\zeta = 1$ ($r_1 = r_2 = -1/\tau$; **Real and equal roots**)

In this case because we have a pair of repeated roots, we cannot use Eq. (6.35) directly. Laplace inversion of the appropriately modified version of Eq. (6.34):

$$y(s) \; = \; \frac{AK/\tau^2}{s \, (s-r)^2}$$

gives the required response:

$$y(t) \; = \; AK\left[1 - \left(1 + \frac{t}{\tau} \right) e^{-t/\tau} \right] \tag{6.39}$$

since $r = -1/\tau$. The time behavior indicated by this equation is an exponential approach to the ultimate value of AK. Note that just as in Case 1, as $t \rightarrow \infty$, $y(t) \rightarrow AK$.

CASE 3: $\zeta > 1$ (r_1 and r_2 **Real and distinct**)

With β as defined in Eq. (6.38a), the roots are now given by:

$$r_1, r_2 \; = \; -\frac{\zeta}{\tau} \pm \frac{\beta}{\tau}$$

and the response is given by:

$$y(t) \; = \; AK\left[1 - e^{-\zeta t/\tau} \left(\cosh \frac{\beta}{\tau} t + \frac{\zeta}{\beta} \sinh \frac{\beta}{\tau} t \right) \right] \tag{6.40}$$

where the hyperbolic functions sinh, cosh are defined as:

$$\sinh \theta \equiv \frac{1}{2}\left(e^{\theta} - e^{-\theta} \right)$$

$$\cosh \theta \equiv \frac{1}{2}\left(e^{\theta} + e^{-\theta} \right)$$

The indicated time response is another exponential approach to the ultimate value of AK. Although perhaps not immediately obvious from Eq. (6.40), it is true, however, that this particular exponential approach is somewhat slower than the one indicated in Eq. (6.39).

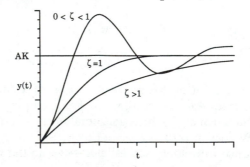

Figure 6.8.   Step responses of the second-order system.

These three responses are sketched in Figure 6.8.

1.   The Case 1 response (when $0 < \zeta < 1$) is oscillatory and is said to be *underdamped*.

2.   The Case 2 response (when $\zeta = 1$) is said to be *critically damped*. It offers the most rapid approach to the final value without oscillation.

3.   The Case 3 response (when $\zeta > 1$) is sluggish and is said to be *overdamped*.

We thus see that whether the second-order response is underdamped, overdamped, or critically damped is determined solely by one parameter ζ. This is why it is referred to as the damping coefficient.

Let us now return to the underdamped response in Eq. (6.37) and consider what happens when ζ is set equal to zero. This represents the situation in which there is no damping at all. The resulting response in this case is:

$$y(t) = AK\left[1 - \sin\left(\frac{1}{\tau}t + \frac{\pi}{2}\right)\right] \tag{6.41}$$

a pure, undamped sine wave, with frequency $1/\tau$. This is referred to as the natural frequency of oscillation ω_n. Observe that its reciprocal, the natural *period* of oscillation, is τ, establishing the reason for the name given to this parameter.

Having established that there are three categories of second-order systems (underdamped, critically damped, and overdamped) we are now interested in finding out under what category the ensemble of two first-order systems in series falls. The following example is used to resolve this issue.

Example 6.2   DAMPING CHARACTERISTICS OF TWO FIRST-ORDER SYSTEMS IN SERIES.

Having shown that two first-order systems in series (be they interacting or otherwise) constitute a second-order system, the following questions are to be answered.

1.   What type of second-order system (underdamped, overdamped, or critically damped) is the noninteracting system?

2.   Compare the damping characteristics of the interacting and the noninteracting configurations and hence determine the type of second-order system that describes the interacting system.

Solution:

1.   As earlier demonstrated, the damping characteristics of any second-order system are determined solely by the value of the parameter ζ. For the noninteracting arrangement, we had earlier shown that this parameter is given by:

$$\zeta = \frac{1}{2} \frac{(\tau_1 + \tau_2)}{\sqrt{\tau_1 \tau_2}} \tag{6.42}$$

and our task is now to find out whether this quantity is greater than, less than, or equal to 1.

First observe that Eq. (6.42) is a ratio of the arithmetic mean and the geometric mean of the two time constants τ_1 and τ_2. We may thus use the argument that the arithmetic mean of two positive quantities can never be smaller than their geometric mean to establish that $\zeta \geq 1$.

Alternatively, let us assume that the converse is true: that is $\zeta < 1$. This leads us to conclude from Eq. (6.42) that:

$$\tau_1 + \tau_2 < 2\sqrt{\tau_1 \tau_2} \tag{6.43}$$

Since the quantities involved in this inequality are all positive we may square both sides without altering the inequality; the result is:

$$\tau_1^2 + 2\tau_1 \tau_2 + \tau_2^2 < 4\tau_1 \tau_2$$

which simplifies to:

$$\tau_1^2 - 2\tau_1 \tau_2 + \tau_2^2 < 0$$

or

$$(\tau_1 - \tau_2)^2 < 0 \tag{6.44}$$

Observe, however, that regardless of the actual values of τ_1 or τ_2, a squared quantity can never be less than zero, thus proving false the initial assumption of $\zeta < 1$. We therefore conclude that $\zeta \geq 1$.

Note that when $\tau_1 = \tau_2$ Eq. (6.42) indicates that $\zeta = 1$.
The conclusion is therefore that the noninteracting arrangement of two first-order systems in series is either critically damped (when the two time constants are identical) or overdamped; it can *never* be underdamped.

2.   For the interacting case, the damping coefficient was earlier given as:

$$\zeta = \frac{1}{2} \frac{(\tau_1 + \tau_2 + K_2 \tau_1)}{\sqrt{\tau_1 \tau_2}} \tag{6.45}$$

A comparison of this expression with Eq. (6.42) shows that unless K_2 is negative (which is not the case with the two physical tanks of Figure 6.2)) the damping coefficient for the interacting system takes on a value that is even greater than that for the noninteracting system.

Thus if the noninteracting system is overdamped, then the interacting system is even more so. (As an exercise, the reader should show that the interacting system can never be critically damped.)

Thus we conclude that the interacting system is also overdamped; exhibiting "heavier" damping characteristics than the noninteracting system. This implies the interacting system will show more sluggish response characteristics than the noninteracting system; this, of course, had already been indicated earlier on in Example 6.1.

One of the most important conclusions from this example is that two first-order systems in series can *never* exhibit underdamped characteristics.

There are a number of other physical systems that do exhibit underdamped behavior. For example, the U-tube manometer of Section 6.2.1 will exhibit underdamped behavior for its most common configuration with water or mercury. Another physical system that can exhibit underdamped response is the spring-shock absorber system of the last section. Let us illustrate with the following example.

Example 6.3   THE DYNAMIC BEHAVIOR OF A SUSPENSION SYSTEM.

A suspension system similar to the one shown in Figure 6.7 has been designed as part of a special cart to carry delicate instrumentation around the plant for special tests and measurements. Unfortunately the shock absorbers tend to wear out with use, and must be replaced. In order to determine the performance of the system with various types of shock absorbers and to construct a model for the system, some experiments are to be performed with the following equipment:

1. New heavy duty shocks: $c = 1600$ newtons/(m/s)
2. New regular shocks: $c = 1200$ newtons/(m/s)
3. Worn regular shocks: $c = 400$ newtons/(m/s)

The parameters of the suspension system are:

spring constant, $k = 2000$ newtons/m
mass of load, $m = 500$ kg
amplitude of test force, $A = 50$ newtons

Using the model developed in Section 6.2.1 for this system we see that the deviation from the desired position, y, is given by:

$$m\frac{d^2y}{dt^2} = -ky - c\frac{dy}{dt} + f(t)$$

which can be put in the standard form (6.25) with $u = f(t)$:

$$\tau = \sqrt{\frac{m}{k}}, \qquad K = \frac{1}{k}, \qquad \zeta = \sqrt{\frac{c^2}{4mk}}$$

Using the parameters given above for this suspension system we obtain

$$\tau = \sqrt{\frac{500}{2000}} = 0.5 \text{ s}, \qquad K = \frac{1}{2000} = 0.0005 \text{ m/newtons},$$

$$\zeta = \sqrt{\frac{c^2}{4 \times 10^6}} = \frac{c}{2000}$$

with a natural frequency of $\omega_n = 1/\tau = 2$ radians/s.

Thus for new heavy duty shocks, the damping factor $\zeta = 0.8$, for new regular shocks, $\zeta = 0.6$, and for worn regular shocks, $\zeta = 0.2$. Before beginning the experiment, the model is used to predict the dynamic response. To apply a *step input*, $u(t) = AH(t)$, one simply places a force of 50 newtons on the system for $t \leq 0$, defines this as the origin, and then suddenly releases this force at $t = 0$. The resulting responses from the model

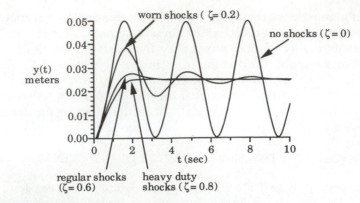

Figure 6.9.   Response of suspension system to step change in force of 50 newtons.

for the three shock absorbers tested are shown in Figure 6.9. Note that both new shocks give good response with the heavy duty ones giving less overshoot. The worn shocks oscillate for a long time before coming to rest, while removal of the shock absorbers ($c = 0$) gives sustained oscillations with no damping.

   Note that for this example, the characteristic time τ, or equivalently, the natural period of oscillation $\omega_n = 1/\tau$, is determined by the mass of the load and the spring constant k. The damping factor ζ depends on these factors as well as the shock absorber viscous damping coefficient c.

Underdamped behavior in chemical process systems often arises as a result of combining a feedback controller *and* the process itself. Since the issue of feedback control is not to be addressed until later, we will at this point settle for satisfying ourselves about this issue with the following example.

**Example 6.4   UNDERDAMPED BEHAVIOR ARISING FROM THE
                  COMBINATION OF A FIRST-ORDER SYSTEM AND A
                  PROPORTIONAL + INTEGRAL CONTROLLER.**

As we know, the differential equation model for a first-order system with time constant τ_1 and steady-state gain K is given as:

$$\tau_1 \frac{dy}{dt} + y = K u(t) \tag{6.46}$$

This system is now to be operated in conjunction with a certain type of controller known as a *Proportional + Integral controller*. The implication of this arrangement is that the input variable, $u(t)$, is now determined by the controller according to the control law:

$$u(t) = K_c \left[(y_d - y) + \frac{1}{\tau_I} \int_0^t (y_d - y) \, dt' \right] \tag{6.47}$$

where the constants K_c and $1/\tau_I$ are controller parameters; y_d is the desired set-point value for y. (We should point out that the name given to this controller derives from the fact that Eq. (6.47) indicates control action *proportional* to the error $(y_d - y)$ as well as involving the *integral* of this error.) Investigate the dynamic characteristics of this combination, first in general, and then for the specific case in which the system's steady-state gain and time constant are respectively given by: $K = 2$, and $\tau_1 = 4$; and the controller parameters are: $K_c = 0.5$, $\tau_I = 0.25$.

Solution:

By introducing Eq. (6.47) into Eq. (6.46), we obtain the integro-differential equation:

$$\tau_1 \frac{dy}{dt} + y = KK_c \left[(y_d - y) + \frac{1}{\tau_I} \int_0^t (y_d - y)\, d\,t' \right] \tag{6.48}$$

where now y_d has become the input.

To proceed from here, we consider, for simplicity, that whatever changes scheduled to occur in y_d will take place at $t = 0$, and are such that y_d is constant for $t > 0$. This is to enable the vanishing of the derivative of y_d for $t > 0$.

Thus for $t > 0$, differentiating Eq. (6.48) gives, upon some simple rearrangement:

$$\tau_1 \frac{d^2y}{dt^2} + (1 + KK_c)\frac{dy}{dt} + \frac{KK_c}{\tau_I} y = \frac{KK_c}{\tau_I} y_d \tag{6.49}$$

This second-order differential equation may now be rearranged into the "standard form" to give:

$$\frac{\tau_1 \tau_I}{KK_c} \frac{d^2y}{dt^2} + \frac{(1 + KK_c)}{KK_c}\tau_I \frac{dy}{dt} + y = y_d \tag{6.50}$$

so that now the characteristic second-order system parameters are seen to be given by:

$$\tau = \sqrt{\frac{\tau_1 \tau_I}{KK_c}} \; ; \; \zeta = \frac{(1 + KK_c)}{2} \sqrt{\frac{\tau_I}{\tau_1 KK_c}} \tag{6.51}$$

Thus, whenever:

$$(1 + KK_c) \sqrt{\frac{\tau_I}{\tau_1 KK_c}} < 2$$

the combined (process + feedback controller) system will exhibit underdamped behavior.

For the specific situation where $K = 2$, and $\tau_1 = 4$; and $K_c = 0.5$, and $\tau_I = 0.25$, we deduce from Eq. (6.51) that $\tau = 1$, and $\zeta = 0.25$ and the response to a step change in y_d will exhibit underdamped behavior, since $\zeta < 1$ in this case.

In light of this example, we now state that the dynamic behavior of many systems under feedback control is very similar to the underdamped second-order response. For this reason, the underdamped response takes on special significance in process control practice, and as a result, some special terminology (which will now be defined) has been developed to characterize this behavior.

Characteristics of the Underdamped Response

Using Figure 6.10 as reference, the following are the terms used to characterize the underdamped response.

1. Rise Time, t_r: Time to reach the ultimate value for the first time; it can be shown that, for a second-order system:

$$t_r = \frac{\tau}{\beta}[\pi - \phi] \qquad (6.52)$$

where β and ϕ have been defined as:

$$\beta = |\zeta^2 - 1|^{1/2} \qquad (6.38a)$$

$$\phi = \tan^{-1}(\beta/\zeta) \qquad (6.38b)$$

2. Overshoot: The maximum amount by which the transient exceeds the ultimate value AK; expressed as a fraction of this ultimate value, i.e., the ratio a_1/AK in the diagram.

For a second-order system, the maximum value attained by this response is given by:

$$y_{max} = AK\left[1 + \exp\left(\frac{-\pi\zeta}{\beta}\right)\right] \qquad (6.53)$$

so that the overshoot is now given by:

$$\text{Overshoot} = \frac{a_1}{AK} = \exp\left(\frac{-\pi\zeta}{\beta}\right) \qquad (6.54)$$

The time to achieve this maximum value is:

$$t_{max} = \frac{\pi}{\omega} \qquad (6.55)$$

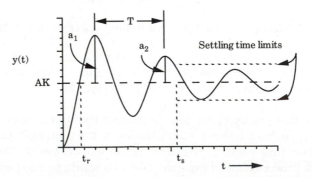

Figure 6.10.   Characteristics of the underdamped response.

3. Period of Oscillation: From Eq. (6.37) observe that the radian frequency of oscillation is:

$$\omega = \frac{\beta}{\tau}$$

and since the period (in times/cycle) is given by $2\pi/\omega$ we therefore have:

$$T = \frac{2\pi\tau}{\beta} \tag{6.56}$$

4. Decay Ratio: A measure of the rate at which oscillations are decaying, expressed as the ratio a_2/a_1 in the diagram:

$$\text{Decay Ratio} = \frac{a_2}{a_1} = \exp\left(\frac{-2\pi\zeta}{\beta}\right) = (\text{Overshoot})^2 \tag{6.57}$$

5. Settling Time, t_s: A somewhat arbitrary quantity defined as the time for the process response to settle to within some small neighborhood of the ultimate value; usually taken to be within $\pm 5\%$.

6.3.2 Impulse Response

The response of the second-order system to the input function $u(t) = A\,\delta(t)$ is obtained from:

$$y(s) = \frac{K}{\tau^2 s^2 + 2\zeta\tau s + 1} A \tag{6.58}$$

Again, depending on the value of ζ, there are three possible types of responses, and they are summarized below.

Case 1: $\zeta < 1$ **(Underdamped Response)**

$$y(t) = \frac{AK}{\beta\tau} e^{-\zeta t/\tau} \sin\left(\frac{\beta t}{\tau}\right) \tag{6.59}$$

Case 2: $\zeta = 1$ **(Critically Damped Response)**

$$y(t) = \frac{AK}{\tau^2} t\, e^{-t/\tau} \tag{6.60}$$

Case 3: $\zeta > 1$ **(Overdamped Response)**

$$y(t) = \frac{AK}{\beta\tau} e^{-\zeta t/\tau} \sinh\left(\frac{\beta t}{\tau}\right) \tag{6.61}$$

where the constant β is as earlier defined in Eq. (6.38).
    These responses are shown in Figure 6.11.

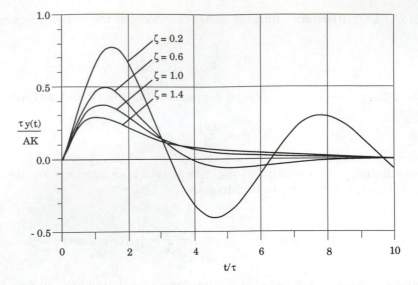

Figure 6.11.   Impulse response of a second-order system.

6.3.3   Sinusoidal Response

The response of the second-order system to the sine wave input of Eq. (3.51) is obtained by combining Eq. (6.26) with Eq. (3.52) to form:

$$y(s) = \frac{K}{\tau^2 s^2 + 2\zeta\tau s + 1} \frac{A\omega}{s^2 + \omega^2} \tag{6.62}$$

This may be factored to read:

$$y(s) = \frac{AK\omega/\tau^2}{(s + j\omega)(s - j\omega)(s - r_1)(s - r_2)} \tag{6.63}$$

where r_1 and r_2 are the second-order transfer function poles, and j is the imaginary number $\sqrt{-1}$.

Once again, we observe that upon partial fraction expansion, three types of responses can be identified. It can be shown that these responses are given as follows:

Case 1: $\zeta < 1$ **(all 4 roots are imaginary)**

$$y(t) = P_1\cos\omega t + P_2\sin\omega t + e^{-\zeta t/\tau}\left[P_3 \cos\left(\frac{\beta t}{\tau}\right) + P_4 \sin\left(\frac{\beta t}{\tau}\right)\right] \tag{6.64}$$

Case 2: $\zeta = 1$ **(2 imaginary roots, 2 repeated roots)**

$$y(t) = Q_1 \cos\omega t + Q_2 \sin\omega t + Q_3 e^{-t/\tau}\left(1 + \frac{t}{\tau}\right) \tag{6.65}$$

Case 3: $\zeta > 1$ (2 imaginary roots, 2 real, distinct roots)

$$y(t) = R_1 \cos\omega t + R_2 \sin\omega t + e^{-\zeta t/\tau}\left[R_3 \cosh\left(\frac{\beta t}{\tau}\right) + R_4 \sinh\left(\frac{\beta t}{\tau}\right)\right] \tag{6.66}$$

The constants P_1, P_2 ; Q_1, Q_2 ; and R_1, R_2 are obtained by partial fractions.

We now point out some salient features of these responses.

1. It can be shown that $P_1 = Q_1 = R_1$; and $P_2 = Q_2 = R_2$.

2. As $t \to \infty$ the first two terms in each case persist; the other terms vanish by virtue of the exponential terms becoming zero.

Thus in each case, the ultimate response is also sinusoidal and is of the form:

$$y(t)\big|_{t \to \infty} = C_1 \cos\omega t + C_2 \sin\omega t$$

and now if we evaluate the constants C_1 and C_2, and apply the usual trigonometric identities to combine the sine and cosine terms into one sine term, we obtain the ultimate periodic response for the second-order system as:

$$y(t)\big|_{t \to \infty} = \frac{AK}{\sqrt{(1 - \omega^2\tau^2)^2 + (2\zeta\omega\tau)^2}} \sin(\omega t + \phi) \tag{6.67}$$

with ϕ given by:

$$\phi = -\tan^{-1}\left[\frac{2\zeta\omega\tau}{(1 - \omega^2\tau^2)}\right] \tag{6.68}$$

Frequency Response

A comparison of the input sine wave, $A\sin\omega t$, with the ultimate output shown in Eq. (6.67) indicates that the frequency response of the second-order system has amplitude ratio given by:

$$AR = \frac{K}{\sqrt{(1 - \omega^2\tau^2)^2 + (2\zeta\omega\tau)^2}} \tag{6.69}$$

and phase angle ϕ given by Eq. (6.68).

There are several very interesting characteristics of this frequency response, a full-scale discussion of which is deferred till Chapter 9.

6.4   *N* FIRST-ORDER SYSTEMS IN SERIES

Consider the situation shown in block diagramatic form in Figure 6.12, in which *N* noninteracting, first-order systems are connected in series. The arrangement is such that the output of the *i*th system is the input of the (*i*+1)th system, with *u*, the process input, as the input to the first system in the series (see Figure 6.12).

The analysis of this system's dynamic behavior is actually a straightforward extension of the discussion in Section 6.1.1. Firstly, because each of the *N* systems in the ensemble is first order, the transfer function relationship for each component is given by:

$$y_1(s) = \frac{K_1}{\tau_1 s + 1} u(s)$$

$$y_2(s) = \frac{K_2}{\tau_2 s + 1} y_1(s)$$

$$\cdots \quad \cdots \quad \cdots$$

$$y_i(s) = \frac{K_i}{\tau_i s + 1} y_{i-1}(s)$$

$$\cdots \quad \cdots \quad \cdots$$

$$y_{N-1}(s) = \frac{K_{N-1}}{\tau_{N-1} s + 1} y_{N-2}(s)$$

$$y_N(s) = \frac{K_N}{\tau_N s + 1} y_{N-1}(s)$$

Next, by a series of cascaded substitutions (an extension of the procedure by which Eq. (6.9) was obtained from Eqs. (6.5) and (6.6)) we obtain the fact that the overall transfer function for this system is simply a product of the contributing transfer functions — a generalization of our earlier result for two first-order systems in series, i.e.:

$$y_N(s) = \left(\prod_{i=1}^{N} \frac{K_i}{\tau_i s + 1} \right) u(s) \tag{6.70}$$

Just as two first-order systems in series turned out to be a second-order system, the ensemble of *N* first-order systems in series is an *N*th-order system.

Characteristic Parameters

The characteristic parameters for this system are easily identified. They are:

- $K = \prod_{i=1}^{N} K_i$ the combined steady-state gain, a product of the *N*

  individual, contributing system steady-state gains,

- *N* time constants, τ_i, $i = 1, 2, ..., N$.

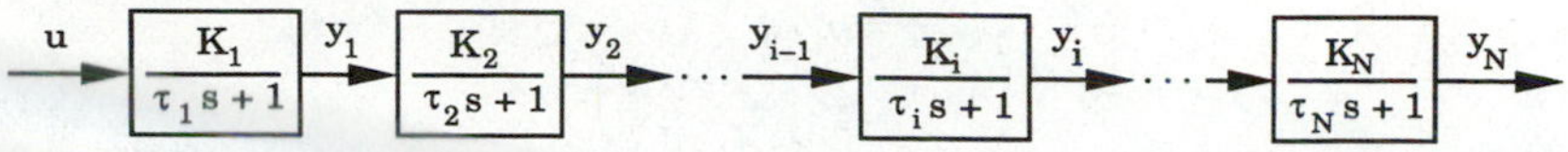

Figure 6.12. Block diagramatic representation of N noninteracting systems in series.

The transfer function for this process, extracted from Eq. (6.70), is:

$$g(s) = \left(\prod_{i=1}^{N} \frac{K_i}{\tau_i s + 1} \right) \tag{6.71a}$$

The N factored terms in the denominator expand out to give an Nth-order polynomial in s so that Eq. (6.71a) may be expressed as:

$$g(s) = \frac{K}{a_N s^N + a_{N-1} s^{N-1} + \dots + a_2 s^2 + a_1 s + a_0} \tag{6.71b}$$

where it is easy to see that the first N coefficients of the Nth-order denominator polynomial are directly related to the time constants shown in Eq. (6.71a), and $a_0 = 1$.

Thus an ensemble of N first-order systems in series behaves like an Nth-order system. This is a generalization of our earlier statement regarding the fact that two first-order systems in series constitute a second-order (overdamped or critically damped) system.

Observe that the transfer function for N first-order systems in series has N poles located at $s = -1/\tau_i$; $i = 1, 2, \dots, N$ and no zeros.

6.4.1 Unit Step Response

The response of this system to a unit step change in the input is obtainable from Eq. (6.70) with $u(s) = 1/s$, i.e:

$$y_N(s) = \left(\prod_{i=1}^{N} \frac{K_i}{\tau_i s + 1} \right) \frac{1}{s} \tag{6.72}$$

which, by partial fraction expansion becomes:

$$y_N(s) = K \left(\frac{A_0}{s} + \sum_{i=1}^{N} \frac{A_i}{\tau_i s + 1} \right) \tag{6.73}$$

where the constants A_i (obtained from the partial fraction expansion) are given by:

$$A_0 = 1; \quad A_i = \lim_{s \to (-1/\tau_i)} \frac{[(\tau_i s + 1) g(s)]}{Ks} \tag{6.74}$$

when the poles of the process transfer function $g(s)$ given in Eq. (6.81) are all distinct.

By inverting Eq. (6.73), the unit step response is obtained as:

$$y_N(t) = K\left(1 + \sum_{i=1}^{N} \frac{A_i}{\tau_i} e^{-t/\tau_i}\right) \tag{6.75}$$

Concerning the nature of this response, let us note the following points.

1. We recall that the sluggishness introduced into the response of a first-order system by adding a second one in series was demonstrated in Figure 6.4. It turns out that subsequent addition of more first-order systems actually causes the sluggishness to become more pronounced. This fact is illustrated in Figure 6.13, a plot of different unit step responses for N first-order systems in series, when N takes on various values.

2. It can be shown that, as long as $N > 1$, the slope of the unit step response at the origin will be zero, i.e.:

$$\left.\frac{dy_N}{dt}\right|_{t=0} = 0 \tag{6.76}$$

3. The "inherent reluctance" factor, obtained by evaluating the following integral where $y^* = K$ and $y_N(t)$, is as given in Eq. (6.75):

$$J_N = \int_0^\infty [y^* - y(t)]\, dt$$

(analogous to J_2 obtained in Eq. (6.15) for two first-order systems in series), can be shown to be given in this case by:

$$J_N = K\left(\sum_{i=1}^{N} \tau_i\right) \tag{6.77}$$

once more, a mere generalization of Eq. (6.15).

The response of this system to various other inputs can be derived along the same lines as illustrated with the unit step response. This is left as an exercise

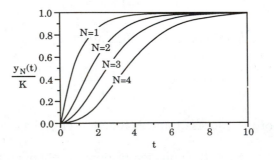

Figure 6.13.   Unit step response of several noninteracting first-order systems in series.

to the reader. The sinusoidal response, and the subsequent frequency response derived from the ultimate, purely periodic portion of this response, are of special significance and will be discussed in Chapter 9.

6.4.2 A Special Limiting Case

To round up our discussion of N first-order systems in series, we wish to consider a very special case of such systems. The situation under consideration is as follows:

- All the N steady-state gains involved are equal to 1
- All the time constants are identical, and
- Each time constant is equal to α/N

The implication of this third point is that each time the number of systems in the ensemble increases, the time constant of each of these systems is decreased.

The transfer function for such a system is given by:

$$g_N(s) = \frac{1}{\left(\dfrac{\alpha}{N} s + 1\right)^N} \tag{6.78}$$

We are particularly interested in what happens in the limit as $N \to \infty$, when there are infinitely many such systems in this ensemble. (We should mention that our interest at this point is purely qualitative, primarily to set the stage for the quantitative discussion to be found in an upcoming chapter.)

Observe that as $N \to \infty$, each contributing "time constant" goes to zero. The immediate consequence is that we now have an *infinite* sequence of individual systems, each of which behaves essentially like a pure gain process.

Let us now address the issue of how this system will respond to a unit step input. From the description given above, we are led to believe that a unit step change in the input will result in each component system transferring this unit step instantaneously (for that is what a pure gain system does) to the immediate neighboring system. Theoretically, this instantaneous, stage-to-stage transfer of the step function goes on through the entire *infinite* sequence of systems. However, we may imagine that after a sufficiently large number of stages, we obtain the final output from the "final" stage.

Keep in mind that so far, all we know is that the (limiting) response at each stage is a step function; it is still far from obvious what the response at the "final" stage (which is the overall system output) will actually be. However, from Eq. (6.77) we can obtain what the "inherent reluctance" factor is in this case; it is given by:

$$\lim_{N \to \infty} J_N = \lim_{N \to \infty} \left(\sum_{i=1}^{N} \tau_i\right) \tag{6.79}$$

and since all the time constants are equal, each taking the value α/N, the sum of these time constants is simply equal to α, regardless of the number of systems in the sequence. Thus, Eq. (6.79) becomes

$$\lim_{N \to \infty} J_N = \alpha \tag{6.80}$$

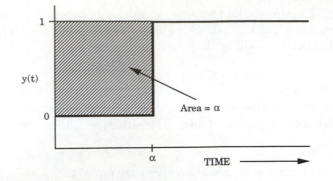

Figure 6.14.   An intuitive deduction of the unit step response of a special infinite sequence of first-order systems in series.

This is a very helpful result because it implies that the area under the curve between the instantaneous unit response $y^* = 1$ and the actual unit step response of this infinite sequence of systems is equal to α.

When we combine this with the fact that the response expected from the "final" stage in this infinite sequence is a step function, we are led to deduce that the overall unit step response from this system can only be as shown in Figure 6.14, a unit step function delayed by α time units.

What we have deduced from such intuitive, qualitative arguments is in fact what actually obtains, but we will have to wait until Chapter 8 for a rigorous confirmation of this statement.

This special infinite sequence of first-order systems behaves like a system that does not in any way alter the form of the input function: it merely introduces a *delay*, making the output appear as the same function as the input, only shifted by a certain amount of time units. Such systems are of importance in process control studies and will be discussed in Chapter 8.

6.5   THE GENERAL Nth-ORDER SYSTEM

Before proceeding to the next stage, perhaps a few statements are in order here about the general Nth-order system.

Rather than starting with N first-order transfer functions and combining them to give the transfer function in Eq. (6.71a), and consequently the one in Eq. (6.71b), if we had started instead with Eq. (6.71b), we would be dealing with a *general* Nth-order system. And just like a general second-order system can exhibit dynamic behavior that differs from that of two first-order systems in series, the general Nth-order system can also exhibit dynamic behavior that differs from that of N first-order systems in series.

Now, it is clear, following from the discussion of the general second-order system, that the behavior of the general Nth-order system will depend on the nature of the N roots of the denominator polynomial in Eq. (6.71b); i.e., the N system poles.

Observe, however, that these roots are either real or they are complex; and when complex, they occur in pairs, as complex conjugates. We therefore have only two situations:

1.   The roots are all real (either distinct or multiple), or
2.   Some roots occur as complex conjugates while the others are real

In the first instance, the behavior will be exactly like that of N first-order systems in series, and what we have discussed in the preceding section holds. In the second instance, we only need to observe that each pair of complex conjugate roots contributes the behavior of an *underdamped* second-order system to the overall system behavior; the real roots contribute first-order system behavior. Thus what we have in this case is an ensemble of several first-order systems connected in series with as many underdamped second-order systems as we have *pairs* of complex conjugate roots.

The final conclusion is therefore the following:

The general Nth-order system behavior is composed either entirely of the behavior of N first-order systems in series, or a combination of several underdamped second-order systems included in the series of first-order systems.

Although in practice the dynamic behavior of most industrial processes tends to be higher order and overdamped, we should note that a general treatment only requires an inclusion of first-order systems plus underdamped second-order systems into the series of systems composing the higher order systems. The basic principles of decomposing the high-order system into several systems of lower order remains the same.

6.6   HIGHER ORDER SYSTEMS WITH ZEROS

A discussion of systems that exhibit higher order dynamics would be incomplete without considering that class of systems whose transfer functions involve zeros.

Such systems have the general transfer function form:

$$g(s) = \frac{K\left(\xi_1 s + 1\right)\left(\xi_2 s + 1\right) \cdots \left(\xi_q s + 1\right)}{\left(\tau_1 s + 1\right)\left(\tau_2 s + 1\right) \cdots \left(\tau_p s + 1\right)} \tag{6.81}$$

where for an underdamped response, we may consider a pair of τ_i to be complex conjugates. This model may be written more compactly as:

$$g(s) = \frac{K \prod_{i=1}^{q}\left(\xi_i s + 1\right)}{\prod_{i=1}^{p}\left(\tau_i s + 1\right)} \tag{6.82}$$

To facilitate our study of such systems we now introduce the following notation:

We shall denote a (p,q)-order system as one whose transfer function has p poles and q zeros, or equivalently, whose transfer function has a pth-order denominator polynomial and a qth-order numerator polynomial.

Thus the characteristic parameters of the general, (p,q)-order system whose transfer function is shown in Eq. (6.81) are:

- K, the steady-state gain,
- p poles, located at $s = -1/\tau_i$, $i = 1,2, \dots p$
- q zeros, located at $s = -1/\xi_i$, $i = 1,2, \dots q$

When q is strictly less than p, the model is termed *strictly proper*, and when $q = p$, it is called *semiproper*. For realistic process models, q will always be less than or equal to p. The first in this class of systems is therefore seen to be the $(2,1)$-order system; its dynamic behavior will be investigated first before generalizing.

6.6.1   Unit Step Response of the (2,1)-Order System

The transfer function for the $(2,1)$-order system is given by:

$$g(s) = \frac{K\left(\xi_1 s + 1\right)}{\left(\tau_1 s + 1\right)\left(\tau_2 s + 1\right)} \tag{6.83}$$

At this point we should note that the difference between this system's transfer function and that for the system composed of two first-order systems in series (see Eq. (6.10)) is the presence of the first-order lead term with the lead time constant ξ_1 in Eq. (6.83), contributing the only transfer function zero. Thus we may expect that any difference observed between the response of the $(2,1)$-order system and the pure second-order system will be attributable directly to the presence of this single zero.

The response of the $(2,1)$-order system to a unit step input is obtained from:

$$y(s) = \frac{K\left(\xi_1 s + 1\right)}{\left(\tau_1 s + 1\right)\left(\tau_2 s + 1\right)} \frac{1}{s} \tag{6.84}$$

By partial fraction expansion, Eq. (6.84) becomes:

$$y(s) = K\left[\frac{A_0}{s} + \frac{A_1}{\left(\tau_1 s + 1\right)} + \frac{A_2}{\left(\tau_2 s + 1\right)}\right] \tag{6.85}$$

where the constants are given by:

$$A_0 = 1\,; \quad A_1 = \frac{-\tau_1\left(\tau_1 - \xi_1\right)}{\left(\tau_1 - \tau_2\right)}\,; \quad A_2 = \frac{-\tau_2\left(\tau_2 - \xi_1\right)}{\left(\tau_2 - \tau_1\right)} \tag{6.86}$$

and Laplace inversion now gives:

$$y(t) = K\left[1 - \left(\frac{\tau_1 - \xi_1}{\tau_1 - \tau_2}\right)e^{-t/\tau_1} - \left(\frac{\tau_2 - \xi_1}{\tau_2 - \tau_1}\right)e^{-t/\tau_2}\right] \tag{6.87}$$

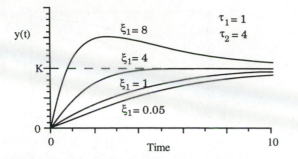

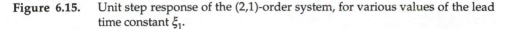

Figure 6.15.   Unit step response of the (2,1)-order system, for various values of the lead time constant ξ_1.

If we now compare this with the corresponding expression for the step response of the second-order system with no zero:

$$y_2(t) = K\left[1 - \left(\frac{\tau_1}{\tau_1 - \tau_2}\right)e^{-t/\tau_1} - \left(\frac{\tau_2}{\tau_2 - \tau_1}\right)e^{-t/\tau_2}\right] \qquad (6.12)$$

we can immediately observe, very clearly, the influence of the zero.

Firstly, let the lag time constants be ordered such that $\tau_1 < \tau_2$. (That this imposes no restriction whatsoever is obvious from the fact that Eq. (6.87) remains absolutely unchanged when τ_1 and τ_2 are interchanged.) If this be the case, then it can be shown that for *positive* values of ξ_1, the response in Eq. (6.87) can take several different forms summarized below:

Case 1: $\xi_1 > \tau_2$

In this case it can be shown (see problems at the end of the chapter) that the response overshoots the ultimate value K.

This brings out a very important difference between the system *with* a zero (shown in Eq. (6.83)) and the one without (see Eq. (6.10)). Observe that while the system of two first-order systems in series can never overshoot the final value, the system shown in Eq. (6.83) can exhibit overshoot if the lead time constant is large enough.

Case 2: $\xi_1 = \tau_2$ or τ_1

Here the zero cancels one of the poles, and the response becomes *identical* to that of a first-order system with time constant equal to the surviving time constant.

Case 3: $0 < \xi_1 < \tau_2$

The response in this case does not show any overshoot, and it actually resembles that of a first-order system until ξ_1 becomes quite small compared to τ_1 and τ_2.

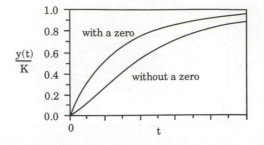

Figure 6.16.   Comparison of the unit step responses of the second-order system, with and without a zero.

These facts are demonstrated in Figure 6.15, where the unit step response of a (2,1)-order system with $\tau_1 = 1$, $\tau_2 = 4$, is shown for four different values of ξ_1 covering all the cases indicated above:

1. $\xi_1 = 0.05$, Case 3,
2. $\xi_1 = 1$ and $\xi_1 = 4$, both Case 2 conditions, and
3. $\xi_1 = 8$, Case 1.

Figure 6.16 shows a comparison of the (nonovershooting) step response of the (2,1)-order system and the second-order system response of Eq. (6.12) to demonstrate what the presence (and the absence) of the zero does to the unit step responses for each of these systems.

Let us now note the following important points about the (2,1)-order system's step response in general.

1. Unlike the pure second-order response, the response for the system with a zero is swifter. We therefore see clearly that the effect of the lead term is to "speed up" the process response.

2. It is possible to speed up the response to the point where overshoot occurs. This happens when the lead time constant is larger than the larger of the two lag time constants.

3. It can be shown that the response of the system with the zero has a nonzero slope at the origin. In particular, either directly from Eq. (6.97), or by using the initial value theorem of Laplace transforms, we obtain for this (2,1)-order system that:

$$\frac{dy}{dt}\bigg|_{t=0} = \frac{K\xi_1}{\tau_1 \tau_2} \tag{6.88}$$

which is nonzero. (Observe that the slope becomes zero when $\xi_1 = 0$, when the lead term in the system transfer function disappears.)

It should also be noted that the slope is positive when the signs of K and ξ_1 are the same, and negative when K and ξ_1 have *opposite* signs. This fact will become useful in the next chapter.

4.   The "inherent reluctance" factor in this case is easily shown to be given by:

$$J_{(2,1)} = \int_0^\infty [y^* - y(t)]dt = K\left(\tau_1 + \tau_2 - \xi_1\right) \tag{6.89}$$

which provides interesting insight into the effect of zeros on the "inherent reluctance" of the process. Compared to Eq. (6.15) we see from Eq. (6.89) that the (2,1)-order system is less reluctant (more "willing") to respond than the pure second-order counterpart by virtue of the presence of the zero. In general, therefore, the tendency is for the lead term in a transfer function to reduce the inertia inherent within higher order systems. However, when ξ_1 becomes too large, the possibility exists that the overshoot area may overwhelm the area before the rise time, thus allowing negative values of $J_{(2,1)}$. An interesting situation arises when $\xi_1 = \tau_1 + \tau_2$. The resulting zero reluctance factor in this case is *not* indicative of instantaneous response; it is due to the equality of the overshoot area above the final steady-state line *and* the area below the line. Since these areas counterbalance each other, the net area is "computed" as zero.

We will now generalize the discussion given above.

6.6.2 Unit Step Response of the General (*p,q*)-Order System

The unit step response of the general (*p,q*)-order system is obtained using Eq. (6.81):

$$y(s) = \frac{K\left(\xi_1 s + 1\right)\left(\xi_2 s + 1\right) \cdots \left(\xi_q s + 1\right)}{\left(\tau_1 s + 1\right)\left(\tau_2 s + 1\right) \cdots \left(\tau_p s + 1\right)} \frac{1}{s} \tag{6.90}$$

Once again, by partial fraction expansion (assuming distinct poles, τ_i) Eq. (6.90) becomes:

$$y(s) = K\left(\frac{A_0}{s} + \sum_{i=1}^{p} \frac{A_i}{\tau_i s + 1}\right) \tag{6.91}$$

just like in Eq. (6.73), and in fact the constants A_i are also given by the same expression Eq. (6.74), i.e.:

$$A_0 = 1; \quad A_i = \lim_{s \to (-1/\tau_i)} \frac{[(\tau_i s + 1)g(s)]}{Ks} \tag{6.74}$$

However, the actual values obtained for this set of constants will be different because the transfer function $g(s)$ in this case is given by Eq. (6.81) (as opposed to Eq. (6.71)) and the presence of the zeros will influence the values obtained. (For example, see Eq. (6.86) for the case with $p=2$, and $q=1$.)

From Eq. (6.91) the step response is given by:

$$y(t) = K \left(1 + \sum_{i=1}^{p} \frac{A_i}{\tau_i} e^{-t/\tau_i} \right) \tag{6.92}$$

The points to note about this response are as follows:

1. When $p - q = 1$ (i.e., when we have *exactly* one more pole than we have zeros — referred to as a pole-zero excess of 1) regardless of the actual values of p and q, as long as $q \geq 1$, it can be shown that the initial slope of this response will be nonzero. In particular:

$$\left. \frac{dy}{dt} \right|_{t=0} = \frac{K \prod_{i=1}^{q} \xi_i}{\prod_{i=1}^{p} \tau_i} \tag{6.93}$$

and the response is rapid, not exhibiting the typical sigmoidal characteristics of higher order systems. Note that Eq. (6.88) is a special case of the more general Eq. (6.93).

2. When $p - q > 1$ the initial slope of the response is zero, and the shape of the response curve is sigmoidal (see Figure 6.17).

3. In general, the response of the system with the higher value of q (more zeros) will be swifter than the response of a corresponding system with a lower value of q (fewer zeros). This is in keeping with the conclusion that lead terms tend to "speed up" process responses; and we expect that the higher the number of lead terms, the more pronounced will be the "speeding up" effect.

This fact is illustrated in Figure 6.17 for a third-order system with no zero, and a third-order system with one zero; i.e., $p = 3, q = 1$. Observe that because $p - q = 2$ (which is greater than 1), the response for the system with the zero is sigmoidal, but it is still less sluggish than the response for the pure third-order system with no zero.

In closing, let us return once more to the (2,1)-order system. One of the most important aspects of this system's step response is that unlike the pure second-order system's response, the slope at the origin is nonzero, and the actual value is given in Eq. (6.88) as $K\xi_1 / \tau_1 \tau_2$.

We did specify that for this system ξ_1 is positive (as such, the system's transfer function zero, $s = -1/\xi_1$, is *negative*), and we assumed that the system gain K is also positive. As a result, this slope is positive, since K and ξ_1 have the same positive sign.

When the value of ξ_1 is negative, so that the transfer function zero is *positive* (with the steady-state gain still positive), observe that in this case, the slope of the step response at the origin will now be *negative*. The implication here is as follows: the system response starts out heading "downwards" by virtue of the negative initial slope, turning around sometime later to head in the direction of the steady-state value dictated by K (remember, this is positive.)

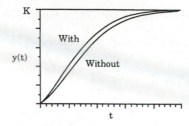

Figure 6.17.   Effect of a single zero on the unit step response of a third-order system.

Such surprising responses are sometimes encountered in actual practice, and in processes of some importance. The dynamic behavior of such processes are so unusual that they will be discussed separately in the next chapter.

6.7   SUMMARY

Processes whose dynamic behavior are represented by higher (but finite) order models have been studied in this chapter. We have found these processes to be somewhat more complex in nature than the low-order systems treated in Chapter 5. The transfer function models for these processes have at least two poles, and may or may not have zeros.

When two first-order systems are connected in series, whether they are connected in an interacting or noninteracting format, we have found that the behavior exhibited is that of a second-order system. In general, N first-order systems connected in series constitute an Nth-order system. The step response of such systems (indicative of typical dynamic behavior) are in general more sluggish than the step response of the first-order system. Furthermore, the higher the order, the more sluggish the response.

The study of the behavior of the general second-order system confronted us with the first major departure from low-order system behavior. We found out that unlike the first-order system, there are actually three different kinds of second-order systems: overdamped, underdamped, and critically damped; and that two first-order systems in series are of the overdamped (or at best the critically damped) type.

When zeros are present in a process transfer function model, we have seen that the general tendency is for the responses to be swifter than what would be observed in the corresponding systems with no zeros. Zeros tend to "speed up" process response. In particular we saw that for the (2,1)-order system (i.e., one with two poles and one zero) the presence of the zero can speed up the response to such an extent that overshooting the final value is experienced, whereas it was impossible for its counterpart without the zero to exhibit an overshoot.

Our treatment of higher (but finite) order systems is not yet complete. There is a special class of higher order systems with zeros that exhibit what can only be considered as unusual dynamic behavior. These will be dealt with next, in Chapter 7.

Also, higher order systems of infinite order have not yet been discussed, only hinted at. As it turns out, one of the most prominent aspects of industrial process control systems is the almost ubiquitous presence of time delays, and time-delay systems are actually infinite-order systems. This fact will be established in Chapter 8 where we will undertake a study of the dynamic behavior of time-delay systems.

REFERENCES AND SUGGESTED FURTHER READING

1.   Bird, R.B., W.E. Stewart, and E.N. Lightfoot, *Transport Phenomena*, J. Wiley, New York (1960)
2.   Stephanopoulos, G., *Chemical Process Control: An Introduction to Theory and Practice*, Prentice-Hall, Englewood Cliffs, NJ (1985)

REVIEW QUESTIONS

1.   An ensemble of two first-order systems in series is equivalent to a system of what order? In general, what will the resulting system order be for an ensemble of N first-order systems in series?

2.   What is the rationale behind the terminology adopted for ζ and τ, two of the characteristic parameters of the general second-order system?

3.   What range of values of the parameter ζ will give rise to:

     (a) Overdamped response
     (b) Underdamped response
     (c) Critically damped response
     (d) Pure *undamped* response?

4.   Why is it impossible for two first-order systems in series to exhibit underdamped behavior?

5.   Under what condition will the step response of a second-order system with a single zero exhibit overshoot?

6.   In general, what effect does a lead term have on the theoretical process response?

7.   Is it possible to distinguish between the step response of a first-order system and that of a second-order system with a single zero, whose characteristic parameters are related according to: $\tau_1 < \xi_1 < \tau_2$?

8.   If a second-order system with a single zero now has characteristic parameters related according to: $\xi_1 \ll \tau_1 < \tau_2$ (i.e., the lead term is almost negligible), how can we distinguish between its step response and that of a first-order system?

9.   How can you distinguish between the step response of an underdamped second-order system (with no zero) and that of a second-order system with a single zero that exhibits overshoot only because of the presence — and location — of its zero?

10. What does the term "pole-zero excess" mean, and what effect does it have on the nature of the initial portion of the system step response?

PROBLEMS

6.1   (a) By taking Laplace transforms of the expression in Eq. (6.8) in the main text, obtain the transfer function relating y_2 and u, and show that it is identical to that given in Eq. (6.10).
(b) By differentiating the unit step response given in Eq. (6.12) and setting $t = 0$, establish that the initial slope of the second-order system's step response is zero when $\tau_1 \neq \tau_2$. Repeat this exercise using the expression given in Eq. (6.14) and establish that even when $\tau_1 = \tau_2$, the initial slope of the second-order response is still zero.

6.2 By using the initial value theorem of Laplace transforms, or otherwise, show that for the general system whose transfer function is given in factored form as in Eqs. (6.81) and (6.82), i.e.:

$$g(s) = \frac{K\left(\xi_1 s + 1\right)\left(\xi_2 s + 1\right) \cdots \left(\xi_q s + 1\right)}{\left(\tau_1 s + 1\right)\left(\tau_2 s + 1\right) \cdots \left(\tau_p s + 1\right)} \tag{6.81}$$

$$g(s) = \frac{K \prod\limits_{i=1}^{q}\left(\xi_i s + 1\right)}{\prod\limits_{i=1}^{p}\left(\tau_i s + 1\right)} \tag{6.82}$$

the initial slope of the step response is zero whenever $p - q > 1$, and nonzero only when $p - q = 1$.

Also show that when the pole-zero excess is precisely 1 (i.e., $p - q = 1$) — *all other things being equal* — the initial portion of the step response becomes steeper as the values of ξ_i , $i = 1, 2, ..., q$, the lead time constants, increase; whereas, as the value of the lag time constants τ_i , $i = 1, 2, ..., p$, increase, the initial response becomes less steep.

6.3 (a) Obtain the impulse response of two first-order systems in series directly by using the transfer function in Eq. (6.10), with $K = K_1 K_2$.
(b) Differentiate the step response given in Eq. (6.12) with respect to time and compare the result with the impulse response obtained in (a); hence establish that, at least for this system, one can obtain the impulse response by differentiating the step response.
(c) Starting from the general, arbitrary linear system model:

$$y(s) = g(s)\, u(s)$$

with no particular form specified for $g(s)$, by introducing appropriate expressions for $u(s)$, establish that, in general:

$$\frac{d}{dt}\left(y_{\text{step}}\right) = \left(y_{\text{impulse}}\right) \tag{P6.1}$$

where $\left(y_{\text{step}}\right)$ is the step response, and $\left(y_{\text{impulse}}\right)$ is the impulse response.

6.4 A *parallel* arrangement of two first-order systems, each with individual transfer functions given by:

$$g_1(s) = \frac{K_1}{\tau_1 s + 1}$$

$$g_2(s) = \frac{K_2}{\tau_2 s + 1}$$

gives rise to an overall composite system with transfer function given by:

$$g(s) = g_1(s) + g_2(s)$$

(a) You are now required to show that, unlike the *series* arrangement discussed in the text, this $g(s)$ has the following properties:

1. Its gain is $K_1 + K_2$, not K_1K_2;
2. *It has a single zero* whose location is determined by the *lead* time constant, ξ_1, a parameter whose value depends on all the four contributing system parameters. Give an explicit expression for the dependence of this *lead* time constant on the contributing system parameters. (The series arrangement, of course, gives rise to no zero.)
3. For the situation in which both K_1 and K_2 are positive, the lead time constant ξ_1 is related to the contributing *lag* time constants according to: $\tau_1 < \xi_1 < \tau_2$, if $\tau_1 < \tau_2$, or, according to: $\tau_2 < \xi_1 < \tau_1$ if the converse is the case.

Hence, or otherwise, establish that a parallel arrangement of two first-order systems, each with a *positive* steady-state gain, *will never exhibit overshoot* in response to a step input.

(b) If now K_1 is *positive* and K_2 is *negative*, but with $|K_1| > |K_2|$, under what condition will the *lead* time constant of the system resulting from this parallel arrangement be negative? (Note that both τ_1 and τ_2 are strictly positive.)

6.5   For the interacting system discussed in Section 6.1.3, introduce the following deviation variables:

$$x_1 = h_1 - h_{1s} \,;\ x_2 = h_2 - h_{2s} \,;\ u = F_0 - F_{0s}$$

and let $y = x_2$; further, introduce the parameters, τ_1, τ_2, K_1, and K_2, precisely as defined for the noninteracting system discussed in Section 6.1.1 and thereby combine the expressions given in Eqs. (6.18) and (6.19) into a vector-matrix differential equation model of the form:

$$\dot{\mathbf{x}} = \mathbf{Ax} + \mathbf{Bu}$$
$$y = \mathbf{c}^T\mathbf{x}$$

(a) What are the elements of **A**, **B**, **c**, in this case?

(b) Compare the **A** matrix for this interacting system with that given in Eq. (6.29) and hence identify the factor responsible for the "interaction."

6.6   (a) Show that the expression for $y(t)$ in Eq. (6.87) attains a maximum only if $\xi > \tau_1$ and τ_2, and hence establish that a (2,1)-order system will *not* exhibit overshoot unless the lead time constant is greater than both lag time constants; i.e., $\xi > \tau_1$ and τ_2.

(b) Under conditions guaranteeing the existence of a maximum in Eq. (6.87) derive an expression for the time t_{max} at which this maximum occurs; also derive an expression for the time t_1 when the ultimate value K is first attained.

6.7   Using whichever control system simulation software is available to you (e.g., CONSYD, CC, Matlab, etc.) obtain *unit* step response plots for systems whose transfer functions are given below:

(a) $g(s) = \dfrac{2(3s + 1)}{(2s + 1)(s + 1)}$

(b) $g(s) = \dfrac{1.5(4s + 1)}{(2s + 1)(s + 1)}$

From Eq. (6.89) compute $J_{(2,1)}$, the inherent reluctance factor for each system and interpret your results in light of the plots and of what $J_{(2,1)}$ is supposed to measure.

6.8   An industrial distillation column of an unusual design is the first part of a complex monomer/solvent recovery system used to separate the components in the product stream of a polymerization reactor. After a carefully conducted procedure for process identification (see Chapter 13), the dynamic behavior of the most critical portion of the column was found to be reasonably well characterized by the transfer function:

$$g(s) = \frac{6s^2 + s + 1}{(6s + 1)(s^2 + 3s + 1)} \qquad \text{(P6.2)}$$

(a) How many poles and how many zeros does this system transfer function have and where are they located?
(b) Use a control system simulation package to obtain the system's unit step response; what do you think might be responsible for this somewhat unusual response?
(c) Expand $g(s)$ in the form:

$$g(s) = \frac{As + B}{s^2 + 3s + 1} + \frac{C}{6s + 1} \qquad \text{(P6.3)}$$

What insight does such an expansion give in terms of an "empirical mechanism" for explaining the observed system behavior?

6.9 In modeling the process of drug ingestion, distribution, and subsequent metabolism in an individual, a simplified mathematical model was given in Problem 4.4 at the end of Chapter 4 as:

$$\frac{dx_1}{dt} = -k_1 x_1 + u \qquad \text{(P4.3)}$$

$$\frac{dx_2}{dt} = k_1 x_1 - k_2 x_2 \qquad \text{(P4.4)}$$

$$y = x_2 \qquad \text{(P4.5)}$$

where, as indicated in that problem statement, x_1, x_2, and u respectively represent (as *deviations* from initial steady-state values) the drug mass in the Gastrointestinal tract (GIT), the drug mass in the Bloodstream (BS), and the drug ingestion rate (see Problem 4.4).

For a given small mammal used in a toxicology laboratory, the values of the effective "rate constants" associated with drug distribution *into* the bloodstream, and with drug elimination *from* the bloodstream, were found respectively to be: $k_1 = 5.63$ (min)$^{-1}$ and $k_2 = 12.62$ (min)$^{-1}$.

Using a transfer function version of the process model, or otherwise,
(a) Obtain the response of the drug mass in the bloodstream to a drug ingestion program which, as shown below in Figure P6.1, calls for a 10 mg/min ingestion rate started instantaneously at $t = 0$, sustained for precisely 5 min, and discontinued thereafter. What is the maximum value attained by the drug mass in the bloodstream?
(b) Repeat (a) for the situation in which:

$$u(t) = \begin{cases} 0; & t < 0 \\ 10e^{-k_2 t}; & t \geq 0 \end{cases} \qquad \text{(P6.4)}$$

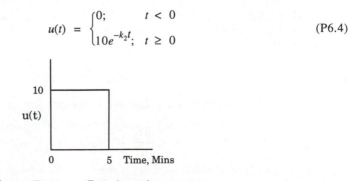

Figure P6.1.   Drug ingestion program

a program which starts out with a drug ingestion of 10 mg/min and decays exponentially at a rate characterized by k_2, the same "rate constant" associated with drug elimination from the bloodstream.

(c) Repeat (a) for the situation in which:

$$u(t) = \begin{cases} 0; & t < 0 \\ 10 \sin 2t; & t \geq 0 \end{cases} \qquad \text{(P6.5)}$$

a periodic drug ingestion program with a 10 mg/min amplitude and a frequency of 2 radians/min.

6.10 An experimental biotechnology process utilizes a microorganism M_1 for the batchwise manufacture of a vaccine. After each production cycle, most of the original mass of M_1 is recovered and reused, but after several cycles, a fresh supply of M_1 must be introduced and the "spent" batch of M_1 discarded. A special pond, cultivated as the natural habitat of a different microorganism M_2 is used for the disposal of M_1 because it is known that the former microorganism thrives and reproduces by feeding exclusively on M_1. However, if M_1 is left by itself in the pond, it will mutate somewhat and then begin to thrive and reproduce by feeding on lower forms present in the pond.

In monitoring the pond for the amount of M_1 and M_2 it contains, it is found, however, that contrary to initial expectations, the amount of M_1 does not decrease steadily with time; rather, it fluctuates in a distinctly periodic fashion. The same analysis uncovers another equally unsettling fact: that the amount of M_2 itself also fluctuates essentially with the same period as the M_1 fluctuations.

As a first step in understanding the observed phenomenon, the following mathematical model was developed for the dynamic behavior of the mass concentration of M_1 and M_2 in the pond:

$$\frac{d\xi_1}{dt} = a_1\xi_1 - a_{12}\xi_1\xi_2 + b\mu \qquad \text{(P6.6a)}$$

$$\frac{d\xi_2}{dt} = -a_2\xi_2 + a_{21}\xi_1\xi_2 \qquad \text{(P6.6b)}$$

where:

ξ_1 is the mass concentration of M_1 (microgram/liter)
ξ_2 is the mass concentration of M_2 (microgram/liter)
μ is the rate of introduction of M_1 into the pond (assumed steady, for simplicity) (microgram/hr)
$a_1, a_{12}, a_2, a_{21}, b$, are all parameters characteristic of the specific pond in question

This nonlinear model may be *linearized* around ξ^*_1, ξ^*_2, and μ^*, some appropriate reference values of the variables ξ_1, ξ_2, and μ respectively (see Chapter 10), and by defining the following deviation variables:

$$x_1 = \xi_1 - \xi^*_1; \qquad x_2 = \xi_2 - \xi^*_2; \qquad u = \mu - \mu^*$$

we obtain the following approximate, linear model:

$$\frac{dx_1}{dt} = \alpha_{11}x_1 - \alpha_{12}x_2 + bu \qquad \text{(P6.7a)}$$

$$\frac{dx_2}{dt} = \alpha_{21}x_1 - \alpha_{22}x_2 \qquad \text{(P6.7b)}$$

For the specific pond in question, the appropriate parameter values are estimated to be: $\alpha_{11} = 1.0$; $\alpha_{12} = 1.5$; $\alpha_{21} = 2.0$; $\alpha_{22} = 2.0$; and $b = 0.25$.
Reduce the two equations to a single second-order differential equation in x_2 of the form:

$$\beta_2 \frac{d^2 x_2}{dt^2} + \beta_1 \frac{dx_2}{dt} + \beta_0 x_2 = \phi u(t) \tag{P6.8}$$

and give the values of the new parameters β_0, β_1, β_2, and ϕ. From your knowledge of second-order systems, what do these parameters indicate about the type of behavior predicted for x_2 by the approximate model? Finally, obtain and plot the *actual* unit step response of x_2. Comment on whether or not the predictions of this simplified, linearized model are consistent with the dynamic behavior observed via the pond monitoring system.

6.11 Revisit Problem 6.10. By taking Laplace transforms of the modeling equations (P6.7a,b), solve the two equations simultaneously for $x_1(s)$ and $x_2(s)$ and obtain transfer function models of the form:

$$x_1(s) = g_1(s)\, u(s) \tag{P6.9a}$$
$$x_2(s) = g_2(s)\, u(s) \tag{P6.9b}$$

(a) Write explicit expressions for the two transfer functions $g_1(s)$ and $g_2(s)$; first compare and contrast them, then confirm that the expression obtained for x_2 in Problem 6.10 is equivalent to that obtained here in Eq. (P6.9b).
(b) Using the transfer function representation in Eq. (P6.9a), obtain and plot the unit step response of x_1, the mass concentration of M_1 in the pond. Compare this plot with that obtained in Problem (6.10) for the mass concentration of M_2.

6.12 The following transfer function was obtained by correlating input/output data collected from a series ensemble of three CSTR's used for the sulphonation reaction in a detergent manufacturing process:

$$g(s) = \frac{3.6}{(12s + 1)(15s + 1)(20s + 1)} \tag{P6.10}$$

The input to the process u is the mole fraction of the main reactant in the feedstream to the first reactor; the output y is the normalized (dimensionless) weight percent of the product at the third reactor's outlet. The time units are in minutes.
(a) Obtain an expression for the system's response to a change of 0.1 in the system input, and use a control system simulation package to obtain a plot of this response.
(b) The final reactor, having been identified in isolation from the rest as the one with a time constant of 20 min, and with a gain of 1.5, is to be taken out of commission temporarily. All other things being equal, what will be the transfer function of the surviving two-reactor ensemble? Obtain and plot the response of the smaller ensemble to the same input as that in part (a); compare this plot with that obtained in part (a).

6.13 A stream containing an essentially constant concentration of pollutant flows steadily at a rate $F = 65.2$ km^3/year into Lake Superior, a lake whose volume, $V = 12.2 \times 10^3$ km^3, may be assumed constant (cf. R. H. Rainey, "Natural Displacement of Pollution from the Great Lakes," *Science*, **155**, 1242–1243 (1967)). (If, for simplicity, the effects of volumetric changes due to water evaporation or due to precipitation are ignored, it may be assumed that water seepage through the lake's porous bed is at the same rate of 65.2 km^3/year; V is therefore considered constant.)

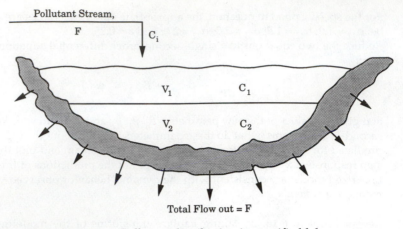

Figure P6.2.   Pollutant distribution in stratified lake.

The effect of stratification brought about by temperature differences along the depth of such deep lakes is clearly significant in determining the dispersion of pollutants; also the vastness of the extent of the lake, and the general absence of any reasonably vigorous mechanism for agitation, all conspire to make for a nonuniform dispersion of pollutants within the lake. Thus, to assume that the pollutants are distributed evenly within the entire body of water only leads to the following grossly oversimplified model for C, the concentration of pollutants in the lake:

$$V\frac{dC}{dt} = FC_i - FC \tag{P6.11}$$

with C_i as the pollutant concentration in the inlet stream.

(a) Let us now consider that the stratifying effect of temperature results in the lake being made up of *two* distinct (idealized) layers, a top and a bottom half, each of equal volume, and with respective pollutant concentration C_1 and C_2. If all the other assumptions leading to Eq. (P6.11) are still retained; if the flow across the stratification is assumed constant and equal to F; and if each stratum is again assumed to be well mixed and uniform, but distinct one from the other, then an alternative model consistent with these new considerations will be:

$$V_1\frac{dC_1}{dt} = FC_i - FC_1 \tag{P6.12a}$$

$$V_2\frac{dC_2}{dt} = FC_1 - FC_2 \tag{P6.12b}$$

with

$$V_1 = V_2 = V/2$$

Using the Lake Superior parameters given above, obtain the time behavior of the pollutant concentrations C_1 and C_2 in response to a *change* in C_i of 5 kg mole/km^3. Plot these responses and compare them with the response predicted by the simpler model in Eq. (P6.11).

(b) Were the lake now assumed to be stratified into *three* distinct layers of equal volume $V_1 = V_2 = V_3$, and each with respective pollutant concentration C_1, C_2, and C_3, write an appropriate model, the equivalent of the one in Eq. (P6.12); state your assumptions. Plot the time responses of C_1, C_2, and C_3 to the same change in C_i used in part (a) above. Compare with the plots in part (a).

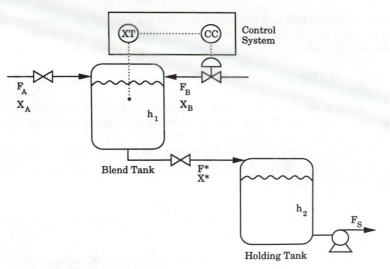

Figure P6.3.   Simplified schematic diagram of a gasoline blending process.

6.14   Figure P6.3 is a simplified schematic diagram of a gasoline blending process. Stream A of gasoline with octane rating X_A comes directly from the production facility at a volumetric flowrate F_A gal/min; stream B of gasoline with *uniform* octane rating X_B is drawn from a storage tank, and its flowrate, F_B gal/min, is determined by a very effective control system designed to maintain the octane rating of the gasoline stream exiting the blend tank at the desired value, X^*. The blended gasoline flows into the holding tank at a flowrate F^* determined by the level in the blend tank. A metering pump draws the blended gasoline out of the holding tank at a constant, supply flowrate F_s gal/min.

To simplify the problem at hand, we shall assume the following:

1. X_A, the octane rating of the gasoline in stream A is constant; only the flowrate F_A varies. (In reality, both change.)
2.   The gasoline density is essentially constant at ρ regardless of the octane rating.
3.   The control system works perfectly, so that the octane rating of the blend tank gasoline is not only uniform, it is also constantly held at X^*; the blend level h_1, however, changes because of the control moves made on F_B.
4.   The cross-sectional areas of the blend tank A_1 and that of the holding tank A_2 are uniform and constant.
5.   The flowrate F^* is *directly* proportional to the level h_1.

(a) Obtain a reasonable mathematical model which describes the dynamic behavior of the levels h_1, and h_2 of the blended gasoline respectively in the blend tank, and the holding tank. Define the deviation variables:

$$u = F_B - F_{Bs}\,;\ x_1 = h_1 - h_{1s}\,;\ x_2 = h_2 - h_{2s}\,;\ d = F_A - F_{As}$$

cast your model in terms of these variables and obtain a transfer function model relating u to x_2.
(b) Obtain an expression for the response of the holding tank level (x_2 in deviation variables) to a rectangular pulse of magnitude δ in flowrate F_A, lasting for θ time units. To what value does x_2 settle ultimately?

6.15 Data on the monthly before-tax earnings of a high-tech corporation over the lifetime of its main product were found to be reasonably well explained by the simple econometric model:

$$y(k) = 1.15y(k-1) - 0.49y(k-2) + 0.95u(k-1) + 0.75u(k-2) \qquad \text{(P6.13)}$$

where $y(k)$ is earnings (in millions of dollars), and $u(k)$ is the *total* number of new versions of the product in the market; k is the month-to-month time index (i.e., $k = 1$ for the first month, 2 for the second month, etc.).

It can be shown (cf. Box and Jenkins, *Time Series Analysis: Forecasting and Control*, Holden-Day, San Francisco, 361 (1976)) that a continuous time approximation of this model is given by:

$$2\frac{d^2 y}{dt^2} + 1.414\frac{dy}{dt} + y = 5u(t) \qquad \text{(P6.14)}$$

(with the time units still in months), an expression that indicates a second-order dynamic behavior for the company's earnings.

(a) What type of second-order behavior will this economic system exhibit?

(b) By obtaining the response to $u(t) = 1$, derive an expression for the time history of the company's earnings as a result of introducing one new version of the main product to the market. Plot this response.

(c) From this plot, or otherwise, determine *the ultimate value to which the earnings settle*, *the maximum earnings* made on this product, and *when* (in months after inception) this maximum occurs.

(d) If this new version of the product is discontinued after 12 months, what are the *total* earnings realized over the product's lifetime? [Hint: Earnings are obtained *and computed* on a monthly basis.]

6.16 In a study demonstrating the application of control systems theory to biological systems, Grodins et al.[†] modeled the respiratory process as consisting of two essentially constant-volume reservoirs connected by the circulating blood, as shown in Figure P6.4

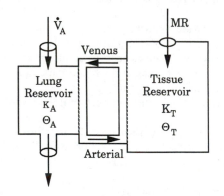

Figure P6.4.   Lung, blood, tissue CO_2 exchange system.

[†] F. S. Grodins, J. S. Gray, K. R. Schroeder, A. L. Norins, and R. W. Jones, "Respiratory responses to CO_2 inhalation: a theoretical study of a nonlinear biological regulator," *J. Appl. Physiol.*, **7**, 283 (1954).

By making several simplifying assumptions carefully discussed in the reference they obtained the following set of two second-order differential equations for the Lung, Blood, Tissue CO_2 exchange system:

$$\frac{K_A K_T}{Q\dot{V}_A} \ddot{\theta}_T + \left(\frac{K_A}{\dot{V}_A} + \frac{K_T BA_s}{\dot{V}_A} + \frac{K_T}{Q} \right) \dot{\theta}_T + \theta_T$$

$$= BA_s F_{CO_2}^I(t) + \frac{BA_s MR}{\dot{V}_A} + \frac{MR}{Q} + A_i \qquad (P6.15)$$

$$\frac{K_A K_T}{Q\dot{V}_A} \ddot{\theta}_A + \left(\frac{K_A}{\dot{V}_A} + \frac{K_T BA_s}{\dot{V}_A} + \frac{K_T}{Q} \right) \dot{\theta}_A + \theta_A$$

$$= F_{CO_2}^I(t) + \frac{K_T}{Q} \dot{F}_{CO_2}^I(t) + \frac{MR}{\dot{V}_A} \qquad (P6.16)$$

where each ($\bullet$) indicates a derivative with respect to time. The variables in these equations are defined as follows:

K_A = Volume of lung (alveolar) reservoir
K_T = Volume of tissue reservoir
θ_A = Alveolar concentration of CO_2
$\dot{\theta}_A$ = Rate of change of alveolar concentration of CO_2
θ_T = Tissue concentration of CO_2
$\dot{\theta}_T$ = Rate of change of tissue concentration of CO_2
$\dot{V}_A$ = Ventilation rate
Q = Cardiac output
B = Barometric pressure
A_s, A_i = Parameters of an empirical absorption relation
MR = Rate of metabolic CO_2 production
$F_{CO_2}^I$ = Fraction of CO_2 in inspired gas
(with the implication that $F_{CO_2}^I \dot{V}_A$ represents the rate of CO_2 intake via inspiration.)

By doing the following:

1. Assuming that $B, A_s, A_i, MR, \dot{V}_A$, and Q are constants;
2. Introducing the consolidated parameters ξ_1, ξ_2, ξ_{12} defined as:

$$\xi_1 = \frac{K_A}{\dot{V}_A}; \qquad \xi_2 = \frac{K_T}{Q}; \qquad \xi_{12} = \frac{K_T}{\dot{V}_A} BA_s$$

3. Defining "effective time constants" τ_1, and τ_2 as:

$$\tau_1 \tau_2 = \xi_1 \xi_2; \qquad \tau_1 + \tau_2 = \xi_1 + \xi_{12} + \xi_2$$

hence, or otherwise, recast the original set of modeling equations in terms of deviation variables:

$$y_1 = \theta_T - \theta_T^*; \qquad y_2 = \theta_A - \theta_A^*; \qquad u = F_{CO_2}^I - (F_{CO_2}^I)^*$$

take Laplace transforms and show that the equivalent transfer function model for this biological system is the following set of equations:

$$y_1(s) = \frac{(\xi_2 s + 1)}{(\tau_1 s + 1)(\tau_2 s + 1)} u(s) \qquad \text{(P6.17a)}$$

$$y_2(s) = \frac{K}{(\tau_1 s + 1)(\tau_2 s + 1)} u(s) \qquad \text{(P6.17b)}$$

and hence that:

$$y_2(s) = \frac{K}{(\xi_2 s + 1)} y_1(s)$$

or, as presented in Grodins *et al.* (1954):

$$y_1(s) = \frac{(\xi_2 s + 1)}{K} y_2(s)$$

6.17 (a) Revisit Problem 6.16. An inspection of the transfer function representing the behavior of the alveolar concentration of CO_2 indicates the presence of a zero. From the definition of ξ_2 in terms of actual physiological parameters, and in terms of its relation to the "effective time constants" τ_1 and τ_2 introduced in Problem 6.15, will the step response of the alveolar CO_2 concentration ever show an overshoot?
(b) For the specific situation in which the following parameters are given (cf. Grodins *et al.* (1954) cited in Problem 6.16):

$K_T = 40$ liters; $Q = 6$ liters/min; $K_A = 3$ liters; $\dot{V}_A = 5$ liters/min; $B = 760$ mm Hg; $A_s = 0.00425$ liters/liter/mm Hg

obtain and plot the change observed in the tissue CO_2 concentration (y_1) and the alveolar CO_2 concentration (y_2), in response to a change of 0.1 in the inspired fraction of CO_2, $F_{CO_2}{}^I$.

6.18 In the 1959 paper, "Integrative cardiovascular physiology: a mathematical synthesis of cardiac and blood vessel hemodynamics" (*Quart. Rev. Biol.*, **34**, 93 (1959)) F. S. Grodins developed a fairly complex control system model for the cardiovascular regulator.

Of the complete overall cardiovascular system, an important subsystem of considerable physiological interest involves the effect of hemorrhage on the systemic arterial load pressure. If Q_H represents the *rate* of hemorrhage (defined as the rate of blood flow *from* the system *to* the external environment), and if P_{AS} represents the systemic arterial load pressure, then the Grodins model for this subsystem may be reduced to the block diagrammatic representation shown below in Figure P6.5,

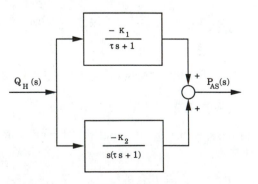

Figure P6.5.

where τ, the systemic arterial system's effective time constant, and the gains K_1, K_2, are defined in terms of physiological parameters as:

$$\tau = \frac{R_s C_{As} C_{Vs}}{C_s}; \qquad K_1 = \frac{R_s C_{Vs}}{C_s}; \qquad K_2 = \frac{1}{C_s}$$

with:

R_s =   a linear Poisseuille resistance (arising from the hydraulic analog on which the full-scale model is based),

C_{As} =   systemic arterial compliance,

C_{Vs} =   systemic venal compliance,

C_s = $(C_{As} + C_{Vs})$, the *total* compliance.

(Compliance is defined as a ratio of volume to pressure.)

(a) Consolidate the subsystem model to obtain:

$$P_{AS}(s) = g(s) Q_H(s)$$

and identify the poles and zeros of the consolidated transfer function, leaving these in terms of the given parameters, τ, K_1, and K_2. If the model were to be used instead to study the effect of the *rate of transfusion* on systemic arterial pressure, where transfusion is defined as blood flow *to* the system *from* the external environment (precisely the *reverse* of hemorrhage), what changes, if any, need to be made in the model?

(b) Again, in terms of the parameters τ, K_1, and K_2, obtain an expression for the response of P_{AS} to a unit step change in Q_H. From your result, deduce the long-term effect of sustained hemorrhaging on the systemic arterial load pressure.

(c) Given the following parameters: $R_s = 0.0167$ mm Hg; $C_{Vs} = 500$ cc/mm Hg, (cf. Grodins 1959), obtain and plot the response of P_{AS} to a rectangular pulse in Q_H of magnitude 20 cc/s, sustained for 30 s, for two different values of $C_{As} = 250$ cc/mm Hg, $C_{As} = 750$ cc/mm Hg. To what ultimate value does P_{AS} settle in each case?

CHAPTER

7

INVERSE-RESPONSE SYSTEMS

When the initial direction of a process system's step response is *opposite* to the direction of the final steady state, it is said to exhibit *inverse response*. There is an initial inversion in the response of such processes that starts the process output heading *away* from its ultimate value; this is later reversed, and the process output eventually heads in the direction of the final steady state.

Unusual as this phenomenon may appear, it nevertheless occurs in enough real processes to warrant a closer look. In doing so, we will see that systems that exhibit such inverse-response characteristics actually belong in the same class as the higher order systems with zeros studied in Chapter 6. However, while the zeros of the previously studied systems have all been negative, we will find that inverse-response behavior is associated with the presence of positive zeros; i.e., zeros located in the right-half of the complex s plane.

7.1   INTRODUCTION

Consider a system whose transfer function, $g(s)$, is composed of two parts as shown below:

$$g(s) = g_1(s) - g_2(s) \qquad (7.1)$$

and because the second "mode," $g_2(s)$, appears to be in "opposition" to the first "mode," $g_1(s)$ (by virtue of the negative sign), we refer to $g_1(s)$ as representing the "main mode" and to $g_2(s)$ as the "opposition mode" of the system.

Let us consider in particular the situation where we have a system composed of two opposing first-order modes (see Figure 7.1), i.e.:

$$g(s) = \frac{K_1}{\tau_1 s + 1} - \frac{K_2}{\tau_2 s + 1} \qquad (7.2)$$

where both steady-state gains are positive, and by definition, we consider the "main" steady-state gain K_1 to be larger than the "opposition" steady-state gain K_2.

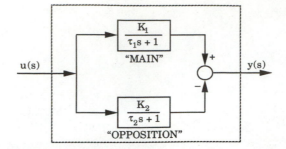

Figure 7.1. System composed of two opposing first-order systems in parallel.

The dynamic behavior of such a system, obtained from the usual transfer function model:

$$y(s) = g(s)\,u(s)$$

will in this case be given by:

$$y(s) = g_1(s)u(s) - g_2(s)u(s) \tag{7.3a}$$

or

$$y(s) = y_1(s) - y_2(s) \tag{7.3b}$$

where $y_1(s)$ is the response of the "main" first-order process, and $y_2(s)$ is the response of the "opposition" first-order process.

Let us now investigate the qualitative nature of this system's unit step response.

Unit Step Response

It is clear from Eq. (7.3b) that the overall unit step response of this composite system will be the difference between the unit step response of system 1 and that of system 2.

Steady-State Considerations

From our knowledge of the dynamic behavior of first-order systems, we know that the unit step response of system 1 has a steady-state value of K_1, and that the corresponding value for system 2 is K_2. Thus the overall steady-state value attained by the composite system's unit step response is:

$$y(\infty) = K_1 - K_2 \tag{7.4}$$

and since by definition $K_1 > K_2$, this quantity is *positive*.

Initial Slope

Once again, from Eq. (7.3b), for all $t \geq 0$, we know that:

$$y(t) = y_1(t) - y_2(t)$$

therefore upon differentiation we have that, for all $t \geq 0$:

$$\frac{dy}{dt} = \frac{dy_1}{dt} - \frac{dy_2}{dt} \tag{7.5}$$

The application of Eq. (7.5) in particular for $t = 0$ now assures us that the net initial slope of the composite system response is, once again, the difference between the initial slope of the "main" system's response and that of the "opposition" system.

Recalling once more our results for first-order step responses, we immediately obtain from Eq. (7.5) that the initial slope of the overall system is given by:

$$\left.\frac{dy}{dt}\right|_{t=0} = \frac{K_1}{\tau_1} - \frac{K_2}{\tau_2} \tag{7.6}$$

Let us now make an interesting proposition: suppose that the natures of the "main" and "opposition" systems are such that:

$$\frac{K_2}{\tau_2} > \frac{K_1}{\tau_1} \tag{7.7}$$

i.e., the initial slope of the opposing system response is greater than the initial slope of the "main" system response; then the net initial slope of the overall system response as given in Eq. (7.6) will be *negative*.

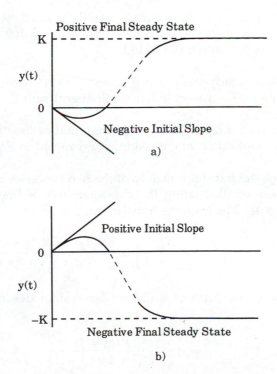

a)

b)

Figure 7.2.   Initial and final aspects of the behavior of the composite system of Figure 7.1.

When we combine the steady-state considerations with the initial slope results, we find that the step response for this composite system, under conditions indicated in Eq. (7.7), will have a *negative* slope at the origin — indicating an initial downward trend — and a *positive* final steady-state value. As indicated in Figure 7.2(a), the only way it is going to be possible for the response to start out heading downward, but end up being ultimately positive, is if there is a "turn around" somewhere between the initial point, $t = 0$, and when steady state is achieved.

The opposite behavior shown in Figure 7.2(b) can be obtained if $|K_1| > |K_2|$, but both K_1 and K_2 are *negative*. In this case the ultimate response given by Eq. (7.4) will be *negative*, and the initial slope satisfying Eqs. (7.6) and (7.7) will be *positive*. Thus the system will begin to respond in a positive direction and then reverse itself and ultimately reach a negative steady state as shown in Figure 7.2(b).

Systems whose step responses are characterized by such an initial inversion (starting out the process in the "wrong" direction which is later reversed so that the process response eventually heads in the "right" direction) are said to exhibit inverse response.

The main point here is the following:

Inverse response occurs as the net effect of (at least two) opposing dynamic modes of different magnitudes, operating on different characteristic timescales. The faster mode, which must have the smaller magnitude, is responsible for the initial, "wrong way" response; this is eventually overwhelmed by the slower mode having the larger magnitude.

Observe that in the specific example we have used, with competition between two first-order modes, we have been able to show that inverse response is possible only when the "opposition" mode

1. Has a lower steady-state gain than the "main" mode (recall that the condition $K_2 < K_1$ was an initial requirement), and

2. Responds with a faster initial slope than that of the "main" mode, for this is the implication of the condition expressed in Eq. (7.7).

Let us now combine the transfer functions of the two processes in Figure 7.1 into one for the purpose of illustrating the characteristics of linear systems that show inverse response. The resulting transfer function is:

$$g(s) = \left(\frac{K_1}{\tau_1 s + 1} - \frac{K_2}{\tau_2 s + 1} \right) = \frac{(K_1 \tau_2 - K_2 \tau_1)s + (K_1 - K_2)}{(\tau_1 s + 1)(\tau_2 s + 1)}$$

Note that this is the form of a (2,1)-order system described in the last chapter:

$$g(s) = \frac{K(\xi s + 1)}{(\tau_1 s + 1)(\tau_2 s + 1)} \qquad (6.83)$$

with

$$K = K_1 - K_2 \qquad (7.8a)$$

$$\xi = \frac{K_1\tau_2 - K_2\tau_1}{(K_1 - K_2)} = \left(\frac{K_1}{\tau_1} - \frac{K_2}{\tau_2}\right)\frac{\tau_1\tau_2}{(K_1 - K_2)} \qquad (7.8b)$$

To obtain the inverse response shown in Figure 7.2(a), we required both K_1, K_2 positive, $K_1 > K_2$, and also $K_2/\tau_2 > K_1/\tau_1$ (see Eq. (7.7)). This means that in Eqs. (6.83) and (7.8), $K > 0$, $\xi < 0$.

Alternatively, the opposite response in Figure 7.2(b) occurs if $|K_1| > |K_2|$ with both K_1, K_2 negative and $|K_2/\tau_2| > |K_1/\tau_1|$. This results in $K < 0$ and $\xi < 0$ in Eqs. (6.83) and (7.8).

Thus we see that while the overall process gain can be of either sign, $\xi < 0$ is a requirement for inverse response of $g(s)$ in Eq. (6.93); i.e., $g(s)$ must have a positive (right-half plane) zero.

With this as background, we will now look into some physical processes that exhibit inverse-response characteristics. We should now be in a position to understand, on purely physical grounds, why such behavior is inevitable in such processes. Once we can identify two opposing mechanisms as being responsible for the dynamic behavior of a physical process, especially if one of them exercises greater influence on the process output than the other, it should no longer come as a surprise if we find that such a process exhibits inverse response.

7.2 INVERSE RESPONSE IN PHYSICAL PROCESSES

1. The Drum Boiler

The drum boiler shown schematically in Figure 7.3 is the process currently under consideration. It is typically used for producing steam as a plant utility. The boiler contains boiling liquid, vapor bubbles, and steam, and the level of nongaseous liquid material accumulated in the drum is the key variable of interest desired to be kept under control.

The observed level of material in the drum is due to *both* the liquid and the

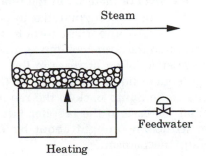

Figure 7.3. Schematic diagram of the drum boiler process.

entrained vapor bubbles that, because of their higher specific volumes, rise through the boiling liquid and temporarily "settle" on the boiling liquid surface. This fact is central to the dynamic behavior of the level in the drum boiler.

Let us now consider how the level in the drum boiler will respond to a positive step change in the input; i.e., an increase in the cold feedwater flowrate. First, we note that in the long run, the addition of more feed material into the drum boiler must ultimately result in an increase in the process inventory (the heat supply to the reboiler has not been changed). However, the same action of adding more cold feedwater sets in motion, simultaneously, an opposing phenomenon: for observe that the entrance of cold water *immediately* results in a drop in drum liquid temperature, causing vapor bubbles to collapse, thus reducing the observed level in the boiler drum.

The overall result is that shown in Figure 7.2(a), where a step change in the cold feedwater rate initially causes a drop in the boiler level before we ultimately see the expected increase, a typical inverse response.

The two opposing dynamic modes in the boiler drum process are now fairly easy to identify:

1. The ultimate material increase in the boiler drum as a result of material input into the process,

2. The initial reduction in the volume of entrained vapor bubbles as a result of the decrease in temperature brought about by the addition of colder material into the drum.

Observe carefully that the influence exerted on the drum boiler level by the first mode is by far the greater, in the long run; but this influence is exerted on a much slower timescale than the less influential second mode. However, despite the fact that the influence of collapsing bubbles on the boiler drum level is, in the long run, obviously far less significant, it is nevertheless exerted on a much swifter timescale. These factors combine to make the drum boiler process a prime candidate for inverse-response behavior as is well known (see also Shinskey [1] and Smith and Corripio [2]).

2.   The Reboiler Section of a Distillation Column

In operating a distillation column, it is well known that without intervention from a level controller, an increase in the reboiler steam duty will result ultimately in a decrease in the level of material in the reboiler. This is because such an action will cause more low boiling material to be boiled off from the inventory in the reboiler. It is also important to note that there will be an increase in the vapor rate through the entire column as a result.

However, the immediate effect of an increase in the vapor rate is to cause increased frothing and subsequent liquid spillage from the trays immediately above the reboiler, sending more liquid back to the reboiler from these trays. This will result in an initial increase in the reboiler liquid level prior to the ultimate, and inevitable, reduction brought about by the more substantial influence of the vapor boil-up mechanism.

Again we see two opposing mechanisms in action here: the more substantial boil up, steadily (but *slowly*) causing a reduction in the inventory in the reboiler, pitted against the frothing and subsequent liquid spillage (caused by the same action responsible for the boil-up) responsible for the less substantial, but rapidly executed, initial increase in the reboiler material. The first mode exerts the more substantial influence but on a much slower timescale; the influence of the second mode is less substantial in the long run, but it is exerted on a much swifter timescale. The inevitable result of such a combination is inverse response of the form shown in Figure 7.2(b). (See also Luyben[3].)

3. An Exothermic Tubular Catalytic Reactor

When the temperature of the reactants in the feed stream to a tubular catalytic reactor is increased, the temperature of product material at the reactor exit exhibits inverse-response behavior if the reaction is exothermic. The reason for this observation is as follows.

The increase in the feed temperature has the initial effect of increasing the conversion in the region near the entrance of the reactor, thereby depleting, for a short period, the amount of reactants that will reach the region around the exit. This results in a temporary reduction in the exit temperature.

However, the catalyst in the bed begins to increase in temperature so that eventually the reactants reaching the exit region do so at a higher temperature, promoting an increase in the reaction rates in this region, and ultimately resulting in an increase in the exit temperature.

Once more, we see the initial reduction in the amount of unreacted material reaching the exit region as the less significant, yet fast mode, standing in opposition against the increase in reactant temperature, the mode that has the more significant influence on exit temperature but whose influence, even though steady, is on a slower timescale. Thus the reactor exit temperature will respond as shown in Figure 7.2(a). Reference [2] (p. 391) reports an example of another chemical reactor that exhibits inverse-response characteristics.

7.3   DYNAMIC BEHAVIOR OF SYSTEMS WITH SINGLE, RIGHT-HALF PLANE ZEROS

Recall from Section 7.1 that for first-order processes in opposition, inverse response required that the overall transfer function $g(s)$ have a right-half plane zero. Let us now consider the dynamic behavior of the process whose transfer function is as shown below:

$$g(s) \; = \; \frac{K \, (-\eta s + 1)}{(\tau_1 s + 1)(\tau_2 s + 1)} \tag{7.9}$$

where $\eta > 0$. Note also that this is just the (2,1)-order system discussed in Chapter 6, with $\xi = -\eta$.

Observe that this transfer function has two poles located at $s = -1/\tau_1$ and $s = -1/\tau_2$, but its only zero is located as $s = +1/\eta$; i.e., it is a *positive* zero. Because of the location of the zero in the right-half of the complex s plane, such zeros are commonly referred to as *right-half plane* (RHP) zeros.

7.3.1   Unit Step Response

The unit step response for this system may be obtained using Eq. (7.9) and the usual transfer function model with, $u(s) = 1/s$, i.e.:

$$y(s) = \frac{K\,(-\eta s + 1)}{(\tau_1 s + 1)(\tau_2 s + 1)} \, \frac{1}{s} \tag{7.10}$$

We find it more instructive first to investigate certain limiting behavior of this response before obtaining the entire $y(t)$.

Ultimate Response

Employing the final-value theorem of Laplace transform:

$$\lim_{t \to \infty} y(t) = \lim_{s \to 0} [s\,y(s)]$$

we have in this case from Eq. (7.10) that:

$$\lim_{t \to \infty} y(t) = \lim_{s \to 0} \frac{K\,(-\eta s + 1)}{(\tau_1 s + 1)(\tau_2 s + 1)} = K \tag{7.11}$$

that is the process gain, K; i.e., the ultimate response is positive if K is positive and negative if K is negative.

Initial Slope

Let $dy(t)/dt$ represent the derivative of the unit step response function. We may now employ the initial value theorem of Laplace transforms to obtain the initial value of $dy(t)/dt$, the value of the initial response slope at the origin. In this case, we have:

$$\lim_{t \to 0} \frac{dy(t)}{dt} = \lim_{s \to \infty} \left[sL \left\{ \frac{dy}{dt} \right\} \right] = \lim_{s \to \infty} [s^2 y(s)] \tag{7.12}$$

Using Eq. (7.10) in Eq. (7.12) now gives:

$$\lim_{t \to 0} \frac{dy(t)}{dt} = \lim_{s \to \infty} \left[\frac{s\,K\,(-\eta s + 1)}{(\tau_1 s + 1)(\tau_2 s + 1)} \right]$$

The right-hand side limit is easier to evaluate if the numerator and denominator terms are divided by s^2; i.e.:

$$\lim_{t \to 0} \frac{dy(t)}{dt} = \lim_{s \to \infty} \left[\frac{K\left(-\eta + \dfrac{1}{s}\right)}{\left(\tau_1 + \dfrac{1}{s}\right)\left(\tau_2 + \dfrac{1}{s}\right)} \right] \tag{7.13}$$

finally giving:

$$\lim_{t \to 0} \frac{dy(t)}{dt} = \frac{-K\eta}{\tau_1 \tau_2} \tag{7.14}$$

Since η, τ_1, τ_2 are all positive, then the initial slope takes the opposite sign of the process gain. This fact combined with the results of Eq. (7.11) prove that the system Eq. (7.9) will always show an inverse response.

We may now obtain the complete unit step response from Eq. (7.10) in the usual manner, yielding:

$$y(t) = K\left[1 - \left(\frac{\tau_1 + \eta}{\tau_1 - \tau_2}\right) e^{-t/\tau_1} - \left(\frac{\tau_2 + \eta}{\tau_2 - \tau_1}\right) e^{-t/\tau_2} \right] \tag{7.15}$$

This is similar in every respect to the solution to Eq. (6.84) but with the singular and most important exception that the substitution $\xi = -\eta < 0$ has been made in Eq. (6.87).

Figure 7.4 shows this response for the system with $K = 1$, $\tau_1 = 2$ min, $\tau_2 = 5$ min, and $\eta = 3$ min; i.e., the system whose transfer function is given by:

$$g(s) = \frac{(-3s + 1)}{(2s + 1)(5s + 1)} \tag{7.16}$$

The plotted response is:

$$y(t) = \left(1 + \frac{5}{3} e^{-t/2} - \frac{8}{3} e^{-t/5} \right) \tag{7.17}$$

The response of this system to other input functions can be found using the same principles and is left as an exercise for the reader (also see problems at the end of the chapter).

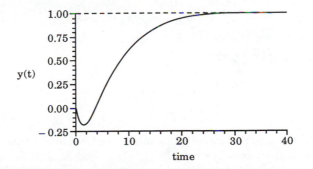

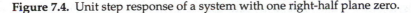

Figure 7.4. Unit step response of a system with one right-half plane zero.

7.3.2 Additional Characteristics of the Unit Step Response

The most prominent feature of the unit step response of the inverse-response system is its initial inversion and subsequent reversal, necessitating that the response pass through a minimum at some time $t_m > 0$. By differentiating Eq. (7.15), setting the result to zero, and solving the expression for t, we find that the response passes through a minimum at $t = t_m$ given by:

$$t_m = \frac{\tau_1 \tau_2}{(\tau_2 - \tau_1)} \ln \left(\frac{\mu_1}{\mu_2} \right) \qquad (7.18)$$

where

$$\mu_i = \frac{\tau_i + \eta}{\tau_i} \; ; \quad i = 1,2 \qquad (7.19)$$

A second differentiation of Eq. (7.15) establishes that this is truly a minimum point.

Example 7.1   CHARACTERIZING THE UNIT STEP RESPONSE OF AN
               INVERSE-RESPONSE SYSTEM.

For the system whose transfer function is given in Eq. (7.16) and whose step response is given in Eq. (7.17) show from this step-response expression that this is an inverse-responding system and find the time at which the response experiences "turn around."

Solution:

The first characteristic of the response obtained from systems of the type indicated in Eq. (7.6) is that the initial slope possesses a different sign from that of the ultimate value. From Eq. (7.17) we see immediately that:

$$y(\infty) = 1$$

which is positive.
Differentiating Eq. (7.17) gives:

$$\frac{dy}{dt} = -\frac{5}{6} e^{-t/2} + \frac{8}{15} e^{-t/5} \qquad (7.20)$$

which, upon setting $t = 0$ gives:

$$\left. \frac{dy}{dt} \right|_{t=0} = -0.3$$

so that with the negative initial slope and a positive final value, the response is identified as an inverse response.
The process response experiences "turn around" at the point where the response passes through a minimum. By setting the expression in Eq. (7.20) to zero, we observe

that this occurs when:

$$\frac{5}{6} e^{-t/2} = \frac{8}{15} e^{-t/5}$$

which when solved for t gives the time of the minimum t_m:

$$t_m = \frac{10}{3}\left[\ln\left(\frac{5}{6}\right) - \ln\left(\frac{8}{15}\right)\right]$$

or

$$t_m = 1.488 \text{ min}$$

Alternatively, one could substitute directly into the general result Eqs. (7.18) and (7.19) to obtain t_m.

7.3.3 Effect of Additional Poles and Zeros

We have shown that the system in Eq. (7.9), whose transfer function has a single RHP zero and two poles, exhibits inverse-response behavior. Let us now consider the more general system whose transfer function is given by:

$$g(s) = \frac{K(-\eta s + 1)(\xi_1 s + 1)(\xi_2 s + 1) \cdots (\xi_q s + 1)}{(\tau_1 s + 1)(\tau_2 s + 1) \cdots (\tau_p s + 1)} \tag{7.21}$$

or, more compactly:

$$g(s) = \frac{K(-\eta s + 1)\prod_{i=1}^{q}(\xi_i s + 1)}{\prod_{i=1}^{p}(\tau_i s + 1)} \tag{7.22}$$

which is a system with p poles, q regular, negative zeros, and a single RHP zero (with $p > q + 1$; i.e., total number of zeros less than the number of poles). It can be shown readily that this system will also exhibit inverse response. The issue now is: how will the presence of the additional poles and zeros influence the dynamic behavior of such systems?

In our previous study of the dynamic behavior of higher order systems (in Chapter 6), we discovered that the addition of lag terms (which contribute the poles) has the net effect of slowing down the response; on the other hand, the addition of lead terms (contributing the zeros) has the net effect of speeding up the system response. The same is true for the inverse-response system whose transfer function has a single RHP zero: additional poles will tend to slow the response down while additional zeros will tend to speed up the response.

To illustrate this point, the unit step responses of higher order systems (which we shall call Systems B and C) are compared with the unit step response of the system given in Eq. (7.16), which we shall call System A, i.e.:

• **System A:**

$$g(s) = \frac{(-3s + 1)}{(2s + 1)(5s + 1)} \tag{7.16}$$

the system with one RHP zero and two poles, whose unit step response was presented in Eq. (7.17) and in Figure 7.4.

• **System B:**

$$g(s) = \frac{(-3s + 1)}{(2s + 1)(5s + 1)(4s + 1)} \tag{7.23}$$

a system with one RHP zero and *three* poles, one more than System A.

• **System C:**

$$g(s) = \frac{(-3s + 1)(s + 1)}{(2s + 1)(5s + 1)(4s + 1)} \tag{7.24}$$

a system with one RHP zero, one regular, negative zero, and three poles, one zero more than System B, and one pole and one zero more than System A.

Figure 7.5 shows the unit step responses of Systems A and B. Right away, we clearly see that the influence of the additional pole possessed by System B is to slow down its response in comparison to that of System A. Observe also that as a result of the increased sluggishness of System B, the inverse response it exhibits is less pronounced than that of System A. Thus the effect of an additional pole is to slow down the overall response while reducing the strength of the initial inverse response.

When one zero is added to the System B transfer function, the result is System C. A comparison of the unit step responses of these two systems is shown in Figure 7.6. As expected, the additional zero causes the System C response to be somewhat swifter than that of System B. The effect of a simultaneous addition of a pole *and* a zero to the transfer function of System A, to produce System C, is demonstrated by the unit step responses shown in Figure 7.7.

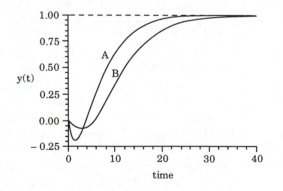

Figure 7.5. Unit step responses of systems A and B, demonstrating the effect of an additional pole.

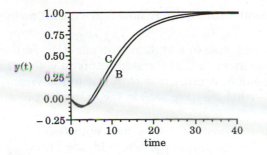

Figure 7.6. Unit step responses of systems B and C, demonstrating the effect of an additional zero.

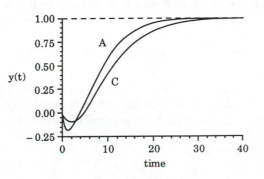

Figure 7.7. Unit step responses of systems A and C, demonstrating the effect of an additional zero *and* pole.

Because the additional lag term, $(4s + 1)$, contributing the pole has a stronger effect than the added lead term, $(s + 1)$, contributing the zero, the overall effect is biased towards that due to the added pole; the system response is slowed down, and the strength of the initial inverse response is reduced.

7.4 DYNAMIC BEHAVIOR OF SYSTEMS WITH MULTIPLE, RIGHT-HALF PLANE ZEROS

The discussion in the preceding section centered around the influence of additional poles and zeros on the dynamic behavior of an inverse-response system, with the understanding that the inverse-response characteristics are due to the presence of a *single* RHP zero. Note carefully that the additional zeros have been regular, negative zeros, so that in the midst of all the multiple zeros contained in the system transfer function, we still find only one RHP zero.

At this point, we might well ask the question: *what effect would the addition of another RHP zero have on the dynamic behavior of a system already exhibiting inverse response?* Or, more generally: what are the main characteristics of the dynamic behavior of a system with multiple RHP zeros?

In this regard, we have the following results:

The initial portion of the step response of a system with multiple RHP zeros will exhibit as many inversions as there are RHP zeros.

From this general result, we may now make the following deductions:

1. The step response of a system with *one* RHP zero exhibits (as we now know) one inversion. If the step response is therefore to end up in the right direction, it must first start out in the wrong direction; this, of course, is what we have seen in the examples presented thus far.

2. The step response of the system with *two* RHP zeros exhibits two inversions in the initial portion, with the following consequences: for this response to end in the direction of the final steady state (as it must) after executing two inversions, it must start out heading in the right direction first, take a first inversion and temporarily head in the wrong direction, and return to the right direction upon the second and final inversion.

3. The step response of the system with *three* RHP zeros exhibits three inversions in the initial portion. Consequently, we observe once again that such a response must start out in a direction opposite to that of the final steady state. The first inversion temporarily turns the response in the right direction, and the second inversion reverses the trend; the third and final inversion is what puts the response in the direction of the final steady state.

In taking such arguments as these to their logical conclusions, we may thus deduce the qualitative nature of the step responses upon consecutive additions of RHP zeros. The following is a summary of these conclusions:

*The system with an **odd** number of RHP zeros exhibits **true** inverse response in the sense that the initial direction of the step response will always be **opposite** to the direction of the final steady state, regardless of the number of inversions involved in this response.*

*On the other hand, the initial portion of the step response of a system with an **even** number of RHP zeros exhibits the same even number of inversions before heading in the direction of the final steady state, but the initial direction is always the same as the direction of the final steady state.*

To demonstrate these results, the unit step responses of the following systems are plotted in Figures 7.8, 7.9, and 7.10.

• **System D:**

$$g(s) = \frac{(-3s + 1)(-s + 1)}{(2s + 1)(5s + 1)(4s + 1)} \tag{7.25}$$

a system with *two* RHP zeros (Figure 7.8);

• **System E:**

$$g(s) = \frac{(-3s + 1)(-s + 1)(-2.5s + 1)}{(2s + 1)(5s + 1)(4s + 1)(3.5s + 1)} \qquad (7.26)$$

a system with *three* RHP zeros (Figure 7.9); and finally

• **System F:**

$$g(s) = \frac{(-3s + 1)(-s + 1)(-2.5s + 1)(-6s + 1)}{(2s + 1)(5s + 1)(4s + 1)(3.5s + 1)(1 + 7s)} \qquad (7.27)$$

a system with *four* RHP zeros (Figure 7.10).

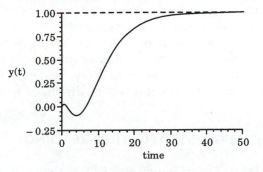

Figure 7.8. Dynamic behavior of System D with two RHP zeros.

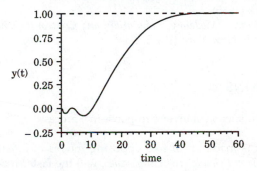

Figure 7.9. Dynamic behavior of System E with three RHP zeros.

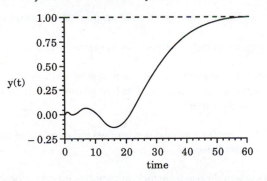

Figure 7.10. Dynamic behavior of System F with four RHP zeros.

7.5   SUMMARY

We have been concerned in this chapter with investigating the dynamic behavior of systems exhibiting inverse-response behavior. We have found that such responses are possible whenever there are two (or more) *opposing* mechanisms (operating on two different timescales) responsible for the overall system behavior. Several examples of real processes that exhibit these unusual dynamic characteristics were presented.

It was shown that the transfer functions of inverse-response systems are usually characterized by the presence of a RHP zero; the presence of additional poles and regular, negative zeros may modify the dynamic behavior, but the inverse-response characteristics will still be in evidence. The presence of multiple RHP zeros, however, introduces some more interesting characteristics, bringing about multiple inversions in the initial portions of the step responses, a phenomenon that is rarely observed in practice.

With the material covered in this chapter, we have now completed the discussion of the dynamic behavior of higher (but finite) order systems with zeros. The dynamic behavior of *time-delay systems*, perhaps the most important of the infinite-order systems will be discussed in the next chapter.

REFERENCES AND SUGGESTED FURTHER READING

1.   Shinskey, F.G., *Process Control Systems* (2nd ed.), McGraw-Hill, New York (1979)
2.   Smith, C.A. and A.B. Corripio, *Principles and Practice of Automatic Process Control*, J. Wiley, New York (1985)
3.   Luyben, W.L., *Process Modeling, Simulation, and Control for Chemical Engineers* (2nd ed.), McGraw-Hill, New York (1991)

REVIEW QUESTIONS

1.   Describe the phenomenon of inverse response in a process.

2.   When a system is made up of two opposing modes of *different magnitudes*, each operating on *different characteristic timescales*, will the faster mode have to be smaller or larger in magnitude for the system to exhibit inverse response?

3.   What characterizes the transfer function of an inverse-response system?

4.   Can a *series* arrangement of two first-order systems exhibit inverse response?

5.   If two inverse-response systems A and B are otherwise identical, except that the system A transfer function has an additional (left-half plane) *pole*, how will one system's step response differ from the other?

6.   What aspect of the transfer function determines the number of inversions exhibited by a system's initial step response?

7.   Why is it that a system whose transfer function has an *even* number of right-half plane zeros is, *strictly speaking*, not really an "inverse-response" system in the strictest possible sense of the term "inverse response"?

PROBLEMS

7.1 (a) Derive individual expressions for the response obtained when the system whose transfer function is given in Eq. (7.16) is subjected to the following inputs:

1. $u(t) = \delta(t)$; a unit impulse function

$$2.\ u(t) = \begin{cases} 0; & t < 0 \\ A; & 0 < t < b; \\ 0; & t \geq b \end{cases} \quad \text{a pulse input of magnitude } A \text{ and duration } b$$

(b) For the two distinct situations in which the magnitude and duration of the pulse input investigated above are given as:

1. $A = 1/8$, $b = 8$, and
2. $A = 1/2$, $b = 2$

obtain plots of the resulting pulse responses and compare them with a plot of the unit impulse response.

(c) Compare the unit impulse response obtained in part (a) with that of the system whose transfer function is given as:

$$g(s) = \frac{(3s + 1)}{(2s + 1)(s + 1)} \tag{P7.1}$$

7.2 From the expression given in Eq. (7.15), show that the inherent reluctance factor for this inverse-response system — defined in the usual fashion as:

$$J = \int_0^\infty [y^* - y(t)]\,dt$$

(where $y^* = K$ in this case) — is given by:

$$J_{(IR)} = K\left(\tau_1 + \tau_2 + \eta\right) \tag{P7.2}$$

Compare this with the expression given in Chapter 6 for $J_{(2,1)}$, for the corresponding system with a *left-half plane* zero.

7.3 While the dynamic behavior of a system with a "strong" left-half plane zero is expected to be different from that of a system with a "strong" *right*-half plane zero, the nature of such differences are, in themselves, quite informative. Plot the unit step responses of the two systems whose transfer functions are given as follows:

$$g_1(s) = \frac{(3s + 1)}{(2s + 1)(s + 1)}\ ;\ \text{and}\ g_2(s) = \frac{(-3s + 1)}{(2s + 1)(s + 1)}$$

Compare and contrast the two responses.

7.4 Two systems, each with individual transfer functions $g_1(s)$ and $g_2(s)$ are connected in a *parallel* arrangement such that the resulting composite system has a transfer function given by:

$$g(s) = g_1(s) + g_2(s)$$

In the specific case with:

$$g_1(s) = \frac{K_1(-\eta s + 1)}{(\tau_1 s + 1)(\tau_2 s + 1)} \tag{P7.3}$$

$$g_2(s) = \frac{K_2}{\lambda s + 1} \tag{P7.4}$$

(a) How many poles and how many zeros will the resulting composite system possess?

(b) Assuming that both K_1 and K_2 are positive, what conditions must be satisfied by K_2, and λ so that $g(s)$ will no longer be an inverse-response system? (You may find it useful to know that: *A necessary and sufficient condition for the quadratic equation $ax^2 + bx + c = 0$ to have no roots with positive real parts is that all its coefficients, a, b, and c be strictly positive.*)

(c) It has been proposed that the inverse response exhibited by the system in Eq. (7.16) be "compensated" by augmenting it with the following first-order system in a parallel arrangement:

$$g_2(s) = \frac{1}{3s + 1} \tag{P7.5}$$

i.e., the parallel arrangement of this first-order system and the inverse-response system in Eq. (7.16) will result in a composite system that does not exhibit inverse response. Confirm or refute this proposition.

7.5 (a) A system consisting of a first-order mode in opposition to the main pure capacity mode, with respective transfer functions $g_2(s) = K_2/(\tau_2 s + 1)$ and $g_1(s) = K_1/s$, can exhibit inverse response only if $K_1 < K_2/\tau_2$. Establish this result.

(b) The level in a certain drum boiler process is known to exhibit inverse response. In identifying a dynamic model for this process, input/output data were collected in accordance with a planned experimental procedure supposedly designed so that the two opposing modes of the overall process can be identified independently. After carrying out the appropriate data analysis, the following candidate transfer function model was obtained:

$$g(s) = g_1(s) - g_2(s) \tag{P7.6}$$

where

$$g_1(s) = \frac{5.3}{s} \tag{P7.7}$$

and

$$g_2(s) = \frac{10.1}{2.2s + 1} \tag{P7.8}$$

It is known that the experimental procedure is much better at identifying the ultimate behavior, characterized mostly by $g_1(s)$, than at identifying the initial transient behavior, characterized mostly by $g_2(s)$. In fact, an independent researcher has suggested that this model identification procedure tends to *overestimate* the first-order time constant in each typical $g_2(s)$ by as much as 20%. In light of the identified transfer functions given above, provide your own assessment of the identification procedure.

7.6 Show that a system composed of a critically damped second-order system in opposition to a first-order system, whose transfer function is given as:

$$g(s) = \frac{K_1}{\tau_1 s + 1} - \frac{K_2}{\left(\tau_2 s + 1\right)^2} \tag{P7.9}$$

will exhibit inverse response only if:

$$\frac{K_2}{\tau_2} > 2 \left(\frac{K_1}{\tau_1}\right) \tag{P7.10}$$

given that:

$$K_1 > K_2 \tag{P7.11}$$

Compare this result with Eq. (7.7).

7.7   In the science of immunology, it is known that the process of vaccination results in responses from two subsystems in opposition: the "disease-causing" mode promotes the growth and replication of the intruding "disease-causing" cells; the immune system, on the other hand, produces the appropriate antibodies required to fight off the intruding cells. The antibodies produced by the *healthy* immune system ultimately overwhelm and completely destroy the disease-causing cells, and then remain behind in the bloodstream as prophylactic sentries against future attacks. A dynamic systems approach to the analysis of this process shows that immunization in the human body is an inverse-response process.

Let x_1 represent the "disease-causing cell count" (in total number of cells per standard sample volume); in general its "isolated" response to the amount of injected vaccine u (in cm^3) is reasonably well approximated by:

$$x_1(s) = \frac{K_1}{s(\tau_1 s + 1)} u(s) \tag{P7.12}$$

The corresponding "isolated" response of the "antibodies cell count," say, x_2, is reasonably well approximated by:

$$x_2(s) = \frac{K_2}{s(\tau_2 s + 1)(\tau_3 s + 1)} u(s) \tag{P7.13}$$

Define a normalized "extent-of-immunization index" y as a weighted average of the disease-causing cell count and the antibodies cell count according to:

$$y = \omega_1 x_1 + \omega_2 x_2 \tag{P7.14}$$

with the weights given by:

$$\omega_1 = \frac{-1}{(K_2 - K_1)} \tag{P7.15a}$$

$$\omega_2 = \frac{1}{(K_2 - K_1)} \tag{P7.15b}$$

With regards to the approximate models given above, the following two facts are pertinent to the analysis:

1.   For the healthy immune system:

$$K_2 > K_1 \tag{P7.16}$$

2.   For effective vaccines:

$$\frac{\tau_1}{\tau_2 + \tau_3} < \frac{K_1}{K_2} \qquad (P7.17)$$

(a) Show that under the conditions represented in Eqs. (P7.16) and (P7.17), the transfer function relating y to u has precisely *one* — not two — right-half plane zeros and therefore that the response of the "extent-of-immunization index" y to the amount of injected vaccine u will exhibit true inverse response.

(b) For the specific situation in which a 5 cm³ ampoule of a certain vaccine was injected into an individual for whom:

$$K_1 = 0.5 \, K_2 \qquad (P7.18a)$$

$$\tau_1 = 2 \text{ hr}, \tau_2 = 3 \text{ hr}, \tau_3 = 5 \text{ hr} \qquad (P7.18b)$$

first establish that this is a situation in which the individual's immune system is healthy and the vaccine is effective; then approximate this vaccination input as $u(t) = 5\delta(t)$ and obtain a plot of the response of the "extent-of-immunization index" y.

(c) Provide your own interpretation of what the "extent-of-immunization index" y measures. Explain carefully what you think a value of $y = 0$, a value of $y < 0$, and a value of $y = 1$ mean.

7.8 Revisit Problem 7.7. Establish from the general expression obtained there for the overall transfer function relating y to u that:

(a) For the *unhealthy* immune system (with $K_1 > K_2$) into which a unit impulse $u(t) = \delta(t)$ of an otherwise effective vaccine has been introduced, the "extent-of-immunization index" y will settle ultimately to a value of –1 (as opposed to the desired value of +1) *without exhibiting inverse response*.

(b) When a healthy immune system receives a unit impulse $u(t) = \delta(t)$ of a vaccine that is considered *less effective* because:

$$\frac{K_1}{K_2}\left(\tau_2 + \tau_3\right) < \tau_1, \qquad (P7.19)$$

as compared to the condition indicated in Eq. (P7.17), the "extent-of-immunization index" y will still settle ultimately to the desired value of +1, and still exhibit inverse response.

(c) When an *unhealthy* immune system receives a unit impulse $u(t) = \delta(t)$ of the *less effective* vaccine considered in part (b), the "extent-of-immunization index" y will still settle ultimately to the undesired value of –1, and its initial response will show *two* inversions.

(d) Given the following three specific cases in which the parameters are given as:

1. $K_1 = 2K_2$; $\tau_1 = 2 \text{ hr}, \tau_2 = 3 \text{ hr}, \tau_3 = 5 \text{ hr}$

2. $K_1 = 0.5K_2$; $\tau_1 = 4 \text{ hr}, \tau_2 = 3 \text{ hr}, \tau_3 = 2 \text{ hr}$

3. $K_1 = 3K_2$; $\tau_1 = 7 \text{ hr}, \tau_2 = 1 \text{ hr}, \tau_3 = 1/3 \text{ hr}$

In each case determine the "health" of the immune system and the "effectiveness" of the vaccine, and then obtain a plot of the response of the "extent-of-immunization index" y to an idealized unit impulse input of vaccine, $u(t) = \delta(t)$. In terms of the two components of y, provide a qualitative explanation for the observed response in each case particularly with respect to the immune system's state, and the effectiveness of the vaccine.

8

TIME-DELAY SYSTEMS

As diverse as the various classes of systems we have studied so far have been, they do have one characteristic in common: upon the application of an input change, their outputs, even though sluggish in some cases, begin to respond without delay.

In the process industries, one often finds systems in which there is a noticeable delay between the instant the input change is implemented and when the effect is observed, with the process output displaying an initial period of no response. Such systems are aptly referred to as time-delay systems, and their importance is underscored by the fact that a substantial number of processes exhibit these delay characteristics.

This chapter is devoted to studying time-delay systems: how they occur, why they are classifiable as infinite-order systems, their dynamic behavior, and how they are related to other higher order systems.

8.1 AN INTRODUCTORY EXAMPLE

Consider the process shown in Figure 8.1, involving the flow of an incompressible fluid, in plug flow, at a constant velocity v, through a perfectly insulated pipe of length L. The fluid is initially at a constant temperature T_0, uniformly, throughout the entire pipe length. At time $t = 0$, the temperature of the fluid coming in at the pipe inlet ($z = 0$) is changed to $T_i(t)$. We are now interested in investigating how the temperature at the pipe outlet ($z = L$) responds to such an input change.

8.1.1 Physical Considerations

On purely physical grounds, it is easy to see that any temperature changes implemented at the pipe inlet will not be registered instantaneously at the outlet. This is because it will take a finite amount of time for an individual fluid element at the new temperature, to traverse the distance L from the inlet to the outlet so that its new temperature can be observed.

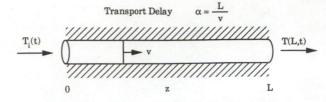

Figure 8.1. Flow through an insulated pipe.

Since the fluid velocity is v, and it is assumed constant, the time for each fluid element to traverse the required distance from the pipe inlet to the outlet is L/v. Since the pipe is also assumed to be perfectly insulated, there will be no heat losses, or any other changes for that matter, experienced by each fluid element in the process. Thus, any changes implemented in the inlet will be preserved intact, to be observed at the outlet after the time required to traverse the entire pipe length has elapsed.

From such physical arguments we have arrived at the conclusion that in response to changes in the process input (i.e., the inlet temperature) the process output (i.e., the exit temperature) will have exactly the same form as the input, only delayed by L/v time units.

Let us now show, mathematically, that this is in fact what obtains.

8.1.2   Mathematical Model and Analysis

Consider an element of fluid of thickness Δz whose boundaries are arbitrarily located at the points z and $z + \Delta z$ along the length of the pipe (see Figure (8.2)).

For a pipe of constant, uniform cross-sectional area A, an energy balance over such an element gives:

$$\rho A C_p \Delta z \frac{\partial T}{\partial t} \;=\; \rho A v C_p (T - T^*)\Big|_z - \rho A v C_p (T - T^*)\Big|_{z+\Delta z} \qquad (8.1)$$

where ρ, C_p, are, respectively, the fluid density, and specific heat capacity, and T^* is the usual, reference temperature (see Chapter 4); the notation $|_z$ is used to indicate that the quantity in question is evaluated at z.

By dividing through by Δz, and taking the limits as $\Delta z \rightarrow 0$, Eq. (8.1) becomes:

$$\frac{\partial T(z,t)}{\partial t} \;+\; v\frac{\partial T(z,t)}{\partial z} \;=\; 0 \; ; \qquad 0 < z < L \qquad (8.2)$$

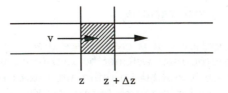

Figure 8.2.   A control volume in the pipe of length Δz.

a partial differential equation (PDE). When we pause to consider the fact that the fluid temperature within the pipe varies not only just along the length of the pipe but also with time at each point along the pipe, it should not come as a surprise that this process is modeled by a PDE.

As earlier alluded to, the system whose dynamic behavior is represented by PDE's is known as a distributed parameter system. Thus the process currently under consideration is one of such systems.

Let us now define the deviation variable:

$$y(z,t) = T(z,t) - T(0) \tag{8.3}$$

where we recall that $T(0) = T_0$ is the initial fluid temperature at $t = 0$ along the pipe length. In terms of this deviation variable, the process model in Eq. (8.2) is:

$$\frac{\partial y(z,t)}{\partial t} + v\frac{\partial y(z,t)}{\partial z} = 0 \tag{8.4}$$

To develop a transfer function model for this process, we must now take Laplace transforms of the expression in Eq. (8.4). From the definition of the Laplace transform, we have:

$$\int_0^\infty e^{-st}\frac{\partial y(z,t)}{\partial t}\,dt + v\int_0^\infty e^{-st}\frac{\partial y(z,t)}{\partial z}\,dt = 0$$

or

$$sy(z,s) + v\int_0^\infty e^{-st}\frac{\partial y(z,t)}{\partial z}\,dt = 0 \tag{8.5}$$

since the first term is simply the Laplace transform (with respect to time) of the indicated time derivative.

For the second term in Eq. (8.5), interchanging the order of integration and differentiation (an exercise which can be shown to be valid in this case) gives:

$$\int_0^\infty e^{-st}\frac{\partial y(z,t)}{\partial z}\,dt = \frac{d}{dz}\int_0^\infty e^{-st}y(z,t)\,dt \tag{8.6}$$

$$= \frac{d}{dz}y(z,s)$$

so that Eq. (8.5) becomes:

$$sy(z,s) + v\frac{d}{dz}y(z,s) = 0 \tag{8.7}$$

a first-order, linear, ordinary differential equation in $y(z, s)$. (It is perhaps of interest at this point to note that while Laplace transformation converts an ordinary differential equation (ODE) to an algebraic equation, it converts the PDE to an ODE).

The solution of the ODE in Eq. (8.7) is easily obtained as:

$$y(z,s) = C_1 e^{-\frac{z}{v}s}$$ (8.8)

We may evaluate the arbitrary constant C_1 by using the entrance condition at $z = 0$ that:

$$y(z,s) = y(0,s)$$

so that Eq. (8.8) becomes

$$y(z,s) = y(0,s)\, e^{-\frac{z}{v}s}$$ (8.9)

Observe now that Eq. (8.9) represents the relationship between the Laplace transform of the inlet temperature and the Laplace transform of the temperature at any other point z along the length of the pipe. Observe further that the term z/v is the ratio of the distance traveled in arriving at this point, to the flow velocity, a direct measure of the time taken to traverse the indicated distance.

We are particularly concerned with the situation at the pipe exit; i.e., at $z = L$. In this case, Eq. (8.9) becomes:

$$y(L,s) = y(0,s)\, e^{-\left(\frac{L}{v}\right)s}$$

and if we define as α the time it takes for a fluid element to traverse the entire length of the pipe, i.e.:

$$\alpha = \frac{L}{v}$$ (8.10)

then we have:

$$y(L,s) = e^{-\alpha s}\, y(0,s)$$ (8.11)

Thus the transfer function relating the process input (changes in the inlet temperature, $y(0,s)$) to the process output (changes in the exit temperature, $y(L,s)$) is clearly seen from Eq. (8.11) to be given by:

$$g(s) = e^{-\alpha s}$$ (8.12)

We may now use Eq. (8.12) to investigate how this process will respond to changes in the input function $u(t) = y(0,t)$.

This task is seen to be very easy once we recall the effect of the translation function of Laplace transforms. We have, for any input $u(t)$ that:

$$y(L,s) = e^{-\alpha s}\, u(s)$$ (8.13)

which, when converted back to the time domain gives the result:

$$y(L,t) = u(t - \alpha)$$ (8.14)

indicating that the output is exactly the same as the input, only delayed by α time units, which, from Eq. (8.10) is the time it takes a fluid element traveling at a velocity v to traverse the pipe length. This, of course, is in perfect agreement with the conclusions we arrived at earlier from purely physical grounds.

The process we have been analyzing is perhaps the most common example of the *pure time-delay* process. It illustrates the well known fact that the transportation of material and/or energy from one part of a plant to the other is usually accompanied by the phenomenon known variously as transportation lag, dead time, distance-velocity lag, or simply time delay. Thus, recycle loops, flow through connecting pipes, etc., all contribute to the existence of the phenomenon of time delays in the process industries.

In measuring fluid concentrations using such equipment as gas or liquid chromatographs, the finite period it takes for the fluid to flow through the equipment and for the analysis to be completed constitutes another source of time delays in physical processes.

As we will see later, time delays also appear in approximate linear models of complicated processes. However, in such cases the delay is not associated with pipes or other transportation processes; it is simply a more economical means of approximating higher order dynamics.

Let us now formally discuss the dynamic behavior of the pure time-delay process.

8.2 THE PURE TIME-DELAY PROCESS

The process whose transfer function is given by:

$$g(s) = e^{-\alpha s} \tag{8.15}$$

is known as a pure time-delay process because its response to any input is merely a delayed version of the input.

It is important to note that the processes we have considered up until now have all had transfer functions that are ratios of (finite-order) polynomials in s. Obviously, this is not the case with the pure time-delay process whose transfer function, as indicated in Eq. (8.15), is a transcendental function.

An attempt to obtain a rational version of the transfer function in Eq. (8.15) by recasting it as:

$$g(s) = \frac{1}{e^{\alpha s}}$$

and carrying out a Taylor series expansion for $e^{\alpha s}$ results in:

$$g(s) = \frac{1}{1 + \alpha s + \dfrac{\alpha^2 s^2}{2!} + \dfrac{\alpha^3 s^3}{3!} + \ldots + \dfrac{\alpha^n s^n}{n!} + \ldots} \tag{8.16}$$

with an *infinite*-order denominator polynomial. In accordance with our classification of system order, therefore, Eq. (8.16) identifies the pure time-delay system as an *infinite-order* system.

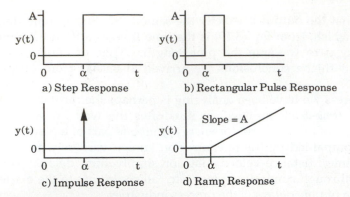

Figure 8.3.   Response of the pure time-delay process to various inputs.

8.2.1   Response to Various Inputs

The nature of the transfer function given in Eq. (8.15) makes it particularly easy to determine the response of the pure time-delay process to any given input. Observe that for any input $u(t) = f(t)$, where $f(t)$ is any arbitrary function, the response is given by:

$$y(t) \ = \ f(t - \alpha) \tag{8.17}$$

The response to some of the standard input functions we have been using thus far are summarized in Figure 8.3.

Of special note is the sinusoidal response. In this case, the input is $A \sin \omega t$ and from Eq. (8.17), the output is given by:

$$y(t) \ = \ A \sin \left[\omega(t - \alpha) \right] \tag{8.18}$$

When compared to the input function, Eq. (8.18) indicates that the frequency response of the pure time-delay system has the following amplitude ratio and phase lag characteristics:

$$\mathrm{AR} \ = \ 1 \tag{8.19a}$$

$$\phi \ = \ -\alpha \omega \tag{8.19b}$$

Again, we postpone a detailed discussion of frequency response until Chapter 9.

8.2.2   Limiting Behavior of First-Order Systems in Series

In Section 6.4.2 of Chapter 6, we considered a special case of N first-order systems in series in which the individual time constants were all equal to α/N, and the steady-state gains were all unity. Recall from Eq. (6.78) that the transfer function for this process is given by:

$$g_N(s) \ = \ \cfrac{1}{\left(\dfrac{\alpha}{N} s \ + \ 1 \right)^N} \tag{6.78}$$

Let us now formalize the heuristic arguments with which we earlier attempted to deduce the dynamic behavior of this system in the limit as $N \to \infty$.

From the expression in Eq. (6.78) we obtain the required limits as:

$$\lim_{N \to \infty} g_N(s) = \lim_{N \to \infty} \frac{1}{\left(\dfrac{\alpha}{N} s + 1\right)^N} \tag{8.20}$$

To evaluate the limits on the RHS of Eq. (8.20) requires the following arguments:

The definition of the exponential number e is:

$$e = \lim_{m \to \infty} \left(1 + \frac{1}{m}\right)^m$$

so that:

$$e^{-\alpha} = \lim_{m \to \infty} \left(1 + \frac{1}{m}\right)^{-\alpha m}$$

$$= \lim_{m \to \infty} \frac{1}{\left(1 + \dfrac{1}{m}\right)^{\alpha m}}$$

If we now let $N = \alpha m$, we have:

$$e^{-\alpha} = \lim_{N \to \infty} \frac{1}{\left(1 + \dfrac{\alpha}{N}\right)^N} \tag{8.21}$$

from where it becomes very clear that the limit in Eq. (8.20) is:

$$\lim_{N \to \infty} g_N(s) = e^{-\alpha s} \tag{8.22}$$

which is exactly the same as the transfer function given for the pure time-delay process given in Eq. (8.15).

We now have the following result:

The pure time-delay process with time delay α is obtained in the limiting case of a sequence of N identical first-order systems (each with time constant α/N) connected in series, as $N \to \infty$.

The most significant implication of this result is that the pure time-delay system, identified in the more revealing transfer function form in Eq. (8.16) as an infinite-order system, is conceptually equivalent to a system composed of an infinite sequence of first-order systems in series. Of more practical importance is the fact that a high-order system (for example, a large number of first-order processes in series) will therefore tend to behave very much like a pure time-delay system.

8.3   DYNAMIC BEHAVIOR OF SYSTEMS WITH TIME DELAYS

8.3.1   A Physical Example

Consider the system shown in Figure 8.4. This is the same isothermal CSTR discussed in Chapter 5, but this time the feed stream is shown as being transported to the reactor, at a constant velocity v, through a long cylindrical pipe of length L, and uniform cross-sectional area.

The feed is assumed to contain species A diluted with water to achieve a concentration c_0. We assume the feed is cold so that no reaction takes place in the pipe; the reaction starts as soon as feed enters the reactor; the concentration of A at the reactor inlet is c_1.

We now make the following observations about the dynamic behavior of this process:

1. Its dynamic behavior is that of two processes connected in series: the long pipe, and the reactor; the output of the first system is the input to the second system.

2. From the discussion in Section 8.1, we know that the long pipe behaves like a pure time-delay process, only in this case, it is material, not energy, which is being transported. The relationship between the pipe inlet concentration c_0 and the pipe outlet concentration c_1 is therefore given by:

$$c_1(t) = c_0(t - \alpha) \tag{8.23a}$$

where $\alpha = L/v$. In terms of the deviation variables $w(t) = c_1(t) - c_{1s}(t)$ and $u(t) = c_0(t) - c_{0s}(t)$ this relationship becomes:

$$w(t) = u(t - \alpha) \tag{8.23b}$$

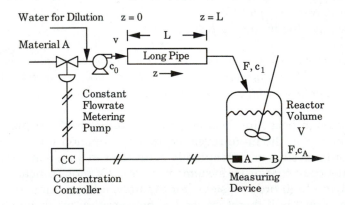

Figure 8.4.   A chemical reactor where reactants are mixed a long distance upstream.

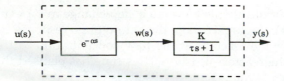

Figure 8.5. Block diagrammatic representation of system in Figure 8.4.

and upon taking Laplace transforms, we obtain the transfer function equivalent:

$$w(s) = e^{-\alpha s} u(s) \tag{8.24}$$

3. From Eq. (5.9) and the discussion in Section 5.1 of Chapter 5, the reactor was identified as a first-order system; i.e., if the deviation of the reactor outlet concentration $c_A(t)$ from its steady-state value c_{As} is designated $y(t)$, the transfer function relationship between $w(s)$ (which is the reactor input) and $y(s)$ is given by:

$$y(s) = \frac{K}{\tau s + 1} w(s) \tag{8.25}$$

In block diagrammatic form, we see that this system is represented as shown in Figure 8.5.

The transfer function relationship between the overall system input $u(s)$ and the output $y(s)$ is obtained by substituting Eq. (8.24) for $w(s)$ in Eq. (8.25), giving:

$$y(s) = \frac{K\, e^{-\alpha s}}{\tau s + 1} u(s) \tag{8.26}$$

This is the transfer function representation of a first-order system with delay, sometimes referred to as a first-order-plus-time-delay system.

Response to Various Inputs

The issue of how this system will respond to changes in the input $u(t)$ is now easily answered. The key lies in recognizing that the effect of the pipe is merely to delay the input function. The overall response will therefore be exactly like that of the first-order system as already discussed in Chapter 5, only shifted by α time units.

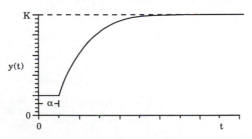

Figure 8.6. Response of system in Figure 8.4 to a unit step change in pipe inlet concentration.

Thus, for example, the response to a step change in pipe inlet concentration will be as shown in Figure 8.6; here the period of no response corresponds to the time required for the reactant at the new concentration to travel through the pipe and enter the reactor. It is after this time has elapsed that the first-order dynamic behavior of the reactor (by itself) is observed.

8.3.2 General Considerations

The ideas illustrated with the aid of the example system considered above may now be generalized.

A time-delay system is a system whose transfer function contains the exponential term $e^{-\alpha s}$, where α is the associated time delay; conceptually, it is composed of a pure time-delay process connected in series with a dynamic system represented by rational polynomials of s in the transform domain.

Thus if the transfer function of the nondelay part of the system is designated, in general, as $g^*(s)$, then a general transfer function for the time-delay system is given by:

$$g(s) = g^*(s)e^{-\alpha s} \tag{8.27}$$

and represented by the block diagram in Figure 8.7.

Thus, for example, the transfer function of a first-order system with delay is as shown in Eq. (8.26); for a second-order system with time delay, the transfer function is:

$$g(s) = \frac{K\,e^{-\alpha s}}{\tau^2 s^2 + 2\zeta\tau s + 1} \tag{8.28}$$

and in general, for the (p,q)-order system with associated time delay α, the transfer function is:

$$g(s) = \frac{K\left(\xi_1 s + 1\right)\left(\xi_2 s + 1\right)\cdots\left(\xi_q s + 1\right)e^{-\alpha s}}{\left(\tau_1 s + 1\right)\left(\tau_2 s + 1\right)\cdots\left(\tau_p s + 1\right)} \tag{8.29}$$

or, more compactly:

$$g(s) = \frac{K\displaystyle\prod_{i=1}^{q}\left(\xi_i s + 1\right)e^{-\alpha s}}{\displaystyle\prod_{i=1}^{p}\left(\tau_i s + 1\right)} \tag{8.30}$$

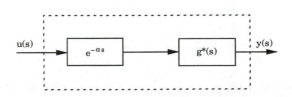

Figure 8.7.   Block diagrammatic representation of the general time-delay system.

In terms of the response of the general time-delay system to various inputs, what is true for the example system we have used above is true in general:

The response of the general time-delay system whose transfer function is represented as $g(s) = g*(s)e^{-\alpha s}$ is exactly the same as the response of the undelayed portion, $g*(s)$, shifted by α time units.

We note that it is possible to have a transfer function involving time-delay processes in parallel, for example:

$$g(s) = g*_1 e^{-\alpha_1 s} + g*_2 e^{-\alpha_2 s} + \dots + g*_n e^{-\alpha_n s} \qquad (8.31)$$

or even more complex forms. The response of such a process is simply a summation of the individual contributing responses and therefore poses no additional complication.

The following example is used to illustrate how to obtain the response of a time-delay system.

Example 8.1   UNIT STEP RESPONSE OF A FIRST-ORDER SYSTEM WITH TIME DELAY.

The dynamic behavior of a certain process is well represented as a first-order system with steady-state gain 0.66, time constant 6.7 minutes, and a time delay of 2.6 minutes.

1.   What is the transfer function representation of this process?
2.   Find the unit step response of this system.

Solution:

1.   From the given information, the system transfer function is :

$$g(s) = \frac{0.66 \, e^{-2.6s}}{6.7s + 1} \qquad (8.32)$$

2.   From our earlier results on the response of first-order systems (see Eq. (5.13)) the unit step response of the nondelay portion of this process is given by:

$$y(t) = 0.66(1 - e^{-t/6.7})$$

If we now combine with this the effect of the time delay, which is to shift the response

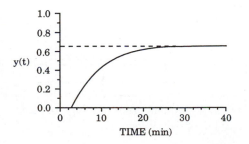

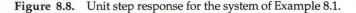

Figure 8.8.   Unit step response for the system of Example 8.1.

by 2.6 minutes, the time-delay system step response is now given by:

$$y(t) = \begin{cases} 0; & t < 2.6 \\ 0.66(1 - e^{-(t-2.6)/6.7}); & t > 2.6 \end{cases} \qquad (8.33)$$

This response is shown in Figure 8.8.

8.3.3 Approximate Behavior of Very High Order Systems

Our earlier result that the pure time-delay system is equivalent to an infinite sequence of identical first-order systems in series has some useful implications. If the behavior of N first-order systems in series approaches that of a time-delay system in the limit as N becomes very large, it seems plausible to attempt to approximate the behavior of very high-order systems as if they were time-delay systems.

To illustrate this concept, Figure 8.9 shows the step responses of a fourth-order system (consisting of four first-order systems in series) and that of a first-order system with time delay.

The advantage in adopting such an approximation strategy lies in the fact that fewer parameters are required to characterize the approximate model than the full Nth-order model. Observe that the Nth-order system is characterized by $N+1$ parameters: N time constants and one steady-state gain, whereas the first-order system with time delay is characterized by only three parameters.

In addition, we need to keep in mind that all models are really idealizations of true process behavior; and no one model is capable of perfectly representing reality. It is in this sense that an approximate but simpler model may be preferable.

Approximating the behavior of very high-order systems with that of low-order systems with delays is a common strategy adopted in Process Control practice. As we will see later, this strategy is not only useful for dynamic analysis, it is sometimes an important aspect of controller design.

One system of importance in the process industries whose time-delay characteristics are explicitly due to the physical presence of a flow time delay is the heat exchanger. Let us now briefly examine its dynamic behavior.

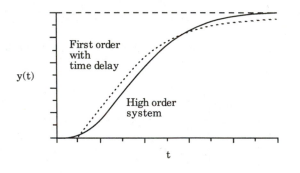

Figure 8.9.   Comparison of step responses of a high-order system with that of a first-order-plus-time-delay system.

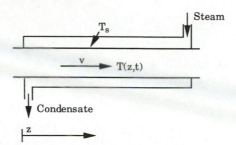

Figure 8.10.   The steam-heated heat exchanger process.

8.4   THE STEAM-HEATED HEAT EXCHANGER

As already mentioned in Chapter 4, we do not intend to include a full-scale discussion of the dynamic behavior of distributed parameter systems (DPS) in this book; however, we will present a discussion of some simple processes that fall under this category.

We have already identified the pure time-delay system as one such system (by virtue of its PDE model of Eqs. (8.2) and (8.4)), and its dynamic behavior has been discussed. Another DPS whose importance is based on the fact that it is a very common feature in the process industries is the steam-heated heat exchanger shown schematically in Figure 8.10.

Observe that the difference between this system and that of Figure 8.1 is that the fluid within the pipe is now being heated from the outside by the condensing steam. Once again, we assume that the fluid is flowing at a constant velocity v in plug flow. The other system parameters are indicated in the diagram.

The mathematical model for this process (derived in Chapter 12) is:

$$\frac{\partial T(z,t)}{\partial t} + v \frac{\partial T(z,t)}{\partial z} = \beta \left[T_{st}(t) - T(z,t) \right] ; \qquad 0 < z < L \tag{8.34}$$

where β is a heat transfer parameter (see Chapter 12). In terms of the usual deviation variables, the process model is:

$$\frac{\partial y(z,t)}{\partial t} + v \frac{\partial y(z,t)}{\partial z} = \beta \left[u(t) - y(z,t) \right] ; \qquad 0 < z < L \tag{8.35}$$

a first-order, hyperbolic PDE.

To develop the transfer function model for this process, once again we must take Laplace transforms of Eq. (8.35). Using the same strategy as before, the result is:

$$s y(z,s) + v \frac{d}{dz} y(z,s) = \beta \{ [u(s) - y(z,s)] \}$$

which rearranges to give:

$$\frac{dy(z,s)}{dz} = -\frac{(s + \beta)}{v} y(z,s) + \frac{\beta}{v} u(s) \tag{8.36}$$

a linear, first-order ODE easily solved for $y(z, s)$.

Using the entrance condition that at $z = 0$, $y(z,s) = y(0, s)$, it is possible to obtain the solution to Eq. (8.36) as:

$$y(z,s) = e^{-\left(\frac{s+\beta}{v}\right)z} y(0,s) + \left(\frac{\beta}{s + \beta}\right)\left[1 - e^{-\left(\frac{s+\beta}{v}\right)z}\right]u(s) \tag{8.37}$$

and in particular for $z = L$ (heat exchanger exit conditions) the transfer function model is:

$$y(L,s) = e^{-\left(\frac{s+\beta}{v}\right)L} y(0,s) + \left(\frac{\beta}{s + \beta}\right)\left[1 - e^{-\left(\frac{s+\beta}{v}\right)L}\right]u(s) \tag{8.38}$$

It is important to now note that this is of the form:

$$y(s) = g(s)\, u(s) + g_d(s)d(s) \tag{4.37}$$

so that for this process,

- The disturbance is $y(0,s)$, the change in the inlet temperature
- The input is $u(s)$, the steam temperature
- The output is $y(L,s)$, the heat exchanger outlet temperature

and the associated transfer functions are:

$$g(s) = \frac{\beta}{s + \beta}\left(1 - K_1 e^{-\alpha s}\right) \tag{8.39}$$

and

$$g_d(s) = K_1 e^{-\alpha s} \tag{8.40}$$

where $\alpha = L/v$, the delay due to fluid transportation. Here:

$$K_1 = e^{-\alpha\beta} \tag{8.41}$$

is constant, since α and β are known constants of the process.

It should be noted that Eq. (8.39) is an example of time-delay process in parallel having the general form of Eq. (8.31). Thus the main process transfer function $g(s)$ consists of two parts:

$$g(s) = g_1(s) - g_2(s) \tag{8.42}$$

where

$$g_1(s) = \frac{\beta}{s + \beta}$$

$$= \frac{1}{(1/\beta)s + 1} \tag{8.43a}$$

and

$$g_2(s) = \frac{K_1 e^{-\alpha s}}{(1/\beta)s + 1} \tag{8.43b}$$

the first being a pure first-order system with steady-state gain 1, and time constant $1/\beta$, and the second, a first-order-plus-time-delay system with steady-state gain K_1, same time constant $1/\beta$, and time delay α.

The disturbance transfer function $g_d(s)$ is that of a pure time-delay system with nonunity gain K_1. The reader should be convinced from Eq. (8.41) that this gain is positive, and less than 1.

These facts significantly simplify the dynamic analysis of an otherwise complex system. In fact, contrary to popular expectations (see Coughanowr and Koppel, Ref. [1], pp. 356, 357) the dynamic response of this system is now very easy to obtain, as illustrated by the following examples.

Example 8.2   UNIT STEP RESPONSE OF THE STEAM-HEATED HEAT EXCHANGER PROCESS.

Obtain an expression for the dynamic behavior of the temperature of the fluid leaving the heat exchanger of Figure 8.10 in response to a unit step change in the steam temperature, assuming the inlet temperature $T(0,t)$ has remained constant.

Solution:

When the inlet temperature is kept constant, the dynamic behavior of the heat exchanger outlet temperature is obtained from:

$$y(s) = g(s) u(s)$$

From Eq. (8.42) and the expressions in Eq. (8.43) we observe that $y(t)$, the unit step response we seek, is given by:

$$y(t) = y_1(t) - y_2(t) \tag{8.44}$$

where $y_1(t)$ is the unit step response of the first-order system whose transfer function is given in Eq. (8.43a), and $y_2(t)$ is the unit step response of the system indicated in Eq. (8.43b).

We may now call upon our previous knowledge of the dynamic behavior of such systems; this enables us to write down immediately the expressions for $y_1(t)$ and $y_2(t)$ as:

$$y_1(t) = 1 - e^{-\beta t} \tag{8.45}$$

and

$$y_2(t) = \begin{cases} 0; & t < \alpha \\ K_1\left(1 - e^{-\beta(t-\alpha)}\right); & t > \alpha \end{cases}$$

$$= \begin{cases} 0; & t < \alpha \\ K_1 - K_1 e^{\alpha\beta} e^{-\beta t}; & t > \alpha \end{cases} \qquad (8.46)$$

and now, using Eq. (8.41), Eq. (8.46) becomes:

$$y_2(t) = \begin{cases} 0; & t < \alpha \\ K_1 - e^{-\beta t}; & t > \alpha \end{cases} \qquad (8.47)$$

If we now subtract Eq. (8.47) from Eq. (8.45), as indicated in Eq. (8.44) we obtain the required unit step response as:

$$y(t) = \begin{cases} 1 - e^{-\beta t}; & t < \alpha \\ 1 - e^{-\alpha\beta}; & t > \alpha \end{cases} \qquad (8.48)$$

where we have made use of Eq. (8.41) in place of the surviving K_1 term in the second half of the expression.

The expression in Eq. (8.48) indicates that the *change* observed in the temperature of the fluid at the heat exchanger outlet (as a deviation from the initial steady-state value) starts out increasing exponentially, with no delay, and after α time units have elapsed (the time required for any incoming fluid element to travel through the entire inner pipe) the temperature change takes on a constant value, $1 - e^{-\alpha\beta}$. The response is shown in Figure 8.11. (It is left as an exercise to the reader to confirm that, on purely physical grounds, this response is indeed what we should expect.)

Example 8.3   RESPONSE OF THE STEAM-HEATED HEAT EXCHANGER PROCESS TO A UNIT STEP CHANGE IN THE DISTURBANCE.

Obtain the response of the temperature of the fluid leaving the heat exchanger to a unit step change in the inlet temperature $T(0,t)$ when the steam temperature is kept constant.

Solution:

From Eq. (8.38) (or the simplified, general form in Eq. (4.37)) when $u = 0$ (for this is the implication of keeping the steam temperature constant) the response to this disturbance input is obtained from:

$$y(s) = g_d(s)d(s)$$

From Eq. (8.40) we easily see that the response will be that of a pure time-delay system with gain K_1; i.e., instead of a repetition of the unit step function shifted in time, we have:

$$y(t) = \begin{cases} 0; & t < \alpha \\ K_1; & t > \alpha \end{cases}$$

or

$$y(t) = \begin{cases} 0; & t < \alpha \\ e^{-\alpha\beta}; & t > \alpha \end{cases} \qquad (8.49)$$

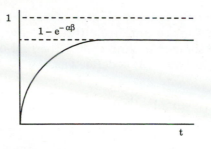

Figure 8.11.   Response of the heat exchanger exit temperature to a unit step change in
                steam temperature.

8.5   RATIONAL TRANSFER FUNCTION
         APPROXIMATIONS

While it is true that from a certain perspective, it is more advantageous to
approximate the dynamic behavior of a high-order system by a lower order
system plus a time delay, this exercise, however, replaces a rational transfer
function with one that contains the transcendental function $e^{-\alpha s}$ associated with
time-delay systems.

As we shall soon see, because of certain analysis problems which are made
easier if the transfer function were completely rational, and because of certain
process analysis tools that absolutely require rational transfer functions, the
transcendental exponential term is sometimes approximated by rational
functions.

8.5.1. Padé Approximations

Padé approximations have been developed to allow $e^{-\alpha s}$ to be represented by
the ratio of low-order polynomials in s. The most general form is:

$$e^{-\alpha s} \cong \frac{a_p(\alpha s)^p + a_{p-1}(\alpha s)^{p-1} + \dots + a_0}{b_q(\alpha s)^q + b_{q-1}(\alpha s)^{q-1} + \dots + b_0} \qquad (8.50)$$

where the orders of the polynomials, p and q, as well as the coefficients a_i, b_j,
are chosen to have the power series expansion for the polynomial ratio
(Eq. (8.50)) match the power series expansion for $e^{-\alpha s}$ as closely as possible.
Recall that a Taylor series expansion of $e^{-\alpha s}$ is:

$$e^{-\alpha s} = 1 - \alpha s + \frac{(\alpha s)^2}{2!} - \frac{(\alpha s)^3}{3!} + \dots\dots \qquad (8.51)$$

However, a large number of terms are required to give a good approximation,
leading to very high-order systems. A better approach is to use Eq. (8.50),
which retains relatively low-order terms, but which can be made equivalent to
the leading terms in Eq. (8.51). The coefficient set a_i, b_j which causes
Eqs. (8.50) and (8.51) to be equivalent for the maximum number of terms
Eq. (8.51) (for a specific order p,q) is termed a "Padé approximant."

The most commonly employed approximations are the first-, second-, or third-order Padé approximant although even higher order Padé approximations are sometimes employed. The first-order Padé approximation $(p = q = 1)$ (valid up to $|\alpha s| \sim 1$) is defined as:

$$e^{-\alpha s} \approx \frac{1 - \frac{\alpha}{2} s}{1 + \frac{\alpha}{2} s} \tag{8.52}$$

Similarly, the second-order approximation $(p = q = 2)$ (valid up to $|\alpha s| \sim 2$) is:

$$e^{-\alpha s} \approx \frac{1 - \frac{\alpha}{2} s + \frac{\alpha^2}{12} s^2}{1 + \frac{\alpha}{2} s + \frac{\alpha^2}{12} s^2}; \tag{8.53}$$

and the third-order approximation $(p = q = 3)$ (valid up to $|\alpha s| \sim 3$) is:

$$e^{-\alpha s} \approx \frac{1 - \frac{\alpha}{2} s + \frac{\alpha^2}{10} s^2 - \frac{\alpha^3}{120} s^3}{1 + \frac{\alpha}{2} s + \frac{\alpha^2}{10} s^2 + \frac{\alpha^3}{120} s^3} \tag{8.54}$$

Obviously even higher order approximations may be used. As is to be expected, the higher the order the better the approximation.

The step responses of the first-, second-, and third-order Padé approximations are shown in Figure 8.12. Note that the number of right-half plane zeros increases with the order of approximation so that the time delay is being approximated by many RHP zeros.

The frequency response of the various Padé approximations is illustrated by a Bode plot in Figure 8.13. Note that the amplitude ratio of the Padé approximation is always exact, but the phase lag is poorly approximated at higher frequencies.

8.5.2 High-Order System Approximation

Recall from Eqs. (8.20) and (8.22) that one can consider the time delay as the limit of a large number of first-order processes in series:

$$e^{-\alpha s} = \lim_{N \to \infty} \frac{1}{\left(\frac{\alpha s}{N} + 1\right)^N} \tag{8.55}$$

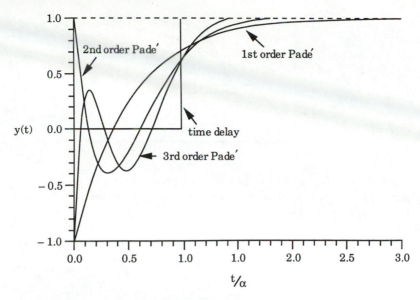

Figure 8.12.   Unit step response of a pure time-delay system using first-, second-, and third-order Padé approximations.

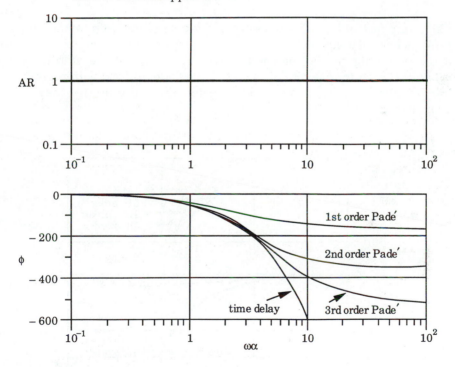

Figure 8.13.   Frequency response of a pure time delay and the first-, second-, and third-order Padé approximations.

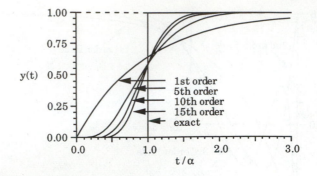

Figure 8.14. Unit step response of a pure time delay using the N first-order system approximation.

Therefore it is reasonable to consider approximating the time delay by a finite number of first-order processes:

$$e^{-\alpha s} \cong \frac{1}{\left(\dfrac{\alpha s}{N} + 1\right)^N} \tag{8.56}$$

The step response of Eq. (8.56) is shown in Figure 8.14 for N up to the 15th order. The frequency response of this approximation is shown in Figure 8.15. Note that the phase lag can be matched reasonably well, but the amplitude ratio fails badly at high frequencies. In general, the Padé approximation gives much better results for the same model order than the high-order system approximation.

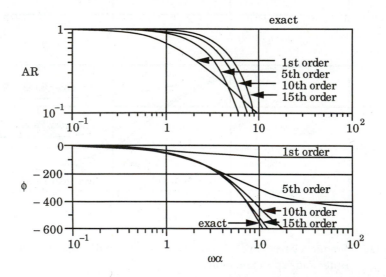

Figure 8.15. Frequency response of a pure time-delay using the N first-order system approximation.

8.6   MODEL EQUATIONS FOR SYSTEMS CONTAINING TIME DELAYS

Recall that in Chapter 4 we presented general equations for models in each of the model forms:

1. State space
2. Transform domain
3. Frequency response
4. Impulse response

However, we did not deal explicitly with models containing time delays in that earlier discussion. Now that we have introduced the time delay and reviewed some of its properties, let us indicate how time delays are introduced into each of the linear model forms discussed earlier. We shall consider only continuous-time systems here; discrete-time system models with time delays are discussed in Chapters 23–26.

8.6.1   State-Space Models

Recall that single-input/single-output (SISO) state-space models *without delays* take the form:

$$\frac{d\mathbf{x}}{dt} = \mathbf{A}\mathbf{x}(t) + \mathbf{b}u(t) + \gamma d(t)$$
$$y(t) = \mathbf{c}^T\mathbf{x}(t) \tag{4.29}$$

When there is an input delay α, a state delay β, an output delay η, and a disturbance delay δ, then the time-domain model is of the form:

$$\frac{d\mathbf{x}(t)}{dt} = \mathbf{A}_0\mathbf{x}(t) + \mathbf{A}_1\mathbf{x}(t - \beta) + \mathbf{b}_0u(t) + \mathbf{b}_1u(t - \alpha) + \gamma_0 d(t) + \gamma_1 d(t - \delta)$$
$$y(t) = \mathbf{c}_0^T\mathbf{x}(t) + \mathbf{c}_1^T\mathbf{x}(t - \eta) \tag{8.57}$$

Note that if there are multiple delays, then additional terms will appear for each delay.

Usually for SISO systems, the model has only a single delay that can be considered an input delay α.

In the case of multiple-input/multiple-output (MIMO) time delays, the equations are similar except that there will be multiple delays and the quantities u, d, y, will become vectors while $\mathbf{b}_0, \mathbf{b}_1, \gamma_0, \gamma_1, \mathbf{c}_0, \mathbf{c}_1$ will be matrices. These systems will be considered in more detail in Chapters 20–22.

8.6.2   Transform-Domain Models

In the transform domain, the transfer function model:

$$y(s) = g(s)u(s) + g_d(s)d(s) \tag{4.37}$$

will have the same form as without delays except that $g(s)$, $g_d(s)$ will have terms $e^{-\alpha s}$, $e^{-\beta s}$, $e^{-\eta s}$, $e^{-\delta s}$ to represent the delays in Eq. (8.57). Thus a general form would be:

$$g(s) \;=\; \frac{\displaystyle\sum_{n=1}^{N} b_n(s) e^{-\alpha_n s}}{\displaystyle\sum_{m=1}^{M} a_m(s) e^{-\beta_n s}} \tag{8.58}$$

where the $a_m(s), b_n(s)$ are rational polynomials in s. The more usual form that arises when there is only an input or output delay is:

$$g(s) \;=\; g^{*}(s) e^{-\alpha s} \tag{8.59}$$

where $g^*(s)$ is a rational transfer function. We have seen many examples of these forms in this chapter.

For multiple-input/multiple-output (MIMO) systems with time delays, each element of the transfer function matrix $G(s)$, $G_d(s)$ in:

$$\mathbf{y}(s) \;=\; \mathbf{G}(s)\mathbf{u}(s) + \mathbf{G_d}(s)\mathbf{d}(s) \tag{4.38}$$

would have the form of Eqs. (8.58) or (8.59). A fuller discussion of this will be in Chapters 20–22.

8.6.3   Frequency-Response Models

There are no special complications introduced by time delays for this class of models because if one substitutes $s = jw$ in Eqs. (8.58) and (8.59) one obtains:

$$g(j\omega) \;=\; \frac{\displaystyle\sum_{n=1}^{N} b_n(j\omega) e^{-j\alpha_n \omega}}{\displaystyle\sum_{m=1}^{M} a_m(j\omega) e^{-j\beta_m \omega}} \tag{8.60}$$

and

$$g(j\omega) \;=\; g^{*}(j\omega) e^{-j\alpha\omega} \tag{8.61}$$

When one uses the identity:
$$e^{-j\phi} \;=\; \cos\phi - j\sin\phi$$

one sees that the frequency response of Eqs. (8.60) or (8.61) is straightforward to determine.

8.6.4   Impulse-Response Models

Recall that the impulse-response model for a single-input/single-output system takes the form:

$$y(t) \;=\; \int_0^t g(t-\sigma)u(\sigma)d\sigma + \int_0^t g_d(t-\sigma)d(\sigma)d\sigma \tag{4.21}$$

while for a multiple-input/multiple-output system one has:

$$\mathbf{y}(t) = \int_0^t \mathbf{G}(t - \sigma)\mathbf{u}(\sigma)d\sigma + \int_0^t \mathbf{G}_d(t - \sigma)\mathbf{d}(\sigma)d\sigma \qquad (4.46)$$

where $\mathbf{G}(t)$, $\mathbf{G}_d(t)$ are matrices of impulse-response functions. Since each of these requires the impulse-response function $g(t)$ either from data or from the inverse Laplace transform of $g(s)$, there are no special difficulties when time delays are present.

To illustrate, we recall that for a first-order system with transform-domain transfer function:

$$g(s) = \frac{K}{\tau s + 1} \qquad (4.60)$$

and the impulse response is:

$$g(t) = \frac{K}{\tau} e^{-t/\tau} \qquad (4.76)$$

If we add a time delay to Eq. (4.60) so that:

$$g(s) = \frac{Ke^{-\alpha s}}{\tau s + 1} \qquad (8.62)$$

The resulting impulse response is:

$$g(t) = \begin{cases} 0 & 0 < t < \alpha \\ \dfrac{K}{\tau} e^{-(t - \alpha)/\tau} & t > \alpha \end{cases} \qquad (8.63)$$

This expression for $g(t)$ can just as easily be inserted into Eq. (4.21) as the expression in Eq. (4.76). Thus time delays introduce no special problems for impulse-response models.

To summarize, we see that time delays can be introduced into each of the model forms without any special problems.

8.7 SUMMARY

We have been mainly concerned in this chapter with studying the characteristics of time-delay systems: those systems for which there is a delay period between when an input change is implemented and when the response is observed. The sources of such behavior in physical processes have been identified, the characterizing factor in their mathematical model identified, and the most salient features of their responses discussed.

We have seen that simple time-delay systems are composed of regular dynamic systems connected in series with pure time-delay elements. Thus their responses are exactly like those of "normal," undelayed systems, only shifted in time. The transfer functions of time-delay systems involve the exponential term $e^{-\alpha s}$, which accounts for the observed time shift in the process responses.

A very important aspect of the discussion in this chapter is the fact that the dynamic behavior of very high-order systems are often conveniently approximated by lower order models with time delays. This has certain implications as far as controller design is concerned which we will have cause to refer to later on.

When the transcendental nature of the time-delay system's transfer function model is a problem, rational approximations of different types may be used. Both Padé approximations and high-order system approximations have been discussed and their properties analyzed.

We have reviewed each of the linear model forms introduced in Chapter 4 and indicated how these are modified to accommodate the presence of time delays. In each case we saw that no special complications arise.

With this we have completed the discussion of the dynamic behavior of linear systems, be they of low or high-order, finite or infinite. The next logical step will be a discussion of *nonlinear* systems. However, before undertaking this task, we will examine the issue of frequency response analysis of linear systems more closely in Chapter 9.

REFERENCES AND SUGGESTED FURTHER READING

1. Coughanowr, D. R. and L. B. Koppel, *Process Systems Analysis and Control*, McGraw-Hill, New York (1965) (2nd ed., D. R. Coughanowr, McGraw-Hill, New York (1991))

REVIEW QUESTIONS

1. What is a time-delay system?

2. What are some aspects of a typical processing plant that will contribute to the existence of time delays?

3. What characterizes the transfer function of a time-delay system?

4. The pure time-delay process may be represented as an infinite sequence of what type of system?

5. What effect does a pure time-delay element have on the response of a dynamic system?

6. What advantage is typically gained by approximating the dynamic behavior of a *very high*-order system with a low-order model and a time-delay element?

7. What is a Padé approximation?

8. What distinguishes the state-space model of a time-delay system from that of a system with no delay?

9. What is peculiar about the impulse-response function of time-delay system when compared with that of a corresponding system with no delay?

PROBLEMS

8.1 Solve Eq. (8.36) and hence confirm Eq. (8.37).

8.2 The transfer function model of the top portion of a distillation column is given as follows:

$$y(s) = \left(\frac{12.8e^{-s}}{16.7s + 1}\right)u(s) + \left(\frac{3.8e^{-8s}}{14.9s + 1}\right)d(s) \qquad (P8.1)$$

which is of the form:

$$y(s) = g(s)u(s) + g_d(s)d(s)$$

(a) Obtain the response of the output, $y(t)$, to a *unit* step change in the disturbance, $d(t)$, assuming that the input, $u(t) = 0$.
(b) Given now that the input u is to be determined according to the expression:

$$u(s) = g_{FF}(s)\, d(s)$$

with $g_{FF}(s)$ given by:

$$g_{FF}(s) = \frac{-g_d(s)}{g(s)}$$

for this specific distillation column, obtain an expression for the dynamic behavior observed in $u(t)$ as a result of a unit step change in the disturbance $d(t)$ under such a control strategy. *Sketch* this response showing clearly all important distinguishing factors.

8.3 Revisit Problem 5.6. First obtain $u(s)$, the Laplace transform of the input "staircase" function of Eq. (P5.7) and Figure P5.3, then use the given transfer function model:

$$y(s) = \frac{0.32}{15.5s + 1}u(s) \qquad (P5.6)$$

to obtain the reactor temperature response, $y(t)$.

8.4 An industrial furnace studied by Endtz *et al.* (1957)[†] has as its input the fuel supply pressure (in psi), with the effluent stream temperature (in °C), as its output. In terms of deviation variables, the transfer function model relating these process variables was given as:

$$y(s) = \frac{40e^{-20s}}{(900s + 1)(25s + 1)}u(s) \qquad (P8.2)$$

The units of time are in seconds.
(a) Use your favorite control system simulation package to obtain the unit step response of this process. If the smaller time constant of 25 s were ignored (since it seems negligible compared to the other time constant of 900 s) obtain the resulting unit step response and compare it with that obtained for the original system.

[†]J. Endtz, J. M. Janssen, and J. C. Vermeulen, *Plant and Process Dynamic Characteristics*, Butterworths, London (1957), p. 176.

(b) The authors also noted that the process could equally well have been described by the transfer function model:

$$y(s) = \frac{40}{(900s + 1)(13s + 1)^3} u(s) \qquad \text{(P8.3)}$$

Obtain the unit step response of this new system model and compare it with the two obtained earlier.

8.5   The production end of a certain polymer manufacturing facility consists primarily of the ensemble shown in Figure P8.1: a premixer and a reactor connected in series by a pipe whose length is *not* negligible.

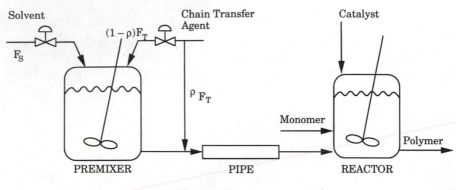

Figure P8.1.

The premixer is a stirred tank in which a large stream of inert solvent **S** (flowrate F_S) is mixed with a small stream of chain transfer agent **T** (flowrate F_T). By design, a fraction ρ of the chain transfer agent stream is fed directly to the outlet of the premixer. The monomer **M** and catalyst **C** are fed into the continuous stirred tank reactor along with the diluted stream of the chain transfer agent from the premixer. Polymer product is withdrawn from the reactor outlet and separated from the unreacted monomer and solvent in another section which is of no interest at the present time.

The main objective is to control the *normalized* melt viscosity of the polymer product by adjusting the ratio of F_T to F_S. The pertinent process variables are defined as follows, *in terms of deviations from appropriate steady-state values*:

u	=	Ratio of the transfer agent flow to inert solvent flow (F_T/F_S)
v	=	Concentration of transfer agent measured at the inlet of the connecting pipe
w	=	Concentration of transfer agent measured at the reactor inlet
y	=	Normalized melt viscosity of polymer product at reactor exit

Given that the following transfer functions are reasonably accurate for each portion of the ensemble as indicated:

PREMIXER: $\qquad g_1(s) \quad = \quad \dfrac{5s + 1}{10s + 1}$

CONNECTING PIPE: $g_2(s) \quad = \quad e^{-6s}$

REACTOR: $\qquad g_3(s) \quad = \quad \dfrac{25}{12s + 1}$

(where the time unit is in *minutes* and all the other units are normalized units conforming with industry standards for this product):

(a) What is the overall transfer function (for the entire polymerization process) that relates — in terms of deviation variables — the process input (the (F_T/F_S) ratio) to the process output, the normalized polymer melt viscosity?

(b) Obtain individual mathematical expressions for the response of $v(t)$, $w(t)$, and $y(t)$ to a step input of size 0.02 in $u(t)$. *Sketch* each of these responses separately, showing clearly *all* important distinguishing characteristics.

(c) If a modification to the process has the apparent effect of shifting the location of the single zero of $g_1(s)$ from its original location in the complex plane to $s = -0.05$, sketch the new response of $v(t)$, $w(t)$, and $y(t)$ to the same step input of size 0.02 in $u(t)$, assuming that only the premixer dynamics have been affected by the modification. Show clearly all the important distinguishing characteristics of each response.

8.6 After linearizing the original set of *six* coupled, nonlinear, ordinary differential equations that describe the dynamics of an industrial separations process, further simplification and Laplace transformation gave rise to the following transfer function relating the input to the output:

$$y(s) = \frac{10.3}{(1.5s + 1)^5(15.2s + 1)} u(s) \qquad \text{(P8.4)}$$

Use your favorite control system simulation package to obtain the unit step response of this process and compare it to the unit step response of the process whose transfer function is given as:

$$g(s) = \frac{10.3e^{-7.5s}}{(15.2s + 1)} \qquad \text{(P8.5)}$$

How is this transfer function related to the one indicated in Eq. (P8.4)? Assess the reasonableness, or otherwise, of using this transfer function in place of the "more complicated" one in Eq. (P8.4).

8.7 A chemical reactor with recycle is modeled by the following differential equation:

$$\frac{dy}{dt} = ay(t) + a_1y(t - \alpha_0) + bu(t) + \gamma d(t) \qquad \text{(P8.6)}$$

when the feed comes from usual feed tank A. When the alternate feed tank B provides the feed, the reactor model is altered slightly to:

$$\frac{dy}{dt} = ay(t) + a_1y(t - \alpha_0) + bu(t - \alpha_1) + \gamma d(t) \qquad \text{(P8.7)}$$

(a) With no additional information given, provide your own speculations as to what might be the difference between tank A and tank B; what u, the input variable to the reactor, might be; and what y, the reactor output variable, might be.

(b) Obtain the transfer functions $g(s)$ and $g_d(s)$ for each of the two models. How do they compare to the transfer functions of the time-delay systems of Section 8.3?

(c) Given a specific situation in which $a = -2$, $a_1 = 0.5$, $b = 3.0$, $\gamma = 0.25$, with $\alpha_0 = 2$ and $\alpha_1 = 5$, use a control systems simulation package to obtain the process response to a unit step change in $u(t)$ when the feed is taken from each of the two feedtanks and compare these two responses. If you had a choice, which feed tank would you rather use? Support your answer as adequately as you can.

8.8   The one-dimensional rod shown in Figure P8.2 is to be heated as indicated using a steam chest.

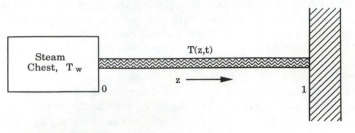

Figure P8.2.

The length coordinate z has been normalized so that the total length is 1; heat is added at $z = 0$, and the $z = 1$ end is assumed to be perfectly insulated. By defining the following deviation variables:

$$y(z,t) \;=\; T - T^* \tag{P8.8a}$$

$$u(t) \;=\; T_w - T_w^* \tag{P8.8b}$$

with (*) representing initial steady-state values, and assuming that heat loss to the atmosphere is negligible, it is possible to show that an appropriate model for this process is given by:[†]

$$\frac{\partial y(z,t)}{\partial t} \;=\; \frac{\partial^2 y(z,t)}{\partial z^2} \tag{P8.9}$$

along with the following boundary conditions:

$$z = 0 \qquad\qquad \frac{\partial y}{\partial z} \;=\; \beta(y - u) \tag{P8.10}$$

$$z = 1 \qquad\qquad \frac{\partial y}{\partial z} \;=\; 0 \tag{P8.11}$$

By taking Laplace transform (with respect to time t) show that the transfer function relating $y(z,s)$ to $u(s)$ *for any z along the length of the rod* is given by:

$$g(z,s) \;=\; \frac{A(z,s)}{B_1(s) + B_2(s)} \tag{P8.12a}$$

where:

$$A(z,s) \;=\; e^{-2\sqrt{s}}\, e^{z\sqrt{s}} - e^{-z\sqrt{s}} \tag{P8.12b}$$

$$B_1(s) \;=\; \left(1 + e^{-2\sqrt{s}}\right) \tag{P8.12c}$$

$$B_2(s) \;=\; \frac{\sqrt{s}}{\beta}\left(1 + e^{-2\sqrt{s}}\right) \tag{P8.12d}$$

[†] Ray, W. H. *Advanced Process Control*, Butterworths, Boston (1989), p. 147.

and that, in particular, for z = 0, the transfer function is:

$$g(0,s) = \frac{1}{1 + \dfrac{\sqrt{s}}{\beta} \tanh \sqrt{s}}$$

(P8.13)

8.9   Use the first-order Padé approximation:

$$e^{-\alpha s} \approx \frac{1 - \dfrac{\alpha}{2}s}{1 + \dfrac{\alpha}{2}s}$$

in reverse to approximate:

$$g(s) = \frac{(1 - 3s)}{(2s + 1)(5s + 1)}$$

(P8.14)

with another transfer function which no longer involves a right-half plane zero but a time delay. Obtain a plot of the unit step response of the inverse-response system represented by the original transfer function and compare it with the step response obtained from the approximating time-delay system obtained by using the Padé approximation *in reverse*.

8.10   By using arguments similar to those employed in Section 8.2.2, or otherwise, show that the time-delay term $e^{-\beta s}$ in a transfer function may also be represented as:

$$e^{-\beta s} = \lim_{N \to \infty} \left(1 + \frac{-\beta s}{N}\right)^N$$

(P8.15)

i.e., as an infinite set of right-half plane zeros. By writing $e^{-\alpha s}$ as:

$$e^{-\alpha s} = \frac{e^{-\frac{\alpha s}{2}}}{e^{+\frac{\alpha s}{2}}}$$

now show from Eq. (P8.15) that it is *also* possible to represent the time-delay term as:

$$e^{-\alpha s} = \lim_{N \to \infty} \left[\frac{\left(1 + \dfrac{-\alpha s}{2N}\right)}{\left(1 + \dfrac{\alpha s}{2N}\right)}\right]^N$$

(P8.16)

i.e., as an infinite set of poles and an infinite set of right-half plane zeros.

8.11   (a) Given the following transfer function model for a heat exchanger:

$$y(s) = \frac{(1 - 0.5e^{-10s})}{(40s + 1)(15s + 1)} u(s)$$

(P8.17)

Obtain the equivalent differential equation model, as well as the equivalent *impulse-response* model.

8.12 Show that the system whose transfer function is composed as follows:

$$g(s) = \frac{K_1 e^{-\alpha s}}{\tau s + 1} - \frac{K_2}{\tau s + 1} \tag{P8.18}$$

has a single zero located at:

$$s = \frac{-\ln\left(\dfrac{K_2}{K_1}\right)}{\alpha} \tag{P8.19}$$

and hence establish that the system will exhibit inverse response only if:

$$\left|K_2\right| < \left|K_1\right| \tag{P8.20}$$

8.13 (a) Two first-order-plus-time-delay systems in parallel have the composite transfer function given by:

$$g(s) = \frac{K_1 e^{-\alpha_1 s}}{(\tau_1 s + 1)} - \frac{K_2 e^{-\alpha_2 s}}{(\tau_2 s + 1)} \tag{P8.21}$$

Why is it that under conditions indicated by $\alpha_1 > \alpha_2$ and $K_1 > K_2$, the system will exhibit inverse response regardless of the relative values of τ_1 and τ_2? Given the specific situation with $\alpha_1 = 10$, $\alpha_2 = 3$; $K_1 = 5$, $K_2 = 2$; with $\tau_1 = 7$ and $\tau_2 = 6$, obtain a plot of the composite system's unit step response.

(b) Under the situation in which $\tau_1 = \tau_2 = \tau$, show that the composite transfer function shown in Eq. (P8.21) has a single zero located at:

$$s = \frac{\ln K_1 - \ln K_2}{\alpha_1 - \alpha_2} \tag{P8.22}$$

an expression that is entirely independent of τ. From here, deduce the conditions to be satisfied for the composite system to exhibit inverse response when $\tau_1 = \tau_2$.

CHAPTER

9

FREQUENCY-RESPONSE ANALYSIS

Since our study of the dynamic behavior of various processes began in Chapter 5, we have had brief encounters with how these linear systems respond to sinusoidal inputs. Such responses have been identified as consisting of two parts: the transient part (which dies out as $t \to \infty$) and the part that persists, itself a sinusoidal function, and therefore called the ultimate periodic response.

Studying the functional relationship between the input and output sinusoids is the central theme of frequency-response analysis, and as we shall see, such studies provide crucial information about the character of the process in question. The objectives in this chapter are first to consolidate, and then to generalize, the frequency-response results encountered earlier; the final task of presenting new frequency-response results will be much facilitated after this.

9.1   INTRODUCTION

As has been shown with isolated examples at various stages of our discussion thus far, the response of a linear system to a sinusoidal input, say, $A \sin \omega t$, given enough time for all the transients to die out, is also a sine wave. As illustrated in Figure 9.1, the output sinusoid has the form $A_o \sin (\omega t + \phi)$; i.e., it has the same frequency as the input but a different amplitude A_o; and it is out of phase by ϕ degrees (or radians). The ratio of output and input amplitudes, A_o/A, is known as the amplitude ratio, AR; ϕ is called the phase angle.

For any given system, the AR and ϕ values vary with ω, the frequency of the input signal. The results we have obtained so far bear out this fact (see Eq. (5.29) for the first-order system, Eq. (5.54) for the pure capacity system, Eqs. (6.78) and (6.79) for the second-order system, and Eq. (8.19) for the pure time-delay system). The investigation of how these quantities vary with ω, for any particular system, is the main concern of frequency-response analysis.

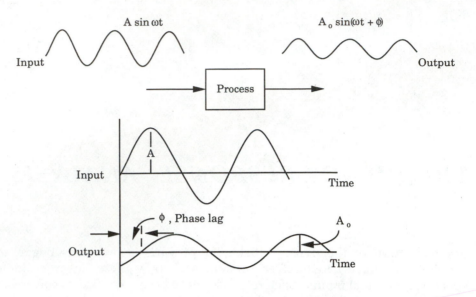

Figure 9.1. Frequency response of a linear process.

Our interest in this sort of analysis is justified in part by the fact that the dynamics of a linear process can be *entirely* characterized by the AR and ϕ behavior over a range of ω values. Since we have earlier introduced the Bode diagram as the most common method of presenting the AR and ϕ behavior as a function of ω, the implication here is that the dynamic behavior of a linear process system is entirely characterized by its Bode diagram.

As we demonstrated in Chapter 4, the frequency-response transfer function $g(j\omega)$ can be obtained from the other model forms. In this chapter we show that the dynamic information contained in the frequency response (e.g., AR and ϕ from a Bode plot) is precisely what constitutes the frequency-response transfer function $g(j\omega)$. Thus, the frequency response of a system provides direct information about the *entire* transfer function — $g(j\omega)$; no other response (step response, impulse response, etc.) provides such complete information directly. As a result, frequency response plays an important role in the analysis and design of feedback control systems. This aspect will be picked up again in Chapter 15; for now we will concentrate on the dynamic analysis aspect of frequency response.

The method of obtaining the frequency response of a general process from its Laplace domain transfer function will first be presented. After this, the main characteristics of the frequency response for various types of linear systems will be discussed, and their individual Bode diagrams presented. In anticipation of the most significant application of frequency-response analysis — in the design of feedback controllers — we will also present frequency-response results (including Bode diagrams) for these feedback controllers, even though they are not to be formally introduced until Chapter 15.

9.2   A GENERAL TREATMENT

9.2.1   The Fundamental Frequency-Response Result

Before giving the main frequency-response result, we need to familiarize ourselves with the following preliminary facts about a complex number (see Appendix B for more details).

Given a complex number z whose real part is a and whose imaginary part is b, i.e.:

$$z = a + bj \tag{9.1}$$

located in the complex plane as shown in Figure 9.2, we define its magnitude $|z|$ and argument arg (z), or $\angle z$, as:

$$|z| = \sqrt{a^2 + b^2} \tag{9.2}$$

and

$$\arg(z), \text{ or } \angle z = \tan^{-1}\left(\frac{b}{a}\right) \tag{9.3}$$

The main result is now given as follows:

Given a system whose frequency-response transfer function is $g(j\omega)$, the amplitude ratio AR and phase angle ϕ for this system are given by:

$$\text{AR} = |g(j\omega)| \tag{9.4a}$$

and

$$\phi = \arg(g(j\omega)), \text{ or } \angle g(j\omega) \tag{9.4b}$$

where we recall from Chapter 4 that

$$g(j\omega) = g(s)\big|_{s=j\omega} \tag{9.4c}$$

The relationship given in Eq. (9.4c) which allows $g(j\omega)$ (and thus the frequency response) to be obtained directly from $g(s)$ is a very powerful result, and establishing it requires deriving the frequency response for the general

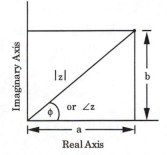

Figure 9.2.   The complex number.

system whose transform-domain transfer function is given simply as $g(s)$. The reader willing to accept this result on faith can skip the somewhat lengthy derivation we are about to present.

Frequency Response of the General Linear System

For the general linear system whose transform-domain transfer function model is given as:

$$y(s) \; = \; g(s) \, u(s)$$

the response to an input sine wave $A \sin \omega t$ is obtained from:

$$y(s) \; = \; g(s) \frac{A \omega}{s^2 + \omega^2} \tag{9.5}$$

The frequency response of this system may now be obtained by carrying out the following steps:

1. Partial Fraction Expansion

Upon partial fraction expansion in the usual manner, Eq. (9.5) may be rewritten as:

$$y(s) \; = \; \sum_{i=1}^{n} \left[\frac{A_i}{(s - r_i)} \right] + \frac{B_1}{(s - j\omega)} + \frac{B_2}{(s + j\omega)} \tag{9.6}$$

where the r_i's are the n (possibly complex) roots of the nth-order denominator polynomial of $g(s)$, i.e., the system poles; and, of course, $+j\omega$ and $-j\omega$ are recognized as the complex conjugate roots of $(s^2 + \omega^2)$.

2. Laplace Inversion

Equation (9.6) becomes:

$$y(t) \; = \; \sum_{i=1}^{n} A_i \, e^{r_i t} + B_1 e^{j\omega t} + B_2 e^{-j\omega t} \tag{9.7}$$

3. Asymptotic *Behavior*

If each of the r_i is negative (or has a negative real part), we observe that the terms involving $e^{r_i t}$ in Eq. (9.7) all die away with time. Furthermore, if we recall the Euler identities:

$$e^{j\theta} \; = \; \cos\theta + j \sin\theta \tag{9.8a}$$

$$e^{-j\theta} \; = \; \cos\theta - j \sin\theta \tag{9.8b}$$

we see that the other two terms in Eq. (9.7) persist, since they are, in fact, a sum of sines and cosines. Thus, the ultimate periodic response (UPR) is given by:

$$y(t)\Big|_{t \to \infty} = B_1 e^{j\omega t} + B_2 e^{-j\omega t} \tag{9.9}$$

The necessity for imposing the condition of negativity on the r_i stems from the following fact: if just one of these roots is positive (or has a positive real part), the exponential term associated with it will grow indefinitely with time, making nonsense of the concept of an ultimate response that is strictly periodic; there is, in fact, no longer a *bounded* function to which such a response ultimately settles. The importance of this statement will be developed further in Chapter 11.

4. Evaluate Constants B_1 and B_2

The constants B_1 and B_2 in Eq. (9.7) are given by:

$$B_1 = \lim_{s \to (j\omega)} (s - j\omega) \left[g(s) \frac{A\omega}{(s - j\omega)(s + j\omega)} \right]$$

$$= \lim_{s \to (j\omega)} \left[g(s) \frac{A\omega}{(s + j\omega)} \right] \tag{9.10}$$

or

$$B_1 = \frac{Ag(j\omega)}{2j} \tag{9.11}$$

and a similar procedure gives:

$$B_2 = \frac{Ag(-j\omega)}{-2j} \tag{9.12}$$

5. Convert from Exponentials to Sines and Cosines

Making use of the Euler identities, and the values given in Eqs. (9.11) and (9.12) for the constants, Eq. (9.9) becomes:

$$y(t)\Big|_{t \to \infty} = A \left[\frac{g(j\omega) - g(-j\omega)}{2j} \right] \cos\omega t + A \left[\frac{g(j\omega) + g(-j\omega)}{2} \right] \sin\omega t \tag{9.13}$$

6. Write $g(j\omega)$ as a Complex Variable and Simplify

A crucial step. Recognizing that $g(j\omega)$ is a complex number with real part $\text{Re}(\omega)$ and imaginary part $\text{Im}(\omega)$, we obtain:

$$g(j\omega) = \text{Re}(\omega) + j \, \text{Im}(\omega) \tag{9.14a}$$

which implies that $g(-j\omega)$ is given by:

$$g(-j\omega) = \text{Re}(\omega) - j \, \text{Im}(\omega) \tag{9.14b}$$

If we now substitute these quantities in Eq. (9.13), the result is:

$$y(t)\Big|_{t \to \infty} = A\,(\text{Re}(\omega)\,\sin\omega t + \text{Im}(\omega)\,\cos(\omega t)) \qquad (9.15)$$

We now need to recall the following general trigonometric identity:

$$p\,\sin\theta + q\,\cos\theta = \rho\,\sin(\theta + \psi) \qquad (9.16a)$$

where

$$\rho = \sqrt{p^2 + q^2} \qquad (9.16b)$$

and

$$\psi = \tan^{-1}\left(\frac{q}{p}\right) \qquad (9.16c)$$

This identity enables one to consolidate a sum of sines and cosines into one sine function as in Eq. (9.16a); the relationship between the "newly introduced" parameters ρ and ψ, and the original ones, p and q, are explicitly given in Eqs. (9.16b,c).

If we now let:

$$\theta = \omega t$$
$$p = \text{Re}(\omega)$$
$$q = \text{Im}(\omega)$$

then Eq. (9.16) allows one to consolidate the RHS expression in Eq. (9.15) to give:

$$y(t)\Big|_{t \to \infty} = A\,[\rho\,\sin(\omega t + \psi)] \qquad (9.17)$$

where, in this specific case, the parameters ρ and ψ are related to $\text{Re}(\omega)$ and $\text{Im}(\omega)$ according to:

$$\rho = \sqrt{\text{Re}(\omega)^2 + \text{Im}(\omega)^2} \qquad (9.18a)$$

and

$$\psi = \tan^{-1}\left[\frac{\text{Im}(\omega)}{\text{Re}(\omega)}\right] \qquad (9.18b)$$

Thus the amplitude ratio, of this general response (which we recall is the magnitude of the output Eq. (9.17) divided by the input $A\,\sin\omega t$), is given by ρ in Eq. (9.18a), and the phase angle is given by ψ in Eq. (9.18b).

7. Relate to the Magnitude and Argument of $g(j\omega)$

We now note from Eq. (9.14a) that the magnitude and the argument of the complex number $g(j\omega)$ are given by:

$$|g(j\omega)| = \sqrt{\text{Re}(\omega)^2 + \text{Im}(\omega)^2} \qquad (9.19a)$$

and

$$\arg[g(j\omega)], \text{ or } \angle g(j\omega) = \tan^{-1}\left[\frac{\text{Im}(\omega)}{\text{Re}(\omega)}\right] \tag{9.19b}$$

which are precisely the same corresponding expressions obtained in Eq. (9.18) as the required amplitude ratio and phase angle, thus establishing the result:

The frequency response of a process represented by g(s) may be found by letting s = jω and using the results:

$$\text{AR} = |g(j\omega)| \tag{9.4a}$$

$$\phi = \arg[g(j\omega)], \text{ or } \angle g(j\omega) \tag{9.4b}$$

as defined in Eq. (9.19).

In closing, let us note that the transfer function $g(j\omega)$, being complex, may be represented either in the Cartesian form:

$$g(j\omega) = \text{Re}(\omega) + j\,\text{Im}(\omega) \tag{9.14a}$$

or in the equivalent polar form:

$$g(j\omega) = |g(j\omega)|\,e^{j(\angle g(j\omega))}$$

which, from Eq. (9.4) is:

$$g(j\omega) = \text{AR}e^{j\phi}$$

with AR, the amplitude ratio, and ϕ, the phase angle, given by:

$$\text{AR} = \sqrt{\text{Re}(\omega)^2 + \text{Im}(\omega)^2} \tag{9.20a}$$

$$\phi = \tan^{-1}\left[\frac{\text{Im}(\omega)}{\text{Re}(\omega)}\right] \tag{9.20b}$$

9.2.2   General Procedure

In light of the fundamental frequency-response result we have just presented, the procedure for finding the frequency response of *any* system represented by its transform-domain transfer function $g(s)$ is as follows (provided the poles of $g(s)$ all have negative real parts):

1. Substitute $j\omega$ for s in the transfer function expression to obtain the corresponding frequency-response transfer function, $g(j\omega)$.

2. Owing to the nature of $g(s)$, the obtained function $g(j\omega)$ will be of the form $g(j\omega) = N(j\omega)/D(j\omega)$; *rationalize* this expression to obtain the Cartesian form:

$$g(j\omega) = \text{Re}(\omega) + j\,\text{Im}(\omega) \tag{9.14a}$$

Clearly identify $\text{Re}(\omega)$ and $\text{Im}(\omega)$ respectively, the real and imaginary parts.

3. Having done this, the required AR and ϕ are given by:

$$\text{AR} = \sqrt{\text{Re}(\omega)^2 + \text{Im}(\omega)^2} \tag{9.20a}$$

and

$$\phi = \tan^{-1}\left[\frac{\text{Im}(\omega)}{\text{Re}(\omega)}\right] \tag{9.20b}$$

4. Alternatively, $g(j\omega)$ could be represented in the polar form:

$$g(j\omega) = \text{AR}\, e^{j\phi} \tag{9.21}$$

from which the required AR and ϕ are extracted directly.

To accomplish this, we can deal individually with the numerator and denominator of $g(j\omega) = N(j\omega)/D(j\omega)$ by representing both $N(j\omega)$ and $D(j\omega)$ in polar form. Thus $g(j\omega)$ will take the form:

$$g(j\omega) = \frac{|N(j\omega)|e^{j\angle N(j\omega)}}{|D(j\omega)|e^{j\angle D(j\omega)}}$$

so that the polar form for $g(j\omega)$ will be:

$$g(j\omega) = \left|\frac{N(j\omega)}{D(j\omega)}\right| e^{j\angle N(j\omega) - \angle D(j\omega)}$$

From Eq. (9.4), therefore, the required AR and ϕ will now be given by

$$\text{AR} = \left|\frac{N(j\omega)}{D(j\omega)}\right| \tag{9.22a}$$

$$\phi = \angle N(j\omega) - \angle D(j\omega). \tag{9.22b}$$

Let us illustrate this procedure with the following examples.

Example 9.1   FREQUENCY RESPONSE OF A FIRST-ORDER SYSTEM.

Use the procedure given above to obtain the frequency-response of a first-order system and compare this result with that obtained earlier in Eq. (5.29).

Solution:

For the first-order system:

$$g(s) = \frac{K}{\tau s + 1}$$

and therefore, substituting $s = j\omega$ gives

$$g(j\omega) = K\left(\frac{1}{j\omega\tau + 1}\right) \tag{9.23}$$

Next we rationalize this expression to eliminate the quantity j from the denominator; this is done by multiplying numerator and denominator terms by the complex conjugate of the denominator; i.e.:

$$g(j\omega) = K\left(\frac{1}{j\omega\tau + 1}\right)\frac{1 - j\omega\tau}{1 - j\omega\tau}$$

giving

$$g(j\omega) = K\left(\frac{1 - j\omega\tau}{1 + \omega^2\tau^2}\right)$$

This may now be put in the Eq. (9.14a) Cartesian form as:

$$g(j\omega) = K\left(\frac{1}{1 + \omega^2\tau^2} - \frac{\omega\tau}{1 + \omega^2\tau^2} j\right)$$

with the immediate implication that:

$$\text{Re}(\omega) = K\left(\frac{1}{1 + \omega^2\tau^2}\right) \tag{9.24a}$$

and

$$\text{Im}(\omega) = -K\left(\frac{\omega\tau}{1 + \omega^2\tau^2}\right) \tag{9.24b}$$

We therefore have:

$$\text{AR} = |g(j\omega)| = K\sqrt{\frac{1 + \omega^2\tau^2}{(1 + \omega^2\tau^2)^2}}$$

$$= \frac{K}{\sqrt{1 + \omega^2\tau^2}} \tag{9.25a}$$

and

$$\phi = \arg(g(j\omega)), \text{ or } \angle g(j\omega) = \tan^{-1}(-\omega\tau) = -\tan^{-1}(\omega\tau) \tag{9.25b}$$

We may now compare these with the expressions obtained earlier in Eq. (5.29) and observe that they are exactly the same.

Alternatively, with the polar form approach, we recognize from Eq. (9.23) that:

$$g(j\omega) = \frac{N(j\omega)}{D(j\omega)} = \frac{K}{1 + j\omega\tau}$$

with

$$N(j\omega) = K,$$

$$D(j\omega) = 1 + j\omega\tau.$$

Thus

$$|N(j\omega)| = K; \quad \angle N(j\omega) = 0$$

$$|D(j\omega)| = \sqrt{1 + \omega^2\tau^2}; \quad \angle D(j\omega) = \tan^{-1}(\omega\tau)$$

with the immediate result that, according to Eqs. (9.22a,b):

$$AR = \frac{|N(j\omega)|}{|D(j\omega)|} = \frac{K}{\sqrt{1 + \omega^2\tau^2}}$$

$$\phi = \angle N(j\omega) - \angle D(j\omega) = -\tan^{-1}(\omega\tau)$$

as in Eqs. (9.25a,b).

Example 9.2   **FREQUENCY RESPONSE OF A SECOND-ORDER SYSTEM.**

Use the same procedure of Example 9.1 to obtain the frequency response of a general second-order system and compare this result with that obtained earlier in Eqs. (6.78) and (6.79).

Solution:

For the second-order system:

$$g(s) = \frac{K}{\tau^2 s^2 + 2\zeta\tau s + 1} \tag{6.26}$$

and, therefore, substituting $s = j\omega$ gives, upon some simplification:

$$g(j\omega) = \frac{K}{(1 - \omega^2\tau^2) + 2\zeta\omega\tau j} \tag{9.26}$$

Upon rationalization in the usual manner, we have:

$$g(j\omega) = K\left[\frac{(1 - \omega^2\tau^2) - 2\zeta\omega\tau j}{(1 - \omega^2\tau^2)^2 + (2\zeta\omega\tau)^2}\right]$$

so that the AR and ϕ are now obtained as:

$$AR = |g(j\omega)| = \frac{K}{\sqrt{(1 - \omega^2\tau^2)^2 + (2\zeta\omega\tau)^2}} \tag{9.27a}$$

and

$$\phi = \arg [g(j\omega)], \text{ or } \angle g(j\omega) = \tan^{-1}\left(\frac{-2\zeta\omega\tau}{1 - \omega^2\tau^2}\right) \qquad (9.27b)$$

which are exactly the same as the expressions given earlier in Eqs. (6.78) and (6.79) for the AR and the phase angle of the second-order system.

With the polar form approach, we recognize that in Eq. (9.26):

$$N(j\omega) = K$$

$$D(j\omega) = (1 - \omega^2\tau^2) + (2\zeta\omega\tau)j$$

so that

$$|N(j\omega)| = K \; ; \; \angle N(j\omega) = 0$$

$$|D(j\omega)| = \sqrt{(1 - \omega^2\tau^2)^2 + (2\zeta\omega\tau)^2} \; ; \; \angle D(jw) = \tan^{-1}\left(\frac{2\zeta\omega\tau}{1 - \omega^2\tau^2}\right)$$

We may now use Eq. (9.22) to obtain the required AR and ϕ directly:

$$AR = \frac{|N(j\omega)|}{|D(j\omega)|} = \frac{K}{\sqrt{(1 - \omega^2\tau^2)^2 + (2\zeta\omega\tau)^2}}$$

and

$$\phi = \angle N(j\omega) - \angle D(j\omega) = -\tan^{-1}\left(\frac{2\zeta\omega\tau}{1 - \omega^2\tau^2}\right)$$

as in Eq. (9.27a,b).

One important point to note from these examples is that applying the general frequency-response result greatly facilitates the determination of a

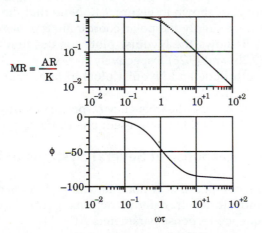

Figure 9.3. Bode diagram for a first-order system.

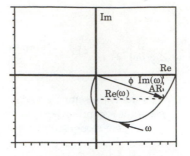

Figure 9.4.   A Nyquist plot of $g(j\omega)$.

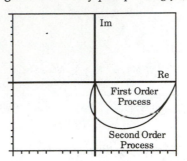

Figure 9.5.   Nyquist plots of first- and second-order process $g(j\omega)$ given by Eqs. (9.23) and (9.26).

system's frequency response. The method presented earlier, which involves first finding the complete sinusoidal response and then dropping off the transient terms, is clearly more tedious.

As we showed in earlier chapters, a natural way to display the frequency response graphically is through the use of a *Bode plot* in which we plot AR and ϕ vs. frequency as shown in Figure 9.3. These plots will be used throughout this book to display and analyze the frequency response of various types of model systems.

There is also another plot which is often used to show the frequency response of a process — the *Nyquist plot*. The Nyquist plot is simply a graph of the real and imaginary parts of $g(j\omega)$ directly in the complex plane as a function of frequency, as shown in Figure 9.4. Note that the phase angle ϕ and amplitude ratio AR are simply the polar coordinate representation of $g(j\omega)$. To illustrate, Figure 9.5 shows the Nyquist plots for the first- and second-order processes in Eqs. (9.25) and (9.26) respectively.

We shall use both Bode and Nyquist plots in later chapters for the design of process control systems.

Before going on to discuss the frequency response of various process systems, we find it profitable, at this point, to present one more very pertinent result.

9.2.3   Frequency Response of Several Systems in Series

Consider the system composed of N noninteracting systems connected in series, each one with individual transfer functions $g_1(s)$, $g_2(s)$, ... , $g_N(s)$, and corresponding frequency-response parameters AR_1, AR_2, ... , AR_N, and ϕ_1, ϕ_2, ... , ϕ_N. The frequency response of this composite system is given by:

$$(AR)_{\text{overall}} = (AR_1)(AR_2) \dots (AR_N) \tag{9.28}$$

a *product* of the individual amplitude ratios of the contributing systems; and:

$$(\phi)_{\text{overall}} = \phi_1 + \phi_2 + \dots + \phi_N \tag{9.29}$$

a *sum* of the individual, contributing phase angles.

This result is most easily established by using the polar form for representing a complex variable. If we recall that the complex variable z whose magnitude and argument are given as $|z|$ and ϕ has the exponential polar form:

$$z = |z| \, e^{j\phi}$$

then establishing Eqs. (9.28) and (9.29) from Eqs. (9.1) and (9.3) and Figure 9.2 is straightforward, and it is left as an exercise to the reader.

As we already know from Chapter 6, $g(s)$, the overall transfer function for the composite system, is given by:

$$g(s) = g_1(s)g_2(s) \dots g_N(s)$$

Substituting $s = j\omega$ now gives:

$$g(j\omega) = g_1(j\omega)g_2(j\omega) \dots g_N(j\omega) \tag{9.30}$$

from where writing each $g_i(j\omega)$ in polar form gives:

$$g(j\omega) = \left| g_1(j\omega) \right| e^{j\phi_1} \left| g_2(j\omega) \right| e^{j\phi_2} \dots \left| g_N(j\omega) \right| e^{j\phi_N}$$

which rearranges to:

$$g(j\omega) = \left| g_1(j\omega) \right| \left| g_2(j\omega) \right| \dots \left| g_N(j\omega) \right| e^{j(\phi_1 + \phi_2 + \dots + \phi_N)} \tag{9.31}$$

If we now write the overall transfer function in its polar form, i.e.:

$$g(j\omega) = |g(j\omega)| \, e^{j\phi} \tag{9.32}$$

a direct comparison of Eqs. (9.32) and (9.31) gives the required result.

It is now important to point out that taking logarithms in Eq. (9.28) gives the following alternative, but equivalent form which is particularly useful in constructing Bode diagrams:

$$\log (AR)_{\text{overall}} = \log AR_1 + \log AR_2 + \dots + \log AR_N \tag{9.33}$$

implying that the logarithm of the overall amplitude ratio is a sum of the logarithms of the contributing system amplitude ratios.

These results will be useful in our subsequent discussions as we start to look at the frequency response of the various classes of process systems. First, let us summarize the implications of these results on the construction of Bode diagrams from complex transfer functions.

Implications for Construction of Bode Diagrams

Any complex transfer function $g(s)$ can be broken down into simpler individual components $g_1(s), g_2(s), \ldots, g_N(s)$, whose frequency-response parameters, AR_1, $AR_2, \ldots, AR_N$, and $\phi_1, \phi_2, \ldots, \phi_N$ can be easily found. As a result of Eqs. (9.28), (9.29) and (9.33), the following are the *classical* rules for constructing the Bode diagram corresponding to the complex transfer function, using the frequency-response characteristics of the simpler components:

1. The steady-state gain of the complex transfer function is extracted and used as a scaling factor for the overall AR plot.
2. The logarithm of the (scaled) overall AR is obtained by summing up the logarithms of the individual contributing AR's.
3. The overall phase angle is obtained by summing up the individual contributing phase angles.

The emphasis on the word *classical* as used above is deliberate: it is meant to convey the fact that the construction of Bode diagrams is hardly ever done by hand any longer; the current practice is to employ computer software packages for this purpose. These rules are therefore primarily of classical interest; but they are presented here for pedagogical reasons also, since they aid the understanding of the frequency-response characteristics of complex systems.

9.3   LOW-ORDER SYSTEMS

9.3.1   First-Order System

As shown earlier, the frequency-response characteristics of a first-order system are given by:

$$AR = \frac{K}{\sqrt{1 + \omega^2 \tau^2}} \tag{9.34a}$$

and

$$\phi = \tan^{-1}(-\omega\tau) \tag{9.34b}$$

Taking logarithms in Eq. (9.34) we have:

$$\log \frac{AR}{K} = -\frac{1}{2}\log(1 + \omega^2\tau^2) \tag{9.35}$$

and for purposes of the Bode diagram, we shall retain K as a scaling factor for AR, and use the *magnitude ratio*, MR $= AR/K$, as a scaled amplitude ratio. We will also use $\omega\tau$ as the abscissa (i.e., use τ as a scaling factor for the frequency axis).

Asymptotic Considerations

Amplitude Ratio

Concerning the AR expressions, we make the following observations about its asymptotic behavior:

1. As $\omega\tau \to 0$, $AR/K \to 1$, and $\log AR/K \to \log 1$
    Thus, at very low frequencies, the log-log plot of MR $=$ AR/K vs. $\omega\tau$ tends to the horizontal line represented by:

$$MR = \frac{AR}{K} = 1 \qquad (9.36)$$

    This line is called the *low-frequency asymptote* for this reason.

2. As $\omega\tau \to \infty$, $\log AR/K \to - \log \omega\tau$ and the log-log plot of MR $=$ AR/K vs. $\omega\tau$ tends to the line represented by:

$$\log \frac{AR}{K} = - \log \omega\tau \qquad (9.37)$$

    a straight line with slope –1. Because this situation arises at very high frequencies, this line is known as the *high-frequency asymptote.*

One final point: the high-frequency asymptote intersects the low-frequency asymptote at the point where $\omega\tau$ takes on the value of 1 (since the low-frequency asymptote, for which $AR/K = 1$, must have this point in common with the high-frequency asymptote, for which $\log AR/K = - \log \omega\tau$). The value of the frequency at this point is ω_n, known as the *corner frequency* or the natural frequency. Thus, since:

$$\omega_n\tau = 1$$

we have the expression for the corner frequency:

$$\omega_n = \frac{1}{\tau} \qquad (9.38)$$

the reciprocal of the system time constant. This is illustrated in Figure 9.3.

Phase Angle

The important aspects of the asymptotic behavior of the phase angle are as follows:

1. As $\omega\tau \to 0$, $\phi \to 0$, and
2. As $\omega\tau \to \infty$, $\phi \to - 90°$
3. At the corner frequency, $\omega\tau = 1$, and $\phi = - \tan^{-1}(1) = - 45°$

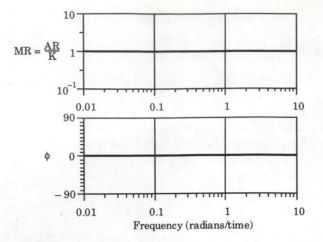

Figure 9.6.   Bode diagram for a pure gain system.

These facts tell us that the phase angle for the first-order system becomes asymptotic to –90° as frequency increases, never falling below this value at any time.

All these points are illustrated in the Bode diagram given in Figure 9.3.

9.3.2   Pure Gain System

The transfer function for this process, as we know, is:

$$g(s) \ = \ K \tag{9.39a}$$

so that

$$g(j\omega) \ = \ K \tag{9.39b}$$

and in terms of the form in Eq. (9.14a), we have Re = K and Im = 0, with the immediate implication that:

$$AR \ = \ K \tag{9.40a}$$

and

$$\phi \ = \ 0 \tag{9.40b}$$

Thus the Bode diagram for this system has the simple form shown in Figure 9.6. The important point to note here is how the frequency response of the pure gain system corresponds to that of the first-order system at low frequency.

9.3.3   Pure Capacity System

Since the transfer function in this case is:

$$g(s) \ = \ \frac{K^*}{s} \tag{5.44}$$

we have:

$$g(j\omega) = \frac{K^*}{j\omega}$$

Rationalization now gives:

$$g(j\omega) = -\frac{K^* j}{\omega} \tag{9.41}$$

so that in this case, $\text{Re}(\omega) = 0$ and $\text{Im}(\omega) = -K^*/\omega$; therefore:

$$AR = \frac{K^*}{\omega} \tag{9.42a}$$

and

$$\phi = -\tan^{-1}(\infty)$$

giving

$$\phi = -90° \tag{9.42b}$$

It is important to note that these results are exactly the same as earlier obtained in Eq. (5.54).

Observe from Eq. (9.42a) that:

$$\log\frac{AR}{K^*} = -\log\omega \tag{9.43}$$

Thus the log-log plot of the magnitude ratio vs. ω for the pure capacity system will be a straight line with slope −1; and the phase angle plot will be a horizontal line fixed at $\phi = -90°$. The actual Bode diagram is shown in Figure 9.7.

It is important to note that both the MR and ϕ behavior correspond to the asymptotic behavior of a first-order system at high frequency.

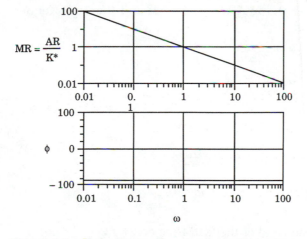

Figure 9.7. Bode diagram for a pure capacity system.

9.3.4 First-Order Lead System

The first-order lead process has the transfer function form:

$$g(s) = K(\xi s + 1) \tag{9.44a}$$

so that:

$$g(j\omega) = K(\xi\omega j + 1) \tag{9.44b}$$

Finding the magnitude and argument of this complex number whose real and imaginary parts are directly available without rationalization is a very easy exercise; the results are:

$$AR = K\sqrt{1 + \omega^2\xi^2} \tag{9.45a}$$

and

$$\phi = \tan^{-1}(\omega\xi) \tag{9.45b}$$

Asymptotic Considerations

Amplitude Ratio

The AR expression has the following asymptotic behavior:

1. As $\omega\xi \to 0$, $AR/K \to 1$, and $\log AR/K \to \log 1$

    Thus, at very low frequencies, the log-log plot of MR = AR/K vs. $\omega\xi$ tends to the horizontal line:

    $$MR = \frac{AR}{K} = 1$$

    Thus the *low-frequency asymptote* for this system is the same as that for the first-order system.

2. As $\omega\xi \to \infty$, $\log AR/K \to \log \omega\xi$ and the high-frequency asymptote is the line:

    $$\log\frac{AR}{K} = \log \omega\xi \tag{9.46}$$

    a straight line with slope +1. This is the exact opposite of the case with the first-order system.

3. The corner frequency in this case is:

    $$\omega_c = \frac{1}{\xi} \tag{9.47}$$

    the reciprocal of the lead time constant.

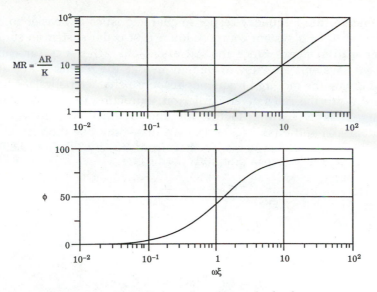

Figure 9.8. Bode diagram for a first-order lead system.

Phase Angle

The phase angle asymptotic behavior is as follows:

1. As $\omega\xi \to 0$, $\phi \to 0$
2. As $\omega\xi \to \infty$, $\phi \to +90°$
3. At the corner frequency, $\omega\xi = 1$ and $\phi = \tan^{-1}(1) = +45°$

The Bode diagram for this system is shown in Figure 9.8.
There are some very important points to note about this frequency response:

1. The magnitude ratio for the first-order lead process is greater than or equal to 1; with the first-order system, MR is *less* than or equal to 1.

2. The phase angle is always greater than or equal to zero, with the implication that the output sine wave actually leads the input sine wave. This is the basis for the term "lead system."

It is interesting to note that just as the transfer function for the first-order lead is like the inverse of the transfer function for a first-order system, the Bode diagrams for these systems are mirror images of each other; for the magnitude ratio plots the reflection is about the line MR = 1, while for the phase angle plots, the reflection is about the line $\phi = 0$.

9.3.5 Lead/Lag System

The lead/lag system has a Laplace domain transfer function given by:

$$g(s) = K\frac{(\xi s + 1)}{(\tau s + 1)}$$ (5.55)

For this system, the frequency-response characteristics are easier to obtain by considering it to be a system composed of a first-order system in series with a first-order lead system. Since the AR and ϕ characteristics for each of the "component" systems are known, we may invoke the results in Eqs. (9.28) and (9.29) and obtain the required frequency-response characteristics at once.

According to the rules given earlier on, extracting the steady-state gain K (to be used as a scaling factor for the overall AR) leaves us with $g_1(s) = 1/(\tau s + 1)$ and $g_2(s) = (\xi s + 1)$. We may now use the results obtained earlier for each of these systems, so that, from Eqs. (9.28) and (9.29), the AR and ϕ for the lead/lag system are given by:

$$\frac{AR}{K} = \frac{\sqrt{1 + \omega^2\xi^2}}{\sqrt{1 + \omega^2\tau^2}} \tag{9.48a}$$

and

$$\phi = \tan^{-1}(\omega\xi) - \tan^{-1}(\omega\tau) \tag{9.48b}$$

and from Eqs. (9.48a) and (9.33) we have:

$$\log\frac{AR}{K} = \frac{1}{2}\log(1 + \omega^2\xi^2) - \frac{1}{2}\log(1 + \omega^2\tau^2) \tag{9.49}$$

The asymptotic characteristics of this frequency response are a combination of the asymptotic characteristics of the lead system and those of the first-order system. It is left as an exercise to the reader to confirm that:

1. As $\omega \to 0$, $AR/K \to 1$, and as $\omega \to \infty$, $AR/K \to \xi/\tau = \rho$
2. As $\omega \to 0$, $\phi \to 0$, and as $\omega \to \infty$, $\phi \to 0$

Thus for the lead/lag system, the high- and low-frequency asymptotes are the same for the phase angle plot, and *similar* in character for the AR plot (similar in the sense that both the low- and high-frequency asymptotes are horizontal straight lines).

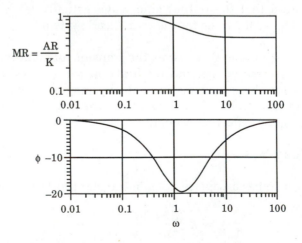

Figure 9.9. Bode diagram for the lead/lag system when $\xi < \tau$; ($\xi = 0.5$, $\tau = 1$).

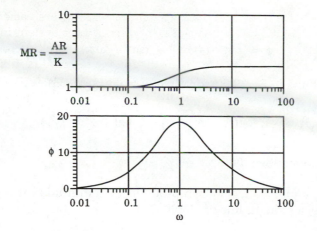

Figure 9.10. Bode diagram for the lead/lag system $\xi > \tau$, $(\xi = 2, \tau = 1)$.

It is easy to see, once again, that there are three possible types of behavior arising from the lead/lag system, depending on the relationship between the lead and lag time constants.

Eliminating the trivial case when $\xi = \tau$ (a pure gain system), the two possible Bode diagrams, when $\xi < \tau$ and when $\xi > \tau$, are shown in Figures 9.9 and 9.10 respectively for the specific case when $\tau = 1$ and ξ takes on the values 0.5 and 2. Note the high-frequency behavior when the lag term dominates (Figure 9.9) and when the lead term dominates (Figure 9.10). The asymptotic value of MR = AR/K lies either above the line MR = 1, or below depending on whether $\rho = \xi/\tau$ is larger or smaller than 1.

9.4 HIGHER ORDER SYSTEMS

One of the main points emerging from the discussion in Chapter 6 is the fact that, with the exception of the underdamped second-order system, higher order systems are composed of combinations of lower order systems. This being the case, the issue of obtaining the frequency-response characteristics of such systems is greatly simplified by the application of the results (and rules) given in Section 9.2. Let us illustrate with some higher order systems that frequently arise.

9.4.1 Two First-Order Systems in Series

For two first-order systems in series, the transfer function is given as:

$$g(s) = \frac{K_1 K_2}{(\tau_1 s + 1)(\tau_2 s + 1)} \tag{6.10}$$

and for $K = K_1 K_2$, we may apply the results of Section 9.2 to immediately obtain the following:

$$\frac{AR}{K} = \frac{1}{\sqrt{1 + \omega^2 \tau_1^2}} \frac{1}{\sqrt{1 + \omega^2 \tau_2^2}} \tag{9.50a}$$

and

$$\phi = \tan^{-1}(-\omega\tau_1) + \tan^{-1}(-\omega\tau_2) \qquad (9.50b)$$

Taking logarithms in Eq. (9.50), or directly from Eq. (9.33), we have:

$$\log\frac{AR}{K} = -\frac{1}{2}\log\left(1 + \omega^2\tau_1^2\right) - \frac{1}{2}\log\left(1 + \omega^2\tau_2^2\right) \qquad (9.51)$$

It is now easy to see that:

1.   The high-frequency asymptote of the amplitude ratio plot is, in this case, a straight line of slope –2;

2.   The low-frequency asymptote is the same as for the first-order system;

3.   The phase angle approaches –180° at high frequency, and approaches 0° at low frequency.

To illustrate how the Bode diagram is generated by combining the AR and ϕ of two first-order elements, we present Figure 9.11, for the system:

$$g(s) = \frac{1}{(s + 1)(5s + 1)} \qquad (9.52)$$

It is important to emphasize again that, as far as the construction of Bode diagrams go, employing the methods illustrated by this example are of historical, and pedagogical interest only; the typical computer generated plot will show only the overall AR and ϕ curves.

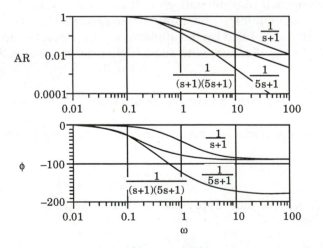

Figure 9.11.   Bode diagram for two first-order systems in series.

9.4.2   Second-Order System

For the general second-order system, from the earlier results given in Eqs. (6.78) and (6.79) (and confirmed in Example 8.2), we have:

$$\frac{AR}{K} = \frac{1}{\sqrt{(1 - \omega^2\tau^2)^2 + (2\zeta\omega\tau)^2}} \qquad (9.53a)$$

and

$$\phi = \tan^{-1}\left(\frac{-2\zeta\omega\tau}{1 - \omega^2\tau^2}\right) \qquad (9.53b)$$

Asymptotic Considerations

Considering the expressions in Eq. (9.53) as functions of $\omega\tau$, with ζ as a parameter, the following asymptotic characteristics are worthy of note:

Amplitude Ratio

1. As $\omega\tau \to 0$, $AR/K \to 1$, and $\log AR/K \to \log 1$
    Thus, the low-frequency asymptote is the horizontal line:

$$\frac{AR}{K} = 1 \qquad (9.54)$$

2. As $\omega\tau \to \infty$, $\log AR/K \to -2\log \omega\tau$ so that the high-frequency asymptote will be the line:

$$\log \frac{AR}{K} = -2\log \omega\tau \qquad (9.55)$$

    a straight line with slope –2.

It is important to point out that the high-frequency asymptote intersects the low-frequency asymptote when $\omega\tau = 1$. However, the parameter τ for the second-order system is defined as the natural period, or the reciprocal of the natural frequency. Thus from this definition, it is easy to see that value of the corner frequency at this point is the same as the natural frequency of oscillation ω_n.

Phase Angle

1. As $\omega\tau$ becomes much less than 1:

$$\tan \phi \approx -2\zeta\omega\tau$$

    so that as $\omega\tau \to 0$, $\phi \to 0$.

2. When $\omega\tau = 1$, regardless of the value of ζ, $\tan \phi = -\infty$, so that $\phi = -90°$.

3. As $\omega\tau \rightarrow \infty$,

$$\phi \rightarrow \tan^{-1}\left(\frac{2\zeta}{\omega\tau}\right) = \tan^{-1}(0) = -180°$$

implying that the phase angle for the second-order system becomes asymptotic to $-180°$ as frequency increases, never falling below this value at any time.

Note also that these facts imply that the various phase angle plots, obtained for different values of the parameter ζ pass through the *same* point at $\phi = -90°$ when $\omega\tau = 1$. The Bode diagram is shown in Figure 9.12.

One peculiar characteristic of the frequency response of a second-order system is the fact that the AR plot shows a "hump" for certain values of ζ. By differentiating Eq. (9.53a) with respect to $\omega\tau$, and setting to zero, we may use elementary calculus to determine that this phenomenon, known as *resonance*, occurs for:

$$\zeta < \frac{1}{\sqrt{2}} = 0.707 \tag{9.56}$$

The frequency at the point where the AR plot passes through this maximum, called the *resonance frequency*, ω_r, is:

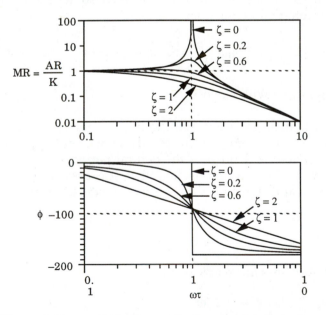

Figure 9.12.   Bode diagram for the second-order system.

$$\omega_r = \frac{1}{\tau}\sqrt{1 - 2\zeta^2} \tag{9.57}$$

This resonance phenomenon leads to maximum values of MR > 1, with increasingly pronounced MR peaks as the damping coefficient ζ gets smaller. In the undamped case, $\zeta = 0$, the MR peak actually becomes infinite.

Note finally from Eq. (9.57) that the resonance frequency gets closer and closer to the natural frequency of oscillation as ζ gets smaller.

9.4.3 Other Higher Order Systems

It is now a fairly easy matter to generalize the results obtained thus far. We have that:

The Bode diagram for the Nth-order system has for the amplitude ratio high-frequency asymptote a straight line of slope $-N$, and for the low-frequency asymptote, the horizontal line: AR/K = 1. The phase angle plot starts at 0° at zero frequency and asymptotically approaches $-(90 \times N)°$ at very high frequencies.

The complete Bode diagram for an Nth-order system will have the properties of the lower order systems of which it is composed, and will therefore not be much different in character from the ones we have already encountered.

Systems with Zeros

In much the same way that the step responses of systems with zeros exhibit interesting characteristics, so also do their frequency responses.

Rather than give a general treatment, we will illustrate some of the several possibilities by presenting some example Bode diagrams for the (2,1)-order system:

$$g(s) = \frac{K\left(\xi_1 s + 1\right)}{\left(\tau_1 s + 1\right)\left(\tau_2 s + 1\right)} \tag{6.83}$$

The Bode diagram in Figure 9.13 is for the system:

$$g(s) = \frac{(0.5s + 1)}{(s + 1)(4s + 1)} \tag{9.58}$$

for which $\xi_1 < \tau_1 < \tau_2$.

Some of the important aspects to note here are:

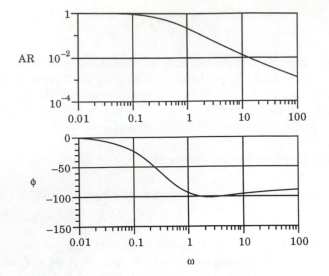

Figure 9.13. Bode diagram for the (2,1)-order system with $\xi_1 < \tau_1 < \tau_2$.

1. The high-frequency amplitude ratio asymptote is a straight line of slope -1; and the phase angle tends asymptotically to $-90°$ at high frequencies. It is left as an exercise to the reader to confirm that this is indeed what should be expected when two lag terms are combined with a lead term.

2. The phase angle plot shows an "undershoot," with the value of ϕ falling below the ultimate value of $-90°$ over a certain range of frequency values. It can be shown that whenever $\xi_1 < \tau_1 < \tau_2$, i.e., the lag time constants are all larger than the lead time constant (as is the case here), this phenomenon will be in evidence.

The Bode diagram for the system:

$$g(s) = \frac{(2s + 1)}{(s + 1)(4s + 1)} \tag{9.59}$$

for which $\tau_1 < \xi_1 < \tau_2$ is shown in Figure 9.14.

The most significant characteristic of this Bode diagram is that it can easily be mistaken for that of a purely first-order system. This will always be the case whenever the value of the lead time constant lies in between the values of the two lag time constants.

When the lead time constant is larger than both lag time constants, as with the following system:

$$g(s) = \frac{(8s + 1)}{(s + 1)(4s + 1)} \tag{9.60}$$

we recall first that, as was shown in Figure 6.15 (for this same system), the step response overshoots the final steady-state value.

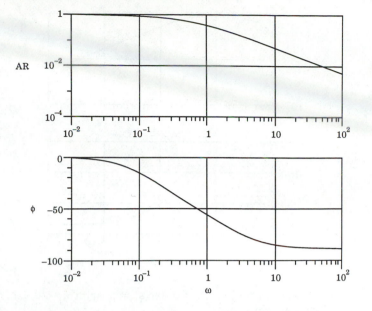

Figure 9.14. Bode diagram for the (2,1)-order system with $\tau_1 < \xi_1 < \tau_2$.

The Bode diagram for this system is shown in Figure 9.15. It is interesting to note that both the AR and ϕ aspects of this Bode diagram exhibit "overshoot." The AR plot overshoots the value of 1 over a certain range of frequency values, and the phase angle is actually positive for some low-frequency values. This is, of course, due to the fact that the influence of the lead term is stronger in this case than in the earlier two cases.

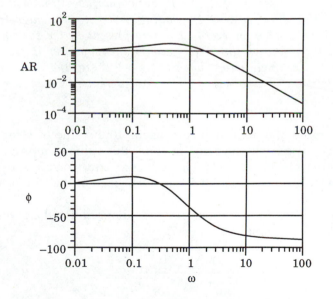

Figure 9.15. Bode diagram for the (2,1)-order system with a stronger lead influence; $\tau_1 < \tau_2 < \xi_1$.

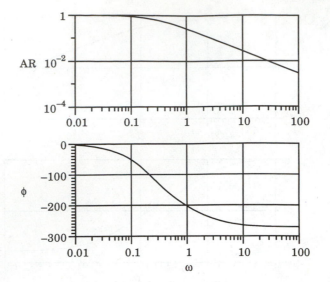

Figure 9.16. Bode diagram for the (2,1)-order system with $\xi_1 < 0$; i.e., an inverse-response system.

Inverse-Response Systems

When the zero present in the system transfer function is a RHP zero, i.e., $\xi_1 < 0 < \tau_1 < \tau_2$, we already know that the system will exhibit inverse response. In terms of the effect of the RHP zero on the frequency response, the main point of interest is summarized below:

> *Whereas the presence of a LHP zero improves the phase angle characteristics of a system by adding +90° to the high-frequency asymptotic value of φ for every LHP zero present in the system transfer function, the effect of the RHP zero is the opposite: it worsens the phase angle characteristics by adding −90° to the high-frequency asymptotic value of φ for every RHP zero present in the system transfer function.*

Thus, for example, the high-frequency asymptotic value of φ for the second-order system by itself, with no zeros, is −180°; this value is improved to −90° for the second-order system with a single LHP zero, and worsened to −270° for the corresponding system with one RHP zero. The sign of the zero has no effect in the amplitude ratio.

For the purpose of illustration, the Bode diagram for the inverse-response system used as an example in Chapter 7:

$$g(s) = \frac{(-3s + 1)}{(2s + 1)(5s + 1)} \tag{7.16}$$

is shown in Figure 9.16. Note the asymptotic value of the phase angle at high frequencies tends towards −270°.

9.4.4   Time-Delay Systems

For the pure time-delay system:

$$g(s) \; = \; e^{-\alpha s} \tag{8.15}$$

with frequency-response transfer function given by:

$$g(j\omega) \; = \; e^{-j\alpha\omega} \tag{9.61}$$

which, when compared with the polar form:

$$g(j\omega) \; = \; |g(j\omega)| \, e^{\,j\phi}$$

establishes at once the following frequency-response characteristics:

$$|g(j\omega)| = AR \; = \; 1 \tag{9.62a}$$

and

$$\phi \; = \; -\alpha\omega \tag{9.62b}$$

once again confirming the results given earlier in Eq. (8.19). The Bode diagram is shown in Figure 9.17.

The most important features are the constant, unity AR characteristics, and the perpetually increasing nature of the phase lag. This is in contrast to each of the systems we have considered so far, where there has always been a limiting value to which the phase angle tends at very high frequency: the pure time-delay system is unique in that its phase angle decreases indefinitely as frequency increases.

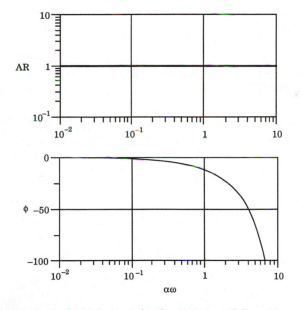

Figure 9.17 Bode diagram for the pure time-delay system.

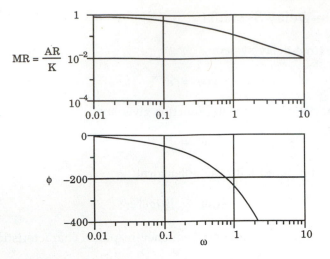

Figure 9.18.   Bode diagram for a first-order-plus-time-delay system.

For any system with an associated delay, we note from Eq. (9.62a) that the logarithm of the AR contributed by the time-delay element is zero; we therefore have the following result:

If AR* is the amplitude ratio of the undelayed portion of a time-delay system, with the corresponding phase angle of ϕ^*, then the frequency response of the system including the delay element $e^{-\alpha s}$ is:

$$AR = AR^* \tag{9.63a}$$

and

$$\phi = \phi^* - \alpha\omega \tag{9.63b}$$

In other words, the AR characteristics of a system are unaffected by the presence of a time delay, but the phase angle characteristics are considerably altered.

For the purposes of illustration, Figure 9.18 shows the Bode diagram for the first-order-plus-time-delay system used in Example 8.1:

$$g(s) = \frac{0.66\, e^{-2.6s}}{6.7s + 1} \tag{8.32}$$

Observe that the first-order AR characteristics are still retained; but the phase angle no longer has the −90° limiting value. The fact that the phase angle no longer has a limiting value is due directly to the presence of the time-delay element.

9.5 FREQUENCY RESPONSE OF FEEDBACK CONTROLLERS

Listed below with their corresponding transfer functions are the four types of feedback controllers we will deal with later on in Part IV:

1. Proportional (P) Controller:

$$g_c(s) = K_c \qquad (9.64)$$

2. Proportional + Integral (PI) Controller:

$$g_c(s) = K_c \left(1 + \frac{1}{\tau_I s} \right) \qquad (9.65)$$

3. Proportional + Derivative (PD) Controller:

$$g_c(s) = K_c \left(1 + \tau_D s \right) \qquad (9.66)$$

4. Proportional + Integral + Derivative (PID) Controller:

$$g_c(s) = K_c \left(1 + \frac{1}{\tau_I s} + \tau_D s \right) \qquad (9.67)$$

Close observation will reveal that the P controller has identical characteristics with the pure gain system, while the PD controller has identical characteristics with the first-order lead system Eq. (9.42), which have all been previously discussed.

It is left as an exercise to the reader to use the methods we have discussed above to obtain the frequency-response characteristics of the PI and the PID controllers, and confirm that the Bode diagrams are as given in Figures 9.19(a) and (b). (See also problems at the end of the chapter.)

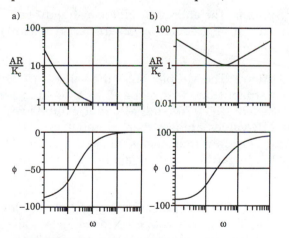

Figure 9.19.   Bode diagrams for (a) the PI controller and (b) the PID controller.

9.6   SUMMARY

We have studied the frequency-response characteristics of the various classes of processes in this chapter, deriving the individual AR and ϕ expressions, and studying their respective Bode diagrams. We have also introduced the Nyquist plot as an alternative method of displaying the frequency response.

Looking back, we should now be able to see why it is true that the fundamental features of these process systems are fully characterized by their frequency responses. Firstly, we have shown that the frequency-response characteristics are immediately obtained from the magnitude and argument of $g(j\omega)$, the process frequency-response transfer function. We have seen that the frequency response of the process, represented by $g(j\omega)$, fully characterizes a linear process.

Even though we have presented several technical pieces of information useful in the construction of Bode diagrams (especially for systems with complex transfer functions) it should be kept in mind that the standard practice is to employ computer software packages for generating such diagrams.

Since Bode diagrams are just as characteristic of individual classes of linear systems as their transform-domain transfer functions, it is important that the essential aspects of these diagrams for each class of systems be well known. It is perhaps helpful to summarize these salient features here:

First-Order System: The MR plot is 1 at very low frequencies and approaches a straight line with slope –1 as $\omega \to \infty$. The phase angle plot starts out at 0° at very low frequencies, takes the value $-45°$ at the corner frequency $\omega_n = 1/\tau$, and approaches a limiting value of $-90°$ as $\omega \to \infty$.

Pure Gain System: Constant MR = 1 and $\phi = 0$ at all frequencies, characteristics similar to those of the first-order system as $\omega \to 0$.

Pure Capacity System: The entire MR plot is a straight line with slope –1 and ϕ is constant at –90° at all frequencies, characteristics similar to those of the first-order system as $\omega \to \infty$.

Second-Order System: The MR plot takes the value 1 at very low frequencies and approaches a straight line with slope –2 as $\omega \to \infty$; in between, for values of damping coefficient $\zeta < 0.707$, the MR plot exhibits resonance; for other values of ζ, the MR values are less than or equal to 1. The phase angle plot starts out at 0° at very low frequencies, takes the value $-90°$ at the natural frequency $\omega_n = 1/\tau$ (regardless of the value of ζ), and approaches a limiting value of $-180°$ as $\omega \to \infty$.

Nth-Order System: The MR plot takes the value 1 at very low frequencies and approaches a straight line with slope $-N$ as $\omega \to \infty$. The phase angle plot starts out at 0° at very low frequencies and approaches a limiting value of $-(90 \times N)°$ as $\omega \to \infty$.

First-Order Lead System: The MR value is greater than or equal to 1 at all frequencies, and the phase angle is always positive. The MR plot is 1 at very low frequencies and approaches a straight line with slope + 1 as $\omega \to \infty$. The phase angle plot starts out at 0° at very low frequencies and approaches a limiting value of $+ 90°$ as $\omega \to \infty$. The Bode plot for this system looks like a mirror image of that for the first-order (lag) system.

Time Delay: MR = 1 for all frequencies while ϕ has monotonically decreasing phase angle characteristics.

It is also advantageous to be familiar with the effect of system zeros on the frequency response; LHP zeros in general improve the phase angle characteristics, while these characteristics are worsened by RHP zeros.

We should mention, finally, that this chapter has actually raised some issues that have very significant bearings on controller design for process systems. We will be recalling several of these issues in Chapter 15, where the application of frequency-response analysis to conventional feedback controller design will be discussed.

REVIEW QUESTIONS

1. The complete response of a linear system to a sinusoidal input consists of two parts: what are they and which part is studied in "Frequency-Response Analysis"?

2. What is the main concern of frequency-response analysis?

3. What is the fundamental frequency-response analysis result, and what is its main benefit?

4. What is a Bode plot and how is it different from a Nyquist plot?

5. How are the amplitude ratio and phase angle characteristics of the individual elements of a composite system related to the corresponding overall system characteristics?

6. Why is the reciprocal of the first-order system time constant referred to as the "corner frequency"?

7. Which portion of the frequency response of a first-order system is related to the frequency response of the pure capacity system? Which portion is related to the frequency response of the pure gain system?

8. From a frequency-response perspective, why is the transfer function in Eq. (9.44) referred to as a "lead element" and the transfer function for a first-order system (with $K = 1$) referred to as a "lag element"?

9. What key parameter determines the shape of the lead/lag system's Bode diagram?

10. What conditions must the damping coefficient of a second-order system satisfy in order that the phenomenon of resonance be evident in its Bode diagram?

11. A common misconception about underdamped second-order systems is that they *all* exhibit resonance. What do you think is responsible for this misconception? Give a range of values of the damping coefficient for which the second-order system will be underdamped *without* exhibiting resonance.

12. What is the difference between the effect of a left-half plane zero and that of a right-half plane zero on the phase angle characteristics of a process?

13. How can you distinguish the frequency response of a system with a strong lead from that of a system that exhibits resonance?

14. What is unique about the frequency-response behavior of a time-delay system?

PROBLEMS

9.1 Establish that the complex number:

$$z = a + bj$$

may be expressed in the exponential polar form:

$$z = |z|e^{j\phi} \tag{P9.1}$$

where

$$|z| = \sqrt{a^2 + b^2} \tag{P9.2a}$$

$$\phi = \tan^{-1}\left(\frac{b}{a}\right) \tag{P9.2b}$$

9.2 The derivation of the fundamental frequency-response analysis result given in the main text, from Eq. (9.5) through Eq. (9.19), was for the case when the transfer function did not involve a time delay. Consider the situation in which:

$$g(s) = g^*(s)\,e^{-\alpha s}$$

and rederive these results for the time-delay system and show that the expressions given for AR and ϕ in Eq. (9.4) hold regardless of the presence of time delays.

9.3 (a) Establish Eqs. (9.56) and (9.57).
(b) Plot the Nyquist diagrams for the systems whose transfer functions are given as follows:

$$g_1(s) = \frac{1}{s^2 + 4s + 1}$$

$$g_2(s) = \frac{1}{s^2 + 0.2s + 1}$$

Is the effect of the resonance exhibited by System 2 obvious?

9.4 Show that the response of a pure capacity system with transfer function:

$$g(s) = \frac{K}{s}$$

to a sinusoidal input $u(t) = A \sin \omega t$ is exactly equivalent to the unit step response of an *undamped* second-order system; specify the undamped system gain.

9.5 Obtain Bode plots for the system whose transfer function is given by:

$$g(s) = \frac{1}{bs}\left(1 - e^{-bs}\right) \tag{P9.3}$$

for $b = 1, 0.1, 0.01$ and compare with the Bode plot for $g(s) = 1$. How can these results help you in assessing the effectiveness of approximating the ideal impulse function with a rectangular pulse function?

9.6 (a) *Sketch* the Bode diagram for each of the systems whose transfer functions are given below. Be sure to identify clearly on each sketch all the important distinguishing characteristics:

$$g_1(s) = \frac{K}{as^3 + bs^2 + cs + d} \tag{P9.4}$$

$$g_2(s) = \frac{Ke^{-\alpha s}}{s^2 + s + 1} \tag{P9.5}$$

(b) A pulse test identification experiment performed on a process gave rise to the Bode diagram shown below in Figure P9.1.

Postulate a possible transfer function form that you believe is consistent with the features displayed by this Bode diagram. *Support your postulate adequately.*

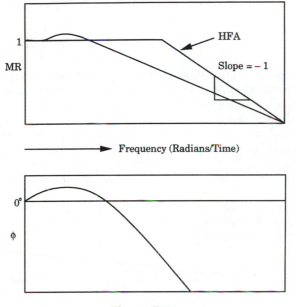

Figure P9.1.

9.7 The following differential equations were given in Problem 8.7 (Chapter 8) for a chemical reactor operating under two different feed conditions:

$$\frac{dy}{dt} = -2y(t) + 0.5_1 y(t - 2) + 3u(t) \tag{P9.6}$$

and

$$\frac{dy}{dt} = -2y(t) + 0.5_1 y(t - 2) + 3u(t - 5) \tag{P9.7}$$

assuming the disturbance is zero. Obtain the transfer functions in each case and from these obtain the corresponding Bode diagrams and compare them. How does the

Bode diagram for the system described by the model in Eq. (P9.6) differ from the standard first-order system Bode diagram?

9.8 The transfer function for a steam-heated heat exchanger is given as:

$$g(s) = \frac{(1 - 0.5e^{-10s})}{(40s + 1)(15s + 1)} \qquad (P9.8)$$

Obtain the corresponding Bode diagram. How does this differ from the Bode diagram of a typical time-delay system of the type considered in Section 9.4.4?

9.9 (a) In each case, by obtaining first, $g(j\omega)$, or otherwise, derive the expressions for the AR and ϕ for the PI and PID controllers whose transfer functions are given in Eqs. (9.65) and (9.67).
(b) Derive expressions for the asymptotic behavior of the AR and ϕ for these controllers and confirm that Figure 9.19 is, in fact, consistent with such results.

9.10 An anaerobic digester used in a small town's waste management facility, which is approximately modeled by the following transfer function:

$$g(s) = \frac{0.52}{(45.5s + 1)(12.2s + 1)} \qquad (P9.9)$$

is subjected to periodic loading which, for simplicity, will be approximated as a pure sine wave: $u(t) = A \sin \omega t$. The process output is the "normalized" digester biomass, and the input is the sludge flowrate into the facility, in "normalized" mass flow units; the unit of time is minutes.

Find the amplitude of the digester output when the periodic loading amplitude is 5 units, at a frequency of $\omega = 0.2$ radians/min. If the frequency were to go up by an order of magnitude to 2 radians/min., with the amplitude remaining constant, by how much will the amplitude of the digester output change?

9.11 For each of the following pairs of transfer functions, plot and compare the Bode diagrams, noting which of the two has the *minimum* phase angle behavior. (These results will be useful in Chapter 18.)

(a) $g_1(s) = \dfrac{5}{6s + 1}$; $\qquad\qquad g_2(s) = \dfrac{5e^{-2s}}{6s + 1}$

(b) $g_1(s) = \dfrac{5(4s + 1)}{(2s + 1)(6s + 1)}$; $\qquad g_2(s) = \dfrac{5(-4s + 1)}{(2s + 1)(6s + 1)}$

(c) $g_1(s) = \dfrac{5}{2s + 1}$; $\qquad\qquad g_2(s) = \dfrac{5}{2s - 1}$

9.12 Repeat Problem 9.11 for Nyquist diagrams.

CHAPTER
10

NONLINEAR SYSTEMS

The processes we have studied so far have all been represented by linear models. Because this entitles them to be classified as linear systems (and linear systems, as we know, are fully characterized by their transfer functions) we have been able to obtain fairly general results by taking full advantage of the convenience offered by the transfer function model form.

However, the truth of the matter is that, to varying degrees of severity, all physical processes exhibit some nonlinear behavior. Furthermore, when a process shows strong nonlinear behavior, a linear model may be inadequate; a nonlinear model will be more realistic. Unfortunately, this improved closeness to reality is attained at a cost: the convenience and simplicity offered by the linear model is sacrificed. To be sure, many processes exhibit only mildly nonlinear dynamic behavior and therefore can be reasonably approximated by linear models. However, even for such systems, the behavior over a wide range of operating conditions will be noticeably nonlinear, and no single linear model is able to represent such behavior adequately.

The issue of analyzing the dynamic behavior of processes represented by nonlinear models is the main concern in this chapter; this is motivated by the fact that situations do arise when it is undesirable to neglect the inherent nonlinearities of a process, be they mild or otherwise. The objective here is to provide an overview of how we can expect to carry out dynamic analysis when the process models are nonlinear.

The nature of nonlinear problems (in all areas of analytical endeavor, not just in Process Dynamics and Control alone) is such that they are not as easily amenable to the general, complete treatment possible with linear systems. To keep nonlinear dynamic analysis on a realistic and practical level therefore usually involves trade-offs between accuracy and simplicity. Keeping this in mind will help us maintain a proper perspective throughout this chapter.

10.1   INTRODUCTION: LINEAR AND NONLINEAR BEHAVIOR IN PROCESS DYNAMICS

There are two basic properties that characterize the behavior of a linear system:

1. The Principle of Superposition

If the response of a process to input I_1 is R_1, and the response to I_2 is R_2, then according to the principle of superposition, the response to $(I_1 + I_2)$ is $(R_1 + R_2)$ *if the system is linear*.

In general, this principle states that the response of a linear system to a sum of N inputs is the same as the sum of the responses to the individual inputs.

As a result of this principle, for example, the response of a linear system to a step change of magnitude A is exactly the same as A times the system's *unit* step response. The same is true for all other types of input functions.

2. Independence of Dynamic Response Character and Process Conditions

The dynamic character of the process response to an input change is independent of the specific operating conditions at the time of implementing the input change, *if the system is linear*. In other words, *identical* input changes implemented at *different* operating steady-state conditions will give rise to output changes of *identical* magnitude and dynamic character (see Figure 10.1).

The principle of superposition also means that from the same initial steady-state conditions, the output change observed in response to a certain positive input change will be a perfect mirror image of the output observed in response to a negative input change of the same magnitude (see Figure 10.2). According to this property, the step response of a linear system, for example, is the same regardless of the actual initial value of the output or input variable, or, for that matter, the direction of the input change, the "step down" giving rise to a perfect mirror image of the "step up" response.

It is because of these two characterizing properties that we have been able to use the linear transfer function representation for carrying out dynamic analysis in the last five chapters.

A nonlinear system does not exhibit any of these properties; the response to a sum of inputs is *not* equal to a sum of the individual responses; the response to a step change of magnitude A is *not* equal to A times the unit step response; the magnitude and dynamic character of the step response *are* dependent on the initial steady-state operating conditions; a "step down" response is *not* a mirror image of a "step up" response; and what is true for the step input function is true

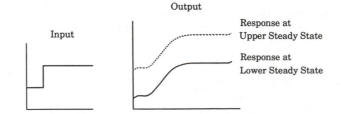

Figure 10.1.   Step response of a truly linear system at two different steady states.

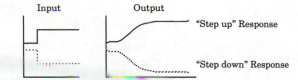

Figure 10.2. "Step down" and "step up" response of a linear system.

for all the other input functions. In particular, the ultimate response of the nonlinear system to a sinusoidal input function is *not* a perfect sinusoidal output function.

The model equations that will represent the dynamic behavior of such systems adequately must necessarily contain nonlinear terms; and immediately, the convenience of being able to completely characterize the system's dynamic behavior by the linear transfer function is lost.

That nonlinear dynamics are inevitable in most chemical processes is a fact that needs very little justification; one only has to call to mind, for example, the equations of chemical phase equilibria when Henry's Law doesn't apply; heat flux expressions for heat transfer by radiation; chemical kinetic rate expressions, especially the well-known Arrhenius expression, all of which are hardly linear. A survey discussing nonlinear processes that occur in the chemical industry may be found in Ref. [3].

To buttress this point, we will now provide a sample of some very simple, very common process systems and their process models.

10.2   SOME NONLINEAR MODELS

1. Outflow from a Tank

Regardless of tank geometry, the outlet flowrate through an orifice at the bottom of a tank is proportional to the square root of the liquid level in the tank. Thus, for the liquid level process considered in Chapter 1 (Figure 1.19), and used as the first example in Chapter 5, a more realistic model is:

$$A \frac{dh}{dt} = F_i - c \sqrt{h} \tag{10.1}$$

more realistic, that is, than Eq. (1.13) which was based on the simplifying approximation that outflow varied linearly with liquid level.

The nonlinearity of this model is, of course, due to the presence of the square root term.

2. Isothermal Continuous Stirred Tank Reactor (CSTR)

If in the isothermal CSTR of Chapter 5 an nth-order reaction were taking place, the mathematical model would read:

$$\frac{d}{dt}(c_A) = \frac{1}{\tau} c_{Af} - \frac{1}{\tau} c_A - k c_A{}^n \tag{10.2}$$

which is nonlinear when $n \neq 1$.

3. The Nonisothermal CSTR

Perhaps the most popular example of a classic nonlinear chemical process is the nonisothermal CSTR.   As we had earlier shown in Chapter 4, the mathematical model for this process system is:

$$\frac{dc_A}{dt} = -\frac{1}{\theta}c_A - k_0\, e^{-\{E/RT\}}c_A + \frac{1}{\theta}c_{Af} \tag{4.26}$$

$$\frac{dT}{dt} = -\frac{1}{\theta}T + \beta k_0\, e^{-\{E/RT\}}c_A + \frac{1}{\theta}T_f - \chi \tag{4.27}$$

Observe that the nonlinearities in this model are due to two factors:

1.   The nonlinear functions of temperature — involving the exponential of the reciprocal.
2.   The product functions — involving products of c_A and a function of T.

The point here is that if such simple, common systems have nonlinear mathematical models, we can imagine the extent of nonlinearities to be encountered in the models for more complex systems, in which more complicated chemical and physical mechanisms are responsible for the observed system behavior. A good catalog of such system models is available, for example, in Refs. [1, 2].

10.3   METHODS OF DYNAMIC ANALYSIS OF NONLINEAR SYSTEMS

Coming to grips with the fact that realistic process models are almost always nonlinear, the next major question we wish to ask ourselves is the following:

How do we analyze the dynamic behavior of processes when they are represented by nonlinear models?

Assuming, for the moment, that we are able to answer this question satisfactorily, there is yet another important question, raised primarily because of our experience with linear systems analysis:

Is there a technique for analyzing the dynamic behavior of nonlinear systems that can be generalized to cover a wide variety of such systems?

The answers to these questions form the basis of the material covered in this chapter.

To analyze the dynamic behavior of a system represented by nonlinear equations, there are usually four alternative strategies to adopt:

1. The rigorous analytical approach
2. Exact linearization by variable transformation
3. Numerical analysis
4. Approximate linearization

We shall consider each of these in turn, with greater emphasis laid on the third and fourth strategies for reasons that will become clear as the discussion progresses.

10.3.1 Rigorous Analytical Techniques

It is well known that very few nonlinear problems can be solved formally, primarily because there is no general theory for the analytical, closed-form solution of nonlinear equations. This means that, unlike linear systems where analytical techniques abound for the general solution of the modeling equations, nonlinear systems face a more restrictive situation.

Nevertheless, there are a few analytical methods for analyzing the dynamic behavior of nonlinear systems; for example, classical methods due to Poincaré, Krylov, and Bogolyubov (see Friedly Ref. [2]), and the techniques of Liapunov, Hopf, Golubitsky, and Arnold (see Refs. [4–7]). These methods will allow one to characterize qualitatively the dynamics of nonlinear systems.

However, these approaches find limited use in process control practice; thus a detailed discussion of their salient features has been excluded from the intended scope of this textbook. Perhaps, with further research, the analytical treatment of nonlinear systems will evolve to a state where it will be more useful for process control.

10.3.2 Variable Transformation

On the basis of the fact that linear systems are easier to analyze than nonlinear ones, it seems attractive to consider the following proposition:

If a system is nonlinear in its original variables (say, x, y, and u) is it not possible that it will be linear in some transformation of these original variables?

If this were possible, the general strategy for dynamic analysis would then involve first carrying out this transformation, then performing dynamic analysis on the linear, transformed version, and finally transforming the result back to the original variables.

Let us illustrate with an example.

Example 10.1 **VARIABLE TRANSFORMATION FOR THE ISOTHERMAL CSTR WITH A SECOND-ORDER REACTION.**

The dynamic behavior of this process is given by:

$$\tau \frac{dc_A}{dt} = -\left(c_A + k\tau c_A{}^2\right) + c_{Af} \tag{10.3}$$

which is nonlinear because of the square term.

Let us now define a new state variable:

$$z = \frac{c_A}{1 + k\tau c_A} \tag{10.4}$$

Since:

$$\frac{dz}{dt} = \frac{dz}{dc_A}\frac{dc_A}{dt}$$

then, according to this definition Eq. (10.4), and from Eq. (10.3) we have:

$$\frac{dz}{dt} = \frac{1}{\left(1 + k\tau c_A\right)^2}\left[-\frac{1}{\tau}(c_A + k\tau c_A^{\,2}) + \frac{1}{\tau}c_{Af}\right] \tag{10.5}$$

which simplifies to:

$$\frac{dz}{dt} = -\frac{1}{\tau}\frac{c_A}{1 + k\tau c_A} + \frac{1}{\tau}\frac{c_{Af}}{(1 + k\tau c_A)^2} \tag{10.6}$$

If we now define another new input variable v as:

$$v = \frac{c_{Af}}{(1 + k\tau c_A)^2} \tag{10.7}$$

and recall Eq. (10.4), Eq. (10.6) becomes:

$$\frac{dz}{dt} = -\frac{1}{\tau}z + \frac{1}{\tau}v \tag{10.8}$$

which is now linear in the new variables z and v defined in terms of the old variables as shown in Eqs. (10.4) and (10.7).
In principle, we may now use Eq. (10.8) to investigate the dynamic behavior of this system and relate it to the original system variables via the expressions in Eqs. (10.4) and (10.7).

As promising as this idea of variable transformations may seem, there are some important questions which indicate that this strategy might be of limited practical utility:

1. How was the transformation in Eq. (10.4) found?
2. Is there a general procedure by which such transformations can always be found, for *all* nonlinear systems?
3. How easily can these transformations be found?

There is, in fact, no general procedure for finding such transformations for all nonlinear systems; and for the restricted classes of systems for which a general procedure exists, it is not easy to apply (cf. Ref. [8]).

There are other related methods based on differential geometry, but these also suffer from similar limitations (cf. Ref. [9]). A good overview of such transformation methods may be found in Ref. [10].

10.3.3 Numerical Solutions and Computer Simulation

Let us remind ourselves that the first step in the analysis of a system's dynamic behavior is the solution of the modeling equations; and the models, as we recall, are usually in the form of differential equations (ordinary or partial). For linear systems, analytical solutions are possible, while for nonlinear systems, analytical solutions are usually not possible; however, there is a fallback position: we can obtain the solution in *numerical* form.

Thus, instead of having an analytical expression that represents the behavior of the process output over time, we make use of a scheme that provides numerical values for this process output at specific points in time.

General Procedure

The general procedure for analyzing the system's behavior numerically is as follows:

1. Discretize the Process Model

This involves taking the original system model, consisting of one, or possibly several, equations of the type:

$$\frac{dx_i}{dt} = f_i(x_1, x_2, ..., x_n, u, d, t) \tag{10.9}$$

and appropriately discretizing it to obtain:

$$x_i(k+1) = f_i^d(x_1(k), x_2(k), ..., x_n(k), u(k), d(k), k, \Delta t) \tag{10.10}$$

A variety of techniques are available for carrying out this step.

2. Recursive (or Iterative) Solution of the Discretized Model

If we are now given the initial values $x_i(0)$ for each of the variables, and the value of the forcing functions, these equations can now be solved recursively for the value taken by each $x_i(k)$ in the next time interval.

Owing to the recursive nature of this aspect of the problem, the digital computer is the perfect tool for doing the job. When the model for a process system is thus solved with the aid of a computer, we are said to have obtained a computer simulation of the process behavior.

There are many specialized techniques for obtaining numerical solutions to differential equation models and these are typically covered in courses on numerical methods and analysis (see also Appendix B). These techniques usually differ in the discretization strategy employed in taking the model from the continuous form in Eq. (10.9) to the discrete form in Eq. (10.10).

In addition, computer packages are available that offer a wide variety of these numerical techniques, thereby facilitating the otherwise potentially tedious task of numerical dynamic analysis.

For the purpose of illustrating the principles involved in numerically obtaining a process response, we now present the following example.

Example 10.2   NUMERICAL STEP RESPONSE OF THE LIQUID LEVEL IN A TANK.

The dynamic behavior of the liquid level in a tank has a model given as:

$$A \frac{dh}{dt} = F_i - c\sqrt{h} \tag{10.1}$$

with a cross-sectional area A of 0.25 m^2 and was initially at steady state with inlet flowrate F_i at 0.3 m^3/min, while the level in the tank h was 0.36 m. It is required to obtain the response of this system when the inlet flowrate is increased to 0.4 m^3/min. Derive an expression to be used for obtaining the numerical value of the level in the tank system at intervals of 0.1 min using the *explicit Euler method* for discretization. The constant c has the value 0.5 m$^{5/2}$/min.

Solution:

Using the given data, the model for this specific process is:

$$\frac{dh}{dt} = 4F_i - 2\sqrt{h} \tag{10.11}$$

Now, with the explicit Euler method (cf. Appendix B), a first derivative is represented in the following discrete form:

$$\frac{dy}{dt} \approx \frac{y(t + \Delta t) - y(t)}{\Delta t}$$

or, taking time t as the kth time interval, with Δt as the width of each of these time intervals:

$$\frac{dy}{dt} \approx \frac{y(k + 1) - y(k)}{\Delta t} \tag{10.12}$$

According to this scheme therefore, Eq. (10.11) becomes:

$$\frac{h(k + 1) - h(k)}{\Delta t} = 4F_i(k) - 2\sqrt{h(k)} \tag{10.13}$$

or

$$h(k + 1) = h(k) + \Delta t \left[4F_i(k) - 2\sqrt{h(k)} \right]$$

and for $\Delta t = 0.1$ as given, the required expression is:

$$h(k + 1) = h(k) + 0.1 \left[4F_i(k) - 2\sqrt{h(k)} \right] \tag{10.14}$$

We may now use this equation along with the given initial condition $h(0) = 0.36$, and set $F_i(0)$ to 0.4 to obtain the response of the liquid level to this change in the input variable. For $k = 0$, for example, we have:

$$h(1) = 0.36 + 0.1(1.6 - 1.2) = 0.4$$

implying that 0.1 minutes after the implementation of the step change in the inlet flowrate, the liquid level will be 0.4 m. A recursive application of the expression:

$$h(k + 1) = h(k) + 0.1\left[1.6 - 2\sqrt{h(k)}\right] \qquad (10.15)$$

gives the value of the liquid level at intervals of 0.1 minutes; a compilation of these values constitute the required numerical step response.

Note that the recursive calculations required by this procedure are best carried out with the aid of a digital computer. There are many special subroutines available for the numerical integration of differential equations, so that details of this procedure are usually handled by the software.

Let us now give a quick recapitulation of the three techniques we have so far discussed for nonlinear dynamic analysis. The first two are clearly of limited practical application and need no further comment; the technique of numerical analysis, even though straightforward, provides limited information because it is usually impossible to infer anything of a general form from the numerical response of a process system. Numerical analysis provides specific *numerical* answers to specific problems; it is not possible to obtain general solutions in terms of arbitrary parameters and unspecified inputs in order to understand the process behavior in a more general fashion, as was the case with linear systems.

The question now is: can we ever achieve this desirable objective of being able to analyze, in a general fashion, the behavior of an arbitrary nonlinear system, in response to unspecified inputs? The answer is *yes*: but at a cost. For observe that the nonlinear process model can be *linearized* around a particular steady-state value, and with this linearized approximation of the true nonlinear model, we can obtain fairly general results regarding the process behavior, but these will only be approximate, never completely representing true behavior accurately.

When faced with the problem of analyzing the behavior of a nonlinear process, therefore, we have a choice to make: if accuracy is more important, then numerical analysis is the recommended approach; but general inferences are ruled out. When the emphasis is more on simplicity, and it is also important to be able to make some general statements about the dynamic behavior, one will then have to sacrifice accuracy and use the method of linearization.

Because the compromise involved in the latter strategy is often considered more advantageous, linearization is the most commonly utilized technique in the analysis of nonlinear systems.

10.4   LINEARIZATION

In the context of process dynamics and control, *linearization* is a term used in general for the process by which a nonlinear system is approximated by a linear one; i.e., the nonlinear process model is somehow approximated by a linear process model.

The most popular technique for obtaining these linear approximations is based on Taylor series expansions of the nonlinear aspects of the process model, as demonstrated below.

Any function $f(x)$ (under certain conditions which are always satisfied by the models of realistic processes) can be expressed in the following power series around the point $x = x_s$:

$$f(x) = f(x_s) + \left(\frac{df}{dx}\right)_{x = x_s} (x - x_s) + \left(\frac{d^2f}{dx^2}\right)_{x = x_s} \frac{(x - x_s)^2}{2!} + \cdots$$

$$+ \left(\frac{d^nf}{dx^n}\right)_{x = x_s} \frac{(x - x_s)^n}{n!} + \cdots \tag{10.16}$$

If we now ignore second- and higher order terms, Eq. (10.16) becomes:

$$f(x) \approx f(x_s) + f'(x_s)(x - x_s) \tag{10.17}$$

which is now linear.

Observe that if $(x - x_s)$ is very small, the higher order terms will be even smaller and the linear approximation is satisfactory. The implication of this linearization process is depicted in Figure 10.3.

The approximation is exact at the point $x = x_s$, remains satisfactory as long as x is close to x_s, and deteriorates as x moves farther away from x_s — the rate of this deterioration being directly related to the severity of the nonlinearity of the function $f(x)$.

This is easily extended to a function of two variables $f(x_1, x_2)$, expanded in a Taylor series around the point (x_{1s}, x_{2s}); the result is:

$$f(x_1, x_2) = f(x_{1s}, x_{2s}) + \left(\frac{\partial f}{\partial x_1}\right)_{\substack{x_1 = x_{1s} \\ x_2 = x_{2s}}} (x_1 - x_{1s}) + \left(\frac{\partial f}{\partial x_2}\right)_{\substack{x_1 = x_{1s} \\ x_2 = x_{2s}}} (x_2 - x_{2s})$$

$$+ \ *higher order terms* \tag{10.18}$$

This purely mathematical idea will now be applied to the problem of nonlinear dynamic analysis.

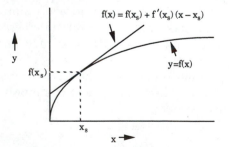

Figure 10.3.   Nonlinear function and linearized approximation.

The General Linearization Problem

Consider the general nonlinear process model:

$$\frac{dx}{dt} = f(x,u) \tag{10.19a}$$

$$y = h(x) \tag{10.19b}$$

where $f(\bullet, \bullet)$ is an arbitrary nonlinear function of the two variables, x, the process state variable, and u the process input; $h(\bullet)$ is another nonlinear function relating the process output, y to the process state variable x.

The linearized approximation of this very general nonlinear model may now be obtained by carrying out a Taylor series expansion of the nonlinear functions around the point (x_s, u_s). This gives:

$$\frac{dx}{dt} = f(x_s, u_s) + \left(\frac{\partial f}{\partial x}\right)_{(x_s, u_s)} (x - x_s) + \left(\frac{\partial f}{\partial u}\right)_{(x_s, u_s)} (u - u_s)$$
$$+ \textit{higher order terms} \tag{10.20a}$$

$$y = h(x_s) + \left(\frac{\partial h}{\partial x}\right)_{x_s} (x - x_s) + \textit{higher order terms} \tag{10.20b}$$

Ignoring the higher order terms now gives the linear approximation:

$$\frac{dx}{dt} = f(x_s, u_s) + a(x_s, u_s)(x - x_s) + b(x_s, u_s)(u - u_s) \tag{10.21a}$$

$$y = h(x_s) + c(x_s)(x - x_s) \tag{10.21b}$$

where

$$a(\bullet, \bullet) = \left(\frac{\partial f}{\partial x}\right)_{(x_s, u_s)}$$

$$b(\bullet, \bullet) = \left(\frac{\partial f}{\partial u}\right)_{(x_s, u_s)}$$

$$c(\bullet) = \left(\frac{dh}{dx}\right)_{x_s}$$

It is customary to express Eqs. (10.21a,b) in terms of deviation variables:

$$\tilde{x} = x - x_s$$

$$\tilde{u} = u - u_s$$

$$\tilde{y} = y - y_s$$

If in addition to this, the linearization point (x_s, u_s) is chosen to be a steady-state operating condition, then observe from the definition of a steady state and from Eq. (10.19) that both dx_s/dt and $f(x_s, u_s)$ will be zero. Equations (10.21a,b) then become:

$$\frac{d\tilde{x}}{dt} = a\tilde{x} + b\tilde{u} \tag{10.22a}$$

$$\tilde{y} = c\tilde{x} \tag{10.22b}$$

where, for simplicity, the arguments have been dropped from a, b, and c. A transform-domain transfer function model corresponding to Eq. (10.22) may now be obtained by the usual procedure; the result is:

$$\tilde{y}(s) = \left[\frac{c(x_s)\, b(x_s, u_s)}{s - a(x_s, u_s)}\right] \tilde{u}(s) \tag{10.23}$$

with the transfer function as indicated in the square brackets. This transfer function should provide an approximate linear model valid in a region close to (x_s, u_s).

The principles involved in obtaining approximate linear models by linearization may now be summarized as follows:

1. Identify the functions responsible for the nonlinearity in the system model.
2. Expand the nonlinear function as a Taylor series around a steady state, and truncate after the first-order term.
3. Reintroduce the linearized function into the model; simplify, and express the resulting model in terms of deviation variables.

The application of these principles to specific problems is usually much simpler than one would think at first, as we now demonstrate with the following examples.

Example 10.3   LINEARIZATION OF A NONLINEAR MODEL INVOLVING A NONLINEAR FUNCTION OF A SINGLE VARIABLE.

Obtain an approximate linear model for the liquid level in a tank whose nonlinear model was given as:

$$A\frac{dh}{dt} = F_i - c\sqrt{h} \tag{10.1}$$

where F_i, the inlet flowrate, is the manipulated variable.

Solution:

The nonlinear function in this case is $f(h) = h^{1/2}$, a function of a single variable. A Taylor series expansion of this function around the steady state h_s gives:

$$h^{1/2} = h_s^{1/2} + \frac{1}{2} h_s^{-1/2} (h - h_s) + \textit{higher order terms} \tag{10.24}$$

Ignoring the higher order terms and introducing the linear approximation into Eq. (10.1) for $h^{1/2}$ now gives:

$$A\frac{dh}{dt} = F_i - c\left[h_s^{1/2} + \frac{1}{2}h_s^{-1/2}(h - h_s)\right] \tag{10.25}$$

The deviation variable $y = (h - h_s)$ naturally presents itself; if we now add the variable $u = (F_i - F_{is})$, realizing that under steady-state conditions $(dh/dt = 0)$, Eq. (10.1) implies that $F_{is} = c h_s^{1/2}$, Eq. (10.25) reduces at once to:

$$\tau \frac{dy}{dt} = Ku - y \tag{10.26}$$

where we have defined:

$$\tau = A\frac{2(h_s)^{1/2}}{c}, \quad K = \frac{2h_s^{1/2}}{c} \tag{10.27}$$

From this approximate linear model, we may now take Laplace transforms in the usual manner to obtain the following approximate transform-domain transfer function model:

$$y(s) = \frac{K}{\tau s + 1}u(s) \tag{10.28}$$

a first-order transfer function with steady-state gain and a time constant given by Eq. (10.27). It is now very important to note that both K and τ are functions of h_s, so that the apparent steady-state gain and time constant exhibited by this system will be different at each operating steady-state condition.

Example 10.4 LINEARIZATION OF A NONLINEAR MODEL INVOLVING A NONLINEAR FUNCTION OF TWO VARIABLES.

The dynamic behavior of the liquid level h in the conical storage tank system shown in Figure 10.4 can be shown to be represented by Eq. 10.1, where now the cross section area of the tank is given by:

$$A = \pi r^2 = \pi \left(\frac{Rh}{H}\right)^2 \tag{10.29}$$

so that the tank model becomes:

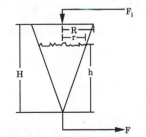

Figure 10.4.   The conical tank system.

$$\frac{dh}{dt} = \frac{\alpha F_i}{h^2} - \beta h^{-3/2} \tag{10.30}$$

where α and β are parameters defined by:

$$\alpha = \frac{1}{\pi}\left(\frac{H}{R}\right)^2, \qquad \beta = c\alpha$$

Here, F_i is, again, the inlet flowrate, the manipulated variable. Obtain an approximate linear model for this system.

Solution:

The process model has two types of nonlinear functions: $F_i h^{-2}$, a product of two functions, and $h^{-3/2}$. We shall have to linearize each of these functions separately around the steady state (h_s, F_{is}).

1.   The linearization of $f(h, F_i) = F_i h^{-2}$ proceeds according to Eq. (10.18) as follows:

$$f(h, F_i) = f(h_s, F_{is}) + \left(\frac{\partial f}{\partial h}\right)_{(h_s, F_{is})} (h - h_s) + \left(\frac{\partial f}{\partial F_i}\right)_{(h_s, F_{is})} (F_i - F_{is})$$
$$+ \ higher \ order \ terms$$

whereupon carrying out the indicated operations now gives:

$$f(h, F_i) \approx f(h_s, F_{is}) - 2F_{is}h_s^{-3}(h - h_s) + h_s^{-2}(F_i - F_{is}) \tag{10.31}$$

if we ignore the higher order terms.

2.   The steps involved in linearizing the second nonlinear term are no different from those illustrated in the previous example; the result is:

$$h^{-3/2} \approx h_s^{-3/2} - \frac{3}{2}h_s^{-5/2}(h - h_s) \tag{10.32}$$

We may now introduce these expressions in place of the corresponding nonlinear terms in Eq. (10.30). Recalling that under steady-state conditions $\alpha F_{is} = \beta h_s^{1/2}$, and introducing the deviation variables $y = (h - h_s)$ and $u = (F_i - F_{is})$, the approximate linear model is obtained upon further simplification as:

$$\tau \frac{dy}{dt} + y = Ku \tag{10.26}$$

where the steady-state gain, and time constant associated with this approximate linear model, are given by:

$$K = \frac{2\alpha}{\beta}h_s^{1/2} = \frac{2h_s^{1/2}}{c} \tag{10.33a}$$

and

$$\tau = \frac{2h_s^{5/2}}{\beta} \tag{10.33b}$$

If we desire an approximate transform-domain transfer function model, Laplace transformation in Eq. (10.26) gives:

$$y(s) = \frac{K}{\tau s + 1} u(s) \tag{10.34}$$

Note that the approximate linear model for the conical tank has the same process gain, K, as for the cylindrical tank; but the time constant is a much stronger function of the liquid level h_s.

Some important points to note about the results of these two illustrative examples are the following:

1. In each case, the approximate transfer function model is of the first-order kind, with the time constants, and steady-state gain values dependent on the specific steady state around which the system model was linearized.

2. Using such approximate transfer function models will give approximate results which are good in a small neighborhood around the initial steady state; farther away from this steady state, the accuracy of the approximate results becomes poorer.

3. Using such approximate models, we are able to get a general (even if not 100% accurate) idea about the speed and magnitude of response to expect from any *arbitrary* nonlinear system. Numerical analysis can give more accurate information, but only about the specific response to a specific input, starting from a specific operating condition for a specific set of parameters.

In Chapter 13, where we discuss model building from experimental data, we will introduce another approach to the problem of creating approximate linear models for nonlinear processes — through fitting approximate linear models to data.

Let us conclude with a comparison of the responses of the nonlinear model and an approximate linear model for the conical tank of Example 10.4. Suppose that $\alpha = 2, \beta = 1, c = 0.5$ so that the nonlinear model is:

$$\frac{dh}{dt} = 2F_i h^{-2} - h^{-3/2} \tag{10.35}$$

while the approximate linear model has the transfer function

$$y(s) = \frac{4h_s^{1/2}}{(2h_s^{5/2})s + 1} u(s) \tag{10.36}$$

which in the time domain is

$$\frac{dh(t)}{dt} = \frac{1}{\left(2h_s^{5/2}\right)} \left\{ h_s - h(t) + 4h_s^{1/2} \left[F_i(t) - F_{is} \right] \right\} \tag{10.37}$$

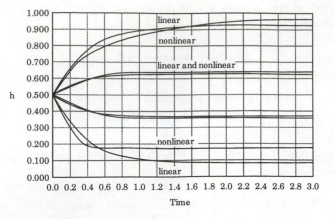
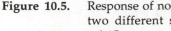

Figure 10.5.   Response of nonlinear model Eq. (10.35) and linear model Eq. (10.37) for two different steps in feed flowrate up and down $\Delta F_i = \pm 0.05$ and ± 0.15.

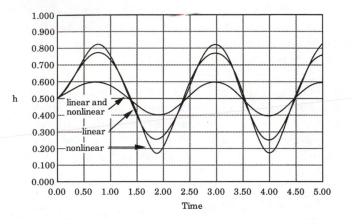

Figure 10.6.   Response of nonlinear model Eq. (10.35) and linear model Eq. (10.37) to sine wave in inlet flowrate with two different amplitudes $A = 0.05, 0.15$.

The step response (up and down) for these models is shown in Figure 10.5 for steps of two different magnitudes. The response to a sine wave input of two different amplitudes is shown in Figure 10.6. Note that for small magnitude step inputs and small amplitude sine waves, the linear model is a good approximation of the nonlinear model; however, for the large amplitude inputs the linear model predictions deviate somewhat from the nonlinear model behavior. Also note the distinct differences in nonlinear response for a step up and step down.

10.5   SUMMARY

Our discussion in this chapter has been motivated by the fact that it is indeed the rare process that is absolutely linear in behavior. Understandably, the analysis of the dynamic behavior of nonlinear systems is not as easily, or conveniently, carried out as the analysis of linear systems; but it is important to know the various options available to us when we cannot ignore the inherent nonlinearities in a process model.

We have identified four techniques by which nonlinear systems may be analyzed, passing relatively quickly over two of them because they are of limited practical application, and focusing mostly on the two most viable options — numerical analysis and model linearization.

It has become clear that with nonlinear analysis, it is not possible to have accuracy and generality at the same time, unlike the situation with linear analysis. When accuracy is the more important objective, the numerical approach is the appropriate choice. Apart from requiring computer assistance, the other drawback associated with this approach is that it makes available only simulation results for a specific set of parameters.

The model linearization technique is the most popular, primarily because it allows the use of the convenient methods of linear analysis and it is quite general. It is generally perceived that this advantage outweighs the disadvantage of potentially poor accuracy, especially when analysis is required to cover a wide range of operating conditions. All in all, when the limitations of linearization are kept in proper perspective, it can prove to be a very powerful tool for analyzing the dynamic behavior of nonlinear systems. (This last statement takes on even greater significance in the next chapter where we consider system stability.)

We must now draw a most important conclusion from the proceedings in this chapter, enabling us to justify why so much time is invested on linear systems analysis when real process systems are almost always nonlinear to some extent.

The objective of process control is the design of effective control systems that will hold the process conditions close to its desired steady-state value. Even though the system is inherently nonlinear, the influence of effective regulatory control is to ensure that deviations from this steady state will be small, in which case the behavior will be essentially indistinguishable from that of a linear system. It is in this sense that the process system is primarily considered to be approximately linear, and the linear methods we have studied in considerable detail are applicable.

Before leaving the main subject of Process Dynamics, there remains one final issue to be discussed, that of system stability. This is the subject of the next chapter, the last chapter of Part II.

REFERENCES AND SUGGESTED FURTHER READING

1. Luyben, W. L., Process Modeling, Simulation, and Control for Chemical Engineers (2nd ed.), McGraw-Hill, New York(1991)
2. Friedly, J. C., Dynamic Behavior of Processes, Prentice-Hall, Englewood Cliffs, NJ (1972)
3. Longwell, E., "The Industrial Importance of Nonlinear Control," Chemical Process Control IV (Y. Arkun and W.H. Ray, Ed.), AIChE, (1991)
4. Iooss, G. and D. D. Joseph, Elementary Stability and Bifurcation Theory, Springer-Verlag, New York (1981)
5. Guckenheimer, J. and P. Holmes, Nonlinear Oscillations, Dynamical Systems, and Bifurcation of Vector Fields, Springer-Verlag, New York (1983)
6. Golubitsky, M. and D. G. Schaeffer, Singularities and Groups in Bifurcation Theory, Vol. 1, Springer-Verlag, New York (1985)

7. Arnold, V. I., Geometrical Methods in the Theory of Ordinary Differential Equations, Springer-Verlag, New York (1988)
8. Hunt, L. R., R. Su, and G. Meyer, "Global Transformations of Nonlinear Systems," IEEE Trans. Auto Control, **28**, 24 (1983)
9. Isidori, A., Nonlinear Control Systems, Springer-Verlag, New York (1989)
10. Kravaris, C. and Y. Arkun, "The Differential Geometry Approach to Nonlinear Control," Chemical Process Control IV (Y. Arkun and W.H. Ray, Ed.), AIChE (1991)

REVIEW QUESTIONS

1. What two properties characterize the behavior of linear systems, and how do they assist in recognizing nonlinear behavior?

2. What are some of the most common sources of nonlinearities in chemical process models?

3. When confronted with the analysis of nonlinear systems dynamic behavior, what are some of the approaches available for consideration?

4. Why is exact linearization by variable transformation currently not very practical?

5. What are some of the main advantages and disadvantages of the numerical approach to nonlinear dynamic analysis?

6. What is the main advantage of the model linearization approach to nonlinear dynamic analysis? What is the main disadvantage of this approach?

7. Which approach makes it possible to draw *general* but approximate inferences about the behavior of a nonlinear system, and which approach makes it possible to draw only *specific* but accurate inferences?

8. Why is it necessary to truncate a Taylor series expansion of a nonlinear function (of a single variable) after the first term if a linear approximation is desired?

9. If chemical processes are almost universally nonlinear in nature, so that their mathematical models are almost always nonlinear, why are linear approximations found to be adequate in a substantial number of cases, especially when feedback control is involved?

PROBLEMS

10.1 The following is a sample catalog of some common constitutive relations of chemical engineering and the nonlinear functions they introduce into process models:

CHEMICAL KINETICS:

(a)*Temperature dependence of reaction rate "constant": The Arrhenius expression*:

$$k(T) = k_0 e^{-E/RT} \qquad \text{(P10.1)}$$

with k_0, E, R, as constants.

EQUATIONS OF STATE:

(b)*(Empirical) Temperature dependence of enthalpy*:

$$H(T) = b_0 + b_1 T + b_2 T^2 + b_3 T^3 + b_4 T^4 \qquad \text{(P10.2)}$$

with b_0, b_1, b_2, b_3, b_4 as constants.

(c) *Antoine's equation: Temperature dependence of vapor pressure:*

$$p^0(T) = e^{[A - B/(T + C)]} \qquad \text{(P10.3)}$$

with A, B, C as constants.

PHASE EQUILIBRIUM:

(d) *Vapor-liquid equilibrium relation for binary mixture:*

$$y(x) = \frac{\alpha x}{1 + (\alpha - 1)x} \qquad \text{(P10.4)}$$

with α, the relative volatility, as a constant.

HEAT TRANSFER:

(e) *Radiation heat transfer:*

$$q(T) = \varepsilon \sigma A T^4 \qquad \text{(P10.5)}$$

with ε, σ, A as constants.

Linearize each of these five expressions, expressing your results in terms of deviation variables which you should clearly define.

10.2 The differential equation model for a first-order bilinear system:

$$\frac{dY(t)}{dt} = aY(t) + nY(t)U(t) + bU(t) \qquad \text{(P10.6)}$$

differs from the differential equation model for the first-order system:

$$\frac{dY(t)}{dt} = aY(t) + bU(t) \qquad \text{(P10.7)}$$

only by the bilinear $nY(t)U(t)$ term.

(a) Linearize the bilinear system model around the steady-state values, Y_s and U_s, define deviation variables, $y = Y - Y_s$ and $u = U - U_s$, and obtain the approximate, transfer function relating $y(s)$ to $u(s)$.

(b) Recast Eq. (P10.7) in terms of the deviation variables of part (a) and obtain the first-order system transfer function. Compare this transfer function with the approximate transfer function obtained for the linearized bilinear system model. What effect does the presence of the bilinear term have on the steady-state gain and the time constant of a first-order system?

10.3 The dynamic behavior of the liquid level in a certain cylindrical storage tank is modeled by:

$$A \frac{dh}{dt} = F_i - ch^{2/3} \qquad \text{(P10.8)}$$

where A is the uniform cross-sectional area (m²); F_i is the inlet flowrate (m³/min); h is the liquid level (m); and c is a constant.

(a) Linearize this process model around a steady-state level h_s, and obtain an approximate transfer function model relating the *deviation variables* $y = h - h_s$ and $u = F_i - F_{is}$, where F_{is} is the steady-state inlet flowrate corresponding to h_s. Explicitly specify what the steady-state gain and the time constant of the approximate transfer functions are.

(b) The specific tank with $c = 0.5$ (m⁷ᐟ³/min), $A = 0.25$ m², and whose total height is 1 m, had a steady-state liquid level of 0.512 m when operating at an initial steady-state flowrate of 0.32 m^3/min. It is believed that if the flowrate were to increase suddenly from this initial value to a new value of 0.52 m^3/min, and remain there, the approximate, linearized model will mislead us into concluding that the liquid level will rise to a new steady-state level within the limits of the total tank height, and the tank will *not* overflow; the more accurate nonlinear model, on the other hand, predicts that this change will, in fact, cause the tank to overflow. Confirm or refute this statement.

10.4 A mathematical model for the level of gasoline in a hemispherical storage tank is given as:

$$\frac{dh}{dt} = \frac{1}{\pi} \frac{1}{(2Rh - h^2)} \left(F_i - ch^{1/2} \right) \tag{P10.9}$$

where R is the radius of the hemispherical tank, and F_i is the flowrate of gasoline into the storage tank. Show that by introducing the following transformations:

$$z = \exp\left[h^{3/2} \left(\frac{4R}{3} - \frac{2h}{5} \right) \right] \tag{P10.10}$$

and

$$v = \frac{F_i z}{h^{1/2}} \tag{P10.11}$$

the original nonlinear model equation is converted to the first-order linear differential equation:

$$\frac{dz}{dt} = az + bv \tag{P10.12}$$

Show explicitly what the constants a and b are.

10.5 Revisit Problem 10.4. Given a specific tank with parameters $R = 2$ m, and $c = 1.5$ (m$^{5/2}$/min.), operating at initial steady-state conditions: $F_{is} = 1.35$ m^3/min, $h_s = 0.81$ m;
(a) Linearize the original mathematical model around this steady state, using a first-order Taylor series expansion; obtain an approximate transfer function model, and use this approximate model to obtain a response of the gasoline level to a step change in F_i from the initial value to 1.8 m^3/min. Plot this response.
(b) Using a numerical integration software of your choice, obtain a more accurate response of the gasoline level (to the same change in F_i investigated in part (a)) by solving the original modeling equation, incorporating the given parameters. Superimpose a plot of this response on the approximate response plot obtained in part (a) and comment on the "goodness" of the linearized, approximate transfer function model.

10.6 In Example 10.4, an approximate linear first-order model was derived for the conical tank by linearizing the original nonlinear model. The following problem (Olson 1992)[†] shows an alternative method for doing the same thing.
Starting from the original material balance equation:

$$\frac{dV}{dt} = F_i - ch^{1/2} \tag{P10.13}$$

given initial steady-state values V_s for the tank volume, h_s for the tank level, and F_{is} for the inlet flowrate, define the following new variables:

$$V = V/V_s$$

[†] J. H. Olson, *private communications*, March 1992.

$$H = h/h_s$$
$$F_i = F_i/F_{is}$$

respectively, the *relative* volume in the tank, the *relative* tank level, and the *relative* inlet flowrate. Show that with this definition, the material balance equation becomes:

$$\tau_s \frac{dV}{dt} = F_i - H^{1/2} \tag{P10.14}$$

with $\tau_s = V_s/F_{is}$, the residence time of the tank at the current operating steady state. Now introduce the deviation variable $\tilde{V} = V - V_s = V - 1$, with the others, $\tilde{H}$, and $\tilde{F}_i$ similarly defined, and show that upon linearization, Eq. (P10.14) becomes:

$$6\tau_s \frac{d\tilde{H}}{dt} = 2\tilde{F}_i - \tilde{H} \tag{P10.15}$$

an equation that is easier to use than Eqs. (10.26) and (10.33) since no constitutive relation is needed to find the time constant and the gain.

10.7 In Problem 6.10 the model for an experimental biotechnology process was given as:

$$\frac{d\xi_1}{dt} = a_1\xi_1 - a_{12}\xi_1\xi_2 + b\mu \tag{P6.6a}$$

$$\frac{d\xi_2}{dt} = -a_2\xi_2 + a_{21}\xi_1\xi_2 \tag{P6.6b}$$

where:
  ξ_1 is the mass concentration of microorganism 1, M_1 (microgram/liter)
  ξ_2 is the mass concentration of microorganism 2, M_2 (microgram/liter)
  μ is the rate of introduction of M_1 into the pond (microgram/hr)
  $a_1, a_{12}, a_2, a_{21}, b$, are fixed parameters
  Linearize this nonlinear model around the reference values ξ^*_1, ξ^*_2, and μ^*, defining the following deviation variables:

$$x_1 = \xi_1 - \xi^*_1; \qquad x_2 = \xi_2 - \xi^*_2; \qquad u = \mu - \mu^*$$

Show that the resulting linearized model is of the form given in Eqs. (P6.7a,b):

$$\frac{dx_1}{dt} = \alpha_{11}x_1 - \alpha_{12}x_2 + bu \tag{P6.7a}$$

$$\frac{dx_2}{dt} = \alpha_{21}x_1 - \alpha_{22}x_2 \tag{P6.7b}$$

Provide explicit expressions for the new parameters $\alpha_{11}, \alpha_{12}, \alpha_{21}, \alpha_{22}$, and b in terms of the original model parameters and the reference values around which the model was linearized.

10.8 Perhaps the most popular nonlinear chemical engineering system is the nonisothermal CSTR in which the single, irreversible reaction $\mathbf{A} \to \mathbf{B}$ is taking place. The model for this reactor is given in Chapters 4 and 12 as:

$$\frac{dc_A}{dt} = -\frac{1}{\theta}c_A - k_0e^{-(E/RT)}c_A + \frac{1}{\theta}c_{Af} \tag{4.26}$$

$$\frac{dT}{dt} = -\frac{1}{\theta}T + \beta k_0e^{-(E/RT)}c_A + \frac{1}{\theta}T_f - \chi \tag{4.27}$$

where, as previously noted: c_A is the concentration of reactant A in the reactor while T is reactor temperature, $\theta = V/F$ is the mean residence time in the reactor, $k = k_0 e^{-(E/RT)}$ is the reaction rate constant, $\beta = (-\Delta H)/\rho C_p$ is a heat of reaction parameter, and χ is the heat removed by the tubular cooling coil:

$$\chi = \frac{hA}{\rho C_p V}(T - \bar{T}_c) \tag{12.3}$$

with h as the coil heat transfer coefficient, A the total coil heat transfer area, and $\bar{T}_c$ is the average temperature along the length of the coil. Linearize this set of equations around the steady states c_{As} and T_s.

10.9 A detailed mathematical model describing the behavior of a jacketed CSTR used for the free-radical polymerization of methylmethacrylate (MMA) using azo-bis-isobutyronitrile (AIBN) as initiator, and toluene as solvent, was presented by Congalidis *et al.* (1989)[†] and has been used in several other studies. Recently, Doyle *et al.* (1992),[††] by making some simplifying assumptions, and introducing the published kinetic and other reactor parameters, reduced this reactor model to the following set of equations:

$$\frac{d\xi_1}{dt} = 3(6 - \xi_1) - 1.36594\xi_1\sqrt{\xi_2} \tag{P10.16}$$

$$\frac{d\xi_2}{dt} = 60\mu - 3.03161\xi_2 \tag{P10.17}$$

$$\frac{d\xi_3}{dt} = -0.00134106\xi_1\sqrt{\xi_2} + 0.034677\,\xi_2 - 10\xi_3 \tag{P10.18}$$

$$\frac{d\xi_4}{dt} = 136.758\xi_1\sqrt{\xi_2} - 10\xi_4 \tag{P10.19}$$

along with:

$$\eta = \frac{\xi_4}{\xi_3} \tag{P10.20}$$

Here, ξ_1, ξ_2, ξ_3, and ξ_4 respectively represent the dimensionless monomer concentration, initiator concentration, bulk zeroth moment, and bulk first moment; η is the number average molecular weight; and μ is the volumetric flowrate of the initiator.

(a) Define the deviation variables: $x_i = \xi_i - \xi_i^*$; $i = 1, 2, 3, 4$; $u = \mu - \mu^*$; and $y = \eta - \eta^*$ and linearize this set of modeling equations around the steady state indicated by ξ_i^*; $i = 1, 2, 3, 4$; μ^*; and η^*.

(b) Given the following specific steady-state operating conditions:

$\xi_1^* = 5.50677$, $\xi_2^* = 0.132906$, $\xi_3^* = 0.001975$, $\xi_4^* = 49.3818$, and $\mu^* = 0.01678$,

take Laplace transforms of the linearized equations and obtain a *single* transfer function that relates $y(s)$ to $u(s)$.

[†] J. P. Congalidis, J. R. Richards, and W. H. Ray, "Feedforward and Feedback Control of a Copolymerization Reactor" *AIChEJ*, 35—891 (1989).

[††] F. J. Doyle, III, B. A. Ogunnaike, and R. K. Pearson, "Nonlinear Model Predictive Control Using Second Order Volterra Models," Preprints of the Annual A.I.Ch.E. meeting, Miami Beach, Florida, Nov. 1992.

CHAPTER

11

STABILITY

A very important aspect of the dynamic behavior of a process that so far has not received any formal attention has to do with the following issue: when a (finite) input change is implemented on a physical system, does the resulting transient response ultimately settle to a new steady state, or does it grow indefinitely? This issue did not arise in all our previous discussions because of a deliberate attempt on our part to defer it until all other aspects of dynamic analysis have been discussed.

If it is indeed possible for a process not to settle eventually to another steady state, but have its output grow indefinitely when disturbed from an initial steady state, what characteristics of the process determine whether or not such behavior will occur? This chapter is primarily concerned with investigating those issues having to do with the *stability* properties of a dynamic process as represented by its model.

11.1 INTRODUCTORY CONCEPTS OF STABILITY

When a process is "disturbed" from an initial steady state, say, by implementing a change in the input forcing function, it will, in general, respond in one of three ways:

- **Response 1:** Proceed to a steady state and remain there, or
- **Response 2:** Fail to attain to steady-state conditions because its output grows indefinitely, or
- **Response 3:** Fail to attain steady-state conditions because the process oscillates indefinitely with a *constant* amplitude.

These possible patterns of behavior are illustrated in Figure 11.1.

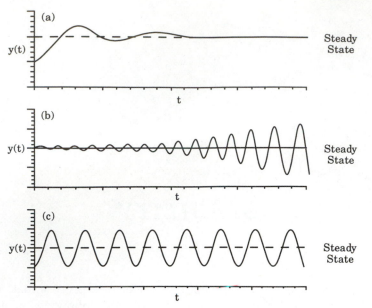

Figure 11.1.   The three possible responses of a dynamic system to a disturbance:
(a) returns to a steady state; (b) output grows indefinitely; (c) output
oscillates indefinitely with a constant amplitude.

It now seems perfectly logical to consider the process that ultimately
settles down after having been disturbed as being "stable" and the one that
fails to settle down as being "unstable." A classic illustration of this concept is
provided by the simple pendulum, consisting of a small bob of mass m connected
to a fixed point pivot by a rod of length R, considered to be essentially rigid and
of negligible mass.

It can be shown (and we will do this later on) that there are two possible
steady-state (equilibrium) positions for the pendulum. These are shown in
Figure 11.2(a), where the pivot is above the bob, so that the pendulum is
hanging straight down, and in Figure 11.3, where the bob is directly above the
pivot.

The nature of the pendulum is such that when disturbed from an equilibrium
position, the bob will swing in a circular arc about the pivot. Its position at any
time during this period is indicated by θ, the angle of deviation from the
straight-down equilibrium position. The equilibrium position represented in
Figure 11.2(a) therefore corresponds to $\theta = 0$, while the one in Figure 11.3
corresponds to $\theta = \pi$.

Now, if in Figure 11.2(a) the bob is displaced to the left by an angle θ, as
shown in Figure 11.2(b), and then released, we find that this disturbance causes
the pendulum to swing back and forth for a while and eventually settle back to
the initial equilibrium position. We will then consider Figure 11.2(a) to be a
stable configuration for the pendulum system; the point $\theta = 0$ representing a
stable equilibrium position or a *stable steady state*.

In contrast, for the configuration in Figure 11.3 where the bob is above the
pivot, it is easy to see that any perturbation in the position of the bob will
cause a change in the pendulum position is such a way that the initial
equilibrium position is no longer attainable. This configuration will then

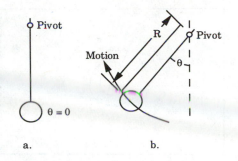

a.   b.

Figure 11.2.   A stable configuration for the pendulum.

be considered as being *unstable*; that is, $\theta = \pi$ is an *unstable steady state*. Note that the upward pointing configuration is indeed a steady state because all the forces are balanced when $\theta = \pi$ exactly. However, this steady state is in fact unattainable by the system unless some external force is used to hold the pendulum in this position.

11.1.1 Definitions of Stability

Although there are many slightly different mathematical definitions of the terms *stable* and *unstable*, we shall choose to use the concept of stability in the sense of Liapunov (see Refs. [1,2]). In this section we shall introduce several rather precise definitions that will aid in understanding the concepts of stability; we conclude this section with a more restricted definition of stability which is used most frequently in process control.

For a given process steady state, we have the following definitions.

Definition 1:

> A system steady state x_s is said to be **stable** if for each possible region of radius $\varepsilon > 0$ around the steady state, there is an initial state x_0 at $t = t_0$ falling within a radius $\delta > 0$ around the steady state that causes the dynamic trajectory to stay within the region $|(x - x_s)| < \varepsilon$ for all times $t > t_0$.

This is illustrated in Figure 11.4 where we see that Definition 1 simply says that for *all* possible regions ε (large or small), there is a corresponding region of initial conditions bounded by δ so that the dynamic system stays within ε. This, of course, implies that $\delta < \varepsilon$ and that even for a very small region ε near the steady state, there must exist an initial radius $\delta > 0$ that keeps the system within ε of x_s.

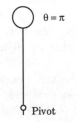

$\theta = \pi$

Pivot

Figure 11.3.   An unstable configuration for the pendulum.

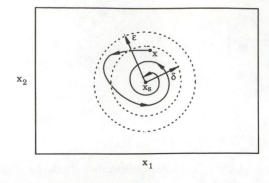

Figure 11.4.   Regions bounded by ε and δ in the definition of stability.

Definition 2:

> *A system steady state* $\mathbf{x}_s$ *is said to be* **asymptotically stable** *if it is both stable and in addition there exists a region of initial conditions of radius* $\delta_0 > 0$ *around* $\mathbf{x}_s$ *for which the system approaches* $\mathbf{x}_s$ *as* $t \to \infty$.

Thus Definition 2 means that any steady state that is *asymptotically stable* will return to the steady state $\mathbf{x}_s$ as $t \to \infty$ after some initial perturbation $\mathbf{x}_0$. Usually for process dynamics and control problems there is no practical difference between steady states that are stable and asymptotically stable.

Definition 3:

> *A system steady state is said to be* **unstable** *if it is not stable.*

This definition needs no explanation.

Note that these precise definitions refer to the steady states of the system rather than to the system itself. If we wish to discuss the stability of the system itself we may do so in some domain of the space of state variables $\mathbf{x}$.

Definition 4:

> *A dynamic system is said to be* **stable** *on some domain M if for every region of radius* ε *in the domain there is a region of radius* δ *in the domain such that the system trajectories are bounded by* ε *if the initial condition is bounded by* δ.

Note that this is just the definition of stability (Definition 1) but limited to some domain M. If we allow the domain M to be the whole space, then the system is said to be *globally stable*. This concept of system stability on a domain is important in interpreting the behavior of nonlinear systems where there may be multiple steady states or domains in the state space that are unstable.

The precise definitions above are necessary when we consider all the possibilities arising from nonlinear systems; however, the simpler concept of Bounded-Input/Bounded-Output (BIBO) stability is often sufficient for linear systems:

Definition 5 (Bounded-Input/Bounded-Output Stability):

If in response to a bounded *input, the dynamic trajectory of a system remains* bounded *as t → ∞, then the system is said to be* **stable**; *otherwise, it is said to be* **unstable**.

For the study of linear systems in process control, it is normally the concept of *BIBO stability* that we shall use.

Based on these definitions, we can now classify the types of behavior shown in Figure 11.1. Obviously, the steady state in Figure 11.1(a) is stable and probably asymptotically stable, while the steady state in Figure 11.1(b) is clearly unstable. The sustained oscillatory trajectory in Figure 11.1(c) implies that the steady state is unstable because it violates Definition 1; however, it satisfies Definition 5 for BIBO stability of the system. Thus, in this case the *steady state* is unstable while the *system* has *BIBO stability*. It is for situations such as this that we have been so careful in our definitions of stability.

If we analyze the pendulum shown in Figures 11.2 – 11.3, we see that the downward-pointing configuration ($\theta_s = 0$) is stable (Definition 1) and the upward-pointing configuration ($\theta_s = \pi$) is unstable. Thus the system is stable on a domain M that excludes the unstable steady state (Definition 4). However, the system remains bounded, so that it has BIBO stability.

11.1.2 Practical Issues in Process Dynamics and Control

There are several practical questions concerning system stability that arise in process dynamics and control. Some of the most important are:

- When is a linear system stable?
- When is a nonlinear system stable?
- Can a system that is unstable by itself (i.e., a "naturally unstable" system) be made stable by addition of a control system?
- Can a "naturally stable" system be made unstable by the addition of a control system?

We will apply the stability concepts just discussed in order to answer these questions in this chapter.

11.2 STABILITY OF LINEAR SYSTEMS

For the purpose of establishing conditions for the stability of a linear system, we shall consider the response to bounded inputs such as the unit step. We shall begin by dealing with systems *without time delays* because these are amenable to exact analytical treatment. Methods for analyzing the stability of systems having time delays are introduced in Chapters 15 and 17. Let us only state here that the *open-loop*[†] stability of a linear transfer function $g^*(s)$ is *not* affected by the presence of a time delay $e^{-\alpha s}$. Thus the open-loop stability character of $g^*(s)$ is the same as that of $g^*(s)e^{-\alpha s}$.

[†] Here the term *open-loop* refers to a system without a feedback controller.

Let us represent the linear system (without time delays) by the transform-domain transfer function model:

$$y(s) = g(s) u(s) \tag{11.1}$$

where $g(s)$ is the system transfer function that consists of a qth-order numerator polynomial and a pth-order denominator polynomial:

$$g(s) = \frac{b_q s^q + b_{q-1} s^{q-1} + \ldots + b_0}{a_p s^p + a_{p-1} s^{p-1} + \ldots + a_0} = \frac{B(s)}{A(s)} \tag{11.2}$$

Recall that the polynomials in $g(s)$ can be factored into the form:

$$g(s) = \frac{K(s - z_1)(s - z_2) \ldots (s - z_q)}{(s - r_1)(s - r_2) \ldots (s - r_p)} \tag{11.3}$$

so that $z_k, k = 1, 2, \ldots q$ the roots of the numerator polynomial, $B(s)$, are the zeros of $g(s)$. Similarly, $r_i, i = 1, 2, \ldots p$, the roots of the denominator polynomial, $A(s)$, are the poles of $g(s)$.

To illustrate the response to a bounded input, consider the unit step response given by:

$$y(s) = g(s) \frac{1}{s} \tag{11.4}$$

Observe that upon partial fraction expansion in the usual manner, Eq. (11.4) may be rewritten as:

$$y(s) = \frac{A_0}{s} + \sum_{i=1}^{p} \left[\frac{A_i}{(s - r_i)} \right] \tag{11.5}$$

where the r_i's are the p system poles. Recall that the r_i can be either real or complex numbers, $r_i = \gamma_i + j\eta_i$, where γ_i is the real part and η_i is the imaginary part of the ith system pole.

Laplace inversion here immediately yields:

$$y(t) = A_0 + \sum_{i=1}^{p} \left(A_i e^{r_i t} \right) \tag{11.6}$$

We now note that the stability characteristics of this response is determined entirely by the p poles, $r_1, r_2, \ldots r_p$: the response will remain bounded if and only if all these poles have *negative* real parts ($\gamma_i < 0$). If just one of these poles has a *positive* real part, the exponential terms associated with it will grow indefinitely and the response becomes unbounded. The special case when some of the γ_i are exactly equal to zero is a singularity that is unlikely to occur for a real process. Thus these are only of interest as the point of transition from stability to instability. We therefore have the following stability test:

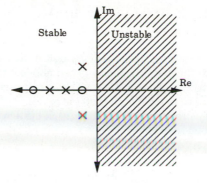

Figure 11.5. The complex plane showing the location of system poles (×) and system zeros (O).

A linear system is stable if and only if all its poles have negative real parts so that they lie in the left half of the complex plane; otherwise it is unstable.

This is a very powerful result because it clearly states that the stability of a linear system is dependent only on its poles; *zeros do not affect system stability*. Thus the issue of determining the stability character of a linear system is no more than that of determining the nature of its poles; it is not at all necessary to determine the details of the dynamic process response. Figure 11.5 shows the complex plane with some poles and zeros plotted. Note that from the stability test above, if any pole is in the right-half plane, the system is unstable. Let us illustrate this with an example.

Example 11.1 STABILITY OF TWO LINEAR SYSTEMS: THE EFFECT OF POLES AND ZEROS.

Investigate the stability properties of the systems whose transfer functions are given as:

$$g_1(s) = \frac{5}{s^2 + s - 6} \tag{11.7}$$

$$g_2(s) = \frac{5(s + 1)}{s^2 + s - 6} \tag{11.8}$$

where the only difference between System 1 and System 2 is that the latter has a zero while the former has none; they both have identical denominator polynomials, i.e., identical poles.
Solution:

The denominator polynomials for each of these systems can be factored to obtain the roots, in which case the transfer functions can be represented as:

$$g_1(s) = \frac{5}{(s + 3)(s - 2)}$$

$$g_2(s) = \frac{5(s + 1)}{(s + 3)(s - 2)}$$

Observe that both systems have identical poles: two each, located at $s = -3$ and $s = +2$. According to the stability result given above, therefore, both systems are *unstable* by virtue of the pole at $s = +2$. The zero possessed by System 2 exerts absolutely no influence as to whether it is stable or not.

Let us confirm this further by obtaining each system's unit step response. For System 1, this is obtained from:

$$y_1(s) = \frac{5}{(s + 3)(s - 2)} \frac{1}{s}$$

while for System 2, it is obtained from:

$$y_2(s) = \frac{5(s + 1)}{(s + 3)(s - 2)} \frac{1}{s}$$

Partial fraction expansion of each of these expressions now gives:

$$y_1(s) = \frac{-5/6}{s} + \frac{1/3}{s + 3} + \frac{1/2}{s - 2}$$

and

$$y_2(s) = \frac{-5/6}{s} + \frac{-2/3}{s + 3} + \frac{3/2}{s - 2}$$

which upon inversion become:

$$y_1(t) = -\frac{5}{6} + \frac{1}{3}e^{-3t} + \frac{1}{2}e^{2t}$$

and

$$y_2(s) = -\frac{5}{6} - \frac{2}{3}e^{-3t} + \frac{3}{2}e^{2t}$$

It is important to observe the following regarding these responses:

- They each have the same set of exponential functions, $e^{r_i t}$.
- The constant coefficients of each of the corresponding exponential functions are different for each response.

Now, each of these responses grows indefinitely with time by virtue of the e^{2t} terms, and therefore both systems are in fact *unstable*, as we had earlier concluded.

Thus, the answer to the question "when is a linear system stable?" is: when all the transfer function *poles* (the roots of the denominator polynomial) have negative real parts; i.e., when they lie in the left half of the complex s-plane.

If the linear system in Eq. (11.2) were expressed in terms of a pth-order differential equation model of the type:

$$a_p \frac{d^p y}{dt^p} + a_{p-1} \frac{d^{p-1} y}{dt^{p-1}} + \ldots + a_2 \frac{d^2 y}{dt^2} + a_1 \frac{dy}{dt} + a_0 y = f[u(t)] \qquad (11.9)$$

where $f[u(t)]$ is some forcing function (which could involve derivatives of $u(t)$), then it is easy to show (cf. Appendix B) that the solution of Eq. (11.9) is of the same form as Eq. (11.6):

$$y(t) = A_0 + \sum_{i=1}^{p} \left(A_i e^{m_i t} \right) \tag{11.10}$$

where the p *characteristic values*, m_i, are found by using the coefficients of the derivatives in Eq. (11.9) to form the characteristic equation:

$$a_p m^p + a_{p-1} m^{p-1} + \ldots + a_2 m^2 + a_1 m + a_0 = 0 \tag{11.11}$$

Thus for models in the form of Eq. (11.9) we have the following stability result:

A linear system in the form of Eq. (11.9) is stable if and only if all its characteristic values m_i have negative real parts so that they lie in the left half of the complex plane.

Recall from Chapter 4 that we could also represent the transform-domain transfer function model Eqs. (11.1) and (11.2) or the high-order equation (11.9) in the state-space:

$$\begin{aligned} \dot{\mathbf{x}} &= \mathbf{A}\,\mathbf{x} + \mathbf{b}\,u \\ y &= \mathbf{c}^T \mathbf{x} \end{aligned} \tag{11.12}$$

where $\mathbf{x}$ is a p vector of states defined by:

$$\dot{x}_1 = x_2$$
$$\dot{x}_2 = x_3$$
$$\vdots$$
$$\dot{x}_{p-1} = x_p$$
$$\dot{x}_p = \frac{1}{a_p}\left(-a_{p-1} x_p - a_{p-2} x_{p-1} \ldots -a_0 x_1 + K u \right)$$

with $x_1 = y$

(Here we consider $f = Ku(t)$ in Eq. (11.9) for simplicity.)
In this case the quantities $\mathbf{A}, \mathbf{b}, \mathbf{c}$ are given by:

$$\mathbf{A} = \begin{bmatrix} 0 & 1 & 0 & \ldots & \ldots & 0 \\ 0 & 0 & 1 & \ldots & \ldots & \ldots \\ \ldots & \ldots & \ldots & \ldots & \ldots & \ldots \\ \ldots & \ldots & \ldots & \ldots & \ldots & \ldots \\ 0 & 0 & 0 & \ldots & \ldots & 1 \\ \dfrac{-a_0}{a_p} & \dfrac{-a_1}{a_p} & \ldots & \ldots & \ldots & \dfrac{-a_{p-1}}{a_p} \end{bmatrix}$$

$$
\mathbf{b} = \begin{bmatrix} 0 \\ 0 \\ \cdot \\ \cdot \\ \cdot \\ \cdot \\ \dfrac{K}{a_p} \end{bmatrix} \qquad \mathbf{c} = \begin{bmatrix} 1 \\ 0 \\ 0 \\ \cdot \\ \cdot \\ 0 \end{bmatrix} \tag{11.13}
$$

It is straightforward to show (cf. Appendices B and D) that the system Eq. (11.12) has a solution of the form of Eqs. (11.6) and (11.10); that is:

$$
y(t) = A_0 + \sum_{i=1}^{p} \left(A_i \, e^{\lambda_i t} \right) \tag{11.14}
$$

Here the quantities λ_i, $i = 1, 2, \ldots, p$ are the *eigenvalues* of the matrix $\mathbf{A}$ in Eq. (11.12) found from the roots of the polynomial defined by:

$$
|\mathbf{A} - \lambda \mathbf{I}| = 0 \tag{11.15}
$$

which when applied to $\mathbf{A}$ in Eq. (11.13) results in the characteristic equation:

$$
a_p \, \lambda^p + a_{p-1} \, \lambda^{p-1} + a_{p-2} \, \lambda^{p-2} + \ldots + a_0 = 0 \tag{11.16}
$$

Thus for models of the form of Eq. (11.12), stability is determined as follows:

> *A linear system in the form of Eq. (11.12) is stable if and only if all of its eigenvalues λ_i have negative real parts so that they lie in the left half of the complex plane.*

Note that the polynomials in Eqs. (11.11) and (11.16) are identical to the denominator polynomial in Eq. (11.2) so that all three equations have the same roots; i.e., the poles of the transfer function $g(s)$ in Eq. (11.2) are the same as the characteristic values of the high-order Eq. (11.9) and the same as the eigenvalues of the matrix $\mathbf{A}$ in Eq. (11.12). Thus $r_i = m_i = \lambda_i$; $i = 1, 2, \ldots, p$.

From this similarity we know that the stability of a linear system is determined by the sign of the real parts of the poles, characteristic values, or eigenvalues, depending on the system model form we use. These all result from the same polynomial equations.

It is important at this point to mention that the sign of the real parts of the roots of these polynomials can be determined without explicitly solving the equations. Straightforward and valuable methods for doing this are presented in Chapter 14.

Example 11.2.   STABILITY OF A LINEAR SYSTEM IN THE TIME DOMAIN.

Using the transfer function $g_1(s)$ from Example 11.1, transform the model:

$$
y(s) = \left[\frac{5}{(s^2 + s - 6)} \right] u(s) \tag{11.17}
$$

into the time domain and investigate the stability character of the time-domain model.

Solution:

Transforming into the time domain, one obtains the model:

$$\frac{d^2y(t)}{dt^2} + \frac{dy(t)}{dt} - 6y(t) = 5u(t) \tag{11.18}$$

or equivalently letting:

$$\frac{dx_1}{dt} = x_2(t)$$

$$\frac{dx_2}{dt} = -x_2(t) + 6x_1(t) + 5u(t) \tag{11.19}$$

$$y = x_1$$

we have the state-space form of Eq. (11.12) where:

$$\mathbf{A} = \begin{bmatrix} 0 & 1 \\ 6 & -1 \end{bmatrix}, \qquad \mathbf{b} = \begin{bmatrix} 0 \\ 5 \end{bmatrix}, \qquad \mathbf{c} = \begin{bmatrix} 1 \\ 0 \end{bmatrix}$$

Now we see that the three model representations Eqs. (11.17), (11.18) and (11.19) produce the set of characteristic equations:

$$s^2 + s - 6 = 0 \tag{from 11.2}$$

$$m^2 + m - 6 = 0 \tag{from 11.11}$$

$$\lambda^2 + \lambda - 6 = 0 \tag{from 11.16}$$

Thus $s_1 = m_1 = \lambda_1 = -3$, $s_2 = m_2 = \lambda_2 = +2$, and the system is unstable however it is analyzed.

11.3 STABILITY OF NONLINEAR SYSTEMS

11.3.1 Some Preliminary Considerations

In order to have a comprehensive analysis of the stability of nonlinear systems, we must use much more sophisticated tools than are required for linear systems because nonlinear systems offer such a wide range of dynamic phenomena. For an introduction to these deep waters, see Refs. [3, 4].

Fortunately for us, there are relatively simple results that allow us to make strong statements about the stability of the steady states of nonlinear systems. To demonstrate this, let us recall the rather general nonlinear model from Chapter 10 for the n state variables x_i, $i = 1, 2, \ldots n$:

$$\frac{dx_i}{dt} = f_i(x_1, x_2, \ldots, x_n, u, d, t) \tag{10.9}$$

Let us neglect the inputs u, d, t, since the variables have nothing to do with the inherent stability properties of the system, and put Eq. (10.9) into vector notation:

$$\frac{d\mathbf{x}}{dt} = \mathbf{f}(\mathbf{x}) \tag{11.20}$$

where $\mathbf{x}$ is a time-dependent vector of the n states and $\mathbf{f}$ is a vector function:

$$
\mathbf{x} = \begin{bmatrix} x_1 \\ x_2 \\ \cdot \\ \cdot \\ \cdot \\ x_n \end{bmatrix}, \qquad
\mathbf{f} = \begin{bmatrix} f_1(\mathbf{x}) \\ f_2(\mathbf{x}) \\ \cdot \\ \cdot \\ \cdot \\ f_n(\mathbf{x}) \end{bmatrix}
\tag{11.21}
$$

If we take the Taylor series expansion of each element of $\mathbf{f}$, i.e., $f_i(x)$, $i = 1, 2, ..., n$, around the steady state $\mathbf{x}_s$, as was done in Eq. (10.16), and neglect all terms higher than first order, we obtain the linearized model:

$$
\frac{dx_i(t)}{dt} = f_i(\mathbf{x}_s) + \sum_{j=1}^{n} \left[\frac{\partial f_i(\mathbf{x})}{\partial x_j} \right]_{\mathbf{x} = \mathbf{x}_s} [x_j(t) - x_{js}];
\tag{11.22}
$$
$$
i = 1, 2, ..., n
$$

Note that $(\partial f_i(\mathbf{x})/\partial x_j)_{\mathbf{x} = \mathbf{x}_s}$ is the partial derivative of f_i with respect to the jth state x_j, evaluated at the steady state $\mathbf{x}_s$. If we define the deviation variable:

$$
\tilde{x}_i(t) = x_i(t) - x_{is}
\tag{11.23}
$$

and recall that at steady state

$$
f_i(\mathbf{x}_s) = 0; \qquad i = 1, 2, ..., n
\tag{11.24}
$$

we obtain the linearized model in terms of deviation variables:

$$
\frac{d\tilde{x}_i(t)}{dt} = \sum_{j=1}^{n} \left[\frac{\partial f_i(\mathbf{x})}{\partial x_j} \right]_{\mathbf{x} = \mathbf{x}_s} \tilde{x}_j(t)
\tag{11.25}
$$
$$
i = 1, 2, ..., n
$$

This can also be put into vector-matrix notation:

$$
\frac{d\tilde{\mathbf{x}}}{dt} = \mathbf{A}\tilde{\mathbf{x}}
\tag{11.26}
$$

where we define:

$$
\tilde{\mathbf{x}} = \begin{bmatrix} \tilde{x}_1 \\ \tilde{x}_2 \\ \cdot \\ \cdot \\ \cdot \\ \tilde{x}_n \end{bmatrix}, \qquad
\mathbf{A} = \begin{bmatrix}
\dfrac{\partial f_1}{\partial x_1} & \dfrac{\partial f_1}{\partial x_2} & \cdot\cdot & \cdot\cdot & \cdot\cdot & \dfrac{\partial f_1}{\partial x_n} \\[2mm]
\dfrac{\partial f_2}{\partial x_1} & \dfrac{\partial f_2}{\partial x_2} & \cdot\cdot & \cdot\cdot & \cdot\cdot & \dfrac{\partial f_2}{\partial x_n} \\[2mm]
\cdot\cdot & & & & & \\[2mm]
\dfrac{\partial f_n}{\partial x_1} & \dfrac{\partial f_n}{\partial x_2} & \cdot\cdot & \cdot\cdot & \cdot\cdot & \dfrac{\partial f_n}{\partial x_n}
\end{bmatrix}_{\mathbf{x} = \mathbf{x}_s}
\tag{11.27}
$$

Now, the approximate linear model for the nonlinear system will be adequate in representing the dynamic behavior as long as the system remains in some domain Ω surrounding the steady state x_s (see Figure 11.6). The bounds of this domain Ω could be, for example, when the maximum error is 1%, or 5%, or some other bound. One thing is clear, however: as the domain shrinks in size, the linear model becomes better and better until there is some region very close to x_s where the error is infinitesimal. This fact is illustrated most clearly in Figure 10.3 of the last chapter for the case of $n = 1$. Thus, analysis of a linearized model for a nonlinear system will give exact results regarding the stability or instability of the steady state x_s. This result, originally due to Liapunov [1, 2], is the basis of all steady-state stability analysis of nonlinear systems.

11.3.2 Main Stability Result

The main result is now stated as follows:

> *To investigate whether a steady state of a nonlinear process is stable or unstable, linearize the process model around the steady state and analyze the linearized model for stability. If the **linearized** model indicates stability of the steady state, then the corresponding nonlinear system will also be stable in the vicinity of this point. If the linearized model indicates instability of the steady state, then the steady state of the original nonlinear system will be unstable.*

Note that this result only refers to system behavior in a small region around the steady state in question. The stability behavior of the nonlinear system for perturbations outside this region must be investigated by other means. Therefore, one must be careful in drawing conclusions for strongly nonlinear systems because the domain for system stability for the steady state in question might be quite small.

11.3.3 Some Applications

The application of the nonlinear stability criterion to some example systems will not only help illustrate how it is used, it will also help illustrate its authenticity. We will consider two examples: the pendulum and the liquid level of the tank process from Examples 10.2 and 10.3.

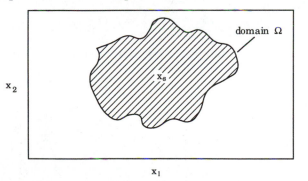

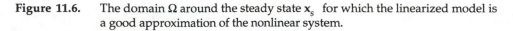

Figure 11.6.   The domain Ω around the steady state x_s for which the linearized model is a good approximation of the nonlinear system.

The Pendulum

As is shown in several texts on classical dynamics, as well as in Ref. [5], the mathematical model for this system, obtained on the basis of Newton's law of motion, is:

$$\frac{d^2\theta}{dt^2} + \frac{k}{m}\frac{d\theta}{dt} + \frac{g}{R}\sin\theta = 0 \tag{11.28}$$

where k is some *positive* constant of proportionality; m and R are respectively the mass of the bob and the length of the connecting rod; and g is the acceleration due to gravity. This model is nonlinear because of the presence of the $\sin\theta$ term.

Steady States (or Equilibrium Points)

Since all time derivatives vanish at steady state, we have from Eq. (11.28) that:

$$\sin\theta_s = 0 \tag{11.29}$$

an equation that holds true for:

$$\theta_s = \pm n\pi; \quad n = 0, 1, 2, \ldots \tag{11.30}$$

Observe, however, that on physical grounds only two of the positions represented by this equation — the ones corresponding to $\theta_s = 0$ and $\theta_s = \pi$ — are truly distinct. Note that these correspond to the "straight down" position of Figure 11.2(a) and the "straight up" position of Figure 11.3, as we had earlier asserted.

Stability Analysis

Let us now apply the stability results of this chapter to determine the stability of this system close to the two steady states, $\theta_s = 0$ and $\theta_s = \pi$.

First let us reduce the second-order model Eq. (11.28) to a set of two first-order equations of the form of Eq. (10.9) by letting $x_1 = \theta$, $x_2 = d\theta/dt$ so that:

$$\frac{dx_1}{dt} = x_2 = f_1(x_1, x_2)$$

$$\frac{dx_2}{dt} = -\frac{k}{m}x_2 - \frac{g}{R}\sin x_1 = f_2(x_1, x_2) \tag{11.31}$$

We can use this form for stability analysis by defining the deviation variables:

$$\tilde{x}_1 = x_1 - x_{1s}$$

$$\tilde{x}_2 = x_2 - x_{2s}$$

and invoking Eq. (11.25):

$$\frac{d\tilde{x}_1}{dt} = \tilde{x}_2 \tag{11.32}$$

$$\frac{d\tilde{x}_2}{dt} = -\frac{g}{R}(\cos x_1)_{x_1 = x_{1s}}\tilde{x}_1 - \frac{k}{m}\tilde{x}_2$$

where the two possible steady states are now represented as:

Steady state 1: $x_{1s} = 0, \; x_{2s} = 0$
Steady state 2: $x_{1s} = \pi, \; x_{2s} = 0$

If we now put Eq. (11.32) in the vector-matrix form of Eq. (11.26), we have:

$$\frac{d\widetilde{\mathbf{x}}}{dt} = \mathbf{A}\widetilde{\mathbf{x}} \tag{11.26}$$

where

$$\widetilde{\mathbf{x}} = \begin{bmatrix} \widetilde{x}_1 \\ \widetilde{x}_2 \end{bmatrix}, \qquad \mathbf{A} = \begin{bmatrix} 0 & 1 \\ -\dfrac{g}{R}(\cos x_{1s}) & -\dfrac{k}{m} \end{bmatrix} \tag{11.33}$$

Observe that in the linearized form (either Eqs. (11.32) or (11.26) and (11.33)), the coefficients of $\widetilde{x}_i$ can contain the steady state-values x_{is} as parameters. This allows the linearization to be valid for all possible steady states.

Recall that the condition for stability of a system of linear first-order equations of the form Eq. (11.26) is that all the eigenvalues λ_i of the matrix $\mathbf{A}$ have negative real parts. The eigenvalues of $\mathbf{A}$ arise from the characteristic polynomial:

$$|\mathbf{A} - \lambda \mathbf{I}| = \left| \begin{bmatrix} 0 & 1 \\ -\dfrac{g}{R}(\cos x_{1s}) & -\dfrac{k}{m} \end{bmatrix} - \begin{bmatrix} \lambda & 0 \\ 0 & \lambda \end{bmatrix} \right| = 0 \tag{11.34}$$

Evaluating the determinant yields:

$$\lambda^2 + \left(\frac{k}{m}\right)\lambda + \left(\frac{g}{R}\right)(\cos x_{1s}) = 0 \tag{11.35}$$

which has roots:

$$\lambda_1, \lambda_2 = -\frac{k}{2m} \pm \frac{1}{2}\sqrt{\left(\frac{k}{m}\right)^2 - \frac{4g}{R}\cos x_{1s}} \tag{11.36}$$

Let us now analyze the stability of each steady state in turn.

Steady state 1, $x_{1s} = \theta_s = 0$ (downward position)

If we substitute $x_{1s} = 0$ in our general linearized model (Eqs. (11.32), (11.33), (11.36)) we see that $\cos(0) = 1$ and the quantity under the square root sign will be certainly less than $(k/m)^2$ so that both λ_1, λ_2 have negative real parts. Thus, steady state 1 is stable.

Steady state 2, $x_{1s} = \theta_s = \pi$ (upward position)

For this steady state, we see that $\cos \pi = -1$ in the linearized model, and the quantity under the square root sign will always be positive and greater than $(k/m)^2$. Thus we will have the situation where $\lambda_1 > 0$, $\lambda_2 < 0$, and the steady state will be unstable.

Note that these results confirm our physical intuition and our earlier conclusions.

Finally, it should be noted that it is also possible simply to linearize the nonlinear model of Eq. (11.28) and analyze the linearized second-order differential equation for stability directly without first converting to the matrix form (see Problem 11.2).

Liquid Level Dynamics in a Tank

The liquid level in a tank has been modeled by Eq. (10.1), i.e.:

$$A \frac{dh}{dt} = F_i - c \sqrt{h} \tag{10.1}$$

whose stability characteristics we now wish to investigate.

A linearization of this model around some arbitrary steady-state level was carried out in Example 10.3 and the linearized transfer function model can be obtained from Eqs. (10.27) and (10.28) as:

$$y(s) = \frac{2h_s^{1/2}/c}{\left(A \frac{2h_s^{1/2}}{c} s + 1 \right)} u(s)$$

indicating that the linearized model has a single pole located at:

$$s = - \frac{c/A}{2h_s^{1/2}} \tag{11.37}$$

We now observe that there is *no* h_s value for which this pole will lie in the right half of the complex plane; we therefore conclude, on the strength of the stability result given above, that this system is always stable!

It is truly remarkable that we can come to such a strong conclusion on the basis of a linearized model. It is, however, easy to see from physical grounds that this conclusion is in fact true.

The nature of the system is such that the flowrate out of the tank is proportional to $\sqrt{h}$, and when inflow balances outflow, we have a steady state. Now, if there is a disturbance of this equilibrium, say, for example, the inflow is increased, the immediate effect is an increase in the liquid level in the tank; but this increased level now causes the outflow to increase, albeit as a nonlinear function.

When the level has increased to such a point where the resulting increase in outflow *balances* the initial inflow increase responsible for the disturbance, we attain another steady state; and the system has regulated itself. Such systems are, therefore, called *self-regulating* systems.

Note that this fact holds regardless of the initial steady state in question; it is only the *rate* at which the new steady state is attained that depends on the actual liquid level at the initial steady state. It is left as an exercise for the reader to confirm that the system attains to a new steady state faster when starting from lower values of the steady-state liquid level.

11.4   DYNAMIC BEHAVIOR OF OPEN-LOOP UNSTABLE SYSTEMS

A system which is inherently unstable is called an open-loop-unstable system in the sense that by itself, without the interference from a feedback controller in the *closed loop*, the system is unstable.

Such systems are represented by linear (or linearized) transfer function models that contain at least one right-half plane (RHP) pole. Thus, for example, the system whose model is given by:

$$g(s) = \frac{K}{\tau s - 1} \tag{11.38}$$

is open-loop unstable because its only pole, located at $s = +1/\tau$, is a RHP pole.

Thus while the presence of a RHP *zero* indicates inverse response, the presence of a RHP pole indicates instability.

11.4.1 Exponential Instability

The unit step response of the system whose transfer function is shown in Eq. (11.38) is easily obtained as:

$$y(t) = K\left(-1 + e^{t/\tau}\right) \tag{11.39}$$

which grows exponentially, indefinitely. In general, whenever the RHP pole responsible for the instability is real, the system step response shows an exponential increase, a situation classified as *exponential instability*.
For the purpose of illustration, Figure 11.7 shows the unit step response of the system whose transfer function is given as:

$$g(s) = \frac{2}{5s - 1} \tag{11.40}$$

Effect of Additional Poles and Zeros

Regardless of whether a system is open-loop stable or not, additional poles (provided they are regular, LHP poles) have the effect of slowing down the system response. For the open-loop-unstable system, having other LHP poles in addition to its RHP pole may slow down the response, but it will not eliminate the instability.

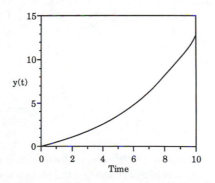

Figure 11.7.   Unit step response of an open-loop-unstable system exhibiting exponential instability.

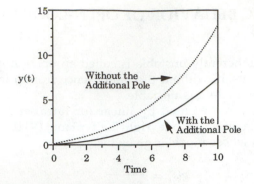

Figure 11.8.   Effect of an additional LHP pole on the unit step response of an open-loop-unstable system exhibiting exponential instability.

This fact is demonstrated in Figure 11.8 which compares the step response of the system in Eq. (11.40) (in the dashed line) with the response (in the solid line) of the system whose transfer function is given by:

$$g(s) = \frac{2}{(5s - 1)(3s + 1)} \tag{11.41}$$

As we already know, zeros tend to speed up the system response. For open-loop-unstable systems, additional zeros therefore tend to quicken the pace of the observed instability.

The unit step response of the system whose transfer function is given by:

$$g(s) = \frac{2(4s + 1)}{(5s - 1)} \tag{11.42}$$

is plotted in the solid line in Figure 11.9 and compared with the original unit step response for the system in Eq. (11.40), and we can clearly observe the acceleration effect of the zero.

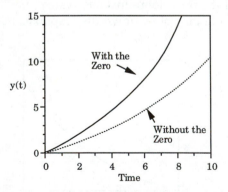

Figure 11.9.   Effect of an additional LHP zero on the unit step response of an open-loop-unstable system exhibiting exponential instability.

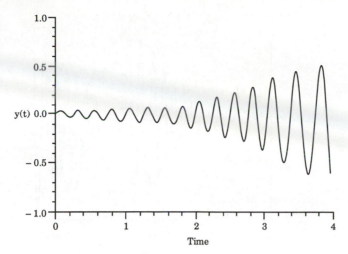

Figure 11.10. Unit step response of an open-loop-unstable system exhibiting oscillatory instability.

11.4.2 Oscillatory Instability

When the open-loop-unstable system poles occur as complex conjugates having positive real parts, the step response will be sinusoidal with increasing amplitude; thus we have oscillatory instability, as distinct from exponential instability in which no oscillations are involved.   We illustrate this phenomenon in Figure 11.10, with the unit step response of the system whose transfer function is given as:

$$g(s) = \frac{10}{(s^2 - 2s + 629)} \tag{11.43}$$

The retardation effect of an additional pole on the step response tends to reduce the amplitude of the oscillations somewhat, but it cannot completely

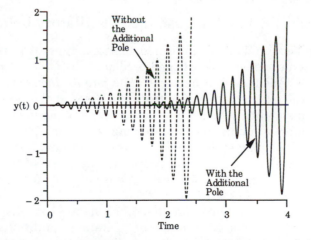

Figure 11.11. Effect of an additional LHP pole on the unit step response of an open-loop-unstable system exhibiting oscillatory instability.

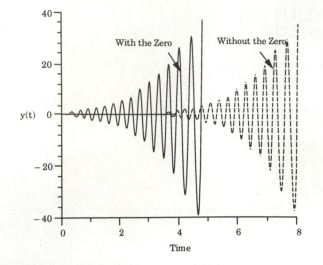

Figure 11.12. Effect of an additional LHP zero on the unit step response of an open-loop-unstable system exhibiting oscillatory instability.

arrest the unstable behavior. This is illustrated in Figure 11.11 (in the solid line) by the response of the system:

$$g(s) = \frac{10}{(s^2 - 2s + 629)(s + 1)}$$ (11.44)

compared to the response of the system of Eq. (11.43) (dashed line).

The unit step response of the system:

$$g(s) = \frac{10(s + 1)}{s^2 - 2s + 629}$$ (11.45)

is shown in Figure 11.12 (solid line), to demonstrate the effect of the additional zero. As expected, the acceleration effect brought about by the zero causes the system instability to become apparent much sooner than in the case of the system Eq. (11.43) without the zero.

11.4.3 Systems Displaying Sustained Oscillatory Behavior

Recall that in Section 11.1 we indicated that one type of dynamic behavior observed for some systems is sustained oscillatory behavior (see Figure 11.1(c) as an example). This type of behavior is quite rare for *linear* systems not under feedback control. It can result only from processes at the boundary between stability and instability when the system poles (or eigenvalues) are purely imaginary complex conjugates (i.e., having zero real parts). This would correspond to a pair of system poles on the imaginary axis in Figure 11.5. However, as we shall see in the example of the next section, many linear systems under feedback control can exhibit this behavior for controller gains at the point of instability of the process.

Many *nonlinear* systems, on the other hand, can display sustained oscillatory behavior for a wide range of operating conditions even without feedback control. Chemical reactors are perhaps the most common processes where such oscillations are seen. The conditions for the onset of these oscillations are quite complex and will not be dealt with in this book. The reader is referred to Refs. [3, 4] for more details.

11.5   STABILITY OF DYNAMIC SYSTEMS UNDER FEEDBACK CONTROL

As we shall see in Part IVA, one of the more important applications of the concept of system stability arises from the fact that a system that is stable by itself can be made unstable by the addition of feedback controllers. Thus a device that is supposed to improve the dynamic behavior of a process has the potential to be detrimental. One of the objectives of control systems design is, therefore, to design control systems that do not jeopardize process stability.

Not to preempt a fuller discussion in Chapters 14 and 15, we only illustrate these ideas here with an example.

Let us consider the problem of controlling the level in Tank 3 of the three tank process shown in Figure 11.13. To accomplish this, the inlet flowrate to the first tank, F_0, will be manipulated based on a signal from a proportional feedback controller. Recall from Chapter 5 that a *linear* model for the liquid level h_i of each tank will take the form:

$$A_1 \frac{dh_1}{dt} = F_0 - c_1 h_1$$

$$A_2 \frac{dh_2}{dt} = c_1 h_1 - c_2 h_2 \qquad (11.46)$$

$$A_3 \frac{dh_3}{dt} = c_2 h_2 - c_3 h_3$$

where A_j, c_j represent the cross-section area and the outlet valve discharge constant for each tank. In terms of deviation variables $y_i = h_i - h_{is}, u = F_0 - F_{0s}$. This becomes:

$$\tau_1 \frac{dy_1}{dt} = -y_1 + K_1 u$$

$$\tau_2 \frac{dy_2}{dt} = K_2 y_1 - y_2 \qquad (11.47)$$

$$\tau_3 \frac{dy_3}{dt} = K_3 y_2 - y_3$$

$$\tau_j = \frac{A_j}{c_j}, \ K_1 = \frac{1}{c_1}, \ K_2 = \frac{c_1}{c_2}, \ K_3 = \frac{c_2}{c_3} \qquad (11.48)$$

Taking the Laplace transform of Eq. (11.47) and relating y_3 to u, one obtains the transform-domain transfer function:

$$y_3(s) = \frac{K}{(\tau_1 s + 1)(\tau_2 s + 1)(\tau_3 s + 1)} u(s) \qquad (11.49)$$

where $K = K_1 K_2 K_3$. From Figure 11.13 we see that y_3 is measured and u is adjusted by the controller. Let us assume for the purpose of our example that each of the tanks is different, and:

$$\tau_1 = 2, \ \tau_2 = 4, \ \tau_3 = 6, \ K = 6$$

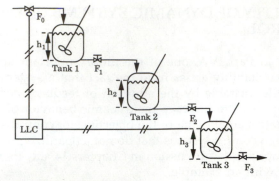

Figure 11.13.   Three noninteracting tanks in series with proportional feedback control.

so that:

$$g(s) = \frac{6}{(2s + 1)(4s + 1)(6s + 1)} \tag{11.50}$$

For the proportional controller with gain K_c, i.e., $u(s) = K_c (y_d (s) - y_3 (s))$ we can see in Figure 11.14 the effect of increasing the controller K_c on the dynamic

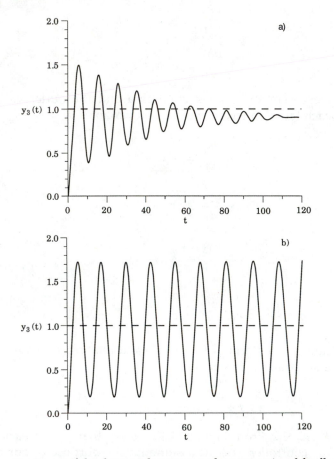

Figure 11.14.   Response of the three-tank system under proportional feedback control to a unit step change in set point; (a) $K_c = 1.0$; (b) $K_c = 1.67$.

response for a unit step change in set point for y_3. Note that for $K_c = 1.0$, the response is stable, while for $K_c = 1.67$ the response is purely oscillatory because the system is at the transition between stability and instability. For $K_c > 1.67$ the process becomes unstable. This example clearly shows that it is possible for this process to become unstable as a result of overly aggressive control action (i.e., by having the controller gain too large).

We note, finally, that just as it is possible for an inherently stable system to be made unstable by feedback control, it is also possible to stabilize an inherently unstable system by feedback control. We will discuss these issues further in Chapter 17.

11.6 SUMMARY

We have focused attention in this chapter on the issue of system stability, and we have provided methods for determining the stability character of both linear and nonlinear systems.

For a linear system to be stable, we have found that the poles of the transfer function must all lie in the left half of the complex plane. This means that an open-loop-unstable system is one whose transfer function possesses at least one RHP pole.

Equivalent time-domain criteria for linear system stability were presented with the result that the eigenvalues of the system coefficient matrix must lie in the left half of the complex plane. Shortcut methods for determining the location of the poles or eigenvalues of the system from the polynomial characteristic equation are presented in Chapter 14.

It comes as a pleasant surprise to find that we can establish the stability of the steady states of a nonlinear system by investigating the behavior of the *linearized* model in the vicinity of the steady state. A few examples were used to illustrate the procedure for carrying out nonlinear stability analysis.

We have also investigated the dynamic behavior of inherently unstable systems, pointing out that they exhibit two kinds of unstable responses: exponential (nonoscillatory) instability, when the RHP poles are real, and oscillatory instability, when the RHP poles occur as complex conjugates. The possibility of destabilization of a process by feedback control was briefly illustrated, awaiting full treatment in Chapters 14 and 15.

REFERENCES AND SUGGESTED FURTHER READING

1. Hahn, W., Theory and Application of Liapunov's Direct Method, Prentice-Hall, New York (1963)
2. Hahn, W., Stability of Motion, Springer-Verlag, New York (1967)
3. Iooss, G. and D. D. Joseph, Elementary Stability and Bifurcation Theory, Springer-Verlag, New York (1981)
4. Guckenheimer, J. and P. Holmes, Nonlinear Oscillations, Dynamical Systems, and Bifurcation of Vector Fields, Springer-Verlag, New York (1983)
5. Coughanowr, D. R. and L. B. Koppel, Process Systems Analysis and Control, McGraw-Hill, New York (1965) (2nd ed., D. R. Coughanowr, McGraw-Hill, New York (1991))

REVIEW QUESTIONS

1. When a process is "disturbed" from an initial steady state, what are the various ways in which it can respond?

2. Why is it necessary to distinguish between the stability of a system *steady state* and the stability of the system itself?

3. What is the difference between the stability definitions 1, 2, and 4 given in the main text?

4. When is a system considered globally stable?

5. Define "Bounded-Input-Bounded-Output" stability. Is this definition applicable to all systems?

6. What are some of the practical issues concerning system stability that arise in process dynamics and control?

7. Is the stability of a linear system steady-state dependent? That is, can a *linear* system be stable at one steady state and be unstable at another?

8. What about the stability of a nonlinear system: is it steady-state dependent?

9. When is a linear system stable? When is it unstable?

10. How does one investigate the stability of a nonlinear system steady state?

11. What characterizes the transfer function of an open-loop unstable system?

12. What influence, if any, do the zeros of a linear system's transfer function exert on its stability?

13. When a linear system exhibits oscillatory instability, what can be said about the nature of the system transfer function poles?

14. Which system is more likely to exhibit sustained oscillatory behavior by itself (without a controller): a linear or a nonlinear system?

15. Is it possible for an otherwise open-loop stable system to be made unstable by the introduction of feedback control?

PROBLEMS

11.1   Given the following system transfer functions:

(a)
$$g(s) = \frac{K}{(\tau s + 1)(s^2 + \omega^2)} \tag{P11.1}$$

(b)
$$g(s) = \frac{K}{(s^2 + 4s + 3)} \tag{P11.2}$$

(c)
$$g(s) = \frac{K}{(s^2 + 2s - 3)} \tag{P11.3}$$

(d)
$$g(s) = \frac{K(s + 1)}{s(s^2 + 4s + 3)} \qquad \text{(P11.4)}$$

investigate each one for bounded-input-bounded-output stability and asymptotic stability.

11.2   Linearize the model in Eq. (11.28) around *each* of the steady states indicated by Eqs. (11.29) and (11.30), and analyze the stability of these steady states directly from the linearized differential equation without using matrix methods.

11.3   The mathematical model of the most important aspect of a pilot scale bioreactor has been reduced to the following bilinear equation:

$$\frac{dY(t)}{dt} = -2.5Y(t) + 0.5Y(t)U(t) + 1.5U(t) \qquad \text{(P11.5)}$$

where $U(t)$ and $Y(t)$ are, respectively, transformed versions of the dimensionless dilution rate, and the dimensionless reactor biomass.
(a) Obtain an expression for the relationship between Y^* and U^*, the respective steady-state values of Y and U. Plot Y^* versus U^*.
(b) A laboratory experiment calls for running the bioreactor at the transformed dilution rate of $U^* = 3.0$. Investigate the stability of the bioreactor at this operating condition.
(c) It has been reported that the reactor has never been successfully operated at the transformed dilution rate of $U^* = 6.0$. Establish that the reason for this is that the conditions correspond to an infeasible steady state, and that even if the steady state were feasible, the reactor will be unstable at such an operating point.

11.4   In Problem 10.9, the model for a free-radical polymerization reactor was given as:

$$\frac{d\xi_1}{dt} = 3(6 - \xi_1) - 1.36594\xi_1\sqrt{\xi_2} \qquad \text{(P10.16)}$$

$$\frac{d\xi_2}{dt} = 60\mu - 3.03161\xi_2 \qquad \text{(P10.17)}$$

$$\frac{d\xi_3}{dt} = -0.00134106\xi_1\sqrt{\xi_2} + 0.034677\,\xi_2 - 10\xi_3 \qquad \text{(P10.18)}$$

$$\frac{d\xi_4}{dt} = 136.758\xi_1\sqrt{\xi_2} - 10\xi_4 \qquad \text{(P10.19)}$$

$$\eta = \frac{\xi_4}{\xi_3} \qquad \text{(P10.20)}$$

where ξ_1, ξ_2, ξ_3, and ξ_4 respectively represent the dimensionless monomer concentration, initiator concentration, bulk zeroth moment, and bulk first moment; η is the number average molecular weight; and μ is the volumetric flowrate of the initiator. Investigate the stability of the following steady-state conditions:

$\xi_1^* = 5.50677$, $\xi_2^* = 0.132906$, $\xi_3^* = 0.001975$, $\xi_4^* = 49.3818$, and $\mu^* = 0.01678$.

11.5   For the nonisothermal CSTR, assuming, for simplicity, that the concentration c_A is being held constant, by analyzing the energy balance equation:

$$\frac{dT}{dt} = -\frac{1}{\theta}T + \beta k_0 e^{-(E/RT)} c_A + \frac{1}{\theta}T_f - \frac{hA}{\rho C_p V}(T - \bar{T}_c)$$

derive the conditions to be satisfied if the reactor is to be stable under these conditions. You may assume that $\bar{T}_c$, the average coil temperature, will be used to control the reactor temperature T.

11.6   Investigate the steady states of the hemispherical gasoline storage tank of Problem 10.4 for stability. The mathematical model was given as:

$$\frac{dh}{dt} = \frac{1}{\pi(2Rh - h^2)}\left(F_i - ch^{1/2}\right) \qquad (P10.9)$$

where R is the radius of the hemispherical tank, and F_i is the flowrate of gasoline into the storage tank. Is the system stable at all steady states? Provide a physical justification of your answer.

11.7   The mathematical model for an exothermic chemical reactor has been reduced by some elementary transformations to:

$$\frac{dY(t)}{dt} = -Y(t) + 2.6U(t) - 2.0U(t)^2 + F(t) \qquad (P11.6)$$

where $Y(t)$ is a transformed reactor conversion, $U(t)$ is the transformed, dimensionless reactor temperature, and $F(t)$ is a transformed, dimensionless reactant inlet concentration (which will be assumed constant, for simplicity).
(a) Linearize this model around the steady state Y^*, U^*, define deviation variables $y = Y - Y^*$ and $u = U - U^*$, obtain a transfer function relating $y(s)$ to $u(s)$, and note the gain and time constant. What happens to the steady-state gain when $U^* = 0.65$?
(b) Obtain a plot of the steady-state relationship between Y^* and U^*; use this plot to explain what happens to the steady-state gain at the operating point corresponding to $U^* = 0.65$.
(c) Investigate the stability of this reactor at every steady-state operating point *excluding the point $U^* = 0.65$*.

11.8   Consider the situation in which the transfer function of a dynamic system, $g^+(s)$, is generated according to:

$$g^+(s) = \frac{g(s)\, K}{1 + g(s)K} \qquad (P11.7)$$

(a) For $g(s)$ given as:

$$g(s) = \frac{6}{(2s + 1)} \qquad (P11.8)$$

show that there is no value of $K > 0$ for which the system represented by $g^+(s)$ will be unstable.
(b) Show also that for $g(s)$ given as:

$$g(s) = \frac{6}{(2s + 1)(4s + 1)} \qquad (P11.9)$$

once again, there is no value of $K > 0$ for which the system represented by $g^+(s)$ will be unstable.

(c) When $g(s)$ is given as:

$$g(s) = \frac{6}{(2s + 1)(4s + 1)(6s + 1)} \tag{P11.10}$$

determine the stability of $g^+(s)$ for $K = 1$ and $K = 2$. You may find it useful to know the following result:

The cubic equation:

$$a_0 s^3 + a_1 s^2 + a_2 s + a_3 = 0$$

will have no roots in the right half of the complex plane if and only if:

$$a_0, a_1, a_2, a_3 > 0; \text{ and}$$
$$a_1 a_2 > a_0 a_3$$

11.9 (a) When an open-loop-unstable system with the transfer function:

$$g(s) = \frac{4.5}{(3s - 1)} \tag{P11.11}$$

is under proportional-only feedback control, it is easy to show (see Chapters 14, 15, and 17) that the transfer function for the resulting "closed-loop" system is given by:

$$g_{CL}(s) = \frac{gK_c}{1 + gK_c} \tag{P11.12}$$

where K_c is the proportional controller gain. Evaluate $g_{CL}(s)$ in this case and determine the conditions that K_c must satisfy so that the closed-loop system is now stable; show that this condition is *entirely independent* of the location of the open-loop-unstable system's right-half plane pole.
(b) Consider now the case that the open-loop-unstable system is given by:

$$g(s) = \frac{4.5}{(3s - 1)(\tau_2 s + 1)} \tag{P11.13}$$

which differs from the transfer function given in Eq. (P11.11) only by the presence of the additional pole: show that under proportional-only control, the conditions for closed-loop stability in this case now depend on the location of the additional left-half plane pole relative to the location of the unstable pole in the right half of the complex plane.

part III

PROCESS MODELING
AND
IDENTIFICATION

Three organically related elements are essential for effective control systems analysis and design: the *inputs* to the process, the *process model* as an abstraction of the process itself, and the *outputs* from the process. Once any two have been specified, the third element may then in principle be derived.

When the *input* is specified along with the *model*, and the resulting process *output* is to be derived, this is the *Process Dynamics* problem, a rather comprehensive treatment of which has just been completed in Part II for various processes represented by diverse models. When the desired process *output* is specified along with the process *model*, and the *input* required to produce such an output is to be derived, this is the *Process Control* problem; it will be discussed in Part IV. Deriving the process model from process input and output information (and perhaps some knowledge of the fundamental laws governing process operation) is the *Process Modeling* problem, quite often the most important (and most difficult) problem of all; it will be the focus here in Part III consisting of Chapters 12 and 13.

Chapter 12 is devoted to the development of theoretical process models from *first principles*, using material and energy balances (and fundamental physical/chemical laws) to determine the model equation structure, and estimating unknown model parameters from the process input and output data. Chapter 13 is devoted to *Process Identification*: the alternative approach of developing purely empirical models, in which the model form, and all the model parameters, are obtained strictly from input and output data acquired from the operating process. The relative merits of each approach will be discussed in the context of examples illustrating model development for distillation columns, chemical reactors, etc.

part III

PROCESS MODELING AND IDENTIFICATION

"...the Science of Deduction and Analys
is one which can only be acquir
by long and patient stud
nor is life long enough to allow any mort
to attain the highest possible perfection in it

Sherlock Holmes, *A Study in Scar*
(Sir Arthur Conan Doyl

"Aristotle could have avoided the mista
of thinking that women had fewer teeth than m
by the simple device
asking Mrs Aristotle to open her mouth

Bertrand Russell (1872–197

CHAPTER

12

THEORETICAL PROCESS MODELING

In Chapter 4, we had our first contact with the process model primarily as a tool of dynamic analysis. There the various forms of process models were presented and the interrelationships between them were discussed. In the rest of the book, the model is used as a tool for process analysis and for the design of control systems. *The question of how one obtains good process models will now be dealt with in this chapter.* In this regard, there are three fundamental issues that will occupy our attention: model formulation, parameter estimation, and model validation.

Regarding model formulation, the principles involved will be illustrated by several examples covering a spectrum of chemical processes; regarding the other two issues, even though the scope of coverage will be limited by space, we will examine enough of the principles involved to enable the reader to obtain a flavor for what is required in estimating the unknown parameters that show up in a process model, and how to check the adequacy of process models before they are used for process control or other applications.

12.1   INTRODUCTION

Inasmuch as entire books have been written on process modeling (cf. Refs. [1–3]), our objective in this chapter is not to present a comprehensive treatment of theoretical modeling. Further, since mathematical modeling is recognized as a curious mixture of "art" and science, the *subjective* and the *objective*, neither is it our objective to present a rigid, cookbook procedure for developing theoretical models. Instead, our aim is to present first the philosophy and principles involved in modeling the behavior of process systems, and then to illustrate the application of these principles with several examples.

Process Modeling is not a rigid science that can be practiced in only one "right" way, but neither is it amorphous and unstructured: there is some technique involved, and the procedure, even if not rigid, must be systematic if the modeling exercise is to be carried out successfully and efficiently.

12.1.1 Modeling the Behavior of Dynamic Processes

In modeling the behavior of dynamic processes, the main objective may be reduced to that of obtaining a functional relationship between the various process variables which explains the observed process behavior. The underlying belief is that a true — but unknown — fundamental relationship, dictated by natural laws, exists between the process variables, and our modeling exercise is directed towards discovering this relationship.

It is also understood right at the outset that it is impossible to portray perfectly all the details of the actual process behavior in mathematical form. Therefore, *theoretical modeling* seeks to arrive at a fundamental modeling relationship through systematic application of the laws of nature to the *most important phenomena* assumed to determine the behavior of the process. The end result is known as a theoretical process model.

On the other hand, *empirical modeling* seeks to approximate this unknown functional relationship by some (usually simple) mathematical functions, using information gathered experimentally from the system. Because they depend entirely on experimental information and experience, the approximating mathematical functions obtained are known as *empirical* models.

It is typical to adopt the theoretical modeling approach when the underlying mechanisms by which a process operates are reasonably well understood. When the process is too complicated for the theoretical approach (usually because very little information about the fundamental nature of the process is available, or the theoretical model equations are enormously complex) the empirical approach is the appropriate choice. The entire subject of constructing a process model from purely experimental information (with no recourse to any theoretical bases for the observed behavior) is known as *Process Identification*; it is the subject of Chapter 13.

Although these two approaches to process modeling appear to be mutually independent, we will soon see that theoretical modeling cannot be completed without some experimentation; in Chapter 13, we will also find that some theoretical considerations are required for empirical modeling to be successful.

12.1.2 Procedure for Theoretical Process Model Development

To start from a physical system and ultimately abstract from it a process model that can be used as its surrogate is a process that consists of several stages:

Stage 1:   Problem Definition

There are several factors that make it very important to have defined very clearly the scope of the problem we wish to solve; some of these are listed below:

1. It is impossible to represent all aspects of the physical process; we can only hope to capture those aspects that are most relevant to the problem at hand.

2. The behavior of a dynamic process can be interpreted mathematically in several different ways, the various phenomena explained to varying degrees of detail: the result is that several different models are

possible for any given process, all of which might even attempt to capture the same aspect of the process, but from various angles, and to varying degrees of complexity.

3. A process model is as useful as the tools available for obtaining solutions to its equations; some modeling equations can be solved analytically, while some can only be solved by numerical methods using a computer.

As a result, before embarking on the actual task of developing a mathematical model for a physical process, a number of questions — which are by no means independent of one another — should first be answered:

1. What do we intend to use the model for?
2. How simple, or complex, will the model have to be?
3. Which aspects of the process do we consider the most relevant and therefore should be contained in such a process model?
4. To what extent are the fundamental principles underlying the operation of this aspect of the process known?
5. How can we test the adequacy of the model?
6. How much time do we have for the modeling exercise?

.

.

.

The answer to these questions will enable us decide, for example, whether to use the theoretical approach or the alternative empirical approach.

Stage 2:   *Model Formulation*

What this stage entails for theoretical model development was briefly introduced in Section 4.2 of Chapter 4 and will be discussed in more detail below. This stage in empirical model development (or Process Identification) will be discussed in Chapter 13.

Stage 3:   *Parameter Estimation*

In developing a model for a physical process (whether by theoretical or empirical means), certain parameters appear whose values must be specified before the model can be used to predict process behavior. For example, the theoretical model obtained for the nonisothermal CSTR in Eqs. (4.26) and (4.27):

$$\frac{dc_A}{dt} = -\frac{1}{\theta}c_A - k_0\, e^{-(E/RT)}c_A + \frac{1}{\theta}c_{Af} \tag{4.26}$$

$$\frac{dT}{dt} = -\frac{1}{\theta}T + \beta k_0\, e^{-(E/RT)}c_A + \frac{1}{\theta}T_f - \chi \tag{4.27}$$

contains the following parameters:

θ, the reactor residence time
$k = k_0\, e^{-(E/RT)}$, the reaction rate constant
$\beta = (-\Delta H)/\rho C_p$, a ratio made up of the heat of reaction divided by the density and specific heat capacity of the reactor content.

For any process model a subset of the parameters may be known *a priori*, available in the literature, or estimated from independent experiments performed, for example, on a pilot plant, or in the laboratory; the remaining unknown parameters must be determined through other means. It is customary to obtain such parameter estimates using experimental data obtained directly from the physical process.

Stage 4:   Model Validation

Before proceeding to use the model, it is essential to evaluate how closely it predicts the behavior of the physical system it is supposed to represent. This is usually accomplished by testing the model against additional process data in the context of our process experience.

Let us now consider, in order, the issues involved with Model Formulation, Parameter Estimation, and Model Validation for the theoretical modeling approach; a similar discussion for the empirical approach will be presented in Chapter 13.

12.2   DEVELOPMENT OF THEORETICAL PROCESS MODELS

The detailed steps in Theoretical Process Modeling are illustrated in Figure 12.1. The physical laws governing the process are analysed to produce modeling equations, which are usually complex and nonlinear, and may involve partial differential or integral equations. This model is usually simplified to a form that is easily handled, but that still retains the essential features of the process. At this point the equation structure of the model is fixed. Next, those model parameters that are completely unknown or not known precisely enough are estimated using process data taken from dynamic input/output experiments performed on the actual process. As indicated in the figure, various types of errors are introduced at each stage of the model development:

1. *Model Formulation Errors:* In postulating the physical phenomena and applying their laws, approximations are made to the model equations that introduce errors.
2. *Model Simplification Errors:* In simplifying the model to make it mathematically and numerically tractable, approximations are made that introduce errors.
3. *Experimental Measurement Errors:* In collecting the process data, both systematic calibration errors and random experimental errors occur. In addition, there can be instrument or human errors resulting in erroneous or missing data.
4. *Parameter Estimation Errors:* In addition to the effects of experimental measurement errors on estimated parameters, there are other sources of parameter estimation errors. For example, inadequate experimental design may produce data sets that intrinsically lack the essential information necessary for obtaining reasonable parameter estimates. Choosing a parameter estimation algorithm that is inappropriate for the data set may also result in poor parameter estimates.

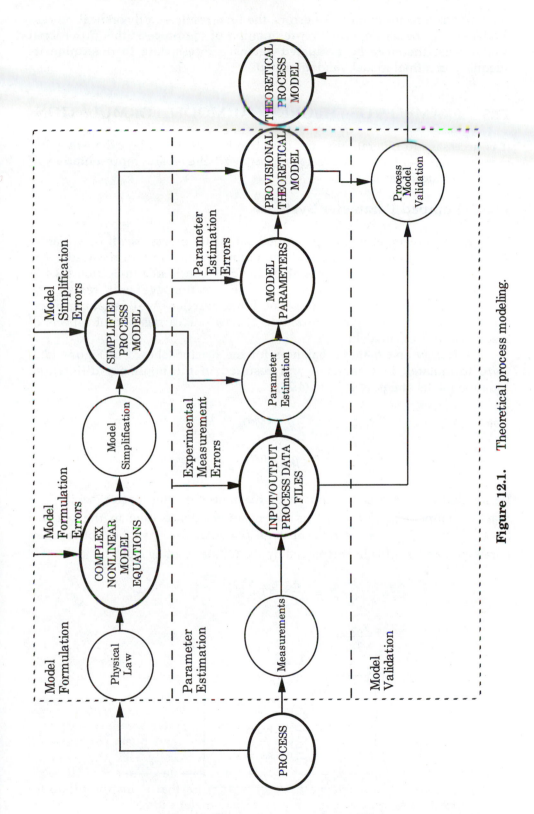

Figure 12.1. Theoretical process modeling.

Primarily as a result of all these errors, the final result — a theoretical process model — will be an imperfect representation of the process; this provisional model must therefore be evaluated against process data to determine its adequacy in a final model validation step.

12.3   EXAMPLES OF THEORETICAL MODEL FORMULATION

It is necessary to reiterate that the objective here is not to be exhaustive, but merely to present illustrative examples we believe are representative of typical chemical processes.

12.3.1 Lumped Parameter Systems

In Chapter 4 several examples of lumped parameter processes whose variables are essentially uniform throughout the entire system were introduced. The stirred heating tank and the nonisothermal CSTR considered in Section 4.2.4 of Chapter 4 are typical examples. In this section we will revisit the nonisothermal CSTR in which the exothermic reaction A→B is taking place. We will use this process as an example of how one carries out all aspects of theoretical process modeling.

The energy and material balances for the reactor (shown in Figure 12.2) were formulated in Chapter 4 and resulted in the nonlinear differential equation model of Eqs. (4.26) and (4.27):

$$\frac{dc_A}{dt} = -\frac{1}{\theta}c_A - k_0\,e^{-(E/RT)}c_A + \frac{1}{\theta}c_{Af} \tag{12.1}$$

$$\frac{dT}{dt} = -\frac{1}{\theta}T + \beta k_0\,e^{-(E/RT)}c_A + \frac{1}{\theta}T_f - \chi \tag{12.2}$$

Recall that c_A is the concentration of reactant A in the reactor, while T is reactor temperature, $\theta = V/F$ is the mean residence time in the reactor, $k = k_0 e^{-\{E/RT\}}$ is the reaction rate constant, $\beta = (-\Delta H)/\rho C_p$ is a heat of reaction parameter, and χ is the heat removed by the tubular cooling coil:

$$\chi = \frac{hA}{\rho C_p V}(T - \bar{T}_c) \tag{12.3}$$

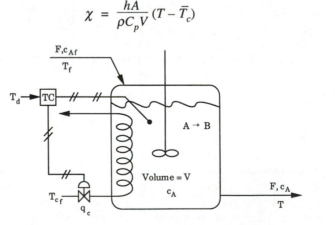

Figure 12.2.   The nonisothermal CSTR.

where h is the coil heat transfer coefficient, A the total coil heat transfer area, and:

$$\bar{T}_c = \frac{1}{L} \int_0^L T_c(z)dz \tag{12.4}$$

the average temperature along the length of the coil $0 < z < L$. It is known (Aris [4]) that χ depends nonlinearly on the coolant flowrate q_c according to:

$$\chi = U q_c \left(1 - e^{-(\alpha/q_c)}\right) (T - T_{cf}) \tag{12.5}$$

where T_{cf} is the coolant inlet temperature, and $\rho_c C_{pc}$ is the volumetric heat capacity of the coolant. In addition, the parameters U, α are given by:

$$U = \frac{\rho_c \, C_{pc}}{\rho C_p \, V} \tag{12.6}$$

$$\alpha = \frac{hA}{\rho_c \, C_{pc}} \tag{12.7}$$

Observe that our model has five parameters, β, U, k_0, E, α; two of these (β and U) are typically available in the literature; and the other three are to be determined. In addition there are five variables c_{Af}, θ, T_f, T_{cf}, q_c determined by the reactor operating conditions. In subsequent sections of this chapter, we shall show how the unknown parameters are estimated and how the final CSTR model is validated against process data.

12.3.2   Stagewise Processes

There are many processes in chemical engineering that are carried out in stages: gas absorption, extraction, and distillation, for example, are carried out on several individual plates enclosed in a column, or alternately, carried out in packed columns with "equivalent stages" equal to a specified length of packing in the column. Leaching tanks, and multistage chemical reactor trains, are also examples of stagewise processes.

These processes are characterized by the occurrence of *stepwise changes* in the value taken by the process variables from stage to stage. We will illustrate how such processes are modeled, and show the typical form their theoretical models take, using an example of the binary distillation column — one of the most important chemical processes.

Following the distillation column model proposed by Luyben [5], let us consider the schematic diagram of a typical distillation column, shown in Figure 12.3, where there is a total of N trays, numbered from the bottom to the top. The distillation column is used to separate a binary mixture of two components, A and B, the former being the lighter component.

The feed, a saturated liquid containing x_F mole fraction of component A, is fed to the column at molar flowrate F (mole/time); the point of entry into the column is tray f, the feed tray.

The overhead vapor is condensed by the indicated heat exchanger arrangement, using cooling water; the condensed liquid flows into the reflux

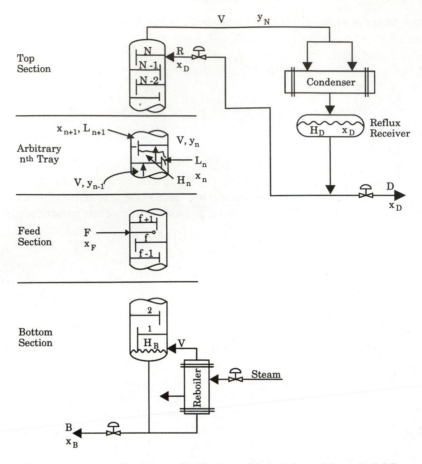

Figure 12.3.   The binary distillation column (adapted from Ref. [5]).

receiver from where it is returned as reflux to the column, at molar flowrate R; and overhead distillate product is withdrawn at molar flowrate D. The content of the reflux receiver is assumed to have a uniform composition x_D, and the liquid holdup is H_D. Note that the composition of the distillate product stream and the reflux stream will be the same as x_D.

The vapor flowing through the column at the molar rate V (mole/time) is generated at the base of the column through the steam-heated, thermo-siphon reboiler arrangement. Bottoms liquid, of composition x_B, is withdrawn as product at the molar flowrate B, while the rest is sent through the reboiler. The liquid holdup at the bottom of the column is H_B.

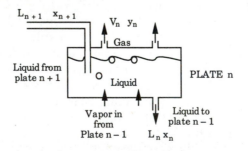

Figure 12.4.   Plate n of a binary distillation column.

The activities taking place on a typical tray n are shown in Figure 12.4. At each step a vapor flowing upward is brought into contact with the liquid at that stage and achieves thermodynamic equilibrium. The liquid from the nth stage flows downward to the next lowest stage in the column. Here:

x_n = mole fraction of material A in the liquid phase
y_n = mole fraction of material A in the vapor phase
L_n = liquid flowrate *out of* plate n
V_n = vapor flowrate *out of* plate n

If the liquid hold up (total volume of accumulated liquid) on plate n is H_n, and the vapor holdup is h_n, assuming there are no heat effects, a theoretical model for this process may now be obtained by taking a component A mass balance on tray n as follows:

Component A Mass Balance

Rate of material *input* with
liquid coming in from plate $n + 1$ $= L_{n+1}\, x_{n+1}$

Rate of material *input* with
vapor coming in from plate $n - 1$ $= V_{n-1}\, y_{n-1}$

Rate of material *output* with
liquid leaving plate n $= L_n\, x_n$
(for plate $n - 1$)

Rate of material *output* with
vapor leaving plate n $= V_n\, y_n$
(for plate $n + 1$)

Rate of accumulation of A
in the liquid phase on plate n $= \dfrac{d}{dt}\left(H_n\, x_n \right)$

Rate of accumulation of A
in the vapor phase on plate n $= \dfrac{d}{dt}\left(h_n\, y_n \right)$

The material balance equations for the typical nth tray are therefore given by:

$$\frac{d}{dt}\left(H_n\, x_n \right) + \frac{d}{dt}\left(h_n\, y_n \right) = L_{n+1}\, x_{n+1} + V_{n-1}\, y_{n-1} - L_n\, x_n - V_n\, y_n \qquad (12.8)$$

$$\frac{dH_n}{dt} = L_{n+1} - L_n \qquad (12.9)$$

$$\frac{dh_n}{dt} = V_{n-1} - V_n \qquad (12.10)$$

These form the basic building blocks for modeling the entire distillation column composed of several such stages.

Overall Problem Definition and Assumptions

In constructing a model for the complete multistage distillation column we must represent the dynamic behavior of the mole fraction of the lighter component A on each tray, as well as in the overhead and bottom product streams. (Frequently, it is the latter two compositions that are really of importance, but as we shall see, these depend on the compositions of the other trays.) Thus, the complete theoretical model for this column will involve overall balance equations written for each of the N trays, as well as for the condenser/receiver, and the reboiler.

It is now convenient to divide the column up into distinct sections, each with its own distinguishing characteristics as follows:

1. The condenser/receiver,
2. The top tray (Tray N), different from all the other trays because it is the point of entry of the reflux stream),
3. An arbitrary tray n representative of all the other trays like it,
4. The feed tray (Tray f), different from the others because it is the point of entry of the feed stream,
5. The first tray,
6. The reboiler and column base.

The following assumptions are typically used to simplify the task of formulating the balance equations for each section of the distillation column:

1. The liquid on each tray is perfectly mixed and of uniform composition x_n.
2. There are negligible heat losses from the column to the atmosphere.
3. The vapor holdup h_n on each tray is negligible.
4. The molal heat of vaporization of components A and B are approximately equal, with the implication that for each mole of vapor that condenses, exactly enough heat required to vaporize one mole of liquid is released (this is the *equimolal overflow* assumption).
5. The relative volatility α is assumed constant throughout the column; this implies that the effect of temperature on the vapor pressures of A and B is approximately the same.
6. Plate efficiencies of 100% are assumed, implying that the vapor leaving each tray is in equilibrium with the liquid accumulated on the tray. (This means that a column with lower plate efficiencies is modeled by fewer trays at 100% efficiency.)
7. The dynamics of the reboiler and condenser heat exchangers may be ignored.

Assumption 1 allows us to carry out overall balances around each tray; assumptions 2, 3, and 4 imply that the vapor rate through the column is the same from tray to tray, i.e.:

$$V = V_n; \quad n = 1, 2, \ldots, N$$

and also that the temperature on each tray is determined from thermodynamic equilibrium, making an energy balance unnecessary.

Assumptions 5 and 6 provide us with one of the constitutive relations needed for this system: for each tray, the mole fraction of A in the vapor phase is related to the liquid phase mole fraction by:

$$y_n = \frac{\alpha x_n}{1 + (\alpha - 1)x_n} \tag{12.11}$$

Assumption 7 allows us to treat the heat exchangers used as the condenser and the reboiler as units which perform their tasks instantaneously. This is obviously not entirely realistic, but neither is it completely unreasonable: the dynamic responses of these heat exchangers occur on a much faster timescale than the response of most industrial distillation columns. Nevertheless, we recognize that these heat exchangers are processing units in their own rights, and the dynamic models for such units will be obtained in the next section.

We are now in a position to obtain the mass balance equations that constitute the theoretical model for this system, starting from the top and working our way to the bottom of the distillation column.

CONDENSER AND REFLUX RECEIVER

- *Component A Mass Balance:*

$$\frac{d}{dt}\left(H_D x_D\right) = V y_N - (R + D) x_D \tag{12.12}$$

- *Total Mass Balance:*

$$\frac{dH_D}{dt} = V - R - D \tag{12.13}$$

TOP TRAY (TRAY N)

- *Component A Mass Balance:*

$$\frac{d}{dt}\left(H_N x_N\right) = V y_{N-1} + R x_D - V y_N - L_N x_N \tag{12.14}$$

- *Total Mass Balance:*

$$\frac{dH_N}{dt} = R - L_N \tag{12.15}$$

ARBITRARY TRAY n ($n = 2, 3, \ldots, N - 1; n \neq f$)

- *Component A Mass Balance:*

$$\frac{d}{dt}\left(H_n x_n\right) = V y_{n-1} + L_{n+1} x_{n+1} - V y_n - L_n x_n \tag{12.16}$$

- *Total Mass Balance:*

$$\frac{dH_n}{dt} = L_{n+1} - L_n \tag{12.17}$$

FEED TRAY

- *Component A Mass Balance:*

$$\frac{d}{dt}\left(H_f x_f\right) = V y_{f-1} + L_{f+1} x_{f+1} - V y_f - L_f x_f + F x_F \tag{12.18}$$

- *Total Mass Balance:*

$$\frac{dH_f}{dt} = L_{f+1} - L_f + F \tag{12.19}$$

FIRST TRAY

- *Component A Mass Balance:*

$$\frac{d}{dt}\left(H_1 x_1\right) = V y_B + L_2 x_2 - V y_1 - L_1 x_1 \tag{12.20}$$

- *Total Mass Bulance:*

$$\frac{dH_1}{dt} = L_2 - L_1 \tag{12.21}$$

COLUMN BASE AND REBOILER

- *Component A Mass Balance:*

$$\frac{d}{dt}\left(H_B x_B\right) = L_1 x_1 - V y_B - B x_B \tag{12.22}$$

- *Total Mass Balance:*

$$\frac{dH_B}{dt} = L_1 - V - B \tag{12.23}$$

Constitutive Relations

For each of the component balance equations given above, the mole fraction of A in the vapor phase can be related to the liquid phase mole fraction by using Eq. (12.11).

With the exception of the holdup in the reflux receiver and the column base (which are usually kept under feedback control) it is possible to relate the rate of liquid flow from each tray to the holdup on the tray using the Francis weir formula:

$$L_n = 3.33 l \left(\delta_n\right)^{3/2} \tag{12.24}$$

where l is the length of the weir (ft); δ_n is the height of the liquid above the weir (ft) on tray n. From a knowledge of the tray geometry, and the (average) density of material on each tray, this expression could be used to relate the liquid holdup H_n and flowrate L_n.

The equations may now be made more compact and rearranged in one of the standard model forms of Chapter 4. Substituting the total material balances (12.13, 12.15, 12.17, 12.19, 12.21, 12.23) into the component mass balances gives the equations:

$$H_D \frac{dx_D}{dt} = V(y_N - x_D) \tag{12.25}$$

$$H_N \frac{dx_N}{dt} = V(y_{N-1} - y_N) + R(x_D - x_N) \tag{12.26}$$

$$H_n \frac{dx_n}{dt} = V(y_{n-1} - y_n) + L_{n+1}(x_{n+1} - x_n) \tag{12.27}$$

$$n = 2, 3, \ldots, N-1; n \neq f$$

$$H_f \frac{dx_f}{dt} = V(y_{f-1} - y_f) + L_{f+1}(x_{f+1} - x_f) + F(x_F - x_f) \tag{12.28}$$

$$H_1 \frac{dx_1}{dt} = V(y_B - y_1) + L_2(x_2 - x_1) \tag{12.29}$$

$$H_B \frac{dx_B}{dt} = V(x_B - y_B) + L_1(x_1 - x_B) \tag{12.30}$$

Note that we have $(N + 2)$ differential equations for the liquid phase compositions, and $(N + 2)$ additional differential equations for the liquid holdups; and the equations are all nonlinear and coupled. The vapor phase compositions can be eliminated by using the equilibrium relation Eq. (12.11), while the liquid flowrates L_n can be related to liquid holdups, H_n, by Eq. (12.24).

The manipulated (control) variables are reboiler vapor rate V, resulting from adjusting the steam to the reboiler, and the reflux rate R (or alternatively the distillate product flowrate D). The disturbances are feedrate F, and feed composition x_f. The measured output variables are typically the composition of the overhead and bottoms products, x_D, x_B. Thus the nonlinear model would take the form of Eq. (4.31):

$$\frac{d\mathbf{x}(t)}{dt} = \mathbf{f}(\mathbf{x}(t), \mathbf{u}(t), \mathbf{d}(t); \theta)$$

$$\mathbf{y}(t) = \mathbf{h}(\mathbf{x}(t))$$

where the state vector $\mathbf{x}$ is of dimension $2N + 4$ and consists of all liquid compositions and liquid holdups, while the output vector $\mathbf{y}$ is of dimension 2, the disturbance vector $\mathbf{d}$ has dimension 2, and the manipulated variable vector $\mathbf{u}$ also has dimension 2. This is seen explicitly in Eq. (12.31):

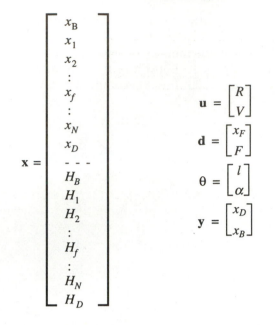

$$\mathbf{x} = \begin{bmatrix} x_B \\ x_1 \\ x_2 \\ \vdots \\ x_f \\ \vdots \\ x_N \\ x_D \\ --- \\ H_B \\ H_1 \\ H_2 \\ \vdots \\ H_f \\ \vdots \\ H_N \\ H_D \end{bmatrix} \qquad \begin{aligned} \mathbf{u} &= \begin{bmatrix} R \\ V \end{bmatrix} \\ \mathbf{d} &= \begin{bmatrix} x_F \\ F \end{bmatrix} \\ \mathbf{\theta} &= \begin{bmatrix} l \\ \alpha \end{bmatrix} \\ \mathbf{y} &= \begin{bmatrix} x_D \\ x_B \end{bmatrix} \end{aligned} \qquad (12.31)$$

Observe that if there are 50 trays in the column, the model would consist of 104 nonlinear differential equations even though we have only two outputs and two control variables. In Chapter 13 we demonstrate that in some cases, simple empirical models often do a fine job of representing this two-input/two-output system with much less work.

In conclusion, we note that the situation with most industrial distillation columns is more complicated: the feed mixtures are usually multicomponent; the trays are not 100% efficient; the vapor holdups on the trays are not negligible; there are multiple feed entry points and multiple sidestream draw-offs; and the condenser and reboiler heat exchanger dynamics may be important.

However, the basic principles involved in developing the more complicated model (which takes all these factors into consideration) are exactly the same as those illustrated with the relatively simpler situation we have just dealt with. Such a complex theoretical distillation column model has been derived in Ref. [5].

12.3.3   Distributed Parameter Systems

As we have seen, the process variables of a lumped parameter system can be assumed to be homogeneous within the process, so that spatial variations can be considered as negligible. The resulting theoretical models therefore involve ordinary differential equations expressing time variations of these process variables.

At the next level are the stagewise processes for which, even though we can no longer assume total homogeneity within the entire process, the spatial variations have been localized to finite stages; but each of these stages is considered as a lumped parameter system in its own right. The resulting theoretical models involve sequences of differential equations *jointly* expressing the time variation of the process variable *on each stage*, as well as the stage-to-stage variations.

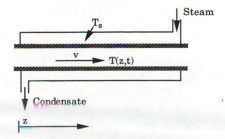

Figure 12.5.   The steam-heated shell-and-tube heat exchanger.

We are now interested in systems in which the changes in spatial position cannot be quantized into stages as with stagewise processes; the theoretical modeling equations will therefore have to be partial differential equations, if the variations in both space and time are to be represented properly.

Shell-and-Tube Heat Exchanger

The process in question, first introduced in Chapter 8, is shown in Figure 12.5.

The Process

A fluid with constant density ρ, and specific heat capacity C_p is flowing through the tube of a shell-and-tube heat exchanger with velocity v, as shown in Figure 12.5; it enters at a temperature T_0 and is heated from the shell side by condensing steam. The tube has a uniform cross-sectional area A, and the surface area available for heat transfer *per unit length* is A_s.

It is clear that the fluid temperature will vary along the length of the heat exchanger, and also with time at any particular point. Thus fluid temperature is both a function of time as well as spatial position along the length of the tube.

Problem Definition and Assumptions

We wish to develop a theoretical model that will predict the variation of the fluid temperature both with time and with position.

To do this we will consider changes taking place during the infinitesimally small time interval $[t, t + \Delta t]$, in the space contained within the infinitesimally sized elemental ring section shown in Figure 12.6, whose boundaries are at z and $z + \Delta z$. The following assumptions are used in conjunction with this *microscopic* description of the process:

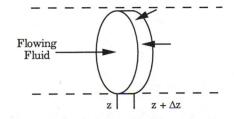

Figure 12.6.   Shell balance for heat exchanger.

1. The material within the element is at a uniform temperature T, but the temperature at the boundaries of the element are $T|_z$ and $T|_{z+\Delta z}$.

2. The fluid is flowing through with a flat velocity profile: this eliminates the need to model variations in the radial direction.

3. Physical fluid properties are assumed constant; and the cross-sectional area of the heat exchanger is assumed uniform.

4. The dynamics of the tube and shell walls are assumed small enough to be negligible: thus any accumulation of energy within the elemental ring is due entirely to the fluid occupying the element, not the heat exchanger wall.

5. The steam on the shell side is assumed to be at a constant temperature T_{st} and heat exchange across the tube wall is assumed to follow the Newton's "law" of cooling.

Microscopic Energy Balance

Amount of energy *in* at z: $\rho A v C_p \, \Delta t \, (T - T^*)|_z$
with flowing fluid,
over the time interval $[t, t + \Delta t]$

Amount of energy *out* at $z + \Delta z$: $\rho A v C_p \Delta t \, (T - T^*)|_{z+\Delta z}$
with flowing fluid,
over the time interval $[t, t + \Delta t]$

Amount of energy in through: $h A_s \Delta z \Delta t \left(T_{st} - T\right)$
the heat exchanger wall
over the time interval $[t, t + \Delta t]$

(Note that the total heat transfer area on this elemental ring is $A_s \, \Delta z$.)

Amount of energy accumulated: $\rho A C_p \, \Delta z \left(T|_{t+\Delta t} - T|_t\right)$
over the time interval $[t, t + \Delta t]$

An energy balance now demands that the total amount of energy accumulated in the element, over the time interval $[t, t + \Delta t]$, be equal to the total amount coming into the element less the total amount leaving within the same time interval. Mathematically, this means:

$$\rho A C_p \, \Delta z \left(T|_{t+\Delta t} - T|_t\right) = \rho A v C_p \, \Delta t \, (T - T^*)|_z - \rho A v C_p \, \Delta t \, (T - T^*)|_{z+\Delta z}$$
$$+ \, h A_s \, \Delta z \Delta t \left(T_{st} - T\right)$$

Dividing through by $\rho A C_p \Delta z \Delta t$ gives:

$$\frac{\left(T|_{t+\Delta t} - T|_t\right)}{\Delta t} = \frac{\left(v T|_z - v T|_{z+\Delta z}\right)}{\Delta z} + \frac{h A_s}{\rho A C_p}\left(T_{st} - T\right) \qquad (12.32)$$

and now, if we take the limits as Δz and Δt simultaneously tend to zero, recalling the definition of the partial derivative of a function, Eq. (12.32) becomes:

$$\frac{\partial T(z, t)}{\partial t} + v \frac{\partial T(z, t)}{\partial z} = \beta \left[T_{st} - T(z, t) \right] \qquad (12.33)$$

a linear, first-order PDE, where:

$$\beta = \frac{hA_s}{\rho A C_p} \qquad (12.34)$$

The following are some important points regarding this model:

1. Presented as in Eq. (12.33), the model has two parameters, β and v. In principle, the fluid velocity v is known so that only β needs to be estimated before the model can be used to predict temperature profiles in the heat exchanger.

2. If the temperature variation on the shell side were important, a similar procedure — this time based on a balance on a microscopic element on the shell side — can be used to obtain the necessary equations; if the wall dynamics were to be taken into consideration, energy balances could also be written for both the tube as well as the shell walls. Such a modeling exercise has, in fact, been carried out in Friedly [6], to which the reader is referred. The complete model will then contain four equations: one for each temperature profile on the tube side and the shell side, and one each to represent the dynamics of the two walls.

3. Upon inspecting Eq. (12.33) carefully, we see that the main difference between this type of model and the others we have obtained for lumped systems is that there is an additional term on the LHS, involving the fluid velocity and a temperature gradient in the spatial coordinate z; this is known as a *convective* term, indicating the rate of heat transfer due to fluid motion, or *convection*. The term on the RHS is related to the rate of heat transfer by *conduction* across the walls of the heat exchanger.

4. Distributed parameter system models are usually incomplete without the specification of boundary conditions; in other words, the boundary conditions that go with the PDE's are just as much a part of the process model as the PDE's themselves. In this particular case, since the PDE is first order in time and also first order in the spatial coordinate z, the model requires *one* initial condition (for the time derivative) and *one* boundary condition (for the z derivative); the boundary condition in this case is usually specified at $z = 0$ since the inlet conditions are usually known, i.e.:

$$\text{At } z = 0, \qquad T(0, t) = T_0(t) \qquad (12.35)$$

The initial condition is usually some starting temperature profile such as:

$$\text{At } t = 0, \qquad T(z, 0) = T^*(z) \qquad\qquad (12.36)$$

The model for the heat exchanger is linear and can therefore be reduced to a transfer function between the steam temperature T_{st} and the exchanger outlet temperature $T(L, t)$. This was made explicit in Chapter 8 where an example may be found.

12.4   PARAMETER ESTIMATION IN THEORETICAL MODELS

As stated earlier — and as has been demonstrated with concrete examples in the preceding section — the typical theoretical model for a process consists of a differential equation (ordinary, or partial; linear or nonlinear) in the process state, input, and output variables (e.g., concentrations, temperatures, pressures, liquid levels, flowrates, etc.), as functions of time, and possibly spatial position.

In addition, however, these model equations also contain *parameters*: reaction rate constants, activation energies, overall heat transfer coefficients, thermal conductivities, diffusion coefficients, relative volatilities, specific heat capacities, densities, and the like, which are usually unknown.

The process model becomes useful as a surrogate for the real process when we can generate from it predictions of the state and output variable behavior over time and possibly space, given specific input conditions. This is only possible if the specific values of the unknown parameters have been determined for the specific process in question.

It is important to note in this regard that a process model might be a very accurate representation of the process model *in form*; but without the right values for the unknown parameters, the model prediction will be inaccurate, and may be quite useless. For example, the differential equation that will accurately represent the behavior of the reactant concentration in an isothermal, batch reactor, in which a well-known, and well-characterized, first-order reaction is taking place, is easy to write down; if however, the reaction rate constant is not accurately known, despite the accuracy of the model form, the model prediction will be inaccurate *on account of the inaccurate reaction rate constant*.

It should be noted very carefully that entire books have been written on the issue of parameter estimation (cf. Beck and Arnold [7], Bard [8]), and major portions of Stewart and Caracotsios [9] and Seinfeld and Lapidus [10] are devoted to parameter estimation. In the amount of space available in this section, therefore, we can only hope to provide a flavor for the essential ingredients, and a summary of the important techniques of parameter estimation, as far as process modeling is concerned. Pertinent references will be inserted at appropriate points during the discussion.

Determining Unknown Model Parameters

There are three ways by which the value of unknown parameters in a process model may be determined:

1. Extracting needed parameters from the published literature (for example, physical properties such as densities, heat capacities, thermal conductivities, etc., are readily available in the literature, as are thermodynamic properties of certain liquids and gases, and mixtures thereof).

2. Carrying out independent experiments to determine fundamental model parameters; for example, laboratory experiments for determining kinetic rate constants to be used in a model for a full-scale industrial reactor.

3. Performing experiments on the particular physical system of interest and determining the unknown parameter values that produce model predictions that are the closest fit to the experimentally observed data.

Thus, when the required parameters are unavailable in the literature, the only other alternative is to estimate them from experimental data.

12.4.1   Basic Principles of Parameter Estimation

For parameter estimation purposes, the theoretical model for any process can be represented in the following form:

$$\eta = f(z, \theta) \tag{12.37}$$

where:

η is an n-dimensional vector of actual process outputs that can be measured in an experiment.

z is an m-dimensional vector of "independent" variables that can be specified arbitrarily for each experiment (e.g., reactant flowrates, heat input to the reboiler, or other such process *input* variables), or variables that, for each experimental observation, are known precisely (e.g., time, or the spatial coordinate variable).

θ is a p-dimensional vector of unknown parameters.

f is some functional relationship, between these variables; it may be an explicit functional relationship as in an algebraic model, or an analytical solution to a differential equation model; it may also be an implicit function, as in the differential equation itself; *but the form that f takes is known.*

Let us illustrate with an example.

Example 12.1   EXPLICIT FUNCTIONAL FORM FOR THE MODEL OF AN ISOTHERMAL BATCH REACTOR.

A batch reactor in which the isothermal, first-order conversion of reactant A to product is taking place is modeled by:

$$\frac{dc_A}{dt} = -kc_A \tag{12.38}$$

where k is the reaction rate constant. This is a linear, first-order, ODE whose solution is easily written as:

$$c_A = c_{A_0} e^{-kt} \tag{12.39}$$

where c_{A_0} is the initial concentration of the reactant. Observe now that Eq. (12.39) is in the form of Eq. (12.37) with:

$$\eta = c_A; \quad \mathbf{z} = [c_{A_0}, t]; \quad \theta = k$$

i.e.

$$\eta = z_1 e^{-\theta z_2} \tag{12.40}$$

Thus in this case, the measurable output is the reactant concentration; we assume that the initial reactant composition is an independent variable at our disposal to specify arbitrarily; and the reaction time is known precisely. Hence the only unknown parameter is the reaction rate constant.

Let us now state the objective of parameter estimation:

Parameter estimation is concerned with **the determination** *from experimental data the best set of values for the unknown parameters* θ *in a process model of known form.*

Conceptually, this involves matching the process model solution to experimentally obtained data, for various values of the unknown model parameters; the set of values of the unknown parameters for which the process model prediction matches the experimental data the "closest" is then chosen as the "best" set of parameter estimates.

Linear or Nonlinear Parameter Estimation

The parameter estimation problem is said to be linear or nonlinear with regards to the vector of parameters, and not with regards to the state variables. If the function $\mathbf{f}$ in Eq. (12.37) is linear *with respect to the vector* θ, then the model is said to be linear *in the parameters*, and it gives rise to a *linear* parameter estimation problem; when the parameters enter the model in a nonlinear fashion, we have a nonlinear parameter estimation problem.

Thus to check the parameter estimation problem presented in Eq. (12.37) for linearity, obtain $\partial \mathbf{f} / \partial \theta_i$; if the result is independent of θ_i, then the problem is linear in θ_i; otherwise it is nonlinear in that parameter.

This classification of the parameter estimation problem is, unfortunately, a potential source of confusion, because a model could be *linear* in terms of the *process variables* (this is the sense in which we have defined linearity in process models up until the issue of parameter estimation arose) but may be *nonlinear* in terms of the parameters. A good example is the batch reactor of Example 12.1: its model, in Eq. (12.38), is a *linear* first-order ODE in terms of c_A; but the solution of this model, written in the form of Eq. (12.39), is as shown in Eq. (12.40), and it contains a *nonlinear* function of the reaction rate constant, the unknown parameter.

However, it is worthy of note that in most cases, even when the process model is linear in the state variables, it is almost always nonlinear in the parameters. Nevertheless, the distinction is well worth keeping in mind.

Elements of Parameter Estimation

Experimental Data

Each single experiment involves measuring all the n output variables η_1, η_2, ..., η_n, for a specified set of values for the independent variables, z_1, z_2, ..., z_m. In order to be able to determine the p parameters θ_1, θ_2, ..., θ_p, independently, it is necessary to perform *at least* p such experiments. As we might expect, all things being equal, the greater the number of experiments performed, the better our estimates will be.

The result of *each* individual experiment is a set of vectors η and z that may be related to the process model; in particular, for the kth experiment, we have:

$$\eta(k) = \mathbf{f}(\mathbf{z}(k), \theta); \; k = 1, 2, ..., N \qquad (12.41)$$

if N separate experiments have been performed ($N \geq p$).

We now note that the experimental measurements of $\eta(k)$ will not be exactly equal to its true value because of measurement error. By this statement we mean, for example, that a thermocouple measurement of an equilibrium mixture of ice and water will *not* be exactly 0°C, but some other value close to it. We therefore differentiate the actual process output $\eta(k)$ from its experimentally observed *measurement*, denoted by $\mathbf{y}(k)$.

And now, because of the unavoidable measurement error, as well as the inaccuracies in the model formulation, the functional form on the RHS of Eq. (12.41) will *not* exactly fit the observed data set $\mathbf{y}(k)$, rather:

$$\mathbf{y}(k) = \mathbf{f}(\mathbf{z}(k), \theta) + \varepsilon(k); \quad k = 1, 2, ..., N \qquad (12.42)$$

where $\varepsilon(k)$ is the vector of errors between the model prediction and the actual data.

Parameter estimation is now involved with finding a specific *set* of parameter values $\hat{\theta}$ such that some scalar function S of the error vector — known as the objective function and usually represented as $S(\theta)$, since it depends on the θ parameter value — is minimized; i.e., over the entire range of possible values of θ, the smallest value of $S(\theta)$ is given by $S(\hat{\theta})$.

Note now that the type of parameter estimates we obtain will depend on the error criterion we use in the objective function to be minimized.

Error Criterion

Several criteria are used to obtain optimal estimates of unknown model parameters, but the most widely used is the least square criterion. In this case, the sum of squares of the errors is the function to be minimized, i.e., $S(\theta)$ is defined as:

$$S(\theta) = \sum_{k=1}^{N} [\varepsilon(k)]^T [\varepsilon(k)] \qquad (12.43)$$

or, from Eq. (12.42):

$$S(\theta) = \sum_{k=1}^{N} \{ \mathbf{y}(k) - \mathbf{f}(\mathbf{z}(k), \theta) \}^T \{ \mathbf{y}(k) - \mathbf{f}(\mathbf{z}(k), \theta) \} \qquad (12.44)$$

where the summation is over all the data points.

Sometimes it might be necessary to assign more weight to more precise measurements, and less weight to the others; this is accomplished by introducing a weighting matrix as follows:

$$S(\theta) = \sum_{k=1}^{N} \{ \mathbf{y}(k) - \mathbf{f}(\mathbf{z}(k), \theta) \}^{T} \mathbf{W}(k) \{ \mathbf{y}(k) - \mathbf{f}(\mathbf{z}(k), \theta) \} \qquad (12.45)$$

and $\mathbf{W}(k)$ is the $n \times n$ weighting matrix whose elements reflect the relative precision of the various measurements (Seinfeld and Lapidus [10]). The criterion in Eq. (12.45) is known by the self-explanatory term: *weighted least squares*.

There are several other criteria used for the optimal determination of model parameters: for example, the method of maximum likelihood, the method of moments, and decision theoretic methods, of which Bayesian estimation is the most popular. We will not discuss any of these methods here; the interested reader is referred to Refs. [7] – [10] for further information.

Optimization

Focusing attention now on just the least square criterion, we see from Eq. (12.44) that parameter estimation in this context boils down to finding the $\hat{\theta}$ that gives the minimum S. Thus, in essence, the parameter estimation problem is really an optimization problem.

Observe that for each trial value of the vector θ, we will obtain a corresponding value for $S(\theta)$, as indicated in Eq. (12.44). This value of $S(\theta)$ represents how far the model prediction is from the experimental data; it provides a means of assessing each trial value of θ objectively as a possible choice for the "best estimate." The collection of all such values of $S(\theta)$ as a function of θ yields a sum-of-squares surface in the p-dimensional parameter space. The objective of parameter estimation is to locate the global minimum of this surface; the θ value at which this minimum occurs is designated $\hat{\theta}$, and called the *least squares estimate* of θ.

Formulated in this sense — as an optimization problem — least squares parameter estimation may be carried out using various types of optimization techniques. The specific situation usually determines which technique is most profitable.

Let us now indicate how the foregoing principles can be applied in estimating parameters from process models that occur in terms of differential equations.

12.4.2   Parameter Estimation in Differential Equation Models with Analytical Solutions

When the process model is such that it possesses an analytical solution, the parameter estimation problem is made that much easier, since an explicit expression of the form given in Eq. (12.41) is obtained directly from the analytical solution.

As previously noted, we may now distinguish two separate cases: the first, in which the function $\mathbf{f}(\mathbf{z},\theta)$ is *linear* in θ; the second in which $\mathbf{f}(\mathbf{z},\theta)$ is *nonlinear* in θ.

Linear Estimation

When the analytical solution of a process model is obtained in the form:

$$\eta = \mathbf{f}(\mathbf{z}, \theta)$$

then, from Eq. (12.42), we see that the experimentally obtained measurements of the process output (which we have referred to as $y(k)$), obtained for various values of the vector of independent variables, $\mathbf{z}(k)$, are related to the process model according to:

$$
\begin{bmatrix}
\mathbf{y}(1) \\
\mathbf{y}(2) \\
\cdots \\
\mathbf{y}(k) \\
\cdots \\
\mathbf{y}(N)
\end{bmatrix}
=
\begin{bmatrix}
\mathbf{f}(\mathbf{z}(1), \theta) \\
\mathbf{f}(\mathbf{z}(2), \theta) \\
\cdots \\
\mathbf{f}(\mathbf{z}(k), \theta) \\
\cdots \\
\mathbf{f}(\mathbf{z}(N), \theta)
\end{bmatrix}
+
\begin{bmatrix}
\varepsilon(1) \\
\varepsilon(2) \\
\cdots \\
\varepsilon(k) \\
\cdots \\
\varepsilon(N)
\end{bmatrix}
\qquad (12.46)
$$

If $\mathbf{f}(\mathbf{z}, \theta)$ is linear in the parameters θ, then it can be shown that Eq. (12.46) becomes:

$$\mathbf{Y} = \mathbf{X}\theta + \mathbf{E} \qquad (12.47)$$

where $\mathbf{X}$ is the complete matrix of independent variables (and functions thereof) compiled for each data set; $\mathbf{Y}$ is the entire collection of the experimentally obtained data; and $\mathbf{E}$ is the collection of errors. We will demonstrate this fact shortly with an example process control system, but we wish to carry the linear parameter estimation problem to its logical conclusion first.

The linear (least squares) parameter estimation problem is now to find the vector $\hat{\theta}$ for which the squared deviation of the data set $\mathbf{Y}$ from the model $\mathbf{X}\theta$ (i.e., the magnitude of the vector $(\mathbf{Y} - \mathbf{X}\theta)$), is minimized. There is an analytical solution to this problem (see Appendix D), and it is given by:

$$\hat{\theta} = (\mathbf{X}^T\mathbf{X})^{-1}\mathbf{X}^T\mathbf{Y} \qquad (12.48)$$

Let us illustrate these concepts with a concrete example process control system, the pure capacity system.

Example 12.2   CONSIDERATIONS FOR LINEAR PARAMETER ESTIMATION IN A PURE CAPACITY SYSTEM.

The theoretical model for a storage tank was given in Chapter 5 as:

$$\frac{dh}{dt} = K^*\left(F_i - F\right) \qquad (12.49)$$

where F_i and F respectively represent the inlet stream flowrate, and the *metered* outlet flowrate. The only process parameter, K^*, is to be estimated from experimental data.

Show that for the situation in which a unit step increase is made in the inlet flowrate F_i,

(a) The analytical solution to this modeling equation can be represented in the form given in Eq. (12.37),

(b) Show that it is linear in the parameter K^*, and

(c) Show that experimental data collected on the level in the tank at time points $t_1, t_2, ..., t_N$, when compiled and related to the analytical solution of the process model give rise to a matrix equation of the type in Eq. (12.47).

Solution:

(a) For the situation in which the system was initially at steady state (i.e., $F_i = F$) before the unit step increase in F_i was implemented, the modeling equation becomes:

$$\frac{dh}{dt} = K^* \tag{12.50}$$

since now, $\left(F_i - F\right) = 1$. The solution to Eq. (12.50) is easily obtained as:

$$h = h_0 + K^* t \tag{12.51}$$

where h_0 is the initial steady-state level, which will be known.

If we now let:

$$\eta_1 = h - h_0 \quad \text{the only process output,}$$
$$z_1 = t \quad \text{time, the only independent variable,}$$
$$\theta_1 = K^* \quad \text{the only parameter,}$$

then it is easy to see that Eq. (12.51) could be written in the form:

$$\eta_1 = z_1 \theta_1 \tag{12.52}$$

which is of the form:

$$\eta = \mathbf{f}(\mathbf{z}, \theta) \tag{12.53}$$

(b) By inspection, we see that Eq. (12.51), the analytical solution to the process model, contains only a linear function of the process parameter K^*; equivalently, we may observe that the alternative representation in Eq. (12.52) is a linear function of θ_1.

(c) If the level measurements at times $t_1, t_2, ..., t_N$, are, respectively $h_1, h_2, ..., h_N$, and in particular if the *deviation* of these measurements from the initial steady-state value h_0 is represented, respectively as $y_1, y_2, ..., y_N$, then these measurements could be related to the process model in Eq. (12.51) as follows:

$$y_1 = h_1 - h_0 = K^* t_1 + \varepsilon_1$$
$$y_2 = h_2 - h_0 = K^* t_2 + \varepsilon_2$$
$$... \quad ... \quad ...$$
$$y_N = h_N - h_0 = K^* t_N + \varepsilon_N$$

which may be written in vector form as:

$$\begin{bmatrix} y_1 \\ y_2 \\ ... \\ y_N \end{bmatrix} = \begin{bmatrix} t_1 \\ t_2 \\ ... \\ t_N \end{bmatrix} K^* + \begin{bmatrix} \varepsilon_1 \\ \varepsilon_2 \\ ... \\ \varepsilon_N \end{bmatrix} \tag{12.54}$$

which is of the form:

$$Y = X\theta + E$$

From here it is straightforward to employ Eq. (12.48) in finding a least square estimate of the *integrator gain* K^* as the next example demonstrates.

Example 12.3   LEAST SQUARES ESTIMATE OF THE INTEGRATOR GAIN OF A PURE CAPACITY SYSTEM.

For the storage tank in Example 12.2, obtain an expression for the least squares parameter estimate of its integrator gain K^*, in terms of experimental level measurements (denoted as y_k, deviations from the initial steady-state value) and time t_k.

Solution:

Since the experimental data can be related to the analytical form of the process model written as in Eq. (12.47), we can use Eq. (12.48) to obtain the required least square estimate of K^*. In this particular case, from the results in Example 12.2, by comparing Eq. (12.54) with Eq. (12.47) we see that:

$$X = \begin{bmatrix} t_1 \\ t_2 \\ \dots \\ t_N \end{bmatrix}; \quad \text{and} \quad Y = \begin{bmatrix} y_1 \\ y_2 \\ \dots \\ y_N \end{bmatrix}$$

If we therefore apply Eq. (12.48) using these specific vectors, we obtain that $\hat{K}^*$, the least squares estimate of the integrator gain K^*, is given by:

$$\hat{K}^* = \frac{\sum_{k=1}^{N} t_k y_k}{\sum_{k=1}^{N} t_k^2} \tag{12.55}$$

since, in this relatively simple case,

$$X^T X = \sum_{k=1}^{N} t_k^2 \tag{12.56a}$$

and

$$X^T Y = \sum_{k=1}^{N} t_k y_k \tag{12.56b}$$

The implication of Eq. (12.55) is therefore that the "best" estimate of the integrator gain for the storage tank may be obtained as follows:

1. Perform a step-response experiment and record the level measurements in terms of *deviations* from the initial steady-state level at time instants t_k; $k = 1, 2, \dots, N$; let these experimental measurements be designated as y_k.
2. Take the product of the measurement y_k and the time at which the measurement was observed t_k, and sum up these products; also take a sum of squares of the time at which each observation was made. As indicated in Eq. (12.55) the ratio of the crossproduct sum to the sum of squares of the times gives the best estimate of the integrator gain (in the least squares sense).

A somewhat more involved situation, but which is based on exactly the same principles, is illustrated with the following example. The problem is the characterization of the *steady-state* temperature profile obtained in the control of a cylindrical ingot being heated in a furnace.

Example 12.4   CHARACTERIZATION AND PARAMETER ESTIMATION FOR A CYLINDRICAL INGOT BEING HEATED IN A FURNACE.

The equation describing the temperature profile in a cylindrical ingot being heated is a furnace is (see Ray [11]):

$$\frac{\partial T}{\partial t} = \frac{\alpha}{r} \frac{\partial}{\partial r}\left(r \frac{\partial T}{\partial r}\right) \tag{12.57}$$

where α is the metal's thermal diffusivity.

(a) Show that an analytical expression (derived from this model) for the steady-state temperature profile in the ingot is:

$$T(r) = C_1 \ln r + C_2 \tag{12.58}$$

with two unknown parameters, C_1 and C_2. These parameters characterize the steady-state temperature profile.

(b) Because Eq. (12.58) is linear in these unknown parameters, they are relatively easy to estimate. The following experimental data were obtained from a particular system (from temperature probes placed at various positions along the radial direction of the ingot) after a unit step change was implemented in the quantity of heat applied to the boundary of the ingot by the furnace. Use the data to obtain a least squares estimate of the two characterizing parameters C_1 and C_2.

r (cms)	T (°C)
0.5	18
1.0	21
1.5	23
2.0	24
2.5	26

Solution:

(a) The model for steady-state conditions is extracted from the general model given in Eq. (12.57) by setting the time derivative to zero, leaving:

$$0 = \frac{\partial}{\partial r}\left(r \frac{\partial T}{\partial r}\right) \tag{12.59}$$

which can be solved in a number of different ways. Perhaps the most straightforward solution is obtained by noting that if the derivative of a quantity is zero, then that quantity must itself be constant; for Eq. (12.59) this means that:

$$\left(r \frac{\partial T}{\partial r}\right) = C_1; \text{ a constant}$$

or

$$\frac{dT}{dr} = \frac{C_1}{r} \qquad (12.60)$$

where we have now switched to total derivatives since only derivatives in r are involved. This equation is now easily solved (e.g., by direct integration) to give:

$$T = C_1 \ln r + C_2 \qquad (12.61)$$

as required; C_2 is an integration constant.

(b) Observe now that even though this expression is *nonlinear* in the radial position variable r, it is linear in the unknown parameters C_1, and C_2; and since we are now concerned with a parameter estimation problem, we have a linear problem on our hands.

Observe again that given data on temperature measurements at various radial positions, these data could be related to the model in Eq. (12.61) according to:

$$T_1 = C_1 \ln r_1 + C_2 + \varepsilon_1$$
$$T_2 = C_1 \ln r_2 + C_2 + \varepsilon_2$$
$$\dots \ \dots \ \dots$$
$$T_N = C_1 \ln r_N + C_2 + \varepsilon_N$$

If we now write this in vector-matrix form we will obtain:

$$\begin{bmatrix} T_1 \\ T_2 \\ \dots \\ T_N \end{bmatrix} = \begin{bmatrix} \ln r_1 & 1 \\ \ln r_2 & 1 \\ \dots & \dots \\ \ln r_N & 1 \end{bmatrix} \begin{bmatrix} C_1 \\ C_2 \end{bmatrix} + \begin{bmatrix} \varepsilon_1 \\ \varepsilon_2 \\ \dots \\ \varepsilon_N \end{bmatrix} \qquad (12.62)$$

which is now of the form:

$$\mathbf{Y} = \mathbf{X}\theta + \mathbf{E}$$

where the correspondence is obvious.

Using the supplied data, we have, upon taking natural logarithms of the supplied r values:

$$\begin{bmatrix} 18 \\ 21 \\ 23 \\ 24 \\ 26 \end{bmatrix} = \begin{bmatrix} -0.693 & 1 \\ 0.000 & 1 \\ 0.405 & 1 \\ 0.693 & 1 \\ 0.916 & 1 \end{bmatrix} \begin{bmatrix} C_1 \\ C_2 \end{bmatrix} + \begin{bmatrix} \varepsilon_1 \\ \varepsilon_2 \\ \dots \\ \varepsilon_N \end{bmatrix} \qquad (12.63)$$

and now using the expression in Eq. (12.48) upon carrying out the indicated matrix operations, we obtain the least squares estimate for these characterizing parameters C_1 and C_2 as:

$$\begin{bmatrix} \hat{C}_1 \\ \hat{C}_2 \end{bmatrix} = \begin{bmatrix} 5.0 \\ 21.0 \end{bmatrix} \qquad (12.64)$$

Nonlinear Estimation

The more common situation with process models for which analytical solutions exist, is for the functional form of $f(\mathbf{z}, \theta)$ to be *nonlinear* in θ. For example, for any first-order system, as we recall from Chapter 5, the process model, is:

$$\tau \frac{dy}{dt} + y = Ku(t) \tag{5.2}$$

and the general solution to this model, for a unit step input in $u(t)$, is:

$$y(t) = K \left(1 - e^{-t/\tau}\right) \tag{12.65}$$

which is linear in K but *nonlinear* in τ. The same is true for other models for systems of low order, which are linear in the *process variables*; the resulting analytical solutions (many of which were obtained in Chapters 5 and 6) are, however, nonlinear in the *parameters* usually because of the exponential function.

Nonlinear estimation problems are handled, in general, by numerical techniques, and we will discuss these shortly. However, primarily because of the elegance — and ready availability — of the analytical least square solution to the linear estimation problem, it is quite popular to seek methods by which the process models can be rearranged or transformed into forms which are linear in the parameters.

Perhaps the most familiar example of this involves the estimation of the first-order reaction rate constant of the isothermal batch reactor of Example 12.1. Recall from Eq. (12.39) that the process model may be represented as:

$$\xi = e^{-kt} \tag{12.66}$$

where ξ is c_A/c_{A_0}. Observe now that a logarithmic transformation converts Eq. (12.66) to:

$$\ln \xi = -kt \tag{12.67}$$

which is now linear in the parameter k. Another such linearizing transformation frequently used to "convert" models that are nonlinear in the parameters to linear ones is illustrated in the following example.

Example 12.5   **LEAST SQUARE ESTIMATE OF THE OVERALL HEAT TRANSFER COEFFICIENT FOR HEAT LOSS FROM A HEATED PLATE.**

The heated plate shown below in Figure 12.7 is heated up to an initial temperature T_0 and allowed to lose heat to the atmosphere.

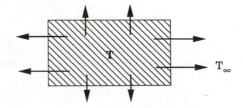

Figure 12.7.   Heat transfer from a heated plate.

Assuming that the plate temperature T is uniform over the entire surface, that the surrounding air is calm, and that the atmospheric temperature remains constant at T_∞, the model representing the change in plate temperature due to heat loss to the atmosphere is:

$$mC_p \frac{dT}{dt} = -UA(T - T_\infty) \qquad (12.68)$$

where m, C_p, and A are respectively the total mass, specific heat capacity, and heat transfer area of the plate.

If we now let:

$$\alpha = \frac{UA}{mC_p} \qquad (12.69)$$

the solution to this first-order differential modeling equation is:

$$T = T_\infty + (T_0 - T_\infty) e^{-\alpha t} \qquad (12.70)$$

which is nonlinear in the parameter α.

However, we note that Eq. (12.70) can be rearranged to give:

$$\xi = \frac{T - T_\infty}{T_0 - T_\infty} = e^{-\alpha t} \qquad (12.71)$$

and the logarithmic transformation converts Eq. (12.71) to:

$$\ln \xi = -\alpha t \qquad (12.72)$$

Now, from temperature data taken at various points in time during the course of experimentation on this plate, the corresponding value of the variable ξ can be calculated; from the logarithm of this derived variable, using Eq. (12.72) and the linear least square technique, we can obtain an estimate for the parameter α. From Eq. (12.69), given the plate area, and other physical properties, we may then deduce the overall heat transfer coefficient U from this estimate.

There are two problems with this approach:

1. It is not applicable in all cases; there are several cases of importance for which no such linearizing transformation can be found.

2. The whole idea of model transformation can be tricky if the measurement error is included; for example, for the batch reactor discussed earlier, if the concentration C_A is measured, and each measurement is referred to as y_i, relating this to the model equation now gives:

$$y_i = C_{A_0} e^{-kt_i} + \varepsilon_i \qquad (12.73)$$

from where it now becomes clear that we can no longer carry out the logarithmic transformation because of the additional error term. Nevertheless, even though it might be questionable from a strict, statistical point of view, it is possible to get around this problem by specifying a different error structure; for example, the batch reactor case might be presented as:

$$\ln\left(\frac{y_i}{C_{A_0}}\right) = -kt_i + \varepsilon_i \qquad (12.74)$$

In other words, the transformation is taken *before* the error is added, implying that the "observed" error is an error in the transformed — not the original — variable. The expression in Eq. (12.74) is now eligible for linear least squares estimation where Eq. (12.73) is not.

In general, nonlinear estimation will be required in most of the cases in which the analytical solution to the differential equation model is a nonlinear function. As noted earlier, this involves employing numerical optimization techniques to find the parameter vector $\hat{\theta}$ which minimizes the objective function $S(\theta)$, in Eqs. (12.44) and (12.45). However, in many cases even such an analytical solution may not be possible, so that we must perform nonlinear parameter estimation on the differential equations themselves.

In this case numerical methods will be required both to solve the process model as well as to find the "best" parameter estimates. To be very general, then, we note that nonlinear parameter estimation can be applied to differential equation models, whether or not analytical solutions are possible. Because this procedure is applicable to *all* process models occurring as differential equations, it will now be briefly discussed. The reader interested in more detailed discussion about the theory and practice of nonlinear estimation is referred to Bard [8] and to Seinfeld and Lapidus [10].

12.4.3   Parameter Estimation in Differential Equation Models: Numerical Methods

The mathematical models for many realistic systems contain enough complexity that they are amenable only to numerical methods of solution. For such systems the following is the typical procedure for estimating the set of unknown parameters contained in the model.

We start by assuming that the model is of the form:

$$\frac{d\eta}{dt} = \mathbf{f}(\eta, \mathbf{z}, \theta, t) \qquad (12.75)$$

where $\mathbf{f}(\bullet)$ is some nonlinear function of the indicated arguments, and, as usual, θ represents the set of unknown parameters to be estimated.

It is also assumed that experiments have been performed on the physical process which have yielded N *independent* measurements of the *vector* of process outputs η, for various settings of the input variables $\mathbf{z}$ and time. These measurements are represented as $\mathbf{y}_k, k = 1, 2, ..., N$.

General Algorithm for Numerical Parameter Estimation

1. Start with an initial guess for the value of the vector of parameters; denote this as $\theta^{(0)}$.

2. For the same conditions under which the experiments were performed, integrate the process model (numerically) to produce predicted model output η_k, *given that the value of the unknown parameter vector is* $\theta^{(0)}$.

3. Evaluate the sum-of-squares function by subtracting the model prediction η_k from $\mathbf{y}_k$ (the actual data it is supposed to predict) and finding the *norm* of the resulting discrepancy vector, i.e.:

$$S_j = S(\theta^{(j)}) = \sum_{k=1}^{N} \left[\mathbf{y}_k - \eta_k(\theta^{(j)})\right]^T \left[\mathbf{y}_k - \eta_k(\theta^{(j)})\right] \qquad (12.76)$$

4. *Update the estimate* $\theta^{(j)}$; This is where specific methods for nonlinear parameter estimation in nonlinear process models differ. We will later summarize some of the more popular methods for carrying out this step. Let the updated estimate be $\theta^{(j+1)}$.

5. Return to Step 2 and iterate; obtain S_{j+1} using the updated estimate.

6. Continue until $\left| S_{j+1} - S_j \right|$ is less than some predetermined tolerance. The final updated estimate is the required least squares estimate, $\hat{\theta}$.

Thus we see that the general procedure is based on searching a p-dimensional space of the parameter vector θ, to locate the global minimum of the sum-of-squares surface $S(\theta)$; the point at which this minimum occurs corresponds to the desired least squares estimate.

It is fairly straightforward to program the computer to carry out the procedure indicated above; we only need to specify a technique for carrying out the critical Step 4, where the parameter estimates are successively corrected, starting with an initial guess.

The following is a summary of some of the techniques that are used for this purpose.

Gradient Methods

The general parameter updating expression for a host of methods known by this name is:

$$\theta^{(j+1)} = \theta^{(j)} - \lambda \Omega \mathbf{g} \qquad (12.77)$$

where: λ is some scalar, Ω is a matrix, and $\mathbf{g}$ is the gradient vector of the sum-of-squares surface S, i.e., each element g_i of this vector $\mathbf{g}$ is given by:

$$g_i = \frac{\partial S}{\partial \theta_i} \qquad (12.78)$$

There are several methods that operate according to Eq. (12.77), and they acquired their collective name because they require the computation of the indicated gradient. The choice of matrix Ω, and scalar λ, is what differentiates one gradient method from the other.

1. For $\Omega = \mathbf{I}$, we have the *method of steepest descent*; it converges too slowly in most practical problems.

2. The *Newton-Raphson method* uses for Ω, the *inverse* of the Hessian matrix $\mathbf{H}$, defined as:

$$\mathbf{H} = \left[\frac{\partial^2 S}{\partial \theta_i \, \partial \theta_j}\right] \qquad (12.79)$$

i.e., $\Omega = \mathbf{H}^{-1}$.

It works well close to the minimum, but in general does not guarantee that at each step $S(\theta^{(j+1)}) < S(\theta^{(j)})$; besides, its requirement of the computation of second derivatives can prove very tedious.

3.   The *Levenberg-Marquardt method* uses instead:

$$\Omega = (\mathbf{H} + K\mathbf{I})^{-1} \tag{12.80}$$

It is usually more efficient than the Newton-Raphson method.

For all these methods the value of λ is set at 1, but with the Levenberg-Marquardt method, the value of the scalar K can be altered during the course of computation, depending on certain circumstances (see Seinfeld and Lapidus [10]).

There are other methods, for example, the *direct search* methods, which do not require computation of the gradient of the sum-of-squares surface, $\partial S/\partial \theta$; they perform well on well-conditioned problems but may not be as efficient as the gradient methods. These methods and others, such as the method of *quasilinearization*, are discussed in Ref. [10], to which the interested reader is referred.

It is important to note that computer packages that incorporate several of these estimation techniques are now routinely available; it is therefore no longer necessary to write one's own parameter estimation program (see, for example, Stewart and Caracotsios [9]).

Final Comments

Partial Differential Equation Models

The techniques we have highlighted above are also applicable to parameter estimation in partial differential equation models; the computational burden might be heavier, but the principles are essentially the same (see Ref. [10], 401–406).

Issues of Convergence

Since these algorithms are iterative, it is essential to be sure that, first, they will converge; and second, that they converge to the true minimum. While it is possible to guarantee convergence, if the problem is carefully defined, there is no procedure by which one can absolutely guarantee that the technique will achieve convergence to a *global* minimum. It is customary to employ a trial-and-error procedure for "probing" for a global minimum after convergence to some minimum has been achieved.

Another important issue is that some problems might be such that the sum-of-squares surface is flat; in this case, large changes in θ values lead to only small changes in $S(\theta)$. The problem here is that the parameter estimates can no longer be very reliable since several, obviously different, combinations of parameter values would give equally small values of the objective function to be minimized; the true minimum is therefore difficult to identify, and tends to get masked even more by numerical round-off errors.

Such problems are said to be ill-posed; and in some cases, by redesigning the experiment, this problem may be alleviated; in other cases nothing much can be done. It will be essential to verify the parameter estimates visually by plotting, where possible, the model prediction against the experimental data.

Further discussions about parameter estimation with regard to the modeling of process control systems are available in Ref. [10].

12.4.4   A Detailed Example: The Nonisothermal CSTR

Let us recall the CSTR process of Section 12.3.1 shown in Figure 12.2. The model equations may be written:

$$\frac{dc_A}{dt} = \frac{1}{\theta}(c_{Af} - c_A) - k_0 \, e^{\,(-E/RT)} c_A \tag{12.81}$$

$$\frac{dT}{dt} = \frac{1}{\theta}(T_f - T) + \beta k_0 \, e^{\,(-E/RT)} c_A - Uq_c\,(1 - e^{\,(-\alpha/q_c)})(T - T_{cf}) \tag{12.82}$$

Note that these contain three unknown parameters k_0, E, α, which must be estimated; the first two are kinetic parameters, while the third is a heat transfer parameter.

There are two distinct philosophies of approach to the current problem of estimating these unknown parameters:

1. The parameters could be estimated in classical fashion using data from a set of *steady-state* experiments specially designed for independent determination of reactor parameters; this usually involves the use of process knowledge.

2. They could be estimated from one or more *dynamic* experiments: by solving the differential equation model and simply choosing the parameters to match the model prediction to the data collected in time.

The Steady-State Approach

Here, we use our process understanding to design two types of steady-state experiments: one for estimating the kinetic parameters, the other for estimating the heat transfer parameter.

Experiment 1: Kinetic Parameters

The kinetic parameters, k_0, E, will be estimated from Experiment 1 which consists of a series of steady-state experiments at three different temperatures T_1, T_2, T_3. The feedback controller shown in Figure 12.2 will be used to hold the reactor temperature constant at these steady states. Note that the specific values of three temperatures T_1, T_2, T_3 is not important; they need only be sufficiently different to allow the activation energy E to be determined. By using the feedback controller, the energy balance Eq. (12.82) is removed from consideration, and we need only consider the steady-state material balance on A at each temperature. Thus the steady-state version of Eq. (12.81) can be rearranged in the form of Eq. (12.47):

$$y = \left(\frac{c_{Af}}{c_{As}} - 1\right) = k\theta + \varepsilon \tag{12.83}$$

where c_{As} is the steady-state value of c_A and ε is the experimental error, an equation is linear in the parameter k, making it easy to determine. However, k is given by:

$$k = k_0\, e^{(-E/RT)} \tag{12.84}$$

(the familiar Arrhenius expression from chemical kinetics). Thus, by running several steady-state experiments at different values of mean residence time θ, good estimates of the rate constant k are first obtained from Eq. (12.83); when this is done at all three temperatures, T_1, T_2, T_3, the resulting three values, $k(T_1)$, $k(T_2)$, $k(T_3)$ can be used to estimate the parameters k_0, E using the following version of Eq. (12.84):

$$\ln k\,(T_i) = \ln k_0 - \frac{E}{RT_i}, \; i = 1, 2, 3 \tag{12.85}$$

obtained via a logarithmic transformation.

To illustrate explicitly how the parameter estimation is accomplished, we use the data in Table 12.1 to estimate the values of $k(T_i)$ recorded in Table 12.2. A plot of Eq. (12.85) is shown in Figure 12.8.

The least squares estimates are:

$$k_0 = 158{,}300 \; (\text{hr}^{-1})$$
$$E = 7996 \; (\text{cal/gm-mole})$$

TABLE 12.1

$y = \dfrac{c_{Af}}{c_{As}} - 1$	Mean Residence Time θ (hr)	Temperature (°C)
0.445	1.00	40.0
0.910	2.00	40.0
1.340	3.00	40.0
0.490	0.50	60.0
0.960	1.00	60.0
1.920	2.00	60.0
0.490	0.25	80.0
0.950	0.50	80.0
2.860	1.50	80.0

TABLE 12.2

Temperature (°C)	$k(\text{hr}^{-1})$
40.0	0.45
60.0	0.96
80.0	1.91

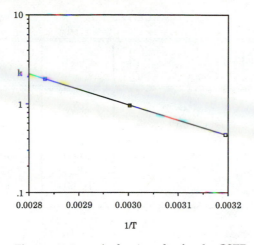

Figure 12.8.   Arrhenius plot for the CSTR.

It is important to realize that the error structure is altered by rearranging the raw c_{As} data to c_{Af}/c_{As} so that if the experimental errors are large, better estimates may sometimes be obtained by leaving the measurements c_{As} in the original form:

$$y = c_{As} = \frac{c_{Af}}{\left(1 + \theta k_0 \, e^{\,(-E/RT)}\right)} + \varepsilon \qquad (12.86)$$

and using nonlinear techniques to estimate both parameters k_0 and E simultaneously.

Experiment 2: Heat Transfer Parameter

The heat transfer parameter, $\alpha = hA/\rho_c C_{pc}$, is composed of four parameters that are all known except for the heat transfer coefficient h. Thus we need to design experiments to estimate h. One approach is to use steady-state experiments without reaction with a liquid for which the values of density, heat capacity, viscosity, etc., match those of the reaction medium. In this case, the steady-state energy balance Eq. (12.82) becomes:

$$T_s = \frac{T_f + \theta U q_c (1 - e^{-\alpha/q_c}) T_{cf}}{1 + \theta U q_c (1 - e^{-\alpha/q_c})} + \varepsilon$$

and can be used with steady-state data in which T_s varies with parameters q_c, T_f, T_{cf}, θ to estimate the parameter α using nonlinear estimation. Alternatively the steady-state equation could be transformed to become linear in the parameter α, i.e.:

$$y = \ln\left[1 + \frac{1}{\theta U q_c}\left(\frac{T_s - T_f}{T_s - T_{cf}}\right)\right] = -\frac{\alpha}{q_c} + \varepsilon \qquad (12.87)$$

Thus experiments in which q_c and T_f vary independently would allow estimation of α by linear regression. To illustrate, consider the data in Table 12.3 taken for $\theta = 0.5$ hr, $T_{cf} = 30°C$, $U = 0.5$:

TABLE 12.3

q_c (m³/hr)	$\left[1 + \dfrac{1}{\theta U q_c}\left(\dfrac{T_s - T_f}{T_s - T_{cf}}\right)\right]$	T_f (°C)	T_s (°C)
1	0.37	85.0	77.5
3	0.70	85.0	75.3
5	0.81	85.0	74.8
1	0.36	100.0	90.4
3	0.71	100.0	87.7
5	0.82	100.0	87.0

Using the linear least squares technique to fit Eq. (12.87) to the data in Table 12.3, the parameter α is estimated as:

$$\alpha = 1.0 \ (m^3/hr)$$

Now we have determined all the parameters for our CSTR model.

The Dynamic Approach

An alternative to the use of carefully designed steady-state experiments for parameter estimation as illustrated above is to use dynamic experiments in which data are taken at intervals over the course of time. In the case where the model has analytical solutions such as in Examples 12.1, 12.2, 12.3, and 12.5, the model reduces to algebraic equations and the estimation is relatively simple. Let us illustrate this for the CSTR process of Eqs. (12.81) and (12.82).

Experiment 1: Kinetic Parameters

To estimate the kinetic parameter k_o, E, we could conduct dynamic step test experiments where the reactant feed concentration c_{Af} is increased by Δc_{Af} at $t = 0$. If we employ a temperature controller to keep the reactor at a fixed temperature during the step test, then Eq. (12.81) can be integrated analytically to yield the dynamic response to the step input Δc_{Af}:

$$y = \left(\frac{c_A - c_{As}}{c_{As}}\right) = \left(\frac{\Delta c_{Af}}{c_{As}}\right)\frac{1}{(1 + k\theta)}\left[1 - e^{-\left(\frac{1+k\theta}{\theta}\right)t}\right] + \varepsilon \qquad (12.88)$$

Here c_{As} is the reactant concentration at the steady state before the step test begins. By conducting this step test at three different temperatures, T_1, T_2, T_3, we would obtain three dynamic data sets at a fixed mean residence time θ, and could then estimate the Arrhenius parameters, k_o, E, in Eq. (12.84). Note that the parameter k appears nonlinearly in Eq. (12.88) so that nonlinear estimation involving the numerical optimization methods of Section 12.4.2 will be required. We would have two alternative ways to proceed. One choice would be to substitute Eq. (12.84) into Eq. (12.88) and estimate k_o,E simultaneously by finding the overall least squares fit to the three data sets. A second alternative would be to perform a least squares fit of k to each data set, and then to use the three values of k in an Arrhenius plot similar to Figure 12.8 in order to obtain k_o, E . Both methods would work.

Experiment 2: Heat Transfer Parameter

As in the steady-state approach, the heat transfer parameter, $\alpha = hA/\rho_c C_{pc}$, could also be determined from experiments under nonreactive conditions; however, this time the experiments are dynamic. To illustrate, consider the CSTR without chemical reaction.   By carrying out dynamic step test experiments where the reactor feed temperature is increased by an amount ΔT_f, at a fixed coolant flowrate, q_c, we can generate data records useful for estimating the heat transfer parameter α. To obtain the corresponding dynamic model we can drop the chemical reaction term from the energy balance Eq. (12.82), and solve the resulting linear equation for response to a step input ΔT_f:

$$
y = \left(\frac{T - T_s}{T_s}\right) = \left(\frac{\Delta T_f}{T_s}\right) \frac{\left[1 - \exp\left(-\left\{\frac{1}{\theta} + Uq_c\left[1 - \exp(-\alpha/q_c)\right]\right\}t\right)\right]}{\left\{1 + \theta Uq_c\left[1 - \exp(-\alpha/q_c)\right]\right\}} + \varepsilon
$$

Here T_s is the tank temperature at the steady state before the step test begins. By conducting this step test at three different coolant flowrates, we would have the dynamic data sets necessary to estimate the heat transfer parameter α. Because the parameter α appears nonlinearly in the model, numerical methods for nonlinear estimation must be employed (cf. Section 12.4.2).

Although carefully designed steady-state or dynamic experiments are preferred in carrying parameter estimation, it sometimes happens that only a limited amount of dynamic data is available from plant records and there is no opportunity for experimental design. In this case, the numerical methods of Section 12.4.3 must be applied in order to carry out nonlinear parameter estimation. For example, for the CSTR of the present example this might correspond to open-loop dynamic experiments in which both temperature and composition vary with time. If one is fortunate enough to have the composition, temperature, and coolant flowrate vary over a sufficiently wide range in these experiments, then reasonable values of the parameters k_o, E, α could be estimated from this data.

12.5   VALIDATION OF THEORETICAL MODELS

After a theoretical model has been formulated, and the unknown parameters estimated, it is still important to check that the model provides an adequate representation of the physical process.

Even though there are several techniques for assessing model adequacy, whose bases are firmly rooted in theoretical statistics, the most satisfying technique — from the point of view of modeling process control systems — involves superimposing a plot of the theoretical model prediction on experimental data obtained from the physical process.   Such a visual evaluation at least provides a first impression about how well the model represents the data; and this is usually enough to enable us decide whether to improve on the model, or use it as it is.

To illustrate this point, we will use the nonisothermal CSTR for which we estimated model parameters in the last section. The model equations are given by Eqs. (12.81) and (12.82) with the following parameters:

$$U = 0.5 \ (m^3)^{-1}$$

$$\beta = 50 \left(\frac{°C \ liter}{gm - mole} \right)$$

$$k_0 = 160,000 \ (hr^{-1})$$

$$E = 8,000 \ (cal/gm - mole)$$

$$\alpha = 1 \ (m^3/hr)$$

In order to validate the model, it must be tested against process experimental data — including data not used in estimating parameters — in order to determine its ability to represent the actual process. In the present example, we have determined the parameters with special experiments in which only one of the two balances Eqs. (12.81) (12.82) was active. Now we can test the model with new experiments involving no feedback control and with chemical reaction. Figure 12.9 shows new steady-state data compared to model predictions and illustrates the final proof of model quality at least for steady-state behavior. To test the dynamic performance, one must test the model against new dynamic data to ensure that the model is adequate for predicting the process behavior for data sets not used in estimating the parameters.

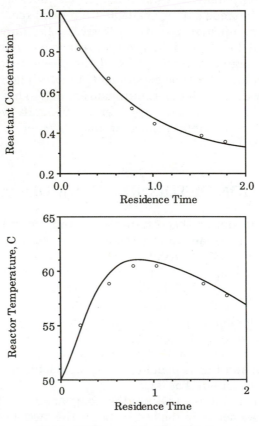

Figure 12.9.   Model validation.

12.6 SUMMARY

We have confronted the issue of theoretical modeling fully in this chapter, extending the horizons of ideas which were only rather briefly sketched in Chapter 4. Yet, despite its length, only the most pertinent aspects of modeling have been treated in this chapter; and some aspects received only sketchy outlines. Nevertheless, the material covered and the extent of coverage have been in keeping with the main objective: to alert the reader that process modeling does *not* end with the formulation of a model; the unknown parameters have to be estimated, and the model has to be validated. Only the theoretical approach to model formulation has been discussed; the parallel discussion for empirical modeling is reserved until the next chapter.

The principles of parameter estimation in theoretical models — using essentially the least squares approach — have been summarized, both for cases where analytical solutions exist and where they do not exist. A summary of typical algorithms used for nonlinear parameter estimation was presented, while emphasizing the point that since efficient parameter estimation programs are now commercially available, it is hardly ever necessary for one to write one's own routine.

The method of model validation advocated by this chapter is the time-honored comparison of model prediction and experimental data; an example was used to illustrate the situation in which a nonlinear model for a chemical reactor was evaluated and found adequate.

To complete the discussion on how to develop useful models for all classes of process systems, we must now consider the situation in which it is impossible and/or impractical to employ the theoretical approach. This is the subject of the next chapter.

REFERENCES AND SUGGESTED FURTHER READING

1. Franks, R. G. E., *Mathematical Modeling in Chemical Engineering*, J. Wiley, New York (1967)
2. Ranz, W., *Describing Chemical Engineering Systems*, McGraw-Hill, New York (1970)
3. Himmelblau, D. M. and K. B. Bischoff, *Process Analysis and Simulation*, J. Wiley, New York (1968)
4. Aris, R., *Elementary Chemical Reactor Analysis*, Prentice-Hall, Englewood Cliffs, NJ (1969) (reprinted in paperback by Butterworths, Boston (1989))
5. Luyben, W. L., *Process Modeling, Simulation, and Control for Chemical Engineers*, McGraw-Hill, New York (1973)
6. Friedly, J. C., *Dynamic Behavior of Processes*, Prentice-Hall, Englewood Cliffs, NJ (1972)
7. Beck, J. V. and K. J. Arnold, *Parameter Estimation in Engineering and Science*, J. Wiley, New York (1977)
8. Bard, Y., *Nonlinear Parameter Estimation*, Academic Press, New York (1974)
9. Stewart, W.E. and M. Caracotsios, *Computer-Aided Modeling of Reactive Systems* (in press)
10. Seinfeld, J. H. and L. Lapidus, *Mathematical Methods In Chemical Engineering, Vol. 3: Process Modeling, Estimation and Identification*, Prentice-Hall, Englewood Cliffs, NJ (1974)
11. Ray, W. H., *Advanced Process Control*, Butterworths, Boston (1989)(paperback reprint of original 1981 book)

REVIEW QUESTIONS

1.   What are the three fundamental issues of concern in process modeling?

2.   What is the main objective of process modeling?

3.   What is the fundamental difference between "theoretical modeling" and "empirical modeling"?

4.   When is it reasonable to adopt the theoretical modeling approach?

5.   When is the empirical modeling approach to be preferred over the theoretical modeling approach?

6.   Why is it necessary to start a process modeling exercise with a "problem definition"?

7.   What is involved in the "problem definition" stage of the procedure for process model development?

8.   What is involved in "model formulation" for theoretical modeling?

9.   What is "parameter estimation"?

10.   What is implied by "model validation"?

11.   What are some of the various classes of errors introduced in the process of developing a theoretical model?

12.   When the theoretical modeling approach is employed in developing a model for a *lumped parameter system*, the modeling equations naturally occur in what form?

13.   What differentiates a stagewise process from a lumped parameter system on the one hand, and a distributed parameter system on the other?

14.   When the theoretical modeling approach is employed in developing a model for a *distributed parameter system*, the modeling equations naturally occur in what form?

15.   To complete the model for a distributed parameter system, what must accompany the basic partial differential modeling equations?

16.   What is the objective of parameter estimation?

17.   What characterizes a linear parameter estimation problem?

18.   Carefully explain why it is possible for a *linear process model* to give rise to a *nonlinear* parameter estimation problem.

19.   How do "experimental data," "error criteria," and "optimization" feature in parameter estimation?

20.   Is it *always possible* to convert a nonlinear parameter estimation problem into one that is linear by appropriate nonlinear variable transformation?

21.   What are the steps involved in the numerical estimation of the parameter in a differential equation model?

22. How are the optimization methods:

    (a) Steepest descent,
    (b) Newton-Raphson, and
    (c) Levenberg-Marquardt
    different from one another?

23. What is the usual procedure for validating theoretical models?

PROBLEMS

12.1 (a) When a mercury-in-glass thermometer is introduced suddenly into a surrounding fluid, the mass m of mercury in the bulb prevents instantaneous attainment of the surrounding temperature.

Given that the specific heat capacity C_p of mercury is constant, and that the rate at which energy is transferred from the surrounding fluid to the thermometer is governed by "Newton's law of cooling," obtain a mathematical model for the transient behavior of T, the temperature of the mercury in the bulb, if the surrounding fluid is at a constant temperature T_s. You may assume:

1. A constant effective area of heat transfer A between the bulb and the surrounding fluid; and a constant heat transfer coefficient h,
2. No significant energy loss from the thermometer.

(b) Table P12.1 below contains the temperature readings obtained from a thermometer plunged into a bath of liquid at temperature 139.8°F. Use these data along with the model obtained in (a) to estimate the composite thermometer parameter defined as:

$$\tau = \frac{mC_p}{hA} \qquad \text{(P12.1)}$$

TABLE P12.1
Temperature Response Data from a Thermometer

Time (s)	Temperature (°F)
0.0	78
10.0	88
20.0	102
30.0	113
40.0	120
50.0	126
60.0	130
70.0	132
80.0	135
90.0	136
100.0	138
120.0	139

12.2 The water heater shown below in Figure P12.1 is a well-mixed, constant volume (V_l), tank through which fluid flows at a constant *mass* flowrate w (or $\rho_l F$ if F is the volumetric flowrate, and ρ_l is the liquid density); the specific heat capacity of the

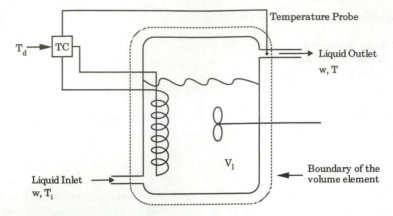

Figure P12.1.   Water heater with temperature controller.

fluid is C_{p_l}. Because the incoming fluid temperature T_i is subject to fluctuations, an electric coil to which a simple *proportional controller* is attached is used to *regulate* the temperature of the liquid in the heater. To facilitate the design of this controller, it is desired to obtain a theoretical process model.

The electrical coil is made of a metal whose specific heat capacity is C_{p_c}, with a surface area A, and mass m_c; the overall heat transfer coefficient is given as U. The temperature controller is provided with information about the tank temperature T, through a thermocouple probe; it is designed to supply energy to the coil at the rate:

$$Q_c = K_c \left(T_d - T \right) \tag{P12.2}$$

where K_c is a predetermined constant, and T_d is the desired tank temperature. Develop a theoretical model for the water heater. You may wish to make use of the following assumptions:

1. The agitation is assumed perfect, so that the temperature within the heater may be considered uniform.
2. The coil temperature T_c is also assumed uniform at any instant, but different from the boiler liquid temperature T.
3. The physical properties of all the components of the process are assumed constant.
4. There are no heat losses to the atmosphere.

Cast your model in terms of deviation variables, and present it in the state-space form.

12.3 The process shown below in Figure P12.2 is a continuous stirred mixing tank used to produce F liters/min of brine solution of mass concentration C_B gm/liter. The raw

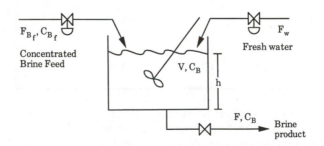

Figure P12.2. Stirred mixing tank for brine production.

materials are fresh water, supplied at a flowrate of F_w liters/min, and a highly concentrated brine solution (mass concentration C_B gm/liter) supplied at a flowrate of F_B liters/min. The volume of material in the tank is V liters, the liquid level in the tank is h meters, and the tank's cross-sectional area, $A(m^2)$, is assumed constant.

(a) Assuming that the tank is well mixed, so that the brine concentration in the tank is uniformly equal to C_B, and that the flowrate out of the tank is proportional to the square root of the liquid level, obtain a mathematical model for this process.

(b) If the process is redesigned to operate at *constant volume*, obtain a new process model and compare it with the one obtained in part (a).

12.4 The continuous anaerobic digester used for treating municipal waste sludge is essentially like any other CSTR except that it involves multiphase biological reactions. Material essentially consisting of organisms of concentration C_{x0} mole/m³, and substrate of concentration C_{s0} mole/m³, is fed to a certain anaerobic digester at F m³/day. Using the following additional assumptions:

1. That the reactions of interest take place only in the liquid phase and that the reactor (liquid) volume V (m³) is constant,
2. That the organism concentration in the digester C_x is uniform, as is the substrate concentration C_s,

(a) Obtain a model for this digester process by carrying out organism and substrate balances and using the following constitutive relationships:

1. The organism growth rate (in the reactor) is given by:

$$\frac{dC_x}{dt} = \mu C_x \tag{P12.3a}$$

where μ, the specific growth rate, is itself given by the *Monod function*:

$$\mu = \mu_0 \left(\frac{C_s}{K_s + C_s} \right) \tag{P12.3b}$$

Here μ_0 is the maximum specific growth rate, and K_s is the saturation constant.

2. The yield (the rate of growth of organism ratioed to the rate of consumption of substrate) in the reactor is given by:

$$\frac{dC_x}{dt} = -S_{xs} \frac{dC_s}{dt} \tag{P12.4}$$

(b) A specific digester with the following parameters: $K_s = 3.33 \times 10^{-3}$ mole/m³, $\mu_0 = 0.4$ day^{-1}, $S_{xs} = 0.02$, $V = 10$ m³, is being fed at the constant rate of 1 m³/day with material containing organism and substrate concentration of 0.3 mole/m³ and 167 mole/m³ respectively. Assuming that the concentrations of both organism and substrate in the digester were initially zero at the commencement of the operation, use the model obtained in (a) (along with a differential equation solver) to obtain, and plot, the behavior of μ, C_x, and C_s in the digester as a function of time.

12.5 The countercurrent liquid/liquid extraction process shown below in Figure P12.3 consists of N separate stages used to contact a liquid mixture containing transferable material A with the raffinate solvent. In each stage, n $(n = 1, 2, \dots, N)$,

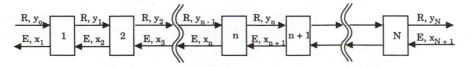

Figure P12.3. Countercurrent liquid/liquid extraction process.

R moles of raffinate solvent from stage $n - 1$, containing y_{n-1} moles of A per mole of solvent, is charged in along with E moles of extract solvent from stage $n + 1$, containing x_{n-1} moles of A per mole of solvent. Extract and raffinate are taken to be immiscible, and the residence time in each tank is assumed to be such that equilibrium is attained; i.e.:

$$x_n = k y_n \qquad \qquad \text{(P12.5)}$$

After the attainment of equilibrium in stage n, the raffinate is sent to stage $(n + 1)$ and extract to stage $(n - 1)$, etc. Initially, R moles of solvent with composition y_0 (moles of A per mole of solvent) is charged into stage 1, and the extract E moles of solvent with composition x_{N+1} (moles of A per mole of solvent) is charged into stage N.

Derive a dynamic model for this process, stating clearly any assumptions you have made.

12.6   Consider the packed bed tubular reactor whose schematic diagram is shown below (Figure P12.4). It is a hollow cylindrical tube of uniform cross-sectional area A, packed with solid catalyst pellets, in which the exothermic reaction $A \rightarrow B$ is taking place.

The packing is such that the ratio of void space to the total reactor volume — the *void fraction* — is known; let its value be represented by ε. Reactant flows in at one end at constant velocity v, and reaction takes place within the reactor. Obtain a theoretical model that will represent the variation in the reactant concentration C and reactor temperature T as a function of time and spatial position z. Consider that the temperature on the surface of the catalyst pellets T_s is different from the temperature of the reacting fluid, and that its variation with time and position is also to be modeled.

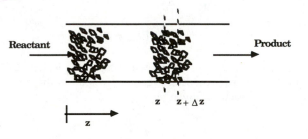

Figure P12.4.   The packed bed tubular reactor.

You may make use of the following assumptions:

1.   The reacting fluid flows through the reactor with a flat velocity profile (implying hat there are no temperature, concentration, or velocity variations in the radial direction).
2.   Transfer of material and energy in the axial (i.e., z) direction is by convective forces only; there is no diffusion in this direction.
3.   The rate of reaction is first order with respect to the reactant concentration, with a rate constant k. It does not depend explicitly on position along the reactor; it only depends on reactant concentration and temperature. The heat of reaction may be taken as $(-\Delta H)$.
4.   The heat transfer from the solid catalyst to the fluid is assumed to follow Newton's "law" of cooling; the area over which this heat transfer takes place is assumed to be a known constant A_s.

5. The fluid density ρ and specific heat capacity C_p are constant; the solid catalyst particles are assumed to be all identical, with identical density ρ_s, and identical specific heat capacity C_{ps}.
6. The catalyst packing is assumed to be uniform, so that across any cross section of the reactor, the number and arrangement of particles are identical. (This allows the use of an arbitrarily located microscopic element for developing the model.)
7. There are no heat losses to the atmosphere.

If you need to make any additional assumptions, state them clearly.

12.7   The following reaction scheme is carried out in an isothermal continuous stirred tank reactor:

where the reactants A and B react to a common product C. In addition there is a limited solubility of the feed material at the feed temperature so that $\alpha\, c_{Af} + c_{Bf} = c_f$. Under these conditions the material balances on A and B are:

$$\frac{dc_A}{dt} = \frac{1}{\theta}\left(c_{Af} - c_A\right) - k_1 c_A$$

$$\frac{dc_B}{dt} = \frac{1}{\theta}\left[(c_f - \alpha c_{Af}) - c_B\right] - k_2 c_B$$

and the product concentration is given by $c_C = c_T - c_A - c_B$. If we define the deviation variables: $x_1 = c_A - c_{As}$, $x_2 = c_B - c_{Bs}$, $u = c_{Af} - c_{Afs}$, $y = c_C - c_{Cs}$, show that the model can be put in the form

$$\frac{dx_1}{dt} = a_1 x_1 + b_1 u$$

$$\frac{dx_2}{dt} = a_2 x_2 + b_2 u$$

where the deviation variable for the product concentration is

$$y = -x_1 - x_2$$

Clearly state the physical meaning of the parameters a_1, a_2, b_1, b_2.
The table below contains the result of a carefully planned dynamic experiment performed for the purpose of estimating the four unknown parameters in this model from measurements of y, the product concentration in the reactor.
Given that initially the nominal operating reactor conditions were such that $x_1 = x_2 = 0$, with $u = 0$, and that the experiment involved implementing the change $u = -1$ at $t = 0$:
(a) Use the data to estimate the parameters $a_1, a_2, b_1,$ and b_2 to the best of your ability.
(b) Plot the *theoretical response* of the model above using the parameters estimated in part (a) and compare it with the experimental data. Evaluate the model fit.

TABLE P12.2
Reactor Experimental Data

Time t (min)	Normalized Product Conc, y (Deviation from nominal value)
0.0	0.00
1.0	− 0.160
2.0	− 0.175
3.0	− 0.090
4.0	0.025
5.0	0.150
6.0	0.259
7.0	0.401
8.0	0.490
9.0	0.555
10.0	0.650
11.0	0.725
12.0	0.750
13.0	0.801
14.0	0.852
16.0	0.901
18.0	0.922
20.0	0.947
22.0	0.965
24.0	0.971
26.0	0.985
28.0	0.990
30.0	0.992

PROCESS IDENTIFICATION: EMPIRICAL PROCESS MODELING

Important situations arise in which a dynamic process must be treated like a "black box" in order to obtain a reasonable mathematical description. In such cases, the most important characteristics of the system are "identified" entirely from its response to known input forcing functions; no attempt is made at utilizing any mechanistic information regarding the fundamental nature of the process. This approach to modeling is known as *Process Identification*; it results in empirical process models.

Just as Chapter 12 was concerned with discussing the main issues involved in the theoretical approach to modeling, this chapter is concerned with process modeling when the theoretical approach is either impossible, impractical or inconvenient. The objectives here are:

- To discuss, in general terms, what *Process Identification* is,
- To compare and contrast it with theoretical modeling, and
- To provide a summary of what is involved in "identifying" a useful model for a dynamic process strictly from input/output data.

13.1 INTRODUCTION AND MOTIVATION

There are many processes of importance that are not very well understood, in the sense that the fundamental laws responsible for their dynamic behavior are poorly known, if at all. An attempt to model such systems using the theoretical approach is not likely to be particularly successful; an alternative approach is clearly needed.

There are certain other situations in which the issue is not our knowledge of the process; enough is known about these processes to permit reasonable success with the theoretical modeling approach. The problem is that the resulting theoretical models are too complicated to be useful, especially for controller design purposes.

A case in point is the simple binary distillation column of the type discussed in Chapter 12, for which a theoretical model will consist of *at least*

($2N + 4$) nonlinear differential equations, if the total number of trays is N. Since a typical industrial distillation control problem may require the control of the mole fraction of lighter material only at the top *and* at the bottom of the column, a simpler model may, in fact, be more useful.

In each of these cases — the first in which theoretical modeling is impossible, the second in which it is both impractical and inconvenient — useful models can be obtained by process identification.

13.1.1 What Is Process Identification

Consider the situation depicted in Figure 13.1 below, in which the "unknown" process is subjected to a forcing function, $u(t)$, and the response, $y(t)$, is measured. Even though the basic principles responsible for the observed stimulus/response behavior may not be known, it is still possible to develop a model for this system by directly correlating the input and output data; the unknown system is then said to have been "identified." We therefore have the following definition:

> *Process identification involves constructing a process model strictly from experimentally obtained input/output data, with no recourse to any laws concerning the fundamental nature and properties of the system.*

No *a priori* knowledge about the process is necessary (although it can be helpful): the system is treated like a "black box," and the experimental information gathered from its response to external stimuli is used to infer (or "identify") what goes on within the "box."

Linear or Nonlinear Process Identification

It would seem natural that for continuous systems, the result of input/output process identification shown in Figure 13.1 would be one of the linear transfer function models introduced in Chapter 4. In fact, in process control practice, this is indeed the case. However, it is also possible to obtain nonlinear models for nonlinear systems by process identification (cf., e.g., Ref. [1]).

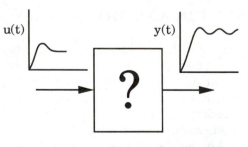

Figure 13.1.   Process identification.

The basic assumption in process control practice is that the process to be identified is approximately linear in some neighborhood of the operating steady state; as such, the linear transfer function model identified for such a process is considered adequate. We will therefore be concerned in this chapter exclusively with *linear* process identification; the reader interested in the nonlinear aspects of process identification should consult the growing literature in this area (cf., e.g., Chapter 5 of Ref.[6], and Ref. [7]).

"Off-line" or "On-line" Identification

Since process identification involves model construction from experimental data, it can, in principle, be carried out during normal process operation, using data obtained under these conditions. The usual practice, however, is to suspend "normal" process operation and perform experiments specifically designed for empirical model development. In the former case process identification is said to be done "on-line," while in the latter case, it is done "off-line."

We will be concerned in this chapter with "off-line" identification, because it is more frequently practiced. It should be noted that even though practiced less frequently, "on-line" process identification is an integral part of most adaptive control schemes (cf. Chapter 18).

Continuous or Discrete Model Identification

Process measurements are almost always available as discrete data obtained at discrete points in time, particularly when acquired using the digital computer. If such data are to be used as the sole basis for developing process models, it is therefore valid to inquire whether only discrete models can be identified from such data.

It is in fact possible to use such "discrete" data to identify *both* continuous as well as discrete models. In "continuous" model identification, the data set is regarded appropriately as sampled information from an otherwise continuous signal: the process identification task is therefore one of identifying an appropriate *continuous* empirical process model most likely to have produced a *continuous* signal that when sampled matches closely the observed *discrete* data.

In "discrete" model identification on the other hand, the discrete data set is taken at "face value" and utilized directly to identify a discrete empirical process model.

We will restrict ourselves in this chapter to "continuous" model identification; a discussion of "discrete" model identification is postponed until after the distinguishing characteristics of discrete processes have been properly introduced in Chapter 23.

13.1.2 Theoretical Modeling versus Process Identification

To put each modeling approach in proper perspective, the following table compares and contrasts theoretical modeling and process identification.

THEORETICAL MODELING	PROCESS IDENTIFICATION
1. Usually involves fewer measurements; requires experimentation only for the estimation of unknown model parameters.	Requires extensive measurements, because it relies entirely on experimentation for the model development.
2. Provides information about the internal state of the process.	Provides information only about that portion of the process which can be influenced by control action (i.e., the controllable part), and that portion can be measured (i.e., the observable part).
3. Promotes fundamental understanding of the internal workings of the process.	Treats the process like a "black box."
4. Requires fairly accurate and complete process knowledge .	Requires no such detailed knowledge; only that output data be obtainable in response to input changes.
5. Not particularly useful for poorly understood and/or complex processes.	Quite often proves to be the only alternative for modeling the behavior of poorly understood and/or complex processes.
6. Naturally produces both linear and nonlinear process models.	Requires special methods to produce nonlinear models.

13.2   PRINCIPLES OF EMPIRICAL MODELING

The basic principle behind the development of empirical models by process identification may be deduced from the diagramatic representation shown in Figure 13.1. In the context of our discussion in this chapter, it is given as follows:

> *Given $u(t)$, the input to a dynamic process, and $y(t)$, the process response to this input function, what is $g(\bullet)$, the process transfer function model?*

Observe that this is a converse of the process dynamics problem which, as we recall, may be stated thus: given $g(\bullet)$ and $u(t)$, obtain $y(t)$. The process dynamics problem is obviously more straightforward than the process identification problem; however, the fact that one is a converse of the other will eventually prove useful.

In somewhat of a parallel to the discussion given in Chapter 12 for theoretical modeling, we note that the process of constructing a working empirical model by process identification also consists of the following stages:

1. Problem Definition
2. Model Formulation
3. Parameter Estimation
4. Model Validation

The difference lies in the specific issues involved at each stage. Let us now consider these issues one by one.

13.2.1 Problem Definition

The considerations listed under this heading in Section 12.1 of Chapter 12 are pertinent regardless of which modeling approach is to be eventually adopted: in fact, it is such considerations that enable us to choose rationally one approach over the other.

13.2.2 Model Formulation

In process identification, this step entails *postulating* models (usually simple ones) which are to be considered as possible candidates. Whereas model formulation in theoretical modeling involves the application of fundamental conservation laws and constitutive relations, in process identification, it involves appropriate analysis of the input/output data for the sole purpose of detecting which model form is likely to be capable of explaining the observed behavior.

Process Data Files

Process data files are to process identification what constitutive relations are to theoretical modeling: they each provide the specific bases for model development for specific processes.

Since the model obtained by process identification is to be based entirely on experimental data, an important fact to keep in mind is that information not contained in the data *cannot* magically appear in the model; it is just as absurd as expecting an unspecified constitutive relation to show up miraculously in the final theoretical process model.

It is therefore important to collect as much data — in the proper range of operating conditions — as is necessary to "identify" those aspects of the process we consider important, much in the same way that in theoretical modeling, all the constitutive relations necessary to capture what we consider important in the process should be specified.

To illustrate this point, note, for example, that to identify properly the steady-state gain of a process from *step-response* data, it is advantageous to collect data until the process is well established at the new steady state; if the data collection is stopped *before* the new steady state is achieved, the steady-state gain will be difficult to identify accurately.

Choice of Input Functions

Central to the issue of process identification is the choice of input functions used to elicit the process response collected as output data. Observe, first, that the totality of the experimental data used for process identification consists of *both* the input as well as the output data. Over and above this, however, is the fact that the useful process information content of the output data is very much dependent on the nature of the input function. It is therefore equally important to choose an input function capable of providing output that is rich in useful process information. It is an added advantage if such information is also easily extracted.

The typical input functions employed in process identification are:

- Step
- Impulse
- Pulse (rectangular or arbitrary)
- Sine waves
- White noise
- Pseudo random binary sequences

The factors that influence the choice of input functions, and other related issues, will be mentioned at the appropriate places in the subsequent discussion.

The Role of Process Dynamics

In process dynamics studies, the fundamental problem is that of generating the responses of a system to various types of inputs, using a theoretical process model; by studying such responses we are then able to characterize these systems. In process identification we are faced with a reverse problem: that of identifying a process model from its response to certain input functions. It is therefore easy to see how results of process dynamics studies will be very useful.

Observe, therefore, that a familiarity with the qualitative shape of the response of various classes of process systems will make it easy to postulate candidate process models from input/output data records. For example, because the *ideal* step response of a first-order system is well known, it is possible to tell if a physical process may be approximately modeled as a first-order system or not, simply by examining its experimental step response.

Thus, process dynamics results — theoretical step, impulse, pulse, and frequency responses — provide the rational bases for postulating candidate models in process identification. This is one of the most important roles of process dynamics in process identification.

Typical Candidate Process Models

For most chemical processes, *unless something else in the observed data indicates otherwise*, the following is the collection of typical "continuous-time" candidate models (represented in the form of transform-domain transfer functions):

1. First-order-plus-time-delay

$$g(s) = \frac{Ke^{-\alpha s}}{\tau s + 1} \tag{13.1}$$

2. Second-order-plus-time-delay

$$g(s) = \frac{Ke^{-\alpha s}}{(\tau_1 s + 1)(\tau_2 s + 1)} \tag{13.2}$$

3. Single zero, two poles, plus time delay

$$g(s) = \frac{K\,(\xi s + 1)\,e^{-\alpha s}}{(\tau_1 s + 1)(\tau_2 s + 1)} \tag{13.3}$$

It should be mentioned that in a number of industrial situations, the process identification problem is often reduced to that of simply fitting the experimental data to one particular, prespecified type of model; in other words, the model form is determined *a priori*, regardless of what the data might indicate. For example, for controller tuning purposes, it is sometimes recommended to fit a first-order-plus-time-delay model to process data even though the response usually indicates a slightly nonlinear, possibly higher order, behavior (see Chapter 15). In such cases, the process identification problem reduces to finding the first-order-plus-time-delay model that provides the "best" fit to the experimental data.

Nevertheless, even in such cases, it is important to note that the *a priori* decision on the model type is usually based on several years of experience with such processes, and does not completely nullify the need for careful examination of what the data themselves might suggest in terms of which candidate models to consider.

13.2.3 Parameter Estimation

Once the candidate model has been postulated, the next task is that of estimating its unknown parameters, in much the same fashion as with theoretical modeling. In classical process identification, this is in fact the main task: that of fitting the experimental data to a predetermined model form by finding the parameter values which provide the "best" fit.

Estimating unknown parameters in an empirical process model can be done in the time domain, or in the frequency domain.

13.2.4 Model Validation

As with theoretical modeling, the final step in process identification involves checking how the empirical model fits the data it is supposed to represent. Model validation is usually done by comparing model predictions with additional process data, and evaluating the fit. This can also be carried out either in the time domain, or in the frequency domain.

The overall steps in the process of Process Identification are illustrated in Figure 13.2.

In the next section we discuss some of the approaches used in identifying empirical process models.

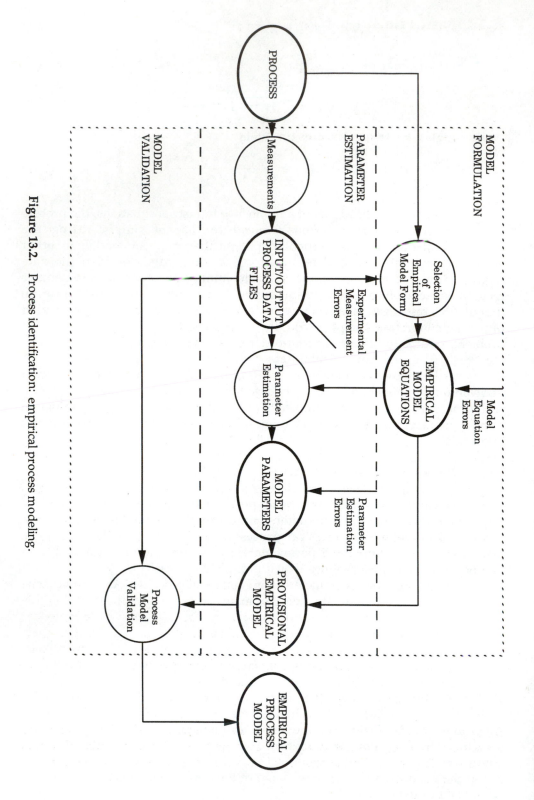

Figure 13.2. Process identification: empirical process modeling.

13.3   STEP-RESPONSE IDENTIFICATION

13.3.1 Basic Principles

The main strategy of step-response identification is to fit *theoretical* step-response functions to experimental step-response data. The experimental step-response is obtained by implementing a step change in the input variable of the physical process and recording the observed change in the output variable. The experimental procedure of introducing a step input and gathering the step-response data from a process, for the sole purpose of identifying an empirical process model, is known as *step testing*.

Model formulation consists of:

1. Inspecting the step-response data and proposing potential candidate models, and
2. Obtaining, from the proposed candidate models, the theoretical step-response functions, as was done in Part II.

Parameter estimation involves obtaining values for the unknown model parameters (in the theoretical step functions) that provide the best fit to the data. *Model validation* involves plotting the theoretical step response of each identified model and evaluating the fit to experimental data.

The main advantages of step-response identification are:

1. Theoretical step-response functions are easy to derive for typical candidate models of interest.
2. Experimental step-response data are relatively easy to obtain in practice.

The main disadvantages are:

1. Proposing the "right" candidate model by inspecting step-response data is complicated by the fact that fundamentally different systems sometimes have qualitatively similar step responses. For example, the step response obtained for the system in Eq. (13.1) will be virtually indistinguishable from that obtained for the system in Eq. (13.3) whenever $\tau_1 < \xi < \tau_2$ (cf. Section 6.6.1 of Chapter 6).
2. Step tests fundamentally involve perturbing the (usually nonlinear) process from one steady-state operating condition to another: they are therefore more likely to produce data containing nonlinear effects that cannot be modeled by any linear candidate model. The farther apart the initial and final steady-state conditions are, the more significant the effects of nonlinearities contained in the step test data.

13.3.2 Parameter Estimation

Once step test data files have been obtained, the main task of step-response identification is to take the theoretical step response of the postulated model as the working model and estimate the unknown parameters by fitting it to the experimental data.

Theoretical Step-Response Expressions

The theoretical step responses for the typical candidate models given above are now listed below along with the parameters that must be estimated. In each case, the magnitude of the step size is A.

1. First-order-plus-time-delay

With transfer function as given in Eq. (13.1):

$$y(t) = \begin{cases} 0; & t < \alpha \\ AK(1 - e^{-(t-\alpha)/\tau}); & t \geq \alpha \end{cases} \tag{13.4}$$

Parameters: K, τ, α

2. Second-order-plus-time-delay

$$y(t) = \begin{cases} 0; & t < \alpha \\ AK\left[1 - \left(\dfrac{\tau_1}{\tau_1 - \tau_2}\right)e^{-(t-\alpha)/\tau_1} - \left(\dfrac{\tau_2}{\tau_2 - \tau_1}\right)e^{-(t-\alpha)/\tau_2}\right]; & t \geq \alpha \end{cases}$$

$$\tag{13.5}$$

Parameters: $K, \tau_1, \tau_2, \alpha$

3. Single zero, two poles, plus time delay

$$y(t) = \begin{cases} 0; & t < \alpha \\ AK\left[1 - \left(\dfrac{\tau_1 - \xi}{\tau_1 - \tau_2}\right)e^{-(t-\alpha)/\tau_1} - \left(\dfrac{\tau_2 - \xi}{\tau_2 - \tau_1}\right)e^{-(t-\alpha)/\tau_2}\right]; & t \geq \alpha \end{cases}$$

$$\tag{13.6}$$

Parameters: $K, \tau_1, \tau_2, \xi, \alpha$

The time delays, time constants, and steady-state gains in these models are estimated by matching these expressions to the experimentally observed $y(t)$ data. With the exception of the first-order-plus-time-delay model form, numerical procedures of the type described in Chapter 12 are usually required. An alternative procedure involving exact discretization of the theoretical step responses is illustrated with Problem 13.1.

Because step testing is used frequently to fit first-order-plus-time-delay models to process data, we now describe how the three parameters, K, τ, and α, contained in this model may be estimated.

First-Order-Plus-Time-Delay Model Form

The theoretical step response in Eq. (13.4) for times greater than the time delay may be rewritten as:

$$y(t) = y_\infty(1 - e^{-(t-\alpha)/\tau}) \tag{13.7}$$

where we have introduced the new variable y_∞ because it represents the ultimate value of the response.

Given data $y(k)$, the value of the process output observed at t_k, then, from Eq. (13.7), we have:

$$y(k) \approx y_\infty\left(1 - e^{-(t_k - \alpha)/\tau}\right) \tag{13.8}$$

A straightforward technique for estimating the unknown parameters is as follows:

Steady-State Gain, K

From the ultimate value of the step-response data y_∞ and the magnitude of the implemented input step function A (which is known), we see that:

$$y_\infty = AK \tag{13.9}$$

then, $\hat{K}$, an estimate of the steady-state gain, is:

$$\hat{K} = \frac{y_\infty}{A} \tag{13.10}$$

which makes sense, if we recall the definition of the process steady-state gain (see Chapter 5).

Time Constant τ and Time Delay α

The expression in Eq. (13.7) is nonlinear in the parameters τ and α, but it can be rearranged as follows:

$$\frac{y_\infty - y}{y_\infty} = e^{-(t - \alpha)/\tau} \tag{13.11}$$

from which we observe that:

$$\ln\left(\frac{y_\infty - y}{y_\infty}\right) = \frac{\alpha}{\tau} - \frac{t}{\tau} \tag{13.12}$$

Thus, a plot of the transformation of the step-response data indicated on the LHS of Eq. (13.12) against time t should give a straight line with a slope of $-1/\tau$, and an intercept on the ordinate of α/τ. Also, it is easy to see that this straight line will meet the t-axis (the abscissa) at the point $t = \alpha$. These facts are summarized in Figure 13.3. We may therefore obtain estimates of these parameters graphically. However, more precise estimates could be obtained by linear regression from Eq. (13.12); and since computer programs abound for carrying out the computations involved in linear regression, this approach is to be preferred in practice.

Let us illustrate the procedure of process identification by step testing with the following example.

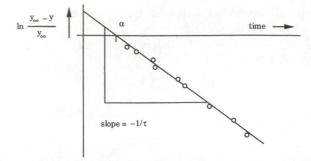

Figure 13.3.   Estimating time constant and time delay from step-response data.

Example 13.1   **IDENTIFICATION OF AN INDUSTRIAL DISTILLATION COLUMN.**

The temperature on Tray #16 of a certain industrial distillation column is regulated by the flowrate of the underflow reflux to the tray just below it. The typical steady-state operating values for these variables are 380°F and 25,000 lb/hr respectively. The result of a step test carried out on this column is presented below in Table 13.1.

TABLE 13.1

Time (hr)	Reflux Flowrate (R lb/hr)	Tray #16 Temp. (T °F)
$t < 0$	25,000	380.0
0.0	27,500	380.0
0.5	27,500	380.9
1.0	27,500	383.1
1.5	...	386.1
2.0	...	387.8
2.5	...	389.1
3.0	...	391.2
3.5	...	391.7
4.0	...	392.4
4.5	...	393.0
5.0	...	393.4
5.5	...	393.6
6.0	...	394.0
6.5	...	394.3
7.0	...	394.4
7.5	...	394.5
8.0	...	394.7
9.0	...	394.8
10.0	...	394.9
11.0	...	395.0
12.0	...	394.9
13.0	...	395.1
14.0	27,500	395.0

A candidate model being entertained for this process is:

$$y(s) = \frac{Ke^{-\alpha s}}{\tau s + 1} u(s) \tag{13.13}$$

Use the provided information to estimate the unknown parameters.

Solution:

We first note that the data are given in terms of actual process variables; they must be recast in terms of deviations from initial steady-state values. Thus, from the supplied table, we obtain output $y(t)$ as $T - 380°$, and input $u(t)$ as $R - 25,000$.

A careful inspection of the step-response data indicates that the ultimate *change* in the Tray #16 temperature as a result of the implemented reflux flow change is, *on the average*, 15.0°F, i.e.:

$$y_\infty = 15.0°F \tag{13.14}$$

Observe therefore that since the step input is of magnitude 2,500 (lb/hr), (27,500 − 25,000), the steady-state gain is estimated as:

$$\hat{K} = \frac{15}{2500} = 6 \times 10^{-3} \text{ °F/(lb/hr)} \tag{13.15}$$

We next obtain the variable $(y_\infty - y)/y_\infty$. Table 13.2 shows the data set up in the form required for estimating the model parameters.

TABLE 13.2

Time (hr)	u	y	$\dfrac{y_\infty - y}{y_\infty}$
$t < 0$	0	0.0	1.000
0.0	2,500	0.0	1.000
0.5	2,500	0.9	0.940
1.0	2,500	3.1	0.793
1.5	...	6.1	0.593
2.0	...	7.8	0.480
2.5	...	9.1	0.393
3.0	...	11.2	0.253
3.5	...	11.7	0.22
4.0	...	12.4	0.173
4.5	...	13.0	0.133
5.0	...	13.4	0.107
5.5	...	13.6	0.093
6.0	...	14.0	0.067
6.5	...	14.3	0.047
7.0	...	14.4	0.040
7.5	...	14.5	0.033
8.0	...	14.7	0.020
9.0	...	14.8	0.013
10.0	...	14.9	0.007
11.0	...	15.0	0.00
12.0	...	14.9	0.007
13.0	...	15.1	−0.007
14.0	2,500	15.0	0.00

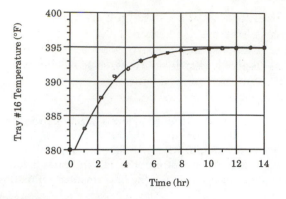

Figure 13.4.   A comparison of model and data for the distillation column of Example 13.1.

Applying linear regression on these data against the expression in Eq. (13.12) we obtain the following estimates for the time delay and time constant:

$$\hat{\alpha} = 0.56 \text{ hr} \tag{13.16}$$

$$\hat{\tau} = 1.91 \text{ hr} \tag{13.17}$$

A graphical display of the data and the identified model is shown in Figure 13.4.

There are quicker, strictly graphical, methods for estimating K, τ, and α directly from experimental step-response plots; but they yield much cruder estimates. There are, however, certain applications that do not require extremely accurate estimates for these model parameters. In such cases, these quicker graphical methods are quite useful. One such application is in the tuning of feedback controllers when no fundamental process model is available. The use of such graphical techniques will be discussed in that context in Chapter 15.

It is also possible to estimate K, τ, and α directly from Eq. (13.4) using nonlinear regression. Such estimates may be more accurate in the presence of substantial experimental error, but the increased accuracy will be obtained at the cost of increased procedural complexity. However, again, we note that many useful computer packages are available for handling the substantial computational burdens usually associated with nonlinear regression.

13.4   IMPULSE-RESPONSE IDENTIFICATION

13.4.1 Basic Principles

The main strategy of *impulse-response identification* is to obtain process transfer function models directly from experimental *impulse-response* data. This is possible because, as we shall soon show, there is a *direct* relationship between a system's transfer function and the *moments* of its theoretical impulse response. Technically, no formal parameter estimation (by data fitting in the usual sense) is required.

Process identification via impulse responses apparently suffers from one major disadvantage: the impulse function is difficult to implement in practice. In reality, however, this is an easily surmountable obstacle: the usual practice is to *derive* impulse-response data (using techniques to be discussed) from other response data that are more easily obtained, such as step, or rectangular pulse responses, or for that matter, from the response to any *arbitrary* input function.

13.4.2 An Introductory Example: The First-Order System

The *unit* impulse response of the first-order system having the standard transfer function:

$$g(s) = \frac{K}{\tau s + 1}$$

is given by:

$$y(t) = \frac{K}{\tau} e^{-t/\tau} \tag{13.18}$$

Because this expression is identical to the inverse Laplace transform of the process transfer function $g(s)$, let us refer to it as $g(t)$. (Recall the definition of the impulse-response function in Chapter 4.)

Let us now note the following about this particular impulse-response function, $g(t)$:

$$\int_0^\infty g(t)\, dt = \frac{K}{\tau} \int_0^\infty e^{-t/\tau}\, dt = K \tag{13.19a}$$

and

$$\int_0^\infty t\, g(t)\, dt = \frac{K}{\tau} \int_0^\infty t e^{-t/\tau}\, dt = K\tau \tag{13.19b}$$

Since Eq. (13.19a) is the definition of the *zeroth* moment of the function $g(t)$ about the origin, and Eq. (13.19b) defines its *first* moment, we may now observe that the steady-state gain K and the time constant τ of a first-order system may be obtained from these moments of the impulse-response function.

In general, if we define the jth moment of the function $g(t)$ as:

$$m_j = \int_0^\infty t^j g(t)\, dt \tag{13.20}$$

then, we have, for the first-order system, that:

$$K = m_0 \tag{13.21a}$$

and

$$\tau = m_1/m_0 \tag{13.21b}$$

The implication now is that given experimental impulse-response data, $g(k)$, $k = 0, 1, 2, \ldots, N$ (which is $g(t)$ observed at various discrete points $t_0, t_1, t_2, \ldots, t_N$ in time, at equally spaced time intervals Δt), then if the data came from a first-order system — or one which is approximately so — the system gain and time constant are estimated from the expressions:

$$\hat{K} = \int_0^\infty g(t)\, dt \tag{13.22a}$$

and

$$\hat{\tau} = \frac{\displaystyle\int_0^\infty t g(t)\, dt}{\displaystyle\int_0^\infty g(t)} \tag{13.22b}$$

where we may evaluate the integrals using quadratures with the discrete experimental $g(k)$ data.

For example, using Simpson's rule (where N must be even), the expressions in Eq. (13.22) take the explicit form:

$$\hat{K} = \int_0^{t_N} g(t)\, dt$$

$$\approx \frac{\Delta t}{3}\left[g(0) + 4\sum_{i=1}^{N/2} g(2i-1) + 2\sum_{i=1}^{N/2-1} g(2i) + g(N) \right] \tag{13.23a}$$

and

$$\hat{\tau} = \frac{\displaystyle\int_0^{t_N} t g(t)\, dt}{\displaystyle\int_0^{t_N} g(t)\, dt}$$

$$\approx \frac{\left[4\sum_{i=1}^{N/2} t_{2i-1}\, g(2i-1) + 2\sum_{i=1}^{N/2-1} t_{2i}\, g(2i) + t_N\, g(N) \right]}{\left[g(0) + 4\sum_{i=1}^{N/2} g(2i-1) + 2\sum_{i=1}^{N/2-1} g(2i) + g(N) \right]} \tag{13.23b}$$

What we have just illustrated for a first-order system can be generalized to any transfer function, of any order.

13.4.3 A General Theory of Process Identification from Impulse Responses

Let $g(s)$ be the transform-domain transfer function of an arbitrary system. The inverse Laplace transform of this transfer function is identical to the system's *theoretical* impulse-response function $g(t)$. By definition of the Laplace transform of a function, therefore:

$$g(s) = \int_0^\infty e^{-st} g(t)\, dt \qquad (13.24)$$

Now, the power series expansion for e^{-st} is:

$$e^{-st} = \sum_{j=0}^\infty \frac{(-st)^j}{j!} \qquad (13.25)$$

so that Eq. (13.24) becomes:

$$g(s) = \int_0^\infty \sum_{j=0}^\infty \frac{(-st)^j}{j!} g(t)\, dt \qquad (13.26)$$

or

$$g(s) = \int_0^\infty \sum_{j=0}^\infty (-1)^j \frac{s^j t^j}{j!} g(t)\, dt \qquad (13.27)$$

Since we started out assuming that the transfer function exists for the process in question, then the Laplace transform indicated in Eq. (13.24) exists, and therefore both the integral, and the summation in Eq. (13.26) converge. This permits us to reverse the order of integration and summation; we may therefore rearrange Eq. (13.27) to give:

$$g(s) = \sum_{j=0}^\infty (-1)^j \frac{(s)^j}{j!} \int_0^\infty t^j g(t)\, dt \qquad (13.28)$$

or

$$g(s) = \sum_{j=0}^\infty (-1)^j \frac{(s)^j}{j!} m_j \qquad (13.29)$$

where m_j is the jth moment of the impulse-response function $g(t)$, defined as:

$$m_j = \int_0^\infty t^j g(t)\, dt \qquad (13.30)$$

The expression in Eq. (13.29) indicates that the transfer function of an arbitrary system can be related directly to the moments of its impulse-response function. A transfer function model can therefore be "identified" for any system from its experimentally obtained impulse response, through the moments of the experimental data. Let us now consider how this is actually done.

Implications for Process Identification

By definition, the steady-state gain for a system whose transfer function is $g(s)$ is given by:

$$K = \lim_{s \to 0} g(s) \qquad (13.31)$$

if the limit exists. (Systems for which this limit fails to exist — e.g., pure capacity systems — have no steady-state gains.)

If we now introduce Eq. (13.24) into Eq. (13.31) for $g(s)$, we have:

$$K = \lim_{s \to 0} \int_0^\infty e^{-st} g(t) \, dt \qquad (13.32)$$

which immediately gives the important result:

$$K = \int_0^\infty g(t) \, dt = m_0 \qquad (13.33)$$

The zeroth moment of the impulse response of any arbitrary system is equal to its steady-state gain,

a generalization of the result given in Eq. (13.19) for the first-order system.

If we now consider using the steady-state gain to normalize the transfer function, we have:

$$g\#(s) = \frac{g(s)}{K} \qquad (13.34)$$

a "normalized" transfer function whose steady-state gain is now 1. Furthermore, if we define the normalized moments of the impulse response as:

$$\mu_j = \frac{m_j}{m_0} \qquad (13.35)$$

then, by dividing through in Eq. (13.29) by m_0, and recalling that $m_0 = K$, we have:

$$g\#(s) = \sum_{j=0}^\infty (-1)^j \frac{(s)^j}{j!} \mu_j \qquad (13.36)$$

or

$$g\#(s) = 1 - \mu_1 s + \frac{\mu_2}{2} s^2 - \frac{\mu_3}{6} s^3 + \ldots \qquad (13.37)$$

The normalized moments μ_j of a system's impulse response can be obtained from experimental data; they are therefore known quantities in Eq. (13.37) which can be related to the elements of the "unknown" transfer function.

Thus, for example, for a system suspected to be an nth-order system, whose transfer function is:

$$g(s) = \frac{K}{a_n s^n + a_{n-1} s^{n-1} + \ldots + a_2 s^2 + a_1 s + 1} \qquad (13.38)$$

observe that the normalized transfer function is:

$$g\#(s) = \frac{1}{a_n s^n + a_{n-1} s^{n-1} + \ldots + a_2 s^2 + a_1 s + 1} \qquad (13.39)$$

After deriving the normalized moments of the impulse response μ_j from experimental data, from Eq. (13.37) we now have:

$$\frac{1}{a_n s^n + a_{n-1} s^{n-1} + \ldots + a_2 s^2 + a_1 s + 1} = 1 - \mu_1 s + \frac{\mu_2}{2} s^2 - \frac{\mu_3}{6} s^3 + \ldots$$

or

$$1 = \left(a_n s^n + a_{n-1} s^{n-1} + \ldots + a_1 s + 1 \right) \left(1 - \mu_1 s + \frac{\mu_2}{2} s^2 + \ldots \right) \qquad (13.40)$$

from where, by carrying out the indicated multiplication and equating coefficients of powers of s on both sides of the equation, we obtain the following expressions that explicitly relate the moments and the coefficients of the transfer function model:

$$a_1 = \mu_1 \qquad (13.41)$$

$$a_2 = a_1 \mu_1 - \frac{\mu_2}{2} = \mu_1^2 - \frac{\mu_2}{2} \qquad (13.42)$$

$$a_3 = a_2 \mu_1 - a_1 \frac{\mu_2}{2} + \frac{\mu_3}{6} \qquad (13.43)$$

$$\ldots \qquad \ldots$$

etc.

13.4.4 Identification of Simple Model Forms

Let us recall that the main objective of process identification is to fit an approximate model, usually of simple form, to experimentally observed data. The procedure for doing this, using the impulse-response approach, is now summarized.

1. Experimentally obtain the system's impulse response.
2. From the impulse-response data, calculate the normalized moments μ_j.
3. Postulate an approximate transfer function model.
4. Estimate the unknown parameters in the postulated, transfer function model by relating the transfer function to the moments.

The relationship between the parameters of the transfer function model and the normalized moments of the impulse response will now be derived for some typical candidate transfer function models commonly used in process identification. In all the cases, the steady-state gain is always given by the zeroth moment m_0 and will therefore not be included explicitly.

First-Order Model

In this case:

$$g\#(s) = \frac{1}{\tau s + 1}$$

and, from Eq. (13.40):

$$1 = (\tau s + 1)\left(1 - \mu_1 s + \frac{\mu_2}{2} s^2 + \ldots\right) \tag{13.44}$$

from where we now obtain the following relations by equating coefficients of powers of s on either side of the equation:

$$\mu_1 = \tau \tag{13.45}$$
$$\mu_2 = 2\tau^2 \tag{13.46}$$
$$\mu_3 = 6\tau^3, \text{ etc.} \tag{13.47}$$

or, in general:

$$\mu_n = n!\tau^n \tag{13.48}$$

from which we can obtain several estimates of the time constant τ. Because the computation of higher moments from data is usually subject to more error, it is advisable to restrict attention to no more than the first three moments.

It should be kept in mind, of course, that the entire data set is used in computing each moment; thus if τ were estimated solely from the first moment as indicated in Eq. (13.45), the result will still be based on the *entire* data set. The estimates obtained from the higher moments could then be considered as providing diagnostic information about the "closeness" of the actual process dynamic behavior to that of an ideal first-order system.

Second-Order Model

In this case:

$$g\#(s) = \frac{1}{a_2 s^2 + a_1 s + 1}$$

and, once more, from Eq. (13.40):

$$1 = \left(a_2 s^2 + a_1 s + 1\right)\left(1 - \mu_1 s + \frac{\mu_2}{2} s^2 + \ldots\right) \qquad (13.49)$$

Again, upon equating the coefficients of powers of s on either side of the equation, we obtain:

$$a_1 = \mu_1 \qquad (13.50)$$

$$a_2 = \mu_1^2 - \frac{\mu_2}{2} \qquad (13.51)$$

which are sufficient for the determination of these unknown constants.

Additional moments can be used to obtain additional estimates of these same constants. These may then be used to verify how truly close the observed system response is to second-order behavior. Again, the degree of discrepancy between the estimates obtained from Eqs. (13.50) and (13.51), and the additional estimates obtained using the higher moments, provides an indication of how good or bad an approximation the postulated second-order model is to true system behavior.

Second-Order-Plus-Single-Zero Model

The normalized transfer function in this case is:

$$g\#(s) = \frac{\xi s + 1}{a_2 s^2 + a_1 s + 1}$$

and this time, from Eq. (13.40), we have:

$$(\xi s + 1) = \left(a_2 s^2 + a_1 s + 1\right)\left(1 - \mu_1 s + \frac{\mu_2}{2} s^2 + \ldots\right) \qquad (13.52)$$

There are three unknown parameters in Eq. (13.52), requiring three equations for unique determination. The three equations are obtained by equating the coefficients of s, s^2, and s^3, i.e:

$$a_1 - \mu_1 = \xi \qquad (13.53a)$$

$$\frac{\mu_2}{2} - a_1 \mu_1 + a_2 = 0 \qquad (13.53b)$$

$$\frac{\mu_3}{6} - a_1 \frac{\mu_2}{2} + a_2 \mu_1 = 0 \qquad (13.53b)$$

When these equations are solved simultaneously for the unknown transfer function parameters, we obtain:

$$a_1 = \frac{3\mu_1\mu_2 - \mu_3}{3(2\mu_1^2 - \mu_2)} \tag{13.54}$$

$$a_2 = a_1\mu_1 - \frac{\mu_2}{2} \tag{13.55}$$

and

$$\xi = a_1 - \mu_1 \tag{13.56}$$

First-Order-Plus-Time-Delay Model

In this case:

$$g\#(s) = \frac{e^{-\alpha s}}{\tau s + 1}$$

and, we recall from Eq. (13.40) that in this case we will now have:

$$e^{-\alpha s} = (\tau s + 1)\left(1 - \mu_1 s + \frac{\mu_2}{2}s^2 + \ldots\right) \tag{13.57}$$

and upon introducing the infinite series representation for the exponential function, the result is:

$$\left(1 - \alpha s + \frac{\alpha^2 s^2}{2!} - \frac{\alpha^3 s^3}{3!} + \ldots\right) = (\tau s + 1)\left(1 - \mu_1 s + \frac{\mu_2}{2}s^2 + \ldots\right) \tag{13.58}$$

and, again, by matching coefficients of s, and of s^2, we may write the two equations needed for unique determination of the two unknown parameters:

$$-\alpha = \tau - \mu_1 \tag{13.59}$$

$$\frac{\alpha^2}{2} = -\tau\mu_1 + \frac{\mu_2}{2} \tag{13.60}$$

These two equations may be solved simultaneously for α and τ by rearranging to give:

$$\mu_g = \mu_1 = \tau + \alpha \tag{13.61}$$

and

$$\sigma_g^2 = \mu_2 - (\mu_1)^2 = \tau^2 \tag{13.62}$$

from which independent estimates of τ and α are easily obtained. It is interesting to note what these equations indicate: since μ_1 is the mean of the normalized impulse-response function, $g\#(t)$, and since the "variance" of $g\#(t)$ is $\mu_2 - (\mu_1)^2$, we then have that:

The normalized impulse-response function for a first-order-plus-time-delay process has a mean value of $\tau + \alpha$ and a variance of τ^2.

This provides a very simple algorithm for determining the parameters of a first-order-plus-time-delay process model from impulse-response data $g(t_k)$ as follows:

1.   Obtain the zeroth moment m_0 by appropriately integrating the data; this is an estimate of the steady-state gain K say, $\hat{K}$ (cf. Eq. (13.33)).
2.   Normalize the data $g(t_k)$ with $\hat{K}$ obtained from 1 to obtain

$$g\#(t_k) \;=\; \frac{g(t_k)}{\hat{K}}$$

3.   Obtain the mean value of $g\#(t_k)$.
4.   Obtain the standard deviation of $g\#(t_k)$; this is an estimate of τ, say, $\hat{\tau}$.
5.   Obtain the estimate of the time delay α by subtracting $\hat{\tau}$ from the mean.

Obtaining Moments from Experimental Data

Experimental impulse-response data are usually obtained at discrete points in time, usually at equally spaced time intervals Δt, and therefore the integral in Eq. (13.30) will have to be approximated by quadratures or other efficient numerical integration schemes. Thus, for example, using Simpson's rule we have:

$$m_j \;=\; \frac{\Delta t}{3}\left[(t_0)^j g(0) \;+\; 4\sum_{i=1}^{N/2}(t_{2i-1})^j g(2i-1) \right.$$
$$\left. +\; 2\sum_{i=1}^{N/2-1}(t_{2i})^j g(2i) \;+\; (t_N)^j g(N)\right]; \quad j = 1, 2, \ldots \qquad (13.63)$$

The steady-state gain is therefore obtained from:

$$m_0 \;=\; \frac{\Delta t}{3}\left[g(0) \;+\; 4\sum_{i=1}^{N/2}g(2i-1) \;+\; 2\sum_{i=1}^{N/2-1}g(2i) \;+\; g(N)\right] \qquad (13.64)$$

exactly as indicated for the first-order system in Eq. (13.23a). The normalized moments are given by introducing Eqs. (13.63) and (13.64) for m_j and m_0 in:

$$\mu_j \;=\; \frac{m_j}{m_0} \qquad (13.65)$$

The indicated summations will be sufficiently close to the integrals provided Δt is small enough.

Let us illustrate the ideas we have presented in this section with an example.

Example 13.2   IDENTIFICATION OF A FIRST-ORDER SYSTEM FROM THE MOMENTS OF ITS IMPULSE RESPONSE.

The impulse response of a system strongly suspected to be first order is given in the table below. From the supplied data, find the steady-state gain and the time constant for this system. How good is the postulate that the system is first order?

Solution:

We observe first that no value is available for $g(0)$. By linear extrapolation using $g(0.5)$ and $g(1)$ the approximate value $\hat{g}(0) = 4.22$ is substituted. Using the data, with $\Delta t = 0.5$ min, and the substituted value for $g(0)$, we obtain the moments from Eq. (13.63) as follows:

$$m_0 = 4.86$$
$$m_1 = 4.96$$
$$m_2 = 9.72$$

The first two normalized moments are now obtained as:

Time t_k (min)	Impulse Response $g(t_k)$
0.5	3.03
1.0	1.84
1.5	1.12
2.0	0.68
2.5	0.41
3.0	0.25
3.5	0.15
4.0	0.09
4.5	0.06
5.0	0.03
5.5	0.02
6.0	0.01
6.5	0.008
7.0	0.005

$$\mu_1 = \frac{4.96}{4.86}$$
$$= 1.03$$

along with:

$$\mu_2 = \frac{9.72}{4.86}$$
$$= 2.00$$

Thus, from these data, we obtain the following estimate of the steady-state gain:

$$\hat{K} = 4.86 \qquad (13.66)$$

Recalling Eqs. (13.45) and (13.46), two possible estimates of the time constant for the postulated first-order model may be obtained; from the first moment:

$$\hat{\tau} = 1.03 \text{ min} \qquad (13.67)$$

and, from the second moment:

$$\hat{\tau} = \sqrt{\frac{2.00}{2}} = 1.00 \text{ min} \qquad (13.68)$$

We may now observe that these two estimates are sufficiently close that we have reason to believe that the system must truly be first order.

It is pertinent at this point to disclose that the data set was generated from the continuous function:

$$g(t) = 5e^{-t} \qquad (13.69)$$

which is recognizable as the impulse response of a first-order system with gain 5, and time constant, 1 min. We may now observe that:
1. The estimates (particularly of the time constant) are fairly accurate,
2. The steady-state gain estimate, as expected, is less accurate. This is due to errors introduced by the absence of $g(t_0)$ from the data (making it necessary to use an approximate extrapolant), the finite sized time interval between experimental observations, and the finite number of data points.

13.4.5 Impulse-Response Data from Other Responses

In practice, it may not always be possible to implement an input function that is close enough to an ideal impulse; it is possible, however, to generate impulse-response data from step responses, rectangular pulse responses, or the response to any arbitrary input function.

From Experimental Step-Response Data

Since the first derivative of the theoretical step response gives the theoretical impulse response (cf. Eq. (5.21)), we may generate impulse-response data from step-response data by differentiation. Thus if $\beta(t_k)$ ($k = 0, 1, 2, ..., N$) represents *unit* step-response data observed at N discrete points t_k in time, the corresponding impulse-response data are given by (recall Eq. (4.55)):

$$g(t_k) = \beta(t_k) - \beta(t_{k-1}) \qquad (13.70)$$

However, this procedure is recommended only if the original step-response data are reasonably noise free; numerical differentiation is notorious for its tendency to amplify the noise component of data.

From Arbitrary Input/Output Data

Consider a data record consisting of the arbitrary input $u(t_k)$ (or $u(k)$ for simplicity) and the corresponding process response $y(t_k)$ (or $y(k)$ for simplicity). So long as the system is linear (or approximately so), then this process response is related to the arbitrary input by no other than the *unit* impulse-response function we seek. Recall that according to Eq. (4.51), for N impulse-response coefficients:

$$y(k) = \sum_{i=1}^{N} g(i)\, u(k-i) \tag{4.51b}$$

For each observed experimental data point, we may now write, using this expression:

$$y(1) = g(1)\, u(0)$$

$$y(2) = g(1)\, u(1) + g(2)\, u(0)$$

$$y(3) = g(1)\, u(2) + g(2)\, u(1) + g(3)\, u(0)$$

$$\dots \qquad \dots$$

etc.
which may be written together as:

$$
\begin{bmatrix} y(1) \\ y(2) \\ y(3) \\ \dots \\ y(M) \end{bmatrix}
=
\begin{bmatrix}
u(0) & 0 & \dots & \dots & 0 \\
u(1) & u(0) & \dots & \dots & 0 \\
u(2) & u(1) & u(0) & \dots & 0 \\
\dots & \dots & \dots & \dots & \dots \\
u(M-1) & u(M-2) & \dots & \dots & u(M-N)
\end{bmatrix}
\begin{bmatrix} g(1) \\ g(2) \\ g(3) \\ \dots \\ g(N) \end{bmatrix}
\tag{13.71a}
$$

or in the form:

$$\mathbf{y} = \mathbf{U}\mathbf{g} \tag{13.71b}$$

Here N is the total number of impulse-response "data" $g(k)$ to be derived from the input/output data record, $u_k, y_k; k = 1, 2, \dots, M$ where $N \leq M$. Observe the implication: that the response to an arbitrary input is merely a transformation of the unit impulse response via the "input matrix" $\mathbf{U}$.

If $M = N$ and $u(0) \neq 0$, the matrix $\mathbf{U}$ is square and nonsingular; hence the solution to Eq. (13.71b) is given by:

$$\mathbf{g} = \mathbf{U}^{-1}\mathbf{y} \tag{13.72}$$

This is the transformation required for deriving impulse-response "data" $g(k)$, $k = 1, 2, \dots, N$, from the process response to any arbitrary input. Because of the special nature of the matrix $\mathbf{U}$ (see Eq. (13.71a) above), the inversion implied in Eq. (13.72) is easily carried out, as will soon be illustrated with an example.

Example 13.3 DERIVING IMPULSE-RESPONSE DATA FROM ARBITRARY INPUT/OUTPUT DATA.

The response of a particular system to the indicated input sequence is given in the table below. From the supplied data, derive the corresponding impulse-response data.

Solution:

We observe first that the expanded version of Eq. (13.71a) may be solved one at a time with the first equation yielding:

$$g(1) = y(1)/u(0)$$

which may be used in the second to obtain:

$$g(2) = \frac{1}{u(0)}[y(2) - g(1)u(1)]$$

We may thus proceed recursively to obtain, in general:

$$g(M) = \frac{1}{u(0)}[y(M) - g(1)u(M-1) - g(2)u(M-2) - \dots - g(M-1)u(1)]$$

Time t_k (min)	Input $u(t_k)$	Response $y(t_k)$
0.0	2.0	0
0.5	2.0	9,80
1.0	1.0	15.86
1.5	0.0	14.64
2.0	0.0	8.95
2.5	0.0	5.44
3.0	0.0	3.30
3.5	0.0	2.00
4.0	0.0	1.21
4.5	0.0	0.73
5.0	0.0	0.45
5.5	0.0	0.27
6.0	0.0	0.16
6.5	0.0	0.09
7.0	0.0	0.0056
7.5	0.0	0.0036

If we now apply this to the given data, we easily obtain $g(0) = 4.9$; $g(1) = 3.03$; $g(2) = 1.84$; $g(3) = 1.12$; $g(4) = 0.68$; $g(5) = 0.41$; $g(6) = 0.25$; $g(7) = 0.15$; $g(8) = 0.09$; $g(9) = 0.06$; $g(10) = 0.03$; $g(11) = 0.02$; $g(12) = 0.01$; $g(13) = 0.008$; and $g(14) = 0.005$ as the required corresponding impulse-response data.

Finally, we note that by following a procedure to be discussed fully in the next section, it is also possible to "Fourier-transform" arbitrary input/output data records to frequency-response data in the form of $g(jw)$, the frequency-domain transfer function (cf. Eq. (13.78a) and the subsequent discussion).

Converting this frequency-response data to the desired impulse-response data is then achieved by inverse Fourier transformation (cf. Eq. (13.76), Ref. [2], Chapter 8 of Ref. [5]; also see Eq. (4.81) and Figure 4.4). As we will soon see, however, it is more common to utilize frequency-response data for process identification in a different (but related) manner.

13.5 FREQUENCY-RESPONSE IDENTIFICATION

13.5.1 Basic Principles

As with step-response identification, the main strategy of frequency-response identification is to fit *theoretical frequency-response functions* to experimental frequency-response data. The theoretical frequency-response expressions are obtained for any given candidate model as discussed in Chapter 9, and even though it is possible to obtain frequency-response data by direct sine wave testing, as with impulse-response testing, it is customary to *derive* such data from more conveniently implemented input responses.

13.5.2 Direct Sine Wave Testing

One of the most important conclusions of our discussion in Chapter 9 is that the fundamental features of any linear process are fully characterized by its frequency response. Thus the Bode diagrams for various linear systems have peculiar, distinguishing characteristics that differentiate one system from another. The following method therefore suggests itself as a means of identifying a process model:

1. Vary the system input $u(t)$ sinusoidally, with a specific amplitude A and frequency ω.

2. After the transients have died out, measure the output data.

3. From the observed output data, deduce the amplitude ratio (AR), and the phase angle ϕ; this provides a *single* point on each half of the experimental Bode diagram.

4. Change the frequency of the input function and go back to Step 2; repeat until enough frequencies have been covered.

5. Compile the data in the form of an experimental Bode diagram.

6. Compare the experimental Bode diagram with theoretical ones; postulate plausible model forms on this basis; estimate parameters by matching theoretical frequency response with the experimental data.

Let us now recall that in Chapter 9, we determined a system's ultimate response to an input sinusoidal function directly from $g(j\omega)$, given its transform-domain transfer function, $g(s)$. Observe that the procedure outlined above is for the reverse process of determining $g(s)$ from an experimentally obtained $g(j\omega)$.

As intuitively appealing as this method appears to be, it suffers from two major disadvantages:

1. Sinusoidal inputs are often extremely difficult to generate experimentally.

2. It is too tedious and too time consuming. Testing at each frequency requires sustaining the sinusoidal input long enough to permit the output transients to die out; and in each case, the result is only one point on the experimental Bode diagram. This procedure must be repeated several times in order to have enough data points on the Bode diagram.

It is therefore not surprising that direct sine wave testing is hardly used in practice. *Pulse testing*, a procedure which we shall now discuss, retains all the benefits of direct sine wave testing while avoiding the two serious disadvantages noted above.

13.5.3 Pulse Testing

The pulse testing procedure involves performing a *single* experiment in which an *arbitrary* pulse input is implemented on the process, and the resulting process response is measured. The input function is not required to take any specific mathematical form; it is only required to start from zero, and end at zero (see Figure 13.5). The reason for requiring that the input function go to zero after a period of time will soon be pointed out.

The main attractive feature of pulse testing is that complete frequency-response information can be obtained by properly processing the input/output data obtained from this single experiment.

Frequency Response through Pulse Testing

By definition, the transfer function for a single input, single output process is:

$$g(s) = \frac{y(s)}{u(s)} \tag{13.73}$$

and since the frequency response can be obtained by setting $s = j\omega$, we have, from Eq. (13.73):

$$g(j\omega) = \frac{y(j\omega)}{u(j\omega)} \tag{13.74}$$

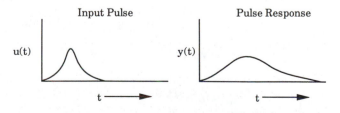

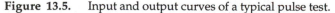

Figure 13.5.   Input and output curves of a typical pulse test.

Observe, therefore, that if $y(j\omega)$ and $u(j\omega)$ can be obtained from respective time data, $y(t)$, and $u(t)$, Eq. (13.74) guarantees that $g(j\omega)$ can be obtained from a ratio of these two frequency quantities.

It can be shown that *any function $f(t)$ which goes to zero sufficiently fast as $t \rightarrow \pm \infty$* can be written as:

$$f(t) = \frac{1}{2\pi} \int_{-\infty}^{+\infty} \hat{f}(j\omega)\, e^{j\omega t}\, d\omega \tag{13.75a}$$

where $\hat{f}(j\omega)$ is the *Fourier Transform* of $f(t)$, defined by:

$$\hat{f}(j\omega) = \int_{-\infty}^{+\infty} f(t)\, e^{-j\omega t}\, dt \tag{13.75b}$$

These equations, in which the symmetry between $f(t)$ and $\hat{f}(j\omega)$ is evident, constitute what is known as a *Fourier transform pair* (see Appendix C). The Fourier transform function $\hat{f}(j\omega)$ is entirely equivalent to $f(t)$, since when the former is known, the latter is determined from Eq. (13.75a).

The main implication of these results is that, in effect, we have two equivalent ways of representing any function: $f(t)$ in the time domain, and $\hat{f}(j\omega)$ in the frequency domain.

Thus, by Fourier transformation, we can obtain the *entirely* equivalent frequency-domain representation of our time data, enabling us to utilize Eq. (13.74).

To understand why it is possible to obtain frequency information from non-periodic data, we now rewrite Eq. (13.75a) by introducing Euler's identities (recall Eq. (9.8)), and approximate the integral by a sum, giving:

$$f(t) = \frac{1}{2\pi} \Delta\omega \sum_{n=-\infty}^{\infty} \hat{f}(j\omega_n) \left(\cos \omega_n t + j \sin \omega_n t \right) \tag{13.76}$$

which shows one of the most significant implications of Fourier's theorem, that:

> *Any function $f(t)$ can be written as a sum of harmonic (or sinusoidal) functions, with amplitude $\hat{f}(j\omega_n)$ and frequency ω_n.*

How strongly a certain frequency ω_n is represented in the sum in Eq. (13.76) is clearly indicated by $\left| \hat{f}(j\omega_n) \right|$; a small value indicates $f(t)$ contains very little of the particular frequency in question, while a large value indicates strong representation.

We may now use these results to obtain experimental Bode diagrams from pulse test results as follows:

Equation (13.74) may now be written as:

$$g(j\omega) = \frac{\hat{y}(j\omega)}{\hat{u}(j\omega)} \tag{13.77}$$

where $\hat{y}(j\omega)$ and $\hat{u}(j\omega)$, are, respectively, the Fourier transformations of the input and output pulse test data, i.e.:

$$g(j\omega) = \frac{\displaystyle\int_{-\infty}^{\infty} y(t)\, e^{-j\omega t} dt}{\displaystyle\int_{-\infty}^{+\infty} u(t)\, e^{-j\omega t} dt} \tag{13.78a}$$

Now, it is assumed that the process is initially at steady state prior to the commencement of the pulse test experiment, at time $t = 0$; thus in these integrals, only the portion from $t = 0$ to $t = \infty$ is relevant. Furthermore, we had made it a requirement of the pulse test that the input should become zero after some finite time T_u. For all realistic systems (with the exception of pure capacity systems), this input condition also essentially forces the output to become zero after some time T_y.

The implication is that Eq. (13.78a) may now be represented as:

$$g(j\omega) = \frac{\displaystyle\int_{0}^{T_y} y(t)\, e^{-j\omega t} dt}{\displaystyle\int_{0}^{T_u} u(t)\, e^{-j\omega t} dt} \tag{13.78b}$$

or, upon introducing the Euler identities:

$$g(j\omega) = \frac{\displaystyle\int_{0}^{T_y} y(t)\cos \omega t\, dt - j \int_{0}^{T_y} y(t)\sin \omega t\, dt}{\displaystyle\int_{0}^{T_u} u(t)\cos \omega t\, dt - j \int_{0}^{T_u} u(t)\sin \omega t\, dt} \tag{13.79a}$$

or

$$g(j\omega) = \frac{A_1(\omega) - jA_2(\omega)}{B_1(\omega) - jB_2(\omega)} \tag{13.79b}$$

where, now, for the output:

$$A_1(\omega) = \int_{0}^{T_y} y(t)\cos \omega t\, dt \tag{13.80a}$$

$$A_2(\omega) = \int_{0}^{T_y} y(t)\sin \omega t\, dt \tag{13.80b}$$

and, for the input:

$$B_1(\omega) = \int_0^{T_u} u(t)\cos \omega t\, dt \qquad (13.81\text{a})$$

$$B_2(\omega) = \int_0^{T_u} u(t)\sin \omega t\, dt \qquad (13.81\text{b})$$

The procedure for obtaining the Bode diagram is now outlined below:

1. Examine the output data and determine T_y, the time beyond which the output may be considered as having essentially returned to zero; T_u the corresponding time for the input pulse is predetermined as part of the experimental design.

2. Choose a range of frequencies, ω_{min} and ω_{max}, over which the Bode plot is to be obtained; for different values of ω in this range, evaluate each of the four integrals in Eqs. (13.80) and (13.81); this gives $g(j\omega)$ as a function of ω.

3. The amplitude ratio and phase angle for each of these frequencies are now obtained from the expression in Eq. (13.79):

$$AR(\omega) = \sqrt{\frac{A_1^2 + A_2^2}{B_1^2 + B_2^2}} \qquad (13.82)$$

and

$$\phi(\omega) = \tan^{-1}\left\{\frac{\text{Im } [g(j\omega)]}{\text{Re } [g(j\omega)]}\right\} \qquad (13.83)$$

where the real part of $g(jw)$ is given by:

$$\text{Re } [g(j\omega)] = \frac{A_1 B_1 + A_2 B_2}{B_1^2 + B_2^2} \qquad (13.84\text{a})$$

and the imaginary part is given as:

$$\text{Im } [g(j\omega)] = \frac{A_1 B_2 - A_2 B_1}{B_1^2 + B_2^2} \qquad (13.84\text{b})$$

Thus, compared to direct sine wave testing, we see that pulse testing:

*   Gives the same results (experimental Bode diagrams), but it does so more conveniently.
*   Requires considerable computational effort, but this is a burden which the computer bears with considerable ease.

Pulse testing also has the additional advantage that, unlike step testing, it is less likely to produce data containing significant nonlinear effects: the input returns to its initial starting value, and therefore the process is not *permanently* perturbed away from its initial.

As with direct sine wave testing, a process model is obtained, complete with parameter estimates, by fitting a theoretical frequency response to the experimentally obtained data. This essentially involves an optimization problem in which we attempt to match ideal model information to experimental data information; the parameter set which provides the closest fit is accepted as the "best" estimates.

In principle, once the plausible models have been postulated by inspection of the experimentally obtained Bode diagram, parameter estimation can be carried out either in the frequency domain — using the frequency data — or directly in the time domain, using the raw time interval data.

Frequency Spectra

We reiterate the importance of keeping in mind that in process identification, the model is based entirely on experimental data; and therefore whatever information is *not* contained in the data *cannot* magically manifest itself in the model.

What this means for pulse testing is that if certain frequencies are not strongly represented in the input and output data, we should not expect the experimentally derived Bode diagram to contain useful information at these frequencies.

Thus, it is important to analyze the input and output data for their "frequency content," by inspecting the magnitudes of the functions $\hat{u}(j\omega)$ and $\hat{y}(j\omega)$ at various frequencies. As noted earlier, this provides us with information regarding how strongly any particular frequency is represented in the input or output data.

From Eqs. (13.77) to (13.81) it is easy to see that for the input:

$$\left|\hat{u}(j\omega)\right| = \sqrt{B_1^2(\omega) + B_2^2(\omega)} \tag{13.85}$$

and for the output:

$$\left|\hat{y}(j\omega)\right| = \sqrt{A_1^2(\omega) + A_2^2(\omega)} \tag{13.86}$$

A plot of Eq. (13.85) against frequency gives the input *frequency spectrum*; the corresponding plot of Eq. (13.86) gives the output *frequency spectrum*. From these frequency spectra, we are able to determine what portions of the experimental Bode diagram contains useful information.

Figure 13.6 shows an example plot of theoretical frequency spectra for a rectangular pulse function of height 1, and various durations of 90 sec, 60 sec, and 30 sec. Observe that the shorter the pulse width, the wider the frequency range represented with enough significance in the input signal. The pulses with longer widths tend to represent only a relatively narrow range of low frequencies with considerable significance; they contain very little of higher frequency components.

It can be shown that the pulse input guaranteed to contain all possible frequency components with equally significant magnitude is the impulse

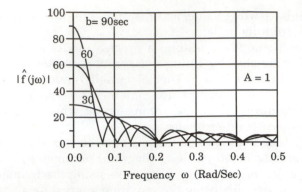

Figure 13.6.   Theoretical frequency spectra of a rectangular pulse function of different duration.

function — a pulse function with infinite height and zero width. As this is physically impossible to realize, it is best (in this context) to aim for an input pulse whose width is as small as possible while assuring an adequate amplitude of the excitation.

A Practical Application

Let us now illustrate the preceding discussion with an example application.

The System

The system in question is a pilot scale distillation column used for instructional purposes at the Operations and Process Laboratories of the University of Wisconsin–Madison. It is a 19–plate, 12–inch diameter, copper column, having variable feed and sidestream draw-off locations; it is used for separating a binary mixture of ethanol and water.

Thermocouples are available on every tray, and in the overhead, reflux, and feed lines, to provide temperature measurements. A density meter is available for on-line composition measurement; a minicomputer is available for data acquisition, and ultimately for computer control.

The Problem

The ultimate objective is to design a control system for controlling the composition of ethanol in the overhead stream, one sidestream, and at the bottom. The reflux flowrate, sidestream draw-off rate, and the reboiler steam pressure are the main input variables that can be used to control the three compositions; the feed flowrate and feed temperature constitute disturbances. A schematic diagram of this system is shown in Figure 13.7.

To design an effective control system, however, a mathematical model of the distillation column is required; this is the immediate problem.

Observe that the configuration of this system is similar to that of the ideal binary distillation column discussed in Chapter 12; we could therefore apply the same principles to obtain a theoretical model for our experimental column.

On the basis of the same assumptions enumerated in Chapter 12, such a theoretical model will:

1. Consist of 42 nonlinear, coupled, ordinary differential equations.

2. Have unknown parameters, such as relative volatility and other thermodynamic properties of ethanol/water mixtures, to be estimated independently.

3. Require numerical methods for its solution — a solution that will provide ethanol compositions on each of the 19 trays, in the condenser, and at the bottom of the column.

However, the model is needed simply for predicting how the compositions in the overhead, sidestream, and bottom of the column will respond (a) to changes in the reflux flowrate, sidestream draw-off rate, and reboiler steam pressure; and (b) to changes in the feed flowrate and the feed temperature. Keep in mind that the modeling exercise is not an end in itself; it is a means to an end.

Thus, not only will the theoretical model be too tedious to deal with, it will provide us with far more information than we need, at a price we may not be able to afford. Observe therefore that this is a typical example of a situation in which theoretical modeling, even though possible, is both impractical as well as inconvenient: process identification by empirical modeling will yield more useful results.

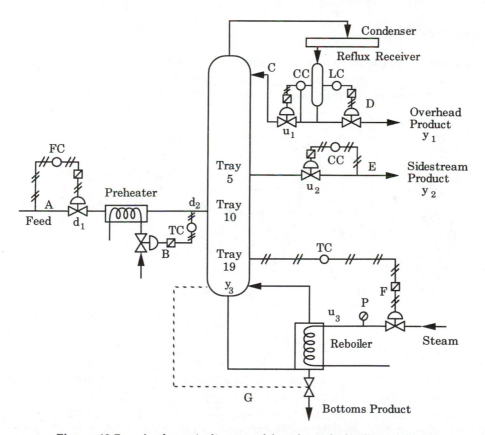

Figure 13.7.   A schematic diagram of the pilot scale distillation column.

Pulse Testing

Because of the nonlinearities inherent in the process, the pulse testing approach was chosen over the step testing approach; transfer function models were identified that represented how each of the three noted outputs respond to each of the three inputs, and the two disturbance variables.

The identification procedure employed is as follows:

1.   Rectangular pulse changes were implemented in each of the input, and disturbance variables, one at a time, and the corresponding output data were recorded.

2.   A computer program was used to convert the time data to the frequency domain via Fourier transformation; frequency spectra of the input and output information were generated, along with experimental Bode plots.

3.   By inspecting and interpreting these Bode plots, candidate models were postulated, and the model parameters estimated numerically using standard optimization routines.

4.   The identified model was validated by superimposing the theoretical pulse response obtained from these models on the experimental data.

Some Results

The response of the overhead composition to a pulse of 15 minutes duration in the reflux rate is shown in triangles in Figure 13.8; the arrow indicates the point in time when the pulse was initiated.

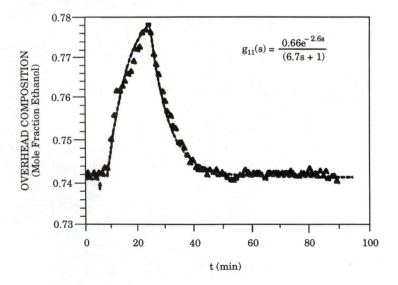

$$g_{11}(s) = \frac{0.66e^{-2.6s}}{(6.7s + 1)}$$

Figure 13.8.   Overhead composition response to a pulse of 15 minutes duration in the reflux rate.

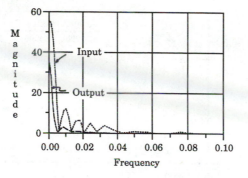

Figure 13.9.   Frequency spectra of the input and output data of Figure 13.8.

Ignoring the dashed lines for now, a comparison of the general shape indicated by the data with that shown in Figure 5.5 — the theoretical pulse response of a first-order system — indicates that a first-order-plus-time-delay model will very likely provide an adequate fit to this data set. It is important to note that even though the data were collected at 1 min intervals, we are fitting a continuous-time model from such discrete data.

The frequency spectra of the input and output data are shown in Figure 13.9, and the experimental Bode diagram in Figure 13.10.

From the output frequency spectrum, we observe that this data set contains useful information only up to frequencies in the neighborhood of 6×10^{-3} radians/s; the Bode diagram should therefore be interpreted with this in mind.

Thus, we see that an interpretation of the relevant portion of Figure 13.10 appears to support the candidacy of a first-order-plus-time-delay model.

Estimates of the steady-state gain, time constant, and time delay that provided the best fit to the experimental pulse response data are indicated in the transfer function model included in Figure 13.8; the dashed line represents the theoretical response of this model to the same pulse input used for the experiment. The agreement between the model and the data is particularly notable.

A response of a different kind was obtained from the sidestream composition when a rectangular pulse change of 15 minutes duration was implemented in the feed temperature; this is shown in Figure 13.11.

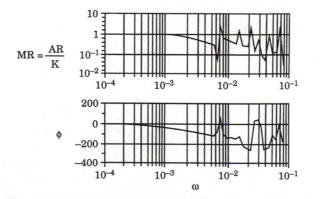

Figure 13.10.   Bode diagram derived from the data of Figure 13.8.

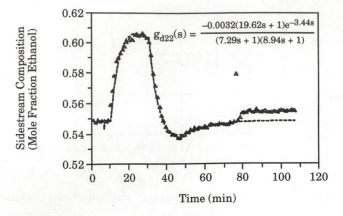

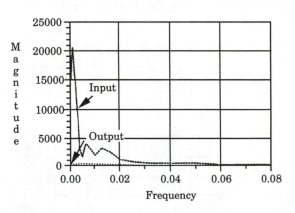

Figure 13.11. Sidestream composition response to a pulse of 15 minutes duration in feed temperature.

Again, ignoring the dashed line for now, we see that because of the "undershoot," a first-order-plus-time-delay model will not adequately represent the observed behavior; such a behavior indicates the presence of a strong zero (recall Section 6.6 of Chapter 6). The frequency spectra and the Bode diagram for this data set are shown respectively in Figures 13.12 and 13.13.

The Bode diagram is particularly revealing. The observed "hump" could not be attributed to resonance in a second-order system: if that were the case, the phase angle plot will *not* show a similar "hump." Such a behavior can only be due to the presence of at least one zero. (Note that the frequency spectra confirms that the observed "hump" is real, since it occurs within the frequency range containing authentic information.)

This leads us to postulate a model of the type given in Eq. (13.3), with one zero, two poles, and a time delay. The estimates of the model parameters that provided the best fit to the experimental data are shown in the transfer function model included in Figure 13.11. As with the first case, the dashed line

Figure 13.12. Frequency spectra of the input and output data of Figure 13.11.

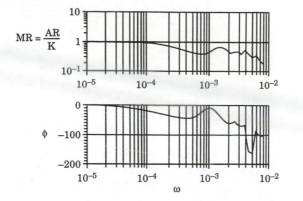

$$MR = \frac{AR}{K}$$

ϕ

Figure 13.13. Bode diagram derived from the data of Figure 13.11.

represents the theoretical pulse response of this model. Once more, the agreement between the model and the data is notable. In particular, note that the estimate of the lead time constant is almost double the value of the larger of the two lag time constants; such a strong zero is needed if the observed undershoot is to be possible.

The complete data set and the remaining results of the modeling exercise (Bode diagrams, frequency spectra, transfer function models, model vs. data plots) are available in Refs.[3, 4]; the sample we have presented above is enough to illustrate the practical application of process identification via frequency response.

13.6   ISSUES FOR MULTIVARIABLE SYSTEMS

The example distillation column we have just now used to illustrate the application of pulse testing is a typical example of a multivariable system — a system with many inputs and many outputs. A significant number of industrial processes are of this nature (see Part IVB).

In identifying process models for such systems, the result is usually a collection of several transfer function elements, relating each input to each output. To be sure, these transfer functions may be identified one at a time, using the strategy just discussed in the previous section. This is an acceptable, but clearly time-consuming, strategy.

As we now illustrate by an example below, it is also possible to estimate all the parameters in the entire transfer function matrix *simultaneously* by carefully designed experiments in which all inputs are varied simultaneously. The input function considered most suitable for such identification is the *pseudorandom binary sequence* (PRBS) (Ref. [5], Chapter 10). It is important to emphasize that there are computer programs (such as MIDENT, a module of the control system design package CONSYD, or the process identification toolbox in MATLAB) that are dedicated to multivariable identification; it is customary to employ such programs since they typically have provision for efficient data analysis and parameter estimation.

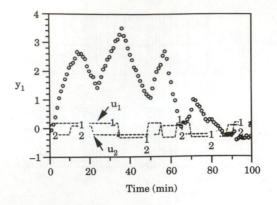

Figure 13.14   Input/output data for multivariable system; output variable 1.

An Illustrative Example

To illustrate these ideas, let us now consider, as an example, the problem of identifying a model for a multivariable system having two inputs, u_1, u_2, and two outputs, y_1, y_2.

The input/output data record shown in Figure 13.14 (for y_1) and Figure 13.15 (for y_2) were obtained by simulating the response of a system represented by (see Chapter 20):

$$\begin{bmatrix} y_1 \\ y_2 \end{bmatrix} = \begin{bmatrix} \dfrac{12.8e^{-s}}{16.7s + 1} & \dfrac{-18.9e^{-3s}}{21.0s + 1} \\ \dfrac{6.6e^{-7s}}{10.9s + 1} & \dfrac{-19.4e^{-3s}}{14.4s + 1} \end{bmatrix} \begin{bmatrix} u_1 \\ u_2 \end{bmatrix} \tag{13.87}$$

to pseudorandom binary sequences in u_1 and u_2, as indicated in the dashed lines in each of these figures. A stochastic disturbance of nonzero average value was added to the simulated output to emulate reality. The model used to generate this data was considered "unknown."

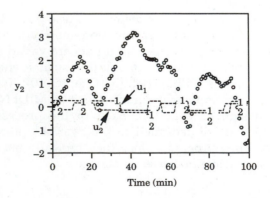

Figure 13.15.   Input/output data for multivariable system; output variable 2.

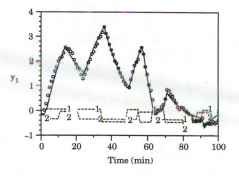

Figure 13.16. Model *vs.* experimental data for multivariable system; output variable 1.

The input/output data record was presented to the program MIDENT, and, as a starting point for the identification, a first-order-plus-time-delay model was postulated for each of the four transfer functions in the multivariable process model.

On the basis of the supplied input/output data record, and the postulated model forms, MIDENT returned the following estimated matrix of transfer functions for the process:

$$\hat{\mathbf{G}}(s) = \begin{bmatrix} \dfrac{11.8e^{-1.00s}}{14.9s + 1} & \dfrac{-18.2e^{-2.76s}}{18.0s + 1} \\ \dfrac{5.58e^{-6.97s}}{10.2s + 1} & \dfrac{-17.5e^{-3s}}{11.8s + 1} \end{bmatrix} \tag{13.88}$$

A comparison of this transfer function matrix with the "actual" one in Eq. (13.87) indicates the effect of the bias introduced by the added stochastic disturbance and the added difficulty in estimating large numbers of parameters from data containing several inputs varying simultaneously.

Since in reality it is impossible to have an "actual" model to compare with, the experimentally determined model is usually evaluated by comparison with the data, typically in a plot. Such plots for this example system are shown in Figures 13.16 and 13.17; the model response to the same input sequence in u_1 and u_2 is shown in the solid lines. Observe that the model fits the data well enough.

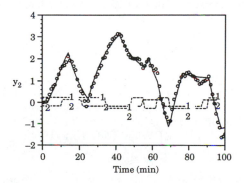

Figure 13.17.   Model *vs.* experimental data for multivariable system; output variable 2.

13.7   SUMMARY

This chapter has focused attention on how to develop a mathematical model for a process when any one of the following situations of practical importance arises:

1. When we know little or nothing about the system, ruling out theoretical modeling.

2. When we know something about the system — perhaps even enough to develop an adequate theoretical model — but we wish to place emphasis on the utility of the model rather than rigor, or completeness.

Under these circumstances, the system is treated like a "black box" and stimulated by changes in its input, so that the response can be observed and analyzed. The process of fitting mathematical models to such experimental input/output data is known as process identification.

The main issues involved in successful process identification have been discussed; and the various methods in the time domain (using step and impulse responses) as well as in the frequency domain have been presented. The most popular methods are step testing for time-domain identification, and pulse testing for frequency-domain identification.

It should not be surprising that process identification draws heavily from our past knowledge of process dynamics: the process identification problem is a converse of the process dynamics problem. A good grasp of process dynamics is therefore important in successful process identification.

In the most general sense, the identification of multivariable systems is not significantly different from that of SISO systems; therefore not much in the way of new material was discussed about this issue. A detailed treatment of this problem lies outside the intended scope of this chapter, and the interested reader is referred, for example, to Refs. [5] and [6]. Nevertheless, an example was presented to illustrate the application of a multivariable identification computer program.

The discussion in this chapter focused exclusively on continuous systems. The issues involved with the identification of discrete-time systems is reserved until Chapter 23 where such a discussion will be more appropriate.

REFERENCES AND SUGGESTED FURTHER READING

1. Seinfeld, J. H. and L. Lapidus, Mathematical Methods in Chemical Engineering, Vol. 3: Process Modeling, Estimation and Identification, Prentice-Hall, Englewood Cliffs, NJ (1974)

2. Bracewell, R. N., The Fourier Transform and Its Applications (2nd ed.), McGraw-Hill, New York (1986)

3. Ogunnaike, B. A., Control Systems Design for Multivariable Systems with Multiple Time Delays, Ph.D. Thesis, University of Wisconsin–Madison (1981)

4. Ogunnaike, B. A., J. P. Lemaire, M. Morari, and W. H. Ray, "Advanced Multivariable Control of a Pilot Plant Distillation Column," AIChE J., **29**, 632 (1983)

5. Eykhoff, P. System Identification, J. Wiley, New York (1974)

6. Ljung, L., System Identification: Theory for the User, Prentice-Hall, Englewood Cliffs, NJ (1989)

7. Bendat, J. S., Nonlinear System Analysis and Identification from Random Data, Wiley-Interscience, New York (1990)

REVIEW QUESTIONS

1. When is "empirical modeling" the only reasonable approach to employ in developing a process model?

2. What is "Process Identification"?

3. What is the difference between "off-line" and "on-line" identification, and which is more frequently employed in practice?

4. Can you think of some potential problems likely to be associated with "on-line" identification?

5. Compare and contrast "Theoretical modeling " and "Process identification."

6. In process identification, what is involved at the model formulation stage?

7. Why is the choice of input functions used in process identification important?

8. What is the role of "Process Dynamics" in "Process Identification"?

9. What is "step testing," and what is the general principle behind identifying process models by this technique?

10. What are some of the major advantages and disadvantages of step testing?

11. From what major disadvantage does "impulse-response identification" suffer, and what is the usual strategy for avoiding this problem?

12. What is the procedure for process identification using the impulse-response approach?

13. From what two major disadvantages does the direct sine wave testing approach to "frequency-response identification" suffer?

14. In generating frequency-response information what is the main attractive feature of the pulse testing technique when compared with the direct sine wave testing?

15. What role does the Fourier transform play in identifying process models via pulse testing?

16. What information is provided by the input and output frequency spectra, and in what way is such information useful in interpreting experimental Bode diagrams?

17. Can you think of some potential problems you are likely to face when trying to identify models for multivariable systems?

PROBLEMS

13.1 Process identification data are usually available at discrete points $t_0, t_1, t_2, ..., t_N$ in time. The value of the *theoretical model response* at each of these same discrete points in time, $y(t_k)$, $k = 1, 2, ..., N$, (or $y(k)$ for simplicity), can be shown to be given *precisely* by:

$$y(k) = a_1 y(k-1) + a_2 y(k-2) + b_1 u(k-m-1) + b_2 u(k-m-2) \qquad (P13.1)$$

for the process whose transfer function is given in Eq. (13.3):

$$g(s) = \frac{K\,(\xi s + 1)e^{-\alpha s}}{(\tau_1 s + 1)(\tau_2 s + 1)} \tag{13.3}$$

where

$$a_1 = \left(e^{-\Delta t/\tau_1} + e^{-\Delta t/\tau_2}\right) \tag{P13.2a}$$

$$a_2 = -\left(e^{-\Delta t/\tau_1}\,e^{-\Delta t/\tau_2}\right) \tag{P13.2b}$$

and

$$b_1 = K\left[1 - \left(\frac{\tau_1 - \xi}{\tau_1 - \tau_2}\right)e^{-\Delta t/\tau_1} - \left(\frac{\tau_2 - \xi}{\tau_2 - \tau_1}\right)e^{-\Delta t/\tau_2}\right] \tag{P13.3a}$$

$$b_2 = K\left[e^{-\Delta t/\tau_1}\,e^{-\Delta t/\tau_2} - \left(\frac{\tau_1 - \xi}{\tau_1 - \tau_2}\right)e^{-\Delta t/\tau_2} - \left(\frac{\tau_2 - \xi}{\tau_2 - \tau_1}\right)e^{-\Delta t/\tau_1}\right] \tag{P13.3b}$$

along with:

$$m = \frac{\alpha}{\Delta t} \tag{P13.4}$$

(a) From this information alone, derive the corresponding expressions for obtaining $y(k)$ for the processes whose transfer functions are given in Eqs. (13.1) and (13.2):

$$g(s) = \frac{Ke^{-\alpha s}}{\tau s + 1} \tag{13.1}$$

$$g(s) = \frac{Ke^{-\alpha s}}{(\tau_1 s + 1)(\tau_2 s + 1)} \tag{13.2}$$

(b) If for simplicity we assume that the delay term m in the expression in Eq. (P13.1) is known independently, given discrete input/output data sets $u(k)$ and $y(k)$, the remaining parameters, $a_1, a_2, b_1,$ and b_2 are relatively easy to estimate *using linear regression*. Derive explicit expressions for determining the original transfer function parameters $K, \tau_1, \tau_2,$ and ξ from estimates of $a_1, a_2, b_1,$ and b_2.

13.2   A laboratory scale stirred mixing tank is strongly believed to behave like a first-order system subject to some delay. The process was subjected to a step *change* of 0.2 gallons/minute in the hot water flowrate; the recorded response is shown in Figure P13.1.

Strictly from the provided response curve, determine (to the best of your ability) estimates of the parameters $K, \tau,$ and α which best fit the stated postulate. Briefly provide the rationale behind whatever technique you have chosen to employ.

13.3   Even though rare, freak accidents sometimes occur in industrial practice that can cause blocks of computer data files to be irrecoverably lost. Under such circumstances, crucial information may have to be extracted instead from strip-chart recordings.

Assuming that such is the case with the process in Problem 13.2, and that Figure P13.1 is the only surviving strip-chart recording of the step test, extract to the best of

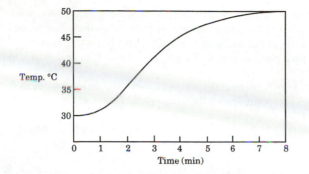

Figure P13.1.   Step response of a stirred mixing tank.

your ability *at least* 15 data points from the given curve and use them to estimate K, τ, and α for a postulated first-order-plus-time-delay model. Compare these estimates with those obtained in Problem 13.2.

TABLE P13.1
SO₂ Scrubber Step Test Data

Time (s)	Water Flowrate (gpm)	Outlet SO₂ Conc (ppm)
$t < 0$	500	25.00
0.0	400	25.00
10.0	400	25.00
20.0	400	25.01
30.0	...	25.00
40.0	...	25.06
50.0	...	25.15
60.0	...	25.30
70.0	...	25.39
80.0	...	25.45
90.0	...	25.53
100.0	...	25.60
110.0	...	25.63
120.0	...	25.67
130.0	...	25.74
140.0	...	25.78
150.0	...	25.82
160.0	...	25.85
170.0	...	25.88
180.0	...	25.89
190.0	...	25.90
200.0	...	25.89
220.0	...	25.90
240.0	400	25.90

13.4   The result of a step test performed on an SO_2 scrubber is shown in Table P13.1. The input variable is the inlet water flowrate (in gallons per minute) and the output variable is the outlet concentration of SO_2 (in parts per million).
(a)Postulate a simple model to represent these data and estimate the parameters of your model.
(b) Obtain the theoretical response of your identified model to the indicated input, and plot and compare it with the given data. Comment on the model fit.

13.5   Solve Eqs. (13.53a,b,c) simultaneously and confirm Eqs. (13.54), (13.55), and (13.56) as the required solutions.

13.6   Given the step-response data for the distillation column used in Example 13.1, generate the corresponding (*unit*) impulse-response data. Reestimate the model parameters using the derived impulse-response data. Compare your estimates with those obtained in the example.

13.7   Table P13.2 contains impulse-response data gathered from a process whose dynamic behavior is to be approximated by a first-order-plus-time-delay transfer function model. First determine the appropriate values for the model parameters; then plot the *theoretical unit impulse response* of your identified model along with the given data. Evaluate the model fit.

TABLE P13.2
Impulse Response Data

Time t_k (min)	Impulse Response $g(t_k)$
0.0	0.0
0.5	0.0
1.0	0.0
1.5	0.0
2.0	10.10
2.5	6.36
3.0	3.86
3.5	2.36
4.0	1.43
4.5	0.86
5.0	0.53
5.5	0.31
6.0	0.19
6.5	0.13
7.0	0.06
7.5	0.042
8.0	0.021
8.5	0.017
9.0	0.011
9.5	0.006
10.0	0.003

TABLE P13.3
Reactor Impulse-Response Data

Time t_k (min)	Impulse Response $g(t_k)$ (Normalized °C)	Time t_k (min)	Impulse Response $g(t_k)$ (Normalized °C)
0.0	0.00	8.0	0.10
0.5	0.13	8.5	0.09
1.0	0.22	9.0	0.08
1.5	0.27	9.5	0.07
2.0	0.28	10.0	0.05
2.5	0.30	11.0	0.04
3.0	0.29	12.0	0.03
3.5	0.28	13.0	0.02
4.0	0.26	14.0	0.017
4.5	0.23	15.0	0.012
5.0	0.22	16.0	0.009
5.5	0.19	17.0	0/006
6.0	0.17	18.0	0.005
6.5	0.15	19.0	0.003
7.0	0.13	20.0	0.002
7.5	0.12		

TABLE P13.4
Frequency-Response Data

Frequency, ω (radians/sec)	Phase Angle, ϕ (Degrees)	Amplitude Ratio (Dimensionless)
10^{-4}	-0.15	0.71
10^{-3}	-0.30	0.70
10^{-2}	-2.75	0.71
2×10^{-2}	-5.70	0.69
4×10^{-2}	-11.05	0.68
6×10^{-2}	-16.50	0.67
10^{-1}	-26.50	0.63
0.2	-44.50	0.50
0.4	-63.45	0.35
0.6	-70.50	0.20
1.0	-78.50	0.15
2.0	-84.25	0.07
4.0	-84.00	0.025
6.0	-88.00	0.02
10.0	-88.90	0.015

13.8 The unit impulse response of the normalized temperature of a chemical reactor is given in Table P13.3 (the input variable is the jacket cooling water flowrate, in gallons/min).

(a) Plot this response. What does the plot suggest about the candidacy of a first-order model? Explain your answer briefly but adequately.

(b) If it is desired to obtain a second-order model, estimate the model parameters. [**Hint:** *Be careful; data record is not evenly spaced.*]

13.9 Table P13.4 contains frequency-response data derived from a pulse test performed on an approximately linear process.

(a) Plot the data appropriately and postulate a candidate transfer function model. Justify your postulate.

(b) Estimate the parameters in your model. [**Hint:** *It may not be necessary to estimate all the paramters simultaneously.*]

(c) Obtain a plot of your identified model's theoretical frequency response and compare it with the data. Comment on the model fit.

13.10 *(This problem is somewhat open ended, and requires a little more effort than usual; it is based on an actual industrial process, and the data record is based on actual data obtained from the process.)*

The mole fraction X_T of a chain transfer agent in a solution polymerization reactor is to be controlled by manipulating F_T, the chain transfer agent flowrate into the reactor. To facilitate the controller design, a simple, low-order model is to be determined from identification experiments. The raw process data file from one such experiment is shown in Table P13.5.

(a) Postulate a model and estimate the parameters.

(b) Validate the model by simulating the theoretical response of your identified model to the indicated input and comparing it to the actual experimental data. Comment on the fit.

(c) If the model obtained in (a) is shown in (b) to be inadequate, postulate another model and repeat the exercise.

TABLE P13.5
Polymer Reactor Identification Data

Time (Hr:Min)	F_T (lb/hr)	X_T (ppm)	Time (Hr:Min)	F_T (lb/hr)	X_T (ppm)
16:01	0.04132	97.27	19:01	0.3894	107.0
16:06	0.04132	97.75	19:06	0.3491	107.6
16:11	0.04132	97.50	19:11	0.3898	107.6
16:16	1.3400	97.11	19:16	0.4132	107.7
16:21	2.8800	96.72	19:21	0.3792	106.7
16:26	4.6400	96.17	19:26	0.3948	107.9
16:31	1.3700	97.81	19:31	0.3560	105.1
16:36	2.8610	107.9	19:36	0.4049	106.6
16:41	1.1070	121.7	19:41	0.4149	107.2
16:46	0.4340	126.0	19:46	0.3980	107.5
16:51	0.4185	131.3	19:51	0.4093	107.3
16:56	0.3559	131.7	19:56	0.4258	108.1
17:01	0.3580	130.4	20:01	0.4546	107.9
17:06	0.3558	131.4	20:06	0.3743	108.1
17:11	0.2949	129.9	20:11	0.4474	109.2
17:16	0.2334	127.3	20:16	0.3923	110.9
17:21	0.2165	124.2	20:21	0.3841	109.5
17:26	0.2497	122.9	20:26	0.4348	109.1
17:31	0.2042	118.6	20:31	0.4531	108.8
17:36	0.3357	107.9	20:36	0.4094	108.4
17:41	0.3730	116.3	20:41	0.3986	108.3
17:46	0.3479	114.4	20:46	0.3646	107.5
17:51	0.2324	112.2	20:51	0.4817	108.5
17:56	0.2399	111.1	20:56	0.4133	109.2
18:01	0.2521	111.4	21:01	0.4376	108.0
18:06	0.2713	109.9	21:06	0.3666	109.4
18:11	0.2487	109.4	21:11	0.4404	109.5
18:16	0.1885	108.3	21:16	0.4832	111.0
18:21	0.3206	108.5	21:21	0.3966	109.2
18:26	0.3021	108.8	21:26	0.4616	108.7
18:31	0.2901	108.3	21:31	0.4091	108.9
18:36	0.2865	108.4	21:36	0.4327	108.9
18:41	0.3412	108.0	21:41	0.4935	109.4
18:46	0.3367	106.3	21:46	0.4088	107.9
18:51	0.3861	106.5	21:51	0.4321	108.8
18:56	0.3736	107.5	21:56	0.4618	109.3

part IV

PROCESS CONTROL

The fundamental premise of process control is that the natural response of all dynamic processes can be modified by the influence of a controller. The objective is therefore to design and implement the controller so that the dynamic response of the process is modified to some desired form. However, the extent to which the natural process response can be modified will usually be determined by our knowledge of the intrinsic process characteristics, by the versatility of the hardware elements available for implementing the controller, and by the nature of inherent process limitations.

Providing the platform for understanding intrinsic process characteristics was the focus in the previous parts of this book; utilizing this knowledge in conjunction with design concepts for the rational design of process controllers is the focus here in Part IV. So that our discussions might proceed along clear lines demarcated on the one hand by intrinsic process characteristics, and on the other by controller hardware, Part IV is broken into three parts as follows:

- *Part IVA: Single-Loop Control* introduces the topic of control system design by presenting design concepts and methods for systems characterized by a single manipulated variable and a single-output variable.

- *Part IVB: Multivariable Process Control* extends controller design concepts to systems having multiple manipulated variables and multiple output variables that interact with each other.

- *Part IVC: Computer Process Control* reconsiders controller design from the digital perspective where the digital computer is the controller hardware element that accepts measurements and provides control action only at discrete points in time.

part IV

PROCESS CONTROL

> "...but I keep under my body
> and keep it under control:
> lest that by any chance
> when I have preached to others,
> I myself should be a castaway."
>
> St. Paul, I Corinthians 9:27

part *IVA*

SINGLE LOOP CONTROL

"Before turning to those moral
mental aspects of the ma
which present the greatest difficul
let the inquirer begin by maste
the more elementary probl

Sherlock Holmes, *A Study in S*
(Sir Arthur Conan D

CHAPTER

14

FEEDBACK CONTROL SYSTEMS

Our study of the control of process systems begins in this chapter with an introduction to perhaps the most common (definitely the oldest) control strategy: feedback control. The main objective in this chapter is to acquaint us with feedback control systems, what they are and how they behave; the issue of *how* to design feedback control systems comes in the next chapter. However, since this is meant to be a fairly comprehensive introduction to the subject, the intended scope of this chapter is quite broad: we will introduce the concept of feedback control and the individual elements that make up the feedback control system; develop the ideas of closed-loop block diagrams and transfer functions; and use these to investigate the dynamic behavior of feedback control systems, leading finally to the important issue of closed-loop system stability.

14.1 THE CONCEPT OF FEEDBACK CONTROL

Let us introduce the concept of feedback control by revisiting a now familiar example process system, the stirred heating tank first introduced in Chapter 1 (cf. Figure 14.1), and recalled several times since then at various places in Chapters 3, 4, and 5.

Let us recall from our various encounters with this process that it has:

- One output variable, the tank temperature T, referred to in Chapter 4 in terms of deviations from its steady-state value T_s as $y = T - T_s$,
- One input variable Q, the steam flowrate through the coils (in terms of deviations from the steady-state value Q_s as $u = Q - Q_s$),
- One disturbance variable T_i, the temperature of inflowing liquid (d is the corresponding deviation from steady state, T_{is}, given by $d = T_i - T_{is}$).

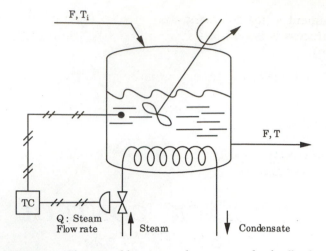

Figure 14.1.   The stirred heating tank system under feedback control.

Let us consider that this time, the process is to be operated under automatic control with the objective of maintaining the tank temperature at a desired value. To accomplish this a temperature controller (denoted by the box labeled TC in Figure 14.1) is provided. In terms of the deviation variables and standard process control terminology, the objective is to maintain y at the desired set-point y_d, given by $y_d = T_d - T_{ds}$ where T_{ds} is the steady-state set-point temperature.

It is pertinent to remind ourselves once again that y may deviate from its desired set-point (and thus precipitate the need for control) either as a result of the effect of changes in the disturbance variable d, or because the set-point itself changes. The problem of maintaining y at its set-point in the face of variations in d (with y_d held constant) is the regulatory control problem; the other problem of making y track y_d as it changes (with d held constant) is the servo problem, also known as *set-point tracking*.

14.1.1 The Feedback Control Strategy

Regardless of whether we have a set-point tracking or a regulatory problem, the feedback control strategy for carrying out the task of maintaining y at the desired set-point y_d is as follows:

1. "Measure" y (i.e., measure T to produce T_m) using a *measuring device* (say, a thermocouple); determine the deviation variable $y_m = T_m - T_s$; the result, y_m, may or may not have the same value as y.

2. *Compare* the measured value y_m, with the desired set-point value y_d to obtain the deviation as an error signal:

$$\varepsilon = y_d - y_m \tag{14.1}$$

This is known as the "feedback" error signal because it is generated from measurement that has been "fed back" from the process.

3. Supply the *controller* with the feedback error signal ε, to be used in obtaining the value of u (the change in the steam flowrate) which will be implemented on the *process*.

4. Implement u (the command to implement the calculated value of u on the process is issued to a *final control element*, in this case, the steam valve).

5. Measure y again, and repeat the entire procedure.

Let us now look more closely at the elements involved in implementing this feedback control strategy.

14.1.2 Elements of the Feedback Loop and Block Diagram Representation

The combination of the physical process and the feedback controller, as shown in Figure 14.1, is known as a *feedback control system*. This feedback control system is seen to consist of the following elements:

- The process itself
- The measuring device
- A comparator (comparing y_m and y_d)
- The controller
- The final control element

Observe that each of these elements is a physical system in its own right, with clearly identifiable inputs and outputs; hence they can each be characterized by a transfer function. It is also important to notice how, according to the feedback control strategy we have just introduced, *information flows from one element to the other in the form of a loop*. Such a view of the feedback control system is reinforced by the block diagram representation shown in Figure 14.2, where the transfer functions of each of these elements are shown as individual blocks whose input and output signals are linked appropriately to indicate the flow of information around the feedback loop.

As we shall see, the block diagram format is a very convenient, very intuitive way of representing control systems. Let us now consider how to develop the block diagram representation of a physical control system.

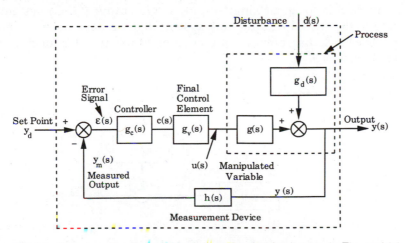

Figure 14.2.   Equivalent block diagram for the process in Figure 14.1.

14.2   BLOCK DIAGRAM DEVELOPMENT

Block diagram representations of control systems are developed by:

1. Identifying the individual elements of the control system (as was done in the previous section for the process in Figure 14.1),

2. Identifying the input and output for each element,

3. Representing the individual input/output transfer function relationship for each element in the block diagram form introduced in Chapter 4,

4. Finally, combining the individual block diagrams for each element to obtain the overall block diagram.

We will now continue with our example process in Figure 14.1 and show how the block diagram in Figure 14.2 was developed.

14.2.1 Individual Elements and Transfer Functions

The individual elements for the process as identified in Section 14.1 have the following characteristics.

The Process

Inputs: u, d, the process input and process disturbance
Output: y

As obtained earlier in Eq. (4.16), the transform-domain transfer function model for the process is:

$$y(s) \; = \; g(s) \, u(s) + g_d(s) \, d(s) \tag{14.2}$$

As we will recall from the discussion in Chapter 4, this model form is, of course, generally applicable to linear single-input, single-output (SISO) systems, not just specific to the stirred heating tank system that we have used merely as a backdrop. The block diagramatic representation of this transfer function model is shown in Figure 14.3.

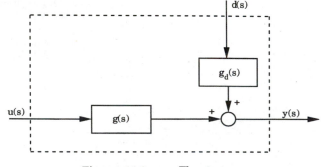

Figure 14.3.   The process.

The Measuring Device (or Sensor)

Input: y
Output: y_m

In the specific case of the stirred heating tank system, using a thermocouple as a measuring device, y represents the temperature, and y_m the voltage output of the thermocouple. Other types of sensors were mentioned in Section 1.4 of Chapter 1.

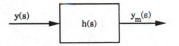

Figure 14.4.   The measuring device.

Designating the transfer function for the measuring device in general as $h(s)$, we have, in terms of the Laplace transforms of the input and output to this system:

$$y_m(s) = h(s)y(s) \tag{14.3}$$

The Comparator (Part of the Controller Mechanism)

Inputs: y_m, y_d
Output: ε

related according to:

$$\varepsilon(s) = y_d(s) - y_m(s) \tag{14.1}$$

Figure 14.5.   The comparator.

The Controller (The "Heart" of the Feedback Control System)

Input: ε
Output: c

The transfer function for the controller is usually designated g_c; thus, we have:

$$c(s) = g_c(s)\,\varepsilon(s) \tag{14.4}$$

As we might expect, $g_c(s)$ can take various forms; and the form it takes obviously determines the nature of $c(s)$. We shall examine the different forms of $g_c(s)$ below.

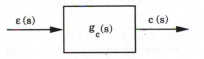

Figure 14.6.   The controller.

The Final Control Element

Input: c, command signal from the controller
Output: u, actual value of the process input variable

Since the final control element is most often a valve (see Section 1.4 for other

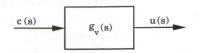

Figure 14.7.   The final control element.

types of final control elements), its transfer function is usually designated g_v; we
therefore have:

$$u(s) = g_v(s)\, c(s) \tag{14.5}$$

Typically, g_v is of the form:

$$g_v = \frac{K_v}{\tau_v s + 1} \tag{14.6}$$

and τ_v is usually quite small compared to the dynamics of the process itself. It
is therefore customary to assume that the dynamics of the valve can be ignored,
and $u(s)$ is taken to be the output of the controller, directly. Alternatively, it is
also possible to lump the dynamics of the valve along with that of the process;
i.e., the valve is treated as an integral part of the process.

14.2.2 The Closed-Loop System Block Diagram

The flow of information into and out of each of the elements of the feedback
control loop is aptly represented by the individual block diagrams in Figures
14.3 through 14.7. If we now combine these individual blocks, linking each of
the indicated signals, we end up with the overall block diagram as shown in
Figure 14.2. The relationship between each component in this feedback control
loop is now easier to visualize, and as we might suspect, subsequent analysis of
the overall system is also greatly facilitated.

This block diagram is typical of all standard feedback control systems,
regardless of the specific process, controller type, sensor, or final control
element in any particular case. Even though more complicated variations on
the same theme are sometimes possible, the basic principles underlying the
development of such diagrams and the relationships between the components
are always the same.

Such a block diagram is said to be for a *closed-loop system* for the simple
reason that the control loop is "closed": the controller is connected with the
process so that information flows to the process from the controller, and also
from the process *back* to the controller, continuously, in one unbroken (hence
"closed") loop.

These closed-loop block diagrams are particularly useful for the analysis of
closed-loop systems; but before taking a closer look at the issues involved, we
wish to consider first the various forms that feedback controllers take.

14.3   CLASSICAL FEEDBACK CONTROLLERS

As noted above, the feedback controller is that component of the feedback loop that issues control commands to the process (via the final control element) based on ε, the deviation of the process measurement from its desired set-point value. How the controller calculates what command signal to issue for any given value of ε is what differentiates one controller from another.

The name of each classical feedback controller, along with the equations that describe how they each relate $\varepsilon(t)$ to $c(t)$, will now be given.

14.3.1 The Proportional Controller

Let $p(t)$ be the actual output signal from the controller (a pressure signal in the case of a pneumatic controller; an electrical signal in the case of an electronic controller), and let p_s be its (constant) value when $\varepsilon(t)$ is zero. Then the *proportional feedback controller* operates according to:

$$p(t) = K_c \varepsilon(t) + p_s \tag{14.7}$$

The deviation of p from its "zero error" value p_s is the control command signal, i.e.:

$$c(t) = p(t) - p_s = K_c \varepsilon(t) \tag{14.8}$$

Thus the command signal from this controller is scheduled to be directly proportional to the observed error, hence its name.

Laplace transformation of Eq. (14.8) leads immediately to:

$$c(s) = K_c \varepsilon(s) \tag{14.9}$$

identifying the transfer function for the proportional controller as:

$$g_c(s) = K_c \tag{14.10}$$

The characteristic parameter of this controller K_c is called the *proportional gain*. Sometimes in commercial controller hardware, the parameter in the proportional controller is represented as the *proportional band, PB*, defined as:

$$PB = \frac{100 \cdot (\text{max range of controller output})}{K_c \cdot (\text{max range of measured variable})}$$

Thus:

$$PB \propto \frac{1}{K_c}$$

14.3.2 The Proportional + Integral (PI) Controller

In this case, $p(t)$ is given by:

$$p(t) = K_c \left[\varepsilon(t) + \frac{1}{\tau_I} \int_0^t \varepsilon(t) \, dt \right] + p_s \tag{14.11}$$

which contains a proportional portion as well as an integral of the error over time. Upon taking the Laplace transform, we obtain:

$$c(s) = K_c \left(1 + \frac{1}{\tau_I \, s} \right) \varepsilon(s) \qquad\qquad (14.12)$$

so that the transfer function for the PI controller is:

$$g_c(s) = K_c \left(1 + \frac{1}{\tau_I s} \right) \qquad\qquad (14.13)$$

The additional parameter τ_I is called the integral time, or the reset time. Often the reciprocal, $1/\tau_I$ is the preferred parameter; this is known as the *reset rate*.

14.3.3 The Proportional + Integral + Derivative (PID) Controller

Here the operating equation is:

$$p(t) = K_c \left[\varepsilon(t) + \frac{1}{\tau_I} \int_0^t \varepsilon(t) \, dt + \tau_D \frac{d\varepsilon}{dt} \right] + p_s \qquad\qquad (14.14)$$

and Laplace transformation gives:

$$c(s) = K_c \left(1 + \frac{1}{\tau_I \, s} + \tau_D \, s \right) \varepsilon(s) \qquad\qquad (14.15)$$

The transfer function for the PID controller is therefore given by:

$$g_c(s) = K_c \left(1 + \frac{1}{\tau_I \, s} + \tau_D \, s \right) \qquad\qquad (14.16)$$

The parameter τ_D is called the derivative time constant.

14.3.4 The Proportional + Derivative (PD) Controller

This is the controller obtained by leaving out the integral action mode from the PID controller, and as such its transfer function is given by:

$$g_c(s) = K_c \left(1 + \tau_D \, s \right) \qquad\qquad (14.17)$$

14.4   CLOSED-LOOP TRANSFER FUNCTIONS

In terms of the outer box in Figure 14.2, observe that the overall closed-loop system in reality just has two inputs, y_d and d, and one output y. This indicates that the individual transfer functions for the separate components of this feedback control system may be consolidated to give composite, overall closed-

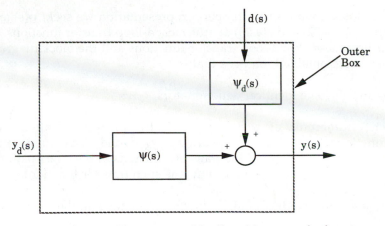

Figure 14.8. The concept of the closed-loop transfer function.

loop transfer functions: one relating y to y_d, and another relating y to d, as illustrated in Figure 14.8.

Let us now consolidate the transfer functions for the closed-loop system by analyzing Figure 14.2 and its component parts.

Even though the process by itself is given by Eq. (14.2), the input u is now obtained according to Eq. (14.5); i.e., $u(s) = g_v(s)c(s)$. However, c itself is obtained according to Eq. (14.4); so that:

$$u = g_v g_c \varepsilon \tag{14.18}$$

Using Eq. (14.1) for ε and Eq. (14.3) for y_m, Eq. (14.18) becomes:

$$u = g_v g_c \left(y_d - hy \right) \tag{14.19}$$

Introducing Eq. (14.19) into the process equation for u now completes the consolidation exercise, having eliminated every other variable except y, y_d, and d; the result is:

$$y = g\left[g_v g_c \left(y_d - hy \right) \right] + g_d d \tag{14.20}$$

which, when solved for y, now gives:

$$y = \frac{g g_v g_c}{1 + g g_v g_c h} y_d + \frac{g_d}{1 + g g_v g_c h} d \tag{14.21}$$

and by defining:

$$\psi(s) = \frac{g g_v g_c}{1 + g g_v g_c h} \tag{14.22}$$

and

$$\psi_d(s) = \frac{g_d}{1 + g g_v g_c h} \tag{14.23}$$

we obtain:

$$y(s) = \psi(s) y_d(s) + \psi_d(s) d(s) \tag{14.24}$$

This is the closed-loop transfer function representation we seek; $\psi(s)$ and $\psi_d(s)$ defined in Eqs. (14.22) and (14.23) are the closed-loop transfer functions.

Regarding these closed-loop transfer functions and the block diagram from which they were obtained, observe that:

1. ψ and ψ_d have identical denominators.

2.   The numerator of ψ, i.e., gg_vg_c, is the product of all the transfer functions in the *direct* path from y to y_d ; while g_d, the numerator of ψ_d is also the "product" of the transfer functions in the *direct* path between y and d (in this case, there is only one such transfer function).

3. gg_vg_ch is the product of all the transfer functions in the entire loop.

These results may now be generalized as follows:

For single-loop feedback systems (such as that in Figure 14.2) the closed-loop transfer functions (CLTF) are given by the general expression:

$$\text{CLTF} = \frac{\pi_f}{1 + \pi_L} \tag{14.25}$$

where
   π_f = product of all the transfer functions in the *direct* path between the output y and the input (y_d or d).

   π_L = product of all the transfer functions in the entire loop.

Let us illustrate the application of these results with the following examples.

Example 14.1   CLOSED-LOOP TRANSFER FUNCTION FROM SINGLE LOOP, FEEDBACK CONTROL SYSTEM BLOCK DIAGRAM.

For the feedback control system whose block diagram is shown in Figure 14.9, first find the closed-loop transfer function relating
(a) y and y_d, (b) y and d_1, and (c) y and d_2,
and then obtain the overall closed-loop transfer function expression relating the output to all the inputs.

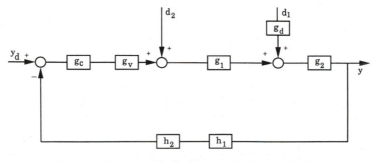

Figure 14.9.   Block diagram for Example 14.1.

Solution:

(a) The product of the transfer functions in the direct path between y and y_d is $g_2 g_1 g_v g_c$; and the product of the transfer functions in the entire loop is $g_2 g_1 g_v g_c h_2 h_1$. Thus the closed-loop transfer function relation between y and y_d is:

$$y = \frac{g_2 g_1 g_v g_c}{1 + g_2 g_1 g_v g_c h_2 h_1} y_d \qquad (14.26)$$

(b) The product of the transfer functions in the direct path between y and d_1 is $g_2 g_d$; thus the closed-loop transfer function relating these two quantities will be given by:

$$y = \frac{g_2 g_d}{1 + g_2 g_1 g_v g_c h_2 h_1} d_1 \qquad (14.27)$$

(c) In the direct path between y and d_2, we have the following product of transfer functions $g_2 g_1$. The closed-loop transfer function in this case will therefore be given by:

$$y = \frac{g_2 g_1}{1 + g_2 g_1 g_v g_c h_2 h_1} d_2 \qquad (14.28)$$

Combining all three expressions, we now obtain the overall transfer function relating all the inputs of this feedback control system to the output as:

$$y = \frac{g_2 g_1 g_v g_c}{1 + g_2 g_1 g_v g_c h_2 h_1} y_d + \frac{g_2 g_d}{1 + g_2 g_1 g_v g_c h_2 h_1} d_1$$

$$+ \frac{g_2 g_1}{1 + g_2 g_1 g_v g_c h_2 h_1} d_2 \qquad (14.29)$$

As we will see in Chapter 16, it is possible to have a SISO feedback control system for which the overall scheme involves multiple control loops. In such cases, the strategy for finding the overall closed-loop transfer function is to simplify the multiloop block diagram by first reducing the inner loops to single blocks. Such an exercise reduces the multiple loops to a single loop which is then easily handled by the technique we have discussed.

It is best to illustrate this concept with an example.

Example 14.2   **CLOSED-LOOP TRANSFER FUNCTION FROM MULTILOOP FEEDBACK CONTROL SYSTEM BLOCK DIAGRAM.**

Find the overall closed-loop transfer function for the multiloop feedback control system whose block diagram is shown in Figure 14.10. (As we will see later in Chapter 16, this is a *cascade control system* configuration.)

Solution:

Let us start by considering the inner loop by itself. It has two inputs, designated m_2 and d_2, and one output m_1. The closed-loop transfer functions relating these variables are easily obtained using the approach illustrated in Example 14.1; the result is:

$$m_1 = \frac{g_{c2} g_v}{1 + g_{c2} g_v h_2} m_2 + \frac{1}{1 + g_{c2} g_v h_2} d_2 \qquad (14.30)$$

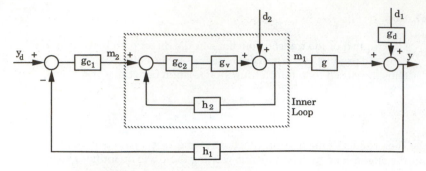

Figure 14.10. Block diagram for Example 14.2.

If we represent these as:

$$m_1 = g_a m_2 + g_b d_2 \tag{14.31}$$

then the multiloop block diagram in Figure 14.10 is simplified to the single-loop one shown in Figure 14.11.

It is now easy from here to obtain the overall transfer function expression for the complete system; it is given by:

$$y = \frac{g g_a g_{c1}}{1 + g g_a g_{c1} h_1} y_d + \frac{g_d}{1 + g g_a g_{c1} h_1} d_1 + \frac{g g_b}{1 + g g_a g_{c1} h_1} d_2 \tag{14.32}$$

If necessary, the newly introduced transfer functions g_a and g_b may be replaced by their explicit expressions obtainable from Eqs. (14.30) and (14.31).

14.5   CLOSED-LOOP TRANSIENT RESPONSE

14.5.1 Preliminary Considerations

In Part II, we studied the dynamic behavior of processes (in the open loop) using their transfer function representations to investigate their responses to various forcing functions. Observe now that for the closed-loop system, we have obtained the overall closed-loop transfer function representation as:

$$y(s) = \psi(s) y_d(s) + \psi_d(s) d(s) \tag{14.33}$$

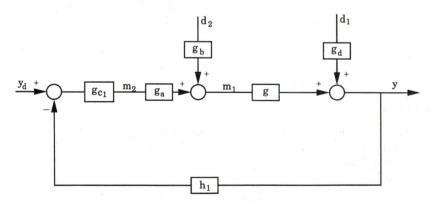

Figure 14.11.   Single-loop equivalent of the multiloop block diagram of Figure 14.10.

The inputs for the closed-loop system are y_d, the set-point, and d the disturbance. In precisely the same manner as with open-loop dynamic analysis, we may now use Eq. (14.33) to investigate the dynamic response of closed-loop systems to changes in the set-point (the servo, or set-point tracking problem), or the response to input disturbances (the regulatory problem).

To be sure, the transfer functions involved in such closed-loop dynamic analyses are more complicated than those used for open-loop analysis, but the principles are exactly the same in each case.

Let us now combine all the blocks in the direct path between y and y_d (excepting the controller, g_c) in a general feedback control system block diagram and refer to this product as simply g; and let g_d represent the product of all the transfer functions in the direct path between y and d; further let the product of all the blocks in the feedback path be referred to as h, giving rise to the generic closed-loop system shown in Figure 14.12. (It is important to keep in mind what g and h stand for here; we have introduced these "combined" transfer functions as a shorthand notation only; by no means are we restricting the following discussion to only those systems having one g and one h block.)

For this generic system, the closed-loop transfer functions are given by:

$$\psi = \frac{gg_c}{1 + gg_ch} \tag{14.34a}$$

and

$$\psi_d = \frac{g_d}{1 + gg_ch} \tag{14.34b}$$

and the transfer function representation for the closed-loop system is:

$$y = \frac{gg_c}{1 + gg_ch}y_d + \frac{g_d}{1 + gg_ch}d \tag{14.35}$$

These will now be used to investigate the transient behavior of closed-loop systems in response to input changes. It is of prime importance to note that both closed-loop transfer functions, ψ and ψ_d, have identical denominators, i.e.: $1 + gg_ch$.

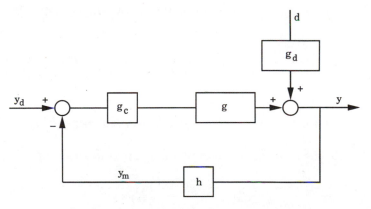

Figure 14.12. A generic closed-loop system.

14.5.2 Transient Response of a First-Order System under Proportional Feedback Control

The stirred heating tank system under feedback control (Figure 14.1) gave rise to the block diagram in Figure 14.2. Let us now investigate the transient response of this system to changes in the set-point and the disturbance (sometimes called "load").

To do this we need explicit expressions for the transfer functions represented in the block diagram. For the process itself, we recall from Chapter 4 that its transfer functions are given by:

$$g(s) = \frac{\beta\theta}{\theta s + 1} \qquad\qquad\qquad (4.15a)$$

$$g_d(s) = \frac{1}{\theta s + 1} \qquad\qquad\qquad (4.15b)$$

where $\beta = \lambda/\rho V C_p$ and θ is the residence time, the ratio of the tank volume V to the throughput F. Thus this is a first-order system with steady-state gain and time constant given by:

$$K = \beta\theta \qquad\qquad\qquad (14.36a)$$

and

$$\tau = \theta \qquad\qquad\qquad (14.36b)$$

Let us assume, for simplicity, that valve dynamics are negligible and that temperature measurements obtained from the thermocouple are perfect and instantaneous, implying:

$$g_v = 1; \; h = 1$$

Under proportional only control, with proportional gain K_c, the controller transfer function is given by:

$$g_c = K_c \qquad\qquad\qquad (14.37)$$

We will now use this process information to investigate the response to two inputs:

1. A unit step change in set-point (servo response; with $y_d = 1$ and $d = 0$)
2. A unit step change in disturbance (regulatory response; with $d = 1$ and $y_d = 0$)

Response to Unit Step Change in Set-point

From Eq. (14.35) the transfer function relationship in this case is:

$$y(s) = \frac{gg_c}{1 + gg_c} \, y_d(s) \qquad\qquad\qquad (14.38)$$

which, upon introducing the provided information becomes:

$$y(s) = \frac{\left(\dfrac{KK_c}{\theta s + 1}\right)}{1 + \left(\dfrac{KK_c}{\theta s + 1}\right)} \frac{1}{s} \qquad (14.39)$$

(with K as given in Eq. (14.36a)). This expression simplifies to:

$$y(s) = \frac{KK_c}{\theta s + 1 + KK_c} \frac{1}{s}$$

which we now put in the more revealing form:

$$y(s) = \frac{K^*}{\theta^* s + 1} \frac{1}{s} \qquad (14.40)$$

where the "closed-loop gain," K^* is:

$$K^* = \frac{KK_c}{1 + KK_c} \qquad (14.41a)$$

and the "closed-loop time constant," θ^* is:

$$\theta^* = \frac{\theta}{1 + KK_c} \qquad (14.41b)$$

In order to solve for the transient response of Eq. (14.40) we observe that it looks like the unit step response of a first-order system with steady-state gain K^* and time constant θ^*. In recognition of this fact therefore, we may immediately write down the required transient response:

$$y(t) = K^* (1 - e^{-t/\theta^*}) \qquad (14.42)$$

This equation represents how the entire control system (the heating tank, valve, and proportional controller) responds to a unit step change in the set-point; i.e., how y changes with time in response to a change in y_d from 0 to 1. In physical terms, it shows how the *actual* tank temperature will respond when its desired value is changed by 1° with the controller trying to meet this demand.

Let us note the following from Eq. (14.42):

1.   As $t \to \infty$, $y \to K^*$.

2.   The approach of y to the ultimate value K^* is exponential.

3.   However, from the definition in Eq. (14.41a) we see that K^* will always be less than 1, for all finite values of K_c.

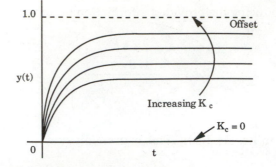

Figure 14.13. Response of a first-order system under proportional feedback control to a unit step change in the set-point.

4.   Because the final value attained by y is not the same as the desired value y_d, there is said to be a *steady-state offset* defined by:

$$\text{offset} = y_d - y(\infty) \tag{14.43}$$

which, in this specific case, is given by:

$$\text{offset} = 1 - \frac{KK_c}{1 + KK_c} = \frac{1}{1 + KK_c} \tag{14.44}$$

The inability to eliminate offset is one of the limitations of proportional-only controllers.

5.   Note that as the controller gain, $K_c \to \infty$, the offset tends to zero.

A plot of this transient behavior is shown in Figure 14.13.

Response to Unit Step Change in Disturbance

For this situation, the transfer function relationship is:

$$y(s) = \frac{g_d}{1 + gg_c} d(s) \tag{14.45}$$

and from the expressions for g, g_d, g_c, we have:

$$y(s) = \frac{\left(\dfrac{1}{\theta s + 1}\right)}{1 + \left(\dfrac{KK_c}{\theta s + 1}\right)} \frac{1}{s} \tag{14.46}$$

which again may be rearranged to give the familiar form:

$$y(s) = \frac{K_d^*}{\theta^* s + 1} \frac{1}{s} \tag{14.47}$$

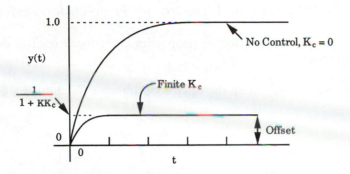

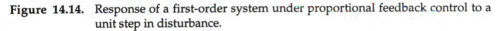

Figure 14.14. Response of a first-order system under proportional feedback control to a unit step in disturbance.

where the closed-loop disturbance gain is:

$$K_d^* = \frac{1}{1 + KK_c} \tag{14.48}$$

and θ^* is as given earlier in Eq. (14.41b).

Once again the transient response is easily obtained as:

$$y(t) = K_d^* (1 - e^{-t/\theta^*}) \tag{14.49}$$

representing how the tank temperature changes in response to a unit step increase in the inlet stream temperature as the controller tries to handle this disturbance.

Once again we note from Eq. (14.49) that:

1. As $t \to \infty$, $y \to K_d^*$ and that the approach is exponential.

2. The controller is supposed to keep y at zero in the presence of the disturbance when $y_d = 0$; however, from Eq. (14.48) we see that this is not the case, since:

$$K_d^* = \frac{1}{1 + KK_c} \neq 0$$

The offset in this case (from Eq. (14.43)) is:

$$\text{offset} = 0 - \frac{1}{1 + KK_c} = -\frac{1}{1 + KK_c} \tag{14.50}$$

3. In the absence of control, i.e., with $K_c = 0$, K_d^* will be 1, θ^* will be θ and the response to this disturbance will be given by:

$$y(t) = (1 - e^{-t/\theta}) \tag{14.51}$$

Figure 14.14 shows the closed-loop response, Eq. (14.49), along with the open-loop response, Eq. (14.51). Note that using the proportional controller reduces the offset, and that the amount of offset decreases as the value of K_c increases.

Effect of Proportional Control on First-Order Systems

Let us now summarize the effect of proportional feedback control on a *general* first-order system.

The first-order system with the open-loop transfer function:

$$y(s) = \left(\frac{K}{\tau s + 1} \right) u(s) \tag{14.52}$$

under proportional feedback control is transformed to a system with the *closed-loop* transfer function representation:

$$y(s) = \left(\frac{K^*}{\tau^* s + 1} \right) y_d(s) \tag{14.53}$$

(assuming $g_v = 1$, $h = 1$) where:

$$K^* = \frac{KK_c}{1 + KK_c} \tag{14.54a}$$

and

$$\tau^* = \frac{\tau}{1 + KK_c} \tag{14.54b}$$

Thus proportional feedback control modifies the dynamic behavior of the first-order system, retaining the order but modifying the values of the characteristic parameters.

It is particularly important to note the effect of proportional control on the value of the system time constant. Observe that so long as $KK_c > 0$, then $\tau^* < \tau$ (i.e., the time constant in the closed loop is smaller than in the open loop) so that feedback control has the effect of speeding up the process response.

It is left as an exercise for the reader to establish that the same is true for the first-order disturbance transfer function: that proportional feedback control makes the closed-loop transfer function time constant smaller than the value in the open loop, resulting in a speeding up of the closed-loop response.

14.5.3 Transient Response of a First-Order System under Proportional + Integral (PI) Feedback Control

In this case the controller transfer function is given in Eq. (14.13), and if we introduce this into the general process transfer function given in Eq. (14.35), then the closed-loop transfer function relation for set-point changes (when $h = 1$) is:

$$y(s) = \frac{\dfrac{K}{\tau s + 1} K_c \left(1 + \dfrac{1}{\tau_I s} \right)}{1 + \dfrac{K}{\tau s + 1} K_c \left(1 + \dfrac{1}{\tau_I s} \right)} y_d(s) \tag{14.55}$$

which rearranges to give:

$$y(s) = \frac{KK_c \, (\tau_I s + 1)}{\tau\tau_I \, s^2 + \tau_I \, (1 + KK_c) \, s + KK_c} \, y_d(s) \tag{14.56}$$

We now note the following:

1. The closed-loop transfer function has a first-order numerator polynomial and a second-order denominator polynomial; i.e., it has one zero and two poles.

2. Under P control alone, the closed-loop transfer function was first order, with one pole and no zero; the effect of P control therefore was merely to change the characteristic parameters. The effect of PI control, on the other hand, is to add a pole and a zero to the process transfer function.

The response of this closed-loop system to a unit step change in set-point can be obtained using Eq. (14.56) and $1/s$ for $y_d(s)$, i.e.:

$$y(s) = \frac{KK_c \, (\tau_I s + 1)}{\tau\tau_I \, s^2 + \tau_I \, (1 + KK_c) \, s + KK_c} \, \frac{1}{s} \tag{14.57}$$

and if r_1 and r_2 are the roots of the denominator polynomial, Eq. (14.57) becomes:

$$y(s) = \frac{KK_c \, (\tau_I s + 1)}{\tau\tau_I \, (s - r_1) \, (s - r_2)} \, \frac{1}{s} \tag{14.58}$$

Using partial fraction expansion, of course, we can invert Eq. (14.58) in the usual manner to obtain:

$$y(t) = A_0 + A_1 \, e^{r_1 t} + A_2 \, e^{r_2 t} \tag{14.59}$$

Qualitative Nature of the Response

On the basis of our earlier discussion of the dynamic behavior of higher order systems in Chapter 6, we can say a few things about the *qualitative* nature of this response. Note now that:

1. Because of the zero involved in Eq. (14.58), it is possible for this response to exhibit overshoot.

2. If r_1 and r_2 are real and negative, the response approaches its ultimate value exponentially.

3. If r_1 and r_2 are complex conjugates, with negative real parts, the response will be damped oscillatory.

4. If at least one of these roots is positive (or in the case of complex conjugates both roots have a positive real part), then the response will grow indefinitely, and fail to settle down to a final value.

Ultimate Value of the Response

Provided the circumstances are such that the roots r_1 and r_2 both have negative real parts, it is possible to use the final-value theorem of Laplace transforms and Eq. (14.57) to obtain the final value to which this response settles.

Recall that:

$$\lim_{t \to \infty} y(t) = \lim_{s \to 0} [s\, y(s)]$$

so that from Eq. (14.57) we have:

$$\lim_{t \to \infty} y(t) = \lim_{s \to 0} \left[\frac{KK_c (\tau_I s + 1)}{\tau\tau_I s^2 + \tau_I (1 + KK_c) s + KK_c} \right] = 1 \qquad (14.60)$$

and there will be *no offset* since the desired value and the ultimate value to which the process output settles are identical.

We can go through the same procedure for the regulatory problem; we will arrive at the same conclusion: with PI control, steady-state offsets are completely eliminated.

14.5.4 Transient Response of a First-Order System under Other Feedback Control Strategies

It is now an easy exercise (which the reader will be asked to carry out in one of the problems at the end of the chapter) to show that:

1. *For a first-order system under* **PD feedback control**:

    *   *The closed-loop transfer function has a first-order numerator polynomial and also a first-order denominator polynomial; i.e., it has one zero and one pole, and therefore the transient response will be that of a lead/lag system.*

    *   *The transient response will exhibit steady-state offset, as was the case with proportional control.*

2. *For a first-order system under* **PID feedback control**:

    *   *The closed-loop transfer function has a second-order numerator polynomial and also a second-order denominator polynomial; i.e.,it has two zeros and two poles.*
    *   *The transient response will* **not** *exhibit any steady-state offset.*

What has been done in this section for the first-order system will now be generalized to any arbitrary linear system.

14.5.5 A General Treatment of Transient Responses

Let us consider, for simplicity, that the process and the measuring device have been lumped together, so that the output obtained from the measuring device is considered as the process output; further we shall refer to their combined transfer function simply as g. In what follows we will assume no specific form for g other than that it has a steady-state gain of K.

The closed-loop transfer function representation for the general servo (set-point tracking) problem is:

$$y(s) = \frac{g g_c}{1 + g g_c} y_d(s) \tag{14.61}$$

and for the regulatory problem:

$$y(s) = \frac{g_d}{1 + g g_c} d(s) \tag{14.62}$$

Transient Response under Proportional Control

Under proportional only control, for the servo problem, Eq. (14.61) becomes:

$$y(s) = \frac{g K_c}{1 + g K_c} y_d(s) \tag{14.63}$$

and the response to a unit step change in set-point will be obtained from:

$$y(s) = \frac{g K_c}{1 + g K_c} \frac{1}{s} \tag{14.64}$$

where factorization of the denominator polynomial into its roots, and partial fraction expansion, followed by Laplace inversion will give the required transient response.

However, we are particularly interested in investigating the ultimate value to which this response settles, prompted by the fact that with the first-order system, we had *offsets*.

Provided that the conditions of the roots of the denominator polynomial in Eq. (14.64) are such that the response indeed tends to a finite limiting value (a condition we shall revisit quite soon), we may now apply the final-value theorem of Laplace transforms to Eq. (14.64) to obtain the ultimate value of this transient response.

We thus have:

$$\lim_{t \to \infty} y(t) = \lim_{s \to 0} \left(\frac{g K_c}{1 + g K_c} \right) \tag{14.65}$$

and since the only term involving s in the RHS of Eq. (14.65) is g, and we know, by definition of its steady-state gain that:

$$\lim_{s \to 0} (g) = K$$

we now have:

$$\lim_{t \to \infty} y(t) = \left(\frac{K K_c}{1 + K K_c} \right) \tag{14.66}$$

which is always less than 1 for all finite values of K_c.

It is easy to show, going through precisely the same procedure that the same result holds for the regulatory problem under proportional-only feedback control: there will always be an offset.

We have thus established a well-known result in feedback control:

The feedback control system using a proportional controller will always exhibit a nonzero steady-state offset.

There is an exception to this rule: when the system in question is a pure capacity system (or when the system contains a pure capacity element) it *does not* exhibit steady-state offset under proportional feedback control.

Observe that if $g(s) = K^*/s$ in Eq. (14.65), the result is:

$$\lim_{t \to \infty} y(t) = \lim_{s \to 0} \left(\frac{\dfrac{K^* K_c}{s}}{1 + \dfrac{K^* K_c}{s}} \right) \tag{14.67}$$

or

$$\lim_{t \to \infty} y(t) = \lim_{s \to 0} \left(\frac{K^* K_c}{s + K^* K_c} \right) = 1 \tag{14.68}$$

Thus only the pure capacity system does not exhibit steady-state offset under proportional-only feedback control. The fundamental reason for this will be shown later.

Transient Response under PI Control

Under PI control, from Eq. (14.61), the closed-loop transfer function for the servo problem is:

$$y(s) = \frac{g\, K_c \left(1 + \dfrac{1}{\tau_I s} \right)}{1 + g\, K_c \left(1 + \dfrac{1}{\tau_I s} \right)} y_d(s) \tag{14.69}$$

and upon rearrangement, we find that the response to a unit step change in set-point will be obtained from:

$$y(s) = \frac{g K_c \left(\tau_I s + 1 \right)}{\tau_I s + g K_c \left(\tau_I s + 1 \right)} \frac{1}{s} \tag{14.70}$$

Once again, for any specified $g(s)$, we can factorize the denominator polynomial into its roots, and in the usual manner, by partial fraction expansion followed by Laplace inversion, we will obtain the required transient response. The specific nature of the transient response will, of course, depend on the nature of the process as represented by $g(s)$, and the values of the controller parameters; but we can say something general about the ultimate value to which this transient response will settle, particularly as regards whether or not there will be steady-state offset.

Again, provided that the conditions of the roots of the denominator polynomial in Eq. (14.70) are such that the response tends to a finite limiting value, we may apply the final-value theorem of Laplace transforms to obtain:

$$\lim_{t \to \infty} y(t) = \lim_{s \to 0} \left[\frac{gK_c \left(\tau_I s + 1 \right)}{\tau_I s + gK_c \left(\tau_I s + 1 \right)} \right] \qquad (14.71)$$

and since, as before:

$$\lim_{s \to 0} (g) = K$$

we have:

$$\lim_{t \to \infty} y(t) = \lim_{s \to 0} \left[\frac{KK_c \left(\tau_I s + 1 \right)}{\tau_I s + KK_c \left(\tau_I s + 1 \right)} \right] = 1 \qquad (14.72)$$

for all values of K and K_c, implying that there will be no steady-state offset. By going through precisely the same procedure, we can also arrive at the same result for the regulatory problem under PI feedback control.

Thus we have established the following well-known result in feedback control:

The closed-loop system will never exhibit steady-state offset under PI control.

Similar arguments can be presented for the feedback control system operating under PD, and under PID control; we will find that there *will be* steady-state offset under PD control, but none under PID control. This finally leads us to the following statement:

The presence of integral action in a feedback control system will eliminate any steady-state offset from the closed-loop transient response; steady-state offset occurs as a result of the absence of integral action.

Thus, the pure capacity system under proportional-only feedback control exhibits no steady-state offset because the system has its own built-in integral action (recall that the system is called a pure *integrator* by virtue of its transfer function form); other systems not so endowed *will exhibit steady-state offset* under proportional control.

Before leaving the subject of transient responses of closed-loop systems, let us emphasize the point that the analytical approach we have presented here is for pedagogical reasons only, to provide general insight into the fundamentals of closed-loop dynamics of feedback systems; this is *not* the customary approach to take in practice. The availability of a variety of computer packages for control systems design and analysis makes it unnecessary to carry out such analysis by hand.

Even though computer assisted analysis is currently the common practice, it should be noted, however, that the analytical approach is still the preferred means by which to obtain *general* insight into the dynamic behavior of closed-loop systems. Computer assisted analysis is most useful in investigating *specific* system behavior. We will give one illustrative example.

Example 14.3   CLOSED-LOOP TRANSIENT RESPONSE OF A THIRD-ORDER SYSTEM UNDER PROPORTIONAL FEEDBACK CONTROL.

Recall the three tanks in series process of Section 11.5 represented by a third-order linear model whose transfer function is given below. We propose to use proportional-only feedback control of the liquid level in Tank 3 with $K_c = 1$. Find the closed-loop transient response to a unit step change in set-point. What is the final steady-state value to which the system settles, and what is the steady-state offset? The transfer function relating Tank 3 level to Tank 1 inlet flowrate is:

$$g(s) = \frac{6}{(2s + 1)(4s + 1)(6s + 1)} \tag{14.73}$$

Solution:

The closed-loop transfer function relation for this system under proportional feedback control is:

$$y(s) = \frac{\dfrac{6K_c}{(2s + 1)(4s + 1)(6s + 1)}}{1 + \dfrac{6K_c}{(2s + 1)(4s + 1)(6s + 1)}} y_d(s)$$

which, for $K_c = 1$ simplifies to:

$$y(s) = \frac{6}{(2s + 1)(4s + 1)(6s + 1) + 6} \frac{1}{s} \tag{14.74}$$

In principle, we could now expand the cubic denominator polynomial, perform the indicated algebra and factor it into its three roots; combine this with the input term $(1/s)$ and carry out partial fraction expansions; and after Laplace inversion we will obtain the required analytical expression for the transient response. This is clearly a very tedious exercise.

However, by entering this information into any control system analysis and design package (in this particular case, we used one known as CONSYD) the generated transient response is shown below in Figure 14.15. Note the indicated offset.

It is easy to calculate the actual value to which the transient response settles at steady state; it is given by:

$$y(\infty) = \frac{KK_c}{1 + KK_c}$$

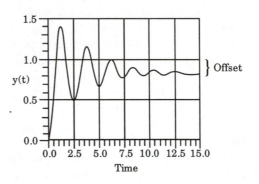

Figure 14.15.   Transient response of the example system under proportional feedback control to a unit step change in set-point.

which in this case is:

$$y(\infty) = \frac{6}{7} \text{ or } 0.857$$

The steady-state offset is obtained by subtracting this from the desired value of 1; the result is:

$$\text{offset} = 0.143$$

The reader may want to verify that this is accurately represented on the computer-generated transient response of Figure 14.15.

14.5.6 Feedback Controllers and Steady-State Offset

What we have succeeded in doing as yet, as far as feedback controllers and steady-state offset are concerned, is to show which ones will result in steady-state offset and which ones will not. It should be interesting to provide intuitive explanations for why these controllers exhibit such characteristics.

Why Does Proportional Feedback Control Result in Steady-State Offset?

The proportional controller operates according to:

$$u(t) = K_c \varepsilon(t) \tag{14.75}$$

so that in the transient state, while things are still changing, the rate at which the process input is being changed is given by:

$$\frac{du(t)}{dt} = K_c \frac{d\varepsilon(t)}{dt} \tag{14.76}$$

Of course, at steady state, all the derivatives will vanish, but the real issue is now the following: is it possible for $d\varepsilon(t)/dt$ to be zero without $\varepsilon(t)$ itself being zero? The answer, of course, is *yes*; as soon as $\varepsilon(t)$ becomes constant, whether at zero, or at a nonzero value, steady state is achieved. Observe from Eq. (14.75), however, that for $\varepsilon(t)$ to be zero, $u(t)$ must also be 0; but from the process model we immediately see that $u(t)$ will never be zero for nonzero set-point y_d or nonzero disturbance $d(t)$. Thus there will always be steady-state offset.

Why Does PI Control Not Result in Steady-State Offset?

For the PI controller, control action is determined according to:

$$u(t) = K_c \left[\varepsilon(t) + \frac{1}{\tau_I} \int_0^t \varepsilon(t) \, dt \right] \tag{14.77}$$

and under transient conditions, the rate at which $u(t)$ changes will be given by:

$$\frac{du(t)}{dt} = K_c \left[\frac{d\varepsilon(t)}{dt} + \frac{1}{\tau_I} \varepsilon(t) \right] \tag{14.78}$$

obtained by differentiating Eq. (14.77). Let us now observe that regardless of controller parameters, or the specific process in question, whenever steady state is achieved, all the derivatives in Eq. (14.78) will vanish, and $\varepsilon(t)$ will always be zero, and there will be *no* steady-state offset.

It is left as an exercise to the reader to establish similar results for the PD and the PID controllers.

14.6   CLOSED-LOOP STABILITY

14.6.1 Preliminary Considerations

Let us return once again to the issue of investigating transient responses of closed-loop systems. The starting point is the closed-loop transfer function representation given in Eq. (14.24):

$$y(s) = \psi(s)\, y_d(s) + \psi_d(s)\, d(s) \tag{14.24}$$

where, if g represents the product of all the transfer functions in the forward path apart from the controller, and h is the composite transfer function in the feedback path, then the closed-loop transfer functions are given by:

$$\psi(s) = \frac{gg_c}{1 + gg_ch} \tag{14.79}$$

and

$$\psi_d(s) = \frac{g_d}{1 + gg_ch} \tag{14.80}$$

It is worth mentioning again that these two transfer functions have identical denominator polynomials.

Characteristic Equation

As demonstrated in the preceding section, the transient response to a unit step change in the set-point is obtained from:

$$y(s) = \frac{gg_c}{1 + gg_ch} \frac{1}{s} \tag{14.81}$$

which in general inverts to give:

$$y(t) = A_0 + A_1\, e^{r_1 t} + A_2\, e^{r_2 t} + \dots + A_n\, e^{r_n t} \tag{14.82}$$

where $r_1, r_2, \dots, r_n$ are the roots of the denominator polynomial in Eq. (14.81). As we will soon show, these are exactly the same as the roots of the equation:

$$1 + gg_ch = 0 \tag{14.83}$$

As we recall from Section 11.2, the denominator polynomial Eq. (14.83) is termed the "characteristic equation." In this case it is the characteristic equation of the closed-loop response Eq. (14.81).

Observe now that in response to a unit step change in the set-point (which is a bounded input), if all the r_i's in Eq. (14.82) have negative real parts, the output will ultimately tend to a finite, constant value; if *just one* of the r_i's has a positive real part, the transient response will grow indefinitely with time.

In terms of the ideas of bounded-input-bounded-output stability which we introduced in Chapter 11, we observe therefore that this closed-loop system will be stable if and only if all the roots of Eq. (14.83) have negative real parts; otherwise, it is unstable.

It is of course quite clear, since both $\psi(s)$ and $\psi_d(s)$ have identical denominator polynomials, that the same considerations as those given above apply for the regulatory problem.

We may now state the following important facts:

Stability analysis for any closed-loop system is carried out by investigating the nature of the roots of the system's characteristic equation. A system is stable if and only if all the roots of the characteristic equation have negative real parts (or, equivalently, lie in the left half of the complex plane); otherwise it is unstable.

Why Equation (14.83) Is the Characteristic Equation

The r_i's in Eq. (14.82), as we know, are the roots of the denominator polynomial of:

$$\psi(s) = \frac{gg_c}{1 + gg_c h}$$

To obtain this denominator however, requires some rearrangement since $gg_c h$ is itself a ratio of polynomials. However, we will show that these roots are exactly the same as the roots of Eq. (14.83).

The most general form that gg_c could take is:

$$gg_c = \frac{N_1(s)\, e^{-\alpha_1 s}}{D_1(s)} \tag{14.84}$$

when time delays are involved, and h can also take the form:

$$h = \frac{N_2(s)\, e^{-\alpha_2 s}}{D_2(s)} \tag{14.85}$$

where we use $N_i(s)$ to represent numerator polynomials and $D_i(s)$ to represent denominator polynomials. Thus, Eq. (14.83) becomes:

$$1 + \frac{N_1(s)\, N_2(s)\, e^{-\alpha s}}{D_1(s)\, D_2(s)} = 0 \tag{14.86}$$

where $\alpha = \alpha_1 + \alpha_2$. This simplifies to:

$$D_1(s)\, D_2(s) + N_1(s)\, N_2(s)\, e^{-\alpha s} = 0 \tag{14.87}$$

Now, from the expression for the closed-loop transfer function, and from Eqs. (14.84) and (14.85), we have that:

$$\psi(s) = \frac{gg_c}{1 + gg_c h} = \frac{\dfrac{N_1(s) \, e^{-\alpha_1 s}}{D_1(s)}}{1 + \dfrac{N_1(s) \, N_2(s) \, e^{-\alpha s}}{D_1(s) \, D_2(s)}} \tag{14.88}$$

which simplifies to:

$$\psi(s) = \frac{N_1(s) \, D_2(s) \, e^{-\alpha_1 s}}{D_1(s) \, D_2(s) + N_1(s) \, N_2(s) \, e^{-\alpha s}} \tag{14.89}$$

and since the roots of the denominator polynomial are determined from the equation obtained by setting the denominator "polynomial" to zero, we immediately see that this gives us the same result as Eq. (14.87), thus establishing that Eq. (14.83) indeed represents the general closed-loop system characteristic equation.

14.6.2 Stability Analysis

From the foregoing, we see that stability analysis for a closed-loop system boils down to investigating the roots of the characteristic equation. This characteristic equation is a polynomial in s whose order is, of course, determined by the order of the contributing transfer functions.

It is interesting to note that the roots of the characteristic equation are identical to the closed-loop system poles. Thus closed-loop stability requires that all closed-loop system poles lie in the LHP just as open-loop stability also requires all open-loop system poles to lie in the LHP.

Let us illustrate the general principles of closed-loop stability analysis with the following example.

Example 14.4   **CLOSED-LOOP STABILITY ANALYSIS OF A FIRST-ORDER SYSTEM UNDER PI CONTROL.**

A first-order system with a steady-state gain of 2, and time constant of 3 is under PI control with $K_c = 4$ and $1/\tau_I = 3/4$. Is this closed-loop system stable or not?

Solution:

The characteristic equation for this system is given by:

$$1 + gg_c = 0$$

or

$$1 + \frac{K}{\tau s + 1} K_c \left(1 + \frac{1}{\tau_I s} \right) = 0$$

which becomes, upon introducing the given parameters, and rearranging:

$$(s^2 + 3s + 2) = 0$$

a quadratic whose two roots are easily obtained as:

$$s = -1; \; s = -2$$

and since these roots are all negative, we conclude that the system is stable.

It should not be too difficult to see from this example that for low-order polynomial characteristic equations, the roots are easy to find, and stability analysis is therefore easy to carry out. For higher order polynomials, however, finding these roots explicitly can be quite tedious. Thus less tedious methods of stability analysis, which cleverly sidetrack the issue of actually calculating these roots, are to be preferred.

Let us now consider some of these stability tests.

14.6.3 Routh's Stability Test

This method for testing for system stability is due to E. J. Routh [1], and it has the following features:

- *It is a purely algebraic test that tells us how many roots of a polynomial have positive real parts without calculating the actual values of these roots.*

- *It is limited to polynomial equations only.*

This means that the Routh test can only be applied to systems with polynomial characteristic equations, immediately excluding time-delay systems whose characteristic equations contain the transcendental $e^{-\alpha s}$ term. (This is one of the situations alluded to in Chapter 8 in which rational approximations for $e^{-\alpha s}$ are sometimes utilized.)

The following is the procedure for applying the Routh test:

1. **Write the characteristic equation in the "standard" form:**

$$a_0 s^n + a_1 s^{n-1} + \ldots + a_{n-2} s^2 + a_{n-1} s + a_n = 0 \qquad (14.90)$$

where the leading term, a_0 is *positive*.

- If a_0 is originally negative, multiply the characteristic equation by -1.

- Inspect the coefficients $a_0, a_1, a_2, \ldots, a_n$; it is necessary that they all be positive; if just one is negative, at least one root of the characteristic equation lies in the RHP, and as far as control systems are concerned, the system is therefore definitely unstable.

- If all the coefficients are positive, we do not have conclusive evidence that there are no roots in the RHP; i.e., we cannot tell whether the system is stable, or unstable. Proceed with the actual Routh test.

2. **Generate the Routh Array:**

This array consists of $n + 1$ rows for an nth-order polynomial, and it is generated entirely from the coefficients of this polynomial as shown below:

Row 1 a_0 a_2 a_4 ... ...

Row 2 a_1 a_3 a_5 ... ...

Row 3 b_1 b_2 b_3 ... ...

Row 4 c_1 c_2 c_3 ... ...

...   .

Row $(n + 1)$ φ_1

- The first two rows are obtained directly from the coefficients of the polynomial characteristic equation.
- The elements in the succeeding rows 3, 4, ... are calculated as follows:

The elements of Row 3

$$b_1 = \frac{a_1 a_2 - a_0 a_3}{a_1} \; ; \; b_2 = \frac{a_1 a_4 - a_0 a_5}{a_1} \; ; \; ... \tag{14.91}$$

The elements of Row 4

$$c_1 = \frac{b_1 a_3 - a_1 b_2}{b_1} \; ; \; c_2 = \frac{b_1 a_5 - a_1 b_3}{b_1} \; ; \; ... \tag{14.92}$$

etc.

- The elements in each row are always obtained from the elements of the two preceding rows.

3. **Analyze the resulting array using the following theorems of the Routh test:**

   - All the roots of the polynomial have negative real parts (i.e., lie in the LHP) if, and only if, all the elements in the first *column* are positive and nonzero.

   - If any element of this first column is negative, there is *at least* one root in the RHP. The number of such RHP roots is equal to the number of sign changes in the first column.

   - The situation in which one pair of roots are on the imaginary axis (all others being in the LHP) manifests itself with the vanishing of all the elements in the nth row. We can locate this pair of purely imaginary roots exactly by solving the equation:

$$p_1 s^2 + p_2 = 0 \tag{14.93}$$

   where p_1 and p_2 are the elements in row $(n - 1)$.

Let us illustrate the application of the Routh test with the following examples.

Example 14.5 ROUTH TEST FOR THE ROOTS OF A POLYNOMIAL EQUATION.

Find how many roots of the fourth-order polynomial:

$$s^4 + 2s^3 + 5s^2 + 6s + 1 = 0 \tag{14.94}$$

lie in the RHP.

Solution:

The equation is already arranged in the standard form; the leading coefficient is positive, as are all the other coefficients. Thus we are unable to say conclusively, just by inspection, whether or not there are any roots of this polynomial in the RHP.

We must now proceed to the Routh test.

Since we have a fourth-order polynomial, the Routh array will have five rows. Following the procedure given above, the following Routh array is generated:

```
Row 1   1   5   1
Row 2   2   6
Row 3   2   1
Row 4   5
Row 5   1
```

And now, since no elements in the first column is negative, the number of roots in the RHP is zero. Thus if this were the characteristic equation for a control system, we will conclude that the system is stable.

The most important application of this test is in evaluating the range of controller parameter values required for a control system to be stable. Let us illustrate this using the three-tank process we introduced earlier in Section 11.5 of Chapter 11 and which we met again in Example 14.3. Recall that for this system, we found, purely by simulation, that for $K_c = 1$ the system is stable; for $K_c = 1.67$ the system is on the verge of instability, and it goes completely unstable for $K_c = 2$. Let us now establish the precise range of K_c for which the system is stable using the Routh test.

Example 14.6 CLOSED-LOOP STABILITY LIMITS FOR A THIRD-ORDER SYSTEM UNDER P CONTROL.

The third-order system whose transfer function was given in Eq. (14.73) is under P control. Find the values of K_c for which the closed-loop system will remain stable.

Solution:

The characteristic equation for this system is:

$$1 + \frac{6K_c}{(2s + 1)(4s + 1)(6s + 1)} = 0 \tag{14.95}$$

which simplifies to:

$$48s^3 + 44s^2 + 12s + \left(1 + 6K_c\right) = 0 \tag{14.96}$$

Note:

- The roots of this characteristic equation depend on what value K_c takes.

- The characteristic equation is in the standard form with the leading term positive; also all coefficients are positive, for values of $K_c > -1/6$.

- We cannot decide conclusively whether or not the system is stable by inspection; we must proceed to the Routh test.

Because we have a third-order polynomial, the Routh array in this case has four rows and the specific entries are shown below:

$$\begin{array}{lcc} \text{Row 1} & 48 & 12 \\ \text{Row 2} & 44 & 1 + 6K_c \\ \text{Row 3} & \dfrac{120}{11} - \dfrac{72}{11}K_c & 0 \\ \text{Row 4} & 1 + 6K_c & \end{array}$$

Analysis:

The first column will contain only positive elements if:

1. $\dfrac{120}{11} - \dfrac{72}{11}K_c > 0 \quad \Rightarrow \quad K_c < \dfrac{5}{3}$

2. $1 + 6K_c > 0 \quad \Rightarrow \quad K_c > -1/6$

The conclusion therefore is that closed-loop stability requires that the proportional controller gain satisfy the condition: $-1/6 < K_c < 5/3$; exactly what we had observed in Chapter 11. Note that while most controllers are designed so that $|KK_c| > 0$, it is not necessary for $K_c > 0$ for closed-loop stability.

The following example illustrates what happens when there is a pair of purely imaginary roots. From the last example we know that the system is on the verge of instability when $K_c = 5/3$. Let us investigate the nature of the roots of the characteristic equation under these conditions.

Example 14.7   ROUTH STABILITY ANALYSIS FOR A CLOSED SYSTEM HAVING PURELY IMAGINARY ROOTS.

Investigate the stability characteristics of the system whose closed-loop characteristic equation is given by:

$$48s^3 + 44s^2 + 12s + 11 = 0 \tag{14.97}$$

Solution:

Since we cannot tell anything about this system's stability by inspection of the characteristic equation, we proceed to the Routh test.

Either by recognizing that this is the Example 14.6 system with $K_c = 5/3$, or by generating the Routh array from scratch, we obtain:

$$\begin{array}{lcc} \text{Row 1} & 48 & 12 \\ \text{Row 2} & 44 & 11 \\ \text{Row 3} & 0 & 0 \\ \text{Row 4} & ? & \end{array}$$

Of course, the vanishing of all the elements in the third row indicates that we are unable to evaluate the final element in the fourth row. It is not necessary, however, to do this. All the other elements in the first column are positive, and therefore we conclude that we have a situation in which a pair of roots lie on the imaginary axis.

The exact location of these roots can now be found, according to Eq. (14.93), by solving:

$$44s^2 + 11 = 0 \tag{14.98}$$

from where we find that the roots are located at:

$$s = \pm\frac{1}{2}j \tag{14.99}$$

This example raises a very interesting question regarding what happens when some coefficients of the characteristic equation are missing, especially those that show up in the denominator in determining the values of entries in subsequent rows. The following example illustrates what to do under such circumstances.

Example 14.8   ROUTH STABILITY ANALYSIS WHEN ONE COEFFICIENT IS ZERO IN THE CHARACTERISTIC EQUATION.

Determine how many roots of the characteristic equation given below lie in the RHP:

$$s^3 - 3s + 2 = 0 \tag{14.100}$$

Solution:

Observe that the coefficient of s^2 is zero. The first step therefore is to replace the missing coefficient with a small *positive* number ε, which we shall use in our subsequent calculations of the elements of the Routh array. Afterwards we will allow this quantity to go to zero and investigate the behavior of the Routh array in the limiting situation.

The "working" characteristic equation becomes:

$$s^3 + \varepsilon s^2 - 3s + 2 = 0 \tag{14.101}$$

Let us note right away that regardless of the value of ε, the system is unstable by virtue of the negative coefficient of s. Thus our objective is not so much to determine whether the system is stable or not as it is to determine the number of roots of the characteristic equation are in the RHP.

The four rows of the Routh array are generated below in terms of the parameter ε:

Row 1	1	-3
Row 2	ε	2
Row 3	$\dfrac{-3\varepsilon - 2}{\varepsilon}$	0
Row 4	2	

Analysis:

In the limit as $\varepsilon \to 0$, we note that the term in Row 3, when written as: $(-3 - 2/\varepsilon)$ will clearly be negative.

Thus we have a negative entry in the first column showing up at the third row; this is not much of a surprise since we already know that the system is unstable.

However, since technically ε is a positive quantity (more precisely, it approaches zero from the right-hand side of the real line) we observe that in going down the first column, we encounter the first sign change from row 2 to row 3, and another sign change from row 3 to row 4.

Since there are now two sign changes, we conclude that the equation has two roots in the RHP.

It turns out that we can actually factorize Eq. (14.100) as $(s-1)^2(s+2)$; from where we see that our Routh analysis is indeed correct: the polynomial has 2 repeated roots at $s = 1$.

As noted earlier, when dealing with time-delay systems, the Routh array is inapplicable. We can do one of two things: obtain approximate results by using the rational polynomial approximations of Chapter 8, or use other methods which tolerate nonrational elements. The frequency response methods which will be dealt with in the next chapter are the most convenient. Another method that is also applicable will now be presented, but the treatment will be general, not just for time-delay systems.

14.6.4 Stability Limits by Direct Substitution

It bears repeating again that one of the most important applications of stability analysis in feedback control is in helping us ascertain what range of controller parameter values guarantees closed-loop stability of the process.

Apart from the Routh test, such stability limits can be obtained by the method of direct substitution. The basis of this method is as follows:

When a closed-loop system is on the verge of instability, at least one pair of roots of the characteristic equation must lie on the imaginary axis of the complex plane; in this case these marginally stable roots, $s = \pm j\omega$, will satisfy the characteristic equation and can be substituted directly. Solving the resulting equation for the unknown controller parameters will establish the required stability limits.

This is best illustrated by an example; and once again, we choose the example three-tank system of Section 11.5. We will recall that for this system we established independently in Examples 14.6 and 14.7 that it will be on the verge of instability when $K_c = 5/3$, and that the purely imaginary roots are located at $s = \pm 1/2\,j$. We will reconfirm these results by the method of direct substitution.

Example 14.9   CLOSED-LOOP STABILITY LIMITS FOR A THIRD-ORDER SYSTEM UNDER P CONTROL BY DIRECT SUBSTITUTION.

Using the method of direct substitution, find the limiting value of K_c for which the closed-loop system whose characteristic equation is given below will remain stable:

$$48s^3 + 44s^2 + 12s + \left(1 + 6K_c\right) = 0 \qquad (14.96)$$

Solution:

By substituting $s = j\omega$ into this equation, we have:

$$-48j\omega^3 - 44\omega^2 + 12j\omega + \left(1 + 6K_c\right) = 0$$

which upon rearranging and collecting like terms gives:

$$(-48\omega^3 + 12\omega)j + \left[(1 + 6K_c) - 44\omega^2\right] = 0 \tag{14.102}$$

Now this equation can be satisfied if and only if both the real and imaginary parts of the LHS are identically zero. This leads to the conditions:

$$-48\omega^3 + 12\omega = 0 \tag{14.103a}$$

and

$$1 + 6K_c - 44\omega^2 = 0 \tag{14.103b}$$

Solving both Eqs. (14.103a) and (14.103b) simultaneously for w and K_c yields two solutions:

$$\omega = 0, \ K_c = -1/6$$

or

$$\omega = \pm 1/2, \ K_c = 5/3$$

which confirm the earlier results that $-1/6 < K_c < 5/3$ is the stable range of controller gains.

When time delays are involved, the solution of the simultaneous equations such as Eq. (14.103) can require some iteration. Let us illustrate this with an example.

Example 14.10 CLOSED-LOOP STABILITY LIMITS BY DIRECT SUBSTITUTION FOR A FIRST-ORDER SYSTEM WITH TIME DELAY.

Consider a first-order-plus-time-delay process under proportional-only control with the following characteristics:

$$g(s) = \frac{e^{-5s}}{(10s + 1)}$$

$$h(s) = 1; \ \ g_c = K_c$$

so that:

$$gg_c h = \frac{K_c}{(10s + 1)} e^{-5s} \tag{14.104}$$

The closed-loop characteristic equation then becomes:

$$(10s + 1) + K_c e^{-5s} = 0 \tag{14.105}$$

Now substituting $s = j\omega$ to find the values of K_c, and ω when the system is at the verge of instability we obtain:

$$(10\omega j + 1) + K_c e^{-5j\omega} = 0 \tag{14.106}$$

Recalling the Euler relation:

$$e^{-5\omega j} = \cos 5\omega - j \sin 5\omega$$

Equation (14.106) becomes:

$$(1 + K_c \cos 5\omega) + j (10\omega - K_c \sin 5\omega) = 0$$

This requires that both the real and imaginary parts vanish simultaneously so that:

$$1 + K_c \cos 5\omega = 0 \tag{14.107a}$$

$$10\omega - K_c \sin 5\omega = 0 \tag{14.107b}$$

must be solved for K_c, ω. Because of the sine and cosine functions in Eq. (14.107), there are an infinite number of solutions. Some of these are provided in the table below.

K_c	ω
-1	0
3.81	± 0.367
15.87	± 1.583
28.36	± 2.834
.	.
.	.

We can ignore all but the first two solutions because the others are due to the periodic nature of the time delay; i.e., all solutions with $\omega > 2\pi/5$ in Eq. (14.107) are higher harmonics. Thus the system in Eq. (14.104) will be stable under proportional control for controller gains in the range $-1.0 < K_c < 3.81$.

Let us observe in closing that the roots of the characteristic Eq. (14.83) will take on different values as the value of K_c changes. We are sometimes interested in more than just the critical value of K_c (i.e., the stability limit); we want to know the entire history of how changes in K_c values affect the roots of the characteristic equation.

A plot of the entire spectrum of values taken by each root of this characteristic equation as K_c varies from 0 to ∞ is known as a *root locus*. Because it offers insight into the closed-loop behavior of feedback systems, we will close this chapter with an introduction to the root locus diagram.

14.7   THE ROOT LOCUS DIAGRAM

14.7.1 General Concepts

The root locus diagram is a plot in the complex plane of all of the roots, r_1, r_2, $\ldots r_n$, as the control gain K_c is varied over some range $0 < K_c < K_{c\max}$. To illustrate, let us consider the three-tank process of Section 11.5 which was analyzed earlier in Examples 14.3, 14.6, 14.7, and 14.9. Recall that the process transfer function is:

$$g(s) = \frac{6}{(2s + 1)(4s + 1)(6s + 1)} \tag{14.73}$$

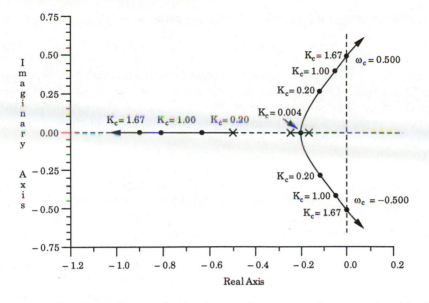

Figure 14.16. Root locus diagram for the three-tank process under proportional feedback control.

Under proportional feedback control, the characteristic equation for this system (assuming $h = 1$) is:

$$48s^3 + 44s^2 + 12s + \left(1 + 6K_c\right) = 0 \qquad (14.96)$$

This is a cubic equation with roots r_1, r_2, r_3 which may be calculated for each value of controller gain, K_c. Using a simple numerical polynomial root finder, one may generate the following table for $0 \le K_c \le 3.0$.

The *Root Locus Diagram*, a plot of these roots in the complex plane, with K_c as a parameter, is shown in Figure 14.16.

K_c	r_1	r_2	r_3
0	-0.167	-0.250	-0.500
0.0040	-0.205	-0.205	-0.506
0.05	$-0.182 + 0.126j$	$-0.182 - 0.126j$	-0.553
0.20	$-0.140 + 0.229j$	$-0.140 - 0.229j$	-0.637
0.50	$-0.093 + 0.325j$	$-0.093 - 0.325j$	-0.731
1.00	$-0.045 + 0.417j$	$-0.045 - 0.417j$	-0.828
1.67	$0.500j$	$-0.500j$	-0.917
3.00	$0.062 + 0.613j$	$0.062 - 0.613j$	-1.04

Let us recall that the closed-loop transient response is given by:

$$y(t) = A_0 + A_1 e^{r_1 t} + A_2 e^{r_2 t} + \dots + A_n e^{r_n t} \tag{14.82}$$

where $r_1, r_2, \dots, r_n$ are the roots of the characteristic equation plotted in the root locus diagram. Thus, the qualitative nature of the closed-loop transient response is entirely determined by the roots of the characteristic equation. Observe that the root locus diagram is a collection of what these roots will be for any given controller parameter value.

The most important implication therefore is that the root locus diagram presents us with information particularly useful for predicting closed-loop transient responses. By the same token, the stability characteristics of a closed-loop system can also be determined from an inspection of the root locus diagram.

For example, from Figure 14.16, we observe that:

- For $K_c = 0$, there are three real roots: $-0.167, -0.250, -0.500$ which are the open-loop poles of g indicated by ×'s in Figure 14.16.
- For $0 \leq K_c < 0.004$, all the roots are real and there is no oscillatory response.
- For $0.004 < K_c < 1.67$, two of the roots are complex with negative real parts so that the response is damped oscillatory.
- For $K_c > 1.67$, two of the roots are complex with positive real parts so that the response is oscillatory with increasing amplitude and unstable.
- Since the frequency of the sinusoidal portion of a system's oscillatory response is determined by the imaginary part of the characteristic equation roots, the oscillatory response takes on higher and higher frequency as K_c increases.
- Since the third root, r_3, stays real and becomes more negative as K_c increases, the transient response arising from the term $A_3 e^{r_3 t}$ in Eq. (14.82) becomes faster as K_c increases.

Let us now generalize the results from this example.

14.7.2 General Features of a Root Locus Diagram

Earlier generations of control engineers were required to master the techniques for efficiently constructing root locus diagrams. However, this is no longer necessary. Computer programs that generate such diagrams with a great deal of facility are now routinely available, thereby allowing the student or practicing engineer to focus on the task of learning how to *interpret* these diagrams, not how to *draw* them.

The following therefore are the general features exhibited by the root locus diagram, a knowledge of which will facilitate the task of extracting, and properly interpreting, the information contained in the diagram.

1. The root locus diagram will show as many roots as the poles of $gg_c h$.

This is easily established from the fact that the root locus diagram is simply a plot of the roots of the characteristic equation:

$$1 + gg_c h = 0 \tag{14.83}$$

with the controller gain K_c as a parameter. When there are no time delays involved, gg_ch can be represented as the following ratio of polynomials:

$$gg_ch = \frac{K_pK_c N(s)}{D(s)} \tag{14.108}$$

where K_p is a combination of all the loop gains, $N(s)$ is a polynomial of order r, and $D(s)$ is a polynomial of order n; and for realistic closed-loop systems, $r \leq n$. Thus gg_ch has r zeros and n poles.

The characteristic equation is therefore given by:

$$1 + \frac{K_pK_c N(s)}{D(s)} = 0$$

or

$$D(s) + K_pK_c N(s) = 0 \tag{14.109}$$

which will be an nth-order polynomial (because $r \leq n$) and will therefore have as many roots as the poles of gg_ch.

In case there is a time delay α in the open-loop process, gg_ch will take the form:

$$gg_ch = \frac{K_pK_c N'(s) e^{-\alpha s}}{D'(s)} \tag{14.110}$$

Since $e^{-\alpha s}$ is periodic, it will introduce an infinite number of roots (recall Example 14.10). However, all of the higher order harmonics are not of interest so that if we represent $e^{-\alpha s}$ by a high-order Padé approximate (say, mth-order), we would have:

$$e^{-\alpha s} = \frac{N_\alpha(s)}{D_\alpha(s)} \tag{14.111}$$

where both $N_\alpha(s)$, $D_\alpha(s)$ are mth-order polynomials. Then if we let:

$$N(s) = N'(s) N_\alpha(s) \tag{14.112}$$

$$D(s) = D'(s) D_\alpha(s) \tag{14.113}$$

we may use the arguments above even in the case of time delays. It can be shown that even though delays introduce additional higher harmonic loci (branches) on the root locus diagram, only those close to the origin need be considered.

Since the loci of the root locus diagram show the trajectory of each root of the characteristic equation, it follows therefore that:

The root locus diagram has as many branches (or loci) as the poles of gg_ch.

2. The root locus diagram is symmetrical about the real axis.

This is simply because of the fact that "branching" off the real axis occurs when there are complex roots, and such roots only occur in mirror image pairs, as complex conjugates.

3. The root loci commence at the poles of $gg_c h$ and terminate at its zeros.

The key to establishing this is the representation of the characteristic equation given in Eq. (14.109). The root loci commence when $K_c = 0$, and terminate when $K_c = \infty$. Setting $K_c = 0$ in Eq. (14.109) leaves:

$$D(s) = 0 \tag{14.114}$$

which, of course, gives rise to roots identical to the *poles* of $gg_c h$ in Eq. (14.108).

To obtain the terminating point of the root loci, we divide through in Eq. (14.109) by K_c first, to obtain:

$$\frac{D(s)}{K_c} + K_p N(s) = 0 \tag{14.115}$$

and now, as $K_c \to \infty$, Eq. (14.115) becomes:

$$K_p N(s) = 0 \tag{14.116}$$

which gives rise to roots identical to the *zeros* of $gg_c h$ in Eq. (14.108).

It is in this sense therefore that we may consider the root locus diagram as the "history of the migration of the poles of $gg_c h$ to its zeros" as a function of the controller gain.

This raises an interesting question, however: what happens in the situation where $gg_c h$ has more poles than zeros? The answer is that the remaining roots go to infinity. We therefore have the result:

The root locus has n branches, r of which terminate at the zeros of $gg_c h$,
while the remaining (n − r) branches go to infinity as K_c increases.

Note that the system with the root locus diagram shown in Figure 14.16 has no zeros so that all three branches go to infinity.

4. The magnitude of $gg_c h$ is 1 for all values of s.

By rearranging the characteristic Eq. (14.83) we see that:

$$gg_c h = -1$$

and taking the absolute value of each side, we obtain:

$$|gg_c h| = 1 \tag{14.117}$$

The zeros and poles of $gg_c h$ are easily identified on the root locus diagram, but not the process gain. This fourth property may be used to evaluate the total process gain K_p from the root locus diagram. Observe that from Eq. (14.108) that this property implies:

$$\left| gg_c h \right| = \left| K_p K_c \right| \frac{|N(s)|}{|D(s)|} = 1 \qquad (14.118)$$

from which K_p may be obtained.

The following three properties have to do with extracting transient response information from the root locus diagram, and, on the basis of our knowledge of the relationship between characteristic equation roots and transient response, they are self-evident.

5.   The closed-loop system response is nonoscillatory for points on the loci that lie completely on the real axis.

6.   When the loci branch away from the real axis, the closed-loop system response becomes oscillatory.

7.   The closed-loop system is on the verge of instability at the point where one of the branches first intersects the imaginary axis in going from the left half to the right half of the complex plane.

It should be quite obvious by now that $gg_c h$ plays a very important role in issues concerning the root locus and in the determination of the characteristics of the closed-loop system in general. This quantity is referred to as the *open-loop transfer function* of the closed-loop system. This is because if the feedback loop is "opened" right at the point where the feedback information y_m is passed on to the comparator (see Figure 14.12), the transfer function relation between this output and the set-point y_d is:

$$y_m = gg_c h \, y_d \qquad (14.119)$$

This is the transfer function we obtain by "opening" the otherwise closed feedback loop.

Let us now use two examples to illustrate the general features of the root locus which we have just presented. It should be mentioned that the notational standard is to represent the poles of the open-loop transfer function by the symbol $\times$ while the zeros are represented on the root locus diagram by $\circ$.

Example 14.11   ANALYSIS OF CLOSED-LOOP SYSTEM BEHAVIOR FROM A ROOT LOCUS DIAGRAM.

The root locus diagram shown in Figure 14.17 is for the process with:

$$gg_c h = \frac{2K_c(s+1)(-s+1)}{(0.5s+1)(2s+1)(4s+1)} \qquad (14.120)$$

(a)   Confirm that the poles and zeros shown in the root locus plot correspond to those in the transfer function Eq. (14.120)

(b)   What will the nature of the transient response be when $K_c = 1$?

(c)   At what K_c value is the system on the verge of instability?

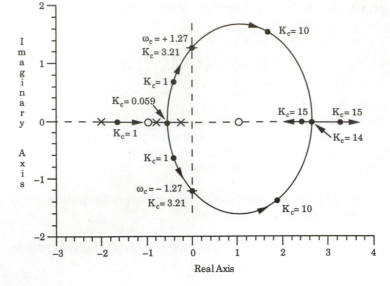

Figure 14.17.   Root locus diagram for process modeled by Eq. (14.120).

Solution:

(a)   The root locus diagram indicates three poles (located at $s = -2, -0.50, -0.25$) and two zeros at $s = -1, +1$. This is consistent with the original model. Observe that the root locus has three branches, two of which go to system zeros and one of which goes to positive infinity.

(b)   From the root locus diagram, when $K_c = 1$, we see that two of the roots have already branched out well into the complex plane. The conclusion is therefore that the transient response will be damped oscillatory.

(c)   When $K_c = 3.21$, the root locus diagram shows that two of the branches are on the imaginary axis and about to cross into the RHP. The conclusion is therefore that this is the controller gain value for which the system is on the verge of instability.

Example 14.12   ANALYSIS OF CLOSED-LOOP SYSTEM BEHAVIOR FROM A ROOT LOCUS DIAGRAM WHEN THERE ARE TIME DELAYS.

As was indicated above, the presence of a time delay adds additional branches to the root locus plot. Figure 14.18 shows the root locus diagram of the first-order process with a time delay and a proportional-only controller considered earlier in Example 14.10:

$$gg_ch = \frac{K_c}{(10s + 1)} e^{-5s} \tag{14.104}$$

The time delay introduces an infinite number of almost horizontal harmonic branches at regular intervals up and down the imaginary axis (we show only three of these). However, only the branch coming in from $-\infty$ on the real axis is of importance to stability. This root meets the root arising from the single pole at -0.1 and branches from the real axis when $K_c = 0.446$. Thus beyond this point the process dynamics are oscillatory. The root locus diagram predicts that for controller gains $K_c > 3.81$ the process will be unstable. These results can be verified by simulations of the closed-loop behavior shown in Figure 14.19.

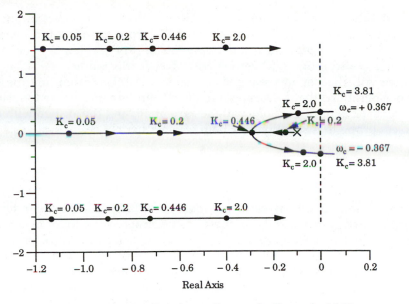

Figure 14.18. Root locus diagram for Example 14.12.

Before leaving the issue of the root locus, it is enlightening to think about the influence of the open-loop transfer function zeros on the stability properties of the closed-loop system, as K_c varies. (The root locus diagram happens to offer a unique insight into this issue.)

We have shown that the roots of the closed-loop system characteristic equation start from the poles of the open-loop transfer function $gg_c h$, and end at its zeros. Thus as K_c increases, the roots of the characteristic equation are attracted towards the zeros. If the open-loop transfer function of a system has zeros that all lie in the LHP, then we see that as K_c increases, the tendency is for such a system to be more stable than the system with no zeros to attract the poles.

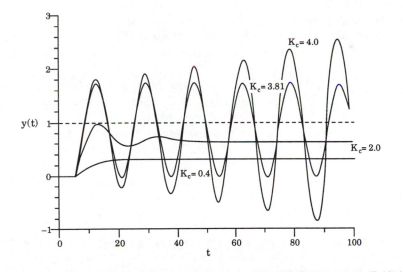

Figure 14.19. Simulation of the closed-loop response for the system in Eq.(14.104).

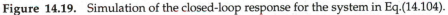

On the other hand, the presence of a RHP zero has the effect of attracting the poles into the RHP as K_c increases and therefore encouraging instability. Thus while a RHP zero does not imply instability (remember, only poles determine stability), the RHP zero is a catalyst for instability.

As we will discuss in greater detail later, it is clear from just these root locus considerations that a system with a RHP zero is more prone to instability under feedback control than one with no zero at all; and the system with a LHP zero is less prone to instability than both of the other classes of systems.

14.8   SUMMARY

We have concerned ourselves in this chapter with getting acquainted with feedback control systems, and our quest has taken us over a lot of material. We started with an introduction to the concept of feedback as a control strategy and followed with the idea of block diagramatic representation for feedback systems. After discussing closed-loop transfer functions and how they are obtained, the rest of the chapter was devoted to various aspects of the analysis of feedback control systems, all of which were dependent on the closed-loop transfer functions.

The presentation of the analysis of feedback control systems in this chapter is somewhat parallel to the discussion of the analysis of open-loop dynamic behavior given in Part II: we started with the ideas of block diagrammatic representations for feedback control systems, and derived the resulting, overall, (closed-loop) *transfer functions*; this was followed by utilizing these transfer functions to analyze the transient responses of feedback control systems; and this in turn led inexorably to the issue of stability.

The importance of the issue of closed-loop stability is only beginning to emerge in this chapter; it will take on greater dimensions as we progress into the issue of the design of feedback control systems.

We have found that the issue of closed-loop stability analysis boils down to investigating the nature of the roots of the characteristic equation. We have also seen that beyond the very low order polynomial equations, evaluating these roots for purposes of stability analysis can be quite tedious. As a result, we presented two stability analysis techniques that circumvent the tedium of polynomial root extraction: the Routh test and the method of direct substitution.

The graphical root locus method was also introduced as a means of investigating closed-loop stability, and also determining qualitative characteristics of closed-loop transient responses.

With the material covered in this chapter as background, we are now in a position to embark on the central issue of Part IV: the design of effective controllers. This endeavor, which will take us through the rest of the book, begins with the design of the relatively simple conventional feedback control systems in the next chapter.

REFERENCES AND SUGGESTED FURTHER READING

1.   Routh, E. J., Dynamics of a System of Rigid Bodies, Part II, Macmillan, London (1907)

REVIEW QUESTIONS

1. What is feedback control?

2. What are the typical elements of a feedback control loop?

3. What steps are involved in developing a block diagram for a physical control system?

4. What is a closed-loop transfer function?

5. Which of the classical controllers, P, PI, PD, and PID, usually gives rise to transient responses with nonzero offset when used in a feedback loop?

6. What is the role of the characteristic equation in analyzing a closed-loop system for stability?

7. The Routh test is an algebraic test for determining what? How is it used in feedback control and what are its limitations?

8. When a closed-loop system is on the verge of instability, what happens to its characteristic equation? How is this concept utilized in determining stability limits for feedback control systems?

9. What is a root locus diagram?

10. What is the "open-loop" transfer function of a closed-loop feedback control system, and why is it so called?

11. How is the "open-loop" transfer function of a feedback control system related to the closed-loop transfer function and the characteristic equation?

12. In what ways do the poles and zeros of the "open-loop" transfer function of a feedback control system determine the general features of its root locus diagram?

13. Why is a root locus diagram symmetric about the real axis?

14. What distinguishes the root locus diagram of a feedback system which contains a time-delay element?

PROBLEMS

14.1   The crude oil preheater furnace first introduced in Chapter 1 (see Fig. P14.1) is used to preheat the crude oil feed material to a fractionator.

As presented previously in Chapter 1, the crude oil flowrate F and temperature T_i at the inlet of the furnace tend to fluctuate substantially. The crude oil flowrate out of the furnace F_o may vary, but it is desired to control the exit temperature T.

(a) Given the following transfer function relations (where (*) indicates nominal steady-state operating values):

$$(T - T^*)(s) = g_1(s)\left(Q_F - Q_F^*\right)(s) \tag{P14.1a}$$

$$(T - T^*)(s) = g_2(s)\left(T_i - T_i^*\right)(s) \tag{P14.1b}$$

$$(T - T^*)(s) = g_3(s)\left(F - F^*\right)(s) \tag{P14.1c}$$

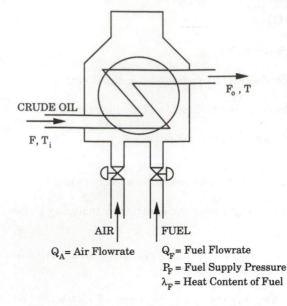

Figure P14.1.   Crude oil preheater furnace.

draw a block diagram for the overall feedback control system which employs a controller (with the transfer function $g_c(s)$) to regulate the furnace exit temperature by adjusting the fuel flowrate; the feedback measurements are acquired via a temperature measuring device with transfer function $h(s)$. The fuel line valve dynamics are represented by the transfer function $g_v(s)$. Include all the given transfer functions and assume that the fuel supply pressure and the heat content of the heating fuel remain constant.

(b) When the fuel supply pressure varies and its effect on the fuel flowrate is represented by:

$$\left(Q_F - Q_F^* \right)(s) \ = \ g_4(s)\left(P_F - P_F^* \right)(s) \tag{P14.2}$$

modify the block diagram appropriately.

14.2   Figure P14.2 shows a feedback scheme for controlling the exit temperature of an industrial heat exchanger that utilizes chilled brine to cool down a hot process stream.

Assuming that T_B, the brine temperature (in °C), remains constant, and given the following transfer function relations:

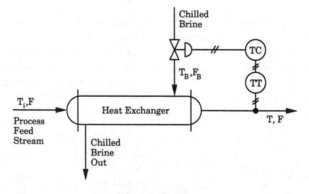

Figure P14.2.

$$(T - T^*)(°C) = \frac{-(1 - 0.5e^{-10s})}{(40s + 1)(15s + 1)}\left(F_B - F_B^*\right) \text{ (kg/s)} \tag{P14.3a}$$

$$(T - T^*) \text{ (°C)} = 0.5e^{-10s}\left(T_i - T_i^*\right) \text{ (°C)} \tag{P14.3b}$$

and given that the controller is a pure proportional controller operating according to:

$$\left(F_B - F_B^*\right) \text{ (kg/s)} = -K_c \left(T_{\text{setpoint}} - T\right) \text{ (°C)} \tag{P14.4}$$

and that the measuring device has negligible dynamics:

(a) Draw a block diagram for this feedback control system.

(b) Using the supplied information, obtain the closed-loop transfer function relationships for this system and write the characteristic equation.

(c) For a 2°C step increase in the process feedstream temperature, what is the steady-state offset observed in the exit temperature, as a function of the controller gain K_c?

14.3 The level control scheme on the reflux receiver of an industrial low boiler column is shown in Figure P14.3. The receiver is an essentially cylindrical vessel 2 ft in diameter.

(a) Ignoring for the moment the level transmitter and the level controller, develop a mathematical model for this process, assuming that F_c, the flowrate of material from the condenser, is unmeasured and varies substantially with time, but that F_R, the reflux flowrate, is held constant; the distillate "product" flowrate F_D is to be used to control the drum level. Cast your model in terms of deviation variables and obtain a transfer function model for this process. The flowrates are in lb/hr, and the density of the material may be assumed constant at 30 lb/ft^3.

(b) Given that the valve on the distillate "product" line has dynamics represented by:

$$g_v(s) = \frac{0.02}{0.2s + 1}$$

(with the time constant in minutes) so that:

$$\left(F_D - F_D^*\right) \text{ (lb/hr)} = g_v (P_v - P_v^*) \text{ (\%)} \tag{P14.5}$$

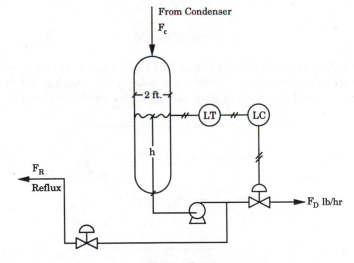

Figure P14.3.

where P_v represents the valve position (in %), and (*) indicates nominal steady-state operating values, develop a block diagram for this process when the controller is a proportional controller operating according to:

$$\left(P_v - P_v{}^*\right)(\%) \;=\; -K_c\,(h_{\text{setpoint}} - h)\ (\text{ft}) \tag{P14.6}$$

You may assume that the measuring device has negligible dynamics.

(c) Given that $K_c = 10$, obtain an expression for the dynamic response of the *closed-loop system* to a step *change* of 2.5 lb/hr in F_c. Plot this response.

14.4 (a) Find the closed-loop transfer function equation for the control system whose block diagram is shown below in Figure P14.4(a).

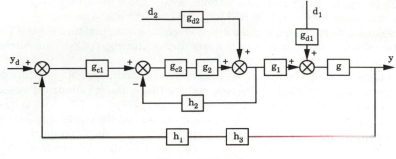

Figure P14.4(a).

(b) Find the closed-loop transfer function equation for the control system whose block diagram is shown below in Figure P14.4(b), and show that with g_k as indicated, the system characteristic equation does not contain the time-delay element. Compare this with the characteristic equation for the situation with $g_k = 0$.

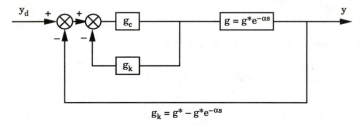

$$g_k = g^* - g^* e^{-\alpha s}$$

Figure P14.4(b).

14.5 The block diagram shown in Figure P14.5 is for the so-called "parallel control" structure in which *two* input variables, u_1 and u_2, are adjusted by *two* independent controllers for the purpose of controlling a *single* output variable y.

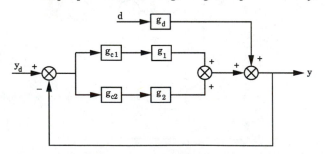

Figure P14.5. The parallel control structure.

Obtain the closed-loop transfer function equation for this system. Given that one of the controllers, say, g_{c2}, is a pure integral controller, show that under this condition, no steady-state offset will result if the other controller g_{c1} is specified as a pure proportional controller.

14.6 When the human eye is exposed to an external light flux, a simplified model of the feedback control mechanism used to regulate the *retinal light flux* by opening the human iris is shown in the block diagram of Figure P14. 6.

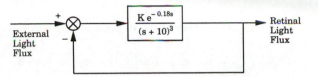

External
Light
Flux

$$\frac{K e^{-0.18s}}{(s + 10)^3}$$

Retinal
Light
Flux

Figure P14.6. Simplified block diagram of the human pupil servo mechanism.

(a) By using a first-order Padé approximation for the time-delay term, find the value of K for which the closed-loop system will exhibit sustained oscillations. What will the frequency of these oscillations be? (Adapted from Nise (1991)[†]) (Note that it is not unusual that the retinal light flux will "exhibit oscillations." The opening of the iris directly determines the amount of retinal light flux: thus, retinal light flux oscillations imply oscillations of the iris.)

(b) By using the method of direct substitution, or otherwise, repeat part (a) without using an approximation for the time-delay element. Compare your results with those obtained in part (a) and thereby assess the adequacy of the approximation used.

14.7 (a) Given the control system in Figure P14.7:

with the transfer functions given by:

$$g(s) = \frac{5e^{-8s}}{15s + 1}$$

$$g_v(s) = \frac{1}{3s + 1}$$

$$g_c(s) = K_c(1 + 3s)$$

determine the largest value of K_c for which the control system is stable.

(b) For the same control system indicated in Figure P14.7, if now, $g_v(s) = 1$, and the plant, given as a second-order system, with the usual transfer function:

$$g(s) = \frac{K}{\tau^2 s^2 + 2\zeta\tau s + 1}$$

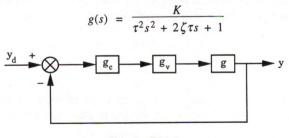

Figure P14.7.

[†] Norman S. Nise, *Control Systems Engineering*, Benjamin/Cummings Publishing Co., (1991), p. 410.

is to be controlled with a PI controller:

$$g_c(s) = K_c\left(1 + \frac{1}{\tau_I s}\right)$$

determine the inequality (or inequalities) that define the range of values for the controller gain, K_c, and the integral time constant, τ_I, for which the closed-loop system will be stable.

14.8   Consider the control system whose block diagram is shown in Figure P14.8:

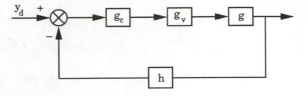

Figure P14.8.

(a) First show that when $g_c(s) = K_c$, $h(s) = 1$, $g_v(s) = 1$, and:

$$g(s) = \frac{5}{(2s + 1)}$$

the closed-loop system is stable for all values of $K_c > 0$.

(b) Investigate the *destabilizing* effect of time delays in a feedback loop by considering $h(s) = e^{-s}$; use a first-order Padé approximation for this irrational function and find the range of K_c values for which the closed-loop system is stable. Compare this with the result in part(a).

(c) Investigate the *stabilizing* effect of derivative control action by considering a PD controller, $g_c(s) = K_c(1 + 0.2s)$, for the system in (b). Find the new range of K_c values required for closed-loop stability and compare this with the result in part (b). Also, show analytically that this system suffers from steady-state offset irrespective of the value of the derivative time constant or the proportional gain.

14.9   Repeat parts (b) and (c) of Problem 14.8, using the method of direct substitution, or otherwise, to obtain more accurate stability limits by dropping the Padé approximation, and using the exact e^{-s} function. Compare these limits with the approximate limits found in Problem 14.8.

14.10 A closed-loop system consists of a second-order process, with steady-state gain 1, and two time constants, 1 and 0.5; a measuring device, with steady-state gain 1, and time constant 1/3; and a pure proportional controller. Determine the frequency of sustained oscillation when the closed-loop system is on the verge of instability. If a new measuring device with negligible dynamics were introduced, with what frequency will the new closed-loop system now oscillate when on the verge of instability?

14.11 Consider the control system whose block diagram is shown in Figure P14.9 where the controller is proportional-only, and the process transfer function is given by:

$$g(s) = \frac{(1 - 5s)}{(5s + 1)(3s + 1)}$$

(a) Show that with $g'(s) = 0$, the closed-loop system will be stable under proportional feedback control only for K_c values less than 1.6.

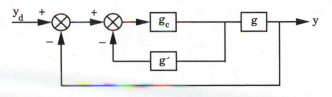

Figure P14.9.

(b) When $g'(s)$ is now given as:

$$g'(s) = \frac{10s}{(5s + 1)(3s + 1)}$$

obtain the new stability limits for the closed-loop system and compare with the results in (a). Comment on the effect of the minor loop, incorporating the given $g'(s)$ on the overall system stability.

14.12 A certain feedback control system consists of a process with transfer function:

$$g(s) = \frac{1}{(20s + 1)(10s + 1)}$$

valve dynamics represented by:

$$g_V(s) = \frac{0.25}{0.5s + 1}$$

and a controller of *unknown* structure. The root locus diagram for the control system is, however, known to have the following features:

Two zeros located at: $s = -0.5$, and $s = -1.0$;
Four poles located at: $s = 0$, $s = -0.05$, and $s = -0.1$, and $s = -2.0$

From the given information, deduce the controller *type* and controller parameter values. (Note that K_c varies for each point on the diagram so that it is *not* one of the controller parameters you are to deduce.)

14.13 An unstable first-order process, with transfer function:

$$g(s) = \frac{2}{(s - 4)}$$

is to be controlled using a PI controller with $\tau_I = 1$.
(a) Use a root locus plotting program to obtain a root locus diagram for the closed-loop system. From this diagram, specify the range of K_c values for which the closed-loop system is stable, and the frequency of oscillation when the system is on the verge of instability.
(b) Confirm analytically the graphical results obtained in part (a).

14.14 A feedback control system has the following components in its loop:

Controller:

$$g_c(s) = K_c\left(1 + \frac{1}{s}\right)$$

Process:

$$g(s) = \frac{2}{(s^2 + 1)}$$

Measuring device:

$$h(s) = \frac{1}{(s + 2)}$$

(a) Use a root locus plotting program to obtain a root locus diagram for the closed-loop system.

(b) From the root locus diagram, draw *qualitative sketches* of the expected closed-loop responses of the system to a unit step change in set-point for $K_c = 0$, $K_c = 1$, $K_c = 10$, and $K_c = 100$. *It is important to do this part before proceeding to part (c), otherwise you will miss the point of the exercise.*

(c) Use a control system simulation package to verify your sketches in part (b).

CHAPTER 15

CONVENTIONAL FEEDBACK CONTROLLER DESIGN

Our studies in Part II showed how the dynamic behavior of processes, in the open loop, are determined solely by their individual, intrinsic characteristics. In Chapter 14, however, we saw how such dynamic behavior can be influenced by the controller. Reduced to the most basic terms, the controller is simply a device for *modifying* the dynamic behavior of a process. The central issues in controller design are, therefore, how to choose what controller type to use, and what values to choose for the controller parameters such that the closed-loop system behavior is thereby *modified appropriately*.

This chapter is the first of several chapters that will be concerned with the overall issue of controller design for all types of processes; it is devoted to the design of conventional single-loop feedback controllers, the typical starting point for studies in controller design methodology.

15.1 PRELIMINARY CONSIDERATIONS

As noted in Chapter 1, the main objective of the control system is to maintain the process output at its desired set-point. To perform this task adequately, the control system must be set up to:

- Monitor the process output,
- Get this information to the controller,
- Have the controller make appropriate decisions about the current situation and issue a control command accordingly,
- Pass the control command to the process.

It is clear that even though we may be able to monitor the process perfectly, and convey information back and forth between the process and the controller perfectly, the success of the entire control scheme obviously depends, to a great extent, on how well the controller performs its task. It is in this sense that the design of the feedback controller is considered the central issue in conventional control systems design.

However it is also clear that successful design of the controller is not sufficient — by itself — to guarantee the success of the overall feedback system. There are some other preliminary issues with significant bearing on the success of a feedback control system; they will be discussed now before we go on to the main task at hand.

The complete control system design problem includes, apart from the obvious design of the controller itself, such considerations as:

- Choosing sensors and transmitters
- Selecting control valves or other final control elements
- Familiarization with the commercial controllers

It is important to note that traditionally these issues have often been taken care of at the process/plant design stage; they are typically discussed in greater detail in books on control system instrumentation. In the past, process control textbooks have been more concerned with the problem of designing the controller, given that all these other elements have been properly designed and selected, and are already in place. However, such preliminary choices of sensors, control valves, etc., if made without an eye to process dynamics can often lead to poor control system performance. Thus modern industry practice is to bring these preliminary issues into the overall design of the control system.

15.1.1 Choice of Sensors, Transmitters, and Final Control Elements

Let us begin with all of the links between the controller and the process.

Sensors and Transmitters

As we will recall, sensors are instruments used to measure process variables; they typically operate by producing electrical signals, which can be related directly or indirectly to the value of the actual process variable. What the sensor produces is usually converted into a more appropriate form before being conveyed to the controller by the transmitter.

Concerning sensors and transmitters, the most important points to note are:

1. The *range* of the sensor instrument must cover the range of the process variable it is to measure.

2. The signal generated by the transmitter must be compatible with the input range of whatever is on the receiving end of the transmission, be it the controller (in case of transmission *from* the process) or the final control element (in case of transmission *to* the process).

More detailed information about types, design, and guidelines for selection of sensors and transmitters for various applications can be obtained, for example, from Considine [1].

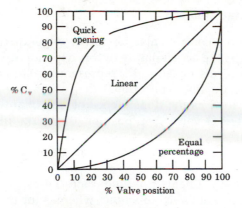

Figure 15.1. Control valve flow characteristics.

Control Valves

Concerning the selection of control valves, the main points of interest are summarized below.

Control Valve Action

The pneumatic control valves are so called because they are activated (their valve stems are moved up or down, causing more, or less valve opening) by air pressure signals. If adding more air causes the valve opening to increase, the valve action is said to be "air-to-open"; the other obvious alternative is the "air-to-close" action, in which the addition of more air causes a reduction in valve opening.

A very important aspect of control valve action is what happens when the air supply fails. The "air-to-open" valves require air to open them and will therefore "fail close"; i.e., these valves will automatically close upon failure of air supply, since they require air supply to keep them open. On the other hand, the "air-to-close" valve will go fully open upon failure of air supply; i.e., they will "fail open."

The "fail position" of a valve is very important from the safety point of view. For example, an "air-to-open" valve might be chosen for the steam valve in our example system of Figure 14.1 because, if ever there is a failure of air supply, we prefer to have the steam valve completely shut off than completely open.

Control Valve Sizing

Since control valves come in different sizes, it is important to have a means by which to determine what valve size is required for what application.

It is standard practice to use the valve discharge coefficient C_v for control valve sizing. By definition, C_v is the flowrate of water through the valve (in U.S. gallons/min) when it is wide open, and the pressure drop across it is 1 psi. It is obvious how, given the maximum C_v value for a valve, we can decide whether or not it is adequate for our specific application.

Control Valve Characteristics

The flowrate through a control valve obviously depends on the valve opening; what is not obvious is the actual nature of this dependence. If we quantify flow through a valve in terms of $\%C_v$ (0% representing *no flow* conditions; 100% representing maximum flow conditions) then the valve characteristics for a control valve is defined as the dependence of this quantity $\%C_v$ on the valve opening (quantified as % valve stem position, 0% being fully closed, with 100% being fully opened).

There are three most common control valve flow characteristics, and they are shown in Figure 15.1.

For a valve with linear flow characteristics, the flow is directly proportional to the percent valve opening; whereas for the quick opening flow characteristics, larger flows are obtained at the lower valve opening positions and increasing the valve opening produces smaller flow increments.

The "equal percentage" valve allows a very small flowrate through at very small percent valve openings which increases rapidly as the valve opening approaches the full open position. This characteristic allows an $x\%$ increase in the flowrate for a commensurate $x\%$ increase in the valve position, regardless of the initial valve opening or flow, hence the "equal percentage" tag.

It is important to choose valves with the right characteristics to match the overall behavior we expect from the control system. Various rules of thumb are available in books on control system instrumentation that offer guidance in this regard. It should be mentioned that perhaps the most commonly used valves are the "equal percentage" type.

Control valves also have time constants for movement that vary from *negligibly small* (for small valves or slow processes) to *important* (for large valves or fast processes). These dynamics must be considered in the control system design.

Other Final Control Elements

Although control valves make up the great majority of final control elements, there are a number of other devices that find use; for example:

- Electrical heaters
- Variable speed motors (as in pumps, compressors, extruders, etc.)
- Stepping motors (as in motor driven valves, positioners, etc.)

Each of these has characteristics (usually gains and dynamics) that can influence the control system design.

15.1.2 Commercial Controllers

Before discussing the characteristics of commercial controllers in detail, we must distinguish between the types one might find in practice:

1. Electronic or digital controllers designed for continuous-time operation
2. Digital controllers designed for discrete-time operation

In this section and in the controller designs discussed in Chapters 16–22, we shall assume we are dealing with controllers of the first type; i.e., continuous-time controllers. Controllers of the second type will be dealt with in Chapters 23–26.

It is also pertinent at this stage to point out several important features of commercially available feedback controllers before going into the main subject of feedback controller design.

Automatic/Manual Modes

Commercial controllers are equipped with the ability to operate either in the automatic or manual modes. Choosing which option the controller operates on is accomplished by means of a switch.

When in the automatic mode of operation, the controller itself makes the control decisions of how to manipulate the process input to maintain the process output at its desired set-point value. When the controller is in the manual mode, the controller output can be changed only manually by an operator; the controller ceases to be in charge of making control decisions.

This feature is advantageous in situations where it is necessary to override the automatic controller. For the most part, however, if the controller is well designed, it usually operates in the automatic mode with little or no intervention necessary from the operator.

Direct/Reverse Action

As we recall from Chapter 14, the proportional controller responds as:

$$c(t) = K_c \, \varepsilon(t) \tag{14.8}$$

i.e., to a positive increase in the error signal, it responds by demanding a proportionate increase in the process input. For the stirred tank process shown in Figure 14.1, for example, if, for some reason the temperature of the liquid in the tank falls below the desired set-point value, the controller receives a signal $\varepsilon(t)$ that will be positive and the controller signal to the control valve will be positive; for an air-to-open valve, this decision will translate to an increase the steam rate. Of course, this is the appropriate decision to take, since such action will have the effect of raising the value of the tank temperature.

However, consider the level control system we discussed in Chapter 1, the configuration shown in Figure 1.17. If the liquid level in this tank falls below the desired set-point value, the control action required to rectify the situation should call for a *reduction* in the outflow from the tank. As presented in Eq. (14.8), the controller will call for an increase in the outflow, and this will be inappropriate action.

The problem is not with the definition of the controller; the issue is the difference between these two processes. Observe that the process in Figure 14.1 has a *positive* steady-state gain (an increase in the input variable causes the output variable to settle to a higher value at the new steady state), while the process in Figure 1.17 has a *negative* steady-state gain (an increase in the input variable causes a reduction in the value of the output variable).

To allow the controller to make the appropriate decision regardless of the sign of the process steady-state gain, commercial controllers are equipped with a feature that makes it possible to change the sign of the controller gain K_c.

For systems with positive steady-state gains, such as the process in Figure 14.1, the controller action as depicted in Eq. (14.8) is appropriate, and there is no need to change the sign of the controller gain. The controller acts in *direct action* mode for such systems. For systems with negative steady-state gains, such as the liquid level system referred to above, a reversal of the sign of the controller gain K_c is necessary for appropriate action. The controller is then said to operate in *reverse action* mode.

Perhaps the most important point to note is that good controller performance requires that the controller gain should take on a sign that ensures that the overall process gain (the product of the process steady-state gain and the controller gain) is positive.

In what follows, we will assume that the overall process gain is positive.

Commercial Controller Forms and Parameters

The equations that describe the operation of feedback controllers as given by various commercial manufacturers are not always in the same form. The most important difference lies in the fact that many commercial manufacturers use the term *proportional band* (PB) (or, more precisely, *percent proportional band*) in place of the controller gain K_c. The important point to note is the relationship between these two quantities:

$$\text{PB} = \frac{100 \text{ (maximum range of controller output)}}{K_c \text{ (maximum range of measured variable)}} \tag{15.1}$$

and the fact that a *low* proportional gain corresponds to a *wide* proportional band, and conversely, a *high* proportional gain corresponds to a *narrow* proportional band.

To prove the relationship in Eq. (15.1), let us define some variables:

$\Delta\varepsilon^* \equiv$ the range in the measured variable error causing the final control element to go from zero to its maximum value with controller gain K_c

$\Delta c^* \equiv$ the range of controller output causing the final control element to go from zero to its maximum value

$\Delta y^* \equiv$ the maximum range of the measured variable

Now the fundamental definition of Proportional Band is:

$$PB = 100 \cdot \frac{(\Delta\varepsilon^*)}{(\Delta y^*)} \tag{15.2}$$

Also from Eq. (14.8) we see that:

$$K_c = \frac{(\Delta c^*)}{(\Delta\varepsilon^*)} \tag{15.3}$$

Substituting Eq. (15.3) into Eq. (15.2) yields:

$$PB = \frac{100 \, (\Delta c^*)}{K_c \, (\Delta y^*)} \tag{15.4}$$

which is equivalent to Eq. (15.1).

There are also other important differences between controllers offered by different manufacturers. For example, a Honeywell PI controller operates according to:

$$p(t) = K_c \, \varepsilon(t) + K_c \tau_R \int_0^t \varepsilon(t) \, dt + p_s \tag{15.5}$$

where τ_R (referred to as the *reset rate*, defined as the reciprocal of the integral time τ_I) and proportional controller gain K_c is used. On the other hand, the Foxboro PI controller operates according to:

$$p(t) = \frac{100}{PB} \, \varepsilon(t) + \frac{100}{PB \, \tau_I} \int_0^t \varepsilon(t) \, dt + p_s \tag{15.6}$$

where PB and τ_I are the specified parameters.

There is also a very important point to note about the commercial implementation of the PID controller, precipitated by the fact that to implement the PID controller as given in Eq. (14.16) requires obtaining the derivative of the error signal. The problem arises because it is impossible to obtain the exact derivative of a signal with a physical device. Thus the commercial PID controller usually approximates the "ideal" PID controller of Eq. (14.16) by:

$$g_c(s) = K_c \left(1 + \frac{1}{\tau_I s} \right) \left(\frac{1 + \tau_D s}{1 + \alpha \tau_D s} \right) \tag{15.7}$$

where α is a small quantity, typically less than 0.1. The essence of the approximation is that the derivative aspect in Eq. (14.16) is approximated by a lead/lag term with a very small lag time constant.

It is also important to note that, for commercial controllers configured in the form shown in Eq. (15.7), the derivative term "interacts" with the other two terms because they multiply each other; whereas for the ideal PID in Eq. (14.16) the integral and derivative terms are distinct, and do not interact. Another problem with the ideal PID controller of Eq. (14.16) is the fact that when a step change is made in the set-point, the derivative action makes a sudden change in controller output — called "derivative kick." To avoid this kick, commercial controllers often replace the derivative term, $d\varepsilon/dt$, with dy_m/dt.

15.2 CONTROLLER DESIGN PRINCIPLES

The objective of maintaining the output of a process at the desired set-point value can be carried out by employing any of the feedback controllers we introduced in Chapter 14. As we recall from the equations governing the operation of these controllers, each one of them will respond differently to the same error information, $\varepsilon(t)$. The first issue therefore has to do with which controller type (i.e., P, PI, PID, or PD) to use for which process.

The next design issue is to choose the magnitude of the controller parameters K_c, τ_I, and τ_D, so as to achieve the best possible overall closed-loop system response. However, to evaluate which parameter set is "best," we must consider specific criteria for judging overall control system performance, as this will now enable objective discrimination among competing, alternative controller parameter combinations.

The controller design problem is thus seen to boil down essentially to the following task:

> *Choose the controller type, and the parameters for the chosen controller, so that some objective criteria of closed-loop performance are satisfied.*

It is thus necessary to decide first by which criteria the closed-loop system performance will be judged. This often provides sufficient information to make the controller type decision. The final step, and from a practical viewpoint, the most important, will be that of choosing the controller parameters so that the system response resulting from this choice meets the specified performance criteria as closely as possible.

Let us consider some standardized performance criteria.

15.2.1 Performance Criteria

Even though specific details of what is considered acceptable performance will vary from individual to individual, there are some general principles that apply universally. An effective closed-loop system, for example, is expected to be *stable*, and to be capable of causing the system output ultimately to attain its desired set-point value. In addition, the approach of this system output to the desired set-point should neither be too sluggish, nor too oscillatory.

A careful examination of what we have just said will reveal three types of criteria by which closed-loop system performance may be assessed in general:

1. Stability Criteria
2. Steady-State Criteria
3. Dynamic Response Criteria

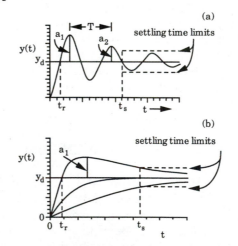

Figure 15.2. Some possible closed-loop responses to a set-point change: (a) oscillatory response with overshoot; (b) some nonoscillatory responses.

Of these, the first two are very easy to specify: the only reasonable specification is that the system must be stable; also, the key steady-state criterion is that there be little or no steady-state offset (i.e., the error ε is brought close to zero at steady state). The third class of criteria specifies *how* the system responds under closed-loop control. As indicated below, these criteria must be chosen according to the problem at hand.

Figure 15.2 illustrates several types of closed-loop responses that can arise depending on the process being controlled, choice of controller type, and the controller parameters selected. Figure 15.2(a) shows an underdamped oscillatory response, while 15.2(b) illustrates several possible types of nonoscillatory responses. The control system designer must decide which type of response is the "best" for any particular problem. To help in this choice, some precise mathematical criteria may be specified. Some of these are:

- Minimum rise time t_r
- Minimum settling time t_s
- Specified maximum overshoot, $a_1 < a_{1spec}$
- For oscillatory responses, a specified maximum decay ratio, $a_2/a_1 < d_r$, to ensure sufficiently rapid settling of oscillations

Obviously there may be some trade-offs with these criteria. Let us use an example to illustrate.

Example 15.1 THE EFFECT OF CONTROLLER TUNING FOR THE THREE WATER TANK SYSTEM.

Let us recall the three water tank system of Section 11.5 which had the open-loop transfer function relating the level in tank 3 to the feedrate of tank 1 as:

$$g(s) = \frac{6}{(2s + 1)(4s + 1)(6s + 1)} \tag{15.8}$$

If one uses a PID controller to control this system, the closed-loop set-point response is given by:

$$y(s) = \frac{6K_c(1 + \dfrac{1}{\tau_I s} + \tau_D s)}{6K_c (1 + \dfrac{1}{\tau_I s} + \tau_D s) + (2s + 1)(4s + 1)(6s + 1)} y_d \tag{15.9}$$

For only PI action determine the effect of tuning parameters K_c, τ_I on the closed loop response.

Solution:

Recalling that derivative action is not recommended for level control, let us set $\tau_D = 0$ and rearrange Eq. (15.9) to:

$$y(s) = \frac{6K_c (\tau_I s + 1)}{6K_c (\tau_I s + 1) + \tau_I s (2s + 1)(4s + 1)(6s + 1)} y_d \tag{15.10}$$

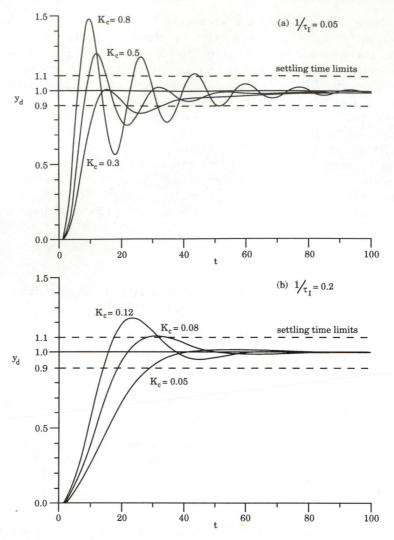

Figure 15.3. Closed-loop responses of the three-tank system under PI control.

The effect of τ_I, K_c on each of the control system performance criteria for this process can be seen in Figure 15.3. Note that for $1/\tau_I$ small, one must use higher controller gains to eliminate offset in a timely fashion. This provides short rise times, but poor decay ratio performance (poorly damped oscillations) and significant overshoot. For larger values of $1/\tau_I$, smaller controller gains may be used, and this results in larger rise times, but less overshoot. It is generally difficult to have very short rise times and small overshoot. However, if settling time was the only criterion (e.g., time to stay with ± 10% of the set-point change), then a wide range of controller parameters with different types of responses would provide settling times of about 30 minutes.

There are also "time-integral" performance criteria that can be used to optimize the choice of controller parameters. The most common of these involve minimizing time-integral functions of $\varepsilon(t) = y_d - y$; for example, the minimization of the following:

1. **Integral Absolute Error (IAE)**

$$IAE = \int_0^\infty |\varepsilon(t)|\, dt \qquad (15.11a)$$

2. **Integral Squared Error (ISE)**

$$ISE = \int_0^\infty \varepsilon^2(t)\, dt \qquad (15.11b)$$

which selectively penalizes large errors.

3. **Integral Time-weighted Absolute Error (ITAE)**

$$ITAE = \int_0^\infty t\, |\varepsilon(t)|\, dt \qquad (15.11c)$$

which more heavily penalizes errors at long times.

4. **Integral Time-weighted Squared Error (ITSE)**

$$ITSE = \int_0^\infty t\, \varepsilon^2(t)\, dt \qquad (15.11d)$$

which more heavily penalizes large errors at long times.
Typical responses resulting from applying these criteria will be illustrated with an example in the next section.

15.2.2 The Controller-Type Decision

Even though we could advance various reasons and criteria for selecting which controller type will be adequate for which application, it is generally agreed that selections made on the basis of the general characteristics of the different feedback controllers are the most practical.

Based on detailed theoretical analyses of closed-loop transient responses (as illustrated in Section 14.4 of Chapter 14), the following is a summary of the most salient characteristics of the classical feedback controllers:

1. **Proportional Controller:** Accelerates the control system response but leaves a *nonzero steady-state offset* for all processes except for the pure capacity process.

2. **Proportional + Integral Controller:** Eliminates offsets but the system response becomes more oscillatory; the added integral action tends to increase the propensity towards instability as K_c increases.

3. **Proportional + Derivative Controller:** Enjoys the *anticipatory* and *stabilizing* effect of derivative action but still leaves a nonzero steady-state offset except for the pure capacity process.

4. **Proportional + Integral + Derivative Controller:** Integral action eliminates offsets, and the oscillations normally introduced as a result may be curbed somewhat by the derivative action; the presence of derivative action tends to amplify the noise components in noisy signals.

In light of these characteristics, the following guidelines may be used in selecting the most suitable controller type:

1. When steady-state offsets are unimportant and can therefore be tolerated, or when the process possesses a natural integrator (as is the case with pure capacity systems), use a P controller. Many liquid level control loops, for example, are on P control.

2. When offsets cannot be tolerated, use a PI controller. A large proportion of feedback controllers in a typical plant are of the PI type.

3. When it is important to compensate for some natural sluggishness in the overall system, and the process signals are relatively noise-free, use a PID controller. For example, temperature control loops are sometimes under PID control; the effect of the "lag" usually introduced by the measuring devices are compensated for by the derivative action. In contrast, flow loops are seldom on PID control; the signals are more susceptible to noise, and the processes are usually not sluggish.

4. PD controllers are seldom used; however, when used, they make it possible for the control system to withstand higher controller gain values and still remain stable, so that smaller steady-state offsets are therefore achievable with PD controllers than are normally possible with P controllers.

15.2.3 Choosing Controller Parameters

This crucial aspect of the overall scheme of controller design is also known as "controller tuning," since, in effect, it involves adjusting feedback controller parameters in order that the performance objective specified for the closed-loop system is achieved.

The process of selecting controller parameters (whether from "scratch," or by adjusting existing controller settings) can be carried out in several ways and to varying degrees of sophistication. However, one subtle fact remains: all controller tuning methods are based, explicitly or implicitly, on some form of *a priori* process characterization. By this we mean that even the "instrument mechanic" who is well accustomed to tuning the controller "by feel" is obviously drawing upon his experience; in reality, an informal, implicit, and often fairly accurate, characterization of the process.

We will thus consider controller tuning in two categories: when explicit process models are available, and when they are not. In the following three sections we consider controller tuning with various types of process models. In Section 15.6 we consider the case when models are not available.

15.3 CONTROLLER TUNING WITH FUNDAMENTAL PROCESS MODELS

A number of controller tuning strategies can be employed when a fundamental process model is available. These can be classified into several types that we now examine:

15.3.1 Optimization of Time-Integral Criteria

It is possible, given any arbitrary process model, to use standard computer programs (e.g., the expert system, CONSYDEX [2]) to find the optimal tuning parameters K_c, τ_I, τ_D for any mathematical objective such as shown in Eq. (15.11). The procedure simply involves optimizing over the controller tuning parameters in order to minimize the chosen error criterion. To illustrate, the following example shows the results of this type of design for a three-tank system under feedback level control.

Example 15.2 **CONTROLLER DESIGN BASED ON TIME-INTEGRAL OBJECTIVES.**

For the three-tank system of Example 15.1, determine the optimal PI controller settings and the optimal closed-loop responses for each of the integral-time criteria in Eq. (15.11).

Solution:

By using the expert system program CONSYDEX to perform the required optimization, the following parameters were determined to minimize the indicated time-integral criteria:

IAE: $K_c = 0.314$; $1/\tau_I = 0.769$;
ISE: $K_c = 0.537$; $1/\tau_I = 0.0432$;
ITAE: $K_c = 0.207$; $1/\tau_I = 0.0996$;
ITSE: $K_c = 0.343$; $1/\tau_I = 0.0708$;

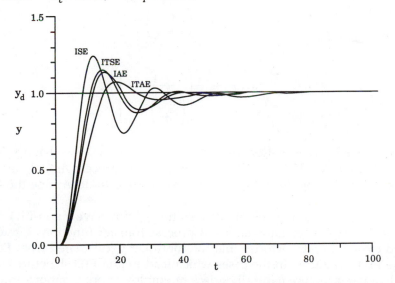

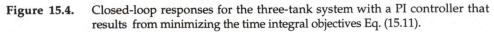

Figure 15.4. Closed-loop responses for the three-tank system with a PI controller that results from minimizing the time integral objectives Eq. (15.11).

The various closed-loop responses obtained using these settings are shown in Figure 15.4. The results indicate that ISE and ITSE produce shorter rise times than IAE and ITAE but larger overshoot. Also the time weighted objectives ITSE and ITAE have longer rise times because deviations at short times are not penalized heavily. All criteria provide rather good settling times.

15.3.2 Model Following PID Controller Designs

There are a number of model following tuning procedures that work on the principle of finding the PID controller parameters that cause the actual closed-loop system to behave in a prescribed fashion. In some cases, what is prescribed is an entire closed-loop trajectory which the actual process is required to track; in others, it is the location of the closed-loop poles, rather than the entire trajectory, which is prescribed.

Direct Synthesis Tuning

The direct synthesis approach (to be discussed fully in Chapter 19) seeks to find the feedback controller g_c required to produce a prespecified closed-loop response. Recall that the closed-loop transfer function for a process under feedback control is:

$$y(s) = \frac{gg_c}{1 + gg_c h} y_d(s) \tag{15.12a}$$

if we assume for the moment that $h = 1$, and require that the closed-loop response follow a desired trajectory represented by the transfer function $q(s)$, i.e.:

$$y(s) = q(s) y_d(s) \tag{15.12b}$$

then as is shown in Chapter 19, by equating Eqs. (15.12a) and (15.12b) we obtain:

$$g_c = \frac{1}{g}\left(\frac{q}{1-q}\right)$$

or

$$g_c = \frac{1}{g}\left[\frac{1}{\left(\dfrac{1}{q}\right) - 1}\right] \tag{15.13}$$

as the feedback controller that will result in a closed-loop behavior represented by Eq. (15.12b). Thus, given $g(s)$, the process model, and $q(s)$ the specified desired trajectory, Eq. (15.13) may be used to determine the feedback controller g_c.

We note here that in general, the resulting g_c may have a non-PID structure; in fact it is only for certain classes of process transfer functions g and certain desired closed-loop trajectories that the controller g_c calculated from Eq. (15.13) has the PID structure. In the cases which lead to non-PID structures, one may use whatever structure naturally arises or employ approximations to bring the controller into the PID form (cf. Chapter 19).

For the purpose of illustration, suppose the desired approach to a new setpoint is modeled by a reference trajectory:

$$\frac{y(s)}{y_d(s)} = q(s) = \frac{1}{\tau_r^2 s^2 + 2\tau_r \zeta_r s + 1} \tag{15.14}$$

let us find the controller tuning parameters that bring the actual process response Eq. (15.12a) as close as possible to the specified response in Eq. (15.14).

Recall from Chapter 6 that Eq. (15.14) is the general model for a second-order system that can have either overdamped (no overshoot) or underdamped response (cf. Figure 6.8). Thus Eq. (15.14) can approximate most of the actual closed-loop responses possible.

Let us now illustrate the design of direct synthesis controllers using as an example a third-order process with $g(s)$ of the form:

$$g(s) = \frac{K}{(\tau_1 s + 1)(\tau_2 s + 1)(\tau_3 s + 1)} \tag{15.15}$$

then Eq. (15.13) takes the form:

$$g_c(s) = \frac{(\tau_1 s + 1)(\tau_2 s + 1)(\tau_3 s + 1)}{K\tau_r s (\tau_r s + 2\zeta_r)}$$

which can be rearranged to:

$$g_c(s) = \frac{(\tau_1 s + 1)(\tau_2 s + 1)(\tau_3 s + 1)}{\beta s (\phi s + 1)} \tag{15.16}$$

where

$$\phi = \frac{\tau_r}{2\zeta_r}$$
$$\beta = 2K \tau_r \zeta_r = 4K(\zeta_r)^2 \phi \tag{15.17}$$

Now if we choose ϕ to be equal to one of the time constants (say, τ_3), we obtain:

$$g_c(s) = \frac{(\tau_1 s + 1)(\tau_2 s + 1)}{\beta s} \tag{15.18}$$

which when compared to the PID controller transfer function:

$$g_c(s) = K_c \left(1 + \frac{1}{\tau_I s} + \tau_D s\right) = \frac{K_c(1 + \tau_I s + \tau_I \tau_D s^2)}{\tau_I s} \tag{15.19}$$

yields the PID tuning parameters:

$$K_c = \frac{(\tau_1 + \tau_2)}{\beta}$$
$$\tau_I = (\tau_1 + \tau_2) \tag{15.20}$$
$$\tau_D = \left(\frac{\tau_1 \tau_2}{\tau_1 + \tau_2}\right)$$

Using the same arguments, if $g(s)$ were a second-order system:

$$g(s) = \frac{K}{(\tau_1 s + 1)(\tau_2 s + 1)} \qquad (15.21)$$

then selecting $\phi = \tau_2$ yields:

$$g_c(s) = \frac{(\tau_1 s + 1)}{\beta s} = \frac{\tau_1}{\beta}\left(1 + \frac{1}{\tau_1 s}\right) \qquad (15.22a)$$

which, when compared with:

$$g_c(s) = K_c\left(1 + \frac{1}{\tau_I s}\right) \qquad (15.22b)$$

shows that the controller would be a PI controller with:

$$K_c = \frac{\tau_1}{\beta}$$

$$\tau_I = \tau_1 \qquad (15.23)$$

Let us illustrate the tuning procedure with a concrete example.

Example 15.3   DIRECT SYNTHESIS CONTROLLER TUNING FOR THE MULTIPLE-TANK PROCESS.

Use the direct synthesis tuning procedure to obtain PID controller parameters for the three-tank system of Example 15.1. Recall that for this system:

$$g(s) = \frac{6}{(2s + 1)(4s + 1)(6s + 1)} \qquad (14.73)$$

Assume that a reasonable closed-loop response given by Eq. (15.14) is desired. Repeat the design for the situation in which there are only two tanks so that:

$$g(s) = \frac{6}{(4s + 1)(6s + 1)} \qquad (15.24)$$

Solution:

Using Eq. (15.20) for the tuning parameters, let us find a good reference trajectory for which ϕ defined by Eq. (15.17) is close to one of the three time constants in Eq. (14.73). Let us choose $\tau_r = 4.0$, $\zeta_r = 0.5$ in the reference trajectory Eq. (15.14) to yield:

$$\frac{y(s)}{y_d} = q(s) = \frac{1}{16s^2 + 4s + 1} \qquad (15.25)$$

which produces the underdamped trajectory shown in Figure 15.5. Note that from Eq. (15.17) this yields $\phi = 4$, $\beta = 24$. In this case $\tau_3 = 4.0$ is canceled exactly, and the controller tuning parameters from Eq. (15.20) become $K_c = 0.333$, $1/\tau_I = 0.125$, $\tau_D = 1.50$ for this PID controller. The desired reference trajectory given in Eq. (15.25) is achieved exactly as illustrated in Figure 15.5.

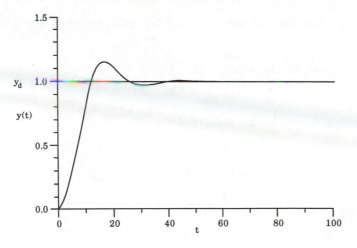

Figure 15.5.   Results of direct synthesis tuning of the three-tank (PID) and two-tank (PI) processes. The reference and feedback trajectories are identical here as are the resulting trajectories for the two problems.

If we apply the same type of design to the two-tank system with the model given by Eq. (15.24), we see that we may choose $\tau_r = 4.0$, $\zeta_r = 0.5$ again yielding the reference trajectory in Eq. (15.25) with $\phi = 4$, $\beta = 24$. The resulting PI controller parameters are now given by $K_c = 0.250$, $1/\tau_I = 0.1667$. With this controller, the desired reference trajectory is achieved exactly — this time by a PI controller for the two-tank system. Note that because we chose the same reference trajectory for each problem, the closed-loop responses for both problems are identical and shown in Figure 15.5.

IMC Tuning

Rivera *et al.* [4] use the ideas of Internal Model Control (IMC), a control strategy discussed in Chapter 19, to produce equivalent PID tuning rules. Again, as with the direct synthesis approach, there are only certain classes of process transfer functions g and desired reference trajectories (y/y_d) that lead to the PID structure and for which IMC tuning can be directly applied. In addition, their controller structure requires a damping filter multiplying the PID controller. Later, in Section 15.4, we will present the IMC-based PID tuning rules for a class of approximate models.

Pole Placement Designs

Another way of specifying desired closed-loop behavior is to require that the closed-loop poles of the system be placed at specified locations. Recall that the closed-loop poles of a feedback control system are the roots of the characteristic equation:

$$1 + g g_c h = 0 \tag{15.26}$$

If we replace g_c by a PID controller, we may then select the tuning parameters in order to cause the roots of Eq. (15.26) to be located where we wish. Let us illustrate for the case of a first-order system:

$$g(s) = \frac{K}{(\tau s + 1)} \tag{15.27}$$

when $h = 1$. Substituting for g_c from Eq. (15.19) gives the characteristic equation:

$$\tau_I \frac{(KK_c\tau_D + \tau)}{KK_c} s^2 + \frac{(1 + KK_c)}{KK_c} \tau_I s + 1 = 0 \qquad (15.28)$$

The two closed-loop poles are given by:

$$r_1, r_2 = \frac{-(1 + KK_c)}{2(KK_c\tau_D + \tau)}$$

$$\pm \left[\frac{KK_c}{2(KK_c\tau_D + \tau)\tau_I} \right] \sqrt{\left[\frac{(1 + KK_c)\tau_I}{KK_c} \right]^2 - \frac{4(KK_c\tau_D + \tau)\tau_I}{KK_c}} \qquad (15.29)$$

and by adjusting K_c, τ_I, τ_D we can locate r_1, r_2 wherever we want in the complex plane.

One difficulty with pole placement designs is that one must know where to place the poles to obtain good closed-loop response. Thus it might be advisable to use a reference trajectory such as Eq. (15.14) as a guide in the choice of desired pole locations. For example, if we wished to have the same poles as in the response shown in Figure 15.5, we can require the closed-loop response poles to be the same as the reference trajectory in Eq. (15.25). Comparing Eq. (15.28) with Eq. (15.25) requires the adjustment of K_c, τ_I, τ_D to achieve:

$$\frac{\tau_I (KK_c\tau_D + \tau)}{KK_c} = 16$$

$$\frac{(1 + KK_c)\tau_I}{KK_c} = 4$$

Note however, that we do not achieve the same closed-loop response as shown in Figure 15.5 because with the PID controller, the actual closed-loop transfer function contains poles as well as zeros, *but only the poles are chosen in pole placement design*; the effect of the zeros on the actual closed-loop response function are not taken into consideration.

It should be clear that pole placement designs do allow good closed-loop response, but require some experience in order to decide which pole locations are the best for any particular problem.

15.3.3 Use of Stability Margins

A less ambitious, but often effective, means of controller tuning is to use the process model to determine the values of controller parameters that bring the closed-loop process to the verge of instability (for example, by simulation, or by using the Routh criterion or the direct substitution methods of Chapter 14). Rules of thumb are then used to reduce the controller gain, to provide a margin of safety away from closed-loop instability. The specific values of K_c, τ_I, τ_D obviously depend on the specific process model; however, Ziegler and Nichols in their classic paper [3], suggest approximate general tuning rules based on a knowledge of the stability limit and a requirement that the controller provide a *quarter decay ratio*; i.e., the oscillation amplitudes in Figure 15.2 dampen such that $a_2/a_1 < 0.25$. Their rules are quite simple:

1. Using proportional-only control, determine the value of controller gain, $K_c = K_{cu}$ (ultimate controller gain) which causes the system to be at the verge of instability. Denote the period of the sustained oscillation at this controller gain as P_u (ultimate period of oscillation).

2. To provide good controller tuning parameters for each type of controller, use the relations in Table 15.1.

Table 15.1
Ziegler-Nichols Stability Margin Controller Tuning Parameters, Ref. [3]

Controller Type	K_c	τ_I	τ_D
P	$0.5K_{cu}$	—	—
PI	$0.45K_{cu}$	$P_u/1.2$	—
PID	$0.6K_{cu}$	$P_u/2$	$P_u/8$

Let us illustrate this procedure with an example.

Example 15.4 ZIEGLER-NICHOLS TUNING USING A FUNDAMENTAL PROCESS MODEL.

Design P, PI, and PID controllers for the three-tank system of Example 15.1 using Ziegler-Nichols tuning rules.

Solution:

From Example 14.9 we see from direct substitution that the proportional controller gain which brings the closed-loop response of the system to the verge of instability is:

$$K_{cu} = 5/3 = 1.667$$

and the frequency of sustained oscillation at that point is:

$$\omega_c = 0.5 \text{ rad/min}$$

From this the ultimate period is obtained as:

$$P_u = \frac{2\pi}{\omega_c} = 4\pi = 12.56$$

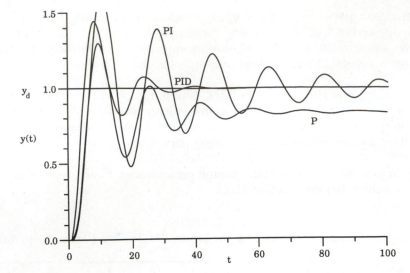

Figure 15.6.   Closed-loop response of the three-tank process under P, PI, and PID control
using Ziegler-Nichols settings.

Thus the Ziegler-Nichols settings for the various controllers are then:

P: $K_c = 0.8333$
PI: $K_c = 0.75, 1/\tau_I = 0.0955$
PID: $K_c = 1.0, 1/\tau_I = 0.159, \tau_D = 1.57$

The response to a set-point change for each type of controller is seen in Figure 15.6.
Note that the P controller has some offset while the PI controller has large overshoot
and slowly damped oscillations. By contrast, the PID controller provides a rather
good response except that there is about 40 percent overshoot. Clearly, for the PI
controller, the earlier, more rigorous controller designs are much better than the
Ziegler-Nichols designs based on rules of thumb.

15.4   CONTROLLER TUNING USING APPROXIMATE PROCESS MODELS

As indicated in Chapters 12–13, rarely is a fundamental, first-principles model
available for most processes. It is more usual for approximate process models to
be constructed from available process data. It is, however, important to stress
that when the model is needed strictly for controller tuning and for nothing
else, the modeling objectives differ drastically from those espoused in
Chapter 13.

The objective in Chapter 13 was to construct models that purport to explain
the correlation between the input and output data for the express purpose of
predicting process behavior, particularly over the operating region covered by
the data. For this reason, care must be taken in postulating candidate models
and in estimating the parameters. While such models can be useful for
controller tuning, if they are readily available, such detail is not absolutely
crucial to the success of controller tuning. Constructing approximate models *for
the sole purpose of feedback controller tuning* requires only very rough estimates
of a relatively small number of parameters in a very simple model; none of the

sophisticated identification techniques of Chapter 13 are necessary in this case.

For feedback controller tuning purposes, the approximate model most commonly employed has the first-order-plus-time-delay transfer function:

$$g = \frac{Ke^{-\alpha s}}{\tau s + 1} \tag{15.30}$$

This model may be obtained by many procedures, but in the interest of simplicity, a particularly popular procedure utilizes open-loop step-response data. This involves making a step change in the input signal to the process in order to elicit a response in the measured variable y_m; a record of this response with time is known as the *process reaction curve*. It must be noted here that the observed response is due to the combined dynamics of the final control element, the process itself, and the measuring device; the approximate model will therefore be for this combined process ensemble.

There are two aspects to the tuning techniques based on these approximate models:

1. Determining the approximate model parameters that best characterize the process response, and
2. Using these parameters in some specified formulas to obtain recommendations of PID controller parameter values.

We will consider each aspect in order.

15.4.1 Process Characterization

Cohen and Coon [5] first noted that the process reaction curve for most controlled processes may be reasonably well approximated by the step response of a first-order system with time delay (see Figure 15.7), requiring only the three parameters in Eq. (15.30) to characterize the process. All that need be determined are the values of K, τ, and α which give a first-order-plus-time-delay step response that most closely approximates the actual process reaction curve. In other words, the process is characterized by a *steady-state gain*, an *effective time constant*, and an *effective time delay*.

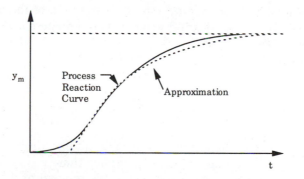

Figure 15.7. The process reaction curve and the first-order-plus-time-delay approximation.

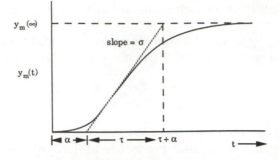

Figure 15.8.   Fitting model parameters to the process reaction curve.

The following is a simple procedure for obtaining the approximate model Eq. (15.30) from the process reaction curve:

1. *Obtain the process reaction curve*: Disconnect the controller from the controlled process; implement a step change of magnitude A in the input to the controlled process via the final control element (e.g., open or close the control valve by a certain fixed amount). Record the resulting transient response of y_m, the measured output; this gives the process reaction curve, an example of which is shown in Figure 15.8.

2. Estimate the characterizing parameters.

   (a) *Estimate the effective time delay,* α:
       Draw a tangent line at the inflexion point on the process reaction curve, where this line intersects the time axis gives an estimate of the effective time delay, α (see Figure 15.8).

   (b) *Estimate the effective gain,* K.
       By definition of the steady-state gain:

   $$K = \frac{\text{ultimate value of the output}}{\text{magnitude of the step input}}$$

   In this case, from Figure 15.8, we have:

   $$K = \frac{y_m(\infty)}{A} \qquad\qquad (15.31)$$

   It must be kept in mind that y_m is in terms of the output deviation from the initial starting value; it is not an *absolute* measurement of the process output.

   (c) *Estimate the effective time constant,* τ.
       A useful but rough estimate of the time constant τ is obtained as:

   $$\tau = \frac{y_m(\infty)}{\sigma} \qquad\qquad (15.32)$$

   where σ is the slope of the tangent line drawn at the inflexion point. Note that Eq. (15.32) implies that $(\tau + \alpha)$ can be picked off the time axis at the point where the tangent at the inflection point intersects the ultimate output value, $y_m(\infty)$.

We now illustrate this procedure with an example.

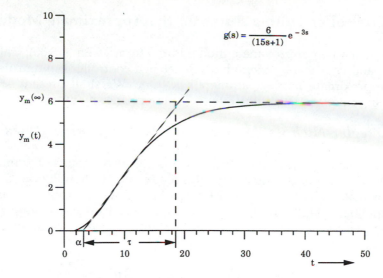

Figure 15.9. Process reaction curve for the three-tank system and the graphically determined approximate model.

Example 15.5   FITTING PARAMETERS TO THE PROCESS REACTION CURVE FOR THE THREE-TANK SYSTEM.

Let us apply this approximate modeling procedure to the three-tank system of Example 15.1 (which we recall has the actual third-order transfer function given by Eq. (14.73)). If we introduce a unit step change ($A = 1$) in the process input, the resulting process reaction curve is given in Figure 15.9. Applying the graphical procedure outlined above gives the approximate model:

$$g(s) \approx \frac{6e^{-3s}}{(15s + 1)} \tag{15.33}$$

Figure 15.10 shows a comparison of the responses of the exact and the approximate models to a unit step input for the three-tank system.

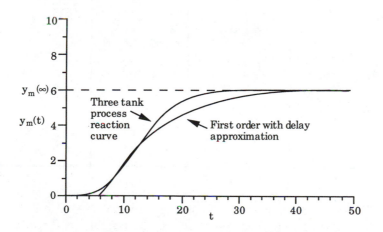

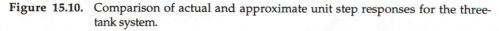

Figure 15.10. Comparison of actual and approximate unit step responses for the three-tank system.

15.4.2 Controller Tuning Rules for the Approximate Model

Having obtained an approximate model in the form of Eq. (15.30), Cohen and Coon [5] (and later many others) have developed controller tuning rules based on the approximate model parameters K, τ, α. We shall present several of these tuning rules and then illustrate their performance by example.

Ziegler-Nichols Approximate Model Tuning Rules

In their classic paper, Ziegler and Nichols [3] also proposed tuning rules designed to provide a quarter decay ratio (i.e., $a_2/a_1 \leq 0.25$ in Figure 15.2) when an approximate model of the form Eq. (15.30) is known. Table 15.2 provides their guidelines. Smith and Corripio [6] point out that these tuning rules are very sensitive to the time delay/time constant ratio; they recommend these rules be limited to models with $0.1 < (\alpha/\tau) < 1.0$.

Cohen-Coon Tuning Rules

Cohen and Coon [5] also developed tuning rules based on the concept of achieving a quarter decay ratio. However, because this criterion does not produce a unique set of tuning parameters, the Cohen-Coon rules given in Table 15.3 are somewhat different from those of Ziegler-Nichols. Again, these should be applied only for a limited range $0.1 < (\alpha/\tau) < 1.0$.

Table 15.2
Ziegler-Nichols Approximate Model PID Tuning Rules [3]
(as reinterpreted by Smith and Corripio [6])

Controller Type	K_c	τ_I	τ_D
P	$\dfrac{1}{K}\left(\dfrac{\tau}{\alpha}\right)$	—	—
PI	$\dfrac{0.9}{K}\left(\dfrac{\tau}{\alpha}\right)$	3.33α	—
PID	$\dfrac{1.2}{K}\left(\dfrac{\tau}{\alpha}\right)$	2.0α	0.5α

Table 15.3
Cohen-Coon Approximate Model PID Tuning Rules [5]

Controller Type	K_c	τ_I	τ_D
P	$\dfrac{1}{K}\left(\dfrac{\tau}{\alpha}\right)\left[1 + \dfrac{1}{3}\left(\dfrac{\alpha}{\tau}\right)\right]$	—	—
PI	$\dfrac{1}{K}\left(\dfrac{\tau}{\alpha}\right)\left[0.9 + \dfrac{1}{12}\left(\dfrac{\alpha}{\tau}\right)\right]$	$\alpha\left[\dfrac{30 + 3\left(\dfrac{\alpha}{\tau}\right)}{9 + 20\left(\dfrac{\alpha}{\tau}\right)}\right]$	—
PD	$\dfrac{1}{K}\left(\dfrac{\tau}{\alpha}\right)\left[\dfrac{5}{4} + \dfrac{1}{6}\left(\dfrac{\alpha}{\tau}\right)\right]$	—	$\alpha\left[\dfrac{6 - 2\left(\dfrac{\alpha}{\tau}\right)}{22 + 3\left(\dfrac{\alpha}{\tau}\right)}\right]$
PID	$\dfrac{1}{K}\left(\dfrac{\tau}{\alpha}\right)\left[\dfrac{4}{3} + \dfrac{1}{4}\left(\dfrac{\alpha}{\tau}\right)\right]$	$\alpha\left[\dfrac{32 + 6\left(\dfrac{\alpha}{\tau}\right)}{13 + 8\left(\dfrac{\alpha}{\tau}\right)}\right]$	$\alpha\left[\dfrac{4}{11 + 2\left(\dfrac{\alpha}{\tau}\right)}\right]$

Time-Integral Tuning Rules

Smith, Murrill and coworkers [7–9] have developed correlations for PID tuning parameters which minimize integral-time objectives such as IAE, ISE, ITAE, and ITSE. Their correlations distinguish between set-point changes and disturbance-rejection responses. A rather complete set of results are summarized in Ref. [6]; however, we will choose to present only the ITAE results in Table 15.4 because these are often the best solution in a practical sense (e.g., see Figure 15.4).
Again, these should only be applied in the approximate range $0.1 < (\alpha/\tau) < 1.0$.

It is particularly interesting to note now how these different tuning rules based on somewhat different criteria *all* depend on the time-constant–to–time-delay ratio. It is therefore easy to see why many practical industrial control engineers consider this ratio to be a very crucial process parameter when it comes to tuning feedback controllers.

Table 15.4
Minimum ITAE Approximate Model Controller Tuning Rules [7–9]

Controller Type	Type of Response	K_c	τ_I	τ_D
P	Disturb.	$\dfrac{0.49}{K}\left(\dfrac{\tau}{\alpha}\right)^{1.084}$	—	—
PI	Set-point	$\dfrac{0.586}{K}\left(\dfrac{\tau}{\alpha}\right)^{0.916}$	$\dfrac{\tau}{\left[1.03 - 0.165\left(\dfrac{\alpha}{\tau}\right)\right]}$	—
PI	Disturb.	$\dfrac{0.859}{K}\left(\dfrac{\tau}{\alpha}\right)^{0.977}$	$\dfrac{\tau}{0.674}\left(\dfrac{\alpha}{\tau}\right)^{0.680}$	—
PID	Set-point	$\dfrac{0.965}{K}\left(\dfrac{\tau}{\alpha}\right)^{0.855}$	$\dfrac{\tau}{\left[0.796 - 0.147\left(\dfrac{\alpha}{\tau}\right)\right]}$	$0.308\tau\left(\dfrac{\alpha}{\tau}\right)^{0.929}$
PID	Disturb.	$\dfrac{1.357}{K}\left(\dfrac{\tau}{\alpha}\right)^{0.947}$	$\dfrac{\tau}{0.842}\left(\dfrac{\alpha}{\tau}\right)^{0.738}$	$0.381\tau\left(\dfrac{\alpha}{\tau}\right)^{0.995}$

Direct Synthesis Approximate Model Tuning Rules

As shown in Chapter 19, given an approximate model as in Eq. (15.30), it is possible to develop tuning rules for PID controllers based upon requiring a closed-loop trajectory $q(s)$ which is first-order-plus-time-delay with a time constant indicated by the parameter τ_r. This determines the desired speed of the closed-loop response, so that for small τ_r, the controller will respond more quickly than for larger τ_r. The tuning rules are given in Table 15.5.

Table 15.5
Direct Synthesis Approximate Model Controller Tuning Rules (cf. Chapter 19)

Controller Type	K_c	τ_I	τ_D	Additional First-Order Filter Parameter $\theta*$
PI	$\dfrac{\tau}{K(\tau_r + \alpha)}$	τ	—	—
PID	$\dfrac{2\tau + \alpha}{2K(\tau_r + \alpha)}$	$\tau + \dfrac{\alpha}{2}$	$\dfrac{\tau\alpha}{2\tau + \alpha}$	$\dfrac{\alpha\tau_r}{2(\tau_r + \alpha)}$

IMC Approximate Model Tuning Rules

As noted in Section 15.3, Rivera and Morari *et al.* [4, 10] have proposed tuning rules for PID controllers based on the Internal Model Control (IMC) methodology. In addition to the model parameters, the designer must select a "filter parameter," λ, which amounts to the desired closed-loop time constant for a first-order response. Thus for λ small, the controller will respond more quickly than if λ is larger. The tuning rules are given in Table 15.6.

Table 15.6
IMC Approximate Model Controller Tuning Rules [4,10]

Controller Type	K_c	τ_I	τ_D	Recommended Choice of λ ($\lambda > 0.2\ \tau$ Always)
PI	$\dfrac{\tau}{\lambda K}$	τ	—	$\dfrac{\lambda}{\alpha} > 1.7$
"Improved" PI	$\dfrac{2\tau + \alpha}{2\lambda K}$	$\tau + \dfrac{\alpha}{2}$	—	$\dfrac{\lambda}{\alpha} > 1.7$
PID	$\dfrac{2\tau + \alpha}{2K(\lambda + \alpha)}$	$\tau + \dfrac{\alpha}{2}$	$\dfrac{\tau\alpha}{2\tau + \alpha}$	$\dfrac{\lambda}{\alpha} > 0.25$

Note the similarity between the direct synthesis tuning rules and those based on IMC. We shall have more to say about this in Chapter 19.

A number of other approximate model tuning rules are available, but the ones presented here are certainly sufficient to provide a wide choice. Let us now compare the performance of the various tuning rules with an example.

Example 15.6.   CONTROLLER TUNING USING AN APPROXIMATE PROCESS MODEL.

Consider again the three-tank process of Example 15.1 and now design PI and PID controllers based on the approximate model Eq. (15.33) determined from the process reaction curve in Example 15.5. Compare the responses to a set-point change using different tuning guidelines.

Solution:

Based on all of the tuning rules discussed above, the controller tuning parameters for PI and PID control are given in the following table (where we recall, for the approximate model $K = 6$, $\tau = 15$, $\alpha = 3$).

Tuning Rules	PI		PID		
	K_c	$\dfrac{1}{\tau_I}$	K_c	$\dfrac{1}{\tau_I}$	τ_D
Ziegler-Nichols (Table 15.2)	0.75	0.100	1.00	0.167	1.50
Cohen-Coon (Table 15.3)	0.764	0.142	1.15	0.147	1.05
ITAE (Table (15.4)	0.427	0.066	0.637	0.051	1.04
Direct Synthesis (Table 15.5) $\tau_r = 3$ for PI $\tau_r = 2$ for PID	0.417	0.067	0.55	0.061	1.36; ($\theta^* = 0.6$)
IMC (Table 15.6) $\lambda = 6$ for PI $\lambda = 2$ for PID	0.417	0.067	0.55	0.061	1.36
	0.458 (Improved PI)	0.061			

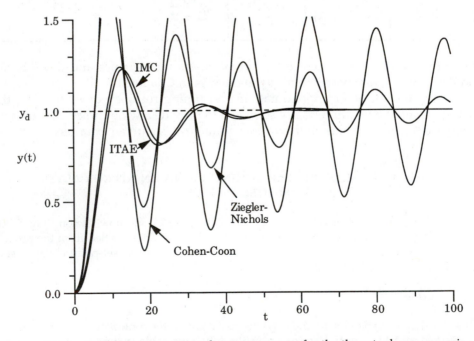

Figure 15.11. Closed-loop set-point change responses for the three-tank process using a PI controller designed with various controller tuning rules (the IMC and direct synthesis results are identical).

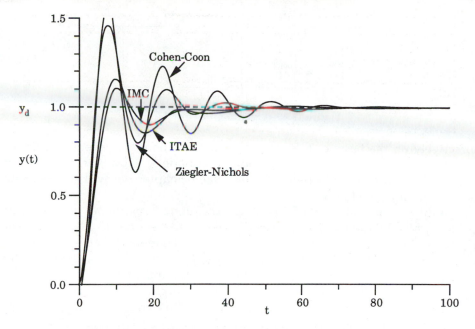

Figure 15.12. Closed-loop set-point change responses for the three-tank process using a PID controller designed with various tuning rules (the IMC and direct synthesis results are identical).

The set-point responses of the true third-order tank process for PI and PID controller designs using each tuning rule are shown in Figures 15.11 and 15.12 respectively. Note that for PI control, both Ziegler-Nichols and Cohen-Coon are much too close to the stability limit to provide good responses. The IMC, improved IMC, Direct Synthesis, and ITAE designs have nearly identical settings and provide nearly identical responses. (In fact the Direct Synthesis, and the IMC settings are, for both PI and PID, identical except for the additional filter time constant of 0.6 in the PID case which is so small as to be negligible when compared with the effective process time constant of 15.)

The PID responses, shown in Figure 15.12, have similar trends. The Ziegler-Nichols and Cohen-Coon settings are somewhat overly aggressive, but the ITAE, Direct Synthesis and IMC tunings are quite good.

Based on this example and other general experience, the ITAE and IMC/Direct synthesis settings appear to be the best choice with the approximate model approach.

15.5 CONTROLLER TUNING USING FREQUENCY-RESPONSE MODELS

As we saw in Chapter 13, process identification sometimes produces frequency-response models in the form (cf. Chapter 4):

$$y(j\omega) = g(j\omega)u(j\omega) + g_d(j\omega)d(j\omega) \tag{4.18}$$

where $g(j\omega)$ is a complex function having a real and a complex part:

$$g(j\omega) = \text{Re}(\omega) + j\,\text{Im}(\omega) \tag{4.41}$$

These transfer functions are often represented in the form of Bode or Nyquist plots (see Chapter 9). Thus it is useful to consider how to tune controllers

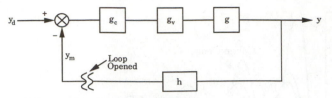

Figure 15.13. Opening the closed loop.

effectively on the basis of this class of models. There are several specialized tuning methods based on frequency-domain models. However, here we will focus on stability margin methods; in a later chapter we will discuss more advanced tuning methods.

Stability margin methods for frequency-response models are philosophically identical to those discussed in Section 15.3.3. The basic principle involves determining the controller parameters that will bring the closed-loop system to the verge of instability (i.e., when the closed-loop system is at the border between stability and instability) and then backing off on the tuning parameters to provide some stability margin. Therefore the Ziegler-Nichols stability margin tuning rules of Table 15.1 can also be applied in this case.

In order to apply stability margin methods, we need a means of determining the point of closed-loop instability from frequency-response models. Consider the feedback control system shown in Figure 15.13. If the feedback loop were to be opened after the measuring device, as indicated in the diagram, the response of y_m to changes in y_d will be given by:

$$y_m = g g_v g_c h \, y_d \tag{15.34}$$

or

$$y_m = g_L y_d$$

where

$$g_L = g g_v g_c h \tag{15.35}$$

Here, g_L is referred to as the open-loop transfer function for this closed-loop system because it relates the closed-loop system's input and output variables *when the loop is opened*. Now we recall that the closed-loop response is given by:

$$y = \frac{g g_v g_c}{1 + g g_v g_c h} y_d$$

which, from Eq. (15.35) may be rewritten as:

$$y = \frac{g g_v g_c}{1 + g_L} y_d \tag{15.36}$$

so that the closed-loop system stability is determined from the roots of the closed-loop system characteristic equation given by:

$$1 + g_L = 0 \tag{15.37}$$

Thus, g_L is seen to play a key role in determining the stability of a closed-loop system.

In the frequency domain, $g_L(j\omega)$ is given by:

$$g_L(j\omega) = g(j\omega) \, g_v(j\omega) \, g_c(j\omega) \, h(j\omega)$$ (15.38)

Let us now determine the closed-loop stability of the system by analyzing the characteristic Eq. (15.37) in the frequency domain using Eq. (15.38). There are two equivalent graphical criteria which can be readily applied — one based on a Nyquist plot, and the other based on a Bode plot.

15.5.1 Nyquist Stability Criterion

Recall from Chapter 9 that the Nyquist plot of any process is simply a plot of the transfer function in the complex plane. Thus the open-loop transfer function, g_L, defined by Eq. (15.38) can be written two ways (cf. Appendix B) in cartesian form:

$$g_L(j\omega) = \text{Re}(\omega) + j \, \text{Im}(\omega)$$ (15.39)

or in polar form:

$$g_L(j\omega) = \text{AR}(\omega) \, e^{-j\phi(\omega)}$$ (15.40)

where the amplitude ratio, AR, and phase angle, ϕ, are given by

$$\text{AR}(\omega) = |g_L(j\omega)| = \sqrt{\text{Re}(\omega)^2 + \text{Im}(\omega)^2}$$ (15.41)

$$\phi(\omega) = \angle g_L(j\omega) = \tan^{-1}\left[\frac{\text{Im}(\omega)}{\text{Re}(\omega)}\right]$$ (15.42)

If necessary, the reader may wish to refer back to Chapter 9 and Appendix B for more discussion of these relationships.

The relations (15.39) – (15.42) can then be represented in the complex plane by a Nyquist plot as shown in Figure 15.14.

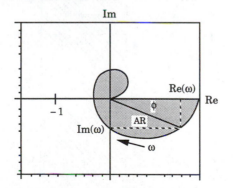

Figure 15.14.   A Nyquist plot of $g_L(j\omega)$.

Note that the curve traced by $g_L(j\omega)$ sweeps out a path as frequency ω is varied. As we recall from Chapter 9, depending on the order of the system, the curve rotates through 90° (first order), 180° (second order), 270° (third order), etc., or rotates an infinite number of times if there is a time delay. Note also that the AR (the distance of the curve from the origin) tends to decrease with increasing frequency. Also note that the AR is proportional to the controller gain because of the definitions Eqs. (15.38) and (15.41). For example, by doubling the controller gain we cause the Nyquist plot to have twice the value of the AR at every frequency.

The Nyquist Map of the Cartesian Complex Plane

To use the Nyquist plot in determining the closed-loop system stability, we must first carefully translate our current understanding of stability into the "new language" of the Nyquist diagram. Our current understanding of stability involves investigating where in the Cartesian complex plane the roots of the characteristic equation are located; we must now reconsider what this means in terms of the Nyquist diagram.

   Let us therefore consider all the possible locations of the roots, $r_i = \alpha_i + j\,\beta_i$ of the characteristic Eq. (15.37) in the complex plane. As shown in Figure 15.15, we recall that roots that cause the process to be unstable will be in the right-half plane while those which allow stable process behavior will be in the left-half plane.

   Let us now translate this into the new setting of the Nyquist diagram. For this purpose, we consider the open-loop transfer function $g_L(s)$ as a function that takes values of the complex variable s (say, s_i) that can have locations in Figure 15.15 and produces another complex number, (say, f_i) given by:

$$f_i = g_L(s_i)$$

which can have locations on the Nyquist plot, Figure 15.14.

   Mathematically, we then say that $g_L(s)$ *maps* the chosen value of s into the resulting value of f. For example, by this "mapping," all the values $s = j\omega$ (for all values of ω) — corresponding to the upper portion of the imaginary axis in the complex plane — give rise to:

$$f = g_L(j\omega)$$

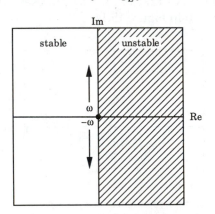

Figure 15.15. Plot of possible locations of the roots, $r_i = \alpha_i + j\,\beta_i$, of characteristic Eq. (15.37).

But recall that the Nyquist diagram is nothing more than a plot of $g_L(j\omega)$ in the complex plane. Observe therefore that this mapping actually corresponds to the very curve traced out by the Nyquist diagram. Thus we say that *the upper portion of the imaginary axis is mapped onto the curve $g_L(j\omega)$ of the Nyquist plot.*

It is easy to show that the lower portion of the imaginary axis, corresponding to the values $s = -j\omega$ (for all ω), gives rise to:

$$f = g_L^*(j\omega)$$

where $g_L^*(j\omega)$ is the complex conjugate of $g_L(j\omega)$. Thus, *the lower portion of the imaginary axis is mapped onto the mirror image of the curve $g_L(j\omega)$ of the Nyquist plot.*

It is now possible to show that all the values $s = \alpha + j\beta$ for which $\alpha > 0$, i.e., all the values in the right half of the complex plane, are mapped by $g_L(s)$ into the shaded region *inside* the curve $g_L(j\omega)$ in the Nyquist plot of Figure 15.14.

Thus we may now conclude:

1. The imaginary axis in Figure 15.15, the stability boundary in the Cartesian coordinate representation of the complex variable s, translates to the curve $g_L(j\omega)$ — and its mirror image — in the Nyquist plot.

2. The location of all the potentially unstable roots of the characteristic equation (the entire right-half plane (RHP) in the Cartesian coordinate representation) translates into the shaded portion in Figure 15.14, the interior of the region swept out by $g_L(j\omega)$ as ω goes from 0 to ∞.

It is now a relatively simple matter to use the Nyquist plot to investigate stability. For any value of s to be a root of the characteristic equation:

$$1 + g_L(s) = 0 \tag{15.37}$$

it must obviously satisfy the condition that $g_L(s) = -1$; i.e., the complex number $g_L(s)$ has a real part $= -1$, and an imaginary part $= 0$. Observe therefore that the shaded region in Figure 15.14 *must* include (or encompass) the point $(-1,0)$ if there are any roots of the characteristic equation in the right half of the complex plane. By the same token, if there are no values of s in the RHP that satisfy the characteristic equation (indicating that the characteristic equation has no roots in the RHP) the shaded region will exclude the point $(-1,0)$. This leads us to the *Nyquist Stability Criterion*:

In the Nyquist plot, if N is the net number of times that the curve $g_L(s)$ encircles the point $(-1,0)$ in the clockwise direction, and P is the number of unstable open-loop poles of $g_L(s)$ (i.e., $P > 0$ implies an open-loop unstable process), then the number of unstable roots of the closed-loop characteristic Eq. (15.37) (i.e., roots that lie in the right-half plane) is $U = N + P$.

There are some immediate implications of this result.

1. If the open-loop process $g_L(s)$ is stable, then $P = 0$ and the closed-loop process is unstable only if there are net clockwise encirclements of $(-1,0)$.

2. If the open-loop process is unstable, then $P > 0$ and there must be at least P *counterclockwise* encirclements of $(-1,0)$ (i.e., $N = -P$) for the controller to stabilize the unstable process.

3. Since the controller gain, K_c, controls the size of AR in Figure 15.14, all of these conclusions depend critically on the choice of K_c (and obviously the other controller parameters as well). The Nyquist plot may therefore be used in conjunction with the Nyquist stability criterion to determine stability limits on the controller parameter values.

Determining Stability Limits

There are several ways by which stability limits on controller parameter values may be determined from the Nyquist diagram. One approach is outlined below:

1. *Obtain $g_L(s)$ for the process in question.*
   The controller parameters will occur as unknown variables; insert provisional values for all but K_c, the proportional gain.

2. *Obtain the Nyquist plot for $g_L(s)$ resulting from Step 1, using the value of $K_c = 1$.*
   This gives what we may refer to as the "gain neutral" Nyquist plot. The premise is that for any other arbitrary value of K_c, say, K_{ca}, the AR at each frequency will be precisely K_{ca} times the value on the "gain neutral" Nyquist plot. (Computer programs are available for generating such plots.)

3. *Determine the critical point at which the "gain neutral" Nyquist plot intersects the real axis.*
   Let this point be denoted as $(-\gamma, 0)$.

4. *Determine the value of the ultimate gain K_{cu} and stability limits, from the critical point determined in Step 3.*
   From the Nyquist stability criterion, we know that the ultimate gain K_{cu} is the value of K_c required to cause the Nyquist plot to intersect the real axis at $(-1,0)$. However, when $K_c = K_{cu}$, the previous "gain neutral" Nyquist plot will have AR values multiplied by K_{cu}, so that the Nyquist plot for this gain value will now intersect the real axis at $(-K_{cu}\gamma, 0)$. The value required to make this coincide with $(-1,0)$ is then easily determined:

$$K_{cu} = \frac{1}{\gamma} \qquad (15.43a)$$

For simple Nyquist curves, $g_L(j\omega)$ with only one intersection of the real axis, the system is then unstable for:

$$K_c > \frac{1}{\gamma} \qquad (15.43b)$$

However as we shall see later, for multiple intersections of the real axis, the situation is more complex.

Let us illustrate this procedure with an example:

Example 15.7 NYQUIST STABILITY TEST OF THREE-TANK SYSTEM UNDER PID CONTROL.

Consider the three-tank system investigated earlier, now operating with a PID controller, a perfect measuring device ($h = 1$), and a perfectly responding control valve ($g_v = 1$). In this case the open-loop transfer function $g_L(s)$ will be given by:

$$g_L(s) = \frac{6K_c (1 + \tau_I s + \tau_D \tau_I s^2)}{\tau_I s (s + 2)(s + 4)(s + 6)} \tag{15.44}$$

Determine the range of stable controller gains when:

(a) $\tau_D = 0, \dfrac{1}{\tau_I} = 0$ (P control)

(b) $\tau_D = 0, \dfrac{1}{\tau_I} = 0.05$ (PI control)

(c) $\tau_D = 10, \dfrac{1}{\tau_I} = 1.33$ (PID control)

Solution:

The "gain neutral" Nyquist plots for $K_c = 1$ (obtained using the program CONSYD) are given in Figure 15.16. Note that for P control (Figure 15.16(a)), the system is stable for $K_c = 1$ and the Nyquist curve intersects the real axis at $(-0.60, 0)$, indicating that $\gamma = 0.6$. Thus for all controller gains $K_c > 1/0.60 = 1.667$, the system is unstable because the curve will then encircle $(-1,0)$ once; by the same token the system is stable for all $K_c < 1/0.60 = 1.667$.

For PI control (Figure 15.16(b)) the "gain neutral" Nyquist plot (for $K_c = 1$) begins at $-\infty$ on the imaginary axis (because of the $1/s$ factor in $g_L(s)$) and finally intersects the real axis at $(-0.732, 0)$. Thus the system is unstable for all controller gains $K_c > 1/0.732 = 1.37$.

For the PID controller, the situation is a bit more complex. As indicated in Figure 15.16(c), the "gain neutral" Nyquist curve begins at $-\infty$ on the imaginary axis, but intersects the real axis at two places: $(-19.8,0)$ and $(-1.48,0)$ for $K_c = 1$. Note however that the net number of encirclements is 0 because the point $(-1,0)$ is never encircled for $K_c = 1$.

Observe now that:

1. As K_c is increased beyond 1, the amplification will still not cause the resulting Nyquist curve to encircle the point $(-1,0)$, and the process remains stable.
2. However, in *decreasing* K_c the resulting curve shifts to the right, and the first encirclement of $(-1,0)$ will occur when:

$$K_c < \frac{1}{1.48} = 0.674$$

Subsequently, as we continue to decrease K_c, for K_c values between 0.674 and $(1/19.8) = 0.051$, the system remains *unstable*. Beyond this point, for:

$$K_c < \frac{1}{19.8} = 0.051$$

the entire curve shifts beyond the point $(-1,0)$ and the system becomes stable again.

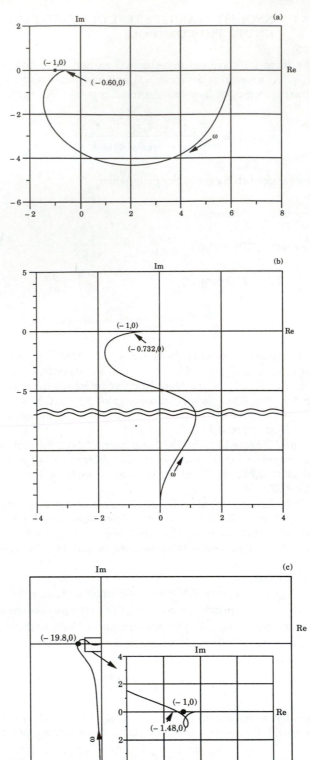

Figure 15.16. Nyquist plots of the three-tank system under (a) P control, (b) PI control $(\tau_I = 20)$, (c) PID control $(\tau_I = 0.75, \tau_D = 10)$.

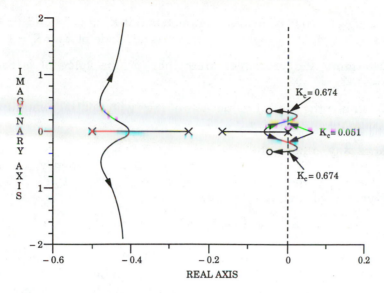

Figure 15.17. Root locus diagram for the three-tank system under PID control $(1/\tau_I = 1.33, \tau_D = 10)$. Compare with Nyquist plot (Figure 15.16(c)) and Bode plot (Figure 15.20).

Thus with the PID controller, the system is "conditionally stable," i.e., for $K_c < 0.051$ and $K_c > 0.674$ the system is stable, but in the range $0.051 < K_c < 0.674$, the system is unstable.

The explanation for this type of behavior may be made more explicit by the Root Locus diagram of Figure 15.17. There it is seen that a pair of closed-loop poles cross into the RHP when $K_c = 0.051$, but turn around and cross back into the LHP when $K_c = 0.674$.

15.5.2 Bode Stability Criterion

An alternate form of frequency domain representation is the Bode plot, discussed in some detail in Chapter 9. Because it is closely related to the Nyquist diagram, it can also be used for determining closed-loop stability from open-loop frequency-response data.

Putting the Nyquist stability criterion in the context of the Bode diagram, we observe first that the critical point $(-1,0)$ corresponds to a situation in which AR = 1 for a phase angle $\phi = -180$. The frequency at which this occurs (shown explicitly on the Bode diagram but not on the Nyquist plot) is known as the *crossover frequency*, ω_c. Thus the encirclement condition of the Nyquist criterion may be reinterpreted in the context of the Bode diagram leading to the *Bode Stability Criterion*:

> *If the AR of the open-loop transfer function, $g_L(s)$, is greater than 1 when $\phi = -180°$, the closed-loop system is unstable.*

Determining Stability Limits

Stability limits for controller parameters may now be determined from the Bode diagram, using the Bode stability criterion. The following is a procedure for doing this:

1. Specify all other controller parameters but K_c in $g_L(s)$ and then obtain (using a computer program) a "gain neutral" Bode plot for $K_c = 1$.

2. Determine the crossover frequency ω_c the value of ω for which $\phi = -180°$.

3. Determine the critical value AR_c of the "gain neutral" amplitude ratio at this crossover frequency.

4. Since any other arbitrary value of K_c, say, K_{ca}, will produce an amplitude ratio of $K_{ca}AR_c$ at this crossover frequency, the ultimate value K_{cu} is then that value for which $K_{cu}AR_c = 1$ at $\omega = \omega_c$. K_{cu} is therefore obtained as:

$$K_{cu} = \frac{1}{AR_c} \qquad (15.45)$$

As illustrated by the example in Figure 15.16(c), we must hasten to add the restriction that the Bode stability criterion only applies to systems that cross $\phi = -180°$ once. For multiple crossings one must use the Nyquist criterion.

Let us now illustrate the application of the Bode stability criterion with an example.

Example 15.8   BODE STABILITY TEST OF THREE-TANK PROCESS UNDER PID CONTROL.

For the PID controller applied to the three-tank process discussed in Example 15.7, determine the range of stable controller gains using the Bode stability criterion.

Solution:
The "gain neutral" Bode plots for the control system (when $K_c = 1$) under:

a)   P control
b)   PI control $(1/\tau_I = 0.05)$
c)   PID control $(1/\tau_I = 1.33, \tau_D = 10)$

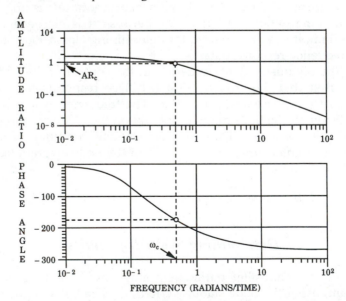

Figure 15.18.   Open-loop Bode plot for three-tank process with P control; $K_c = 1$.

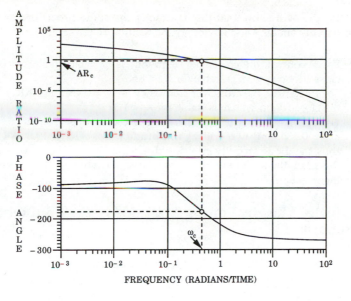

Figure 15.19. Open-loop Bode plot for three-tank process with PI control; $K_c = 1, 1/\tau_I = 0.05$.

are shown in Figures 15.18 – 15.20. Note that the same procedure is applied in each figure:

1. Determine the crossover frequency, ω_c, where $\phi = -180°$.
2. Determine the value of the critical amplitude ratio, AR_c, at the crossover frequency ω_c.
3. Calculate the stability limit on K_c from $K_c < 1/AR_c$.

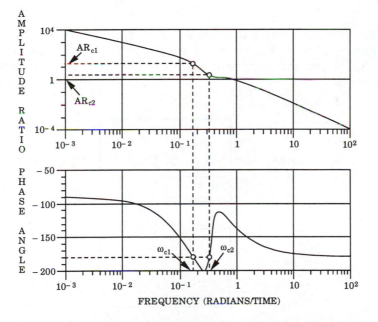

Figure 15.20. Open-loop Bode plot for three-tank process with PID control; $K_c = 1, 1/\tau_I = 1.33, \tau_D = 10$.

A little thought will show that this is exactly the same procedure as followed for Nyquist stability analysis and obviously the results are the same:

(a) For P control, the crossover frequency is $\omega_c = 0.5$ and $AR_c = 0.60$; thus the closed-loop system is stable for $K_c < 1.667$.

(b) For PI control, the crossover frequency is $\omega_c = 0.457$ and $AR_c = 0.732$ so that the closed-loop system is stable for $K_c < 1.37$.

(c) For PID control, note that there are two crossover frequencies, $\omega_{c_1} = 0.172$, $\omega_{c_2} = 0.338$, and two critical amplitude ratios $AR_{c_1} = 19.8$, $AR_{c_2} = 1.48$, and thus the Bode stability criterion does not apply. If one were to apply it in this case, the system would be declared unstable for $K_c = 1$ which is clearly incorrect.

15.5.3 Stability Margin Tuning Methods in the Frequency Domain

Now that we have stability criteria which we can apply to frequency domain models, let us illustrate some stability margin tuning strategies that have found widespread use in practice.

Gain and Phase Margin Tuning

Consider the "gain neutral" Bode diagram shown in Figure 15.21 for open-loop transfer function $g_L(s)$ (i.e., with $K_c = 1$). Every "gain neutral" Bode diagram of this type has the following important characteristics:

1. Two critical points, one on each plot: the phase plot has ω_c, the frequency for which $\phi = -180°$, and the AR plot has the point where $AR = 1$.

2. The corresponding values of AR and ϕ at each of the noted critical points: AR_c is the value of AR at the crossover frequency ω_c; ϕ_1 is the value of ϕ at the point when $AR = 1$, the value of the frequency here being ω_1.

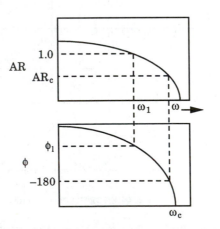

Figure 15.21. Bode diagram for the open-loop transfer function, $g_L(s)$.

According to the Bode stability criterion, if $AR_c > 1$ then the system is unstable; conversely, the system is stable if $AR_c < 1$. Furthermore, we observe that an intuitively appealing measure of *how nearly unstable* a system is may be defined as how far AR_c is from the critical value of 1; or how far ϕ_1 is from $-180°$.

We therefore define a *gain margin*, GM, as:

$$GM = \frac{1}{AR_c} \qquad (15.46)$$

and a *phase margin*, PM, as:

$$PM = \phi_1 - (-180°) \qquad (15.47)$$

These are quantitative measures of stability margins — measures of how far the system is from being on the verge of instability. (Note that for a stable system, $GM > 1$ and $PM > 0$.) Obviously these values depend on the actual value chosen for K_c and other controller parameters.

Conceptually, we could now use this notion of gain and phase margins for controller design. Observe that we could conceive of a scheme by which a controller g_c is chosen for a process, the Bode diagram for the resulting open-loop transfer function g_L obtained, and the critical points of the Bode diagram investigated first for absolute stability, then for stability margins if the system is stable.

Regarding stability margins, it is typical for design purposes to choose the controller parameters to give a GM of about 1.7, or a PM of about 30°. If the chosen g_c fails to give a control system that meets the noted stability margin specifications, another choice is made, and the resulting control system reinvestigated until a g_c is found for which the specifications are met.

Strictly speaking, this concept of controller design using gain and phase margin specifications is most useful for selecting K_c values for proportional-only controllers. For selecting the additional parameters required for other controller types, a time-consuming, trial-and-error procedure must be followed. A more useful, and more popular method for obtaining controller settings from Bode diagrams is now discussed.

Ziegler-Nichols Stability Margin Tuning

From previous discussions in this section, we know that two critical pieces of information concerning closed-loop system stability can be extracted from the Bode diagram: the value of K_c at which the system will be on the verge of instability, and the frequency at this point; the crossover frequency ω_c.

In Section 15.3.3 we discussed a method due to Ziegler and Nichols [3] by which appropriate controller parameter values can be derived from these two parameters. The following is the procedure for obtaining Ziegler-Nichols controller settings from Bode diagrams:

1. Choosing g_c as a proportional-only controller, with $K_c = 1$, obtain the "gain neutral" Bode diagram for the open-loop transfer function g_L of the given system.

2. From this plot, obtain the critical value of AR (i.e., AR_c) at the crossover frequency ω_c (see Figure 15.22).

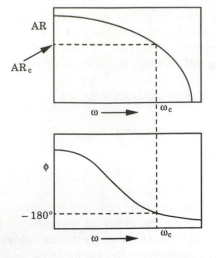

Figure 15.22.   Controller tuning from a Bode diagram.

Observe from here therefore that as explained in Section 15.5.2, the ultimate controller gain K_{cu} (which will make $K_{cu} AR_c = 1$) is given by:

$$K_{cu} = \frac{1}{AR_c} \tag{15.48}$$

The immediate implication, according to the Bode stability criterion, is now that the system will be on the verge of instability when $K_c = K_{cu}$.

3. It can be shown that when the system is on the verge of instability (when $K_c = K_{cu}$), the system output will oscillate perpetually as an undamped sinusoid with frequency ω_c. The period of this oscillation, known as the *ultimate period* P_u, is given by:

$$P_u = \frac{2\pi}{\omega_c} \tag{15.49}$$

4. With the values of the ultimate gain and period thus determined, the Ziegler-Nichols stability margin settings for different controller types may be determined from Table 15.1 in Section 15.3.3.

Clearly, any other tuning rules which depend only on a knowledge of the ultimate gain and ultimate period could be used as well. Alternatively, there are tuning rules that depend on other aspects of the frequency response; these will be discussed in a later chapter.

Example 15.9   ZIEGLER-NICHOLS STABILITY MARGIN TUNING FROM A BODE PLOT.

Consider the three-tank system with P control having the Bode plot of Figure 15.18. Determine the Ziegler-Nichols settings for P, PI, PID control.
Solution:
    From Figure 15.18, we see that $\omega_c = 0.5$, $K_{cu} = 1.667$ with proportional control. Thus the ultimate period is $P_u = 12.56$ and the Ziegler-Nichols P, PI, PID settings are as given in Example 15.4.

15.6   CONTROLLER TUNING WITHOUT A MODEL

In many practical situations involving simple loops, it may not be necessary to obtain a model (fundamental or approximate) in order to tune the controller. In other cases, time or other limitations might prevent the development of any form of model, even if there was the need for one. Under these circumstances, it is possible to use simple, unobtrusive on-line experiments, in conjunction with tuning rules such as those found in Table 15.1, to provide good initial controller parameter values. Let us consider two experimental approaches for implementing this strategy.

15.6.1 Experimental Determination of Closed-Loop Stability Boundaries

This approach involves using a proportional controller and experimentally turning up the controller gain until the closed-loop system begins to oscillate at constant amplitude with period P_u. This is an experimental determination of K_{cu}, P_u for a proportional controller.

   With these values determined, the tuning guidelines such as the Ziegler-Nichols stability margin rules (shown in Table 15.1) may then be used to configure a PI or PID controller. For simple, rapidly responding loops (such as local flow or temperature loops), this approach is easy and quick to implement. Obviously, for large processes with slow closed-loop response, this approach is much less convenient and may be unsafe because one is bringing the entire process to the point of unstable behavior. Thus this approach must be used with caution.

15.6.2 "Auto Tuning" with Relay Controller to Determine Stability Boundaries

Åström and Hagglund [11] show how a simple relay controller may be used to determine the ultimate gain and ultimate period from small amplitude experiments. The controller is a simple on-off controller such as is found in the home furnace. The system block diagram is shown in Figure 15.23. The relay controller has a specified amplitude h and may have a small dead zone δ. It can be created easily by using a very high gain proportional controller with constrained controller action $- h \leq u \leq h$. An example of the resulting closed-loop response is shown in Figure 15.24. Åström and Hagglund have shown that to a very good approximation, the period of the closed-loop response is the ultimate period P_u, and the ultimate proportional controller gain is given by:

$$K_{cu} = \frac{4h}{\pi A} \tag{15.50}$$

where h is the control amplitude and A is the resulting output amplitude.

   This approach, termed "auto tuning," allows one to determine the essential stability character of the closed-loop system with only one experiment involving relatively low-amplitude inputs, (with magnitude h), and outputs (with magnitude A). This introduces only small disturbances into the normal process operation and does not require the entire process to be brought to the

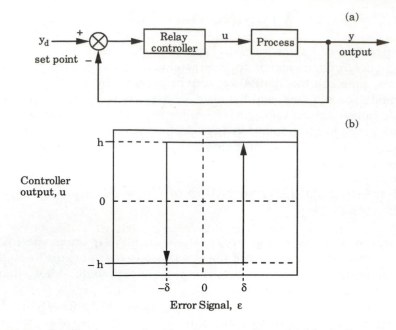

Figure 15.23. Feedback control with relay controller: (a) block diagram, (b) relay controller with amplitude h and dead zone δ.

verge of instability. Thus this approach is much more convenient and safer than the first approach. The disadvantage is that we must have available (in hardware or software) a relay controller with which to perform the tuning. In industrial practice, some vendors provide auto tuning cards for electronic controllers, while for distributed control systems, these relay controllers are provided as software.

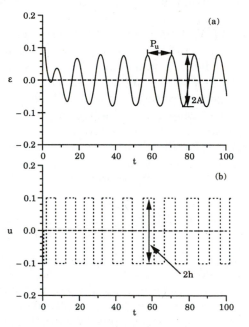

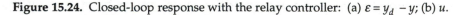

Figure 15.24. Closed-loop response with the relay controller: (a) $\varepsilon = y_d - y$; (b) u.

Figure 15.24 illustrates the performance of this auto tuning approach applied to the three-tank process. Here the relay controller is a high-gain proportional controller (e.g., $K_c = 100$) with constrained small amplitude control action (e.g., $h = 0.1$). The closed-loop response of this relay controller to an initial disturbance leads to period $P_u = 12.6$, and the output amplitude is $A = 0.076$. Thus the ultimate controller gain, calculated from Eq. (15.50), is $K_{cu} = 1.67$.

Semino [12] shows how these ideas can be applied for tuning unstable and nonlinear processes.

15.7 SUMMARY

We have taken the first step in what is to be an extensive study of controller design for chemical processes by undertaking in this chapter a rather comprehensive treatment of the design of conventional, single-loop, feedback controllers. We have introduced the basic principles of single-loop feedback controller design, identifying the main issues as: *the controller type decision*, and *choosing controller parameters* (also known as controller tuning).

Various concepts, methods, criteria, and techniques typically used for making rational decisions about what controller type to use, and what parameters to choose for these controllers have been introduced.

The nature of a substantial number of industrial processes is such that they can be adequately controlled by straightforward, conventional, single-loop, feedback controllers. This chapter is therefore directly relevant for such systems. However, important situations arise in which the conventional single-loop feedback control scheme proves inadequate. The remaining chapters in Part IVA deal with the issue of controller design (for single-input, single-output systems) under circumstances in which more is required than the conventional scheme can offer. The issue of controller design for multivariable systems is discussed later, in Part IVB.

REFERENCES AND SUGGESTED FURTHER READING

1. Considine, D. M., *Process Instruments and Controls Handbook*, McGraw-Hill, New York (1974)
2. Smith, R. E. and W. H. Ray, *CONSYDEX — An Expert System for Computer-Aided Control System Design* (in press)
3. Ziegler, J. G. and N. B. Nichols, "Optimum Settings for Automatic Controllers," *Trans. ASME*, **64**, 759 (1942)
4. Rivera, D. E., M. Morari, and S. Skogestad, "Internal Model Control 4. PID Controller Design," *I&C Process Design Development*, **25**, 252 (1986)
5. Cohen, G. H. and G. A. Coon, "Theoretical Considerations of Retarded Control," *Trans. ASME*, **75**, 827 (1953)
6. Smith, C. A. and A. B. Corripio, *Principles and Practice of Automatic Process Control*, J. Wiley, New York (1985)
7. Murrill, P. W., *Automatic Control of Processes*, International Textbook Co., Scranton, PA (1967)
8. Lopez, A. M., P. W. Murrill, and C. L. Smith, "Controller Tuning Relationships Based on Integral Performance Criteria," *Instr. Tech.*, **14**, 11, 57 (1967)
9. Rovira, Alberto H., PhD. Thesis, Louisiana State University (1981)

10. Morari, M. and E. Zafirou, *Robust Process Control*, Prentice-Hall, Englewood Cliffs, NJ (1989)

11. Åström, K. J. and T. Hagglund, "Automatic Tuning of Simple Regulators with Specifications on Phase and Amplitude Margins," *Automatica*, **20**, 645 (1984)

12. Semino, D., "Automatic Tuning of PID Controllers for Unstable Processes," *Proceedings ADCHEM '94*, Kyoto, Japan (1994)

REVIEW QUESTIONS

1. Apart from the design of the controller itself, what other considerations are important in control systems design?

2. What are the most important points about sensors and transmitters?

3. What is the difference between an "air-to-open" and an "air-to-close" valve ?

4. What are the three most common control valve flow characteristics?

5. Why is the "equal" percentage valve so called?

6. Apart from the control valve, what are some other final control elements?

7. Why is it necessary for a controller to be equipped with a direct/reverse action selection feature?

8. What is "proportional band" and how is it related to the proportional gain?

9. What is the main essence of the controller design problem?

10. What are the three types of criteria by which closed-loop system performance may be assessed in general?

11. What are the most salient characteristics of:
    (a) a P controller
    (b) a PI controller
    (c) a PD controller
    (d) a PID controller?

12. Under what conditions should the use of a P controller be avoided?

13. Why is it that a large number of feedback controllers in a typical plant are of the PI type?

14. When is it advisable to use a PID controller as opposed to a PI controller? When is this inadvisable?

15. Why do you think a PD controller is seldom used? When it is used what advantage accrues?

16. What is "controller tuning"?

17. What are some of the strategies available for controller tuning *when a fundamental model is available*?

18. What is the fundamental principle behind the use of stability margins for controller design when process models are available?

19. Explain the philosophical difference between the models developed via the principles of process identification discussed in Chapter 13, and the models constructed for the sole purpose of feedback controller tuning.

20. For feedback controller tuning purposes, what is the most commonly employed approximate process model form?

21. What is the process reaction curve, and how is it used for feedback controller tuning?

22. What is the Nyquist stability criterion, and how is it used for controller tuning?

23. What is the Bode stability criterion, and how is it used for controller tuning?

24. Within the context of controller design using frequency-response models, what is a "gain margin" and how is it different from a "phase margin"?

25. What strategies are available for controller design without the use of a "formal" model?

26. What is the fundamental principle underlying the "auto tuning" technique using relays?

PROBLEMS

15.1 The dynamic behavior of a pilot scale absorption process has been approximated by the transfer function model:

$$y(s) = \frac{-0.025}{1.5s + 1} u(s) \tag{P15.1}$$

where y is the deviation of SO_2 concentration from its nominal operating value of 100 ppm, measured at the absorber's outlet; and u is the absorber water flowrate, in deviation from its nominal operating value of 250 gpm; time is in minutes.

(a) It is desired to design a PI controller for this process (assuming for now that the dynamics of the sensor as well as those of the valve are negligible) by assigning the closed-loop system poles to the same location as the roots of the quadratic equation:

$$s^2 + s + 1 = 0 \tag{P15.2}$$

Find the values of K_c and $1/\tau_I$ required to achieve this pole assignment, and identify the location of the resultant closed-loop zero which this technique is unable to place arbitrarily. [Hint: *Pay attention to the sign which K_c must have for the controller's action to be reasonable.*]

(b) Implement this PI controller on the process and obtain a simulation of the closed-loop response to a step change of −10 ppm in the set-point for the absorber's outlet SO_2 concentration. Use the $g(s)$ given in Eq. (P15.1) above for the process model.

(c) To investigate the effect of plant/model mismatch and unmodeled dynamics on control system performance, consider now that:

   1. the "true" process model has the same first-order structure, but with a gain of −0.03, and a time constant of 1.0 min;
   2. the unmodeled dynamics ignored in part (a) are given by:

Composition Analyzer Dynamics:

$$h(s) = \frac{e^{-0.2s}}{0.2s + 1} \tag{P15.3a}$$

Water Line Valve Dynamics:

$$g_v(s) = \frac{1}{0.25s + 1} \tag{P15.3b}$$

Introduce this information into a control system simulation package, implement the controller of part (a) on this as the "true" process, and compare the closed-loop response to the same –10 ppm set-point change investigated in part (b). Comment briefly on whether or not the difference in performance is significant enough to warrant a redesign of the controller.

15.2 Revisit Problem 15.1. Reduce the K_c value obtained in part (a) by a factor of 10, and implement this new controller first on the process as indicated in part (b), and then implement it on the process as modified in part (c). Compare these two closed-loop responses.

By evaluating this set of responses in light of the corresponding set of closed-loop responses obtained in Problem 15.1, comment briefly on what effect (if any) the use of a smaller controller gain has on the control system's "sensitivity" to the effect of plant/model mismatch and unmodeled dynamics.

15.3 A conventional controller is to be used in the scheme shown below in Figure P15.1 for controlling the exit temperature of the industrial heat exchanger first introduced in Problem 14.2.

Assuming that T_B, the brine temperature (in °C), remains constant, the transfer function relations for the process have been previously given as:

$$(T - T^*)(°C) = \frac{-(1 - 0.5e^{-10s})}{(40s + 1)(15s + 1)}\left(F_B - F_B{}^*\right) \text{ (kg/s)} \tag{P14.3a}$$

$$(T - T^*) \text{ (°C)} = 0.5e^{-10s}\left(T_i - T_i{}^*\right) \text{ (°C)} \tag{P14.3b}$$

Assume that the valve dynamics are negligible.

(a) Use the provided model in a control system simulation package to obtain a *process reaction curve*; use this process reaction curve to design a PI controller for this process based on:

1. The Ziegler-Nichols *approximate model* tuning rules
2. The Cohen-Coon tuning rules
3. The IMC approximate model tuning rules (Improved PI)

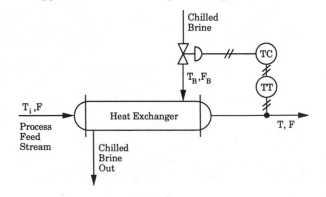

Figure P15.1.

(b) Using each of the PI controllers designed in part (a) in turn, obtain the overall closed-loop response of the heat exchanger system (using the actual transfer function models given in Eq. (P14.3)) to:

1. a + 1°C exit temperature set-point change
2. a + 2°C change in the inlet temperature

In each case briefly evaluate the performance of each of the three PI controllers. If you had to recommend one controller for use on the actual process, which one would you recommend?

15.4 The process shown in Figure P15.2 consists of an insulated hollow pipe of uniform cross-sectional area A m², and length L (m), through which liquid is flowing at a constant velocity, v (m/s).

It is desired to maintain the temperature of the liquid stream flowing through the pipe at some reference value, T_r°C, and a control system is to be designed to achieve this objective. A temperature controller (TC) has been set up to receive T, a temperature measurement, and use this information to adjust the quantity of heat, Q (kW.hr) supplied to the inlet stream by the electrical heater. The electric heater is so efficient that the temperature at the pipe inlet T_i responds *instantaneously* to changes in Q; however, the only thermocouple available for temperature measurement has been installed right at the pipe outlet (see diagram). Thus, if we define the following deviation variables:

$$T_i - T_{is} = x ; \quad T - T_s = y ; \quad Q - Q_s = u$$

assuming the thermocouple has negligible dynamics, and that it is equipped with built in calibration so that its output is identical to the measured temperature, in °C, then the equations describing the behavior of this system are:

$$x(s) = K u(s) \tag{P15.4}$$

where K is the heater constant (units $\equiv$ °C/kW.hr), and:

$$y(s) = e^{-\alpha s} x(s) \tag{P15.5}$$

(a) Draw a block diagram for the entire feedback control system; be careful to label *every* signal and identify *every* block.

(b) Consider the specific situation in which a *pure proportional controller* is used on a pipe whose length is 1 m, and whose cross-sectional area is 0.05 m²; liquid is flowing through this pipe at a *volumetric* rate of 0.2 m³/min. If the heater constant for this system is 0.4 °C/kW.hr, what is the ultimate gain, K_u?

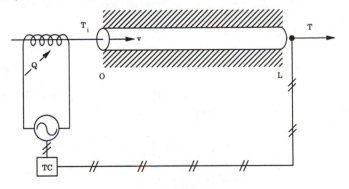

Figure P15.2.

(c) Use the Ziegler-Nichols stability margin tuning rules to design a PID controller for the system described in part (b). Implement this controller and obtain the closed-loop system response to a step change of +2°C in the fluid temperature set-point.

15.5 The liquid level in a drum boiler used in the utilities section of a research laboratory complex is controlled by the feedwater flowrate as indicated in Figure P15.3. An approximate transfer function model representing this process is given by:

$$y(s) = \frac{0.1(1 - 2s)}{s(5s + 1)} u(s) \tag{P15.6}$$

where, in terms of deviation variables, y is the drum level (in inches) and u is the feedwater flowrate (in lb/hr), time is in minutes.

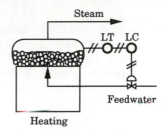

Figure P15.3.

It is known that there are no significant sensor dynamics, but that the valve dynamics are represented by:

$$g_v(s) = \frac{1.2}{(0.8s + 1)} \tag{P15.7}$$

(a) Design a proportional controller for this process by generating an appropriate Bode diagram and using the Bode stability criterion in conjunction with a gain margin of 1.7. Redesign the proportional controller using a phase margin of 30°. Which controller is more conservative?

(b) Using a control system simulation package, implement the two controllers obtained in (a) and obtain a simulation of the closed-loop system response to a step change of 5 inches in the drum level set-point. Plot both the y and u response in each case. Briefly evaluate the performance of each controller.

15.6 For a pilot scale, binary (Ethanol/Water) distillation column, the process reaction curve, representing the experimentally recorded response of the overhead mole fraction of ethanol to a step change in the overhead reflux flowrate, has been approximately modeled by the transfer function:

$$g(s) = \frac{0.66e^{-2.6s}}{6.7s + 1} \tag{P15.8}$$

(a) From an appropriate Bode diagram for this process, obtain the parameters necessary for designing a PI controller using the Ziegler-Nichols stability margin tuning rules (Table 15.1).

(b) It has been suggested by Fuentes (1989),[†] that given a process reaction curve represented by the transfer function:

$$g(s) = \frac{Ke^{-\alpha s}}{\tau s + 1} \tag{P15.9}$$

[†] Y. O. Fuentes, *personal communications*, March 1989.

it is, in fact, possible to use the Ziegler-Nichols stability margin tuning rules to design a controller *without having to obtain the required parameters from a Bode diagram*. The analytical procedure proceeds as follows:

For the given $g(s)$, recall from Chapter 9 that the frequency-response characteristics are given by:

$$AR = \frac{K}{\sqrt{1 + \tau^2\omega^2}} \tag{P15.10}$$

$$\phi = -\tan^{-1}(\omega\tau) - \omega\alpha \tag{P15.11}$$

Determining the crossover frequency ω_c now requires solving the transcendental equation:

$$-\pi = -\tan^{-1}(\omega_c\tau) - \omega_c\alpha \tag{P15.12}$$

for which no analytical, closed-form solution exists. However, Fuentes has shown that under certain reasonable conditions, the solution is well approximated by the expression:

$$\zeta = \frac{1}{2} + \frac{\pi}{2}\lambda \tag{P15.13}$$

where $\zeta = \omega_c\tau$, and $\lambda = \tau/\alpha$.

Now use the Fuentes technique to obtain PI controller tuning parameters for the process in part (a), using the Ziegler-Nichols stability margin tuning rules in Table 15.1. Compare your results with what was obtained in part (a) using the Bode diagram. Comment briefly on the accuracy (and usefulness) of the Fuentes approach.

15.7 In a series of experiments aimed at studying the characteristics of blood pressure control in mammals when external intervention is required, the drug *Nitroprusside* was infused into the bloodstream of a dog, and the mean arterial pressure (MAP) of the dog was monitored. Even though the observed responses were nonlinear, the approximate linear model:

$$y(s) = \frac{-2.5}{(5.0s + 1)^2(2.9s + 1)} u(s) \tag{P15.14}$$

was found to be adequate. Here, y is the *deviation* if the MAP (in mm Hg) from the average value observed prior to the administration of the drug; u is the *Nitroprusside* infusion rate (in ml/min); time is in minutes.

Since it could be fatal to deviate significantly from the prescribed value of the MAP in either direction, it is desired to design a feedback control system which will exhibit little or no oscillatory behavior.

(a) Assuming that this model is accurate enough, and assuming that no other dynamics are involved in the "process," use the stability margin approach and the Ziegler-Nichols tuning rules in Table 15.1 to design P, PI, and PID controllers for this "process."

(b) Evaluate all three controllers by obtaining a simulation of the closed-loop system response to a demand for a –20 mm Hg step change in the dog's MAP. Plot both y and u for each of the three controllers.

In light of the requirement stipulated above, are any of the three controllers acceptable? If not, how would you modify the controller parameters to make them acceptable?

15.8 The cardiac and blood vessel hemodynamics model developed for the cardiovascular regulator by Grodins (1959),[†] includes the subsystem shown in Figure P15.4, which may be used to study the effect of hemorrhaging (or the effect of the rate of blood transfusion) on the human systemic arterial pressure.

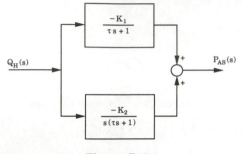

Figure P15.4.

Here, as first introduced in Problem 6.18, Q_H represents the *rate* of hemorrhage (defined as the rate of blood flow *from* the system *to* the external environment), and P_{AS} represents the systemic arterial load pressure. When the model is to be used to study the effect of blood transfusion, Q_H represents the *rate of transfusion* (defined as the rate of blood flow *to* the system *from* the external environment), and the signs on the gains indicated in the transfer functions in Figure P15.4 must be switched.

The model parameters τ, the systemic arterial system's effective time constant, and the gains K_1, K_2, are defined in terms of physiological parameters as:

$$\tau = \frac{R_s C_{As} C_{Vs}}{C_s}$$

$$K_1 = \frac{R_s C_{Vs}}{C_s}$$

$$K_2 = \frac{1}{C_s}$$

with:
R_s = a linear Poisseuille resistance (arising from the hydraulic analog on which the full-scale model is based)
C_{As} = systemic arterial compliance
C_{Vs} = systemic venal compliance
C_s = $(C_{As} + C_{Vs})$; the *total* compliance
(Compliance is defined as a ratio of volume to pressure.)
(a) For the specific subject for whom the following parameters apply: $R_s = 0.0167$ mm Hg, $C_{Vs} = 500$ cc/mm Hg, $C_{As} = 500$ cc/mm Hg (cf. Grodins 1959), design a P controller that will adjust Q_H, now representing the *blood transfusion rate*, using feedback measurements of P_{AS}. Use the pole assignment method, in conjunction with a desired closed-loop characteristic equation that is second order:

$$s^2 + 0.9\,s + 1 = 0$$

Will this controller leave a steady-state offset? Briefly justify your answer.
(b) Implement this controller on the "process" and obtain a simulation of the closed-loop response to a step change of +15 mm Hg in the P_{AS} set-point. Plot both the P_{AS} response as well as that of the blood transfusion rate, Q_H.

[†] F. S. Grodins, "Integrative Cardiovascular Physiology: A Mathematical Synthesis of Cardiac and Blood Vessel Hemodynamics," *Quart. Rev. Biol.* **34**, 93, (1959).

CHAPTER

16

DESIGN OF MORE COMPLEX CONTROL STRUCTURES

The control problems posed by a process that is subject to significant disturbances cannot be effectively handled by conventional feedback control. The same is true of the problems posed when multiple outputs are to be controlled with a *single* input, or when multiple inputs are to be used to control a single output. In such situations, controllers designed by the techniques discussed in Chapter 15 will not perform acceptably, and it becomes necessary to employ more complex control system structures. This chapter is devoted to a discussion of the design of these more complex control systems.

16.1   PROCESSES WITH SIGNIFICANT DISTURBANCES

Consider the conventional feedback control system shown in Figure 16.1, in which the temperature at the bottom of the distillation column is controlled by adjusting the steam flowrate to the reboiler. According to this scheme, the controller determines the steam flowrate required to achieve good temperature regulation, but this command is actually executed in form of an appropriate valve opening that supposedly corresponds to the desired steam flowrate.

Observe, however, that for a given valve opening, a change in the steam supply pressure will alter the actual steam flowrate delivered to the reboiler, ultimately causing an unwanted alteration in the column temperature. If such fluctuations in steam supply pressure are frequent and substantial, the performance of the conventional scheme will deteriorate. Under these circumstances, a more effective control scheme is required for improved performance.

The block diagram of the distillation column bottom's temperature control problem of Figure 16.1 is shown in Figure 16.2 where y is the bottom's temperature, u is the steam flowrate, and d_2 represents the steam air header pressure fluctuation. Here d_1 represents other disturbances to the bottom's temperature not related to the steam flowrate.

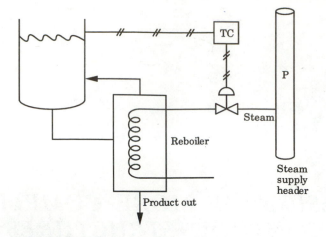

Figure 16.1.   Conventional feedback control of a distillation column's bottom temperature.

Consider the second example system shown in Figure 16.3, the familiar stirred heating tank under conventional feedback control (as earlier shown in Figure 14.1). As we recall, the objective is to regulate T, the temperature within the tank, by adjusting the rate of steam flow through the coil. According to the conventional scheme, any changes in T_i must first be registered as an upset in the value of T before the controller can take corrective action. However, where the fluctuations in T_i are frequent and substantial, it appears as if an alternative scheme which will detect changes in the disturbance T_i and implement preventive control action *before* an upset is registered in T will be far more effective.

The block diagram for the stirred heating tank control problem may also be represented by Figure 16.2. However, in this case, disturbance d_1 represents the inlet temperature variations and d_2 the disturbances related to the coil steam flow rate.

These two examples are illustrative of processes with significant disturbances. There are two common controller structures that are often used for this type of problem: *cascade control* and *feedforward control*. Cascade control is required when the significant disturbances affect the *input*; feedforward control is required when the significant disturbances affect the *output*. In the next two sections we will discuss the design of these two types of control structures.

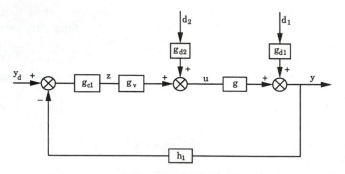

Figure 16.2.   Block diagram for a feedback control system with two disturbances.

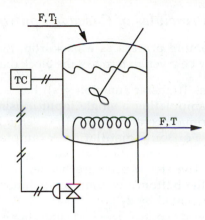

Figure 16.3.   The stirred heating tank under conventional feedback control.

16.1.1 Cascade Control

Let us return to the distillation column temperature control problem shown in Figure 16.1. Suppose we now install a flow controller (FC) in between the temperature controller (TC), and the control valve, for the sole purpose of overseeing the flow conditions, to ensure that, irrespective of fluctuations in the steam supply pressure, the desired steam flowrate is delivered to the reboiler. The temperature controller now needs only to adjust the set-point of the flow controller. Such an arrangement (see Figure 16.4) constitutes a *cascade control* configuration. Observe that under this scheme, the output of the TC (the *primary* controller) is the set-point for the FC (the *secondary* controller). With proper cascade controller design, the effect of supply pressure disturbances on the steam flow conditions can therefore be significantly reduced, if not completely eliminated.

This illustrates the fundamental characteristics of cascade control: two controllers configured such that the output of one is the set-point of the other.

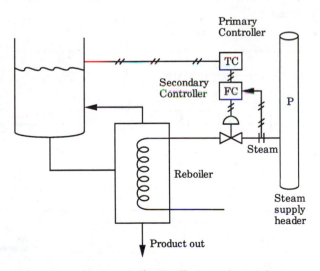

Figure 16.4.   Cascade control of a distillation column's bottom temperature.

Basic Principles of Cascade Control

With the foregoing example process as a backdrop, the general structure of cascade control may now be given in terms of the block diagram of Figure 16.5:

- The main process (transfer function $g(s)$) has a single output y to be controlled by manipulating a single input u using controller g_{c1}. This constitutes the primary control loop. However, u is subject to significant disturbances. For our distillation example, y is the bottom's temperature, u is the steam flowrate, and $g(s)$ is the transfer function between them. The steam pressure header disturbance is d_2. Other disturbances to the bottom's column temperature not related to steam flowrate are represented by d_1.
- To implement cascade control a second controller is set up to regulate u by adjusting the final control element with transfer function $g_v(s)$; this constitutes the inner (secondary) control loop. For the distillation example, $g_v(s)$ is the dynamic relation between the controller signal and the control valve stem position, while u_d is the desired steam flowrate.
- The signal from the primary controller does not go directly to the main process $g(s)$; it becomes the set-point for the secondary controller whose job is to ensure that the main process receives the input u as originally intended by the primary controller.

The two cascade controllers are sometime referred to as *master* (primary) and *slave* (secondary) controllers, for the obvious reason that the one issues the commands which the other is required to implement.

There are two essential requirements in order to implement cascade control successfully:

1. The disturbance to be regulated (e.g., d_2 in Figure 16.5) must be *within* the inner loop, so that it is controllable by the secondary controller.

2. The inner control loop must respond much more quickly that the outer control loop in order to allow u to be regulated before it disturbs the entire process.

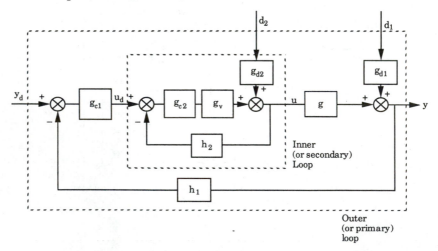

Figure 16.5.   Typical block diagram of a cascade control system.

For our distillation column example this means that a steam flow measurement device must be installed. The other conditions are satisfied by the physics of the problem.

Closed-Loop Characteristics of the Cascade Control System

One of the most crucial facts to emerge from the last section is that even though we have a single-input, single-output (SISO) system, the control of the single-output variable is achieved, under cascade control, with *two* control loops, not one. In other words, the cascade control system is a *multiloop*, SISO control system.

To analyze the overall behavior of the cascade control system, first we need to consolidate the block diagram of Figure 16.5, by dealing first with the inner loop.

As illustrated in Example 14.2, it is easy to see that the closed-loop relationship for the inner-loop elements is:

$$u = \frac{g_{c2}\, g_v}{1 + g_{c2}\, g_v h_2}\, u_d + \frac{g_{d2}}{1 + g_{c2}\, g_v h_2}\, d_2 \tag{16.1}$$

or

$$u = g_1{}^*\, u_d + g_2{}^*\, d_2 \tag{16.2}$$

where $g_1{}^*$ and $g_2{}^*$ are obtained directly from Eq. (16.1). The block diagram is therefore consolidated as shown in Figure 16.6.

From here, the overall closed-loop transfer function representation is now immediately obtained as:

$$y = \frac{g g_1{}^*g_{c1}}{1 + g g_1{}^*g_{c1}h_1}\, y_d + \frac{g g_2{}^*}{1 + g g_1{}^*g_{c1}h_1}\, d_2 + \frac{g_{d1}}{1 + g g_1{}^*g_{c1}h_1}\, d_1 \tag{16.3}$$

At this point, it is important to investigate what the system will be like under conventional feedback control, without the additional cascade control

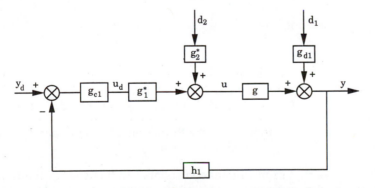

Figure 16.6.   Consolidated block diagram for the cascade control system.

loop. In this case the block diagramatic representation is given in Figure 16.2, for which the closed-loop transfer function representation is easily obtained as:

$$y = \frac{g g_v g_{c1}}{1 + g g_v g_{c1} h_1} y_d + \frac{g g_{d2}}{1 + g g_v g_{c1} h_1} d_2 + \frac{g_{d1}}{1 + g g_v g_{c1} h_1} d_1 \qquad (16.4)$$

A careful comparison of Eqs. (16.3) and (16.4) now reveals the effects of cascade control on the closed-loop performance of such systems:

$$g_v \text{ is replaced by: } g_1^* = \frac{g_{c2} g_v}{1 + g_{c2} g_v h_2}$$

$$g_{d2} \text{ is replaced by: } g_2^* = \frac{g_{d2}}{1 + g_{c2} g_v h_2}$$

and when $h_2 = 1$, we now note the following important facts:

- As the magnitude of g_{c2} increases, $g_1^* \to 1$ and $g_2^* \to 0$
- And, as a result, from Eq. (16.2), $u \to u_d$

Thus with high-performance inner-loop control, the overall effect of the disturbance d_2 on u (and hence on the process system) is totally eliminated. The conclusion is therefore that with a fast-acting inner loop (this is the real implication of having a high magnitude for g_{c2}), the overall cascade system becomes much less vulnerable to the effects of the fluctuations in d_2 than the equivalent system under conventional feedback control.

Because the inner loop of a cascade controller often responds orders of magnitude faster than the outer loop, the steady-state values of g_1^*, and g_2^* can often be used in Eq. (16.3). This amounts to a "quasi-steady-state approximation" for the inner-loop dynamics. Consideration of the distillation column bottom temperature under cascade control (cf. Figure 16.4) illustrates the point. The inner loop (which needs only to move the steam valve to cause an almost instantaneous adjustment in steam flowrate) has a timescale of a few seconds; by contrast the outer control loop (which includes the very slow dynamics of the steam heating the entire liquid contents of the column bottoms) would have a timescale of many minutes.

Cascade Controller Tuning

It is clear that the controller tuning exercise for a cascade control system must proceed in two stages:

1. The inner loop is usually tuned "very tightly" (i.e., as high a proportional gain value as feasible) to give a fast inner-loop response required for effective cascade control. This must be done first.

2. The outer loop is then tuned, with the inner loop in operation, using any of the methods discussed in Chapter 15.

It is useful to note that proportional controllers can be used for the inner loop of a cascade controller because any offset in y can be corrected by using integral action in the outer loop.

16.1.2 Feedforward Control

Let us use the stirred heating tank of Figure 16.3 to illustrate the concepts of feedforward control. The process under conventional feedback control suffers from frequent disturbances in feed temperature, and the controller must wait until these have upset the process before feedback action can be taken. Note that cascade control is not a viable solution because the disturbance is not associated with the manipulated variable. However, if we measure the feed temperature T_i, then we can configure a feedforward controller as shown in Figure 16.7. The feedforward controller then makes adjustments in the steam flowrate to the heating coil in response to observed changes in T_i.

It is important to note the difference between this scheme and conventional feedback. With feedforward control, compensation for the effect of a measured disturbance is possible *before* the process is affected; with conventional feedback control, corrective action can only be taken *after* the process has registered the effect of the disturbance. The immediate implication is therefore that perfect disturbance rejection — an absolute impossibility with feedback control — is possible (at least theoretically) with feedforward control. Note however that the process output itself is never measured in feedforward control.

Design of Feedforward Controllers

As we might suspect, feedforward controllers are designed with the aid of process models. Consider, therefore, the following familiar general transfer function model for a process relating manipulated variable and disturbance to process output, with no controller yet attached:

$$y(s) = g(s) u(s) + g_d(s) d(s) \tag{16.5}$$

The objective is for $y(s)$ to equal its set-point value, $y_d(s)$, at all times, regardless of changes in $d(s)$. This raises a design problem that translates simply to:

Derive the $u(s)$ required to make $y(s) = y_d(s)$ for all time.

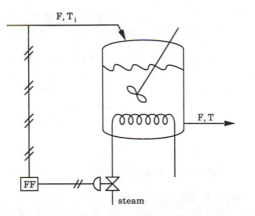

F, T$_i$

F, T

FF

steam

Figure 16.7.   The stirred mixing tank under feedforward control.

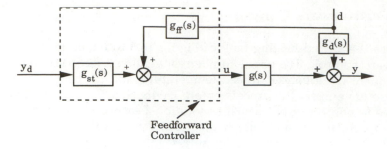

Figure 16.8.   The feedforward controller.

Observe now that this translates, algebraically, to the problem of solving Eq. (16.5) for u upon setting $y = y_d$; doing this gives:

$$u(s) = \left[\frac{1}{g(s)} y_d - \frac{g_d(s)}{g(s)} d(s) \right] \qquad (16.6)$$

or

$$u(s) = g_{st}(s) y_d + g_{ff}(s) d(s) \qquad (16.7)$$

the controller equation for the feedforward scheme. Here $g_{st}(s) = 1/g(s)$ is the set-point tracking controller and $g_{ff}(s) = -g_d(s)/g(s)$ is the disturbance rejection controller. In block diagramatic form, what Eq. (16.7) represents is shown in Figure 16.8.

Some important points to note from Eqs. (16.6) and (16.7) about the feedforward scheme are as follows:

1. The scheme can, in principle, be used both for set-point tracking (the servo problem), and disturbance rejection (the regulatory problem). Observe that for servo problems ($d = 0$), the controller equation is:

$$u(s) = g_{st}(s) y_d \qquad (16.8)$$

while for the regulator problem ($y_d = 0$), we have:

$$u(s) = g_{ff}(s) d(s) ; \qquad (16.9)$$

The feedforward scheme is, however, almost exclusively associated with disturbance rejection; it is hardly ever used for set-point tracking.

2. In either case (set-point tracking or disturbance rejection), the controller has *no* access to measurements of y, the real variable to be controlled; attention is focused instead on the potential cause of the upset (*viz.*, changes in y_d or d), while relying on the accuracy of the process transfer function models to ensure that the desired behavior is obtained, even though y itself is not monitored.

Thus observe that feedforward control is possible when:

*   The disturbance d is measured, and
*   A process model which includes the effect of d on the output y is available.

Before commenting on the merits and demerits of such a scheme, let us first consider the following examples.

Example 16.1 DESIGN OF A FEEDFORWARD CONTROLLER FOR THE STIRRED MIXING TANK.

Design a feedforward controller for the stirred heating tank of Figure 16.7.

Solution:

The first task is to monitor the inlet temperature T_i, so as to have a measurement of the disturbance d. We recall from our numerous encounters with this process that its transfer function model is of the form in Eq. (16.5) with:

$$g(s) = \left(\frac{\beta\theta}{\theta s + 1} \right) \tag{16.10a}$$

$$g_d(s) = \left(\frac{1}{\theta s + 1} \right) \tag{16.10b}$$

where the parameters β and θ have been previously given (see Chapters 4 and 14, Eqs. (4.11) through (4.15)).

According to the design procedure given above, the equation for the feedforward controller therefore is:

$$u(s) = \frac{\theta s + 1}{\beta\theta} y_d(s) - \frac{1}{\beta\theta} d(s) \tag{16.11}$$

The performance of the above-derived feedforward controller is investigated with the next example.

Example 16.2 SET-POINT TRACKING PROPERTIES OF THE FEEDFORWARD CONTROLLER DESIGNED FOR THE STIRRED HEATING TANK.

Investigate the controller performance and the overall behavior of the stirred mixing tank system in response to a step change of magnitude A in the set-point y_d, while under pure feedforward control as designed in Example 16.1.

Solution:

According to Eq. (16.11) the feedforward controller equation for set-point tracking is:

$$u(s) = \frac{1}{\beta\theta} (1 + \theta s) y_d(s) \tag{16.12}$$

and since, in this case, $y_d(s) = A/s$, we have:

$$u(s) = \frac{A}{\beta\theta s} + \frac{A}{\beta} \tag{16.13}$$

The overall system response is now obtained by using Eq. (16.13) for $u(s)$ in the process model:

$$y(s) = \frac{\beta\theta}{\theta s + 1} u(s);$$

(16.14)

immediately giving:

$$y(s) = \frac{A}{\theta s + 1} \left(\frac{1}{s} + \theta \right)$$

which rearranges straightforwardly to give:

$$y(s) = \frac{A}{s}$$

which, when inverted back to time, gives the result:

$$y(t) = A$$

(16.15)

The implication of this result is that the system output, $y(t)$, tracks the desired set-point perfectly. This should not come as a surprise; so long as the process model matches the actual process behavior, it is easy to show, in general, that regardless of the particular form of y_d, this feedforward scheme always gives the result that $y = y_d$ for all time; recall that the controller was designed specifically with this condition as the basis.

Of course, more realistically, the process model will typically not match the process behavior exactly and such perfect set-point tracking will no longer be possible and steady-state offsets will be inevitable.

The disturbance rejection properties of this controller are investigated in the next example.

Example 16.3 **DISTURBANCE REJECTION PROPERTIES OF THE FEEDFORWARD CONTROLLER DESIGNED FOR THE STIRRED MIXING TANK.**

According to Eq. (16.11) the feedforward controller equation for disturbance rejection is simply:

$$u(s) = -\frac{1}{\beta\theta} d(s)$$

(16.16)

which is inverted back to time straightforwardly to give:

$$u(t) = -\frac{1}{\beta\theta} d(t)$$

(16.17)

Recalling from Chapter 4 the physical quantities that β and θ represent, Eq. (16.17) becomes:

$$u(t) = -\frac{F\rho C_p}{\lambda} d(t)$$

(16.18)

Let us now consider the situation in which $d(t)$ increases from its initial value of 0 to a new value δ (i.e., T_i increases from T_{is} to $T_{is} + \delta$); the controller in Eq. (16.18) recommends that the steam flowrate be reduced by the factor $(F\rho C_p/\lambda)\delta$. Now, since the quantity of heat carried by steam at flowrate Q is λQ (recall that λ represents the latent heat of vaporization), then the indicated control action calls for the energy input to the process to be reduced by the factor $F\rho C_p \delta$.

Observe now that since F is the volumetric flowrate of the inlet stream, $F\rho C_p \delta$ is precisely that amount of additional energy brought into the system as a result of the increase in T_i. The result, therefore, is an exact counterbalancing of the disturbance effect by the prescribed control action. This leaves the tank temperature entirely unaffected by whatever fluctuations the inlet stream temperature T_i may experience, a situation of perfect disturbance rejection.

Practical Considerations

As attractive as pure feedforward control may appear to be, it suffers from some significant drawbacks:

1. It is not useful if d cannot be measured;

2. Even when d is measurable, the scheme requires a perfect process model to achieve its objectives;

3. Even in the unrealistic event that a perfect process model exists, the derived feedforward controller transfer functions — both g_{st} and g_{ff} — involve the reciprocal of the process transfer function g, and may therefore present any of the following problems:
   (a) They may be too complicated to be realizable in time.
   (b) If g has a time delay element, $e^{-\alpha s}$, then its reciprocal will contain the term $e^{\alpha s}$, implying that implementing g_{st} in real time will require a prediction.
   (c) If both g and g_d have time-delay elements, and the time delay associated with g_d is *less* than that associated with g, then g_{ff} will contain a term in $e^{\gamma s}$ where $\gamma > 0$; again, as in (b), prediction is required to implement such a controller in real time.
   Under the conditions specified in (b) and (c), the controllers are said to be *physically unrealizable* since, at time t, each of these controllers will require knowledge of the value that $y_d(t)$, or $d(t)$ will take at some time in the future; in (b) $y_d(t + \alpha)$ is required, in (c), $d(t + \gamma)$ is required. It is, of course, impossible to predict these quantities exactly.

All these facts notwithstanding, feedforward control has proved to be a very powerful process control scheme, especially for disturbance rejection. For one reason, many process disturbances are indeed measurable; for another, even when the available process models are imperfect, and/or the derived g_{ff} is too complicated, it is known that the lead/lag unit (whose dynamic behavior, as we may recall, was studied in Chapter 5) provides reasonably good performance as a feedforward controller.

This last statement deserves a closer look.

Application of Lead/Lag Units in Feedforward Control

Let us justify why the lead/lag unit will make a good feedforward controller. We start by recalling the Cohen and Coon process reaction curve technique for approximate process dynamic characterization, in which a general approximate representation for many g and g_d elements is:

$$g(s) \approx \frac{Ke^{-\alpha s}}{\tau s + 1}$$

and

$$g_d(s) \approx \frac{K_d \, e^{-\beta s}}{\tau_d s + 1}$$

leading to a g_{ff} having the following generalized form:

$$g_{ff}(s) \approx -\frac{K_d \, e^{-\beta s}}{\tau_d s + 1} \frac{\tau s + 1}{Ke^{-\alpha s}} \tag{16.19}$$

This, of course, simplifies to:

$$g_{ff}(s) \approx -\frac{K_{ff} \, (\tau s + 1)}{\tau_d s + 1} \, e^{-\gamma s} \tag{16.20}$$

which is a lead/lag unit with a time delay. In the event that α and β are approximately equal Eq. (16.20) is a pure lead/lag element.

This offers an objective justification for the effectiveness of lead/lag units in the implementation of feedforward control schemes even in situations when the dynamic behavior of the processes in question are poorly modeled.

As noted in Chapter 5, lead/lag units are commercially available as analog units; they can also be implemented on the digital computer. As indicated in Eq. (16.20), by adjusting the values for the gain term, as well as the lead and lag time constants on the equipment, it is possible to obtain acceptable approximations to the exact feedforward controller derived in Eq. (16.6).

Tuning the Lead/Lag Unit

In light of the discussion leading to the Eq. (16.20), the following procedure for tuning the lead/lag unit for feedforward control applications immediately suggests itself:

1. *Obtain approximate characterizing process parameters*
    Obtain process reaction curves for
    (a) The response of the process output to a step change in the input u (with $d = 0$), and
    (b) The corresponding process response to a step change in the disturbance d (with $u = 0$). Characterize each response by their respective steady-state gains, and effective time constants, and time delays; i.e., obtain estimates for $K, \tau,$ and α, for the effect of u on the process; and $K_d, \tau_d,$ and β, for the effect of d.

2. *Estimate lead/lag unit parameters*
    From the six parameters obtained from the two process reaction curves,
    we obtain the lead/lag unit parameters as follows:

Gain term: $K_{ff} = K_d/K$
Lead time constant: τ
Lag time constant: τ_d
Associated delay: $\gamma = \beta - \alpha$

Feedback-Augmented Feedforward Control

Our discussion to this point has pointed out the main disadvantages of
feedforward control, chief among which is the unrealistic requirement of very
accurate process models. However, its potential ability to compensate for the
effect of a disturbance *before* the process is affected remains a very attractive
advantage.

The design of a feedback control scheme, on the other hand, does not require
a highly accurate process model. This makes it a very "rugged" control scheme,
but by the same token, it must wait for the disturbance to affect the process
output before corrective action can be taken.

It thus appears that feedforward control is weak on points where feedback
control is strong, and conversely, where feedback control is weak, there the
strength of feedforward control lies. The fact that these two schemes appear to
be somewhat complementary — at least as far as disturbance rejection is
concerned — suggests that a combination of the two might in fact prove to be
quite advantageous. In practice, it is common in the chemical process industries
to augment a feedforward control scheme with feedback compensation, the
basic premise being to complement the excellent disturbance rejection properties
of the feedforward controller with the "ruggedness" of the feedback controller.
We thus have a situation in which feedforward control is used to contain the
effects of substantial, and frequent, disturbances while the "feedback trim" is
used to take care of the   errors introduced as a result of the inevitable
imperfections of feedforward control and/or other unmeasured disturbances.

The block diagram for this feedforward/feedback scheme is shown in
Figure 16.9. The dual objective here is to choose g_{ff} for perfect disturbance
rejection (within limits of process model accuracy, of course) and g_c so that the
closed-loop system is stable, and the other conventional feedback control
objectives, as enumerated in Chapter 15, are met.

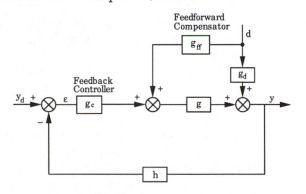

Figure 16.9.   Block diagram for the feedforward/feedback control scheme.

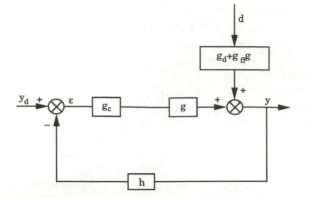

Figure 16.10. Equivalent block diagram for system in Figure 16.9.

Upon simple block diagram manipulations on Figure 16.9, we obtain the equivalent block diagram of Figure 16.10.

The closed-loop transfer functions are easily obtained from Figure 16.10 as:

$$y = \frac{gg_c}{1 + gg_ch}y_d + \frac{g_d + g_{ff}g}{1 + gg_ch}d \tag{16.21}$$

from where we observe immediately that if we now choose g_{ff} as:

$$g_{ff} = -\frac{g_d(s)}{g(s)} \tag{16.22}$$

(see the expressions in Eqs. (16.6) and (16.7) given earlier) then the effect of d on y is theoretically eliminated.

The following are some important points to note about the feedforward/feedback scheme:

1. The overall system stability is determined by the roots of the closed-loop system's characteristic equation, which from Eq. (16.21) is:

   $$1 + gg_ch = 0 \tag{16.23}$$

   Observe carefully that this has nothing to do with g_{ff}. The implication is that the stability of the feedback loop remains unaltered by including feedforward compensation. This is a very significant point in favor of the feedforward control scheme; in terms of closed-loop stability, we have nothing to lose by applying it.

2. When g_d and g are perfectly known, perfect compensation is possible; in the more realistic case where this is not the case, and/or the disturbances are (partially or completely) unmeasurable, the feedback loop picks up the residual error and eliminates it with time.

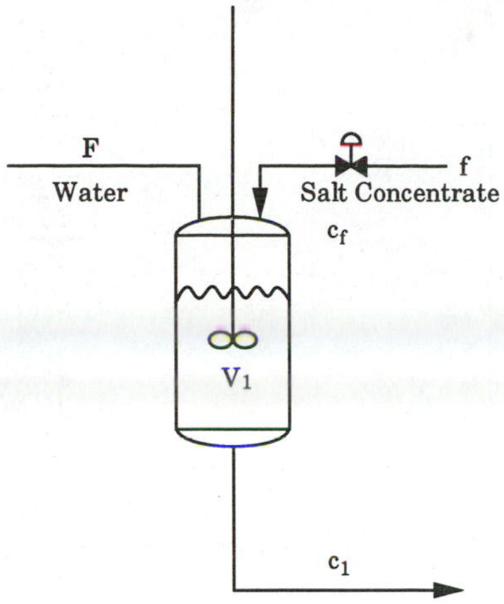

Figure 16.11. The salt solution mixing tank.

16.1.3 Ratio Control

As we have seen in the discussion above, feedforward control involves compensating for the effect of a measurable disturbance d by measuring it directly, and using such information to adjust the manipulated variable.

Certain situations arise in which the requirement is simply that the value observed for d is to be a certain percentage of another process input variable (be it a control variable or another disturbance variable). In this case, the objective is therefore simply to maintain a constant ratio between d and this other variable. Thus as d changes, commensurate changes are made in this other variable in order to maintain the desired ratio. Such a control scheme is called *ratio control* for obvious reasons, and it is easy to see that it is a special form of feedforward control.

Process Flow Applications

By its very nature, ratio control finds application mostly in systems that involve mixing of two or more process streams. In such process flow applications, there are typically two streams whose flowrates are to be maintained at a certain prespecified proportion. While both flowrates are measurable, usually the flowrate of one stream can be controlled while the other flowrate cannot be controlled. The stream whose flowrate cannot be controlled is referred to by the rather descriptive term: *wild stream.*

To illustrate the concept of ratio control and how it is carried out, consider the process shown in Figure 16.11; a salt solution mixer for which the salt concentrate flowrate can be regulated but the water flowrate cannot: making the water stream the "wild stream."

To maintain uniform salt concentration within the mixing tank, it is necessary to maintain a constant ratio of water and salt concentrate flowrates. A ratio control scheme to achieve this objective can be set up in one of the two ways shown in Figure 16.12.

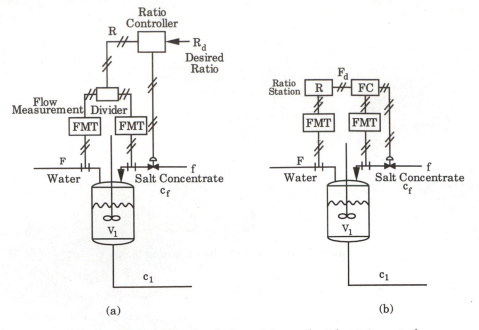

Figure 16.12. The salt solution mixing tank under ratio control.

Under the strategy indicated in Figure 16.12(a),

- Both flowrates are measured and their current ratio obtained by an electronic divider.
- The observed ratio is transmitted to the ratio controller where this is compared to the desired ratio set-point; the error is used to set the flowrate required for the salt concentrate stream.

The actual, current ratio of the flowstreams is the information obtained from the process and passed on to the controller; the controller will now have to use this information to calculate, and implement, the change needed in the salt concentrate flowrate to attain to the desired ratio.

Under the strategy indicated in Figure 16.2(b) on the other hand,

- The "wild stream" flowrate is measured and multiplied at the "ratio station" by the desired ratio: this produces the value of the flowrate at which the other stream must be set to maintain the desired ratio.
- The output of the ratio station is the set-point for the flow controller; it is compared with the actual flowrate of the salt concentrate stream, and the controller acts on the basis of the observed deviation.

In this case there is a flow controller that directly regulates the flowrate of the salt concentrate stream; the information it receives from the ratio station is its set-point.

Regardless of the particular application, configuration (b) is usually preferable; for one thing, it is easier to design, and more intuitively appealing; for another, it can be shown that some nonlinearities are involved with configuration (a) which are absent in configuration (b).

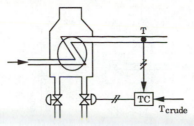

Figure 16.13. Crude oil preheater furnace under feedback control.

16.2 PROCESSES WITH MULTIPLE OUTPUTS CONTROLLED BY A SINGLE INPUT

Special types of control structures, termed *selective control*, have been developed to deal with a process for which there is only one manipulated variable and several outputs to be controlled. Depending on the application there are several approaches to *selective controller design*.

16.2.1 Override Controllers

Override controllers are often used to handle the situation in which one wishes to use the single manipulated variable to regulate one output under normal operation, but in abnormal operation the manipulated variable is brought into play to control another output variable. This usually arises when the process would be unsafe under abnormal operation without the "override" to the second output variable. Let us illustrate the events with an example.

Consider the crude oil preheater furnace discussed in Chapter 1. The feedback control system shown in Figure 16.13 adjusts the furnace fuel flowrate to maintain the desired exit temperature of the heated crude oil stream. Recall from Chapter 1 that both feedforward and cascade control could be added to better compensate for fuel header pressure fluctuations and crude oil feedrate disturbances. However, there is another problem that is characteristic of such heating furnaces: the outside metal surface temperature of the tubes containing the crude oil must remain below the metallurgical limit of the tube material. Thus override controllers are often used to maintain the tube surface temperature below this critical value. The override control scheme is shown in Figure 16.14. Note that optical pyrometers are used to measure the tube surface temperature at several locations along the tube. Let us assume that, based on

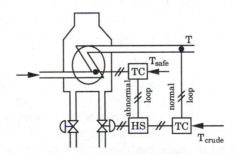

Figure 16.14. Crude oil preheater furnace with override control.

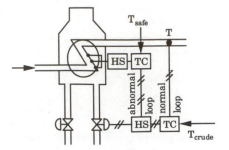

Figure 16.15. Crude oil preheater furnace with override and auctioneering control.

experience, we use one of these tube temperature measurements and send it to a temperature controller and *High Selector* (HS) switch. When the measured tube temperature exceeds a preset critical value, the HS switches from the "normal" loop to the "abnormal" loop in which the fuel flowrate is adjusted to control tube skin temperature at a preset safe maximum. When the "normal" loop again calls for less fuel than required to maintain the maximum tube surface temperature, the selector will switch back to the "normal" loop. In this way the crude oil exit temperature from the furnace will be maintained at its set-point except when the tube surface temperature is exceeded. During "abnormal" operation, the crude oil exit temperature will be somewhat below its set-point, but as close as possible while ensuring safe operation of the furnace.

16.2.2 Auctioneering Control

A second type of selective control is so-called "auctioneering" control. This is where there is a fixed control loop, but the output variable to be used may change. This may be illustrated by the furnace example of the previous section. Recall from Figure 16.14 that the tube surface temperature was measured at several locations. Obviously for the "abnormal" loop we are interested in the location with the highest tube surface temperature. Thus, we can add a HS switch to choose the largest temperature of those measured in regulating the tube surface temperature. This ensures that the highest measured temperature is maintained at a safe level. This auctioneering control scheme has been added to the override controller and shown in Figure 16.15.

Other applications of auctioneering control include controlling the hot spot temperature of packed bed reactors with multiple measurement locations, and the refrigeration of a storage area for temperature sensitive materials to ensure that all parts of the cooler are below a material degradation temperature.

16.3   PROCESSES WITH A SINGLE OUTPUT CONTROLLED BY MULTIPLE INPUTS

In contrast to the last section, here we wish to design control systems for which there are two or more manipulated variables available for the control of a single-output variable. We shall discuss two specific cases and control structures for this class of problems, depending on whether the multiple manipulated variables are required to control the output over the entire range of possible set-points, or whether the multiple input variables are employed for closer control of the speed of response.

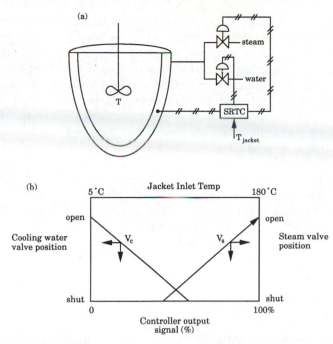

Figure 16.16.   A nonisothermal batch reactor under split-range control; (a) controller configuration; (b) split-range schedule.

16.3.1 Split-Range Control

Split-range control is the usual solution when one requires multiple manipulated variables in order to span the range of possible set-points. The concepts may be explained with a simple, but common example. Let us suppose that we have a nonisothermal batch reactor with a certain specified temperature program required to produce the product. However, the temperature program covers the range from 15°C at the beginning of the batch to 100°C at the end. In the plant the cooling water is at 5°C and the low-pressure steam is at 180°C. Obviously we will require both cooling water and steam in order to span the temperature range of interest. Split-range control allows both cooling water and steam to be used as shown in Figure 16.16. Note that the schedule allows the appropriate fraction of cooling water and steam to be used for each value of the controller output signal and reactor jacket inlet temperature required. Usually the design calls for overlap in the range of each valve for more precise control.

Split-range control is often employed for broad span temperature control of small plant or pilot scale processes, and for year-round heating and cooling of office buildings.

16.3.2 Multiple Inputs for Improved Dynamics

Sometimes multiple inputs are employed in order to speed up the dynamic response of a process during serious upsets or transitions between set-points (i.e., to improve servo behavior). To illustrate the basic principles, consider the stirred heating tank of Figure 16.3, but now assume that there is an additional auxiliary electrical heater and cooler that can be used to control the tank temperature as shown in Figure 16.17. The heating tank has several temperature set-points depending on the product being made in a downstream reactor. The normal regulatory operation at each of these set-points can be readily handled with the process steam going to the tank heating coil; however, it is important for the downstream reactor that set-point changes be made rapidly. Unfortunately, the stirred heating tank temperature responds rather slowly even with the steam valve full open (to go to a higher temperature) or full shut (to go to a lower temperature). Thus for an increase in temperature set-point the auxiliary heater helps speed the move to the new set-point. Similarly when the temperature set-point change is down, the auxiliary cooler helps the tank quickly achieve the new set-point. The drawback is that this auxiliary heater and cooler require expensive electrical energy while the steam to the coil is very cheap because it is produced as a by-product from another process unit. Thus energy efficiency requires that the auxiliary heater and cooler be used only during set-point changes. This could be accomplished with three controllers as shown in Figure 16.17 with a schedule defining when each controller is active. For example, one could choose the following schedule to achieve the control objective.

CONTROLLER SCHEDULE

Deviation from Set-point (°C)	Controllers Active
$\Delta T < -10°C$	TC3 (Aux. heater) and TC1 (steam coil)
$-10°C < \Delta T < +10°C$	TC1 (steam coil alone)
$\Delta T > 10°C$	TC2 (Aux. cooler) and TC1 (steam coil)

The controllers TC2 and TC3 could be on-off controllers or proportional controllers with suitable gain. The controller for the steam coil would have to have integral action and be tuned to reject disturbances.

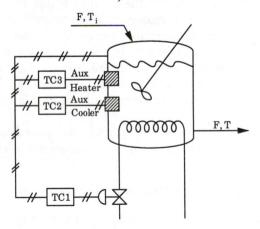

Figure 16.17. The stirred heating tank under multiple-input control.

16.4 ANTIRESET WINDUP

Consider the situation in which a feedback controller receives an error signal ε that is so large as to call for corrective control action beyond the saturation limit of the actuator. For example, a valve for reactor coolant may become full open. Under these circumstances, there will be a difference between the controller output, say, c and u, the actual control action implemented by the actuator on the process. Since the appropriate control action has been "abbreviated" as a result of the actuator saturation, the feedback error will persist much longer than usual. Note that this situation is equivalent to a *temporary* disconnection of the feedback loop since the actuator output remains fixed despite the fact that the control output may continue to change. For as long as this condition remains, the integral action of the controller will simply continue to integrate the persistent error, thereby compounding the problem. When the error is finally reduced, it may take a long time for the "wound up" integral error to return to a more normal level. This phenomenon is known as *reset* (or integral) *windup* (cf., for example, Shinskey 1979; Hanus *et al.* 1987; Åström and Wittenmark 1990).

Conceptually, reset windup can be prevented by stopping the error integral update whenever the actuator is saturated. One example of the many schemes for achieving antireset windup is shown in Figure 16.18. The scheme involves an additional feedback path through the integrator mode of the PID controller, with the primary purpose of making the indicated "actuator error" ε_a (the difference between the actual actuator output u and the desired control action c) equal to zero. When this is achieved, the integral mode of the PID controller is reset, and the controller output will remain at the saturation limit of the actuator, and windup is prevented. The rate at which the integral is reset is determined by the so-called *tracking time constant* τ_t, indicated in the block diagram. Of course if the actuator is not saturated, $\varepsilon_a = 0$, the additional loop will be inactive, and regular PID control is recovered.

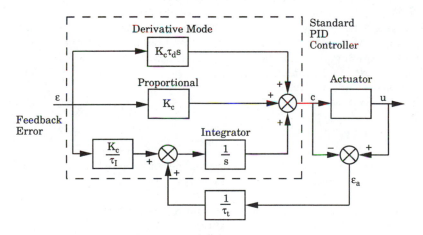

Figure 16.18. PID controller with antireset windup scheme.

16.5   SUMMARY

In this chapter we have introduced some special types of control systems that are required for those situations in which the performance of the conventional feedback controller, by itself, will be inadequate. We have shown that when the process input variable is itself subject to disturbances, the cascade control scheme is recommended. An inner control loop, which is tuned to respond faster than the main outer control loop, is used to contain the influence of the external disturbance on the process input variable.

When the disturbances that directly affect the process output are significant and occur frequently, the feedforward control scheme is able to detect such disturbances and compensate for them before they affect the process output. The design of ideal feedforward controllers and how lead/lag units are useful in the practical implementation of the feedforward control scheme have been discussed. Since perfect disturbance rejection by feedforward control is not possible in practice, it is usual to augment this with the addition of "feedback trim." The issues involved in such feedforward/feedback control schemes have been analyzed.

The concept of ratio control was introduced, and the various alternative configurations for this special form of the feedforward controller illustrated.

We have also discussed the application of selective control schemes such as override and auctioneering control when there are several output variables controlled by a single input. Conversely, we also discussed the design of split range controllers and scheduled controllers where several manipulated variables are used to control a single output. These control schemes have been illustrated by practical examples.

Finally the concept of reset windup was introduced, and a popular scheme for preventing it was presented.

REFERENCES AND SUGGESTED FURTHER READING

1.   Shinskey, F. G., *Process Control Systems* (2nd ed.), McGraw-Hill, New York (1979)
2.   Hanus, R., M. Kinnaert, and L. Henrotte, "Conditioning Technique: A General Anti-windup and Bumpless Transfer Method," *Automatica*, **23**, 729 , (1987)
3.   Åström, K. J. and B. Wittenmark, *Computer-Controlled Systems* (2nd ed.), Prentice-Hall, Englewood Cliffs, NJ (1990)

REVIEW QUESTIONS

1.   What control strategy is needed for improved performance when the *input* of a process is affected by significant disturbances?

2.   What control strategy is needed for improved performance when significant disturbances affect the process *output* directly?

3.   What is the fundamental characteristic structure of cascade control?

4.   What are the essential requirements for the successful implementation of cascade control?

5. What is the main purpose of the secondary loop in a cascade control structure?

6. Of the two loops in a standard cascade control structure — the inner (secondary) loop, and the outer (primary) loop — which needs to be faster, and why?

7. What is the typical controller tuning procedure for a cascade control system?

8. Why is it acceptable to use a proportional-only controller for the inner loop of a cascade control system?

9. What is the fundamental difference between feedforward control and conventional feedback control?

10. Why is a process model required for feedforward controller design?

11. Is feedforward control possible if the disturbance affecting the process output is not measurable?

12. What are the main drawbacks of feedforward control?

13. Why are lead/lag units useful for implementing feedforward controllers?

14. What properties of feedforward control and of conventional feedback control make them mutually complementary?

15. Why is it that ratio control can be considered a special form of feedforward control?

16. To what aspect of the process does the term "wild stream" refer in process flow applications of ratio control?

17. Under what conditions will a process require the use of selective control structures?

18. When is override control the appropriate selective control structure to use?

19. What is auctioneering control and when is it the appropriate selective control structure to use?

20. To what process control problem is *split-range control* an appropriate solution?

21. What is reset windup and how can it be prevented?

PROBLEMS

16.1 (a) Figure P16.1(a) shows a feedback control scheme for a blending process that is subject to significant disturbances. It is desired to augment this scheme with feedforward control. Of the four potential locations for the feedforward measurement indicated in the diagram, A, B, C, or D, which would be the best? (The θ values noted in the diagram correspond to the average residence time in the indicated hollow pipes.) Justify your answer briefly.

(b) Figure 16.1(b) shows a schematic diagram for a heat exchanger exit temperature control problem. The exit temperature of the process stream is influenced by the steam flowrate into the heat exchanger shell, and by the erratically varying inlet temperature of the process stream. The steam flowrate itself is also subject to erratic fluctuations induced by unsteady, unpredictable, and unmeasured steam supply pressure.

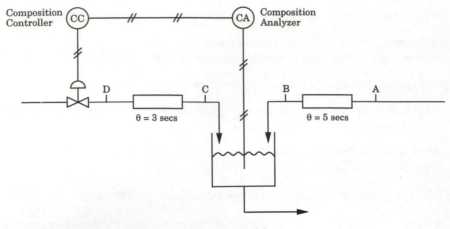

Figure P16.1(a).

Temperature measurements are provided by two temperature sensors and accompanying transmitters indicated as TT1 and TT2 in the diagram; and steam flow measurements are provided by a flowmeter and the accompanying flow transmitter, FT, also indicated in the diagram.

A flow controller, FC, a temperature controller, TC, and a feedforward controller, FFC (which can be configured any way you wish), have been made available to you. You are required to place these controllers such that the heat exchanger system is configured for:

1. Feedforward control only,

2. Feedback control with cascade control,

3. Feedforward control augmented with feedback,

4. Feedforward, feedback, and cascade control.

For each case, draw a process diagram similar to that given in Figure P16.1(b), showing clearly the connections between each of the controllers in question and the sensor/transmitter from which it gets its information.

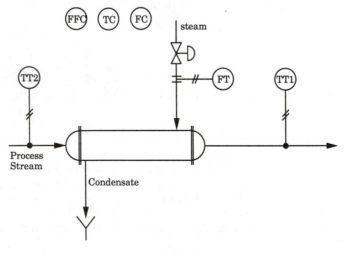

Figure P16.1(b).

16.2 In an industrial polymer process, the weight average molecular weight of the product is controlled by manipulating the chain transfer agent flowrate. However, a certain impurity that enters the process via the solvent stream also acts as a chain transfer agent. Since the impurity concentration in the solvent stream (available via on-line chromatographic analysis) varies unpredictably, the process has proven very difficult to control by the usual feedback control technique.

The following is an approximate model available for the reactor:

$$y(s) = \frac{-550.0}{11.0s + 1} u(s) + \frac{-4.5}{(5.0s + 1)(25.0s + 1)} d(s) \qquad (P16.1)$$

where, in deviation variables, y is the product's weight average molecular weight, u is the chain transfer agent flowrate (in lb/hr), d is the impurities concentration in the solvent stream (in ppm); time is in minutes.

(a) If a feedforward control strategy *alone* is to be used for this process, design this controller and show in a block diagram how it will be implemented. Assume that the y and d measurements are both available instantaneously, with no measuring device dynamics. Is this feedforward controller realizable? Justify your answer briefly.

(b) Given now that a feedback controller with transfer function:

$$g_c(s) = -0.004 \left(1 + \frac{1}{12s} \right)$$

is to be used on the process, and that in reality, the "true" process gain is actually −600.0, and that the disturbance gain is − 4.0, but the time constants were correctly estimated in Eq. (P16.1), implement this feedback controller *alone* on the "true" process, and obtain the closed-loop response to a step change of 20 ppm in the impurities concentration. Plot this response.

(c) For the "true" process as given in part (b), implement the feedforward controller designed in part (a); obtain and plot the control system response to the same step change of 20 ppm in the impurities concentration. Comment on the performance of this control scheme. What effect, if any, does the modeling error have on the performance?

(d) Now implement both the feedback controller of part (b) and the feedforward controller of (a) on the process; obtain the overall system response to the same step change of 20 ppm in the impurities concentration. Compare this response with the ones obtained in parts (c) and (d), and with reference to this specific example, briefly comment on the strengths and weaknesses of each control of the schemes investigated: feedback alone, feedforward alone, and feedforward/feedback.

16.3 The temperature T, on a critical tray in a pyrolysis fractionator, is to be controlled as indicated in the following schematic diagram (Figure P16.2), employing a *feedforward controller* that manipulates the underflow reflux flowrate R on the basis of feed flowrate measurements F. By experimental testing, the following individual transfer function models were obtained for the response of the critical tray temperature (1) to feed rate changes, and (2) to reflux flowrate changes:

1. $$(T - T^*) = \frac{20e^{-15s}}{30s + 1} (F - F^*) \qquad (P16.2)$$

and

2. $$(T - T^*) = \frac{-5e^{-2s}}{6s + 1} (R - R^*) \qquad (P16.3)$$

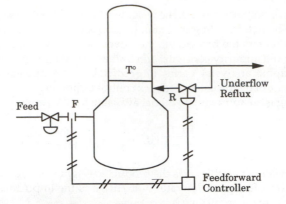

Figure P16.2. The pyrolysis fractionator.

the timescale is in minutes, the flowrates and temperature are, respectively, in Mlb/hr (thousands of pounds/hour) and °F, and the (*) indicates steady-state conditions.

(a) Rewrite the model in the form:

$$y(s) = g(s)\,u(s) + g_d(s)\,d(s)$$

stating explicitly what the variables y, u, and d represent in terms of the original process variables, and clearly identifying $g(s)$ and $g_d(s)$.

(b) Assuming that the specified process model is accurate, design a feedforward controller to achieve "perfect" feed disturbance rejection by giving an explicit expression that indicates how the reflux flowrate is made to change with observed variations in the feedrate. Is this feedforward controller realizable? Provide a physical interpretation for why, or why not.

(c) For the situation in which the process is operating under the indicated feedforward control strategy, obtain an expression for the reflux rate response to the step change of 2 Mlb/hr introduced in part (a). Plot the response.

16.4   The bottoms temperature of a distillation column, in °C (represented as y in deviation variables) is controlled by manipulating the steam flowrate to the reboiler, in lb/hr (represented as u_1 in deviation variables). This purely feedback

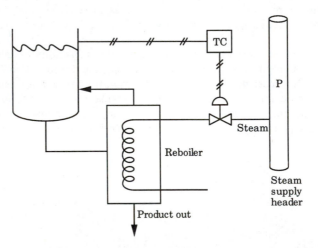

Figure P16.3. Feedback control of a distillation column's bottom temperature.

control strategy is shown in Figure P16.3. An approximate transfer function model for this process is given as:

$$y(s) = \frac{0.25}{10s + 1} u_1(s) \tag{P16.4}$$

However, the steam flowrate itself depends on the percent valve opening (represented as u_2 in deviation variables), and the steam supply pressure, in psi (represented as d_1 in deviation variables) which is known to fluctuate in an unpredictable, but measurable, fashion. These process variables are related according the following approximate model:

$$u_1(s) = \frac{2.2}{2s + 1} u_2(s) + \frac{1.5}{0.5s + 1} d_1(s) \tag{P16.5}$$

(a) As discussed in the main text, cascade control is a popular strategy for dealing with the control problems created by such steam pressure fluctuations. Draw a block diagram for this process under such a cascade control strategy; include all the given transfer functions (and the controllers), and label all the signals.

(b) Because the steam supply pressure is measured, it is also possible to configure this process for feedforward/feedback control *instead* of cascade control. Reconfigure the process for this new control structure, draw a block diagram showing this new configuration, and obtain an expression for the feedforward controller to be implemented. Briefly comment on which scheme (the cascade scheme in part (a), or the feedforward/feedback scheme) you think will be more effective and why. Under what condition will you choose one over the other?

16.5 Figure P16.4(a) shows the control scheme for controlling the effluent temperature of a CSTR in which an *endothermic* reaction is taking place. Because the temperature controller manipulates the preheater feed valve directly, the significant variations in T_i, the temperature of the reactant feed to the preheater, adversely affect the performance of this temperature control scheme.

In modifying this scheme for cascade control, the following block diagram shown in Figure P16.4(b) was obtained.

(a) In terms of the *physical system* variables, state carefully what the following quantities in the block diagram represent: $y, y_d, g_{c1}, u_{p1}, g_{c2}, g_v, g_2, d_1, g_{d1}, u,$ and g.

(b) Obtain the closed-loop transfer function relation between y and y_d, and between y and d_1. What is the closed-loop characteristic equation?

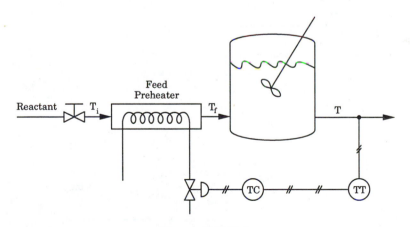

Figure P16.4(a).

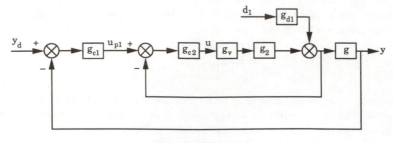

Figure P16.4(b).

(c) Given that both g_{c1} and g_{c2} are proportional controllers with gains K_{c1} and K_{c2} respectively, and that:

$$g(s) = \frac{2e^{-4s}}{3s + 1}$$

$$g_2(s)g_v(s) = \frac{2.5}{2s + 1}$$

$$g_{d1}(s) = \frac{e^{-2s}}{0.2s + 1}$$

for $K_{c2} = 2$, find the K_{c1} range over which the cascade system is stable. Use a first-order Padé approximation for the time delay involved.

(d) Consider the simpler case in which the reactor effluent temperature controller communicates directly with the preheater valve (as in Figure P16.4(a)). Redraw the process block diagram, and for the same process information as in part (c) (except that now there is no longer a *second controller*), find the K_{c1} range over which the simpler system is stable. Compare your result with that obtained in part (c).

16.6 The objective in the water heating tank process shown in Figure P16.5 is to maintain the outlet temperature at some desired set-point T_d by adjusting q, the quantity of heat sent to the tank via steam heating. The steam delivery pressure is P, and it experiences variations over which we have no control, but which affect q.
The following deviation variables are given:

$$T_i - T_i^* = d_1; \quad P - P^* = d_2; \quad T - T^* = y; \quad T_d - T_d^* = y_d; \quad q - q^* = u$$

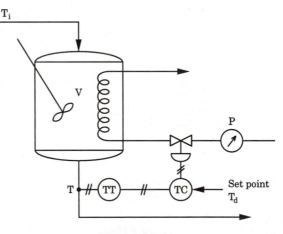

Figure P16.5.

The relevant transfer functions for the components of this closed-loop system are also given as follows: $g_1(s)$, process; $g_{d1}(s)$, disturbance (inlet stream temperature); $h(s)$, thermocouple; $g_c(s)$, feedback controller; $g_v(s)$, control valve; $g_{d2}(s)$, disturbance (steam pressure).

(a) Develop a block diagram for the closed-loop system showing all the components, and obtain the closed-loop transfer function equation.

(b) For the specific situation in which the following transfer functions are given:

$$g_1(s) = \frac{0.5}{5s + 1} \qquad\qquad g_{d1}(s) = \frac{1}{3s + 1}$$

$$g_v(s) = \frac{2}{0.5s + 1} \qquad\qquad g_{d2}(s) = \frac{0.15}{15s + 1}$$

$$h(s) = \frac{1}{0.2s + 1}$$

under PI control with $\tau_I = 0.2$, find the range of K_c values for which the system remains stable.

(c) If a cascade control strategy is to be employed, how should the second control loop be configured (i.e., what will it manipulate, and what measurement will it accept)? Draw the block diagram for the process under this strategy. If the new controller is specified as a proportional controller with gain $K_{c2} = 5.0$, with the primary controller as given in part (b), find the range of K_c values for which the cascade control system will be stable.

16.7 Revisit Problem 16.6, part (c).

(a) Take the secondary controller as the pure proportional controller specified in that problem, and use the stability margin technique, and the corresponding Ziegler-Nichols tuning rules, to obtain acceptable controller parameters if a PID controller is to be used in the primary loop.

(b) Evaluate the performance of the entire control system as designed in part (a) by obtaining the system response to simultaneous *unit* step changes in d_1, and d_2.

16.8 The primary control objective for a certain industrial polymer reactor is to control the polymer product melt viscosity; and two input variables, the reactor temperature, and the mole fraction of the chain transfer agent in the reactor, are available for this purpose. However, because the latter affects the melt viscosity on a much slower timescale than the former, reactor temperature is used as the primary control variable when close to steady-state operating conditions. An approximate model for this aspect of the process under such conditions is given by:

$$y(s) = \left(\frac{3.5}{5.0s + 1}\right)u_1(s) + \left(\frac{0.5}{20.0s + 1}\right)d_1(s) \qquad (P16.6)$$

where, in terms of deviation variables, y is the *normalized* melt viscosity (in dimensionless, industry standard form); u_1 is the reactor temperature, in °C; and d_1 is the reactor mole fraction of the chain transfer agent, in ppm.

The reactor temperature itself is not controlled directly; the viscosity controller merely adjusts the jacket cooling water valve, as indicated in Figure P16.6(a).

It has been observed, however, that the reactor temperature is influenced not only by the jacket cooling water flowrate, but also by variations in the jacket cooling water supply temperature, according to the following expression:

$$u_1(s) = \left(\frac{-1.5}{3.0s + 1}\right)u_2(s) + \left(\frac{0.3}{0.5s + 1}\right)d_2(s) \qquad (P16.7)$$

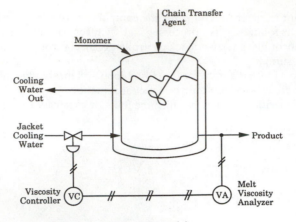

Figure P16.6(a).

where, again, in terms of deviation variables, u_2 is the cooling water flowrate, in gpm, and d_2 is the cooling water supply temperature, in °C.

Furthermore, the valve dynamics, and the (relatively infrequent, but not unimportant) fluctuations in the jacket cooling water supply pressure affect the cooling water flowrate according to the following expression:

$$u_2(s) = \left(\frac{2.3}{0.02s + 1}\right)u_3(s) + 1.5\, d_3(s) \tag{P16.8}$$

where, in terms of deviation variables, u_3 is the dimensionless jacket cooling percent valve opening, and d_3 is the cooling water supply pressure, in psi.

All the foregoing information notwithstanding, the conventional feedback control scheme indicated in Figure P16.6(a) is used; the viscosity controller has the yet unspecified transfer function, $g_c(s)$.

(a) Draw a block diagram for this process under this control strategy, taking care to label each essential element.

(b) Obtain a closed-loop transfer function expression, and use the Ziegler-Nichols stability margin tuning rules to design a PI controller for this process. Implement this controller and obtain the closed-loop system response to simultaneous step changes of +3°C in the cooling water supply temperature, and +2 psi in the supply pressure. Plot this response.

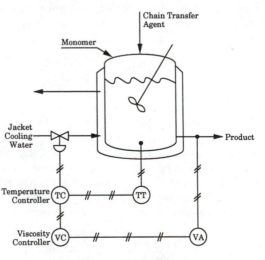

Figure P16.6(b).

(c) The cascade control scheme indicated in Figure P16.6(b) calls for a temperature controller to be interposed between the viscosity controller and the jacket cooling water valve. Redraw the block diagram for the modified process. Given that the temperature controller is a proportional controller with a gain of −4.0, and retaining the viscosity controller designed in part (b), obtain the response of this new closed-loop system to the same input disturbances investigated in part (b). Plot this response and compare it to that obtained in part (b).

16.9 This problem illustrates the physical interpretation, analysis, and design of multilayered cascade control loop systems.

Revisit the polymer process in Problem 16.8, and consider the introduction of yet another cascade control loop, as indicated in Figure P16.7; here a flow controller is now interposed between the temperature controller and the jacket cooling water valve.

(a) Redraw the block diagram for the modified process, and obtain a closed-loop transfer function expression.

(b) Given that the third controller is also a proportional controller, this time with gain 0.5, and retaining the two controllers as previously specified, obtain the response of this double cascaded control system to the same input disturbances investigated in Problem 16.8 parts (b) and (c). Plot this response and compare it to those obtained in Problem 16.8.

(c) In retrospect, by revisiting the description of the process given in Problem 16.8, explain on purely physical grounds why the single cascade system will be better than the pure feedback control system, and why the double cascaded system will be even better, whenever the cooling water supply temperature and pressure variations are significant.

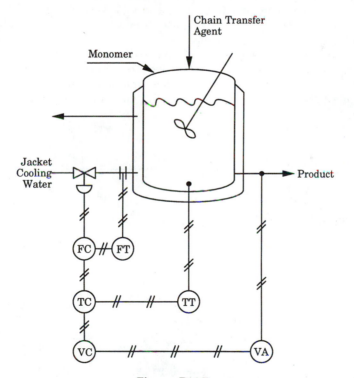

Figure P16.7.

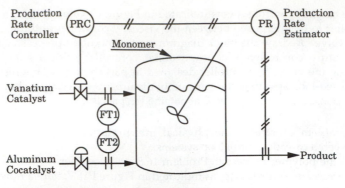

Figure P16.8.

16.10 A polymerization process requires the use of a Vanadium catalyst and an aluminum cocatalyst in a fixed ratio. However, the catalyst flowrate is used to control the production rate and therefore varies independently as shown in Figure P16.8.
Two flow transmitters are available for measuring the catalyst and cocatalyst flowrates. Briefly describe how you would go about designing a ratio control scheme for this process. Modify the process diagram is Figure P16.8 to show your ratio control scheme.

16.11 Figure P16.9 shows a typical application of a parallel control structure to obtain better exit temperature control in a heat exchanger (cf. Balchen and Mummé 1988[†]). The stream of cold liquid feed to be heated by the heat exchanger is divided into two: the main portion, of fixed flowrate, passes through the heat exchanger; the other portion, with an adjustable flowrate, bypasses the heat exchanger, and is later mixed with the heated stream, as indicated in Figure P16.9. It is desired to control the final, mixed stream, temperature T, with *both* the steam control valve on the primary side of the heat exchanger, as well as the heat exchanger bypass stream control valve, available for manipulation.
Because the effect of steam flowrate changes on T is on a slower timescale than the effect of the bypass stream flowrate, the *parallel* control scheme indicated in the figure is to be used. The temperature controller TC1, *a proportional-only controller,*

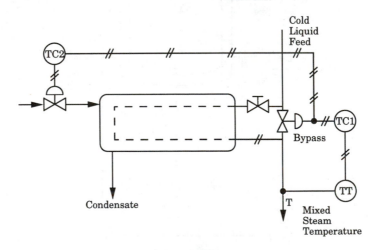

Figure P16.9.

[†] J. G. Balchen and K. I. Mummé, *Process Control: Structures and Applications*, Van Nostrand Reinhold (1988), pp. 33–35.

adjusts the bypass stream valve on the basis of the mixed stream temperature measurements. However, the *output* of this controller is fed to the second temperature controller, TC2, a PI controller, which uses this signal to adjust the steam valve.

(a) Draw a block diagram for this process, assuming that the transfer functions representing the effect of the bypass stream flow on T is $g_1(s)$, and that representing the effect of steam flowrate on T is $g_2(s)$.

(b) Provide a *qualitative* discussion of how such a control scheme will respond to a disturbance of a unit step decrease in the cold liquid feed temperature, assuming the bypass valve was initially 50% open just before this event. To what value will the bypass valve opening ultimately settle? Contrast this with a "standard" feedback scheme in which TC2 alone is used to control T. Which scheme do you think will be more effective, and why?

16.12 An opportunity for investigating the "parallel" control structure arises with the following temperature control problem for an exothermic reactor subject to feed temperature disturbances.

The reactor is traditionally cooled by jacket cooling water, and an approximate mathematical model for the process (in the vicinity of the normal operating conditions) has been obtained as:

$$y(s) = \frac{-4.0e^{-0.1s}}{15.0s + 1} u_1(s) + \frac{1.5e^{-0.1s}}{2.5s + 1} d(s) \tag{P16.9}$$

with y as the reactor temperature (°F), u_1 as the jacket cooling water flowrate (gpm), d as the reactant feed temperature (°F), all in terms of deviation variables. The $e^{-0.1s}$ is the approximate dynamics introduced by the temperature sensor.

In order to increase the cooling capacity, a condenser was installed as shown in Figure P16.10, with the objective of using condenser cooling water flowrate as an

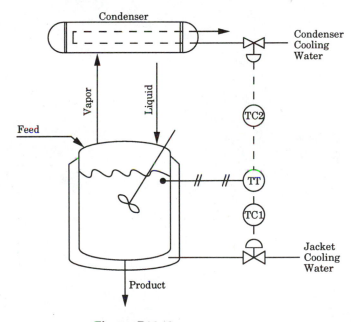

Figure P16.10.

additional control variable. An approximate model for the effect of condenser cooling water on reactor temperature is:

$$y(s) = \frac{-0.5e^{-0.1s}}{2s + 1} u_2(s) \qquad (P16.10)$$

(a) Assuming negligible valve dynamics, design a PI controller for the traditional process (i.e., with jacket cooling water to control reactor temperature) using the Ziegler-Nichols approximate model tuning rules (Table 15.2). Implement this controller and obtain the closed-loop response to a +1°F change in the feed temperature.

(b) Repeat part (a) for the situation in which the condenser cooling water *alone* is used to control the reactor temperature.

(c) Now consider the situation in which both the condenser and the jacket cooling water flows are to be used *simultaneously* in a parallel control arrangement. Let the "primary" controller, TC1, with a transfer function $g_{c1}(s)$, manipulate the jacket cooling water valve, and the "secondary" controller, TC2, with a transfer function $g_{c2}(s)$, manipulate the condenser cooling water valve.

Let TC1 be the PI controller obtained in part (a), while TC2 is a proportional controller obtained by dropping the integral portion of the PI controller of part (b). Implement this parallel control scheme, feeding the same feedback error to both controllers; obtain the closed-loop response to the same input disturbance investigated in parts (a) and (b). Plot the y response as well as both the u_1 and u_2 responses. Briefly comment on the nature of the observed control action.

CHAPTER 17

CONTROLLER DESIGN FOR PROCESSES WITH DIFFICULT DYNAMICS

The open-loop dynamic behavior of certain processes are so different from what one normally observes for most other processes that they pose special controller design problems. If we cast our minds back to our studies of the dynamic behavior of various processes in Part II, we will recall that processes with inverse response, as well as those with time delays, exhibit dynamic behavior that is qualitatively different from what we would expect of "normal" dynamic systems. Also in this class are processes that show open-loop instability.

As we might expect, the control problems created by these processes can be quite difficult. As a result, controller design strategies that will produce effective controllers for such processes must necessarily take into consideration the unusual open-loop dynamic behavior they exhibit. This chapter is thus concerned with controller design for processes having these difficult types of dynamics: *time delays, inverse response,* and *open-loop instability.*

17.1 DIFFICULT PROCESS DYNAMICS

So far in our discussion on control system design, we have tacitly assumed that, by itself in the open loop (before the controller is attached), the process to be controlled exhibits no especially difficult dynamic behavior that might require special attention. For such a system, if the input variable were, for example, increased from an initial steady-state value of u_1 to a new value of, say, $u_1 + \Delta u$, the dynamic behavior is considered *normal* if the output variable responds qualitatively as one of the responses depicted in Figures 17.1(a) and (b); i.e., the output satisfies the following conditions:

1. It begins to respond quickly without significant delay.
2. It heads *directly* for a new steady-state value without first taking an excursion in the opposite direction.
3. It finally settles to a new steady-state value.

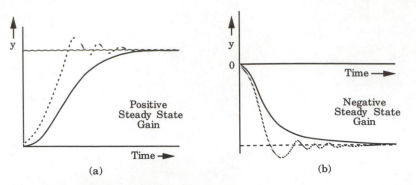

Figure 17.1.   Step responses for processes with "normal" open-loop dynamics.

The control system structures discussed up to this point do an adequate job of controlling processes with "normal" dynamics as we have seen from many previous examples.

By contrast, conventional control systems often do a poor job of controlling the three classes of processes with difficult dynamics. Thus we need to analyze the key features of these difficult processes and provide new control system designs for improved closed-loop performance of these processes.

17.1.1 Characteristics of Difficult Process Dynamics

From our studies in Part II, we know that the presence of any of these difficult characteristics in the dynamic behavior of a process will be identified in the process model as indicated below:

- **Time delay**   transfer function has the $e^{-\alpha s}$ term
- **Inverse response**   transfer function has a RHP *zero*
- **Open-loop instability**   transfer function has a RHP *pole*

Processes that exhibit "normal" behavior do not have any of these terms in their transfer function models.

As we recall from Chapter 8, when the response of a process is as depicted in Figure 17.2(a), i.e., it exhibits an initial, distinct period of no response, we have a time-delay (or dead-time) process.

A process with *time delay* thus violates the first condition noted above for normal dynamic behavior: it does not respond instantaneously to input changes. Unfortunately, a substantial number of chemical processes exhibit time-delay (or time-delay-like) behavior.

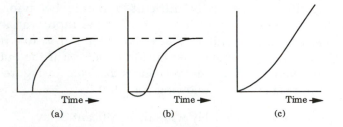

Figure 17.2.   Step responses for processes with difficult dynamics: (a) time delay, (b) inverse response, (c) open-loop unstable.

A process with *inverse response* violates the second condition for normal dynamic behavior in that, even though its step response eventually ends up heading in the direction of the new steady state, it starts out initially heading in the opposite direction, away from the new steady state, changing direction somewhere during the course of time. Such a response is depicted in Figure 17.2(b). Examples of physical processes that exhibit this unusual dynamic behavior were given in Chapter 7.

As we saw in Chapter 11, a process for which the step response is unbounded, i.e., the output increases (or decreases) indefinitely with time, as depicted in Figure 17.2(c) (or, also as in Figure 11.1(b)), is said to be *open-loop unstable* for obvious reasons. An open-loop unstable process thus violates the third condition noted above; its output fails to settle to a new steady-state value in response to a step change in the input. The most well-known chemical processes that exhibits open-loop instability is the exothermic CSTR.

Before we analyze these three classes of processes in more mathematical detail, it is useful to consider physical reasons why they would cause difficulty for a conventional feedback controller:

1. Processes with time delays have a significant delay before they respond to control action (cf. Figure 17.2(a)) so that controllers with aggressive action (high controller gain) will tend to overcompensate and become unstable. Thus there is a limit on the controller gain that can be used for a process with time delay.

2. Processes with inverse response will initially move in the wrong direction as they respond to control action (cf. Figure 17.2(b)). Thus if the controller is tuned too tightly (high controller gain) it will attempt to correct for the movement in the wrong direction and overcompensate. Again there is a limit on the controller gain that can be used for a process having inverse response.

3. Processes that are open-loop unstable will "run away" without control, so most of the controller tuning procedures of Chapter 15 cannot be applied. In addition, open-loop unstable processes can be unstable for various reasons so that simple PI control may not be enough to stabilize them.

These difficult dynamics translate into unusual phase behavior as we see next.

17.1.2 Nonminimum Phase (NMP) systems

Processes with time delays and with inverse response are sometimes collectively referred to as *nonminimum phase* (NMP) systems, a term first introduced by Bode [1]. Let us explain the meaning of this terminology.

A process with a general, "normal" transfer function, $g_1(s)$ has identical amplitude ratio (AR) characteristics as the process whose transfer function is $g_2(s) = g_1(s) e^{-\alpha s}$, but the phase angle characteristics are very different: that much we know from Chapter 9. If $g_1(s)$ has n poles and m zeros, as we may recall, its phase angle $\phi_1(s)$, approaches $(n - m) \times (-90°)$ asymptotically at high frequencies. The phase angle of $g_2(s)$, on the other hand, is given by:

$$\phi_2(s) = \phi_1(s) - \alpha\omega \qquad (17.1)$$

and decreases monotonically due to the influence of the delay term.

Thus of *all* possible transfer functions that show identical AR characteristics, the one possessing a time delay *cannot* be the one having the minimum possible phase characteristics.

In a similar vein, the two processes having transfer functions:

$$g_1(s) \; = \; g°(s)(1 + \eta s) \tag{17.2}$$

and

$$g_2(s) \; = \; g°(s)(1 - \eta s) \tag{17.3}$$

show identical AR characteristics but the asymptotic limit of the phase angle for $g_2(s)$ is $-180°$ in excess of the corresponding value for $g_1(s)$. Thus, the process $g_2(s)$ does *not* have the smallest possible phase characteristics of all processes with which it shares identical AR characteristics:

> *Thus in general, within a class of processes $\{g_1(s), g_2(s), ..., g_n(s)\}$ having amplitude ratio characteristics $\{AR_1, AR_2, ..., AR_n\}$, all identical (i.e., $AR_1 = AR_2 = ... = AR_n$), and phase angles $\{\phi_1, \phi_2, ..., \phi_n\}$ of which the minimum is designated ϕ_{min}, if g_i contains a time-delay element or a RHP zero, then $\phi_i \neq \phi_{min}$.*

It is in this sense that such systems are referred to as nonminimum phase systems.

It has been customary to reserve the term *nonminimum phase* for time-delay and inverse-response systems only (see for example [2]). What is not widely acknowledged, however, is the fact that open-loop unstable systems also exhibit nonminimum phase characteristics.

Observe, for example, that the first-order process:

$$g_1(s) \; = \; \frac{K}{\tau s + 1} \tag{17.4}$$

has identical AR behavior as the open-loop unstable process:

$$g_2(s) \; = \; \frac{K}{\tau s - 1} \tag{17.5}$$

(although the AR is meaningless since the unstable process does not have finite output amplitude); however, it can be shown that:

$$\phi_2 \; = \; -180° + \phi_1 \tag{17.6}$$

so that the open-loop unstable process is also nonminimum phase.

In fact, as originally presented by Bode [1], it was actually the RHP *pole* and the RHP *zero* — and, of course, the time delay — and not the systems in whose transfer functions they appear, that were referred to as the nonminimum phase elements. A system which therefore contains any nonminimum phase element, be it a RHP pole, a RHP zero, or a time delay, is a NMP system.

Observe therefore that the systems we have decided to focus attention on in this chapter share a common feature: they all contain nonminimum phase elements. The presence of NMP elements in a process system's transfer function is thus identified as being responsible for its "difficult" dynamic behavior; it is also the source of a considerable amount of difficulty in control systems design.

Another aspect of processes with "difficult" dynamics is presented in Chapter 19; there we show the serious problems which arise when, in order to achieve high performance, the controller must contain an inverse of the process model.

Let us now consider in more detail the special control problems presented by these systems, and the controller design techniques available for handling them.

17.2   TIME-DELAY SYSTEMS

17.2.1 Control Problems

When a time-delay element is present in a feedback control loop, the following obvious control problems arise:

1.   With a delay in the measuring device, control action will be based on delayed, hence obsolete, process information that is usually not representative of the current situation within the process.

2.   If the process itself has an input delay (because of a transport delay or other process feature), then the effect of the control action will not be immediately felt by the process, compounding the problem even further.

It is easy to see how such a situation could provoke instability.

To provide a quantitative sense of how much destabilization can be caused by the presence of a delay element in a process transfer function, recall Figure 9.3, the Bode diagram of a "normal" first-order system; observe that the phase angle asymptotically approaches a limiting value of $-90°$. Thus from our discussion in Chapter 15, this "normal" system can never go unstable under proportional feedback control since the phase angle can never attain the critical value of $-180°$.

By contrast, consider the Bode diagram for the open-loop transfer function of a first-order system with time delay given in Figure 9.18. Now the phase angle characteristics have been radically transformed: where there once was a limiting value of $-90°$, there is now in fact no limit; the phase angle decreases monotonically with frequency. Thus there is now a limiting value of proportional controller gain at which the phase angle crosses $-180°$ and the system becomes unstable. Furthermore, the higher the value of the time delay, the smaller this limiting value of controller gain becomes.

17.2.2   Conventional Feedback Controller Design

Let us consider the general model for a process with time delay as:

$$g(s) = g^*(s)\, e^{-\alpha s} \tag{17.7}$$

where $g^*(s)$ has "normal" dynamics. Under conventional feedback control as shown in Figure 17.3, the closed-loop system has the characteristic equation:

$$1 + g_c g^*(s) e^{-\alpha s} = 0 \tag{17.8}$$

As discussed above, the increased phase lag of the delay term requires, for closed-loop stability, a reduction in the allowable value of the controller gain. The closed-loop system will therefore have to be more sluggish than the corresponding system without delay. Therefore even though conventional controllers can be used for time-delay systems, we usually have to sacrifice speed of response in order to have closed-loop stability.

Classical Techniques

The techniques for designing conventional feedback controllers for time-delay systems are no different from what we have already discussed in Chapter 15. As evidenced by the performance of the conventional feedback controllers designed in Chapter 15 for the example time-delay system:

$$g(s) = \frac{6e^{-3s}}{15s + 1} \tag{17.9}$$

(see Figures 15.11–15.12) acceptable results can be obtained using these classical techniques *provided the time delay is not too large.* In this case, closed-loop stability, and acceptable overall closed-loop performance can be obtained without sacrificing too much of the speed of response.

The major problem with conventional feedback controller design for time-delay systems lies in the fact that for systems with large time delays, considerably smaller controller gains are recommended yielding sluggish response. Under these circumstances, it is customary to employ a scheme that explicitly takes the presence of the time delay into consideration and actively compensates for it. Such an approach is referred to, in general, as *time-delay compensation.*

17.2.3 Time-Delay Compensation

To implement time-delay compensation, we can alter the control scheme shown in Figure 17.3 to the new configuration of Figure 17.4 where a minor feedback loop has been introduced around the conventional controller, as shown. The subscript m denotes a model, while unsubscripted elements are the real process. Thus:

$$y_m(s) = g_m^*(s) e^{-\alpha_m s} u(s) \tag{17.10}$$

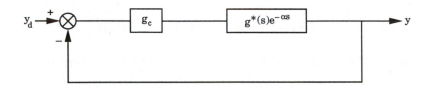

Figure 17.3.   Block diagram of a time-delay system under conventional feedback control.

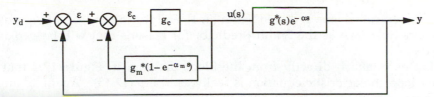

Figure 17.4.   Block diagram incorporating the Smith predictor.

represents the model of the real process. Let us define new variables $y^*(s)$, $y_m^*(s)$ as follows:

$$y^*(s) = g^*(s)\, u(s) \qquad\qquad (17.11)$$

$$y_m^*(s) = g_m^*(s)\, u(s)$$

Since the actual process output y is given by:

$$y(s) = g^*(s)e^{-\alpha s}\, u(s) \qquad\qquad (17.12)$$

then $y^*(s)$ is the output of the "undelayed" version of the process output $y(s)$.

Let us proceed to analyze the time-delay compensator assuming for the moment that there are no model errors (so that $g_m(s) = g(s)$ and $\alpha_m = \alpha$). Observe that the signal reaching the controller, designated as ε_c in the diagram, is a "corrected" error signal given by:

$$\varepsilon_c = y_d - y(s) - [y^*(s) - y(s)] \qquad\qquad (17.13)$$

or

$$\varepsilon_c = y_d - y^*(s) \qquad\qquad (17.14)$$

implying, as a result, that the block diagram of Figure 17.4 is apparently equivalent to that shown in Figure 17.5. If this is so (and we will soon show that it is) then the net result of the introduction of the minor loop is therefore to eliminate the time-delay factor from the feedback loop — where it causes stability problems — and "move" it outside of the loop, where it has no effect on closed-loop system stability. The characteristic equation of the equivalent system in Figure 17.5 is:

$$1 + g_c g^*(s) = 0 \qquad\qquad (17.15)$$

which no longer contains the time-delay element and therefore allows the use of higher controller gains without placing the closed-loop stability in jeopardy.

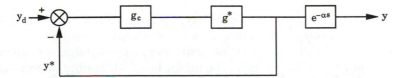

Figure 17.5.   Equivalent block diagram for system incorporating the Smith predictor.

This control scheme is due to Smith [3, 4], and the minor loop of Figure 17.4 is generally known as the Smith predictor for reasons that will be explained shortly.

Let us establish directly from the block diagram in Figure 17.4 that the closed-loop characteristic equation is really as in Eq. (17.15). Again, assuming that there are no model errors, we can consolidate the minor loop around the controller to obtain the transfer-function relationship between ε and u as:

$$u = g_c^* \varepsilon \tag{17.16}$$

where

$$g_c^* = \frac{g_c}{1 + g_c g^* (1 - e^{-\alpha s})} \tag{17.17}$$

The overall closed-loop transfer function is now given by:

$$y = \frac{g^*(s) e^{-\alpha s} g_c^*}{1 + g^*(s) e^{-\alpha s} g_c^*} y_d \tag{17.18}$$

If we now introduce Eq. (17.17) for g_c^*, we obtain:

$$g^*(s) e^{-\alpha s} g_c^* = \frac{g^* g_c e^{-\alpha s}}{1 + g^* g_c - g^* g_c e^{-\alpha s}} \tag{17.19}$$

so that:

$$1 + g^*(s) e^{-\alpha s} g_c^* = \frac{1 + g^* g_c}{1 + g^* g_c - g^* g_c e^{-\alpha s}} \tag{17.20}$$

Simplifying the expression for the closed-loop transfer function in Eq. (17.18) by combining Eqs. (17.19) and (17.20) gives:

$$y = \left(\frac{g^* g_c}{1 + g^* g_c} \right) e^{-\alpha s} y_d \tag{17.21}$$

establishing two things:

1. The Figure 17.5 block diagram is indeed equivalent to that in Figure 17.4, because Eq. (17.21) is the closed-loop transfer function for the Figure 17.5 system.

2. The characteristic equation for the system is as given in Eq. (17.15), by inspection of the denominator Eq. (17.21).

The specialized control scheme we have just considered is known in general as *time-delay compensation* because, as we have seen, the minor loop (which differentiates it from the conventional controller) was introduced to *compensate* for the presence of the time delay. The minor loop is therefore often referred to as a *compensator*.

It is important to note the following points about this time-delay compensation scheme:

1. The effective action of the compensator is to feed the signal y^* to the controller instead of the actual process output y.

2. By the definition in Eqs. (17.11) and (17.12), we may observe that:

$$y^*(s) = e^{\alpha s} y(s)$$

so that:

$$y^*(t) = y(t + \alpha) \qquad (17.22)$$

and it becomes perfectly clear that $y^*(t)$ is a prediction of $y(t)$ exactly α time units ahead, hence the name "Smith predictor" generally associated with the scheme.

3. The scheme will work perfectly as long as the process model is perfectly known; modeling errors will obviously affect its performance. The most significant criticism of the Smith predictor technique is its sensitivity to modeling errors.

4. In those chemical processes for which the time delays are due to transport of material and/or energy through long pipes, the time delays observed for such processes will vary with the fluid flowrate; an increase in flowrate giving rise to lower time delays, and vice-versa. The Smith predictor scheme is designed for constant time delays and may therefore not perform as well for systems with time delays which vary significantly over time.

Design Procedure

The following is the design procedure for using the Smith predictor for time-delay compensation:

1. *Design the Smith Predictor (the minor loop)*
   Recall that the implication of this minor loop is to cause the controller to utilize a "corrected error" signal ε_c instead of the actual error ε; i.e., the control law is:

$$u = g_c \varepsilon_c \qquad (17.23)$$

where

$$\varepsilon_c = \varepsilon - (y^* - y) \qquad (17.24)$$

The design of the minor loop involves setting up a means by which y^* and y are produced from the process model, typically by simulation, using a digital computer; y is obtained directly from the process model, and y^* is obtained from the undelayed version of the process model.

2. *Design g_c*

Since, according to the Smith predictor scheme, the controller "thinks" it is controlling the process $g^*(s)$ for which there is no time delay (see Figure 17.5), we now design the controller on this basis, using any of the previously discussed techniques; i.e., design g_c for the undelayed version of the time-delay system. Observe that the absence of the time delay from the apparent process permits the use of much higher controller gains than would otherwise be allowable. However, since, for perfect delay compensation, the Smith predictor requires a perfect model, and real models are never perfect, we must be cautious in choosing the controller parameters for g_c. In practice, one would choose controller parameters large enough to achieve much better performance than feedback control alone, but not so large as to cause serious deterioration in performance resulting from inevitable plant/model mismatch.

17.3   INVERSE-RESPONSE SYSTEMS

17.3.1 Control Problems

The major problem created by the inverse-response system is the confusing scenario it presents to the automatic controller: observe that having taken the proper action that will eventually yield the desired result, the controller is first given the impression that it, in fact, took the wrong action. The "uninformed" controller, in reacting to this illogical state of affairs, is liable to compound the problem further. It is therefore not difficult to see that such a system's closed-loop stability stands in real jeopardy.

One can obtain a sense of the destabilizing effect caused by the presence of a RHP zero in a system's transfer function by recalling Figure 9.16, the Bode diagram for the process whose open-loop transfer function is:

$$g(s) = \frac{(1 - 3s)}{(2s + 1)(5s + 1)} \tag{17.25}$$

Without the RHP zero in the numerator, the transfer function will be second order, and from our results in Chapter 9, we know that the phase angle will asymptotically approach a limiting value of $-180°$ — implying that the closed-loop system will always be stable, since, theoretically, the phase angle only attains the critical value of $-180°$ at infinite frequency.

With the RHP zero present, observe from Figure 9.16 that the limiting value for the phase angle has been altered to $-270°$ and as such, there is now a finite (crossover) frequency at which $\phi = -180°$, and hence there is a limiting value of K_c above which the system will be unstable.

Typically, inverse-response systems are controlled either using conventional PID control or an inverse-response compensator, similar in nature to the time-delay compensator discussed in the previous section. Let us look at the details of these controller designs more closely.

17.3.2 Conventional Feedback Controller Design

Because the derivative mode of the PID controller endows it with "anticipatory" character, this controller, of all the conventional controllers, is able to cope somewhat with controlling the inverse-response system; it does this by anticipating the "wrong way behavior" and appropriately accommodating it.

It is not difficult to visualize how the PID controller achieves this: at the initial stage of the process response, rather than the error reducing in response to the correct controller action, it actually increases because of the inversion. However, the derivative of the response is *negative* during this period, and when this information is incorporated into the controller equation (as is the case with the PID controller), the result is a net reduction in the magnitude of the control action. On the other hand, immediately after the inversion is over, and the response begins heading in the right direction, the derivative is *positive* and usually quite large; with the PID controller, this translates to a net increase in control action.

Thus, at the initial stage, when the initial, temporary increase in the error would mislead other conventional controllers into increasing the control action (a potentially dangerous situation), the PID controller, with its access to the derivative of the error, actually cuts back on the applied control action; when the initial period of inversion is over, and it is now advantageous to step up the magnitude of control action in order to make up for lost ground, the PID controller is again able to do precisely this, because of the positive derivative of the error in this region.

Classical Techniques: Ziegler-Nichols Designs

As demonstrated by Waller and Nygardas [5], the Ziegler-Nichols settings for PID controllers yield acceptable control of inverse-response systems. (Later in Chapter 19, we will uncover another justification for why, of all conventional feedback controllers, only the PID type can be used to control the inverse-response system with any degree of success.) Let us illustrate with an example.

Example 17.1   DESIGN OF A CONVENTIONAL PID CONTROLLER FOR AN INVERSE-RESPONSE SYSTEM BY THE ZIEGLER-NICHOLS METHOD.

Design a PID controller for the inverse-response system of Eq. (17.25) using the frequency-response Ziegler-Nichols technique.

Solution:

Figure 17.6 shows the Bode diagram for the system under proportional-only control: i.e., the open-loop transfer function is given by:

$$g(s)g_c(s) = \frac{K_c(1-3s)}{(2s+1)(5s+1)} \qquad (17.26)$$

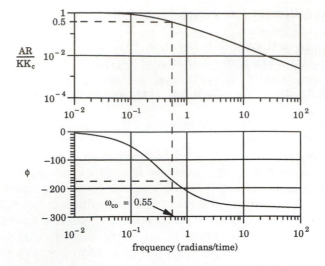

Figure 17.6.   Bode diagram for Example 17.1.

From this diagram, we obtain the following critical pieces of information required for the Ziegler-Nichols design:

- The crossover frequency: ω_{co} = 0.55 radians/time
- The magnitude ratio at this point: MR = AR/KK_c = 0.5

From here, it is easy to see that the ultimate gain and period are given, respectively, by:

$$K_u = 2 \tag{17.27a}$$

and

$$P_u = 11.43 \tag{17.27b}$$

The Ziegler-Nichols recommended values for the PID controller parameters are now obtained as:

$$K_c = 1.2; \qquad \tau_I = 5.7; \qquad \tau_D = 1.4$$

We will evaluate the performance of this controller shortly.

Classical Techniques: Cohen and Coon Designs

It is also possible to use the Cohen and Coon process reaction curve technique for obtaining PID controller designs for inverse-response systems; again the procedure is the same as before:

1. Obtain the process reaction curve (PRC),
2. Characterize the PRC with a steady-state gain, an effective time constant, and an effective time delay,
3. Use the characterizing parameters in the Cohen and Coon PID controller parameter formulas of Chapter 15.

Of all these steps, characterizing the PRC approximately as a first-order system with a time delay is typically the most difficult from a practical standpoint. Naturally, the difficulty lies in deciding on where to draw the tangent line in order to estimate both the effective time delay and the effective time constant. We recommend the following technique for taking the guesswork out of this step.

Recall that at the end of Chapter 8 we introduced Padé approximations for time-delay systems. We see that this particular approximation introduces a RHP *zero* element as well as a mirror image LHP *pole* into the transfer function as an approximation for the time-delay element.

This suggests that, by reversing the Padé approximation, a RHP zero can also be approximated by a time-delay element and a mirror image LHP *zero*; i.e., if:

$$e^{-\alpha s} \approx \frac{1 - \dfrac{\alpha}{2}s}{1 + \dfrac{\alpha}{2}s}$$

then, by the same token:

$$1 - \eta s \approx (1 + \eta s)\, e^{-2\eta s} \tag{17.28}$$

Let us illustrate how to apply this procedure with the following example.

Example 17.2   APPROXIMATE TIME-DELAY MODEL FOR AN INVERSE-RESPONSE SYSTEM USING A REVERSE PADÉ APPROXIMATION.

Obtain an approximate time-delay model for the inverse-response system of Eq. (17.25) using the reverse Padé approximation and compare the unit step responses of both the inverse-response system and its time-delay approximation.

Solution:

Since the original system transfer-function model has a RHP zero for which $\eta = 3$, then according to Eq. (17.28) using the reverse Padé approximation requires replacing this with: $(1 + 3s)\, e^{-6s}$. The approximate time-delay model is therefore given by:

$$\tilde{g}(s) = \frac{(1 + 3s)e^{-6s}}{(2s + 1)(5s + 1)} \tag{17.29}$$

Figure 17.7 shows the unit step response of the original inverse-response system of Eq. (17.25) and the time-delay approximation Eq. (17.29), indicating quite reasonable agreement.

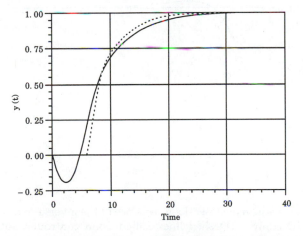

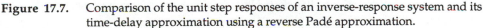

Figure 17.7.   Comparison of the unit step responses of an inverse-response system and its time-delay approximation using a reverse Padé approximation.

Let us now demonstrate the design of a PID controller by the Cohen and Coon process reaction method with the following example.

Example 17.3   DESIGN OF A CONVENTIONAL PID CONTROLLER FOR AN INVERSE-RESPONSE SYSTEM BY THE COHEN AND COON METHOD.

Design a PID controller for the inverse-response system of Example 17.1 (whose transfer function is given in Eq. (17.25)) using the process reaction curve method to obtain the Cohen and Coon controller parameter recommendations. Compare the behavior of the closed system in response to a unit step change in the set-point under this controller with the corresponding behavior under the Ziegler-Nichols PID controller.

Solution:

Recall that the unit step response for this process and the approximate time-delay response is shown in Figure 17.7. From the figure the time delay and steady-state gain are directly available; the effective time constant is easily obtained by taking the slope of the approximate time-delay response at the origin. The values taken by these parameters are:

$$K = 1; \quad \alpha = 6; \quad \tau = 4$$

The Cohen and Coon PID formula from Chapter 15 now recommends the following parameter values:

$$K_c = 0.91; \quad \tau_I = 9.84; \quad \tau_D = 1.71$$

Figure 17.8 shows the closed-loop system response to a unit step change in the set-point under Cohen-Coon PID controller design and under the Ziegler-Nichols PID controller design of Example 17.1. Both controllers perform very well, with the Cohen-Coon PID controller being somewhat more conservative.

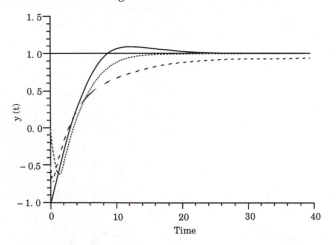

Figure 17.8.   Comparison of closed-loop response of the inverse-response system under PID control. (Dashed line: Cohen-Coon controller; solid line: Ziegler-Nichols Controller; dotted line: inverse-response compensator).

It should be mentioned that the observed conservative nature is actually not a reflection on the Cohen and Coon method which is known for producing less conservative controllers than the Ziegler-Nichols method. The conservative nature of the Cohen-Cohen PID controller in this case stems from the method by which the inverse-response reaction curve has been approximated. In general, the normal graphical tangent method will yield a time-delay factor that will be considerably less than that obtained using the reverse Padé approximation technique, with the net effect that the normal Cohen-Coon controller will be less conservative.

17.3.3 Inverse-Response Compensation

The same principles behind the time-delay compensation technique discussed above have been extended to inverse-response systems by Iinoya and Altpeter [6]. What we now present is a slightly different, and more general version of what is available in Ref. [6].

Consider the block diagram shown in Figure 17.9 for the conventional feedback control of an inverse-response system with transfer-function representation $g(s) = g°(s)(1 - \eta s)$; where $g°(s)$ represents that portion of the transfer function left after factoring out the problematic RHP zero element. For example, for the transfer function in Eq. (17.25):

$$g°(s) = \frac{1}{(2s + 1)(5s + 1)} \tag{17.30}$$

Now consider the situation in Figure 17.10 in which a minor loop is introduced as shown, with the transfer function g' given by:

$$g'(s) = g°(s)\lambda s \tag{17.31}$$

The objective here is to choose the quantity λ such that the signal reaching the controller appears to be from a "normal" system; in this context, one which does not exhibit inverse response.

Let us define the variable y' according to:

$$y'(s) = g'(s)u(s) \tag{17.32}$$

and observe that this is the signal generated by the minor loop. As a result of this minor loop, the signal reaching the controller, designated as ε_c in the diagram, is given by:

$$\varepsilon_c = y_d - y(s) - y'(s) \tag{17.33a}$$

or

$$\varepsilon_c = y_d - [g(s) + g'(s)] \, u(s) \tag{17.33b}$$

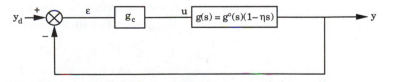

Figure 17.9.   Conventional feedback control of an inverse-response system.

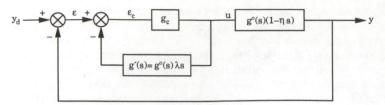

Figure 17.10. Block diagram incorporating the inverse-response compensator.

Now, let:

$$g^*(s) = g(s) + g'(s) \tag{17.34}$$

and

$$y^*(s) = g^*(s)\, u(s) \tag{17.35}$$

then Eq. (17.33b) becomes:

$$\varepsilon_c = y_d - y^* \tag{17.36}$$

As a result of Eq. (17.36), observe that we now have a situation in which, as far as the controller, $g_c(s)$, is concerned, it thinks it is interfaced with the apparent process that generates the signal y^*, and whose transfer function, $g^*(s)$, is as given in Eq. (17.34).

If we now introduce the expressions for g and g' into Eq. (17.34), we obtain:

$$g^*(s) = g°(s)(1 - \eta s) + g°(s)\lambda s \tag{17.37}$$

or

$$g^*(s) = g°(s)[1 + (\lambda - \eta)s] \tag{17.38}$$

Observe that if we now choose λ such that:

$$\lambda \geq \eta \tag{17.39}$$

the transfer function of the apparent process generating y^* no longer contains a RHP zero. Thus the minor loop provides a corrective signal that effectively eliminates the inverse response from the feedback loop. Owing to its similarity to the Smith predictor, the inverse-response compensator also suffers from the same sensitivity to model inaccuracies.

There is some strategy in selecting the exact value of λ satisfying Eq. (17.39). Choosing $\lambda > \eta$ (as opposed to $\lambda = \eta$) has some advantage in case of plant-model mismatch. However, one must be careful not to chose λ too large because then $g^*(s)$ will have a much faster response than $g°(s)$ (see discussion in Chapter 6) and the control loop will be sluggish. It can be shown that the choice:

$$\lambda = 2\eta \tag{17.40}$$

is optimal, providing minimum mean square deviation of plant output from desired set-point.

Let us demonstrate the design of the inverse-response compensator and the influence on closed-loop system stability with the following examples.

Example 17.4   DESIGN OF AN INVERSE-RESPONSE COMPENSATOR.

Design an inverse-response compensator for the inverse-response system of Eq. (17.25).

Solution:

The design of the inverse-response compensator boils down to obtaining the appropriate transfer function $g'(s)$ to use in the minor feedback loop in the block diagram shown in Figure 17.10.

For the process given by Eq. (17.25), $g°(s)$ is given by:

$$g°(s) = \frac{1}{(2s + 1)(5s + 1)} \qquad (17.41)$$

and therefore:

$$g'(s) = \frac{\lambda s}{(2s + 1)(5s + 1)} \qquad (17.42)$$

For this system $\eta = 3$; thus according to Eq. (17.40) a good value of λ to be used in Eq. (17.42) is $\lambda = 6$ so that:

$$g'(s) = \frac{6s}{(2s + 1)(5s + 1)} \qquad (17.43)$$

is the transfer function to use in the inverse-response compensator loop.

Using this transfer function in the inverse-response compensator loop produces the effect of interfacing the controller with the apparent process whose transfer function is given by:

$$g*(s) = \frac{(3s + 1)}{(2s + 1)(5s + 1)} \qquad (17.44)$$

which does not have a RHP zero.

The influence that the inverse-response compensator has on the closed-loop system stability characteristics can be dramatic, as illustrated in the following example.

Example 17.5   CLOSED-LOOP STABILITY OF AN INVERSE-RESPONSE SYSTEM UNDER CONVENTIONAL FEEDBACK CONTROL WITH AND WITHOUT THE COMPENSATOR.

Investigate the closed-loop stability properties of the inverse-response system of Eq. (17.25) under proportional-only feedback control, first without any compensation, and then with the inverse-response compensator designed in Example 17.4.

Solution:

Under conventional, proportional feedback control, the characteristic equation for the closed-loop system is:

$$1 + \frac{K_c(1 - 3s)}{(2s + 1)(5s + 1)} = 0 \qquad (17.45)$$

which rearranges to:

$$10s^2 + \left(7 - 3K_c\right)s + \left(1 + K_c\right) = 0 \qquad (17.46)$$

From here, we observe that the condition for stability is:

$$K_c < 7/3 \qquad (17.47)$$

any value of K_c higher than this will make the system unstable.

With the inverse-response compensator, let us first obtain the overall closed-loop transfer function from the block diagram of Figure 17.10; we will then deduce the characteristic equation from here and use it to investigate the stability properties.

Consolidating the block diagram, dealing first with the minor loop, gives:

$$u = g_c{}^*\varepsilon \qquad (17.48)$$

where

$$g_c{}^* = \frac{g_c}{1 + g_c g'} \qquad (17.49)$$

The overall closed-loop transfer function is now given by:

$$y = \frac{g g_c{}^*}{1 + g g_c{}^*} y_d \qquad (17.50)$$

and the characteristic equation is:

$$1 + g g_c{}^* = 0 \qquad (17.51)$$

Upon introducing into Eq. (17.51) all the appropriate expressions for the indicated transfer-function elements, and rearranging, the characteristic equation becomes:

$$10s^2 + 7s + \left(1 + K_c\right) = 0 \qquad (17.52)$$

which is stable for *all* positive values of K_c; clearly demonstrating the restrictive effect of the RHP zero on closed-loop system stability and the advantages of using the inverse-response compensator.

To illustrate the performance of the inverse-response compensation compared to conventional PID control, the response to a unit set-point change with the inverse-response compensator combined with a PI controller ($K_c = 10$, $1/\tau_I = 0.167$) is shown in Figure 17.8 together with two responses of conventional feedback controllers using PID control. Note that the inverse-response compensator produces the smallest negative deviation and a rapid response without overshoot.

Design Procedure

As with the time-delay compensator, the following is the design procedure for designing an inverse-response compensation scheme:

1. *Design the inverse-response compensator loop*
   As demonstrated in Example 17.4, this simply involves obtaining the appropriate transfer function $g'(s)$ to use in the minor feedback loop in the block diagram shown in Figure 17.10; it specifically entails finding λ which satisfies the conditions in Eq. (17.39) and using this in the expression for $g'(s)$ in Eq. (17.31).

2. *Design* g_c

   Once the compensator has been designed, its effect is to make the controller "think" it is interfaced to the process $g^*(s)$ which has no RHP zero. Again, as with the time-delay compensator, we now simply design the controller for $g^*(s)$; and the absence of the RHP zero permits the use of higher controller gains than would otherwise be possible. However, because of process/model mismatch, one must be careful not to increase the controller gains too aggressively.

17.4   OPEN-LOOP UNSTABLE SYSTEMS

The transfer function that describes the dynamic behavior of an open-loop unstable system has at least one RHP pole. The obvious problem with such a system is that, in its natural state, it is unstable, so that any upset in any direction will result in unstable response. This creates one of the most significant difficulties in control system design; i.e., one cannot use the standard process model identification procedures, but must keep the process under control while carrying out the modeling experiments. If one wishes to obtain only the critical controller gains, we can use the "auto tuning relay controller" of Chapter 15 to determine the upper bound on controller gain and then decrease the gain in closed-loop experiments until the lower bound is determined. If one wishes to obtain an explicit model, one can identify the model under feedback control with known controller parameters using a sequence of set-point changes and disturbances. The open-loop model can then be determined.

This difficult model identification problem is balanced by the fact that a process which is unstable in the open loop can usually be controlled quite well by conventional feedback control *if the controller parameters are carefully chosen*. Let us illustrate with some examples.

Example 17.6   STABILIZATION OF A FIRST-ORDER OPEN-LOOP UNSTABLE SYSTEM BY PROPORTIONAL FEEDBACK CONTROL.

Obtain the range of K_c values required to ensure that the closed-loop system involving the first-order system:

$$g(s) = \frac{K}{\tau s - 1} \tag{17.53}$$

and a proportional controller is stable.

Solution:

The characteristic equation for this closed-loop system is:

$$1 + \frac{KK_c}{\tau s - 1} = 0 \tag{17.54}$$

which rearranges to give:

$$\tau s - 1 + KK_c = 0 \tag{17.55}$$

an equation with only one root located at:

$$s = \frac{1 - KK_c}{\tau} \tag{17.56}$$

Thus, observe that this root will be negative (and the closed-loop system will be stable) as long as:

$$K_c > \frac{1}{K} \tag{17.57}$$

with the immediate implication that any K_c value which satisfies Eq. (17.57) will stabilize the otherwise open-loop unstable system.

Beyond merely stabilizing the system, observe that we may go one step further and obtain the values of controller parameters that will not merely stabilize the closed-loop system, but which will place the closed-loop system poles in prespecified locations in the LHP.

Let us demonstrate this with the following example.

Example 17.7 STABILIZATION OF AN OPEN-LOOP UNSTABLE SYSTEM BY CLOSED-LOOP POLE PLACEMENT

Design a PI controller for the system of Example 17.6 that will stabilize the closed-loop system with the closed-loop system poles located at $s = -2$, and at $s = -4$.

Solution:

The closed-loop characteristic equation in this case becomes, upon some elementary rearrangement:

$$\tau\tau_I s^2 + (KK_c - 1)\,\tau_I s + KK_c = 0 \tag{17.58}$$

The roots of this quadratic are related to its coefficients by:

$$-(r_1 + r_2) = \frac{(KK_c - 1)}{\tau} \tag{17.59}$$

and

$$r_1 r_2 = \frac{KK_c}{\tau\tau_I} \tag{17.60}$$

Introducing the process parameters, $K = 2$, $\tau = 5$ and solving these two equations simultaneously gives the required controller parameter values:

$$K_c = 15.5; \; \tau_I = 0.775 \tag{17.61}$$

It should be mentioned that not all open-loop unstable processes can be stabilized by P control or PI control. It is left as an exercise to the reader (also one of the problems at the end of the chapter) to establish that, using a P or a PI controller, it is *impossible to stabilize* the process whose open-loop transfer function is given as:

$$g(s) = \frac{2}{(2s - 1)(5s + 1)} \tag{17.62}$$

Here either a PD or a PID controller is required for stabilization. This illustrates the fact that open-loop unstable systems can require special types of controllers to stabilize them.

Open-loop unstable systems can also have *conditional* closed-loop stability, where there is only a range of controller gains, $K_{c_l} < K_c < K_{c_u}$, which stabilize the process. This occurs whenever there is an upper limit in controller gain (e.g., when the open-loop phase angle is less than $-180°$ for some frequency) combined with the lower threshold level of controller gain necessary for closed-loop stability of the open-loop unstable system. Let us illustrate with an example.

Example 17.8 CONDITIONAL CLOSED-LOOP STABILITY OF AN OPEN-LOOP UNSTABLE SYSTEM.

Consider the open-loop unstable process given by:

$$g(s) = \frac{2(-s + 1)}{(4s - 1)(2s + 1)} \tag{17.63}$$

under proportional feedback control. Determine the range of controller gains for which the closed-loop system is stable.

Solution:

The characteristic equation for the closed-loop system is given by:

$$8s^2 + 2(1 - K_c)s + (2K_c - 1) = 0 \tag{17.64}$$

and by inspection, the stable range of controller gain is $\frac{1}{2} < K_c < 1$. Thus the system exhibits conditional stability.

17.5 SUMMARY

We have focused attention in this chapter on those processes that exhibit such dynamic behavior in the open loop as to make them quite difficult to control, for the most part, by conventional means. We have seen that there are three classes of such systems: time-delay systems, inverse-response systems, and open-loop unstable systems; we have also seen that they have something in common: they each have *nonminimum phase elements*. The presence of these nonminimum phase elements (the exponential term for the time-delay system; the RHP zero for the inverse-response system; and the RHP pole for the open-loop unstable system) constitutes the main source of difficulty in designing effective controllers for these systems.

Particularly for the time-delay and the inverse-response systems, whenever the influence of the nonminimum phase element is not too strong, conventional feedback controllers can give acceptable results. However, compensators introduced as minor loops around the controllers are recommended when the nonminimum phase elements exert substantial influence; i.e., when the time delay is very large, and for the inverse-response system, when the negative lead time constant is large. Aside from their sensitivity to modeling errors, these compensators can be quite useful in that they enable the use of higher controller gains; thus improving the overall closed-loop performance. Explicit procedures for designing control systems that incorporate these compensators have been given.

The open-loop unstable system was shown to be a comparatively less demanding system to control; it seldom requires more than closed-loop stabilization by conventional feedback control. However, the design procedure can be difficult because the identification of the open-loop model is difficult and the conditional stability properties make controller tuning more complex.

REFERENCES AND SUGGESTED FURTHER READING

1. Bode, H. W., *Network Analysis and Feedback Amplifier Design*, Von Nostrand, Princeton (1985)
2. Morari, M. and E. Zafiriou, *Robust Process Control*, Prentice-Hall, Englewood Cliffs, NJ (1989)
3. Smith, O. J. M., "Close Control of Loops with Dead Time," *Chem. Eng. Prog.*, **53**, (5), 217 (1957)
4. Smith, O. J. M., "A Controller to Overcome Deadtime," *ISA Journal*, **6**, (2), 28 (1959)
5. Waller, K. and C. Nygardas, "On Inverse Response in Process Control," *I&EC Fund.*, **14**, 221 (1975)
6. Iinoya, K. and R. J. Altpeter, "Inverse Response in Process Control," *I&EC*, **54**, (7), 39 (1962)

REVIEW QUESTIONS

1. What are the three classes of processes with difficult dynamics discussed in this chapter, and how does the process transfer-function model indicate the presence of such difficult dynamic characteristics?

2. What is a nonminimum phase system?

3. Why are the time delay, the RHP zero, and the RHP pole considered nonminimum phase elements?

4. In physical terms, how can the presence of a time delay cause problems in conventional feedback control?

5. In terms of phase angle characteristics, how does the presence of a time delay cause problems in conventional feedback control?

6. In applying conventional feedback control for time-delay systems, what is usually sacrificed, and why?

7. When does it become *inadvisable* to apply conventional feedback control for a time-delay system?

8. What is the Smith predictor, and why is it so called?

9. What is the procedure for controller design incorporating a Smith predictor for a time-delay system?

10. How is the minor loop of the Smith predictor scheme usually implemented?

11. Why do you think the performance of the Smith predictor scheme will be sensitive to modeling errors?

12. What is the main problem created by an inverse-response system under conventional feedback control?

13. Of all the conventional feedback controllers, which kind has the potential of coping with the problems created by inverse response? Why is this controller type able to cope with this problem?

14. How is a reverse Padé approximation useful in conventional feedback controller design for inverse-response systems?

15. What is the main objective in the design of an inverse-response compensator?

16. How is the inverse-response compensator similar to the Smith predictor?

17. What is the procedure for controller design incorporating an inverse-response compensator for an inverse-response system?

18. What poses the more significant challenge in dealing with an open-loop unstable system: model identification, or controller design?

19. Why is it not possible to use standard process identification procedures for open-loop unstable processes?

20. Can *any* process that is unstable in the open loop be stabilized by *any* conventional feedback controller?

PROBLEMS

17.1 For the time-delay process with the following transfer-function model:

$$y(s) = \frac{5.0e^{-10.0s}}{15.0s + 1} u(s) \qquad (P17.1)$$

(a) Use the Cohen-Coon tuning rules to design a PI, and a PID, controller; implement each of these controllers and obtain the closed-loop system response to a unit step change in the set-point. Compare the performance of these two controllers.
(b) Repeat part (a) using the appropriate time integral tuning rules to design the PI and PID controllers. Compare this set of closed-loop responses with those obtained in part (a).

17.2 (a) Design a Smith predictor for the time-delay process given in Problem 17.1. by specifying the transfer function to be implemented in the minor loop around the controller yet to be designed.
(b) Assuming the process model given in Eq. (P17.1) is perfect, implement the Smith predictor, using for g_c, a PI controller with gain 0.3 and integral time constant 15.0. Obtain the overall response to a unit step change in the set-point.
(c) Investigate the effect of process gain errors on the performance of the Smith predictor by repeating part (b), retaining the same PI controller, but this time using as the "true" process model:

$$y(s) = \frac{5.5e^{-10.0s}}{15.0s + 1} u(s) \qquad (P17.2)$$

while using the model in Eq. (P17.1) in the Smith predictor's minor loop. Compare this response with the "perfect" Smith predictor response.

(d) Now investigate the effect of time-delay errors on the performance of the Smith predictor by repeating part (c), but this time using as the "true" process model:

$$y(s) = \frac{5.0e^{-11.0s}}{15.0s + 1} u(s) \qquad \text{(P17.3)}$$

Compare this response with the "perfect" Smith predictor response, and with the response obtained in part (c) with the process gain errors.

(e) Finally, investigate the effect of time constant errors on the performance of the Smith predictor by repeating part (c), but this time using as the "true" process model:

$$y(s) = \frac{5.0e^{-10.0s}}{16.5s + 1} u(s) \qquad \text{(P17.4)}$$

Compare this response with the "perfect" Smith predictor response, and with the other responses obtained in parts (c) and (d).

From this specific example, comment on which error causes the most significant performance degradation.

17.3 It is required to control a certain inverse-response system whose transfer-function model is given as:

$$g(s) = \frac{2(1 - 4s)}{(2s + 1)(5s + 1)} \qquad \text{(P17.5)}$$

and several control strategies are to be investigated.

(a) Under pure proportional control, what is the range of values that K_c can take for the closed-loop system to be stable? (Assume there are no other elements in the feedback loop, just the process and the feedback controller.)

(b) When the system operates under pure proportional control with $K_c = 0.5$, in response to a *unit* step change in set-point, what will the steady-state offset in the process output be?

(c) Using an inverse-response compensator in a minor loop around *a pure proportional controller*, if $g'(s)$, the transfer function in this minor loop, is chosen to be:

$$g'(s) = \frac{10s}{(2s + 1)(5s + 1)} \qquad \text{(P17.6)}$$

now find the range of values which K_c can take for the overall system to be stable.

(d) If the pure proportional controller *and* the inverse-response compensator in the minor loop around it are combined into one "apparent controller" block referred to as g_c^*, show that in the limit as $K_c \to \infty$, this apparent controller, g_c^*, in fact tends to a PID controller. What are the values of K_c, τ_I, and τ_D for this controller?

(e) Implement the limiting PID controller derived in part (d) on the process model given in Eq. (P17.5) and obtain the closed-loop system response to a *unit* step change set-point.

17.4 The dynamic behavior of a biological process which consists of two subprocesses in opposition is represented by the following transfer function:

$$g(s) = \frac{3}{(7s + 1)} - \frac{3}{(s + 1)(s + 3)} \qquad \text{(P17.7)}$$

(a) Assume that the process is to be under proportional feedback control and obtain a *root locus diagram* for the closed-loop system, and from it determine the range of K_c

values required for closed-loop stability. What is the frequency of oscillation when the closed-loop system is on the verge of instability?

(b) Use this information to design a PI controller for the process. Implement this controller and obtain the closed-loop response to a unit step change in the set-point. Briefly comment on the control system performance *vis-à-vis* the nature of the process transfer function.

(c) *For Extra Credit Only*. Derive an inverse-response compensator for the class of systems with two right-half plane zeros, whose transfer functions are of the form:

$$g(s) = \frac{K(1 - \eta_1 s)(1 - \eta_2 s)}{(\tau_1 s + 1)(\tau_2 s + 1)(\tau_3 s + 1)} \tag{P17.8}$$

17.5 (a) Find the range of K_c values for which the inverse-response system with the following transfer function will remain stable under proportional feedback control:

$$y(s) = \frac{5(1 - 0.5s)}{(2s + 1)(0.5s + 1)} u(s) \tag{P17.9}$$

(b) Consider now the situation in which the following unorthodox controller is utilized:

$$g_c(s) = K_c \frac{(\tau_z s + 1)}{(\tau_L s + 1)} \tag{P17.10}$$

a proportional controller with a lead-lag element. If now τ_z is chosen to be 0.5, find the τ_L value required to obtain a stability range of $0 < K_c < 4$ for the proportional gain.

(c) Determine whether or not the controller of Eq. (P17.10) will leave a steady-state offset.

17.6 A certain process whose transfer function is given as:

$$g(s) = \frac{2.0}{(5s + 1)(2s - 1)} \tag{P17.11}$$

has proved particularly difficult to control. After careful analysis, the following was discovered about the process:

1. *If either P control or PI control is used on this process, then there are no values of K_c or τ_I for which it can ever be stabilized.*

2. *This process can only be stabilized with a PD or a PID controller.*

3. *Furthermore, for the PID controller, the parameters must satisfy the following three conditions:*

(a) $K_c > \dfrac{1}{2}$

(b) $\tau_D > \dfrac{3}{2K_c}$

(c) $\tau_I > \dfrac{20K_c}{(2K_c \tau_D - 3)(2K_c - 1)}$

Confirm or refute these statements.

17.7   (a) Use the Ziegler-Nichols stability margin rules to design a PI, and a PID, controller for the process whose transfer function is given as:

$$g(s) = \frac{2.0}{(2s + 1)(5s - 1)} \tag{P17.12}$$

Implement these two controllers and obtain the closed-loop responses to a unit step change in the set-point. Which controller appears to be more effective?
(b) If the "true" process transfer function is now given as:

$$g(s) = \frac{2.5}{(3.0s + 1)(4.5s - 1)} \tag{P17.13}$$

instead of Eq. (P17.12), investigate the effect of plant/model mismatch by implementing the controllers designed in (a) on this "true" process. Comment on the effectiveness of each of the two controllers in the face of the modeling errors.

17.8   Prove that when $\lambda = 2\eta$, the inverse-response compensator is *exactly* the same as the Smith predictor with a first-order Padé approximation.

CHAPTER

18

CONTROLLER DESIGN FOR NONLINEAR SYSTEMS

It was in Chapter 10, after having devoted five chapters to the analysis of various classes of linear systems, that we first stated that virtually all physical systems are nonlinear in character. The implications of this statement of fact are just as serious now, after having devoted four chapters to various aspects of linear control systems design.

Traditionally, chemical process control has focused, almost entirely, on the analysis and control of linear systems; and therefore most existing control systems design and analysis techniques are suitable only for linear systems. It is possible to justify this state of affairs by recognizing the fortunate circumstance that many processes are in fact only mildly nonlinear; and that all nonlinear processes take on approximately linear behavior as they approach steady state (see Chapter 10). Thus for such mildly nonlinear processes, and for close regulatory control of arbitrary nonlinear processes near steady state, it is entirely reasonable to use linear control system design methods. Nature has been kind in that the majority of real process control problems fall into these categories.

Nevertheless, for those nonlinear processes whose nonlinearities are strong, and for sufficiently large excursions from steady state, linear controller design techniques often prove inadequate, and more effective alternatives must be considered. The objective in this chapter is to examine the main aspects of the rapidly developing subject of nonlinear process control, and to provide a summary of the various alternative techniques available for designing control systems, including those that explicitly recognize the nonlinearities of the process.

18.1 NONLINEAR CONTROLLER DESIGN PHILOSOPHIES

Controller design may be carried out for nonlinear systems along the lines indicated by any of the following four schemes:

1. *Local Linearization*
    This involves linearizing the modeling equations around a steady-state operating condition and applying linear control systems design results. It is obvious that the controller performance will deteriorate as the process moves further away from the steady state around which the model was linearized, but quite often, adequate controllers can be designed this way.

2. *Local Linearization with Adaptation*
    This scheme seeks to improve on Scheme 1 by recognizing the presence of nonlinearities and consequently providing the controller a means for systematically adapting whenever the adequacy of the linear approximation becomes questionable.

3. *Exact Linearization by Variable Transformations*
    This is a scheme by which a system model that is nonlinear in its original variables is converted into one that is exactly linear in a different set of variables by the use of appropriate nonlinear transformations; controller design may then be carried out for the transformed system with greater facility since it is now linear.

4. *"Special Purpose" Procedures*
    These are usually custom made for specific processes, or specific types of nonlinearities.

Some of the factors that typically determine which of these approach philosophies one adopts are now summarized below:

•   The specific nature of the nonlinear control problem at hand; the nature of the process, and the control system performance objectives,
•   The amount of time available for carrying out the design,
•   The type of hardware available for implementing the controller,
•   The availability and quality of the process model.

We will now discuss these approaches in more detail.

18.2   LINEARIZATION AND THE CLASSICAL APPROACH

It is always possible to obtain an approximate linear model for any nonlinear system, either by linearizing the nonlinear modeling equations (as discussed in Chapter 10), or by correlating input/output data, on the assumption that within the range covered by the experimental data, the process is approximately linear. Either way, the result is an approximate linear model, usually in the transfer function form.

The classical approach to controller design for nonlinear systems involves using the approximate linear model as the basis for the design and applying standard linear control systems design strategies. The following example illustrates how this may be done when a nonlinear process model is available.

Example 18.1 PI CONTROLLER DESIGN FOR THE LIQUID LEVEL PROCESS OF EXAMPLE 10.2.

Design a PI controller for the liquid level system of Example 10.2 using an approximate linear model. Choose the controller parameters K_c and τ_I such that the closed-loop poles for the approximate linear system are located at $r_1, r_2 = -2 \pm 2j$.

Solution:

A linear approximation for this system's nonlinear model was obtained earlier on in Example 10.3; it is the first-order transfer function model given below:

$$y(s) = \frac{2h_s^{1/2}/c}{\left(A\dfrac{2h_s^{1/2}}{c}s + 1\right)} u(s) \tag{18.1}$$

Making use of the specific process parameters given in Example 10.2, the approximate model, for the indicated steady-state conditions, has an apparent steady-state gain $K = 2.4\ m^{-2}$ min, and apparent time constant $\tau = 0.6$ min.

Using the results given in Chapter 15 on controller design by closed-loop pole specification (in particular, Eq. (15.28) for first-order systems under PI control), we obtain the following equations to be solved for the PI controller parameters required for placing the approximate system's closed-loop poles in the indicated positions:

$$-(r_1 + r_2) = 4 = \frac{1}{0.6} + \frac{K_c}{0.25} \tag{18.2}$$

$$r_1, r_2 = 8 = \frac{K_c}{0.25\tau_I} \tag{18.3}$$

which when solved simultaneously yield:

$$K_c = 0.58; \quad \tau_I = 0.29$$

When a nonlinear process model is not available, the identification methods of Chapter 13 may be applied. The magnitude of the test inputs used should span the range of expected process operation so that the approximate linear model is valid in the operating regime of interest.

The classical approach of linearization and subsequent linear control systems design may therefore be used with success when:

- The system is only mildly nonlinear,
- The performance objectives are not too stringent, so that very tight control is not required,
- Large deviations from nominal steady-state operating conditions (e.g., for regulatory control) are not expected.

With strong linearities, the classical scheme should *not* be used for carrying out large set-point changes, which, by definition, require moving the nonlinear process from one steady state to another, quite different, steady state.

18.3   ADAPTIVE CONTROL PRINCIPLES

A controller designed for the system in Example 18.1 using the approximate transfer function model given in Eq. (18.1) will function acceptably as long as the level, h, is maintained at, or close to, the steady-state value h_s. Under these conditions $(h - h_s)$ will be small enough to make the linear approximation adequate. However, if the level is required to change over a wide range, the farther away from h_s the level gets, the poorer the linear approximation.

Observe also from the approximate transfer function model in Eq. (18.1) that the apparent steady-state gain and time constant are dependent on the steady state around which the linearization was carried out. This implies that in going from one steady state to the other, the approximate process parameters will change.

The main problem with applying the classical approach under these circumstances is that it ignores the fact that the characteristics of the approximate model *must* change as the process moves away from h_s, if the approximate model is to remain reasonably accurate. The immediate implication is that any controller designed on the basis of this changing approximate model must also have its parameters adjusted if it is to remain effective.

In the adaptive control scheme, the controller parameters are adjusted (in an automatic fashion) to keep up with the changes in the process characteristics. We know intuitively that, if properly designed, this scheme will be a significant improvement over the classical scheme.

There are various types of adaptive control schemes, differing mainly in the way the controller parameters are adjusted. The three most popular schemes are: scheduled adaptive control, model reference adaptive control, and self-tuning controllers. We shall discuss each of these in the following sections.

18.3.1 Scheduled Adaptive Control

A *scheduled adaptive control* scheme is one in which, as a result of *a priori* knowledge and easy quantification of what is responsible for the changes in the process characteristics, the commensurate changes required in the controller parameters are programmed (or scheduled) ahead of time. This type of adaptive control, sometimes referred to as *gain scheduling*, is illustrated by the block diagram in Figure 18.1. As the input and output variables of the process

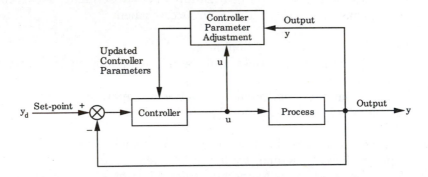

Figure 18.1.   Scheduled adaptive control scheme.

change significantly, this information is sent to the controller, and its parameters are adjusted according to the preprogrammed adjustment schedule. In practice this often reduces to a table look-up procedure.

Let us illustrate scheduled adaptive control with an example.

Example 18.2   PROGRAMMED ADAPTIVE CONTROL OF A CATALYTIC REACTOR.

Let us consider the situation in which the effect of catalyst decay on the dynamic behavior of a catalytic reactor control system is to change the effective steady-state gain of the entire process (including the measuring device, and final control element) according to the equation:

$$K(t) = K_0 f(t) \qquad (18.4)$$

where K_0 is the process gain with fresh catalyst, t is the age of the catalyst in days, and $f(t)$ is some unknown catalyst decay function.

If K_c is the proportional gain of the feedback controller attached to this process, then the overall gain in the control loop will be KK_c. If we now wish to keep this overall gain constant at its initial value of $K_0 K_{co}$ (since, for example, closed-loop stability analysis was carried out on this basis), then observe that this would require K_c to change according to:

$$K_c(t) = \frac{K_0 K_{co}}{K(t)} \qquad (18.5)$$

The actual process gain, $K(t)$, can be determined at any time from measurements of $u(t)$, $y(t)$ as shown in Figure 18.1. Thus the controller parameter adjustment would involve calculating $K(t)$ and using Eq. (18.5) to adjust $K_c(t)$.

Another common type of scheduled adaptive control involves predetermining a set of controller tuning parameters, and possibly a process model, for each region of operating space (e.g., as indicated by u and y) and then applying this schedule of appropriate controller parameters for each part of phase space. Such preprogrammed adaptive control schemes can be very effective when there is a large amount of prior knowledge about the process; however, in other situations in which the process is not so well known as to permit the use of scheduled adaptive control, there are more *self-adaptive* schemes, in which the measured value of the process inputs and outputs are used to deduce the required controller parameter adaptation by using some learning mechanism. These will be discussed next.

18.3.2 Model Reference Adaptive Control

The first of these self-adaptive schemes is called the model reference adaptive controller (sometimes abbreviated MRAC) and is illustrated by the block diagram of Figure 18.2. The key component of the MRAC scheme is the reference model that consists of a reasonable closed-loop model of how the process should respond to a set-point change. This could be as simple as a reference trajectory, or it could be a more detailed closed-loop model. The reference model output is compared with the actual process output and the observed error ε_m is used to drive some adaptation scheme to cause the controller parameters to be adjusted so as to reduce ε_m to zero. The adaptation scheme

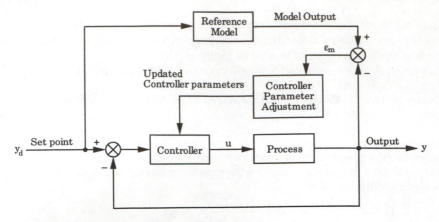

Figure 18.2.   The model reference adaptive control scheme.

could be some control parameter optimization algorithm that reduces the integral squared value of ε_m or some other procedure.

Let us illustrate a simple MRAC with the following example.

Example 18.3   **MODEL REFERENCE ADAPTIVE CONTROL OF A CATALYTIC REACTOR.**

Let us revisit the reactor of Example 18.2 in which the catalyst decays in an unknown way over time. In this case we wish to design a very simple MRAC in order to maintain good control system response over the entire lifetime of the catalyst. Let us assume that we have a PI controller and wish to update the controller parameters by using an optimization algorithm to choose K_c, τ_I in order to minimize:

$$ISE = \int_0^{t_f} \varepsilon_m^2(t)\, dt$$

for a fixed test sequence of small set-point changes. The test sequence is repeated with optimization until the ISE reaches a specific low level. As the catalyst decays, this optimization scheme is applied periodically to ensure that the desired control system response to the test sequence is maintained.

A discussion of the great variety of methods for the design and implementation of MRAC schemes fall outside the intended scope of this book; such information may be found, for example, in Refs. [1, 2].

18.3.3 Self-Tuning Adaptive Control

Self-tuning adaptive controllers such as the self-tuning regulator differ from the model-reference adaptive controller in basic principle. The self-tuning controller, illustrated by the block diagram in Figure 18.3, makes use of the process input and output to estimate recursively, on-line, the parameters of an approximate process model. Thus as the actual nonlinear process changes operating region or changes with time, an approximate linear model is continuously updated with new parameters. The updated model is then used in a prespecified control system design procedure to generate updated controller parameters. The controller could be a PID controller or some of the model-based controllers discussed in Chapter 16, 19, and 26.

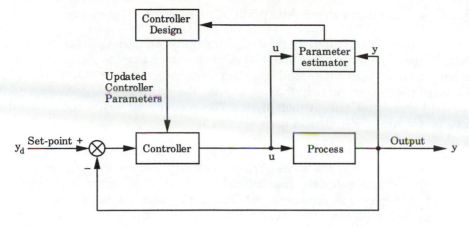

Figure 18.3. The self-tuning adaptive control scheme.

Since the model estimated determines the effectiveness of the controller, the most essential feature of the self-tuning controller is reliable and robust model identification. This requires a good parameter estimation algorithm and procedures for ensuring adequate dynamic experimental design. More details on the design and implementation of self-tuning controllers may be found in Refs. [1, 3, 4].

Let us illustrate the general principles with a simple example.

Example 18.4   SELF-TUNING CONTROL OF A CATALYTIC REACTOR.

Again, let us revisit the catalytic reactor of Example 18.2. In order to implement a self-tuning controller for the reactor, we would need to update our process model as the catalyst decays. Let us assume that we will use an approximate process model of the form:

$$g(s) = \frac{Ke^{-\alpha s}}{\tau s + 1} \tag{18.6}$$

and we will implement a PI controller based on the minimum ITAE tuning rules of Table 15.4:

$$K_c = \frac{0.586}{K}\left(\frac{\tau}{\alpha}\right)^{0.916}; \qquad \frac{1}{\tau_I} = \frac{1.03 - 0.165\left(\dfrac{\alpha}{\tau}\right)}{\tau} \tag{18.7}$$

Thus if we have good current estimates of the model parameters, the controller parameters are determined. The implementation would involve using a test sequence of small set-point changes to provide data on u and y in order to estimate K, α, and τ in the process model, Eq. (18.6). This would provide the initial process model. The model parameters could be updated continuously if there is enough dynamic movement in u and y for proper estimation. Otherwise, the estimation should only be turned on periodically when the test sequence of small set-point changes is applied to update the model parameters. Leaving the estimator on without dynamic movement in u and y is very dangerous because the process model will degrade and the resulting controller based on Eq. (18.7) could be very bad.

18.3.4 A Comparison of Adaptive Controllers

There is clear differentiation in the philosophy behind each of the three adaptive controllers discussed above. Let us use an analogy to illustrate. When parents send a child off to college, they usually try to provide them with "adaptive controllers" to help them deal with the perils of unknown situations they will encounter out on their own. The reader will easily recognize the types of adaptive controllers in use:

1. *Scheduled Adaptive Control* is used by parents who use a list of "if this occurs, do this" instances. This means they assume they can anticipate and enumerate all the situations that can occur and specify preprogrammed remedies. No thinking or decision making is expected of their children. Thus scheduled adaptive control will work when all the possibilities are known in advance.

2. *Model-Reference Adaptive Control* is employed by parents when they have a good role model (such as an older sibling) available. They simply tell their child to emulate the behavior of the role model; i.e., "do whatever they do in the same situation." In this case there is some limited decision making on the part of the child, but it is constrained to deciding how best to emulate their role model. So long as such emulation is possible, then MRAC works well.

3. *Self-Tuning Adaptive Control* is the strategy recommended by parents who have a little more confidence in the thinking ability of their children. While they do not recommend complete independence of action, they ask that the child carefully evaluate the situation based on the data at hand (model building) and then based on their judgment of the situation, take action based on fixed principles. Self-tuning adaptive control allows the most freedom to the controller and thus provides the broadest range of adaptation to unknown situations; however, it is also the most dangerous, potentially unstable, form of adaptive control because poor model identification (poor assessment of situations) can lead to unstable systems and process runaway.

All of these types of adaptive controllers are commercially available and have enjoyed a certain level of industrial acceptance. By their very nature, adaptive controllers can be applied to any system for which the process parameters vary with time; they are also applicable to other systems for which these parameter variations are not due to nonlinearities per se, but to other physical phenomena such as fouling in heat exchangers, catalyst decay in catalytic reactors, etc. More information about the applications of adaptive control may be found in Ref. [5].

18.4 VARIABLE TRANSFORMATIONS

As was briefly noted in Chapter 10, it is sometimes possible to find transformations that convert the nonlinear modeling equations of a process into one that is *exactly* linear; i.e., the process that is nonlinear in its original variables is transformed to one that is linear in a new set of variables.

The idea of variable transformation can be exploited for nonlinear controller design in the following manner:

1. Obtain an equivalent linear system by variable transformation.
2. Design a linear controller for the transformed system, in terms of the new variables.
3. Implement the controller in terms of the original variables of the nonlinear system (such a controller will, of course, be nonlinear).

Several different methods have been reported in the literature for carrying out such variable transformations for nonlinear systems and utilizing the result for controller design (see Refs. [6–10]). The following is a simple illustration of the general principles involved (cf. Ref. [6]).

18.4.1 Transformation of Nonlinear Models

To illustrate the essential idea of transformation methods consider the SISO nonlinear system whose model is given by:

$$\frac{dx}{dt} = f(x,u) \tag{18.8}$$

where $f(\cdot, \cdot)$ is an arbitrary nonlinear function of x, the system state variable, and u the input (control) variable.

Now, it is always possible to separate out the part of this function that is a function of x, and x alone; let this be, say, $c_1 f_1(x)$. Then $f(x,u)$ may always be represented as:

$$f(x,u) = c_1 f_1(x) + c_2 f_2(x,u) \tag{18.9}$$

where $c_2 f_2(x,u)$ is what is left after separating out that portion of $f(x,u)$ that is a function of x alone. Both f_1 and f_2 are taken to be nonlinear, and no restrictions are placed on their functional forms; c_1 and c_2 are constants.

Thus the process model can be written as:

$$\frac{dx}{dt} = c_1 f_1(x) + c_2 f_2(x,u) \tag{18.10}$$

Now consider the transformation:

$$z = g(x) \tag{18.11}$$

where $g(\cdot)$ is a function of x alone, to be determined such that the process model Eqs. (18.8) or (18.10), which is nonlinear in the original variables x and u, is transformed to, for example:

$$\frac{dz}{dt} = az + bv \tag{18.12}$$

or

$$\frac{dz}{dt} = a + bv \tag{18.13}$$

models that are linear in z and v.

By differentiating Eq. (18.11) with respect to t, we obtain:

$$\frac{dz}{dt} = \frac{dz}{dx}\frac{dx}{dt} \tag{18.14}$$

and introducing Eq. (18.10) into Eq. (18.14), we have:

$$\frac{dz}{dt} = c_1 f_1(x)\frac{dz}{dx} + c_2 f_2(x,u)\frac{dz}{dx} \tag{18.15}$$

Observe now that if the function $z = g(x)$ were to be chosen such that:

$$\frac{dz}{dx} = \frac{z}{f_1(x)} \tag{18.16}$$

with

$$f_2(x,u)\frac{dz}{dx} = v \tag{18.17}$$

then Eq. (18.15) will become identical to Eq. (18.12), with constants $a = c_1$, and $b = c_2$.

The required transformation is therefore obtained from Eq. (18.16), i.e.:

$$\frac{dz}{z} = \frac{dx}{f_1(x)}$$

or

$$\ln z = \int \frac{dx}{f_1(x)}$$

which finally gives:

$$z = exp\left[\int \frac{dx}{f_1(x)}\right] \tag{18.18}$$

an expression reminiscent of the integrating factor used in solving first-order, linear differential equations.

The control equation relating the old variable u to the new v, is obtained from Eq. (18.17), which, with the aid of Eq. (18.16) becomes:

$$f_2(x,u) = \frac{v f_1(x)}{z} \tag{18.19}$$

To find an alternate transformation leading to a linear model, one may follow a similar procedure to establish that:

$$z = \int \frac{dx}{f_1(x)} \tag{18.20}$$

along with:

$$f_2(x,u) = v f_1(x) \tag{18.21}$$

will transform the nonlinear Eq. (18.15) to the linear Eq. (18.13) with constants $a = c_1$, and $b = c_2$.

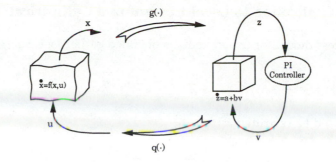

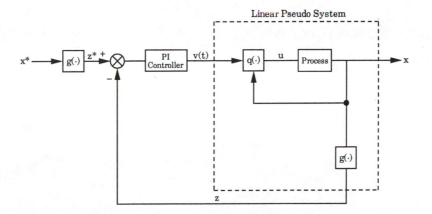

Figure 18.4.   (Top) Conceptual configuration of the transformation controller. (Bottom) Block diagram configuration of the nonlinear transformation controller (adapted from Ref. [6]).

Once the specific forms of the indicated functions in Eqs. (18.19) or (18.21) are given, these equations can be solved for u (explicitly, or if necessary, numerically) as a function of x and v to obtain:

$$u \;=\; q(x,v) \tag{18.22}$$

Figure 18.4 shows the conceptual configuration, and the block diagram of the control system based on these transformation ideas. The control system is implemented as follows:

1. A conventional (e.g., a PI) controller is designed for the transformed system in Eqs. (18.12) or (18.13).

2. As far as the controller is concerned, it is interfaced with a linear system whose state variable is z and whose input variable is v.

3. The controller receives information about z (obtained by transforming the actual system variable x according to Eq. (18.11)) and its desired set-point, and calculates the input v required to adequately control this linear *pseudo-system*.

4. The control action implemented on the real, nonlinear system is u, obtained from Eq. (18.22).

The procedure will be made clearer by considering a specific application.

18.4.2 Application to Level Control in a Cylindrical Tank

The process model for this system, as obtained earlier (see Eq. (10.1)), is:

$$A \frac{dh}{dt} = F_i - c \sqrt{h} \tag{18.23}$$

which could be rewritten as:

$$\frac{dx}{dt} = c_1 x^{1/2} + c_2 u \tag{18.24}$$

where x and u have respectively replaced h and F_i, and the constants are given by:

$$c_1 = -\frac{c}{A}; \quad c_2 = \frac{1}{A}$$

Thus in accordance with Eq. (18.9) we have, for this specific system that:

$$f_1(x) = x^{1/2} \tag{18.25}$$

and

$$f_2(x,u) = u \tag{18.26}$$

If, for example, we wish to transform this system to the linear type shown in Eq. (18.13), then from Eq. (18.20) the new variable z is obtained from:

$$z = \int \frac{dx}{x^{1/2}}$$

or

$$z = 2x^{1/2} \tag{18.27}$$

and from Eq. (18.21) in conjunction with Eqs. (18.25) and (18.26), the control equation is given by:

$$u = vx^{1/2} \tag{18.28}$$

Controller Design and Simulation Results

We now wish to apply these results to a specific flow system shown in Figure 18.5, a cylindrical pipe for which $A = 22.062$ cm^2 (I.D. is 5.3 cm) and, from previous experimentation, it is known that the dependence of the outflowrate on the liquid level in the pipe is given by:

$$F_{out} = A \times 0.1176 \, h^{1/2} \text{ cm}^3/\text{s}$$

(These are actual specifications for a level control system used for instructional purposes in the Process Control laboratory at the University of Wisconsin–Madison.)

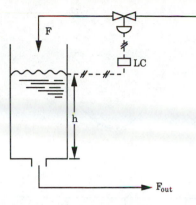

Figure 18.5.   The liquid level system.

The control system is implemented as follows:

1. Obtain the actual measurement x (in this case h) from the process.

2. Calculate z from Eq. (18.27); i.e., take the square root, and multiply by 2 (the constant multiplication is actually unnecessary, as it merely scales the variable).

3. From the same expression, given the desired value for x, obtain the desired value for z; i.e., $z_d = 2\sqrt{x_d}$.

4. The PI controller (tuned for the linear, transformed system Eq. (18.13)) receives z and z_d, and prescribes $v(t)$, the control action required for the linear *pseudo-system*.

5. The actual flowrate implemented on the real, nonlinear flow system is obtained from Eq. (18.28); a control law nonlinear in the original process variable x.

It is interesting to note that the transformation controller bases the prescribed inflowrate on a product of the control action for the linear pseudo-system and the *square root of the liquid level* in the tank. The physical implications of this are particularly important: the nonlinearity in the flow system is caused by the fact that the outflow is proportional to the square root of the liquid level in the tank; the transformation controller is seen to take advantage of this fact by adopting the indicated strategy. The conventional PI control strategy will base the inlet flowrate on the actual liquid level in the tank; the transformation controller bases its action on the product of the output of a linear controller, and the *square root of the liquid level*.

Using controller parameters $K_c = 90$ and $K_I = K_c / \tau_I = 0.2$ (for the PI controller tuned for the linear *pseudo-system*), the response of the example flow system to a set-point change in the liquid level from 40 to 75 cm, operating under the nonlinear transformation controller, is shown in the dashed line in Figure 18.6.

A conventional PI controller using the same controller parameters resulted in an almost unstable system. Upon further reducing the controller settings to

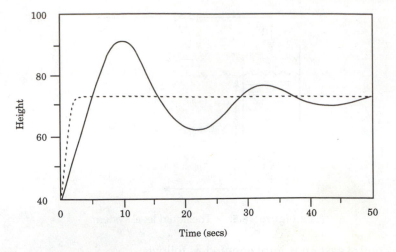

Figure 18.6.   Flow system response to set-point change; dashed line: nonlinear
controller; solid line: conventional PI controller.

$K_c = 9$ and $K_I = K_c / \tau_I = 1$, the system response to the same set-point change under
conventional PI control (i.e., decisions on the inlet flowrate are taken using
direct level measurements) is shown in the solid line. As we would expect, the
transformation controller performs better than the conventional PI controller.

Application of the nonlinear transformation strategy is restricted by the
following considerations:

- The nonlinear transformations are not always easy to find (see Refs.
  [8, 9] for some general approaches).
- The success of the controller design is dependent on having a good
  nonlinear process model — something that is unusual in practice.

Thus it is not a universally applicable control system design procedure.

18.5   SUMMARY

What we have presented in this chapter is no more than a somewhat brief
overview of what is available for designing more effective controllers for
nonlinear processes. We have indicated the conditions under which the
classical approach of linearization followed by linear controller design will
yield acceptable behavior and provided strong reasons for why this approach
is still very pervasive in practice despite the fact that virtually all process
systems of interest are nonlinear.

When the nonlinearities of a process are mild, the classical approach is
recommended, as there is no real advantage in using any of the more
complicated approaches. When the nonlinearities are more pronounced, the
classical controller will not be as effective, and alternative approaches are
called for. The various adaptive control schemes will usually work well under
these conditions.

The technique of variable transformations was also presented for those
situations in which the nonlinearities in the process are to be recognized
explicitly and taken into account in the design of effective controllers. Like

most other model-based controller design techniques, it is expected to be susceptible to modeling errors.

Research in nonlinear process control is rapidly developing and useful results are continuing to appear. Thus in the future there will be more to be learned about nonlinear control system designs.

REFERENCES AND SUGGESTED FURTHER READING

1. Sastry, S. and M. Bodson, *Adaptive Control*, Prentice-Hall, Englewood Cliffs, NJ (1989)
2. Landau, I. D., *Adaptive Control—The Model Reference Approach*, Marcel-Dekker, New York (1979)
3. Åström, K. J. and B. Wittenmark, *Adaptive Control*, Addison-Wesley, New York (1989)
4. Åström, K. J. and B. Wittenmark, "On Self-Tuning Regulators," *Automatica*, **9**, 185 (1973)
5. Narendra, K. S. and R. V. Monopoli, *Applications of Adaptive Control*, Academic Press, New York (1980)
6. Ogunnaike, B. A., "Controller Design for Nonlinear Processes via Variable Transformations," *I.&EC Proc. Des. Dev.*, **25** (1), 241 (1986)
7. Hoo, K. and J. C. Kantor, "An Exothermic Reactor is Feedback Equivalent to a Linear System," *Chem. Eng. Comm.*, **37**, 1 (1985)
8. Hunt, L. R., R. Su, and G. Meyer, "Design for Multi-Input Nonlinear Systems," *Differential Geometry Control Theory* (R.W. Brockett, R.S. Millman, H.J. Sussmann, ed.), Birkhäuser (1983)
9. Su, R., "On Linear Equivalents of Nonlinear Systems," *Systems and Control Letters*, **2** (1), 48 (1982)
10. Reboulet, C. and C. Champetier, "A New Method for Linearizing Nonlinear Systems: The Pseudolinearization," *Int. J. Control*, **40** (4), 631 (1984)

REVIEW QUESTIONS

1. Why is it that despite the fact that virtually all physical systems are nonlinear in character, much of chemical process control focuses on linear systems?

2. What are the different ways by which controller design may be carried out for nonlinear systems?

3. What are some of the important factors to consider in deciding on which nonlinear controller design strategy to adopt?

4. In dealing with nonlinear systems, when is it appropriate to use the popular classical approach of linearizing and subsequent linear controller design?

5. When is it inadvisable to adopt the classical approach to nonlinear controller design?

6. What are the three most popular adaptive control schemes, and what differentiates one from the other?

7. What is required for a typical scheduled adaptive control scheme to be effective?

8. What are the basic principles and the key components of model reference adaptive control?

9. What are the basic principles of self-tuning adaptive control?

10. What is the single most essential feature of the self-tuning controller?

11. What is the fundamental premise of the variable transformation approach to nonlinear controller design?

12. Why is the variable transformation approach not always very practical?

PROBLEMS

18.1 A reactor similar to the one discussed in Doyle, Ogunnaike, and Pearson (1992),[†] used for the free-radical polymerization of methylmethacrylate (MMA) using azo-bis-isobutyronitrile (AIBN) as initiator, and toluene as solvent, is modeled by the following set of equations:

$$\frac{d\xi_1}{dt} = 3(6 - \xi_1) - 1.4\xi_1\sqrt{\xi_2} \tag{P18.1}$$

$$\frac{d\xi_2}{dt} = 50\mu - 3\xi_2 \tag{P18.2}$$

$$\frac{d\xi_3}{dt} = -0.001\xi_1\sqrt{\xi_2} + 0.035\,\xi_2 - 10\xi_3 \tag{P18.3}$$

$$\frac{d\xi_4}{dt} = 130\xi_1\sqrt{\xi_2} - 10\xi_4 \tag{P18.4}$$

along with:

$$\eta = \frac{\xi_4}{\xi_3} \tag{P18.5}$$

As previously noted in Problem 10.9, ξ_1, ξ_2, ξ_3, and ξ_4, respectively represent the dimensionless monomer concentration, initiator concentration, bulk zeroth moment, and bulk first moment; η is the number average molecular weight; and μ is the volumetric flowrate of the initiator.

(a) Given an operating steady-state value, $\mu^* = 0.02$, obtain from the modeling equations the corresponding steady-state values for ξ_1^*, ξ_2^*, ξ_3^*, ξ_4^*, and η^*.

(b) Starting from these steady-state conditions, obtain a process reaction curve *directly* from these nonlinear set of equations (using a differential equation solver), for a 25% increase in the nominal value of μ. Design a PID controller from this process reaction curve, using the time-integral tuning rules (for set-point changes), and design another PID controller using the Ziegler-Nichols tuning rules. Which controller is more conservative?

(c) Evaluate the performance of the Ziegler-Nichols controller (on the nonlinear model equations) for a step change of 2,000 in the number average molecular weight set-point. Plot the closed-loop system response

18.2 (a) Revisit Problem 18.1. This time, linearize the model equations around the steady-state values obtained in part (a), define the deviation variables $x_1 = \xi_1 - \xi_1^*$, $i = 1, 2, 3, 4$; $y = \eta - \eta^*$; $u = \mu - \mu^*$; and obtain an approximate transfer function model relating y to u. Using the Bode stability criterion and the Ziegler-Nichols tuning

---

[†] F. J. Doyle, III, B. A. Ogunnaike, and R. K. Pearson, "Nonlinear Model Predictive Control Using Second-Order Volterra Models," Preprints of the annual A.I.Ch.E. meeting, Miami Beach, FL, Nov. 1992.

rules, design a PID controller for the polymerization reactor with the aid of this approximate transfer function. Implement the controller and evaluate its performance for the same step change of 2,000 in the number average molecular weight set-point investigated in Problem 18.1(c); compare the performance of the two controllers.

(b) If confronted with this polymer reactor control problem *in practice*, under what conditions would the controller design procedure illustrated in Problem 18.1 be preferable over the one used in part (a); and under what conditions will the preference be reversed?

18.3 The dynamic behavior of the isothermal series/parallel Van de Vusse reaction taking place in a CSTR may be modeled by the following equations (cf. E. Hernandez (1992), Ph. D. thesis, Georgia Institute of Technology, Chapter 4) :

$$\frac{d\xi_1}{dt} = -50\,\xi_1 - 10\xi_1{}^2 + (10 - \xi_1)\,\mu \qquad\qquad (\text{P18.6})$$

$$\frac{d\xi_2}{dt} = 50\,\xi_1 - 100\xi_2 - \xi_2\,\mu \qquad\qquad (\text{P18.7})$$

along with:

$$\eta = \xi_2 \qquad\qquad (\text{P18.8})$$

where ξ_1, ξ_2, are, respectively, the dimensionless reactant and product concentration in the reactor, μ is the dimensionless dilution rate, and η is a measurement of the dimensionless product concentration.

(a) Given an operating steady-state value of $\mu^* = 17.5$ for the dilution rate, obtain from the modeling equations the corresponding steady-state values for ξ_1^*, ξ_2^* (and hence η^*). Linearize the model around this steady state, define the deviation variables $x_1 = \xi_1 - \xi_1^*$; $i = 1, 2$; $y = \eta - \eta^*$; $u = \mu - \mu^*$; and obtain an approximate transfer function model relating y to u for this process. What does the transfer function reveal about the character of this process at this particular steady state?

(b) Using a nonlinear differential equation solver, obtain a simulation of the process response to a step change in dilution rate from 17.5 to 30.0. Obtain the commensurate step response using the linearized model derived in part (a). Compare these two responses and comment on whether or not the approximate linear model captures the qualitative process behavior adequately.

(c) Using the Bode stability criterion and the Ziegler-Nichols tuning rules, design a PI controller for this process with the aid of the approximate transfer function. Implement this controller on the nonlinear process and evaluate its performance if a step change in the value of ξ_2 from the initial steady-state value obtained in part (a) to a new value of 1.0 is desired.

(d) Again, using the Bode stability criterion and the Ziegler-Nichols tuning rules, design a PID controller for the process and repeat part (c). Compare the performance of the PI and PID controllers. It is normally expected that for such processes as this, the PID controller should provide the superior performance. Why should this be the case, and does this particular PID controller live up to these expectations?

18.4 Revisit Problem 18.3. With the aid of the approximate transfer function obtained in part (a), design an appropriate compensator for this process, and implement this along with the PI controller designed in part (c) on the nonlinear system. Evaluate the performance for the same step change investigated in Problem 18.3(c). Compare this performances first with that of the PI controller *without* the compensator, and then with the performance of the PID controller investigated in Problem 18.3(d).

18.5   The process model for the liquid level control system in a cylindrical tank has been given as:

$$A\frac{dh}{dt} = F_i - c\sqrt{h} \tag{18.23}$$

(a) Revisit Example 18.1, where, using the approximate transfer function given in Eq. (18.1), a static PI controller was designed for this process by pole placement. The process parameters taken from Example 10.2 are: $A = 0.25$ m^2, $c = 0.5$ m$^{5/2}$/min, $h_s = 0.36$ m. Implement this controller on the nonlinear system, and evaluate its performance for a step change in the level set-point from 0.36 to 0.5 m. Plot this response.

(b) Design a scheduled adaptive PI controller for this process if, with the controller used in part (a) as the basis, the objective is to maintain *constant* the product of the controller gain and the apparent process gain. Write an explicit expression for how the controller gain should change with the level in the tank. Implement this scheduled adaptive PI controller on the nonlinear process, using the same static integral time constant as in part (a), and evaluate the performance for the same step change in the level set-point. Compare the responses.

18.6   The mathematical model for the level of gasoline in a hemispherical storage tank was given in Problem 10.4 as:

$$\frac{dh}{dt} = \frac{1}{\pi(2Rh - h^2)}\left(F_i - ch^{1/2}\right) \tag{P10.9}$$

where R is the radius of the hemispherical tank, and F_i is the flowrate of gasoline into the storage tank. Follow the procedure of Section 18.4.1 and derive the expressions for the transformations required to convert this model to the form given in Eq. (18.12):

$$\frac{dz}{dt} = az + bv \tag{18.12}$$

If instead it is desired to convert the model to the form in Eq. (18.13):

$$\frac{dz}{dt} = a + bv \tag{18.13}$$

what nonlinear transformation will be required?

18.7   A mathematical model for the isolated population dynamics of a typical autoregenerative biological system is given by:

$$\frac{d\xi}{dt} = \alpha\xi(\mu - \xi) \tag{P18.9}$$

where

　　　　ξ is the instantaneous population variable
　　　　α is a characteristic constant of the biological system
　　　　μ is the independent "source" term

(a) For any given constant value $\mu = \mu^*$, solve the model Eq. (P18.9) explicitly, given the initial population ξ_0 at $t = 0$, and show that the time behavior of ξ is given by:

$$\xi = \frac{\mu^*}{1 + \left(\dfrac{\mu^*}{\xi_0} - 1\right)e^{-\alpha\mu^* t}} \tag{P18.10}$$

and hence establish that at steady state, as $t \to \infty$, $\xi \to \mu^*$. Also establish from the nonlinear model in Eq. (P18.9) the steady-state relationship between the nontrivial equilibrium points ξ^* and μ^*. What does this indicate about the process steady-state gain?

(b) Linearize this model around an arbitrarily specified equilibrium point indicated by ξ^* and μ^*, define the deviation variables $x = \xi - \xi^*$ and $u = \mu - \mu^*$, obtain an approximate transfer function relating x to u, and show that the steady-state gain indicated by this approximate transfer function is in fact entirely independent of the actual equilibrium point around which the model was linearized. What is the indicated apparent time constant?

(c) For a particular system with characteristic constant $\alpha = 2.5$, and an initial population variable $\xi_0 = 2$, obtain the response to a change in the source term from its initial value to $\mu = 5.0$ using the nonlinear model. Obtain the commensurate response using the approximate transfer function. Plot and compare these two responses.

(d) Given that the population variable ξ can only be measured with a device that introduces a time delay of 5 units, i.e.:

$$\eta = \xi(t - 5) \tag{P18.11}$$

introduce the deviation variable, $y = \eta - \eta^*$, and obtain the transfer function relating y to u. With the aid of this transfer function, design a PID controller for the process using the time integral tuning criteria (for set-point changes). Implement this controller on the nonlinear process and evaluate its performance for a unit step change in the desired set-point of the measured value of ξ from its initial value of 2. Plot this response and comment on the controller performance.

CHAPTER

19

MODEL-BASED CONTROL

So far, feedback controller design has been approached from the classical viewpoint in which the controller structure choice is restricted to the PID form, and design guidelines are used merely to choose appropriate parameters for this prespecified class of controllers.

There is an alternative approach in which what is specified *a priori* is the dynamic behavior desired of the control system — not the controller structure; the design procedure is then invoked to *derive* the controller (the structure as well as the parameters) that will achieve the prespecified desired objectives. This completely different controller design paradigm involves the *explicit* use of a process model directly; it is therefore often referred to as *model-based control*. The objective of this chapter is to introduce the concepts of model-based control, and then to discuss some of the most common model-based control techniques.

19.1 INTRODUCTION

In introducing the subject matter of Part III, it was stated that the three constituent elements of process analysis are: the *inputs* to the process, the *process model* (as an abstraction of the process itself), and the *outputs* from the process. By specifying *any* two of these elements, and requiring that the unspecified third be determined from the supplied information, we obtain the three fundamental problems of systems analysis illustrated in Figure 19.1:

1. **The "Process Dynamics" problem**: in which the input u is specified along with the model, say, in the form of a transfer function g and the resulting output is to be determined. This was the subject of Part II; it is the most straightforward problem.

2. **The "Process Modeling/Identification" problem**: in which the process input u and output y are provided (sometimes along with some knowledge of the fundamental laws governing the process operation),

and it is required to determine the process model g. This was the subject of Part III; it is the most difficult problem.

3.   **The "Process Control" problem**: in which the *desired* output y^* is specified along with the model g, and the input u required to produce such an output is to be determined. This is the subject of Part IV; it is the most important problem.

19.1.1 Solving the Process Control Problem

In solving the "process control" problem in practice, there is the important issue of what physical device will be used to generate the process input u, and also how this device is to be configured in relation to the physical process. When constrained to use specific hardware elements to implement the controller, this necessarily imposes limitations on our freedom to schedule input adjustments. At the same time, by insisting on specifying the desired output behavior explicitly and arbitrarily, the resulting controller requirement may be impossible to realize because of hardware limitations.

Observe therefore that there are two, *apparently* mutually exclusive, approaches to solving the posed process control problem: one in which the *primary basis* is the hardware element available for implementing the controller, with the desired output behavior only of secondary concern; the other in which the *primary basis* is the desired output behavior, with the hardware element for controller implementation of secondary concern.

The former is the situation with conventional feedback control: with the prespecified hardware element (the PID controller), the desired output behavior must be defined somewhat more loosely, so that feasible solutions can be found within the constraints imposed by the prespecified controller structure.

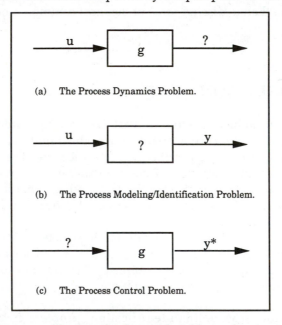

(a)   The Process Dynamics Problem.

(b)   The Process Modeling/Identification Problem.

(c)   The Process Control Problem.

Figure 19.1.   The three fundamental problems of systems analysis.

The PID controller parameters thus determined, the resulting output behavior must be evaluated (typically by simulation) for acceptability.

The latter is the situation with model-based control: the desired output behavior is precisely, and explicitly, specified, and, *using the process model explicitly*, the controller required to obtain the prespecified behavior is derived. In most cases, a microprocessor-based hardware element may be required to implement the resulting controller, but as we shall soon see, the resulting controllers sometimes take the familiar PID form.

19.1.2 Model-Based Approaches

There are two approaches to model-based control system design:

1. **The "Direct Synthesis" Approach**: in which the desired output behavior is specified in the form of a trajectory, and the process model is used *directly* to synthesize the controller required to cause the process output to follow this trajectory *exactly*. Some of the particular techniques falling under this category, and to be discussed in this chapter, are:

   - Direct Synthesis Control
   - Internal Model Control [1]
   - Generic Model Control [2]

2. **The "Optimization" Approach**: in which the desired output behavior is specified in the form of an objective function (which may or may not involve an output trajectory explicitly), and the process model is used to derive the controller required to *minimize* (or maximize) this objective. It is also possible to include some known operating constraints in the optimization objective. Apart from "Optimal Control" which we will review briefly in this chapter, most of the other techniques falling under this category deal with discrete, single-variable or multivariable systems; we will consequently discuss them at the appropriate places in later chapters.

It should be noted that we have already seen some types of model-based controllers in earlier chapters: the feedforward controller of Chapter 16, the Smith predictor, and the inverse-response compensator of Chapter 17 are all model-based controllers. They all use the process model explicitly in deriving the resulting controllers; and without making the issue of the hardware required for implementation the foremost concern, the primary basis for each technique is a clearly defined output behavior objective: For feedforward control it is "perfect" disturbance rejection, while for the Smith predictor and the inverse-response compensator it is elimination of the time-delay term, and the right-half plane zero, respectively, from the closed-loop transfer function.

Let us now begin our discussion of the other model-based control techniques by starting with the one closest in form to the feedback controllers with which we are already familiar.

19.2   CONTROLLER DESIGN BY DIRECT SYNTHESIS

19.2.1 The Controller Synthesis Problem

Consider the block diagram of Figure 19.2, where g represents the consolidated transfer function for the process + measurement device + final control element. In this case the closed-loop transfer function is given by:

$$y(s) = \frac{gg_c}{1 + gg_c} y_d(s) \tag{19.1}$$

Suppose now that we require the closed-loop behavior to be:

$$y(s) = q(s)\, y_d(s) \tag{19.2}$$

where $q(s)$ is a specific, predetermined transfer function of our choice. The choice of the reference trajectory $q(s)$ depends on the type of closed-loop response desired and on the possible responses from the process. Some particular choices for the reference trajectory are shown in Figure 19.3. Note that for systems with no time delays or inverse response, reference trajectories of type (a) or (b) would be appropriate. However, if the process has a significant time delay, then a reference trajectory of type (c) is required because no controller can overcome the initial delay in the closed-loop response. Similarly, if the process has an inverse response, then the reference trajectory must allow inverse response.

Having defined an appropriate form for the reference trajectory, the controller synthesis problem is now posed as follows:

What is the form of the controller g_c required to produce in the process the closed-loop behavior represented by the reference trajectory $q(s)$?

In other words, given q and g, find the g_c required to make the process closed-loop transfer function exactly equal to q. Noting that the closed-loop behavior (as represented in Eq. (19.1)) will be as we desire it to be in Eq. (19.2) if, and only if:

$$q(s) = \frac{gg_c}{1 + gg_c} \tag{19.3}$$

then the controller synthesis problem reduces to solving Eq. (19.3) for g_c to yield:

$$g_c = \frac{1}{g}\left(\frac{q}{1 - q}\right) \tag{19.4}$$

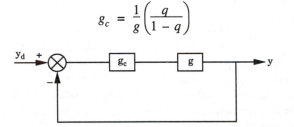

Figure 19.2.   Block diagram of a feedback control system.

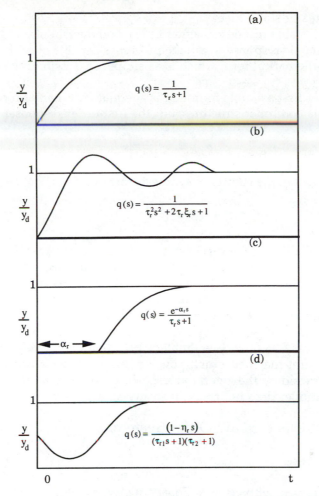

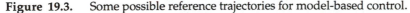

Figure 19.3.   Some possible reference trajectories for model-based control.

This is the *controller synthesis formula* from which, given the process model g, we may derive the controller g_c required for obtaining *any* desired closed-loop behavior represented by q.

Observe that there are no restrictions on the form this controller can take; for this reason, we may also expect that there will be no guarantees that the controller will be implementable in all cases. However, we shall see that if q is chosen appropriately, the synthesized controller g_c will take on practical forms. In particular, we will show that in some important cases, the synthesized controller actually takes the familiar form of conventional PID controllers.

Specifying The Desired Closed-Loop Behavior

The type of controller synthesized using Eq. (19.4) clearly depends on the nature of the transfer function form chosen for q, the desired closed-loop response. To guide us in our choice, let us recall the earlier discussion of Figure 19.3 and supplement this by specifying some of the most important features we would normally expect to find in acceptable closed-loop reference trajectories:

1. No steady-state offset,
2. Quick, stable response, having little or no overshoot,
3. A dynamic response which the process is capable of achieving, and
4. A mathematical form which is as simple as possible.

Let us consider first a simple form that is adequate for many processes and that leads to simple controller structures; i.e., the first-order response:

$$q(s) = \frac{1}{\tau_r s + 1} \tag{19.5}$$

For this choice of q, the controller synthesis formula Eq. (19.4) simplifies to:

$$g_c = \frac{1}{\tau_r s} \frac{1}{g} \tag{19.6}$$

Before investigating the various types of controllers prescribed by Eq. (19.6) for various types of process models, it is important to note that this direct synthesis controller involves the inverse of the process model. This feature, common to all model-based controllers, will have significant implications later on, particularly when we consider the issue of synthesis for nonminimum phase systems.

Also, it is easy to show (see Problem 19.1) that even when the actual plant dynamics are not matched exactly by the model transfer function g, used in Eq. (19.6), provided the overall closed-loop system is stable, this direct synthesis controller does not result in steady-state offset.

19.2.2 Synthesis for Low-Order Systems

Pure Gain Processes

For these processes, with $g(s) = K$, Eq. (19.6) gives rise to:

$$g_c = \frac{1}{K \tau_r s} \tag{19.7}$$

a *pure integral* controller with integral time constant $\tau_I = K \tau_r$.

Pure Capacity Processes

Since in this case $g(s) = K/s$, Eq. (19.6) gives rise to:

$$g_c = \frac{1}{K \tau_r} \tag{19.8}$$

a *pure proportional* controller, with gain $K_c = 1/K \tau_r$.

First-Order Processes

In this case, with the process transfer function:

$$g(s) = \frac{K}{\tau s + 1} \tag{19.9}$$

Eq. (19.6) gives the required synthesized controller as:

$$g_c = \frac{\tau s + 1}{K \tau_r s} \tag{19.10}$$

which may be rearranged to take the more familiar form:

$$g_c = \frac{\tau}{K \tau_r}\left(1 + \frac{1}{\tau s}\right) \tag{19.11}$$

immediately recognizable as a PI controller with $K_c = \tau/K \tau_r$ and $\tau_I = \tau$.

We may summarize the most important implications of the foregoing as follows:

1.   Once the desired speed of the closed-loop response is specified (i.e., a value is chosen for τ_r), for any of the simple low-order processes we have considered, both the *controller type* and the *controller parameters* required to achieve the closed-loop objectives specified by Eq. (19.5) are immediately given in the derived controller transfer functions; furthermore, these controller parameters are given in terms of the process model parameters and τ_r.

2.   For these simple low-order processes, we observe that the direct synthesis approach prescribes nothing outside of the familiar collection of classical feedback controllers; the advantage with this approach is that it automatically prescribes the controller parameters to go with each controller type.

Let us now consider an illustrative example.

Example 19.1   DIRECT SYNTHESIS CONTROLLER FOR A FIRST-ORDER SYSTEM.

Design a controller for the following first-order system:

$$g(s) = \frac{0.66}{6.7s + 1} \tag{19.12}$$

using the direct synthesis approach, and given that the desired closed-loop behavior is as in Eq. (19.5), with $\tau_r = 5$. Compare this controller with that resulting from choosing $\tau_r = 1$.

Solution:

Let us observe first that this is the system used in Problem 15.6 without the time delay.

Either from first principles, or directly from Eq. (19.11), we obtain that the required direct synthesis controller (for $\tau_r = 5$) is:

$$g_c = 2.03\left(1 + \frac{1}{6.7s}\right) \tag{19.13}$$

a PI controller with $K_c = 2.03$ and $\tau_I = 6.7$.

The controller obtained with the choice $\tau_r = 1$ (the faster closed-loop trajectory) is:

$$g_c = 10.15\left(1 + \frac{1}{6.7s}\right) \tag{19.14}$$

and the only difference between these controllers is that the latter has a proportional gain value which is *five times* that of the former; the integral times are identical. As we would expect, a controller with a higher gain value is required to provide the faster closed-loop response indicated by the smaller value chosen for τ_r. Further, it should be noted that a reduction in the value of τ_r by a factor of 5 resulted in a fivefold increase in the value of K_c.

19.2.3 Synthesis for Higher Order Systems

For the second-order system:

$$g(s) = \frac{K}{\tau^2 s^2 + 2\zeta\tau s + 1} \tag{19.15}$$

it is still possible to require the closed-loop response to be first order with $q(s)$ given by Eq. (19.5). In this case the required controller is:

$$g_c = \frac{\tau^2 s^2 + 2\zeta\tau s + 1}{K\tau_r s}$$

which, upon rearrangement into a more familiar form, gives:

$$g_c = \frac{2\zeta\tau}{K\tau_r}\left(1 + \frac{1}{2\zeta\tau s} + \frac{\tau}{2\zeta}s\right) \tag{19.16}$$

immediately recognizable as a PID controller with the indicated parameters.

Example 19.2 DIRECT SYNTHESIS CONTROLLER FOR A SECOND-ORDER SYSTEM.

Design a controller for the following second-order system:

$$g(s) = \frac{1}{(2s + 1)(5s + 1)} \tag{19.17}$$

using the direct synthesis approach, given that the desired closed-loop behavior is as in Eq. (19.5), with $\tau_r = 5$. Also compare this controller with that resulting from choosing $\tau_r = 1$.

Solution:

Once more we observe that this is the example system used in Section 17.3 of Chapter 17 (see Eq. (17.25)) without the RHP zero.

Either from first principles, or by applying Eq. (19.16) directly (being careful to extract the parameters ζ and τ from the given model parameters correctly), we obtain that the required direct synthesis controller (for $\tau_r = 5$) is:

$$g_c = 1.4\left(1 + \frac{1}{7s} + 1.43\ s\right) \tag{19.18}$$

a PID controller with $K_c = 1.4$, $\tau_I = 7$, and $\tau_D = 1.43$.

Upon reducing the value of τ_r to 1, we obtain the following controller:

$$g_c = 7\left(1 + \frac{1}{7s} + 1.43\ s\right) \tag{19.19}$$

a PID controller which has the same integral and derivative times as the former controller, but whose proportional gain is five times larger.

We therefore observe that for this form of the direct synthesis controller, changing the value of τ_r only affects the value of the proportional gain; all other controller parameters are unaffected.

It is easy to show (see Problem 19.2) that for a second-order system with a single zero, the direct synthesis controller resulting from Eq. (19.6) takes the form of a PID controller with a first-order filter. Examples of direct synthesis controller designs for second-order reference trajectories were presented in Chapter 15 where we saw that they also took the form of a PID controller in many cases.

For processes having models which are third order or higher, it is easy to see that the controllers synthesized using Eq. (19.6) are bound to take increasingly more esoteric forms. We will not pursue this aspect further since it is not of pressing practical interest. What is of practical interest is the synthesis of such controllers for processes with more complex dynamics: time delays, inverse response, and open-loop instability.

19.2.4 Synthesis for Processes with Complex Dynamics

The fact that the general controller synthesis formula in Eq. (19.4) involves the inverse of the process model should make us cautious in applying it for time delay as well as inverse-response systems: the $e^{-\alpha s}$ term in the time-delay system's transfer function will become $e^{+\alpha s}$ in the controller, thereby requiring prediction; and for the inverse-response system, the relatively harmless right-half plane zero translates to a right-half plane pole for the controller, making it unstable. The primary reason for caution in applying Eq. (19.4) for open-loop unstable systems is not as obvious, but it is equally critical, as we shall soon show.

Synthesis for Time-Delay Systems

In order to focus attention on essentials, we shall consider that the time-delay system in question is represented by the simple transfer function:

$$g(s) = \frac{Ke^{-\alpha s}}{\tau s + 1} \tag{19.20}$$

even though the discussion to follow generalizes to other forms of time-delay systems. If the first-order reference trajectory given in Eq. (19.5) is still retained as the desired closed-loop behavior, then the direct synthesis controller that will achieve this objective, as derived from Eq. (19.6), is:

$$g_c = \frac{(\tau s + 1)e^{+\alpha s}}{K\tau_r s} \tag{19.21}$$

which is unrealizable, since it requires a prediction. The problem, of course is that the choice of reference trajectory in Eq. (19.5) fails to consider the most critical dynamic characteristic of time-delay systems: their inherent inability to respond instantaneously. Thus one must choose a more realistic form of the reference trajectory:

$$q(s) = \frac{e^{-\alpha_r s}}{\tau_r s + 1} \tag{19.22}$$

shown in Figure 19.3. If we redefine the controller equation using Eq. (19.22) for q in Eq. (19.4) we obtain:

$$g_c = \frac{(\tau s + 1)}{K} \left[\frac{e^{-(\alpha_r - \alpha)s}}{(\tau_r s + 1) - e^{-\alpha_r s}} \right] \tag{19.23}$$

which is realizable in practice so long as the delay in the reference trajectory is at least as long as the delay in the process, i.e., $\alpha_r \geq \alpha$. This condition is necessary to prevent any predictions, $e^{+\alpha s}$, in the controller equation. In the special case when we choose $\alpha_r = \alpha$, the controller in Eq. (19.23) reduces to:

$$g_c = \frac{(\tau s + 1)}{K} \left(\frac{1}{\tau_r s + 1 - e^{-\alpha s}} \right) \tag{19.24}$$

which we will show is *identical* to the Smith predictor (Chapter 17), plus a PI controller. We will also show that by using various approximations for the exponential term, this controller simplifies to the classical PI and PID forms.

The Direct Synthesis Controller in the Smith Predictor Form

To implement the controller as given in Eq. (19.24) implies that control action $u(s)$ will be determined using $u(s) = g_c(s)\, \varepsilon(s)$:

$$u(s) = \frac{(\tau s + 1)}{K} \left(\frac{1}{\tau_r s + 1 - e^{-\alpha s}} \right) \varepsilon(s)$$

This equation may now be rearranged to:

$$K\,(\tau_r s + 1 - e^{-\alpha s})u(s) = (\tau s + 1)\,\varepsilon(s)$$

or

$$\frac{K\tau_r s}{(\tau s + 1)} u(s) + \frac{K}{\tau s + 1} u(s) = \frac{Ke^{-\alpha s}}{\tau s + 1} u(s) + \varepsilon(s) \tag{19.25}$$

We now recall an earlier discussion in Section 17.2.3 that for:

$$y(s) = \frac{Ke^{-\alpha s}}{\tau s + 1} u(s)$$

we defined as $y^*(s)$, the output of the equivalent system without the delay, i.e.:

$$y^*(s) = \frac{K}{\tau s + 1} u(s)$$

This being the case, Eq. (19.25) then becomes:

$$\frac{K\tau_r s}{(\tau s + 1)} u(s) + y^*(s) = y(s) + \varepsilon(s)$$

so that $u(s)$ is now given by:

$$u(s) = \left(\frac{\tau s + 1}{K\tau_r s}\right)\{\varepsilon(s) - [y^*(s) - y(s)]\}$$

or, finally:

$$u(s) = \frac{\tau}{K\tau_r}\left(1 + \frac{1}{\tau s}\right)\{\varepsilon(s) - [y^*(s) - y(s)]\} \qquad (19.26)$$

Observe that this is a PI controller operating on a signal composed of $\varepsilon(s)$, the regular feedback error signal, augmented by the signal $y^* - y$, recognizable as the signal from the minor loop of the Smith predictor strategy (see Figure 17.4). With the direct synthesis controller however, the parameters required for the PI controller are provided. By contrast with the Smith predictor strategy as presented in Chapter 17, neither the controller type nor the controller parameters are specified; the main "conventional" controller must be designed separately. Thus the direct synthesis approach has some advantages.

Approximate Classical Controller Forms

By using the simple first-order Taylor series approximation:

$$e^{-\alpha s} \approx 1 - \alpha s$$

in Eq. (19.24), the direct synthesis controller equation simplifies to:

$$g_c = \frac{\tau}{K(\tau_r + \alpha)}\left(1 + \frac{1}{\tau s}\right) \qquad (19.27)$$

a PI controller with the indicated parameters.
   A much better approximation for $e^{-\alpha s}$ is obtained with the first-order Padé approximation:

$$e^{-\alpha s} \approx \frac{1 - \frac{\alpha}{2}s}{1 + \frac{\alpha}{2}s}$$

Upon introducing this into Eq. (19.24), the result may be rearranged to give:

$$g_c = \frac{\tau}{K(\tau_r + \alpha)}\left(1 + \frac{1}{\tau s}\right)\left(\frac{1 + \frac{\alpha}{2}s}{1 + \tau^* s}\right) \qquad (19.28a)$$

with

$$\tau^* = \frac{\alpha\tau_r}{2(\alpha + \tau_r)} \tag{19.28b}$$

This is recognizable as the commercial PID controller form (see Eq. (15.7)), with the controller parameters as indicated. Alternatively, the same controller could also be rearranged to give:

$$g_c = \frac{\tau + \dfrac{\alpha}{2}}{K(\tau_r + \alpha)}\left[1 + \frac{1}{\left(\tau + \dfrac{\alpha}{2}\right)s} + \left(\frac{\alpha\dfrac{\tau}{2}}{\tau + \dfrac{\alpha}{2}}\right)s\right]\left(\frac{1}{\tau^*s + 1}\right) \tag{19.29}$$

which is in the form of a "standard" PID controller, in conjunction with a first-order filter having a filter time constant τ^*.

Example 19.3   DIRECT SYNTHESIS CONTROLLER FOR A TIME-DELAY SYSTEM.

Obtain approximate PI and PID controllers for the following first-order-plus-time-delay system:

$$g(s) = \frac{0.66e^{-2.6s}}{6.7s + 1}$$

using the direct synthesis approach, and given the reference trajectory of Eq. (19.22), with $\tau_r = 5$. Compare this PI controller with that obtained (using $\tau_r = 5$) for the system in Example 19.1 which is identical in all respects to this system except for the presence of the time delay.

Solution:

From Eq. (19.27) the required PI controller is:

$$g_c = 1.34\left(1 + \frac{1}{6.7s}\right)$$

a PI controller with $K_c = 1.34$ and $\tau_I = 6.7$.

From Eq. (19.29), the required PID-type controller is:

$$g_c = 1.59\left(1 + \frac{1}{8s} + 1.09\ s\ \right)\left(\frac{1}{0.86s + 1}\right)$$

and if we now neglect the 0.86 filter time constant (considering it as small compared to the system time constant and time delay) we obtain an approximate PID controller with parameters:

$$K_c = 1.59,\ \tau_I = 8,\ \tau_D = 1.09$$

For the equivalent system without the time delay treated in Example 19.1, the PI controller recommended in that case has exactly the same integral time, but a higher proportional gain value (2.03 as opposed to 1.34).

Thus in designing a PI controller for the time-delay system, the direct synthesis method has recommended a reduction in the proportional controller gain value as compared to the value recommended for the equivalent system without the delay.

Observe that the controller gain value recommended for the PID controller is slightly higher than that for the PI controller. This is true in general for all types of controller design techniques: the presence of derivative action typically allows the use of higher controller gain values.

Example 19.4 AN EXACT DIRECT SYNTHESIS CONTROLLER FOR A TIME-DELAY SYSTEM.

Obtain the exact direct synthesis controller for the first-order-plus-time-delay system of Example 19.3.

Solution:

From the preceding discussion, we observe that the exact direct synthesis controller for the Example 19.3 system:

$$g(s) = \frac{0.66e^{-2.6s}}{6.7s + 1}$$

is the Smith predictor *in conjunction with* the PI controller:

$$g_c = 2.03 \left(1 + \frac{1}{6.7s} \right)$$

This is, of course, the same controller obtained for the equivalent system without the time delay in Example 19.1, but the signal operated upon by this controller is *not* $\varepsilon(s)$; as a result of its association with the Smith predictor, it is :

$$\left\{ \varepsilon(s) - [y^*(s) - y(s)] \right\}$$

where $\varepsilon(s)$ is the regular feedback error, with $y^*(s)$ and $y(s)$ obtained *by simulation*, respectively from:

$$y^*(s) = \frac{0.66}{6.7s + 1} u(s)$$

and

$$y(s) = \frac{0.66e^{-2.6s}}{6.7s + 1} u(s)$$

Synthesis for Inverse-Response Systems

Once more, to focus attention on essentials, we shall consider the inverse-response system in question to have the transfer function form:

$$g(s) = \frac{K (1 - \eta s)}{(\tau_1 s + 1)(\tau_2 s + 1)} \tag{19.30}$$

Let us now choose as the reference trajectory:

$$q(s) = \frac{(1 - \eta_r s)}{(\tau_{r_1} s + 1)} \frac{1}{\tau_{r_2} s + 1} \tag{19.31}$$

which is shown in Figure 19.3(d) and recognizes the inevitability of inverse-response behavior in the closed-loop response. It is known from Chapter 15 and elsewhere that a good response for an inverse-response system is obtained when a closed-loop pole is placed symmetric with the right-half plane zero. Thus if we choose $\eta_r = \eta$, $\tau_{r_1} = \tau_r$, $\tau_{r_2} = \eta$ in Eq. (19.31), we obtain a direct synthesis controller that could be arranged in two forms:

$$g_c = \frac{\tau_1}{K(2\eta + \tau_r)}\left(1 + \frac{1}{\tau_1 s}\right)\left(\frac{\tau_2 s + 1}{\tau^* s + 1}\right) \tag{19.32}$$

i.e., the commercial PID controller form (with the parameters as indicated), or, alternatively:

$$g_c = \frac{(\tau_1 + \tau_2)}{K(2\eta + \tau_r)}\left[1 + \frac{1}{(\tau_1 + \tau_2)s} + \frac{\tau_1 \tau_2}{(\tau_1 + \tau_2)}s\right]\left(\frac{1}{(\tau^* s + 1)}\right) \tag{19.33}$$

the standard PID form with the additional first-order filter. In each case:

$$\tau^* = \frac{\eta \tau_r}{2\eta + \tau_r} \tag{19.34}$$

Recall that in Section 17.3 of Chapter 17, we had indicated that of all the classical feedback controllers, the PID type is the only one that has been found useful in controlling inverse-response systems; this direct synthesis controller provides another justification for this fact.

Example 19.5   DIRECT SYNTHESIS CONTROLLER FOR AN INVERSE-RESPONSE SYSTEM.

Design a controller for the following inverse-response system:

$$g(s) = \frac{(1 - 3s)}{(2s + 1)(5s + 1)}$$

using the direct synthesis approach, with Eq. (19.31) as the desired closed-loop behavior, with $\tau_r = 1$. Compare the resulting controller with that obtained (using $\tau_r = 1$) for the Example 19.2 second-order system which, but for the absence of the RHP zero, is identical in all other respects to this system.

Solution:

Directly from Eq. (19.33), upon introducing all the appropriate parameters we obtain the direct synthesis controller as:

$$g_c = 1\left(1 + \frac{1}{7s} + 1.43\ s\right)\left(\frac{1}{0.43s + 1}\right)$$

If we now observe that the filter time constant 0.43 is small compared with all the other process time constants, dropping this term simplifies the direct synthesis controller to a PID controller with $K_c = 1$, $\tau_I = 7$, and $\tau_D = 1.43$.

In Example 19.2, for the second-order system having identical time constants but no RHP zero, we obtained a PID controller with identical integral and derivative times but whose proportional gain value was *seven times* larger. We therefore see how the direct synthesis technique recognizes the presence of inverse-response behavior (and its potential problems) and prescribes a substantially lower value of the controller gain.

Synthesis for Open-Loop Unstable Systems

Consider the first-order open-loop unstable system whose transfer function model is given as:

$$g(s) = \frac{K}{\tau s - 1} \qquad (19.35)$$

with the indicated single RHP pole; Eq. (19.6) gives the direct synthesis controller required to provide the closed-loop behavior indicated by q in Eq. (19.5) as:

$$g_c = \frac{\tau}{K\tau_r}\left(1 - \frac{1}{\tau s}\right) \qquad (19.36)$$

which looks like a PI controller with $K_c = \tau/K\tau_r$, but with the singular, and very important, difference that its integral time is *negative*; i.e., $\tau_I = -\tau$.

The problem with this controller is that it will only function properly if all the process parameters are exactly equal to the model parameters; if this is not the case the overall closed-loop system will be unstable. In other words, the controller in Eq. (19.36) will give rise to a stable closed-loop system only under the obviously unrealistic situation that the model is an exact representation of the process; *the controller therefore has a severe robust stability problem.* Let us first establish this important fact before providing a solution.

Robust Stability Problems

Consider the situation in which the actual open-loop unstable process has the transfer function:

$$g_p(s) = \frac{K_p}{\tau_p s - 1} \qquad (19.37)$$

for which Eq. (19.35) is now only an approximate model. When this system is operating in the closed loop with the controller in Eq. (19.36) derived on the basis of the approximate model of Eq. (19.35), the closed-loop characteristic equation:

$$1 + g_p g_c = 0 \qquad (19.38)$$

becomes

$$K\tau_r \tau_p s^2 + \left(K_p \tau - K\tau_r\right) s - K_p = 0 \qquad (19.39)$$

where $K_p \neq K$ and $\tau_p \neq \tau$.

Observe carefully now that because of the negative coefficient of K_p, according to the Routh stability test, the system with this characteristic equation will *always* be unstable. (If the plant and model parameters had been identical, cancellation would have been possible in Eq. (19.38), and the characteristic equation would have indicated a single, closed-loop pole located at $s = -1/\tau_r$).

The direct synthesis philosophy, as embodied in requiring the closed-loop response to be as in Eq. (19.5), is responsible for this problem. According to Eq. (19.5), the single, closed-loop system pole is to be located at $s = -1/\tau_r$; to achieve this requires a controller that will do two things simultaneously: cancel the RHP pole, and replace it with the desired LHP pole (the one located at $s = -1/\tau_r$). Such perfect pole cancellation is, of course, only possible when the location of the RHP pole is *exactly* known.

Robust Controller Synthesis

The indicated robustness problem may be solved by considering the following not-so-obvious modification to q:

$$q(s) = \frac{(\eta_r s + 1)}{(\tau_{r_2} s + 1)} \frac{1}{\tau_{r_1} s + 1} \tag{19.40}$$

With this modification:

$$\frac{q}{1 - q} = \frac{\eta_r s + 1}{(\tau_{r_1} + \tau_{r_2} - \eta_r)s \left(\dfrac{\tau_{r_1} \tau_{r_2}}{\tau_{r_1} + \tau_{r_2} - \eta_r} s + 1 \right)} \tag{19.41}$$

If we now choose the coefficient of s in the denominator term inside the large brackets to be equal to $-\tau$; i.e.:

$$\frac{\tau_{r_1} \tau_{r_2}}{\tau_{r_1} + \tau_{r_2} - \eta_r} = -\tau$$

rearrangement gives:

$$\eta_r = \tau_{r_1} + \tau_{r_2} + \frac{\tau_{r_1} \tau_{r_2}}{\tau} \tag{19.42}$$

so that Eq. (19.41) becomes:

$$\frac{q}{1 - q} = \frac{\eta_r s + 1}{\dfrac{\tau_{r_1} \tau_{r_2}}{\tau} s (\tau s - 1)} \tag{19.43}$$

Using this result, the direct synthesis controller based on Eqs. (19.43) and (19.4) is:

$$g_c = \frac{\tau \eta_r}{K \tau_{r_1} \tau_{r_2}} \left(1 + \frac{1}{\eta_r s} \right) \tag{19.44}$$

recognizable as a normal PI controller with $K_c = \tau\eta_r / K\tau_{r_1}\tau_{r_2}$ and $\tau_I = \eta_r$.

It can be shown that the closed-loop characteristic equation for this system, when there is plant/model mismatch, is given by:

$$K\tau_{r_1}\tau_{r_2}\tau_p \, s^2 + (K_p\tau\eta_r - K\tau_{r_1}\tau_{r_2}) \, s + K_p\tau = 0 \qquad (19.45)$$

Thus if the parameters τ_{r_1}, τ_{r_2} are chosen such that:

$$K\tau_{r_1}\tau_{r_2} < K_p\tau\eta_r \qquad (19.46)$$

the closed-loop system will be stable even though the parameters of the plant and its model are not exactly the same.

Note:

1. By examining the controller gain indicated in Eq. (19.44), we see that the robust stability condition indicated in Eq. (19.46) is exactly equivalent to requiring that:

$$K_c > \frac{1}{K_p} \qquad (19.47)$$

    which is the same condition obtained in Chapter 17 for closed-loop stabilization of the open-loop unstable system by conventional feedback control.

2. The condition indicated in Eq. (19.46) is used to choose the parameters that characterize the desired closed-loop response q. Only two of the three parameters can be chosen independently (Eq. (19.42) determines the dependence of η_r on τ_{r_1} and τ_{r_2}). Thus, Eq. (19.46) can be rearranged to give:

$$\frac{K_p}{K}\left(\frac{\tau}{\tau_{r_1}} + \frac{\tau}{\tau_{r_2}} + 1\right) > 1 \qquad (19.48)$$

    and the two closed-loop time constants τ_{r_1} and τ_{r_2} are now chosen such that Eq. (19.48) is satisfied.

For the second-order open-loop unstable system with a single RHP pole:

$$g(s) = \frac{K}{(\sigma s + 1)(\tau s - 1)} \qquad (19.49)$$

it is left as an exercise to the reader to show that for the reference trajectory in Eq. (19.40), and with the aid of the same choice of parameters as indicated in Eq. (19.42), the direct synthesis controller is obtained as:

$$g_c = \frac{(\sigma + \eta_r)\tau}{K\tau_{r_1}\tau_{r_2}}\left[1 + \frac{1}{(\sigma + \eta_r)s} + \frac{\sigma\eta_r}{(\sigma + \eta_r)}s\right] \qquad (19.50)$$

which is a PID controller with the indicated parameters. The accompanying robust stability conditions for this process, akin to those given in Eq. (19.46) for the first-order open-loop unstable system, are also easily derived. Recall the results given for the system in Eq. (17.62) in Chapter 17, and compare with Eq. (19.50). It is by no means obvious that for such a system, only a PD or a PID controller will stabilize the process; the direct synthesis controller, however, indicates this fact directly in Eq. (19.50).

Table 19.1

Some Formulas for Direct Synthesis Tuning of Processes with Simple Dynamics

Process Model	Reference Trajectory Model	
	$\dfrac{1}{\tau_r s + 1}$	$\dfrac{1}{\tau_r^2 s^2 + 2\zeta\tau_r s + 1}$
$\dfrac{K}{\tau s + 1}$	PI Controller: $K_c = \dfrac{\tau}{K\tau_r}$ $\tau_I = \tau$	
$\dfrac{K}{\tau^2 s^2 + 2\zeta\tau s + 1}$ or $\dfrac{K}{(\tau_1 s + 1)(\tau_2 s + 1)}$	PID Controller: $K_c = \dfrac{2\zeta\tau}{K\tau_r};\left(\text{or }\dfrac{\tau_1 + \tau_2}{K\tau_r}\right)$ $\tau_I = 2\zeta\tau;(\text{or }\tau_1 + \tau_2)$ $\tau_D = \dfrac{\tau}{2\zeta};\left(\text{or }\dfrac{\tau_1\tau_2}{\tau_1 + \tau_2}\right)$	PI Controller with $\dfrac{\tau_r}{2\zeta_r} = \tau_2;\ \beta = 2K\tau_r\zeta_r$ $K_c = \tau_1/\beta$ $\tau_I = \tau_1$
$\dfrac{K}{(\tau_1 s + 1)(\tau_2 s + 1)(\tau_3 s + 1)}$		PID Controller with $\dfrac{\tau_r}{2\zeta_r} = \tau_3;\ \beta = 2K\tau_r\zeta_r$ $K_c = \dfrac{\tau_1 + \tau_2}{\beta}$ $\tau_I = \tau_1 + \tau_2$ $\tau_D = \left(\dfrac{\tau_1\tau_2}{\tau_1 + \tau_2}\right)$

Table 19.2
Some Formulas for Direct Synthesis Tuning of Processes with Complex Dynamics

Process Model: $\dfrac{K\,e^{-\alpha s}}{\tau s + 1}$	**Reference Trajectory:** $\dfrac{e^{-\alpha_r s}}{\tau_r s + 1}$

Three alternate designs for $\alpha = \alpha_r$:

  a. Smith Predictor with PI Controller:

$$K_c = \frac{\tau}{K\tau_r} \;\; ; \;\; \tau_I = \tau$$

  b. Commercial PID Controller:

$$K_c = \frac{\tau}{K\,(\tau_r + \alpha)} \; ; \; \tau_I = \tau \; ; \; \tau_D = \frac{\alpha}{2} \; ; \; \tau^* = \frac{\alpha}{2}\left(\frac{\tau_r}{\alpha + \tau_r}\right)$$

  c. Ideal PID Controller with first-order filter:

$$K_c = \frac{\tau + \dfrac{\alpha}{2}}{K(\tau_r + \alpha)} \; ; \; \tau_I = \tau + \frac{\alpha}{2} \; ; \; \tau_D = \frac{\dfrac{\alpha}{2}\,\tau}{\tau + \dfrac{\alpha}{2}} \; ; \; \tau^* = \frac{\alpha}{2}\left(\frac{\tau_r}{\alpha + \tau_r}\right)$$

Process Model: $\dfrac{K\,(1 - \eta s)}{(\tau_1 s + 1)\,(\tau_2 s + 1)}$	**Reference Trajectory:** $\dfrac{(1 - \eta_r s)}{(\tau_{r_1} s + 1)\,(\tau_{r_2} s + 1)}$

Two alternate designs for $\eta_r = \eta, \;\; \tau_{r_2} = \eta$:

  a. Commercial PID Controller:

$$K_c = \frac{\tau_1}{K(2\eta + \tau_{r_1})} \; ; \; \tau_I = \tau_1 \; ; \; \tau_D = \tau_2 \; ; \; \tau^* = \frac{\eta\tau_{r_1}}{2\eta + \tau_{r_1}}$$

  b. Ideal PID Controller with first-order filter:

$$K_c = \frac{\tau_1 + \tau_2}{K\,(2\eta + \tau_{r_1})} \; ; \; \tau_I = \tau_1 + \tau_2 \; ; \; \tau_D = \frac{\tau_1\,\tau_2}{\tau_1 + \tau_2} \; ; \; \tau^* = \frac{\eta\tau_{r_1}}{2\eta + \tau_{r_1}}$$

Table 19.3
Some Formulas for Direct Synthesis Tuning of
Unstable Processes

Process Model: $\dfrac{K}{\tau s - 1}$	Reference Trajectory: $\dfrac{(1 - \eta_r s)}{(\tau_{r_1} s + 1)(\tau_{r_2} s + 1)}$

PI Controller *with* $\eta_r = \tau_{r_1} + \tau_{r_2} + \dfrac{\tau_{r_1} \tau_{r_2}}{\tau}$:

$$K_c = \frac{\tau \, \eta_r}{K \tau_{r_1} \tau_{r_2}} \;;\quad \tau_I = \eta_r$$

Process Model: $\dfrac{K}{(\tau_1 s - 1)(\tau_2 s + 1)}$	Reference Trajectory: $\dfrac{(1 - \eta_r s)}{(\tau_{r_1} s + 1)(\tau_{r_2} + 1)}$

PID Controller *with* $\eta_r = \tau_{r_1} + \tau_{r_2} + \dfrac{\tau_{r_1} \tau_{r_2}}{\tau_1}$:

$$K_c = \frac{(\tau_2 + \eta_r)}{K \tau_{r_1} \tau_{r_2}} \;;\quad \tau_I = \tau_2 + \eta_r \;;\quad \tau_D = \frac{\tau_2 \, \eta_r}{\tau_2 + \eta_r}$$

19.2.5 Summary of Direct Synthesis Formulas

Although the discussion above would allow the reader to derive the direct synthesis tuning parameters for any class of models and choice of reference trajectories, it is useful to provide a summary of some common cases. Tables 19.1 – 19.3 provide formulas for PI and PID controller tuning based on the most common classes of linear models and for appropriate choices of reference trajectories. These formulas provide an excellent set of tuning parameters for a wide range of problems when an adequate linear model is available for the process.

19.2.6 Direct Synthesis for Nonlinear Systems

Since by its very nature the method of direct synthesis is based on deriving the controller required to make the process output achieve some prespecified objective, it is, in principle, not restricted to linear systems. However, since nonlinear systems are *usually* not represented in the transform domain by transfer function models, applications to nonlinear systems will have to be carried out in the time domain. A technique particularly suited to this, *Generic Model Control*, will be discussed in Section 19.4.

19.3 INTERNAL MODEL CONTROL

19.3.1 Motivation

Consider the process whose dynamic behavior is represented by:

$$y(s) = g(s) u(s) + d \tag{19.51}$$

with the block diagram shown in Figure 19.4 Here d represents the collective effect of *unmeasured* disturbances on the process output y. If it is desired to have "perfect" control, in which the output tracks the desired set-point y_d perfectly, the control action required to achieve this objective is easily obtained by substituting $y = y_d$ in Eq. (19.51):

$$y_d(s) = g(s) u(s) + d \tag{19.52}$$

and then solving for $u(s)$ to obtain:

$$u(s) = \frac{1}{g(s)} \left[y_d(s) - d \right] \tag{19.53}$$

The implication is that if both d and $g(s)$ are known, then for any given y_d, Eq. (19.53) provides the controller that will achieve perfect control. Since in reality, d is unmeasured, and hence unknown, and since g may not always be known perfectly, but only modeled approximately by, say, $\tilde{g}(s)$, we may now adopt the following strategy for implementing Eq. (19.53):

1. Assuming that $\tilde{g}$ is our best estimate of the plant dynamics g, then our best estimate of d is obtained by subtracting the model prediction, $\tilde{g}(s)u(s)$, from the actual plant output y to yield the estimate:

$$\hat{d} = y - \tilde{g}(s) u(s) \tag{19.54}$$

2. Let us choose the notation:

$$c(s) = \frac{1}{\tilde{g}(s)} \tag{19.55}$$

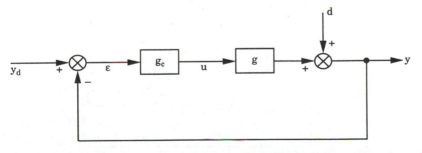

Figure 19.4.   A process under feedback control with unmeasured disturbances d.

and rewrite Eq. (19.53) as:

$$u(s) = c(s)[y_d(s) - \hat{d}(s)] \qquad (19.56)$$

where $\hat{d}$ is the estimate of d given by Eq. (19.54). A block diagrammatic representation of Eqs. (19.54) and (19.56) takes the form shown in Figure 19.5. This structure, first suggested by Frank in 1974 [3], is now known as the "Internal Model Control" structure [1].

If we wish to extend consideration to both measured disturbances d_1, and unmeasured disturbances d_2, we can let:

$$d(s) = g_d(s)\, d_1(s) + d_2(s)$$

in Figure 19.5 and in Eqs. (19.54) and (19.56) so that the "perfect" controller for both set-point changes and measured disturbance rejection is:

$$u(s) = \frac{1}{g(s)}\left[y_d(s) - d_2\right] - \frac{g_d(s)}{g(s)}\, d_1(s) \qquad (19.57)$$

Observe now that the additional term in Eq. (19.57) is no more than what has already been derived in Chapter 16 for the feedforward controller used for cancelling out the effect of measured disturbances.

19.3.2 Basic Features and Properties of IMC

By rearranging the block diagram of Figure 19.5 as indicated in Figure 19.6(a) and consolidating the inner loop into the single block indicated in Figure 19.6(b), it is possible to compare the IMC structure with an equivalent conventional feedback control structure. The following relationships between the conventional feedback controller, $g_c(s)$, and the internal model controller, $c(s)$, are easy to establish:

$$g_c(s) = \frac{c(s)}{1 - c(s)\, \tilde{g}(s)} \qquad (19.58a)$$

$$c(s) = \frac{g_c(s)}{1 + g_c(s)\, \tilde{g}(s)} \qquad (19.58b)$$

These will be useful in the discussion to follow.

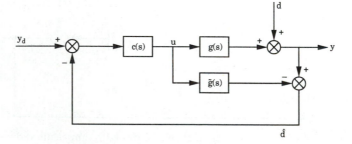

Figure 19.5. The internal model control structure.

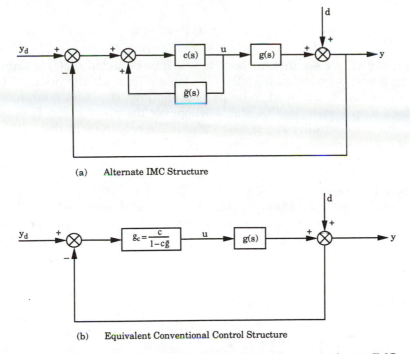

(a)   Alternate IMC Structure

(b)   Equivalent Conventional Control Structure

Figure 19.6.   The conventional control structure equivalent to IMC.

Closed-Loop Transfer Function Relations and Controller Properties

From Figure 19.5, it is a straightforward matter to derive the following closed-loop transfer function relations:

Relation of u to y_d and d:

$$u = \frac{c}{1 + c(g - \tilde{g})} (y_d - d) \tag{19.59}$$

Relation of y to y_d and d:

$$y = d + \frac{gc}{1 + c(g - \tilde{g})} (y_d - d) \tag{19.60a}$$

or

$$y = \frac{gc}{1 + c(g - \tilde{g})} y_d + \frac{1 - \tilde{g}c}{1 + c(g - \tilde{g})} d \tag{19.60b}$$

From these transfer function relations, we may now establish the following three basic properties of the IMC scheme:

1.   Nominal Closed-Loop Stability

In the nominal case, where the model is perfect so that $\tilde{g} = g$, Eqs. (19.59) and (19.60a) become, respectively:

$$u = c (y_d - d) \tag{19.61}$$

and

$$y = gc\,(y_d - d) + d \qquad\qquad (19.62)$$

From Eq. (19.61) we see that the control u will be bounded if $c(s)$ is stable, and the output $y(s)$ is bounded if $g(s)c(s)$ is stable. Thus for the nominal case in which the model is assumed to be perfect, the IMC structure guarantees that the closed-loop system will be stable whenever the process $g(s)$ is open-loop stable, and the controller $c(s)$ is also stable. The requirement that $g(s)$ also be stable is not so stringent, but to require $c(s) = 1/g(s)$ to be stable is often difficult to satisfy in practice.

2. "Perfect" Control

If $g = \tilde{g}$, and $c = g^{-1}$, then from Eq. (19.60), $y = y_d$ for all time and for all d. Thus the IMC structure provides "perfect" control. However, in practice, one has limited controller power so that requiring $c(s) = g^{-1}$ can seldom be satisfied in reality.

3. "Offset-free" Control

So long as $c(s)$ is chosen such that:

$$\lim_{s \to 0} c(s) = \frac{1}{\tilde{g}(0)}$$

then regardless of the fact that $g \neq \tilde{g}$, Eq. (19.60) shows that $y = y_d$ at steady state. Thus so long as the steady-state gain of the controller is the same as the reciprocal of the gain of the *process model*, the IMC structure guarantees offset-free control.

The IMC structure is an idealization that can seldom be realized in practice; however, it is a useful ideal for purposes of discussion, and by adding some approximations and additional constraints, IMC can be implemented in some cases.

19.3.3 Design and Implementation of Internal Model Controllers

In implementing the IMC scheme, the following practical issues must be taken into consideration:

1. $\tilde{g}$ is never equal to g in practice.

2.   Implementing $c(s)$ as $\tilde{g}^{-1}$ is rarely feasible because the transfer function is "improper," and there may be time delays or right-half plane zeros in the model.

3.   Even in the absence of such "noninvertible" dynamics, observe from Eq. (19.58a) that implementing $c(s)$ as $\tilde{g}^{-1}$ will often require controllers with infinitely large control action, which is both inadvisable and impractical.

As a result, the IMC design procedure has to be modified as follows:

1. **Process model factorization:**
   The process model is separated into two parts:

   $$\tilde{g} = \tilde{g}_+ \tilde{g}_- \tag{19.63}$$

   where $\tilde{g}_+$ contains all the noninvertible aspects (time delays, right-half plane zeros), with a steady-state gain of 1, with $\tilde{g}_-$ as the remaining, invertible part.

2. **Controller Specification and Filter Design:**
   The controller is specified as:

   $$c(s) = \frac{1}{\tilde{g}_-} f(s) \tag{19.64}$$

   where $f(s)$ is a filter usually of the form:

   $$f(s) = \frac{1}{(\lambda s + 1)^n} \tag{19.65}$$

   with parameters λ and n chosen to ensure that $c(s)$ is proper (i.e., the numerator order is less than, or at most equal to, the denominator order).

3. **Equivalent Conventional Controller form:**
   If necessary, the IMC controller, $c(s)$, may be converted to the conventional form, $g_c(s)$, for implementation. This is accomplished by using Eq. (19.58a).

Example 19.6   IMC DESIGN FOR A FIRST-ORDER PROCESS.

Design a controller for the first-order process whose transfer function is given by:

$$g(s) = \frac{5.0}{8s + 1} \tag{19.66}$$

using the IMC strategy. Convert this controller to the conventional feedback form.

Solution:
Observing that the transfer function in Eq. (19.66) is completely invertible, we obtain:

$$\frac{1}{g(s)} = \frac{8s + 1}{5.0}$$

which requires only a first-order filter ($n = 1$) in order that $c(s) = f/g$ be proper. Thus, from Eq. (19.64), we have, in this case:

$$c(s) = \frac{1}{g(s)} f(s) = \frac{1}{5.0}\left(\frac{8s + 1}{\lambda s + 1}\right) \tag{19.67}$$

which can be implemented using a lead/lag element.

Introducing Eqs. (19.67) and (19.66) into Eq. (19.58a), we obtain:

$$g_c(s) = \frac{8}{5\lambda}\left(1 + \frac{1}{8s}\right)$$

(19.68)

as the equivalent conventional feedback form, a PI controller whose gain depends on the filter parameter λ.

If the resulting PI controller in Eq. (19.68) appears reminiscent of the direct synthesis controller result of Section 19.2.2, it is because the IMC approach and the direct synthesis approach produce *identical* controllers under certain conditions.

Consider the situation for which $\tilde{g}$ is invertible so that:

$$c(s) = \frac{f(s)}{\tilde{g}(s)}$$

then under these conditions, Eq. (19.58a) becomes:

$$g_c(s) = \frac{f(s)/\tilde{g}(s)}{1 - f(s)}$$

or

$$g_c(s) = \frac{1}{\tilde{g}(s)}\left[\frac{f(s)}{1 - f(s)}\right]$$

(19.69)

which is identical to Eq. (19.4) with $f(s) = q(s)$. Thus for minimum phase systems, the filter of the IMC scheme and the reference trajectory of the direct synthesis scheme have the same interpretation. It is easy to show that even when $\tilde{g}$ is not invertible, the IMC scheme and the direct synthesis scheme still give rise to identical controllers, only this time, $f(s)$ and $q(s)$ take on different forms.

It can also be shown that the equivalence between the IMC controller and the direct synthesis controller is not limited to the nominal case; even in the presence of modeling errors, the direct synthesis controller and the IMC controller also give rise to identical closed-loop responses.

Example 19.7   IMC DESIGN FOR A FIRST-ORDER PROCESS WITH TIME DELAY.

Design a controller for the first-order-plus-time-delay process whose transfer function is given by:

$$g(s) = \frac{5.0e^{-3s}}{8s + 1}$$

(19.70)

using the IMC strategy.

Solution:

Observing that the presence of the time delay makes the transfer function not completely invertible in this case, we proceed to factor the transfer function as follows: Let:

$$g_+(s) = e^{-3s}$$

so that:

$$g_-(s) = \frac{5.0}{8s + 1}$$

The desired controller is then easily determined from Eq. (19.64) as:

$$c(s) = \frac{1}{g_-(s)} f(s) = \frac{1}{5.0}\left(\frac{8s + 1}{\lambda s + 1}\right) \qquad (19.71)$$

where we have chosen a first-order filter.

Note that this controller is precisely the same as the one obtained in Example 19.6 for the first-order system without the delay. It is left as an exercise to the reader, however, to show that the corresponding conventional feedback forms are in fact *not identical*. In this regard, the reader may want to keep in mind that in this case, $c(s)$ is based on a "truncated" form of the model, so that the denominator of Eq. (19.58a) will be more complicated.

Problem (19.9) illustrates how the IMC approach is used to obtain PID tuning rules for processes whose dynamics are represented by first-order-plus-time-delay transfer functions. These results were presented in Chapter 15, in Table 15.6.

For additional details about the IMC scheme, particularly, regarding how model uncertainty information is used to design a filter that will ensure robustness, the interested reader is referred, for example, to Morari and Zafiriou [4].

19.4 GENERIC MODEL CONTROL

Consider the situation in which the dynamics of a process are represented by the first-order differential equation:

$$\frac{dy}{dt} = f(y, u, d) \qquad (19.72)$$

where y is the process output, u is the input, d is the disturbance, and $f(\bullet, \bullet, \bullet)$ can be a nonlinear function. Further, consider that we wish for the process output to approach its desired value of y_d along a reference trajectory, y_r, determined by the equation:

$$\left(\frac{dy_r}{dt}\right) = K_1\left(y_d - y\right) + K_2 \int_0^t \left(y_d - y\right) dt \qquad (19.73)$$

This can be related to $q(s)$ in the Laplace domain by taking the Laplace transforms of Eq. (19.73) and letting $y = y_r$, to yield:

$$y_r(s) = \frac{K_1 s + K_2}{s^2 + K_1 s + K_2} y_d$$

or $y_r = q(s)y_d$ where:

$$q(s) = \frac{K_1 s + K_2}{s^2 + K_1 s + K_2}$$

The process model Eq. (19.72) may be used directly to obtain the controller required to cause the process to follow this trajectory by equating Eqs. (19.72) and (19.73) to obtain:

$$f(y, u, d) = K_1 \left(y_d - y \right) + K_2 \int_0^t \left(y_d - y \right) dt \tag{19.74}$$

and solving for u. In general, since it is impossible to model the process perfectly, our best approximation, $\hat{f}(y, u, d)$, is used instead, and the resulting implicit, nonlinear equation:

$$\hat{f}(y, u, d) - K_1 \left(y_d - y \right) - K_2 \int_0^t \left(y_d - y \right) dt = 0 \tag{19.75}$$

is the control law to be solved for u. Although it may not be obvious, it should be pointed out that solving Eq. (19.75) for u involves an *implicit* inversion of the process model.

This technique, known as "Generic Model Control," was first presented by Lee and Sullivan [2]. It has the following properties:

1.   Any available (possibly nonlinear) process is employed *directly* in the controller.
2.   The integral term provides *additional* compensation for inevitable modeling errors.
3.   The control law has only two parameters, K_1 and K_2, whose values determine the nature of the desired reference trajectory in a well-characterized form.

Relationship between GMC and Other Model-Based Schemes

Lee and Sullivan [2] have discussed the similarities between this control scheme and IMC. We will now show that for $K_2 = 0$, when the process model is linear, first-order, the closed-form solution of the resulting control law yields a PI controller identical to the direct synthesis controller.

For the first-order system whose transfer function is given in Eq. (19.9), the differential equation model is:

$$\tau \frac{dy}{dt} = -y + Ku(t) \tag{19.76}$$

Setting $K_2 = 0$ in Eq. (19.73) implies a reference trajectory described by:

$$\left(\frac{dy_r}{dt}\right) = K_1 \left(y_d - y \right)$$

or

$$\left(\frac{dy_r}{dt}\right) = K_1 \, \varepsilon(t) \tag{19.77}$$

where $\varepsilon(t)$ is the usual feedback error term. We may now integrate Eq. (19.77) directly to obtain:

$$y_r(t) = K_1 \int_0^t \varepsilon(t)dt \qquad (19.78)$$

The control action required to cause the process modeled by Eq. (19.76) to follow the trajectory described by Eqs. (19.77) and (19.78) may now be obtained straightforwardly by substituting the desired trajectory expressions into the model equation and solving for $u(t)$; the result is:

$$u(t) = \frac{\tau K_1}{K} \left[\varepsilon(t) + \frac{1}{\tau} \int_0^t \varepsilon(t)dt \right] \qquad (19.79)$$

which, of course, is a PI controller. But beyond this fact, observe that this is precisely the same PI controller obtained for the direct synthesis controller in Eq. (19.11) with $K_1 = 1/\tau_r$. Note that if the process output y is set equal to the reference trajectory y_r in Eq. (19.73), a Laplace transform of the resulting expression gives:

$$y_r = \frac{1}{(1/K_1) s + 1} y_d$$

and only the substitution $K_1 = 1/\tau_r$ is required to establish that this desired trajectory is precisely the one indicated by $q(s)$ given in Eq. (19.5) and used by the direct synthesis controller.

GMC for Higher Order Systems

For processes modeled by the general, nth-order, nonlinear differential equation:

$$f\left(y; \frac{dy}{dt}; \frac{d^2y}{dt^2}; \ldots; \frac{d^ny}{dt^n}; u; d \right) = 0 \qquad (19.80)$$

applying the GMC scheme requires no more than obtaining the required higher order derivatives of the desired trajectory, introducing these into the model in Eq. (19.80) and solving for u. Thus, control action will be obtained by solving:

$$f\left(y; \frac{dy_r}{dt}; \frac{d^2y_r}{dt^2}; \ldots; \frac{d^ny_r}{dt^n}; u; d \right) = 0 \qquad (19.81)$$

for u.

For the purpose of illustration, consider the design of a GMC controller for the second-order process modeled by:

$$\tau^2 \frac{d^2y}{dt^2} + 2\zeta\tau\frac{dy}{dt} + y = Ku \qquad (19.82)$$

and suppose that the desired trajectory is obtained by setting $K_2 = 0$ in Eq. (19.73). Then as obtained previously, the first derivative of the desired trajectory is as given in Eq. (19.77), from which we easily obtain the second derivative as:

$$\frac{d^2 y_r}{dt^2} = K_1 \frac{d\varepsilon(t)}{dt} \tag{19.83}$$

And now, by introducing Eqs. (19.77), (19.78), and (19.83) into the process model in Eq. (19.82) and solving for u, we obtain:

$$u(t) = \frac{2\zeta\tau K_1}{K} \left\{ \varepsilon(t) + \frac{1}{2\zeta\tau} \int_0^t \varepsilon(t)dt + \frac{\tau}{2\zeta} \left[\frac{d\varepsilon(t)}{dt} \right] \right\} \tag{19.84}$$

the same PID controller obtained in Eq. (19.16) using the direct synthesis approach.

For additional information about the GMC technique, including discussions about its applications on several experimental case studies, the reader is referred to Lee and coworkers [2, 5, 6].

19.5   OPTIMIZATION APPROACHES

Let us return, once again, to the process whose dynamics are represented by the differential equation:

$$\frac{dy}{dt} = f(y, u, d) \tag{19.72}$$

which was used to introduce the GMC technique in Section 19.4; and instead of requiring that the process output approach its desired value of y_d along a specific reference trajectory, y_r, we now wish to approach the control problem differently. We start by setting up an *objective functional* of the general form:

$$\Phi = G\left(y\left(t_f\right)\right) + \int_0^{t_f} F(y, u, d) \, dt \tag{19.85}$$

where $y(t_f)$ is the value taken by the process output at the final time $t = t_f$; and $G(\cdot)$ and $F(\cdot, \cdot, \cdot)$ are functions chosen by the control system designer to reflect the aspect of the process behavior to be optimized. The idea is now to find the control policy $u(t)$ that *minimizes* (or maximizes) Φ, perhaps even subject to such operating constraints as:

$$g(y, u) \leq 0 \tag{19.86a}$$

$$h(y, u) = 0 \tag{19.86b}$$

The philosophical differences between this technique known as "Optimal Control," and GMC are as follows:

1.   When control action is determined according to the optimal control strategy, the trajectory followed by y is *not prespecified*; rather, the "best" possible trajectory, consistent with the process model and the objective embedded in Eqs. (19.85) and (19.86), will be found, and this will be followed by the process output. The prespecified trajectory of the GMC technique (or any other approach not based on optimization) is arbitrary and may quite possibly be unattainable.

2.   The optimal control strategy is automatically equipped to consider process operating constraints; the other strategies must incorporate additional considerations when process constraints are involved.

3.   The *general* optimal control problem as posed above cannot be solved in closed form, or even implicitly, as with the GMC in Eq. (19.72); quite often, intensive computation is required.

Optimal control is not limited to the first-order nonlinear case presented above; in fact, the original development was for nonlinear multivariable systems of arbitrary order. The resulting optimal control problem, however, simplifies considerably when specific special cases of the general form are considered. These and other aspects of controller design by the optimization approach are discussed in Ray [7]. The application of these optimization techniques to the design of discrete-time multivariable controllers for industrial processes has led to the evolution of a class of control schemes known as *Model Predictive Control*; this will be discussed in greater detail in Chapter 27.

19.6   SUMMARY

The approach to controller design that we have introduced in this chapter is radically different from anything we have discussed up until now. These *model-based* approaches allow the control system designer to specify the closed-loop behavior desired of the process, and from this, to derive the controller required to obtain the specified closed-loop behavior directly, using the process model. Even though the two classes of model-based control techniques were identified, our discussion focussed more on the direct synthesis approaches (Direct Synthesis Control, Internal Model Control, and Generic Model Control) while passing more lightly over the optimization approaches. This latter class of techniques are much more naturally suited to multivariable systems and will therefore be discussed more fully later when it will be more appropriate to do so.

We showed that model-based controllers all share in common the concept of using the process model inverse — in one form or another — as part of the controller; we also showed that even though philosophically different from conventional feedback control, model-based control strategies often give rise to controllers that may be rearranged to take on the familiar forms of these classical controllers. The special considerations required for dealing with those processes that exhibit more complex dynamics were carefully discussed, and we found that the derived controllers then take on special forms, some of which we had encountered earlier. Upon some simplification even these special forms reduce to classical feedback controllers.

The principles underlying model-based control make it applicable for the rational design of controllers for nonlinear systems. The Generic Model Control scheme was introduced as a strategy most suited for such applications, particularly when a nonlinear differential equation process model is available.

With this, we conclude our discussion of controller design for single-input, single-output systems. Next, in Part IVB, we will devote our attention to the issue of control system analysis and design when the process has multiple input and output variables.

REFERENCES AND SUGGESTED FURTHER READING

1. Garcia, C. E. and M. Morari, "Internal Model Control: 1 — A Unifying Review and Some New Results," *Ind. Eng. Chem. Proc. Res. Dev.*, **21**, 308 (1982)
2. Lee, P. L. and G. R. Sullivan, "Generic Model Control (GMC)," *Computers and Chemical Engineering*, **12**, (6), 573 (1988)
3. Frank, P. M., *Entwurf von Regelkreisen mit vorgeschriebenem Verhalten*, G. Braun, Karlsruhe (1974)
4. Morari, M. and E. Zafiriou, *Robust Process Control*, Prentice-Hall, Englewood Cliffs, NJ (1989)
5. Lee, P. L., G. R. Sullivan, and W. Zhou, "Process/Model Mismatch Compensation for Model-Based Controllers," *Chem. Eng. Comm.*, **80**, 33 (1989)
6. Newell, R. B. and P. L. Lee, *Applied Process Control: A Case Study*, Prentice-Hall, Englewood Cliffs, NJ (1989)
7. Ray, W. H., *Advanced Process Control*, Butterworths, Boston (1989)

REVIEW QUESTIONS

1. What are the three fundamental problems of systems analysis?

2. What is model-based control, and what differentiates it from conventional feedback control?

3. What are the two approaches to model-based control system design?

4. What are some of the model-based controllers encountered in earlier chapters?

5. How is the controller design problem posed in the direct synthesis approach?

6. What are some of the most important features of an acceptable closed-loop reference trajectory?

7. What are some model forms and closed-loop reference trajectories that give rise to direct synthesis controllers in the form of the classical feedback controllers?

8. What is the advantage of using the direct synthesis approach to design feedback controllers for low-order processes?

9. What restrictions are placed on admissible reference trajectories for direct synthesis controller design when time-delay and inverse-response elements are present in process models? Why are these restrictions necessary?

10. What additional approximations are needed in order to obtain classical feedback controllers from a direct synthesis controller designed for a first-order-plus-time-delay system?

11. The direct synthesis controller designed for a first-order open-loop unstable system suffers from what problem if a first-order closed-loop reference trajectory is used? How is the problem remedied?

12. Why is it rarely feasible to use the model inverse directly as the controller in the IMC strategy?

13. What is the typical procedure for the IMC design?

14. What is the purpose of the IMC filter?

15. What is the relationship between the IMC filter and the transfer function for the direct synthesis reference trajectory?

16. What are some of the main properties of Generic Model Control?

17. What is the fundamental philosophy of the optimization approaches?

18. What are some of the philosophical differences between optimal control and GMC?

PROBLEMS

19.1 Suppose that the direct synthesis controller given in Eq. (19.2) is implemented as in Figure 19.2, but the true process transfer function is g_p rather than g. Obtain the closed-loop transfer function between y_d and y, and show that, despite the fact that $g_p \neq g$, so long as the overall system is stable, this direct synthesis controller guarantees that at steady state, $y = y_d$, and there will be no offset.

19.2 Obtain the direct synthesis controller for the process whose transfer function is given as:

$$g(s) = \frac{K(\xi s + 1)}{(\tau_1 s + 1)(\tau_2 s + 1)}$$

if the closed-loop reference trajectory in Eq. (19.5) is desired. Show that this controller may be arranged in two ways: first, as:

$$g_c = \frac{(\tau_1 + \tau_2)}{K\tau_r}\left[1 + \frac{1}{(\tau_1 + \tau_2)s} + \frac{\tau_1\tau_2}{(\tau_1 + \tau_2)}s\right]\left[\frac{1}{(\xi s + 1)}\right] \quad (P19.1)$$

which is a standard PID controller with an additional term; and then as:

$$g_c = \frac{\tau_1}{K\tau_r}\left(1 + \frac{1}{\tau_1 s}\right)\left(\frac{\tau_2 s + 1}{\xi s + 1}\right)$$

which is the commercial PID controller form.

19.3 Establish the result presented in Eq. (19.50) for the open-loop unstable system modeled by the transfer function in Eq. (19.49).

19.4 *The following problems have to do with the theoretical robustness stability properties of the direct synthesis controller.*
(a) Consider a first-order process whose *true* transfer function:

$$g_p(s) = \frac{K_p}{\tau_p s + 1} \qquad (P19.2)$$

is approximated by the model transfer function:

$$g(s) = \frac{K}{\tau s + 1} \qquad (P19.3)$$

and suppose that this approximate transfer function is used to design a direct synthesis controller to achieve a first-order desired closed-loop trajectory in Eq. (19.5). Show that when the resulting controller is implemented on the "real" process in Eq. (P19.2), the overall closed-loop system will *always* be stable regardless of the value of the true process parameters, K_p and τ_p, and for all values of τ_r, the reference trajectory time constant, *provided K_p and its estimate, K, have the same sign.*

(b) Suppose now that the true process transfer function is second order, i.e.:

$$g_p(s) = \frac{K_p}{(\tau_p s + 1)(\tau_n s + 1)} \qquad (P19.4)$$

containing an additional time constant, τ_n, which has been neglected by the model transfer function in Eq. (P19.3). Furthermore, suppose that the direct synthesis controller has been designed on the basis of the first-order desired trajectory, using the Eq. (P19.3) model. Establish that in implementing such a controller on the real second-order process, closed-loop stability will be guaranteed provided:

$$\tau_r > \rho_p \left(\frac{\tau_p \tau_n}{\tau_p + \tau_n} - \tau \right) \qquad (P19.5)$$

where ρ_p is the multiplicative uncertainty in the process gain estimate, defined by:

$$K_p = \rho_p K \qquad (P19.6)$$

(c) To illustrate how relatively weak the restriction indicated in Eq. (P19.5) is, consider the situation in which the true process transfer function is:

$$g_p(s) = \frac{4}{(10s + 1)(6s + 1)} \qquad (P19.7)$$

and that the model not only ignores the smaller time constant, but in addition grossly underestimates the dominant time constant as 3.0, and the gain as 2.0; i.e.:

$$g(s) = \frac{2.0}{3.0s + 1} \qquad (P19.8)$$

is the model used to design the direct synthesis controller. First show that, in this case, guaranteeing closed-loop stability merely requires that $\tau_r > 1.5$. Next obtain a PI controller for this process by direct synthesis, using the model in Eq. (P19.8) and a first-order desired reference trajectory with $\tau_r = 2.5$; simulate and plot the response of the closed-loop system to a unit step change in the set-point.

19.5 The following second-order transfer function was obtained via an identification experiment performed in a small neighborhood of a particular steady-state operating condition of a nonlinear process:

$$g(s) = \frac{K}{(3.0s + 1)(5.0s + 1)} \tag{P19.9}$$

While the time constants (in minutes) are relatively easy to estimate, the gain is very difficult to estimate; it is therefore left indeterminate as a free design parameter.

(a) Use this model along with a first-order desired trajectory, with $\tau_r = 4.0$, to design a controller for this process. Leave your controller in terms of the parameter K.

(b) Suppose now that the true process has a transfer function of the form:

$$g_p(s) = \frac{K_p(-\xi s + 1)}{(3.0s + 1)(5.0s + 1)} \tag{P19.10}$$

where, as a consequence of the nonlinearity, the process sometimes passes through a region of operation in which it exhibits inverse-response behavior not captured by the approximate linear model in Eq. (P19.9). The time constants are assumed to be essentially perfectly known.

If the controller designed in part (a) is now to be implemented on this process (assuming for simplicity that the parameters in Eq. (P19.10), even though unknown, are fixed), and if it is known that in the worst case $K_p = 2K$, (i.e., in the worst case the process gain is underestimated by a factor of 2), find the maximum value of the unknown right-half plane zero parameter ξ which the closed-loop system can tolerate without going unstable.

(c) If it is now known that for the process in part (b), ξ takes the value of 3.5 in the worst case, and that the worst case for the gain estimate remains as given in part (b), what value of τ_r is required for guaranteeing closed-loop system stability in the worst case?

19.6 Revisit Problem 19.5 for the specific situation in which the true process transfer function is given as:

$$g_p(s) = \frac{3.0(-1.5s + 1)}{(3.0s + 1)(5.0s + 1)}$$

but the model transfer function is as in Eq. (P19.9) with K chosen as 2.0.

(a) Use the model transfer function along with a first-order desired trajectory, with $\tau_r = 4.0$, to design a direct synthesis controller; implement the controller on the true process in Eq. (P19.10) and plot the closed-loop system response to a unit step change in set-point.

(b) Repeat part (a) for the choice of $\tau_r = 6.0$. Compare the performance of this controller with that obtained in part (a) and comment on the effect of larger values of τ_r on closed-loop performance in the face of plant/model mismatch.

19.7 The *nominal* transfer function for an open-loop unstable process was obtained via closed-loop identification experimentation as:

$$g(s) = \frac{-0.55}{(-5.2s + 1)} \tag{P19.11}$$

and the true process gain and time constant are known to lie in the following limits:

$$-0.45 \le K_p \le -0.65 \tag{P19.12a}$$

$$-4.9 \le \tau_p \le -5.5 \tag{P19.12b}$$

Choose an appropriate desired closed-loop transfer function and design a direct synthesis controller for this process. Implement your controller on a process whose transfer function is given as:

$$g_p(s) = \frac{-0.6(-1.5s + 1)}{(-5.0s + 1)} \tag{P19.13}$$

and plot the closed-loop system response to a unit step change in set-point.

19.8 The following sixth-order transfer function was obtained by linearizing the original set of ordinary differential equations used to model an industrial chemical reactor:

$$g(s) = \frac{10.3}{(1.5s + 1)^5(15s + 1)} \tag{P19.14}$$

The indicated time constants are in minutes. For the purposes of designing a practical controller, this model was further simplified to:

$$g_1(s) = \frac{10.3e^{-7.5s}}{15s + 1} \tag{P19.15}$$

(a) Design a controller for this process by direct synthesis, using the transfer function $g_1(s)$, for $\tau_r = 5.0$ min. Introduce a first-order Padé approximation for the delay element *in the controller*. What further approximation will be required in order to reduce the result to a standard PID controller? Can you justify such an approximation?

(b) Implement the PID controller obtained in part (a) on the sixth-order process and obtain the closed-loop response to a step change of 0.5 in the output set-point.

(c) Repeat part (a) but this time introduce the Padé approximation for the delay element *in the model* before carrying out the controller design, and then design a direct synthesis controller for the resulting inverse-response system, again using $\tau_r = 5.0$ mins. Compare the resulting controller with the one obtained in part (a).

19.9 Use the IMC approach to design a classical controller for the process whose dynamics are represented by the general first-order-plus-time-delay transfer function:

$$y(s) = \frac{K\,e^{-\alpha s}}{\tau s + 1}$$

by following this procedure:

    1.   Introduce a first-order Padé approximation for the delay.
    2.   Factor the resulting transfer function into $g_-(s)\,g_+(s)$ as required.
    3.   Choose a first-order filter with time constant λ and obtain the IMC controller.
    4.   Convert from the IMC form to the standard feedback form. The result should be a PID controller.

What values does this IMC procedure recommend for K_c, τ_I, and τ_D?

19.10 Revisit Problem 19.8 and design a PID controller for the sixth-order process in Eq. (P19.14), based on the transfer function in Eq. (P19.15) and the IMC procedure of Problem 19.9, with the specific choice of $\lambda = 5.0$ min. Compare this controller to that obtained in part (a) of Problem 19.8. What does this example illustrate about the relationship between the IMC strategy and the direct synthesis strategy for PID controller tuning for systems with time delays?

19.11 [*This problem is open ended*]
Via frequency response identification, the following model was obtained for a heat exchanger:

$$y(s) = \frac{K e^{-\alpha s}}{(\tau_1 s + 1)(\tau_2 s + 1)} u(s) + K_d e^{-\beta s} d(s) \qquad (P19.16)$$

with parameter values estimated as:

$$
\begin{aligned}
K &= 2.07 \pm 0.8; \ (°F/lb/hr) \\
K_d &= 0.478 \pm 0.1; \ (°F/°F) \\
\tau_1 &= 7.60 \pm 0.5; \ (s) \\
\tau_2 &= 30.40 \pm 1.2; \ (s) \\
\alpha &= 7.6 \pm 1.3; \ (s) \\
\beta &= 25.3 \pm 5.5; \ (s)
\end{aligned}
$$

It may be assumed that the bounds given above for the parameter values cover the entire range of possible values taken by the true process parameters.

In terms of deviation from steady-state values, y is the exit temperature (°F); u is the steam rate (lb/hr); d is the inlet temperature of the process stream.

(a) Assume that $d(s)$ is measured and design an IMC controller for this process using the supplied model. Justify all the design choices made.

(b) Implement the controller designed in part (a) on a "process" represented by:

$$g_p(s) = \frac{1.3 e^{-8s}}{(8s + 1)(29s + 1)} + 0.55 e^{-21s} \qquad (P19.17)$$

and plot the control system response to a change of 2.0°F in the inlet temperature of the process stream. Evaluate the performance of your controller. If it is necessary to redesign your controller, explain what needs to be done, do it, and then implement the redesigned controller for the same disturbance.

19.12 The following model has been given for a CSTR in which an isothermal second-order reaction is taking place:

$$\frac{dC_A}{dt} = \frac{F}{V}(C_{A0} - C_A) - k C_A^2 \qquad (P19.18)$$

Here F, the volumetric flowrate of reactant A into the reactor, is the manipulated variable; C_{A0}, the inlet concentration of pure reactant in the feed, is the disturbance variable; and C_A, the reactor concentration of A, is the output variable to be controlled; V is the reactor volume, assumed constant. The nominal process operating conditions are given as:

$$
\begin{aligned}
F &= 7.0 \times 10^{-3} \ \text{m}^3/\text{s} \\
V &= 7.0 \ \text{m}^3/\text{s} \\
k &= 1.5 \times 10^{-3} \ \text{m}^3/\text{kg mole-s} \\
C_{A0} &= 2.5 \ \text{kg mole/m}^3 \\
C_A &= 1.0 \ \text{kg mole/m}^3
\end{aligned}
$$

(a) Use this information to design a GMC controller for the reactor, leaving as unspecified the controller parameters K_1 and K_2. Describe how such a controller would be implemented.

(b) Assume now that the gas chromatograph measurement of C_{A0} made available to the controller is permanently biased such that the true measurement is always $C_{A0} + 0.25$ kg mole/m^3. Under these circumstances, implement the GMC controller with $K_1 = 1.0 \times 10^{-3}$ and $K_2 = 2.0 \times 10^{-3}$ and plot the control system response to a change in the actual inlet concentration from 2.5 to 2.0 kg mole/m^3. Assume all the other model parameters are as given.

part IVB

MULTIVARIABLE PROCESS CONTROL

"...how they have increased that trouble
many are they that rise up against m

Psalm

"It is quite a three pipe probl
and I beg that you won't speak to
for fifty minut

Sherlock Holmes, *The Red-Headed Lea*
(Sir Arthur Conan Do

CHAPTER

20

INTRODUCTION TO MULTIVARIABLE SYSTEMS

When a process has only one input variable to be used in controlling one output variable, the controller design problem can be handled, fairly conveniently, by the methods discussed in Part IVA. In several important situations, however, the system under consideration has multiple inputs and multiple outputs, making it multivariable in nature. In actual fact, the most important chemical processes are often multivariable in nature. Many nontrivial issues, not encountered hitherto in our discussion of single-input, single-output (SISO) systems, are now raised when controllers are to be designed for these multivariable systems.

Our study of multivariable control systems analysis and design begins in this chapter with an introduction to multivariable systems and the unique control problems they present. This is where we lay the foundation required for a thorough treatment of the controller design issues to be found in later chapters.

20.1 THE NATURE OF MULTIVARIABLE SYSTEMS

A multivariable process is one with multiple inputs, $u_1, u_2, u_3, \ldots, u_m$, and multiple outputs, $y_1, y_2, y_3, \ldots, y_n$, where m is not necessarily equal to n; it could be a single process, such as the stirred mixing tank shown in Figure 20.1, or it could be an aggregate of many process units constituting part of an entire plant, or it could be the entire plant itself.

Input/Output Pairing

Up until now, because we have had to deal with only SISO systems, the question of what input variable to use in controlling what output variable did

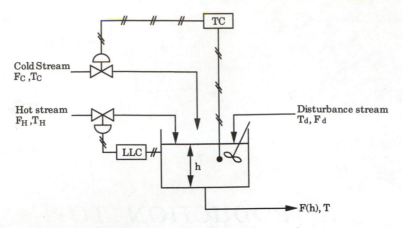

Figure 20.1.   Stirred mixing tank requiring level and temperature control.

not arise, since there was just one of each. But now, with m inputs and n outputs, we are faced with a new problem:

> *Which input variable should be used in controlling which output variable?*

This is referred to as the *input/output pairing* problem.

One of the consequences of having several input and output variables is that such a control system can be configured several different ways depending on which input variable is paired with which output variable. Each u_i to y_j "pairing" constitutes a control configuration, and for a two-input, two-output system (often abbreviated as a 2×2 system) we have two such configurations:

Configuration 1   **Configuration 2**

$u_1 - y_1$ $u_1 - y_2$

and

$u_2 - y_2$ $u_2 - y_1$

What this means in a physical sense is illustrated by the following example.

Example 20.1   CONTROL CONFIGURATIONS FOR THE STIRRED MIXING TANK.

The stirred mixing tank shown in Figure 20.1 has two input variables, the cold stream flowrate, and the hot stream flowrate, to be used in controlling two output variables, the temperature of the liquid in the tank, and the liquid level (see Ref. [1]). Enumerate the different ways the control system can be configured.

Solution:

There are *two* possible configurations:
1. Use hot stream flowrate to control liquid level, and use cold stream to control liquid temperature (this is the configuration shown in Figure 20.1).
2. Use cold stream flowrate to control liquid level, and use hot stream to control liquid temperature.

Note that these two configurations are *mutually exclusive*, i.e., only one of them can be used at any particular time. ·

It can be shown that for a 3×3 system, there are six such configurations; for a 4×4 system there are twenty-four; and in general, for an $n \times n$ system, there are $n!$ possible input-output pairing configurations.

We know, of course, that our controllers can be set up according to *only one* of these configurations. Furthermore, we would intuitively expect one of the configurations to yield "better overall control system performance" than the others: how, then, to choose among these possibilities?

At the simplest level, therefore, the first problem in the analysis and design of multivariable control systems is that of deciding on what input variable to pair with what output variable; and the problem is by no means trivial.

Notation

Because we will be engaged with this issue of input/output pairing quite a bit, we now find it necessary to introduce the following notation, to avoid the confusion that usually attends this exercise:

*We shall refer to the input (or manipulated) variables of a process in their **free** i.e., "unpaired" form as m_1, m_2, m_3, ... , m_m; the output variables will still be y_1, y_2, y_3, ... , y_n.*

After the input variables have been paired with the output variables, we shall reserve the notation u_j for the input variable paired with the jth output, y_j.

Thus, for example, if the third input variable m_3 is paired with the first output variable y_1, then m_3 becomes u_1. In this sense, the subscripts on the m's are merely for counting purposes; the subscripts on the u's actually indicate control loop assignment.

Let us illustrate again with the stirred mixing tank. The input variables are:
1.   The cold steam flowrate
2.   The hot stream flowrate
We may then refer to these respectively as m_1 and m_2, in their "unpaired state."

The output variables are:
1.   Liquid level
2.   Liquid temperature
and let these be y_1 and y_2 respectively.

Now, as configured in Figure 20.1, the hot stream m_2 is paired with the liquid level y_1; and the cold stream m_1 is paired with the liquid temperature y_2. Thus, in terms of keeping track of how many input variables we have, the cold stream is m_1, and the hot stream is m_2; but in terms of *which output they have been paired with*, the hot stream is u_1 (because of its association with the "first" output variable, y_1), and the cold stream is u_2.

If the configuration changes, so will the assignment of the u's; this will ensure that u_j is always paired with y_j. The advantage of this notational convenience will become particularly important in Chapter 21.

Interactions

Another important consideration that did not feature in discussions involving SISO systems but that is of prime importance with multivariable systems is the following: after somehow arriving at the conclusion that input, m_i, is best used to control output y_j, we must now answer the very pertinent question:

> In addition to affecting y_j, its assigned output variable, will m_i affect any other output variable? That is, can m_i control y_j in isolation, or will the control effort of m_i influence other outputs apart from y_j?

Unfortunately, in virtually all cases, a particular m_i will influence several outputs. For example, even though the hot stream flowrate is paired with the level in the stirred mixing process, changes in this input variable will obviously affect the liquid temperature, in addition to affecting its assigned output variable, the liquid level.

This is the *interaction problem*, recognized as the other main problem with the control of multivariable systems.

There is one final set of related issues that do not come into consideration with SISO systems but that are sometimes important with multivariable systems; these will now be introduced briefly.

Controllability and Observability

To illustrate these concepts we introduce another example multivariable system: the two-tank network shown in Figure 20.2. For this process, liquid streams flow into Tanks 1 and 2 at respective volumetric rates F_1, and F_2; the outflow from each tank is assumed (for simplicity) to be proportional to the respective liquid levels h_1 and h_2 in each tank. The liquid leaving Tank 2 is split into two with a fraction F, exiting, and the remainder R pumped back to the first tank (see Ref. [1]).

Thus this is a two-input, two-output (or 2×2) system, with the flowrates of the two inlet streams as the two inputs, and the liquid level in each tank as the two output variables.

We now wish to investigate (for now only in a qualitative sense) how this process will operate under the following conditions:

1. As shown in Figure 20.2, with both inlet streams, and the recycle pump in operation;

2. The recycle pump is turned off, and only stream F_1 is operational; stream F_2 is set at constant flow which *cannot* be changed;

3. The recycle pump is turned off, but this time stream F_1 is now set at constant flow that *cannot* be changed; only stream F_2 is operational.

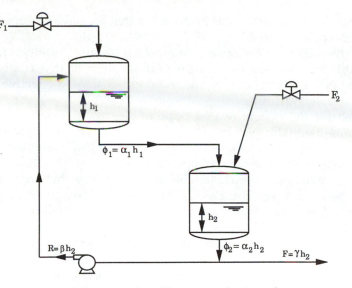

Figure 20.2.   The two-tank network.

In particular, we are interested in answering the following question:

With whatever input variables we have available, can we **simultaneously** *influence the liquid level in* **both** *tanks? Or, put more formally, can we "drive" the liquid level in* **both** *tanks from a given initial value to any other arbitrarily set desired value, in finite time?*

Upon some reflection, we can answer this question for each of the three situations noted above as follows:

1. Under Condition 1, since both inlet streams are available, we can clearly influence the level in both tanks simultaneously.

2. Under Condition 2, observe that with stream F_2 no longer available for control, we have lost our source of *direct* influence on h_2; however, we still have *indirect influence* on it through the outflow from Tank 1.
    Thus, even though we do not have as much control over the system variables (particularly h_2) as we would under Condition 1, we have not lost total control; h_2 can still be influenced by the only available input variable, the stream F_1 flowrate.

3. Under Condition 3, it is clear that we have *absolutely* no control over h_1; the stream F_1 flowrate cannot be changed, and the recycle stream is no longer in operation; thus we can only control h_2. (Note that if the recycle stream were still in operation, through it we would have indirect influence on the level in Tank 1, but this is not the case.)

Under Conditions 1 and 2, the system is said to be *controllable* because, loosely speaking, we can influence each of the two process output variables with whatever input variables we have available; under Condition 3, the system is said to be *uncontrollable* in the sense that we have no control over one of the process output variables.

From this simple illustrative example, it should be clear that not all multivariable systems are necessarily *controllable*, and that controllability is an essential factor to take into consideration in designing effective controllers for multivariable systems.

To illustrate the closely related issue of *observability*, let us imagine that with the same two-tank system of Figure 20.2, we can now only measure the liquid level in *one* tank; along with this single-level measurement restriction, the recycle loop is also taken out of operation.

If the only measurement available is h_2, observe that since its value is *influenced* by h_1 through the Tank 1 outflow, we can, in principle, extract information about the value h_1 given only h_2, and a fairly accurate process model.

If we turn the situation around, and consider the situation in which the only available measurement is h_1, it is clear from the physics of the process that *without the recycle loop in operation*, there is *absolutely* no way we can infer anything about the value of h_2 given only information about h_1.

In the former case, the process is said to be *observable*, while it is *unobservable* in the latter case.

Thus while controllability has to do with the intrinsic property of the system which makes it possible to control all the process state or output variables with the available input variables, observability has to do with the ability to infer *all* the process state variables from available measurements.

Formal definitions of controllability and observability, and quantitative methods for determining the controllability and observability of a process are available, for example, in Ref. [1].

On the basis of the foregoing discussions, it is clear, therefore, that in dealing with multivariable systems, much more so than SISO systems, the need for systematic methods of analysis cannot be overemphasized. The main characteristic problems of multivariable systems (some of which have been highlighted above) typically have very significant implications on control system performance; and by their very nature, these problems are such that they cannot be handled any other way but by systematic analysis.

The same basic tools of systems analysis (e.g., mathematical models, block diagrams, etc.) are still used for multivariable systems, but because of increased dimensionality, additional considerations have to be introduced. The rest of this chapter is devoted to developing the framework for multivariable control systems analysis and design, mostly by extending pertinent ideas we are already familiar with from our encounter with single-variable systems, and introducing new ones which have arisen to handle problems peculiar to multivariable systems.

20.2   MULTIVARIABLE PROCESS MODELS

Models used to represent the dynamic behavior of a multivariable process occur most commonly either in the *state-space* form, or in the *transfer function* form.

Owing to the multivariable nature of such modeling equations they are more conveniently written in vector-matrix form.

State-Space Model Form

As was briefly introduced in Section 4.3 in Chapter 4, the linear, multivariable system is represented in the state space as:

$$\dot{x}(t) = Ax(t) + Bu(t) + \Gamma d(t) \qquad (20.1)$$

$$y(t) = Cx(t) \qquad (20.2)$$

where the dot ($\bullet$) represents differentiation with respect to time, and

$$
\begin{aligned}
x &\equiv l\text{-dimensional vector of state variables} \\
u &\equiv m\text{-dimensional vector of input (control) variables} \\
y &\equiv n\text{-dimensional vector of output variables} \\
d &\equiv k\text{-dimensional vector of disturbance variables}
\end{aligned}
$$

with A, B, C, and Γ as appropriately dimensioned system matrices of conformable order with the respective multiplying vectors.

Nonlinear multivariable systems are typically represented by several nonlinear differential equations, which might be represented in general as given earlier in Eq. (4.31), i.e.:

$$\frac{dx(t)}{dt} = f(x, u, d) \qquad (20.3)$$
$$y(t) = h(x(t))$$

where $f(x, u, d)$, and $h(x(t))$ are usually vectors of nonlinear functions.

Transfer Function Model Form

Analogous to the situation with SISO systems, the multivariable transfer function model form relates the Laplace transform of the vector of input variables to that of the output variables, as given earlier in Eq. (4.38); i.e.:

$$y(s) = G(s)\,u(s) + G_d(s)\,d(s) \qquad (20.4)$$

where, for an m-input, n-output system, $G(s)$ is an $m \times n$ transfer function matrix with elements:

$$
G(s) = \begin{bmatrix}
g_{11}(s) & g_{12}(s) & \cdots & g_{1m}(s) \\
g_{21}(s) & g_{22}(s) & \cdots & g_{2m}(s) \\
\cdots & \cdots & \cdots & \cdots \\
g_{n1}(s) & g_{n2}(s) & \cdots & g_{nm}(s)
\end{bmatrix} \qquad (20.5)
$$

and each typical $g_{ij}(s)$ transfer function element is exactly like the familiar SISO transfer functions. $G_d(s)$ is an $n \times k$ transfer function matrix consisting of similar transfer function elements.

Interrelationships between Model Forms

As with SISO systems, it is possible to convert the state-space model representation to the transfer function form by taking Laplace transforms and rearranging.

Taking Laplace transforms in Eqs. (20.1) and (20.2), assuming that the matrices involved are constant, and that we have zero initial conditions, we obtain:

$$s\mathbf{x}(s) = \mathbf{A}\mathbf{x}(s) + \mathbf{B}\mathbf{u}(s) + \mathbf{\Gamma}\mathbf{d}(s) \tag{20.6}$$

$$\mathbf{y}(s) = \mathbf{C}\mathbf{x}(s) \tag{20.7}$$

which can be rearranged (paying due respect to the rules of matrix algebra) to give:

$$\mathbf{y}(s) = \left[\mathbf{C}(s\mathbf{I} - \mathbf{A})^{-1}\mathbf{B}\right]\mathbf{u}(s) + \left[\mathbf{C}(s\mathbf{I} - \mathbf{A})^{-1}\mathbf{\Gamma}\right]\mathbf{d}(s) \tag{20.8}$$

Hence, by comparison with Eq. (20.4), we obtain the relationship between the transfer function matrices and the state-space matrices:

$$\mathbf{G}(s) = \left[\mathbf{C}(s\mathbf{I} - \mathbf{A})^{-1}\mathbf{B}\right] \tag{20.9}$$

and

$$\mathbf{G_d}(s) = \left[\mathbf{C}(s\mathbf{I} - \mathbf{A})^{-1}\mathbf{\Gamma}\right] \tag{20.10}$$

The reverse problem of deducing the equivalent state-space model from the transfer function model — typically known as the *realization* problem — is not by any means trivial.

For one thing, no realization is *unique*, in the sense that there are several equivalent sets of differential equations that, upon Laplace transformation, will yield the *same* transfer function matrix. Further, there are several different ways of obtaining these different, but equivalent, realizations. As a result it is often necessary to utilize a method that provides the so-called "minimal realization."

We consider a discussion of how to obtain minimal realizations as being outside the intended scope of this book; the interested reader is referred to Ref. [1] for the important details. A simple procedure for obtaining state-space realizations will be illustrated with a problem at the end of the chapter.

It is important to note that most computer-aided control systems design packages with multivariable control capabilities (cf. Appendix E) have routines for obtaining time-domain realizations; this step is necessary when carrying out process simulation using transfer function matrix models.

Let us consider some examples of multivariable process models.

Example 20.2   MATHEMATICAL MODEL FOR THE TWO-TANK SYSTEM OF FIGURE 20.2.

Assuming uniform cross-sectional areas A_1, and A_2, respectively for Tanks 1 and 2, material balances over each of the tanks yield the following modeling equations:

$$A_1\frac{dh_1}{dt} = -\alpha_1 h_1 + \beta h_2 + F_1 \tag{20.11}$$

$$A_2 \frac{dh_2}{dt} = -\alpha_2 h_2 + \alpha_1 h_1 + F_2 \tag{20.12}$$

and if we let $\alpha_2 = \beta + \gamma$ and define the following deviation variables:

$$x_1 = h_1 - h_{1s}; \qquad x_2 = h_2 - h_{2s}$$
$$u_1 = F_1 - F_{1s}; \qquad u_2 = F_2 - F_{2s}$$

then these modeling equations become:

$$\frac{dx_1}{dt} = -\frac{\alpha_1}{A_1} x_1 + \frac{\beta}{A_1} x_2 + \frac{1}{A_1} u_1 \tag{20.13}$$

$$\frac{dx_2}{dt} = \frac{\alpha_1}{A_2} x_1 - \frac{(\beta + \gamma)}{A_2} x_2 + \frac{1}{A_2} u_2 \tag{20.14}$$

If we introduce the variables y_1 and y_2 to represent the *measurements* of the liquid levels (as deviations from steady-state values) we have, in the situation where *both* are measured:

$$y_1 = x_1 \tag{20.15a}$$

$$y_2 = x_2 \tag{20.15b}$$

Observe now that these equations, (20.13), (20.14), and (20.15), are in the form given in Eqs. (20.1) and (20.2); the state-space vectors and matrices are easily seen to be:

$$\mathbf{x} = \begin{bmatrix} x_1 \\ x_2 \end{bmatrix}; \quad \mathbf{y} = \begin{bmatrix} y_1 \\ y_2 \end{bmatrix}; \quad \mathbf{u} = \begin{bmatrix} u_1 \\ u_2 \end{bmatrix}$$

$$\mathbf{A} = \begin{bmatrix} -\dfrac{\alpha_1}{A_1} & \dfrac{\beta}{A_1} \\ \dfrac{\alpha_1}{A_2} & -\dfrac{(\beta + \gamma)}{A_2} \end{bmatrix}; \quad \mathbf{B} = \begin{bmatrix} \dfrac{1}{A_1} & 0 \\ 0 & \dfrac{1}{A_2} \end{bmatrix}$$

with $\mathbf{C} = \mathbf{I}$, and $\boldsymbol{\Gamma} = 0$.

Observe that, in this case, this is a 2×2 system.

As the next example demonstrates, the stirred mixing tank of Figure 20.1 is a nonlinear multivariable system; it is modeled by two differential equations consisting of nonlinear functions.

Example 20.3   **MATHEMATICAL MODEL FOR THE STIRRED MIXING TANK OF FIGURE 20.1.**

Assuming uniform cross-sectional area A_c for the tank, constant liquid physical properties ρ and C_p, and $F = K\sqrt{h}$, we obtain the following from material and energy balances over the tank:

$$A_c \frac{dh}{dt} = F_H + F_C + F_d - K\sqrt{h} \tag{20.16}$$

$$\rho C_p A_c \frac{d(hT)}{dt} = \rho C_p \left(F_H T_H + F_C T_C + F_d T_d - K \sqrt{hT} \right) \tag{20.17}$$

which are nonlinear because of the presence of the square root term, and the product functions of h and T.

Example 20.4   APPROXIMATE LINEAR MODEL FOR THE STIRRED MIXING TANK OF FIGURE 20.1.

The modeling equations obtained in Example 20.3 can be linearized around some steady-state operating point, h_s, T_s along precisely the same lines indicated in Chapter 10. If in addition we define the following deviation variables:

$$
\begin{aligned}
x_1 &= h - h_s; & x_2 &= T - T_s \\
u_1 &= F_H - F_{Hs}; & u_2 &= F_C - F_{Cs} \\
d_1 &= F_d - F_{ds}; & d_2 &= T_d - T_{ds}
\end{aligned}
$$

then the linearized modeling equations become:

$$\frac{dx_1}{dt} = \frac{1}{A_c} \left(-\frac{K}{2\sqrt{h_s}} x_1 + u_1 + u_2 + d_1 \right) \tag{20.18}$$

$$
\begin{aligned}
\frac{dx_2}{dt} = \frac{1}{A_c h_s} \Big[&-K\sqrt{h_s}\, x_2 + (T_H - T_s)\, u_1 + (T_C - T_s)\, u_2 \\
&+ (T_{ds} - T_s)\, d_1 + F_{ds}\, d_2 \Big]
\end{aligned} \tag{20.19}
$$

These equations can now be expressed in the vector-matrix form given in Eqs. (20.1) and (20.2); the resulting state-space matrices are:

$$\mathbf{A} = \begin{bmatrix} -\dfrac{K}{2A_c\sqrt{h_s}} & 0 \\[3ex] 0 & -\dfrac{K}{A_c\sqrt{h_s}} \end{bmatrix} \tag{20.20}$$

$$\mathbf{B} = \begin{bmatrix} \dfrac{1}{A_c} & \dfrac{1}{A_c} \\[3ex] \dfrac{(T_H - T_s)}{A_c h_s} & \dfrac{(T_C - T_s)}{A_c h_s} \end{bmatrix} \tag{20.21}$$

with

$$\Gamma = \begin{bmatrix} \dfrac{1}{A_c} & 0 \\[3ex] \dfrac{(T_{ds} - T_s)}{A_c h_s} & \dfrac{F_{ds}}{A_c h_s} \end{bmatrix} \tag{20.22}$$

If both level and temperature are measured, then $\mathbf{y} = \mathbf{x}$ and $\mathbf{C} = \mathbf{I}$, the identity matrix.

The next example illustrates how to obtain transfer function matrices given the state-space model.

Example 20.5   **APPROXIMATE TRANSFER FUNCTION MATRICES FOR THE STIRRED MIXING TANK OF FIGURE 20.1, USING THE STATE-SPACE MODEL OF EXAMPLE 20.4.**

From the approximate state-space model obtained for the stirred mixing tank system in Example 20.4, we can use the relationships in Eqs. (20.9) and (20.10) to deduce the corresponding transfer function matrices for this process.

From Eq. (20.9), and recalling the state-space matrices of Eqs. (20.20), (20.21), and (20.22) we see that $\mathbf{G}(s)$ is given by:

$$\mathbf{G}(s) = \begin{bmatrix} 1 & 0 \\ 0 & 1 \end{bmatrix} \begin{bmatrix} s + \dfrac{K}{2A_c\sqrt{h_s}} & 0 \\ 0 & s + \dfrac{K}{A_c\sqrt{h_s}} \end{bmatrix}^{-1} \begin{bmatrix} \dfrac{1}{A_c} & \dfrac{1}{A_c} \\ \dfrac{(T_H - T_s)}{A_c h_s} & \dfrac{(T_C - T_s)}{A_c h_s} \end{bmatrix}$$

which upon carrying out the indicated algebra finally gives:

$$\mathbf{G}(s) = \begin{bmatrix} \dfrac{1}{A_c\left(s + a_{11}\right)} & \dfrac{1}{A_c\left(s + a_{11}\right)} \\ \dfrac{(T_H - T_s)}{A_c h_s\left(s + a_{22}\right)} & \dfrac{(T_C - T_s)}{A_c h_s\left(s + a_{22}\right)} \end{bmatrix} \qquad (20.23)$$

where:

$$a_{11} = \frac{K}{2A_c\sqrt{h_s}} \qquad (20.24a)$$

$$a_{22} = \frac{K}{A_c\sqrt{h_s}} \qquad (20.24b)$$

and we also obtain, in similar fashion, that:

$$\mathbf{G}_d(s) = \begin{bmatrix} \dfrac{1}{A_c\left(s + a_{11}\right)} & 0 \\ \dfrac{(T_{ds} - T_s)}{A_c h_s\left(s + a_{22}\right)} & \dfrac{F_{ds}}{A_c h_s\left(s + a_{22}\right)} \end{bmatrix} \qquad (20.25)$$

State space models are usually obtained from theoretical modeling exercises, and, as shown in the last example, they can be used to derive transfer function models when the modeling equations are linear. Often, when a multivariable process is modeled by experimental means — by correlating input/output data — the model is constructed in the transfer function matrix form as illustrated in the following example.

**Example 20.6   TRANSFER FUNCTION MODEL FOR A BINARY
DISTILLATION COLUMN.**

The model for a distillation column used in separating methanol and water was
reported in Ref. [2] as being of the form Eq. (20.4), with two output variables, two
input variables, and one disturbance variable defined (in terms of deviation variables)
as:

$$y_1 = \text{overhead mole fraction methanol}$$
$$y_2 = \text{bottoms mole fraction methanol}$$
$$u_1 = \text{overhead reflux flowrate}$$
$$u_2 = \text{bottoms steam flowrate}$$
$$d = \text{column feed flowrate}$$

The transfer function matrices are:

$$\mathbf{G}(s) = \begin{bmatrix} \dfrac{12.8e^{-s}}{16.7s + 1} & \dfrac{-18.9e^{-3s}}{21.0s + 1} \\[3mm] \dfrac{6.6e^{-7s}}{10.9s + 1} & \dfrac{-19.4e^{-3s}}{14.4s + 1} \end{bmatrix} \tag{20.26}$$

and

$$\mathbf{G}_d(s) = \begin{bmatrix} \dfrac{3.8e^{-8.1s}}{14.9s + 1} \\[3mm] \dfrac{4.9e^{-3.4s}}{13.2s + 1} \end{bmatrix} \tag{20.27}$$

Later on, we will use this process model to evaluate various control system designs.

20.3   OPEN-LOOP DYNAMIC ANALYSIS IN STATE SPACE

Conceptually, open-loop dynamic analysis is carried out for multivariable
systems in precisely the same fashion as for SISO systems; it involves solving
for the process output $\mathbf{y}(t)$ (now a vector!), given the process model, and specific
forms of forcing functions $\mathbf{u}(t)$. The computational effort involved will, of
course, be greater, but the principles are essentially the same.

The transient responses can be obtained using the state-space model form, or
the transfer function form; depending on the problem at hand, one form might be
easier, or more appropriate, to use than another. In this section we will use
state-space models and then demonstrate the use of transfer function models in
the next section.

For linear systems, carrying out dynamic analysis in the state space
involves no more than solving Eqs. (20.1) and (20.2) for any given form of forcing
function, and investigating the nature of the resulting solution.

For nonlinear multivariable systems, it is clear that rigorous dynamic
analysis is only possible in the state space; however, such analysis requires
mathematical tools far beyond the intended scope of this book. We will
assume, therefore, that approximate linear analysis, using either linearized
state-space models or equivalent transfer function matrices, will suffice for our
purposes. (This, of course, does not preclude numerical analysis using the
complete nonlinear model, but as was pointed out in Chapter 10, such analysis is
too specific to be of any general use.)

20.3.1 Analytical Solutions

As it turns out, because the state-space model Eq. (20.1) is linear with constant coefficients, and because of the striking resemblance it bears with the scalar, first-order ODE, it has a general, analytical, solution that can be written down immediately, in closed form:

$$\mathbf{x}(t) = \int_0^t e^{\mathbf{A}(t - \sigma)} \left[\mathbf{B}\mathbf{u}(\sigma) + \mathbf{\Gamma}\mathbf{d}(\sigma) \right] d\sigma \qquad (20.28)$$

and the output $\mathbf{y}(t)$ is obtained simply by multiplying Eq. (20.28) by $\mathbf{C}$, to give:

$$\mathbf{y}(t) = \mathbf{C} \int_0^t e^{\mathbf{A}(t - \sigma)} \left[\mathbf{B}\mathbf{u}(\sigma) + \mathbf{\Gamma}\mathbf{d}(\sigma) \right] d\sigma \qquad (20.29)$$

The indicated integration is then carried out given a specific form for the input variables, $\mathbf{u}(t)$ and $\mathbf{d}(t)$.

20.3.2 Unit Step Responses

In particular, for *unit* step changes in the *entire* input vector $\mathbf{u}(t)$, (i.e., $\mathbf{u}(t) = \mathbf{1}$, a vector whose elements are all 1's) with $\mathbf{d}(t) = 0$, Eq. (20.28) reduces to:

$$\mathbf{x}(t) = \int_0^t e^{\mathbf{A}(t - \sigma)} \mathbf{B} \bullet \mathbf{1} \, d\sigma \qquad (20.30)$$

and *provided* $\mathbf{A}$ *is square, and nonsingular*, we can now carry out the indicated integration (again, paying due attention to the laws of matrix algebra) and obtain:

$$\mathbf{x}(t) = \mathbf{A}^{-1}(e^{\mathbf{A}t} - \mathbf{I})\mathbf{w} \qquad (20.31a)$$

where $\mathbf{w}$ is a vector whose ith element is given by:

$$\mathbf{w}_i = \sum_j b_{ij} \qquad (20.31b)$$

and the output is obtained by premultiplying Eq. (20.31(a)) by $\mathbf{C}$, giving:

$$\mathbf{y}(t) = \mathbf{C}\mathbf{A}^{-1}(e^{\mathbf{A}t} - \mathbf{I})\mathbf{w} \qquad (20.32)$$

Eqs. (20.31) and (20.32) give the multivariable process response to *simultaneous* unit step changes in *all* the process inputs. Often, we are more interested in how the process responds to a unit step change in only *one* of the input variables while the others are set to zero. (As we will see later, this provides greater insight into the nature of the multivariable system than the response to simultaneous unit step changes *all* of the input variables.)

In this case, partitioning $\mathbf{B}$ as a matrix composed of n column vectors $\mathbf{b}_1$, $\mathbf{b}_2$, $\mathbf{b}_3$, ..., $\mathbf{b}_n$, as follows:

$$\mathbf{B} = \left[\mathbf{b}_1. \mathbf{b}_2. \mathbf{b}_3. ... \mathbf{b}_n \right] \tag{20.33}$$

then, for a unit step change in u_i alone, with all the other inputs remaining at zero, the response is:

$$\mathbf{x}(t) = \mathbf{A}^{-1}(e^{\mathbf{A}t} - \mathbf{I})\mathbf{b}_i \tag{20.34a}$$

or

$$\mathbf{y}(t) = \mathbf{C}\mathbf{A}^{-1}(e^{\mathbf{A}t} - \mathbf{I})\mathbf{b}_i \tag{20.34b}$$

We will refer to these as the *individual input* unit step responses.

The main obstacle to be surmounted in applying these equations lies in the evaluation of the matrix exponential $e^{\mathbf{A}t}$, especially when $\mathbf{A}$ is a nondiagonal matrix. As shown in Appendix D, this matrix exponential is evaluated with the aid of the eigenvalues and eigenvectors of $\mathbf{A}$, using the following expression:

$$e^{\mathbf{A}t} = \mathbf{M}e^{\Lambda t}\mathbf{M}^{-1} \tag{20.35}$$

where Λ is the diagonal matrix of the eigenvalues of $\mathbf{A}$:

$$\Lambda = \begin{bmatrix} \lambda_1 & 0 & ... & ... & 0 \\ 0 & \lambda_2 & 0 & ... & 0 \\ ... & ... & ... & ... & ... \\ 0 & 0 & ... & ... & \lambda_n \end{bmatrix}$$

and $\mathbf{M}$ is the modal matrix of corresponding eigenvectors.

Thus the *individual input* unit step-response equations become:

$$\mathbf{x}(t) = \mathbf{A}^{-1}\left(\mathbf{M}e^{\Lambda t}\mathbf{M}^{-1} - \mathbf{I}\right)\mathbf{b}_i \tag{20.36a}$$

or

$$\mathbf{y}(t) = \mathbf{C}\mathbf{A}^{-1}\left(\mathbf{M}e^{\Lambda t}\mathbf{M}^{-1} - \mathbf{I}\right)\mathbf{b}_i \tag{20.36b}$$

where the matrix exponential of the diagonal matrix Λt is given by:

$$e^{\Lambda t} = \begin{bmatrix} e^{\lambda_1 t} & 0 & ... & ... & 0 \\ 0 & e^{\lambda_2 t} & 0 & ... & 0 \\ ... & ... & ... & ... & ... \\ 0 & 0 & ... & ... & e^{\lambda_n t} \end{bmatrix}$$

20.3.3 Stability

Following similar arguments as those presented in Chapter 11 for the SISO system, we note from the step-response equations in Eq. (20.36a,b) that these responses will remain bounded if all the elements of the (diagonal) matrix Λ have negative real parts; if just one of them has a positive real part the process response will grow indefinitely with time. If we recall that the elements of the matrix Λ are the eigenvalues of the system matrix $\mathbf{A}$, and since the unit step input is a *bounded input*, we now state the following result for Bounded-Input, Bounded-Output (BIBO) stability of multivariable systems:

A multivariable process whose model is given in the state space as in Eqs. (20.1) and (20.2) is open-loop stable if and only if all the eigenvalues of the matrix $\mathbf{A}$ *have negative real parts; otherwise it is unstable.*

It is important to note the following points:

1. The open-loop stability of the multivariable system depends only on the matrix $\mathbf{A}$, and on nothing else. Observe that in Eq. (20.36), boundedness or otherwise of the response is ascertained regardless of i, the index on the $\mathbf{b}$ vector (i.e., regardless of which of the n inputs we used to introduce the test step function) or, for that matter, the vector itself.

2. The stability conditions hold regardless of whether we are investigating the response in the *state* vector, $\mathbf{x}$, or in the *output* vector, $\mathbf{y}$.

3. For nonlinear systems, upon linearization of the nonlinear model around some steady state in order to obtain a linearized model of the form in Eqs. (20.1) and (20.2), it is customary to refer to the resulting $\mathbf{A}$ matrix as the *Jacobian* of the nonlinear system. As was the case with nonlinear SISO systems, it is also true of multivariable systems that: *the stability of the nonlinear multivariable system near the steady state is determined by the eigenvalues of its Jacobian; if these eigenvalues all have negative real parts, the system is stable; otherwise the system is unstable.*

4. As a reminder, it should be noted that the eigenvalues of any matrix $\mathbf{A}$ are determined by finding the roots of the polynomial characteristic equation:

$$|(\mathbf{A} - \lambda \mathbf{I})| = 0 \tag{20.37}$$

This will be important when we consider multivariable system stability using transfer function matrices.

Let us illustrate the principles involved in dynamic analysis in the state space with the following examples, all based on the two-tank process of Figure 20.2.

Example 20.7 ANALYSIS OF THE DYNAMIC BEHAVIOR OF THE TWO-TANK PROCESS: 1. *STABILITY.*

Consider the two-tank process of Figure 20.2 under the situation in which Tank 1 is a 500 liter tank, while Tank 2 has a capacity of 200 liters; the respective cross-sectional areas are 0.5 m^2, and 0.2 m^2.

The flowrate out of each tank is assumed proportional to the liquid level in each tank; the constants of proportionality are identical and are given as: $\alpha_1 = \alpha_2 = 0.1$ m^2/s; 40% of the amount of liquid leaving Tank 2 is recycled back to Tank 1, implying that $\beta = 0.04$ m^2/s, and $\gamma = 0.06$ m^2/s.

The entire process is initially at steady state at the following conditions:

Liquid levels: $h_{1s} = 0.7$ m ; $h_{2s} = 0.5$ m
Inlet flowrates: $F_{1s} = 0.05$ m^3/s ; $F_{2s} = 0.05$ m^3/s

Using the general results in Example 20.2, evaluate the model parameters for this specific process, under the specified conditions and investigate whether or not it is open-loop stable.

Solution:

Introducing the specific process parameters into the model obtained in Example 20.2, we find that the system matrices are given by:

$$\mathbf{A} = \begin{bmatrix} -0.2 & 0.08 \\ 0.5 & -0.5 \end{bmatrix} ; \quad \mathbf{B} = \begin{bmatrix} 2 & 0 \\ 0 & 5 \end{bmatrix} \tag{20.38}$$

with $\mathbf{C} = \mathbf{I}$.

To investigate the stability of this system, all that is required is to evaluate the eigenvalues of $\mathbf{A}$; these are obtained from:

$$\left| (\mathbf{A} - \lambda\mathbf{I}) \right| = 0$$

or, in this case:

$$\begin{vmatrix} -0.2 - \lambda & 0.08 \\ 0.5 & -0.5 - \lambda \end{vmatrix} = 0 \tag{20.39}$$

which, upon evaluating the indicated determinant, gives rise to the characteristic equation:

$$\lambda^2 + 0.7\,\lambda + 0.06 = 0 \tag{20.40}$$

a quadratic whose two roots, the eigenvalues of the matrix $\mathbf{A}$, are given by:

$$\lambda_1 = -0.1 \tag{20.41a}$$

$$\lambda_2 = -0.6 \tag{20.41b}$$

and since both of these eigenvalues are *negative*, we conclude that the system is open-loop stable.

The matrix of eigenvalues for this system is therefore seen to be:

$$\Lambda = \begin{bmatrix} -0.1 & 0 \\ 0 & -0.6 \end{bmatrix} \tag{20.42}$$

With the following two examples, we investigate the response of the two-tank process to step changes in the inlet flowrate to Tank 1 first, and then Tank 2. Two important facts are to be illustrated by these examples:

1. That there really *are* interactions between the process variables, in the sense that a change in the Tank 1 inlet flowrate will affect the liquid level in *both* tanks, and not just the level in Tank 1; just as a change in Tank 2 inlet flowrate will not only affect the level in Tank 2 but that in Tank 1 also.

2. The interaction effects of input variable 1 on output variable 2 are not necessarily the same as the interaction effects of input variable 2 on output variable 1; in other words, within a multivariable process, the interaction effects exerted on the various output variables differ for the different input variables. Such information will prove useful, in a qualitative sense, when we need to choose input/output pairs.

Example 20.8   ANALYSIS OF THE DYNAMIC BEHAVIOR OF THE TWO-TANK PROCESS: 2. *RESPONSE TO STEP CHANGE IN TANK 1 INFLOW.*

Obtain expressions for how the liquid level in each of the two tanks will respond to a step increase in the Tank 1 inlet flowrate from its initial value of 0.05 m^3/s to a new value of 0.065 m^3/s (while the Tank 2 inflow is maintained constant at its initial steady-state value). Comment on the *interactive* influence of this input variable on the level in Tank 2.

Solution:

We will first find the *unit* step response in terms of deviation variables, multiply this by 0.015, and finally express it in terms of actual process variables.

Under the conditions stipulated above, observe that the unit step response (with $u_1 = 1$, and $u_2 = 0$) will be given by:

$$\mathbf{x}(t) = \mathbf{A}^{-1}\left(\mathbf{M}\,e^{\Lambda t}\,\mathbf{M}^{-1} - \mathbf{I}\right)\mathbf{b}_1 \tag{20.43}$$

where, from Eq. (20.38):

$$\mathbf{b}_1 = \begin{bmatrix} 2 \\ 0 \end{bmatrix}$$

The inverse of the **A** matrix given in Eq. (20.38) is easily evaluated as:

$$\mathbf{A}^{-1} = \begin{bmatrix} -\dfrac{25}{3} & -\dfrac{4}{3} \\[2mm] -\dfrac{25}{3} & -\dfrac{10}{3} \end{bmatrix} \tag{20.44}$$

As shown in Appendix D, the eigenvectors corresponding to each of the two eigenvalues calculated in Example 20.7 are obtained by choosing *any* nonzero column of the adjoint of the matrix $(\mathbf{A} - \lambda_i \mathbf{I})$ (scaled by any constant term). In this fashion, we obtain the eigenvector corresponding to $\lambda_1 = -0.1$ as:

$$\mathbf{m}_1 = \begin{bmatrix} 4 \\ 5 \end{bmatrix}$$

and the other eigenvector, corresponding to $\lambda_2 = -0.6$, is similarly obtained as:

$$\mathbf{m}_2 = \begin{bmatrix} 1 \\ -5 \end{bmatrix}$$

upon putting these together, we obtain the modal matrix as:

$$\mathbf{M} = \begin{bmatrix} 4 & 1 \\ 5 & -5 \end{bmatrix} \tag{20.45}$$

the inverse of which is now easily obtained, in the usual manner, giving:

$$\mathbf{M}^{-1} = \begin{bmatrix} 0.2 & 0.04 \\ 0.2 & -0.16 \end{bmatrix} \tag{20.46}$$

What is now left is to introduce the appropriate matrices into Eq. (20.43). In doing this, it is perhaps of interest to see what the matrix exponential turns out to be in this case. From Eq. (20.35) upon introducing Eqs. (20.42), (20.45), and (20.46), and simplifying, we obtain:

$$e^{\mathbf{A}t} = \mathbf{M}e^{\Lambda t}\mathbf{M}^{-1}$$

$$= \begin{bmatrix} \left(\dfrac{4}{5}e^{-0.1t} + \dfrac{1}{5}e^{-0.6t} \right) & \left(\dfrac{4}{25}e^{-0.1t} - \dfrac{4}{25}e^{-0.6t} \right) \\[2ex] \left(e^{-0.1t} - e^{-0.6t} \right) & \left(\dfrac{1}{5}e^{-0.1t} + \dfrac{4}{5}e^{-0.6t} \right) \end{bmatrix}$$

After completing the remaining algebra indicated in Eq. (20.43), and simplifying, the final result is:

$$\mathbf{x}(t) = \begin{bmatrix} \dfrac{50}{3}\left(1 - 0.96e^{-0.1t} - 0.04e^{-0.6t} \right) \\[2ex] \dfrac{50}{3}\left(1 - 1.2e^{-0.1t} + 0.2e^{-0.6t} \right) \end{bmatrix} \tag{20.47}$$

From here, multiplying by 0.015, and introducing the actual process variables in place of the deviation variables, we obtain the following explicit expressions for the responses of the liquid level in the two tanks to a step change of the indicated magnitude in the inlet flowrate to Tank 1:

$$h_1 = 0.7 + 0.25\left(1 - 0.96e^{-0.1t} - 0.04e^{-0.6t} \right) \tag{20.48}$$

and

$$h_2 = 0.5 + 0.25\left(1 - 1.2e^{-0.1t} + 0.2e^{-0.6t} \right) \tag{20.49}$$

Comments

1. These step-response expressions indicate exponential increases in both h_1 and h_2.
2. There is a net increase of 0.25 m in the liquid levels in each tank after steady state has been achieved.
3. Despite the fact that only the inlet flow to Tank 1 was changed, we see that this change affects the level in *both* tanks (that the effect is identical in both tanks is purely coincidental; see next example).
4. Had there been no recycle loop, observe from Eq. (20.11) or Eq. (20.13) that Tank 1 would have exhibited *first-order dynamics* in response to changes in F_1; the fact that the response indicated in Eq. (20.48) is not of the first-order type is another indication of the influence of process interactions; changes in F_1 affect the level in Tank 2, but, through the recycle loop, the resultant change in Tank 2 level returns to affect the level in Tank 1. Thus the two tanks are seen to be *mutually* interacting.

   To conclude, we note that if there were no interactions between the process variables, in response to changes in F_1, h_2 would have remained at its initial steady-state value, and h_1 would have increased in a first-order fashion.

Having investigated the effect of changes in F_1, particularly on h_2, we now investigate, in the next example, the effect of changes in F_2. It should prove interesting to see if such a change will show a greater, or lesser effect on h_1, than that shown by F_1 on h_2.

Example 20.9   **ANALYSIS OF THE DYNAMIC BEHAVIOR OF THE TWO-TANK PROCESS: 3. *RESPONSE TO STEP CHANGE IN TANK 2 INFLOW.***

In the same manner as in Example 20.8 obtain expressions for how the liquid level in each of the two tanks in Figure 20.2 will respond to a step increase of the same magnitude (0.015 m^3/s) in F_2, the Tank 2 inlet flowrate (i.e., from its initial value of 0.05 m^3/s to a new value of 0.065 m^3/s) while maintaining the Tank 1 inflow at its initial steady-state value.

Solution:

This time, since the input change is in u_2, the unit step response is now obtained using:

$$x(t) = A^{-1} \left(M e^{At} M^{-1} - I \right) b_2$$

where, from Eq. (20.38),

$$b_2 = \begin{bmatrix} 0 \\ 5 \end{bmatrix}$$

Everything else is exactly as in Example 20.8, and therefore following exactly the same procedure, we obtain the following results:

$$h_1 = 0.7 + 0.10 \left(1 - 1.2e^{-0.1t} + 0.2e^{-0.6t} \right) \tag{20.50}$$

and

$$h_2 = 0.5 + 0.25 \left(1 - 0.6e^{-0.1t} - 0.4e^{-0.6t} \right) \tag{20.51}$$

We may now note that the net increase in h_1 caused by the step change in F_2 is 0.1 m; for the same step change in F_1, the net increase we observed in h_2 was 0.25 m.

Thus, *the net interaction effect of F_2 on h_1 is **less** than that of F_1 on h_2.*

It is easy to see why this will be the case from purely physical grounds: from Figure 20.2 we see that the conditions in Tank 1 affect Tank 2 *directly* via the outflow from Tank 1; all the material leaving Tank 1 flows into Tank 2. On the other hand, conditions in Tank 1 are affected by Tank 2 conditions via the recycle stream only; and this is but a fraction (in this particular example, only 40%) of the outflow from Tank 2;

20.4   MULTIVARIABLE TRANSFER FUNCTIONS AND OPEN-LOOP DYNAMIC ANALYSIS

In much the same way as in Section 4.8 where we first formally introduced the transfer function, we note that the multivariable transfer function is a *matrix* that relates the Laplace transforms of the input and output vectors, $\mathbf{u}(t)$, and $\mathbf{y}(t)$.

This transfer function matrix (sometimes abbreviated to *transfer matrix*) enjoys the same linearity property noted in Section 4.8 for its scalar counterpart, and also has poles and zeros. As we would expect, these poles and zeros by and large determine the nature of the multivariable system response; what is far from obvious is how these poles and zeros are defined. These and other important characteristics of the multivariable transfer function will now be discussed.

20.4.1 Poles and Zeros

The SISO transfer function, represented as in:

$$y(s) = g(s)\,u(s)$$

and its multivariable counterpart:

$$\mathbf{y}(s) = \mathbf{G}(s)\,\mathbf{u}(s)$$

look very similar until we try to define the poles, and, in particular, the zeros of the latter.

We recall from Section 4.8 that for the SISO system whose transfer function is given by:

$$g(s) = \frac{N(s)}{D(s)} \tag{20.52}$$

the poles are the roots of the denominator polynomial, $D(s)$, while the zeros are the roots of the numerator polynomial, $N(s)$.

As noted in Section 20.2 of this chapter (in Eq. (20.5)), the transfer function matrix consists of an array of several transfer function elements of the type in Eq. (20.52); this raises the following question: *since each transfer function element making up the multivariable transfer function matrix has its own set of poles and zeros, how do we define the poles and zeros of the entire system?* This question is easier to answer for poles than it is for zeros.

Multivariable System Poles

It is customary to define the poles of a multivariable system as the collection of *all* the poles of the individual transfer function elements of the transfer function matrix.

The rationale behind such a definition is that the transfer function elements can always be expressed as different numerator polynomials all divided by the *same* denominator polynomial, and this common denominator polynomial, will consist of *all* the poles of the individual transfer function elements. For example, the transfer function matrix:

$$\mathbf{G}(s) = \begin{bmatrix} \dfrac{1}{s+1} & \dfrac{1}{s+3} \\[2mm] \dfrac{1}{s+4} & \dfrac{1}{s+2} \end{bmatrix} \tag{20.53}$$

can be rearranged to give:

$$\mathbf{G}(s) = \frac{\begin{bmatrix} (s+2)(s+3)(s+4) & (s+1)(s+2)(s+4) \\[1mm] (s+1)(s+2)(s+3) & (s+1)(s+3)(s+4) \end{bmatrix}}{(s+1)(s+2)(s+3)(s+4)} \tag{20.54}$$

Observe that the common denominator polynomial is made up of the totality of the poles of all the individual transfer function elements.

Returning now to the general definition of a pole, observe therefore that the *entire* transfer function matrix becomes indeterminate at each one of the individual poles.

In the specific example transfer function matrix of Eq. (20.53), note that the individual elements have single poles at $s = -1$, $s = -2$, $s = -3$, and $s = -4$; this entire collection therefore constitutes the poles of the transfer function matrix. From the equivalent representation in Eq. (20.54), observe that $\mathbf{G}(s)$ becomes indeterminate at any one of these values.

It should be mentioned that if the transfer function matrix was obtained from the state-space representation by Laplace transformation, it will be as indicated in Eq. (20.9), and from the definition of the inverse of a matrix — as the adjoint divided by the determinant — we have, from Eq. (20.9) that:

$$\mathbf{G}(s) = \frac{[\mathbf{C} \, \text{Adj}(s\mathbf{I} - \mathbf{A}) \, \mathbf{B}]}{|(s\mathbf{I} - \mathbf{A})|} \tag{20.55}$$

where we now note that each element of the transfer function will have, as a common denominator, the polynomial resulting from evaluating the determinant $|(s\mathbf{I} - \mathbf{A})|$, with the immediate implication that the poles of the entire multivariable system are the roots of:

$$|(s\mathbf{I} - \mathbf{A})| = 0 \tag{20.56}$$

If we compare this with Eq. (20.37), the equation used to evaluate the eigenvalues of the matrix $\mathbf{A}$, we immediately see that they are the same equation and will therefore have the same roots. This leads us to the following result:

> The poles of a transfer function matrix, and the eigenvalues of the equivalent system matrix **A** in the state-space form, are one and the same.

The obvious implications of these results on how to investigate the open-loop stability of a multivariable system from its transfer function matrix will be deferred until later.

Multivariable System Zeros

It is very tempting to want to borrow from the definition of the multivariable system poles, and thus define the zeros as the totality of the zeros of the individual transfer function elements; but this is incorrect.

Observe that if a single element of a matrix becomes indeterminate at any particular point, the entire matrix becomes indeterminate; thus even from this viewpoint, it is reasonable to define the multivariable system poles in terms of the poles of the individual transfer function elements, as we have done. However, the entire transfer function matrix, **G**(s), *does not* become zero at a zero of the individual transfer function elements; that particular element becomes zero, but one zero element does not make the entire matrix into a zero matrix.

In actual fact, it makes no sense at all to search for values of s that make the entire matrix **G**(s) become zero, and to define these as the system zeros; for then only in the extremely rare case in which all the individual transfer function elements in the matrix have a common numerator factor will the multivariable system have a zero.

The appropriate definition of multivariable system zeros can be obtained with proper interpretation of the zeros of a SISO system: *as the poles of the inverse system*; that is, from Eq. (20.52), the zeros of this SISO system are interpreted as the *poles* of $1/g(s)$, since:

$$\frac{1}{g(s)} = \frac{D(s)}{N(s)} \tag{20.57}$$

Since we have already agreed on a reasonable definition of the *poles* of a multivariable system, we should be able to proceed from here.

Thus, we tentatively define as the zeros of a multivariable system, the *poles* of the inverse of the transfer function. Let us now investigate this a little more closely.

By the definition of a matrix inverse, we know that the inverse of the transfer function matrix **G**(s), *if the matrix is square* and nonsingular, is given by:

$$\mathbf{G}^{-1}(s) = \frac{\text{Adj } \mathbf{G}(s)}{|\mathbf{G}(s)|} \tag{20.58}$$

from where we see clearly that this inverse matrix has *poles* at the roots of the common denominator polynomial resulting from evaluating the indicated determinant. Thus the roots of the equation:

$$|G(s)| = 0 \qquad\qquad (20.59)$$

are therefore the *zeros* of the (square) transfer function matrix $G(s)$. We now state the following definition of the zero of a multivariable system:

The zeros of a (square) multivariable system with transfer function matrix $G(s)$ are the zeros of the determinant of this transfer function matrix, i.e., $|G(s)|$.

When the transfer function matrix is not square (indicating that the number of inputs and outputs are different) it is no longer possible to define the zeros in terms of the determinant. We will be concerned mostly with square systems, but for completeness, we note that a generalization of the idea of the determinant of a square matrix leads us to the following more general definition of the zeros of a (possibly nonsquare) multivariable system:

Those values of s for which the rank of the transfer function matrix $G(s)$ is reduced are called the zeros of $G(s)$.

The next example illustrates how to evaluate the zeros of a square, multivariable system, by simply looking for the zeros of the determinant of the transfer function matrix.

Example 20.10   THE ZEROS OF A MULTIVARIABLE SYSTEM.

Find the zero(s) of the multivariable system whose transfer function is given in Eq. (20.53).

Solution:

The determinant of this transfer function matrix is easily obtained as:

$$|G(s)| = \frac{1}{(s+1)(s+2)} - \frac{1}{(s+3)(s+4)} \qquad\qquad (20.60)$$

which simplifies to:

$$|G(s)| = \frac{4s+10}{(s+1)(s+2)(s+3)(s+4)} \qquad\qquad (20.61)$$

with the implication that the multivariable system has a single zero at $s = -10/4$.

Apart from the mechanics of evaluating multivariable zeros, there are three major points that this example illustrates:

1. The zeros of a multivariable system are very subtle characteristics of the system that are not as obvious to discern as the poles. For example, there is no way of knowing, by inspection alone, what the system zero is, or for that matter whether or not the example system even has a zero. (We will give some examples shortly of systems whose transfer functions are very similar to that in Eq. (20.53) but for which the zeros are radically different.)

2. We could have made the following point in a general fashion, but it is
    more instructive to observe it from the example:

    *With the exception of time-delay systems, the determinant of a
    multivariable system's transfer function matrix is a ratio of two
    polynomials; the roots of the numerator polynomial are the
    multivariable system **zeros**, and the roots of the denominator
    polynomial are the multivariable system **poles**.*

    Observe from Eq. (20.61) that the poles we had earlier obtained for the
    system in Eq. (20.53) have shown up in the denominator of the
    expression for the determinant of the transfer function matrix.

3. It is possible for a multivariable system to have poles and zeros in the
    same location. In evaluating the zeros of a multivariable system from
    the determinant of its transfer function matrix, it is therefore of utmost
    importance to ensure that the denominator polynomial contains *all* the
    system poles, i.e., ensure that there has been no pole-zero cancellation
    when forming the determinant (see Morari and Zafiriou [3]).

In light of the foregoing, we see that in terms of the intrinsic nature of the
multivariable system, as characterized by the system poles and zeros, the role
played by the single transfer function in SISO systems analysis is now played
by the determinant of the transfer function matrix in multivariable systems
analysis. We will have more to say about this later.

For now, we leave the following exercises to the reader: to show that the
system:

$$\mathbf{G}(s) = \begin{bmatrix} \dfrac{1}{s+1} & \dfrac{1}{s+2} \\[2mm] \dfrac{1}{s+3} & \dfrac{1}{s+4} \end{bmatrix} \tag{20.62}$$

which is just a reshuffling of the same elements in Eq. (20.53), has *no zero*.

Also, the reader should show that the system:

$$\mathbf{G}(s) = \begin{bmatrix} \dfrac{1}{s+1} & \dfrac{1}{s+3} \\[2mm] \dfrac{1}{s+4} & \dfrac{1}{s+7} \end{bmatrix} \tag{20.63}$$

which differs from the one in Eq. (20.53) by the single fact that the $(s+2)$ term
has been replaced by $(s+7)$, has a RHP zero at $s = +5$.

In terms of the system matrices of the equivalent state-space model, it can
be shown from Eq. (20.55) that the system zeros are the roots of the equation:

$$|\mathbf{C} \; \text{Adj}(s\mathbf{I} - \mathbf{A}) \; \mathbf{B}| = 0 \tag{20.64}$$

Thus, system zeros can be computed readily from the state-space model
formulation as well.

20.4.2 A Quantitative Measure of Singularity — Singular Values

One of the most important facts to emerge from our recent discussion is the significance of the *determinant* of a transfer function matrix and the conditions governing when this determinant becomes zero, for as we have just seen, these conditions identify the zeros of the multivariable system, and these zeros play an important role in characterizing the system behavior.

Recall from matrix algebra that when the determinant of a matrix is zero, the matrix is said to be singular. As we shall see later, how close a transfer function matrix is to being singular also provides other related information about how easy, or how difficult, it is to control such a system.

To be sure, simply evaluating the determinant of a matrix and inspecting how close it is to zero is *one* way of finding out how near to being singular the matrix is; however, this approach is not always very reliable.

The classic example for illustrating how poor the determinant is as an indicator of incipient singularity is the following matrix:

$$\mathbf{H}(\varepsilon) = \begin{bmatrix} 1 & \varepsilon \\ \dfrac{1}{\varepsilon} & 1 \end{bmatrix} \qquad (20.65)$$

which is clearly singular, regardless of the value of ε. However, consider the situation in which ε takes on very small values; if, as a result, we neglect the small ε value at the top right hand corner in Eq. (20.65), we obtain:

$$\mathbf{H}_1 = \begin{bmatrix} 1 & 0 \\ \dfrac{1}{\varepsilon} & 1 \end{bmatrix} \qquad (20.66)$$

However, whereas $\mathbf{H}$ is singular, the determinant $\left| \mathbf{H}_1 \right|$ is always equal to 1, regardless of the value of ε. Thus, even though we know, by construction, that as ε gets smaller, the matrix $\mathbf{H}_1$ approaches singularity, its determinant gives no such indication.

It is known that the determinant of a matrix is equal to the product of its eigenvalues; we might therefore be led to think that the eigenvalues might contain more information than the determinant. While this is true in general, when it comes to indicating the onset of singularity, the eigenvalues are not much better than the determinant. For example, the two eigenvalues of the matrix $\mathbf{H}_1$ are both equal to 1, and provide no better indication than the determinant regarding how nearly singular the matrix is. Thus we need something better as a quantitative measure of near singularity.

Singular Values

It turns out that a good measure of the near singularity of a matrix can be found from quantities known as the *singular values* of the matrix. For any general, possibly complex, $n \times m$ matrix $\mathbf{H}$, the singular values are defined (cf. Appendix D) as:

$$\sigma_i = (\lambda_i (\mathbf{H}^* \mathbf{H}))^{1/2} \qquad i = 1, 2, \ldots, \text{n} \qquad (20.67)$$

where $\mathbf{H}^*$ is the transpose of the complex conjugate of $\mathbf{H}$ (simply the transpose if $\mathbf{H}$ is real) and $\lambda_i (\mathbf{H}^* \mathbf{H})$ represents the ith eigenvalue of the matrix $(\mathbf{H}^* \mathbf{H})$.

For our example matrix, $\mathbf{H}_1$:

$$\mathbf{H}_1^T \mathbf{H}_1 = \begin{bmatrix} 1 + \left(\dfrac{1}{\varepsilon}\right)^2 & \dfrac{1}{\varepsilon} \\ \dfrac{1}{\varepsilon} & 1 \end{bmatrix} \tag{20.68}$$

and the eigenvalues may be obtained as the two roots of:

$$\lambda^2 - \left(2 + \frac{1}{\varepsilon^2}\right)\lambda + 1 = 0 \tag{20.69}$$

i.e.:

$$\lambda_1 = 1 + \frac{1}{2\varepsilon^2} + \frac{\sqrt{1 + 4\varepsilon^2}}{2\varepsilon^2} \tag{20.70a}$$

$$\lambda_2 = 1 + \frac{1}{2\varepsilon^2} - \frac{\sqrt{1 + 4\varepsilon^2}}{2\varepsilon^2} \tag{20.70b}$$

The singular values are therefore given by:

$$\sigma_1 = \sqrt{\lambda_1} = \sqrt{1 + \frac{1}{2\varepsilon^2} + \frac{\sqrt{1 + 4\varepsilon^2}}{2\varepsilon^2}} \tag{20.71a}$$

$$\sigma_2 = \sqrt{\lambda_2} = \sqrt{1 + \frac{1}{2\varepsilon^2} - \frac{\sqrt{1 + 4\varepsilon^2}}{2\varepsilon^2}} \tag{20.71b}$$

We may now observe that as $\varepsilon \to 0$, the larger singular value, σ_1, goes to infinity, and the smaller, σ_2, goes to unity; this widening "separation" between the two singular values provides a reliable indication of the onset of singularity, or "ill-conditioning." In fact, the ratio of the largest and smallest singular value:

$$\kappa = \frac{\sigma_{max}}{\sigma_{min}} \tag{20.72}$$

is known as the "condition number" of a matrix. Thus for our example, $\kappa \to \infty$ smoothly as $\varepsilon \to 0$ and provides a good measure of the progress towards singularity.

In general, given the transfer function matrix $\mathbf{G}(s)$, the singular values are obtained, as a function of frequency, by substituting $s = j\omega$, and evaluating the square root of the eigenvalues of the matrix $\mathbf{S}(j\omega)$ given by:

$$\mathbf{S}(j\omega) = \mathbf{G}^T(-j\omega)\, \mathbf{G}(j\omega) \tag{20.73}$$

since the complex conjugate of $\mathbf{G}(j\omega)$ is $\mathbf{G}(-j\omega)$. Plots of these singular values and the condition number κ as a function of frequency provide useful design information for multivariable systems (see Ref. [3]). Singular value analysis has proved to be a useful tool in the analysis and design of multivariable control systems, but a detailed discussion lies outside the intended scope of this book.

With the foregoing characteristics of the multivariable transfer function in mind, we now proceed to the more mechanical aspects of how to use transfer function matrices for obtaining transient responses. The principles involved are mostly straightforward extensions of what we already know for SISO systems.

20.4.3 Dynamic Analysis

Consider the general $n \times m$ multivariable system whose transfer function model is given by:

$$y(s) = G(s) u(s) \tag{20.74}$$

where $G(s)$ is an $n \times m$ transfer function matrix whose elements are as given in Eq. (20.5). Writing Eq. (20.74) more explicitly gives:

$$y_1(s) = g_{11}(s) u_1(s) + g_{12}(s) u_2(s) + \ldots + g_{1m}(s) u_m(s)$$

$$y_2(s) = g_{21}(s) u_1(s) + g_{22}(s) u_2(s) + \ldots + g_{2m}(s) u_m(s)$$

$$\ldots \quad \ldots \ldots \quad \ldots \quad \ldots \quad \ldots$$

$$y_n(s) = g_{n1}(s) u_1(s) + g_{n2}(s) u_2(s) + \ldots + g_{nm}(s) u_m(s)$$

Let us now define as x_{ij} the product of the ijth transfer function element and the jth input, i.e.:

$$x_{ij}(s) = g_{ij}(s) u_j(s) \tag{20.75}$$

Observe now that this is exactly like the representation for a SISO system; we can therefore obtain the response of x_{ij} to changes in u_j in precisely the same manner as with the SISO system.

If, therefore, each transfer function element $g_{ij}(s)$ has p_{ij} poles, then upon carrying out the usual partial fraction expansion for this transfer function element, and inverting, we will obtain:

$$x_{ij}(t) = \sum_{q=1}^{p_{ij}} \left(A_{ijq} e^{r_{ijq} t} \right) \tag{20.76}$$

where the coefficients A_{ijq} are obtained from the partial fraction expansion exercise, and $r_{ijq}, q = 1, 2, \ldots, p_{ij}$ are the roots of the denominator polynomial of $g_{ij}(s)$, i.e., the transfer function poles.

The actual multivariable system outputs are obtained by taking the appropriate sums of these x_{ij} responses, i.e.:

$$y_i = \sum_{j=1}^{m} x_{ij}(t) \tag{20.77}$$

or, from Eq. (20.76):

$$y_i = \sum_{j=1}^{m} \sum_{q=1}^{p_{ij}} \left(A_{ijq} \, e^{r_{ijq}t} \right) \tag{20.78}$$

Stability

Observe that if any of the r_{ijq} terms in Eq. (20.78) has a positive real part, then the output associated with this term will grow indefinitely, and even though all the other outputs might not suffer the same fate, the entire multivariable system is said to be unstable. Of course, we know that the r_{ijq} terms are the transfer function poles, and therefore, just as with the SISO system, the stability of the multivariable system is determined by the nature of the transfer function poles. We now have the following result:

> *A multivariable system is stable if **all** the poles of the transfer function matrix lie in the left-half plane (LHP); otherwise it is unstable.*

The multivariable system poles have earlier been identified as the totality of the poles of each transfer function element. Thus for a multivariable system to be stable, *none* of the individual transfer function elements should have a right-half plane (RHP) pole.

In relation to the equivalent state-space form, the multivariable system poles have been identified in Eq. (20.56) as the roots of the polynomial resulting from evaluating the determinant indicated in this equation, i.e.:

$$|(s\mathbf{I} - \mathbf{A})| = 0$$

Thus from the transfer function matrix point of view, stability requires that the roots of this equation all lie in the LHP.

Upon comparing this equation with the equation given in Eq. (20.37) for the determination of the eigenvalues of the matrix $\mathbf{A}$, we immediately see that the condition that *all* the eigenvalues of $\mathbf{A}$ lie in the LHP is exactly identical to the condition requiring the poles of the transfer function matrix $\mathbf{G}(s)$ to lie in the LHP.

Thus, it matters not whether we determine the stability of a multivariable system in terms of its transfer function matrix or its state-space model; the results are identical.

Transient Responses

The actual mechanics of obtaining multivariable transient responses using transfer functions can be quite tedious (except in some unusually simple cases) even though the principles are trivial extensions of what we already know from SISO systems analysis; this makes computer assistance inevitable.

For most computer packages with multivariable systems analysis and controller design capabilities, the user provides the transfer function matrix and the desired input forcing function, and the transient responses will be calculated and the time behavior plotted.

The following example illustrates the analytical aspects of obtaining multivariable transient responses using the methanol/water distillation column introduced in Example 20.6.

Example 20.11 **STEP RESPONSES OF THE BINARY DISTILLATION COLUMN.**

Obtain the response of the binary distillation column of Example 20.6 to a unit step change in u_1, the overhead reflux flowrate; also obtain the response to a unit step change in u_2, the bottoms steam flowrate. Comment on the interactive influence of u_1 on y_2 in comparison to its "main" effect on y_1. For the second unit step response, also comment of the interactive influence of u_2 on y_1 in comparison to its "main" effect on y_2.

Solution:

Response to unit step change in u_1.

From the transfer functions in Eq. (20.26), given that $u_1 = 1$ and $u_2 = 0$, then the transient responses are obtained from:

$$y_1(s) = \frac{12.8e^{-s}}{16.7s + 1}\frac{1}{s} \tag{20.79}$$

and

$$y_2(s) = \frac{6.6e^{-7s}}{10.9s + 1}\frac{1}{s} \tag{20.80}$$

and recalling Example 8.1, where the unit step response of a time-delay system was first illustrated, we obtain:

$$y_1(t) = \begin{cases} 0; & t < 1 \\ 12.8\left(1 - e^{-(t-1)/16.7}\right); & t \geq 1 \end{cases} \tag{20.81}$$

and

$$y_2(t) = \begin{cases} 0; & t < 7 \\ 6.6\left(1 - e^{-(t-7)/10.9}\right); & t \geq 7 \end{cases} \tag{20.82}$$

Observe that these equations are similar to the step response of any other first-order-plus-time-delay system.

Note that the "interaction" effect of u_1 on y_2 is significant with a net change of 6.6 in y_2 at steady state.

Response to unit step change in u_2.

In this case, setting $u_2 = 1$ and $u_1 = 0$, we find that the transient responses are obtained from:

$$y_1(s) = \frac{-18.9e^{-3s}}{21.0s + 1}\frac{1}{s} \tag{20.83}$$

and

$$y_2(s) = \frac{-19.4e^{-3s}}{14.4s + 1}\frac{1}{s} \tag{20.84}$$

which result in the unit step responses:

$$y_1(t) = \begin{cases} 0; & t < 3 \\ -18.9\left(1 - e^{-(t-3)/21.0}\right); & t \geq 3 \end{cases} \tag{20.85}$$

and

$$y_2(t) = \begin{cases} 0; & t < 3 \\ -19.4\left(1 - e^{-(t-3)/14.4}\right); & t \geq 3 \end{cases} \qquad (20.86)$$

again similar to the step response of any other first-order-plus-time-delay system.

Again there is a significant "interaction effect" of u_2 on y_1 with a steady-state change in y_1 of -18.9.

20.5   CLOSED-LOOP DYNAMIC ANALYSIS

As with SISO systems, so it is with feedback control of multivariable systems: all the output variables are measured, and the information is sent to the appropriate controllers assigned the task of regulating the behavior of each output. These controllers in turn pass on their control decisions to the process through the appropriate input variables. Thus such arrangement gives rise to *multiple* feedback control loops, with each individual loop resembling the SISO control loop we investigated in Chapter 14.

The analysis of the behavior of multivariable systems in the feedback control loop follow similar lines as the SISO counterpart, with just a few additional considerations, as we now show.

20.5.1 Multivariable Block Diagrams

In the same vein as with SISO systems, multivariable systems may be represented with block diagrams. Figure 20.3(a) shows an example block diagram for an m-input, n-output (k-disturbance) system where the multiple lines are used to indicate the multiplicity of the variables involved.

Such a block diagram is often "simplified" to the type shown in Figure 20.3(b), where it is enough that the blocks contain transfer function matrices, and that the signals are designated as vectors, to set it apart from the SISO counterpart to which it bears potentially confusing similarities. Such a block diagram is quite useful for overall closed-loop analysis.

Alternatively, block diagrams can be "expanded" to emphasize the individual effects of *each* element of $\mathbf{G}(s)$ on each output. For example, a 2×2 system may be represented as in Figure 20.3(c) which shows, quite explicitly, the interactive effects of u_2 on y_1, and of u_1 on y_2. This block diagram type is useful for interaction analysis and the design of interaction compensators, as we shall see later.

20.5.2 Closed-Loop Behavior

Closed-Loop Transfer Functions

From the more compact block diagram (Figure 20.3(b)), and from Eq. (20.4) we have:

$$\mathbf{y}(s) = \mathbf{G}(s)\,\mathbf{u}(s) + \mathbf{G_d}(s)\,\mathbf{d}(s) \qquad (20.87)$$

but

$$\mathbf{u}(s) = \mathbf{G}_c(s)\,\varepsilon(s)$$

and

$$\varepsilon(s) = \mathbf{y}_d - \mathbf{y}(s)$$

so that Eq. (20.87) becomes:

$$\mathbf{y}(s) = \mathbf{G}\mathbf{G}_c\left[\mathbf{y}_d - \mathbf{y}(s)\right] + \mathbf{G}_d\,\mathbf{d} \qquad (20.88)$$

We may now rearrange this to give:

$$\left(\mathbf{I} + \mathbf{G}\mathbf{G}_c\right)\mathbf{y}(s) = \mathbf{G}\mathbf{G}_c\,\mathbf{y}_d + \mathbf{G}_d\,\mathbf{d}$$

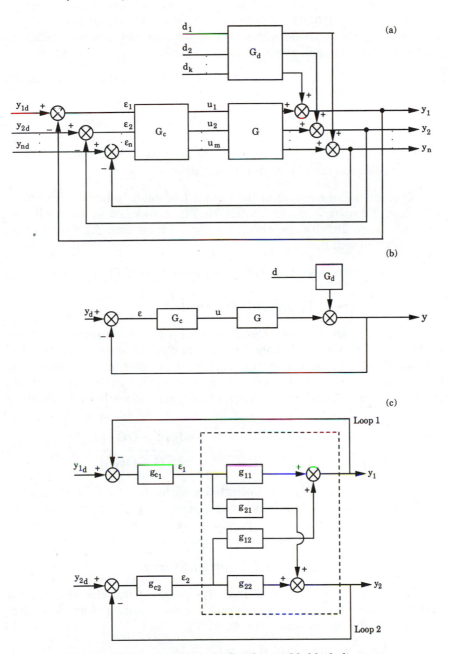

Figure 20.3. Three types of multivariable block diagrams.

and by premultiplying with the inverse of the matrix $(I + GG_c)$, we finally obtain the closed-loop response of the outputs **y** to set-point changes $\mathbf{y}_d$ and disturbances **d**:

$$\mathbf{y}(s) = \left(\mathbf{I} + \mathbf{GG}_c\right)^{-1}\mathbf{GG}_c\mathbf{y}_d + \left(\mathbf{I} + \mathbf{GG}_c\right)^{-1}\mathbf{G_d}\,\mathbf{d} \qquad (20.89)$$

The following points should be noted:

1. If this expression is compared with the equivalent closed-loop transfer function expression for the SISO system obtained in Chapter 14, we see that the only difference is that, since we are now dealing with matrices, Eq. (20.89) involves multiplication by the inverse of $(\mathbf{I} + \mathbf{GG}_c)$, whereas the equivalent closed-loop transfer function for a SISO system contains a division by the scalar $(1 + gg_c)$.

2. Both closed-loop transfer function matrices indicated in Eq. (20.89) contain the inverse of $(\mathbf{I} + \mathbf{GG}_c)$.

3. This matrix $(\mathbf{I} + \mathbf{GG}_c)$ is often referred to as the *return difference matrix*, and in taking the analogy with the SISO system farther, we may anticipate that it will play a key role in the determination of the closed-loop stability of multivariable systems.

4. It is left as an exercise to the reader to establish that if a measurement device transfer function matrix $\mathbf{H}(s)$ is introduced in the feedback path in Figure 20.3(b), the resulting closed-loop transfer function matrix equation will then be:

$$\mathbf{y}(s) = \left(\mathbf{I} + \mathbf{GG}_c\mathbf{H}\right)^{-1}\mathbf{GG}_c\mathbf{y}_d + \left(\mathbf{I} + \mathbf{GG}_c\mathbf{H}\right)^{-1}\mathbf{G_d}\,\mathbf{d} \qquad (20.90)$$

20.5.3 Closed-Loop Transient Responses

In principle, we can now use the closed-loop transfer function expressions Eqs. (20.89) or (20.90) to calculate the response of the multivariable system in the closed loop to changes in the vector of set-points $\mathbf{y}_d$ or the vector of disturbances **d** in precisely the same fashion indicated in Section 14.4 for SISO systems. From the definition of a matrix inverse, observe that Eq. (20.89) can be rewritten as:

$$\mathbf{y}(s) = \frac{\text{Adj}\left(\mathbf{I} + \mathbf{GG}_c\right)\mathbf{GG}_c}{\left|(\mathbf{I} + \mathbf{GG}_c)\right|}\,\mathbf{y}_d + \frac{\text{Adj}\left(\mathbf{I} + \mathbf{GG}_c\right)\mathbf{G_d}}{\left|(\mathbf{I} + \mathbf{GG}_c)\right|}\,\mathbf{d} \qquad (20.91)$$

and each element of each of the indicated transfer function matrices will have *identical* denominator polynomials corresponding to the determinant of the return difference matrix, and from previous experience, we know that these will determine the nature of the closed-loop responses.

Closed-Loop Stability

It becomes very clear from here that the poles of the closed-loop transfer function matrices are the roots of the polynomial resulting from evaluating the determinant of the return difference matrix:

$$\left|(\mathbf{I} + \mathbf{GG}_c)\right| = 0 \qquad (20.92a)$$

or from

$$|(\mathbf{I} + \mathbf{G}\mathbf{G}_c\mathbf{H})| = 0 \qquad (20.92b)$$

when the measuring devices are separately represented in the feedback loop poles, and these determine whether the multivariable system is stable or not.

Thus, Eq. (20.92) becomes the characteristic equation for the multivariable system for which stability is to be investigated. Observe that the conditions for stability are that *all* the roots of this characteristic equation must lie in the LHP.

A comparison of Eq. (20.92) with Eq. (14.83), the characteristic equation for the SISO system, establishes the analogy between SISO systems analysis and its multivariable counterpart; the denominator of the scalar closed-loop transfer function has been replaced by the *determinant* of the return difference matrix.

A comparison of the conditions required for stability in a multivariable system and the stability conditions for individual single-loop systems is illustrated by the following example.

Example 20.12   STABILITY OF A 2 × 2 SYSTEM.

The general 2 × 2 system whose transfer function is given as:

$$\mathbf{G}(s) = \begin{bmatrix} g_{11}(s) & g_{12}(s) \\ g_{21}(s) & g_{22}(s) \end{bmatrix} \qquad (20.93)$$

is to be controlled by two independent feedback controllers, so that the controller transfer function matrix is:

$$\mathbf{G}_c(s) = \begin{bmatrix} g_{c1}(s) & 0 \\ 0 & g_{c2}(s) \end{bmatrix} \qquad (20.94)$$

Obtain the characteristic equations to be used in investigating the closed-loop stability of this system under the following three conditions:

1.   Only Loop 1 is closed; Loop 2 is open.
2.   Only Loop 2 is closed; Loop 1 is open.
3.   The truly multivariable situation in which both loops are closed.

Solution:

Under Conditions 1 and 2, we essentially have the system operating as a SISO system since only *one loop* is closed in each case.

The characteristic equations are:

$$1 + g_{11}g_{c1} = 0 \qquad (20.95)$$

for Condition 1, and:

$$1 + g_{22}g_{c2} = 0 \qquad (20.96)$$

for Condition 2.

For the complete multivariable system, with both loops closed (Condition 3), the characteristic equation is obtained from:

$$\begin{vmatrix} 1 + g_{11} \, g_{c1} & g_{12} \, g_{c2} \\ g_{21} \, g_{c1} & 1 + g_{22} \, g_{c2} \end{vmatrix} = 0 \qquad (20.97)$$

and upon evaluating the determinant, we obtain:

$$\left(1 + g_{11} \, g_{c1}\right)\left(1 + g_{22} \, g_{c2}\right) - g_{12} \, g_{21} \, g_{c1} \, g_{c2} = 0 \qquad (20.98)$$

The main point is to observe that the conditions which will guarantee stability for each individual loop when operating in isolation are completely different from the conditions that will guarantee stability when both loops are closed. However, one should note that:

1. If *either* g_{12} or g_{21} is zero, then the conditions indicated in Eq. (20.95) along with Eq. (20.96) are identical to those indicated in Eq. (20.98) and the individual loops can be considered independently in analyzing stability.

2. If either g_{12} or g_{21} (but not both) are zero, there are still interactions between the loops, but these do not influence the stability condition. In this case (called *one-way interaction*), the interaction appears as a disturbance in the affected loop.

From this example we see that one of the implications of interactions in multivariable systems is that guaranteeing stability for individual loops operating by themselves provides no guarantee that when all the loops are closed, and are operating simultaneously, the overall system will be stable. Thus, contrary to what we might be led to think, it is not always possible to design a control system for a multivariable system by treating it like several, independent SISO systems.

20.6   SUMMARY

This chapter has been concerned with providing an introduction to multivariable systems, emphasizing the peculiar characteristics which sets them apart from SISO systems, but where necessary also pointing out the similarities they bear with their SISO counterparts. Our objective has been to provide a framework for the discussion of effective controller design techniques for multivariable systems in the following two chapters, rather than a detailed analysis of multivariable system structure.

We have pointed out the problems of input/output pairing, interactions, and controllability and observability, and have shown that they arise as a result of the presence of more than one input and/or output variable. These problems are therefore unique to multivariable systems; they do not arise for SISO systems.

We have introduced multivariable process model forms and how they are used for both open- and closed-loop dynamic analyses; in this connection, the important properties of a multivariable system — poles, zeros, and singular values — that influence dynamic behavior were also introduced.

With the material presented in this chapter as background, we are now in a position to consider the issue of controller design for multivariable systems. This will be the subject of the next two chapters.

REFERENCES AND SUGGESTED FURTHER READING

1. Ray, W. H., *Advanced Process Control*, McGraw-Hill, New York (1981), Butterworths, Boston (1989)

2. Wood, R. K. and M. W. Berry, "Terminal Composition Control of a Binary Distillation Column," *Chem. Eng. Sci.*, **29**, 1808 (1973)

3. Morari, M. and E. Zafiriou, *Robust Process Control*, Prentice-Hall, Englewood Cliffs, NJ (1989)

4. Luyben, W. L. and C. D. Vinante, "Experimental Studies in Distillation Decoupling," *Kemian Teollisauus*, **29**, 499 (1972)

REVIEW QUESTIONS

1. What is a multivariable system?

2. What are some of the major issues that differentiate a multivariable system from a single-input/single-output (SISO) system?

3. In very general terms, what do the concepts of "controllability" and "observability" mean?

4. Is open-loop dynamic analysis of multivariable systems *conceptually* different from open-loop dynamic analysis of SISO systems?

5. What determines Bounded-Input-Bounded-Output (BIBO) stability of a linear multivariable system whose model is given in the state-space form?

6. What are the poles of a transfer function matrix? Can they be determined *by mere inspection*?

7. Given a multivariable system's transfer function model and the equivalent state-space model, what is the relationship between the transfer function matrix poles and the system matrix **A**?

8. Is it possible, by mere inspection alone, to determine the zeros of a transfer function matrix?

9. How are the zeros of a square transfer function matrix determined?

10. What are the singular values of a matrix?

11. Which of the following properties of a multivariable system matrix provides the most reliable indication of incipient singularity: the eigenvalues, the determinant, or the singular values?

12. When is a multivariable system modeled by a linear transfer function matrix stable?

13. How is closed-loop stability determined for a linear multivariable feedback system?

14. When a multivariable control system consists of several individual single loops, is it possible to guarantee stability for the overall system (with all loops operating *simultaneously*) by guaranteeing stability for each individual loop independently?

PROBLEMS

20.1 A multivariable system has the following state-space model:

$$\dot{x} = Ax + Bu$$
$$y = Cx \qquad\qquad\qquad (P20.1)$$

where the matrices and vectors are given by:

$$x = \begin{bmatrix} x_1 \\ x_2 \end{bmatrix}; \qquad y = \begin{bmatrix} y_1 \\ y_2 \end{bmatrix}; \qquad u = \begin{bmatrix} u_1 \\ u_2 \end{bmatrix};$$

$$A = \begin{bmatrix} -3 & 2 \\ 1 & -4 \end{bmatrix}; \quad B = \begin{bmatrix} 2 & 0 \\ 0 & 1 \end{bmatrix}; \quad C = I, \text{ the identity matrix.}$$

(a) Find the eigenvalues of the matrix A, the corresponding matrix of eigenvectors M, and confirm that for this system:

$$A = M\Lambda M^{-1} \qquad\qquad (P20.2)$$

where Λ is the *diagonal* matrix of the eigenvalues of A.
(b) Obtain the corresponding transfer function matrix model for this system.

20.2 Given that the impulse response model form for a system whose state-space model is as given in Eq. (P20.1) is:

$$y(t) = \int_0^t G(t - \sigma)u(\sigma)\, d\sigma \qquad (P20.3)$$

with the impulse response *matrix* given by:

$$G(t) = Ce^{At}B \qquad\qquad (P20.4)$$

(a) Obtain the equivalent impulse response model for the system in Problem 20.1. [**Hint:** *You may wish to use the results obtained in part (a) of Problem 20.1; see also Eq. (20.35).*]
(b) Using the impulse-response model form or otherwise, obtain an analytical expression for the response of the system in Problem 20.1 to a *unit* step change in u_1 only (i.e., at $t > 0$ $u_1 = 1$ but $u_2 = 0$).

20.3 The complex theoretical model for a certain novel bioreactor has been linearized around a proposed steady-state operating point, resulting in the following approximate *linear* state-space model:

$$\dot{x} = Ax + Bu$$

with

$$A = \begin{bmatrix} 4 & -1 \\ 5 & -2 \end{bmatrix}; \quad B = \begin{bmatrix} 1 & 0 \\ 0 & 2 \end{bmatrix}; \quad x = \begin{bmatrix} x_1 \\ x_2 \end{bmatrix}; \quad u = \begin{bmatrix} u_1 \\ u_2 \end{bmatrix}$$

For simplicity we shall assume that the two state variables are measured directly so that:

$$y = x$$

(a) Show that the bioreactor is *open-loop unstable* at this operating condition.
(b) Obtain the corresponding transfer function matrix $\mathbf{G}(s)$ for the bioreactor at this operating point; find the system poles and zeros.
(c) Certain economic considerations dictate that the bioreactor is most profitably operated at this unstable steady state. Under proportional-only feedback control, using two single-loop controllers with gains K_{c_1} and K_{c_2}, with:

$$\mathbf{u} = -\mathbf{K}_c \mathbf{x} \tag{P20.5}$$

where

$$\mathbf{K}_c = \begin{bmatrix} K_{c_1} & 0 \\ 0 & K_{c_2} \end{bmatrix}$$

given that $K_{c_2} = 2$, find the range of values which K_{c_1} must take for the closed-loop system to be stable.

20.4 A 2×2 system is modeled approximately by:

$$\begin{bmatrix} y_1 \\ y_2 \end{bmatrix} = \begin{bmatrix} \dfrac{2}{s+3} & \dfrac{-1}{s+1} \\ \dfrac{-1}{s+7} & \dfrac{1/2}{s+4} \end{bmatrix} \begin{bmatrix} m_1 \\ m_2 \end{bmatrix} \tag{P20.6}$$

(a) Find the poles and zeros of the transfer function matrix.
(b) Obtain explicit expressions for the system outputs $y_1(t)$, $y_2(t)$ in response to inputs $m_1(t) = m_2(t) = 1$ at $t > 0$. Plot these responses.

20.5 Luyben and Vinante [4] studied a 24-tray distillation column used for separating methanol and water. Considering the temperatures on trays 17 and 4 as the most important output variables, they studied their responses to changes in the reflux flowrate, and steam flowrate, and obtained the following transfer function matrix model:

$$\begin{bmatrix} y_1 \\ y_2 \end{bmatrix} = \begin{bmatrix} \dfrac{-2.16e^{-s}}{8.5s+1} & \dfrac{1.26e^{-0.3s}}{7.05s+1} \\ \dfrac{-2.75e^{-1.8s}}{8.2s+1} & \dfrac{4.28e^{-0.35s}}{9.0s+1} \end{bmatrix} \begin{bmatrix} m_1 \\ m_2 \end{bmatrix} \tag{P20.7}$$

where, in terms of deviations from their respective steady states, the input and output variables are:

y_1 = Tray #17 temperature (°C)
y_2 = Tray #4 temperature (°C)
m_1 = Reflux flowrate (kg/hr)
m_2 = Steam flowrate (kg/hr)

(a) By defining four new "state" variables x_1, x_2, x_3, and x_4 as follows:

$$x_1(s) = \left(\frac{-2.16e^{-s}}{8.5s+1} \right) m_1(s) \tag{P20.8}$$

$$x_2(s) = \left(\frac{1.26e^{-0.3s}}{7.05s+1} \right) m_2(s) \tag{P20.9}$$

$$x_3(s) = \left(\frac{-2.75e^{-1.8s}}{8.2s + 1}\right)m_1(s) \tag{P20.10}$$

$$x_4(s) = \left(\frac{4.28e^{-0.35s}}{9.0s + 1}\right)m_2(s) \tag{P20.11}$$

write the *differential equations* to be solved *in time* to simulate the dynamic behavior of this process. Such a set of equations are called a *realization* of the original transfer function model.

(b) Using your favorite simulation package (or otherwise) obtain the response of y_1 (the tray #17 temperature) to unit step changes implemented *simultaneously* in *both* the reflux flowrate and the steam flowrate. By examining the given transfer function model carefully, explain the nature of the response you have obtained.

20.6 Obtain the equivalent impulse response matrix model for the Luyben and Vinante column of Problem 20.5.

[**Hint**: *The impulse response matrix* $\mathbf{G}(t)$ *has as its elements* $g_{ij}(t)$, *the inverse Laplace transform of* $g_{ij}(s)$, *the individual elements of the transfer function matrix. Do not forget the presence of the time delays.*]

20.7 For the *multivariable* system whose block diagram is shown below in Figure P20.1, find the closed-loop transfer function *matrices* relating $\mathbf{y}$ to both $\mathbf{y}_d$ and $\mathbf{d}$. What is the characteristic equation?

20.8 The transfer function of a 2×2 system is given as:

$$\mathbf{G}(s) = \begin{bmatrix} \dfrac{1}{5s + 1} & \dfrac{0.0001}{0.2s + 1} \\[2ex] \dfrac{100}{3s + 1} & \dfrac{1}{0.5s + 1} \end{bmatrix} \tag{P20.12}$$

(a) Set $s = 0$ and obtain the steady-state gain matrix $\mathbf{K} = \mathbf{G}(0)$.

(b) First find the eigenvalues of $\mathbf{K}$, then find its *singular* values. Obtain the ratio of the larger to the smaller singular values (the "condition number"; see Appendix D) and compare it to the corresponding ratio of the eigenvalues. From your knowledge of what the condition number of a matrix indicates, what does the value you have obtained imply about the "conditioning" of this process?

(c) By examining the steady-state gain matrix for this system carefully, comment on the relative effectiveness of the two process input variables in affecting the two output variables.

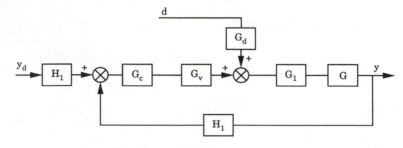

Figure P20.1.

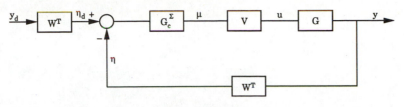

Figure P20.2.

20.9 The block diagram shown in Figure P20.2 (where all the indicated blocks contain transfer function *matrices*) is for a certain multivariable control system to be discussed in Chapter 22.
(a) Find the *closed-loop* transfer function between **y** and $\mathbf{y}_d$.
(b) What is the characteristic equation for this system?
(c) Given that the following relationship exists between **G**, the process transfer function matrix, and the other transfer function matrices **V** and **W**:

$$\mathbf{G} \ = \ \mathbf{W} \, \Sigma \, \mathbf{V}^{\mathrm{T}}$$

where Σ is a *diagonal* matrix, and the matrices **V** and **W** are *unitary* matrices (see Appendix D), show that the characteristic equation for this system reduces to:

$$\left| \mathbf{I} + \Sigma \mathbf{G}_c^{\Sigma} \right| \ = \ 0 \qquad\qquad (\text{P20.13})$$

where $\mathbf{G}_c^{\Sigma}$ is the indicated (diagonal) matrix of controller transfer functions and **I** is the identity matrix.

[**Hint:** *You may find the following determinant relations useful:*

1. $|\mathbf{ABC}| \ = \ | \, \mathbf{A}| \, |\mathbf{B}| \, |\mathbf{C}| \ = \ | \, \mathbf{B}| \, |\mathbf{C}| \, |\mathbf{A}| \ = \ | \, \mathbf{C}| \, |\mathbf{A}| \, |\mathbf{B}|$, etc., for nonsingular, square matrices **A**, **B**, and **C** of identical order.
2. $|\mathbf{A}| = 1/|\mathbf{A}^{-1}|$.]

20.10 A two-input/two-output system whose transfer function matrix model is as shown below:

$$\begin{bmatrix} y_1 \\ y_2 \end{bmatrix} = \begin{bmatrix} \dfrac{2}{s+1} & \dfrac{1}{3s+1} \\[2mm] \dfrac{4}{s+1} & \dfrac{1}{2s+1} \end{bmatrix} \begin{bmatrix} m_1 \\ m_2 \end{bmatrix} \qquad (\text{P20.14})$$

is under feedback control, using the y_1–m_1/y_2–m_2 pairing, so that the process transfer function matrix, $\mathbf{G}(s)$, is as indicated above; the controller, $\mathbf{G}_c$ has been chosen as two pure proportional controllers, i.e.:

$$\mathbf{G}_c \ = \ \begin{bmatrix} K_{c_1} & 0 \\ 0 & K_{c_2} \end{bmatrix}$$

and it is desired to use the following specific controller parameters: $K_{c_1} = 1$, $K_{c_2} = 5$.
(a) First show that with the given controller parameters, the closed-loop system will be *unstable*; then
(b) Show that with the same controller parameters, *if the loop pairing is switched* (i.e., changing to a $y_1 - m_2/y_2 - m_1$ configuration) the system *will be stable*. [**Hint:** *Find the appropriate transfer function matrix first, since the transfer function matrix in this case is different from the one used in part (a).*]

21

INTERACTION ANALYSIS AND MULTIPLE SINGLE-LOOP DESIGNS

In designing controllers for multivariable systems, a typical starting point is the use of multiple, *independent* single-loop controllers, with each controller using one input variable to control a preassigned output variable. Since such a control system will consist of several individual, single-loop feedback systems of the type we studied in Part IVA, it has the obvious advantage of building on familiar material. However, as shown in Chapter 20, primarily because of the interactions among the process variables, multivariable systems *cannot*, in general, be treated like multiple, independent, single-loop systems. Thus, a multiple single-loop control strategy for multivariable systems must take these interactions into consideration, and should be applied with care.

This chapter is concerned with discussing how interaction analysis is carried out and how the results of such analysis are used for multiple single-loop controller designs. The main questions regarding *when* to use this control strategy and *how* to design the multiple single-loop controllers will be answered in this chapter; what to do to obtain improved performance when this strategy proves ineffective will be discussed in the next chapter.

21.1   INTRODUCTION

As introduced in Chapter 20, a multivariable system has multiple inputs and multiple outputs. When confronted with the task of using the multiple inputs to control the multiple output variables of such processes, it seems natural to start out by considering the possibility of pairing the input and output variables and assigning *one* feedback controller to link each input/output variable pair, resulting in several feedback control loops, as illustrated in Figure 20.3(c).

With this approach, the issues to be resolved are twofold:

- How to pair the input and output variables
- How to design the individual single-loop controllers

Consider, as an example, the transfer function model for a 2×2 system, which in its expanded form is given by the following equations:

$$y_1(s) = g_{11}(s)\, u_1(s) + g_{12}(s)\, u_2(s) \tag{21.1}$$

$$y_2(s) = g_{21}(s)\, u_1(s) + g_{22}(s)\, u_2(s) \tag{21.2}$$

where we clearly see how each input variable is capable of influencing each output variable.

Here, we have a situation in which if u_1 has been assigned to y_1 (and, by default, u_2 to y_2), each of these control loops will experience interactions from the other loop; Loop 1 (the $u_1 - y_1$ loop) will experience interactions coming from Loop 2 (the $u_2 - y_2$ loop) via the g_{12} element, since it is by this that y_1 is influenced by u_2. In the same manner, Loop 2 experiences interactions coming from Loop 1 via the g_{21} element. This situation is clearly illustrated, once again, in Figure 20.3(c).

If the input/output pairing were switched — giving a second, alternative configuration — each loop will still be subject to interactions from the other loop, but this time, the interactions will come about via the g_{11} and g_{22} elements in Eqs. (21.1) and (21.2).

It seems reasonable that, as long as we intend to use two single-loop controllers for this 2×2 system, we should pair the input and output variables such that the resulting interaction is "minimum"; in other words, since there are two possible input/output pairing configurations for this particular system, it would be to our advantage to choose that configuration that results in smaller net interaction among the variables. This strategy works well for the great majority of processes; however, one should be aware that there are more complex control system designs that try to use the interactions rather than eliminate them.

To proceed logically to determine which input/output pairing configuration to use for the multiple, single-loop control strategy, we must have a means of discriminating between the $N!$ different configurations possible with an $N \times N$ system. This means we must be able to quantify the interactions experienced by the process under each configuration.

21.2 PRELIMINARY CONSIDERATIONS OF INTERACTION ANALYSIS AND LOOP PAIRING

21.2.1 A Measure of Control Loop Interactions

Consider an example 2×2 system (see Figure 21.1) with two output variables, y_1 and y_2, and two input variables that we will now refer to as m_1 and m_2, because we do not yet know which one will be paired with which output variable.

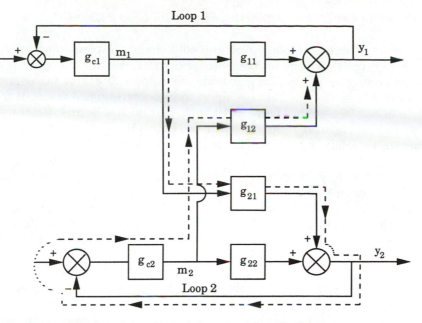

Figure 21.1.   Loop interactions for a 2 x 2 system.

Note that each loop can be opened or closed. When both loops are open, then m_1 and m_2 can be manipulated independently and the effect of each of these inputs on each of the outputs is represented by the following transfer function model:

$$y_1(s) = g_{11}(s)\, m_1(s) + g_{12}(s)\, m_2(s) \tag{21.3}$$

$$y_2(s) = g_{21}(s)\, m_1(s) + g_{22}(s)\, m_2(s) \tag{21.4}$$

where each transfer function element consists of an unspecified dynamic portion, and a steady-state gain term K_{ij}.

Let us now consider m_1 as a candidate input variable to pair with y_1. In order to evaluate this choice against the alternative of using m_2 instead, we will perform two "experiments" on the system.

• **Experiment 1:** *Unit step change in m_1 with all loops open.*

When a unit step change is made in the input variable m_1, with all the loops open, the output variable y_1 will change, and so will y_2, but for now we are primarily interested in the response in y_1.

After steady state has been achieved, let the change observed in y_1 as a result of this change in m_1 be referred to as Δy_{1m}; this represents the *main effect* of m_1 on y_1. Observe from Eq. (21.3) that:

$$\Delta y_{1m} = K_{11} \tag{21.5}$$

Keep in mind that no other input variable apart from m_1 has been changed, and all the control loops are opened so that there is no feedback control involved.

• **Experiment 2:** *Unit step change in m_1 with Loop 2 closed.*

In this "experiment," the same unit step change implemented in m_1 in the first "experiment" will be implemented; however, this time, Loop 2 is closed, and the controller g_{c2} is charged with the responsibility of using the other input variable m_2 to ward off any upset in y_2 occurring as a result of this step change in m_1. Note that Figure 21.1 shows that m_1 has both a direct influence on y_1 and an indirect influence with the path given by the dotted line. In particular the following things happen to the process due to the change in m_1:

1.   Obviously y_1 changes because of g_{11}, but because of interactions via the g_{21} element, so does y_2.

2.   Under feedback control, Loop 2 wards off this interaction effect on y_2 by manipulating m_2 until y_2 is restored to its initial state before the occurrence of the "disturbance."

3.   The changes in m_2 now return to affect y_1 via the g_{12} transfer function element.

Thus, the changes observed in y_1 are from two different sources: (1) the direct influence of m_1 on y_1 (which we already know from "experiment" 1 to be Δy_{1m}), and (2) the additional, *indirect* influence, precipitated by the retaliatory action from the Loop 2 controller in warding off the interaction effect of m_1 on y_2, say, Δy_{1r}. Note that this quantity, Δy_{1r}, is indicative of how much interaction m_1 is able to provoke from the other control loop in its attempt to control y_2.

After the dynamic transients die away and steady state has been achieved, there will be a net change observed in y_1, Δy_1^* given by:

$$\Delta y_1^* \ = \ \Delta y_{1m} + \Delta y_{1r}$$

a sum of the *main effect* of m_1 on y_1 and that of the interactive effect provoked by m_1 interacting with the other loop.

As we will show below, the quantity Δy_1^* will be given by, say:

$$\Delta y_1^* \ = \ K_{11}\left(1 - \frac{K_{12}\,K_{21}}{K_{11}\,K_{22}}\right) = K_{11}^* \tag{21.6}$$

Observe now that a good measure of how well the process can be controlled if m_1 is used to control y_1 is:

$$\lambda_{11} = \frac{\Delta y_{1m}}{\Delta y_1^*}$$

or

$$\lambda_{11} = \frac{\Delta y_{1m}}{\Delta y_{1m} + \Delta y_{1r}} \tag{21.7}$$

This quantity is a measure of the main effect of m_1 on y_1 compared to the total effect including the effect it provokes from the other controller, since it cannot control y_1 without upsetting y_2. Thus the quantity, λ_{11}, provides a measure of the extent of interaction in using m_1 to control y_1 while using m_2 to control y_2.

Observe that we can now perform a similar set of "experiments" that, this time, investigate the candidacy of m_2 as the input variable to use in controlling y_1. By properly interpreting the measures calculated from Eq. (21.7), we can quantify the degree of steady-state interaction involved with each control configuration, and thus determine which configuration minimizes steady-state interaction.

21.2.2 Loop Pairing on the Basis of Interaction Analysis

Let us return to the results of the "experiments" performed on the example 2×2 system of the previous subsection and now interpret the quantity λ_{11} as a measure of loop interaction. We will do this by evaluating the consequences of having various values λ_{11}.

1. The case when $\lambda_{11} = 1$

This can happen only if Δy_{1r} is zero; in physical terms, it means that the main effect of m_1 on y_1, measured when all the loops are opened, and the total effect, measured when the other loop is closed, *are identical*.

This will be the case only if:

- m_1 does not affect y_2, in which case there is no retaliatory control action from m_2, or
- m_1 *does* affect y_2, but the retaliatory control action from m_2 does not cause any changes in y_1 because m_2 does not affect y_1.

Under such circumstances, observe that m_1 is the perfect input variable to use in controlling y_1, because we will then have *no* interaction problems.

2. The case when $\lambda_{11} = 0$

This indicates a condition in which m_1 has no effect on y_1, for only then will Δy_{1m} be zero in response to a change in m_1. Observe that under these circumstances, m_1 is the perfect input variable for controlling *not* y_1 but y_2: since m_1 does not affect y_1, we can therefore control y_2 with m_1 without interacting with y_1 at all.

3. The case when $0 < \lambda_{11} < 1$

This corresponds to the condition in which the direction of the interaction effect is the same as that of the main effect. In this case the total effect is greater than the main effect. For $\lambda_{11} > 0.5$, the main effect contributes more to the total effect than the interaction effect, and as the main effect contribution increases, λ_{11} becomes closer to 1. For $\lambda_{11} < 0.5$, the contribution from the interaction effect dominates, and as this contribution increases, λ_{11} moves closer to zero. For $\lambda_{11} = 0.5$, the contributions from the main effect and the interactive effect are exactly equal.

4. The case when $\lambda_{11} > 1$

This corresponds to the condition where Δy_{1r} (the interaction effect) is opposite in sign to Δy_{1m} (the main effect), but smaller in absolute value. In this case the total effect, Δy_1^*, is less than the main effect, Δy_{1m}, and thus a larger controller action m_1 is required to achieve a given change in y_1 in the closed loop than in the open loop. Obviously for λ_{11} very large and positive, the interactive effect almost cancels the main effect and closed-loop control of y_1 by m_1 will be very difficult to achieve.

5. The case when $\lambda_{11} < 0$

This case arises when Δy_{1r} (the interaction effect) is not only opposite in sign to Δy_{1m} (the main effect), but is larger in absolute value. The pairing of m_1 with y_1 in this case is not very desirable because the direction of the effect of m_1 on y_1 in the open loop is opposite to the direction in the closed loop. The consequences of using such a pairing could be catastrophic as we shall demonstrate below.

 With these preliminary ideas in mind, it is possible to generalize the interaction analysis to multivariable systems of arbitrary dimension.

21.3   THE RELATIVE GAIN ARRAY (RGA)

The quantity λ_{11} introduced in the last section is known as the *relative gain* between output y_1 and input m_1, and as we have seen, it provides a measure of the extent of the influence of process interactions when m_1 is used to control y_1. Even though we introduced this quantity in reference to a 2×2 system, it can be generalized to any other multivariable system of arbitrary dimension.

 Let us define λ_{ij}, the relative gain between output variable y_i and input variable m_j, as the ratio of two steady-state gains:

$$\lambda_{ij} = \frac{\left(\dfrac{\partial y_i}{\partial m_j}\right)_{\text{all loops open}}}{\left(\dfrac{\partial y_i}{\partial m_j}\right)_{\text{all loops closed except for the } m_j \text{ loop}}} \qquad (21.8)$$

$$= \left(\frac{\text{open-loop gain}}{\text{closed-loop gain}}\right) \text{for loop } i \text{ under the control of } m_j$$

When the relative gain is calculated for all the input/output combinations of a multivariable system, and the results are presented in an array of the form shown below:

$$\Lambda = \begin{bmatrix} \lambda_{11} & \lambda_{12} & \cdots & \lambda_{1n} \\ \lambda_{21} & \lambda_{22} & \cdots & \lambda_{2n} \\ \cdots & \cdots & \cdots & \cdots \\ \lambda_{n1} & \lambda_{n2} & \cdots & \lambda_{nn} \end{bmatrix} \qquad (21.9)$$

the result is what is known as the *relative gain array* (RGA) or the *Bristol array*. The RGA was first introduced by Bristol [1] and has become the most widely used measure of interaction.

21.3.1 Properties of the RGA

The following are some of the most important properties of the RGA; they provide the bases for its utility in studying and quantifying interactions among the control loops of a multivariable system.

1. The elements of the RGA across any row, or down any column, sum up to 1, i.e.:

$$\sum_{i=1}^{n}\lambda_{ij} = \sum_{j=1}^{n}\lambda_{ij} = 1 \qquad (21.10)$$

2. λ_{ij} is dimensionless; therefore, neither the units, nor the absolute values actually taken by the variables m_j, or y_i, affect it.

3. The value of λ_{ij} is a measure of the *steady-state* interaction expected in the ith loop of the multivariable system if its output y_i is paired with m_j; in particular, $\lambda_{ij} = 1$ implies that m_j affects y_i without interacting with, and/or eliciting interaction from, the other control loops. By the same token, $\lambda_{ij} = 0$ implies that m_j has absolutely no effect on y_i.

4. Let K_{ij}^{*} represent the loop i steady-state gain when all the other loops — excepting this one — are closed, (i.e., a generalization of Eq. (21.6) for the example 2×2 system), whereas K_{ij} represents the normal, *open-loop* gain. By the definition of λ_{ij}, observe that:

$$K_{ij}^{*} = \frac{1}{\lambda_{ij}} K_{ij} \qquad (21.11)$$

with the following, very important implication: $1/\lambda_{ij}$ tells us by what factor the *open-loop gain* between output y_i and input m_j will be altered when the other loops are closed.

5. When λ_{ij} is negative, it is indicative of a situation in which loop i, with all loops open, will produce a change in y_i in response to a change in m_j totally opposite in direction to that when the other loops are closed. Such input/output pairings are potentially unstable and should be avoided.

Before discussing the general principles of how the RGA is actually used for loop pairing, it seems appropriate first to discuss the methods by which it is computed.

21.3.2 Calculating the RGA

There are two basic methods used for calculating the RGA for a square, linear system, whose transfer function matrix $\mathbf{G}(s)$ is available: (1) the "first principles" method, and (2) the matrix method.

Calculating RGA's from First Principles

We will illustrate this procedure with the 2×2 system of Eqs. (21.3) and (21.4).

First, we observe that, by definition, the RGA is concerned with steady-state conditions, thus we only need the steady-state form of this model, which is:

$$y_1 = K_{11} m_1 + K_{12} m_2 \tag{21.12}$$

$$y_2 = K_{21} m_1 + K_{22} m_2 \tag{21.13}$$

To obtain λ_{11} from the definition in Eq. (21.8), we need to evaluate both of the indicated partial derivatives from Eqs. (21.12) and (21.13) as follows: Eqs. (21.12) and (21.13) represent steady-state, open-loop conditions; therefore, the numerator partial derivative in Eq. (21.8) is obtained from Eq. (21.12) by straightforward differentiation:

$$\left(\frac{\partial y_1}{\partial m_1} \right)_{\text{all loops open}} = K_{11} \tag{21.14}$$

The second partial derivative calls for Loop 2 to be closed, so that in response to changes in m_1, the second control variable m_2 can be used ultimately to restore y_2 to its initial value of 0. Thus, to obtain the second partial derivative, we first find, from Eq. (21.13), the value m_2 must take to keep $y_2 = 0$ in the face of changes in m_1; what effect this will have on y_1 is deduced by substituting this value for m_2 in Eq. (21.12).

Setting $y_2 = 0$ and solving for m_2 in Eq. (21.13) gives:

$$m_2 = -\frac{K_{21}}{K_{22}} m_1 \tag{21.15}$$

and substituting this into Eq. (21.12) gives:

$$y_1 = K_{11} m_1 - \frac{K_{12} K_{21}}{K_{22}} m_1 \tag{21.16}$$

Having thus "eliminated" m_2 we may then differentiate, so that now:

$$\left(\frac{\partial y_1}{\partial m_1} \right)_{\text{loop 2 closed}} = K_{11} \left(1 - \frac{K_{12} K_{21}}{K_{11} K_{22}} \right) \tag{21.17}$$

and we obtain, finally, that:

$$\lambda_{11} = \frac{1}{1 - \zeta} \tag{21.18}$$

where

$$\zeta = \frac{K_{12} K_{21}}{K_{11} K_{22}} \tag{21.19}$$

As an exercise, the reader is encouraged to follow the illustrated procedure and confirm that the other relative gains, λ_{12}, λ_{21}, and λ_{22}, are given by:

$$\lambda_{12} = \lambda_{21} = \frac{-\zeta}{1 - \zeta} \tag{21.20}$$

and

$$\lambda_{22} = \lambda_{11} = \frac{1}{1 - \zeta} \tag{21.21}$$

Thus the RGA for the 2×2 system is given by:

$$\Lambda = \begin{bmatrix} \dfrac{1}{1 - \zeta} & \dfrac{-\zeta}{1 - \zeta} \\ \dfrac{-\zeta}{1 - \zeta} & \dfrac{1}{1 - \zeta} \end{bmatrix} \tag{21.22}$$

Note carefully that the elements add up to unity across any row as well as down any column.

It is useful to observe that if we define:

$$\lambda = \lambda_{11} = \frac{1}{1 - \zeta} \tag{21.23}$$

for this 2×2 system, then the RGA may be written:

$$\Lambda = \begin{bmatrix} \lambda & 1 - \lambda \\ 1 - \lambda & \lambda \end{bmatrix} \tag{21.24}$$

The immediate consequence is that the single element λ is sufficient to determine the RGA for a 2×2 system. Thus for 2×2 systems, λ is often called the relative gain parameter.

If we realistically assess this first-principles method of obtaining the RGA, it will not be difficult to come to the conclusion that it will be too tedious to apply in cases where the system is of higher dimension than 2×2. The problem is not with the numerator partial derivatives — these are always easy to obtain. The problem is with the denominator partial derivatives: the procedure for their evaluation in higher dimensional systems will involve solving several systems of linear algebraic equations.

The fact that linear algebraic equations are quite conveniently solved by matrix methods should then alert us to the possibility of using matrix methods to simplify the tedium involved in calculating RGA's, especially for higher dimensional systems.

The Matrix Method for Calculating RGA's

Let $\mathbf{K}$ be the matrix of steady-state gains of the transfer function matrix $\mathbf{G}(s)$, i.e.:

$$\lim_{s \to 0} \mathbf{G}(s) = \mathbf{K} \tag{21.25}$$

whose elements are K_{ij}; further, let $\mathbf{R}$ be the transpose of the inverse of this steady-state gain matrix, i.e.:

$$\mathbf{R} = \left(\mathbf{K}^{-1}\right)^{\mathrm{T}}$$

with elements r_{ij}. Then, it is possible to show that the elements of the RGA can be obtained from the elements of these two matrices according to:

$$\lambda_{ij} = K_{ij}\, r_{ij} \tag{21.26}$$

It is important to note that the above equation indicates an element-by-element multiplication of the corresponding elements of the two matrices $\mathbf{K}$ and $\mathbf{R}$; it has nothing to do with the standard matrix product.

Let us illustrate this procedure with the following examples.

Example 21.1   RGA FOR A GENERAL 2 × 2 SYSTEM USING MATRIX METHODS.

Find the RGA for the 2 × 2 system whose transfer function matrix model is given in Eqs. (21.3) and (21.4) and compare the results with those obtained earlier in Eqs. (21.18) through (21.22).

Solution:

For this system, the steady-state gain matrix is given by:

$$\mathbf{K} = \begin{bmatrix} K_{11} & K_{12} \\ K_{21} & K_{22} \end{bmatrix} \tag{21.27}$$

and from the definition of the inverse of a matrix, we have that:

$$\mathbf{K}^{-1} = \frac{1}{|\mathbf{K}|} \begin{bmatrix} K_{22} & -K_{12} \\ -K_{21} & K_{11} \end{bmatrix}$$

where $|\mathbf{K}|$, the determinant of $\mathbf{K}$, is given by:

$$|\mathbf{K}| = K_{11}K_{22} - K_{12}K_{21} \tag{21.28}$$

and taking the transpose of the matrix given in Eq. (21.27), we obtain:

$$\mathbf{R} = \left(\mathbf{K}^{-1}\right)^{\mathrm{T}} = \frac{1}{|\mathbf{K}|} \begin{bmatrix} K_{22} & -K_{21} \\ -K_{12} & K_{11} \end{bmatrix} \tag{21.29}$$

If we now carry out a term-by-term multiplication of the elements of the matrices $\mathbf{K}$ (in Eq. (21.26)) and $\mathbf{R}$ (in Eq. (21.29)) we obtain:

$$\lambda_{11} = \frac{K_{11}K_{22}}{|\mathbf{K}|} \tag{21.30}$$

or

$$\lambda_{11} = \frac{K_{11}K_{22}}{K_{11}K_{22} - K_{12}K_{21}} \tag{21.31}$$

which, for $K_{11}K_{22} \neq 0$ simplifies to:

$$\lambda_{11} = \frac{1}{1 - \zeta} \tag{21.32}$$

with

$$\zeta = \frac{K_{12}K_{21}}{K_{11}K_{22}} \tag{21.33}$$

exactly as we earlier obtained in Eqs. (21.18) and (21.19). By considering each element of the **K** and **R** matrices, we can generate an expression for each of the other λ_{ij}.

The following are examples of how to calculate the RGA for some specific systems.

Example 21.2   RGA FOR THE WOOD AND BERRY DISTILLATION COLUMN OF EXAMPLE 20.6 USING MATRIX METHODS.

Find the RGA for the Wood and Berry binary distillation column whose transfer function matrix is as given in Eq. (20.26).

Solution:

For this system, the steady-state gain matrix is easily extracted from the transfer function matrix by setting $s = 0$, giving:

$$\mathbf{K} = \mathbf{G}(0) = \begin{bmatrix} 12.8 & -18.9 \\ 6.6 & -19.4 \end{bmatrix} \tag{21.34}$$

and its inverse is obtained, in the usual manner, as:

$$\mathbf{K}^{-1} = \begin{bmatrix} 0.157 & -0.153 \\ 0.053 & -0.104 \end{bmatrix} \tag{21.35}$$

from where, upon taking the transpose of this matrix, we obtain:

$$\mathbf{R} = \left(\mathbf{K}^{-1}\right)^{\mathrm{T}} = \begin{bmatrix} 0.157 & 0.053 \\ -0.153 & -0.104 \end{bmatrix} \tag{21.36}$$

and a term-by-term multiplication of K_{ij} and r_{ij} now gives the RGA as:

$$\Lambda = \begin{bmatrix} 2.0 & -1.0 \\ -1.0 & 2.0 \end{bmatrix} \tag{21.37}$$

The following example is computationally more burdensome because it involves a 3×3 system, but the basic principles are the same.

**Example 21.3   RGA FOR A BINARY DISTILLATION COLUMN WITH A
SIDESTREAM DRAW-OFF USING MATRIX METHODS.**

In Ref. [2], the transfer function model for a pilot scale, binary distillation column used to separate ethanol and water was given as in Eq. (21.38) below, where the process variables are (in terms of deviations from their respective steady-state values):

y_1 = overhead mole fraction ethanol
y_2 = mole fraction of ethanol in the side stream
y_3 = temperature on Tray #19
u_1 = overhead reflux flowrate
u_2 = side stream draw-off rate
u_3 = reboiler steam pressure

$$\begin{bmatrix} y_1 \\ y_2 \\ y_3 \end{bmatrix} = \begin{bmatrix} \dfrac{0.66e^{-2.6s}}{6.7s+1} & \dfrac{-0.61e^{-3.5s}}{8.64s+1} & \dfrac{-0.0049e^{-s}}{9.06s+1} \\[2ex] \dfrac{1.11e^{-6.5s}}{3.25s+1} & \dfrac{-2.36e^{-3s}}{5s+1} & \dfrac{-0.012e^{-1.2s}}{7.09s+1} \\[2ex] \dfrac{33.68e^{9.2s}}{8.15s+1} & \dfrac{46.2e^{-9.4s}}{10.9s+1} & \dfrac{0.87(11.61s+1)e^{-s}}{(3.89s+1)(18.8s+1)} \end{bmatrix} \begin{bmatrix} u_1 \\ u_2 \\ u_3 \end{bmatrix} \quad (21.38)$$

Find the RGA for this 3×3 system.

Solution:

The steady-state gain matrix in this case is given by:

$$\mathbf{K} = \mathbf{G}(0) = \begin{bmatrix} 0.66 & -0.61 & -0.0049 \\ 1.11 & -2.36 & -0.012 \\ -33.68 & 46.2 & 0.87 \end{bmatrix} \quad (21.39)$$

The procedure for obtaining the inverse of this matrix is long and tedious: first the determinant is found, by using any of the methods indicated in Appendix D (the value is −0.5037), and then all the *nine* cofactors for this matrix are found; since the adjoint is the transpose of the matrix of cofactors, the inverse is easily found thereafter. The result is:

$$\mathbf{K}^{-1} = \begin{bmatrix} 2.955 & -0.6 & 0.0083 \\ 1.066 & -0.797 & -0.005 \\ 61.157 & 18.405 & 1.748 \end{bmatrix} \quad (21.40)$$

so that:

$$\mathbf{R} = (\mathbf{K}^{-1})^{\mathrm{T}} = \begin{bmatrix} 2.955 & 1.066 & 61.157 \\ -0.6 & -0.797 & 18.405 \\ 0.0083 & -0.005 & 1.748 \end{bmatrix} \quad (21.41)$$

From here, a term-by-term multiplication of the elements in Eqs. (21.39) and (21.41), as indicated in Eq. (21.26), now gives the complete RGA for this system as:

$$\Lambda = \begin{bmatrix} 1.95 & -0.65 & -0.3 \\ -0.66 & 1.88 & -0.22 \\ -0.29 & -0.23 & 1.52 \end{bmatrix} \tag{21.42}$$

Observe carefully that the elements sum to unity across any row, and down any column.

21.4   LOOP PAIRING USING THE RGA

Having established the RGA as a reasonable means of quantifying control loop interactions at steady state, and having illustrated how it is usually obtained, we will now consider how it may be used as a guide for pairing input and output variables in order to obtain the control configuration with minimal loop interaction.

21.4.1 Interpreting the RGA Elements

It is important to know how to interpret the RGA elements properly before we can use the RGA for loop pairing.

For the purpose of properly interpreting the RGA element λ_{ij}, it is convenient to classify all the possible values it can take into five categories, and in each case, investigate the implications for loop interaction, and what such implications suggest regarding input/output pairing.

1. $\lambda_{ij} = 1$, indicating that the open-loop gain between y_i and m_j is *identical* to the closed-loop gain:

   - *Implication for loop interactions*: loop i will *not* be subject to retaliatory action from other control loops when they are closed, therefore m_j can control y_i without interference from the other control loops. If, however, any of the g_{ik} elements in the transfer function matrix are nonzero, then the ith loop will experience some disturbances from control actions taken in the other loops, but these are not provoked by control actions taken in the ith loop.

   - *Pairing recommendation*: pairing y_i with m_j will therefore be ideal.

2. $\lambda_{ij} = 0$, indicating that the open-loop gain between y_i and m_j is *zero* :

   - *Implication for loop interactions*: m_j has no direct influence on y_i (even though it may affect other output variables).

   - *Pairing recommendation*: do *not* pair y_i with m_j; pairing m_j with some other output will however be advantageous, since we are sure that at least y_i will be immune to interaction from this loop.

3. $0 < \lambda_{ij} < 1$, indicating that the open-loop gain between y_i and m_j is *smaller* than the closed-loop gain:

- *Implication for loop interactions*: since the closed-loop gain is the sum of the open-loop gain *and* the retaliatory effect from the other loops:

    (a) The loops are definitely interacting, but

    (b) They do so in such a way that the retaliatory effect from the other loops is in the same direction as the main effect of m_j on y_i.

    Thus the loop interactions "assist" m_j in controlling y_i. The extent of the "assistance" from other loops is indicated by how close λ_{ij} is to 0.5.

    When $\lambda_{ij} = 0.5$, the main effect of m_j on y_i is exactly identical to the retaliatory (but complementary) effect it provokes from other loops; when $\lambda_{ij} > 0.5$ (but still less than 1) this retaliatory effect from the other interacting loops is lower than the main effect of m_j on y_i; if, however, $0 < \lambda_{ij} < 0.5$, then the retaliatory effect is actually more substantial than the main effect.

- *Pairing recommendation*: If possible, avoid pairing y_i with m_j whenever $\lambda_{ij} \leq 0.5$.

4. $\lambda_{ij} > 1$, indicating that the open-loop gain between y_i and m_j is *larger* than the closed-loop gain:

- *Implication for loop interactions*: the loops interact, and the retaliatory effect from the other loops acts in *opposition* to the main effect of m_j on y_i (thus reducing the loop gain when the other loops are closed); but the main effect is still dominant, otherwise λ_{ij} will be negative. For large λ_{ij} values, the controller gain for loop i will have to be chosen much larger than when all the other loops are open. This could cause loop i to become unstable when the other loops are open.

- *Pairing recommendation*: The higher the value of λ_{ij}, the greater the opposition m_j experiences from the other control loops in trying to control y_i; therefore, where possible, do not pair m_j with y_i if λ_{ij} takes a very high value.

5. $\lambda_{ij} < 0$; indicating that the open-loop and closed-loop gains between y_i and m_j have opposite signs:

- *Implication for loop interactions*: the loops interact, and the retaliatory effect from the other loops is not only in *opposition* to the main effect of m_j on y_i, it is also the more dominant of the two effects. This is a potentially dangerous situation because opening the other loops will likely cause loop i to become unstable.

- *Pairing recommendation*: because the retaliatory effect that m_j provokes from the other loops acts in opposition to, and in fact dominates, its main effect on y_i, avoid pairing m_j with y_i.

21.4.2 Basic Loop Pairing Rules

In light of the foregoing discussion, we are led first to the conclusion that the ideal situation occurs when the RGA takes the form:

$$\Lambda = \begin{bmatrix} 1 & 0 & 0 & \dots & 0 \\ 0 & 1 & 0 & \dots & 0 \\ 0 & 0 & 1 & \dots & 0 \\ \dots & \dots & \dots & \dots & \dots \\ 0 & 0 & 0 & \dots & 1 \end{bmatrix}$$

in which each row, and each column, contains one and only one nonzero element whose value is unity. Such an RGA is obtained when the process transfer function matrix is diagonal, or triangular; the first indicates *no* interaction among the loops, while the latter indicates *one-way* interaction (see example below and problems at the end of the chapter). It is easy to pair the input and output variables under these circumstances.

The case of *one-way* interaction requires some further discussion. Consider the 2×2 system with the following lower triangular transfer function:

$$\mathbf{G}(s) = \begin{bmatrix} \dfrac{1}{s + 1} & 0 \\ \dfrac{3}{3s + 1} & \dfrac{4}{4s + 1} \end{bmatrix} \tag{21.43}$$

Calculation of the RGA yields:

$$\Lambda = \begin{bmatrix} 1 & 0 \\ 0 & 1 \end{bmatrix} \tag{21.44}$$

Note that because the element $g_{12}(s)$ is zero, the input m_2 has no effect on output y_1 and thus any control action in Loop 2 has no effect at all on Loop 1 (cf. Figure 21.1). However, the input m_1 does influence the output y_2 through the nonzero $g_{21}(s)$ element. Thus upsets in Loop 1 requiring control action by m_1 would appear as a disturbance, $g_{21}(s) m_1(s)$, which would have to be handled by the feedback controller of Loop 2. Hence, even though the RGA of Eq. (21.44) is ideal, Loop 2 would be at a disadvantage. Thus in deciding on loop pairing one should distinguish between loop pairings having ideal RGA's resulting from diagonal transfer functions and those arising from triangular transfer functions. The latter case will always have poorer performance under feedback control.

In the more common nonideal cases, the following pairing rule originally published by Bristol [1] should be used:

• **Rule 1**

Pair input and output variables that have positive RGA elements that are closest to 1.0.

Thus, for a 2×2 system with output variables y_1 and y_2, to be paired with input variables m_1 and m_2, if the RGA is:

$$\Lambda = \begin{bmatrix} 0.8 & 0.2 \\ 0.2 & 0.8 \end{bmatrix}$$

then it is recommended pair to pair y_1 with m_1, and y_2 with m_2, quite often referred to as the 1-1/2-2 pairing. On the other hand, for the 2×2 system whose RGA is:

$$\Lambda = \begin{bmatrix} 0.3 & 0.7 \\ 0.7 & 0.3 \end{bmatrix}$$

the recommended pairing is y_1 with m_2 and y_2 with m_1, or the 1-2/2-1 pairing, while the following RGA:

$$\Lambda = \begin{bmatrix} 1.5 & -0.5 \\ -0.5 & 1.5 \end{bmatrix}$$

recommends the 1-1/2-2 pairing in order to avoid pairing on a negative RGA element. (Despite the fact that pairing on RGA values greater than 1 is not very desirable, to pair on a negative RGA value is worse still.)

Thus, for the 2×2 Wood and Berry distillation column whose RGA was calculated in Example 21.2 as:

$$\Lambda = \begin{bmatrix} 2.0 & -1.0 \\ -1.0 & 2.0 \end{bmatrix}$$

the 1-1/2-2 pairing is recommended; in terms of the actual process variables, (see Example 20.6) this translates to using the overhead reflux flowrate to control overhead mole fraction of methanol ($y_1 - m_1$), and the bottoms steam flowrate to control the bottoms mole fraction methanol ($y_2 - m_2$).

For the 3×3 distillation column of Example 21.3 with the RGA:

$$\Lambda = \begin{bmatrix} 1.95 & -0.65 & -0.3 \\ -0.66 & 1.88 & -0.22 \\ -0.29 & -0.23 & 1.52 \end{bmatrix}$$

the 1-1/2-2/3-3 pairing is recommended. In physical terms, this gives rise to a control configuration which uses the overhead reflux flowrate to control y_1, the overhead mole fraction ethanol; the side stream draw-off rate to control y_2, the mole fraction of ethanol in the side stream, and the reboiler steam pressure to control y_3, the temperature on Tray #19.

Niederlinski Index

Even though pairing rule 1 is usually sufficient in most cases, it is often necessary (especially with 3×3 and higher dimensional systems) to use this rule in conjunction with stability considerations provided by the following theorem originally due to Niederlinski (Ref. [3]), and later modified by Grosdidier et al. (Ref. [4]):

> *Consider the n × n multivariable system whose input and output variables have been paired as follows: $y_1 - u_1, y_2 - u_2, ..., y_n - u_n$, resulting in a transfer function model of the form:*

$$\mathbf{y}(s) = \mathbf{G}(s)\,\mathbf{u}(s)$$

Further let each element of $\mathbf{G}(s)$, $g_{ij}(s)$, be

1. *Rational, and*
2. *Open-loop stable,*

and let n individual feedback controllers (which have integral action) be designed for each loop so that each one of the resulting n feedback control loops is stable when all the other $n - 1$ loops are open.

> *Under closed-loop conditions in all n loops, the multiloop system will be unstable for all possible values of controller parameters (i.e., it will be "structurally monotonic unstable") if the Niederlinski index N defined below is negative, i.e.:*

$$N \equiv \frac{|\mathbf{G}(0)|}{\prod\limits_{i=1}^{n} g_{ii}(0)} < 0 \qquad (21.45)$$

The following important points help us utilize this result properly:

1. This result is both necessary and sufficient *only* for 2 × 2 systems; for higher dimensional systems, it provides only sufficient conditions: i.e., *if* Eq. (21.45) holds then the system is definitely unstable; however, if Eq. (21.45) does *not* hold, the system may, or may not be unstable: the stability, will, in this case, depend on the values taken by the controller parameters.

2. For 2 × 2 systems the Niederlinski index becomes:

$$N = 1 - \zeta$$

where ζ is given by Eq. (21.19). Thus for a 2 × 2 with a negative relative gain, $\zeta > 1$, the Niederlinski index is always negative; *hence 2 × 2 systems paired with negative relative gains are always structurally unstable.*

3. The theorem is for systems with rational transfer function elements, thereby *technically* excluding time-delay systems. However, since Eq. (21.45) depends only on steady-state gains (which are independent of time delays), the results of this theorem provide useful information about time-delay systems also, but the analysis is no longer rigorous (see Ref. [4].) Thus Eq. (21.45) should therefore be applied with caution when time delays are involved.

This leads to loop pairing rule 2:

- **Rule 2**

    Any loop pairing is unacceptable if it leads to a control system configuration for which the Niederlinski index is negative.

Thus, the strategy for using the RGA for loop pairing may be summarized as follows:

1. Given the transfer function matrix $G(s)$, obtain the steady-state gain matrix $K = G(0)$, and from this, obtain the RGA, Λ; also obtain the determinant of K, and the product of the elements on its main diagonal.

2. Use Rule #1 to obtain tentative loop pairing suggestions from the RGA, by pairing on *positive* elements which are closest to 1.0.

3. Use Niederlinski's condition of Eq. (21.45) to verify the stability status of the control configuration resulting from 2; if the pairing is unacceptable, select another.

21.4.3 Some Applications of the Loop Pairing Rules

As we have shown above, for a 2×2 system, the Niederlinski stability condition is exactly equivalent to requiring that we pair only on positive RGA elements. Thus as long as we are dealing with 2×2 systems, Rule #1 is sufficient for obtaining appropriate loop pairing.

This is, however, not the case for 3×3 and higher dimensional systems; it is possible to pair on positive RGA elements and still have a structurally unstable system, as the following example illustrates.

Example 21.4   LOOP PAIRING FOR A 3×3 SYSTEM.

Calculate the RGA for the system whose steady-state gain matrix is given below, and investigate the loop pairing suggested upon applying Rule #1 (see Koppel, Ref. [5]):

$$
\mathbf{K} \; = \; \mathbf{G}(0) \; = \;
\begin{bmatrix}
\dfrac{5}{3} & 1 & 1 \\[2mm]
1 & \dfrac{1}{3} & 1 \\[2mm]
1 & 1 & \dfrac{1}{3}
\end{bmatrix}
\tag{21.46}
$$

Solution:

By taking the inverse, and calculating the RGA in the usual manner as illustrated in Example 21.3 we obtain the following RGA:

$$
\Lambda \; = \;
\begin{bmatrix}
10 & -4.5 & -4.5 \\
-4.5 & 1 & 4.5 \\
-4.5 & 4.5 & 1
\end{bmatrix}
\tag{21.47}
$$

from which, by Rule #1, the 1-1/2-2/3-3 pairing is recommended.

According to this pairing however, the steady-state gain matrix for the system has determinant given by:

$$|\mathbf{K}| = -0.148 \tag{21.48}$$

and the product of the diagonal elements of $\mathbf{K}$ is:

$$\prod_{i=1}^{3} K_{ii} = \left(\frac{5}{3}\right)\left(\frac{1}{3}\right)\left(\frac{1}{3}\right) = \frac{5}{27} \tag{21.49}$$

Thus, for this system, according to Eq. (21.45) this pairing will lead to a negative Niederlinski index and an unstable configuration.

This example provides a situation in which Rule #2 disqualifies a loop pairing suggested by Rule #1; thus an alternative loop pairing must now be considered. While the available alternatives are not especially attractive, a possible pairing would be 1-1/2-3/3-2 which has the rearranged relative gain array:

$$\Lambda = \begin{bmatrix} 10 & -4.5 & -4.5 \\ -4.5 & 4.5 & 1 \\ -4.5 & 1 & 4.5 \end{bmatrix}$$

and new steady-state gain matrix:

$$\mathbf{K} = \begin{bmatrix} \dfrac{5}{3} & 1 & 1 \\ 1 & 1 & 1/3 \\ 1 & 1/3 & 1 \end{bmatrix}$$

with determinant $|\mathbf{K}| = \dfrac{4}{27}$ and Niederlinski index:

$$N = \frac{4/27}{5/3} = \frac{4}{45}$$

so that with this pairing the system is no longer structurally unstable.

In higher dimensional systems, it is possible to pair on negative RGA values and still have a stable system, something of an impossibility in a 2×2 system. However, for such systems, it has been shown (see also Gagnepain and Seborg [6]) that if the loop which is paired on the negative RGA element is opened, the lower dimensional subsystem will be unstable.

An example system that illustrates this (see Koppel [5] and McAvoy [7]) has steady-state gain matrix:

$$\mathbf{K} = \mathbf{G}(0) = \begin{bmatrix} 1 & 1 & -0.1 \\ 0.1 & 2 & -1 \\ -2 & -3 & 1 \end{bmatrix} \tag{21.50}$$

with a determinant, $|\mathbf{K}| = 0.53$. The RGA for this system is:

$$\Lambda = \begin{bmatrix} -1.89 & 3.59 & -0.7 \\ -0.13 & 3.02 & -1.89 \\ 3.02 & -5.61 & 3.59 \end{bmatrix} \tag{21.51}$$

from which one finds that the only feasible pairing has to involve a negative RGA element for the first loop because the other pairing configurations are disqualified by Rule #2.

The suggested 1-1/2-2/3-3 pairing gives rise to a configuration for which the system, according to Niederlinski's theorem, is *not* structurally unstable despite pairing on the negative (1,1) RGA element.

However, if the first loop is opened (i.e., the $y_1 - m_1$ variables are dropped from the process model) the resulting subsystem will have a steady-state gain matrix relating the remaining two output variables y_2 and y_3, to the remaining two input variables, m_2 and m_3, given by:

$$\tilde{\mathbf{K}} = \begin{bmatrix} 2 & -1 \\ -3 & 1 \end{bmatrix} \tag{21.52}$$

and it is easy to verify that the $y_2 - m_2$, $y_3 - m_3$ pairing will violate the Niederlinski theorem and thus give rise to an unstable system.

Such a system that is stable when all loops are closed, but that goes unstable should one of them become open, is said to have a *low degree of integrity*.

Thus, in the final analysis, after we put all the foregoing factors into consideration, we come to the following conclusion about using the RGA for loop pairing:

> *Always pair on positive RGA elements that are closest to 1.0 in value; thereafter use Niederlinski's condition to check the resulting configuration for structural instability. Wherever possible, avoid pairing on negative RGA elements; for 2 × 2 systems such pairings always lead to an unstable configuration, while for higher dimensional systems, they lead to a configuration which at best has a low degree of integrity.*

21.5   LOOP PAIRING FOR NONLINEAR SYSTEMS

So far, we have discussed how to obtain the RGA essentially from transfer function models, which, as we know, represent only linear systems. By virtue of the fact that most chemical processes of importance are nonlinear, it is pertinent at this stage to consider whether Bristol's RGA technique can be used for pairing the input/output variables of nonlinear systems.

The reader should have noted by now that interaction analysis by means of the RGA is based only on steady-state information. While this is often considered a limitation (see Section 21.8), this particular aspect is in fact what makes the RGA technique easily applicable to nonlinear systems:

Assuming that a process model is available, any of the following two approaches will enable us obtain RGA's for nonlinear systems:

1. From first principles (i.e., taking partial derivatives), using the steady-state version of the nonlinear model, because only steady-state models are involved, it is sometimes possible to obtain analytical expressions.

2. By linearizing the nonlinear model around a specific steady state and using the approximate **K** matrix to obtain the RGA.

The second approach is usually less tedious than the first one, and is therefore more popular.

It should not come as a surprise that RGA's for nonlinear systems are distinguished by their dependence on specific steady-state process conditions. As we may recall from Chapter 10, it is the nature of nonlinear process systems for characteristic process parameters to be different at different steady-state operating conditions. Thus, while RGA's for linear systems consist only of constant elements, the RGA's for nonlinear systems are functions of process steady-state operating conditions and will therefore change as these operating conditions change.

We shall now present examples to illustrate these two procedures for obtaining RGA's for nonlinear systems, and how such RGA's are used for loop pairing.

Example 21.5   RGA AND LOOP PAIRING FOR NONLINEAR SYSTEMS:
1. THE BLENDING PROCESS.

The process shown in Figure 21.2 below is used to blend two process streams containing pure material A and pure material B, with respective (molal) flowrates, F_A and F_B. The objective is to control *both* the total product flowrate F, and the product composition, represented as x, and calculated in terms of mole fraction of material A in the blend.

The mathematical model for the process, assuming negligible mixing dynamics, is obtained from total and component mass balances as follows:

Total Mass Balance:

$$F = F_A + F_B \tag{21.53}$$

Component A Mass Balance:

$$x = \frac{F_A}{F_A + F_B} \tag{21.54}$$

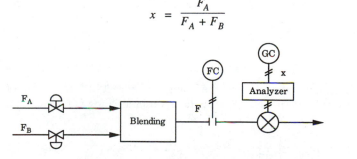

Figure 21.2.   The blending process.

Obtain the RGA for this process and suggest which input variable to pair with which output variable to achieve good control.

Solution:

Observe that for this process, the two output variables are F and x, and the input (or control) variables are F_A and F_B; let us refer to these input variables as m_1 and m_2 respectively. The process modeling equations then become:

$$F = m_1 + m_2 \qquad (21.55)$$

which is linear, and:

$$x = \frac{m_1}{m_1 + m_2} \qquad (21.56)$$

which is nonlinear. Since this is a 2×2 system, we only need to obtain the $(1,1)$ element of the RGA given by:

$$\lambda = \frac{\left(\dfrac{\partial F}{\partial m_1}\right)_{\text{both loops open}}}{\left(\dfrac{\partial F}{\partial m_1}\right)_{\text{second loop closed}}} \qquad (21.57)$$

From Eq. (21.55), with both loops open:

$$\left(\frac{\partial F}{\partial m_1}\right)_{\text{both loops open}} = 1 \qquad (21.58)$$

Now, upon closing the second loop, the question now is: what value will m_2 have to take, in order that, for any change in m_1, x is restored to its desired steady-state value, say x^*? To answer this question calls for setting $x = x^*$ in Eq. (21.56) and solving for m_2 in terms of m_1 and x^*; the result is:

$$m_2 = \frac{m_1}{x^*} - m_1 \qquad (21.59)$$

Thus, when the second loop is closed, to perform its task of maintaining the mole fraction of material A at x^*, m_2 will respond to any changes in m_1 as indicated in Eq. (21.59). If we now introduce this into Eq. (21.55) we will have:

$$F = m_1 + \frac{m_1}{x^*} - m_1$$

or

$$F = \frac{m_1}{x^*} \qquad (21.60)$$

an expression that represents the changes to expect in F as a result of changes in m_1, *when the second loop is closed*. From here, differentiating with respect to m_1 now gives:

$$\left(\frac{\partial F}{\partial m_1}\right)_{\text{second loop closed}} = \frac{1}{x^*} \qquad (21.61)$$

with the final result that:

$$\lambda = \frac{1}{1/x^*} = x^* \tag{21.62}$$

and therefore, the RGA for this blending system is:

$$\Lambda = \begin{bmatrix} x^* & 1 - x^* \\ 1 - x^* & x^* \end{bmatrix} \tag{21.63}$$

where x^* is the desired mole fraction of species A in the blend.

It is now important to observe the following about this result:

1. The RGA *is* dependent on the particular steady-state value x^* desired for the composition of the blend; it is not constant as was the case with the linear systems we have dealt with so far.

2. The implication here is that the recommended loop pairing will depend on our steady-state operating point.

3. Because x^* is a mole fraction, it is bounded between 0 and 1, i.e., $0 \le x^* \le 1$. Thus, none of the elements of the RGA will ever be negative. The implication here is that the worst we can do is to have severe interactions if the input and output variables are not properly paired; the system cannot be unstable as a result of poor pairing. A little reflection on the nature of the physical blending system will confirm this.

The loop pairing strategy can now be developed as follows:

1. If x^* is close to 1, the first implication, from Eq. (21.56), is that m_1 (which represents the flowrate of pure material A) is larger than m_2, the other flowrate. From Eq. (21.63) the recommended pairing in this case is $F - m_1$ and $x - m_2$, i.e., the *larger* flowrate is to be used in controlling total flow, while the smaller flowrate is used to control the composition.

2. That this is the most reasonable pairing can be seen by examining the physical system: when the product composition is close to 1, we have almost pure A in the blend, and the total flowrate can be modified quite easily using the flowrate of A, without affecting the composition too drastically; similarly, if we wish to alter the composition, the addition of small amounts of material B can cause significant changes in the composition without altering the flowrate significantly. Thus, the flow controller will not interact strongly with the composition controller if this $F - m_1$ and $x - m_2$ pairing is used. By contrast, if the alternate pairing configuration is used, the interactions between the two control loops will be quite severe.

3. When the steady-state product composition is closer to 0, the RGA suggests that we switch loop pairings, and use m_2 to control F, and m_1 to control x. Again, the physics of the problem indicates that this is the best course of action.

4. An interesting situation arises when $x^* = 0.5$; in this case, it matters not which input variable is used to control which output variable; the observed interactions will be significant in either case.

Nonlinear system models are not always as easy to deal with as the preceding example appears to indicate. In most cases, the nonlinear steady-state model can be quite tedious to handle, especially in obtaining the denominator partial derivative that almost always requires analytical solutions of several nonlinear algebraic equations.

The technique of utilizing an approximate linear model (obtained by linearization around an operating steady state) is not only more convenient in such cases, it is also consistent with other aspects of nonlinear control system analysis and design we have presented so far.

The following example illustrates this technique.

Example 21.6   RGA AND LOOP PAIRING FOR NONLINEAR SYSTEMS:
**   2. THE STIRRED MIXING TANK.**

The stirred mixing tank introduced in Chapter 20 (see Figure 20.1), is a multivariable system in which the cold and hot stream flowrates are to be used in controlling the liquid level and tank temperature.

The nonlinear modeling equations for this system were obtained in Example 20.3, and an approximate transfer function, obtained by linearizing around some nominal steady-state liquid level h_s and temperature T_s, has also been obtained in Example 20.5, and presented in Eqs. (20.23) and (20.24).

Obtain the RGA for this system and use it to recommend which of the input variables should be used to control the liquid level and which should be used to control the tank temperature.

Solution:

From the approximate transfer function matrix given in Example 20.5, we obtain the steady-state gain matrix for the process as:

$$\mathbf{K} = \mathbf{G}(0) = \frac{1}{K} \begin{bmatrix} 2\sqrt{h_s} & 2\sqrt{h_s} \\ \dfrac{(T_H - T_s)}{\sqrt{h_s}} & \dfrac{(T_C - T_s)}{\sqrt{h_s}} \end{bmatrix} \tag{21.64}$$

We recall from our previous encounters with this process that its output variables are represented as follows:

$$y_1 = \text{liquid level;} \qquad\qquad y_2 = \text{tank temperature}$$

and the input variables are represented as:

$$m_1 = \text{hot stream flowrate;} \qquad m_2 = \text{cold stream flowrate}$$

Either by taking advantage of the fact that this is a 2×2 system, and therefore using Eqs. (21.32) and (21.33), (this is recommended), or by using the full matrix method, the RGA for this system is obtained as:

$$\Lambda = \begin{bmatrix} \dfrac{T_C - T_s}{T_C - T_H} & \dfrac{-(T_H - T_s)}{T_C - T_H} \\[3ex] \dfrac{-(T_H - T_s)}{T_C - T_H} & \dfrac{T_C - T_s}{T_C - T_H} \end{bmatrix} \qquad (21.65)$$

where we see again that the RGA depends on the steady-state operating temperature T_s, and as such, we will expect different loop pairing suggestions for different steady-state operating conditions. (It is interesting to note that the RGA is not directly dependent on the steady-state level h_s.)

To illustrate these points, let us introduce numerical values for the parameters of this process as follows:

Hot stream temperature, $T_H = 65°C$
Cold stream temperature, $T_C = 15°C$

We will now investigate loop pairing for this process under four different steady-state operating conditions:

Condition 1: $T_s > 40°C$ (in particular $T_s = 55°C$)

Under these conditions, observe that the operating steady-state temperature is closer to the hot stream temperature. Introducing the numerical values into the RGA gives:

$$\Lambda = \begin{bmatrix} 0.8 & 0.2 \\ 0.2 & 0.8 \end{bmatrix} \qquad (21.66)$$

and the RGA recommends a 1-1/2-2 pairing, i.e.:

Hot stream (m_1) to control liquid level, (y_1), and
Cold stream (m_2) to control temperature, (y_2).

Upon examining the physical process, we immediately see that this is the only reasonable thing to do: the cold stream temperature (15°C) is much farther away from the steady-state operating tank temperature (55°C) than the hot stream temperature (65°C). As a result, small changes are required in the cold stream flowrate in order to produce noticeable changes in the tank temperature, making it possible for the temperature controller to do its job effectively without causing much upset in the liquid level.

Conversely, because the hot stream temperature is closer to the operating steady-state tank temperature, it can be used to control the level without causing significant changes in the tank temperature.

Note therefore that the alternative pairing (hot stream to control temperature, and cold stream to control level) will cause significantly more pronounced problems due to loop interactions, and will therefore not be as effective.

Condition 2: $T_s < 40°C$ (In particular $T_s = 25°C$)

We now have a situation in which the operating steady-state temperature is closer to the cold stream temperature. The RGA in this case, upon introducing the numerical values, is:

$$\Lambda = \begin{bmatrix} 0.2 & 0.8 \\ 0.8 & 0.2 \end{bmatrix} \tag{21.67}$$

and the RGA recommends a pairing which is the reverse of that recommended under condition 1: the 1-2/2-1 pairing, in which:

Liquid level (y_1) is controlled by the cold stream (m_2), and
Temperature (y_2) is controlled by the hot stream (m_1).
Again, from physical considerations, this makes perfect sense, since this is the only configuration that minimizes the steady-state interaction effects among the process variables.

Condition 3: $T_s = 40°C$

In this case, the operating temperature is exactly equidistant from the cold and hot stream temperatures. The RGA in this case is:

$$\Lambda = \begin{bmatrix} 0.5 & 0.5 \\ 0.5 & 0.5 \end{bmatrix} \tag{21.68}$$

and, as we might also expect from the physics of this situation, either pairing is equally bad.

Condition 4: $T_s = T_H$

In this case, the tank is to operate at the same temperature as the hot stream. Observe that the RGA in this case is:

$$\Lambda = \begin{bmatrix} 1 & 0 \\ 0 & 1 \end{bmatrix} \tag{21.69}$$

and it suggests that we can achieve perfect control of the level without interacting with the temperature if we use the hot stream for this purpose. It also indicates the obvious fact that variations in the hot stream flowrate are not able to affect the tank temperature.

These simple examples have therefore demonstrated that the RGA can indeed be used for nonlinear as well as for linear systems. That the RGA sometimes recommends different pairings at different operating conditions confirms to us that even though the analysis has been based on approximate linearized models, this cardinal property of nonlinear systems is not lost.

21.6   LOOP PAIRING FOR SYSTEMS WITH PURE INTEGRATOR MODES

Since interaction analysis via the RGA involves using steady-state information, a particularly interesting situation arises when dealing with processes that contain pure integrator elements, since pure integrator elements show no steady state.

Even though McAvoy [7] has several suggestions as to what might be done in such cases, we use the following example to suggest an alternative way of dealing with systems of this nature.

Example 21.7 RGA FOR AN INDUSTRIAL DE-ETHANIZER.

The transfer function for a 2×2 subsystem extracted from the 3×3 model obtained by Tyreus [8] for an industrial de-ethanizer is given below. Obtain the RGA for this system and use it to recommend loop pairing:

$$\mathbf{G}(s) = \begin{bmatrix} \dfrac{1.318e^{-2.5s}}{20s + 1} & \dfrac{-e^{-4s}}{3s} \\ \dfrac{0.038(182s + 1)}{(27s + 1)(10s + 1)(6.5s + 1)} & \dfrac{0.36}{s} \end{bmatrix} \tag{21.70}$$

Solution:

First we observe that to obtain the steady-state gain matrix $\mathbf{K}$, setting $s = 0$ in Eq. (21.70) will not work since the (1,2) and (2,2) elements contain the pure integrator mode represented by $1/s$.

Let us replace this term by I, i.e.:

$$I = \frac{1}{s}$$

We then have:

$$\mathbf{K} = \lim_{s \to 0} \mathbf{G}(s) = \lim_{I \to \infty} \begin{bmatrix} 1.318 & \dfrac{-I}{3} \\ 0.038 & 0.36I \end{bmatrix} \tag{21.71}$$

Thus the relative gain parameter for this system may be obtained from Eqs. (21.18), (21.19), and (21.71) to yield:

$$\lambda = \lim_{I \to \infty} \left(\cfrac{1}{1 + \cfrac{0.038 \times 0.333I}{1.138 \times 0.36I}} \right) \tag{21.72}$$

Since the variables I cancel, we obtain:

$$\lambda = 0.97 \tag{21.73}$$

and the resulting RGA is:

$$\Lambda = \begin{bmatrix} 0.97 & 0.03 \\ 0.03 & 0.97 \end{bmatrix} \tag{21.74}$$

Obviously 1-1/2-2 pairing is recommended.

The approach of representing the integrator element by I, calculating the RGA with this as a parameter, and finally taking the limit as this quantity tends to infinity will work as illustrated in the example whenever the integrating elements occur in such a way that the I terms cancel. When this is not the case, other approaches suggested by McAvoy [7] can be used.

21.7   LOOP PAIRING FOR NONSQUARE SYSTEMS

Our discussion so far has centered around how to obtain RGA's, and how to use them for input/output pairing when the process has an equal number of input and output variables. Such systems are referred to as *square systems* because their transfer function matrices are square.

It is not always true, however, that a multivariable process *must* have an equal number of input and output variables. Some multivariable systems are *nonsquare systems* in the sense that they have an unequal number of input and output variables, and therefore their transfer function matrices will *not* be square.

The most obvious problem with nonsquare systems is that after input/output pairing, there will always be a residual of unpaired input or output variables, depending on which of these are in excess. Thus, quite apart from the nontrivial issue of deciding which input variable to pair with which output variable, we now have to contend with another problem: which variables should be left redundant, and which ones should take active part in the control scheme? These are the questions to be answered in this section.

21.7.1 Classifying Nonsquare Systems

We start by noting that there are *two* types of nonsquare systems: systems with *fewer* input than output variables, and those with *more* input than output variables. In order to be able to control the output variables arbitrarily, we need *at least* an equal number of input variables; thus systems that have fewer input than output variables will be referred to as *underdefined* systems, while the systems with more input than output variables will be referred to as *overdefined*. The former has a deficiency in input variables; the latter has an excess of input variables. Thus, a multivariable system with n output and m input variables — whose transfer function matrix will therefore be $n \times m$ in dimension — is underdefined if $m < n$, and overdefined if $m > n$.

21.7.2 Underdefined Systems

For underdefined systems, the main issue is that not all the outputs can be controlled, since we do not have enough input variables. The loop pairing decision is, however, much simpler in this case:

> *By economic considerations, or other such means, decide which m of the n output variables are the most important; these will be paired with the m available input variables; the less important n − m output variables will not be under any form of control.*

Let us illustrate this with an example.

> **Example 21.8   RGA AND LOOP PAIRING FOR AN UNDERDEFINED NONSQUARE SYSTEM: A BINARY DISTILLATION COLUMN.**

> Consider the situation in which the binary distillation column used in Example 21.3 has its sidestream draw-off rate set at a fixed amount that *cannot* be changed, and yet all three of its output variables, the overhead mole fraction of ethanol, the sidestream

mole fraction of ethanol, and the Tray #19 temperature are still being monitored, and are to be controlled. Since we have lost one control variable, the process model now becomes:

$$
\begin{bmatrix} y_1 \\ y_2 \\ y_3 \end{bmatrix} = \begin{bmatrix} \dfrac{0.66e^{-2.6s}}{6.7s + 1} & \dfrac{-0.0049e^{-s}}{9.06s + 1} \\[2mm] \dfrac{1.11e^{-6.5s}}{3.25s + 1} & \dfrac{-0.012e^{-1.2s}}{7.09s + 1} \\[2mm] \dfrac{-33.68e^{-9.2s}}{8.15s + 1} & \dfrac{0.87(11.61s + 1)e^{-s}}{(3.89s + 1)(18.8s + 1)} \end{bmatrix} \begin{bmatrix} m_1 \\ m_2 \end{bmatrix}
$$

(21.75)

where the input variables m_1 and m_2 are, respectively, the overhead reflux rate and the reboiler steam pressure.

It is impossible to control all three output variables with only two input variables; however, the sidestream mole fraction of ethanol is deemed the least important of the output variables, leaving two output variables to be controlled with the two input variables. How should these input and output variables be paired?

Solution:

Upon deciding to leave the control of the sidestream composition out of the control scheme, we may now rewrite the model as:

$$
\begin{bmatrix} y_1 \\ y_3 \end{bmatrix} = \begin{bmatrix} \dfrac{0.66e^{-2.6s}}{6.7s + 1} & \dfrac{-0.0049e^{-s}}{9.06s + 1} \\[2mm] \dfrac{-33.68e^{-9.2s}}{8.15s + 1} & \dfrac{0.87(11.61s + 1)e^{-s}}{(3.89s + 1)(18.8s + 1)} \end{bmatrix} \begin{bmatrix} m_1 \\ m_2 \end{bmatrix}
$$

(21.76)

along with the additional relation

$$
y_2 = \frac{1.11e^{-6.5s}}{3.25s + 1} m_1 + \frac{-0.012e^{-1.2s}}{7.09s + 1} m_2
$$

(21.77)

The modified (square) subsystem's steady-state gain matrix is obtained from Eq. (21.76) as:

$$
\tilde{\mathbf{K}} = \begin{bmatrix} 0.66 & -0.0049 \\ -33.68 & 0.87 \end{bmatrix}
$$

(21.78)

and the RGA is:

$$
\Lambda = \begin{bmatrix} 1.4 & -0.4 \\ -0.4 & 1.4 \end{bmatrix}
$$

(21.79)

The recommended input/output pairing scheme is:

$y_1 - m_1$; overhead reflux to control overhead composition
$y_3 - m_2$; reboiler steam pressure to control Tray #19 temperature

Note that according to this control scheme, as indicated in Eq. (21.77), the sidestream composition will drift as the values of m_1, and m_2 change, but that is the nature of underdefined systems: we can only achieve arbitrarily good control of two out of the three output variables and accept the drift in the third one.

This example has shown that for underdefined systems, the strategy is to choose a square subsystem by dropping off the excess number of output variables on the basis of economic importance; the subsequent analysis is the same as for square systems.

21.7.3 Overdefined Systems

The real challenge in deciding on loop pairings for nonsquare systems is presented by overdefined systems. In this case, we have an excess of input variables, and therefore we can actually achieve arbitrary control of the fewer output variables *in more than one way.*

The situation here is as follows: since there are m input variables to n output variables (and $m > n$), there are many more input variables to choose from in pairing the inputs and the outputs, and therefore there will be several different square subsystems from which the pairing possibilities are to be determined; in actual fact, there are exactly $\binom{m}{n}$ possible square subsystems, where we recall that $\binom{m}{n} = m!/n!\,(m-n)!$.

The strategy is therefore:

1. First determine, from the given process model, all the $\binom{m}{n}$ possible square subsystems.
2. Obtain the RGA's for each of these square subsystems.
3. Examine these RGA's and pick the best subsystem on the basis of the overall *character* of its RGA (in terms of how close it is to the ideal RGA).
4. Having thus determined the best subsystem, use its RGA to determine which input variable within this subsystem to pair with which output variable.

This is best illustrated with an example.

Example 21.9   RGA AND LOOP PAIRING FOR AN OVERDEFINED NONSQUARE SYSTEM.

A certain multivariable system has two outputs y_1 and y_2 that can be controlled by any of three available inputs m_1, m_2, and m_3. Through pulse testing, the following transfer function model was obtained:

$$\begin{bmatrix} y_1 \\ y_2 \end{bmatrix} = \begin{bmatrix} \dfrac{0.5e^{-0.2s}}{3s+1} & \dfrac{0.07e^{-0.3s}}{2.5s+1} & \dfrac{0.04e^{-0.03s}}{2.8s+1} \\[3mm] \dfrac{0.004e^{-0.5s}}{1.5s+1} & \dfrac{-0.003e^{-0.2s}}{s+1} & \dfrac{-0.001e^{-0.4s}}{1.6s+1} \end{bmatrix} \begin{bmatrix} m_1 \\ m_2 \\ m_3 \end{bmatrix} \qquad (21.80)$$

Which loop pairing is expected to give the best control?

Solution:

This is a 2×3 system, and as such, we have a situation in which only two of the three candidate input variables will be used for control, while the third input variable will have to be set at a fixed value, and will therefore be redundant.

To determine which variable should be active and which should be redundant, we first obtain all the possible 2×2 subsystems; there are $\binom{3}{2} = 3$ such subsystems that we now list, complete with their corresponding transfer function matrices, all extracted, of course, from Eq. (21.80).

- **Subsystem 1** (Utilizing m_1 and m_2 for control):

$$
\begin{bmatrix} y_1 \\ y_2 \end{bmatrix} =
\begin{bmatrix}
\dfrac{0.5e^{-0.2s}}{3s + 1} & \dfrac{0.07e^{-0.3s}}{2.5s + 1} \\[2ex]
\dfrac{0.004e^{-0.5s}}{1.5s + 1} & \dfrac{-0.003e^{-0.2s}}{s + 1}
\end{bmatrix}
\begin{bmatrix} m_1 \\ m_2 \end{bmatrix}
\tag{21.81}
$$

- **Subsystem 2** (Utilizing m_1 and m_3 for control):

$$
\begin{bmatrix} y_1 \\ y_2 \end{bmatrix} =
\begin{bmatrix}
\dfrac{0.5e^{-0.2s}}{3s + 1} & \dfrac{0.04e^{-0.03s}}{2.8s + 1} \\[2ex]
\dfrac{0.004e^{-0.5s}}{1.5s + 1} & \dfrac{-0.001e^{-0.4s}}{1.6s + 1}
\end{bmatrix}
\begin{bmatrix} m_1 \\ m_3 \end{bmatrix}
\tag{21.82}
$$

- **Subsystem 3** (Utilizing m_2 and m_3 for control):

$$
\begin{bmatrix} y_1 \\ y_2 \end{bmatrix} =
\begin{bmatrix}
\dfrac{0.07e^{-0.3s}}{2.5s + 1} & \dfrac{0.04e^{-0.03s}}{2.8s + 1} \\[2ex]
\dfrac{-0.003e^{-0.2s}}{s + 1} & \dfrac{-0.001e^{-0.4s}}{1.6s + 1}
\end{bmatrix}
\begin{bmatrix} m_2 \\ m_3 \end{bmatrix}
\tag{21.83}
$$

Next we obtain the RGA's for each subsystem; the results are:

For Subsystem 1:

$$
\Lambda_1 =
\begin{bmatrix}
0.843 & 0.157 \\
0.157 & 0.843
\end{bmatrix}
\tag{21.84}
$$

For Subsystem 2:

$$
\Lambda_2 =
\begin{bmatrix}
0.758 & 0.242 \\
0.242 & 0.758
\end{bmatrix}
\tag{21.85}
$$

and for Subsystem 3:

$$
\Lambda_3 =
\begin{bmatrix}
-1.4 & 2.4 \\
2.4 & -1.4
\end{bmatrix}
\tag{21.86}
$$

Based on inspection of these three RGA's, it appears that Subsystem 1 will offer us the best possible control, because its RGA is the closest to the ideal situation; it is somewhat better than Subsystem 2, and far superior to Subsystem 3.

Having thus decided on Subsystem 1, the implication is that the input variables m_1 and m_2 are to be used for control and m_3 is to be redundant. The final task is to pair the variables of this subsystem; an easy exercise, since from Eq. (21.84) we see that the best possible loop pairing within this subsystem is: $y_1 - m_1$ and $y_2 - m_2$.

One must realize, of course, that there may be constraints on some of the manipulated variables or time delays in some of the dynamics or other issues that could be important in choosing the best subsystem or pairing. Thus as will be discussed in more detail below, one must consider these as well as the RGA in the final choice of control system design.

21.8 FINAL COMMENTS ON LOOP PAIRING AND THE RGA

21.8.1 Loop Pairing in the Absence of Process Models

When process models are not available, it is still possible to obtain RGA's from experimental process data. One may adopt either of the following two approaches in such situations:

1. *Experimentally determine the steady-state gain matrix* K, *by implementing step changes in the process input variables, one at a time, and observing the ultimate change in each of the output variables.* Let Δy_{1j} be the observed change in the value of output variable 1 in response to a change of Δm_j in the jth input variable m_j; then, by definition of the steady-state gain:

$$k_{1j} = \frac{\Delta y_{1j}}{\Delta m_j}$$

and, in general, the steady-state gain between the ith output variable and the jth input variable will be given by:

$$k_{ij} = \frac{\Delta y_{ij}}{\Delta m_j} \tag{21.87}$$

The elements of the K matrix are thus obtained. Of course, once this gain matrix is known, we can easily generate the RGA.

2. *It is also possible to determine each element of the RGA directly from experiment.* As we may recall, each RGA element λ_{ij} can be determined upon performing two experiments; the first determines the open-loop steady-state gain by measuring the response of y_i to input m_j, when all the other loops are opened; in the second experiment, all the other loops are closed — using PI controllers to ensure that there will be no steady-state offsets — and the response of y_i to input m_j is redetermined. By definition (see Eq. (21.8)), the ratio of these gains gives us the desired relative gain element.

It should be clear that, of these two experimental procedures, the second one is more time consuming, and involves too many upsets to the process; for these reasons it is not likely to be popular in practice. The first procedure is to be preferred.

21.8.2 Final Comments on the RGA

The following are summary comments about Bristol's relative gain array:

1. The RGA requires only steady-state process information; it is therefore easy to calculate, and easy to use.

2. The RGA provides information only about the *steady-state* interaction inherent within a process system; dynamic factors are not taken into consideration in Bristol's relative gain analysis. This remains the main criticism of the RGA. As a result of this apparent limitation of the RGA, several modifications, and alternatives, which take dynamic factors into consideration have been proposed (See Gagnepain and Seborg [6] and McAvoy [7]).

3. The RGA (or any of its newer dynamic variations) only suggests input/output pairings for which the interaction effects are minimized; it provides no guidance about other features which may influence the pairing. Thus this is the topic of the next subsection.

21.8.3 Other Factors Influencing the Choice of Loop Pairing

The RGA and the Niederlinski index provide guidance on the choice of loop pairing in order to minimize steady-state interactions and avoid structural instability. However, there are other factors that influence the choice of loop pairing. Some important ones are:

1. **Constraints on the input variable:** It may happen that the best RGA pairing results in a choice of input variable for y_i that is severely limited by some constraints (maximum feed concentration, maximum heater power, limited heat duty, etc.) so that it cannot carry out the control task assigned. In this case, another pairing might perform better even with increased steady-state interactions.

2. **The presence of time-delay, inverse-response, or other slow dynamics in the best RGA pairing:** Since the RGA is based only on steady-state considerations, it can easily happen that the best RGA pairing results in very sluggish closed-loop response because of long time delays, significant inverse response, or large time constants in the diagonal elements of the transfer function for the recommended pairing. In this situation, choosing a pairing with inferior RGA properties but without these sluggish diagonal dynamics could greatly improve the control system performance.

3. **Timescale Decoupling of Loop Dynamics:** There are often timescale issues that can influence the choice of loop pairing. For example, in a 2 × 2 system it may be that with a certain pairing, the RGA indicates serious loop interactions. However, if at the same time, one of the loops

responds very much faster than the other, then there can be *timescale decoupling* of the loops. This occurs when the fast loop responds so fast that the effect of the slow loop appears as a constant disturbance; conversely the slow loop does not respond at all to the high-frequency disturbances coming from the fast loop. This means that we can safely pair loops with large differences in closed-loop response times even when the RGA is not favorable.

Let us illustrate these ideas with some examples.

Example 21.10   LOOP PAIRING FOR THE STIRRED MIXING TANK WITH ADDITIONAL HEATER.

Suppose that the stirred mixing tank of Example 20.3 (Figure 20.1) has added an in-tank heater for temperature control with heater power Q, so that the model becomes:

$$A_c \frac{dh}{dt} = F_H + F_C + F_d - K\sqrt{h}$$

$$\rho C_p A_c \frac{d(hT)}{dt} = \rho C_p \left(F_H T_H + F_C T_C + F_d T_d + \frac{Q}{\rho C_p} - K\sqrt{h}\, T \right)$$

Then we have three inputs (F_C, F_H, Q) and only two outputs (h, T). By linearizing the equations as before we obtain a transfer function model slightly modified from Eq. (20.23):

$$\mathbf{G}(s) = \begin{bmatrix} \dfrac{1}{A_c(s + a_{11})} & \dfrac{1}{A_c(s + a_{11})} & 0 \\[3mm] \dfrac{(T_H - T_s)}{A_c h_s(s + a_{22})} & \dfrac{(T_c - T_s)}{A_c h_s(s + a_{22})} & \dfrac{1/\rho C_p}{A_c h s_s(s + a_{22})} \end{bmatrix}$$

where a_{11}, a_{22} are given by Eq. (20.24).

Now the three possible square subsystems of this overdetermined system yield relative gain arrays as follows:

1. For F_H, F_C as manipulate variables, the relative gain array is given in Eq. (21.65).

2. For F_H, Q as manipulated variables:

$$\Lambda = \begin{bmatrix} 1 & 0 \\ 0 & 1 \end{bmatrix}$$

3. F_C, Q as manipulated variables:

$$\Lambda = \begin{bmatrix} 1 & 0 \\ 0 & 1 \end{bmatrix}$$

If we want the steady-state tank temperature to be $T_s = \dfrac{T_H + T_C}{2}$, then for Subsystem 1, the RGA is:

$$\Lambda = \begin{bmatrix} 0.5 & 0.5 \\ 0.5 & 0.5 \end{bmatrix}$$

so that either Subsystem 2 or 3 using the in-tank heater would be preferred based on RGA analysis. However, suppose that the in-tank heater has limited power and can barely achieve the steady state, T_s, at maximum power. Thus it would not be a desirable choice for regulatory temperature control. In this case, Subsystem 1 with the poor RGA pairing would be better than using such a low-power heater.

Suppose that in order to overcome the heater power limitation, a much larger heater was installed in the tank, but because of its massiveness it has a very large time delay between the control signal and the actual power delivery. The poor RGA choice of Subsystem 1 could still be the best choice because of the sluggish closed-loop response with the heater.

Example 21.11   LOOP PAIRING WITH TIMESCALE DECOUPLING.

Consider the in-line blending problem of Example 21.5. Recall that when we wish a blend which contains 20% species A, the RGA is:

$$\Lambda = \begin{bmatrix} 0.2 & 0.8 \\ 0.8 & 0.2 \end{bmatrix}$$

Based on RGA analysis alone, pairing the total flowrate, y_1, to the flowrate of A, m_1, and the blend composition, y_2, to the flowrate of B, m_2, would not be a recommended pairing. However, if the flow measurement and valve dynamics were very rapid, then the closed-loop response of the $y_1 - m_1$ loop would be extremely fast (on the order of seconds). By contrast, if the composition analyser had a significant time constant (on the order of minutes), then the $y_2 - m_2$ loop would be 10 – 100 times slower than the $y_1 - m_1$ loop. In this case, timescale decoupling would mean the control loops could be considered virtually independent of each other, and no serious interactions would occur.

Example 21.12   ANOTHER FORM OF TIMESCALE DECOUPLING.

Consider the 2×2 system with the transfer function:

$$G(s) = \begin{bmatrix} \dfrac{2}{10s + 1} & \dfrac{2}{s + 1} \\ \dfrac{1}{s + 1} & \dfrac{-4}{10s + 1} \end{bmatrix} \qquad (21.88)$$

with RGA given by:

$$\Lambda = \begin{bmatrix} 0.8 & 0.2 \\ 0.2 & 0.8 \end{bmatrix}$$

so that the $y_1 - m_1/y_2 - m_2$ pairing shown in Eq. (21.88) would definitely be recommended. The closed-loop response for a unit set-point change in y_1 using this pairing and a diagonal PI controller ($K_{c_1} = 4$, $\tau_{I_1} = 0.5$; $K_{c_2} = -4$, $\tau_{I_2} = 0.3$) is shown in Figure 21.3. The performance is not too bad considering that the open-loop time constants on the diagonal are 10 minutes.

However, note that in spite of an adverse RGA parameter, $\lambda = 0.2$, the reverse $y_1 - m_2/y_2 - m_1$ pairing could have great advantages because with this configuration,

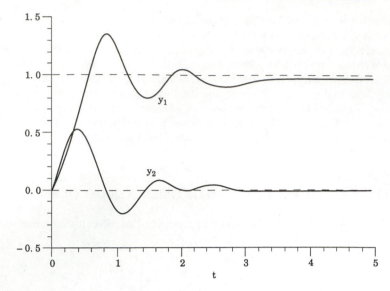

Figure 21.3.   Set-point response for RGA recommended pairing in Example 21.12.

the open-loop time constants on the diagonal are only 1 minute. In fact, with this reverse pairing, the closed-loop response for the same set-point change using a diagonal PI controller ($K_{c_1} = 10$, $\tau_{I_1} = 0.3$; $K_{c_2} = 20$, $\tau_{I_2} = 0.3$) is shown in Figure 21.4. The performance in this case is dramatically better than with the RGA recommended pairing. The reason is that the control loops are able to respond so rapidly that the interactions that appear more slowly are easily dealt with.

The lesson here is that the RGA provides only a guide to steady-state interactions and must be used together with all other engineering considerations in choosing the loop pairing.

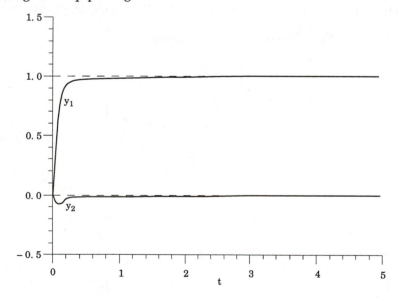

Figure 21.4.   Set-point response for adverse RGA pairing in Example 21.12.

21.9   CONTROLLER DESIGN PROCEDURE

21.9.1   Multiloop Controller Design

The design of multiple single-loop controllers for multivariable systems proceeds in two stages:

1.   Judicious choice of loop pairing
2.   Controller tuning for each individual loop

Since our primary objective in this chapter is really the design of effective multiple single-loop controllers for multivariable systems, the amount of time invested in discussing the issue of input/output pairing should give an indication of how important this aspect is in the overall scheme of multiloop control systems design.   The crucial point is that if the input and output variables making up the control loops are poorly paired, it will be exceedingly difficult or even impossible to design a successful multiloop control system.

Having covered the first half of this design task in earlier portions of this chapter, it now remains to discuss the issue of tuning the individual controllers. Before we do this, however, it is useful to indicate how the RGA analysis can be used to indicate the expected performance of a multiple single-loop control strategy.

It should not be surprising that when the RGA for the process is close to the ideal, (i.e., λ_{ii} very close to 1) that multiloop controllers are likely to function very well if carefully designed.   However, when the RGA indicates strong interactions for the chosen loop pairing and especially when the input and output variables are paired on values of λ_{ij} that are very large or negative, multiloop controllers are not likely to perform well even for the best possible tuning.

If, based on an analysis of the expected extent of loop interaction, the decision is made to proceed with a multiple single-loop control strategy, one must now consider how to tune the individual controllers.

21.9.2   Controller Tuning for Multiloop Systems

The main obstacle to proper controller tuning is presented by the interactions that exist between the control loops of a multiloop system.   This is what makes it risky to adopt the obvious strategy of tuning the controllers individually, in isolation from the others, with the hope that when all the loops are eventually closed, the overall system performance will still be adequate. (Recall from the stability consideration presented in Chapter 20 that the stability of individual control loops of a multiloop system when operating singly does not guarantee the stability of the complete system when all the loops are closed.)

Although several tuning procedures have been proposed (cf. Niederlinski [3] and McAvoy [7]) the procedure that is typically followed in practice is the following:

1. With the other loops on manual control, tune each control loop independently until satisfactory closed-loop performance is obtained.

2. Restore all the controllers to joint operation under automatic control and readjust the tuning parameters until the overall closed-loop performance is satisfactory in all the loops.

When the interactions between the control loops are not too significant, this procedure can be quite useful. It is clear, however, that for systems with significant interactions, the readjustment of the tuning in Step 2 can be difficult and tedious. It is possible to cut down on the amount of guesswork that goes into such a procedure by noting that in almost all cases, the controllers will need to be made more conservative (i.e., the controller gains reduced, and the integral times increased) when all the loops are closed, in comparison to when operating alone, with all the other loops open.

There are various ways of accomplishing this "detuning" (e.g., Ref. [7–10]). The following procedure is recommended by McAvoy [9] for 2×2 problems:

1. Use any of the single-loop tuning rules (Ziegler-Nichols, Cohen and Coon, etc.) to obtain starting values for the individual controllers; let the controller gains be $K_{ci}{}^*$.

2. These gains should be reduced using the following expressions that depend on the relative gain parameter λ:

$$K_{ci} = \begin{cases} \left(\lambda - \sqrt{\lambda^2 - \lambda}\right)K_{ci}{}^* & \lambda > 1.0 \\ \left|\lambda + \sqrt{\lambda^2 - \lambda}\right|K_{ci}{}^* & \lambda < 1.0 \end{cases} \tag{21.89}$$

It may still be necessary to "retune" these controllers after they have been put in operation; however, this will not require as much effort as if one were starting from scratch.

Let us illustrate multiloop controller tuning with the following example.

Example 21.13   MULTIPLE LOOP TUNING FOR THE WOOD AND BERRY DISTILLATION COLUMN.

We wish to design multiloop controllers for the Wood and Berry distillation column of Example 21.2. Recall that this 2×2 system has the transfer function:

$$\mathbf{G}(s) = \begin{bmatrix} \dfrac{12.8e^{-s}}{16.7s + 1} & \dfrac{-18.9e^{-3s}}{21.0s + 1} \\ \dfrac{6.6e^{-7s}}{10.9s + 1} & \dfrac{-19.4e^{-3s}}{14.4s + 1} \end{bmatrix}$$

How do we tune the multiloop controllers?

Solution:

From Example 21.2, the RGA was found to be:

$$\Lambda = \begin{bmatrix} 2.0 & -1.0 \\ -1.0 & 2.0 \end{bmatrix}$$

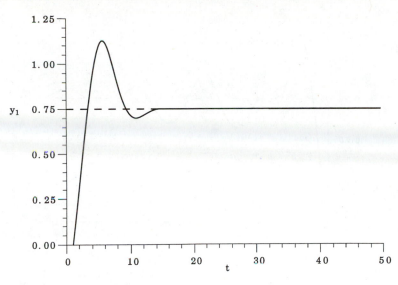

Figure 21.5. Closed-loop response of Loop 1 when Loop 2 is open.

so that $1 - 1/2 - 2$ loop pairing is indicated.

If we make use of the loop tuning procedure suggested above, we begin by tuning the individual loops with the other loops open, i.e., tune PI controllers for the two SISO systems:

$$y_1(s) = \frac{12.8\, e^{-s}}{16.7s + 1}\, u_1(s)$$

$$y_2(s) = \frac{-19.4\, e^{-3s}}{14.4s + 1}\, u_2(s)$$

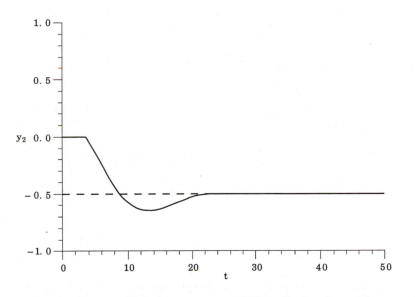

Figure 21.6. Closed-loop response of Loop 2 when Loop 1 is open.

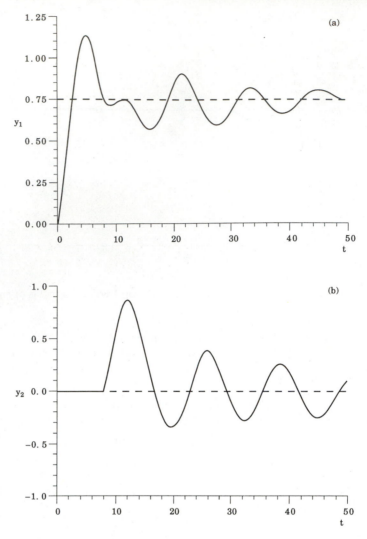

Figure 21.7.   Closed-loop response when both loops are closed simultaneously:
(a) y_1, (b) y_2.

Choosing a PI controller with parameters close to Ziegler-Nichols tuning leads to the
following parameters:

<u>Loop 1</u>   <u>Loop 2</u>

$K_{c_1} = 0.80$ $K_{c_2} = -0.15$

$\tau_{I_1} = 3.26$ $\tau_{I_2} = 9.35$

With the other loop open, these tuning parameters lead to the quite good individual
closed-loop responses to set-point changes shown in Figures 21.5 and 21.6.

If we close both loops and make the same set-point change in y_1 (no set-point
change is desired for y_2) using these tuning parameters, we see that the process is
nearly unstable due to loop interactions and oscillates continuously, as shown in

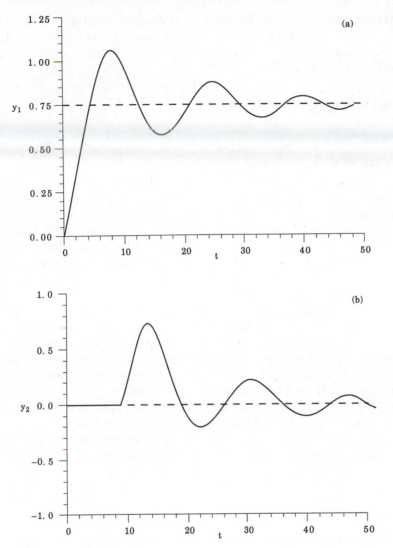

Figure 21.8.   Closed-loop response with both loops closed and "detuned" controllers: (a) y_1, (b) y_2.

Figure 21.7. Thus we must detune both loops. Using the detuning rule in Eq. (21.89), for $\lambda = 2$ we see that the original controller gains must be multiplied by a factor of 0.59 so that the new tuning is:

Loop 1	Loop 2
$K_{c_1} = 0.47$	$K_{c_2} = -0.088$
$\tau_{I_1} = 3.26$	$\tau_{I_2} = 9.35$

The closed-loop performance of the multiloop system with these new parameters is shown in Figure 21.8. Now the process is stable and has better closed-loop response with the new tuning parameters. The response could be further improved by continued adjustment on the tuning parameters.

Other techniques that may be used include optimization procedures that minimize the weighted sum of the integral-squared error for all the loops (recall the discussion of these in Chapter 15) and model-based tuning (to be discussed in Chapter 22).

21.10 SUMMARY

A first approach to the design of effective controllers for multivariable systems has been considered in this chapter through the use of rational loop pairing and multiple single-loop controllers. In adopting the multiple single-loop control strategy, we have shown that the first order of business is to determine the best way to pair the input and output variables, and whether or not such a control strategy is indeed viable. It turns out that the principal loop analysis tool for answering both questions is the Bristol RGA, the most popular means of quantifying process interactions and determining the input/output pairing for which the interaction effects are minimized. We have discussed the RGA in great detail, showing how it is calculated and how it is used for input/output pairing for various classes of systems.

The task of designing multiloop controllers for multivariable systems still remains an art, but some guidelines on to how to systematize this task have been given. These multiple single-loop controllers will work well only in situations in which control loop interactions are not very strong. For multivariable systems whose RGA's indicate very strong loop interactions, closed-loop stability will be attained only by sacrificing speed of response, for, as is well known, it will be necessary to reduce the controller gains in each loop significantly (i.e., "detune" the loops) under such circumstances.

When multiple single-loop controllers cannot be made to perform well, control system performance can be improved by considering other more advanced techniques; these will be discussed in the next chapter.

REFERENCES AND SUGGESTED FURTHER READING

1. Bristol, E. H., "On a New Measure of Interactions for Multivariable Process Control," *IEEE Trans. Auto. Cont.*, AC - 11, 133 (1966)
2. Ogunnaike, B. A., J. P. Lemaire, M. Morari, and W. H. Ray, "Advanced Multivariable Control of a Pilot Plant Distillation Column," *AIChE*, **29**, 632 (1983)
3. Niederlinski, A., "A Heuristic Approach to the Design of Linear Multivariable Interacting Control Systems," *Automatica*, **7**, 691 (1971)
4. Grosdidier, P., M. Morari, and B. R. Holt, "Closed Loop Properties from Steady-State Information," *I&EC Fund.*, **24**, 221 (1985)
5. Koppel, L. B., "Input-Output Pairing in Multivariable Control," *AIChE J.*, (1982)
6. Gagnepain, J.-P. D. and E. Seborg, "Analysis of Process Interactions with Applications to Multiloop Control System Design," *I&EC Process Des. Dev.*, **21**, 5 (1982)
7. McAvoy, T. J., *Interaction Analysis Theory and Application*, ISA, Research Triangle Park, NC (1983)
8. Tyreus, B., "Multivariable Control System Design for an Industrial Distillation Column," *I&EC Proc. Des. Dev.*, **18**, 177 (1979)
9. McAvoy, T. J., "Connection between Relative Gain and Control Loop Stability and Design," *AIChE J.*, **27**, 613 (1981)
10. Marino-Galarraga, M., T. J. McAvoy, and T. E. Marlin, "Short-cut Operability. 2. Estimation of Detuning Parameter for Classical Control Systems," *I&EC Res.*, **26**, 511 (1987)

REVIEW QUESTIONS

1. In using multiple single-loop feedback controllers to control a multivariable system, what are the two issues to be resolved?

2. What is the relative gain array (RGA); what is it used for?

3. What are some of the important properties of the RGA?

4. Why is a single element of the RGA sufficient to determine the entire array for a 2×2 system?

5. Qualitatively, what is the implication for loop interaction when the ijth element of a multivariable system's RGA:
   - Is equal to 1?
   - Is equal to 0?
   - Lies between 0 and 1?
   - Is greater than 1?
   - Is equal to 1?
   - Is negative?

6. What is the ideal RGA, and under what condition can it be obtained?

7. Why is the RGA for a process in which the loops experience *no* interactions identical to the RGA for a process in which the loops experience *one-way* interaction?

8. How is the RGA used for loop pairing?

9. What is the Niederlinski index, and how is it useful for loop pairing?

10. How is loop pairing carried out for nonlinear systems using the RGA?

11. What is peculiar about the RGA's for nonlinear systems?

12. What is a nonsquare system?

13. How is the loop pairing problem for *underdefined* systems fundamentally different from the loop pairing problem for *overdefined* systems?

14. What is the procedure for applying the RGA for pairing the input and output variables of an overdefined system?

15. How can the RGA be obtained and used for loop pairing in the absence of process models?

16. What is the main criticism of the RGA?

17. Even though the RGA is the most widely used loop pairing tool, what are some other important factors that should guide the final choice in this matter?

18. What is the procedure for designing multiple single-loop controllers for multivariable systems?

19. After designing each individual single-loop controller in a multiloop control system *independently*, why is it necessary to detune them when all the loops operate *simultaneously*?

PROBLEMS

21.1 An approximate model identified for an experimental, laboratory scale continuous stirred-tank polymerization reactor is:

$$\begin{bmatrix} y_1 \\ y_2 \end{bmatrix} = \begin{bmatrix} \dfrac{0.81}{2.6s + 1} & 0 \\ \dfrac{0.55}{4.3s + 1} & \dfrac{-0.013}{3.5s + 1} \end{bmatrix} \begin{bmatrix} m_1 \\ m_2 \end{bmatrix}$$ (P21.1)

where, in terms of deviations from their respective steady states, the input and output variables are:

$$
\begin{aligned}
y_1 &= \text{Polymer production rate (gm/min)} \\
y_2 &= \text{Weight average molecular weight} \\
m_1 &= \text{Monomer flowrate (gm/min)} \\
m_2 &= \text{Chain transfer agent flowrate (gm/min)}
\end{aligned}
$$

Not included in the model is the effect of the catalyst flowrate; for the experimental system this was permanently set at a fixed value.

(a) First study the given transfer function matrix; does it show any evidence of interactions between the process variables? Which output variable is expected to be more susceptible to the effect of interactions and why?

(b) Calculate the RGA for this process. What does this suggest in terms of input/output pairing and the loop interactions resulting from such a pairing scheme? Interpret your results in light of your initial assessment in part (a).

21.2 (a) Assuming a y_1–m_1 / y_2–m_2 pairing, design two independent single-loop PI controllers for the polymer reactor of Problem 21.1 using Cohen-Coon settings. What are the recommended controller parameters?

(b) Using your favorite simulation package, obtain the response of the overall closed-loop system to a set-point change of 100 gm/min in the polymer production rate (with no change desired in the weight average molecular weight), using the controllers you designed in part (a). Plot the responses of all the process input and output variables. Evaluate the performance of these controllers.

21.3 A pilot scale distillation column used for separating a binary mixture of ethanol and water is modeled by the following transfer function matrix model:

$$\mathbf{y}(s) = \mathbf{G}(s)\mathbf{m}(s) + \mathbf{G}_d(s)\mathbf{d}(s)$$ (P21.2)

with the matrices $\mathbf{G}(s)$ and $\mathbf{G}_d(s)$ given by:

$$\mathbf{G}(s) = \begin{bmatrix} \dfrac{0.7e^{-2.6s}}{6.7s + 1} & \dfrac{-0.005e^{-s}}{9.1s + 1} \\ \dfrac{-34.7e^{-9.2s}}{8.2s + 1} & \dfrac{0.9(11.6s + 1)e^{-s}}{(3.9s + 1)(18.8s + 1)} \end{bmatrix}$$ (P21.3a)

$$\mathbf{G}_d(s) = \begin{bmatrix} \dfrac{0.14e^{-12s}}{6.2s + 1} & \dfrac{-0.0011(26.3s + 1)e^{-2.7s}}{(7.9s + 1)(14.6s + 1)} \\ \dfrac{-11.5e^{-0.6s}}{7.0s + 1} & \dfrac{0.32e^{-2.6s}}{7.8s + 1} \end{bmatrix}$$ (P21.3b)

The process variables are (in terms of deviations from their respective steady-state values):

$$y_1 = \text{Overhead mole fraction ethanol}$$
$$y_2 = \text{Temperature on Tray \#19}$$
$$m_1 = \text{Overhead reflux flowrate}$$
$$m_2 = \text{Reboiler steam pressure}$$
$$d_1 = \text{Feed flowrate}$$
$$d_2 = \text{Feed temperature}$$

It is desired to design two single-loop feedback controllers for this process with Controller 1 to regulate y_1, and Controller 2 to regulate y_2. Which input variable (m_1, m_2) will be more advantageous to pair with which output variable (y_1, y_2)? Support your decision adequately. Comment on how well you think the *best* set of single-loop controllers designed for this distillation column will perform.

21.4 The "low-boiler column" in a distillation train used to separate the products of an industrial chemical reactor has the following experimentally determined approximate transfer function model :

$$\mathbf{y}(s) = \mathbf{G}(s)\, \mathbf{m}(s) + \mathbf{G}_d(s)\, \mathbf{d}(s)$$

with the matrices $\mathbf{G}(s)$ and $\mathbf{G}_d(s)$ given by:

$$\mathbf{G}(s) = \begin{bmatrix} \dfrac{-0.12}{0.004s+1} & \dfrac{0.011}{0.055s+1} & \dfrac{-0.00175e^{-0.1s}}{0.00525s+1} \\[2mm] \dfrac{-0.013}{0.00325s+1} & \dfrac{0.075}{0.015s+1} & \dfrac{-0.00012e^{-0.25s}}{0.0141s+1} \\[2mm] \dfrac{-0.0043}{s} & \dfrac{0.04}{s} & \dfrac{0.000086}{s} \end{bmatrix} \quad \text{(P21.4a)}$$

$$\mathbf{G}_d(s) = \begin{bmatrix} \dfrac{-0.0041}{0.0041s+1} & \dfrac{0.185}{0.37s+1} \\[2mm] \dfrac{-0.001}{0.000225s+1} & \dfrac{0.16}{0.4s+1} \\[2mm] 0 & 0 \end{bmatrix} \quad \text{(P21.4b)}$$

The distillation column variables are (in terms of deviations from their respective steady-state values):

$$y_1 = T_{36}, \text{ temperature on Tray \#36 (°C)}$$
$$y_2 = \text{Bottoms temperature (°C)}$$
$$y_3 = \text{Reflux receiver level (ft)}$$
$$m_1 = \text{Reflux flowrate (lb/hr)}$$
$$m_2 = \text{Reboiler heat duty (NHDU}^\dagger)$$
$$m_3 = \text{Condenser cooling water flowrate (lb/hr)}$$
$$d_1 = \text{Feed flowrate}$$
$$d_2 = \text{Feed temperature}$$

---

† Normalized Heat Duty Units.

(a) Assuming that it has been predetermined that the bottoms temperature is to be controlled with the reboiler heat duty, use the RGA to determine how the remaining variables should be paired.

(b) Using the $y_1–m_3$ / $y_2–m_2$ / $y_3–m_1$ pairing, and using the following controller parameters:

Loop 1:	(T_{36} loop);	$K_c = -80.0$;	$1/\tau_I = 3.5$
Loop 2:	(Bottoms temperature loop);	$K_c = 2.0$;	$1/\tau_I = 6.0$
Loop 3:	(Reflux level loop);	$K_c = -500.0$;	$1/\tau_I = 2.0$

obtain a simulation of the overall column response to a 600 lb/hr step increase in the feed flowrate.

Now, using a "new configuration" obtained from the alternate pairing: $y_1 – m_1$ / $y_2 – m_2$ / $y_3 – m_3$, along with the *new* controller parameters:

Loop 1:	(T_{36} loop);	$K_c = -500.0$;	$1/\tau_I = 1.0$
Loop 2:	(Bottoms temperature loop);	$K_c = 4.5$;	$1/\tau_I = 2.0$
Loop 3:	(Reflux level loop);	$K_c = 12{,}000$;	$1/\tau_I = 0.8$

obtain the column response to the *same* disturbance (600 lb/hr feed rate step change). Compare, on corresponding plots, the responses under both the old and the new configurations.

(c) Given now that for economic and environmental reasons the temperature T_{36} is the most critical of all the low boiler column variables, and that very tight control is therefore required for this variable, which of the two configurations investigated in part (b) is preferable? Discuss your answer in the light of the RGA results obtained in part (a).

Given that Tray #36 is close to the top of the column, and that the reflux receiver is a 3-ft diameter cylindrical drum into which empties *all* the material from the (essentially total) condenser, from a consideration of the physics of the problem, which would you expect to be the more effective manipulated variable to use in controlling this temperature: the reflux flowrate or the condenser cooling temperature? Support your answer adequately. Offer an explanation for the results given by the RGA.

(d) Repeat part (b) for a step input disturbance of 5°C in the feed temperature. Once again, comment on which configuration provides tighter control on the critical T_{36} variable.

21.5 A certain multivariable system has three outputs $y_1, y_2,$ and y_3 which can be controlled by any of four available inputs $m_1, m_2, m_3,$ and m_4. Through pulse testing, the following 3×4 transfer function matrix model was obtained:

$$\begin{bmatrix} y_1 \\ y_2 \\ y_3 \end{bmatrix} = \begin{bmatrix} \dfrac{0.5e^{-0.2s}}{3s+1} & \dfrac{0.07e^{-0.3s}}{2.5s+1} & \dfrac{0.04e^{-0.3s}}{2.5s+1} & \dfrac{0.01e^{-0.5s}}{10s+1} \\[2ex] 0 & 0 & 0 & \dfrac{0.5e^{-0.1s}}{2.1s+1} \\[2ex] \dfrac{0.004e^{-0.5s}}{1.5s+1} & \dfrac{-0.003e^{-0.2s}}{s+1} & \dfrac{-0.006e^{-0.4s}}{1.6s+1} & 0 \end{bmatrix} \begin{bmatrix} m_1 \\ m_2 \\ m_3 \\ m_4 \end{bmatrix}$$

Which input/output pairing configuration is expected to give the best results for a multiple single-loop control strategy? [**Hint :** *One input/output pair is obvious and can be detected by inspection; eliminate this to reduce the problem to a 2 × 3 loop pairing problem.*]

21.6 The following is a transfer function model representing the dynamic behavior of certain portions of a hot oil fractionator:

$$\begin{bmatrix} y_1 \\ y_2 \end{bmatrix} = \begin{bmatrix} \dfrac{4.05e^{-27s}}{50s+1} & \dfrac{1.77e^{-28s}}{60s+1} & \dfrac{5.88e^{-27s}}{50s+1} & \dfrac{1.44e^{-27s}}{40s+1} \\[3mm] \dfrac{4.38e^{-20s}}{33s+1} & \dfrac{4.42e^{-22s}}{44s+1} & \dfrac{7.2}{19s+1} & \dfrac{1.26}{32s+1} \end{bmatrix} \begin{bmatrix} m_1 \\ m_2 \\ m_3 \\ m_4 \end{bmatrix} \qquad \text{(P21.5)}$$

where, in terms of deviation variables:

y_1 = Top end point; $\qquad$ y_2 = Bottoms reflux temperature;
m_1 = Top draw rate; $\qquad$ m_2 = Side draw rate;
m_3 = Bottoms reflux duty; $\qquad$ m_4 = Upper reflux duty.

Which of these four inputs should be paired with the two outputs?

21.7 (a) The following incomplete RGA was determined *experimentally* (after four very long and tedious experiments) for a 3 × 3 system with outputs y_1, y_2, y_3, to be paired with the inputs m_1, m_2, m_3. Is there enough information to determine an input/output pairing? If so what input/output pairing is recommended?

$$\begin{bmatrix} ? & ? & -0.6 \\ ? & 0.8 & ? \\ -0.2 & -0.2 & ? \end{bmatrix}$$

(b) The same strategy as that used in part (a) was used to generate the following incomplete RGA, this time for a 4 × 4 system with outputs y_1, y_2, y_3, y_4 to be paired with the inputs m_1, m_2, m_3 and m_4. What input/output pairing is recommended?

$$\begin{bmatrix} ? & 0.15 & ? & -0.09 \\ -0.01 & ? & 0.29 & 1.15 \\ ? & 3.31 & 0.27 & ? \\ 0.22 & ? & ? & 1.84 \end{bmatrix}$$

21.8 The following 4 × 5 transfer function matrix model was developed by Congalidis, Richards, and Ray [*Proc. ACC*, 1986, pp. 1779–1793] for a CSTR used for solution copolymerization of methyl-methacrylate (MMA) and vinyl acetate (VA):

$$\mathbf{y}(s) = \mathbf{G}(s)\,\mathbf{m}(s)$$

where, in terms of deviations from their respective steady states, the four output variables are:

y_1 = polymer production rate (kg/hr)
y_2 = mole fraction of monomer A (MMA) in copolymer
y_3 = weight average molecular weight
y_4 = reactor temperature (°K)

the five input variables are:

m_1 = flowrate of monomer B (VA) (kg/hr)
m_2 = flowrate of monomer A (MMA) (kg/hr)
m_3 = flowrate of initiator (kg/hr)
m_4 = flowrate of chain transfer agent (kg/hr)
m_5 = reactor jacket temperature (°K)

and the 4×5 transfer function matrix itself is given by:

$$G(s) = \begin{bmatrix} \dfrac{0.34}{0.85s+1} & \dfrac{0.21}{0.42s+1} & \dfrac{0.50(0.50s+1)}{0.12s^2+0.40s+1} & 0 & \dfrac{6.46(0.90s+1)}{0.07s^2+0.30s+1} \\[3mm] \dfrac{-0.41}{2.41s+1} & \dfrac{0.66}{1.51s+1} & \dfrac{-0.30}{1.45s+1} & 0 & \dfrac{-3.72}{0.80s+1} \\[3mm] \dfrac{0.30}{2.54s+1} & \dfrac{0.49}{1.54s+1} & \dfrac{-0.71}{1.35s+1} & \dfrac{-0.20}{2.71s+1} & \dfrac{-4.71}{0.08s^2+0.41s+1} \\[3mm] 0 & 0 & 0 & 0 & \dfrac{1.03(0.23s+1)}{0.07s^2+0.31s+1} \end{bmatrix}$$

The input and output variables are now to be paired to give the minimum possible (steady-state) interaction. How should this be done? [**Hint**: *Some pairings are obvious upon careful inspection of the transfer function matrix; eliminate these and reduce the problem down before carrying out further analysis.*]

21.9 The steady-state gain matrices for three multivariable (2×2) systems are given as follows:

$$\text{SYSTEM A: } \mathbf{K} = \begin{bmatrix} -10.00 & 0.04 \\ -10.00 & -0.04 \end{bmatrix}$$

$$\text{SYSTEM B: } \mathbf{K} = \begin{bmatrix} -1.0 & 0.4 \\ -1.0 & -0.4 \end{bmatrix}$$

$$\text{SYSTEM C: } \mathbf{K} = \begin{bmatrix} 2.30 & 1.10 \\ 1.14 & 0.55 \end{bmatrix}$$

(a) What do the RGA's for these systems suggest regarding how well we can expect single-loop controllers to work in each case?
(b) Examine the steady-state gain matrix for system C carefully; what clues does it contain about inherent process characteristics which are reflected in the RGA?
(c) Of the two systems A and B, which would you expect to be easier to control? Support your answer adequately.

21.10 The drug *Hydrochlorothiazide* (trade name "Diazide") is sometimes used to regulate blood pressure in elderly patients. Because it is also a diuretic, its main side effect is a substantial increase in urine production, an unpleasant problem for the elderly who are quite often already prone to incontinence. To counter this side effect, the antispasmodic drug *Oxybutynin* (trade name "Ditropan") is often prescribed as well. However, quite apart from the cost of the medication, *Oxybutynin* is known to increase ocular pressure and can therefore be quite dangerous for those patients with glaucoma. As an alternative, a less expensive experimental drug "OGZ" purported to have much milder side effects is being tested under strictly controlled conditions for use in conjunction with *Hydrochlorothiazide*. Preliminary results show that although the experimental "OGZ" can counteract the increased production of urine, in what appears reminiscent of the type of interactions experienced by multivariable chemical processes, it also causes a slight *increase* in blood pressure, perhaps as a result of some yet unidentified chemical interaction with the active ingredient in *Hydrochlorothiazide*.

Tests carried out on a certain patient indicates that the effect of regular doses of *Hydrochlorothiazide* and the drug "OGZ" may be represented by the following approximate transfer function model:

$$\begin{bmatrix} y_1 \\ y_2 \end{bmatrix} = \begin{bmatrix} \dfrac{-0.04e^{-0.1s}}{0.11s + 1} & \dfrac{0.0005e^{-0.15s}}{0.21s + 1} \\ \dfrac{0.22}{0.12s + 1} & \dfrac{-0.02}{0.21s + 1} \end{bmatrix} \begin{bmatrix} u_1 \\ u_2 \end{bmatrix} \qquad (P21.6)$$

where, in terms of deviations from their respective steady states, the four variables are:

y_1 = normalized (dimensionless) blood pressure
y_2 = normalized urine production rate
u_1 = rate of *Hydrochlorothiazide* ingestion (mg/day)
u_2 = rate of "OGZ" ingestion (mg/day)

Note that the unit of time is in days.
(a) What insight can the process control approach to interaction analysis offer in quantifying the degree of seriousness of the "OGZ" interactions with *Hydrochlorothiazide*? Interpret and discuss your results carefully.
(b) A particular test involves administering to the patient a one-time dose of 25 mg/day of *Hydrochlorothiazide*, monitoring the patients normalized blood pressure, inferring urine production rate from some other associated measurements, and administering "OGZ" continuously intravenously, on the basis of the observed urine production rate. Consider the situation in which the "OGZ" flowrate is determined by a proportional-only controller (with $K_c = 0.5$); use your favorite simulation program and the model in Eq. (P21.6) to determine all the responses gathered from this test over the time period of 1 day (24 hr). Plot these responses. You may assume, for simplicity, that the provided model is adequate and that:

1. This particular test program is equivalent to a sustained *step* change $u_1 = 25$ over the entire one-day period.
2. The set-point for the urine production rate is zero (in deviations from normal conditions).
3. Whatever flowrate is called for by the "OGZ" flow controller can be achieved with arbitrarily small resolution.

To what final steady-state values do the normalized blood pressure and the urine production settle? What is the maximum value attained by the urine production rate?
(c) The *Hydrochlorothiazide* ingestion program in part (b) may be more realistically represented as $u_1 = 25\delta(t)$ where $\delta(t)$ is the Dirac delta function. Repeat part (b) with this description of the input function u_1. Compare the two sets of responses obtained in each case.

22

DESIGN OF MULTIVARIABLE CONTROLLERS

In Chapter 21 we discussed the design of controllers for multivariable systems with interactions between the inputs and outputs; however, the strategy employed was to design individual control loops that act independently of each other. By contrast in this chapter, we will discuss the design of true multivariable controllers that utilize *all* the available process output information *jointly* to make decisions on what the complete input vector **u** should be. Thus, each control command issued by the true multivariable controller has been based not just on the status of *one* assigned output variable, but on *all* the output variables. With such controllers, it is possible, in principle, to eliminate completely the effect of interactions between the process variables.

The objective in this chapter is to present some of the principles and techniques for designing multivariable controllers, motivated by the fact that the design of effective multivariable controllers remains one of the more challenging problems faced by industrial process control practitioners. We will begin with the most widely used multivariable control technique: *loop decoupling*. Then the concept of *singular value decomposition* (SVD) will be presented as a means of detecting when it is structurally impossible to apply decoupling to a system, and how it can be useful as a multivariable controller design technique in its own right. Our discussion will end with a summary of some of the other available multivariable controller design techniques.

22.1 INTRODUCTION

In this chapter we will assume that the simple approach of Chapter 21 involving good loop pairing and appropriate tuning of multiple single-loop controllers has not produced a control system that meets the requirements of the process. Thus it is necessary to design a true multivariable controller that acts on data from all the process outputs and makes adjustments to all the process inputs in response to a disturbance or set-point change. To make this more explicit, let us adopt the notation introduced in Chapter 20 and assume that after input/output pairing has been carried out, the input variables become u_i, and that the system has been configured with the $y_1 - u_1, y_2 - u_2, \ldots, y_n - u_n$ pairings.

Under the multiple, independent, single-loop control strategy, each controller g_{c_i} operates according to:

$$u_i = g_{c_i}\left(y_{di} - y_i\right)$$

or

$$u_i = g_{c_i} \varepsilon_i \tag{22.1}$$

On the other hand, a true multivariable controller must decide on u_i, not using only ε_i, but using the entire set, $\varepsilon_1, \varepsilon_2, \ldots, \varepsilon_n$. Thus, the controller actions are obtained from:

$$\left.\begin{array}{rcl}
u_1 & = & f_1(\varepsilon_1, \varepsilon_2, \ldots, \varepsilon_n) \\
u_2 & = & f_2(\varepsilon_1, \varepsilon_2, \ldots, \varepsilon_n) \\
u_3 & = & f_3(\varepsilon_1, \varepsilon_2, \ldots, \varepsilon_n) \\
\ldots & = & \ldots \\
u_n & = & f_n(\varepsilon_1, \varepsilon_2, \ldots, \varepsilon_n)
\end{array}\right\} \tag{22.2}$$

The design problem is therefore that of determining $f_1(.), f_2(.), \ldots, f_n(.)$ so that each of the output errors is driven to zero in an acceptable fashion; this often reduces to designing the controller such that the interaction effects are eliminated, or at least significantly reduced. In a very general sense, the principle behind how the functions indicated in Eq. (22.2) are determined is what distinguishes one multivariable control technique from another. A few representative examples will serve to illustrate this point.

• Decoupling

In *decoupling*, as shown in Figures 22.1 and as discussed in more detail below, additional transfer function blocks are introduced between the single-loop controllers and the process, functioning as links between the otherwise independent controllers. The actual control action experienced by the process will therefore now contain information from all the other controllers; for example, for a 2 × 2 system, whose individual controller outputs are $g_{c_1} \varepsilon_1$ and $g_{c_2} \varepsilon_2$, if the "decoupling blocks" for each loop have transfer functions g_{I_1} and g_{I_2} respectively, then the control equations are given by:

$$u_1 = g_{c_1} \varepsilon_1 + g_{I_1}\left(g_{c_2}\varepsilon_2\right) \tag{22.3}$$

and

$$u_2 = g_{c_2} \varepsilon_2 + g_{I_2}\left(g_{c_1}\varepsilon_1\right) \tag{22.4}$$

which are of the form given in Eq. (22.2) with the specific functional forms taken in this case by $f_1(\varepsilon_1,\varepsilon_2)$, and $f_2(\varepsilon_1,\varepsilon_2)$ shown explicitly.

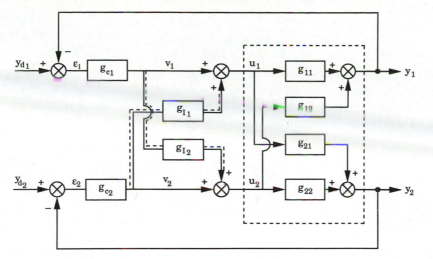

Figure 22.1.   Multivariable 2×2 control system incorporating simplified decoupling.

• Change of Variables

By using transformations to create combinations of the original input and/or output variables of a process experiencing strong interactions, it is possible to obtain an equivalent system whose variables no longer interact; for example, consider a 2×2 system whose transfer function model is given as:

$$y_1 = a(s)\, u_1 + a(s)\, u_2 \tag{22.5}$$

and

$$y_2 = b(s)\, u_1 - b(s)\, u_2 \tag{22.6}$$

(Under certain conditions, the linearized transfer function model for the stirred mixing tank of Examples 20.3 and 21.6 is actually of this form.)

Observe now that according to Eqs. (22.5) and (22.6), the process variables of this system interact with one another; but by defining two new variables μ_1, and μ_2 according to:

$$\mu_1 = u_1 + u_2 \tag{22.7}$$

and

$$\mu_2 = u_1 - u_2 \tag{22.8}$$

the modeling equations become

$$y_1 = a(s)\, \mu_1 \tag{22.9}$$

and

$$y_2 = b(s)\, \mu_2 \tag{22.10}$$

and μ_1 does not interact with y_2; neither does μ_2 interact with y_1. The implication now is that independent controllers can be designed for the transformed system represented in Eqs. (22.9) and (22.10); the control equations will now be:

$$\mu_1 = g_{c_1}\, \varepsilon_1 \tag{22.11}$$

and

$$\mu_2 = g_{c_2} \varepsilon_2 \tag{22.12}$$

where, of course, $\varepsilon_i = (y_{d_i} - y_i)$. Note however, that the new variables μ_1 and μ_2 are not the real process variables; we must now recover u_1 and u_2, the actual process input variables, from Eqs. (22.7) and (22.8), in terms of μ_1, and μ_2. The result of the reverse transformation is:

$$u_1 = \frac{\mu_1 + \mu_2}{2} \tag{22.13a}$$

and

$$u_2 = \frac{\mu_1 - \mu_2}{2} \tag{22.13b}$$

and upon introducing Eqs. (22.11) and (22.12) into these equations, we obtain the actual control equations which, on the basis of the indicated variable transformation, will eliminate interactions from this process:

$$u_1 = \frac{g_{c_1} \varepsilon_1 + g_{c_2} \varepsilon_2}{2} \tag{22.14}$$

and

$$u_2 = \frac{g_{c_1} \varepsilon_1 - g_{c_2} \varepsilon_2}{2} \tag{22.15}$$

again, of the form in Eq. (22.2).

Finding the appropriate combination of process variables that will lead to total elimination of, or, at least a reduction in, the control loop interactions, tends in general to be an "art" (see Refs. [1, 2]); nevertheless, as we will see later, the concept of singular value decomposition can provide guidelines in this regard.

Having thus illustrated the general principles and concepts of a multivariable controller, we now proceed to discuss some of the multivariable controller design techniques in greater detail. In each case we will focus attention on three things:

1. *Principles* behind the technique
2. *Results*
3. *Applications*, including practical considerations for effective utilization

22.2   DECOUPLING

22.2.1 Principles of Decoupling Using Interaction Compensators

From our introduction to multivariable systems in Chapter 20, and our subsequent discussions in Chapter 21, we know that all the input variables of a typical multivariable process are *coupled* to all its output variables; the main $y_1 - u_1, y_2 - u_2, \dots, y_n - u_n$ couplings are desirable (this is what makes it possible to control the process variable in the first instance); it is the $y_i - u_j$ cross-couplings, by which y_i is influenced by u_j (for all i and all j, with $i \neq j$), that are undesirable: they are responsible for the control loop interactions.

It is therefore clear that any technique that eliminates the *effect* of the undesired cross-couplings will improve control system performance. Note that the purpose is *not* to "eliminate" the cross-couplings; that is an impossibility, since to do this will require altering the physical nature of the system. Observe, for example, that it is not possible to stop the hot stream flowrate from affecting the tank temperature in the stirred mixing tank system of Figure 20.1, despite the fact that our main desire is to use this flowrate to control the level; neither can we prevent the cold stream from affecting the level, even though controlling the tank temperature is its main responsibility. We can, however, *compensate* for the noted cross-coupling effects by carefully balancing the hot and cold stream flowrates.

The main objective in decoupling is to *compensate* for the effect of interactions brought about by cross-couplings of the process variables. As depicted in Figure 22.2, this is to be achieved by introducing an additional transfer function "block" (the interaction compensator) between the single-loop controllers, and the process. This interaction compensator, together with the single-loop controllers, now constitute the multivariable decoupling controller.

In the ideal case, the decoupler causes the control loops to act as if totally independent of one another, thereby reducing the controller tuning task to that of tuning several, noninteracting controllers. The main advantage here is that it will now be possible to use SISO controller design techniques.

The decoupler design problem is that of choosing the elements of the compensator, G_I, shown in Figure 22.2, to satisfy usually one of the following objectives:

1. *Dynamic Decoupling*: eliminate interactions from *all* loops, at every instant of time.

2. *Steady-State Decoupling*: eliminate only steady-state interactions from all loops; because dynamic interactions are tolerated, this objective is less ambitious than dynamic decoupling but leads to much simpler decoupler designs.

3. *Partial Decoupling*: eliminate interactions (dynamic, or only steady state) in a subset of the control loops; this objective focuses attention on only the critical loops in which the interactions are strongest, leaving those with weak interactions to act without the aid of decoupling.

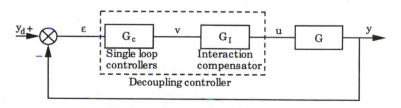

Figure 22.2.   A multivariable decoupling control system incorporating an interaction compensator.

22.2.2 Design of Ideal Decouplers

Even though there are several ways by which decouplers can be designed, the basic principles are all the same. We shall illustrate, first, the principles of *simplified decoupling*, using a 2×2 example and the block diagram shown in Figure 22.1.

Simplified Decoupling

Let us start by observing the following important aspects of the block diagram in Figure 22.1:

1. There are *two* compensator blocks g_{I_1} and g_{I_2}, one for each loop.

2. We have adopted a new notation: the controller outputs are now designated as v_1 and v_2, while the actual control action implemented on the process remains as u_1 and u_2. This distinction becomes necessary because the output of the controllers and the control action to be implemented on the process no longer have to be the same.

3. Without the compensator, $u_1 = v_1$ and $u_2 = v_2$, and the process model remains as:

$$\left. \begin{array}{l} y_1 = g_{11}u_1 + g_{12}u_2 \\ y_2 = g_{21}u_1 + g_{22}u_2 \end{array} \right\} \tag{22.16}$$

so that the interactions persist, since u_2 is still cross-coupled with and therefore affects y_1 through the g_{12} element, and u_1 affects y_2 by cross-coupling through g_{21}.

4. With the interaction compensator, Loop 2 is "informed" of changes in v_1 through g_{I_2}, so that u_2 — what the process actually feels — is adjusted accordingly. The same task is performed for Loop 1 by g_{I_1}, which adjusts u_1 with v_2 information.

The design question is now posed as follows:

What should g_{I_1} and g_{I_2} be if the effects of loop interactions are to be completely neutralized?

We will now answer this question.

First, let us consider Loop 1, in Figure 22.1 where the process model is:

$$y_1 = g_{11}u_1 + g_{12}u_2 \tag{22.17a}$$

$$y_2 = g_{21}u_1 + g_{22}u_2 \tag{22.17b}$$

However, because of the compensators, the equations governing the control action are:

$$u_1 = v_1 + g_{I_1}v_2 \atop u_2 = v_2 + g_{I_2}v_1 \Bigg\} \tag{22.18}$$

Introducing these equations into Eq. (22.17) now gives:

$$y_1 = \left(g_{11} + g_{12}g_{I_2}\right)v_1 + \left(g_{11}g_{I_1} + g_{12}\right)v_2 \tag{22.19a}$$

$$y_2 = \left(g_{21} + g_{22}g_{I_2}\right)v_1 + \left(g_{22} + g_{21}g_{I_1}\right)v_2 \tag{22.19b}$$

Now in order to have only v_1 affect y_1 and to eliminate the effect of v_2 on y_1, it is now enough to choose g_{I_1} such that the coefficient of v_2 in Eq. (22.19a) vanishes, i.e:

$$g_{11}g_{I_1} + g_{12} = 0$$

which requires:

$$g_{I_1} = -\frac{g_{12}}{g_{11}} \tag{22.20}$$

A similar procedure carried out for Loop 2 leads to the condition that to make y_2 in Eq. (22.19b) entirely free of v_1 requires:

$$g_{I_2} = -\frac{g_{21}}{g_{22}} \tag{22.21}$$

These are the transfer functions of the decouplers needed to compensate exactly for the effect of loop interactions in the example 2×2 system of Figure 22.1.

If Eqs. (22.20) and (22.21) are now introduced into Eq. (22.19), we see that the overall system equations become:

$$y_1 = \left(g_{11} - \frac{g_{12}g_{21}}{g_{22}}\right)v_1 \tag{22.22a}$$

and

$$y_2 = \left(g_{22} - \frac{g_{12}g_{21}}{g_{11}}\right)v_2 \tag{22.22b}$$

and the system is completely decoupled, with v_1 affecting only y_1, and v_2 affecting only y_2. The equivalent block diagram is shown in Figure 22.3 where the loops now appear to act independently and can therefore be tuned individually.

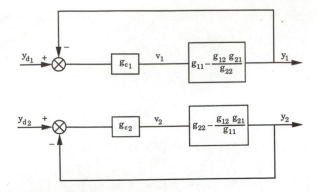

Figure 22.3.   Equivalent block diagram for system in Figure 22.2 with the compensators implemented as in Eqs. (22.20) and (22.21).

It is also interesting to consider what Eqs. (22.22a,b) indicate for the overall closed-loop system behavior *at steady state*. If the steady-state gain for each transfer function element g_{ij} is K_{ij}, observe that at steady state, Eqs. (22.22a,b) become:

$$y_1 = K_{11}\left(1 - \frac{K_{12}K_{21}}{K_{11}K_{22}}\right)v_1 \qquad (22.23a)$$

and

$$y_2 = K_{22}\left(1 - \frac{K_{12}K_{21}}{K_{11}K_{22}}\right)v_2 \qquad (22.23b)$$

If we now recall Eqs. (21.19) and (21.23) as the definition of the relative gain parameter λ for the 2 × 2 system, we immediately see that Eqs. (22.23a,b) become:

$$y_1 = \frac{K_{11}}{\lambda}v_1 \qquad (22.24a)$$

and

$$y_2 = \frac{K_{22}}{\lambda}v_2 \qquad (22.24b)$$

implying that upon implementing simplified decoupling, the effective closed-loop steady-state gain in each loop is the ratio of the open-loop gain and the relative gain parameter. Note that when λ is very large, the effective closed-loop gains become very small, and control system performance may be jeopardized. We will return to this point below.

It is now important to note that when dealing with systems with dimensions larger than 2 × 2, the simplified decoupling approach has the potential of rapidly becoming tedious. For example, the situation for a 3 × 3 system is shown in Figure 22.4 where there are now six compensator blocks to design. More generally, an $N \times N$ system leads to $(N^2 - N)$ compensators. The same principles used for the 2 × 2 system are also applicable, but there is more cumbersome work involved.

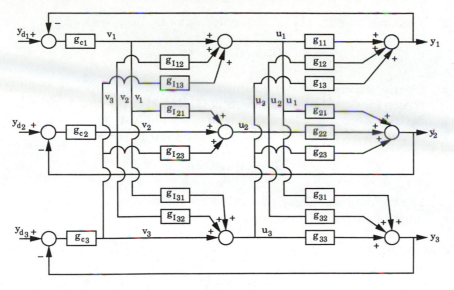

Figure 22.4.   Simplified decoupling for a 3×3 system.

The more general approach to decoupler design employs matrix methods that are particularly convenient for higher dimensional systems.

Generalized Decoupling

A more general procedure for decoupler design may be outlined as follows:

1. From Figure 22.2, we observe that:

$$\mathbf{y} = \mathbf{Gu} \qquad (22.25)$$

$$\mathbf{u} = \mathbf{G}_I \mathbf{v} \qquad (22.26)$$

so that:

$$\mathbf{y} = \mathbf{GG}_I \mathbf{v} \qquad (22.27)$$

2. In order to eliminate interactions absolutely, $\mathbf{y}$ must be related to $\mathbf{v}$ through a diagonal matrix; let us call it $\mathbf{G}_R(s)$. We may now choose $\mathbf{G}_I$ such that:

$$\mathbf{GG}_I = \mathbf{G}_R(s) \qquad (22.28)$$

and the compensated input/output relation becomes:

$$\mathbf{y} = \mathbf{G}_R(s)\mathbf{v}$$

where $\mathbf{G}_R$ represents the equivalent diagonal process that the diagonal controllers $\mathbf{G}_c$ are required to control.

3. Therefore from Eq. (22.28), the compensator $\mathbf{G}_I$ must be given by:

$$\mathbf{G}_I = \mathbf{G}^{-1}\,\mathbf{G}_R \qquad (22.29)$$

4. Clearly, the compensator obtained using Eq. (22.29) depends on what is chosen for $\mathbf{G}_R$. Thus the elements of $\mathbf{G}_R$ should be selected to provide the desired decoupled behavior with the simplest possible decoupler defined in Eq. (22.29). A commonly employed choice is:

$$\mathbf{G}_R = \text{Diag}[\mathbf{G}(s)] \qquad (22.30)$$

i.e., the diagonal elements of $\mathbf{G}(s)$ are retained as the elements of the diagonal matrix $\mathbf{G}_R$; however, other choices may be used.

Relationship between Generalized and Simplified Decoupling

"Generalized" decoupling may be related to simplified decoupling, by noting that for simplified decoupling applied to a 2×2 system, the compensator transfer function matrix is given by:

$$\mathbf{G}_I = \begin{bmatrix} 1 & g_{I_1} \\ g_{I_2} & 1 \end{bmatrix} \qquad (22.31)$$

while for a 3×3 system, the compensator matrix $\mathbf{G}_I$ takes the form:

$$\mathbf{G}_I = \begin{bmatrix} 1 & g_{I_{12}} & g_{I_{13}} \\ g_{I_{21}} & 1 & g_{I_{23}} \\ g_{I_{31}} & g_{I_{32}} & 1 \end{bmatrix} \qquad (22.32)$$

Thus, for simplified decoupling the specific *form* desired for $\mathbf{G}_I$ is specified by Eqs. (22.31) and (22.32), etc., and the task is to find the g_{I_i} required to make $\mathbf{GG}_I$ diagonal. The final diagonal form of $\mathbf{GG}_I$ is a result of the design.

On the other hand with the general decoupling approach, the final diagonal form for $\mathbf{GG}_I$ is specified as $\mathbf{G}_R$, and then the $\mathbf{G}_I$ required to achieve this is derived from Eq. (22.29).

Let us now consider some examples to illustrate the principles of both simplified and generalized decoupler design.

Example 22.1   SIMPLIFIED DECOUPLER DESIGN FOR THE WOOD AND BERRY DISTILLATION COLUMN.

Using the simplified decoupling approach, design a decoupler for the Wood and Berry distillation column whose transfer function model was given in Example 20.6, Eq. (20.26).

Solution:

Recalling the transfer function matrix for this process, we now obtain, with the aid of Eqs. (22.20) and (22.21), that the transfer functions for the simplified decouplers are:

$$g_{I_1} = -\frac{\dfrac{-18.9e^{-3s}}{21.0s + 1}}{\dfrac{12.8e^{-s}}{16.7s + 1}}$$

and

$$g_{I_2} = -\frac{\dfrac{6.6e^{-7s}}{10.9s + 1}}{\dfrac{-19.4e^{-3s}}{14.4s + 1}}$$

which simplify to:

$$g_{I_1} = 1.48\frac{(16.7s + 1)e^{-2s}}{(21.0s + 1)} \tag{22.33}$$

and

$$g_{I_2} = 0.34\frac{(14.4s + 1)e^{-4s}}{(10.9s + 1)} \tag{22.34}$$

Note that these interaction compensator blocks each have the form of a lead/lag system with a gain term and a time-delay term.

In terms of actual implementation, using Eqs. (22.33) and (22.34) in Eq. (22.18) implies that the overhead reflux rate, u_1 will be determined from:

$$u_1 = v_1 + \left[1.48\frac{(16.7s + 1)\,e^{-2s}}{(21.0s + 1)}\right]v_2 \tag{22.35}$$

where, as we may recall, v_1 and v_2 are, respectively, the outputs of the Loop 1 (overhead composition) controller and Loop 2 (bottoms composition) controller. The bottoms steam flowrate will be determined from:

$$u_2 = v_2 + \left[0.34\frac{(14.4s + 1)\,e^{-4s}}{(10.9s + 1)}\right]v_1 \tag{22.36}$$

The next example illustrates the procedure for carrying out generalized decoupling in practice; it also illustrates, rather graphically, the rationale behind referring to the first approach as "simplified" decoupling.

Example 22.2   **A GENERALIZED DECOUPLER FOR THE WOOD AND BERRY DISTILLATION COLUMN.**

Using the generalized decoupling approach, design a decoupler for the Wood and Berry distillation column, the same system of Example 22.1.

Solution:

The transfer function matrix in question is:

$$\mathbf{G}(s) = \begin{bmatrix} \dfrac{12.8e^{-s}}{16.7s + 1} & \dfrac{-18.9e^{-3s}}{21.0s + 1} \\[2ex] \dfrac{6.6e^{-7s}}{10.9s + 1} & \dfrac{-19.4e^{-3s}}{14.4s + 1} \end{bmatrix} \tag{22.37}$$

and if we choose $\mathbf{G}_R$ to be diagonal elements of Eq. (22.37), then:

$$\mathbf{G}_R(s) \; = \; \begin{bmatrix} \dfrac{12.8e^{-s}}{16.7s \, + \, 1} & 0 \\ 0 & \dfrac{-19.4e^{-3s}}{14.4s \, + \, 1} \end{bmatrix} \tag{22.38}$$

To obtain the required decoupler as indicated in Eq. (22.29), we must now evaluate the inverse of the transfer function matrix in Eq. (22.37); this is given by:

$$\mathbf{G}^{-1}(s) \; = \; \frac{1}{\Delta} \begin{bmatrix} \dfrac{-19.4e^{-3s}}{14.4s \, + \, 1} & \dfrac{18.9e^{-3s}}{21.0s \, + \, 1} \\ \dfrac{-6.6e^{-7s}}{10.9s \, + \, 1} & \dfrac{12.8e^{-s}}{16.7s \, + \, 1} \end{bmatrix} \tag{22.39}$$

where Δ is the determinant of the matrix:

$$\Delta = \frac{-248.32(21.0s+1)(10.9s+1)e^{-4s} + 124.74(16.7s+1)(14.4s+1)e^{-10s}}{(16.7s+1)(14.4s+1)(21.0s+1)(10.9s+1)} \tag{22.40}$$

By multiplying the inverse matrix in Eq. (22.39) and $\mathbf{G}_R$ in Eq. (22.38) as indicated in Eq. (22.29) we obtain the generalized decoupler for the Wood and Berry column as:

$$\mathbf{G}_I \; = \; \begin{bmatrix} g_{I_{11}} & g_{I_{12}} \\ g_{I_{21}} & g_{I_{22}} \end{bmatrix} \tag{22.41}$$

with each element given as follows:

$$g_{I_{11}} = \frac{-248.32(21.0s+1)(10.9s+1)}{124.74(16.7s+1)(14.4s+1)e^{-6s} - 248.32(21.0s+1)(10.9s+1)} \tag{22.42}$$

$$g_{I_{12}} = \frac{-366.66(16.7s+1)(10.9s+1)e^{-2s}}{124.74(16.7s+1)(14.4s+1)e^{-6s} - 248.32(21.0s+1)(10.9s+1)} \tag{22.43}$$

$$g_{I_{21}} = \frac{84.48(21.0s+1)(14.4s+1)e^{-4s}}{124.74(16.7s+1)(14.4s+1)e^{-6s} - 248.32(21.0s+1)(10.9s+1)} \tag{22.44}$$

and

$$g_{I_{22}} \; = \; g_{I_{11}} \tag{22.45}$$

   The actual implementation of this decoupler means that given v_1, the output of the Loop 1 (overhead composition) controller, and v_2, the output of the Loop 2 (bottoms composition) controller, the overhead reflux rate, u_1 will be determined from:

$$u_1 \; = \; g_{I_{11}} v_1 \, + \, g_{I_{12}} v_2 \tag{22.46}$$

and the bottoms steam flowrate will be determined from:

$$u_2 \; = \; g_{I_{21}} v_1 \, + \, g_{I_{22}} v_2 \tag{22.47}$$

with the decoupler transfer function elements as given in Eqs. (22.42), (22.43), (22.44), and (22.45).

These two examples clearly illustrate the advantages and disadvantages of simplified versus generalized decoupling:

1.   While both decouplers provide diagonal systems for which SISO controllers can be tuned by SISO methods, the closed-loop response could be completely different in each case. For example, the "equivalent" open-loop decoupled system using simplified decoupling is given by Eq. (22.22) and shown in Figure 22.3 as:

$$y_1 = \left(g_{11} - \frac{g_{12}g_{21}}{g_{22}}\right)v_1 = \left(\frac{12.8e^{-s}}{16.7s + 1} - \frac{(18.9)(6.6)(14.4s + 1)e^{-7s}}{(19.4)(10.9s + 1)(21.0s + 1)}\right)v_1$$

$$y_2 = \left(g_{22} - \frac{g_{12}g_{21}}{g_{11}}\right)v_2 = \left(\frac{-19.4e^{-3s}}{14.4s + 1} + \frac{(18.9)(6.6)(16.7s + 1)e^{-9s}}{(12.8)(10.9s + 1)(21.0s + 1)}\right)v_2$$

which is much more complicated than the $\mathbf{G}_R$ matrix specified in the *Generalized* decoupling design, as, for example, Eq. (22.38). One would expect that the tuning and closed-loop performance for the *Generalized* decoupling case to be much better than for *Simplified* decoupling.

2.   The price we must pay for the improved closed-loop performance in the *Generalized* decoupling case is a more complicated decoupler. This is easily seen if we compare the *Simplified* decoupler given by Eqs. (22.33) and (22.34) with the *Generalized* decoupler in Eqs. (22.42) – (22.45).

22.2.3 Some Limitations to the Application of Decoupling

There are some limitations to the application of decoupling, and these must be kept in mind in order to maintain a proper perspective when designing decouplers.

Perfect decoupling is only possible if the process model is perfect. As this is hardly ever the case, perfect decoupling is unattainable in practice. However, even with imperfect models, decoupling has been applied with notable success on a number of industrial processes [1].

The ideal decouplers derived by the simplified decoupling approach are similar to feedforward controllers (compare Eqs. (22.20) and (22.21) with Eq. (16.22)). Returning to take another look at Figure (22.1) will confirm why this has to be the case: observe that, as far as Loop 1 is concerned, u_2 appears as a "disturbance" passing through the g_{12} transfer function element. Thus designing g_{I_1} to take the form in Eq. (22.20) is identical to the procedure for feedforward controller design. Ideal dynamic decouplers are thus subject to the same realization problems we had enumerated in Chapter 16 (Section 16.1.2) for feedforward controllers, particularly when time delays are involved in the transfer function elements.

Perfect dynamic decouplers (whether simplified or generalized) are based on model inverses. As such, they can only be implemented if such inverses are both *causal* and *stable*. To illustrate, consider the 2 × 2 compensators in

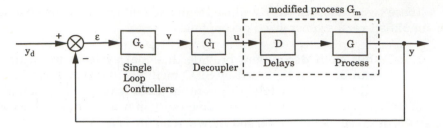

Figure 22.5.   Use of additional time delays to allow dynamic decoupling.

Figure 22.1, where the transfer functions g_{I_1} and g_{I_2} must be causal (no $e^{+\alpha s}$ terms) and stable. To satisfy causality for the 2×2 system, we see from Eqs. (22.20) and (22.21) that any time delays in g_{11} must be smaller than the time delays in g_{12}, and a similar condition must hold for g_{22} and g_{21}. To satisfy stability, a second condition is that g_{11} and g_{22} must not have any RHP zeros and also g_{12} and g_{21} must not have any RHP poles. This leads to the following more general conditions that must be satisfied in order to implement simplified dynamic decoupling for $N \times N$ systems:

1. **Causality:** In order to ensure causality in the compensator transfer functions it is necessary that the time-delay structure in $G(s)$ be such that the smallest delay in each row occurs on the diagonal. For simplified decoupling, this is an absolute requirement; however, it is possible to add delays to the inputs $u_1, u_2, ..., u_n$, in order to satisfy the requirement if the original process G does not comply. This is equivalent to defining a modified process:

$$\mathbf{G}_m = \mathbf{GD} \tag{22.48}$$

    where $\mathbf{D}$ is a diagonal matrix of time delays:

$$\mathbf{D}(s) = \begin{bmatrix} e^{-d_{11}s} & & & 0 \\ & e^{-d_{22}s} & & \\ & & \cdot & \\ & & & \cdot \\ 0 & & & \cdot\ e^{-d_{nn}s} \end{bmatrix} \tag{22.49}$$

    The simplified decoupler is then designed by using the elements of $\mathbf{G}_m$ rather than $\mathbf{G}$, and the matrix $\mathbf{D}$ must be inserted into the control loop as shown in Figure 22.5. An example below illustrates this procedure.
        In the case of generalized decoupling, one may use the modified process $\mathbf{G}_m$ as above, or alternatively adjust the time delays in the diagonal matrix, $\mathbf{G}_R$ in Eq. (22.29), in order that the elements of $\mathbf{G}_I = (\mathbf{GD})^{-1}\mathbf{G}_R$ are causal. This is equivalent to requiring that $\mathbf{G}_R^{-1}\mathbf{GD}$ have the smallest delay in each row on the diagonal.

2. **Stability:** In order to ensure the stability of the compensator transfer functions, it is necessary that the causality condition above be satisfied and that there be no RHP zeros of the process $G(s)$. This is an absolute requirement for simplified decoupling and reduces to the condition that

there be no RHP zeros in the diagonal elements of **G** and that the off-diagonal elements of **G** be stable. For generalized decoupling this may be relaxed by adjusting the dynamics of $\mathbf{G}_R$ in order that the elements of $\mathbf{G}_I = \mathbf{G}^{-1}\mathbf{G}_R$ be stable.

One may quickly verify that the Wood and Berry Column of Examples 22.1 and 22.2 satisfies these dynamic decoupling conditions. However, let us provide an example for which these conditions are not satisfied.

Example 22.3   **DECOUPLING CONTROL OF A DISTILLATION COLUMN WITH DIFFICULT TIME-DELAY STRUCTURE.**

Let us consider a slight modification of the Wood and Berry process where a process change has added a time delay of 3 minutes to the input u_1, so that the process model takes the form:

$$\mathbf{G}(s) = \begin{bmatrix} \dfrac{12.8\ e^{-4s}}{16.7s + 1} & \dfrac{-18.9\ e^{-3s}}{21.0s + 1} \\[3mm] \dfrac{6.6\ e^{-10s}}{10.9s + 1} & \dfrac{-19.4\ e^{-3s}}{14.4s + 1} \end{bmatrix} \qquad (22.50)$$

Note that now the smallest delay in each row is not on the diagonal, and the simplified decoupling compensator Eq. (22.33) becomes:

$$g_{I_1} = 1.48\ \dfrac{(16.7s + 1)\ e^{+s}}{(21.0s + 1)} \qquad (22.51)$$

which is noncausal and cannot be implemented. Thus simplified decoupling cannot be used unless we make use of a delay matrix, **D**(s). Let us design **D**(s) to add a time delay of 1 minute to the input u_2, i.e.:

$$\mathbf{D}(s) = \begin{bmatrix} 1 & 0 \\ 0 & e^{-s} \end{bmatrix} \qquad (22.52)$$

so that the modified process is:

$$\mathbf{G}_m = \mathbf{GD} = \begin{bmatrix} \dfrac{12.8\ e^{-4s}}{16.7s + 1} & \dfrac{-18.9\ e^{-4s}}{21.0s + 1} \\[3mm] \dfrac{6.6\ e^{-10s}}{10.9s + 1} & \dfrac{-19.4\ e^{-4s}}{14.4s + 1} \end{bmatrix} \qquad (22.53)$$

Now we have satisfied the requirement that the smallest delay be on the diagonal and the simplified decoupling compensators become:

$$g_{I_1} = \begin{bmatrix} \dfrac{1.48\ (16.7s + 1)}{21.0s + 1} \end{bmatrix}$$

$$g_{I_2} = \begin{bmatrix} \dfrac{0.34\ (14.4s + 1)\ e^{-6s}}{(10.9s + 1)} \end{bmatrix}$$

and these are implemented together with the delay matrix, $\mathbf{D}(s)$. Note that this decoupling comes at a cost of an additional delay of one minute in Loop 2.

If we use generalized decoupling for the process in Eq. (22.50), all the compensators are causal except for:

$$g_{I_{12}} = \frac{-366.66(16.7s+1)(10.9s+1)e^{+s}}{124.74(16.7s+1)(14.4s+1)e^{-6s} - 248.32(21.0s+1)(10.9s+1)}$$

which cannot be implemented. To remedy this we could use the $\mathbf{D}(s)$ given by Eq. (22.52) and let:

$$\mathbf{G}_I = (\mathbf{GD})^{-1}\,\mathbf{G}_R$$

Alternatively, we could let $\mathbf{D}(s) = \mathbf{I}$ and increase the time delay in the (2,2) element of $\mathbf{G}_R$ to 4 minutes and have the same generalized compensator as in Eqs. (22.42) – (22.45), but now

$$g_{I_{21}} = \frac{84.48(21.0s+1)(14.4s+1)e^{-7s}}{124.74(16.7s+1)(14.4s+1)e^{-6s} - 248.32(21.0s+1)(10.9s+1)}$$

$$g_{I_{22}} = \frac{-248.32(21.0s+1)(10.9s+1)e^{-s}}{124.74(16.7s+1)(14.4s+1)e^{-6s} - 248.32(21.0s+1)(10.9s+1)}$$

The problems at the end of the chapter provide additional examples of dynamic decoupling.

Because of complexities in implementation or the lack of a high-quality dynamic model, it is often necessary to make use of simplifications of the ideal decoupler. For example, steady-state decoupling requires only the matrix of steady-state gains as a model. Alternatively partial decoupling reduces the dimensionality of the problem by requiring decoupling in only a few of the control loops. Often one of these choices is employed as the best balance between complexity and closed-loop controller performance. Let us now discuss some of these simpler types of decoupling.

22.2.4 Simpler Types of Decoupling

When ideal dynamic decoupling is not desirable or possible, useful results may be obtained by using the less ambitious approaches of Partial Decoupling and Steady-State Decoupling.

Partial Decoupling

When some of the loop interactions are weak or if some of the loops need not have high performance, then one may consider partial decoupling. In this case, attention is focused only on a subset of the control loops where the interactions are important and high performance control is required.

It is typical to consider partial decoupling for 3×3 and higher dimensional systems, since the main advantage lies in the reduction of dimensionality; however, partial decoupling is also applicable to 2×2 systems, in which case, one of the compensator blocks in Figure 22.1 is set to zero for the loop that is to be excluded from decoupling.

The following example illustrates the application of partial decoupling for a 3×3 system.

Example 22.4

PARTIAL DYNAMIC DECOUPLING FOR THE HULBERT AND WOODBURN WET-GRINDING CIRCUIT.

Hulbert and Woodburn [3] have reported the control of a wet-grinding circuit whose transfer function model, obtained by experimental means, is given by:

$$\begin{bmatrix} y_1 \\ y_2 \\ y_3 \end{bmatrix} = \begin{bmatrix} \dfrac{119}{217s + 1} & \dfrac{153}{337s + 1} & \dfrac{-21}{10s + 1} \\[2mm] \dfrac{0.00037}{500s + 1} & \dfrac{0.000767}{33s + 1} & \dfrac{-0.00005}{10s + 1} \\[2mm] \dfrac{930}{500s + 1} & \dfrac{-667e^{-320s}}{166s + 1} & \dfrac{-1033}{47s + 1} \end{bmatrix} \begin{bmatrix} u_1 \\ u_2 \\ u_3 \end{bmatrix} \qquad (22.54)$$

The time delay and time constants are in seconds; and the indicated process variables are given in terms of deviations from their respective steady-state values as:

y_1 = Torque required to turn the mill (Nm)
y_2 = Flowrate from the mill (m^3/s)
y_3 = Density of the cyclone feed (kg/m^3)
u_1 = Feedrate of solids to the mill (kg/s)
u_2 = Feedrate of water to the mill (kg/s)
u_3 = Feedrate of water to the sump (kg/s)

A relative gain analysis of this system (see Problem (22.2)) recommends the 1-1/2-2/3-3 input/output pairing configuration; it also indicates that if three independent single-loop feedback controllers are used, the system will experience significant loop interactions. A schematic diagram for this process is shown in Figure 22.6.

Since the least sensitive of the output variables for this system is y_2 (the total flow from the mill), and since from experience, it is known that the greatest amount of interaction is between control Loops 1 and 3, it is desired to design decouplers for only these loops, leaving Loop 2 to operate without decoupling. Obtain the transfer functions required for these decouplers.

Solution:

Under the indicated conditions, the transfer function matrix for the subsystem of interest is now given by:

$$\begin{bmatrix} y_1 \\ y_3 \end{bmatrix} = \begin{bmatrix} \dfrac{119}{217s + 1} & \dfrac{-21}{10s + 1} \\[2mm] \dfrac{930}{500s + 1} & \dfrac{-1033}{47s + 1} \end{bmatrix} \begin{bmatrix} u_1 \\ u_3 \end{bmatrix} \qquad (22.55)$$

and if the decoupler transfer functions are represented as g_{l_1} for Loop 1, and g_{l_3} for Loop 3, then using the simplified decoupling approach, we have, from Eq. (22.55):

$$g_{l_1} = \frac{\dfrac{21}{10s + 1}}{\dfrac{119}{217s + 1}}$$

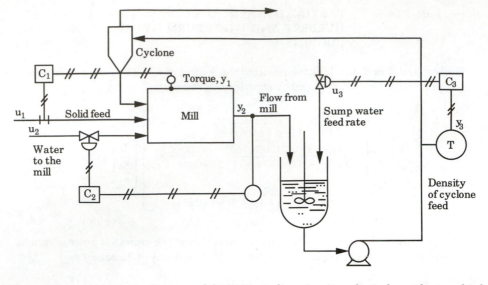

Figure 22.6. Schematic diagram of the wet-grinding circuit under independent multiple single-loop feedback control.

and

$$g_{I_3} = \frac{\dfrac{930}{500s + 1}}{\dfrac{1033}{47s + 1}}$$

which simplify to:

$$g_{I_1} = \frac{0.176(217s + 1)}{(10s + 1)} \tag{22.56}$$

and

$$g_{I_3} = \frac{0.9(47s + 1)}{(500s + 1)} \tag{22.57}$$

each implementable by lead/lag units.

Note that this control scheme is equivalent to letting $g_{I_{13}} = g_{I_1}$ and $g_{I_{31}} = g_{I_3}$ in Figure 22.4 with all the other $g_{I_{ij}} = 0$.

Steady-State Decoupling

The only difference between dynamic decoupling and steady-state decoupling is that while the former uses the complete, dynamic version of each transfer function element in obtaining the decoupler, the latter uses only the steady-state gain portion of these transfer function elements.

Thus, if each transfer function element $g_{ij}(s)$, has a steady-state gain term K_{ij}, and if the matrix of steady-state gains is represented by $\mathbf{K}$, then the steady-state decoupling results equivalent to those obtained in Section 22.2.2 are now summarized:

Simplified steady-state decoupling for a 2 × 2 system

Here:

$$g_{I_1} = -\frac{K_{12}}{K_{11}} \tag{22.58}$$

and

$$g_{I_2} = -\frac{K_{21}}{K_{22}} \tag{22.59}$$

will take simple, constant, numerical values, and will therefore always be realizable, as well as implementable.

Generalized steady-state decoupling

In this case, the decoupler matrix will be given by:

$$\mathbf{G}_I = \mathbf{K}^{-1} \mathbf{K}_R \tag{22.60}$$

where $\mathbf{K}_R$ is the steady-state version of $\mathbf{G}_R(s)$. Since the inversion indicated in Eq. (22.60) now concerns only a matrix of numbers, Eq. (22.60) will always be realizable and easily implemented.

The main advantages of steady-state decoupling are, therefore, that its design involves simple numerical computations, and that the resulting decouplers are always realizable.

Example 22.5 **SIMPLIFIED AND GENERALIZED STEADY-STATE DECOUPLERS FOR THE WOOD AND BERRY DISTILLATION COLUMN.**

Design steady-state decouplers for the Wood and Berry column using both the simplified and the generalized approaches.

Solution:

The simplified steady-state decouplers are easily obtained by retaining only the steady-state gains of the transfer function elements; the results are:

$$g_{I_1} = -\frac{-18.9}{12.8} = 1.48 \tag{22.61}$$

and

$$g_{I_2} = -\frac{6.6}{-19.4} = 0.34 \tag{22.62}$$

and these will be implemented on the real process as:

$$u_1 = v_1 + 1.48 \, v_2 \tag{22.63}$$

and

$$u_2 = 0.34 \, v_1 + v_2 \tag{22.64}$$

For the generalized decoupler on the other hand, we have, upon taking the *inverse* of:

$$\mathbf{K} = \begin{bmatrix} 12.8 & -18.9 \\ 6.6 & -19.4 \end{bmatrix} \tag{22.65}$$

and multiplying by:

$$\mathbf{K}_R = \begin{bmatrix} 12.8 & 0 \\ 0 & -19.4 \end{bmatrix} \tag{22.66}$$

we would obtain:

$$\mathbf{G}_I = \begin{bmatrix} 2.01 & 2.97 \\ 0.68 & 2.01 \end{bmatrix} \tag{22.67}$$

and these will be implemented as:

$$u_1 = 2.01 \, v_1 + 2.97 \, v_2 \tag{22.68}$$

and

$$u_2 = 0.68v_1 + 2.01v_2 \tag{22.69}$$

Note that these same results could have been obtained by setting $s = 0$ in the dynamic decoupler results obtained in Example 22.2.

In principle we would expect the simplified approach and generalized approach to give the same result for steady-state decoupling. That this is true may be seen by dividing Eqs. (22.68) and (22.69) by the common factor 2.01 to obtain Eqs. (22.63) and (22.64). However, the single-loop controller gains in each case will also differ by the factor 2.01.

As the example illustrates, steady-state decouplers are very easy to design and straightforward to implement. This advantage in simplicity can be so compelling that, quite often, in dealing with control loop interaction problems in practice, steady-state decoupling is the first technique to try; only when the dynamic interactions prove to be persistent will dynamic considerations be entertained.

Observe that whenever the dynamic aspects of the transfer function elements in each row of the transfer function matrix are *similar*, the dynamic decoupler will be very close to its steady-state version. It can be shown (and this is left as an exercise to the reader) that for the stirred mixing tank system whose approximate transfer function matrix was given in Eq. (20.23), whether we use the simplified or the generalized approach, the dynamic decoupler obtained in each case is actually identical to the corresponding steady-state decoupler.

Because steady-state decoupling often leads to great improvements in control system performance with very little work or cost, it is the decoupling technique most often applied in practice. Thus in the next section we will discuss the design of steady-state decouplers and their performance in some detail.

22.3   FEASIBILITY OF STEADY-STATE DECOUPLER DESIGN

One would think from the discussion in the last section that design of a steady-state decoupler is trivial. One simply finds the matrix of steady-state process gains, calculates the matrix inverse, and programs these numbers into the control computer. Operationally, this is true — no further compensations are required on-line. However, the *successful* implementation of a steady-state decoupler requires answers to further questions:

1. *Are there constraints on some of the inputs (e.g., valve range, heating or cooling power, etc.) which are reached?* If so the decoupler cannot be used beyond the point of these constraints because the constrained controllers are no longer responding to error signals and the decoupler could even make the feedback control scheme drive the process unstable once an input constraint is reached.

2. *Is the steady-state model of the process sufficiently accurate to permit steady-state decoupling?* Depending on the degree of "ill-conditioning" of the true steady-state gain matrix, an extremely accurate process model or only a rough approximation might be required for successful implementation of decoupling.

3. *Is the steady-state gain structure of the process so "ill-conditioned" that even with a perfect model, successful decoupling is impossible.* As we will demonstrate below, there are "impossible situations" in which the true process model is so ill-conditioned that decoupling control is useless. Unfortunately, such situations are not rare in industrial practice.

Let us now discuss these questions of *feasibility* of the design of steady-state decoupling controllers in more detail.

22.3.1 Effect of Controller Constraints

In process control practice there are always constraints on some of the process input variables because valves cannot go beyond full open or full shut, heaters cannot go beyond full power or zero power, etc. Thus, the important question in evaluating this factor is whether or not one or more of the inputs reaches a constraint during normal operation. If the controller gains are such that none of the constraints come into play, then decoupling control is not affected by these constraints. However, if even one of the inputs reaches a constraint then the control system can no longer send the combination of control actions required to achieve decoupling, and extremely poor (or even unstable) responses could result. Thus more sophisticated controller strategies, which explicitly recognize and take constraints into consideration, are required when input constraints come into play. Let us illustrate with an example.

Example 22.6	**PERFORMANCE OF STEADY-STATE DECOUPLER WITH INPUT CONSTRAINTS.**

Let us consider the implementation of the simplified steady-state decoupler for the Wood and Berry Distillation Column of Example 22.2.

Recall that the decoupler is given by Eqs. (22.61) and (22.62). When this is implemented together with diagonal PI controllers with tuning parameters $K_{c_1} = 0.30$, $1/\tau_{I_1} = 0.307$, $K_{c_2} = -0.05$, $1/\tau_{I_2} = 0.107$, one sees from Figures 22.7–22.8 that the

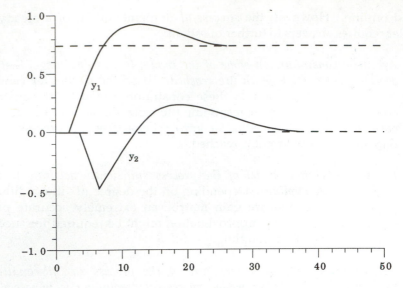

Figure 22.7. Wood and Berry Distillation Column: closed-loop response of y_1 and y_2 under steady-state decoupling control with unconstrained inputs.

closed-loop response is better than what was obtained with multiple single-loop designs (cf. Example 21.13). Thus a steady-state decoupler seems very attractive.

Now suppose that an energy revamp of the column has reduced the size of the reflux drum and reflux control valve so that u_1 has a more limited maximum flowrate, $0 \le u_1 \le 0.15$. When the same decoupling control scheme is applied to the column with the new reflux flowrate constraint, Figures 22.9, and 22.10 show that the closed-loop response is very poor once the reflux valve is full open and the system becomes unstable. This is because u_1 is pegged at its maximum value trying to increase y_1 and Loop 2 is trying to compensate imagined control actions u_1, which are computed, but not implemented.

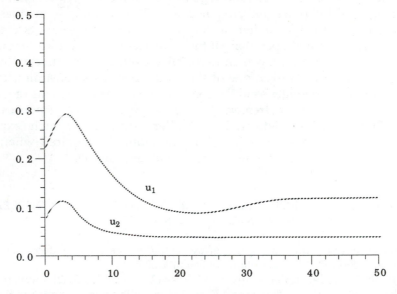

Figure 22.8. Unconstrained manipulated variables, u_1, u_2, for closed-loop response in Figure 22.7.

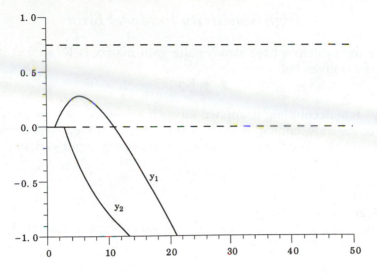

Figure 22.9.   Wood and Berry Distillation Column: closed-loop response of y_1 and y_2 under steady-state decoupling control with constrained u_1, $0 \leq u_1 \leq 0.15$.

Thus steady-state decoupling cannot be applied in this case; however, more sophisticated, constraint following controllers could be implemented (cf. Chapter 27 for more discussion of these).

22.3.2 Decoupling Control of Ill-Conditioned Processes

There are processes whose structure make them poor candidates for decoupling control. Such processes fall under the general category of "ill-conditioned" processes. Although there is a unifying structure to "ill-conditioned" processes, let us begin the discussion by considering the various manifestations of this ill-conditioning.

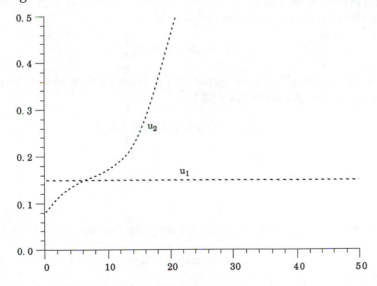

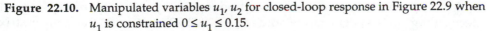

Figure 22.10.   Manipulated variables u_1, u_2 for closed-loop response in Figure 22.9 when u_1 is constrained $0 \leq u_1 \leq 0.15$.

High Sensitivity to Model Error

Consider the system whose steady-state gain matrix is $\mathbf{K}$, so that its steady-state model is given by:

$$\mathbf{y} = \mathbf{K}\,\mathbf{u} \tag{22.70}$$

If steady-state decoupling is applied on such a system, we now know that the outputs of the independent, multiple single-loop controllers, represented by the vector $\mathbf{v}$, will be related to the actual vector of process input variables $\mathbf{u}$ by:

$$\mathbf{u} = \mathbf{G}_I\mathbf{v} \tag{22.71}$$

where $\mathbf{G}_I$ is the decoupler matrix.

For the system in Eq. (22.70), we have established in the previous section that the generalized decoupler is given by:

$$\mathbf{G}_I = \mathbf{K}^{-1}\,\mathbf{K}_R \tag{22.72}$$

where $\mathbf{K}_R$ is the diagonal gain matrix of $\mathbf{G}_R(s)$. Therefore at steady state, *if the model is perfect*, the relationship between $\mathbf{v}$ and the process output will be:

$$\mathbf{y} = \mathbf{K}\,\mathbf{K}^{-1}\,\mathbf{K}_R\mathbf{v} \tag{22.73}$$

or

$$\mathbf{y} = \mathbf{K}_R\mathbf{v} \tag{22.74}$$

and since $\mathbf{K}_R$ is diagonal, $\mathbf{G}_I$ will have achieved decoupling; there will no longer be cross-coupling between the control loops.

We know, however, that no process model (not even the steady-state portion alone) is 100% accurate; let us then consider the situation in which there is an error $\Delta\mathbf{K}$ in the estimate of the steady-state gain matrix. In reality, therefore, the true steady-state model should be:

$$\mathbf{y} = (\mathbf{K} + \Delta\mathbf{K})\,\mathbf{u} \tag{22.75}$$

The decoupler in Eq. (22.72), designed on the basis of $\mathbf{K}$, will now provide the following relationship between $\mathbf{v}$ and $\mathbf{y}$:

$$\mathbf{y} = (\mathbf{K} + \Delta\mathbf{K})\,\mathbf{K}^{-1}\,\mathbf{K}_R\mathbf{v}$$

or

$$\mathbf{y} = \mathbf{K}_R\mathbf{v} + \Delta\mathbf{K}\,\mathbf{K}^{-1}\,\mathbf{K}_R\mathbf{v} \tag{22.76}$$

Observe that there will now be a deviation from the ideal situation indicated in Eq. (22.74):

$$\Delta\mathbf{y} = \Delta\mathbf{K}\,\mathbf{K}^{-1}\,\mathbf{K}_R\mathbf{v} \tag{22.77}$$

which represents the amount of error introduced into $\mathbf{y}$ as a result of carrying out decoupling on a system modeled as in Eq. (22.70), but with the gain matrix subject to modeling error in the amount represented by $\Delta\mathbf{K}$.

From the definition of the inverse of a matrix, Eq. (22.77) becomes:

$$\Delta \mathbf{y} = \frac{\Delta \mathbf{K} \ \text{Adj}(\mathbf{K}) \ \mathbf{K}_R \mathbf{v}}{|\mathbf{K}|} \qquad (22.78)$$

We now observe that if the determinant of the gain matrix, $|\mathbf{K}|$, is very small, its reciprocal will be very large, and:

1. Small modeling errors will be magnified into very large errors in $\mathbf{y}$.

2. Small changes in controller output $\mathbf{v}$ will also result in large errors in $\mathbf{y}$.

Such systems are said to be very *sensitive* to modeling errors, and it is clear that decoupling will be a risky venture in such cases. Furthermore, observe from Eqs. (22.76) and (22.77) that the larger the error $\Delta \mathbf{y}$ the farther away from being decoupled the system is, and thus feedback control has a more difficult time correcting for the error.

We therefore have the first condition under which decoupling will prove difficult:

Whenever the determinant of the process gain matrix is very small, the system will be extremely sensitive to modeling errors and decoupling will be difficult to achieve; in the limit as the determinant completely vanishes, decoupling will be entirely impossible.

Process sensitivity to modeling error can also be deduced from the elements of the RGA of the system because the elements of the RGA are intimately related to the determinant of the process gain matrix. We recall from the matrix method presented in Chapter 21 for calculating RGA elements, that if C_{ij} represents the cofactor of K_{ij} (the i,j th element of the gain matrix $\mathbf{K}$), then the RGA of the system has elements given by:

$$\lambda_{ij} = \frac{K_{ij} C_{ij}}{|\mathbf{K}|} \qquad (22.79)$$

Therefore large sensitivity to modeling errors — heralded by the nearness of $|\mathbf{K}|$ to zero — will be manifested as inordinately large RGA values.

In Bristol's original RGA paper [4] it is stated that in addition to providing a measure of control loop interactions, the RGA also provides a measure of a system's sensitivity to modeling errors. Shinskey [5, 6] has also presented results showing that systems with very large RGA values are very sensitive to decoupler error, with sensitivity increasing with increasing RGA values.

Thus we may conclude that:

Decoupling will be difficult to achieve for a system whose input/output variables are paired on very large RGA values; such a system will also be very sensitive to modeling errors.

Let us illustrate this with an example.

Example 22.7   **PARTIAL STEADY-STATE DECOUPLING OF THE
                          PRETT AND GARCIA HEAVY OIL FRACTIONATOR.**

Prett and Garcia presented in Ref. [7] a 7×5 transfer function model for a heavy oil
fractionator column. Here we will consider the partial decoupling of this system by
focusing on a 2×2 subsystem involving the top end point, and the intermediate reflux
temperature as the two outputs, with top draw rate and the intermediate reflux duty as
the two inputs. This subsystem has the following transfer function model:

$$\mathbf{G}(s) = \begin{bmatrix} \dfrac{4.05e^{-27s}}{50s + 1} & \dfrac{1.20e^{-27s}}{45s + 1} \\ \dfrac{4.06e^{-8s}}{13s + 1} & \dfrac{1.19}{19s + 1} \end{bmatrix} \tag{22.80}$$

The steady-state gains are dimensionless, and the time delays and time constants are in
minutes. Analyze the sensitivity of this system to model errors, and assess the
feasibility of decoupling.

Solution:

The steady-state gain matrix for this subsystem is:

$$\mathbf{K} = \begin{bmatrix} 4.05 & 1.20 \\ 4.06 & 1.19 \end{bmatrix} \tag{22.81}$$

with determinant:

$$|\mathbf{K}| = -0.0525 \tag{22.82}$$

which is very close to zero, indicating that decoupling will be very difficult. The RGA
for this system calculated in the usual manner, is:

$$\Lambda = \begin{bmatrix} -91.8 & 92.8 \\ 92.8 & -91.8 \end{bmatrix} \tag{22.83}$$

Thus, because of the small value of the determinant of the gain matrix, and the
inordinately large values of the RGA elements, we conclude that decoupling will be
extremely difficult for this system.

Below, we will present other reasons why decoupling will in fact be next to
impossible for this system.

Degeneracy

When the determinant of a matrix is close to zero, the matrix is said to be
nearly *singular*; and one of the manifestations of singularity in a matrix is
linear dependence between the columns and/or the rows, or *degeneracy* of the
matrix.

To further illustrate this point, consider the problem of solving the
following system of two algebraic equations:

$$x_1 + 2x_2 = 1 \tag{22.84}$$
$$2x_1 + 4x_2 = 3 \tag{22.85}$$

It is clear that there is no solution, and the reason is that the elements of the LHS of both equations are linearly dependent; one is exactly twice the other. If these two equations are written in matrix form, we would have:

$$\begin{bmatrix} 1 & 2 \\ 2 & 4 \end{bmatrix} \begin{bmatrix} x_1 \\ x_2 \end{bmatrix} = \begin{bmatrix} 1 \\ 3 \end{bmatrix} \tag{22.86}$$

or

$$\mathbf{A}\mathbf{x} = \mathbf{b} \tag{22.87}$$

and ordinarily, the solution to Eq. (22.87), will be:

$$\mathbf{x} = \mathbf{A}^{-1}\mathbf{b} \tag{22.88}$$

Observe, however, that in this case $\mathbf{A}$ is:

$$\mathbf{A} = \begin{bmatrix} 1 & 2 \\ 2 & 4 \end{bmatrix} \tag{22.89}$$

which is singular; hence its inverse does not exist, and there is therefore no solution.

Eqs. (22.84) and (22.85) are called a *degenerate* set of equations; the determinant of the system matrix given in Eq. (22.89) is zero because the two columns (*and* the two rows) are linearly dependent. If now the equations are modified slightly to read:

$$x_1 + 2.01x_2 = 1 \tag{22.90}$$

$$2.01x_1 + 4x_2 = 3 \tag{22.91}$$

so that:

$$\mathbf{A} = \begin{bmatrix} 1 & 2.01 \\ 2.01 & 4 \end{bmatrix} \tag{22.92}$$

in principle, these equations now have a solution; however, small errors in the $\mathbf{b}$ vector will result in very large errors in the solution. These system of equations are now very nearly *degenerate*, and the solutions are not very reliable.

The same issue of *degeneracy* illustrated above applies to decoupling; for observe that what we are trying to do with decoupling is exactly the same as solving the system of linear equations:

$$\mathbf{K}\mathbf{u} = \mathbf{K}_R\mathbf{v} \tag{22.93}$$

to determine the decoupled process input vector $\mathbf{u}$; therefore, if $\mathbf{K}$ is degenerate, it will be extremely difficult to achieve decoupling.

To illustrate this issue of degeneracy in decoupling, and what it means in physical terms, let us return to the Prett and Garcia hot oil fractionator subsystem of Example 22.7; the steady-state model for this system is:

$$\begin{bmatrix} y_1 \\ y_2 \end{bmatrix} = \begin{bmatrix} 4.05 & 1.20 \\ 4.06 & 1.19 \end{bmatrix} \begin{bmatrix} u_1 \\ u_2 \end{bmatrix} \tag{22.94}$$

Note that the rows of the steady gain matrix are almost identical, and if we write out these equations in full to obtain:

$$y_1 = 4.05u_1 + 1.20u_2$$

and

$$y_2 = 4.06u_1 + 1.19u_2$$

we observe that the equation for y_1 is almost identical to y_2. Thus the system is degenerate.

The objective of decoupling is to make it possible for the control system to regulate y_1 and y_2 independently; however, the nature of this system is that *whatever y_1 does, y_2 must do, also*. It is structurally impossible to control these variables independently; thus for all intents and purposes, decoupling is impossible.

The degeneracy in this system has been brought about by the linear dependence of the *output* variables; the next example illustrates degeneracy brought about by near-linear dependence in the input variables.

Example 22.8   **DEGENERACY AND DECOUPLING CONTROL OF THE STIRRED MIXING TANK SYSTEM.**

Let us consider the stirred mixing tank expanded with an additional heater as in Example 21.10. If we fix the heater at a constant power and only consider the hot and cold stream flowrates as inputs, the linearized model is given by Eq. (22.95).

When some specific values are introduced for the physical dimensions of the stirred mixing tank system:

$$\mathbf{G}(s) = \begin{bmatrix} \dfrac{0.261}{43.5s + 1} & \dfrac{0.261}{43.5s + 1} \\ \dfrac{0.01235\,(T_H - T_s)}{21.7s + 1} & \dfrac{0.01235(T_C - T_s)}{21.7s + 1} \end{bmatrix} \qquad (22.95)$$

(These physical dimensions correspond to those for the stirred mixing tank system used for instruction at the University of Wisconsin–Madison.)

Investigate the possible degeneracy for this system under two conditions:

• *Condition 1*

Hot Stream Temperature: $T_H = 53°C$
Cold Stream Temperature: $T_C = 13°C$
Operating steady-state tank temperature: $T_s = 33°C$

• *Condition 2*

Hot Stream Temperature: $T_H = 28.5°C$
Cold Stream Temperature: $T_C = 28°C$
Operating steady-state tank temperature: $T_s = 33°C$

Observe that for Condition 2 the hot stream has cooled off substantially, and the cold stream has heated up significantly so that the desired steady-state is above both feed temperatures. This is only possible because of the auxiliary heater.

Solution:

Under Condition 1, the steady-state gain matrix is given by:

$$\mathbf{K} = \begin{bmatrix} 0.261 & 0.261 \\ 0.247 & -0.247 \end{bmatrix} \tag{22.96}$$

whose determinant is -0.129; the RGA for this system is obtained as:

$$\Lambda = \begin{bmatrix} 0.5 & 0.5 \\ 0.5 & 0.5 \end{bmatrix} \tag{22.97}$$

and even though the indication is that there will be significant interactions, we observe that decoupling is very feasible in this case. (We will show later what is required to achieve decoupling under these conditions.) Note that because of the -0.247 element, the rows and the columns of Eq. (22.96) are, in fact, *not* linearly dependent even though they may appear to be so.

The gain matrix under Condition 2 is:

$$\mathbf{K} = \begin{bmatrix} 0.261 & 0.261 \\ -0.0556 & -0.0618 \end{bmatrix} \tag{22.98}$$

and we observe immediately that if we rounded all the elements off to only the second decimal place, *Column* 1 will be identical to Column 2.

The determinant of this matrix is -0.00162, and the RGA is:

$$\Lambda = \begin{bmatrix} 10 & -9 \\ -9 & 10 \end{bmatrix} \tag{22.99}$$

and it is now clear that the system will be easier to decouple under Condition 1 than under Condition 2.

There is a physical reason for the increased degeneracy under Condition 2: observe that the difference between the hot and cold stream temperatures is only half of a degree. The implication is therefore clear: we no longer have two input variables, we essentially have only one (since the hot stream is almost identical to the cold stream), and as a result we have lost one degree of freedom in controlling the process.

Even though it is not crystal clear from the discussion given above, note that the degeneracy will affect only our control of the temperature; in no way does it affect our control of the level.

Less Obvious Ill-Conditioning

So far, we have seen that one possible measure of the feasibility of decoupling is how small the determinant of the process gain matrix $\mathbf{K}$ is, or, equivalently, how large the RGA elements for the process are. We have also seen that under circumstances in which $|\mathbf{K}| \approx 0$, or λ_{ij}'s are very large, there is usually something wrong with the physical system itself; either the output variables are not independent (as was the case with the Prett and Garcia fractionator) or the input variables are not independent (as with the water tank system of Example 22.8 under Condition 2).

There are however some situations in which the determinant is not too small, neither are the RGA elements too large, and yet decoupling is not feasible.

Consider for example the process whose gain matrix is:

$$\mathbf{K} = \begin{bmatrix} -60 & 0.05 \\ -40 & -0.05 \end{bmatrix} \tag{22.100}$$

The determinant for this matrix is 5; and the RGA is:

$$\Lambda = \begin{bmatrix} 0.6 & 0.4 \\ 0.4 & 0.6 \end{bmatrix} \tag{22.101}$$

neither of which is unusual. Nevertheless, a trained eye will spot the problem with this system upon inspecting Eq. (22.100) closely: observe that if u_1 and u_2 are of the same order of magnitude (i.e., same scaling), this matrix indicates that u_1 is the only effective control variable; the effect of u_2 on the process output variables is negligible in comparison. Thus, we have a situation somewhat similar to that illustrated with the stirred mixing tank in Example 22.8, under Condition 2; the system in Eq. (22.100) essentially has only one input variable; decoupling cannot be achieved in two output variables using only one input variable.

Compare this now with the process whose gain matrix is:

$$\mathbf{K} = \begin{bmatrix} -3 & 1 \\ -2 & -1 \end{bmatrix} \tag{22.102}$$

The two processes have identical determinants and identical RGA's, but while decoupling is not feasible in Eq. (22.100) it is entirely feasible in Eq. (22.102).

It is obvious from this example that neither the determinant nor the RGA is a reliable indicator of ill-conditioning; in this case, what indicates the difference between the two processes in Eqs. (22.100) and (22.102) are the eigenvalues of the two matrices.

The eigenvalues of the matrix in Eq. (22.100) are:

$$\lambda_1 = 59.965 \tag{22.103a}$$
$$\lambda_2 = 0.0835 \tag{22.103b}$$

while those for the other process in Eq. (22.102) are:

$$\lambda_1 = -2 + j \tag{22.104a}$$
$$\lambda_2 = -2 - j \tag{22.104b}$$

We may now note that while the product of each set of eigenvalues is the same (the product of the eigenvalues of a matrix is equal to its determinant) for the difficult-to-decouple system, one eigenvalue is about 720 times larger that the other; for the easy-to-decouple system, the two eigenvalues have *identical* magnitudes.

Unfortunately, eigenvalues are not always reliable indicators of ill-conditioning, as illustrated by the following process having a gain matrix:

$$\mathbf{K} = \begin{bmatrix} 1 & 0.001 \\ 100 & 1 \end{bmatrix} \tag{22.105}$$

The determinant for this matrix is 0.9; and the RGA is:

$$\Lambda = \begin{bmatrix} 1.11 & -0.11 \\ -0.11 & 1.11 \end{bmatrix} \tag{22.106}$$

both of which give absolutely no indication of the serious ill-conditioning problems afflicting the process.

The eigenvalues for this matrix are:

$$\lambda_1 = 1.316 \tag{22.107a}$$
$$\lambda_2 = 0.684 \tag{22.107b}$$

and the larger is less that twice the value of the smaller, again giving no indication of any problems with decoupling.

However, upon critical examination, we see that the problem with this process is that u_1 exerts 1000 times more influence on y_1 than does u_2; and 100 times more influence on y_2. Thus once more, we have a situation in which one input variable is significantly more influential than the other: another case of ill-conditioning.

As mentioned earlier in our introductory discussion in Chapter 20, the most reliable indicators of ill-conditioning in a matrix are its *singular values* defined as the square root of the eigenvalues of the matrix $\mathbf{K}^T\mathbf{K}$. (Since we are now dealing with constant matrices, we do not need to use the more general definition involving the transpose of the complex conjugate.)

For the process in Eq. (22.105) the singular values are:

$$\sigma_1 = 100.01 \tag{22.108a}$$

$$\sigma_2 = 0.009 \tag{22.108b}$$

The ratio of the largest to the smallest singular value of a matrix is called the *condition number*, and it is the single most reliable indicator of the conditioning of a matrix: the larger the condition number, the poorer the conditioning of the matrix.

The condition number for the process currently under investigation is:

$$\kappa = 1.113 \times 10^4$$

which is really quite large, clearly indicating its poor conditioning.

Let us further illustrate the use of singular values to assess the conditioning of a process by referring back to the processes of the earlier examples. First, let us consider the Prett and Garcia fractionator [7]. For this process, recall that the gain matrix is:

$$\mathbf{K} = \begin{bmatrix} 4.05 & 1.20 \\ 4.06 & 1.19 \end{bmatrix} \tag{22.109}$$

which has singular values $\sigma_1 = 5.978$, $\sigma_2 = 0.00878$ and a condition number:

$$\kappa = \frac{5.978}{0.00878} = 680.778$$

This clearly indicates serious ill-conditioning.

Similarly, the stirred mixing tank under Condition 1 stipulated in Example 22.8 has a gain matrix:

$$\mathbf{K} = \begin{bmatrix} 0.261 & 0.261 \\ 0.247 & -0.247 \end{bmatrix} \tag{22.110}$$

which has singular values $\sigma_1 = 0.3695$, $\sigma_2 = 0.3500$ and a condition number:

$$\kappa = \frac{0.3695}{0.35} = 1.06$$

indicating a very well-conditioned system.

However, recall that under Condition 2, the stirred mixing tank has a gain matrix:

$$\mathbf{K} = \begin{bmatrix} 0.261 & 0.261 \\ -0.0556 & -0.0618 \end{bmatrix} \tag{22.111}$$

which has singular values of $\sigma_1 = 0.1432$, $\sigma_2 = 0.000019$ and condition number:

$$\kappa = \frac{0.1432}{0.000019} = 7537$$

indicating that under Condition 2, the mixing tank is a poorly conditioned system.

Note therefore that it is impossible to assess the conditioning of a process *accurately*, and *reliably*, without knowing the singular values of the gain matrix $\mathbf{K}$; the reader is referred to Appendix D for how these singular values are obtained.

Let us now combine all the factors affecting the feasibility of decoupling for any multivariable system. The small determinant of $\mathbf{K}$ and the large RGA values are indicative of sensitivity and degeneracy, and the large condition number indicates ill-conditioning in a way in which neither the determinant, nor the RGA values, nor even the eigenvalues can. However, sensitivity and degeneracy are, in fact, only special cases of ill-conditioning; we therefore have the following as the final conclusion of our discussion in this section:

Feasibility of decoupling is directly related to the conditioning of the process gain matrix. Decoupling is only feasible to the degree that the process is well conditioned; it is virtually impossible to achieve decoupling in a poorly conditioned process.

22.4 STEADY-STATE DECOUPLING BY SINGULAR VALUE DECOMPOSITION

The Singular Value Decomposition (SVD) of the steady-state gain matrix of a process provides another approach to steady-state decoupling. If we recall from Appendix D that the SVD of a process gain matrix $\mathbf{K}$ can be written as:

$$\mathbf{K} = \mathbf{W} \, \Sigma \, \mathbf{V}^T \tag{22.112}$$

then applying the SVD of $\mathbf{K}$, the steady-state model becomes:

$$\mathbf{y} = \mathbf{W} \, \Sigma \, \mathbf{V}^T \mathbf{u} \tag{22.113}$$

and upon premultiplying by $\mathbf{W}^T$ and recalling (from Appendix D) the orthogonality properties of $\mathbf{W}$, we obtain:

$$\mathbf{W}^T\mathbf{y} = \Sigma \, \mathbf{V}^T\mathbf{u} \tag{22.114}$$

From Appendix D, we recall that when $\mathbf{K}$ is square, Σ is a diagonal matrix of singular values. This allows us to define new output variables η and new input variables μ, according to:

$$\eta = \mathbf{W}^T\mathbf{y} \tag{22.115}$$

and

$$\mu = \mathbf{V}^T\mathbf{u} \tag{22.116}$$

so that the process model Eq. (22.109) becomes:

$$\eta = \Sigma \, \mu \tag{22.117}$$

Because Σ is diagonal, Eq. (22.117) indicates a system that is completely decoupled at steady state.

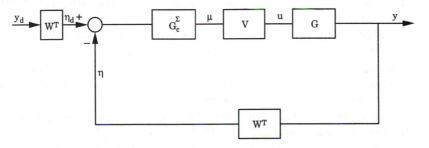

Figure 22.11. Block diagram of the multivariable controller based on the SVD technique.

The implication here is the following: instead of controlling **y** with **u**, the transformation of variables indicated in Eqs. (22.115) and (22.116) will convert our original system (with cross-coupling among the process variables) to the one indicated in Eq. (22.117) in which there is no cross-coupling (at steady state); furthermore, the open-loop gain of each loop in the "transformed" system is indicated clearly by the singular values and the conditioning automatically assessed from the condition number.

Thus a controller can now be designed for the equivalent (steady-state) system indicated in Eq. (22.117) which controls η using μ. If this controller is designated as G_c^{Σ} then the scheme would be implemented as shown in the block diagram in Figure 22.11.

Further details regarding the design of multivariable controllers using this technique are available in Lau, Alvarez, and Jensen [8]; successful applications of the SVD technique, and practical discussions are available in Downs and Moore [9], and in Refs. [10, 11].

22.5   OTHER MODEL-BASED CONTROLLERS FOR MULTIVARIABLE PROCESSES

It should be obvious that all of the controllers discussed so far in this chapter are *model-based multivariable controllers,* because computations from a dynamic or steady-state process model are required in their implementation. It is possible to obtain additional multivariable model-based controllers by extending the ideas of the SISO model-based controllers first developed in Chapter 19 to multivariable processes.   The analysis and mathematical manipulations are more involved, but the underlying concepts are the same as for SISO systems.   Some of the other controllers available are

1. **Multivariable Time-Delay and Inverse-Response Compensators:** Multivariable compensators which will provide model-based controllers for multivariable processes with multiple delays or RHP zeros are available based on the work of Ogunnaike and Jerome *et al.* [12–15].   These are designed in the spirit of the SISO Smith Predictor, but automatically handle dynamic loop interactions as well as provide excellent closed-loop response.

2. **Multivariable IMC and Direct Synthesis Controllers:** These are relatively straightforward extensions of their SISO counterparts in Chapter 19 (cf. the work of Morari *et al.* [16, 17] and others [18, 19]).

3. **Multivariable Frequency Domain Design Techniques such as the Inverse Nyquist Array and Direct Nyquist Array Techniques [19], Characteristic Loci Methods [20], Modal Control [21], and Singular Value and μ Synthesis Techniques [22, 23]:** These design methods use multivariable frequency domain analysis and model-based controller designs to minimize loop interactions and improve robustness to model/process mismatch.

4. **Optimal Control Designs such as Linear Quadratic Controllers, Dynamic Matrix Control, Model Algorithmic Control, and Other Constrained Multivariable Controllers [17, 21, 24, 25]:** These controllers are designed to minimize some specified optimization objective and to deal explicitly with constraints. The most successful of these in process control applications, Dynamic Matrix Control and Model Algorithmic Control, will be discussed in more detail in Chapter 27.

22.6 SUMMARY

The final stage in our discussion of controller design for multivariable systems has been presented in this chapter with a discussion of the principles and techniques of true multivariable control. We spent some time discussing decoupling (by far the most widely used multivariable controller design technique) being careful to differentiate dynamic decoupling from its simplified variations: steady-state decoupling and partial decoupling.

In discussing the issue of feasibility of decoupling (a topic that has important bearing on the application of decoupling but is quite often ignored), we have shown the various physical factors that affect the feasibility of decoupling and that singular value decomposition is the most reliable technique for assessing process decoupleability. In addition, we also discussed how the SVD provides an alternative method for designing steady-state decouplers.

Finally we have provided a directory of other model-based controller design techniques that can be applied to multivariable systems. The most popular of these will be discussed further in Chapter 27.

REFERENCES AND SUGGESTED FURTHER READING

1. McAvoy, T. J., *Interaction Analysis: Principles and Applications*, ISA, Research Triangle Park, NC (1983)
2. McAvoy, T. J., "Steady-State Decoupling of Distillation Columns," *I&EC Fundamentals*, **18**, 269 (1979)
3. Hulbert, D. G. and E. T. Woodburn, "Multivariable Control of a Wet Grinding Circuit," *AIChE J.*, **29**, 186 (1983)
4. Bristol, E. H., "On a New Measure of Interaction for Multivariable Process Control," *IEEE Trans. Autom Cont.*, **AC-11**, 133 (1966)
5. Shinskey, F. G., *Distillation Control for Productivity and Energy Conservation*, McGraw-Hill, New York (1977)
6. Shinskey, F. G., *Process Control Systems*, McGraw-Hill, New York, (1979)
7. Prett, D. M. and C. E. Garcia, *Fundamental Process Control*, Butterworths, Boston (1988)
8. Lau H., J. Alvarez, and K. F. Jensen, "Synthesis of Control Structure by Singular Value Analysis: Dynamic Measures of Sensitivity and Interaction," *AIChE J.*, **31**, 427 (1985)
9. Downs, J. J. and C. F. Moore, "Steady-State Gain Analysis for Azeotropic Distillation," *Proc. JACC*, **WP-7** (1981)

10. Moore, C. F., "Application of SVD for the Design, Analysis, and Control of Industrial Processes," *Proc. ACC*, **TA-1**, 643 (1986)

11. Bruns, D. D. and C. R. Smith, "Singular Value Analysis: A Geometrical Structure for MV Processes," *Preprint of the 1982 AIChE Winter Meeting*, Orlando, FL

12. Ogunnaike, B. A. and W. H. Ray, "Multivariable Controller Design for Linear Systems Having Multiple Time Delays," *AIChE J.*, **25**, 1043 (1979)

13. Ogunnaike, B. A., J. P. Lemaire, M. Morari, and W. H. Ray, "Advanced Multivariable Control of a Pilot Plant Distillation Column," *AIChE J.*, **29**, 632 (1983)

14. Jerome, N. F. and W. H. Ray, "High Performance Multivariable Control Strategies for Systems Having Time Delays," *AIChE J.*, **32**, 914 (1986)

15. Jerome, N. F. and W. H. Ray, "Model-Predictive Control of Linear Multivariable Systems Having Time Delays and RHP Zeros," *Chem. Eng. Sci.*, **47**, 763 (1992)

16. Garcia, C. and M. Morari, "Internal Model Control 2. Design Procedures for Multivariable Systems," *I&EC Proc. Des. Dev.*, **24**, 472 (1985)

17. Garcia, C., D. M. Prett, and M. Morari, "Model Predictive Control: Theory and Practice — a Survey", *Automatica*, **25**, 335 (1989)

18. Ogunnaike, B. A., "Design of Robust Control Systems by Direct Synthesis," *Preprints Annual AIChE Meeting*, paper 21h, San Francisco (1989)

19. Rosenbrock, H. H., *Computer Aided Control System Design*, Academic Press, New York (1974)

20. Maciejowski, J. M., *Multivariable Feedback Control Design*, Addison-Wesley, Reading, MA (1989)

21. Ray, W. H., *Advanced Process Control*, Butterworths, Boston (1989) McGraw-Hill, New York (1980)

22. Doyle, J. C., B. A. Francis, and A. R. Tannenbaum, *Feedback Control Systems*, Macmillan, New York (1992)

23. Morari, M. and E. Zafiriou, *Robust Process Control*, Prentice-Hall, Englewood Cliffs, NJ (1989)

24. Cutler, C. R. and B. L. Ramaker, "Dynamic Matrix Control — A Computer Control Algorithm," *Proc. JACC*, paper WPS-B (1980)

25. Mehra, R. A. and S. Mahmood, "Model Algorithmic Control," in *Distillation Dynamics and Control* (P. B. Desphande, Ed.), ISA, Research Triangle Park, NC (1985)

REVIEW QUESTIONS

1. In terms of how individual output feedback errors are utilized, what is the difference between a true multivariable controller and a set of multiple single-loop controllers?

2. What is the main objective in decoupling?

3. How is decoupling achieved in principle?

4. In terms of controller design and control system performance, what advantages accrue from being able to achieve perfect decoupling?

5. What is dynamic decoupling, and how does it differ from steady-state decoupling and partial decoupling?

6. What is involved in the design of "simplified decouplers"?

7. Upon implementing simplified decoupling for a 2 × 2 system, how are the individual *closed-loop* gains of the decoupled system related to the individual *open-loop* gains?

8.   What is the effect of a large RGA element on the effective closed-loop gains of a 2×2 system which employs two single-loop feedback controllers and simplified decouplers?

9.   Why is the simplified decoupling approach not recommended for higher dimensional systems?

10. What is involved in the design of "generalized decouplers"?

11. In terms of what must be specified *a priori*, and what is obtained as a result of the design, how does the simplified decoupling approach differ from the generalized approach?

12. What are the primary advantages and disadvantages of simplified versus generalized decoupling?

13. In what manner are the simplified ideal decouplers similar to feedforward controllers?

14. What conditions must be satisfed for ideal dynamic decouplers to be implementable?

15. Under what conditions will partial decoupling be an attractive alternative to full, ideal decoupling?

16. What is the main advantage of partial decoupling?

17. What are the main advantages of steady-state decoupling?

18. What further issues are important for the successful implementation of steady-state decouplers?

19. What is the effect of ill-conditioning on the feasibility of decoupling?

20. How can the determinant of a system's gain matrix be used to determine when decoupling will be difficult to achieve? Is this a foolproof means of assessing the feasibility of decoupling?

21. How can the RGA elements on which the input and output variables of a system are paired be used to determine when decoupling will be difficult to achieve? Is this a foolproof means of assessing the feasibility of decoupling?

22. What does it mean in physical terms if a process possesses a nearly degenerate steady-state gain matrix?

23. What is the most reliable indication of ill-conditioning of a process steady-state gain matrix?

24. Why is it virtually impossible to achieve decoupling in a poorly conditioned process?

25. How is SVD used to design steady-state decouplers?

26. What are some other model-based multivariable controller techniques available in the literature but not discussed in detail in this chapter?

PROBLEMS

22.1 (a) Implement the simplified *dynamic* decoupler designed, in Example 22.1, for the Wood and Berry distillation column, on the basis of the transfer function matrix:

$$\mathbf{G}(s) = \begin{bmatrix} \dfrac{12.8e^{-s}}{16.7s + 1} & \dfrac{-18.9e^{-3s}}{21.0s + 1} \\[3mm] \dfrac{6.6e^{-7s}}{10.9s + 1} & \dfrac{-19.4e^{-3s}}{14.4s + 1} \end{bmatrix}$$

Use two single-loop PI controllers with tuning parameters:

$$K_{c1} = 0.3; \quad 1/\tau_{I1} = 0.307;$$
$$K_{c2} = -0.05; \quad 1/\tau_{I2} = 0.107;$$

and compare the closed-loop system response to a step change of 0.75 in the y_1 set-point with that shown in Figure 22.7, for simplified steady-state decoupling.
(b) Now suppose that the "true" process transfer function matrix is:

$$\mathbf{G}(s) = \begin{bmatrix} \dfrac{15e^{-2s}}{15s + 1} & \dfrac{-25e^{-4s}}{16s + 1} \\[3mm] \dfrac{6e^{-5s}}{9s + 1} & \dfrac{-22e^{-4s}}{12s + 1} \end{bmatrix} \qquad \text{(P22.1)}$$

Implement on this "process" the same dynamic decouplers used in part (a), along with the same set of two single-loop controllers; compare the response to the same set-point change in y_1 with that obtained in part (a). Comment on the effect of plant/model mismatch on the performance of this set of dynamic decouplers.

22.2 (a) Carry out the RGA analysis for the Hulbert and Woodburn wet-grinding circuit process presented in Example 22.4, and confirm the input/output configuration implied in Figure 22.6. Interpret your results in terms of the extent of steady-state interaction to expect if three independent single-loop controllers are used to control the process.
(b) Assuming a 1-1/2-2/3-3 pairing, follow the procedure of Section 22.2.2 and design dynamic decouplers for this process using the simplified decoupling approach. State explicitly the transfer functions for the six compensator blocks shown in Figure 22.4.
(c) Implement the partial decouplers obtained in Example 22.4 along with three single-loop PI controllers with tuning parameters:

$$K_{c1} = 0.008; \quad \tau_{I1} = 217;$$
$$K_{c2} = 1303.8; \quad \tau_{I2} = 33;$$
$$K_{c3} = 0.00097; \quad \tau_{I3} = 47;$$

and obtain the closed-loop system response to a step change of $-50\ Nm$ in the torque (i.e., y_1) set-point. Plot all three responses, as well as the corresponding control actions. Comment on the effectiveness of the partial decouplers. Did the response of y_2 (total flow from the mill) suffer significantly as a result of having been left to operate without decoupling?

22.3 (a) Revisit the Hulbert and Woodburn wet-grinding circuit process, and carry out a steady-state SVD analysis; interpret the results. What is the condition number for this process?

(b) If the unit of y_2, the total flow from the mill, is now changed from m^3/s to cm^3/s, obtain the new transfer function matrix and repeat the SVD analysis. What is the new condition number?

(c) Implement on this process a steady-state SVD decoupler as in Figure 22.11, using the SVD matrices obtained in part (b), along with single-loop controllers with the following tuning parameters:

$$K_{c1} = \frac{1}{\sigma_1}; \; \tau_{I1} = 50;$$

$$K_{c2} = \frac{1}{\sigma_2}; \; \tau_{I2} = 50;$$

$$K_{c3} = \frac{1}{\sigma_3}; \; \tau_{I3} = 50;$$

where $\sigma_1, \sigma_2, \sigma_3$ are the computed singular values. Obtain the closed-loop system response to a step change of -50 Nm in the torque (i.e., y_1) set-point and compare the performance of this steady-state SVD decoupler with the partial dynamic decoupler of Problem 22.2 (c). Can you think of a systematic way to select integral time constants for single-loop controllers operating with a steady-state SVD decoupler?

22.4 The transfer function model for a solid-fuel boiler has been given by Johansson and Koivo[†] as:

$$\begin{bmatrix} y_1 \\ y_2 \end{bmatrix} = \begin{bmatrix} \dfrac{-e^{-2s}}{10s + 1} & \dfrac{-1}{10s + 1} \\ 0 & \dfrac{e^{-10s}}{60s + 1} \end{bmatrix} \begin{bmatrix} u_1 \\ u_2 \end{bmatrix}$$ (P22.2)

The unit of time is seconds, and the process variables are (in terms of deviations from their respective steady-state values):

$y_1 = $ Normalized boiler underpressure
$y_2 = $ Percent flue gas O_2 content
$u_1 = $ Damper position (%)
$u_2 = $ Secondary blower motor speed (rpm)

(a) Design a dynamic decoupler for this process first using the *simplified* approach, and then using the *generalized* approach, with G_R chosen as Diag[$G(s)$], and show that the two approaches yield *precisely* the same dynamic decoupler in this case. Comment on whether or not the decoupler is realizable.

(b) Ignore the time delay in the (1,1) element of the transfer function matrix and show in this case, that the resulting *dynamic* decoupler is *precisely* the same as the *steady-state decoupler*. Implement this decoupler on the process, using two single-loop controllers with tuning parameters:

$$K_{c1} = -2.5; \; \tau_{I1} = 12.5;$$
$$K_{c2} = 2.0; \; \tau_{I2} = 57.0;$$

[†] Johansson, L. and H. Koivo, "Inverse Nyquist Array Technique in the Design of a Multivariable Controller for a Solid-Fuel Boiler," *Int. J. Control*, **40**, 1077 (1984).

and obtain the closed-loop system response to a step change of 1% in the flue gas O_2 content set-point.

(c) Now augment the process with a matrix of additional delays:

$$\mathbf{D}(s) = \begin{bmatrix} 1 & 0 \\ 0 & e^{-2s} \end{bmatrix}$$

as illustrated in Figure 22.5, and repeat the design of the dynamic decoupler of part (a). Is this decoupler now realizable?

(d) Implement the decoupler obtained in part (c) on the augmented process using the same two single-loop controllers in part (b); obtain the closed-loop system response to the same 1% step change in the flue gas O_2 content set-point. Compare this closed-loop performance with that obtained in part (b) and comment on the advantages and disadvantages of the additional delay of 2 s in the flue gas O_2 content loop.

22.5 The model for the 2×2 subsystem derived from the heavy oil fractionator column presented by Prett and Garcia [7] was given in Example 22.7 as:

$$\begin{bmatrix} y_1 \\ y_2 \end{bmatrix} = \begin{bmatrix} \dfrac{4.05e^{-27s}}{50s + 1} & \dfrac{1.20e^{-27s}}{45s + 1} \\ \dfrac{4.06e^{-8s}}{13s + 1} & \dfrac{1.19}{19s + 1} \end{bmatrix} \begin{bmatrix} m_1 \\ m_2 \end{bmatrix} \qquad (P22.3)$$

The unit of time is minutes, and the steady-state gains are dimensionless. The subsystem variables are defined as follows:

$$y_1 = \text{Top end point}$$
$$y_2 = \text{Intermediate reflux temperature}$$
$$m_1 = \text{Top draw rate}$$
$$m_2 = \text{Intermediate reflux duty}$$

(a) The RGA for this subsystem (obtained in Example 22.7) indicates a $y_1 - m_2$ and $y_2 - m_1$ pairing to avoid structural instability. Rename the input variable m_2 as u_1 and m_1 as u_2 and rewrite the process model in terms of these new variables. Use this reorganized model to design a *steady-state* decoupler for this process using the generalized matrix approach, with $\mathbf{G}_R(0)$ chosen as Diag[$\mathbf{G}(0)$].

(b) Use the elements on the main diagonal of the reorganized process transfer function matrix, and the IMC "improved PI" tuning rules (Table 15.6, Chapter 15), to obtain two single-loop PI controllers for this process; choose $\lambda = 54$ for Loop 1, and $\lambda = 16$ for Loop 2. Explain why these are reasonable values for λ.

Implement these PI controllers on the process as indicated in the reorganized model, along with the steady-state decoupler obtained in part (a). Obtain the closed-loop system response to a unit step change in the y_1 set-point. Plot the output responses as well as the control actions. Comment on the performance of the closed-loop system.

(c) Repeat part (a) this time using the *simplified* decoupling approach. Implement this new steady-state decoupler along with the same two single-loop PI controllers used in part (b), and obtain the closed-loop system response for the same set-point change in y_1. Compare the closed-loop performance of this simplified steady-state decoupler with that of the generalized steady-state decoupler in part (b). Provide a reason for the observe difference in performance.

22.6 (a) Revisit the heavy oil fractionator of Problem 22.5; multiply the PI controller gains obtained in part (b) by 92.8, the relative gain parameter for the process; implement these new controllers along with the simplified decoupler designed in part (c), and obtain the closed-loop system response for the same set-point change in the y_1. Compare the new closed-loop system performance with the one obtained in Problem 22.5(c).

(b) To investigate the effect of plant/model mismatch on the strategy employed in part (a), implement the same steady-state decoupler of part (a), along with the new PI controllers, but now on a process whose transfer function matrix is:

$$\mathbf{G}(s) = \begin{bmatrix} \dfrac{1.19e^{-27s}}{45s + 1} & \dfrac{4.06e^{-27s}}{50s + 1} \\[2ex] \dfrac{1.20}{19s + 1} & \dfrac{4.05e^{-8s}}{13s + 1} \end{bmatrix} \tag{P22.4}$$

a *very slight* modification of the model used in the design of both the decoupler as well as the controllers. Obtain the closed-loop system response to the same unit step change in the y_1 set-point; plot the output responses as well as the control actions; and comment on the performance of the closed-loop system in this case.

22.7 In a study of temperature-based control of high-purity columns,[†] the following transfer function model was obtained for a methanol-ethanol column from data generated via pseudorandom binary signal (PRBS) testing:

$$\begin{bmatrix} y_1 \\ y_2 \end{bmatrix} = \begin{bmatrix} \dfrac{-33.89}{(98.02s + 1)(0.42s + 1)} & \dfrac{32.63}{(99.60s + 1)(0.35s + 1)} \\[2ex] \dfrac{-18.85}{(75.43s + 1)(0.30s + 1)} & \dfrac{34.84}{(110.50s + 1)(0.03s + 1)} \end{bmatrix} \begin{bmatrix} u_1 \\ u_2 \end{bmatrix} \tag{P22.5}$$

The unit of time is minutes, and the dimensionless process variables are defined as follows:

$$y_1 = \frac{T_{21} - 614.10°R}{30.0°R} \; ; \text{ the normalized tray 21 temperature}$$

$$y_2 = \frac{T_7 - 636.77°R}{30.0°R} \; ; \text{ the normalized tray 7 temperature}$$

$$u_1 = \frac{L - 3.384 \text{ mol/min}}{6.0 \text{ mol/min}} \; ; \text{ the normalized reflux flowrate}$$

$$u_2 = \frac{V - 3.856 \text{ mol/min}}{6.0 \text{ mol/min}} \; ; \text{ the normalized vapor flowrate}$$

(a) Is steady-state decoupling feasible for this process? Justify your answer adequately.

(b) Using the procedure discussed in Chapter 19, design two single-loop direct synthesis (PID) controllers for this distillation column, using the diagonal elements of the transfer function matrix model, with $\tau_r = 50.0$ for each loop. Using the generalized approach, design a steady-state decoupler for this process, with $\mathbf{G}_R(0)$ chosen as Diag$[\mathbf{G}(0)]$. Implement this decoupler with the two PID controllers and obtain the closed-loop system response to a 0.05 step change in the y_2 set-point. Plot the responses and comment on the performance.

[†] Chien, I.-L., and B. A. Ogunnaike, "Modeling and Control of High-Purity Columns," Preprints of the 1992 AIChE Annual Meeting, Paper 2b, Miami, FL, November 1992.

22.8   In a study of anesthesia control using predictive control techniques, Linkens and Mahfouf[†] presented the following transfer function anesthesia model:

$$\begin{bmatrix} y_1 \\ y_2 \end{bmatrix} = \begin{bmatrix} \dfrac{(10.6s + 1)e^{-s}}{(3.08s + 1)(4.81s + 1)(34.4s + 1)} & \dfrac{0.27e^{-s}}{(1.283s + 1)(1.25s + 1)} \\ 0 & \dfrac{-15.0e^{-0.42s}}{(2s + 1)} \end{bmatrix} \begin{bmatrix} u_1 \\ u_2 \end{bmatrix}$$

$$\text{(P22.6)}$$

The unit of time is minutes, and the process variables are (in terms of deviations from their respective steady-state values):

y_1 = Percent paralysis (a measure of muscle relaxation)
y_2 = Mean Arterial Pressure (MAP) (a measure of unconsciousness); (mm Hg)
u_1 = Injection rate of Atracurium (drug for producing muscle relaxation); (ml/min)
u_2 = Injection rate of Isoflurane (drug for inducing unconsciousness) (ml/min)

(a) Obtain a process reaction curve of y_1 (% paralysis) in response to a unit step input in u_1 (Atracurium injection rate); use this information to design a PID controller for the "muscle relaxation loop" using the Cohen and Coon tuning rules (Table 15.3, Chapter 15). Use the g_{22} element directly with the Cohen and Coon tuning rules to obtain the corresponding PID controller parameters for the "unconsciousness loop." Implement these two single-loop controllers and obtain the closed-loop system response to a step change of –10 mm Hg in the mean arterial pressure set-point.
(b) Use the simplified approach to obtain a dynamic decoupler for this process, and with adequate justification, simplify your result down to a lead/lag element. Implement this decoupler along with the single-loop controllers designed in part (a) and obtain the closed-loop system response to the same step change of –10 mm Hg in the mean arterial pressure set-point. Does the decoupler provide an improvement in closed-loop performance?

22.9   In Chapter 21, the transfer function model for a pilot scale, binary distillation column used to separate ethanol and water was given as:

$$\begin{bmatrix} y_1 \\ y_2 \\ y_3 \end{bmatrix} = \begin{bmatrix} \dfrac{0.66e^{-2.6s}}{6.7s + 1} & \dfrac{-0.61e^{-3.5s}}{8.64s + 1} & \dfrac{-0.0049e^{-s}}{9.06s + 1} \\ \dfrac{1.11e^{-6.5s}}{3.25s + 1} & \dfrac{-2.36e^{-3s}}{5s + 1} & \dfrac{-0.012e^{-1.2s}}{7.09s + 1} \\ \dfrac{-33.68e^{-9.2s}}{8.15s + 1} & \dfrac{46.2e^{-9.4s}}{10.9s + 1} & \dfrac{0.87(11.61s + 1)e^{-s}}{(3.89s + 1)(18.8s + 1)} \end{bmatrix} \begin{bmatrix} u_1 \\ u_2 \\ u_3 \end{bmatrix} \quad (21.38)$$

where the process variables are (in terms of deviations from their respective steady–state values):

y_1 = Overhead mole fraction ethanol
y_2 = Mole fraction of ethanol in the side stream
y_3 = Temperature on Tray #19
u_1 = Overhead reflux flowrate
u_2 = Side stream draw-off rate
u_3 = Reboiler steam pressure

[†] Linkens, D. A., and M. Mahfouf, "Supervisory Generalized Predictive Control (GPC) and Fault Detection for Multivariable Anesthesia," *Proceedings of the 2nd European Control Conference*, Groningen, Netherlands, p. 1999 (1993).

Because the RGA analysis indicates the presence of significant steady-state interactions, it is desired to utilize steady-state decoupling along with three single-loop controllers for this process.

(a) Evaluate the feasibility of steady-state decoupling.

(b) Design a steady-state decoupler using the generalized approach, with $G_R(0)$ chosen as $\text{Diag}[G(0)]$, then use a design technique of your choice to obtain appropriate PI controllers. Justify any design choices made. Implement the decoupler and the PI controllers on the process and obtain the closed-loop response to a +5°C change in the y_3 set-point. Plot the responses and comment on the controller performance. If you consider the performance to be unsatisfactory because of design choices you have made, redesign the controllers until you are satisfied with the performance.

22.10 Revisit the pilot scale, binary distillation process of Problem 22.9 and consider a partial decoupling strategy in which the second loop, involving the control of y_2, the sidestream mole fraction of ethanol, with u_2, the sidestream draw-off rate, is to operate without decoupling; dynamic decouplers are to be designed for Loops 1 and 3.

Using the simplified decoupling approach, and making any other decisions you deem necessary, design the decouplers, as well as two PI controllers for each of the two loops. As much as possible, justify any design decisions made.

Design a third controller for Loop 2 which is to operate without decoupling. Implement your design and obtain the closed-loop system response to a +5°C change in the y_3 set-point. Plot the responses and compare the control system performance with that obtained in Problem 22.9.

For this particular system, and on the basis of your design, comment on whether or not partial dynamic decoupling offers any advantages over full-scale *steady-state* decoupling.

22.11 Consider the *multivariable* PI controller described by the following transfer function matrix:

$$G_c(s) = K_c + K_I \frac{1}{s} \qquad (P22.7)$$

where K_c and K_I are full, $n \times n$ *matrices* respectively for the proportional and the integral parts of the controller. The problem with this controller is that there are $2 \times n \times n$ tuning parameters to be specified, and no established tuning methods comparable to those available for its classical single-loop counterpart.

Pullinen and Lieslehto[†] have presented some techniques for choosing the matrices in Eq. (P22.7) and have implemented some designs on the following distillation column model:

$$\begin{bmatrix} y_1 \\ y_2 \end{bmatrix} = \begin{bmatrix} \dfrac{-0.6e^{-1.6s}\%/\%}{(4.75s+1)} & \dfrac{0.28e^{-4.9s}\%/\%}{(12.75s+1)} \\[3mm] \dfrac{-0.27e^{-2.1s}\%/\%}{(4.25s+1)} & \dfrac{0.52\%/\%}{(14s+1)} \end{bmatrix} \begin{bmatrix} u_1 \\ u_2 \end{bmatrix} \qquad (P22.8)$$

The unit of time is minutes, and the dimensionless process variables, in terms of percent of range, are defined as follows:

[†] Pullinen, R. and J. Lieslehto, "Multivariable PI controller in Distillation Column Control," *Proceedings of the 2nd European Control Conference*, Groningen, Netherlands, p. 1747 (1993).

y_1 = Top tray temperature
y_2 = Bottom tray temperature
u_1 = Reflux flow set-point
u_2 = Reboiler steam valve position

(a) Implement on this process the multivariable PI controller with the following parameters:

$$\mathbf{K}_c = \begin{bmatrix} -2.8923 & 1.4336 \\ -3.9732 & 8.4310 \end{bmatrix} ; \quad \mathbf{K}_I = \begin{bmatrix} -0.528 & 0.282 \\ -0.276 & 0.612 \end{bmatrix} \quad \text{(P22.9)}$$

and obtain the closed-loop system response to a change of –2% in the top temperature. Plot the response.

(b) An unpublished alternative tuning method calls for choosing the matrices as follows:

$$\mathbf{K}_c = \Gamma^{-1} \Theta^{-1} \quad \text{(P22.10)}$$

where the elements of Γ, γ_{ij}, are obtained from the process transfer function matrix as:

$$\gamma_{ij} = \frac{K_{ij}}{\tau_{ij}} \quad \text{(P22.11)}$$

the ratio of the steady-state gain to the time constant for each transfer function element; and Θ is a diagonal matrix of tuning constants, one for each loop. The "integral gain" matrix is determined as:

$$\mathbf{K}_I = \mathbf{K}^{-1} \Theta^{-1} \quad \text{(P22.12)}$$

where $\mathbf{K}$ is the matrix of process steady-state gains.

Obtain the matrices $\mathbf{K}_c$ and $\mathbf{K}_I$ for this process using the matrix of tuning constants:

$$\Theta = \begin{bmatrix} 5.0 & 0.0 \\ 0.0 & 10.0 \end{bmatrix}$$

Implement this controller on the process and obtain the closed-loop system response to the same change of – 2% in the top temperature. Plot the responses and compare them with those obtained in part (a).

For extra credit, see if you can improve the performance by choosing different tuning parameters for the diagonal matrix Θ.

22.12 The mathematical model of a Continuous Stirred Tank Reactor (CSTR) in which an exothermic, first-order reaction is taking place, is given below in terms of dimensionless quantities:

$$\frac{dx_1}{dt} = \frac{1}{\tau}\left[-(u_2 + 1)(x_1 - d_2) + Da(1 - x_1)\exp\left(\frac{\gamma x_2}{x_2 + 1}\right) \right] \quad \text{(P22.13)}$$

$$\frac{dx_2}{dt} = \frac{1}{\tau}\left[(u_2 + 1)(d_1 - x_2) + \frac{BDa}{\gamma}(1 - x_1)\exp\left(\frac{\gamma x_2}{x_2 + 1}\right) - B(x_2 - u_1) \right] \quad \text{(P22.14)}$$

$$y_1 = x_1$$
$$y_2 = x_2$$

The output variables, y_1 and y_2, respectively the dimensionless outflow concentration and reactor temperature, are to be controlled with u_1 and u_2, respectively the dimensionless coolant temperature and input feedrate; d_1 and d_2, the disturbances, are respectively the dimensionless feed temperature and feed concentration. The model parameters in the operating steady-state are given as follows:

$$\tau = 1.0; \quad Da = 0.11; \quad \gamma = 20; \quad B = 7.0; \quad \text{and } \beta = 0.5.$$

An approximate linear model obtained for this process has the transfer function:

$$\mathbf{G}(s) = \begin{bmatrix} \dfrac{-1.24e^{-0.745s}}{2.2s + 1} & \dfrac{0.411e^{-0.083s}}{1.8s + 1} \\[3mm] \dfrac{1.124e^{-0.167s}}{1.99s + 1} & \dfrac{-0.244e^{-0.05s}}{1.64s + 1} \end{bmatrix} \tag{P22.15}$$

(a) Use this model and the Ziegler Nichols tuning rules (Table 15.2, Chapter 15) to design two single-loop PI controllers for this reactor; implement the controllers *on the approximate model* in Eq. (P22.15) along with a steady-state decoupler designed using the generalized approach. Obtain the closed-loop system response to a step change of 0.5 in the y_1 set-point.
(b) Implement the controllers and the decoupler of part (a) on the nonlinear system in (P22.13) and (P22.14) and obtain the closed-loop system response to the same step change of 0.5 in the y_1 set-point. Compare this performance with that obtained in part (a). Comment on how pronounced the effect of nonlinearity is on the closed-loop performance.

22.13 Via frequency response identification, the following model was obtained for a multicomponent distillation column:

$$\begin{bmatrix} y_1 \\ y_2 \end{bmatrix} = \begin{bmatrix} \dfrac{-0.6(1 - 3s)\%/\%}{(s + 1)} & \dfrac{0.15(1 - 0.9s)\%/\%}{(s + 1)} \\[3mm] \dfrac{-0.5(1 + 0.3s)\%/\%}{(s + 1)} & \dfrac{0.7(1 + 0.4s)\%/\%}{(s + 1)} \end{bmatrix} \begin{bmatrix} u_1 \\ u_2 \end{bmatrix} \tag{P22.16}$$

The process variables, in terms of percent of range, are defined as follows:

$$y_1 = \text{Top tray temperature}$$
$$y_2 = \text{Bottom tray temperature}$$
$$u_1 = \text{Reflux flowrate}$$
$$u_2 = \text{Steam flowrate}$$

The average liquid holdup time in the rectification section has been used to normalize time.
It has been suggested that we implement a *simplified dynamic* decoupler along with two single-loop PI controllers with the following tuning parameters:

$$K_{c1} = -0.23; \quad 1/\tau_{I1} = 1.0$$
$$K_{c2} = 1.42; \quad 1/\tau_{I2} = 1.0$$

(a) Attempt to design such a simplified decoupler using the provided model. What is wrong with the decoupler?

(b) Use the first-order Padé approximation:

$$e^{-\alpha s} \approx \frac{1 - \dfrac{\alpha}{2} s}{1 + \dfrac{\alpha}{2} s}$$

in *reverse* to approximate each element in the process model that has a right-half plane zero with a time delay and a mirror image left-half plane zero. Now use this modified model, augmented with an appropriate diagonal matrix of additional delays (see Figure 22.5), and design the simplified decoupler.

(c) Implement the decoupler designed in part (b) along with the controllers given, with K_{c2} changed to 0.27, and obtain the column response to a step change of 1% in the top temperature set-point. Was it necessary to reduce the value of K_{c2}? Why, or why not?

22.14 In a lime kiln process used in the paper industry for recovering lime from lime mud, the problem of producing good quality lime often reduces to controlling the *front end temperature* (y_{fet}), and the *back end temperature* (y_{bet}), by manipulating the *fuel flowrate* (u_{fuel}), and the *percent opening of the induced draft damper* (u_{damper}). A series of identification experiments aimed at obtaining an approximate linear model for one such industrial lime kiln yielded the following results:[†]

EXPERIMENT SET # 1
(Percent Opening of induced damper held constant: i.e., $u_{damper} = 0$):

$$y_{fet}(s) = \frac{1.66(°F/SCFM)}{39.0s + 1} u_{fuel}(s) \qquad \text{(P22.17a)}$$

$$y_{bet}(s) = \frac{0.34e^{-s}(°F/SCFM)}{8.9s + 1} u_{fuel}(s) \qquad \text{(P22.17b)}$$

EXPERIMENT SET # 2
(Fuel flowrate held constant: i.e., $u_{fuel} = 0$):

$$y_{fet}(s) = \frac{-1.74e^{-2s}(°F/\%)}{4.4s + 1} u_{damper}(s) \qquad \text{(P22.18a)}$$

$$y_{bet}(s) = \frac{1.40e^{-s}(°F/\%)}{3.8s + 1} u_{damper}(s) \qquad \text{(P22.18b)}$$

The unit of time is minutes.

(a) How should the input and output variables be paired if two single-loop controllers are to be used for this process? Support your recommendation adequately. How severely do you expect the two loops to interact?

(b) Implement multivariable PI controller:

$$G_c(s) = K_c + K_I \frac{1}{s}$$

[†] Charos, G. N., Y. Arkun, and R. A. Taylor, "Model Predictive Control of an Industrial Lime Kiln," *TAPPI Journal*, February 1991.

on this process, choosing the controller matrix elements as recommended in Eqs. (P22.11) and (P22.12) in part (b) of Problem P22.11, with the matrix of tuning parameters as:

$$\Theta = \begin{bmatrix} 30.0 & 0.0 \\ 0.0 & 4.0 \end{bmatrix} \qquad (P22.19)$$

Obtain the closed-loop system response to a change of + 100°F in the front end temperature set-point. Plot the responses of the input as well as the output variables.

(c) *For extra credit*, see if you can improve the control system performance in part (b) by choosing different tuning parameters for the diagonal matrix Θ *but making sure that the following constraints are satisfied*:

$$
\begin{array}{ccccc}
-140°F & \le & y_{fet} & \le & 160°F \\
-75°F & \le & y_{bet} & \le & 55°F \\
-60\ SCFM°F & \le & u_{fuel} & \le & 190\ SCFM \\
-9.5\% & \le & u_{damper} & \le & 18\%
\end{array}
$$

part IVC

COMPUTER
PROCESS
CONTROL

"My mind rebels at stagnati
Give me problems, give me wc
give me the most abstruse cryptogr
or the most intricate analy
and I am in my own proper atmosphe

Sherlock Holmes, *The Sign of I*
(Sir Arthur Conan Dc

INTRODUCTION TO
SAMPLED-DATA SYSTEMS

The universal drive for improved efficiency in the operation of industrial processes has led to more stringent demands on the performance of control systems demands that can no longer be met by relying solely on the traditional, analog control hardware. Fortunately, because of the dramatic changes in computer technology, the digital computer allows almost routine realizations of the most complex control algorithms at very low cost. As a result, the use of the digital computer to implement control schemes has now become so widespread that virtually all new plants are designed to operate under computer control in one form or another.

In our discussions up to this point, we have assumed that the process measurements available to the control system were continuous in time and the controller made continuous changes in control action. By contrast, when the process is being monitored and controlled by a digital computer, data are sampled only at discrete points in time, and changes in control action are taken only at these discrete times. Thus the sampled process and controller are known collectively as a "sampled-data system." Processes operating in this way acquire a new character, completely different from anything we have encountered so far. Our study of computer process control therefore begins in this chapter with an introduction to the characteristics of a process induced by its association with the digital computer.

23.1 INTRODUCTION AND MOTIVATION

23.1.1 The Digital Computer as a Controller

Reduced to its most fundamental essence, a controller is no more than a device that, given information about the error signal ε from a process, computes the control action required to reduce the observed error, using a predetermined control "law."

Viewed in this light, we see that the classical analog feedback controllers — which operate according to the PID control laws discussed earlier — are just *one* possible class of controllers; one can see that there are other devices, operating according to other control laws, that can also function as controllers.

The digital computer is perfectly suited to do the job of a controller with greater flexibility and versatility than the classical controllers because of the following advantages:

1. Data acquisition from the physical process is very easily and rapidly done with the digital computer.

2. A tremendous capacity for mass storage (and rapid retrieval) of the collected process information is readily available.

3. The control "law" for calculating corrective control action no longer has to be restricted to a hardwired analog circuit (e.g., for PID control) because any control law, no matter how unconventional or complicated, can usually be programmed on a computer.

4. The required computations, no matter how complex, can be carried out at very high speeds.

5. The cost of digital computers and ancillary equipment has reduced drastically in the last few years.

All these factors combine to make computer applications to the on-line implementation of process control schemes very attractive.

23.1.2 Idiosyncrasies of Computer Control

If we recall our discussion in Chapter 1, the three main tasks of a control system are:

*   Monitoring process variables by measurement
*   Calculation of required corrective control action
*   Implementation of the calculated control action

For a computer control system, all three tasks are carried out exclusively by the digital computer.

Observe that the use of the computer to carry out the second task is no different from its traditional use as a fast calculator; it is in carrying out the first, and the third, tasks that the unique character of the computer is brought to bear on the physical process in ways that require some attention.

Computers and Data Acquisition

Computers, by nature, deal only in "digitized" entities: integer numbers. The output signals from a process, however, are usually continuous (e.g., voltage signals from a thermocouple). Since the computer can only access information in digital form, computer control applications therefore require that the continuous process output signal be converted to digital form.

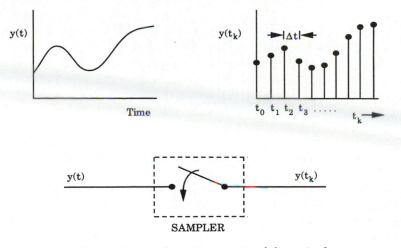

Figure 23.1. A continuous signal discretized.

As we discussed in Chapter 2, this "digitization" of a continuous signal is carried out by an analog-to-digital (A/D) converter. The continuous signal, $y(t)$, is sampled at discrete points in time, $t_0, t_1, t_2, \ldots$, to obtain the "sampled" data, $y(t_k)$, $k = 0, 1, 2, 3, \ldots$, which is then digitized.

As indicated in Figure 23.1 below, the sampler of the A/D converter consists of a switch that closes every Δt time units, theoretically for an infinitesimally small period of time, to produce "samples" of $y(t)$ as $y(t_k)$ "spikes."

Let us therefore note the following about computer control systems:

The use of the computer for data acquisition introduces a new perception of the process: a quantized view, in which the system is no longer seen as giving out continuous information, continuously in time; process information now appears to be available only in discrete form, at discrete points in time.

It is customary to refer to Δt, the length of the time interval between each successive sample, as the *sample time* or *sampling period*; naturally, its reciprocal, $1/\Delta t$, is called the *sampling rate*; and multiplying the sampling rate by 2π gives the sampling frequency.

Computers and Control Action

On the other side of the computer process control coin is the fact that the computer also *gives out* information in digital form, at discrete points in time. This has significant implications on control action implementation, as we now illustrate.

Consider the situation in which the computer gives out control command signals as indicated in Figure 23.2(a); note that the command signal takes on the value $u(t_k)$ at time instants $t_0, t_1, t_2, \ldots$, and is zero in between these sample points. In implementing such a control action sequence on a process whose final control element is, for example, an air-to-open control valve, we will find that the valve opens up by an amount dictated by the value that $u(t_k)$ takes at each

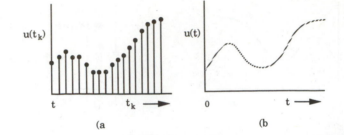

Figure 23.2. A discrete signal from the computer and the intended continuous version.

sampling point, and shuts off completely in between samples. Quite apart from the unusual wear and tear that such control action will have on the valve, it is obvious that the intention is not to have the control valve open up only at sampling points; the intended control action is shown in the dashed line in Figure 23.2(b).

Since the computer is incapable of giving out continuous signals, what is required is a means of reconstructing some form of "continuous" signal from these impulses. The device for doing precisely this is known as a digital-to-analog (D/A) converter. The key elements in D/A converters are called *holds*; they are designed to simply "hold" the previous value of the signal (or its slope, or some other related function) until another sample is made available. We will discuss these further later.

The data acquisition and control action implementation aspects of the computer control systems may be combined to produce Figure 23.3, a typical block diagram for a computer control system — an equivalent of that shown in Figure 14.2 for the analog feedback control system.

Mathematical Descriptions

As a result of the effects of using the computer for data acquisition and control action implementation, computer control systems are often referred to by such terms as *sampled-data* systems, *discrete-time* systems, or *digital control* systems, to indicate explicitly this quantized characteristic bestowed upon it by the digital computer. Such systems must now be described, mathematically, in a manner that reflects their "new" character of giving out, and receiving,

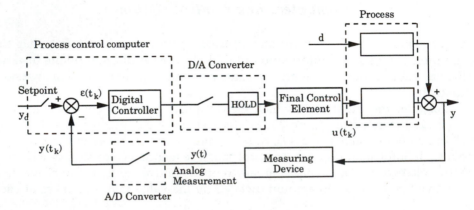

Figure 23.3. Typical block diagram for a computer control system.

information only at discrete points in time. Computer control systems are therefore more conveniently represented by discrete-time models of the type briefly introduced in various sections of Chapter 4.

The three major issues raised by the use of digital computers for control system implementation may therefore be summarized as follows:

1. Sampling (and conditioning) of continuous signals
2. Continuous signal reconstruction
3. Appropriate mathematical description of sampled-data systems

We shall now consider each of these important issues in turn.

23.2 SAMPLING AND CONDITIONING OF CONTINUOUS SIGNALS

23.2.1 Sampling

Sampling a continuous signal is achieved, in practice, with the aid of a *sampler*: a device that, as indicated in Figure 23.1, operates physically like a switch that stays closed for a very short, but finite, period, t_ε seconds; this action is repeated every Δt seconds.

Quantitative Description of Sampled Signals

Consider the arbitrary, continuous function, $f(t)$, sampled every Δt seconds, at specific points in time represented by t_k; $k = 0, 1, 2, \ldots$ It is customary to consider the sampling as commencing at time $t = 0$ such that $t_0 = 0$; $t_1 = \Delta t$; $t_2 = 2\Delta t$; and in general:

$$t_k = k\Delta t \tag{23.1}$$

It is also customary to represent the sampled version of $f(t)$ as $f^*(t)$. Let us now derive an expression for relating these two functions, the continuous $f(t)$, and its sampled version, $f^*(t)$, because this relationship will be required in Chapter 24 when we take up the analysis of sampled-data systems.

Note that the action of the sampler is such that during the kth sampling interval, the sampler output $f^*(t)$ takes the value $f(t = t_k)$ (i.e., $f(t_k)$), for t_ε seconds; immediately afterwards, the sampler output is zero (cf. Figure 23.1). Thus the complete $f^*(t)$ is a train of impulses whose values match that of the continuous function $f(t)$ *only* at the sampling points, and remain at zero elsewhere. If we assume ideal sampler operation in which measurement is provided instantaneously ($t_\varepsilon = \varepsilon \to 0$) then a *single* sampler output at time t_k is most conveniently modeled using the ideal delta (impulse) function as follows:

$$f^*(t_k) = \int_{t_k - \varepsilon}^{t_k + \varepsilon} f(t) \, \delta(t - t_k) \, dt \tag{23.2}$$

The key property of the delta function that makes this description possible is shown below for any arbitrary function $x(t)$:

$$\int_{t-\varepsilon}^{t+\varepsilon} x(t)\, \delta(t-\theta)\, dt = \begin{cases} x(\theta); & t = \theta \\ 0; & t \neq \theta \end{cases} \tag{23.3}$$

That is, the operation of $\delta(t-\theta)$ on any function combined with the indicated integration has the net effect of "wiping out" the original function, leaving a single, surviving "spike" at the point $t = \theta$ whose magnitude coincides with the value of the original function at this single point.

Since what we observe as the sampler output is a sequence of such impulses as that represented in Eq. (23.2), then $f^*(t)$ can be represented as:

$$f^*(t) = \int_0^\infty \sum_{k=0}^\infty f(t)\, \delta(t-t_k)\, dt \tag{23.4a}$$

or

$$f^*(t) = f^*(t_0) + f^*(t_1) + \dots + f^*(t_k) + \dots \tag{23.4b}$$

It is important to recognize that each of the entities indicated in this sum remains permanently at zero until t equals the argument in the parentheses, when it instantaneously takes the indicated value; thereafter, it promptly returns to zero once more. Thus, for example, when $t = t_1$, everything else but $f^*(t_1)$ remains at zero. With this in mind, we may now write Eq. (23.4) more compactly as follows without the risk of confusion:

$$f^*(t) = \sum_{k=0}^\infty f(t_k)\, \delta(t-t_k) \tag{23.5}$$

This is the quantitative representation of the output signal of an ideal impulse sampler whose input is the continuous function $f(t)$.

Choosing Sampling Time, Δt

The process of sampling a continuous signal is, in practice, not as straightforward as it appears at first. The main question to be answered is: *how frequently should we sample?* That is, how should Δt, the sampling interval, be chosen? Surprisingly, this deceptively simple question can be very difficult to answer conclusively.

Let us illustrate the issues involved with choosing Δt by considering the following two extreme situations:

1. **Sampling too rapidly**: The sampled signal, $y(t_k)$, very closely resembles the continuous signal, $y(t)$; but this will obviously require a large number of samples and a large amount of data storage. In addition, this increased work load can limit the number of other loops the computer can service.

2. **Sampling too slowly**: In this case, fewer samples will be taken, and as a result, there will be no unusually heavy burden on the computer. However, the main problem is that $y(t_k)$ may no longer resemble the original, continuous signal, $y(t)$. This problem is known as *aliasing*, and

a classic example may be used to illustrate its effect in Figure 23.4. If the original sinusoidal function, shown in Figure 23.4(a), is sampled four times every three cycles, the result is the sampled signal shown by X's in Figure 23.4(b). If we attempt to reconstruct the original sinusoidal signal from this sampled data, we obtain the apparent sinusoidal signal shown in Figure 23.4(b), whose frequency is clearly seen to be lower. Of course, typical process signals are not sinusoidal, but the example serves to illustrate the problem of capturing high-frequency information when sampling too slowly. The other problem associated with sampling too slowly is that the effectiveness of the control system is often substantially reduced. This stems from the fact that in between samples, the control system actually operates as an open-loop system since no control action is taking place during the entire period in between samples. Thus if the sampling interval is very long the performance will clearly deteriorate.

From such considerations, it is thus clear that the "optimum" sampling time must lie somewhere in between these two extremes. Unfortunately, there are no hard and fast rules for choosing Δt. A prudent choice of Δt is clearly one which is small enough to guarantee that no significant dynamic information is lost in sampling, but which is not so fast that the computer will become overloaded.

If the fastest dynamics one wishes to capture by sampling is known, say, this is frequency ω_{max}, then one must have a *sampling frequency:*

$$\omega_s = \frac{2\pi}{\Delta t} \tag{23.6}$$

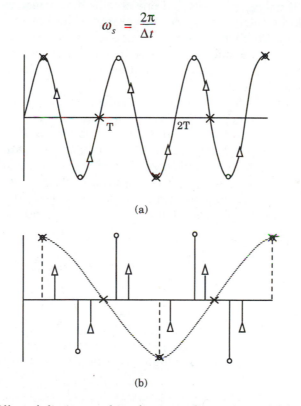

(a)

(b)

Figure 23.4.   Effect of aliasing resulting from sampling a sinusoidal function too slowly:
$\times - \Delta t = 3T/4$; o $- \Delta t = T/2$; $\Delta - \Delta t = T/2$.

which satisfies the *sampling rule:*

$$\omega_s > 2\omega_{max} \tag{23.7}$$

Alternatively, for any given sampling frequency ω_s, the largest frequency that can be extracted from sampled data, called the *Nyquist frequency* ω_N, can be determined from the equality in Eq. (23.7) as:

$$\omega_N = \frac{\omega_s}{2} = \frac{\pi}{\Delta t} \tag{23.8}$$

The validity of the sampling rule in Eq. (23.7) can be seen from Figure 23.4 where we observe that if we want to capture a sine wave, which has period T and frequency, $\omega = 2\pi/T$, then we must sample with frequency:

$$\omega_s > \frac{4\pi}{T} \tag{23.9}$$

By comparing Eqs. (23.8) and (23.9), we see that this means choosing $\Delta t < T/2$ in Figure 23.4. Note that if we were to choose $\Delta t = T/2$ and had our first sample at $T/4$, then we obtain the sequence of samples shown with circles in Figure 23.4, and we have captured the sine wave in sampled form. However, choosing Δt to barely satisfy the sampling rule, $\Delta t = T/2$, does not capture all the details of the sine wave. It only guarantees that one obtains the correct frequency, but the amplitude could be totally incorrect. For example, suppose one sampled with $\Delta t = T/2$, but took the first sample at $T/3$ as shown by triangles in Figure 23.4. In this case the sampled sine wave has reduced amplitude; in fact if one took the first sample at $T/2$, the resulting signal would appear constant because the sampled amplitude is zero.

The lesson here is that one must consider carefully the specification of ω_{max}, the highest frequency one wishes to capture, and allow sufficiently rapid sampling to uncover the details of the signal. Obviously this means that the choice of Δt will very much depend on the particular application at hand. For digital computers that are simply pieces of hardware emulating continuous analog controllers, rapid sampling is important. However, digital control algorithms based on less frequent sampling are usually used. In this case the following rule-of-thumb has been found to provide a good starting choice for sampling time:

For a process with a dominant time constant τ_{dom}, it is recommended that the sample time, Δt, be chosen as:

$$\Delta t \approx 0.2\tau_{dom} \tag{23.10}$$

This rule will ensure, for example, that for a first-order system with time constant τ, no more than 20–25 samples will be required to represent a step response.

23.2.2 Signal Conditioning

The sampled signal is seldom in a condition to be used directly by the computer. For example, temperature measurement signals from a thermocouple are usually in millivolts. Such low-voltage signals are often too weak to be transmitted over some distance to the A/D converters; they will be susceptible to corruption by noise, and will have very low resolution by the time they reach the A/D converters. Therefore, a certain degree of "conditioning" is needed before the sample signals can be useful.

As discussed in Chapter 2, signal conditioning usually involves the following:

- Amplification of weak signals
- Noise suppression by filtering
- Possible multiplexing

Even though the usual practice is for instrumentation engineers to take care of the issue of signal conditioning at the hardware design and selection stage, this only ensures that whatever signal was produced by the measurement device reaches the computer intact. Once the data are in the computer, additional digital filtering may be required to enhance the precision of the measurements.

The simplest, and perhaps the most popular, digital filter is the exponential filter which works according to the following equation:

$$\hat{y}(k) = \beta \hat{y}(k-1) + (1 - \beta)\, y(k) \tag{23.11}$$

where:

$\hat{y}(k)$ is the filtered value of the signal at the sampling instant k (when $t = t_k$)

$y(k)$ is the measured signal value at the sampling instant k

β is the filter constant $0 < \beta < 1$

This filter is classified as a *low-pass* filter because it allows only the low-frequency components of a signal to pass through, the higher frequency components having been substantially reduced. Other types of digital filters are also used in practice.

It is important to note that while filters help suppress the effect of high-frequency fluctuations, they may also introduce additional dynamics into the control system, and hence can affect the dynamic behavior of the overall control system. For example, the exponential filter of Eq. (23.11) can be shown to be the discrete-time version of a first-order system with a steady-state gain of 1, a time constant related to β and Δt, and having as input the "noisy" signal $y(t)$. If such a filter is used in a feedback control system, it will be important to take the additional filter dynamics into consideration if the filter time constant is comparable to the process time constants.

23.3   CONTINUOUS SIGNAL RECONSTRUCTION

The other main consequence of computer control that must be carefully addressed is that, as illustrated in Figure 23.2(a), control commands generated by the computer are in the form of individual impulses, issued at discrete-time points $t_0, t_1, t_2, ..., t_k, ...$ Keep in mind that such impulses take on nonzero values only at the indicated discrete points in time; in between samples, they are zero.

As explained earlier, the objective is clearly not to implement control action in terms of such impulses, but as the continuous function indicated in the dashed line in Figure 23.2(b). Unfortunately, only the sampled versions of the desired continuous function — the indicated "spikes" — are all that are available. The problem is now clearly seen as that of reconstructing a continuous signal, *given only the values it takes at certain discrete points in time*.

Note that this is the reverse of sampling. With sampling, the continuous signal $y(t)$ is available; by choosing an appropriate value for Δt — the time interval between samples — the sampled version $y(t_k)$ is obtained using a sampler. In this case, however, what we have available is the discrete control command signal, $u(t_k)$, from which we must now reconstruct a continuous signal, $u(t)$.

The first point to note about continuous signal reconstruction is that there is no unique solution to the problem: there are many different ways by which one can make a discrete signal take on a continuous form. Thus depending on what strategy is adopted, several different kinds of "continuous" versions are possible for the *same* discrete signal.

Holds

The basic essence of the signal reconstruction problem is really that of connecting discrete points with some curve to give the otherwise discrete signal a semblance of being continuous. From a purely mathematical point of view, it would appear as if the best way to connect a smooth curve through discrete points is by employing splines or other interpolation device. This approach is, however, not very practical, for two reasons:

1.   In practice, the discrete points to be connected by the smooth curve are control commands issued sequentially by the computer. This goes directly against the principles underlying the application of splines, since they are typically used to connect a "static" collection of points, not a set of points made available sequentially.

2.   Even if the process of connecting points with splines could be made sequential, the effort required to make this possible appears disproportionate to any potential benefits to be realized.

A very simple, and intuitively appealing, approach to this problem is to simply *hold* the value of the discrete signal constant at its previous value over the sampling interval until the next sampled value is available; the control variable then takes on this new value that is also held over the next sampling interval, etc.   Mathematically, this means that $u(t)$ and $u(t_k)$ are related according to:

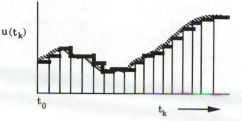

Figure 23.5.   Continuous signal reconstruction using a zero-order hold.

$$u(t) = \begin{cases} u(t_k) ; & t = t_k \\ u(t_k) ; & t_k < t < t_{k+1} \\ u(t_{k+1}) ; & t = t_{k+1} \end{cases} \qquad (23.12)$$

As shown in Figure 23.5, this leads to the "stair-case" function, which, even though not an exact representation of the continuous function in dashed lines, is good enough, and very practical.

Physical devices which make it possible to generate a signal as illustrated above by *holding* the value of the discrete signal in a certain fashion over the sampling interval until new information is available are referred to as *holds*. In particular, what is indicated in Eq. (23.12), and illustrated in Figure 23.5, is achieved by a *zero-order hold* (ZOH), so called because of the "holding" policy of simply maintaining the previous discrete value over the sampling interval.

The *first-order hold*, on the other hand, "holds" at t_k by linearly extrapolating between $u(t_{k-1})$ and $u(t_k)$ until $u(t_{k+1})$ becomes available, i.e.:

$$u(t) = \begin{cases} u(t_k); & t = t_k \\ u(t_k) + \left[\dfrac{u(t_k) - u(t_{k-1})}{\Delta t} \right] (t - t_k); & t_k < t < t_{k+1} \\ u(t_{k+1}); & t = t_{k+1} \end{cases} \qquad (23.13)$$

A reconstruction of the signal in Figure 23.2 using a first-order hold element is shown in Figure 23.6.

Observe that we can extend the foregoing ideas and develop equations for the operation of second, third, and higher order hold elements. However, this

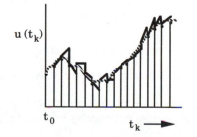

Figure 23.6.   Continuous signal reconstruction using a first-order hold.

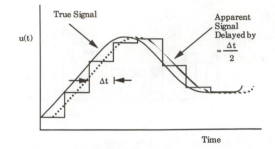

Figure 23.7.   Apparent signal delay induced by the ZOH element.

has the disadvantage that the implementation of a first-order hold requires *two* initial starting values from which to begin the linear extrapolation, and in general, the nth-order hold element will require $n + 1$ initial starting values. Furthermore, it is not always true that the reconstructed signals obtained using higher order holds are better approximations of the actual continuous signal. In fact, there are situations when, for example, the first-order hold will give a poorer approximation than the simple ZOH.

Thus, because higher order holds require more starting values, put more computational burden on the computer, and do not offer much improvement over the simple ZOH, most process control applications use the ZOH element. Unless otherwise stated, when we refer to a hold element in our subsequent discussion, we are referring to the ZOH.

One final point about the ZOH: because of its strategy of holding the value of the discrete signal constant over the sampling interval, the ZOH generates an apparent signal that lags behind the true signal by approximately 1/2 of the sampling time (see Figure 23.7). This fact is sometimes important in the design of digital feedback controllers, especially when Δt is large.

23.4   MATHEMATICAL DESCRIPTION OF DISCRETE-TIME SYSTEMS

From our discussions, it should be clear that it is no longer appropriate, under a sampled-data computer control strategy, to represent a process by the continuous differential equation, or transfer function models used up to now. An appropriate mathematical representation must take cognizance of the fact that the process gives out, and receives, information at discrete points in time, no longer continuously. Thus, as introduced in Chapter 4, discrete-time models for sampled-data systems take the form of *difference equations* (instead of differential equations), pulse transfer functions (instead of Laplace domain transfer functions), and *discrete* impulse-response models involving summations, instead of integrals. The development of such discrete-time representations is the main concern of these final sections of the chapter.

23.4.1 Discrete-Time Difference Equation Models

Discrete-time difference equation models were briefly introduced in Chapter 4; however, here we will generalize the model further to the form:

$$\mathbf{x}(k + 1) = \Phi\mathbf{x}(k) + \beta\mathbf{u}(k) + \sum_{i=1}^{p} \beta_i \mathbf{u}(k - D_i) + \Gamma\mathbf{d}(k) + \sum_{j=1}^{q} \Gamma_j \mathbf{d}(k - M_j) \quad (23.14)$$

$$\mathbf{y}(k) = \mathbf{C}\mathbf{x}(k) \tag{23.15}$$

Here we see that Eqs. (23.14) and (23.15) have terms $\mathbf{u}(k - D_i)$ that allow delays in the control inputs and terms $\mathbf{d}(k - M_j)$ that allow delays in the disturbances. For example, if there were a time delay α in the inputs, this corresponds to $D = \alpha/\Delta t$ being subtracted from k in the input argument. It would also be possible to include terms for delayed states or delayed measurements, but we will not deal with that case here. These models are either derived from the corresponding continuous model or constructed directly from experimental data (see the next two sections). The premise behind Eqs. (23.14) and (23.15) is that they represent the state and output dynamics at sample times $t_k = k\Delta t$ when the controls $\mathbf{u}$ and disturbances $\mathbf{d}$ are applied with a zero-order hold (ZOH). From our discussions in the last section, it is clear that having $\mathbf{u}$ enter the process with a ZOH is straightforward. However, the process disturbances, $\mathbf{d}$, are really continuous variables, and approximating them as constant over a sampling interval, Δt, may not always be appropriate. This point will be discussed in more detail in subsequent chapters.

In order to develop better insight into difference equation models, a few common ones are listed in Table 23.1. These forms are explicitly derived below and in the problems at the end of the chapter.

23.4.2 Discrete-Time Impulse-Response Models

The discrete-time impulse-response model can be found from the difference equation form by letting $u(k) = \delta_{ko}$ (a unit pulse at $k = 0$) in Eqs. (23.14) and (23.15). This leads to $g(k)$, and $g_d(k)$ which can be convoluted with the inputs u and d respectively, to yield the process outputs from:

$$y(k) = \sum_{i=1}^{k} g(k - i)u(i) + \sum_{i=1}^{k} g_d(k - i)d(i) \tag{23.16}$$

for the SISO case or:

$$\mathbf{y}(k) = \sum_{i=1}^{k} \mathbf{G}(k - i)\mathbf{u}(i) + \sum_{i=1}^{k} \mathbf{G}_d(k - i)\mathbf{d}(i) \tag{23.17}$$

in the multivariable case. The convolution identity:

$$\sum_{i=1}^{k} g(k - i)u(i) = \sum_{i=1}^{k} g(i)u(k - i) \tag{23.18}$$

which is valid for any convolution operation may be used to make the computations more convenient. Remember that these impulse-response models include a zero-order hold.

Impulse-response models for some common cases are also shown in Table 23.1.

23.4.3 Discrete-Time Transform-Domain Models

The discrete-time analog to the Laplace transform model is known as the *pulse transfer function model* or the *z-transform model*. These models are discussed in detail in Chapter 24 so that here we only indicate their form for some common cases in Table 23.1.

Table 23.1
Some Common Forms of Discrete-Time Models[*]

Continuous Model	Difference Equation $(t = k\Delta t)$	Impulse Response $g(k)$	z-Transform $g(z)$
First Order: $\tau \dfrac{dy\,(t)}{dt} + y(t) = Ku(t)$ $g(s) = \dfrac{K}{\tau s + 1}$	$y(k+1) = \phi y(k)$ $\quad + K\,(1-\phi)\,u(k)$ $\phi = e^{-\Delta t/\tau}$	$K\phi^{k-1}\,(1-\phi)$	$\dfrac{K\,(1-\phi)z^{-1}}{1 - \phi z^{-1}}$
First Order with Time Delay: $\tau \dfrac{dy}{dt} + y = Ku(t-\delta)$ $g(s) = \dfrac{Ke^{-\delta s}}{\tau s + 1}$	$y(k+1) = \phi y(k)$ $\quad + K\,(1-\phi)\,u(k-m)$ $\phi = e^{-\Delta t/\tau}$ $\delta = m\Delta t$	$K\phi^{k-m-1}\,(1-\phi)$	$\dfrac{K\,(1-\phi)\,z^{-1-m}}{1 - \phi z^{-1}}$
Second Order: $\tau^2 \dfrac{d^2 y(t)}{dt^2} + 2\zeta\tau \dfrac{dy(t)}{dt}$ $\quad + y = Ku(t)$ $g(s) = \dfrac{K}{\tau^2 s^2 + 2\zeta\tau s + 1}$	$y(k+1) =$ $2\alpha\beta y k) - \alpha^2 y\,(k-1) +$ $b_1\,u(k) + b_2\,u\,(k-1)$ where $b_1 = K\left[1 - \alpha\left(\beta + \dfrac{\zeta}{\omega\tau}\gamma\right)\right]$ $b_2 = K\left[\alpha^2 + \alpha\left(\dfrac{\zeta}{\omega\tau}\gamma - \beta\right)\right]$ $\omega\tau = \sqrt{1 - \zeta^2}$ $\alpha = e^{-\zeta\Delta t/\tau}$ $\beta = \cos(\omega\Delta t)$ $\gamma = \sin(\omega\Delta t)$	(cf. Appendix C)	$\dfrac{\left(b_1 z^{-1} + b_2 z^{-2}\right)}{1 - 2\alpha\beta z^{-1} + \alpha^2 z^{-2}}$ where $b_1 = K\Big[1 - \alpha$ $\qquad \times\left(\beta + \dfrac{\zeta}{\omega\tau}\gamma\right)\Big]$ $b_2 = K\Big[\alpha^2 + \alpha$ $\qquad \times\left(\dfrac{\zeta}{\omega\tau}\gamma - \beta\right)\Big]$ $\omega\tau = \sqrt{1 - \zeta^2}$ $\alpha = e^{-\zeta\Delta t/\tau}$ $\beta = \cos(\omega\Delta t)$ $\gamma = \sin(\omega\Delta t)$

[*]These models all incorporate a zero-order hold on the input u.

23.5 THEORETICAL MODELING OF DISCRETE-TIME SYSTEMS

23.5.1 Preliminary Considerations

As with all processes, mathematical models can be developed for discrete-time systems using either the theoretical or the empirical approach. This section is concerned with the theoretical approach to the development of discrete-time models for sampled-data systems.

The theoretical approach to chemical process modeling, as we recall from Chapter 12, gives rise to models in the form of differential equations — models in which the process variables occur as *continuous* functions of time. Thus there is no *direct* way of obtaining, from first principles, theoretical process models that will naturally occur in discrete time. The reason for this state of affairs is that the discrete nature of sampled-data systems is not an intrinsic part of the process; it has been artificially induced by association with the computer. Since the theoretical approach seeks to represent only what is intrinsic to the process, it is impossible for the externally — and artificially — induced discretization of time to be *naturally* manifested in the theoretical model. Hence:

> It is only by discretization of the (continuous-time) theoretical models that "theoretical" discrete-time models are developed for sampled-data systems.

In general, discrete-time models can be obtained from available continuous versions in the following ways:

1. *Discretization by finite-difference approximations.* This is the quickest, but least accurate method of discretization. It is seldom used.

2. *Exact discretization from exact analytical solutions.* This is an exact method which employs the analytical solutions available for linear constant coefficient models. It is the most commonly employed method.

3. *Direct conversion of the continuous Laplace transfer function to the pulse transfer function.* This involves the use of certain tools of dynamic analysis of discrete-time systems to be discussed in Chapter 24.

Our discussion in this chapter will be limited to exact, analytical discretization primarily because theoretical models occur naturally in the form of differential equations and not as transform-domain transfer functions. Chapter 24 is the more appropriate environment for discussing the third method.

23.5.2 Exact Discretization of Continuous Models

By obtaining the exact analytical solution to a continuous-time process model for *piecewise constant* inputs (i.e., when a zero-order hold is employed), it is possible to derive an exact discrete-time model. This approach works well for any linear, constant coefficient, time-domain model.

Let us illustrate the principles involved in exact discretization with a first-order system. The modeling equation in this case is:

$$\tau \frac{dy}{dt} \; + \; y \; = \; Ku$$

which, along with the usual zero initial condition, $y(0) = 0$, has the *exact, analytical* solution:

$$y(t) \; = \; \int_0^t e^{-(t-\sigma)/\tau} \frac{K}{\tau} u(\sigma) \, d\sigma \tag{23.19}$$

If we wish the solution at time $t + \Delta t$, we obtain:

$$y(t + \Delta t) \; = \; \int_0^{t+\Delta t} e^{-(t+\Delta t-\sigma)/\tau} \frac{K}{\tau} u(\sigma) \, d\sigma$$

$$= \; e^{-\Delta t/\tau} \int_0^{t+\Delta t} e^{-(t-\sigma)/\tau} \frac{K}{\tau} u(\sigma) \, d\sigma \tag{23.20}$$

Multiplying Eq. (23.19) by $e^{-\Delta t/\tau}$ and subtracting from Eq. (23.20) yields:

$$y(t + \Delta t) - e^{-\Delta t/\tau} y(t) \; = \; e^{-\Delta t/\tau} \int_t^{t+\Delta t} e^{-(t-\sigma)/\tau} \frac{K}{\tau} u(\sigma) \, d\sigma \tag{23.21}$$

where we have made use of the result from elementary calculus that:

$$\int_t^{t+\Delta t} (\bullet) \; = \; \int_0^{t+\Delta t} (\bullet) \; - \; \int_0^t (\bullet)$$

Note that in Eq. (23.21), if we are modeling a sampled-data system that uses a zero-order hold for input signal reconstruction, $u(\sigma)$ is held constant over the sampling interval $[t, t+\Delta t]$ at the value taken at the beginning of the interval; if this is the kth sampling interval, then $t = t_k$, and this constant input value is $u(t_k)$.

Thus, with current time as t_k (and consequently, $t + \Delta t$ as t_{k+1}), Eq. (23.21) becomes:

$$y(t_{k+1}) - e^{-\Delta t/\tau} y(t_k) \; = \; e^{-\Delta t/\tau} u(t_k) \int_t^{t+\Delta t} e^{-(t-\sigma)/\tau} \frac{K}{\tau} d\sigma \tag{23.22}$$

where we have factored out $u(t_k)$ because it is constant over the interval covered by the integral in Eq. (23.22). If we now evaluate the integral in Eq. (23.22), we see that:

$$e^{-\Delta t/\tau} \int_t^{t+\Delta t} e^{-(t-\sigma)/\tau} \frac{K}{\tau} d\sigma \; = \; K \left(1 - e^{-\Delta t/\tau} \right) \tag{23.23}$$

so that Eq. (23.22) becomes:

$$y(k + 1) = e^{-\Delta t/\tau} y(k) + K \left[1 - e^{-\Delta t/\tau}\right] u(k) \qquad (23.24)$$

Thus if we let $\phi = e^{-\Delta t/\tau}$, we can rewrite the discrete-time first-order model with inputs having a ZOH as:

$$y(k + 1) = \phi y(k) + K (1 - \phi) u(k) \qquad (23.25)$$

In exactly the same way, higher order discrete-time models can also be obtained by exact discretization of the corresponding higher order models. To demonstrate, let us assume the time-domain model is of the form:

$$\dot{\mathbf{x}}(t) = \mathbf{A}\mathbf{x}(t) + \mathbf{B}\mathbf{u}(t) \qquad \mathbf{x}(0) = \mathbf{x}_0 \qquad (23.26)$$

$$\mathbf{y}(t) = \mathbf{C}\mathbf{x}(t) \qquad (23.27)$$

where vector $\mathbf{x}$ represents the process states, $\mathbf{u}$ the control inputs, and $\mathbf{y}$ the process outputs. The matrices $\mathbf{A}, \mathbf{B}, \mathbf{C}$ have constant elements. Recall that the analytical solution to Eq. (23.26) takes the form:

$$\mathbf{x}(t) = e^{\mathbf{A}t} \mathbf{x}_0 + \int_0^t e^{\mathbf{A}(t-\sigma)} \mathbf{B}\mathbf{u}(\sigma) \, d\sigma \qquad (23.28)$$

We may solve for $\mathbf{x}(k\Delta t)$ by repeated application of Eq. (23.28) for each value of k under piecewise constant $\mathbf{u}$ to yield:

$$\mathbf{x}(\Delta t) = e^{\mathbf{A}\Delta t} \mathbf{x}_0 + \left(\int_0^{\Delta t} e^{\mathbf{A}(\Delta t-\sigma)} \, d\sigma\right) \mathbf{B}\mathbf{u}(0)$$

$$\mathbf{x}(2\Delta t) = e^{\mathbf{A}\Delta t} \mathbf{x}(\Delta t) + \left(\int_{\Delta t}^{2\Delta t} e^{\mathbf{A}(2\Delta t-\sigma)} \, d\sigma\right) \mathbf{B}\mathbf{u}(\Delta t) \qquad (23.29)$$

$$\vdots$$

$$\mathbf{x}((k+1)\Delta t) = e^{\mathbf{A}\Delta t} \mathbf{x}(k\Delta t) + \left(\int_{k\Delta t}^{(k+1)\Delta t} e^{\mathbf{A}[(k+1)\Delta t-\sigma]} \, d\sigma\right) \mathbf{B}\mathbf{u}(k\Delta t)$$

Now if we define the constant matrices:

$$\boldsymbol{\Phi} = e^{\mathbf{A}\Delta t} \qquad (23.30)$$

$$\boldsymbol{\beta} = \int_0^{\Delta t} e^{\mathbf{A}(\Delta t-\sigma)} \, d\sigma \, \mathbf{B} \qquad (23.31)$$

then the discrete-time representation of Eqs. (23.26) and (23.27) becomes:

$$\mathbf{x}(k+1) = \boldsymbol{\Phi}\mathbf{x}(k) + \boldsymbol{\beta}\mathbf{u}(k) \qquad (23.32)$$

$$\mathbf{y}(k+1) = \mathbf{C}\mathbf{x}(k+1) \tag{23.33}$$

and this has the "solution" (the analog to Eq. (23.28)):

$$\mathbf{x}(k) = \Phi^k \mathbf{x}(0) + \beta \sum_{i=0}^{k-1} \Phi^{k-1-i} \mathbf{u}(i) \tag{23.34}$$

Although one could evaluate the matrix exponentials required by hand (see Appendix D), fortunately there are standard computer programs available (see Appendix A) for evaluating Eqs. (23.32) and (23.33) given Eqs. (23.26) and (23.27).

23.6 EMPIRICAL MODELING OF DISCRETE-TIME SYSTEMS

Empirical modeling for discrete-time systems involves fitting discrete models to experimental data in much the same way as with continuous-time systems. The steps involved are summarized below:

1. Excite the system with a sequence of discrete inputs $u(k)$, and collect data on $y(k)$, the process response to the input stimuli.
2. Postulate one or more candidate models — usually by examining a plot of the data, or simply from experience.
3. Estimate the model parameters — usually by regression analysis (as illustrated below).
4. Evaluate model adequacy — usually by examining a plot of the model prediction superimposed on the data. (It is also possible to differentiate between competing models in this fashion.)
5. Return to Step 2 if the model does not agree sufficiently well with the data; otherwise the modeling exercise is completed.

Observe that these steps incorporate the same tasks of *Model formulation*, *Parameter estimation*, and *Model validation* discussed in detail for continuous systems in Chapter 13. However, while the *basic principles* of model identification are the same for discrete as well as continuous systems, there are some significant differences in the specific implementation of these principles.

First, we note that for discrete-time systems, difference equation models are often much more convenient to identify than transform-domain (pulse) transfer function models. By contrast, recall that for continuous systems, it is in fact the transform-domain transfer function model that is most conveniently — and most often — identified (cf. Chapter 13). Also, recall from Chapter 13 that the continuous impulse-response function is not easy to identify directly from process data; however, as we shall soon see, the *discrete* impulse-response function is quite conveniently identified even from process data generated using arbitrary inputs.

23.6.1 Difference Equation Model Identification

Because of their recursive structure, difference equation models for linear systems have the most natural form for expressing empirical relations between process variables on the basis of discrete process data. The coefficients in the model are the "unknown parameters," and the primary identification task is parameter estimation.

Parameter Estimation in Difference Equation Models

The nature of linear difference equation models is such that it is particularly convenient to estimate the unknown parameters using linear regression. Rather than give this issue a general treatment, we will illustrate with a simple example.

The most general form of the discrete-time process model for a second-order system is the linear, second-order difference equation:

$$y(k) = a_1 y(k-1) + a_2 y(k-2) + b_1 u(k-1) + b_2 u(k-2) \tag{23.35}$$

Suppose that after having excited the system with the input sequence: $\{u(1), u(2), u(3), ..., u(N)\}$, the process response data $\{y(1), y(2), y(3), ..., y(N), y(N+1)\}$ is measured. Starting from $k = 3$ in Eq. (23.35) we have:

$$y(3) = a_1 y(2) + a_2 y(1) + b_1 u(2) + b_2 u(1)$$

$$y(4) = a_1 y(3) + a_2 y(2) + b_1 u(3) + b_2 u(2)$$

$$\cdots \qquad\qquad \cdots \qquad \cdots \qquad \cdots$$

$$y(N) = a_1 y(N-1) + a_2 y(N-2) + b_1 u(N-1) + b_2 u(N-2)$$

$$y(N+1) = a_1 y(N) + a_2 y(N-1) + b_1 u(N) + b_2 u(N-1)$$

where measurement error terms are not shown explicitly. When these equations are written in vector-matrix form, we obtain:

$$
\begin{bmatrix}
y(3) \\
y(4) \\
\cdots \\
y(N) \\
y(N+1)
\end{bmatrix}
=
\begin{bmatrix}
y(2) & y(1) & u(2) & u(1) \\
y(3) & y(2) & u(3) & u(2) \\
\cdots & \cdots & \cdots & \cdots \\
y(N-1) & y(N-2) & u(N-1) & u(N-2) \\
y(N) & y(N-1) & u(N) & u(N-1)
\end{bmatrix}
\begin{bmatrix}
a_1 \\
a_2 \\
b_1 \\
b_2
\end{bmatrix}
\tag{23.36}
$$

which is of the form:

$$\mathbf{y} = \mathbf{X}\theta \tag{23.37}$$

Since all the elements of the vector $\mathbf{y}$ and the matrix $\mathbf{X}$ are known, we may obtain a least-squares estimate of θ, the vector of unknown parameters, as follows:

$$\hat{\theta} = (\mathbf{X}^T\mathbf{X})^{-1}\mathbf{X}^T\mathbf{y} \tag{23.38}$$

Let us now note the following important points:

1. It is straightforward to generalize the above discussion to a discrete-time system of any order.

2. It is seldom that one actually carries out the indicated regression analysis by hand. As mentioned in Chapters 12 and 13, most computer packages for control systems design and analysis have programs for process identification that contain facilities for computer-assisted parameter estimation.

3. It should be mentioned that estimating discrete-time process model parameters as indicated above actually does transgress certain principles of linear parameter estimation theory. Strictly speaking, the theory requires the elements of the matrix **X** to be *independent* as well as error free; but it is clear from the second-order example in Eq. (23.36) that this will not be so for dynamic discrete-time processes: the nature of the dynamic system dictates that $y(k)$ *must* depend on $y(k-1)$, and because these are measurements, they will, in addition, be subject to measurement error.

To be entirely rigorous — from the statistical point of view — whenever the model (and experimental data) contain variables that are serially correlated in time, parameter estimation should be carried out within the framework of *Time Series Analysis* (see Box and Jenkins [1]). However, in virtually all the cases of interest to us, the simpler — if not entirely rigorous — procedure outlined above gives acceptable results, as we now illustrate with the following example.

Example 23.1   EMPIRICAL DISCRETE-TIME MODEL DEVELOPMENT FOR AN EXPERIMENTAL STIRRED MIXING TANK SYSTEM.

The following is a summary of an actual experiment performed on the stirred mixing tank introduced in Figure 20.1 of Chapter 20. The purpose of the experiment was to model the temperature variation of the tank's contents in response to inlet stream temperature changes. An empirical discrete-time model was considered adequate.

The inlet stream temperature was changed from its initial value of 27°C to 33°C and maintained at this new value while the disturbance stream temperature was kept constant (as much as possible) at its initial steady-state value of 20°C.

The tank temperature, as well as the input temperature, was measured every 2 seconds by a dedicated digital computer. Table 23.2 shows every third point of the original data set. A plot of this data is shown as circles in Figure 23.8.

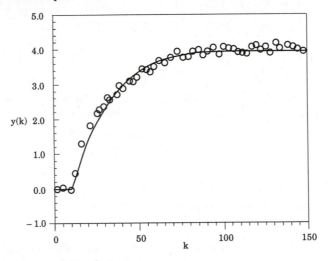

Figure 23.8.   Experimental data and discrete-time model predictions for a stirred mixing tank process.

TABLE 23.2.
Stirred Mixing Tank Data Summary

k	Time (seconds)	$y(k)$	k	Time (seconds)	$y(k)$
0	0	0	76	152	3.98
3	6	0	79	158	3.98
6	12	0	82	164	4.15
10	20	0.35	85	170	4.16
13	26	1.22	88	176	3.98
16	32	1.75	91	182	4.08
19	38	2.10	94	188	4.21
20	40	2.20	97	194	4.02
23	46	2.28	100	200	4.30
26	52	2.55	103	206	4.21
29	58	2.48	106	212	4.19
32	64	2.71	109	218	4.11
35	70	2.98	112	224	4.07
38	76	3.26	115	230	4.02
41	82	3.07	118	236	4.27
44	88	3.30	121	242	4.31
47	94	3.30	124	248	4.17
50	100	3.50	127	254	4.27
53	106	3.70	130	260	4.11
56	112	3.65	133	266	4.40
59	118	3.58	136	272	4.21
62	124	3.74	139	278	4.45
65	130	3.90	142	284	4.35
68	136	3.74	145	290	4.25
71	142	3.95	148	296	4.17
74	148	4.08	151	302	4.25

It is known *a priori* (and a plot of the data also confirmed this) that there is a time delay of 14 seconds (or 7 sample times) in the input; as a result the simple discrete model:

$$y(k) = ay(k-1) + bu(k-8)$$

was postulated. The parameters a and b were estimated by least squares from the *complete* data set, with $u(k) = 6$ for all k values. The result is:

$$a = 0.95; \qquad b = 0.042$$

The prediction of this model:

$$y(k) = 0.95 \, y(k-1) + 0.042 \, u(k-8) \tag{23.39}$$

is shown as the solid line in Figure 23.8. As indicated in the plot the model fits the data reasonably well.

23.6.2 Impulse-Response Model Identification

In general, the output $y(k)$ of a dynamic process observed at discrete points in time can be related to the process input $u(k)$, via its *discrete* impulse-response function as follows:

$$y(k) = \sum_{i=1}^{k} g(i)\, u(k-i) \tag{23.40}$$

(cf. Eq. (4.51) in Chapter 4), the discrete counterpart of the convolution integral of Eq. (4.47). For each observed experimental output data, we may write, using Eq. (23.40):

$$y(1) = g(1)u(0)$$

$$y(2) = g(1)u(1) + g(2)u(0)$$

$$y(3) = g(1)u(2) + g(2)u(1) + g(3)u(0)$$

$$\cdots \qquad \cdots$$

These equations may now be written together as:

$$
\begin{bmatrix} y(1) \\ y(2) \\ y(3) \\ \cdots \\ y(M) \end{bmatrix}
=
\begin{bmatrix}
u(0) & 0 & \cdots & \cdots & 0 \\
u(1) & u(0) & \cdots & \cdots & 0 \\
u(2) & u(1) & u(0) & \cdots & 0 \\
\cdots & \cdots & \cdots & \cdots & \cdots \\
u(M-1) & u(M-2) & \cdots & \cdots & u(M-N)
\end{bmatrix}
\begin{bmatrix} g(1) \\ g(2) \\ g(3) \\ \cdots \\ g(N) \end{bmatrix}
\tag{23.41}
$$

which is of the form:

$$\mathbf{y} = \mathbf{U}\mathbf{g} \tag{23.42}$$

And now, for any arbitrary set of inputs $u(k)$, $k = 0, 1, 2, \ldots, M$, and for N discrete impulse-response elements to be estimated from the data (where $N \leq M$) we obtain from Eq. (23.42), the familiar least square estimate of $\mathbf{g}$ the impulse-response vector, as:

$$\hat{\mathbf{g}} = (\mathbf{U}^T\mathbf{U})^{-1}\mathbf{U}^T\mathbf{y} \tag{23.43}$$

These equations simplify, as we would expect, when the input sequence, $u(k)$, takes on simple, specific forms, such as step, rectangular, or impulse functions.

23.6.3 Frequency-Response Model Identification

To obtain a pulse transfer function model for the discrete-time process specifically by frequency-response identification usually requires that we first fit a *continuous* transform-domain transfer function model to experimental data (as discussed in Chapter 13); the continuous model is then converted to the desired discrete, pulse transfer function form. The procedure for performing such conversions (which in itself is not particularly straightforward) will be discussed in the next chapter because it requires tools yet to be introduced. Nevertheless, we may observe that such a model identification strategy is unnecessarily tortuous; it is therefore not recommended. It is much more direct — and less taxing — to fit experimental data to difference equation models first and then convert these to the transfer function form. The procedure for converting difference equations to pulse transfer functions (by z-transforms) happens to be particularly straightforward and will be discussed in the next chapter.

23.7 SUMMARY

Our objective in this chapter has been simply to acquaint us with the peculiar characteristics of discrete-time systems before fully delving into dynamic analysis and controller design for such systems in the upcoming chapters.

In this regard, the three main issues we addressed were that:

1. Continuous process output information must now be discretized.
2. Discrete control command signals from the computer must be made piecewise continuous.
3. The process model must now be in discrete form.

The means by which a continuous signal is discretized was discussed, emphasizing the point that the appropriate choice of the sampling interval is the main issue. We also discussed how the reverse problem of continuous signal reconstruction is handled with holds of various types, emphasizing the zero-order hold as the most commonly used.

Since the discrete-time system is most appropriately described by discrete-time models, we devoted some time to the development of such discrete-time models. We showed how "theoretical" discrete-time models can be developed by discretizing the theoretical, continuous-time, models, and how empirical models are obtained by fitting discrete-time models to experimental data.

We are now in a position to proceed to the next stage in our study of *Computer Process Control* where we discuss the dynamic analysis of discrete-time systems.

REFERENCES AND SUGGESTED FURTHER READING

1. Box, G. E. P. and G. M. Jenkins, *Time Series Analysis: Forecasting and Control*, Holden-Day, San Francisco (1970).

REVIEW QUESTIONS

1. Reduced to its most fundamental essence, what is a controller?

2. Why is the digital computer perfectly suited to do the job of a controller?

3. In using the computer for data acquisition, what role does the analog-to-digital converter play?

4. What view of the process is induced by using the computer for data acquisition?

5. What key element in the digital-to-analog converter makes it possible to use the computer for implementing control action effectively?

6. Computer control systems more conveniently represented mathematically by what class of models?

7. What are the three major issues raised by the use of the digital computer for control system implementation?

8. How does an ideal sampler operate?

9. In choosing a sampling interval for a sampled-data system, what are the advantages and disadvantages of sampling too rapidly, or too slowly?

10. What is "aliasing"?

11. What is the Nyquist frequency?

12. What is involved in signal conditioning?

13. In what manner is continuous signal reconstruction the reverse of continuous signal sampling?

14. Why is there no unique solution to the problem of continuous signal reconstruction?

15. Why is it impractical, for on-line process control applications, to employ splines (or any other interpolation technique) for continuous signal reconstruction?

16. What is the difference between a zero-order hold and a first-order hold?

17. Why is the zero-order hold preferred over higher order holds for most process control applications?

18. What is the procedure for obtaining discrete-time models from first principles?

19. What specific methods are used for converting continuous models to discrete form? Which is the most commonly employed?

20. What steps are involved in developing empirical discrete models?

21. How does the identification of discrete models differ from the identification of continuous models?

22. In obtaining difference equation models empirically, what is the primary identification task?

23. Why is it not recommended to obtain pulse transfer function models by frequency-response analysis?

PROBLEMS

23.1 (a) Hourly laboratory data on the relative viscosity of a certain commercial elastomer are shown in Table P23.1. Given that the true relative viscosity of the product manufactured during the 25 hour period had remained essentially constant, so that the observed variation is due to laboratory analysis "noise," apply the first-order filter in Eq. (23.11) on the data, using three different values of the parameter $\beta = 0.3$, 0.5, and 0.9. Plot the raw data along with the three sets of filtered data. What would you estimate the true relative viscosity value to be?

(b) One of the primary criticisms of the first-order filter is that it tends to "smear out" abrupt shifts in noisy data. To illustrate this effect, revisit the data in Table P23.1, and starting from entry 13, add 5.0 to each of the last 13 samples; filter the entire data set using the first-order filter using $\beta = 0.3$, and $\beta = 0.9$. Plot the raw data along with the two sets of filtered data and comment on the effect of using high values for the filter parameter β.

TABLE P23.1
Hourly Viscosity Data

Time (hr)	Relative Viscosity (coded units)
1.0	49.763
2.0	50.050
3.0	51.079
4.0	49.310
5.0	49.855
6.0	51.056
7.0	49.887
8.0	49.806
9.0	49.735
10.0	50.799
11.0	50.699
12.0	50.469
13.0	50.007
14.0	50.341
15.0	49.798
16.0	50.767
17.0	48.678
18.0	50.372
19.0	50.782
20.0	49.648
21.0	49.889
22.0	49.669
23.0	49.544
24.0	50.472
25.0	50.837

23.2 The transfer function for the Wood and Berry distillation column used in separating methanol and water was introduced in Chapter 20 as:

$$\mathbf{G}(s) = \begin{bmatrix} \dfrac{12.8e^{-s}}{16.7s + 1} & \dfrac{-18.9e^{-3s}}{21.0s + 1} \\ \dfrac{6.6e^{-7s}}{10.9s + 1} & \dfrac{-19.4e^{-3s}}{14.4s + 1} \end{bmatrix} \tag{P23.1}$$

The time constants and time delays are in minutes, and the input and output variables for the process are:

u_1 = Overhead reflux flowrate
u_2 = Bottoms steam flowrate
y_1 = Overhead mole fraction methanol
y_2 = Bottoms mole fraction methanol

First obtain a differential equation realization of the transfer function model, and then use an exact discretization, with $\Delta t = 1$ min, to obtain a set of equations for the corresponding discrete-time model.

23.3   The data set in Table P23.2 was obtained from an experiment designed for identifying a simple model to be used for computer control of a pilot plant polymer reactor. The output $y(k)$ is the observed change in the "coded" viscosity of the polymer in solvent, as measured every minute by an on-line viscometer; the input $u(k)$ is the *relative change* in the composite manipulated variable C, a fundamentally derived nonlinear function of the initiator flowrate, and the ratio of the catalyst to cocatalyst flowrate, i.e., $u(k) = (C(k) - C^*)/C^*$ where C^* is the typical steady-state operating values for this composite process variable.   (The process exhibits reasonably linear behavior over a wide range of operating conditions as a result of the unusual, composite input variable employed.)

Postulate a discrete first-order model and from the given data estimate the parameters. Evaluate the model fit.

TABLE P23.2
Polymer Reactor Pilot Plant Pulse Test Identification Data

Time (minutes)	Input $u(k)$ (dimensionless)	Output $y(k)$ (coded units)
$t < 0$	0.0	0.0
0.0	1.0	0.0
1.0	2.0	1.766
2.0	1.0	4.458
3.0	0.0	4.414
4.0	0.0	2.447
5.0	...	1.669
6.0	...	1.306
7.0	...	0.882
8.0	...	0.661
9.0	...	0.497
10.0	...	0.486
11.0	...	0.385
12.0	...	0.291
13.0	...	0.197
14.0	...	0.181
15.0	...	0.094
16.0	...	0.165
17.0	...	0.056
18.0	...	0.089
19.0	...	0.073
20.0	...	−0.002
21.0	...	0.014
22.0	...	−0.014
23.0	...	−0.021
24.0	...	0.059
25.0	0.0	0.013

23.4   For the same polymer reactor data of Problem 23.3, postulate a discrete second-order model of the form:

$$y(k) = a_1 y(k - 1) + a_2 y(k - 2) + b_1 u(k - 1) + b_2 u(k - 2) \tag{P23.2}$$

and estimate the parameters. Evaluate the model fit and compare it to the fit obtained in Problem 23.3.

CHAPTER

24

TOOLS OF DISCRETE-TIME SYSTEMS ANALYSIS

In Chapter 23 we introduced the key concepts of discrete-time systems and the types of models employed; in addition we described how discrete-time models can be obtained from first-principles continuous models or directly from experimental data. In Chapter 25 we will begin a detailed discussion of the dynamic behavior of discrete-time systems and how these dynamics can be analyzed. However, first we must introduce some basic tools needed for that discussion. These tools, primarily the z-transform and the pulse transfer function, are direct analogs of the previously encountered tools of continuous-time systems analysis, the Laplace transform, and the Laplace domain transfer function. This analogy will be exploited in our presentation.

24.1   INTRODUCTION

Although we can carry out numerical simulation studies of the difference equations and impulse-response models introduced in the last chapter, such studies are only illustrative and provide no general results. Hence, analytical approaches are required to develop general principles of discrete-time analysis. Recall that in continuous-time systems analysis, we found the Laplace transform to be particularly useful; it provided us with the compact and convenient transfer function representation, from which the indispensable block diagram arose as a natural consequence. In discrete time, the z-transform plays a similar role, providing the *pulse transfer function* (the analog of the Laplace domain transfer function) from which the z-domain block diagrams are built.

Thus, the most important analytical tools for discrete-time systems analysis are:

- The z-transform
- The pulse transfer function
- Linear difference equations

This chapter is devoted to studying the z-transform and the pulse transfer function in the context of their application to discrete systems analysis. We treat linear difference equations and their solution in Appendix B.

It is important to remember that these analytical tools are applicable only when the process model is linear with constant coefficients; for nonlinear process models, dynamic analysis must be carried out numerically, or else *approximate* linear models will have to be substituted for the nonlinear ones.

24.2   THE BASIC CONCEPTS OF z-TRANSFORMS

Like the Laplace transform, the z-transform is a purely mathematical device with its own self-contained body of theory, and just as the Laplace transform is a useful device in the analysis of continuous systems, the z-transform finds valuable application in the analysis of discrete-time systems. What follows is therefore intended to be no more than a brief introduction to those aspects of the z-transform that will be useful for our specific purposes: analyzing the dynamic behavior of discrete-time systems.

24.2.1 Notation

In recognition of what could be potentially confusing subject matter, we find it necessary, before delving into z-transforms proper, to present first a summary of the notation we will employ.

The Time Arguments

t $\equiv$ measure of continuous time, $t \in [0, \infty)$

t_k $\equiv$ a specific, discrete point in time; in particular, the kth sampling point

Δt $\equiv$ amount of time in between samples, i.e., from $t = t_{k-1}$ to $t = t_k$ is precisely Δt units with the usual convention that the "starting" time is $t_0 = 0$, then for a uniform sample time, Δt, $t_k = k\Delta t$)

The Functions

$f(t)$ $\equiv$ the continuous function

$f^*(t)$ $\equiv$ the sampled (discretized) version of the continuous function, $f(t)$; a compact representation of the entire sequence of "spikes" at the various sample points $t = t_0, t_1, ..., t_k$

$f^*(t_k)$ $\equiv$ the specific value of $f^*(t)$ at the kth sample point, $t = t_k$

$f(t_k)$ $\equiv$ the specific value of $f(t)$ at $t = t_k$; by definition of $f(t)$ and $f^*(t)$, it is clear that *in numerical value*, $f(t_k) = f^*(t_k)$

$f(k\Delta t)$ $\equiv$ the value of $f(t)$ at the kth sample point $t = k\Delta t$; same as $f(t_k)$

$f(k)$ $\equiv$ a simpler representation for $f(k\Delta t)$, where it is understood that Δt, the sampling interval, assumed uniform, need not be explicitly indicated

$f(z)$ $\equiv$ the z-transform of the *discrete signal* $f^*(t)$

The z-Transform Arguments

z $\equiv$ the complex z-domain argument; the equivalent of the complex s variable in Laplace transforms

z^{-1} $\equiv$ the reciprocal of the *variable z*

The z-Transform Operators

$\mathcal{Z}\{.\}$ $\equiv$ the z-transform *operator*

$\mathcal{Z}^{-1}\{.\}$ $\equiv$ the *inverse z*-transform *operator*

24.2.2 Definition

The z-transform of the discrete signal $f^*(t)$ made up of the sequence of impulses $f(k\Delta t)$ for $k = 1, 2, \ldots$, is defined as:

$$f(z) = \mathcal{Z}\{f^*(t)\} = \sum_{k=0}^{\infty} f(k\Delta t)z^{-k} \tag{24.1a}$$

or, simplifying by dropping the Δt in the argument:

$$f(z) = \mathcal{Z}\{f^*(t)\} = \sum_{k=0}^{\infty} f(k)z^{-k} \tag{24.1b}$$

*Thus, the **z-transform** computes a function of z, designated f(z), from a **discrete sequence** in k, designated f(k). In computer process control applications, k is the sample time **index**, and f*(t) is the compact, functional representation of the discrete sequence, f(k), k = 1, 2 , ..., ∞.*

Remarks

1. It is important to restate that $f(z)$ is the z-transform of the *discrete* function $f^*(t)$ (equivalently, the *discrete sequence $f(k)$*). It is *incorrect* to consider $f(z)$ as the z-transform of the *continuous* function $f(t)$; *only discrete functions possess z-transforms.*

2. That the z-transform is a discrete analog of the Laplace transform becomes immediately obvious by recalling the definition of the Laplace transform given in Chapter 3:

$$f(s) = \int_0^\infty e^{-st} f(t)\, dt \tag{3.1}$$

and comparing it with Eq. (24.1).

3. Beyond this broad analogy, it is possible to establish a deeper relationship between the z-transform and the Laplace transform as follows:
Recall from Eq. (23.5) that $f^*(t)$ can be represented as:

$$f^*(t) = \sum_{k=0}^{\infty} f(k\Delta t)\, \delta(t - k\Delta t) \tag{24.2}$$

It is now possible to obtain $L\{f^*(t)\}$, the Laplace transform of $f^*(t)$; all we need to do is recall that:

$$L\{\delta(t)\} = 1$$

and also that with the time translation result:

$$L\{\delta(t - k\Delta t)\} = e^{-ks\Delta t}$$

With these two results, taking the Laplace transform of Eq. (24.2) gives:

$$f^*(s) = L\{f^*(t)\} = \sum_{k=0}^{\infty} f(k\Delta t)\, e^{-ks\Delta t} \tag{24.3}$$

from where we now see that by defining:

$$z \equiv e^{s\Delta t} \tag{24.4}$$

we are able to exactly recover the z-transform defined in Eq. (24.1) from the Laplace transform in Eq. (24.3). Thus:

The z-transform is a special case of the Laplace transform in which the variable transformation indicated in Eq. (24.4) has been carried out.

4. Observe from Eq. (24.3) that it would be possible to use Laplace transforms such as $f^*(s)$ for discrete-time systems analysis. However, the results would be very awkward and cumbersome; as it turns out, the z-transform is far more convenient.

Before proceeding farther, let us illustrate how to use Eq. (24.1) to obtain z-transforms with the following examples.

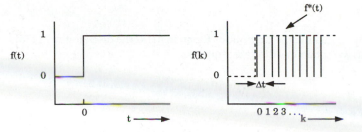

Figure 24.1.   The continuous and the discretized unit step function.

Example 24.1   z-TRANSFORM OF THE DISCRETE UNIT STEP FUNCTION.

Find the z-transform of the *discrete function* $f^*(t)$ obtained by sampling the continuous, unit step function $f(t) = 1$ (for $t \geq 0$) at intervals of Δt time units.
Solution:

The discrete sequence obtained from the simple, unit step function is clearly given by:

$$f(k) \;=\; f(k\Delta t) \;=\; 1; \quad k = 0, 1, 2, \ldots \tag{24.5}$$

obtained by evaluating the continuous function $f(t)$ at the discrete-time points $t = k\Delta t$. The discrete step function may now be represented as:

$$f^*(t) \;=\; \sum_{k=0}^{\infty} 1\; \delta(t - k\Delta t) \tag{24.6}$$

By definition of the z-transform, using Eq. (24.5) in Eq. (24.1b) gives:

$$f(z) = Z\{f^*(t)\} \;=\; \sum_{k=0}^{\infty} 1\; z^{-k} \tag{24.7}$$
$$= \; 1 + z^{-1} + z^{-2} + \ldots$$

an infinite series which, for $|z^{-1}| < 1$ converges to give:

$$f(z) \;=\; \frac{1}{1 - z^{-1}} \tag{24.8}$$

Example 24.2   z-TRANSFORM OF THE DISCRETE EXPONENTIAL FUNCTION.

Find the z-transform of the *discrete function* $f^*(t)$ obtained by sampling the continuous, exponential function $f(t) = e^{-at}$ at intervals of Δt time units.

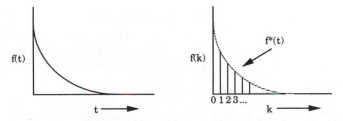

Figure 24.2.   The continuous and the discretized exponential function.

Solution:

By evaluating the continuous function $f(t)$ at the discrete-time points $t = k\Delta t$, we obtain the discrete sequence:

$$f(k) \; = \; f(k\Delta t) \; = \; e^{-ak\Delta t} \; ; \;\; k = 0, 1, 2, \ldots \tag{24.9}$$

and if we let:

$$r \; = \; e^{a\Delta t} \tag{24.10}$$

a constant whose value depends on the sampling interval Δt, then we have the *discrete exponential function*, made up of the discrete sequence:

$$f(k) \; = \; f(k\Delta t) \; = \; r^{-k} \; ; \;\; k = 0, 1, 2, \ldots \tag{24.11a}$$

Alternatively this can be written:

$$f^*(t) \; = \; \sum_{k=0}^{\infty} r^{-k} \, \delta(t - k\Delta t) \tag{24.11b}$$

Taking the required z-transform we obtain:

$$f(z) = Z\{f^*(t)\} \; = \; \sum_{k=0}^{\infty} r^{-k} \, z^{-k} \tag{24.12}$$

which, by comparison with Eq. (24.7), is seen to converge to:

$$f(z) \; = \; \frac{1}{1 - (rz)^{-1}} \tag{24.13}$$

provided that $|(rz)^{-1}| < 1$.

Note that the z-transform in Eq. (24.13) is a function of the variable r, which depends on Δt, the sample time. This illustrates a fundamental fact about z-transforms:

> *With the exception of "uniform" functions such as the unit step function, the z-transform of a sampled signal always depends on the sample time Δt.*

In principle, the z-transform of any discrete function — provided it is z-transformable — can be obtained as illustrated in these examples. In practice, however, it is customary to consult tables of various functions and their corresponding z-transforms, just as one would do with Laplace transforms. (See Appendix C for such a table of z-transforms.)

24.2.3 Some Properties of the z-Transform

As with the Laplace transform, the z-transform possesses certain properties that make it particularly useful in computer process control applications.

z-Transformable Functions

Not all discrete functions, $f^*(t)$, possess z-transforms; the z-transform exists only for those functions — and those values of the complex variable z — for which the infinite sum in Eq. (24.1) exists and is finite. Fortunately, all the functions of practical interest in computer process control are such that they possess z-transforms.

Uniqueness of Transform Pairs

The discrete function, $f^*(t)$, and its corresponding transform, $f(z)$, form a *unique* transform pair, in precisely the same sense that the *continuous* function, $f(t)$, and its Laplace transform, $f(s)$, form a unique transform pair. This property is useful in moving back and forth between the discrete-time domain and the z-domain.

Nonunique Relation with Sampled Continuous Functions

When the discrete function, $f^*(t)$, is obtained by sampling a continuous function, $f(t)$, the unique correspondence between $f^*(t)$ and $f(z)$ *does not* necessarily extend to the continuous function $f(t)$ from which $f^*(t)$ was generated, i.e., given an $f(z)$, there is a unique, corresponding $f^*(t)$, but there may not be a unique corresponding continuous $f(t)$. This situation arises because there is no unique relationship between a continuous function and its sampled version. Observe that depending on how the sampling is done, it is possible to obtain several different discretized functions, $f^*(t)$, from the same $f(t)$ (see Figure 23.4); by the same token, it is also possible to reconstruct many different continuous functions from the same discrete function $f^*(t)$. This is the problem of aliasing introduced in Chapter 23.

This crucial point serves as further warning against committing the blunder of attempting to obtain the z-transform of a continuous function; it is only possible to obtain the z-transform of discretized versions of the continuous function.

Linearity

As with Laplace transforms, the z-transform operation is linear, i.e., given two constants c_1 and c_2 and two (z-transformable) discrete functions $f^*_1(t)$ and $f^*_2(t)$, then:

$$Z\{c_1 f^*_1(t) + c_2 f^*_2(t)\} = c_1 Z\{f^*_1(t)\} + c_2 Z\{f^*_2(t)\} \qquad (24.14a)$$

The validity of this assertion is easily proved from the definition in Eq. (24.1).

It can also be shown that the inverse z-transform operation too is linear, so that:

$$Z^{-1}\{c_1 f_1(z) + c_2 f_2(z)\} = c_1 Z^{-1}\{f_1(z)\} + c_2 Z^{-1}\{f_2(z)\} \qquad (24.14b)$$

Virtually all z-transform-based discrete-time systems analysis makes use of both aspects of this linearity property.

Convolution

If $f_1(k)$ and $f_2(k)$ are two discrete sequences in k, which respectively make up the discrete functions $f^*_1(t)$ and $f^*_2(t)$, and if:

$$Z\{f^*_1(t)\} = f_1(z) \tag{24.15a}$$

and

$$Z\{f^*_2(t)\} = f_2(z) \tag{24.15b}$$

are their corresponding z-transforms, then:

$$Z\left\{\sum_{i=0}^{\infty} f_1(i)\, f_2(k - i)\right\} = f_1(z)\, f_2(z) \tag{24.16}$$

This property is particularly useful for analysis involving transfer functions.

24.2.4 Additional Useful Results

In addition to the properties listed above, the following results are very useful in linear discrete-time systems analysis.

The Inverse Transformation

By drawing on our experience with how the Laplace transform was used for continuous-time systems analysis, it is not difficult to anticipate that the utility of the z-transform in discrete-time systems analysis will also hinge on being able to recover the *discrete* function $f^*(t)$ (or the corresponding discrete sequence, $f(k)$) given only its z-transform, $f(z)$.

The inverse z-transform, which, in principle, provides the means of reversing the process indicated in Eq. (24.1), is given by:

$$f(k) = Z^{-1}\{f(z)\} = \frac{1}{2\pi j} \int_C f(z) z^k \frac{dz}{z} \tag{24.17}$$

a closed complex integral which, not surprisingly, is reminiscent of Eq. (3.10), the inverse of the Laplace transform.

Of course as with the Laplace transform, this inversion formula is hardly ever used in practice. More practical methods for inverting z-transforms will be investigated in the upcoming Section 24.3. However, two very important points must be noted from the expression in Eq. (24.17):

1. The z-transform inverse produces only the *discrete* sequence $f(k)$; it is incapable of providing information about the original continuous function, $f(t)$, from which the sequence could have come.

2. The inverse provides no information about the sampling interval, Δt, for the discrete sequence $f(k)$.

The Transform of Derivatives and Integrals

For the z-transform, *no* unique expressions exist for the transform of derivatives and integrals. This is because:

1. These are continuous-time operations, which, in the strictest possible sense, have no meaning in discrete time.

2. If, however, we were to revert to the discrete-time analogs (or more appropriately, *discrete-time approximations*) of these operations (namely, the *finite difference* for the derivative, and the finite *sum* for the integral) observe carefully that these discrete-time analogs can be obtained in several different ways: the finite difference approximation for a derivative will vary depending on the discretization step; the integral could be approximated by the trapezoidal rule, Simpson's rule, or a host of other rules. Each discrete representation will give rise to a different z-transform.

Shift (or Translation) Properties of the z-Transform

Shift in Time (Real Translation)

If $Z\{f(k)\}$, the z-transform of a discrete sequence, $f(k)$, is $f(z)$, then:

$$Z\{f(k - m)\} = z^{-m} f(z) \tag{24.18}$$

is the z-transform of the sequence delayed by m sampling periods, a result directly analogous to Eq. (3.24) for the Laplace transform.

It is easy to establish Eq. (24.18) from the definition in Eq. (24.1) as follows:

$$Z\{f(k - m)\} = \sum_{k=0}^{\infty} f(k - m)z^{-k} \tag{24.19}$$

and by applying the change of variable: $i = k - m$, we obtain:

$$Z\{f(k - m)\} = \sum_{i=-m}^{\infty} f(i)z^{-i-m} \tag{24.20}$$

For causal functions, $f(i) = 0$ for $i < 0$, hence, Eq. (24.20) simplifies to:

$$Z\{f(k - m)\} = \sum_{i=0}^{\infty} [f(i) \, z^{-i}]z^{-m}$$

$$= z^{-m} \sum_{i=0}^{\infty} [f(i) \, z^{-i}]$$

$$= z^{-m} f(z) \tag{24.21}$$

as required.

This result could also be presented in terms of a *discretized* continuous function as follows:

If $f^*(t)$ is the discrete function obtained by sampling the continuous function $f(t)$ at intervals of Δt time units, and $Z\{f^*(t)\} = f(z)$, then in the presence of a time delay α (assumed to be an integer multiple of the sample time Δt, i.e., $\alpha = m \Delta t$), $Z\{f^*(t - \alpha)\}$ is given by:

$$Z\{f^*(t - \alpha)\} = Z\{f^*(t - m\Delta t)\} = z^{-m} f(z) \tag{24.22}$$

The assumption that the time delay is an integer multiple of the sampling time may not always hold. Under such circumstances, it is possible to make use of the so-called *modified* z-transform (see Refs. [1-3]). However, we will not discuss this in detail here.

Scale "Shift" in z (Complex Translation)

Recall the complex translation theorem of Laplace transforms, which states that if $f(s)$ is the Laplace transform of the continuous function $f(t)$, then:

$$L\{e^{-at}f(t)\} = f(s+a) \tag{24.23}$$

The z-transform equivalent for the discrete sequence $f(k)$ states that:

$$Z\{r^{-k}f(k)\} = f(rz) \tag{24.24}$$

In terms of the discretized version of a continuous function, the same result could be presented as follows:

If $f^*(t)$ is the sampled version of the continuous function $f(t)$, with $Z\{f^*(t)\}$ given as $f(z)$, then:

$$Z\left\{\left[e^{-at}f(t)\right]^*\right\} = f\left(e^{a\Delta t}z\right) \tag{24.25}$$

where [.]* indicates the discrete version of the included function.
Introducing $r = e^{a\Delta t}$, gives:

$$Z\left\{\left[e^{-at}f(t)\right]^*\right\} = f(rz) \tag{24.26}$$

Let us illustrate the application of this result with the following example.

Example 24.3   z-TRANSFORM OF THE SAMPLED, DAMPED SINE FUNCTION.

Find the z-transform of the *discrete function* $f^*(t)$ obtained by sampling the continuous, damped sine function $f(t) = e^{-at} \sin \omega t$ $(t \geq 0)$ at intervals of Δt time units.

Solution:

The solution strategy is first to find the z-transform of the *sampled* sine function, and then to apply the complex translation result.
Thus, we first let:

$$f_1(t) = \sin \omega t \tag{24.27}$$

then, recalling Eq. (24.2), the sampled sine function will be given by:

$$f^*_1(t) = \sum_{k=0}^{\infty} \sin(\omega k\Delta t)\, \delta(t - k\Delta t) \tag{24.28}$$

and the z-transform is given by:

$$f_1(z) = Z\{f*_1(t)\} = \sum_{k=0}^{\infty} \sin(\omega k \Delta t) z^{-k} \tag{24.29}$$

The indicated sum is easier to evaluate if we introduce the Euler identity:

$$\sin \theta = \frac{e^{j\theta} - e^{-j\theta}}{2j} \tag{24.30}$$

so that Eq. (24.29) becomes:

$$f_1(z) = \frac{1}{2j} \sum_{k=0}^{\infty} \left[e^{(j\omega k \Delta t)} - e^{(-j\omega k \Delta t)} \right] z^{-k} \tag{24.31}$$

or

$$f_1(z) = \frac{1}{2j} \left[\sum_{k=0}^{\infty} \left(e^{(j\omega \Delta t)} z^{-1} \right)^k - \sum_{k=0}^{\infty} \left(e^{(-j\omega \Delta t)} z^{-1} \right)^k \right]$$

which, as in Examples 24.1 and 24.2 simplifies to give:

$$f_1(z) = \frac{1}{2j} \left(\frac{1}{1 - e^{j\omega \Delta t} z^{-1}} - \frac{1}{1 - e^{-j\omega \Delta t} z^{-1}} \right) \tag{24.32}$$

Upon further rearrangement, using Eq. (24.30) and the cosine counterpart:

$$\cos \theta = \frac{e^{j\theta} + e^{-j\theta}}{2}$$

Eq. (24.32) simplifies to:

$$f_1(z) = \frac{z^{-1} \sin(\omega \Delta t)}{1 - 2z^{-1} \cos(\omega \Delta t) + z^{-2}} \tag{24.33}$$

We now observe that the desired z-transform, $f(z)$, for the sampled damped sine function may now be obtained from Eq. (24.33) by invoking the complex translation result Eq. (24.26): the result is:

$$f(z) = \frac{(ze^{a\Delta t})^{-1} \sin(\omega \Delta t)}{1 - 2(ze^{a\Delta t})^{-1} \cos(\omega \Delta t) + (ze^{a\Delta t})^{-2}} \tag{24.34}$$

Effect of Sampling Time on z-Transforms

It is clear that the z-transform of the discrete signal obtained by sampling a continuous signal depends on the specific sample time used. However, as we now illustrate with a specific example, this effect could be quite dramatic in certain situations.

By following the procedure of Example 24.3, (see also Problem 24.1) it is possible to deduce that the z-transform of discrete function $f*(t)$ generated by sampling the cosine function $f(t) = \cos(\omega t)$, with a sample time Δt, is given by:

$$f(z) = \frac{1 - z^{-1} \cos(\omega \Delta t)}{1 - 2z^{-1} \cos(\omega \Delta t) + z^{-2}} \tag{24.35}$$

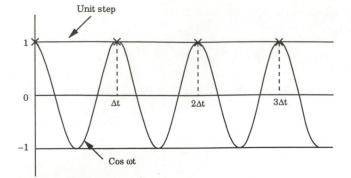

Figure 24.3. Sampling the cosine function with period $\Delta t = 2\pi/\omega$; comparison with the unit step function.

If we now choose the sample time such that:

$$\Delta t = \frac{2n\pi}{\omega} \tag{24.36}$$

for *any* nonzero integer value for n, i.e., $n = 1, 2, 3, \ldots$ observe that Eq. (24.35) immediately reduces to:

$$f(z) = \frac{1 - z^{-1}}{1 - 2z^{-1} + z^{-2}}$$

which, provided $z^{-1} \neq 1$, simplifies to give:

$$f(z) = \frac{1}{1 - z^{-1}} \tag{24.37}$$

and this, we recall from Example 24.1, is identical to the z-transform for the sampled unit step function.

Let us note therefore:

1. That as a result of the influence of the sample time, Δt, it is possible for the sampled versions of radically different continuous functions (such as the unit step function and the cosine function in this example) to have identical z-transforms.

2. Under such circumstances, the sampled functions will obviously be identical in appearance. In this specific case, sampling the cosine function according to Eq. (24.36) (i.e., using a sampling period identical to the natural period of the cosine wave) results in a sampled function that takes the value of 1 at every sample point, a function that is entirely indistinguishable from a sampled, unit step function.

The Initial-Value Theorem

Given $f(z)$, the z-transform of a discrete function $f^*(t)$ (made up of the discrete sequence $f(k)$), the initial value of this function can be evaluated using of the following result:

$$\lim_{t \to 0} [f^*(t)] = \lim_{k \to 0} [f(k)] = \lim_{z \to \infty} [f(z)] \tag{24.38}$$

This result is easily established from the definition in Eq. (24.1) and is left as an exercise to the reader.

The Final-Value Theorem

Provided the "final" value taken by the discrete function $f^*(t)$ (or, equivalently, the "final" value in the discrete sequence $f(k)$) is finite, it can be deduced directly from the z-transform $f(z)$ using the following result:

$$\lim_{t \to \infty} [f^*(t)] = \lim_{k \to \infty} [f(k)] = \lim_{z \to 1} [(1 - z^{-1}) f(z)] \qquad (24.39)$$

24.2.5 Application of z-Transforms to Discrete-Time Process Control

The application of z-transforms to discrete-time control systems analysis and design is analogous to the Laplace transform applications in continuous time. The broad concepts involved are summarized as follows and illustrated in Figure 24.4:

1. The original problem, formulated in the continuous-time domain (or in the Laplace domain) is first discretized with sample time Δt.
2. The discretized problem is transformed to the z-domain.
3. The transformed problem is analyzed in the z-domain to produce the desired result in terms of z-transforms.
4. The desired time domain results are recovered in discrete form from the z-domain solution, through the inverse z-transform operation.

Several examples below will illustrate the details of this procedure. The first two steps in the procedure described above require more detailed discussion. Figure 24.5 illustrates the possible paths one may take in proceeding from a continuous-time formulation to the z-transform. More importantly, the figure also shows:

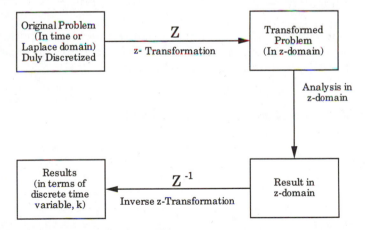

Figure 24.4. Conceptual representation of the application of z-transforms in discrete-time process control.

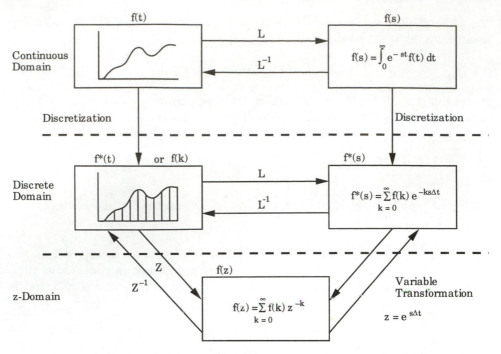

Figure 24.5.   Paths relating the continuous domain to the z-domain via the discrete
domain.

1. The interrelationships between the various functions that can be derived from the original continuous function, $f(t)$, (i.e., $f(s)$, its Laplace transform; $f^*(t)$, its sampled version, whose own Laplace transform and z-transform are, respectively, $f^*(s)$, and $f(z)$).

2. That all paths from the continuous formulation to the z-transform must involve the *irreversible* operation of discretization, making it easy to see why it is impossible to recover the original continuous function from the z-transform of its discretized version.

We have already discussed the actual mechanics involved in performing the operations indicated in Figure 24.5 with the exception of inverting the z-transforms; this is the subject of the next section.

24.3   INVERTING z-TRANSFORMS

The process of recovering the discrete function, $f^*(t)$ — or, equivalently, the discrete sequence, $f(k)$ — from the z-transform, $f(z)$, is known as inverting a z-transform, i.e.:

$$\mathcal{Z}^{-1}\{f(z)\} = f^*(t) \tag{24.40a}$$

or

$$\mathcal{Z}^{-1}\{f(z)\} = f(k) \; ; k = 1, 2, \ldots, \infty \tag{24.40b}$$

It is important to stress once more that the inversion of a z-transform will always yield a *discrete* function of time. Thus even if the discrete function $f^*(t)$ originally resulted from sampling a continuous function, $f(t)$, inverting the z-transform of the discretized function will allow us to recover only $f^*(t)$; *the original continuous function cannot be recovered.*

The standard formula for inverting a z-transform is the closed, complex integral given in Eq. (24.17); however, it is not the most practical means of obtaining z-transform inverses. It is more customary, in practice, to invert a z-transform by one of the following means:

- From tables of z-tranforms
- Expansion in an infinite series (long division)
- Partial fraction expansion

Let us discuss each of these techniques in more detail.

24.3.1 Using Tables of z-Transforms

If the z-transform to be inverted is simple enough, it may be found as an entry in a table of z-transforms, in which case, the process of inversion involves no more than locating the z-transform and its corresponding discrete-time inverse in the table.

For example, given:

$$f(z) = \frac{3}{1 - 0.5z^{-1}} \tag{24.41}$$

we easily find, from any table of z-transforms (such as the one in Appendix C), or from Example 24.2, that this is the z-transform of the *discrete* exponential function, $f(k) = r^{-k}$, with $r = e^{a\Delta t}$, for which:

$$Z\{f(k)\} = Z\{r^{-k}\} = \frac{1}{1 - (rz)^{-1}} \tag{24.42}$$

Thus, by inspection, we immediately observe that the inverse of Eq. (24.41) becomes:

$$f(k) = Z^{-1}\left\{\frac{3}{1 - (2z)^{-1}}\right\} = 3\,(2)^{-k};\ k = 1, 2, \ldots, \infty \tag{24.43a}$$

$$= 3\,(0.5)^k\,;\quad k = 1, 2, \ldots, \infty$$

or

$$= 3\,e^{-0.693k}\,;\quad k = 1, 2, \ldots, \infty \tag{24.43b}$$

since

$$e^{-0.693} = 0.5.$$

Remarks

1. It will be mere speculation on our part to conclude that the continuous function whose sampled version is represented in Eq. (24.43) is in fact an exponential function. Note that this discrete function could equally well have arisen, for example, from sampling a damped cosine function whose natural frequency coincides with the sampling frequency.

2. Even if the true identity of the continuous function were somehow to be determined independently, observe that Eq. (24.43) gives no information regarding Δt, the sampling time.

The most obvious drawback of this method of obtaining z-transform inverses is this: transforms encountered in realistic problems may not always be so simple as to be part of any table of transforms, no matter how extensive. However, there is the prospect of being able to break any complicated z-transform up into a sum of simpler constituents whose inverse transforms may be found in tables. The method of partial fraction expansion to be discussed in Section 24.3.3 is based on precisely such a strategy.

24.3.2 Inversion through Infinite Series Expansion (Long Division)

From the definition of the z-transform in Eq. 24.1 one may write out $f(z)$ as:

$$f(z) = f(0) + f(1)z^{-1} + f(2)z^{-2} + f(3)z^{-3} + \ldots \tag{24.44}$$

recognizable as an infinite series in the variable z^{-1}. Thus for *any* z-transform, $f(z)$, expanded as an infinite series in z^{-1}, say:

$$f(z) = \gamma_0 + \gamma_1 z^{-1} + \gamma_2 z^{-2} + \gamma_3 z^{-3} + \ldots \tag{24.45}$$

its inverse, the discrete sequence, $f(k)$, is easily recovered by comparison of the coefficients of powers of z^{-1} in Eqs. (22.44) and (22.45) to yield

$$f(k) = \gamma_k \; ; \; k = 1, 2, \ldots \tag{24.46}$$

Thus to obtain $f(k)$, we need to show *how* any arbitrary z-transform can be represented as a power series in z^{-1}.

Power Series Expansion for Arbitrary f(z)

In general, the z-transform is usually expressed as a ratio of two polynomials in z^{-1}, i.e.:

$$f(z) = \frac{B(z^{-1})}{A(z^{-1})} \tag{24.47}$$

or

$$f(z) = \frac{b_0 + b_1 z^{-1} + b_2 z^{-2} + \ldots + b_m z^{-m}}{a_0 + a_1 z^{-1} + a_2 z^{-2} + \ldots + a_n z^{-n}} \tag{24.48}$$

Observe that had the transform been expressed instead in terms of polynomials in z, the form in Eq. (24.48) is obtained easily by dividing both numerator and denominator by the appropriate power of z.

In principle, it is possible to carry out explicit "long division" of the numerator polynomial by the denominator polynomial, as indicated in Eq. (24.48); the result will be an infinite series as in Eq. (24.45), with the possibility, of course, that some coefficients will be zero. Obviously, this procedure is tedious.

A less tedious, entirely equivalent procedure is the following: The objective is to represent Eq. (24.48) as in Eq. (24.45); thus if:

$$\Gamma(z^{-1}) = \gamma_0 + \gamma_1 z^{-1} + \gamma_2 z^{-2} + \gamma_3 z^{-3} + \ldots$$

then the required γ_k sequence must satisfy the equation:

$$\Gamma(z^{-1}) = \frac{B(z^{-1})}{A(z^{-1})} \tag{24.49}$$

or

$$\Gamma(z^{-1})A(z^{-1}) = B(z^{-1}) \tag{24.50}$$

which may be expanded to:

$$\left(\gamma_0 + \gamma_1 z^{-1} + \gamma_2 z^{-2} + \gamma_3 z^{-3} + \ldots\right)\left(a_0 + a_1 z^{-1} + a_2 z^{-2} + \ldots + a_n z^{-n}\right)$$

$$= b_0 + b_1 z^{-1} + b_2 z^{-2} + \ldots + b_m z^{-m} \tag{24.51}$$

By carrying out the multiplication indicated on the LHS, collecting like terms, and comparing coefficients of powers of z^{-1}, we obtain the following expressions that can be used recursively to obtain γ_k for $k = 0, 1, 2, \ldots$:

For the constant term:

$$a_0 \gamma_0 = b_0 \tag{24.52a}$$

giving

$$\gamma_0 = \frac{b_0}{a_0} \tag{24.52b}$$

For the coefficient of z^{-1}:

$$a_1 \gamma_0 + a_0 \gamma_1 = b_1 \tag{24.53a}$$

so that:

$$\gamma_1 = \frac{b_1}{a_0} - \frac{a_1}{a_0} \gamma_0 \tag{24.53b}$$

For the coefficient of z^{-2}:

$$a_2 \gamma_0 + a_1 \gamma_1 + a_0 \gamma_2 = b_2 \tag{24.54a}$$

so that:

$$\gamma_2 = \frac{b_2}{a_0} - \frac{a_1}{a_0} \gamma_1 - \frac{a_2}{a_0} \gamma_0 \tag{24.54b}$$

and upon recognizing the emerging pattern, we obtain, for the general expression for finding γ_k given a_k, and b_k:

$$\gamma_k = \frac{1}{a_0} \left[b_k - \sum_{i=1}^{k} a_i \gamma_{k-i} \right] \tag{24.55}$$

which can be used recursively with Eq. (24.52b) as the initial condition.

In summary, to invert a z-transform by infinite series expansion:

1. Represent $f(z)$ as a ratio of polynomials in z^{-1} as indicated in Eq. (24.48); extract the coefficients $a_0, a_1, a_2, ..., a_n$, and $b_0, b_1, b_2, ..., b_m$.
2. With the initial condition $\gamma_0 = b_0/a_0$, use the general expression in Eq. (24.55) recursively to obtain $\gamma_1, \gamma_2, \gamma_3, ...$
3. From Eq. (24.46) and the immediately preceding discussion, we know that the inverse z-transform is given by $f(k) = \gamma_k$, $k = 1, 2, ...$

Let us now illustrate this procedure with an example.

Example 24.4   INVERTING A z-TRANSFORM BY INFINITE SERIES EXPANSION.

Using the technique of infinite series expansion, find the inverse z-transform of

$$f(z) = \frac{C}{1 - \phi z^{-1}} \tag{24.56}$$

Solution:

As $f(z)$ is already in the form required by Eq. (24.48), we simply note that the coefficients are given by:

$$b_0 = C \tag{24.57a}$$
$$b_k = 0 \, ; \quad k = 1, 2, ... \tag{24.57b}$$

and

$$a_0 = 1 \tag{24.58a}$$
$$a_1 = -\phi \tag{24.58b}$$
$$a_k = 0 \, ; \quad k = 2, 3, ... \tag{24.58c}$$

From here we easily obtain:

$$\gamma_0 = C \tag{24.59}$$

and, from Eq. (24.55) for $k = 1$:

$$\gamma_1 = \frac{1}{a_0} \left(b_1 - a_1 \gamma_0 \right)$$

and since $b_1 = 0$, and $a_1 = -\phi$, we immediately have:

$$\gamma_1 = C\phi \tag{24.60}$$

Using both Eqs. (24.59) and (24.60) now in Eq. (24.55) for $k = 2$, remembering that both b_2 and a_2 are zero, we obtain:

$$\gamma_2 = C\phi^2$$

one more such application of Eq. (24.55) gives:

$$\gamma_3 = C\phi^3$$

and we finally see that Eq. (24.55) in this case gives rise to a recursive relation:

$$\gamma_k = -a_1\gamma_{k-1} \tag{24.61}$$

or, that:

$$\gamma_k = C\phi^k \tag{24.62}$$

Thus, we now have that the given $f(z)$ may be expressed as the following infinite series:

$$f(z) = C\{1 + \phi z^{-1} + \phi^2 z^{-2} + \phi^3 z^{-3} + \ldots\}$$

which is, of course, recognizable as the z-transform of the sequence:

$$f(k) = C\phi^k \tag{24.63}$$

The same conclusion could have been arrived at immediately from Eq. (24.62), by equating the γ_k and the $f(k)$ sequences.

Note that this sequence also represents the discrete exponential function, with $\phi = 1/r$, or $\phi = e^{-a\Delta t}$.

24.3.3 Inversion Through Partial Fraction Expansion

In a manner completely analogous to the procedure for obtaining Laplace transform inversions by partial fraction expansion, we will invert a complex z-transform through a two-step procedure:

1. Recast a complicated z-transform, $f(z)$, in terms of a *finite* sum of simpler functions, $f_1(z)$, $f_2(z)$, ..., $f_n(z)$, whose inverse transforms are easily recognizable, i.e.:

$$f(z) = f_1(z) + f_2(z) + \ldots + f_n(z) \tag{24.64}$$

2. Use the linearity property of the inverse z-transform to recover $f(z)$ from

$$Z^{-1}\{f(z)\} = Z^{-1}\{f_1(z)\} + Z^{-1}\{f_2(z)\} + \ldots + Z^{-1}\{f_n(z)\} \tag{24.65}$$

The procedure for breaking the function $f(z)$ up into its simpler *partial fractions* is now outlined:

1. The z-transform to be inverted is represented in the form shown in Eqs. (24.47) and (24.48), as a ratio of two polynomials in z^{-1} of orders m in the numerator, and n in the denominator.

2. Assuming that the denominator polynomial has n distinct, real roots, $z^{-1} = p_i^{-1}$ (or, equivalently $z = p_i$); $i = 1, 2, \ldots, n$, then upon factoring $A(z^{-1})$ into these roots we would have:

$$f(z) = \frac{B(z^{-1})}{a_0 \left(1 - p_1 z^{-1}\right) \left(1 - p_2 z^{-1}\right) \cdots \left(1 - p_n z^{-1}\right)} \tag{24.66}$$

3. By using the methods of partial fraction expansion discussed in Appendix C, Eq. (24.66) is expanded to give:

$$f(z) = \left(\frac{C_1}{1 - p_1 z^{-1}}\right) + \left(\frac{C_2}{1 - p_2 z^{-1}}\right) + \cdots + \left(\frac{C_n}{1 - p_n z^{-1}}\right) \tag{24.67}$$

so that each of the *partial fractions* is of the form:

$$f_i(z) = \frac{C_i}{1 - p_i z^{-1}}; \quad i = 1, 2, \ldots, n \tag{24.68}$$

which is simple enough that its inverse may be obtained directly from z-transform tables. Alternatively we may recall that this is precisely the z-transform we inverted in Example 24.4; thus, we have that for each *partial fraction* in Eq. (24.67):

$$\mathcal{Z}^{-1}\{f_i(z)\} = \mathcal{Z}^{-1}\left\{\frac{C_i}{1 - p_i z^{-1}}\right\}$$

or, that:

$$f_i(k) = C_i \left(p_i\right)^k \tag{24.69}$$

4. We may now take the inverse z-transform in Eq. (24.67), recalling Eq. (24.65) and using Eq. (24.69) to finally obtain:

$$f(k) = C_1 \left(p_1\right)^k + C_2 \left(p_2\right)^k + \cdots + C_n \left(p_n\right)^k \tag{24.70}$$

5. If sample time is known, *and* if for each root, $p_i \geq 0$ so that $\ln p_i$ exists, we may relate each partial fraction inverse to the sampled exponential function by letting:

$$p_i = e^{-q_i \Delta t} \tag{24.71a}$$

this expression can be used to obtain q_i in terms of the roots p_i as:

$$q_i = -\frac{1}{\Delta t} \ln p_i \tag{24.71b}$$

with the result that Eq. (24.70) becomes:

$$f(k) = C_1 e^{-q_1 k \Delta t} + C_2 e^{-q_2 k \Delta t} + \cdots + C_n e^{-q_n k \Delta t} \tag{24.72}$$

Let us illustrate this procedure with an example.

Example 24.5   INVERTING A z-TRANSFORM BY PARTIAL FRACTION EXPANSION.

Find the inverse z-transform of:

$$f(z) = \frac{0.8z^2 - 0.3z}{z^2 - 1.5z + 0.5} \tag{24.73}$$

by the technique of partial fraction expansion.

Solution:

Observe first that $f(z)$ is given in terms of polynomials in z and not in terms of polynomials in z^{-1}. It is important to note that nothing is inherently wrong with dealing with z-transforms presented in this form; however, it is quite often more convenient to deal with polynomials in z^{-1}.

By dividing through by z^2, $f(z)$ is easily converted to the desired form:

$$f(z) = \frac{0.8 - 0.3z^{-1}}{1 - 1.5z^{-1} + 0.5z^{-2}} \tag{24.74}$$

from whence we factor the denominator into its two roots to obtain:

$$f(z) = \frac{0.8 - 0.3z^{-1}}{(1 - z^{-1})(1 - 0.5z^{-1})} \tag{24.75}$$

Upon partial fraction expansion, we see that:

$$f(z) = \frac{1}{1 - z^{-1}} - \frac{0.2}{1 - 0.5z^{-1}} \tag{24.76}$$

The inverse transform is now easily obtained from Eq. (24.69) as:

$$f(k) = 1 - 0.2 \, (0.5)^k \tag{24.77}$$

or, in terms of the discrete exponential function:

$$f(k) = 1 - 0.2 \, e^{-0.693k} \tag{24.78}$$

24.4   PULSE TRANSFER FUNCTIONS

24.4.1 Definition

Analogous to the Laplace domain transfer function, we define a new transfer function, $g(z)$, which relates $u(z)$, the z-transform of a *sampled* (or "pulsed") *input signal*, $u^*(t)$, to $y(z)$, the z-transform of a *sampled* (or "pulsed") *output signal*, $y^*(t)$, according to:

$$y(z) = g(z)\, u(z) \tag{24.79}$$

with one important restriction:

The input and output signals must be sampled synchronously (i.e., at the same time) and also at the same rate.

Figure 24.6.   The concept of a pulse transfer function.

Such a transfer function, represented diagramatically in Figure 24.6, is called a *pulse* transfer function.

24.4.2 Relationship to $g(s)$

As we might expect, the role of the pulse transfer function in sampled-data systems analysis is related to that of the Laplace domain transfer function in classical control systems analysis.

It is important, however, to note the fact that most chemical processes are considered to be inherently continuous, so that process models often occur more naturally in the continuous-time framework, as differential equations, or in the form of the classical transfer functions. One of the main issues of concern for discrete-time systems is therefore how to relate the classical transfer function, $g(s)$, to the pulse transfer function, $g(z)$.

Recall that the impulse response of a continuous system, $g(t)$, represents the inverse Laplace transform of the classical transfer function $g(s)$, i.e.:

$$g(t) = L^{-1}\{g(s)\} \tag{24.80}$$

If we sample $g(t)$ with sample time Δt, then from the definition of the z-transform (Eq. (24.1)), we obtain the pulse transfer function for the sampled process *without a hold:*

$$g_{NH}(z) = \sum_{k=0}^{\infty} g(k\Delta t)z^{-k} \tag{24.81}$$

*The pulse transfer function, $g(z)$, of a sampled-data process **without a hold** is the z-transform of the sampled version of the impulse-response function, $g(t)$, of the continuous process.*

This provides a procedure for generating a pulse transfer function $g_{NH}(z)$ from the classical transfer function $g(s)$ as follows:

1. Obtain the impulse-response function, $g(t)$, by taking inverse Laplace transforms of $g(s)$.
2. Given a sampling time Δt, discretize $g(t)$ to obtain the discrete sequence $g(k\Delta t)$ (or, more simply, $g(k)$).
3. The z-transform of this discrete sequence is the required pulse transfer function.

Let us illustrate with the following example.

Example 24.6   PULSE TRANSFER FUNCTION OF A SAMPLED FIRST-ORDER PROCESS WITHOUT A ZERO-ORDER HOLD.

Find the pulse transfer function for the sampled-data process consisting of a standard first-order process whose inputs and outputs are synchronously sampled at regular intervals of Δt time units. The transfer function for the continuous process without a zero-order hold is given as:

$$g(s) = \frac{K}{\tau s + 1} \tag{24.82}$$

Solution:

Since:

$$L^{-1}\{g(s)\} = g(t) = \frac{K}{\tau} e^{-t/\tau} \tag{24.83}$$

it follows immediately that the pulse transfer function of the first-order process without a zero-order hold, $g_{NH}(z)$, is given by:

$$g_{NH}(z) = \frac{K}{\tau} \sum_{k=0}^{\infty} e^{-k\Delta t/\tau} z^{-k} \tag{24.84}$$

which simplifies to:

$$g_{NH}(z) = \frac{K}{\tau} \left(\frac{1}{1 - e^{-\Delta t/\tau} z^{-1}} \right) \tag{24.85}$$

or

$$g_{NH}(z) = \frac{K}{\tau} \left(\frac{1}{1 - \phi z^{-1}} \right) \tag{24.86}$$

where

$$\phi = e^{-\Delta t/\tau} \tag{24.87}$$

This is the pulse transfer function of a first-order sampled-data process with sampling time Δt and no zero-order hold.

24.4.3 The Zero-Order Hold Element

We might be tempted to assume that in comparing a continuous process system with its sampled-data version that the equivalence indicated in Figure 24.7 applies. Here the continuous process is a composite of a final control element such as a control valve, $g_v(s)$, the process itself, $g_p(s)$, and an output measurement device, $h(s)$, so that $g(s) = g_v(s) g_p(s) h(s)$. The control valve, $g_v(s)$ is expecting a continuous input signal $u(t)$ such as shown in Figure 23.2(b).

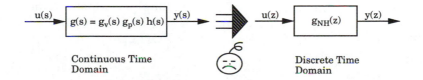

Figure 24.7. A wrong assumption of equivalence between a continuous system and its sampled-data version.

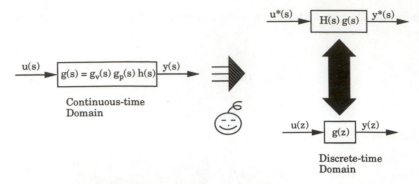

Figure 24.8. Proper equivalence between a continuous system and its sampled-data version achieved with the inclusion of a hold element.

One might think the transfer function $g_{NH}(z)$, such as derived in Example 24.6 for a first-order process, would be the appropriate equivalent transfer function for the discrete-time process. However, this assumption would be wrong! As represented in Figure 24.7, the inputs received by the sampled-data system are pulses of the type indicated in Figure 23.2(a) in Chapter 23; they take the values $u(k\Delta t)$ at the sample point but promptly return to zero thereafter. Thus, as we saw in Chapter 23, the equivalent discrete-time process requires one additional piece of hardware — the zero-order hold, $H(s)$, which acts on the series of spikes of the sampled input, $u(t_k)$ so that it appears as a series of steps. Without this zero-order hold, the discrete-time process cannot function any better than if it had no final control element or any other essential part of the process. Since the pulse transfer function $g_{NH}(z)$ in Eq. (24.86) contains no zero-order hold, it is not very useful for dynamic analysis because it does not represent the true process.

Thus the correctly configured sampled-data system is the one that incorporates a hold element as indicated in Figure 24.8. Note that one first uses the sampled version of $u(s)$, denoted by $u^*(s)$, and sends that to the transfer function $H(s) g(s)$ which includes the zero-order hold element, $H(s)$. This can then be converted to the appropriate pulse transfer function, $g(z)$. Since the zero-order hold is incorporated into essentially all digital control systems, it will be assumed to be present in all of the analysis to follow.

The Pulse Transfer Function g(z)

Let us now derive the expression for the pulse transfer function resulting from a combination of the process transfer function and the ZOH element as shown in Figure 24.8. First, we observe that the operation of the ZOH is to hold whatever value it receives constant over the sampling period Δt; this is precisely the same as "operating" on the pulsed input by a function that takes on a value of 1 at the beginning of the sampling period, and remains at this value over the interval Δt. Such a function is shown in Figure 24.9.

Observe, now, that if $H(t)$ represents the ZOH operation, and if $U(t)$ is the unit step function, then:

$$H(t) = U(t) - U(t - \Delta t) \tag{24.88}$$

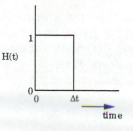

Figure 24.9. A functional representation of the operation of the ZOH element.

with the Laplace transform easily obtained as:

$$H(s) = \frac{1}{s} - \frac{e^{-s\Delta t}}{s} \qquad (24.89)$$

In the Laplace domain, the transfer function of a ZOH element in conjunction with a process with transfer function $g(s)$ is the product $H(s)g(s)$, and from Eq. (24.89) we have:

$$H(s)g(s) = \left(\frac{1}{s} - \frac{e^{-s\Delta t}}{s}\right)g(s) \qquad (24.90)$$

or

$$H(s)g(s) = (1 - e^{-s\Delta t})\frac{g(s)}{s} \qquad (24.91)$$

Since z-transforms are more convenient for discrete-time analysis, what we really seek then is the pulse transfer function in the z-domain, rather than the s-domain expression in Eq. (24.91).

Using the expression in Eq. (24.91), $g(z)$ may be found as:

$$g(z) = Z\{[H(s)g(s)]^*\} = Z\left\{\left[(1 - e^{-s\Delta t})\frac{g(s)}{s}\right]^*\right\} \qquad (24.92)$$

where the notation [.]* is introduced to serve as a reminder that the z-transform operation is performed *not on the Laplace domain function* but on *the discrete version of whatever continuous function gave rise to the indicated function of s*.

The rightmost part of Eq. (24.92) may be simplified further if we regard the indicated operation as that of converting a discrete Laplace transform to the corresponding z-transform. Recalling Eq. (24.4) therefore, and applying it to the exponential term, we have:

$$g(z) = (1 - z^{-1}) Z\left\{\left[\frac{g(s)}{s}\right]^*\right\} \qquad (24.93)$$

The procedure for obtaining the pulse transfer function of a sampled-data process with a ZOH element may now be outlined as follows:

1. Take the inverse Laplace transform to obtain the continuous-time function corresponding to $g(s)/s$; let this function be $\tilde{g}(t)$.
2. Discretize the function $\tilde{g}(t)$ with sample time Δt to yield $\tilde{g}(k\Delta t)$ and then obtain the z-transform $\tilde{g}(z)$.
3. Multiply by $1 - z^{-1}$ as indicated in Eq. (24.93) to obtain $g(z)$, i.e.:

$$g(z) = (1 - z^{-1})\,\tilde{g}(z) \tag{24.94}$$

Let us now illustrate the procedure with an example.

Example 24.7   PULSE TRANSFER FUNCTION OF A FIRST-ORDER SAMPLED-DATA PROCESS INCLUDING A ZERO-ORDER HOLD.

Find the pulse transfer function for a first-order sampled-data process with a ZOH element, given that:

$$g(s) = \frac{K}{(\tau s + 1)} \tag{24.95}$$

Assume that the inputs and outputs are synchronously sampled at regular intervals of Δt time units.

Solution:

From Eq. (24.94) we have:

$$\frac{g(s)}{s} = \frac{K}{s\,(\tau s + 1)} \tag{24.96}$$

and either by recognizing this as the unit step response of a first-order system, or by going through the actual mechanics of partial fraction expansion followed by Laplace inversion, we obtain:

$$\tilde{g}(t) = K\big(1 - e^{-t/\tau}\big) \tag{24.97}$$

Substituting $t = k\Delta t$ and taking the z-transform of this function yields:

$$\tilde{g}(z) = K\left(\frac{1}{1 - z^{-1}} - \frac{1}{1 - e^{-\Delta t/\tau}\,z^{-1}}\right) \tag{24.98}$$

and from Eq. (24.93) or Eq. (24.94), we have finally:

$$g(z) = K\left[\left(1 - z^{-1}\right)\left(\frac{1}{1 - z^{-1}} - \frac{1}{1 - e^{-\Delta t/\tau}\,z^{-1}}\right)\right]$$

which simplifies to give the final desired expression:

$$g(z) = \frac{K\left(1 - e^{-\Delta t/\tau}\right)z^{-1}}{1 - e^{-\Delta t/\tau}\,z^{-1}} \tag{24.99}$$

If we let:

$$\phi = e^{-\Delta t/\tau} \tag{24.100}$$

then Eq. (24.99) becomes:

$$g(z) = \frac{K(1 - \phi)z^{-1}}{1 - \phi z^{-1}} \tag{24.101}$$

A table of some common pulse transfer functions for various sampled continuous processes with ZOH elements can be found in Appendix C.

Let us note in closing that it is tempting to presume that since $g_{NH}(z)$ is relatively easy to obtain from $g(s)$, it should be possible to obtain $g(z)$ by taking the product of $H(z)$, the z-transform of the hold element, and $g_{NH}(z)$. However, this would be wrong, i.e.:

$$g(z) \neq H(z)g_{NH}(z) \tag{24.102}$$

To establish this inequality, we first obtain $H(z)$, say, from Eq. (24.88) as follows:

$$H(z) = \mathcal{Z}\{U^*(t)\} - \mathcal{Z}\{U^*(t - \Delta t)\} \tag{24.103}$$

Recalling the expressions obtained earlier for the discrete unit step function, and also recalling the time-shift result in Eq. (24.18), Eq. (24.103) becomes:

$$H(z) = \frac{1}{1 - z^{-1}} - z^{-1}\left(\frac{1}{1 - z^{-1}}\right) = 1 \tag{24.104}$$

and hence:

$$H(z)g_{NH}(z) = g_{NH}(z) \tag{24.105}$$

which merely reproduces our original *isolated* process pulse transfer function. Thus one must always deal with $g(z)$ rather than $g_{NH}(z)$.

24.4.4 Difference Equations and Pulse Transfer Functions

In Section 4.7 of Chapter 4, we discussed the relationships between various forms of continuous time-based models. Of particular interest throughout most of Part II was the relationship between the differential equation model and the classical transfer function $g(s)$. We will now address the parallel issue in discrete-time analysis, involving difference equations and pulse transfer functions.

From Difference Equations to Pulse Transfer Functions

The general difference equation model for the linear, single-input, single-output process has the form:

$$y(k) + a_1 y(k - 1) + a_2 y(k - 2) + \ldots + a_n y(k - n) = b_0 u(k) + b_1 u(k - 1) +$$
$$b_2 u(k - 2) + \ldots + b_m u(k - m) \tag{24.106}$$

Just as $g(s)$ was obtained in continuous-time analysis by taking Laplace transforms of the differential equation, the pulse transfer function model corresponding to the difference equation model in Eq. (24.106) will be obtained by taking z-transforms.

By invoking the linearity property of the z-transform, as well as recalling the time-shift result of Eq. (24.18), we have for all i:

$$\mathcal{Z}\{y(k-i)\} = z^{-i} y(z) \tag{24.107}$$

$$\mathcal{Z}\{u(k-i)\} = z^{-i} u(z) \tag{24.108}$$

so that taking z-transforms in Eq. (24.106) yields:

$$\left(1 + a_1 z^{-1} + a_2 z^{-2} + \dots + a_n z^{-n}\right)y(z)$$
$$= \left(b_0 + b_1 z^{-1} + b_2 z^{-2} + \dots + b_m z^{-m}\right)u(z) \tag{24.109}$$

or

$$y(z) = \left(\frac{b_0 + b_1 z^{-1} + b_2 z^{-2} + \dots + b_m z^{-m}}{1 + a_1 z^{-1} + a_2 z^{-2} + \dots + a_n z^{-n}}\right) u(z) \tag{24.110}$$

Thus, for this process, the pulse transfer function is:

$$g(z) = \frac{b_0 + b_1 z^{-1} + b_2 z^{-2} + \dots + b_m z^{-m}}{1 + a_1 z^{-1} + a_2 z^{-2} + \dots + a_n z^{-n}} \tag{24.111}$$

Observe that the original difference equation already assumes that the input sequence $u(k)$ is *piecewise constant*, i.e., the input takes a constant value $u(k)$ over the interval $k\Delta t$ and $(k + 1)\Delta t$ indicating that the ZOH element is already incorporated; thus the z-transform automatically produces $g(z)$.

The procedure for converting any difference equation model to a corresponding pulse transfer function model is thus quite straightforward:

1. Take z-transforms of the difference equation, remembering that for any function $f(k–i)$:

$$\mathcal{Z}\{f(k-i)\} = z^{-i} f(z) \tag{24.112}$$

2. Collect and separate terms involving $y(z)$ and $u(z)$ as in Eq. (24.109); rearrange as in Eq. (24.110) to obtain the pulse transfer function.

The next example illustrates these principles.

Example 24.8   PULSE TRANSFER FUNCTION MODEL FROM A FIRST-ORDER DIFFERENCE EQUATION MODEL.

Find the pulse transfer function for the process whose discrete-time model is the following first-order difference equation:

$$y(k) = 0.5\, y(k - 1) + 0.2u(k - 1) \tag{24.113}$$

Solution:

By taking z-transforms we find that:

$$y(z) = 0.5\, z^{-1}\, y(z) + 0.2\, z^{-1}\, u(z)$$

which rearranges to give:

$$y(z) = \left(\frac{0.2\, z^{-1}}{1 - 0.5\, z^{-1}} \right) u(z) \qquad (24.114)$$

as the pulse transfer function model corresponding to the example first-order difference equation model. Thus the pulse transfer function is given by:

$$g(z) = \left(\frac{0.2\, z^{-1}}{1 - 0.5\, z^{-1}} \right) \qquad (24.115)$$

which may now be compared with Eqs. (24.99) or (24.101).

From Pulse Transfer Functions to Difference Equations

The reverse process of obtaining a corresponding difference equation from a pulse transfer function model (i.e., the discrete-time realization problem) is also quite straightforward:

1. Express the given pulse transfer function, $g(z)$ as a ratio of two polynomials in z^{-1}:

$$g(z) = \frac{B(z^{-1})}{A(z^{-1})} \qquad (24.116)$$

    so that the pulse transfer function model will be:

$$y(z) = \frac{B(z^{-1})}{A(z^{-1})}\, u(z) \qquad (24.117)$$

    (Note that if the pulse transfer function model happens to be given in terms of z instead of the more convenient z^{-1}, the desired form may be obtained by dividing numerator and denominator by the appropriate power of z.)

2. Rearrange Eq. (24.117) to give:

$$A(z^{-1})y(z) = B(z^{-1})\, u(z) \qquad (24.118)$$

3. To recover the difference equation model, take inverse z-transforms term by term noting that for $f = y$ or $f = u$:

$$\mathcal{Z}^{-1}\{(z^{-i})\, f(z)\} = f(k - i) \qquad (24.119)$$

Let us illustrate once again with an example.

Example 24.9   DIFFERENCE EQUATION MODEL FROM A PULSE TRANSFER FUNCTION MODEL.

Given the pulse transfer function model:

$$y(z) = \left[\frac{K\left(1 - e^{-\Delta t/\tau}\right) z^{-1}}{1 - e^{-\Delta t/\tau} z^{-1}} \right] u(z) \qquad (24.120)$$

find the corresponding difference equation model.

Solution:

Note that the model is in the form required by Eq. (24.117) and further rearrangement gives:

$$\left(1 - e^{-\Delta t/\tau} z^{-1}\right) y(z) = \left[K\left(1 - e^{-\Delta t/\tau}\right) z^{-1}\right] u(z) \qquad (24.121)$$

If we now take the inverse z-transforms using Eq. (24.119) we immediately obtain:

$$y(k) - \left(e^{-\Delta t/\tau}\right) y(k-1) = K\left(1 - e^{-\Delta t/\tau}\right) u(k-1)$$

which rearranges to:

$$y(k) = \left(e^{-\Delta t/\tau}\right) y(k-1) + K\left(1 - e^{-\Delta t/\tau}\right) u(k-1) \qquad (24.122)$$

This equation is recognizable as the difference equation model for a first-order system obtained in Table 23.1 by exact discretization of a first-order, differential equation model, when the input is taken to be piecewise constant over the sampling interval.

The interrelationships between the most common continuous-time and discrete-time model forms for processes are summarized in Figure 24.10. By following the paths shown one can move from the continuous to the discrete domain and obtain either time- or transform-domain models as desired.

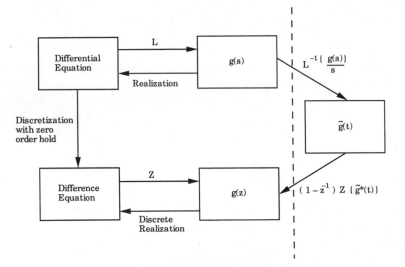

Figure 24.10. Interrelationships between continuous-time and discrete-time process models.

24.5 SUMMARY

This chapter has been concerned primarily with presenting the z-transform and the pulse transfer function as the basic tools of discrete systems analysis. These may be derived from familiar continuous analogs, but the analysis requires great care because the transition from continuous-time to discrete time is fraught with potential pitfalls which the reader can avoid by keeping the essentials in mind:

1. The z-transform does not exist for a continuous function; it exists only for its discretized version.

2. The inverse z-transform provides only the discrete function; the original, continuous function cannot be recovered from the inverse z-transform.

3. The appropriate pulse transfer function, $g(z)$, for a sampled-data system *must* include a hold element along with the process transfer function.

4. Particular care must be taken in obtaining $g(z)$; it *cannot* be obtained directly from $g_{NH}(z)$, the *isolated* process pulse transfer function; it must be obtained as indicated in Figure 24.10, either from a difference equation model, or through $\tilde{g}(t)$, the inverse Laplace transform of $g(s)/s$.

The next logical step in our discussions of computer control systems is that of employing the tools of this chapter to analyze the dynamic behavior of discrete-time systems; this is the subject of the next chapter.

REFERENCES AND SUGGESTED FURTHER READING

Detailed treatment of additional issues involved with digital control systems in general are available, for example, in the following books devoted entirely to this subject matter:

1. Franklin, G. F. and J. D. Powell, *Digital Control of Dynamic Systems*, Addison-Wesley, Reading, MA (1980)
2. Ogata, K., *Discrete-Time Control Systems*, Prentice-Hall, Englewood Cliffs, NJ (1981)
3. Åström, K. J. and B. Wittenmark, *Computer Controlled Systems*, (2nd ed.) Prentice-Hall, Englewood Cliffs, NJ (1992)

REVIEW QUESTIONS

1. What is the z-transform?

2. Why is the z-transform particularly useful in analyzing the behavior of discrete-time systems?

3. In what way is the z-transform a special case of the Laplace transform?

4. Why is it incorrect to assume that one can obtain the z-transform of a continuous function?

5. What are the broad concepts involved in the application of z-transforms to discrete-time control systems analysis?

6. What are the various means by which z-transform inverses can be obtained?

7. What is a pulse transfer function?

8. In relating the transfer function of a continuous process to the corresponding sampled-data version, why is it necessary to include the dynamics of a zero-order hold element?

9. What is the actual procedure for obtaining the pulse transfer function of a sampled-data process with a ZOH element, given its Laplace domain transfer function?

10. What is the procedure for converting a difference equation model to a pulse transfer function model?

11. Why does the pulse transfer function derived from a difference equation process model occur naturally with the ZOH element already incorporated?

12. What is the procedure for obtaining a difference equation model from a pulse transfer function model?

PROBLEMS

24.1 Establish that the z-transform of the discrete function $f^*(t)$ generated by sampling the cosine function $f(t) = \cos(\omega t)$ with a sample time Δt is given by:

$$f(z) = \frac{1 - z^{-1}\cos(\omega \Delta t)}{1 - 2z^{-1}\cos(\omega \Delta t) + z^{-2}} \qquad (P24.1)$$

24.2 For a system whose continuous Laplace domain transfer function is given as:

$$g(s) = \frac{K(\xi s + 1)}{(\tau_1 s + 1)(\tau_2 s + 1)} \qquad (P24.2)$$

show that the pulse transfer function, $g(z)$, for the sampled process (with sample time Δt) incorporating a zero-order hold (ZOH) element, is given by:

$$g(z) = \frac{b_1 z^{-1} + b_2 z^{-2}}{1 + a_1 z^{-1} + a_2 z^{-2}} \qquad (P24.3)$$

where, if we define:

$$\phi_1 = e^{-\Delta t/\tau_1}; \quad \phi_2 = e^{-\Delta t/\tau_2} \qquad (P24.4a)$$

then the indicated coefficients are given by:

$$a_1 = -\left(\phi_1 + \phi_2\right) \qquad (P24.4b)$$

$$a_2 = \phi_1 \phi_2 \qquad \text{(P24.4c)}$$

and

$$b_1 = K\left[1 - \left(\frac{\tau_1 - \xi}{\tau_1 - \tau_2}\right)\phi_1 - \left(\frac{\tau_2 - \xi}{\tau_2 - \tau_1}\right)\phi_2 \right] \qquad \text{(P24.4d)}$$

$$b_2 = K\left[\phi_1 \phi_2 - \left(\frac{\tau_1 - \xi}{\tau_1 - \tau_2}\right)\phi_2 - \left(\frac{\tau_2 - \xi}{\tau_2 - \tau_1}\right)\phi_1 \right] \qquad \text{(P24.4e)}$$

24.3 Given the inverse-response process with the continuous Laplace domain transfer function:

$$g(s) = \frac{2.0(-3s + 1)}{(2s + 1)(5s + 1)} \qquad \text{(P24.5)}$$

with time constants in minutes:
(a) Use the result of Problem 24.2 to obtain a pulse transfer function, $g(z)$, for the sampled process incorporating a zero-order hold (ZOH) element, when the sample time is specified as $\Delta t = 1$ min.
(b) Repeat part (a) for $\Delta t = 7$ min.
(c) From the two pulse transfer functions obtained in parts (a) and (b), derive the corresponding difference equation models for the sampled process operating under the two different sample times; plot the unit step response obtained from each model. In light of what you know about inverse-response systems, comment on the nature of the two responses and the effect of sampling on the *perceived* behavior of continuous processes.

24.4 By linearizing the first-principles model for a CSTR in which a first-order exothermic reaction is taking place, the resulting approximate model for the reactor temperature response to changes in jacket cooling water flowrate is given by the following continuous Laplace domain transfer function:

$$g(s) = \frac{-0.55 \ (°\text{F/gpm})}{(5s - 1)(2s + 1)} \qquad \text{(P24.6)}$$

The time constants are in minutes. Obtain the corresponding pulse transfer function for the sampled process incorporating a ZOH element given that the sample time $\Delta t = 0.5$ min.

24.5 From an open-loop identification experiment, the transfer function model for the level in a thermosiphon reboiler in response to steam rate changes has been obtained as:

$$y(s) = \frac{0.9 \ (0.3s - 1)}{s \ (2.5s + 1)} u(s) \qquad \text{(P24.7)}$$

The time constants are in minutes; and the input and output variables are, respectively, the normalized steam rate, and normalized level, defined in terms of nominal steady-state operating conditions, h^* (in), and F_s^* (lb/hr), as:

$$u = \left[\frac{(F_s - F_s^*)\ (\text{lb/hr})}{F_s^*\ (\text{lb/hr})} \right]$$

$$y = \left[\frac{(h - h^*)\ (\text{in})}{h^*\ (\text{in})} \right]$$

This model is to be used for the design of a digital controller with a sample rate of Δt = 0.1 min; obtain the required pulse transfer function model for the sampled process incorporating a ZOH element.

24.6   In Example 20.3, the nonlinear model for the stirred mixing process of Figure 20.1 was obtained as:

$$A_c \frac{dh}{dt} = F_H + F_C + F_d - K\sqrt{h} \tag{20.16}$$

$$\rho C_p A_c \frac{d(hT)}{dt} = \rho C_p \left(F_H T_H + F_C T_C + F_d T_d - K\sqrt{h}T \right) \tag{20.17}$$

This model was linearized around the steady-state operating point, h_s, T_s in Examples 20.4, and by defining the following deviation variables:

$$\begin{aligned}
y_1 &= x_1 = h - h_s; & y_2 &= x_2 = T - T_s \\
u_1 &= F_H - F_{Hs}; & u_2 &= F_C - F_{Cs} \\
d_1 &= F_d - F_{ds}; & d_2 &= T_d - T_{ds}
\end{aligned}$$

The equivalent Laplace domain transfer function matrix model was obtained in Example 20.5 as:

$$\mathbf{y}(s) = \mathbf{G}(s)\,\mathbf{u}(s) + \mathbf{G}_d(s)\,\mathbf{d}(s)$$

where

$$\mathbf{G}(s) = \begin{bmatrix} \dfrac{1}{A_c(s + a_{11})} & \dfrac{1}{A_c(s + a_{11})} \\[3mm] \dfrac{(T_H - T_s)}{A_c h_s(s + a_{22})} & \dfrac{(T_C - T_s)}{A_c h_s(s + a_{22})} \end{bmatrix} \tag{20.23}$$

and

$$\mathbf{G}_d(s) = \begin{bmatrix} \dfrac{1}{A_c(s + a_{11})} & 0 \\[3mm] \dfrac{(T_{ds} - T_s)}{A_c h_s(s + a_{22})} & \dfrac{F_{ds}}{A_c h_s(s + a_{22})} \end{bmatrix} \tag{20.25}$$

with

$$a_{11} = \frac{K}{2A_c\sqrt{h_s}} = \frac{\beta}{A_c} \tag{20.24a}$$

$$a_{22} = \frac{K}{A_c\sqrt{h_s}} = 2a_{11} \qquad\qquad (20.24b)$$

(a) If a specific process whose physical parameters are given below is sampled with $\Delta t = 10$ sec, obtain the pulse transfer function *matrix* for the sampled process, assuming a ZOH element:

$$A_c = 531.7 \text{ cm}^2$$
$$h_s = 15 \text{ cm}$$
$$\beta = \frac{K}{2\sqrt{h_s}} = 3.25$$
$$T_C = 17°C$$
$$T_H = 72°C$$
$$T_d = 35°C$$
$$T_s = 28°C$$
$$F_s = 103 \text{ cc/s}$$
$$F_{ds} = 10.3 \text{ cc/s}$$

(b) If, as a result of a transportation lag through the water pipes, a time delay of 60 seconds is known to be associated with the effect of the flowrate changes on temperature, how should the pulse transfer function matrix be modified to reflect this fact?

DYNAMIC ANALYSIS
OF
DISCRETE-TIME SYSTEMS

With the necessary tools of Chapter 24 in hand, we are now in a position to begin investigating the dynamic behavior of discrete-time systems. In carrying out this analysis, one has the advantage that purely discrete models involving discrete input and output variables are somewhat easier to handle than their continuous-time counterparts, because differences and sums are less demanding conceptually than derivatives and integrals. However, there are some conceptual complications that require some care. This is primarily due to the fact that in computer process control, the input to the process is discrete, and a discrete measurement results from sampling an otherwise continuous system. On the one hand, this creates the advantageous situation in which our discrete analysis results can be checked against the familiar continuous-time results for consistency; and this will be done from time to time. On the other hand, however, the sampling of a continuous process also creates a very interesting situation in which the resulting discrete system is found embedded in a continuous environment. This state of affairs manifests itself most clearly in the coexistence of both continuous *and* discrete components in the block diagrams for such sampled-data systems. As we shall see, care must be taken in analyzing such block diagrams. A discussion of discrete system stability analysis concludes the chapter, with particular emphasis on the somewhat unexpected stability properties of closed-loop sampled-data control systems.

25.1   OPEN-LOOP RESPONSES

In this opening section, we will be concerned, primarily with addressing the following issue:

Given a discrete-time process model and $u(k)$, $k = 0, 1, 2, ...$, a sequence of discrete values for the process input, investigate the behavior of the process output, $y(k)$, $k = 1, 2, ...$. (For multivariable systems, we would have vectors $\mathbf{u}(k)$ and $\mathbf{y}(k)$ instead of the corresponding scalar sequences.)

This is, of course, a discrete version of the continuous-time dynamic analysis problem already investigated in detail in Part II. As was the case then, unless otherwise stated, we will assume that the system is initially at steady state, and that the model is given in terms of deviation variables.

25.1.1 The First-Order Process

The process whose dynamic behavior is modeled by the first-order, difference equation:

$$y(k) + ay(k-1) = bu(k-1) \tag{25.1}$$

is a discrete version of the (continuous) first-order process whose differential equation model was first given in Eqs. (5.1) and (5.2) (cf. Table 23.1).

The corresponding pulse transfer function model is easily obtained, as outlined in Section 24.4, by taking z-transforms in Eq. (25.1) to produce:

$$y(z) = \frac{bz^{-1}}{1 + az^{-1}} u(z) \tag{25.2}$$

Either of these equivalent model forms could be used to determine the response of the process output $y(k)$ to any input sequence, $u(k)$.

Unit Step Response: Time-Domain Analysis

When $u(k) = 1$ for $k = 0, 1, 2, ...$, we observe from Eq. (25.1) that the response $y(k)$ is given by:

$$y(k) = -ay(k-1) + b \tag{25.3}$$

an expression that can be used recursively to generate the process response. Observe from here that for $k = 1$, if we let:

$$\phi = -a \tag{25.4}$$

we would have:

$$y(1) = \phi y(0) + b$$

and for $k = 2$:

$$y(2) = \phi y(1) + b$$

which becomes:

$$y(2) = \phi^2 y(0) + \phi b + b \tag{25.5}$$

and similarly, for $k = 3$:

$$y(3) = \phi y(2) + b$$

and upon substituting for $y(2)$, this becomes:

$$y(3) = \phi^3 y(0) + \phi^2 b + \phi b + b$$

or

$$y(3) = \phi^3 y(0) + b\,(\phi^2 + \phi + 1) \tag{25.6}$$

It is now easy to see that further recursive evaluation will yield the general result:

$$y(k) = \phi^k y(0) + b\,(\phi^{k-1} + \phi^{k-2} + \ldots + 1) \tag{25.7}$$

Since the model is assumed to be in terms of deviation variables, and since the process is initially at steady state, then $y(0) = 0$; also, it is easily established that the geometric series involving powers of ϕ simplifies to $(1 - \phi^k)/(1 - \phi)$, provided $\phi \neq 1$. Thus the general expression for any value of k becomes:

$$y(k) = \frac{b}{1 - \phi}(1 - \phi^k) \tag{25.8}$$

Remarks:

1. Whether or not a *finite*, final steady-state value exists for this response depends on the parameter a (or, equivalently, ϕ).
2. If $|\phi| < 1$, then as $k \to \infty$, $\phi^k \to 0$, and the response tends to the finite steady-state value, $y(\infty) = b/(1 - \phi)$.
3. If $|\phi| \geq 1$, as $k \to \infty$, Eq. (25.8) ϕ^k becomes infinite, and the response is *unbounded*.
4. The closer $|\phi|$ is to 0, the faster $\phi^k \to 0$, and the faster the approach of $y(k)$ to $y(\infty)$; conversely, the closer $|\phi|$ is to 1, the more sluggish the approach to $y(\infty)$.
5. Thus in terms of the familiar characteristic parameters of the continuous first-order system, ϕ appears to be related to the *time constant*, while $b/(1 - \phi)$ is clearly the steady-state gain. A comparison of Eq. (25.1) with Eq. (24.122) establishes that this is in fact the case.

A sketch of this response (for $|\phi| < 1$) is shown in Figure 25.1.

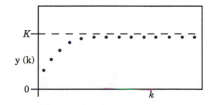

Figure 25.1. Unit step response of a discrete, first-order process.

Unit Step Response: z-Domain Analysis

The same response could be obtained using the z-transfer function representation of Eq. (25.2). To demonstrate this let us recall that for a unit step $u(z) = 1/(1 - z^{-1})$ so that the z-transform for the discrete unit step response gives (with $\phi = -a$):

$$y(z) = \frac{bz^{-1}}{1 - \phi z^{-1}} \frac{1}{1 - z^{-1}}$$

We first simplify by partial fraction expansion, to obtain:

$$y(z) = b\left[\frac{1/(1 - \phi)}{1 - z^{-1}} - \frac{1/(1 - \phi)}{1 - \phi z^{-1}}\right] \tag{25.9}$$

from whence inverse z-transforms are easily obtained, giving the result:

$$y(k) = b\left[\frac{1}{1 - \phi} - \phi^k\left(\frac{1}{1 - \phi}\right)\right]$$

which immediately rearranges to give:

$$y(k) = \frac{b}{1 - \phi}(1 - \phi^k) \tag{25.10}$$

exactly as obtained earlier in Eq. (25.8). We may now recall the corresponding continuous-time result from Chapter 5, (Eq. (5.13) with $A = 1$):

$$y(t) = K(1 - e^{-t/\tau}) \tag{25.11}$$

and observe that Eq. (25.10) is precisely the discrete version of Eq. (25.11) with $b/(1 - \phi) = K$ and $\phi^k = e^{-k\Delta t/\tau}$.

Response to an Arbitrary Input Sequence: Time-Domain Analysis

It is easy to extend the foregoing discussion to cover the situation when we have an arbitrary input sequence for a first-order process. Observe from the time-domain equation in Eq. (25.1) that, with $y(0) = 0$, the general expression for $y(k)$ may be obtained recursively:

for $k = 1$, $y(1) = bu(0)$ \hfill (25.12)

for $k = 2$, and $\phi = -a$, $y(2) = \phi y(1) + bu(1)$

or, using Eq. (25.12) for $y(1)$, $y(2) = \phi bu(0) + bu(1)$

and, for $k = 3$, $y(3) = \phi^2 bu(0) + \phi bu(1) + bu(2)$

$$= b[\phi^2 u(0) + \phi u(1) + u(2)]$$

so that in general: $y(k) = b\left[\sum_{i=1}^{k} \phi^{i-1} u(k-i)\right]$ \hfill (25.13)

Thus, Eq. (25.13) gives the response of the discrete first-order process to any arbitrary input sequence $u(k)$, $k = 0, 1, 2, \ldots$.

Response to an Arbitrary Input Sequence: z-Domain Analysis

The convolution property of the z-transform is particularly useful for this type of analysis. Recall from the convolution result of Eq. (24.16) that:

$$\mathcal{Z}\left\{\sum_{i=0}^{\infty} f_1(i)\, f_2(k - i)\right\} = f_1(z)\, f_2(z) \tag{24.16}$$

This immediately implies that:

$$\mathcal{Z}^{-1}\left\{f_1(z)\, f_2(z)\right\} = \sum_{i=0}^{\infty} f_1(i)\, f_2(k - i) \tag{25.14}$$

Observe now that this inverse z-transform version of the convolution result provides us an immediate, compact representation of the inverse transformation of a product of two z-transforms (compare with Eq. (4.22), the convolution result of Laplace transforms).

The most important implication of this powerful result for open-loop z-domain analysis may be stated as follows:

For *any* discrete process represented by the z-transfer function model:

$$y(z) = g(z)\, u(z) \tag{25.15}$$

Let $\mathcal{Z}^{-1}\{g(z)\} = g(k)$, $k = 0, 1, 2, \ldots$, and $\mathcal{Z}^{-1}\{u(z)\} = u(k)$, $k = 0, 1, 2, \ldots$, then, $y(k)$ is immediately obtained as:

$$y(k) = \sum_{i=0}^{\infty} g(i)u(k - i) \tag{25.16}$$

There are a few important points we must note here:

1. $g(k)$ is the discrete *impulse-response function* for the process whose z-transfer function is $g(z)$. Observe that when $u(k)$ is the *Kronecker delta function* $\delta(k\Delta t)$ defined by:

$$\delta(k\Delta t) = \begin{cases} 1; & k = 0 \\ 0; & k \neq 0 \end{cases} \tag{25.17}$$

    (i.e., the discrete-time equivalent of the continuous-time unit impulse function, $\delta(t)$), it is easy to see that the z-transform, $u(z) = \mathcal{Z}[\delta(k\Delta t)]$ is 1 in this case; hence, $y(z) = g(z)$, and therefore $y(k)$, the response, will be exactly $g(k)$.

2. The expression in Eq. (25.16) indicates that the process response to any other arbitrary sequence of inputs can be obtained by the convolution of this impulse-response function $g(k)$ with the input sequence, $u(k)$.

3. This expression is, in fact, not new; it was encountered earlier in Chapter 4 and in Chapter 23, where it was referred to as the discrete impulse-response model form. Observe, therefore, that Eq. (25.16) is a valid, alternative model form, being entirely equivalent to Eq. (25.2) as we have now established.

Let us now apply Eq. (25.16) to the current problem of finding the response of the first-order process to an arbitrary input sequence. From Eq. (25.2), we observe that $g(z)$ is given by:

$$g(z) = \frac{bz^{-1}}{1 + az^{-1}} = \frac{bz^{-1}}{1 - \phi z^{-1}} \tag{25.18}$$

To obtain the inverse z-transform of this function, first, let us factor out the z^{-1} term in the numerator to obtain:

$$g(z) = \left(\frac{b}{1 - \phi z^{-1}}\right) z^{-1} \tag{25.19}$$

We now recall that:

1. The inverse transform of the terms in winged brackets is $b\phi^k$.

2. From the time-shift property of z-transforms given in Eq. (24.18), the effect of the additional z^{-1} term is merely to introduce a time delay of one sample time.

Thus, we have:

$$g(k) = \begin{cases} 0; & k < 1 \\ b\phi^{k-1}; & k \geq 1 \end{cases} \tag{25.20}$$

and for the general input sequence $u(k)$, we may now use Eq. (25.16) to obtain the desired response as:

$$y(k) = \sum_{i=1}^{\infty} b\phi^{i-1} u(k-i) \tag{25.21}$$

where the index must now start from 1 because, from Eq. (25.20), $g(k)$ is zero until then. By factoring out b, and recalling that $u(k)$ is zero for $k < 0$, so that $u(k-i)$ will vanish for any $i > k$, Eq. (25.21) becomes:

$$y(k) = b\left[\sum_{i=1}^{k} \phi^{i-1} u(k-i)\right] \tag{25.22}$$

which is, of course, identical to Eq. (25.13).

25.1.2 The General, Linear Discrete Process

The models for the general, linear discrete process were given in Chapter 24 as:

$$\begin{aligned} y(k) + a_1 y(k-1) &+ a_2 y(k-2) + \ldots + a_n y(k-n) \\ &= b_0 u(k) + b_1 u(k-1) + b_2 u(k-2) + \ldots + b_m u(k-m) \end{aligned} \tag{24.106}$$

in the time domain, and:

$$y(z) = \left(\frac{b_0 + b_1 z^{-1} + b_2 z^{-2} + \dots + b_m z^{-m}}{1 + a_1 z^{-1} + a_2 z^{-2} + \dots + a_n z^{-n}} \right) u(z) \tag{24.110}$$

in the z-domain.

If all we are interested in is numerical analysis of the behavior of these systems, then Eq. (24.106) is ideal for this purpose, being easily programmed on a digital computer to accept values of the discrete sequence, $u(k)$, and sequentially compute the resultant $y(k)$. To gain insight into the general characteristics of these systems, however, the analysis is somewhat more conveniently carried out in the z-domain.

Unit Step Response

The unit step response of this general, linear system is obtained by introducing $1/(1 - z^{-1})$ for $u(z)$, giving:

$$y(z) = \left(\frac{b_0 + b_1 z^{-1} + b_2 z^{-2} + \dots + b_m z^{-m}}{1 + a_1 z^{-1} + a_2 z^{-2} + \dots + a_n z^{-n}} \right) \frac{1}{1 - z^{-1}} \tag{25.23}$$

The z-transform inversion required for converting Eq. (25.23) to the time domain is best carried out by partial fraction expansion as discussed in Section 24.3 of Chapter 24.

Assuming once again that the denominator polynomial has n distinct, real roots, then recalling Eq. (24.67), we see that by partial fraction expansion, Eq. (25.23) becomes:

$$y(z) = \left(\frac{R_0}{1 - z^{-1}} \right) + \left(\frac{R_1}{1 - p_1 z^{-1}} \right) + \left(\frac{R_2}{1 - p_2 z^{-1}} \right) + \dots + \left(\frac{R_n}{1 - p_n z^{-1}} \right) \tag{25.24}$$

where the first term on the RHS arises because of the additional $1/(1 - z^{-1})$ term. As before, $p_1, p_2, \dots, p_n$ are the roots of the denominator polynomial, and the constants $R_0, R_1, \dots, R_n$ are determined in the usual fashion.

Taking inverse z-transforms in Eq. (25.24) now yields:

$$y(k) = R_0 + R_1 \left(p_1 \right)^k + R_2 \left(p_2 \right)^k + \dots + R_n \left(p_n \right)^k \tag{25.25}$$

It is important to note a few things about this general unit step response:

1. If, for every p_i, $i = 1, 2, \dots, n$ in Eq. (25.25):

$$|p_i| < 1 \tag{25.26}$$

then observe that as $k \to \infty$, all the terms except R_0 vanish and $y(k)$ attains a finite, steady-state value, $y(\infty) = R_0$.

2. If *just one* of the roots p_i fails to satisfy Eq. (25.26), then this portion of the expression in Eq. (25.25) will grow indefinitely as $k \to \infty$, and the response will be unbounded because of this single term.

3. It is therefore clear that the nature of the roots of the denominator polynomial of the z-domain transfer function determines the character of the response. Recall that this was precisely the case with s-domain transfer functions.

Being the roots of the denominator polynomial of the z-transfer function implies that this polynomial will be zero when $z = p_i$, $i = 1, 2, \ldots, n$; therefore, as was the case with s-domain dynamic analysis, these roots p_i are known as the *poles* of the z-transfer function. Later in Section 25.2 we shall have more to say about the poles (and also the *zeros*) of the z-transfer function.

Response to an Arbitrary Input Sequence

The response of the general, linear discrete system to an arbitrary input sequence is best obtained using the convolution result. Since the z-transfer function for this process is:

$$g(z) = \left(\frac{b_0 + b_1 z^{-1} + b_2 z^{-2} + \ldots + b_m z^{-m}}{1 + a_1 z^{-1} + a_2 z^{-2} + \ldots + a_n z^{-n}} \right) \tag{25.27}$$

by partial fraction expansion followed by z-transform inversion (precisely as in Section 24.3.3) we obtain:

$$g(k) = C_1 (p_1)^k + C_2 (p_2)^k + \ldots + C_n (p_n)^k \tag{25.28}$$

where p_i, $i = 1, 2, \ldots, n$ are, once again, the poles of the z-transfer function. If we now use Eq. (25.28) in Eq. (25.16), $y(k)$ may now be obtained as:

$$y(k) = \sum_{i=0}^{\infty} C_1 (p_1)^i u(k-i) + \sum_{i=0}^{\infty} C_2 (p_2)^i u(k-i) + \ldots$$

$$+ \sum_{i=0}^{\infty} C_n (p_n)^i u(k-i) \tag{25.29}$$

which may be further simplified to:

$$y(k) = \sum_{j=1}^{n} \sum_{i=0}^{\infty} C_j (p_j)^i u(k-i) \tag{25.30}$$

The important point to note here is that even in this general case, the ultimate behavior of the response is determined, once again, solely by the nature of the poles of the z-transfer function, a result that is consistent with continuous-time-domain results.

25.1.3 Extension to Multivariable Systems

Extending the foregoing discussion to multivariable systems is fairly straightforward; we simply replace the scalar $u(k)$ and $y(k)$ by appropriately dimensioned vectors $\mathbf{u}(k)$ and $\mathbf{y}(k)$, and the model parameters by appropriate matrices.

The typical multivariable process (in the absence of time delays) is modeled in discrete-time by the equation:

$$\mathbf{y}(k) = \Phi\mathbf{y}(k-1) + \mathbf{B}\mathbf{u}(k-1) \tag{25.31}$$

where $\mathbf{y}(k)$ is a vector consisting of v_1 output variable elements $y_i(k)$; $i = 1, 2, \ldots, v_1$, and $\mathbf{u}(k)$ is a vector consisting of v_2 input variable elements $u_i(k)$; $i = 1, 2, \ldots, v_2$, for a $v_1 \times v_2$ system.

The corresponding z-domain model is easily obtained to be:

$$\mathbf{y}(z) = [\mathbf{I} - z^{-1}\Phi]^{-1}\mathbf{B}z^{-1}\,\mathbf{u}(z) \tag{25.32a}$$

This may be further rearranged to give:

$$\mathbf{y}(z) = [z\mathbf{I} - \Phi]^{-1}\mathbf{B}\mathbf{u}(z) \tag{25.32b}$$

an expression which mimics the continuous-time, s-transfer function matrix form very well.

Example 25.1   DISCRETE-TIME MODEL FOR A 2 × 2 SYSTEM.

Given a two-input, two-output system modeled by the two equations:

$$y_1(k) + a_{11}y_1(k-1) + a_{12}y_2(k-1) = b_{11}u_1(k-1) + b_{12}u_2(k-1) \tag{25.33}$$

$$y_2(k) + a_{21}y_1(k-1) + a_{22}y_2(k-1) = b_{21}u_1(k-1) + b_{22}u_2(k-1) \tag{25.34}$$

observe that by defining:

$$\Phi = \begin{bmatrix} -a_{11} & -a_{12} \\ -a_{21} & -a_{22} \end{bmatrix}; \quad \mathbf{B} = \begin{bmatrix} b_{11} & b_{12} \\ b_{21} & b_{22} \end{bmatrix} \text{ and }$$

$$\mathbf{y}(k) = \begin{bmatrix} y_1(k) \\ y_2(k) \end{bmatrix}; \quad \mathbf{u}(k) = \begin{bmatrix} u_1(k) \\ u_2(k) \end{bmatrix}$$

this model becomes:

$$\mathbf{y}(k) = \Phi\mathbf{y}(k-1) + \mathbf{B}\mathbf{u}(k-1)$$

as given in Eq. (25.31).

By paying careful attention to the laws of matrix algebra, it is easy to show (see Problem 25.4) that the response of the discrete multivariable system in Eq. (25.31) to a unit step in *all* the v_2 inputs (i.e., $u_i(k) = 1$, for all the i and for all discrete-time steps $k = 0, 1, 2, \ldots$) is given by:

$$\mathbf{y}(k) = (\mathbf{I} - \Phi)^{-1}(\mathbf{I} - \Phi^k)\mathbf{B} \tag{25.35}$$

Observe that Eq. (25.35) is merely a matrix version of Eq. (25.8).

It can be shown that provided all the eigenvalues of Φ are strictly less than 1 in absolute value, i.e.:

$$|\lambda(\Phi)| < 1 \tag{25.36}$$

then

$$\lim_{k \to \infty} \Phi^k = 0 \tag{25.37}$$

and the output vector attains to the final, steady-state value:

$$\mathbf{y}(\infty) = (\mathbf{I} - \Phi)^{-1}\mathbf{B} \tag{25.38}$$

The response to the arbitrary vector sequence of inputs, $\mathbf{u}(k)$, is easily shown to be:

$$\mathbf{y}(k) = \left[\sum_{i=1}^{k} \Phi^{i-1} \mathbf{B}\mathbf{u}(k - i) \right] \tag{25.39}$$

which is recognizable as the matrix version of Eq. (25.13).

25.1.4 Effect of Time Delays

So far, we are yet to consider formally the dynamic behavior of discrete systems having time delays. Observe, however, that the presence of a time delay merely causes a time shift in the response of the "undelayed" version of the process; thus, deriving the time response of time-delay systems creates no additional problem.

First-Order System with Time Delay

When a first-order system has an associated delay of D sampling periods, the model in Eq. (25.1) is modified to read:

$$y(k) + ay(k - 1) = bu(k - D - 1) \tag{25.40}$$

and the corresponding pulse transfer function model with $(\phi = -a)$ is:

$$y(z) = \frac{bz^{-1}(z^{-D})}{1 - \phi z^{-1}} u(z) \tag{25.41}$$

It is important to note that the presence of a time delay will always be announced by the appearance of a z^{-D} term common to all elements in the numerator of the z-transfer function. This can therefore be factored out, as indicated in Eq. (25.41).

Deriving the dynamic response of this system should proceed as follows:

1. Recast the z-transfer function as $g_1(z) z^{-D}$, where:

$$g_1(z) = \frac{bz^{-1}}{1 - \phi z^{-1}}$$

the "undelayed" portion of the transfer function.

2. Let $y_1(z) = g_1(z)\, u(z)$, so that:

$$y(z) = y_1(z)\, z^{-D} \tag{25.42}$$

Because $g_1(z)$ has no delays, $y_1(k)$ is obtained as before in Eq. (25.8);

3. Now recall Eq. (24.18) for the effect of time shifts on z-transforms; in particular, its reverse, in Eq. (24.119), is used to derive the desired $y(k)$ from $y_1(k)$, using Eq. (25.42): in this specific case, the result is:

$$y(k) = \begin{cases} 0; & k < D \\[2mm] \dfrac{b\,(1 - \phi^{k-D})}{1 - \phi}; & k \geq D \end{cases} \tag{25.43}$$

General Order System with Time Delay

These ideas are easily extended to the general order system with a time delay. The key lies in representing the general transfer function $g(z)$ as:

$$g(z) = g_1(z)\, z^{-D} \tag{25.44}$$

where $g_1(z)$ is the "undelayed" portion. The "undelayed" response, $y_1(k)$, is obtained, and from the time-shift result, the required time-delayed response will be given by:

$$y(k) = \begin{cases} 0; & k < D \\[2mm] y_1(k - D); & k \geq D \end{cases} \tag{25.45}$$

25.2 CHARACTERISTICS OF OPEN-LOOP PULSE TRANSFER FUNCTIONS

Just as in the case of the dynamic behavior of continuous-time systems in Part II, it is possible to determine quickly the *qualitative* nature of the discrete system dynamic behavior by examining the pulse transfer function. Thus it is worthwhile to discuss the characteristics of pulse transfer functions in more detail.

25.2.1 Equivalent Pulse Transfer Function Forms

The pulse transfer function can be written in one of two different ways, and we would do well to understand each form and how they are related before going on to discuss general characteristics.

The "Backward Shift" Form

The pulse transfer function is often written in the general form:

$$g(z) = \frac{B(z^{-1})\, z^{-D}}{A(z^{-1})} \tag{25.46a}$$

or

$$g(z) = \frac{\left(b_0 + b_1 z^{-1} + b_2 z^{-2} + \dots + b_m z^{-m} \right) z^{-D}}{1 + a_1 z^{-1} + a_2 z^{-2} + \dots + a_n z^{-n}} \qquad (25.46b)$$

where $B(z^{-1})$ is an mth-order polynomial in z^{-1}, and $A(z^{-1})$ is an nth-order polynomial in z^{-1}. For causal systems, $n > m$; and we will represent the polynomial-order difference $(n - m)$ by the variable l.

This form is sometimes known as the "backward shift" form for the following reason: only powers of z^{-1} are involved, and since $z^{-1}f(z)$ in the time domain is $f(k-1)$ (which is $f(k)$ shifted backwards in time by one unit) we may then observe that z^{-1} actually behaves like a unit "backward shift" operator, hence the name.

The "Forward Shift" Form

The pulse transfer function in Eq. (25.46) may be converted to a form involving only positive powers of the variable z as follows:

By factoring out z^{-m} in the numerator and z^{-n} in the denominator, we obtain:

$$g(z) = \frac{z^{-m}\left(b_0 z^m + b_1 z^{m-1} + b_2 z^{m-2} + \dots + b_m \right) z^{-D}}{z^{-n}\left(z^n + a_1 z^{n-1} + a_2 z^{n-2} + \dots + a_n \right)} \qquad (25.47)$$

and introducing $l = n - m$, Eq. (24.47) becomes:

$$g(z) = \frac{z^l\left(b_0 z^m + b_1 z^{m-1} + b_2 z^{m-2} + \dots + b_m \right) z^{-D}}{z^n + a_1 z^{n-1} + a_2 z^{n-2} + \dots + a_n} \qquad (25.48)$$

If we choose not to confound the z^{-D} term, indicative of the presence of a time delay of D sample times, with the z^l term, indicative of the denominator-numerator polynomial-order excess, then it is possible to write Eq. (25.48) as the following ratio of polynomials *involving only z*:

$$g(z) = \frac{z^l\left(b_0 z^m + b_1 z^{m-1} + b_2 z^{m-2} + \dots + b_m \right)}{z^D\left(z^n + a_1 z^{n-1} + a_2 z^{n-2} + \dots + a_n \right)} \qquad (25.49)$$

or

$$g(z) = \frac{z^l B'(z)}{z^D A'(z)} \qquad (25.50)$$

where $B'(z)$ and $A'(z)$ are the numerator and denominator polynomials in z indicated in Eq. (25.49). Since this form is in terms of the variable z (which is the *inverse* of the unit "backward shift" operator z^{-1}), it is sometimes referred to as the "forward shift" form in direct contrast to the "backward shift" form.

If we choose to combine the z^l and the z^D terms, $g(z)$ will have the form:

$$g(z) = \frac{B'(z)}{z^N A'(z)} \qquad (25.51)$$

with

$$N = D - l \qquad (25.52)$$

We must now note the following very crucial points about the relationship between the two forms available for representing the pulse transfer function:

1. The polynomial equations:

    $$A'(z) = z^n + a_1 z^{n-1} + a_2 z^{n-2} + \dots + a_n = 0 \qquad (25.53)$$

    found in Eqs. (25.49) and (25.50), and the corresponding

    $$A(z^{-1}) = 1 + a_1 z^{-1} + a_2 z^{-2} + \dots + a_n z^{-n} = 0 \qquad (25.54)$$

    found in Eq. (25.46) have *identical* sets of roots for z, provided $z \neq 0$, and may therefore be considered equivalent.

2. The same is true for the polynomials $B(z^{-1})$ and $B'(z)$:

    $$B'(z) = b_0 z^m + b_1 z^{m-1} + b_2 z^{m-2} + \dots + b_m = 0 \qquad (25.55)$$

    and

    $$B(z^{-1}) = b_0 + b_1 z^{-1} + b_2 z^{-2} + \dots + b_m z^{-m} = 0 \qquad (25.56)$$

    have identical sets of roots for z.

3. In computer process control, the pulse transfer function almost always occurs either by taking z-transforms of a discrete-time model, or by converting (also via z-transforms) a Laplace transfer function. In either case, since the z-transform naturally occurs in terms of z^{-1}, such pulse transfer functions will *naturally* occur in the "backward shift" form. Thus, the "forward shift" form usually arises from a rearrangement of the "backward shift" form.

4. In the event that a pulse transfer function is given in the form shown in Eq. (25.51) where the time delay D and the numerator-denominator polynomial-order excess l are combined, the value of l can be deduced from the respective orders of $B'(z)$ and $A'(z)$, and hence the actual value of D can be determined from Eq. (25.52).

Before going on to study the characteristics of pulse transfer functions, the following example is used to further clarify the preceding discussion.

Example 25.2 PULSE TRANSFER FUNCTION FORMS FOR A SECOND-ORDER SYSTEM WITH A ZOH.

The pulse transfer function for a second-order system with a ZOH element can be shown to be given by:

$$g(z) = \frac{b_1 z^{-1} + b_2 z^{-2}}{1 + a_1 z^{-1} + a_2 z^{-2}} \qquad (25.57)$$

1.   Express $g(z)$ in the form given in Eq. (25.46); identify D and l.
2.   Convert Eq. (25.57) to a function of z only by multiplying its numerator and denominator by z^2; interpret the resulting function.

Solution:

1.   By observing that there is a common z^{-1} factor in the numerator, Eq. (25.57) may be rearranged as:

$$g(z) = \frac{\left(b_1 + b_2 z^{-1}\right) z^{-1}}{1 + a_1 z^{-1} + a_2 z^{-2}} \tag{25.58}$$

and now, since the order of the remaining numerator polynomial (in parentheses) is 1, and the order of the denominator polynomial is 2, we conclude that Eq. (25.58) is in the form given in Eq. (25.46) with $l = 2 - 1 = 1$, and $D = 1$.

2.   Multiplying by z^2 as required, we obtain:

$$g(z) = \frac{b_1 z + b_2}{z^2 + a_1 z + a_2} \tag{25.59}$$

Let us now note that if we had been given Eq. (25.59) directly as the pulse transfer function for the second-order system with a ZOH, a very common mistake is to inspect Eq. (25.59) — which is entirely identical to Eq. (25.58) — and conclude that there is *no delay involved* in this system. However, since this is a ratio of a first-order and a second-order polynomial, obviously $l = 1$ and $D = 1$.

To summarize, we note that the pulse transfer function may be represented in any of the following entirely equivalent forms:

1.   **The "Backward Shift" Form**

$$g(z) = \frac{B(z^{-1}) z^{-D}}{A(z^{-1})}$$

$$= \frac{\left(b_0 + b_1 z^{-1} + b_2 z^{-2} + \ldots + b_m z^{-m}\right) z^{-D}}{1 + a_1 z^{-1} + a_2 z^{-2} + \ldots + a_n z^{-n}} \tag{25.60}$$

2.   **The "Forward Shift" Form**

$$g(z) = \frac{B'(z)}{z^N A'(z)}$$

$$= \frac{\left(b_0 z^m + b_1 z^{m-1} + b_2 z^{m-2} + \ldots + b_m\right)}{z^N\left(z^n + a_1 z^{n-1} + a_2 z^{n-2} + \ldots + a_n\right)} \tag{25.61}$$

or, when derived from the "backward shift " form, and the delay term D has not been combined with the numerator-denominator polynomial-order excess l:

$$g(z) = \frac{z^l B'(z)}{z^D A'(z)}$$

$$= \frac{z^l \left(b_0 z^m + b_1 z^{m-1} + b_2 z^{m-2} + \ldots + b_m \right)}{z^D \left(z^n + a_1 z^{n-1} + a_2 z^{n-2} + \ldots + a_n \right)} \tag{25.62}$$

We are now in a position to study the characteristics of the pulse transfer function.

25.2.2 Fundamental Characteristics

By analogy with the Laplace transfer function, the pulse transfer function (expressed either in the "backward shift" or the "forward shift" form) is also characterized by its *order*, *time delay* (dead time), *steady-state gain*, *poles*, and *zeros*.

The *order* of a pulse transfer function is the same as the order of the *denominator* polynomial $A(z^{-1})$ in the "backward shift" form, or $A'(z)$ in the "forward shift" form, i.e., n. As previously noted, the time delay is D. Note that the order and associated time delay are more easily identified in the "backward shift" form in Eq. (25.60).

A physical system is said to be of the order of the transfer function that represents its dynamic behavior. Thus the process in Example 25.2 is a second-order system.

Steady-State Gain

Let $S(k)$ represent the unit step response of the system whose pulse transfer function is $g(z)$. Observe that $S(k)$ will be given by:

$$S(k) = \mathcal{Z}^{-1}\{S(z)\} \tag{25.63}$$

where

$$S(z) = g(z)\frac{1}{1 - z^{-1}} \tag{25.64}$$

Provided there is a finite, steady-state value, it may be found by the final value theorem, i.e.:

$$\lim_{k \to \infty} S(k) = \lim_{z \to 1} \left[(1 - z^{-1})S(z) \right] \tag{25.65}$$

and by introducing Eq. (25.64) into Eq. (25.65) we obtain:

$$\lim_{k \to \infty} S(k) = \lim_{z \to 1} \left[g(z) \right] \tag{25.66}$$

If we now recall that the final steady-state value of a system's unit step response is identical to its steady-state gain, we have the final result that K, the system's steady-state gain, is obtained from its pulse transfer function according to:

$$K = \lim_{z \to 1} [g(z)] \tag{25.67}$$

Recall that the corresponding result for the continuous system is:

$$K = \lim_{s \to 0} [g(s)] \tag{25.68}$$

Applying Eq. (25.67) to either representation of the pulse transfer function, we find that *regardless of which representation we choose*, K will be given by:

$$K = \frac{b_0 + b_1 + b_2 + \dots + b_m}{1 + a_1 + a_2 + \dots + a_n} \tag{24.69}$$

Poles and Zeros

We will first give the *formal* definition of the poles and zeros of a pulse transfer function; this requires the use of the "forward shift" form in Eq. (25.61) (cf. for example, Ref. [1], p. 52, and Ref. [2], p. 44). Thereafter we will interpret the results in the context of the "backward shift" form so that poles and zeros may be accurately determined from this more *natural* form.

Given a pulse transfer function expressed in the form shown in Eq. (25.61), by analogy with the Laplace transfer function, the roots of the *denominator* polynomial $A'(z)$ are the *poles* of $g(z)$ while the roots of the numerator polynomial $B'(z)$ are the *zeros* of $g(z)$. The N additional poles at $z = 0$ are contributed by the time delay (cf. Åström and Wittenmark [1], p. 51). (In actual fact, as indicated in Eq. (25.62), the time delay contributes D poles at $z = 0$, but the denominator-numerator polynomial order excess results in l zeros at precisely the same location $z = 0$. By pole-zero cancellation, only $N = D - l$ of the poles at $z = 0$ actually survive.)

Thus, $g(z)$, an *nth-order* pulse transfer function, has:

- n poles located at the roots of the nth-order denominator polynomial $A'(z)$,
- m zeros located at the roots of the mth-order numerator polynomial $B'(z)$, and
- When there is a time delay of D sample times, and a pole-zero excess of $n - m = l$, there will be $N = D - l$ additional poles located at $z = 0$, the origin of the complex z-plane.

However, we had noted earlier that $A'(z)$ in the forward shift form and $A(z^{-1})$ in the backward shift form have identical sets of roots; similarly, $B'(z)$ and $B(z^{-1})$ also have identical sets of roots. The implication for poles and zeros of $g(z)$ in terms of the backward shift representation is now given as follows:

1. **Poles:** The n roots of $A(z^{-1}) = 0$ are the n poles of $g(z)$; (a time delay contributes D additional poles at $z = 0$, the origin of the complex z-plane).

2. **Zeros:** The m roots of $B(z^{-1}) = 0$ are the m zeros of $g(z)$; whenever $n > m$, there will be an excess of $l = (n - m)$ additional zeros at $z = 0$. (If $D > l$ these zeros are canceled with l of the D poles contributed by the time delay).

Let us illustrate these ideas with the following examples.

Example 25.3   CHARACTERISTICS OF SEVERAL PULSE TRANSFER FUNCTIONS.

Obtain the main characteristics (i.e., order, delay, steady-state gain, poles and zeros) of the following pulse transfer functions:

(a) $$g(z) = \frac{b}{1 - az^{-1}} \qquad (25.70)$$

(b) $$g(z) = \frac{bz^{-1}}{1 - az^{-1}} \qquad (25.71)$$

(c) $$g(z) = \frac{bz^{-1-M}}{1 - az^{-1}} \qquad (25.72)$$

(d) $$g(z) = \frac{b_1 z^{-1} + b_2 z^{-2}}{1 + a_1 z^{-1} + a_2 z^{-2}} \qquad (25.73)$$

Solution:

(a) For this transfer function, we observe, first directly from Eq. (25.70), that $n = 1$, $m = 0$, and $D = 0$. Thus $l = 1$ and we now obtain the required characteristics as follows:
- This is a *first*-order transfer function, because $n = 1$; however, there is no ZOH
- The delay is zero, because $D = 0$
- By setting $z = 1$ we obtain $K = b/(1 - a)$
- The transfer function has *one* pole at $z = a$; the only zero is the one at $z = 0$ since $l = 1$ and $m = 0$

Of course, multiplying in Eq. (25.70) by z/z we obtain:

$$g(z) = \frac{bz}{z - a} \qquad (25.74)$$

with a zero at $z = 0$, and a pole at $z = a$. Thus it is much easier to see the poles and zeros in the forward shift form.

(b) In this case, recognizing the presence of the z^{-1} term in the numerator, the transfer function is first "rearranged" as:

$$g(z) = \frac{(b) z^{-1}}{1 - az^{-1}} \qquad (25.75)$$

and we see that this is the first-order transfer function in part (a) with a *unit delay*.
    Thus we have: a first-order transfer function, with delay $D = 1$, steady-state gain $K = b/(1 - a)$, a pole at $z = a$, and *in principle* an excess zero at $z = 0$ because $l = 1$. However, we now observe the pole-zero cancellation phenomenon again: the unit

delay contributes an extra pole at $z = 0$ that *cancels out* the excess zero located at the same point. In fact, the delay associated with this process is only noticed if the backward shift form Eq. (25.75) is used: multiplying both the numerator and denominator by z immediately gives:

$$g(z) = \frac{(b)}{z - a} \tag{25.76}$$

which still shows the pole at $z = a$ but fails to show either the excess zero or the single pole contributed by the delay because they have been canceled out, being opposing terms located at the same point.

(c) In this case, the transfer function clearly has the same form as that in part (b) but with an *additional* delay term; thus it has all the characteristics identified with part (b), but now $D = M + 1$.

We must again note that *one* of the $M + 1$ additional poles contributed at $z = 0$ by the delay term will be canceled out by the *excess* zero at the same location, so that if Eq. (25.72) is put in the forward shift form, it will appear as if it has only a delay of M, i.e.:

$$g(z) = \frac{b}{z^M(z - a)} \tag{25.77}$$

(d) Rearranging the transfer function as follows:

$$g(z) = \frac{\left(b_1 + b_2 z^{-1}\right)z^{-1}}{1 + a_1 z^{-1} + a_2 z^{-2}} \tag{25.78}$$

which we recall from Example 25.2 is a second-order process with a ZOH. To verify this, observe that: $n = 2$, $m = 1$, $D = 1$; so that $l = 1$, and the characteristics of Eq. (25.78) in this case are:

- A second-order transfer function ($n = 2$)
- A *unit* delay ($D = 1$)
- A steady-state gain $K = \left(b_1 + b_2\right)/\left(1 + a_1 + a_2\right)$
- *Two* poles at the roots of the denominator polynomial, i.e., $z = p_1, p_2$, where:

$$p_1, p_2 = \frac{-a_1 \pm \sqrt{a_1{}^2 - 4a_2}}{2} \tag{25.79}$$

- One zero at $z = -(b_2/b_1)$ with an excess zero at $z = 0$ because $l = 1$

Observe again that the excess zero gets canceled out by the additional pole at $z = 0$ contributed by the unit delay. In the forward shift form, Eq. (25.73) reads:

$$g(z) = \frac{b_1 z + b_2}{z^2 + a_1 z + a_2} \tag{25.80}$$

in which the proper zero at $z = -(b_2/b_1)$, and the two poles are preserved, but the excess zero and the additional pole due to the delay have been canceled out.

Recall from Chapter 23 that we indicated that the presence of a ZOH induces a time delay in the process. This is seen in parts (b), (c), and (d) of Example 25.3 as a factor z^{-1}.

25.2.3 Relationship with $g(s)$

Let us keep in mind that in computer process control, the process model may often be available first as a Laplace transfer function; the convenience of the z-transform (over the Laplace transform) in analyzing discrete-time systems is what prompts its replacement with a z-domain pulse transfer function model. Since we are therefore often engaged in converting a Laplace transfer function to its z-domain counterpart, it will, of course, be useful to be able to relate $g(z)$ to the $g(s)$ from which it was derived. It turns out, however, that in going through this exercise our ultimate reward is something more fundamental: we find that we are, in fact, able to relate the *entire* complex s-plane — the familiar arena in which continuous analysis takes place — to the *entire* complex z-plane, the (yet unfamiliar) corresponding arena for discrete analysis.

A Fundamental s-plane ↔ z-plane Relationship

When we sample a continuous process whose Laplace transfer function is $g(s)$, the *Laplace* transform of the *sampled* version, i.e., $g^*(s)$, is given by:

$$g^*(s) = L\{g^*(t)\} = \sum_{k=0}^{\infty} g(k\Delta t)e^{-ks\Delta t} \tag{25.81}$$

where $g^*(t)$ is the sampled version of the continuous impulse response $g(t)$. The corresponding $g(z)$ is obtained, in principle, by taking z-transforms of $g^*(t)$; by definition of the z-transform, therefore, we have:

$$g(z) = Z\{g^*(t)\} = \sum_{k=0}^{\infty} g(k\Delta t)z^{-k} \tag{25.82}$$

and by comparing Eq. (25.82) with Eq. (25.81) we see that the *sampled* s-transfer function is related to the z-transfer function by the relation:

$$z = e^{s\Delta t} \tag{25.83}$$

or, conversely:

$$s = \frac{1}{\Delta t} \ln z \tag{25.84}$$

Thus if one wishes to relate a z-transfer function to the Laplace transfer function from which it arose by "sampling at time intervals of Δt," these results are essential. Eq. (25.83) actually allows us to relate the Laplace transform of any arbitrary discrete function to its z-transform. Thus Eq. (25.83) is, in fact, a fundamental relationship between the complex variables s and z that holds regardless of the specific function whose z-transform is being compared to the corresponding Laplace transform. In particular, Eq. (25.83) can be used to relate *all possible values* of s in the familiar s-plane to the corresponding values of z in the less familiar territory of the z-plane. This will enable us to develop an analogous set of results by which discrete counterparts of continuous processes can be systematically analyzed.

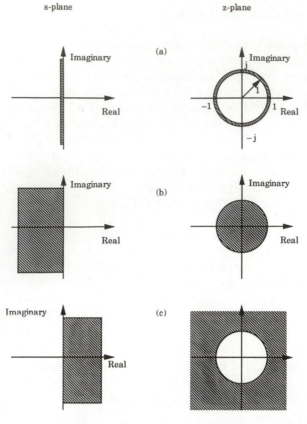

Figure 25.2.   Corresponding regions in the *s*-plane and the *z*-plane.

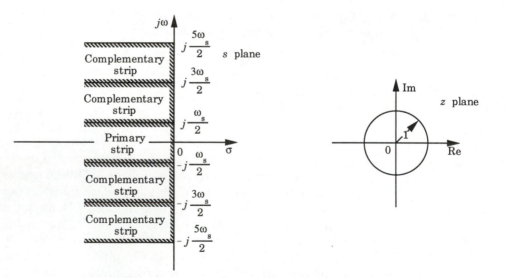

Figure 25.3.   Periodic strips in the *s*-plane that all map into the unit circle in the *z*-plane (adapted from Ref. [2]).

Relating the s-plane and the z-plane

In general, any point in the complex s-plane can be represented by:

$$s = \sigma + j\omega \tag{25.85}$$

where σ is the real part and ω is the imaginary part. From Eq. (25.83) we see that the corresponding point in the z-plane will be given by:

$$z = e^{(\sigma + j\omega)\Delta t} \tag{25.86}$$

or

$$z = r\, e^{j\omega\Delta t} \tag{25.87}$$

where

$$r = e^{\sigma\Delta t} \tag{25.88}$$

We observe immediately that Eq. (25.87) is the polar representation of a complex number in the z-plane whose magnitude, and argument, are given respectively by:

$$|z| = r = e^{\sigma\Delta t} \tag{25.89}$$

and

$$\arg(z) = \angle z = \omega\Delta t \tag{25.90}$$

In order to convert Eq. (25.87) to the form:

$$z = \sigma^* + j\omega^* \tag{25.91}$$

we can use the Euler identity to replace the complex exponential in Eq. (25.87) to yield:

$$z = r\Big[\cos(\omega\Delta t) + j\sin(\omega\Delta t)\Big] \tag{25.92}$$

so that:

$$\sigma^* = r\cos(\omega\Delta t) = e^{\sigma\Delta t}\cos(\omega\Delta t) \tag{25.93}$$

and

$$\omega^* = r\sin(\omega\Delta t) = e^{\sigma\Delta t}\sin(\omega\Delta t) \tag{25.94}$$

Let us now use these relationships to locate points in the z-plane corresponding to several important points in the s-plane (cf. Figures 25.2 and 25.3).

The imaginary axis $(\sigma = 0$ for all $\omega)$

From Eqs. (25.89) and (25.90), we see that these points correspond to the set of complex variables in the z-plane represented by:

$$|z| = r = 1, \text{ for all } \angle z \tag{25.95}$$

the locus of all complex numbers in the z-plane having magnitude 1. This is clearly identified as the unit circle. Thus, we have that:

The imaginary axis in the s-plane corresponds to the unit circle in the z-plane.

The entire left-half plane ($\sigma < 0$ for all ω)

Since σ is negative in this case, we see from Eqs. (25.89) and (25.90), that $e^{\sigma\Delta t}$ will be less than 1. Thus, the corresponding set of points in the z-plane are represented by:

$$|z| = e^{\sigma\Delta t} < 1, \text{ for all } \angle z; \qquad (25.96)$$

i.e., the set of all complex numbers with magnitude less than 1, for all arguments; this clearly represents the points lying in the *interior* of the unit circle. Thus:

The entire left-half plane in the s-plane corresponds to the interior of the unit circle in the z plane.

The entire right-half plane ($\sigma > 0$ for all ω)

From Eqs. (25.89) and (25.90), since σ is *positive* in this case, we easily obtain that the corresponding set of points in the z-plane are now represented by:

$$|z| = e^{\sigma\Delta t} > 1, \text{ for all } \angle z; \qquad (25.97)$$

and this is the set of all complex numbers with magnitude greater than 1; this is clearly the set of points lying in the *exterior* of the unit circle, i.e., the complement of the set represented in Eq. (25.96). Thus:

The entire right-half plane in the s-plane corresponds to the exterior of the unit circle in the z-plane.

Primary and complementary strips in the left-half plane

Even though we have now completely matched the s-plane to the z-plane, there is one last point to be made: we start by considering a specific situation of special significance involving two points in the s-plane represented by:

$$s_1 = \sigma + j\omega_1 \qquad (25.98)$$

and

$$s_2 = \sigma + j\left(\omega_1 \pm \frac{2n\pi}{\Delta t}\right) \qquad (25.99)$$

Observe that these two points have the same real parts, and that their imaginary parts differ by an integral multiple of $2\pi/\Delta t$, which is recognizable as the sampling frequency (recall that a sampling interval of Δt corresponds to a sampling frequency $\omega_s = 2\pi/\Delta t$).

Let us obtain the points in the z-plane corresponding to these two s-plane points. Using Eq. (25.92) we have:

$$z_1 = r\Big[\cos(\omega_1 \Delta t) + j \sin(\omega_1 \Delta t)\Big] \qquad (25.100)$$

and

$$z_2 = r\left[\cos\left(\omega_1 \pm \frac{2n\pi}{\Delta t}\right)\Delta t + j \sin\left(\omega_1 \pm \frac{2n\pi}{\Delta t}\right)\Delta t\right] \qquad (25.101)$$

which simplifies to give:

$$z_2 = r\left[\cos(\omega_1 \Delta t \pm 2n\pi) + j \sin(\omega_1 \Delta t \pm 2n\pi)\right] \qquad (25.102)$$

From the periodicity property of the sin and cosine functions, Eq. (25.102) becomes:

$$z_2 = r\left[\cos(\omega_1 \Delta t) + j \sin(\omega_1 \Delta t)\right] \qquad (25.103)$$

so that points z_1 and z_2 are identical in the z-plane. We therefore have the following result:

If two points in the s-plane have the same real part, but their imaginary parts differ by integral multiples of ω_s, the sampling frequency, they map into the same point in the z-plane.

The following are some of the important consequences of this result:

1. If the s-plane is divided up along the imaginary axis into strips of width ω_s, starting with the "primary" strip represented by $\sigma - j(\omega_s/2) \leq s \leq \sigma + j(\omega_s/2)$, and then "complementary" strips of width $\pm j (2n + 1)\omega_s/2$, $n = 1, 2, ...$, then the points in *each* of these strips map into the interior of the unit circle in the z-plane (see Figure 25.3).

2. In mapping the entire left half of the s-plane into the z-plane, only the points within the primary strip are mapped into unique points within the unit circle in the z-plane; any other points outside the primary strip in the s-plane will map to points in the z-plane that are repetitions of these primary points.

3. The relationship between the z-plane and the s-plane is not unique; a given point in the s-plane corresponds to a single point in the z-plane, but a given point in the z-plane corresponds to an infinite number of points in the s-plane.

There is a very important practical reason for the nonuniqueness in the secondary strips where the frequency in the s-plane is larger than half the sampling frequency, i.e., $\omega_s/2$. Recall from our discussion in Chapter 23 that we defined the *Nyquist frequency*, $\omega_N = \omega_s/2$, as the largest frequency for which information can be extracted from sampled data having sampling frequency $\omega_s = 2\pi/\Delta t$. Thus the *primary strip* shown in Figure 25.3 is the locus of all points in the s-plane with frequency less than the Nyquist frequency and thus that defines those points that transform properly into the discrete z-plane. All the points in the complementary strips have a frequency too high to be contained in the sampled data and in fact may corrupt the sampled-data results by producing aliased data such as seen in Figure 23.4.

Mapping of Transfer Function Poles and Zeros

Having examined the issue of how the s-plane is related to the z-plane in general, we now wish to address a more specific issue: given the poles and zeros of $g(s)$, what can we say about the poles and zeros of the corresponding $g(z)$?

Transfer functions without time delays

In the absence of delays, we recall that the Laplace transfer function occurs as a ratio of two polynomials in s:

$$g(s) = \frac{N(s)}{D(s)} \tag{25.104}$$

and that the n roots of the nth-order denominator polynomial are the poles of $g(s)$, while the m roots of the mth-order numerator polynomial are the zeros.

We also recall that it is possible to decompose Eq. (25.104) into simpler terms by partial fraction expansion, viz.:

$$g(s) = \frac{A_1}{s - s_1} + \frac{A_2}{s - s_2} + \dots + \frac{A_n}{s - s_n} \tag{25.105}$$

where $s_1, s_2, \dots, s_n$ are the n poles of $g(s)$. By Laplace inversion, we obtain, from Eq. (25.105) that:

$$g(t) = A_1 e^{s_1 t} + A_2 e^{s_2 t} + \dots + A_n e^{s_n t} \tag{25.106a}$$

By evaluating this function at discrete points $k\Delta t$ in time, we obtain:

$$g(k\Delta t) = A_1 e^{s_1 k\Delta t} + A_2 e^{s_2 k\Delta t} + \dots + A_n e^{s_n k\Delta t} \tag{25.106b}$$

from which the discrete version of Eq. (25.106a) becomes:

$$g^*(t) = \sum_{k=0}^{\infty} g(k\Delta t) \, \delta(t - k\Delta t) \tag{25.107}$$

By taking z-transforms in Eq. (25.107), using Eq. (25.106b), we obtain the pulse transfer function equivalent of the Laplace transfer function we started with in Eq. (25.104):

$$g(z) = \frac{A_1}{1 - (e^{s_1 \Delta t})z^{-1}} + \frac{A_2}{1 - (e^{s_2 \Delta t})z^{-1}} + \dots + \frac{A_n}{1 - (e^{s_n \Delta t})z^{-1}} \tag{25.108}$$

By recombining these partial fraction terms, we are able to obtain a consolidated pulse transfer function of the form:

$$g(z) = \frac{B(z^{-1})}{\left(1 - p_1 z^{-1}\right)\left(1 - p_2 z^{-1}\right) \dots \left(1 - p_n z^{-1}\right)} \tag{25.109}$$

where each p_i in Eq. (25.109) is given by:

$$p_i = e^{s_i \Delta t} \tag{25.110}$$

Observe that Eq. (25.109) is a pulse transfer function with n poles (located at $z = p_i$, $i = 1, 2, ..., n$) and m zeros determined by the roots of the numerator polynomial $B(z^{-1})$.

We may now summarize the results for $g(s)$ in Eq. (25.104) and its pulse transfer function equivalent, $g(z)$, in Eq. (25.109):

1. While the poles of $g(s)$ are $s_1, s_2, ..., s_n$, the corresponding poles of $g(z)$ are $e^{s_1 \Delta t}, e^{s_2 \Delta t}, ..., e^{s_n \Delta t}$, thus giving the result:

    If $s = s_i$ are the poles of an s-transfer function, the corresponding poles of the z-transfer function are $p_i = e^{s_i \Delta t}$

    thus confirming again that the transfer function poles are mapped from the s-domain to the z-domain in accordance with the fundamental relationship $z = e^{s\Delta t}$ given in Eq. (25.83).

2. The relation in Eq. (25.110) implies that the poles p_i of $g(z)$ could be determined *directly* from the poles s_i of $g(s)$ without explicitly carrying out the z-transform process.

3. Unfortunately, no such clear-cut relationship as Eq. (25.110) exists between the zeros of $g(s)$ and of $g(z)$. Given the zeros of $g(s)$, we can only locate the corresponding zeros of $g(z)$ by explicitly carrying out the z-transform process.

4. Eq. (25.110) emphasizes that the location of the poles of $g(z)$ in the z-plane are strongly influenced by the sampling time Δt. The same is true for the zeros. However, it is only for small sampling times that a zero of $g(s)$ at $s = \zeta_i$ corresponds approximately to a zero of $g(z)$ at $z = e^{\zeta_i \Delta t}$.

Transfer functions with time delays

When time delays are involved, we recall that the Laplace transfer function is no longer completely rational; the pole-zero contribution of the transcendental e^{-as} delay term is not quite as straightforward to analyze precisely. Surprisingly, this is not the case with the z-transfer function. As we had noted earlier, the delay term contributes poles at the origin of the z-plane.

Poles and Zeros of Processes

As far as the mapping of poles and zeros from the s- to the z-plane goes, the important issue in computer process control is really that of finding the relationship between the poles and zeros of the *continuous process* and of the corresponding *discrete process*, with emphasis on the word *process*. The difference between this issue and that of merely mapping transfer function poles and zeros is the crucial point made several times earlier:

> *The discrete process that truly corresponds to the continuous process must incorporate a hold element.*

Thus, for example, given the second-order continuous process with the s-transfer function:

$$g(s) = \frac{K}{(s + \alpha)(s + \beta)} \tag{25.111}$$

by straightforward discretization, we obtain a z-transfer function having the form:

$$g_{NH}(z) = \frac{b}{\left(1 - p_1 z^{-1}\right)\left(1 - p_2 z^{-1}\right)} \tag{25.112}$$

where $p_1 = e^{-\alpha \Delta t}$, and $p_2 = e^{-\beta \Delta t}$. However, upon incorporating a zero-order hold element, it can be shown that the resulting discrete process that truly corresponds to the sampled continuous process in Eq. (25.111) has the pulse transfer function:

$$g(z) = \frac{b_1 z^{-1} + b_2 z^{-2}}{\left(1 - p_1 z^{-1}\right)\left(1 - p_2 z^{-1}\right)} \tag{25.113}$$

We may now observe that for this example, the continuous process has two poles, and both $g_{NH}(z)$ and $g(z)$ also have two poles; the denominator polynomials of $g_{NH}(z)$ and $g(z)$ are the same, but their respective numerator polynomials are different. Also note that $g(s)$ has no zero, while $g(z)$ has a single zero located at $z = -(b_2/b_1)$

We may now make the following statements about the relationship between the poles and zeros of a sampled continuous *process* and the resulting discrete *process* incorporating a ZOH element.

1. A sampled continuous process and the resulting discrete process (incorporating a ZOH) have the same number of poles.

2. The poles s_i of the sampled continuous process and the corresponding poles p_i of the resulting discrete process (incorporating a ZOH) are related according to $p_i = e^{s_i \Delta t}$; thus the incorporation of the ZOH neither affects the *number* nor the *location* of the poles.

3. The number of zeros of the continuous process is, in general, *not* the same as the number of zeros of the equivalent discrete process incorporating a ZOH element. Thus, the incorporation of a ZOH affects the *number* of zeros.

4. Again, unfortunately, it is not possible, given only information about the zeros of a continuous process, to make any general statements about the number or the *exact* location of the zeros of the equivalent discrete process incorporating a ZOH. (See Ref. [1], p. 58 and Ref. [4], p. 61, for heuristic rules giving approximate locations of these zeros.)

Inverse-Response Processes

We recall from Chapter 7 that a continuous process whose transfer function has a zero in the right half of the s-plane will exhibit inverse response. By the same token, a discrete process whose pulse transfer function has a zero outside the unit circle will also exhibit inverse response. However, as we will now show with an example, sampling a continuous, inverse-response process may not always give rise to a discrete-time process that exhibits inverse response, further demonstrating the effect of sampling time on the zeros of sampled processes.

Example 25.4   **THE EFFECT OF SAMPLING TIME ON THE ZERO OF A SAMPLED INVERSE-RESPONSE PROCESS WITH A ZOH ELEMENT.**

For the inverse-response process studied in Chapter 7, with Laplace transfer function given in Eq. (7.16) as:

$$g(s) = \frac{(-3s + 1)}{(2s + 1)(5s + 1)} \tag{7.16}$$

obtain $g(z)$, the hybrid pulse transfer function for the process incorporating a ZOH element. Given that the unit of time is minutes, identify the zero of $g(z)$ and investigate its location with respect to the unit circle when Δt is 1, 2, 6, 7, 8, and 10 min.

Solution:

Following the procedure outlined in Chapter 24, for obtaining $g(z)$ from $g(s)$, we have first that:

$$\tilde{g}(s) = \frac{g(s)}{s} = \frac{(1 - 3s)}{s(2s + 1)(5s + 1)}$$

with inverse Laplace transform:

$$\tilde{g}(t) = 1 + \frac{5}{3} e^{-t/2} - \frac{8}{3} e^{-t/5} \tag{25.114}$$

(compare with Eq. (7.17)). From here, using Eq. (24.92) we obtain $g(z)$ as:

$$g(z) = (1 - z^{-1})\left(\frac{1}{1 - z^{-1}} + \frac{5/3}{1 - p_1 z^{-1}} - \frac{8/3}{1 - p_2 z^{-1}} \right) \tag{25.115}$$

where

$$p_1 = e^{-\Delta t/2} \tag{25.116}$$
$$p_2 = e^{-\Delta t/5} \tag{25.117}$$

Eq. (25.115) rearranges to give:

$$g(z) = \frac{b_1 z + b_2}{z^2 - \left(p_1 + p_2 \right) z + p_1 p_2} \tag{25.118}$$

where

$$b_1 = 1 - \left(\frac{8}{3}\right)p_2 + \left(\frac{5}{3}\right)p_1 \qquad (25.119a)$$

$$b_2 = p_1 p_2 + \left(\frac{5}{3}\right)p_2 - \left(\frac{8}{3}\right)p_1 \qquad (25.119b)$$

Thus the pulse transfer function has a single zero is located at:

$$z_1 = -\left(\frac{b_2}{b_1}\right)$$

or

$$z_1 = \frac{8p_1 - 3p_1 p_2 - 5p_2}{3 - 8p_2 + 5p_1} \qquad (25.120)$$

Upon introducing the given values of Δt and calculating the zero location z_1, we obtain the following:

- For $\Delta t = 1$; $z_1 = 1.414$
- For $\Delta t = 2$; $z_1 = 2.194$
- For $\Delta t = 6$; $z_1 = -1.373$
- For $\Delta t = 7$; $z_1 = -0.8604$
- For $\Delta t = 8$; $z_1 = -0.592$
- For $\Delta t = 10$; $z_1 = -0.32$

Thus we observe that the zero lies outside the unit circle for $\Delta t = 1, 2,$ and 6, but lies within the unit circle for $\Delta t = 7, 8,$ and 10. Therefore the discrete system will exhibit inverse response when the sample time is less than 6 min, but will not exhibit inverse response when sampling time is 7 min or greater. The conclusion is clearly that when a continuous inverse-response system is sampled, whether or not the resulting discrete system (including the ZOH element) exhibits inverse response is influenced by the sampling time.

It is important to understand the physical basis of this curious influence of sample time on zero location for discrete-time processes. Consider the step response of a continuous system with inverse response (such as given by Eq. 7.16)) sketched in Figure 25.4. Let the time at which the "wrong-way" behavior is over be Δt_{min}. Then for sample times $\Delta t > \Delta t_{min}$, the discrete-time system will show no inverse response because the first sample "jumps over" the initial inverse-response period. This can be summarized as follows:

Given a continuous process exhibiting an inverse response lasting a period Δt_{min}, the corresponding discrete-time process will show no inverse response for sample times $\Delta t \geq \Delta t_{min}$.

The samples denoted by ×'s in Figure 25.4 clearly illustrates this effect with $\Delta t = 1.5 \, \Delta t_{min}$. On the other hand, the samples denoted by o's in the figure, with $\Delta t = 0.5 \, \Delta t_{min}$ indicate that the discrete-time process also shows an inverse response.

A similar situation occurs for processes with time delays. Observe that the amount of delay D associated with a discrete-time process depends on how the underlying continuous system is sampled. If the continuous process time delay is α, and the sampling time is Δt, then D is given by:

$$D = \frac{\alpha}{\Delta t} \qquad (25.121)$$

Note then that as Δt increases, the time delay associated with the discrete-time process decreases; when $\Delta t \geq \alpha$, the situation reduces to the relatively benign case of the (unavoidable) unit delay normally associated with the sampling process. Even though this unit delay is now larger than the original delay in the continuous process, the digital control system design does not explicitly have to deal with delays. If we let $\Delta t_{min} = \alpha$ in Figure 25.4, so that the process step response shows a delay in the period $0 < t < \Delta t_{min}$, then the samples denoted by $\bullet$'s, with $\Delta t = \Delta t_{min} = \alpha$, in the figure illustrates the case with time delays as well.

Thus, it is possible to eliminate troublesome nonminimum phase behavior in discrete-time by increasing the sampling time used for generating the sampled processes. However, we will show below that long sample times tend to have a destabilizing effect on closed-loop digital control systems; thus one must exercise care in increasing sample times merely to overcome difficult dynamics.

25.2.4 Effect of Poles and Zeros on Dynamic Response

Although we showed in Section 25.1 how to derive the detailed dynamic response of a process given the discrete-time model and the input function, it is possible (just as with continuous processes) to have a *qualitative* understanding of the dynamic response of a discrete-time process based on a knowledge of the gain, poles, and zeros of the process.

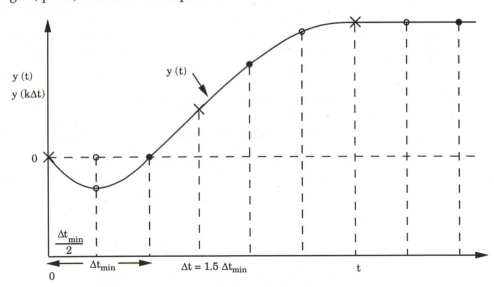

Figure 25.4. Inverse response and time delays for continuous- and discrete-time processes; $\times - \Delta t = 1.5 \,_{min};$ $\circ - \Delta t = 0.5 \,_{min};$ $\bullet - \Delta t = \Delta t_{min}.$

In this section we will investigate the dynamic behavior of discrete processes from the point of view of their pulse transfer function poles and zeros. The dynamic behavior will be characterized by impulse responses, since all other responses can be derived from the impulse response by convolution.

First-Order Process with a ZOH

The Pulse Transfer Function

The pulse transfer function for this process is given in the backward shift form as:

$$g(z) = \frac{bz^{-1}}{1 - pz^{-1}} \qquad (25.122)$$

or, in the forward shift form:

$$g(z) = \frac{b}{z - p} \qquad (25.123)$$

The Transfer Function Characteristics

The transfer function is characterized by a unit delay due to the sample-and-hold process, and a single pole at $z = p$; it has no zero.

The Transient Response Characteristics

Taking the inverse z-transform of $g(z)$ in Eq. (25.122) gives the *impulse response* of this process, $g(k)$, as:

$$g(k) = bp^{k-1}; \text{ for } k \geq 1 \qquad (25.124)$$

The nature of this impulse response is clearly governed by the location of the single pole p which must be a real number. It is possible to identify six qualitatively different types of responses as shown in Figure 25.5:

1. $p > 1$; $g(k)$ grows exponentially with time

2. $p = 1$; $g(k)$ remains constant at b for all k

3. $0 \leq p < 1$; $g(k)$ monotonically decreases to zero

4. $-1 < p < 0$; $g(k)$ is a sequence of decreasing numbers with alternating signs, because p is negative

5. $p = -1$; $g(k)$ is an alternating sequence of $+b$ and $-b$

6. $p < -1$; $g(k)$ is a divergent sequence with alternating signs, because p is both negative and greater than 1 in absolute value

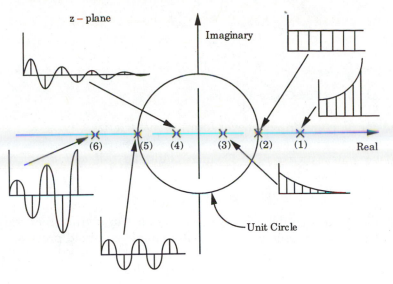

Figure 25.5. Impulse responses of the first-order process with a ZOH as a function of the location of the single pole (adapted from Ref. [4]).

One should realize that for the discrete, first-order process resulting from sampling a real, continuous first-order process, it is physically impossible to obtain the response types 4, 5, and 6; this is because the single pole at $z = p$ arises from the real continuous pole at, say, $s = s_1$ and since these poles are related according to $p = e^{s_1 \Delta t}$, it becomes clear that p cannot be negative. Responses 4–6 are typically associated, not with physical first-order processes, but with certain discrete controllers to be discussed in the next chapter.

Second-Order Process with a ZOH

The Pulse Transfer Function

As indicated in Example 25.2, the pulse transfer function for a second-order process including a ZOH is given in the backward shift, unfactored form as:

$$g(z) = \frac{b_1 z^{-1} + b_2 z^{-2}}{1 + a_1 z^{-1} + a_2 z^{-2}}$$

and if the two roots of the denominator quadratic are p_1 and p_2, then $g(z)$ has the factored form:

$$g(z) = \frac{\left(b_1 + b_2 z^{-1}\right) z^{-1}}{\left(1 - p_1 z^{-1}\right)\left(1 - p_2 z^{-1}\right)} \tag{25.125}$$

In the forward shift form, the transfer function is:

$$g(z) = \frac{b_1 z + b_2}{z^2 + a_1 z + a_2} \tag{25.126}$$

or

$$g(z) = \frac{b_1 z + b_2}{(z - p_1)(z - p_2)} \tag{25.127}$$

The Transfer Function Characteristics

The transfer function (in whichever form) is characterized by:

- Two poles, at $z = p_1, p_2$, which could be *real and distinct*, *real and coincident*, or *complex conjugates*.
- One zero at $z = -b_2/b_1$.

From the backward shift forms, we are also able to deduce that the transfer function has the usual unit delay due to the sample-and-hold process.

The Transient Response Characteristics: Real Poles

When the poles are real and distinct, a partial fraction expansion of Eq. (25.125) yields:

$$g(z) = \frac{A_1 z^{-1}}{1 - p_1 z^{-1}} + \frac{A_2 z^{-1}}{1 - p_2 z^{-1}} \tag{25.128}$$

where the constants A_1 and A_2 are:

$$A_1 = \frac{b_1 p_1 + b_2}{p_1 - p_2} \tag{25.129}$$

and

$$A_2 = \frac{b_1 p_2 + b_2}{p_2 - p_1} \tag{25.130}$$

Thus the transfer function in Eq. (25.128) is a sum of two pulse transfer functions of the type just dealt with for the first-order system, and the contribution of each term to the overall transient response is as illustrated in Figure 25.5. To see this more explicitly, note that taking the inverse of $g(z)$ to produce the impulse response, $g(k)$, gives:

$$g(k) = A_1 \left(p_1\right)^{k-1} + A_2 \left(p_2\right)^{k-1} ; \; k \geq 1 \tag{25.131}$$

when the poles are distinct and:

$$g(k) = b_1 k p^{k-1} + b_2 (k-1) p^{k-2} ; \; k \geq 1 \tag{25.132}$$

when the poles are identical.

A significant difference in the transient response shows up, as we might expect, when we have a pair of complex conjugate poles; this case will be treated below.

   The effect of the zeros are felt only through the constants A_1 and A_2; they affect the *magnitude* of each contributing response, but exercise absolutely no influence on whether the response dies out with time or grows indefinitely: this aspect of the response is solely determined by the poles.

The Transient Response Characteristics: Complex Conjugate Poles

When the poles are complex conjugates, they may be expressed as:

$$p_1, p_2 = a* \pm jb* \qquad (25.133)$$

or, in the polar form as:

$$p_1, p_2 = r* e^{\pm j\omega*} \qquad (25.134)$$

where, as usual:

$$r* = \sqrt{(a*)^2 + (b*)^2} \qquad (25.135)$$

and

$$\omega* = \tan^{-1}\left(\frac{b*}{a*}\right) \qquad (25.136)$$

   Under these circumstances, a partial fraction expansion of the pulse transfer function gives:

$$g(z) = \frac{A_1 z^{-1}}{1 - r* e^{j\omega*} z^{-1}} + \frac{A_2 z^{-1}}{1 - r* e^{-j\omega*} z^{-1}} \qquad (25.137)$$

and upon taking inverse transforms, consolidating the resulting complex exponentials using the Euler identity, and tidying up, the impulse response in this case is obtained as:

$$g(k) = C(r*)^{k-1} \cos\left[\omega*(k-1) + \phi\right]; \text{ for } k \geq 1 \qquad (25.138)$$

where the constants C and ϕ are related to A_1 and A_2 (cf. Appendix C).
   It is clear that the response indicated in Eq. (25.138) will be oscillatory; however, depending on the actual numerical values taken by the real and imaginary portions of these poles (or equivalently, the magnitude $r*$ and argument $\omega*$) it is possible to identify the four different types of *qualitative oscillatory* response shown in Figure 25.6:

1. $r* > 1, \omega* \neq \pi$ (Poles *outside* the unit circle)
    $g(k)$ is oscillatory with increasing amplitude

2. $r* = 1, \omega* \neq \pi$ (Poles *on* the unit circle)
    $g(k)$ is oscillatory with constant amplitude

3. $r* < 1, \omega* \neq \pi$ (Poles *inside* the unit circle)
    $g(k)$ is damped oscillatory

4. $r* < 1, \omega* = \pi$ (Coincident poles *inside* the unit circle)
    $g(k)$ is damped oscillatory; with the
    frequency of oscillation coinciding with
    sampling frequency, ω_s

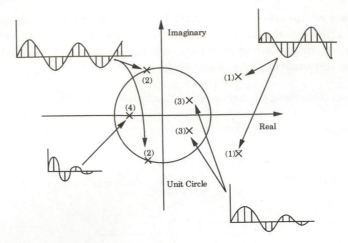

Figure 25.6.   Impulse responses of the second-order process with a ZOH as a function of the location of the complex conjugate poles (adapted from Ref. [4]).

The Type 4 response requires some special mention. Since $\omega^* = \pi$, the two poles are located at $z = r^* e^{+j\pi}$, and $z = r^* e^{-j\pi}$. Using the relation $z = e^{s\Delta t}$, we easily deduce that the corresponding poles in the s-plane have imaginary parts $\pi/\Delta t$ and $-\pi/\Delta t$; i.e., these poles are located exactly on the upper and lower boundaries of the primary strip in the s-plane. Because these imaginary parts are separated by a factor of precisely $2\pi/\Delta t = \omega_s$, we see that they will map into the single pole in the unit circle indicated by the number 4 in Figure 25.6.

Dynamic Response of General Discrete-Time Processes with ZOH

The pulse transfer function of a general process incorporating a ZOH has the form:

$$g(z) = \frac{B(z^{-1})z^{-1}}{A(z^{-1})} \qquad (25.139)$$

where the additional z^{-1} term has occurred as a result of the sampling and hold process.

Let us now observe that upon decomposing Eq. (25.139) into simpler partial fractions, we will obtain only two types of terms: the type in Eq. (25.128) corresponding to those poles of $A(z^{-1})$ that are real, and the type in Eq. (25.137) corresponding to those poles that are complex conjugate pairs. Thus the transient response of *any* general sampled-data process will simply be an appropriate combination of the transient responses cataloged in Figures 25.5 and 25.6, as determined by the nature of the poles of the denominator polynomial. These transient responses therefore constitute the basic building blocks from which the response of any other linear process can be deduced.

Again, as noted in the discussion of the second-order process, the zeros affect only the *magnitude* of each contributing response; they do not affect the ultimate behavior as $k \to \infty$. So long as the zeros all lie within the unit circle, not much else can be said in terms of their influence on transient responses.

Discrete processes with zeros outside the unit circle behave like their continuous counterparts with zeros in the right-half plane: they exhibit inverse-response behavior.

We may now summarize the effect of poles and zeros on the dynamic behavior of discrete systems as follows:

1. Real poles within the unit circle contribute transient responses that ultimately die out with time; if the poles are positive, the contributions decrease monotonically; if the poles are negative, the contributions alternate in sign while decreasing in magnitude, an effect sometimes referred to as "ringing."

2. Real poles on the unit circle contribute transient responses that neither grow nor decrease with time; alternating sign changes accompany the response when the pole is located at $z = -1$.

3. Complex conjugate poles always occur in pairs, and they always contribute oscillatory transient responses.

4. Complex conjugate poles located within the unit circle contribute damped oscillatory behavior; the frequency of oscillation is equal to the imaginary part.

5. Complex conjugate poles whose real parts lie within the unit circle but whose imaginary parts are multiples of $\pm\pi$ appear as a double pole located on the real axis; they contribute damped oscillatory responses with the frequency of oscillation identical to the sampling frequency, ω_s. These types of poles arise from sampling a continuous process whose poles happen to be located at $s = -a \pm \omega_s/2$. (Note the relationship of the continuous system pole to the sampling frequency.)

6. A conjugate pair of poles *on* the unit circle contribute sustained oscillatory responses.

7. Any pole outside the unit circle contributes a response that grows indefinitely with time; the response will be oscillatory with increasing amplitude if the poles are complex conjugate pairs.

8. There is no simple, general way to characterize the effect of zeros that lie within the unit circle: they affect the magnitude of the contributed responses but not the ultimate behavior; zeros that lie outside the unit circle contribute inverse-response transient behavior.

9. Time delays cause a shift in time response; they also contribute poles at the origin of the complex z-plane.

25.3   BLOCK DIAGRAM ANALYSIS FOR SAMPLED-DATA SYSTEMS

What distinguishes a sampled-data control system from its purely continuous counterpart is that it consists of a *discrete* controller — which accepts and gives out only discrete signals — in conjunction with a *continuous* system whose

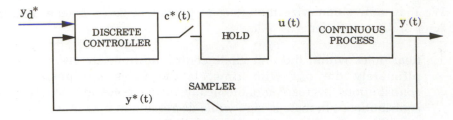

Figure 25.7.   Basic elements of a sampled-data control system.

continuous outputs must now be *sampled*, and for which the *discrete* inputs produced by the discrete controller must be made to appear continuous with a hold element. This concept is illustrated in Figure 25.7.

The block diagramatic representation of such systems will therefore consist of a mixture of continuous as well as sampled signals interconnecting continuous as well as discrete transfer functions.

It is therefore clear that block diagram analysis for sampled-data systems cannot be handled cavalierly; sampled-data control system block diagrams are not just mere discretized versions of the familiar continuous feedback control system block diagrams. However, there are some valuable concepts and principles that help to clarify and simplify sampled-data system block diagram analysis; we will present these next.

25.3.1 Some Useful Concepts and Principles

Notation for Signals and Transfer Functions

A star symbol (*) is used to distinguish sampled signals from continuous signals that carry no such attachment. As usual, an (s) argument indicates the Laplace transform of the indicated signal, while a (z) argument indicates the z-transform of the indicated signal which, by definition, must be discrete.

The "Starred" Laplace Transform Concept

The Laplace transform of a *discrete* (or *sampled*) signal is indicated by $f^*(s)$, the so–called "starred" Laplace transform; it was introduced in Chapter 24, in Eq. (24.3):

$$f^*(s) \;=\; L\{f^*(t)\} \;=\; \sum_{k=0}^{\infty} f(k\Delta t) e^{-ks\Delta t} \qquad (24.3)$$

The transfer function relating the "starred" Laplace transform $u^*(s)$ and $y^*(s)$ is the s^*-transfer function defined as:

$$y^*(s) \;=\; g^*(s)\, u^*(s) \qquad (25.140)$$

It relates the Laplace transform of two *sampled* signals.

Two Basic Problems

We are basically concerned with two problems in the block diagram analysis of sampled-data systems:

1. The combination of *sampled* inputs with *continuous* systems, e.g., $u^*(s)$ with $g(s)$, and
2. The conversion of s-domain transfer function relations to pulse transfer function relations.

The first problem is easily resolved:

By the nature of a continuous process, regardless of what input it receives, its output is always continuous.

Thus, we have that:

$$y(s) = g(s) u^*(s) \tag{25.141}$$

where the absence of a star symbol (*) in $y(s)$ indicates it is the Laplace transform of a continuous signal.

The second problem is that of obtaining the pulse transfer function relation of the type:

$$y(z) = g(z) u(z) \tag{25.142}$$

from expressions of the type in Eq. (25.141), and it consists of three parts: (1) deciding whether or not a zero-order hold needs to be added to $g(s)$; normally this is required; (2) obtaining the discrete version of the continuous signal (i.e., obtaining $y^*(s)$ from Eq. (25.141)); and finally (3) obtaining the z-transform of the discretized signal. If a zero-order hold is added, the resulting pulse transfer function is $g(z)$, whereas if no hold is used, then $g_{NH}(z)$ results. The transformation is achieved by using the following "starred" Laplace transform results.

The Basic "Starred" Laplace Transform Results

Obtaining "Starred" Transforms from a Product of Transforms

Given the continuous signal, $y(s)$, resulting from a sampled input $u^*(s)$:

$$y(s) = g(s) u^*(s) \tag{25.143}$$

then, the sampled output, $y^*(s)$, is given by:

$$y^*(s) = [g(s) u^*(s)]^* = g^*(s) u^*(s) \tag{25.144}$$

This very useful result is central to block diagram analysis for sampled-data systems; its proof may be found in Refs. [2, p. 184–5], and [4, p. 86]

Relating the "Starred" Transform to the z-Transform

The key result here is already familiar to us:

By definition, the "starred" Laplace transform is actually nothing but the z-transform with $e^{s\Delta t}$ replaced by z. Thus the z variable may be considered a more convenient, "shorthand" notation for the (implicit) "starred" transform variable, and the two may therefore be used interchangeably.

Recall that this result was obtained earlier in Chapter 24 and has already been used liberally in this current chapter. The main implication for block diagram analysis is that:

$$y*(s) = g*(s) u*(s)$$

is completely equivalent to:

$$y(z) = g_{NH}(z)u(z) \tag{25.145}$$

The application of these results will become clearer as we now begin our discussion of block diagram analysis.

25.3.2 Open-Loop Block Diagrams

Single Elements

Consider the simple block diagram shown in Figure 25.8 containing a single process element. The objective of this rather simple block diagram analysis is to obtain the z-domain pulse transfer function relating $y(z)$ to $u(z)$.

Clearly from this block diagram:

$$y(s) = g(s) u*(s)$$

Next we obtain $y*(s)$ by taking "starred" transforms; from the result in Eq. (25.144), we get:

$$y*(s) = g*(s) u*(s)$$

from where we immediately obtain that:

$$y(z) = g_{NH}(z)u(z)$$

If we are now given, for example, that:

$$g(s) = \frac{K}{\tau s + 1} \tag{25.146}$$

this block diagram analysis tells us that the corresponding pulse transfer function is:

$$g_{NH}(z) = \frac{K}{\tau}\left(\frac{1}{1 - \phi z^{-1}}\right) \tag{25.147}$$

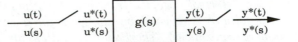

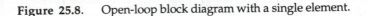

Figure 25.8.   Open-loop block diagram with a single element.

where $\phi = e^{-\Delta t/\tau}$, as obtained earlier in Example 24.6.

Recall from Example 24.7 that *when the zero-order hold is added* to $g(s)$ as is usual practice, the pulse transfer function for the first-order process is:

$$g(z) = \frac{K(1 - \phi)\, z^{-1}}{1 - \phi z^{-1}} \qquad (24.101)$$

Series Elements

Consider now the block diagram shown in Figure 25.9 containing two continuous process elements in series separated by an intermediate sampler. Again, the objective of the block diagram analysis is to obtain the z-domain pulse transfer function relating $y(z)$ to $u(z)$.

The analysis proceeds as follows:

$$y(s) = g_2(s)\, v^*(s) \qquad (25.148)$$

and

$$v(s) = g_1(s)\, u^*(s) \qquad (25.149)$$

We now wish to obtain $v^*(s)$ from Eq. (25.149), introduce the result into Eq. (25.148), and then find $y^*(s)$ from the consolidated expression.

Applying the result from Eq. (25.144) in Eq. (25.149), we obtain:

$$v^*(s) = g^*_1(s)\, u^*(s) \qquad (25.150)$$

which when introduced into Eq. (25.148) gives:

$$y(s) = g_2(s)\, g^*_1(s)\, u^*(s) \qquad (25.151)$$

and upon applying the "starred" transform, we finally obtain:

$$y^*(s) = g^*_2(s)\, g^*_1(s)\, u^*(s) \qquad (25.152)$$

We now straightforwardly obtain the required z-transfer function relation as:

$$y(z) = g_{2NH}(z)\, g_{1NH}(z)\, u(z) \qquad (25.153)$$

where we emphasize that there is no hold on either sampler.

The implication here is that the overall pulse transfer function for the process in Figure 25.9 is the product of the pulse transfer functions of the individual elements. Thus, for example, if:

$$g_1 = \frac{K_1}{\tau_1 s + 1} \qquad (25.154a)$$

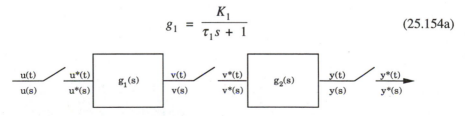

Figure 25.9. Open-loop block diagram for two elements in series with an intermediate sampler.

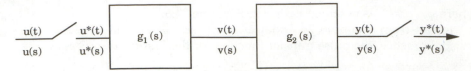

Figure 25.10. Open-loop block diagram for two continuous elements in series with no intermediate sampler.

and

$$g_2 = \frac{K_2}{\tau_2 s + 1} \tag{25.154b}$$

then the overall pulse transfer function will be given in this case by:

$$g_{NH}(z) = g_{1NH}(z) \, g_{2NH}(z)$$

$$= \left[\frac{K_1 K_2}{\tau_1 \tau_2 (1 - e^{-\Delta t/\tau_1} z^{-1}) (1 - e^{-\Delta t/\tau_2} z^{-1})} \right] \tag{25.155}$$

obtained by simply multiplying the individual z-transfer functions for each first-order process.

A completely different situation arises when the two elements in series are *continuous*, with no intermediate sampling, as indicated in Figure 25.10. Let us now proceed to obtain the z-domain pulse transfer function relating $y(z)$ to $u(z)$ for this case.

From the block diagram:

$$y(s) = g_2(s) \, v(s) \tag{25.156a}$$

and

$$v(s) = g_1(s) \, u^*(s) \tag{25.156b}$$

so that:

$$y(s) = g_2(s) \, g_1(s) \, u^*(s) \tag{25.157}$$

We must now observe that what connects $y(s)$ to $u^*(s)$ in Eq. (25.157) (and indeed in the block diagram itself) is a combination of two, *continuous* transfer functions which may be combined into one single — but still continuous — composite transfer function, which we will refer to as $g_2 g_1(s)$ simply to remind us of its individual component elements, i.e., Eq. (25.157) may be written as:

$$y(s) = g_2 g_1(s) \, u^*(s) \tag{25.158}$$

where

$$g_2 g_1(s) = g_2(s) \, g_1(s) \tag{25.159}$$

We may now apply the star transform to Eq. (25.158) to obtain:

$$y^*(s) = \left[g_2 g_1 \right]^*(s) \, u^*(s)$$

and finally:

$$y(z) = g_2 g_{1NH}(z) \, u(z) \tag{25.160}$$

It is important to note that $g_2 g_{1NH}(z)$ represents the z-transform of the discretized version of the composite s-transfer function made up of the product $g_2(s)g_1(s)$; on the other hand, $g_{2NH}(z)g_{1NH}(z)$ is the product of the individual z-transforms of the discretized versions of $g_2(s)$ separately, and $g_1(s)$ separately. In general:

$$g_2 g_{1NH}(z) \neq g_{1NH}(z) \, g_{2NH}(z) \tag{25.161}$$

We therefore have that:

The z-transform of a product of functions is not always the same as the product of their individual z-transforms.

This is a more general form of the expression introduced in Eq. (24.102) when dealing with the ZOH element.

To conclude, as well as to drive home the implication of Eq. (25.161), observe that if, in Figure 25.10, the transfer functions are as given in Eq. (25.154), then, first, we have that:

$$g_2 g_1(s) = g_2(s) \, g_1(s) = \frac{K_2 K_1}{(\tau_2 s + 1)(\tau_1 s + 1)}$$

and the z-transform of the discretized version of this second-order, composite transfer function is given by:

$$g_2 g_{1NH}(z) = \mathcal{Z} \left\{ \left[\frac{K_2 K_1}{(\tau_2 s + 1)(\tau_1 s + 1)} \right]^* \right\}$$

and upon evaluating the indicated z-transform (for $\tau_2 \neq \tau_1$), we then have that the overall pulse transfer function will be given in this case by:

$$g_2 g_{1NH}(z) = \left[\frac{K_1 K_2 (e^{-\Delta t/\tau_2} - e^{-\Delta t/\tau_1}) z^{-1}}{(\tau_2 - \tau_1)(1 - e^{-\Delta t/\tau_1} z^{-1})(1 - e^{-\Delta t/\tau_2} z^{-1})} \right] \tag{25.162}$$

which is *not* the same as that obtained in Eq. (25.155) for the same transfer function elements arranged in the *different* block diagram of Figure 25.9.

Let us repeat the comparison between the situation in Figure 25.9 and in Figure 25.10 except that now we require the input samplers to have zero-order holds. In this case we can use Eq. (24.101) and see that the overall transfer function in Figure 25.9 takes the form:

$$g(z) = g_1(z) \, g_2(z) = \frac{K_1 K_2 (1 - \phi_1)(1 - \phi_2) z^{-2}}{(1 - \phi_1 z^{-1})(1 - \phi_2 z^{-1})} \tag{25.163}$$

where $\phi_i = e^{-\Delta t/\tau_i}$ $i = 1, 2$. This can be compared with Eq. (25.155), the situation without a ZOH.

Following a similar analysis, we see that if the single sampler on the input to $g_1(s)$ in Figure 25.10 has a ZOH, then the overall transfer function in Figure 25.10 becomes (cf. Appendix C):

$$g_2 g_1(z) = \frac{K_1 K_2}{\tau_1 - \tau_2} \times$$ (25.164)

$$\left\{ \frac{\left[\tau_1(1 - \phi_1) - \tau_2(1 - \phi_2)\right] z^{-1} + \left[\tau_2 \phi_1 (1 - \phi_2) - \tau_1 \phi_2 (1 - \phi_1)\right] z^{-2}}{\left(1 - \phi_1 z^{-1}\right)\left(1 - \phi_2 z^{-1}\right)} \right\}$$

In practice, Eqs. (25.163) and (25.164) which include the ZOH are the relevant ones for discrete-time block diagrams of real processes.

The next example illustrates in detail the block diagram with and without a zero-order hold.

Example 25.5   THE EFFECT OF THE ZOH ELEMENT ON SAMPLED-DATA SYSTEMS.

Obtain the *continuous*-time response, $y(t)$, of the process indicated in Figure 25.10, given that the input, $u(t)$, is a unit step function, and that $g_1(s)$ and $g_2(s)$ are first-order processes given by Eq. (25.154). Determine the continuous-time response $y(t)$ for two cases

      1. When $g_1(s)$ includes a zero-order hold
      2. When there is no zero-order hold

Solution:

The unit step response with the zero-order hold can be found easily because the transfer function for the continuous process including the hold is:

$$H(s)\, g_1(s)\, g_2(s) = \left(\frac{1 - e^{-s\Delta t}}{s}\right)\left(\frac{K_1}{\tau_1 s + 1}\right)\left(\frac{K_2}{\tau_2 s + 1}\right)$$ (25.165)

and recall that the sampled unit step input is:

$$u^*(s) = \frac{1}{1 - e^{-s\Delta t}}$$ (25.166)

so that:

$$y(s) = H(s)\, g_1(s)\, g_2(s)\, u^*(s) = \frac{1}{s}\left(\frac{K_1}{\tau_1 s + 1}\right)\left(\frac{K_2}{\tau_2 s + 1}\right)$$ (25.167)

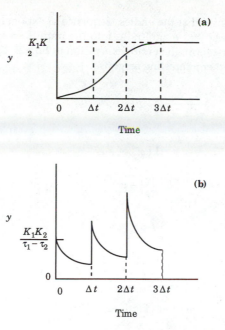

Figure 25.11. Continuous-time response of two first-order processes in series to a sampled unit step function: (a) with the ZOH element; (b) without the ZOH element.

or in the time domain:

$$y(t) \;=\; K_1 K_2 \left\{ 1 \;-\; \frac{e^{-t/\tau_1}}{\left[1 - (\tau_2/\tau_1)\right]} \;-\; \frac{e^{-t/\tau_2}}{\left[1 - (\tau_1/\tau_2)\right]} \right\} \qquad (25.168)$$

which is sketched in Figure 25.11(a).

The response without the zero-order hold can be found from removing $H(s)$ from Eq. (25.167) to yield:

$$y(s) \;=\; g_1(s)\, g_2(s)\, u^*(s) \;=\; \frac{K_1}{\left(\tau_1 s + 1\right)} \frac{K_2}{\left(\tau_2 s + 1\right)} \frac{1}{\left(1 - e^{-s\Delta t}\right)} \qquad (25.169)$$

Recall that $u^*(s)$ can be written:

$$\frac{1}{1 - e^{-s\Delta t}} \;=\; 1 + e^{-s\Delta t} + e^{-2s\Delta t} + \ldots + e^{-ks\Delta t} + \ldots \qquad (25.170)$$

Introducing Eq. (25.170) into Eq. (25.169) now gives:

$$y(s) \;=\; \left(\frac{K_1}{\tau s + 1}\right)\!\left(\frac{K_2}{\tau_2 s + 1}\right)\!\left(1 + e^{-s\Delta t} + e^{-2s\Delta t} + \ldots \right) \qquad (25.171)$$

and if we recognize that the endless sequence of exponentials in s merely introduces delays of the indicated magnitudes into the impulse-response function obtained by inverting the first term, the entire expression in Eq. (25.171) is easily inverted to give the following rather interesting response:

$$y(t) = \begin{cases} \dfrac{K_1 K_2}{\tau_1 - \tau_2} \left(e^{-t/\tau_1} - e^{-t/\tau_2} \right) & 0 < t < \Delta t \\[2em] \dfrac{K_1 K_2}{\tau_1 - \tau_2} \left[\left(e^{-t/\tau_1} - e^{-t/\tau_2} \right) \right. \\[1em] \left. + \left(e^{-(t-\Delta t)/\tau_1} - e^{-(t-\Delta t)/\tau_2} \right) \right] & \Delta t < t < 2\Delta t \\[1em] \bullet \\ \bullet \\ \bullet \\ \bullet \end{cases} \qquad (25.172)$$

with time behavior shown in Figure 25.11(b).

25.3.3 Closed-Loop Block Diagrams

If we refer back to our discussion in the early portions of Chapter 14, we will recall that the components of the feedback loop in a continuous feedback control system were identified as: the process itself, the measuring device, a comparator, the controller, and the final control element.

For the sampled-data control system, the situation is slightly different; some new components have been added, and the character of some of the old components have been modified. If we refer back to Figure 23.3 in which the sampled-data control system was introduced, we may now identify the components of the feedback loop of a sampled-data control system as follows:

1. The process itself
2. The measuring device
3. A sampler for discretizing the continuous measurement, and another for discretizing the set-point signal
4. A comparator (performing its operations on discrete signals)
5. The controller (now completely discrete)
6. The hold element
7. The final control element

Of these components, the third and the sixth are the new additions, not being part of the continuous feedback loop of Part IVA; the fourth and the fifth have acquired new, completely discrete characters.

Transfer Function Relations for the Elements of the Block Diagram

In the same manner as we did for the continuous system in Chapter 14, in Eqs. (14.1) through (14.5), we will now write the transfer function relations for the elements of the sampled-data control system.

The Process:

$$y(s) = g(s) u(s) + g_d(s) d(s) \qquad (25.173)$$

We assume here that the process itself is continuous and the process output can be measured continuously; that the process input can be made continuous; and that the disturbance variable is continuous.

The Measuring device (or sensor):

$$y_m(s) = h(s) y(s) \qquad (25.174)$$

The Sampler:
Performs a discretization operation on y_m; its output is y^*_m.

The Comparator:
Is part of the discrete controller mechanism:

$$\varepsilon^*(s) = y^*_d(s) - y^*_m(s) \qquad (25.175)$$

The Controller:
Being completely discrete, operates according to:

$$c^*(s) = g^*_c(s) \, \varepsilon^*(s) \qquad (25.176)$$

The (Zero-Order) Hold Element:
Converts the discrete control command signal (with Laplace transform $c^*(s)$) to the (piecewise) continuous signal with Laplace transform $c_H(s)$:

$$c_H(s) = H(s) c^*(s) \qquad (25.177)$$

The Final Control Element:
Converts the piecewise continuous control command signal output of the hold element to the actual process input variable:

$$u(s) = g_v(s) c_H(s) \qquad (25.178)$$

If we now put all these elements together, the result is the closed-loop block diagram shown in Figure 25.12.

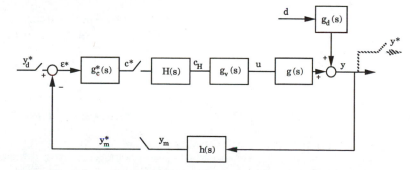

Figure 25.12. Typical block diagram of a digital feedback control system.

Closed-Loop Transfer Function Relations

We are now in a position to use the same principles applied in developing open-loop transfer function relations to analyze the closed-loop system in Figure 25.12. The objective here is to find z-transfer functions that relate the "sampled" process output, y^* to the sampled set-point, y^*_d, and also relate the "sampled" process output, y^* to the disturbance variable d (which we should keep in mind is not a sampled variable).

Transfer Function Relating $y(z)$ to $y_d(z)$

In order to determine the response to set-point changes, let us begin with the process model in Eq. (25.173) and set $d(s) = 0$; the process input, $u(s)$, is determined by Eq. (25.178), but since $c_H(s)$ itself is determined by Eq. (25.177), we have:

$$u(s) = g_v(s) \, H(s) c^*(s)$$

which, when introduced into Eq. (25.173) for $u(s)$ gives:

$$y(s) = g(s) \, g_v(s) \, H(s) c^*(s) \tag{25.179}$$

From this point on, we have two main tasks: (1) determine $y^*(s)$, and (2) obtain a closed-form expression for $y^*(s)$ by eliminating $c^*(s)$.

By taking the usual star transformation in Eq. (25.179), we obtain:

$$y^*(s) = \left[gg_v H \right]^* (s) c^*(s) \tag{25.180}$$

where the product of the three continuous transfer functions $g(s)g_v(s)H(s)$ has been combined into the single, composite, continuous transfer function $gg_v H(s)$, with "starred" Laplace transform indicated in Eq. (25.180).

Next, we observe from Eqs. (25.176) and (25.175) that $c^*(s)$ is given by:

$$c^*(s) = g^*_c(s) \left[y^*_d(s) - y^*_m(s) \right] \tag{25.181}$$

In order to eliminate $y^*_m(s)$, we must now exercise some caution, because the only related expression we have is in Eq. (25.174), and it involves the continuous signal, y_m, not y^*_m. We must first reintroduce Eq. (25.179) for y in Eq. (25.174) to give:

$$y_m(s) = h(s)g(s) \, g_v(s) \, H(s) c^*(s) \tag{25.182}$$

and then take the star transform in order to generate the required expression for $y^*_m(s)$. Here we make the product $h(s)g(s)g_v(s)H(s)$ into the composite $hgg_v H(s)$, and take the star transformation to give:

$$y^*_m(s) = \left[hgg_v H \right]^* (s) c^*(s) \tag{25.183}$$

We may now safely introduce Eq. (25.183) into Eq. (25.181) and solve for $c^*(s)$ to yield:

$$c^*(s) = \left[\frac{g^*_c(s)}{1 + \left[hgg_v H \right]^* (s) g^*_c(s)} \right] y^*_d(s) \tag{25.184}$$

If we now eliminate $c^*(s)$ from Eq. (25.180) by introducing Eq. (25.184) we obtain the starred transfer function relationship:

$$y^*(s) = \left[\frac{[gg_vH]^* (s)g^*_c(s)}{1 + [hgg_vH]^* (s)g^*_c(s)} \right] y^*_d(s) \qquad (25.185)$$

Let us now simplify our notation by introducing the following composite continuous transfer functions, the product of all the continuous transfer functions in the forward part *excluding the controller*:

$$g_f(s) = g(s)g_v(s) H(s) \qquad (25.186a)$$

and

$$g_l(s) = h(s)g(s) g_v(s)H(s) \qquad (25.186b)$$

the product of all continuous transfer functions in the feedback loop (again, excluding the controller); then Eq. (25.185) may be written as:

$$y^*(s) = \left[\frac{g^*_f(s)g^*_c(s)}{1 + g^*_l(s)g^*_c(s)} \right] y^*_d(s) \qquad (25.187)$$

We may now simply recast Eq. (25.187) in the z-transform notation to obtain the required closed-loop transfer function relation:

$$y(z) = \left[\frac{g_f(z)g_c(z)}{1 + g_l(z)g_c(z)} \right] y_d(z) \qquad (25.188)$$

The introduction of the composite transfer functions $g_f(s)$ and $g_l(s)$ is more than just a notational convenience; it permits the development of a simple, error-proof procedure for obtaining closed-loop transfer functions for sampled-data systems without having to go through a lengthy derivation every time. This procedure is now outlined as follows:

1. Obtain the product of all the *continuous* transfer functions in the forward path from the set-point signal to the process output signal. Be careful not to include the controller; it is discrete, and in any event, its transfer function is always presented separately in closed-loop transfer functions even for continuous feedback control systems. Call this consolidated continuous transfer function $g_f(s)$.

2. Obtain $g^*_f(s)$, the starred transform of this consolidated transfer function.

3. Also obtain the product of all the continuous transfer functions in the feedback loop (again, excluding the controller); let this be $g_l(s)$; also obtain the starred transform of this transfer function, $g^*_l(s)$.

4. The starred transform relation for the closed-loop system will be given by Eq. (25.187) from which the z-transfer function relation is easily obtained as given in Eq. (25.188).

Transfer Function Relating $y(z)$ to $d(z)$

The procedure summarized above may now be applied in deriving the transfer function relating the process output to disturbance changes.

The forward path between the disturbance signal and the process output contains only the disturbance transfer function, $g_d(s)$; however, recall that the disturbance signal is also continuous, not sampled, and we must therefore combine it with $g_d(s)$ to obtain the composite $g_d\,d(s)$.

The continuous transfer functions in the feedback loop are precisely as in the last derivation: thus, we have, immediately, that, with respect to disturbance changes:

$$y^*(s) = \frac{[g_d\,d]^*(s)}{1 + g^*_l(s)g^*_c(s)}$$

so that in z-transform notation:

$$y(z) = \frac{g_d\,d(z)}{1 + g_l(z)g_c(z)} \tag{25.189}$$

We may now combine the two closed-loop z-transfer function expressions to obtain:

$$y(z) = \left[\frac{g_f(z)g_c(z)}{1 + g_l(z)g_c(z)}\right] y_d(z) + \frac{g_d\,d(z)}{1 + g_l(z)g_c(z)} \tag{25.190}$$

Observe the striking resemblance between this expression and the corresponding expression obtained in Chapter 14 for continuous systems.

Let us note that:

1. The denominator polynomials in both portions of Eq. (25.190) are identical.

2. Even though we cannot separate out the disturbance signal (because it is continuous), given any specified form for $d(s)$, the composite $g_d\,d(s)$ can be formed, from which $g_d\,d(z)$ can then be obtained.

3. In using Eq. (25.190) to determine closed-loop transient response to either a set-point change, or a change in the input disturbance, the nature of the observed transient response will be determined solely by the roots of the denominator polynomial, i.e.:

$$1 + g_l(z)\,g_c(z) = 0 \tag{25.191}$$

4. Observe that this expression is reminiscent of the characteristic equation obtained for continuous-time closed-loop feedback systems in Chapter 14; it is, in fact, the sampled-data feedback control system characteristic equation, and it plays precisely the same role as its continuous-time counterpart.

We shall illustrate closed-loop analysis with the following example.

Example 25.6   **CLOSED-LOOP BEHAVIOR OF A FIRST-ORDER SYSTEM UNDER PROPORTIONAL DIGITAL CONTROL.**

Given the digital control system whose block diagram is shown in Figure 25.12, find the response to a unit step change in the set-point when the transfer functions are given as follows:

$$g(s) = \frac{K}{\tau s + 1}; \quad g_v(s) = 1; \quad h(s) = 1$$

$$H(s) = \left(\frac{1 - e^{-s\Delta t}}{s}\right) \text{ (the ZOH element)}$$

and the controller is a digital proportional controller, i.e., $g^*_c(s) = K_c$.

Solution:

In this case, the consolidation of the continuous transfer functions in the forward path in this case gives, from Eq. (25.186)

$$g_f(s) = g(s)g_v(s)H(s)$$

$$= \frac{K}{\tau s + 1}\left(\frac{1 - e^{-s\Delta t}}{s}\right) \qquad (25.192)$$

which is the same as $g_l(s)$ since $h(s) = 1$.

Discretizing the composite transfer function in Eq. (25.192) and taking z-transforms, we obtain:

$$g_f(z) = g_l(z) = \left[\frac{K\left(1 - \phi\right)z^{-1}}{1 - \phi\, z^{-1}}\right]$$

where $\phi = e^{-\Delta t/\tau}$. Since $g^*_c(s) = K_c$, we have immediately that:

$$g_c(z) = K_c$$

From Eq. (25.188), the closed-loop transfer function relation we need is:

$$y(z) = \left[\frac{\dfrac{KK_c\left(1 - \phi\right)z^{-1}}{1 - \phi\, z^{-1}}}{1 + \dfrac{KK_c\left(1 - \phi\right)z^{-1}}{1 - \phi\, z^{-1}}}\right]y_d(z)$$

which simplifies to:

$$y(z) = \left\{ \frac{KK_c(1 - \phi)z^{-1}}{1 + \left[KK_c - \phi(1 + KK_c) \right] z^{-1}} \right\} y_d(z) \tag{25.193}$$

To obtain the response to a unit step change in the set-point, we let $y_d(z) = \dfrac{1}{1 - z^{-1}}$ in Eq. (25.193) to obtain:

$$y(z) = \left\{ \frac{KK_c(1 - \phi)z^{-1}}{1 + \left[KK_c - \phi(1 + KK_c) \right] z^{-1}} \right\} \frac{1}{1 - z^{-1}} \tag{25.194}$$

Before carrying out a z-transform inversion it is instructive to simplify the expression in winged brackets first, as follows:

Let:

$$p = \phi\left(1 + KK_c\right) - KK_c \tag{25.195}$$

be the closed-loop transfer function pole in Eq. (25.194); then, solving for ϕ in terms of p gives:

$$\phi = \frac{p + KK_c}{1 + KK_c}$$

so that:

$$1 - \phi = \frac{1 - p}{1 + KK_c} \tag{25.196}$$

Introducing Eqs. (25.196) and (25.195) into the winged brackets in Eq. (25.194) now gives:

$$y(z) = \left[\frac{K_{cl}(1 - p)z^{-1}}{1 - p\, z^{-1}} \right] \frac{1}{1 - z^{-1}} \tag{25.197}$$

where

$$K_{cl} = \frac{KK_c}{1 + KK_c} \tag{25.198}$$

is the closed-loop gain. The z-transform indicated in Eq. (25.197) is now easily inverted (cf. Appendix C) to give:

$$y(k\Delta t) = K_{cl}(1 - p^k) ; \; k \geq 0 \tag{25.199}$$

Some comments about this step response are now in order:
1. Observe now that Eq. (25.199) is analogous to Eq. (14.42) obtained for the continuous system under classical proportional control.
2. Further observe that Eq. (25.198) is identical to Eq. (14.41a); however, Eq. (25.195) does *not* bear the same resemblance to Eq. (14.41b).
3. The fact that Eq. (25.195) is not analogous to its continuous-time counterpart expression is a critical point that provides us a glimpse into the major difference between continuous and sampled-data feedback control systems. As we will soon see, this is a difference with serious implications for the stability of sampled-data control systems.

25.4   STABILITY

A major recurring theme throughout this chapter is that once we have made the modifications required by the special character of discrete-time systems, all the techniques for continuous-time system analysis that we learned earlier carry over nicely for the analysis of discrete-time systems. This is especially true for stability analysis.

As was the case for the continuous system, we are often concerned with whether or not the output sequence of a discrete system remains bounded in response to a bounded sequence of values for the input. While there are other means of defining system stability, this concept of "bounded-input, bounded-output" (BIBO) stability is usually sufficient.

By analogy to the continuous-time definition of BIBO stability, we may now state that

*If in response to a bounded input sequence, u(k), the dynamic output sequence of a system, y(k), remains bounded as k → ∞, then the system is said to be **stable**; otherwise, it is said to be **unstable**.*

It is important to remember that, strictly speaking, this definition has nothing to say about the behavior of the output of sampled-data systems in between sampling points. However, it requires a special pathological situation for an unbounded continuous signal to give rise to a bounded sampled sequence. Therefore, it is standard practice to assume that a bounded sampled sequence implies a bounded continuous output signal.

25.4.1 Conditions for Stability

Recall that a continuous system is considered stable if the roots of its characteristic equation all lie strictly in the left half of the complex *s*-plane. For the open-loop system, this translates to having all the poles of the open-loop transfer function in the left-half plane; for the closed-loop system, it is the roots of the more complicated closed-loop characteristic equation that must lie in the left-half plane. Thus the stability region in the continuous domain is clearly the left half of the *s*-plane, with the imaginary axis as the boundary.

From our discussion in Section 25.2.3, it was established, using the fundamental relationship between the variables *s* and *z*, that the region in the *z*-plane that corresponds to the left half of the *s*-plane is the *interior* of the unit circle (Figure. 25.2(b)). Thus, we are therefore able to deduce that the region of stability in the *z*-plane is the interior of the unit circle, with the circle itself as the boundary. We may therefore state that:

A discrete system is stable if the roots of its discrete characteristic equation all lie strictly within the unit circle in the z-plane.

Let us now discuss what this translates to in the open-loop and closed-loop cases.

Open Loop

For the general linear discrete system whose open-loop transfer function was given earlier as:

$$g(z) = \frac{B(z^{-1}) \, z^{-D}}{A(z^{-1})}$$

(25.46a)

or

$$g(z) = \frac{\left(b_0 + b_1 z^{-1} + b_2 z^{-2} + \dots + b_m z^{-m}\right) z^{-D}}{1 + a_1 z^{-1} + a_2 z^{-2} + \dots + a_n z^{-n}} \qquad (25.46b)$$

the characteristic equation is defined as:

$$A(z^{-1}) = 1 + a_1 z^{-1} + a_2 z^{-2} + \dots + a_n z^{-n} = 0 \qquad (25.200)$$

the roots of which, we may recall, have been previously referred to as the poles of $g(z)$. Thus:

> *The discrete system with an open-loop pulse transfer function $g(z)$ is open-loop stable if all the poles of $g(z)$ lie strictly within the unit circle.*

For the multivariable system whose time-domain model was given earlier as:

$$\mathbf{y}(k) = \mathbf{\Phi}\mathbf{y}(k-1) + \mathbf{B}\mathbf{u}(k-1) \qquad (25.31)$$

it can be shown that the characteristic equation in this case is:

$$|\mathbf{\Phi} - \lambda \mathbf{I}| = 0 \qquad (25.201)$$

which, for an $N \times N$ system, gives rise to an Nth-order polynomial in λ, with roots, λ_i, $i = 1, 2, 3, \dots, N$. The stability condition is that:

$$|\lambda_i| < 1; \, i = 1, 2, \dots, N \qquad (25.202)$$

AN IMPORTANT QUESTION

We must now ask an important question that enables us to relate the stability properties of a sampled-data systems to that of the underlying continuous system:

> *Can the discrete system that arises from sampling a continuous, open-loop stable system ever be open-loop unstable?*

To answer this question, we refer back to our discussion in Section 25.2.3 and recall the earlier statement made that if $s = s_i$, $i = 1, 2, \dots, n$ are the poles of an s-transfer function, the corresponding poles of the z-transfer function are $z = e^{s_i \Delta t}$.

The implications here are as follows: if the continuous system is stable, then $s_i < 0$ for all i; this being the case, observe then that the sampled-data system resulting from sampling this system will have a z-transfer function whose poles will all lie within the unit circle, since $e^{s_i \Delta t} < 1$ if $s_i < 0$. We therefore conclude that:

> *An open-loop stable continuous system can never give rise to an open-loop unstable sampled-data system.*

Observe, however, that while this statement is true, the actual location of the poles within the unit circle will depend on Δt, the sampling time.

Closed Loop

The closed-loop transfer function relation for the typical digital feedback control system was obtained in Eq. (25.190) with the characteristic equation given as:

$$1 + g_I(z)\, g_c(z) \;=\; 0 \tag{25.191}$$

where $g_I(z)$ is the z-transfer function corresponding to the *composite* s-transfer function made up of all the continuous transfer functions in the feedback loop, exclusive of the controller. For the closed-loop system to be stable, all the roots of Eq. (25.191) must lie strictly within the unit circle.

ANOTHER IMPORTANT QUESTION

Let us now ask another important question, related to the previous one:

Can a continuous system which is closed-loop stable under continuous control become closed-loop unstable under digital control with the same controller parameters?

We shall answer this crucial question by considering a very simple, and familiar, example.

Example 25.7   CLOSED-LOOP STABILITY OF A FIRST-ORDER SYSTEM UNDER PROPORTIONAL DIGITAL CONTROL.

Investigate the stability properties of a digital control system involving a first-order process with a ZOH under digital proportional control.

Solution:

In this case, since

$$g(s) \;=\; \frac{K}{\tau s + 1}; \quad \text{and} \quad H(s) \;=\; \left(\frac{1 - e^{-s\Delta t}}{s}\right)$$

are the only other elements in the feedback loop apart from the controller, we have that:

$$g_I(s) \;=\; \left(\frac{1 - e^{-s\Delta t}}{s}\right)\frac{K}{\tau s + 1} \tag{25.203}$$

so that the corresponding z-transfer function is:

$$g_I(z) \;=\; \frac{K(1 - \phi)z^{-1}}{1 - \phi z^{-1}} \tag{25.204}$$

with $\phi = e^{-\Delta t/\tau}$. And now, since $g_c(z) = K_c$, the characteristic equation for this closed-loop system is given by:

$$1 \; + \; \frac{KK_c(1 - \phi)z^{-1}}{1 - \phi z^{-1}} \; = \; 0$$

which simplifies to yield:

$$z - \phi + (1 - \phi)KK_c \; = \; 0 \qquad\qquad (25.205)$$

an equation whose single root is located at:

$$z \; = \; p \; = \; \phi - (1 - \phi)KK_c \qquad\qquad (25.206)$$

(recall Eq. (25.195) in Example 25.7).

Let us observe the following about the nature of this root:

1. When $K_c = 0$ (the open-loop case), the root is located on the real axis at $z = \phi = e^{-\Delta t/\tau}$, and since $e^{-\Delta t/\tau}$ is always less than 1, we therefore have that $0 < \phi < 1$, and the system is stable.
2. As K_c increases, the value of this root shifts to the left along the real axis.
3. At the critical point when $K_c = K^*_c$, where:

$$K^*_c \; = \; \frac{1 + \phi}{K(1 - \phi)} \qquad\qquad (25.207)$$

$z = -1$ in Eq. (25.206), and the root lies on the boundary of the unit circle; for values of $K_c > K^*_c$, the root falls outside the unit circle, and the system becomes unstable.

This last observation is quite significant: recall that a continuous first-order system under continuous proportional control *can never* be unstable; we have just shown that the same system under digital proportional control can only be stable for controller gain values less than the critical value shown in Eq. (25.207).

Let us therefore note very carefully from this example that:

It is possible for a continuous system that is closed-loop stable under continuous control to become closed-loop unstable under digital control, even with the same controller parameters.

It must also be observed that in the simple example just considered, not only is it possible for this system to go unstable, the critical controller gain value required for this system's closed-loop stability depends on the sampling time (through ϕ). The influence of sampling time on the stability of this system is illustrated in the next example.

Example 25.8   EFFECT OF SAMPLING TIME ON THE CLOSED-LOOP STABILITY OF A FIRST-ORDER SYSTEM UNDER PROPORTIONAL DIGITAL CONTROL.

Examine the influence of sampling time on the stability properties of the closed-loop digital control system of Example 25.7.

Solution:

From the expression in Eq. (25.206) for the location of the single root of the characteristic equation, we observe that for a *fixed* value of controller gain K_c, the critical value of ϕ for which the root becomes -1 is given by:

$$\phi_{\text{critical}} = \frac{KK_c - 1}{KK_c + 1} \tag{25.208}$$

Thus, for all values of ϕ for which $\phi < \phi_{\text{critical}}$, the system will be unstable. Since $\phi = e^{-\Delta t/\tau}$, we can determine the critical value for the sampling time $\Delta t_{\text{critical}}$ for which this system is stable for any fixed controller gain:

$$\Delta t_{\text{critical}} = -\tau \ln\left(\frac{KK_c - 1}{KK_c + 1}\right) \tag{25.209}$$

Thus for $\Delta t > \Delta t_{\text{critical}}$ the system will be unstable.

Let us try to understand this result in physical terms. Recall from Chapters 23 and 24 that one of the effects of adding a zero-order hold to a continuous system was effectively to add a unit time delay, which appears in the transfer function as an extra z^{-1} factor. Since the size of this effective time delay is proportional to the sampling time, Δt, and since we recall from Part II that adding a time delay will cause a continuous first-order system to go unstable under proportional feedback control (for sufficiently large K_c), then the results of the last two examples are quite understandable. These results generalize to any order system and serve to warn us that increasing the sampling time means that, for stability, we must "detune" the controllers for discrete-time systems just as we were required to "detune" continuous system controllers in order to preserve stability as time delays were increased.

25.4.2 Stability Tests

We have just established that, as with continuous systems, the determination of the stability of discrete systems involves no more than investigating the nature of the roots of a polynomial equation in the complex variable z. It is clear, however, that it will be particularly inconvenient if one has to explicitly evaluate the roots of characteristic equations higher in order than a quadratic. Thus, as was the case for continuous systems, there are several techniques for testing the stability of a system without explicitly evaluating the roots. Recall that for continuous systems, we used the Routh test, a purely algebraic test that identifies the number of roots of a polynomial with positive real parts. In its application to stability analysis, it was used to test if any roots of the characteristic equation, a polynomial in the variable s, lie in the right-half plane. For the discrete system, because the stability region is the interior of the unit circle, a corresponding test will, of course, be one that can identify roots of the z-domain characteristic equation that lie in the exterior of the unit circle in the z-plane. There are several approaches to this (see Refs. [1–4]); however, here we will focus on two of the most common techniques:

1. The bilinear (Möbius) transformation technique
2. The Jury test

Each of these will be discussed in turn.

The Bilinear (Möbius) Transformation Technique

Consider the following transformation:

$$w = \frac{z + 1}{z - 1} \qquad (25.210)$$

which is known as the bilinear or Möbius transformation. It can be shown that every complex number represented by $|z| < 1$ gets mapped into the set of complex numbers represented by $|w| < 0$, i.e., Eq. (25.210) maps the interior of the unit circle in the z-plane into the left half of the w-plane.

The implication of this is as follows: a polynomial equation in the variable z of the form $P(z) = 0$, with roots $z_1, z_2, z_3, ..., z_n$, can be converted by the bilinear transformation in Eq. (25.210) to a different polynomial equation in the variable w, of the form $\prod(w) = 0$, with roots $w_1, w_2, w_3, ..., w_n$; if the roots $z_1, z_2, z_3, ..., z_n$ lie within the unit circle, then it will also be true that the corresponding roots $w_1, w_2, w_3, ..., w_n$ will lie in the left half of the w-plane. Thus, the Routh test, used previously for continuous systems, can also be used for determining if any of the roots $w_1, w_2, w_3, ..., w_n$ are in the right half of the w-plane, and this will be entirely equivalent to determining if any of the roots $z_1, z_2, z_3, ..., z_n$, is in the exterior of the unit circle in the z-plane.

The transformation in Eq. (25.210), when solved for z in terms of w, gives:

$$z = \frac{w + 1}{w - 1} \qquad (25.211)$$

where the interesting symmetry should be noted. The strategy for using the bilinear transformation for stability analysis is now outlined:

1. By substituting Eq. (25.211) for z in the characteristic equation, $P(z) = 0$, convert it to the equation $\prod(w) = 0$;

2. Use the Routh test on $\prod(w)$; any roots of $\prod(w)$ in the RHP correspond to roots of $P(z)$ in the exterior of the unit circle.

Unfortunately, the conversion of $P(z)$ to $\prod(w)$ frequently requires a substantial amount of algebraic manipulation; this is then followed by the usual amount of computation involved in conducting the Routh test. As a result, this technique is not as popular as other techniques (see below) by which we can tell directly if a polynomial in z has all its roots inside the unit circle or not.

The Jury Test

The Jury test is a truly "discrete-time" version of the "continuous-time" Routh test; it provides the necessary and sufficient conditions for a polynomial equation (with real coefficients) to have roots that are less than 1 in absolute value, (i.e., roots that lie within the unit circle); it is also based on an array, or table, just like the Routh test; and the generation and application of the Jury table bears some remarkable resemblances to the generation and application of the Routh array.

The following is the procedure for applying the Jury test.

1. **Write the characteristic equation in the "forward" form:**

$$P(z) = a_0 z^n + a_1 z^{n-1} + \ldots + a_{n-2} z^2 + a_{n-1} z + a_n = 0 \qquad (25.212)$$

where the leading term, a_0, is *positive*.

2. **Carry out the first part of the Jury test:**

By substituting into Eq. (25.212) determine:

$$P(1) = \sum_{i=0}^{n} a_i \qquad (25.213a)$$

$$P(-1) = \sum_{i=0}^{n} (-1)^{n-i} a_i \qquad (25.213b)$$

and then test if   1. $\left| a_n \right| < a_0$,
2. $P(1) > 0$, and
3. $P(-1) > 0$, for n even, or $P(-1) < 0$, for n odd

If any of these three conditions are *not* satisfied, then we can immediately conclude that $P(z)$ has roots *outside* the unit circle, and it is not necessary to proceed further. The system is unstable. However, if these tests are satisfied, then proceed to Step 3.

3. **Generate the Jury table:**

This is a table consisting of $n + 1$ columns and $n - 1$ *pairs* of rows making a total of $2n - 2$ rows. (It should be noted that in most textbooks, the Jury table is presented with $(2n - 3)$ rows; we have chosen to explicitly include the additional final row for reasons that will soon become clear.)
The elements of the rows in the Jury table are generated from the coefficients of the characteristic polynomial as follows:

1. All the rows are determined in pairs; the elements of the even numbered row of the pair are always the reverse of the elements in the odd numbered row.

2. The elements of first pair of rows (rows 1 and 2) are the coefficients of $P(z)$ arranged in forward and reverse order.

3. The elements of the subsequent pairs of rows are obtained from the elements in the immediately preceding pair of rows according to the following determinant formulas:

THE JURY TABLE

Row 1	a_n	a_{n-1}	a_{n-2}	a_{n-3}	$\cdots$	a_2	a_1	a_0
Row 2	a_0	a_1	a_2	a_3	$\cdots$	a_{n-2}	a_{n-1}	a_n

Row 3	b_{n-1}	b_{n-2}	b_{n-3}	b_{n-4}	$\cdots$	b_1	b_0
Row 4	b_0	b_1	b_2	b_3	$\cdots$	b_{n-2}	b_{n-1}

Row 5	c_{n-2}	c_{n-3}	c_{n-4}	c_{n-5}	$\cdots$	c_0
Row 6	c_0	c_1	c_2	c_3	$\cdots$	c_{n-2}

$$\vdots \quad \vdots$$

Row $2n-5$	p_3	p_2	p_1	p_0
Row $2n-4$	p_0	p_1	p_2	p_3

Row $2n-3$	q_2	q_1	q_0
Row $2n-2$	q_0	q_1	q_2

$$b_k = \begin{vmatrix} a_n & a_{n-1-k} \\ a_0 & a_{k+1} \end{vmatrix} \; ; \; k = 0, 1, 2, \ldots, n-1$$

$$c_k = \begin{vmatrix} b_{n-1} & b_{n-2-k} \\ b_0 & b_{k+1} \end{vmatrix} \; ; \; k = 0, 1, 2, \ldots, n-2$$

$$\vdots$$

$$q_k = \begin{vmatrix} p_3 & p_{2-k} \\ p_0 & p_{k+1} \end{vmatrix} \; ; \; k = 0, 1, 2$$

4. Determine stability from the elements of the first column:

The first necessary condition for stability is that all three tests in Step 2 above are satisfied. If this is true, then for stability it is further necessary that the following conditions relating pairs of *calculated* elements in the *first* column be satisfied:

$$|b_{n-1}| > |b_0|$$

$$|c_{n-2}| > |c_0|$$

$$\bullet$$
$$\bullet$$
$$\bullet$$

$$|q_2| > |q_0|$$

We may now observe that having the additional row $2n - 2$ allows us to consider the stability test as one involving pairs of elements in the first column. The combined tests in Step 2 and Step 4 are the necessary and sufficient conditions that all the roots of $P(z)$ are inside the unit circle and thus determine system stability.

Let us illustrate the Jury test procedure with the following example.

Example 25.9   STABILITY OF A CLOSED-LOOP DIGITAL CONTROL SYSTEM USING THE JURY TEST.

The characteristic equation for a certain closed-loop digital control system is given as:

$$P(z) = 1 + 0.4\,z^{-1} - 0.69z^{-2} - 0.256z^{-3} + 0.032z^{-4} = 0 \qquad (25.214)$$

Determine whether this system is stable or not.

Solution:

The equation is not in the form required by Eq. (25.212), but by multiplying by z^4, we obtain:

$$P(z) = z^4 + 0.4\,z^3 - 0.69z^2 - 0.256z + 0.032 = 0 \qquad (25.215)$$

which is in the required form, with the coefficients given by:

$$a_0 = 1; \quad a_1 = 0.4; \quad a_2 = -0.69; \quad a_3 = -0.256; \quad a_4 = 0.032$$

We may now carry out the first part of the test; keep in mind that $n = 4$ which is even.
1. $|a_4| = 0.032 < 1$
2. $P(1) = 0.486 > 0$
3. $P(-1) = 0.198 > 0$ (because $n = 4$, we apply this condition)

Thus, all the three conditions of the first part of the test are satisfied; we must now proceed to construct the Jury table, to complete the test.

The table will consist of three pairs of rows; the elements of the first pair of rows are already known from the characteristic equation; so the elements we need to determine are b_0, b_1, b_2, b_3, for the second pair of rows, and for the last pair of rows, c_0 and c_2. We may calculate c_1 if we wish, but it is not necessary for the stability test; for completeness its value is included in the table shown below.

After going through the required determinant evaluations, the following Jury table is obtained:

| Row 1 | 0.032 | −0.256 | −0.69 | 0.4 | 1.0 |
| Row 2 | 1.0 | 0.4 | −0.69 | −0.256 | 0.032 |

| Row 3 | −0.9990 | −0.4082 | 0.6679 | 0.2688 |
| Row 4 | 0.2688 | 0.6679 | −0.4082 | −0.9990 |

| Row 5 | 0.9255 | 0.2283 | −0.5575 |
| Row 6 | −0.5575 | 0.2283 | 0.9255 |

Examining the pairs of *calculated* elements in the first column of this table, we find that:

$$0.9990 > 0.2688; \qquad \text{and} \qquad 0.9255 > 0.5575$$

indicating that no roots of the characteristic equation lie outside the unit circle, from which we conclude that the system is stable.

It is, perhaps instructive to note that $P(z)$ may be factored as follows:

$$P(z) = (z - 0.8)(z + 0.5)(z - 0.1)(z + 0.8) \tag{25.216}$$

from where we see that indeed no root lies outside the unit circle.

As with the Routh test, it is possible to conduct the Jury test with unspecified values for controller parameters. The results of such tests will be the range of values of the controller parameters over which the closed-loop system will be stable (see Problem 25.10).

Other Methods

It should be noted that other methods exist for studying the stability properties of discrete-time systems. Some of these other methods include:

1. The Root Locus Method (see Ref. [3])
2. The Nyquist method (see Refs. [1, 3])
3. Lyapunov's direct method (Ref. [1]).

Another obvious choice is to use polynomial root extraction computer programs. When all the parameters in the characteristic equation are specified, and we are only interested in the roots, this is perhaps one of the most efficient ways to investigate closed-loop stability.

25.5 SUMMARY

In this chapter, we have explored in some detail, the dynamic behavior of discrete processes under open-loop as well as closed-loop conditions, using the tools of the z-transform and the pulse transfer function introduced in Chapter 24.

By analogy with the continuous Laplace transfer function, the characteristics of the pulse transfer function were discussed extensively and the relationship to the continuous domain transfer function established. These z-domain transfer functions were used as building blocks in studying the open- and closed-loop behavior of sampled-data processes via block diagram analysis; then the key issues involved in determining the stability of discrete systems were discussed.

The main lessons of this chapter may be summarized as follows:

1. Discrete systems are remarkably similar to their continuous-time-domain counterparts so that once the necessary modifications have been made to account for their special characteristics, *nearly all* of the concepts, techniques, and results which are applicable in continuous-time analysis carry over directly into discrete-time analysis.

2. However, when discrete systems arise by the process of sampling a continuous process, the fact that we have a discrete system existing in a continuous environment requires that analysis be done with great care, as illustrated by several examples.

3. There are certain continuous domain results that *do not* carry over into the discrete domain, especially in the area of stability of closed-loop systems. The most important of these is that while some continuous-time systems exist which, theoretically, can never be closed-loop unstable, this is not the case for sampled-data systems because of the unit sample time delay introduced by the hold element.

The importance of the material covered in this chapter will become even more evident in the next chapter as we take on the issues involved in the design of digital controllers.

REFERENCES AND SUGGESTED FURTHER READING

1. Åström, K. J. and B. Wittenmark, *Computer Controlled Systems*, Prentice-Hall, Englewood Cliffs, NJ (1984)
2. Ogata, K., *Discrete-Time Control Systems*, Prentice-Hall, Englewood Cliffs, NJ (1981)
3. Franklin, G. F. and J. D. Powell, *Digital Control of Dynamic Systems*, Addison-Wesley, Reading MA (1980)
4. Cadzow, J. A. and H. R. Martens, *Discrete-Time and Computer Control Systems*, Prentice-Hall, Englewood Cliffs, NJ (1970)

REVIEW QUESTIONS

1. What determines the order of a pulse transfer function?

2. How are the poles, the zeros, and the steady-state gain of a pulse transfer function determined?

3. What does the time-delay term contribute to the pole-zero distribution of a pulse transfer function?

4. What aspect of a pulse transfer function is responsible for contributing additional zeros at the origin?

5. What is the fundamental relationship between the complex Laplace transform variable s and the z-transform variable?

6. The imaginary axis in the complex s-plane maps into what region in the z-plane?

7. The entire left half of the complex s-plane maps into what region in the z-plane? What does this imply about the region in the z-plane into which the entire right-half plane is mapped?

8. Why is the relationship between the z-plane and the s-plane not unique?

9. How are the poles and zeros of a Laplace domain transfer function related to the poles and zeros of the corresponding z-domain transfer function?

10. Why is the *number* of zeros of a continuous process usually different from the number of zeros of the transfer function of the equivalent discrete process incorporating a ZOH?

11. A discrete process whose pulse transfer function has a *zero* outside of the unit circle will exhibit what type of dynamic behavior?

12. Why does the process of sampling a continuous inverse-response system not always produce a discrete system that exhibits inverse response?

13. How do the following characteristics of a discrete system's pulse transfer function affect its dynamic behavior?

    (a) Positive real poles *within* the unit circle
    (b) Negative real poles *within* the unit circle
    (c) Real poles *on* the unit circle
    (d) Complex conjugate poles *within* the unit circle
    (e) Complex conjugate poles *outside* the unit circle
    (f) Complex conjugate poles whose real parts lie *within* the unit circle but whose imaginary parts are multiples of $\pm \pi$
    (g) Complex conjugate poles *on* the unit circle
    (h) A *pole* outside the unit circle
    (i) A *zero* outside the unit circle
    (j) A time delay

14. What distinguishes the block diagram of a sampled-data feedback control system from that of a continuous feedback control system?

15. What is the "starred" Laplace transform, and what is it used for?

16. Which components of the sampled-data feedback control system are the same as those of the corresponding continuous feedback control system; which are common to both but modified for the sampled-data system; and which are completely unique to the sampled-data system?

17. What is the procedure for obtaining closed-loop transfer functions for sampled-data systems?

18. When is a discrete system stable?

19. Can a continuous, open-loop stable system give rise to an open-loop unstable sampled-data system?

20. Why is it possible for a continuous system that is stable under continuous closed-loop control to become unstable under digital control, even with identical controller parameters?

21. What is the basic principle behind the bilinear transformation technique for determining the stability of a discrete system?

22. What is the procedure for applying the Jury stability test?

PROBLEMS

25.1 Given the following pulse transfer function model for a first-order system incorporating a ZOH element:

$$y(z) = \frac{2.0(1 - \phi)z^{-1}}{1 - \phi z^{-1}} u(z) \qquad (P25.1)$$

Obtain the system's unit step response for the following values of ϕ:

(a) $\phi = 0.5$
(b) $\phi = 0.8$
(c) $\phi = -0.5$
(d) $\phi = 1.5$

Plot these responses and comment on the effect of ϕ on the step response of a first-order system.

25.2 Each of the following pulse transfer functions relates the input u and output y of a sampled process incorporating a ZOH element. For each case:

(a) Evaluate the process steady-state gain
(b) State the location of the poles and zeros
(c) Write the corresponding difference equation process model
(d) Obtain the process unit step response, and
(e) Identify the process *type* (e.g., first-order, inverse-response, etc.)

(1) $\qquad g(z) = \dfrac{0.5z^{-1}}{1 - 0.8z^{-1}}$

(2) $\qquad g(z) = 0.5z^{-1}$

(3) $\qquad g(z) = \dfrac{0.5z^{-1}}{1 - z^{-1}}$

(4) $\qquad g(z) = \dfrac{4.5z^{-1} - 4z^{-2}}{1 - 0.8z^{-1}}$

(5) $\qquad g(z) = \dfrac{6.0z^{-1} - 0.375z^{-2}}{1 + z^{-1} + 1.25z^{-2}}$

(6) $\qquad g(z) = \dfrac{-0.431z^{-1} + 0.6093z^{-2}}{1 - 1.4252z^{-1} + 0.4965z^{-2}}$

(7) $\qquad g(z) = \dfrac{(0.0995z^{-1} + 0.0789z^{-2})z^{-6}}{1 - 1.4252z^{-1} + 0.4965z^{-2}}$

(8) $\qquad g(z) = \dfrac{0.5z^{-1}}{1 - 1.2z^{-1}}$

25.3 In Chapter 7, it was established that a system composed of two first-order systems in opposition, i.e., whose transfer function is given by:

$$g(s) = \frac{K_1}{(\tau_1 s + 1)} - \frac{K_2}{(\tau_2 s + 1)} \qquad \text{(P25.2)}$$

will exhibit inverse response, for $K_1 > K_2$, when:

$$\frac{K_1}{\tau_1} < \frac{K_2}{\tau_2} \qquad \text{(P25.3)}$$

(a) Obtain the equivalent discrete-time result by deriving the conditions under which the system whose pulse transfer function (with ZOH element) is given below will exhibit inverse response:

$$g(z) = \frac{b_1 z^{-1}}{1 - p_1 z^{-1}} - \frac{b_2 z^{-1}}{1 - p_2 z^{-1}} \qquad \text{(P25.4)}$$

(b) By relating Eq. (P25.2) to Eq. (P25.4), show that in the limit as the sampling time Δt tends to zero, the conditions derived in part (a) reduce to the conditions in Eq. (P25.3).

25.4 Establish the result given in Eq. (25.35).

25.5 For each of the block diagrams shown in Figure P25.1, determine the open-loop transfer function between $y(z)$ and $u(z)$. In each case:

$$g_1(s) = \frac{1}{s}$$

$$g_2(s) = \frac{5}{3s + 1}$$

the unit of time is minutes, and the sample time $\Delta t = 0.5$ min.

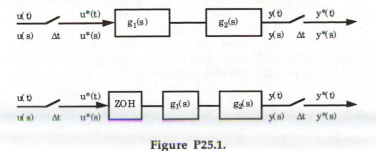

Figure P25.1.

25.6 A process whose transfer function model (including valve and measuring device dynamics) has been obtained as:

$$y(s) = \frac{5.0}{4.5s + 1} u(s) + \frac{1.5}{2.1s + 1} d(s) \qquad (P25.5)$$

is to be controlled using a digital proportional feedback controller, including a ZOH element. Draw a block diagram of the type in Figure 25.11 for this process and obtain the closed-loop transfer function relations.

25.7 Figure P25.2 shows a *digital* cascade control system. Obtain the closed-loop transfer functions between $y(z)$ and $y_d(z)$, and between $y(z)$ and $d_2(z)$.

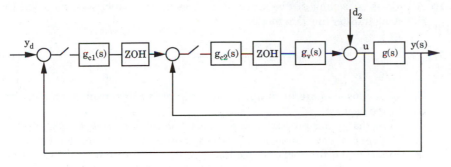

Figure P25.2. A digital cascade control system.

25.8 In Problem 24.5, the transfer function model for a thermosiphon reboiler was given as:

$$y(s) = \frac{0.9\,(0.3s - 1)}{s\,(2.5s + 1)} u(s) \qquad (P24.7)$$

with the input and output variables defined as:

$$u = \frac{(F_S - F_S\,{}^*)\,(lb/hr)}{F_S\,{}^*\,(lb/hr)}$$

$$y = \frac{(h - h^*)\,(ins)}{h^*\,(ins)}$$

Here, h^* (ins) and $F_S{}^*$ (lb/hr) represent nominal steady-state operating conditions. The time constants are in minutes.

Under *digital* proportional feedback control incorporating a ZOH element, with a sample rate of $\Delta t = 0.1$ min, obtain the closed-loop pulse transfer function given that $K_c = -0.5$. From the closed-loop transfer function, write the corresponding difference equation and obtain the process response to a unit step change in the normalized level set-point. Plot this response.

25.9   A proportional-only controller was designed for a first-order process by direct synthesis, in the continuous-time domain; the controller gain has been specified as:

$$K_c = \frac{\tau}{K\tau_r} \tag{P25.6}$$

where the K is the process steady-state gain, τ is the process time constant, and τ_r is the time constant of the desired closed-loop reference trajectory.
(a) In attempting to implement the control system on a digital computer, a young engineer selected $\tau/\tau_r = 3.0$. A more experienced control engineer recommended that with such a selection, the sampling time, Δt, should not exceed 0.7τ, otherwise the closed-loop system might become unstable. Confirm or refute this recommendation.
(b) Because of hardware limitations, it is not possible to sample this process any faster than $\Delta t = \tau$. What is the maximum value of τ/τ_r for which the closed-loop system will remain stable if $\Delta t = \tau$?

25.10 A process consisting of two CSTR's in series is reasonably well modeled by the following transfer function model:

$$y(s) = \frac{1.5}{(12.0s + 1)(5.0s + 1)} u(s) \tag{P25.7}$$

The time constants are in minutes, and the valve and sensor dynamics may be considered negligible.
(a) Show that under proportional feedback control *in continuous time*, the closed-loop system is always stable, for all values of proportional controller gain K_c.
(b) Under digital proportional feedback control, incorporating a ZOH, find the range of K_c values for which the closed-loop system will remain stable if the sample time $\Delta t = 1$ min.
(c) Repeat part (b) for $\Delta t = 5$ min.

25.11 An approximate model for the reactor temperature response to changes in jacket cooling water flowrate was given for a certain reactor in Problem 24.4 as:

$$g(s) = \frac{-0.55 \; (\text{°F/gpm})}{(5s - 1)(2s + 1)} \tag{P24.6}$$

The time constants are in minutes. First show that the closed-loop system is stabilized under proportional feedback control *in continuous time*, with $K_c = -4.0$. Next, investigate the closed-loop stability of the *digital* control system employing the same controller (incorporating a ZOH element) with $\Delta t = 0.5$ min.

25.12 The closed-loop transfer function for a digital feedback control system has been obtained as:

$$g_{cl}(z) = \frac{g_f(z)K_c}{1 + g_f(z)K_c} \tag{P25.8}$$

with $K_c = 2.0$, and:

$$g_f(z) = \frac{0.15z^{-1} - 0.29z^{-2} + 0.32z^{-3}}{1 - 1.6z^{-1} + 0.5z^{-2} - 0.4z^{-3}} \qquad \text{(P25.9)}$$

Investigate the stability of the closed-loop system:

(a) Using the method of bilinear transformations and the Routh array
(b) Using the Jury test

25.13 For the first-order system with the transfer function:

$$g(s) = \frac{0.66}{6.7s + 1} \qquad \text{(P25.10)}$$

a continuous-time-domain design has yielded a PI controller with the following parameters:

$$K_c = \frac{6.7}{0.66\lambda}$$

$$\tau_I = 6.7$$

If this controller is now implemented in digital form, incorporating a ZOH element, and with $\Delta t = 1$, find the range of admissible values of the tuning parameter λ required for closed-loop stability. The transfer function for the digital PI controller is:

$$g_c(z) = K_c \left[1 + \frac{1}{\tau_I(1 - z^{-1})} \right] \qquad \text{(P25.11)}$$

25.14 A closed-loop system consists of a proportional controller, a ZOH element, and first-order-plus-time-delay system. The open-loop pulse transfer function between $y(z)$ and $u(z)$ (incorporating the ZOH) is given by:

$$g(z) = \frac{(bz^{-1})z^{-m}}{1 - \phi z^{-1}} \qquad \text{(P25.12)}$$

(a) What is the characteristic equation for this closed-loop system?
(b) For the specific case in which $b = 0.09$, and $\phi = 0.86$, find the maximum value of K_c for which the closed-loop system remains stable if $m = 0$.
(c) Repeat part (b) for $m = 1$, and for $m = 2$. Comment on the effect of time delays on the range of K_c values required for closed-loop stability.

25.15 The process shown below in Figure P25.3, first introduced in Problem 8.5, is the production end of a certain polymer manufacturing facility.

The main objective is to control the *normalized* melt viscosity of the polymer product by adjusting the ratio of F_T to F_S. The pertinent process variables are defined as follows, *in terms of deviations from appropriate steady-state values*:

u = ratio of the transfer agent flow to inert solvent flow (F_T/F_S)
v = concentration of transfer agent measured at the inlet of the connecting pipe
w = concentration of transfer agent measured at the reactor inlet
y = normalized melt viscosity of polymer product at reactor exit

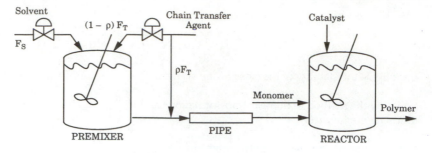

Figure P25.3.

The following transfer functions have been given for each portion of the process:

Premixer: $\qquad g_1(s) \quad = \quad \dfrac{5s + 1}{10s + 1}$

Connecting pipe: $\quad g_2(s) \quad = \quad e^{-6s}$

Reactor: $\qquad g_3(s) \quad = \quad \dfrac{25}{12s + 1}$

the time unit is in *minutes*, and all the other units are normalized units conforming with industry standards for this product.

(a) We want to control this process using a digital controller, $g_c(z)$, incorporating a ZOH. Draw a block diagram of the type in Figure 25.12 for this process and obtain the closed-loop transfer function.

(b) Under proportional control with $\Delta t = 3$ min, find the maximum value of K_c for which the closed-loop system remains stable.

(c) By increasing the sample time to $\Delta t = 6$ min, the delay is reduced to one sample time. Repeat part (b) for this choice of the sample time. In terms of the range of K_c values required for closed-loop stability, is there an advantage to increasing the sample time in this manner in order to reduce the delay time?

26

DESIGN OF
DIGITAL CONTROLLERS

The major distinction between conventional analog and digital computer control systems is that the continuous, conventional controller is restricted by the available hardware elements to take on simple, prespecified forms, such as the P, PI, or PID; by contrast, the digital controller has virtually unlimited flexibility in the possible forms a control algorithm can take.

This chapter is concerned with how digital controllers are designed. We start with an introduction to the digital controller itself, along with a discussion of the two fundamental digital controller design approaches: the *indirect* and the *direct*. The indirect approach involves constructing discrete approximations to continuous controllers obtained by continuous design methodologies; discrete PID controllers and the others that fall under this category are discussed first. With the direct approach, a completely discrete representation is used for the process, and controller design leads directly to discrete controllers. The design techniques in this category usually take full advantage of the latitude offered by the digital computer and therefore quite often result in controllers that are more sophisticated in structure than the discretized classical controllers. The chapter concludes with a discussion of issues involved when designing digital controllers for multivariable processes.

26.1 PRELIMINARY CONSIDERATIONS

As we have done elsewhere in this book, unless otherwise stated, we shall consider that the dynamics of the final control element and the measuring device have all been combined with the dynamics of the process into one overall process transfer function to be designated as $g(s)$. The input and output of this "overall process" shall be designated as $u(t)$, and $y(t)$, respectively, with corresponding Laplace transforms $u(s)$ and $y(s)$. This is usually the case in practice because in identifying a process model by input/output data correlation, it is usually difficult to separate out the contributions of the valve dynamics, and those of the measuring device, from the collected input/output data; thus it is precisely this overall transfer function, $g(s)$, that is identified.

Under digital feedback control, recall that the continuous input, $u(t)$, received by the overall process is actually a discrete signal from the discrete controller that has been "made continuous" by the hold element; in particular, with a ZOH element, $u(t)$ will be the piecewise constant input signal generated from the discrete sequence $u(k)$, the control command signal from the controller. The controller itself is considered as receiving discrete error information, $\varepsilon(k)$, the difference between the sampled, desired set-point value, $y_d(k)$, and the sampled process output, $y(k)$.

This digital feedback control system is shown in Figure 26.1(a). (The sampler at the input of the ZOH element is not explicitly shown because the block represents a combination sample-and-hold device.) Observe that with the exception of $u(t)$ connecting the hold element and the overall process, every other signal indicated in this diagram is sampled. If we now combine the ZOH element and the process, and refer to the composite transfer function as $g_H(s)$, i.e.:

$$g_H(s) = H(s)\, g(s)$$

$$= \left(\frac{1 - e^{-s\Delta t}}{s}\right) g(s) \tag{26.1}$$

then

$$y(s) = g_H(s)\, u*(s) \tag{26.2}$$

and by taking "star" transforms, we obtain:

$$y*(s) = g_H^*(s)\, u*(s) \tag{26.3}$$

and the equivalent z-transform representation is easily obtained:

$$y(z) = g(z)\, u(z) \tag{26.4}$$

With this maneuver, all the explicitly indicated signals will now be discrete, and Figure 26.1(a) can be represented in the completely equivalent form shown in Figure 26.1(b). Once again note that we use $g(z)$ with no subscripts to denote the discrete-time process having a zero-order hold, because this is the *true* equivalent of $g(s)$ in control system design.

Henceforth in our discussion, a reference to the "process" is to be understood to mean that entity represented by g, the overall process (plus valve and measuring device dynamics) incorporating the ZOH element. In a digital feedback control system, the only other component in the loop (as indicated in Figure 26.1(b)) will be the digital controller g_c with discrete input $\varepsilon(k)$ and discrete output $u(k)$.

With these ideas in mind, let us now proceed to the task at hand: the design of digital controllers.

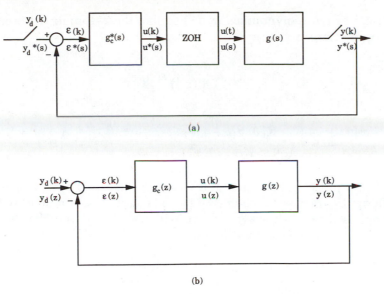

(a)

(b)

Figure 26.1. Two equivalent forms of a digital feedback control system.

26.2   THE DIGITAL CONTROLLER AND ITS DESIGN

We begin our study of the issues involved in the design of digital controllers with an introduction to the digital controller itself, and the philosophies behind its design.

26.2.1 The Digital Controller

A digital controller accepts a discrete sequence of error signals, $\varepsilon(k)$, and computes the discrete sequence of corrective control action, $u(k)$, according to a discrete control law that may be represented in the time domain, or in the z-domain.

z-domain Representation

In the z-domain, the controller has the following pulse transfer function representation (see Figure 26.1(b)):

$$u(z) \;=\; g_c(z)\, \varepsilon(z) \tag{26.5}$$

In its most general form, the digital controller transfer function, $g_c(z)$, has the form:

$$
\begin{aligned}
g_c(z) &= \frac{\Theta(z^{-1})}{\Psi(z^{-1})} \\[1ex]
&= \frac{\theta_0 + \theta_1 z^{-1} + \theta_2 z^{-2} + \ldots + \theta_r z^{-r}}{1 + \psi_1 z^{-1} + \psi_2 z^{-2} + \ldots + \psi_q z^{-q}}
\end{aligned}
\tag{26.6}
$$

(i.e., a ratio of two polynomials in z^{-1}) so that the z-domain discrete control law takes the form:

$$u(z) = \frac{\Theta(z^{-1})}{\Psi(z^{-1})} \, \varepsilon(z) \tag{26.7a}$$

or

$$u(z) = \left(\frac{\theta_0 + \theta_1 z^{-1} + \theta_2 z^{-2} + \ldots + \theta_r z^{-r}}{1 + \psi_1 z^{-1} + \psi_2 z^{-2} + \ldots + \psi_q z^{-q}} \right) \varepsilon(z) \tag{26.7b}$$

As with the pulse transfer functions considered in Chapter 25, the digital controller pulse transfer function could also be written as a ratio of polynomials in z as opposed to z^{-1}; however, as we will see shortly, the "backward" shift form in Eq. (26.6) is the more useful form.

Poles and Zeros of the Digital Controller

From our discussion in Section 25.2 in Chapter 25, we note that the q roots of the polynomial equation:

$$\Psi(z^{-1}) = 1 + \psi_1 z^{-1} + \psi_2 z^{-2} + \ldots + \psi_q z^{-q} = 0 \tag{26.8}$$

are the *poles* of the digital controller, while the r roots of:

$$\Theta(z^{-1}) = \theta_0 + \theta_1 z^{-1} + \theta_2 z^{-2} + \ldots + \theta_r z^{-r} = 0 \tag{26.9}$$

are the *zeros* of the digital controller. Again as with other transfer functions, how the controller output $u(k)$ responds to process error signal $\varepsilon(k)$ is determined by the nature of these poles and zeros.

Realization of the Digital Controller

While z-domain representations might sometimes be convenient for analysis, let us keep in mind that the digital controller must be physically implemented in the time domain. Thus, when the control law is given in the z-domain form as in Eq. (26.7), physical implementation requires that we obtain the equivalent time-domain representation; this is sometimes known as a time-domain *realization*.

The procedure described in Section 24.4.4 of Chapter 24 may be used for deriving the time-domain realization of Eq. (26.7) as follows. First rearrange to obtain:

$$\Psi(z^{-1}) u(z) = \Theta(z^{-1}) \, \varepsilon(z) \tag{26.10}$$

or

$$\left(1 + \psi_1 z^{-1} + \psi_2 z^{-2} + \ldots + \psi_q z^{-q} \right) u(z) =$$
$$\left(\theta_0 + \theta_1 z^{-1} + \theta_2 z^{-2} + \ldots + \theta_r z^{-r} \right) \varepsilon(z) \tag{26.11}$$

Then by taking inverse z-transform in the usual manner, we obtain the time-domain realization:

$$u(k) + \psi_1 u(k-1) + \psi_2 u(k-2) + \ldots + \psi_q u(k-q) =$$
$$\theta_0 \varepsilon(k) + \theta_1 \varepsilon(k-1) + \theta_2 \varepsilon(k-2) + \ldots + \theta_r \varepsilon(k-r) \qquad (26.12)$$

which may also be written in the recursive form:

$$u(k) = -\psi_1 u(k-1) - \psi_2 u(k-2) - \ldots - \psi_q u(k-q) +$$
$$\theta_0 \varepsilon(k) + \theta_1 \varepsilon(k-1) + \theta_2 \varepsilon(k-2) + \ldots + \theta_r \varepsilon(k-r) \qquad (26.13)$$

Let us note the following about this time-domain form of the discrete controller:

1. It is more straightforward to obtain if the controller transfer function is given in the backward shift form.

2. It consists of two distinct parts: the first part involves *past* values of the control sequence, while the second part involves *current* and *past* values of the error sequence.

3. Because the current value, $u(k)$, depends on past values of the control sequence in Eq. (26.13), this portion of the control law may be considered the "autoregressive" part; the second part is a "moving average" of a total of $r + 1$ consecutive values of the error sequence (the current error and r past values); this part may be considered as the "moving average" part. Such terminology is motivated by the striking resemblance that Eqs. (26.10) and (26.13) bear with the ARMA (autoregressive, moving average) statistical models of time series analysis.

4. The contribution from the "autoregressive" part is determined by ψ_i, $i = 1, 2, \ldots, q$, which are directly related to the poles of $g_c(z)$; the contribution from the "moving average" part, on the other hand, is determined by θ_i, $i = 0, 1, 2, \ldots, r$, which are directly related to the zeros of $g_c(z)$. Thus the controller transfer function poles influence the dependence of $u(k)$ on the past values of the control sequence, while the zeros influence its dependence on the error sequence.

Physical Realizability

It has already been observed that as presented in Eq. (26.12), or Eq. (26.13), the digital control law requires the determination of $u(k)$, the current control action, on the basis of only current and past information. Had the control law involved terms containing $\varepsilon(k + i)$, where $i > 0$, this would have required that current control action be calculated on the basis of *future* (and currently unavailable) information. Under such conditions, the controller is said to be physically unrealizable.

The condition of physical realizability may be stated as follows:

*The digital controller whose transfer function is as given in Eq. (26.6) is physically realizable if and only if the highest power of z in the numerator polynomial $\Theta(z^{-1})$ is **less than or equal to** the highest power of z in the denominator polynomial, $\Psi(z^{-1})$.*

Thus, if R_N represents the *highest power of z* in the numerator polynomial of $g_c(z)$, and R_D represents the *highest power of z* in the denominator polynomial, then for $g_c(z)$ to be physically realizable requires that:

$$R_N \leq R_D \tag{26.14}$$

It is very important to note that the realizability condition can be checked directly from the backward shift form; it is not necessary to reexpress $g_c(z)$ in the forward shift form. Let us illustrate with the following examples.

Example 26.1   **PHYSICAL REALIZABILITY OF VARIOUS DIGITAL CONTROLLERS.**

Test the following digital controller pulse transfer functions for physical realizability.

(a) $$g_c(z) = \frac{3 + 2z^{-1} + 3z^{-2}}{1 + 4z^{-1}} \tag{26.15}$$

(b) $$g_c(z) = \frac{z + 3 + 2z^{-1} + 3z^{-2}}{1 + 4z^{-1}} \tag{26.16}$$

(c) $$g_c(z) = \frac{2z^{-1} + 3z^{-2}}{1 + 4z^{-1}} \tag{26.17}$$

(d) $$g_c(z) = \frac{z^2 + z + 3 + 2z^{-1} + 3z^{-2}}{z^2 + 1 + 4z^{-1}} \tag{26.18}$$

Solution

(a)   The highest power of z in the numerator of Eq. (26.15) is 0 (note that the constant 3 is the same as $3z^0$), i.e., $R_N = 0$; the highest power in the denominator is also 0, i.e., $R_D = 0$. Thus, this controller is physically realizable, because $R_N = R_D$. Note that this controller is a special case of the one presented in Eq. (26.7). A time-domain realization of Eq. (26.15) is easily obtained as:

$$u(k) = -4u(k-1) + 3\varepsilon(k) + 2\varepsilon(k-1) + 3\varepsilon(k-2) \tag{26.19}$$

which does not require any future information.

(b)   For the controller in Eq. (26.16), observe that the highest power of z in the numerator is 1, i.e., $R_N = 1$, while for the denominator, $R_D = 0$. Since in this case $R_N > R_D$, the physical realizability condition is violated, and the controller is *not* realizable. To confirm this conclusion, the time-domain realization of Eq. (26.16) may be obtained in the usual manner; the result is:

$$u(k) = -4u(k-1) + \varepsilon(k+1) + 3\varepsilon(k) + 2\varepsilon(k-1) + 3\varepsilon(k-2) \tag{26.20}$$

which is immediately recognized as being unrealizable because of the presence of the $\varepsilon(k+1)$ term.

(c) For the controller in Eq. (26.17), $R_N = -1$ while $R_D = 0$; since in this case $R_N < R_D$, we conclude that the controller is realizable. The time-domain realization in this case is easily obtained as:

$$u(k) = -4u(k-1) + 2\varepsilon(k-1) + 3\varepsilon(k-2) \qquad (26.21)$$

which is, of course, realizable.

(d) For this controller, $R_N = 2$ and also $R_D = 2$ and since $R_N = R_D$, we conclude that the controller is realizable. The time-domain realization in this case might appear confusing at first; observe that upon the usual inversion back into the time domain, we obtain, from Eq. (26.18):

$$u(k+2) + u(k) = -4u(k-1) + \varepsilon(k+2) + \varepsilon(k+1) + 3\varepsilon(k) + 2\varepsilon(k-1) + 3\varepsilon(k-2)$$

which, upon first examination appears to contain terms requiring future information. However, by shifting the time index in each term back by two, we obtain the entirely equivalent form:

$$u(k) + u(k-2) = -4u(k-3) + \varepsilon(k) + \varepsilon(k-1) + 3\varepsilon(k-2) + 2\varepsilon(k-3) + 3\varepsilon(k-4)$$

or

$$u(k) = -u(k-2) - 4u(k-3) + \varepsilon(k) + \varepsilon(k-1) + 3\varepsilon(k-2) + 2\varepsilon(k-3) + 3\varepsilon(k-4) \quad (26.22)$$

which no longer involves terms requiring future information. Note also that multiplying the original pulse transfer function by z^{-2}/z^{-2} would have eliminated the positive powers of z, and the realization in Eq. (26.22) could have been obtained directly.

26.2.2 Two Approaches to Digital Controller Design

Fundamentally, the objectives of digital controller design are no different from those of continuous controller design: in each case the aim is to obtain a controller that satisfies some prespecified performance criteria (typically that the closed-loop system be stable, that there be no steady-state offsets, and that the dynamic response behavior be "acceptable"). In digital controller design, however, the end result is a *discrete* controller.

This discrete controller can be obtained in one of two ways:

1. Carry out the familiar continuous domain design and then discretize the resulting continuous controller, or
2. Carry out the design directly in discrete time, and obtain the discrete controller directly.

The philosophy behind the first approach is as follows: with sampled-data control systems, even though we have a truly discrete controller with a discretized view of the process, the process environment is still fundamentally continuous; thus a useful discrete controller can be obtained by *carefully* constructing the discrete equivalent of a controller designed for the continuous process. The primary focus in this approach is the continuous system; the discrete nature of the controller is secondary, an implementation detail.

This is the *indirect* approach, since the discrete controller is obtained indirectly from a fundamentally continuous design, and has the advantage of relying on controller design principles with which we are already familiar. The main disadvantage, as we shall soon show, lies in the fact that controllers designed this way may not perform as well as the continuous controllers from which they were obtained. These controllers are also limited in form to discretized versions of continuous controllers.

With the second approach, the design perspective is switched, and the primary focus is now the *discrete* controller with its discretized view of the process; the operational representation of the control system is now Figure 26.1(b), where *everything* appears discrete. It is the *direct* approach because the design yields a discrete controller directly. It has two main advantages: all the effects of sampling and discretization are already directly incorporated into the design, and the resulting controllers are only limited in form by realizability conditions. Thus there is broad latitude in the designs that can result from this approach.

The main disadvantage of the direct approach is that the fundamentally continuous nature of the process is not easily taken into account in this strictly discrete-time approach; it is therefore possible to obtain closed-loop system behavior that meets performance objectives at sampling points while exhibiting undesired behavior between samples.

26.2.3 From Continuous Time to Discrete Time

The transition from a continuous closed-loop control system to a sampled-data, computer control system involves certain issues with very important implications for digital controller design.

The Effect of Sampling and Holding

Qualitatively, the application of control action on a piecewise continuous basis, and the intermittent sampling of the process output, affect the control loop in the following manner:

1. In between samples, the control loop actually operates in "open-loop" fashion, with the previously applied control action held constant by the ZOH device until the next sampling point when the input "jumps" to another value at which it is, again, held for the sampling duration. Thus, the behavior of a sampled-data system is, in actual fact, exactly the same as a series of open-loop step responses.

2. No intersample process information is available to the controller. Such loss of process information may become critical if the process output signal, and the frequency at which it is sampled, are such that unacceptable intersample behavior goes largely undetected at sampling points.

Quantitatively, as was discussed in Chapter 23, the effect of the sample-and-hold device on the sampled-data system is approximately equal to adding a time delay of $\Delta t/2$ to the dynamics of the process. To illustrate, let us consider the sinusoidal function shown in solid lines in Figure 26.2; sampling this function at intervals of Δt time units and holding the sampled value constant over the sampling period produces the indicated stepwise continuous function. Observe that the sampled function "appears" as a shifted version of the original continuous function, which may be *approximated* by the dashed line.

This fact may be demonstrated analytically by considering the following approximation to the transfer function of the ZOH device:

$$\frac{1 - e^{-s\Delta t}}{s} \approx \frac{1}{s}\left(1 - 1 + s\Delta t - \frac{(s\Delta t)^2}{2!} + \cdots\right)$$

$$= \Delta t\left(1 - \frac{s\Delta t}{2} + \cdots\right) \tag{26.23}$$

Note that the first two terms in the parenthesis in Eq. (26.23) correspond to a series expansion of $e^{-s\Delta t/2}$, a time delay of magnitude $\Delta t/2$. The implications on closed-loop stability are significant, and may be summarized as follows:

> *The additional, effective time delay contributed to the dynamics of the sampled process by the sample-and-hold device adds $\omega\Delta t/2$ radians (or $180\omega/\omega_s$ degrees) to the process phase lag, thereby reducing the* **phase margin** *by $\omega_{co}\Delta t/2$ radians (or $180\omega_{co}/\omega_s$ degrees), where ω_{co} is the crossover frequency, and ω_s is the sampling frequency.*

This factor is singlehandedly responsible for the closed-loop stability properties observed in Example 25.8 of Chapter 25 for the first-order system under digital proportional control.

The Advantages of Digital Computer Control

One might be tempted from the foregoing to conclude that a closed-loop control system will always perform better under continuous control than under digital control. This is far from true, however. Digital control has one single, overwhelming advantage over conventional control: the computational power and versatility of the digital computer allows the implementation of control strategies that are impossible to implement with conventional analog controllers. Thus, with a proper understanding of the pitfalls associated with

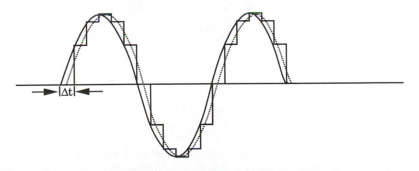

Figure 26.2. The introduction of an effective delay of $\Delta t/2$ by the sample-and-hold operation applied to a continuous signal.

digital process control, it is possible to exploit its advantages of available computing power and versatility in designing sampled-data control systems that will consistently outperform the conventional types.

26.3 DISCRETE PID CONTROLLERS FROM THE CONTINUOUS DOMAIN

The simplest, and the most familiar, digital controllers are the discrete PID controllers. These arise from approximating classical continuous PID controllers. Thus let us begin our study of digital controller design by investigating the form these controllers take, and the strategies available for their design.

26.3.1 Digital Approximation of the Analog PID Controller

The continuous analog PID controller, as we may recall, operates according to the following equation in the time domain:

$$p(t) = K_c \left[\varepsilon(t) + \frac{1}{\tau_I} \int_0^t \varepsilon(t) \, dt + \tau_D \frac{d\varepsilon}{dt} \right] + p_s \qquad (26.24)$$

where p_s is the controller output when $\varepsilon(t)$ is zero. By defining $u(t)$ as the deviation variable $(p(t) - p_s)$, Laplace transformation in Eq. (26.24) gives:

$$u(s) = g_c(s) \, \varepsilon(s) \qquad (26.25a)$$

where the indicated controller transfer function, as we well know, is given by:

$$g_c(s) = K_c \left(1 + \frac{1}{\tau_I \, s} + \tau_D \, s \right) \qquad (26.25b)$$

There are various ways of obtaining digital approximations to this controller. A popular method involves approximating the integral with the rectangular rule for numerical integration, and approximating the derivative with a finite difference, i.e.:

$$\int_0^t \varepsilon(t) \, dt \approx \sum_{i=1}^k \varepsilon(i) \, \Delta t \qquad (26.26a)$$

and

$$\frac{d\varepsilon}{dt} \approx \frac{\varepsilon(k) - \varepsilon(k-1)}{\Delta t} \qquad (26.26b)$$

The digital PID controller may be obtained by discretizing Eq. (26.24), using Eq. (26.26). Depending on how the result is presented, this digital controller can take one of two distinct forms: the *position* form, or the *velocity* form.

The Position Form

By discretizing Eq. (26.24) and substituting Eq. (26.26) directly we have:

$$u(k) = K_c \left\{ \varepsilon(k) + \frac{\Delta t}{\tau_I} \sum_{i=1}^{k} \varepsilon(i) + \frac{\tau_D}{\Delta t} [\varepsilon(k) - \varepsilon(k-1)] \right\} \qquad (26.27)$$

This is known as the *position* form because at each sampling point, this control law gives the actual value (position) of the controller output directly.

The pulse transfer function for this controller may be obtained by taking z-transforms in Eq. (26.27). Observe that this will involve the z-transform of the indicated finite sum; let us employ the following device in simplifying this aspect of the z-transform.

Let $S(k)$ represent the k term error sequence sum, i.e.:

$$S(k) = \sum_{i=1}^{k} \varepsilon(i) \qquad (26.28)$$

then, clearly:

$$S(k-1) = \sum_{i=1}^{k-1} \varepsilon(i) \qquad (26.29)$$

and by subtracting Eq. (26.29) from Eq. (26.28), we obtain straightforwardly, that:

$$S(k) = S(k-1) + \varepsilon(k) \qquad (26.30)$$

If we now take z-transforms, we obtain:

$$S(z) = z^{-1}S(z) + \varepsilon(z)$$

easily rearranging to:

$$S(z) = \left(\frac{1}{1 - z^{-1}} \right) \varepsilon(z) \qquad (26.31)$$

We may now take z-transforms in Eq. (26.27), and immediately obtain:

$$u(z) = K_c \left[\varepsilon(z) + \frac{\Delta t}{\tau_I} \left(\frac{1}{1 - z^{-1}} \right) \varepsilon(z) + \frac{\tau_D}{\Delta t} (1 - z^{-1})\varepsilon(z) \right] \qquad (26.32)$$

and by factoring out the common $\varepsilon(z)$ term, we obtain the pulse transfer function representation:

$$u(z) = g_c(z)\varepsilon(z) \qquad (26.33)$$

with the controller pulse transfer function given by:

$$g_c(z) = K_c \left[1 + \frac{\Delta t}{\tau_I} \left(\frac{1}{1 - z^{-1}} \right) + \frac{\tau_D}{\Delta t} (1 - z^{-1}) \right] \qquad (26.34)$$

COMPARISON WITH THE GENERAL DIGITAL CONTROLLER

The digital PID controller given in Eq. (26.27) may be expanded and rearranged to yield:

$$u(k) = K_c \left(1 + \frac{\tau_D}{\Delta t} + \frac{\Delta t}{\tau_I} \right) \varepsilon(k) + \left(\frac{K_c \Delta t}{\tau_I} - \frac{K_c \tau_D}{\Delta t} \right) \varepsilon(k - 1)$$

$$+ \frac{K_c \Delta t}{\tau_I} [\varepsilon(k - 2) + \varepsilon(k - 3) + \dots + \varepsilon(1)] \qquad (26.35)$$

which takes the general form of a digital controller given in Eq. (26.13), with the following parameters:

$$\psi_i = 0;$$

$$\theta_0 = K_c \left(1 + \frac{\tau_D}{\Delta t} + \frac{\Delta t}{\tau_I} \right); \ \theta_1 = K_c \left(\frac{\Delta t}{\tau_I} - \frac{\tau_D}{\Delta t} \right); \text{ and}$$

$$\theta_i = \frac{K_c \Delta t}{\tau_I}; \ i = 2, 3, \dots, k - 1$$

Thus, the digital PID controller is seen to be a special case of the general digital controller presented earlier.

The Velocity Form

Observe that by shifting the discrete-time index appropriately, we obtain from Eq. (26.27) that $u(k - 1)$ is given by:

$$u(k - 1) = K_c \left\{ \varepsilon(k - 1) + \frac{\Delta t}{\tau_I} \sum_{i=1}^{k-1} \varepsilon(i) + \frac{\tau_D}{\Delta t} [\varepsilon(k - 1) - \varepsilon(k - 2)] \right\} \qquad (26.36)$$

and by subtracting Eq. (26.36) from Eq. (26.27), we obtain:

$$\Delta u(k) = K_c \left\{ [\varepsilon(k) - \varepsilon(k - 1)] + \frac{\Delta t}{\tau_I} \varepsilon(k) + \frac{\tau_D}{\Delta t} [\varepsilon(k) - 2\varepsilon(k - 1) + \varepsilon(k - 2)] \right\}$$

$$(26.37)$$

where

$$\Delta u(k) = u(k) - u(k - 1) \qquad (26.38)$$

This may be further rearranged to give:

$$\Delta u(k) = K_c \left[\left(1 + \frac{\Delta t}{\tau_I} + \frac{\tau_D}{\Delta t} \right) \varepsilon(k) - \left(\frac{2\tau_D}{\Delta t} + 1 \right) \varepsilon(k - 1) + \frac{\tau_D}{\Delta t} \varepsilon(k - 2) \right] \qquad (26.39)$$

This is known as the *velocity* form, since it represents the "change" in the *position* of the control variable.

The pulse transfer function representation of the velocity form is easily obtained by taking z-transforms in Eq. (26.39); the result is:

$$\Delta u(z) = K_c \left[\left(1 + \frac{\Delta t}{\tau_I} + \frac{\tau_D}{\Delta t} \right) - \left(\frac{2\tau_D}{\Delta t} + 1 \right) z^{-1} + \left(\frac{\tau_D}{\Delta t} \right) z^{-2} \right] \varepsilon(z) \qquad (26.40)$$

or, if we wish to represent this in terms of $u(z)$, instead of $\Delta u(z)$, we can take the z-transform of Eq. (26.38) to yield:

$$u(z) = \frac{\Delta u(z)}{1 - z^{-1}}$$

so that:

$$u(z) = g_c(z)\varepsilon(z)$$

with $g_c(z)$ given by:

$$g_c(z) = K_c \left[\frac{\left(1 + \frac{\Delta t}{\tau_I} + \frac{\tau_D}{\Delta t} \right) - \left(\frac{2\tau_D}{\Delta t} + 1 \right) z^{-1} + \left(\frac{\tau_D}{\Delta t} \right) z^{-2}}{(1 - z^{-1})} \right] \qquad (26.41)$$

which is precisely the same expression one obtains by rearranging Eq. (26.34).

COMPARISON WITH THE GENERAL DIGITAL CONTROLLER

If we now compare Eqs. (26.39) or (26.41) with the corresponding expressions Eqs. (26.13) and (26.6), we see that the velocity form for the digital PID controller is a special case of the general digital controller, with the following parameters:

$$\psi_1 = -1; \qquad \psi_i = 0; \ i > 1 \qquad \text{along with:}$$

$$\theta_0 = K_c \left(1 + \frac{\tau_D}{\Delta t} + \frac{\Delta t}{\tau_I} \right); \ \theta_1 = -K_c \left(\frac{2\tau_D}{\Delta t} + 1 \right);$$

$$\theta_2 = K_c \left(\frac{\tau_D}{\Delta t} \right); \ \theta_i = 0; \ i > 2$$

The following points may now be noted about the digital PID controller:

1. It is physically realizable in either the position or the velocity form.

2. If the approximations used for the integral and the derivative in Eq. (26.24) had been different (say, for example, we had used the trapezoidal rule for the integral, and the backward difference for the derivative) completely different expressions would have been obtained for the digital PID controller.

3. The velocity form has certain advantages over the position form; the most obvious is that it eliminates problems of reset windup.

A discussion of several operational aspects of the digital PID controller may be found in Ref. [1], p. 183, and Ref. [2], pp. 616–619.

26.3.2 Digital PID Controller Tuning

The most common approach to tuning digital PID controllers is to employ the same heuristic rules used for tuning continuous PID controllers, i.e., the Cohen and Coon, or the Ziegler-Nichols tuning formulas. The fundamental justification is that for small sampling times, the digital PID controller behaves much like the continuous controller; thus continuous controller settings may be converted for use with the digital version.

However, it is important to remember that the inclusion of the ZOH adds an effective time delay of $\Delta t/2$ to the process time delay. For the situations when Δt is *not* small, it is necessary to adjust the process time delay from α to $(\alpha + \Delta t/2)$ before carrying out the continuous design; this will ensure that when the continuous controller is discretized and implemented in digital form, the destabilizing effect of sampling will be taken into account. It should be noted that if the process reaction curve is generated from the sampled-data process before using the Cohen and Coon tuning method, the effect of the sample-and-hold operation will already be included in the experimentally generated data.

Thus, we have that:

> *In general, digital PID controllers are tuned using the same classical techniques used for tuning continuous PID controllers, but with additional compensation for the effect of sampling time delay.*

Regardless of the method used for tuning, it is often good practice to have a first evaluation of the controller performance via simulation whenever possible.

26.4 OTHER DIGITAL CONTROLLERS BASED ON CONTINUOUS DOMAIN STRATEGIES

The strategy of designing digital controllers for sampled-data systems by converting continuous controllers originally designed for continuous systems is not limited to PID controllers alone. In this section, we take a more general approach to the subject: we pose the problem as that of obtaining digital approximations to any arbitrary s-transfer function; the approximation formulas generated thereby may then be used for the *specific* application of converting continuous controller transfer functions (obtained via continuous domain designs) to digital forms for use with sampled-data systems.

26.4.1 Techniques for Digital Approximation of Continuous Transfer Functions

In Chapter 25, we established that the Laplace transform variable s (on which continuous transfer functions are based) and the z-transform variable (on which discrete pulse transfer functions are based) are related by:

$$z^{-1} = e^{-s\Delta t} \tag{26.42}$$

In using this to obtain the discrete equivalent of an s-transfer function, $g(s)$, it should be recalled that the procedure calls for a Laplace inversion to obtain $g(t)$, followed by discretization, to obtain $g^*(t)$; only after this step can we take z-transforms, a process which we have shown to be exactly equivalent to taking Laplace transforms to obtain $g^*(s)$, and then converting to the z-domain via Eq. (26.42).

Such a procedure provides exact discrete-time equivalents, but it is clearly very tedious. It turns out, however, that by replacing Eq. (26.42) with approximations for $e^{-s\Delta t}$, we obtain relations between s and z that may be used to convert from the s-domain to the z-domain by direct substitution. Three of the most common approximants are now presented below:

1. *Expansion of* $e^{-s\Delta t}$: A series expansion for $e^{-s\Delta t}$ truncated after the first two terms gives:

$$e^{-s\Delta t} \approx 1 - s\Delta t$$

so that, Eq. (26.42) may be approximated as:

$$z^{-1} \approx 1 - s\Delta t$$

which rearranges to give:

$$s \approx \frac{1 - z^{-1}}{\Delta t} \tag{26.43}$$

2. *Expansion of* $1/e^{s\Delta t}$: An alternative approximation may be obtained from a series expansion for the reciprocal function $e^{s\Delta t}$ instead, viz.:

$$e^{s\Delta t} \approx 1 + s\Delta t \tag{26.44}$$

so that $e^{-s\Delta t}$ is approximated by the reciprocal of the right-hand side of Eq. (26.44), with the result that:

$$z^{-1} \approx \frac{1}{1 + s\Delta t}$$

which rearranges to give:

$$s \approx \frac{1 - z^{-1}}{z^{-1}\Delta t} \tag{26.45}$$

3. *Padé Approximation*: A final alternative may be obtained by using a first-order Padé approximation for $e^{-s\Delta t}$, i.e.:

$$e^{-s\Delta t} \approx \frac{1 - s\Delta t/2}{1 + s\Delta t/2}$$

or

$$e^{-s\Delta t} \approx \frac{2 - s\Delta t}{2 + s\Delta t} \tag{26.46}$$

so that:

$$z^{-1} \approx \frac{2 - s\Delta t}{2 + s\Delta t} \tag{26.47}$$

and solving for s in terms of z, we obtain the Tustin approximation [3]:

$$s \approx \frac{2}{\Delta t} \left(\frac{1 - z^{-1}}{1 + z^{-1}} \right)$$

(26.48)

A more rigorous justification for these approximating formulas may be obtained by considering the problem of numerically integrating the differential equation represented by the s-transfer function; if the *backward difference* approach is used, the approximation in Eq. (26.43) is the result; if the *forward difference* approach is used, the approximation in Eq. (26.45) is the result; and if the *trapezoidal numerical integration* approach is used, the approximation in Eq. (26.48) is the result.

The various properties of each of these approximants have been thoroughly investigated in Ogata (Ref. [3], pp. 308–317). We merely note that Tustin's approximation (Eq. (26.48)) is the most accurate; it is also the most complicated to use. The backward difference approximation in Eq. (26.43) is easy to use; and, for a stable $g(s)$, it will always produce a stable pulse transfer function.

The forward difference approximation in Eq. (26.45) (also known as Euler's approximation) is almost as easy to use as Eq. (26.43); however, it can be shown that because it maps certain regions of the left half of the s-plane into the exterior of the unit circle in the z-plane, a stable $g(s)$ might be replaced by an unstable pulse transfer function. (The reader with some familiarity with numerical methods will immediately see the connection with the well-known stability problems associated with Euler's method of solving ordinary differential equations.)

Thus it is advisable to use either the Tustin or the backward difference approximation; for simplicity we will use only the backward difference approximation Eq. (26.43) in what follows.

26.4.2 Translating Analog Controller Designs to Digital Equivalents

We will now use the backward difference conversion formula to translate some typical continuous controllers into digital equivalents.

The PID Controller

Given the PID controller, with the transfer function:

$$g_c(s) = K_c \left(1 + \frac{1}{\tau_I s} + \tau_D s \right)$$

by replacing s according to the expression Eq. (26.43), we obtain the approximate digital equivalent as:

$$g_c(z) = K_c \left[1 + \frac{\Delta t}{\tau_I} \frac{1}{1 - z^{-1}} + \frac{\tau_D}{\Delta t} (1 - z^{-1}) \right]$$

(26.49)

which is, in fact, identical to the transfer function obtained earlier in Eq. (26.34). The fact that Eq. (26.49), a verifiably appropriate approximation to $g_c(s)$, could be obtained so easily from the Laplace transfer function should not be lost on the reader; it is one of the reasons why this approach is very attractive.

The Direct Synthesis Controller

Recall that in Chapter 19, the general formula for generating the direct synthesis controller with a first-order reference trajectory was found to be:

$$g_c(s) = \frac{1}{\tau_r s} \frac{1}{g} \tag{19.6}$$

This is the controller which will cause the process with open-loop transfer function $g(s)$ to follow the closed-loop trajectory represented by a first-order lag with a time constant τ_r.

According to the approximation scheme in Eq. (26.43), the digital equivalent of this controller is:

$$g_c(z) = \frac{\Delta t}{\tau_r} \frac{1}{1 - z^{-1}} \left(\frac{1}{g} \right) \Bigg|_{s = \frac{1-z^{-1}}{\Delta t}} \tag{26.50}$$

To apply this in a specific situation is particularly simple: we take the process transfer function, $g(s)$, substitute $s = (1 - z^{-1})/\Delta t$, and insert the result in Eq. (26.50). The following example illustrates this procedure.

Example 26.2   DIGITAL APPROXIMATION TO THE DIRECT SYNTHESIS CONTROLLER FOR A FIRST-ORDER PROCESS.

Obtain the digital approximation for the continuous direct synthesis controller designed for a first-order system whose transfer function is given by :

$$g(s) = \frac{K}{\tau s + 1}$$

Solution:

Substituting $s = (1 - z^{-1})/\Delta t$ gives, upon some rearrangement:

$$g(s)\Big|_{s = \left(\frac{1-z^{-1}}{\Delta t} \right)} = \frac{\dfrac{K\Delta t}{\tau + \Delta t}}{1 - \left(\dfrac{\tau}{\tau + \Delta t} \right) z^{-1}} \tag{26.51}$$

and, by applying Eq. (26.50), we obtain the desired controller as:

$$g_c(z) = \frac{\tau + \Delta t}{K\tau_r} \left(\frac{1 - b z^{-1}}{1 - z^{-1}} \right) \tag{26.52}$$

where

$$b = \frac{\tau}{\tau + \Delta t}$$

The time-domain realization of this controller is easily obtained as:

$$u(k) = u(k-1) + \left(\frac{\tau + \Delta t}{K\tau_r}\right)\varepsilon(k) - \left(\frac{\tau}{K\tau_r}\right)\varepsilon(k-1) \qquad (26.53)$$

which is recognized as a PI controller in the velocity form; it is, of course, also of the general digital controller form in Eq. (26.13).

The Digital Feedforward Controller

In Chapter 16, we obtained the result that the controller required for feedforward compensation is given by:

$$g_{ff}(s) = -\frac{g_d(s)}{g(s)}$$

where $g_d(s)$ is the disturbance transfer function, and $g(s)$ is the process transfer function. By simply substituting Eq. (26.43) for s in this expression, we would obtain $g_{ff}(z)$, the digital feedforward controller.

In particular, for the purpose of illustration, when $g_d(s)$ and $g(s)$ are assumed to take the first-order-plus-time-delay form, we recall that g_{ff} reduces to a lead/lag element, possibly with a time delay, i.e.:

$$g_{ff}(s) \approx \frac{K_{ff}(\tau s + 1)}{\tau_d s + 1} \, e^{-\gamma s} \qquad (16.20)$$

Assuming that the indicated time delay is an integral multiple of Δt, say, $\gamma = M\Delta t$, then, by observing that the substitution of Eq. (26.43) for $(\tau s + 1)$ results in something of the form $a(1 - bz^{-1})$, it is a trivial exercise to establish that the digital lead/lag form for the feedforward controller will be:

$$g_{ff}(z) \approx \frac{K_{ff} z^{-M}(1 - bz^{-1})}{(1 - cz^{-1})} \qquad (26.54)$$

It is left as an exercise for the reader to deduce what the constants K_{ff}, a, b, and c are. Note that even this digital feedforward controller transfer function is a special case of the general form given in Eq. (26.6).

The Discrete Smith Predictor Algorithm

The Smith predictor, first introduced in Chapter 17, is one of the earliest model-based control schemes. As we recall from the earlier discussion, with the continuous domain control law given by:

$$u(s) = \left[\frac{g_c}{1 + g_c g(1 - e^{-\alpha s})}\right]\varepsilon(s) \qquad (26.55)$$

the Smith predictor improves the closed-loop behavior of processes with transfer function $g\,e^{-\alpha s}$ by eliminating the delay from the closed loop. Here, g_c is the classical controller designed for g, the process *without* the delay.

It should be noted that this control law is not easy to implement in continuous time (cf. Ref. [1], p. 238, and Ref. [4], p. 228). By substituting Eq. (26.43) for s, however, Eq. (26.55) can be converted to a digital control law that is quite easy to implement on the digital computer. To accomplish this, first rearrange Eq. (26.55) to give:

$$u(s)\left[1 + g_c g(1 - e^{-\alpha s})\right] = g_c \varepsilon(s) \tag{26.56}$$

Let us now consider, for the purpose of illustration, the situation in which the process transfer function is of the first-order-plus-time-delay type, so that g, the transfer function *without* the delay, is:

$$g(s) = \frac{K}{\tau s + 1}$$

Further, suppose g_c has been designed for g such that:

$$g g_c = \frac{1}{\tau_r s}$$

(from the discussion in Chapter 19, we will recall that such a g_c will be a PI controller with parameters as indicated in Eq. (19.11)).

Under these conditions, Eq. (26.56) reduces to:

$$u(s)\left[1 + \frac{(1 - e^{-\alpha s})}{\tau_r s}\right] = \frac{1}{g\tau_r s}\,\varepsilon(s)$$

which rearranges to:

$$(\tau_r s + 1)\,u(s) - e^{-\alpha s}u(s) = \left(\frac{\tau s + 1}{K}\right)\varepsilon(s) \tag{26.57}$$

If we now substitute $s = (1 - z^{-1})/\Delta t$ from Eq. (26.43) and assume the time delay $\alpha = D\Delta t$, where D is an integer, the digital approximation to Eq. (26.57) is obtained straightforwardly as:

$$\left(\frac{\tau_r + \Delta t}{\Delta t}\right)u(z) - \frac{\tau_r}{\Delta t}z^{-1}u(z) - z^{-D}u(z) = \left(\frac{\tau + \Delta t}{K\Delta t}\right)\varepsilon(z) - \frac{\tau}{K\Delta t}z^{-1}\varepsilon(z) \tag{26.58}$$

with time-domain realization:

$$u(k) = \left(\frac{\tau_r}{\tau_r + \Delta t}\right)u(k-1) + \left(\frac{\Delta t}{\tau_r + \Delta t}\right)u(k-D) + \left[\frac{\tau + \Delta t}{K(\tau_r + \Delta t)}\right]\varepsilon(k)$$

$$- \left[\frac{\tau}{K(\tau_r + \Delta t)}\right]\varepsilon(k-1) \tag{26.59}$$

Note once again that this control law is of the general form given in Eq. (26.13) and is very easy to implement on the digital computer.

26.5 DIGITAL CONTROLLERS BASED ON DISCRETE DOMAIN STRATEGIES

As noted earlier, the problems with translating analog designs to digital equivalents and employing these for the control of sampled-data systems are twofold: (1) purely continuous designs fail to take the effect of the sample-and-hold operation into consideration; (2) the translation process involves approximations. Thus, these "translated" digital controllers may work well if the sampling time is short because sampling is faster and the approximations involved in the s and z variable substitutions will be more accurate; however, they are not likely to perform as well as the continuous originals.

It is possible to obtain better results by taking the *direct* approach, i.e., carrying out digital controller design within the discrete framework, either in the z-domain, or in the discrete-time domain. Some of the approaches for carrying out such designs will be discussed in this section.

26.5.1 Direct Synthesis in the z-Domain

Rather than approximating the continuous domain results derived in Chapter 19, as was done in the last section, let us borrow the concepts and use them to derive a digital direct synthesis controller from first principles.

Consider the sampled-data control system shown in Figure 26.1(b) which has the closed-loop transfer function between $y(z)$ and $y_d(z)$ as:

$$y(z) = \frac{g(z)\,g_c(z)}{1 + g(z)\,g_c(z)}\,y_d(z) \tag{26.60}$$

If we now represent the desired closed-loop behavior (the reference trajectory) by the pulse transfer function $y(z)/y_d(z) = q(z)$, then by following precisely the same arguments as presented in Section 19.1, we arrive at the conclusion that the digital controller which is required to produce the desired closed-loop response will be given by:

$$g_c(z) = \frac{1}{g(z)}\,\frac{q(z)}{1 - q(z)} \tag{26.61}$$

The choice of the desired closed-loop transfer function $q(z)$ is, of course, restricted by the fact that the resulting controller must be physically realizable.

It may be interesting to inject a historical note at this point. One might conclude from the current literature on model-based controllers that these direct synthesis ideas are relatively new and the concept of using reference trajectories, $q(z)$, and the process model inverse, $1/g(z)$, in designing high-performance controllers is a recent development. *In fact, this is a design procedure that is more than 35 years old*, for one can find the current philosophical discussion and even Eq. (26.61) in the 1958 textbook by Ragazzi and Franklin [5]! Given the state of digital computers in the mid-1950s, it is a wonder the controller designs were so advanced.

There are several different digital controller design techniques that fit into this general direct synthesis scheme; they differ only in their choice of reference trajectory, $q(z)$.

Deadbeat Controller

The deadbeat controller (cf. Ref. [6]) is based on the requirement that, for any set-point change, the closed-loop response should exhibit:

- Minimum rise time
- Zero steady-state offset
- Finite settling time

in other words, the closed-loop system is required to follow the set-point exactly at each sampling point once any period of unavoidable delay is past, i.e.:

$$y(z) = z^{-1}y_d(z) \qquad (26.62)$$

where the unit sampling delay has been incorporated. Thus, observe that this translates to requiring that $q(z)$ take the form:

$$q(z) = z^{-1} \qquad (26.63)$$

and the resulting deadbeat controller will be given by:

$$g_c(z) = \frac{1}{g(z)} \frac{z^{-1}}{1 - z^{-1}} \qquad (26.64)$$

When the process has a time delay of D sample periods, the expression for $q(z)$ is modified to account for the additional time delay to read:

$$q(z) = z^{-D-1} \qquad (26.65)$$

and the deadbeat controller will be:

$$g_c(z) = \frac{1}{g(z)} \frac{z^{-D-1}}{1 - z^{-D-1}} \qquad (26.66)$$

Thus, for example, the deadbeat controller for a first-order system with a ZOH will be obtained by using:

$$g(z) = \frac{K(1 - \phi)z^{-1}}{1 - \phi z^{-1}}$$

in Eq. (26.64), where we recall that $\phi = e^{-\Delta t/\tau}$. This provides the deadbeat controller:

$$g_c(z) = \frac{1 - \phi z^{-1}}{K(1 - \phi)(1 - z^{-1})} \qquad (26.67)$$

The time-domain realization of Eq. (26.67) is obtained in the usual fashion as:

$$u(k) = u(k-1) + \left[\frac{1}{K(1 - \phi)}\right]\varepsilon(k) - \left[\frac{\phi}{K(1 - \phi)}\right]\varepsilon(k-1) \qquad (26.68)$$

which, again has the form given in Eq. (26.13).

A very important point to note about the deadbeat controller (and indeed all other controllers based on the direct synthesis scheme) is that it depends on the inverse of $g(z)$; thus the zeros of $g(z)$ become the controller's poles.

This immediately creates a problem for processes with inverse response (zero outside the unit circle) because the process zero becomes an unstable pole in the controller. Thus deadbeat controllers cannot be used for processes with inverse response.

An interesting situation arises when the process $g(z)$ has a zero located at $z = -\beta$, where $0 < \beta < 1$, so that the zero, even though negative, lies within the unit circle. Under these conditions, this negative process zero becomes a negative controller pole. The contribution of such a pole to the dynamic response of the controller will be of the type 4 discussed in Chapter 25, and illustrated in Figure 25.5; consecutive values of the controller output will alternate in sign, a situation referred to as "ringing."

The effect of controller *ringing* on the process output $y(k)$ is to produce *intersample rippling*; i.e., oscillations in the output between samples. This can be illustrated with the following example.

Example 26.3   DIGITAL DEADBEAT CONTROLLER RINGING AND PROCESS OUTPUT RIPPLING.

Consider the continuous underdamped second-order process:

$$g(s) = \frac{1}{s^2 + s + 1} \tag{26.69}$$

If we convert this to a discrete-time process having a zero-order hold and sample time, $\Delta t = 0.3$, the pulse transfer function is:

$$g(z) = \frac{0.04052\ z^{-1} + 0.03665\ z^{-2}}{1 - 1.664\ z^{-1} + 0.7408\ z^{-2}} \tag{26.70}$$

Note that this has poles $p_i = 0.8318 \pm 0.221\ i$ and a zero at $z_i = -0.9045$.

Design a deadbeat controller and determine the closed-loop response to a unit step set-point change.

Solution:

Since there are no time delays (other than the unit delay due to the hold), the deadbeat controller requires a reference trajectory given by Eq. (26.63). Employing the design Eq. (26.64), one obtains the controller design:

$$g_c(z) = \left(\frac{z^{-1}}{1 - z^{-1}}\right)\left(\frac{1 - 1.664\ z^{-1} + 0.7408\ z^{-2}}{0.04052\ z^{-1} + 0.03665\ z^{-2}}\right) \tag{26.71}$$

which can be realized by inverting the pulse transfer function to:

$$u(k) = 24.68\ [0.0039\ u(k-1) + 0.03665\ u(k-2)$$
$$+ \ \varepsilon(k) - 1.664\ \varepsilon(k-1) + 0.7408\ \varepsilon(k-2)] \tag{26.72}$$

When this controller is applied to the process, the closed-loop response, in terms of the continuous and discrete outputs y, as well as the discrete input u, is shown in Figure 26.3. Note that the controller pole at $z_i = -0.9045$ (coming from the process

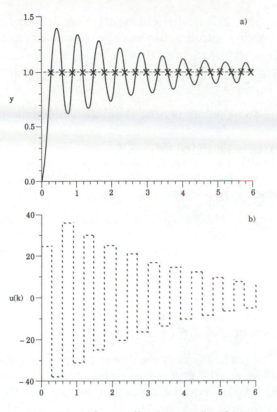

Figure 26.3.   An illustration of controller "ringing" and intersample "rippling."

zero) causes ringing of $u(k)$, and this produces the rippling of the continuous output between samples. Obviously if we did not measure the continuous output and had only the sampled output we would think the controller was outstanding because it caused the sampled process to reach the new set-point in one sample time.

This example illustrates one of the major problems with deadbeat controllers and indeed any other controller with very aggressive controller action and poles close to -1: the sampled process may follow the discrete-time reference trajectory perfectly and yet the actual process is wildly oscillating. See Refs. [3, 6, 7–9] for more discussion of deadbeat controllers.

Dahlin Algorithm

It is possible to improve the performance of the deadbeat controller by choosing a less aggressive reference trajectory; this is the purpose of the Dahlin algorithm. With Dahlin's algorithm, the closed-loop behavior is desired to be of the first-order-plus-time-delay type, with the familiar continuous-time representation:

$$q(s) = \frac{e^{-\gamma s}}{\tau_r s + 1}$$

If the delay is an integral multiple of the sampling time (i.e., $\gamma = M \Delta t$, say), then this translates to a $q(z)$ given by:

$$q(z) = \frac{\left(1 - e^{-\Delta t / \tau_r}\right) z^{-M-1}}{1 - e^{-\Delta t / \tau_r} z^{-1}} \qquad (26.73)$$

which includes the ZOH element with the first-order-plus-time-delay dynamics. The Dahlin controller may now be obtained from Eq. (26.61) as:

$$g_c(z) = \frac{1}{g(z)} \left[\frac{(1 - e^{-\Delta t/\tau_r}) z^{-M-1}}{1 - e^{-\Delta t/\tau_r} z^{-1} - (1 - e^{-\Delta t/\tau_r}) z^{-M-1}} \right] \tag{26.74}$$

The following are some points to note about the Dahlin algorithm:

1. It has two tuning parameters: the time delay M, and the time constant term τ_r in the reference trajectory. M is chosen so that the controller is realizable when there are process time delays; τ_r determines the speed of the closed-loop response.

2. It is clear that with this algorithm, we have the opportunity to avoid the type of excessive control action that is possible with the deadbeat controller since we can choose τ_r to give as conservative a controller as we wish; however, the dependence on the inverse of $g(z)$ means the Dahlin algorithm is still susceptible to controller ringing. It is sometimes possible to minimize the effects of ringing in the Dahlin algorithm by choosing τ_r carefully; other techniques for avoiding ringing may be found, for example, in Ref. [8].

3. Because the controller uses the process inverse, any process zeros outside the unit circle will become unstable poles in the Dahlin controller. Thus the Dahlin controller cannot be used with processes having inverse response.

Let us illustrate the detuning possible with the Dahlin algorithm by considering the following example.

Example 26.4   IMPROVED RESPONSE WITH THE DAHLIN CONTROLLER.

Let us again consider controlling the second-order system of Example 26.3, but this time using the Dahlin algorithm.

Solution:

If we choose a reference trajectory that is only slightly less agressive than that used by the deadbeat controller, i.e., $\gamma = 0$, $\tau_r = 0.3 = \Delta t$, so that Eq. (26.73) becomes:

$$q(z) = \frac{0.6321 \, z^{-1}}{1 - 0.3679 \, z^{-1}} \tag{26.75}$$

then the Dahlin controller is:

$$g_c(z) = \left(\frac{0.6321 \, z^{-1}}{1 - z^{-1}} \right) \left(\frac{1 - 1.664 \, z^{-1} + 0.7408 \, z^{-2}}{0.04052 \, z^{-1} + 0.03665 \, z^{-2}} \right) \tag{26.76}$$

which can be realized as

$$u(k) = 15.60 \, [\varepsilon(k) - 1.664 \, \varepsilon(k - 1) + 0.7408 \, \varepsilon(k - 2) + 0.0039 \, u(k - 1)$$

$$+ 0.03665 \, u(k - 2)] \tag{26.77}$$

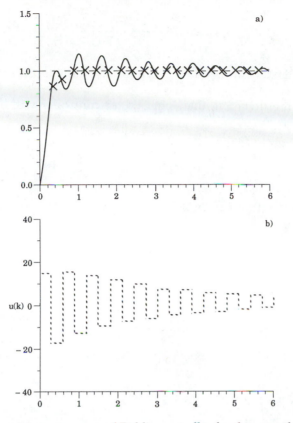

Figure 26.4. Closed-loop response of Dahlin controller for the second-order process of Example 26.3; (a) output response to set-point change, (b) controller action required.

Note that by comparing Eq. (26.77) with the earlier deadbeat controller Eq. (26.72), one sees that with $\tau_r > 0$, the Dahlin controller is only a detuned deadbeat controller.

The closed-loop response of the Dahlin controller in Eq. (26.77) is shown in Figure 26.4. Note that while there is still ringing and rippling, the effects are less pronounced, and the continuous output response is improved. However, the response is still not very good.

It is possible to modify the deadbeat and Dahlin controllers to permit them to handle processes with inverse response and to eliminate ringing. This involves adding detuning filters so that the resulting controller does not strive to cancel process zeros. Vogel and Edgar [8] and later Zafiriou and Morari [9] have proposed specific approaches for the design of these detuning filters. We will discuss only the Vogel and Edgar controller here.

Vogel-Edgar Controller

Vogel and Edgar [8] have proposed modifications to Dahlin's algorithm by removing the model numerator dynamics from the controller. This would mean the controller does not try to cancel the process zeros; thus the Vogel-Edgar controller is not subject to ringing and can handle processes with inverse response. The modification is added to the reference trajectory so that Eq. (26.73) becomes:

$$q(z) = \frac{(1 - e^{-\Delta t/\tau_r})}{(1 - e^{-\Delta t/\tau_r} z^{-1})} \left[\frac{B(z^{-1})}{B(1)} \right] z^{-M-1} \tag{26.78}$$

where $B(z^{-1})$ is the numerator polynomial of the process model given in Eq. (25.46).

With this modification, the Vogel-Edgar controller will be:

$$g_c(z) = \frac{B(z^{-1})}{B(1)\,g(z)} \left[\frac{(1 - e^{-\Delta t/\tau_r})\,z^{-M-1}}{1 - e^{-\Delta t/\tau_r} z^{-1} - \frac{B(z^{-1})}{B(1)}(1 - e^{-\Delta t/\tau_r})z^{-M-1}} \right] \tag{26.79}$$

Let us illustrate with an example.

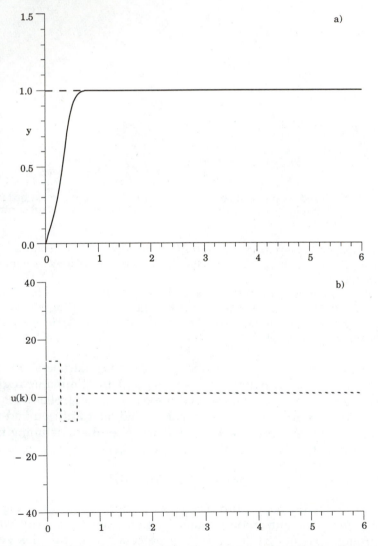

Figure 26.5. Closed-loop response of the Vogel-Edgar controller for the second-order process of Example 26.3; (a) output response to set-point change, (b) controller action required.

Example 26.5   CLOSED-LOOP RESPONSE WITH THE VOGEL-EDGAR CONTROLLER.

Let us again control the second-order system of Example 26.3, but now use the Vogel-Edgar controller.

Solution:

If we modify the reference trajectory by the factor $B(z^{-1})/B(1)$ as indicated in Eq. (26.78), then for $M = 0$, $\tau_r = 0.3$:

$$q(z) = \frac{0.6321\ z^{-1}}{1 - 0.3679\ z^{-1}} \left(\frac{0.04052\ z^{-1} + 0.03665\ z^{-2}}{0.07717} \right) \qquad (26.80)$$

and the Vogel-Edgar controller becomes:

$$g_c(z) = \frac{8.19 - 13.6\ z^{-1} + 6.07\ z^{-2}}{1 - 0.700\ z^{-1} - 0.300\ z^{-2}} \qquad (26.81)$$

which can be realized as:

$$u(k) = 0.700\ u(k-1) + 0.300\ u(k-2)$$
$$+ 8.19\ \varepsilon(k) - 13.6\ \varepsilon(k-1) + 6.07\ \varepsilon(k-2) \qquad (26.82)$$

The closed-loop response to a unit step in set-point (Figure 26.5) shows much improved control action with no ringing and only slightly delayed rise time. Thus for discrete processes with zeros outside the unit circle, or with negative real parts, the Vogel-Edgar controller is far superior to the original deadbeat or Dahlin algorithms.

Zafiriou and Morari [9] provide a review of these and other SISO digital controllers.

26.5.2 Time-Domain Designs

It is possible to design the discrete controller directly in the discrete-time domain, using the discrete-time difference equation model. The general procedure involves starting from the process model:

$$y(k) + a_1 y(k-1) + a_2 y(k-2) + \dots + a_n y(k-n)$$
$$= b_0 u(k) + b_1 u(k-1) + b_2 u(k-2) + \dots + b_m u(k-m)$$

with known parameters, and entertaining a control law of the type given in Eq. (26.13), with yet undetermined controller parameters. This control law is then introduced into the model and the controller parameters required to obtain a prespecified process response derived. For example, if the control law of the type:

$$u(k) = u(k-1) + K_0\ \varepsilon(k) + K_1\ \varepsilon(k-1) \qquad (26.83)$$

is to be used for a first-order system whose model is given by:

$$y(k) = ay(k-1) + bu(k-1) \qquad (26.84)$$

where a and b are given, first we observe, from Eq. (26.84) that:

$$y(k) - y(k-1) = ay(k-1) - ay(k-2) + bu(k-1) - bu(k-2) \qquad (26.85)$$

and upon substituting Eq. (26.83), with its time argument shifted backwards by one, and rearranging, we obtain:

$$y(k) = (1+a)\,y(k-1) - ay(k-2) + bK_0\,\varepsilon(k-1) + bK_1\,\varepsilon(k-2) \qquad (26.86)$$

We may now introduce $\big(y_d(k) - y(k)\big)$ for $\varepsilon(k)$ and rearrange to obtain finally the discrete-time equation for the closed-loop systems as:

$$y(k) = (1 + a - bK_0)\,y(k-1) - (a + bK_1)\,y(k-2) + b(K_0 + K_1)\,y_d \qquad (26.87)$$

where we have assumed that $y_d(k)$ is constant.

Now, we could employ several types of designs here; two of these will be discussed in more detail.

Direct Synthesis Design

Suppose we chose a reference trajectory in the discrete-time domain; for example, by inverting the pulse transfer function for the first order with time-delay response in Eq. (26.73) one obtains:

$$y(k) = \phi_r\,y(k-1) + (1 - \phi_r)\,y_d\,(k - M - 1) \qquad (26.88)$$

where $\phi_r = e^{-\Delta t/\tau_r}$. Thus by comparing Eq. (26.87) with Eq. (26.88) term by term for the case when $M = 0$ one can select K_0 and K_1 to achieve the desired first-order response. In the present case this means selecting:

$$K_0 = \frac{\left(1 + a - \phi_r\right)}{b}$$

$$K_1 = -\frac{a}{b}$$

so that Eqs. (26.87) and (26.88) agree exactly.

Pole Placement Design

If we take the z-transform of Eq. (26.87), one obtains:

$$y(z) = \frac{b\left(K_0 + K_1\right)}{1 - \left(1 + a - b\,K_0\right)z^{-1} + \left(a + b\,K_1\right)z^{-2}}\,y_d \qquad (26.89)$$

indicating a closed-loop pulse transfer function whose poles are roots of the characteristic equation:

$$z^2 - \left(1 + a - b\,K_0\right)z + \left(a + b\,K_1\right) = 0 \qquad (26.90)$$

Note that this characteristic equation can also be obtained directly from the discrete-time Eq. (26.87) by inspecting the coefficients of $y(k)$, $y(k-1)$, $y(k-2)$

to yield Eq. (26.90). Thus one need not calculate the pulse transfer function. Having obtained the closed-loop characteristic equation, it is possible to choose the two controller parameters K_0, K_1 so as to place the closed-loop poles wherever we wish. Since the closed-loop response is determined by the location of these poles, we can place the poles to achieve the desired dynamic behavior of the system under feedback control. This particular technique is referred to as *pole placement* because the controller parameters are determined by "placing" the closed-loop poles in desired locations. A more general discussion of pole placement is available in Ref. [1], Chapter 10, where other design methods are also discussed.

26.5.3 Other Approaches to Digital Controller Design

Several of the other controller design techniques for continuous systems can be modified so that they can be directly applicable for digital controller design. The root locus method, for example, is directly applicable to discrete systems. Discussion of this and other frequency-domain design procedures for digital systems may be found in Ref. [3]. The time-domain techniques of optimal control are also directly applicable; these are discussed in Refs. [1, 3]. Dynamic Matrix Control (DMC), Model Algorithmic Control (MAC), and other model-based digital controller algorithms will be described in Chapter 27.

26.6   DIGITAL MULTIVARIABLE CONTROLLERS

When dealing with multivariable systems, it turns out that with the exception of the most trivial of these systems, any rational controller design procedure leads to controllers that are usually impossible to implement with classical analog controllers. Thus, virtually all multivariable controllers must be implemented in digital form. The issue of multivariable control is therefore almost synonymous with that of designing "digital" controllers for multivariable systems.

As with the single-variable systems, digital controllers may be obtained for multivariable systems either indirectly by discretizing designs carried out in the continuous domain, or directly by carrying out the designs in the discrete domain. With the indirect approach, the same procedure of s, z variable substitution presented earlier in this chapter applies; with the direct approach, however, it is often more common to carry out the design in the discrete-time domain, using the matrix difference equation model as basis. Some of these multivariable, time-domain techniques (discussed in Refs. [1, 3]), have not been very popular with process control practitioners.

Model-based control techniques such as DMC, MAC, and IMC, on the other hand, are applicable to multivariable processes and have enjoyed notable acceptance in industrial applications; the principles involved in designs based on these techniques will be discussed in the next chapter.

26.7   SUMMARY

In this chapter we have introduced the digital controller and presented two approach philosophies for its design. The *indirect* approach, in which continuous designs are "translated" into the digital realm by simple variable substitution, makes it possible to convert to digital form virtually any controller designed by the controller design techniques discussed in Part IVA. By this method, continuous PID controllers, feedforward and cascade controllers, time-delay compensation controllers, direct synthesis controllers, ratio controllers, etc., are all easily digitized, and may be used on the sampled-data system. Despite the convenience, this approach has the disadvantage of requiring special compensation for the effect of sampling if the sample time is large; also, the approximations involved in the digitizing process usually result in some loss of performance.

With the direct approach, the design is carried out directly in the discrete domain, either using transfer functions (as with the deadbeat and the Dahlin algorithms) or the difference equation model in discrete time. While these techniques implicitly take the effect of sampling into consideration, they are unable to deal directly with the fact that the process is inherently continuous; the result, as was demonstrated with the discrete domain direct synthesis techniques, is the possibility of a closed-loop system that achieves the desired behavior at sampling points but that exhibits unacceptable intersample response.

It is important to reiterate, in closing, that the real advantage of digital over analog control is that control algorithms that are impossible to implement because of analog hardware limitations are easily implemented on the digital computer. It is this versatility of the digital computer, more than anything else, which makes the application of advanced control schemes possible.

REFERENCES AND SUGGESTED FURTHER READING

1. Åström, K. J. and B. Wittenmark, *Computer Controlled Systems*, Prentice-Hall, Englewood Cliffs, NJ (1984)
2. Seborg, D. E., T. F. Edgar, and D. A. Mellichamp, *Process Dynamics and Control*, J. Wiley, New York (1989)
3. Ogata, K., *Discrete-Time Control Systems*, Prentice-Hall, Englewood Cliffs, NJ (1981)
4. Deshpande, P. B. and R. H. Ash, *Elements of Computer Process Control with Advanced Control Applications*, ISA, Research Triangle Park, NC (1981)
5. Ragazzi, J. R. and G. F. Franklin, *Sampled-data Control Systems*, McGraw-Hill, New York (1958)
6. Isermann, R., *Digital Control Systems*, Springer-Verlag, Berlin (1981)
7. Kuo, B. C., *Analysis and Synthesis of Sampled-data Control Systems*, Prentice-Hall, Englewood Cliffs, NJ (1963)
8. Vogel, E. F. and T. F. Edgar, "A New Dead Time Compensator for Digital Control," *ISA/80 Proceedings* (1980)
9. Zafiriou, E. and M. Morari, "Digital Controllers for SISO Systems: A Review and a New Algorithm," *Int. J. Control*, **42**, 855 (1985)

REVIEW QUESTIONS

1. What is a digital controller, and what are some of the characteristics that distinguish it from a conventional analog controller?

2. What is the general z-domain form of the digital controller, and how is the corresponding time-domain realization obtained?

3. Why is the concept of "physical realizability" important within the context of digital controller implementation?

4. What does it mean for a digital controller to be physically realizable?

5. What is the fundamental philosophy behind the "indirect" approach to the design of digital controllers? How is it different from the philosophy behind the "direct" approach?

6. What are the advantages and disadvantages of the "indirect" approach to the design of digital controllers?

7. What are the advantages and disadvantages of the "direct" approach to the design of digital controllers?

8. What are the main effects of sampling and holding on the properties of a sampled-data feedback control loop?

9. How does the sample-and-hold device quantitatively affect the closed-loop stability of a sampled-data control system?

10. What is the single most overwhelming advantage of digital control over conventional analog control?

11. What is the position form of the discrete PID controller? How is it different from the velocity form?

12. How are digital PID controllers tuned in general?

13. What is Tutsin's approximation, and what is it used for?

14. Of the two alternatives to Tutsin's approximation discussed in this chapter, which is more useful? Why?

15. What is required in using the "backward difference" approximation to convert a controller transfer function from analog to digital form?

16. What is involved in carrying out controller design by direct synthesis in discrete time? Is it any different from what was discussed in Chapter 19 for direct synthesis design in the continuous-time domain?

17. What are the requirements for a deadbeat controller? Why is this not a practical controller for most applications?

18. What is "ringing," and how does this phenomenon arise?

19. What strategy is employed in Dahlin's algorithm for improving on the performance of the deadbeat controller?

20. What are some of the basic characteristics of the Dahlin controller?

21. What is the Vogel-Edgar modification to Dahlin's algorithm, and in what way does it improve the Dahlin controller's performance?

22. Why is the issue of multivariable control almost synonymous with that of designing "digital" controllers for multivariable systems?

PROBLEMS

26.1 The control system for an industrial gas furnace consists of the following elements:
(a) A digital controller operating according to:

$$u(k) = 0.75u(k-1) + 0.25u(k-2) + 0.01\varepsilon(k) - 0.008\varepsilon(k-1) \qquad (P26.1)$$

(b) A ZOH element;
(c) A valve, with transfer function:

$$g_v(s) = \frac{75.0}{0.01s + 1} \quad \text{(SCFM/psig)} \qquad (P26.2)$$

(d) A transfer line, with transfer function:

$$g_l(s) = \frac{e^{-0.2s}}{0.01s + 1} \quad \text{(SCFM/SCFM)} \qquad (P26.3)$$

(e) The main process, with transfer function:

$$g_f(s) = \frac{2.0}{0.5s + 1} \quad \text{(°F/SCFM)} \qquad (P26.4)$$

The unit of time is seconds.
(a) Obtain the consolidated transfer function $g(s)$ for the process, and combine it with the ZOH element to obtain the process pulse transfer function $g(z)$ given that $\Delta t = 0.1$ sec. Also obtain $g_c(z)$ for the controller.
(b) Draw a block diagram of the type in Figure 26.1(b) for this process; obtain the closed-loop transfer function and the characteristic equation. Investigate the stability of the closed-loop system.

26.2 The relationship between the thick stock flowrate to a paper machine, u, and the basis weight of the paper, y, has been approximated by the following transfer function model:

$$y(s) = \frac{0.55e^{-8s}}{7.5s + 1} u(s) \qquad (P26.5)$$

(a) Use this Laplace domain model, along with the Cohen and Coon tuning rules presented in Chapter 15, to design a PI controller. Discretize the resulting PI controller using $\Delta t = 2$. Implement the digital controller on the continuous process and simulate the closed-loop system response to a unit step change in the basis weight set-point. Plot the response.
(b) If the process is to be operated at the increased sampling rate $\Delta t = 4$, discretize the continuous PI controller using $\Delta t = 4$ and repeat part (a) under these new conditions. Compare the closed-loop responses.
(c) For the situation with $\Delta t = 4$, first add $\Delta t/2$ to the time delay in the process transfer function before using the Cohen and Coon rules to obtain the continuous PI controller. Discretize this new controller (using $\Delta t = 4$) and repeat part (b). Compare the performance of this controller to that in part (b). Is there any significant advantage to the strategy employed in obtaining this last controller?

26.3 The following transfer function model relates the top composition of ethanol, y, in an ethanol/water distillation column to the reflux flowrate, u, and the feed flowrate, d:

$$y(s) = \frac{0.66e^{-2.5s}}{6.7s + 1} u(s) + \frac{0.14e^{-12.0s}}{6.2s + 1} d(s) \qquad \text{(P26.6)}$$

The unit of time is minutes. Observe that the model is of the form:

$$y(s) = g(s) u(s) + g_d(s) d(s)$$

and recall that the transfer function for a feedforward controller designed on the basis of such model is given by:

$$g_{ff}(s) = \frac{-g_d(s)}{g(s)}$$

If such a feedforward controller is to be implemented in digital form with sample time $\Delta t = 0.5$ min, obtain the required digital controller transfer function, $g_{ff}(z)$, from the particular $g_{ff}(s)$ obtained for this process, using:
(a) The forward difference approximation
(b) The backward difference approximation
(c) The Tutsin approximation

26.4   For the general first-order system with the transfer function:

$$g(s) = \frac{K}{\tau s + 1} \qquad \text{(P26.7)}$$

it was shown in Chapter 19 that the continuous direct synthesis controller required to achieve a specified first-order closed-loop reference trajectory is a PI controller. In Example 26.2, the digital version of this direct synthesis controller was obtained using the backward difference approximation.
(a) By comparing Eq. (26.5) with Eq. (26.39) — the expression for the general discrete PID controller — obtain the values for K_c and τ_I for this particular controller and confirm that these values are independent of the sample time, Δt. Why is it desirable to have controller parameters that depend on Δt?
(b) Repeat Example 26.2 using Tutsin's approximation; compare the result with that in Eq. (26.53). Do the controller parameters now depend on Δt?
(c) It has been recommended that for discrete-time systems with long sample times, controller performance can be improved if the design is based instead on the model:

$$g(s) = \frac{K e^{-\alpha s}}{\tau s + 1} \qquad \text{(P26.8)}$$

with $\alpha = D\Delta t$, and then setting $D = 1$ in the resulting controller such as that in Eq. (26.69). (This is equivalent to introducing a sampling delay of 1 time unit as part of the process model.) Let us refer to this as the "modified Smith predictor approach." Show that such an exercise will still result in a PI controller (in velocity form), and compare this controller with the one in Eq. (26.53). Comment on what effect increasing Δt has on the parameters of this alternative controller.

26.5   To illustrate the main points of Problem 26.4 concretely, consider the simple first-order process with transfer function:

$$g(s) = \frac{0.66}{6.7s + 1} \qquad \text{(P26.9)}$$

(a) Obtain a direct synthesis controller for this process for $\tau_r = 2.0$. Implement this in standard digital form on the continuous process, with a sample time $\Delta t = 1.0$, and obtain the closed-loop system response to a unit step change in the process set-point.

(b) Implement the same controller on the same process for the same set-point change, but now with sample time $\Delta t = 4.0$. Compare the closed-loop response obtained in this case with that obtained in part (a). Comment on the effect of increased Δt on the closed-loop system performance. (Recall Examples 25.8 and 25.9 of Chapter 25.)

(c) Use the "modified Smith predictor approach" outlined in Problem 26.4(c) to obtain the digital controller for the case when $\Delta t = 4.0$. Implement this controller on the process (with $\Delta t = 4.0$) and obtain the closed-loop response for the same set-point change. Compare the response with that obtained in part (b).

26.6 [*This problem is open ended*]

The following model for a CSTR in which an isothermal second-order reaction is taking place, was first given in Problem 19.12:

$$\frac{dC_A}{dt} = \frac{F}{V}\left(C_{A0} - C_A\right) - k\,C_A^2 \tag{P19.18}$$

The manipulated variable is F, the volumetric flowrate of reactant A into the reactor; the disturbance variable is C_{A0}, the inlet concentration of pure reactant in the feed; and the output variable to be controlled is C_A, the reactor concentration of A; V is the reactor volume, assumed constant. The nominal process operating conditions are given as:

$$
\begin{aligned}
F &= 7.0 \times 10^{-3} \text{ m}^3/\text{s} \\
V &= 7.0 \text{ m}^3/\text{s} \\
k &= 1.5 \times 10^{-3} \text{ m}^3/\text{kg mole-s} \\
C_{A0} &= 2.5 \text{ kg mole/m}^3 \\
C_A &= 1.0 \text{ kg mole/m}^3
\end{aligned}
$$

Consider the situation in which the gas chromatograph measurement of C_A has a delay of 60 seconds, and that C_{A0} is not available for measurement.

(a) Linearize the model around the provided steady-state operating condition, obtain the approximate transfer function model, and use it to design a classical PID controller, using any tuning method of your choice. Carefully justify any design choices made.

(b) Obtain a discrete version of the controller obtained in part (a); implement it on the original nonlinear system using $\Delta t = 30$ sec, and obtain the closed-loop system response to a change in C_{A0} from 2.5 to 2.0 kg mole/m^3 (remember, C_{A0} is unavailable for measurement).

(c) Use the linearized model to design an IMC controller, justifying any design choices made. Discretize this controller using the approximation technique of your choice, for $\Delta t = 30$ sec, and repeat part (b). Compare the performance with that of the classical controller you designed and implemented in part (b).

26.7 In Problem 24.5, the transfer function model for a thermosiphon reboiler was given as:

$$y(s) = \frac{0.9\,(0.3s - 1)}{s\,(2.5s + 1)}\,u(s) \tag{P24.7}$$

with the input and output variables defined as:

$$u = \left[\frac{(F_s - F_s*)\ (\text{lb/hr})}{F_s*\ (\text{lb/hr})} \right];$$

$$y = \left[\frac{(h - h*)\ (\text{in})}{h*\ (\text{in})} \right]$$

Here, $h*$ (ins), and F_s* (lb/hr), represent nominal steady-state operating conditions. The time constants are in minutes.

It is desired to control this process using the computer to sample the process output and to take control action every 0.1 min.

(a) Use the transfer function model and the referece trajectory:

$$q(s) = \frac{(1 - 0.3s)}{(1 + 0.3s)(\tau_r s + 1)}$$

with $\tau_r = 1.0$ min, to design a direct synthesis controller. What dynamic system can be used to implement this somewhat unusual controller? Also, what is the closest classical controller that can be used as an approximation?

(b) Discretize the controller using the backward difference approach, implement it on the process, and obtain the closed-loop response to a unit step change in the normalized level set-point.

(c) On the chance that the inverse response behavior might be "missed" if the sample time is increased to $\Delta t = 1.0$ min, rederive the discrete version of the direct synthesis controller under these conditions, implement it on the process (using the new sample time), and obtain the closed-loop response to the same unit step change in the normalized level set-point. For this particular process, does the longer sample time provide an advantage or a disadvantage?

26.8 This problem is concerned with comparing Dahlin's digital controller design technique with the continuous direct synthesis technique.

Given the first-order-plus-time-delay system:

$$g(s) = \frac{K\ e^{-\alpha s}}{\tau s + 1} \tag{P26.10}$$

and an appropriate reference trajectory, the continuous direct synthesis controller was given in Chapter 19, in Eq. (19.24).

(a) Discretize this controller and obtain an expression for $g_c(z)$ (Keep in mind that the discrete counterpart of $e^{-\alpha s}$ is z^{-M} where $\alpha = M\Delta t$.)

(b)From the transfer function in Eq. (P26.10), first obtain $g(z)$, the corresponding pulse transfer function incorporating a ZOH; and use this to obtain the Dahlin controller, employing the $q(z)$ given in Eq. (26.79). Compare the Dahlin controller with the discretized version of the continuous direct synthesis controller.

26.9 (a) Design a Dahlin controller for the paper process whose transfer function model was given in Problem 26.2 as:

$$y(s) = \frac{0.55e^{-8s}}{7.5s + 1} u(s) \tag{P26.5}$$

for $\Delta t = 2.0$, $\gamma = 8.0$, and $\tau_r = 3.0$. Write out an explicit expression for the controller. Implement this controller on the "true" process represented by:

$$y(s) = \frac{0.40e^{-10s}}{8.0s + 1} u(s) \tag{P26.11}$$

and obtain the closed-loop system response to a unit step change in the basis weight set-point.

(b) Repeat part (a) with $\tau_r = 6.0$. Compare the closed-loop system response with that obtained in part (a). Comment on the effect of the parameter τ_r on the controller performance in the face of plant/model mismatch.

26.10 [*This problem is open ended*]

Figure P26.1 is a schematic diagram for an industrial heat exchanger that utilizes chilled brine to cool down a hot process stream; it was first introduced in Problem 14.2. Assume that T_B, the brine temperature (in °C), remains constant, that the measuring device has negligible dynamics, and that the following transfer function relations:

$$(T - T^*)(°C) = \frac{-(1 - 0.5e^{-10s})}{(40s + 1)(15s + 1)} \left(F_B - F_B^* \right) \text{ (kg/s)} \tag{P26.12a}$$

$$(T - T^*) \, (°C) = 0.5e^{-10s} \left(T_i - T_i^* \right) \, (°C) \tag{P26.12b}$$

are reasonably accurate.

(a) Design a classical PID controller for this process using a technique of your choice. Justify any design choices made.

(b) Discretize the classical PID controller obtained in part (a) using backward differences, with $\Delta t = 5$ seconds. Implement this controller on the process and obtain the closed-loop system response to a step change of +5°C in the process feedstream temperature.

(c) Obtain a process reaction cubest of your ability, determine from it an approximate first-order-plus-time-delay model. Use this approximate model to obtain a Dahlin controller for $\Delta t = 5$ sec. Justify the design choices made. Implement this controller for the same distrurbance as in part (a). Compare the closed responses.

(d) Which controller would you normally expect to perform better? Did the simulations confirm this expectation? Explain.

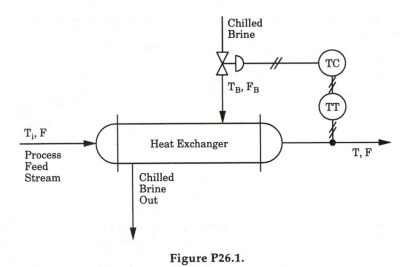

Figure P26.1.

26.11 Given the transfer function for an open-loop unstable process as:

$$g(s) = \frac{-0.55}{5.0s - 1} \qquad \text{(P26.13)}$$

(a) Obtain $g_{Hp}(z)$ the pulse transfer function for the process incoporating a ZOH, for $\Delta t = 1$ min. Write out the coresponding difference equation.
(b) Given that this process is to be controlled using the digital PI controller:

$$u(k) = u(k-1) - 1.2K\varepsilon(k) + K\varepsilon(k-1) \qquad \text{(P26.14)}$$

with the free parameter K, introduce this expression into the difference equation model obtained in part (a), and assuming that the set-point y_d is constant, obtain the closed-loop difference equation relating $y(k)$ to y_d. Obtain the range of values that the controller parameter K can take in order for the closed loop system to be stable.

26.12 A digital controller designed for the polymer process introduced in Problems 8.5 and 25.15 has been given as:

$$u(k) = 0.625u(k-1) + 0.3755u(k-2) + 0.325\varepsilon(k)$$
$$- 0.508\varepsilon(k-1) + 0.199\varepsilon(k-2) \qquad \text{(P26.15)}$$

with the sample time for implementation given as $\Delta t = 3.0$ min. Recall that the composite transfer function for the process is given by:

$$g(s) = \frac{25.0e^{-6s}}{(10s + 1)(12s + 1)} \qquad \text{(P26.16)}$$

(a) Obtain $g(z)$, the pulse transfer function incorporating the ZOH, for $\Delta t = 3.0$ min. Write out the corresponding difference equation.
(b) Introduce the controller expression in Eq. (P26.15) into the difference equation obtained in part (a); assume that the set-point y_d is constant, and rearrange the expression to obtain a difference equation relating $y(k)$ to y_d; use this equation to check for closed-loop stability.
(c) Using the difference equation in part (b), implement a unit step change in y_d; simulate and plot the system response. What type of controller is implied in Eq. (P26.15)?

part V

SPECIAL CONTROL TOPICS

In Parts I–IV we have presented key concepts and many important tools for understanding the dynamics, developing a model, and designing a control system for a process of interest. However, this material is only a fundamental base from which to launch further forays into deeper and more advanced aspects of *Process Dynamics, Modeling, and Control*. Even after the extensive and detailed discussions presented so far, there remain a significant number of important subjects yet to be explored. To sample some of these, we provide in Chapters 27–29 an *introduction* to a broad spectrum of "special topics" and truly advanced process control concepts which the engineer will likely encounter in practice. To avoid undue excursions outside the boundaries imposed by the intended scope of this book, however, the *detailed* treatment of these topics must be left for a more advanced course of study. As a finale, a capstone discussion of several case studies is provided in Chapter 30 to illustrate how appropriate concepts and tools are selected and brought to bear in solving real problems.

part V

SPECIAL CONTROL TOPICS

" 'The time has come,' the Walrus
'To talk of many th
Of shoes — and ships — and sealing w
Of cabbages and kings

Lewis Carroll, *Alice in Wond*

" *And if any man think that he knoweth any*
he knoweth nothing yet as he ought to k

St. Paul, *I Corinthi*

CHAPTER

27

MODEL   PREDICTIVE   CONTROL

In many areas of scientific endeavor, the development of new technology is quite often motivated by practical need. This is certainly true of classical automatic process control where the emergence of the chemical process industries, with its large production volume, and complex interconnection of several unit operations, created the *need* for automation, which, in turn motivated the *development* of automatic process control technology. Furthermore, in process control technology within the field of process control itself, we find that significant advances have also been motivated mostly by practical need. This is particularly so with the model-based control techniques already introduced in Chapters 19, 22, and 26.

In the chemical and allied industries, the universal drive for more consistent attainment of high product quality, more efficient use of energy, and an increasing awareness of environmental responsibility have all combined to impose far stricter demands on control systems than can be met by traditional techniques *alone*. The industry's response to these challenges led to the development of the industrially successful *Dynamic Matrix Control* (DMC) technique [14 – 18, 56] and *Model Algorithmic Control* (MAC) [45, 46, 60, 62 – 65], better known by the name of the commercial computer software used for its implementation — IDCOM. These two control strategies (and subsequent ones of a similar structure) are classified as *Model Predictive Control* (MPC) schemes for reasons we shall soon discuss.

Our primary objective in this chapter is to provide an overview of the principles and practice of *Model Predictive Control* as a family of control schemes; and in this regard we will focus our discussion on its basic principles, some relevant aspects of the underlying theory, and the various commercially available MPC packages. Also, even though the first MPC schemes were developed in industry, insight into the fundamentals of the theory, implementation details, and limitations of MPC have resulted primarily from academic contributions; we will therefore review such contributions and how they have extended the frontier and scope-of-influence of MPC, paying particular attention to the emergence of nonlinear MPC as a distinct and important subject matter in its own right.

27.1 INTRODUCTION

Model Predictive Control is an appropriately descriptive name for a class of computer control schemes that utilize a process model for two central tasks:

1. *Explicit* prediction of future plant behavior, and
2. Computation of appropriate corrective control action required to drive the predicted output as close as possible to the desired target value.

As a class of control schemes, MPC has enjoyed such remarkable industrial success and popularity that according to recent reviews [28, 78] it is currently the most widely utilized of all advanced control methodologies in industrial applications.

Before delving into a full-scale discussion of the principles and practice of MPC, we find it instructive first to place this industrial success and popularity in perspective. We will do this by examining the nature of contemporary industrial processes, and the typical associated control problems; we will then see that MPC in fact addresses a significant number of these problems in a manner that is direct, efficient, and easily understood.

27.1.1 Contemporary Industrial Processes and Associated Control Problems

The typical industrial process is usually complex, having several input, output, and disturbance variables, and processing large volumes of raw materials in continuous, semibatch, or batch mode. Furthermore, the current economic/environmental climate dictates that such processes must operate under conditions of very stringent demands on:

*   Product quality
*   Energy utilization
*   Safety and environmental accountability

In most cases, the key product quality variables by which the product is sold are not available for on-line measurement, and in some cases, even the key indicators of the basic process conditions are not available for measurement frequently enough. In addition, when the same processing units are used to manufacture several different grades of the same basic product, the plant will undergo frequent transitions or start-ups and shutdowns.

Contemporary Industrial Control Problems

As a result of the foregoing, the typical industrial process has associated with it the following control problems:

1. It is almost always *multivariable*, typically with significant interactions between the process variables:
    *   Several process variables ($y_1, y_2, ..., y_n$) to be *controlled*
    *   Several manipulated variables ($u_1, u_2, ..., u_m$) available for control
    *   Several disturbance variables ($d_1, d_2, ..., d_k$), some measurable, some not, over which we have little or no control

2. Difficult dynamic behavior:
    • Time delays caused by transportation lag through pipes, long measurement and analysis delay, approximations to high-order systems, etc.
    • Inverse-response behavior
    • Occasional open-loop instability

3. Inherent nonlinearities (since all chemical processes are nonlinear to varying degrees of severity).

4. Constraints of all kinds, typically as itemized below:
    • Constraints on the absolute values of the input and output variables
    • Constraints on the rates of change of the inputs
    • Other equality and inequality constraints
    • Hard and soft constraints

27.1.2 Anatomy of an "Ideal Controller" and MPC Capabilities

We may now observe that the "ideal controller" required for optimal operation of the contemporary industrial process must have the following attributes:

1. It should be able to handle multivariable process interactions, time delays, and other problematic dynamic behavior, input/output constraints, disturbances (measurable and otherwise), and complications resulting from nonlinear behavior.

2. It should do all the above while optimizing the use of control effort.

3. It should perform well with the barest minimum set of measurements from the plants (i.e., it must have a means of inferring critical, but unavailable, plant information from whatever *is* available).

4. It should remain robust in the face of modeling errors and measurement noise.

5. It should maintain all these attributes during start-ups and shutdowns, as well as during steady-state operation.

While it is clear that no such "ideal" controller exists, it is instructive to summarize now the typical capabilities of model predictive controllers and to compare these with the ideal requirements noted above.

Typical MPC Capabilities

As we will show subsequently, the typical MPC scheme:

1. Is particularly easy to use for multivariable systems and handles process interactions with ease.

2. Handles time delays, inverse response, as well as other difficult process dynamics with ease.

3. Utilizes a process model, but no rigid model form is required (the original MPC schemes are based on so-called "nonparametric" step or impulse-response models).

4. Compensates for the effect of measurable and unmeasurable disturbances (the former directly in a feedforward fashion, the latter by inference in a feedback fashion).

5. Is posed as an optimization problem and is therefore capable of meeting the control objectives by optimizing control effort (if the problem is so posed), and at the same time is capable of handling constraints of all kinds.

Observe now that when these MPC capabilities are compared with the noted "ideal" controller attributes, it is easy to appreciate the industrial success of this class of controllers where others have failed. Let us now give a historical perspective of the evolution of MPC.

27.1.3 A Historical Overview of Model Predictive Control

Even though the current industrial and academic interest in MPC was primarily generated by the 1978 paper of Richalet *et al.* [65], and by the 1979 Shell publications [14, 56], what have since become recognized as the central MPC concepts actually predate these first reports of application by some twenty years.

The textbooks by Newton *et al.* [52], and Horowitz [32], the original dead time compensator (the Smith predictor) [73, 74] discussed in Chapters 17 and 26, and the linear programming approach to optimal control presented by Zadeh and Whalen [83], all contain the essence of the MPC concept.

Apparently, the first publication to introduce, explicitly, the now standard MPC concept of the finite moving prediction horizon seems to be Propoi [59] (cf. Garcia *et al.* [28]). The classic text by Box and Jenkins [6] also contains the fundamental concepts of MPC, complete with a statistical theory for optimal handling of measurement and process noise.

Nevertheless, according to Mehra [45, 46], Model Algorithmic Control (MAC), as the forerunner of the scheme now commercially marketed under the name of IDCOM, was in fact originally conceived in the late 1960s by Adersa/Gerbios, subsequently developed, and successfully applied to several multivariable industrial processes (cf. Richalet *et al.* [62–65]). With the availability of more powerful digital computers in the 1970s, it became more practical and economical to apply MAC to more industrial processes; this led to its evolution into the computer control algorithm now marketed as IDCOM.

The Dynamic Matrix Control (DMC) technique first appeared in the open literature in 1979 (after having been applied with notable success on several Shell processes for a number of years), when Cutler and Ramaker, and Prett and Gillette reported its application to a fluid catalytic cracking unit [14, 56]. A reformulation of the original DMC problem as a quadratic program by Garcia and Morshedi [27] subsequently led to the development of its most popular variation to date: Quadratic Dynamic Matrix Control (QDMC).

The first attempt to put MPC (primarily as embodied by DMC and MAC) within the same framework as the so-called "classical control schemes" was by

Garcia and Morari [23], in which the Internal Model Control (IMC) paradigm was proposed. They showed that the IMC structure (in which, as noted in Chapter 19, an internal model of the plant operates in parallel with the plant, and in which the controller is some appropriate inverse of this plant model) is inherent in all MPC schemes. This and subsequent publications by Morari and coworkers provided insight into the stability, robustness and performance of MPC schemes [25, 26, 70].

It should be noted in this regard that the IMC structure popularized by Garcia and Morari in 1982 [23] was in fact already a central part of the 1974 monograph by Frank [22]. In addition Youla *et al.* [80, 81] and Zames [85] developed the same type of problem formulation. However, the unification of all these concepts and the consolidation of the various structures into the one now known as the IMC structure is due to Garcia and Morari [23]. Several additional MPC techniques have since been developed, and some have become commercial products; these will be reviewed in Section 27.5.

The theory and practice of MPC has since been the principal subject of at least three books [47, 49, 57], two workshops [43, 58], and several sessions at the Fourth Chemical Process Control Conference [1].

Finally, we note that since the publication of the pioneering papers, both IDCOM and DMC as well as other MPC schemes have been applied to many industrial processes. A sample of such processes with the appropriate references is provided in the next section.

27.1.4 A Sample List of Documented MPC Applications

The following list is a representative sample of applications of various MPC schemes that have appeared in the open literature since 1975:

1. Aircraft flight control, jet engine control [45, 60]
2. Superheater, steam generators, utility boiler [34, 75]
3. Transonic wind tunnel [50]
4. Fluid catalytic cracking units [18, 31, 40, 56]
5. Batch reactor [24]
6. Several processes in the pulp and paper industry [41, 67]
7. Hydrocracker reactor [9, 16, 17]
8. Distillation columns [29]
9. Polymer extruder [77]
10. Olefins plant [9, 30]

27.1.5 Process Characteristics and Applicability of MPC

To be sure, not all processes require MPC for effective performance (cf., for example, Ref. [30]). Thus, it is important to identify which processes require MPC for effective control and what characterizes such processes.

A careful study of the published applications of MPC indicates that processes with constrained, multivariable control problems involving difficult dynamics have traditionally benefitted the most from this control methodology. Thus the conventional consensus among the industrial practitioners, as well as both industrial and academic researchers, is that MPC is typically best suited to processes with any combination of the following characteristics:

1. Multiple input and output variables with significant interactions between SISO loops.
2. Either equal or unequal number of inputs and outputs.
3. Complex and unusually problematic dynamics (such as long time delays, and inverse response, or even unusually large time constants).
4. Constraints in inputs and/or output variables.

27.2 GENERAL PRINCIPLES OF MODEL PREDICTIVE CONTROL

Chemical processes often have slow dynamics so that, quite often, it takes a substantial amount of time for the *full effect* of each *single* control action to be completely manifested in the observable process output.   The immediate implication is that it is not possible to obtain the "full picture" of the consequences of previously applied control action from *only* the current process output measurement. In deciding what control action to take at the current time instant therefore, it is useful:

1. To consider first how the process output will behave in the future *if no further control action is taken*,
2. To target control action towards rectifying what remains to be corrected after the full effects of the previously implemented control action have been completely manifested.

This is the fundamental motivation behind Model Predictive Control as a controller design methodology.

Basic Elements of MPC

The fundamental framework of Model Predictive Control consists of *four* elements shared in common by *all* such schemes; what differentiates one specific scheme from the other is the strategy and philosophy underlying how each element is actually implemented. These elements (illustrated in Figure 27.1) may be defined as follows:

1. Reference Trajectory Specification

The first element of MPC is the definition of a desired target trajectory for the process output, $y^*(k)$. This can simply be a step to the new set-point value or more commonly, it can be a desired reference trajectory that is less abrupt than a step.

2. Process Output Prediction

Some appropriate model, $\mathbf{M}$, is used to predict the process output over a predetermined, extended time horizon (with the current time as the prediction origin) *in the absence of further control action*. For discrete-time models this means predicting $\hat{\mathbf{y}}(k + 1)$, $\hat{\mathbf{y}}(k + 2),\dots \hat{\mathbf{y}}(k + i)$ for i sample times into the future based on all actual past control inputs $\mathbf{u}(k)$, $\mathbf{u}(k - 1), \dots \mathbf{u}(k - j)$.

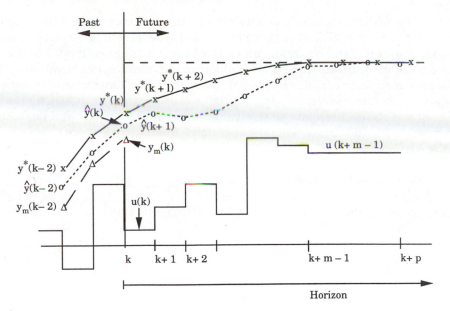

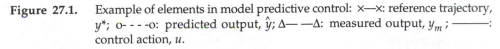

Figure 27.1.   Example of elements in model predictive control: ×—×: reference trajectory, y^*; o- - - -o: predicted output, $\hat{y}$; Δ——Δ: measured output, y_m ; ———: control action, u.

3.   Control Action Sequence Computation

The same model, **M**, is used to calculate a sequence of control moves that will satisfy some specified optimization objective such as:

- Minimizing the predicted deviation of the process output from target over the prediction horizon,
- Minimizing the expenditure of control effort in driving the process output to target,

subject to prespecified operating constraints. This is conceptually equivalent to constructing and utilizing a suitable model inverse, **M***, which is capable of predicting the control input sequence $\mathbf{u}(k)$, ..., $\mathbf{u}(k + m - 1)$ required for achieving the prespecified output behavior p time steps into the future.

4.   Error Prediction Update

In recognition of the fact that no model can constitute a perfect representation of reality, plant measurement, $\mathbf{y}_m(k)$, is compared with the model prediction, $\hat{y}(k)$, and the prediction error, $\hat{\varepsilon}(k) = \mathbf{y}_m(k) - \hat{y}(k)$, thus obtained is used to update future predictions.

Having defined the basic elements of MPC, let us discuss how the underlying process model is developed and applied.

Model Forms for MPC

We note first that standard (conventional) MPC, the main subject of this chapter, utilizes only *linear* models; incorporating nonlinear models of any form within the MPC framework is a relatively recent innovation which will be discussed in Section 27.7. Also, since most MPC schemes are implemented on the digital computer, the discrete, rather than the continuous, model forms are more natural. Thus the three most popular model forms for conventional MPC schemes are the linear, discrete versions of *the finite convolution* (or *impulse-response*), *the state-space*, and *the transform-domain transfer function* models, each of which has been encountered previously in a more general context in Chapters 4, 23, and 24; we will now reexamine these models more closely below in the MPC context.

1.   Finite Convolution Models

There are two entirely equivalent forms of this model type:

The Impulse-Response Model Form:

$$y(k) \ = \ \sum_{i=0}^{k} g(i) \, u \, (k-i) \tag{27.1}$$

with the "parameters" $g(i)$ known as the process *impulse response function*.

The Step-Response Model Form:

$$y(k) \ = \ \sum_{i=0}^{k} \beta \, (i) \, \Delta u(k-i) \tag{27.2}$$

with the "parameters" $\beta(i)$ known as the process *step-response function*; and $\Delta u(k) = u(k) - u(k-1)$.

As we had noted earlier in Chapter 4, it is easily established that the step- and impulse-response functions are related according to:

$$g(i) \ = \ \beta(i) - \beta(i-1)$$

and

$$\beta(i) \ = \ \sum_{j=1}^{i} g(j)$$

It should also be noted that for all real, causal systems, both $g(0)$ and $\beta(0)$ are equal to zero; hence, all realistic discrete-time systems exhibit the mandatory one-step delay.

The impulse (or step) response coefficients are typically obtained either directly from plant data (e.g., Cutler and Yocum [18]), provided the data are reasonably noise free, or they are deduced from other so-called *parametric model forms* such as the other two listed below.

2.   Discrete State-Space Models

These models may be represented by:

$$y(k) \ = \ \sum_{i=0}^{k} a(i) \, y(k-i) \ + \ \sum_{i=0}^{k} b(i) \, u(k-i-m) \tag{27.3}$$

These are also known as ARMAX (for **A**uto**R**egressive, **M**oving **A**verage with e**X**ogenous inputs) because of the similarity in structure with the time series autoregressive, moving average (ARMA) models of Box and Jenkins [6].

The model parameters such as the delay m, and the coefficients $a(i)$, $b(i)$ usually must be obtained by fitting the model to plant data. Again by reason of causality, $a(0) = b(0) = 0$.

3.   Discrete Transfer Function Models

These are models from:

$$y(z) \; = \; \frac{z^{-m}B(z^{-1})}{A(z^{-1})} \, u(z) \tag{27.4}$$

where the indicated transfer function has been represented as a ratio of two polynomials in the z-transform variable, $B(z^{-1})$ and $A(z^{-1})$, along with the associated delay of m sample times (see Chapters 24 and 25 for more discussion of these model forms). The parameters of the transfer function must also be determined from experimental data via process identification.

Other discussions of the various model forms for standard MPC may be found, for example, in Refs. [57, 69]. It is important to remember that it is possible to convert from one model form to another, even though some conversions are more straightforward than others. The details of how to convert one model form to another has been discussed in Chapters 4, 23, and 24; such details may also be found, for example, in Ref. [49].

Obtaining the Model Inverse

In most practical situations, the process of obtaining the "model inverse" is carried out numerically, as the solution of a (possibly constrained) optimization problem. In the unconstrained optimization case, it can be shown (and we shall do so shortly) that the resulting inverses are typically generalized (left, or right) inverses of the mathematical operator that constitutes the plant model.

Updating Model Predictions

Many factors contribute to whatever discrepancy is observed between actual plant measurements and model prediction: the effect of unmodeled, unmeasured disturbances, fundamental errors in the model structure, or errors in the model parameter estimates. As it is often impossible to isolate one cause from another, the standard MPC strategy is simply to attribute the discrepancy entirely to unmeasured disturbances, assume this value will remain constant for all future times over the prediction horizon, and update the model predictions accordingly. The strategy is realized therefore by simply adding the currently observed discrepancy to all the future predictions.

Let us now examine in the next two sections how these general principles are applied specifically in the basic formulation of DMC and MAC, the two schemes considered as the pioneers of the MPC methodology.

27.3 DYNAMIC MATRIX CONTROL

27.3.1 Process Representation

Let $\beta = [\beta(1),\ \beta(2),\ \beta(3),\ ...,\ \beta(n)]^T$ represent the *unit* step-response function obtained for a process, i.e., as illustrated in Figure 27.2, the elements of the vector β represent the *change* observed in the process output at n, consecutive, equally spaced, discrete-time instants after implementing a *unit change* in the input variable.

Also, let the current time instant be k, and let us further assume that, in the absence of further control action, it is predicted that the output of the process in question will take the values:

$$\hat{y}^0(k),\ \hat{y}^0(k+1),\ \hat{y}^0(k+2),\ ...,\ \hat{y}^0(k+p-1)$$

over the p-time step horizon whose origin is the current time instant k. Let this vector be represented as $\mathbf{y}^0(k)$, i.e.:

$$\hat{\mathbf{y}}^0(k)\ =\ [\hat{y}^0(k),\ \hat{y}^0(k+1),\ \hat{y}^0(k+2),\ ...,\ \hat{y}^0(k+p-1)]^T$$

where the argument k in $\hat{\mathbf{y}}^0(k)$ is used to indicate the time origin of the sequential predictions in the vector, and the superscript "0" is used to indicate a prediction conditional on the absence of further control action. (If $k = 0$, for example, this essentially constitutes an initial condition on the predicted plant output; subsequent sequences for $k > 0$ may be obtained recursively using the prediction at the penultimate instant $(k - 1)$). Note that the vector notation for $\hat{\mathbf{y}}^0(k)$ (and for $\mathbf{y}(k+1)$ and $\Delta\mathbf{u}(k)$ below) is special because it denotes a scalar at many different time steps rather than a vector of outputs at one time step as would the usual interpretation.

To proceed, we assume linearity, so that the principle of superposition applies and the step-response model representation Eq. (27.2) indicates that an arbitrary sequence of m control moves, $\Delta u(k)$, $\Delta u(k+1)$, $\Delta u(k+2)$, ..., $\Delta u(k + m - 1)$ will cause the system to change from the noted "initial conditions" $\hat{\mathbf{y}}^0(k)$ to a new state:

$$\hat{\mathbf{y}}\ (k+1)\ =\ [\hat{y}(k+1),\ \hat{y}(k+2),\ \hat{y}(k+3),\ ...,\ \hat{y}(k+p)]^T$$

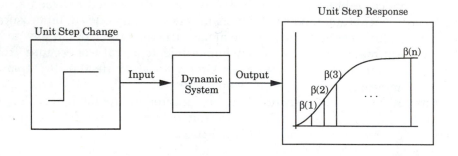

Figure 27.2.   The unit step-response function.

(see Figure 27.3) as given by the following equations, where the appended sequence, $w(k + 1)$, $w(k + 2)$, $w(k + 3)$, ..., $w(k + p)$, represents the collective effect of unmeasured disturbances on the output:

$$\hat{y}(k + 1) = \hat{y}^0(k) + \beta(1)\Delta u(k) + w(k + 1)$$

$$\hat{y}(k + 2) = \hat{y}^0(k + 1) + \beta(2)\Delta u(k) + \beta(1)\Delta u(k + 1) + w(k + 2)$$

$$\hat{y}(k + 3) = \hat{y}^0(k + 2) + \beta(3)\Delta u(k) + \beta(2)\Delta u(k + 1) + \beta(1)\Delta u(k + 2) + w(k + 3)$$

$$... = ... + ...$$

$$\hat{y}(k + m) = \hat{y}^0(k + m - 1) + \beta(m)\Delta u(k) + \beta(m - 1)\Delta u(k + 1) + ... + \beta(1)\Delta u(k + m - 1) + w(k + m)$$

$$\hat{y}(k + m + 1) = \hat{y}^0(k + m) + \beta(m + 1)\Delta u(k) + \beta(m)\Delta u(k + 1) + ... + \beta(2)\Delta u(k + m - 1) + w(k + m + 1)$$

$$... = ... + ...$$

$$\hat{y}(k + p) = \hat{y}^0(k + p - 1) + \beta(p)\Delta u(k) + \beta(p - 1)\Delta u(k + 1) + ... + \beta(p - m + 1)\Delta u(k + m - 1) + w(k + p) \tag{27.5}$$

This may be rewritten as:

$$\hat{\mathbf{y}}(k + 1) = \hat{\mathbf{y}}^0(k) + \mathbf{B}\,\Delta\mathbf{u}(k) + \mathbf{w}(k + 1) \tag{27.6}$$

where $\Delta\mathbf{u}(k)$ is the value of $\Delta\mathbf{u}$ at m points in time beginning at time step k; i.e.:

$$\Delta\mathbf{u}(k) = [\Delta u(k)\ \Delta u(k + 1)\ ...\ \Delta u(k + m - 1)]^T$$

and the matrix $\mathbf{B}$ given by:

$$\mathbf{B} = \begin{bmatrix} \beta(1) & 0 & 0 & ... & 0 \\ \beta(2) & \beta(1) & 0 & ... & 0 \\ \beta(3) & \beta(2) & \beta(1) & ... & 0 \\ ... & ... & ... & ... & ... \\ \beta(m) & \beta(m - 1) & \beta(m - 2) & ... & \beta(1) \\ \beta(m + 1) & \beta(m) & \beta(m - 1) & ... & \beta(2) \\ \beta(p) & \beta(p - 1) & \beta(p - 2) & ... & \beta(p - m + 1) \end{bmatrix}$$

is called the system's "dynamic matrix." Observe that it is made up of m columns of the system's step-response function appropriately shifted down in order. $\Delta\mathbf{u}(k)$ represents the m–dimensional vector of control moves; and $\mathbf{w}(k + 1)$ is the vector of the collective effect of unmodeled disturbances. We are yet to resolve the nontrivial issue of how this latter vector is to be determined ahead of time.

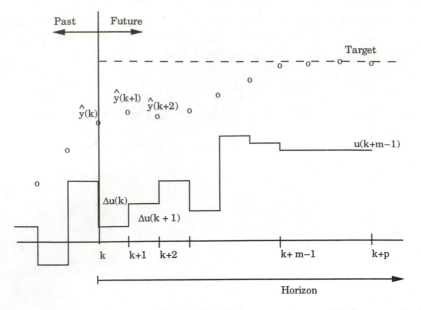

Figure 27.3. The elements of DMC: The "reference trajectory" is the set-point line.

27.3.2 The SISO (Unconstrained) Problem

Unconstrained Problem Formulation

Expressed in the form shown above in Eq. (27.6) the control problem becomes that of judiciously choosing, and implementing, the control sequence represented as $\Delta\mathbf{u}(k)$ such that the *predicted* system output is caused to move to, and remain at, the desired trajectory value:

$$\mathbf{y}^*(k + 1) = [y^*(k + 1), y^*(k + 2), y^*(k + 3), ..., y^*(k + p)]^T$$

(usually set constant at the desired set-point value), i.e., choose $\Delta\mathbf{u}(k)$ such that:

$$\hat{\mathbf{y}}^0(k) + \mathcal{B}\,\Delta\mathbf{u}(k) + \mathbf{w}(k + 1) = \mathbf{y}^*(k + 1) \tag{27.7}$$

If we now define $\mathbf{e}(k + 1)$, the *projected error vector* in DMC parlance, as the difference between the desired output trajectory $\mathbf{y}^*(k + 1)$ and the current output prediction *in the absence of further control action*, $\hat{\mathbf{y}}^0(k)$, corrected for the effect of unmeasured disturbance, $\mathbf{w}(k + 1)$, i.e.:

$$\mathbf{e}(k + 1) \triangleq \mathbf{y}^*(k + 1) - [\hat{\mathbf{y}}^0(k) + \mathbf{w}(k + 1)] \tag{27.8}$$

then the control problem reduces to choosing $\Delta\mathbf{u}(k)$ such that the equation:

$$\mathcal{B}\,\Delta\mathbf{u}(k) = \mathbf{e}(k + 1) \tag{27.9}$$

is satisfied. This is the final process model form for the unconstrained DMC formulation. The set of equations represent, on the RHS, the predicted deviation of plant output from desired set-point *in the absence of further control action*, and on the LHS, the predicted *incremental change* to be observed in the plant output as a result of implementing the sequence of control moves, $\Delta\mathbf{u}(k)$.

Unconstrained Problem Solution

Now, the expression in Eq. (27.9) usually contains an overdetermined set of equations since, in DMC, m, the contemplated number of control moves, is always chosen to be smaller than the prediction horizon length p. Thus, *no exact solution exists* for this set of equations in the sense that there is no vector $\Delta\mathbf{u}(k)$ that will cause the LHS to match the RHS exactly.

The set of equations in Eq. (27.9) may therefore only be "solved" by finding the vector $\Delta\mathbf{u}(k)$ that minimizes some vector norm: $\|(\mathbf{e}(k + 1) - \mathbb{B}\,\Delta\mathbf{u}(k))\|$ (cf. Appendix D), i.e., Eq. (27.9) is solved by minimizing some metric that measures the difference between the vector on the LHS (*what the process can do*) and the one on the RHS (*what the process is required to do*).

In the specific case where the chosen norm is the sum of squares of the elements of $(\mathbf{e}(k + 1) - \mathbb{B}\,\Delta\mathbf{u}(k))$ (i.e., the *Euclidian norm*), this is the same as solving the unconstrained optimization problem:

$$\min_{\Delta\mathbf{u}(k)} \phi \; = \; [\mathbf{e}(k + 1) - \mathbb{B}\,\Delta\mathbf{u}(k)]^T\,[\mathbf{e}(k + 1) - \mathbb{B}\,\Delta\mathbf{u}(k)] \tag{27.10}$$

which, of course, is the classical "least squares" problem with a well-known, analytical, closed-form solution. By differentiating in Eq. (27.10) with respect to $\Delta\mathbf{u}(k)$, we obtain quite straightforwardly:

$$\frac{\partial\phi}{\partial\Delta\mathbf{u}(k)} \; = \; -\mathbb{B}^T[\mathbf{e}(k + 1) - \mathbb{B}\,\Delta\mathbf{u}(k)]$$

and by setting this equal to zero we obtain the so-called normal equations:

$$\mathbb{B}^T\mathbb{B}\,\Delta\mathbf{u}(k) \; = \; \mathbb{B}^T\mathbf{e}(k + 1)$$

and now, provided $\mathbb{B}^T\mathbb{B}$ is nonsingular, we obtain the well-known least squares solution:

$$\Delta\mathbf{u}(k) \; = \; (\mathbb{B}^T\mathbb{B})^{-1}\mathbb{B}^T\mathbf{e}(k + 1) \tag{27.11}$$

or

$$\Delta\mathbf{u}(k) \; = \; \mathbb{B}^+\mathbf{e}(k + 1) \tag{27.12}$$

where

$$\mathbb{B}^+ \; = \; (\mathbb{B}^T\mathbb{B})^{-1}\mathbb{B}^T$$

is known as the "left inverse" of $\mathbb{B}$ because while:

$$\mathbb{B}^+\mathbb{B} \; = \; \mathbf{I}$$

$\mathbb{B}\,\mathbb{B}^+$ on the other hand gives an *idempotent* matrix, and not the *identity* matrix $\mathbf{I}$ (see Appendix D).

Observe therefore that while the model prediction in Eq. (27.9) involves the system's dynamic matrix $\mathbf{B}$, the control action calculation in Eq. (27.11) involves the *left inverse* of $\mathbf{B}$.

It is important to point out here that from the definition in Eq. (27.8), and from the model equation in Eq. (27.6), we see that:

$$\mathbf{e}(k + 1) - \mathbf{B}\,\Delta\mathbf{u}(k) = \mathbf{y}^*(k + 1) - \hat{\mathbf{y}}(k + 1) \tag{27.13}$$

the deviation of the predicted process output from the desired trajectory, which we now define as the "residual error vector" $\varepsilon_r(k + 1)$:

$$\varepsilon_r(k + 1) = \mathbf{y}^*(k + 1) - \hat{\mathbf{y}}(k + 1) \tag{27.14}$$

so that:

$$\mathbf{e}(k + 1) - \mathbf{B}\,\Delta\mathbf{u}(k) = \varepsilon_r(k + 1) \tag{27.15}$$

Thus a control sequence $\Delta\mathbf{u}(k)$ chosen according to Eq. (27.11) minimizes the squared residual error between the predicted process output and the desired trajectory over the p-interval prediction horizon.

The difference between the two error vectors $\mathbf{e}(k + 1)$ and $\varepsilon_r(k + 1)$ in Eq. (27.15) is subtle but extremely important: $\mathbf{e}(k + 1)$ we recall as the "projected error vector" — the predicted output error that will result if no further control action is taken; $\varepsilon_r(k + 1)$ on the other hand is the residual error left *after* the control action aimed at eliminating $\mathbf{e}(k + 1)$ has been implemented. Thus $\mathbf{e}(k + 1)$ is used to compute control action; the result of the computation is conditioned upon minimizing the residual $\varepsilon_r(k + 1)$.

In the *explicit*, unconstrained DMC solution, $\Delta\mathbf{u}(k)$ is determined analytically using $\mathbf{e}(k + 1)$ as shown in Eq. (27.11); in the *implicit* form obtained by solving the optimization problem in Eq. (27.10), $\varepsilon_r(k + 1)$ may be introduced for $\mathbf{e}(k + 1) - \mathbf{B}\,\Delta\mathbf{u}(k)$ and $\Delta\mathbf{u}(k)$ is then determined numerically so that the residual vector is minimized.

In practice, a penalty against excessive control action is often incorporated into the optimization objective to read:

$$\min_{\Delta\mathbf{u}(k)}\phi = [\mathbf{e}(k + 1) - \mathbf{B}\,\Delta\mathbf{u}(k)]^T\,[\mathbf{e}(k + 1) - \mathbf{B}\,\Delta\mathbf{u}(k)] + K^2[\Delta\mathbf{u}(k)^T\Delta\mathbf{u}(k)] \tag{27.16}$$

It is easy to show that in this case the resulting control action is:

$$\Delta\mathbf{u}(k) = \left(\mathbf{B}^T\mathbf{B} + K^2\,\mathbf{I}\right)^{-1}\mathbf{B}^T\mathbf{e}(k + 1) \tag{27.17}$$

a slight modification of Eq. (27.11).

Unmeasured Disturbance Estimation and Prediction Update

The projected error vector $\mathbf{e}(k + 1)$ in Eq. (27.8) requires the vector of *future* values of unmeasured disturbance effects that are unknown at time instant k; they can only be estimated solely on the basis of currently available information.

At the penultimate time instant $(k - 1)$, on the basis of this same step-response model, the value $\hat{y}(k)$ was predicted as the output value to expect at time instant k. With the actual measurement $y(k)$ now available, the discrepancy in the model prediction, i.e.:

$$\delta(k) = y(k) - \hat{y}(k) \tag{27.18}$$

also becomes available.

In DMC, this discrepancy is assumed to be due to the effect of unmodeled disturbances yet unaccounted for by the model; it is further assumed that this is the best estimate of the future values of such disturbance effects. Thus, each element $w(k + i)$, $i = 1, 2, ..., p$ of the vector $\mathbf{w}(k + 1)$ is estimated as:

$$\hat{w}(k + i) = \delta(k) = y(k) - \hat{y}(k); \quad i = 1, 2, ..., p \tag{27.19}$$

and these values introduced for $\mathbf{e}(k + 1)$ in Eq. (27.8).

Implementation Strategy

We now note that because of the inevitable prediction inaccuracies, it is not advisable to implement the entire sequence, $\Delta u(k)$, $\Delta u(k + 1)$, ..., $\Delta u(k + m - 1)$ as calculated from Eq. (27.11) over the next m intervals in automatic succession. This is because despite our best efforts in estimating the vector $\mathbf{w}(k + 1)$, it is impossible to anticipate precisely over the next m time intervals, the unavoidable plant/model mismatch, and other unmodeled disturbances that are bound to cause the actual plant state to differ from the predictions used to compute this sequence of control actions. Also the process output set-point may change at any time over the next m intervals. Such a set of precomputed control moves is therefore inherently incapable of reflecting the changes that would subsequently occur after the computation.

The DMC strategy is therefore to implement only the first move, $\Delta u(k)$, and then:

1. Update the projected error vector $\mathbf{e}(k + 1)$ in Eq. (27.8) at the next time instant, i.e.:
    * Update $\mathbf{y}^*(k + 1)$ with any new desired set-point information.
    * Update $\hat{\mathbf{y}}^0(k)$ by adding the effect of implementing $\Delta u(k)$, and assuming no other control move beyond this will be implemented.
    * Using available system measurement, and the corresponding model prediction, update the $\mathbf{w}(k + 1)$ estimates as indicated in Eqs. (27.18) and (27.19).

2. Shift the origin of the prediction horizon forward to $(k + 1)$ by dropping off the first element in each of the *updated* vectors $\mathbf{y}^*(k + 1)$, $\hat{\mathbf{y}}^0(k)$, $\mathbf{w}(k + 1)$, advancing the other elements in order, so that the second element becomes the first, the third becomes the second, etc., and finally filling the vacant last element by linear extrapolation. The updated and "shifted" vectors now become, respectively, $\mathbf{y}^*(k + 2)$, $\hat{\mathbf{y}}^0(k + 1)$, $\mathbf{w}(k + 2)$.

3. Recompute the control move sequence using these updated and "shifted" vectors, implement the first one, and repeat the cycle.

The discussion given above is easily extended to multivariable systems and to the constrained case. But before demonstrating how this is done, it is important to pause and observe the four MPC elements as utilized in the DMC scheme:

1. The *reference trajectory* is specified simply as a step to the desired set-point or target value.

2. *Process output prediction* is carried out using the step-response model as given in Eq. (27.6), with $\mathbb{B}$, the dynamic matrix as the "main operator"; this is further rearranged and simplified to the equivalent modeling equations in Eq. (27.9).

3. *Control action computation* is carried out using Eq. (27.11), which involves the use of the generalized (in this case, the left) inverse of the dynamic matrix $\mathbb{B}$.

4. *Model prediction update* is achieved by providing estimates for $\mathbf{w}(k + 1)$ as indicated in Eq. (27.8), using the single correction term calculated at time instant k as indicated in Eqs. (27.18) and (27.19).

27.3.3 Extensions to Multivariable Systems

The DMC scheme as discussed above extends quite straightforwardly to systems with multiple inputs and multiple outputs; the basic form of the modeling equation in Eq. (27.9) remains the same, except that the matrices and vectors become larger and appropriately partitioned. For example, for a two-input, two-output system, the DMC equations are given by:

$$\begin{bmatrix} \mathbf{e}_1(k + 1) \\ \cdots \\ \mathbf{e}_2(k + 1) \end{bmatrix} = \begin{bmatrix} \mathbb{B}_{11} & : & \mathbb{B}_{12} \\ \cdots & : & \cdots \\ \mathbb{B}_{21} & : & \mathbb{B}_{22} \end{bmatrix} \begin{bmatrix} \Delta\mathbf{u}_1(k) \\ \cdots \\ \Delta\mathbf{u}_2(k) \end{bmatrix}. \qquad (27.20)$$

where $\mathbf{e}_i(k + 1)$ is the projected error vector for the ith output variable ($i = 1, 2$), $\Delta\mathbf{u}_j(k)$ is the sequence of control moves for the jth input variable ($j = 1, 2$), and $\mathbb{B}_{ij}$ is the dynamic matrix formed from the step-response of the *output variable* i to the *input variable* j ($i = 1, 2; j = 1, 2$).

By simply expanding the definition of the vectors and matrices so that the entire partitioned vector on the LHS is represented as $\mathbf{e}(k + 1)$, the one on the RHS as $\Delta\mathbf{u}(k)$, and the partitioned matrix simply as $\mathbb{B}$, observe therefore that Eq. (27.20) is exactly of the same form as Eq. (27.9), and therefore the least-square control move sequences are also given by Eq. (27.11); the only difference is that the matrices and vectors are now larger in dimension.

Scaling

With multivariable processes, it often happens that a unit change in one output variable is much more important than a unit change in another. A typical example is the two-point distillation control system in which the top product quality (measured in mole fractions) and a tray temperature in the bottoms section (measured in °F) are to be controlled. In the case of the first variable, a unit change represents the entire scale (since mole fractions must be between 0 and 1), while temperature changes of 10°F are not considered unusual. Thus in this particular case, a unit change in temperature may be deemed as important as a change of 0.01 in mole fraction.

DMC recognizes the importance of scaling the error vector $\mathbf{e}(k+1)$ such that equally important changes in the various output variables are treated equally. This is done through the introduction of a weighting matrix Γ as a tuning parameter so that Eq. (27.11) reads:

$$\Delta\mathbf{u}(k) = (\mathfrak{B}^T\Gamma\mathfrak{B})^{-1}\mathfrak{B}^T\Gamma\mathbf{e}(k+1)$$

The role of this matrix and insight into how to chose its elements are best understood via the following considerations that show clearly how it arises fundamentally (cf. Ref. [53]).

Let the elements of the matrix $\mathbf{W}$ represent the importance we wish to attach to a unit change in each variable *relative* to a unit change in the others. Upon introducing this scaling matrix, the DMC Eq. (27.9) becomes:

$$\mathbf{W}\mathfrak{B}\,\Delta\mathbf{u}(k) = \mathbf{W}\mathbf{e}(k+1) \tag{27.21}$$

so that the objective of the least squares optimization problem is now modified to:

$$\min_{\Delta\mathbf{u}(k)} \phi = [\mathbf{e}(k+1) - \mathfrak{B}\,\Delta\mathbf{u}(k)]^T\,\mathbf{W}^T\mathbf{W}[\mathbf{e}(k+1) - \mathfrak{B}\,\Delta\mathbf{u}(k)]$$

or

$$\min_{\Delta\mathbf{u}(k)} \phi = [\mathbf{e}(k+1) - \mathfrak{B}\,\Delta\mathbf{u}(k)]^T\,\Gamma[\mathbf{e}(k+1) - \mathfrak{B}\,\Delta\mathbf{u}(k)] \tag{27.22}$$

with the matrix Γ defined as:

$$\Gamma = \mathbf{W}^T\mathbf{W} \tag{27.23}$$

from which $\partial\phi/\partial\Delta\mathbf{u}(k) = 0$ then yields the required solution:

$$\Delta\mathbf{u}(k) = (\mathfrak{B}^T\Gamma\mathfrak{B})^{-1}\mathfrak{B}^T\Gamma\mathbf{e}(k+1) \tag{27.24}$$

It may be observed therefore that by definition, the weighting matrix Γ is *positive definite*; and it arises from the scaling matrix $\mathbf{W}$ as shown in Eq. (27.23). It is thus best to determine Γ by specifying the elements of $\mathbf{W}$ chosen to achieve the desired scaling effects in the various process outputs.

Let us note in closing that the reader familiar with the statistical theory of parameter estimation will immediately have recognized the striking similarities between least squares parameter estimation and the DMC problem formulation, and how the results shown above are directly related to the ordinary least squares, and the weighted least squares problems in linear statistical parameter estimation. In fact, many concepts which are traditionally associated with various aspects of statistical analysis feature prominently in the structure of DMC. A detailed discussion of these issues are contained in Ref. [53].

27.3.4 The General Constrained Problem

The most general form for specifying the DMC problem may now be expressed as:

$$\min_{(\Delta u(k),...,\Delta u(k+m-1))} \sum_{l=1}^{p} \varepsilon_r(k+l)^T \Gamma \varepsilon_r(k+l) + \sum_{l=1}^{m} [u(k+l-1)^T F u(k+l-1)$$
$$+ \quad \Delta u(k+l-1)^T R \Delta u(k+l-1)] \qquad (27.25)$$

where the elements of the weighting matrices Γ, F, and R, along with the prediction horizon p, and the contemplated number of precomputed control moves m, are the tuning parameters.

When constraints are involved, this problem formulation must be augmented with the constraint equations. In this case, if we define the collective vector of all the control moves as v, i.e.:

$$v = [u(k), u(k+1), ..., u(k+m-1)]^T \qquad (27.26)$$

the optimization problem can be written as the following standard Quadratic Program (QP) (cf., for example, Refs. [28, 84]):

$$\min_{v} \phi(v) = \frac{1}{2} v^T Q v + q^T v \qquad (27.27)$$

subject to:

$$\Lambda^T v \geq \eta \qquad (27.28)$$

where the matrices Q, Λ, are related to the tuning parameters noted above and to some hard constraints; the vectors v and η on the other hand are linear functions of the output prediction vector, $\hat{y}$, the past inputs, $u(k-1)$, $u(k-2)$, ..., $u(k-n)$, as well as being functions of the tuning parameters. There are many efficient algorithms for solving this QP (e.g., Refs. [57, 66, 84]).

It should perhaps be noted, in conclusion, that the original DMC scheme reported in the pioneering papers was not formulated in this fashion; the above is, strictly speaking, the QDMC formulation. As will be noted in Section 27.5, the commercial DMC package is based on optimizing an absolute value (L_1 norm) of the residual error vector rather than the quadratic function (L_2 norm) as noted above.

27.4   MODEL ALGORITHMIC CONTROL (MAC)

As noted earlier, MAC is very similar to DMC *in principle*. However, the fundamental similarities they share notwithstanding, there are some subtle but important distinctions between MAC and DMC. We will therefore focus on such distinctions rather than provide a full treatment comparable to that given DMC in the previous section.

27.4.1 Process Representation

The model form for the MAC scheme:

$$\hat{y}(k+1) = \hat{y}^0(k) + G u(k) + w(k+1) \qquad (27.29)$$

is very similar to that in Eq. (27.6): the main differences are that this is an impulse-response model involving **u**, the actual input, whereas the DMC model in Eq. (27.6) is a step-response model involving Δ**u**, the change in the input; the matrix **G** here is composed of impulse-response coefficients, in precisely the same manner in which the "dynamic matrix" ℬ in Eq. (27.6) is composed of the step-response coefficients.

27.4.2 Problem Formulation, Solution, and Implementation via IDCOM

Problem Formulation

The unconstrained problem formulation may then be posed similarly as in Eq. (27.7), i.e., to choose **u**(*k*) such that:

$$\hat{\mathbf{y}}^0(k) \; + \; \mathbf{G}\mathbf{u}(k) \; + \; \mathbf{w}(k+1) \; = \; \mathbf{y}^*(k+1) \tag{27.30}$$

However, how the elements of the desired trajectory, **y***(*k* + 1) are calculated in this case constitutes what is perhaps the most fundamental difference between DMC and MAC. While with DMC all the elements of the entire vector **y***(*k* + 1) are set equal to the current set-point value, i.e.:

$$y^*(k + i) \; = \; y_{set\text{-}point}(k); \;\; i = 1, 2, \ldots, p \tag{27.31}$$

with MAC, the elements of the desired trajectory are taken as a reference trajectory, $y^*(k)$, based on the current *measured* value, $y(k)$, and the current set-point, $y_{set\text{-}point}(k)$ as follows:

$$y^*(k + 1) \; = \; \alpha y(k) \; + \; (1 - \alpha)y_{set\text{-}point}(k); \;\; 0 < \alpha < 1 \tag{27.32}$$

with subsequent values determined according to:

$$y^*(k + i) \; = \; \alpha y(k + i - 1) \; + \; (1 - \alpha)y_{set\text{-}point}(k); \; i > 1 \tag{27.33}$$

Thus the MAC desired trajectory is a first-order approach to set-point from the current *measured* value. This is indicated in Figure 27.4.

As noted in Ref. [46], it is possible to use higher order trajectories; however, the first-order strategy noted above is used almost exclusively.

It is important to note that the parameter α determines the desired speed of approach to set-point; a small value (close to zero) indicates the desire for a swift approach to set-point, while a larger value (close to 1) indicates the opposite: a more sluggish approach to set-point. This parameter is thus the primary tuning parameter for the MAC controller, and its choice is very closely tied to the robustness of the overall control system (cf. Ref. [54]). As noted in Ref. [28] the parameter α is a more direct, and more intuitive, tuning parameter than the weighting matrices, horizon lengths, etc., utilized in the DMC formulation.

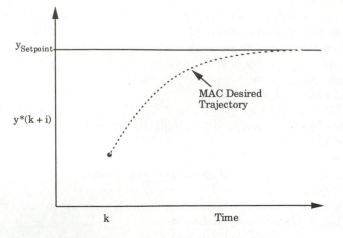

Figure 27.4 .   The MAC desired trajectory.

Finally, we also note that in the DMC formulation, the number of control moves m is deliberately chosen to be less than the prediction horizon length p, and both are available as tuning parameters. With MAC $m = p$ and neither is used as a tuning parameter; p is determined by the truncation order of the finite impulse-response function, i.e., for all $j > p$, $h(j) \approx 0$.

Problem Solution

In a similar vein to the DMC problem, MAC determines the control sequence by minimizing the objective function:

$$\phi = \sum_{i=1}^{p} [y^*(k + i) - \hat{y}(k + i)]^2 \gamma(i) \tag{27.34}$$

or

$$\phi = \sum_{i=1}^{p} [\varepsilon_r(k + i)]^2 \gamma(i)$$

where the $\gamma(i)$ are (possibly) time-varying weights.

Implementation

The model prediction is updated and corrected for the effect of modeling errors and unmodeled disturbances precisely as indicated in Eqs. (27.18) and (27.19) for DMC. The same strategy of implementing only the first element of the computed input sequence and then repeating the optimization at the next time instant is also adopted in MAC.

The constrained problem poses no additional complications; the constraints are simply appended to the objective function and the resulting quadratic program (QP) is solved using a proprietary algorithm resident within the IDCOM software. For obvious reasons, not much has been disclosed about this algorithm; however, it is known to perform a two-step QP when there are more inputs than outputs. Under these conditions, because of the extra degrees of freedom, the objective of the second QP is to minimize the deviation of the input sequence from their respective prespecified ideal resting values in addition to meeting the objectives of the first QP. This ensures that the objectives of the first QP are met with the optimum expenditure of control effort.

27.5 COMMERCIAL MODEL PREDICTIVE CONTROL SCHEMES: A QUALITATIVE REVIEW

A wide variety of MPC schemes are currently available as commercial computer software packages offered by various vendors. The salient features of several of these commercial products are now summarized qualitatively below.

27.5.1 IDCOM

IDCOM — for **ID**entification and **COM**mand — is the commercial package marketed in the United States by Setpoint, Inc. for implementing the Model Algorithmic Control (MAC) scheme. As noted earlier, the original IDCOM controller was developed in France by Adersa/Gerbios; later versions (IDCOM-S for single variable systems, and IDCOM-M for multivariable systems) were developed jointly by Adersa and Setpoint, Inc. in the United States.

Model Form

The process model form utilized by IDCOM is the impulse-response version of the linear discrete convolution model form. The model is usually identified from plant dynamic test data. Rather than identify the model directly in the impulse-response form, however, it is often recommended to identify the ARMAX type model first and then to convert this to the required impulse-response form. This is primarily because the identification of the impulse-response function directly from noisy data usually leads to estimates with large variances — a direct consequence of the fact that the impulse-response coefficients are all fundamentally correlated.

The IDCOM package includes a routine for model identification.

Fundamental Objective

The primary objective of the IDCOM controller is to determine the sequence of control moves that will minimize, over the prediction horizon of interest, the sum of squared deviations of the predicted outputs (Controlled Variables, in IDCOM parlance) from their set-points (or zone limits, when bounds rather than precise target values are specified) subject to *both* absolute, and rate-of-change, constraints on the inputs (Manipulated Variables in IDCOM parlance).

When there are more manipulated variables than controlled variables, incorporated in the IDCOM framework is the concept of Ideal Resting Values for the manipulated variables. In this case, in addition to the primary objective, the IDCOM controller then tries to minimize the sum of squared deviations of the manipulated variables from their respective ideal resting values, but still subject to satisfying the primary objectives regarding the controlled variables.

Thus the IDCOM strategy involves a two-step optimization problem: the primary problem involves choosing the control sequence required to drive controlled variables close to target; the second involves optimizing the use of control effort in achieving the objectives of the primary problem. This two-step optimization problem is solved using a quadratic programming (QP) approach.

Even when there are no excess manipulated variables, the ideal resting values concept is still particularly valuable when, for operational and/or economic reasons, there is some benefit in having a manipulated variable at a specific steady-state value. No other MPC scheme incorporates this concept explicitly.

Tuning

The major tuning parameter in the IDCOM controller is the "reference time" by which the desired closed-loop speed of response is specified. A smaller "reference time" (corresponding to an α value in Eqs. (27.32) and (27.33) close to zero) provides faster closed-loop response, but the manipulated variable is moved more vigorously; a larger "reference time" (corresponding to an α value close to unity) provides a slower but smoother response requiring less vigorous control action. By virtue of this fact, a larger "reference time" also improves the robustness of the overall control system since then the controller becomes less sensitive to plant/model mismatch.

Implementation Platform

Even though a version of IDCOM (IDCOM-B) is currently available for implementation directly on the Bailey DCS, the IDCOM software typically resides in the process control computer and communicates with the distributed control system (DCS) database through some interface. A common hardware implementation strategy recommended by Set-point requires linking IDCOM — resident in a Digital VAX — with the Honeywell TDC 3000 DCS system through Honeywell's CM-50s interface and a generic scanner program, called the Phoenix interface, written by Set-point. Of course, with the appropriate interface, any other available DCS system could be used in place of the TDC 3000.

Example Industrial Applications

Industrial applications of IDCOM include the control of reactors and regenerators of Fluid Catalytic Cracking (FCC) and coking units; cutpoint control of crude distillation units; reactor control in reforming and hydrocracking units; control of C2 splitters in ethylene units; and control of various distillation columns.

27.5.2 DMC

The Dynamic Matrix Control (DMC) scheme was originally developed at Shell and is currently being marketed by Dynamic Matrix Control Corporation. It is also available from other vendors under license from DMC Corporation. As noted earlier, even though they differ in certain details, the basic control philosophies of DMC and IDCOM are similar.

Model Form

DMC utilizes the step-response model form usually obtained from plant dynamic test data. The DMC package includes an identification software package — Dynamic Matrix Identification (DMI) — developed specifically for identifying step-response models for multivariable processes. The strategy employed remains undisclosed for proprietary reasons.

Fundamental Objective

The original version of the DMC algorithm finds the sequence of control moves that minimize the squared deviation of the predicted output from their respective set-points. This is done by first solving the unconstrained optimization problem with the quadratic objective function explicitly, obtaining the least square solution. The constraint handling capability is then added by iteratively checking for constraint violation and resolving the modified problem if violation occurs. In the subsequent Shell versions of DMC, constraints handling is incorporated more directly by formulating the problem as a Quadratic Program. This version, as noted earlier is known as Quadratic Dynamic Matrix Control (QDMC).

The DMC package marketed by DMC Corporation is different from QDMC in that the controller objective function requires the minimization of the *absolute* value of the vector of errors between the predicted process output and their respective set-points or zone limits. This absolute value formulation allows the constrained optimization problem to be solved by the less computationally intensive linear programming (LP) approach.

Since only a single-step optimization is performed, ideal resting value optimization is not an explicit part of the DMC algorithm. In order to keep manipulated variables close to their prespecified economic optimum values (in situations when there are more manipulated variables than controlled variables), additional terms are introduced into the overall objective function. The relative importance of the various aspects of this composite objective function (minimizing prediction error, versus penalty against excessive control action, versus keeping manipulated variables at the economic optimum) is indicated by the values introduced for the various weighting matrices associated with each individual component.

Tuning

The main tuning parameters are known as the *move suppression factors* (the scalar K in Eq. (27.17) or the matrix $\mathbf{R}$ in Eq. (27.25)); they indicate the trade-off between the amount of movement allowed in the manipulated variables and the rate at which the output deviation from set-point is reduced over the prediction horizon. The prediction horizon as well as the number of control moves computed at each optimization cycle are also available for use as tuning parameters.

As earlier explained, the weighting matrix Γ in Eqs. (27.23) and (27.25) is used primarily for scaling in the multivariable case; it permits the assignment of more or less weight to the objective of reducing the predicted error for the individual output variables; the matrix $\mathbf{F}$ in Eq. (27.25) is primarily for penalizing the movement of manipulated variables away from prespecified economic optimum values when such a policy is desired and reasonable.

Implementation Platform

The DMC software will typically reside in the process control computer. Some DMC Corporation applications include cases where DMC resides in the VAX computer and communicates with Honeywell TDC 3000 through CM-50s.

Example Industrial Applications

DMC appears to have the largest industrial application base if the 15 or more years of applications at Shell Oil Company is considered. Nevertheless, DMC applications are very similar to IDCOM's, with the majority in the oil and petrochemical industries. More recent DMC applications now include some chemical and polymerization reactors.

27.5.3 OPC

Optimum Predictive Control (OPC) was developed, and is currently being marketed, by Treiber Controls, Inc. The control concept, the model form, and the fundamental objectives are all similar to those employed by DMC.

Model Form

Like DMC, OPC uses the step-response model for output prediction. However, this model is obtained by employing time-series analysis to identify a transfer function model from plant data; this is then converted to the step-response model. This is in direct contrast to the DMC identification strategy in which the so-called nonparameteric step response is identified *directly* using DMI.

Furthermore the identification module of OPC *only* permits the identification of the component input/output elements of a multivariable transfer function matrix one at a time, whereas with DMI it is possible to identify these elements simultaneously.

Fundamental Objective

The OPC objective function is essentially the same as that of the original DMC; control action is calculated by minimizing a quadratic performance index. However, in order to be able to handle constraints and still run OPC on smaller machines, the optimization problem, which is more naturally cast as the computationally intensive Quadratic Program (QP), had to be reformulated. By the introduction of two artificial variables, the original quadratic objective function is transformed into a linear objective function with one additional equality constraint; this permits the handling of constraints using less computationally intensive Linear Programming (LP) approach, so that OPC can indeed run on smaller machines.

Tuning

The major tuning parameters in OPC are the same as with DMC: the penalty weights on the control moves. Other variables such as the prediction horizon, the number of control moves, etc., are also available as tuning parameters.

Implementation Platform

With OPC the model building, controller design and simulation tasks are all carried out on IBM compatible personal computers. By design, the OPC controller implementation can be carried out on Bailey's MFC03 multifunction controller. This eliminates the need to provide additional communication between the host computer and the DCS system.

Example Industrial Applications

Treiber Control, Inc. is somewhat of a smaller company than Setpoint, Inc. and DMC Corporation. Thus there are not as many OPC applications as there are IDCOM or DMC applications. However, OPC has been applied on hydrocracking units, tar sands extraction cells, high-purity distillation columns, polymer extruders, pulp mills, and multivariable polyethylene reactors.

27.5.4 PC

Predictive Controller (PC) is marketed by Profimatics, Inc. In many ways it appears to combine various aspects of the IDCOM strategy with those of DMC.

Model Form

Its model form is fundamentally similar to that used by IDCOM. Like IDCOM it uses a reference trajectory defined by a filter whose time constant determines the desired speed of closed-loop response.

However, unlike any of the other MPC schemes, the model updating scheme is somewhat more realistic. The model prediction correction strategy is enhanced by including past prediction errors so that observed trends can be incorporated systematically.

Fundamental Objective

The quadratic objective function utilized by PC and its iterative strategy for constraint handling is reminiscent of the original DMC algorithm. However, again, unlike the other MPC schemes in which multiple future control moves are computed at each optimization cycle (even though only the first one is implemented), the PC scheme adopts a single move strategy: each optimization cycle is concerned with computing only *one* control move. This policy may affect controller performance adversely especially when constraints are active.

Tuning

In PC, as in IDCOM, the reference trajectory is used to define the desired closed-loop response; as such the filter time constant for obtaining the reference trajectory is the major tuning parameter.

Implementation Platform

PC is implemented on the same platforms as IDCOM and DMC.

Example Industrial Applications

Published applications of the PC algorithm have been mostly in oil operations, particularly on FCC units.

27.5.5 HMPC

Horizon Predictive Control (HPC) is marketed by Honeywell as a single loop (multi-input, single-output) scheme. The multivariable version, Honeywell Multivariable Predictive Control (HMPC), is relatively new on the market.

Model Form

Both HPC and HMPC are flexible in the model form they can accept for carrying out model prediction. However, the identifier included with these packages identifies parametric ARMAX or transfer function models from plant data.

Fundamental Objective

The objectives of HPC (and of HMPC) are fundamentally different from that of the other MPC schemes. A correction horizon (the main tuning parameter) is chosen by the designer as the time after which the prediction error is to be reduced permanently to zero; as there is an infinite set of control move sequences that will achieve this nominal objective, the HPC objective is therefore to find the sequence that corresponds to a minimum expenditure of control effort.

It should be noted that the quadratic objective function of the DMC algorithm may be reformulated to resemble the HPC objective.

For proprietary reasons, there are several other aspects of the HMPC/HPC algorithm that are currently unavailable.

Tuning

The major tuning parameter is the correction horizon. It is defined as the length of time within which the controller must return the variable to its set-point or nearest limit. It is claimed that even in the multivariable case, all that is required is the specification of this correction horizon for each controlled variable.

Implementation Platform

The model building, controller design, and simulation modules for HMPC/HPC are implemented on the personal computer. Estimation of model parameters from plant dynamic test data is carried out using MATLAB's system identification toolbox. The controller coefficients are transferred from the PC to the Honeywell DCS. The control algorithm is implemented directly in the Honeywell application module so that no process control computer is needed for implementing HMPC/HPC.

Example Industrial Applications

Since HMPC/HPC are relatively new additions to the commercial MPC market, publicly documented industrial applications are still relatively few.

27.6 ACADEMIC AND OTHER CONTRIBUTIONS

Since the publication of the original DMC and MAC papers, the general subject matter of MPC has received considerable attention from academic as well as industrial researchers. The result is that substantial contributions have been made to the advancement of the theory and practice of MPC. What follows is a representative — rather than exhaustive — review of the various aspects of MPC that have received attention.

27.6.1 Fundamental Understanding

As originally conceived and presented in the pioneering publications, DMC and MAC were clearly empirical techniques based on technically sound, but heuristic, arguments. Primarily because a strong theoretical framework was not yet available for these techniques therefore, the remarkable industrial success they enjoyed (especially where others had failed previously) was initially difficult to explain. Garcia and Morari [23] provided some theoretical results, putting these techniques as well as others which, at best appeared only vaguely related, into one unified structure they subsequently referred to as the "Internal Model Control" (IMC) structure (see Figure 27.5; see also the series of papers that followed [25, 26, 70, 20]). More fundamental theoretical results and insight into the stability, robustness, and other properties of MPC appeared later in the papers by Rawlings and coworkers [44, 51, 61].

In the same year that the original Garcia and Morari paper was published, the first paper on the basic theoretical properties of MAC was published by Rouhani and Mehra [71] in which was discussed, among other things, the theoretical basis of how problematic dynamics (particularly inverse response) are handled by MAC.

Seborg and coworkers (e.g., [10, 39]) connected the concepts of optimization via linear programming for constrained multivariable control with MPC; they also provided some fundamental analysis of the concepts involved in the design of both the predictor and the control elements of MPC. In the paper by Maurath *et al.* [42], the concept of principal component analysis was introduced for the design of MPC schemes.

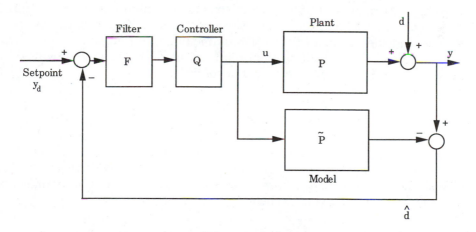

Figure 27.5. The IMC structure.

Brosilow and coworkers [7, 8] also contributed to the initial elucidation of the fundamental structure now known as IMC, as well as to the use of linear programming within the concept of multivariable control.

The quadratic programming approach to MPC was first introduced by Garcia [24] and Ricker [66]. By posing the constrained MPC problem as a Quadratic Program, and by recognizing the similarity between the problem solved at every time instant in standard MPC, it is possible to take advantage of efficient algorithms that already exist for solving such optimization problems.

A statistical interpretation of the DMC strategy was presented by Ogunnaike [53]. This work establishes that within the DMC framework reside the same principles upon which such time-proven statistical concepts as least squares parameter estimation, the Levenberg-Marquart nonlinear regression method, Ridge regression, and Bayesian sequential estimation are based, thereby providing another vehicle for understanding DMC's success.

Using arguments from the statistical theory of time-series analysis, MacGregor [38] established, among other things, the theoretical basis for the first-order filter used to generate the MAC reference trajectory. He showed that if the unmodeled aspects of the plant behavior (due to model errors *and* unmeasured disturbances) can be represented as a random walk process observed with white measurement noise (an assumption which is reasonably close to the observed reality), then the minimum variance predictor of the effect of these unmodeled dynamics, $\hat{w}(k + 1)$ in the language of Section 27.3, is obtained by utilizing a first-order filter in the feedback path, as advocated by the MAC scheme.

MacGregor also established that the output of this filter is easily rearranged to give an exponentially weighted moving average (EWMA) of past values of the disturbance sequence; it then becomes easy to deduce that under conditions in which the process dynamics are ignored, the EWMA control charts used in Statistical Process Control (SPC) become identical to the MAC strategy incorporating this first-order filter. A detailed discussion of this and other related issues may be found in Chapter 28.

27.6.2 Additional Structurally Similar MPC Schemes

Several control methodologies bearing structural similarities with DMC and MAC have since been proposed. Most of the differences between these strategies and the original MPC schemes are subtle, but, in some cases, these subtle differences can lead to important distinctions. The following is a representative selection from this class of control strategies:

1. **Generic Model Control** [35, 36]:
   In which any available process model (linear or nonlinear, dynamic or static) is incorporated *directly* into the algorithm designed to calculate the control action required to cause the process to follow the trajectory prescribed by a reference system. In the SISO case, the reference system has two poles and one zero whose locations constitute the tuning parameters. Lee [37] has compared this scheme with standard MPC (see Chapter 19).

2. **Adaptive Generalized Predictive Control** [11, 12]:
   In which the MPC paradigm is embedded in a self-tuning/adaptive control environment. This (and other similar techniques in this category) are largely limited to SISO cases; extensions to multivariable systems may be conceptually straightforward, but the implementations details are daunting. Clarke [13] presents a review of the issues regarding robustness and parameter estimation within this controller design framework.

3. **Simplified Model Predictive Control** [2–4]:
   In which the computational aspects of the standard MPC are simplified by the device of specifying the nominal open-loop system response as the base case requirement for the closed-loop control system. The obvious comparisons with standard MPC are discussed by Arulanan and Deshpande [2–4]. The original concept may be found in Tu and Tsing [76].

4. **Forward Modeling Control** [21, 55]:
   In which computational simplicity is achieved by requiring that control action be calculated to minimize the error at only *one* single point in the future, as opposed to the multipoint optimization strategy of standard MPC. The strategy is similar to that of Marchetti *et al.* [39].

27.6.3 Robustness Issues

Since the process model plays a central role in MPC, and given that model uncertainty is an unavoidable fact, how MPC schemes perform in the face of plant/model mismatch is of prime importance.

A general analysis of the stability and robustness of *standard* MPC schemes is not possible because of its finite horizon formulation. Certain specific issues concerning robust stability and robust performance *for the IMC technique* (in the unconstrained case) have been discussed by Morari and Zafiriou [47]; particular emphasis was laid on the important issues surrounding robust controller design with explicit incorporation of uncertainty descriptions (see also Ref. [57]). Within the framework of *standard* MPC, Zafiriou [84] discusses the effect of tuning parameters and constraints on the robustness and stability.

Rawlings and coworkers [51, 61] have presented a completely general analysis of the stability and robustness of MPC using an *infinite horizon* representation.

27.6.4 Improvements on Fundamental Elements

The fundamental elements of MPC, as noted in Section 27.2, are *the model, the appropriate model inverse* (explicit as in IMC, or implicit as in standard MPC when posed as an optimization problem), and the *disturbance estimation*, or model prediction updating scheme.

The issues involved in improving the model representation aspects of MPC have been discussed, for example, by Morari and Lee [48], and by Muske and Rawlings [51]; extending MPC to accept nonlinear models will be discussed later.

The constrained optimization approach to implementing the approximate process model inverse as the controller is not without its complications. For multivariable problems with long prediction horizons and requiring computation of a large number of future control moves, the resulting QP can be quite formidable. Strategies for improving this aspect of MPC have been proposed by Brosilow and coworkers [8]. It involves a two-step approach utilizing the less computationally taxing LP formulation.

For good reason, the disturbance estimation strategy of MPC has received the most criticism. Recommendations for posing the MPC problem as a classical state estimation and regulation problem, and taking advantage of the rich heritage of these subject matters have been made by several authors (cf. Refs. [44, 48, 68, 69, and 82]).

27.6.5 Nonlinear Extensions

Since all chemical processes are nonlinear to varying degrees of severity, extending the standard MPC concept to systems described explicitly by nonlinear models has been the subject of intense research. To date there have been essentially three approaches:

1. *Scheduled Linearization* (as exemplified in Ref. [24]) in which the nonlinear model is linearized around several operating steady states and the appropriate linearized model used within the linear MPC framework as the process moves from one operating condition to the other.

2. *Extended Linear MPC* [33] in which the basic linear MPC structure is augmented to reflect the true nonlinearities captured by an explicit nonlinear model; the recommended control moves then become perturbations around the basic linear MPC controller recommendations.

3. *Explicit Nonlinear MPC* in which the fundamental MPC problem is posed as a full-scale nonlinear problem and the nonlinear controller obtained either explicitly by analytical means (as proposed in Refs. [19, 20]) or implicitly as the solution of a nonlinear program (NLP) complete with the full nonlinear model and (possibly nonlinear) constraints (cf., for example, Ref. [5]).

We consider the issues involved with the nonlinear extensions of MPC important enough that they will be discussed separately in the next section.

27.7   NONLINEAR MODEL PREDICTIVE CONTROL

If one considers the basic concepts of MPC, there is nothing in the MPC structure that *fundamentally* forbids the use of a nonlinear model. In fact, the conceptual extension of the intrinsic MPC concept to nonlinear systems has been discussed by Frank [22] in 1974.

However, some very serious practical issues arise that make the nonlinear extension of MPC a nontrivial exercise.

1. *Availability of Nonlinear Models.* It is generally believed that the success of standard MPC is due in part to the relative ease with which step- and impulse-response models can be obtained; nonlinear models, on the other hand, are much more difficult to construct, either from first principles or from input/output data correlation.

2. *Nonlinear Model Inversion.* The procedure for carrying out the model inversion required for the implementation of the standard MPC controller is itself not trivial; the inversion in the nonlinear case can only be expected to be even more complicated.

These factors notwithstanding, there are strong factors motivating the need for nonlinear control in certain chemical process control applications.

27.7.1 Motivation

It is true that in many classical industrial processes, the role of the control system is to hold the process output as close to the operating steady state as possible in the face of disturbances. This is the classic regulatory problem, and by definition, the process will be operating in the vicinity of a steady state, and therefore a linear representation will, in fact, be adequate.

However, there are some very important exceptions to this general conception of chemical processes:

1. For batch and semibatch processes, the entire operation is carried out in transient mode, and there is no such thing as "steady-state operation."

2. Continuous processes that experience frequent start-ups, shutdowns, and transitions spend a substantial period of time away from any identifiable steady-state operating region.

3. There are processes for which the overall nonlinearities (even in the neighborhood of steady states) can be so severe, and so critical to stability, that no linear *time-varying* model will be adequate.

For such processes (which are common in the chemical and polymer industry) linear controllers will not be very effective; nonlinear controllers will be essential for improved performance. A particular example of the latter class of systems is the reactor studied in Ref. [19], for which the nonlinear model exhibits a sign change in the process gain. It was found that there is *no* linear controller (with integral action) which will permit stable operation of this process.

27.7.2 Techniques for Nonlinear MPC

The various techniques that have been proposed for implementing nonlinear versions of the MPC concept will now be briefly reviewed.

Scheduled Linearized MPC

With this technique, a linearized model whose states are updated as the process conditions change is used to generate the step-response coefficients needed to implement the linear QDMC control computation; the model prediction is however carried out by integrating the full-scale nonlinear model on-line, in parallel with the plant. This procedure has worked well for a batch reactor [24].

Extended Linear MPC

The approach here as presented in Hernandez and Arkun [33], is to reformulate the original DMC problem in Eq. (27.6) to read:

$$\hat{\mathbf{y}}^{el}(k+1) = \hat{\mathbf{y}}^0(k) + \mathbf{B} \Delta \mathbf{u}(k) + \mathbf{w}(k+1) + \mathbf{d}^{nl}(k) \qquad (27.35)$$

where everything else but the newly introduced vector $\mathbf{d}^{nl}(k)$ is exactly as described for standard DMC. The elements of the vector $\mathbf{d}^{nl}(k)$ are computed by minimizing the difference between $\hat{\mathbf{y}}^{el}(k + 1)$, the output prediction obtained from the *extended linear* model shown above, and $\hat{\mathbf{y}}(k + 1)$, the prediction obtained from a full-scale nonlinear model of the plant. Thus by this device, the process nonlinearities as captured by the nonlinear model are incorporated directly into the DMC formulation while still preserving the original framework.

The solution to the extended DMC problem so formulated then becomes a perturbation (predicated by the presence of the additional $\mathbf{d}^{nl}(k)$ term) around the standard linear DMC solution. Simminger *et al.* [72], have illustrated via simulation studies the potential effectiveness of this scheme.

Bilinear MPC

As an approximation of the nonlinear plant behavior, a bilinear model can often provide a more accurate representation than a linear model; it can also provide a more tractable representation than a full-scale nonlinear model. An MPC strategy based on bilinear process representations has been presented by Yeo and Williams [79].

The results presented were for SISO systems, and some transformations were employed in order that the resulting algorithm might be explicit in the same vein as the IMC results. Particularly because bilinear models might be easier to obtain than full-scale nonlinear models, this approach seems to hold some promise, especially if it can be extended to multivariable systems relatively easily, and provided it can be formulated to handle constraints.

Nonlinear MPC

Strategies for extending the IMC concept to nonlinear systems have been presented by Economou *et al.* [19, 20]; in particular the issues of generating an appropriate inverse of the nonlinear model and that of implementing it as the (nonlinear) IMC controller were carefully discussed.

As an alternative to the IMC approach, recall that the original MPC problem was posed as a QP (or an LP) primarily in order to take full advantage of these optimization techniques. If the linear model in the standard MPC scheme were then replaced by a full-scale nonlinear model of the type:

$$\frac{d\mathbf{x}}{dt} = \mathbf{f}(\mathbf{x}, \mathbf{u}, \mathbf{d}; \theta) \qquad (27.36)$$

$$y = g(x, u, d) \tag{27.37}$$

and the constraints are permitted to enter in a nonlinear fashion if necessary, then the resulting problem becomes:

$$\min_{u(t)} \phi[u(t), x(t), y(t)] \tag{27.22}$$

subject to:

$$\frac{dx}{dt} - f(x, u, d; \theta) = 0$$

$$y - g(x, u, d) = 0$$

$$h(x, u) = 0$$

$$q(x, u) \leq 0$$

$$x(t_0) = x_0$$

$$t \in [t_0, t_0 + T]$$

Here y is the vector of process outputs, u is the input vector, x is the state vector, d is the disturbance vector, and θ is the vector of model parameters; the current time is indicated at t_0, and the prediction horizon is indicated by T.

Posed this way, the nonlinear MPC problem becomes a nonlinear program (NLP). As a result of the rapid development of efficient nonlinear programming (NLP) algorithms capable of handling large numbers of variables and constraints, it is now possible to solve the NMPC problem as posed above. Biegler and Rawlings [5] provide a very careful review of the work done to date in this regard. Several issues, of course, still remain unresolved; these are also very carefully considered in the reference.

27.8   CLOSING REMARKS

27.8.1 Strengths and Weaknesses of MPC

To conclude our discussion on Model Predictive Control, we now summarize the main strengths and deficiencies of *conventional* MPC schemes. By conventional MPC, we mean those techniques that, in the original sense of DMC and MAC, utilize the (linear) step or impulse-response model for explicit prediction, and calculate control action by solving a (constrained) optimization problem.

Strengths

Conventional MPC has enjoyed unprecedented popularity and success in industry primarily for the following reasons:

1. **Model Form**
    It uses a step (or impulse) response model, a physically intuitive, easily understood, so-called nonparametric model form requiring little or no *a priori* decision about process order, or, for that matter, model structure.

2. **Control Computation**
    Its strategy of utilizing long-range prediction for the computation of
    corrective control action is also intuitively appealing and easy to
    understand. Many otherwise difficult problems created by difficult
    dynamics (dead-time, inverse response, etc.) are handled with remarkable
    facility with this approach.

3. **Constraints**
    It allows the explicit handling of constraints through on-line optimization.
    This aspect, in particular makes conventional MPC the only methodology
    that, in one package, allows the typical performance criteria of importance
    in the process industries to be addressed *directly* and not in the *ad hoc*
    fashion typical of most other schemes.

Deficiencies

Ironically, most of the currently identified deficiencies of MPC are associated
directly or indirectly with the very attributes that constitute its strength.

1. **Model Form**
    There are several industrial processes of importance for which an
    approximate linear model provides a poor representation of the relevant
    plant dynamics. In this case, the attractiveness of the step- (or impulse-)
    response model is supplanted by the unacceptable inaccuracies of the
    naively simple representation.
        Also, the conventional MPC model form is not directly applicable to
    processes with integrating elements; under these conditions special
    modifications noted in Morari and Lee [48] are required.
        The issue of model identification could also be raised here; however,
    model identification is a universal problem in process control and in general
    is not localized to MPC alone.

2. **Tuning**
    The control computation strategy adopted by MPC, even though
    fundamentally appealing, contains a substantial number of tuning
    parameters; and it is not always obvious how these parameters should be
    chosen. In this regard, as noted earlier, MAC is far less cumbersome than
    DMC; but by the same token, the latter has more degrees of freedom
    available for achieving the often conflicting objectives of robustness and
    performance.

3. **Model Prediction Updating**
    The model prediction updating strategy is often extremely inadequate
    because it assumes that the currently observed discrepancy between the
    model prediction and plant measurement is due only to unmodeled
    disturbances; it also assumes further that such discrepancy will remain
    constant over the prediction horizon. This often leads to poor performance,
    especially in disturbance rejection.

4. **Analysis**
    As attractive and as useful as conventional MPC has been, a systematic
    analysis of its stability and robustness properties is not possible in its

original finite horizon formulation. The control law is, in general, time-varying, and especially in the constrained case, cannot be represented in the closed form required for such general analyses. The infinite horizon formulation is more amenable to such analysis (cf. Refs. [51, 61]).

A more detailed discussion of the strengths and weaknesses of MPC along with some recommendations for improvements may be found in Morari and Lee [48].

A natural consequence of recognizing these deficiencies is that the areas of MPC in need of further work are more easily identified. These are summarized next.

27.8.2 Further Work

Despite the fact that significant advances have been made with regard to the theory and practice of Model Predictive Control as a control system design methodology, there is still further work to be done primarily (but not exclusively) in the areas reviewed below.

Model Identification

Model identification is clearly the "Achilles' heel" of MPC (or any other model-based controller design technique for that matter). Developing and evaluating appropriate models still remain the most important and most time-consuming steps in the implementation of MPC; these steps also require more expertise from the user.

Robust model identification methods that will function well in the nonideal industrial environment are needed. In this regard, issues concerning model identification under closed-loop, as opposed to open-loop, conditions (i.e., off-line versus on-line identification); what test signal to use (step, pulse, pseudo-random binary sequence, or others); how to deal effectively with unmeasurable disturbances that unavoidably enter the process during data collection; how to reject process noise but still retain process dynamic information; etc., are all very important and need to be addressed within the framework of MPC.

Unmeasured Disturbance Estimation and Prediction

Most MPC schemes assume that all future model prediction errors remain the same as the current prediction error. Of course, any observed prediction error is composed of components contributed by unmeasurable disturbance, modeling error (parametric as well as structural), and measurement noise.

Were all the contributions due to unmeasured disturbances alone, it would be possible to take advantage of some results from time series analysis and carry out on-line disturbance modeling using the current and past prediction errors. Predicting the future effects of these unmeasured (and hitherto unmodeled) disturbances can then be carried out more systematically and with reduced uncertainty. The closed-loop load response will thus be much improved. If, however, the contributions were primarily due to fundamental errors in the model structure, then such "disturbance modeling" will not be as effective.

On the other hand, if the prediction error was due entirely to measurement noise, then, strictly speaking, no model correction is necessary. Observe therefore that there is a trade-off between closely estimating what is considered as the effect of unmeasured disturbance on the one hand and measurement noise rejection on the other (not to mention the fact that the effect of structural errors in the model may end up being incorrectly ascribed to unmeasured disturbances). How to resolve such issues as these is a subject still under active research.

Systematic Treatment of Modeling Error and Uncertainty

No matter how derived, there will always be some uncertainty associated with any model. The problem of MPC controller design under uncertainty has been studied, but many issues remain to be resolved.

Commercial MPC packages rely on tuning parameters for achieving robustness, but it is not always obvious how these tuning parameters should be selected optimally; besides, an "optimal" set of tuning parameters found for a set of operating conditions is bound to become "suboptimal" or completely inadequate if the operating conditions were to change drastically.

Nonlinear Model Predictive Control

Most of the issues involving the practical implementation of nonlinear MPC are still unresolved; but the most critical remains that of obtaining appropriate nonlinear models.

For complex chemical processes, fairly accurate first principle models are usually difficult and/or very time consuming to obtain. This makes the possibility of identifying empirical nonlinear input/output models from plant data attractive particularly to the industrial practitioner.

Nonlinear input/output models may be obtained using:

- *Dynamic* neural networks
- Polynomial ARMA (autoregressive moving average) models
- Other NARMA (Nonlinear ARMA) models using, for example, radial basis functions
- *Volterra Series* models (a generalization of the linear impulse-response convolution models)

Regarding neural networks, the most critical unresolved issue is how to introduce dynamic elements into networks most effectively in such a way that multistep output prediction, as required by MPC, can be carried out with reasonable confidence.

For the polynomial ARMA models, the major obstacles center around the fact that guidelines for choosing adequate model structures are still not widely available. As for NARMA models using radial basis functions, the underlying concepts are very similar to those on which neural network modeling is based; the two approaches therefore share similar limitations. However, unlike the neural network approach, not as much has been published in terms of developing NARMA models using radial basis functions. The primary concern with Volterra models is that they typically require large number of parameters to represent most nonlinear processes adequately.

27.9   SUMMARY

The current state of Model Predictive Control has been reviewed in some detail in this chapter. The main issues regarding the basic principles of MPC, its historical evolution, and some details of the underlying theory of standard MPC have been discussed. The commercial packages available for the implementation of various MPC schemes have been reviewed, while stressing what differentiates one from the other. The academic contributions to the advancement of MPC as a body of knowledge within the engineering discipline have been reviewed, singling out the critical subject matter of nonlinear MPC as an emerging methodology worthy of separate consideration in its own right. In addition to this, however, a discussion of other areas of MPC in need of further work has also been presented.

It should be clear from this chapter that:

1. Standard MPC as a control methodology has enjoyed remarkable industrial success where others have failed primarily because, to date, it is the only methodology that allows the important performance criteria to be incorporated directly into the controller design formulation.

2. The fundamental principles of MPC are relatively simple to understand, the basic models required for its implementation are reasonably easy to obtain; and it takes full advantage of advances in computer hardware technology as well as the availability of optimization software.

3. The numerous applications available in the open literature attest to the fundamental effectiveness of MPC as a controller design methodology particularly for large-scale, constrained multivariable processes with complex dynamic characteristics.

4. These numerous applications also attest to the fact that the commercial packages available for implementing MPC schemes are all fundamentally adequate; they all have the same fundamental technological base, and the various factors that distinguish one from the other have arisen essentially as a result of the primary motivation and preferences of the developer and/or marketer. However, DMC and IDCOM continue to occupy preeminent positions since, as the pioneering commercial packages, they have been available the longest: they therefore have the largest share of the documented industrial applications.

Yet for all its strengths and successes, not all processes will benefit from the power of MPC. Processes whose variables exhibit little or no interactions, processes with normal dynamic characteristics, and those which are seldom subject to constraints will most likely not benefit significantly from MPC: simpler techniques are just as likely to perform as well.

The main areas where further work is needed MPC are:

- Identification of models from plant dynamic test data,
- Nonlinear extensions,

the former because to a large extent, the performance of the MPC scheme is really only as good as the model used, and the latter because nonlinearities are an integral part of all realistic processes, and they can still constitute significant problems in certain cases.

REFERENCES AND SUGGESTED FURTHER READING

1. Arkun, Y. and W. H. Ray, Eds., *Proceedings of Chemical Process Control – CPC IV*, AIChE (1991)
2. Arulalan, G. R. and P. B. Deshpande, "Synthesizing a Multivariable Prediction Algorithm for Noninteracting Control," *Proc. Am. Control Conf.*, Seattle, WA (1986a)
3. Arulalan, G. R. and P. B. Deshpande, "A New Algorithm for Multivariable Control," *Hydrocarbon Processing*, 51, (1986b)
4. Arulalan, G. R. and P. B. Deshpande, "Simplified Model Predictive Control," *Ind. Eng. Chem. Res.*, **26**, 356 (1987)
5. Biegler, L. T. and J. B. Rawlings, "Optimization Approaches to Nonlinear Model Predictive Control," in *Proceedings of CPC IV*, 543–571 (1991)
6. Box, G. E. P. and G. M. Jenkins, *Time Series Analysis: Forecasting and Control*, Holden-Day, San Francisco (1976)
7. Brosilow, C. B., "The Structure and Design of Smith Predictors from the Viewpoint of Inferential Control," *Proc. Joint Aut. Control Conf.*, Denver (1979)
8. Brosilow, C. B., G. Q. Zhao and K. C. Rao, "A Linear Programming Approach to Constrained Multivariable Control," *Proc. Am. Control Conf.*, San Diego, 667 (1984)
9. Caldwell, J. M. and G. D. Martin, "On-line Analyzer Predictive Control," *Sixth Annual Control Expo Conf.*, Rosemont, IL, 19–21 May (1987)
10. Chang, T. S. and D. E. Seborg, "A Linear Programming Approach to Multivariable Feedback Control with Inequality Constraints," *Int. J. Control*, **37**, 583 (1983)
11. Clarke, D. W. and C. Mohtadi, "Properties of Generalized Predictive Control," *Proc. 10th IFAC World Congress*, Munich, **10**, 63 (1987a)
12. Clarke, D. W., C. Mohtadi and P. S. Tuffs, " Generalized Predictive Control — I and II," *Automatica*, **23**, 137–160 (1987b)
13. Clarke, D. W., "Adaptive Generalized Predictive Control," in *Proc. of CPC IV*, 395–417 (1991)
14. Cutler, C. R. and B. L. Ramaker, "Dynamic Matrix Control — A Computer Control Algorithm," *AIChE National Meeting*, Houston (1979)
15. Cutler, C. R. (1983). Ph. D. Thesis, U. of Houston
16. Cutler, C. R. and R. B. Hawkins, "Constrained Multivariable Control of a Hydrocracker Reactor," *Proc. Am. Control Conf.*, Minneapolis, 1014 (1987)
17. Cutler, C. R. and R. B. Hawkins, "Application of a Large Predictive Multivariable Controller to a Hydrocracker Second Stage Reactor," *Proc. Am. Control Conf.*, Atlanta, 284 (1988)
18. Cutler, C. R. and F. H. Yocum, "Experience with the DMC Inverse for Identification," in *Proc. of CPC IV*, 297–317 (1991)
19. Economou, C. G., M. Morari and B. Palsson, "Internal Model Control — 5. Extension to Nonlinear Systems," *Ind. Eng. Chem. Process Des. Dev.*, **25**, 403 (1986)
20. Economou, C. G. and M. Morari, "Internal Model Control — 6. Multiloop Design," *Ind. Eng. Chem. Process Des. Dev.*, **25**, 411 (1986)
21. Erickson, K. T. and R. E. Otto, "Development of a Multivariable Forward Modeling Controller," *I. and E C Research*, **30**, 482 (1991)
22. Frank, P. M., *Entwurf von Regelkreisen mit Vorgeschriebenem Verhalten*, G. Braun, Karlsruhe (1974)
23. Garcia, C. E. and M. Morari, "Internal Model Control —1. A Unifying Review and Some New Results," *Ind. Eng. Chem. Process Des. Dev.*, **21**, 308 (1982)

24. Garcia, C. E., "Quadratic Dynamic Matrix Control of Nonlinear Processes. An Application to a Batch Reaction Process," *AIChE Annual Mtg*, San Francisco (1984)

25. Garcia, C. E. and M. Morari, "Internal Model Control — 2. Design Procedure for Multivariable Systems," *Ind. Eng. Chem. Process Des. Dev.*, **24**, 472 (1985a)

26. Garcia, C. E. and M. Morari, "Internal Model Control — 3. Multivariable Control Law Computation and Tuning Guidelines," *Ind. Eng. Chem. Process Des. Dev.*, **24**, 484 (1985b)

27. Garcia, C. E. and A. M. Morshedi, "Quadratic Programming Solution of Dynamic Matrix Control (QDMC)," *Chem. Eng. Commun.*, **46**, 73 (1986)

28. Garcia, C. E., D. M. Prett, and M. Morari, "Model Predictive Control: Theory and Practice — A Survey," *Automatica*, **25**, 335 (1989)

29. Georgiou, A., C. Georgakis, and W. L. Luyben, "Nonlinear Dynamic Matrix Control for High-purity Distillation Columns," *AIChE J.*, **34**, 1287 (1988)

30. Gerstle, J. G. and D. A. Hokanson, "Opportunities for Dynamic Matrix Control in Olefins Plants," *AIChE Spring Meeting*, Houston, Texas (1991)

31. Grosdidier, P., A Mason, A. Aitolahti, and V. Vanhamäki, "FCC Unit Reactor-regenerator Control," *submitted for presentation at ACC*, Chicago (1992)

32. Horowitz, I. M., *Synthesis of Feedback Systems*, Academic Press, London (1963)

33. Hernandez, E. and Y. Arkun, "A Nonlinear DMC Controller: Some Modeling and Robustness Considerations," *Proc. Am. Control Conf.*, Boston, 2355 (1991)

34. Lectique J., A. Rault, M. Tessier, and J. L. Testud, "Multivariable Regulation of a Thermal Power Plant Steam Generator," *IFAC World Congress*, Helsinki (1978)

35. Lee, P. L. and G. R. Sullivan, "Generic Model Control (GMC)," *Comp. and Chem. Eng.* **12**, 573 (1988)

36. Lee, P. L., R. B. Newell, and G. R. Sullivan, "Generic Model Control — A Case Study," *Can. J. Chem. Eng.*, **67**, 478 (1989)

37. Lee, P. L., "The Direct Approach to Nonlinear Control," in *Proceedings of CPC IV*, 517–542 (1991)

38. MacGregor, J. F., "On-line Statistical Process Control," *Chem. Eng. Progress*, October, **21** (1988)

39. Marchetti, J. L., D. A. Mellichamp, and D. E. Seborg, "Predictive Control Based on Discrete Convolution Models," *Ind. Eng. Chem. Process Des. Dev.*, **22**, 488 (1983)

40. Martin, G. D., J. M. Caldwell, and T. E. Ayral, "Predictive Control Applications for the Petroleum Refining Industry," *Japan Petroleum Institute — Petroleum Refining Conf.*, Tokyo, Japan, 27–28 October (1986)

41. Matsko, T. N., "Internal Model Control for Chemical Recovery," *Chem. Eng. Progress*, **81**, (12), 46 (1985)

42. Maurath, P. R., D. E. Seborg, and D. A. Mellichamp, "Predictive Controller Design by Principal Components Analysis," *Proc. Am. Control Conf.*, Boston, 1059 (1985)

43. McAvoy, T., Y. Arkun, and E. Zafiriou *Model-based Process Control: Proceeding of the IFAC Workshop*, Pergamon Press, Oxford (1989)

44. Meadows, E. S., K. R. Muske, and J. B. Rawlings, "Constrained State Estimation and Discontinuous Feedback in Model Predictive Control," *Proc. European Control Conference*, Groningen, Netherlands, **4**, 2308–2312 (1993)

45. Mehra, R. K., *et al.* "Model Algorithmic Control Using IDCOM for the F100 Jet Engine Multivariable Control Design Problem," in *Alternatives for Linear Multivariable Control* (ed. M. Sain *et al.*), NEC, Chicago (1978)

46. Mehra, R. K. and S. Mahmood, "Model Algorithmic Control" Chapter 15 in *Distillation Dynamics and Control* (ed. P. B. Deshpande), ISA, Research Triangle Park, NC (1985)

47. Morari, M. and E. Zafiriou, *Robust Process Control*, Prentice-Hall, Englewood Cliffs, NJ (1989)

48. Morari, M. and J. H. Lee, "Model Predictive Control: The Good, the Bad, and the Ugly," in *Proceedings of CPC IV*, 420–444 (1991)

49. Morari, M., C. E. Garcia, J. H. Lee, and D. M. Prett, *Model Predictive Control*, (in press) (1992)

50. Moreau, P. and J. P. Littnaun, "European Transonic Wind Tunnel Dynamics, Simplified Model," Adersa/Gerbios (1978)

51. Muske, K. R. and J. B. Rawlings, "Model Predictive Control with Linear Models," *AIChE J*, **39**, (2), 262 (1993)

52. Newton, G. C., L. A. Gould, and J. F. Kaiser, *Analytical Design of Feedback Controls*, Wiley, New York (1957)

53. Ogunnaike, B. A., "Dynamic Matrix Control: A Non-stochastic Industrial Process Control Technique with Parallels in Applied Statistics," *Ind. Eng. Chem. Fund.*, **25**, 712 (1986)

54. Ohshima, M., I. Hashimoto, T. Takamatsu, and H. Ohno, "Robust Stability of Model Predictive Control," *Int. Chem. Eng.*, **31**, (1), 119 (1991)

55. Otto, R. E., "Forward Modeling Controllers: A Comprehensive SISO Controller," *Annual AIChE Mtg.*, November (1986)

56. Prett, D. M. and R. D. Gillette, "Optimization and Constrained Multivariable Control of a Catalytic Cracking Unit," *AIChE National Mtg.*, Houston, Texas; also *Proc. Joint Aut. Control Conf.*, San Francisco (1979)

57. Prett, D. M. and C. E. Garcia, *Fundamental Process Control*, Butterworths, Stoneham (1988)

58. Prett, D. M., C. E. Garcia, and B. L. Ramaker, *The Second Shell Process Control Workshop*, Butterworths, Boston (1990)

59. Propoi, A. I., "Use of LP Method for Synthesizing Sampled-data Automatic Systems," *Automn. Remote Control*, **24**, 837 (1963)

60. Rault, A., J. Richalet, and P. LeRoux, "Commands Auto Adaptive d'Un Avion," Adersa/Gerbois (1975)

61. Rawlings, J. B. and K. R. Muske, "The Stability of Constrained Receeding Horizon Control," *IEEE Trans. Auto. Cont.* (in Press) (1993)

62. Richalet, J. A. and B. Bimonet, "Identification des Systems Discrets Linearines Mono-variables Par Minimisation d'Une Distance de Sructure," *Elec. Lett.*, **4**, (24) (1968)

63. Richalet, J. A., F. Lecamus, and P. Hummel, "New Trends in Identification, Minimization of a Structural Distance, Weak Topology," *2nd IFAC Symposium on Identification*, Prague (1970)

64. Richalet, J. A., A. Rault, and R. Pouliquen, *Identification des Processus par la Methode du Modele*, Gordon and Breach (1970)

65. Richalet, J. A., A. Rault, J. D. Testud, and J. Papon, "Model Predictive Heuristic Control: Applications to Industrial Processes," *Automatica*, **14**, 413 (1978)

66. Ricker, N. L., "The Use of Quadratic Programming for Constrained Internal Model Control," *Ind. Eng. Chem. Process Des. Dev.*, **24**, 925 (1985)

67. Ricker, N. L., T. Subramanian, and T. Sim, "Case Studies of Model Predictive Control in Pulp and Paper Production," in *Model Based Process Control*, (ed. T. J. McAvoy, Y. Arkun, and E. Zafiriou), Pergamon Press, Oxford (1989)

68. Ricker, N. L., "Model Predictive Control with State Estimation," *Ind. Eng. Chem. Res.*, **29**, 374 (1990)

69. Ricker, N. L., "Model-predictive Control: State of the Art," in *Proceedings of CPC IV*, 271–296 (1991)

70. Rivera, D. E., M. Morari, and S. Skogestad, "Internal Model Control — 4. PID Controller Design," *Ind. Eng. Chem. Process Des. Dev.*, **25**, 252 (1986)

71. Rouhani, R. and R. K. Mehra, "Model Algorithmic Control (MAC): Basic Theoretical Properties," *Automatica*, **18**, (4), 401 (1982)

72. Simminger, J., E. Hernandez, Y. Arkun, and F. J. Schork, "A Constrained Multivariable Nonlinear Model Predictive Controller Based on Iterative QDMC," *Proceedings ADCHEM IFAC*, Toulouse, October (1991)

73. Smith, O. J. M., "Closer Control of Loops with Dead Time," *Chem. Eng. Progress*, **53**, (5), 217 (1957)

74. Smith, O. J. M., "A Controller to Overcome Dead-time," *ISA Journal*, February (1959)

75. Testud, J. L., "Commande Numerique Multivariable du Ballon de Recuperation de Vapeur," Adersa/Gerbios (1979)

76. Tu, F. C. Y., and J. Y. H. Tsing, "Synthesizing a Digital Algorithm for Optimized Control," *In. Tech.*, May (1979)

77. Wassick J. M. and D. T. Camp, "Internal Model Control of an Industrial Extruder," *Proc. ACC*, Atlanta, 2347 (1988)

78. Yamamoto, S. and I. Hashimoto, "Present Status and Future Needs: The View from Japanese Industry," in Proceedings of CPC IV, 1–27 (1991)

79. Yeo, K. Y. and D. C. Williams, "Bilinear Model Predictive Control," *Ind. Eng. Chem. Res.*, **26**, 2267 (1987)

80. Youla, D. C., J. J. Bongiorno, and H. A. Jabr, "Modern Wiener-Hopf Design of Optimal Controllers — I. The Single-input-output Case," *IEEE Trans. Aut. Control*, **AC-21**, 3 (1976a)

81. Youla, D. C., H. A. Jabr, and J. J. Bongiorno, "Modern Wiener-Hopf Design of Optimal Controllers — II. The Multivariable Case," IEEE *Trans. Aut. Control*, **AC-21**, 319 (1976b)

82. Yuan, P. and D. E. Seborg, "Predictive Control Using an Observer for Load Estimation," *Proc. Am. Control Conf.*, Seattle, 669 (1986)

83. Zadeh, L. A. and B. H. Whalen, "On Optimal Control and Linear Programming," *IRE Trans. Auto. Cont.*, **7**, (4), 45 (1961)

84. Zafiriou, E., "On the Robustness of Model Predictive Controllers," in *Proceedings of CPC IV*, 363–393 (1991)

85. Zames, G., "Feedback and Optimal Sensitivity: Model Reference Transformations, Multiplicative Seminorms, and Approximate Inverse," *IEEE Trans. Aut. Control*, **AC-26**, 301 (1981)

REVIEW QUESTIONS

1. In MPC the process model is used for what two central tasks?

2. What are the typical characteristics of those processes most likely to benefit from MPC?

3. What is the relationship between MAC and IDCOM?

4. What four elements constitute the basic framework of all MPC schemes?

5. What is the difference between an impulse-response and a step-response model?

6. What model form is employed in DMC?

7. The dynamic matrix in DMC is made up of what elements?

8. What type of inverse of the dynamic matrix B is used for control action computation when the unconstrained DMC problem is posed as a least squares minimization problem?

9. What is the difference between the "projected error vector" $e(k + 1)$ and the residual error vector $\varepsilon_r(k + 1)$ in DMC?

10. What is the DMC strategy for updating model predictions?

11. Why is it inadvisable to implement the entire sequence of m control moves calculated at time instant k in automatic succession over the next m time intervals?

12. Why is scaling important in using DMC for multivariable control problems posed as least squares optimization problems?

13. What role does the scalar K play in the unconstrained DMC formulation?

14. In the general constrained DMC formulation, explain the roles of the tuning matrices Γ, **B**, and **D**.

15. What model form is employed in the MAC methodology?

16. What are the main differences between DMC and MAC?

17. What role does the parameter α play in MAC?

18. What differentiates the commercial package OPC from DMC?

19. What characteristics set the commercial package PC (marketed by Profimatics) apart from the others?

20. What property makes the Honeywell's HMPC different from the other commercial MPC packages?

21. What is the main tuning parameter in HMPC?

22. What approaches are currently available for nonlinear MPC?

CHAPTER 28

STATISTICAL PROCESS CONTROL

Thus far, we have tacitly assumed that: (1) the manipulated variables of a process can be adjusted to implement control action as frequently as we desire, and that such adjustments are "cost-free"; and (2) the controlled process variables can be measured just as frequently, and that such measurements are available "noise-free" (i.e., with no associated uncertainty, or inherent "variability"). Quite often in industrial practice, however, the most important variables to be controlled are product quality variables whose measurements must be obtained in the laboratory. Such measurements are available only infrequently, and associated with each measurement is an inherent "variability." As such, any observed difference between a controlled process variable and its desired value may simply be due to inherent measurement "variability" — in which case no corrective action is necessary; however, the observed difference may also be due to a real, systematic departure from target — in which case corrective action *may* be necessary. Within this framework, *effective* process control must now consist of two parts:

1. An *objective* analysis of the observed deviation of the process variable from its desired value.
2. A rational decision regarding what to do to minimize such deviation.

At one extreme, with its roots in the Quality Control movement spearheaded by Shewart in the 1920s (cf. Ref. [22]), traditional Statistical Process Control (SPC) focuses almost exclusively on the first part, finding application most appropriately in those manufacturing processes for which the cost for making process adjustments is significant. At the other extreme, standard (or deterministic) Process Control traditionally focuses on the second part, finding application mostly on processes for which little or no cost is associated with taking control action.

The objective in this chapter is to present SPC in its broadest and more modern sense: as the use of statistics for the objective analysis and rational design of effective process control strategies, in the face of inherent process "variability," and taking the cost of control action into consideration. Our discussion will therefore include — but will not be limited to — classical SPC, i.e., the methodology of Quality Control charting.

28.1   INTRODUCTION

Whenever the value of a process variable is observed by measurement at time instant k, say, as $y(k)$, it is generally understood that this measurement consists of two inseparable parts:

$$y(k) = y*(k) + e(k) \qquad (28.1)$$

where $y^*(k)$ is the true, but unknown, value of the process variable, and $e(k)$ is the "measurement error," usually random. Given a sequence of $y(k)$ values, $y^*(k)$ represents the "explainable" part of the data, while $e(k)$ represents the random, "unexplainable" part — the inherent variability associated with the measurement.

In the face of such inherent variability, if we wish to maintain the process variable as close as possible to its desired (target) value, $y_d(k)$, the following two questions must be answered:

1.   Is the observed difference between $y(k)$ and $y_d(k)$ "significant"?
2.   If the difference is "significant," what action should be taken?

Since the first question is concerned with analysis, while the second is concerned with control strategy, we shall refer to the former as the *analysis* problem, and to the latter as the *control strategy* problem.

28.1.1 Common Cause and Special Cause Variation

Any observed differences between $y(k)$ and its desired value $y_d(k)$ is said to be due to "common cause" variation if:

$$y(k) - y_d(k) = e(k) \qquad (28.2)$$

where $e(k)$ is nothing more than the inherent variability due entirely to random measurement errors. In this case, observe that the implication from Eq. (28.1) is that $y^*(k) = y_d(k)$, i.e., the process variable is actually at its desired target value.

More generally, the difference between $y(k)$ and its desired value will be of the form:

$$y(k) - y_d(k) = \delta(k) + e(k) \qquad (28.3)$$

and if $\delta(k)$ is nonzero, then the observed difference is said to be due to "special cause" variation (represented by $\delta(k)$) *in addition to* the inherent common cause variation. The implication is therefore that $y^*(k)$, the true value of the process variable, differs from its desired target value by $\delta(k)$.

The primary issue involved in the *analysis problem* is therefore that of deciding whether the observed difference between $y(k)$ and $y_d(k)$ is due solely to common cause variation, or whether there is an additional special (or "assignable") cause component.

28.1.2 Two Traditional Approaches

The traditional Quality Control (QC) methods were developed to solve the analysis problem systematically by employing statistics and probability theory to detect special cause variations *in the midst of unavoidable common cause variation*. The philosophy of approach to the control strategy problem is more vague: the *source* of any confirmed special cause variation is to be "identified" and "eliminated." The underlying assumption is that the process tends to remain "on-target" for long periods of time, with only occasional (abrupt) shifts "off-target"; that such shifts are caused by infrequent "special events" that can be identified and rectified at source; and that significant costs are incurred whenever adjustments are made in the manipulated variables. Traditional QC schemes therefore naturally place more emphasis on the analysis problem.

On the other hand, standard process control addresses the analysis problem by assuming that common cause variations are almost always negligible compared to special cause variations. The control strategies associated with standard process control are thus fundamentally designed to eliminate *all* deviations from target. The assumption is that the process is always subjected to disturbances of known and unknown sources, and will therefore not remain "on-target" without external intervention from the control system. Furthermore, it is assumed that little or no cost is incurred in implementing control action. Standard process control therefore naturally places more emphasis on the control strategy problem.

We may now observe that while the traditional QC methods deal with the analysis problem systematically, the QC approach to the control strategy problem is overly simplistic and often impractical. The converse is true for standard process control: the analysis problem is dealt with too cavalierly, but the subsequent approach to the control strategy problem is more systematic and usually more realistic. Clearly, neither of these two traditional approaches alone answers *both* questions adequately and simultaneously. Nevertheless, there are many situations of practical importance in which the solution provided by each approach is appropriate, and for some situations, a combination of the two approaches is best.

28.1.3 The Discussion to Follow

The rest of the chapter is organized as follows: the most widely used traditional QC techniques will be reviewed first in Section 28.2; the underlying assumptions will be stated clearly so that the conditions under which they are applicable can be determined objectively. What happens when one of the key assumptions underlying the QC techniques breaks down is discussed in Section 28.3; within this same context, a case is made for the standard process control paradigm, and the philosophy underlying this popular engineering approach is presented. The issue of when traditional SPC is appropriate is discussed explicitly in Section 28.4.

In Section 28.5 we discuss how our knowledge of process dynamics and deterministic process modeling can be combined appropriately with statistics in dealing with *both* the analysis as well as the control strategy problems simultaneously and systematically. The discussion generalizes both the QC methods and standard process control, presenting this framework of "Stochastic Process Control" as a true hybrid between the traditional QC philosophy favored by statisticians, and standard process control favored and practiced routinely by engineers. Section 28.6 provides a brief summary of some of the more advanced multivariate topics.

28.2   TRADITIONAL QUALITY CONTROL METHODS

28.2.1 Preliminary Concepts

Let us start by considering the situation in which the following holds:

1. $y^*(k)$ the true value of a process variable (observed by measurement as $y(k)$) is *constant*.

2. $e(k)$, the measurement error associated with $y(k)$, is a sequence of independent, and identically distributed, *random* variables with mean value of zero, and variance σ^2.

3. The underlying probability distribution from which the observed "sample" $e(k)$, $k = 1, 2, 3, \ldots$, is assumed to have been drawn is the Gaussian (or normal) distribution.

The immediate implication of these assumptions is that the operating model explaining each observation is:

$$y(k) = y^* + e(k) \tag{28.4}$$

i.e., that the observed process measurement $y(k)$ varies around its true, *fixed*, constant, mean value, y^*, in a random fashion that follows the Gaussian probability distribution function. Thus the probability of obtaining a value of y lying in any given interval is obtained from the probability density function:

$$f(y) = \frac{1}{\sigma\sqrt{2\pi}} \exp\left[-\frac{1}{2}\left(\frac{y - y^*}{\sigma}\right)\right] \tag{28.5}$$

One of the many advantages of employing this probability distribution function to describe the process variation is that its very well characterized theoretical properties greatly simplify statistical analysis. It is known, for example, that there is a probability of 0.00135 that a data point $y(k)$ drawn *at random* from such a distribution will exceed the true mean value of y^* by 3σ, i.e.:

$$\Pr[y(k) > (y^* + 3\sigma)] = 0.00135 \tag{28.6}$$

and by the symmetry of this probability distribution function, we also have that:

$$\Pr[y(k) < (y^* - 3\sigma)] = 0.00135 \tag{28.7}$$

Thus the probability that $y(k)$ will fall outside the limits $y^* \pm 3\sigma$ is the sum of the probabilities in Eqs. (28.6) and (28.7), or 0.0027.

With this simple fact, we may now observe that whenever the process is truly "on-target," it is highly unlikely that the discrepancy between the observed value $y(k)$ and its desired target value y_d will exceed the value 3σ, three times the measurement error's standard deviation; any deviation exceeding these bounds is *very likely* (with probability 0.9973) to be due to a "special cause."

Thus the properties of the Gaussian distribution function permits us to *test the hypothesis* that the discrepancy between each observed value $y(k)$ and its desired target value y_d is due solely to inherent measurement variability (commonly known as the *null hypothesis*, H_0), as opposed to the *alternative hypothesis*, H_a, that it is due to a special, assignable cause.

Risks and Errors in Hypothesis Testing

Observe from Eqs. (28.6) and (28.7) that there is a finite, nonzero probability that an observation randomly drawn from a Gaussian distribution will deviate from the true mean by more than $\pm 3\sigma$. Thus in testing the hypothesis that such a deviation is due solely to common cause variation, setting the "acceptance limits" at $\pm 3\sigma$ carries with it a probability of 0.0027 of *falsely* rejecting the hypothesis when in fact it should be accepted.

A Type I error is committed when an otherwise true hypothesis is falsely rejected; the probability of committing such an error is referred to as the α risk. A Type II error, on the other hand, is committed when a false hypothesis is inadvertently accepted; the probability of committing such an error is referred to as the β risk. It should be noted that attempts to lower the α risk are usually accompanied by an increase in the β risk, and vice versa.

28.2.2 The Shewart Chart

What is traditionally referred to as the Shewart Quality Control chart (see Refs. [22, 23]) is a graphical method for testing the hypothesis that the observed process measurement $y(k)$ does not differ *significantly* from its desired target value y_d against the alternative hypothesis that it is significantly different. The theoretical basis is found in Eqs. (28.6) and (28.7): that true, purely random variation of y around y_d can be expected to exceed 3σ only about two or three times in 1,000 observations. The α risk associated with this procedure is therefore 0.0027.

Thus, given y_d the desired target value, and σ, the standard deviation of the inherent measurement error, the Shewart Chart in essence monitors:

$$z = \frac{|y - y_d|}{\sigma} \tag{28.8}$$

(sometimes called the "signal-to-noise" ratio); the process is considered as being "on-target" or "in-control" when:

$$z < 3.0 \tag{28.9}$$

but when:

$$z \geq 3.0 \tag{28.10}$$

a "special event" is declared to have occurred.

Thus the classical Shewart Chart consists of a sequential time plot of process data $y(k)$, on which is superimposed a centerline corresponding to the target value y_d along with upper and lower control limit (UCL, LCL) lines set at $\pm 3\sigma$ from this centerline.

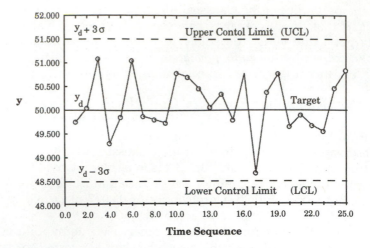

Figure 28.1.   Shewart quality control chart for the Mooney viscosity data of Table 23.1 (a typical Shewart Chart for a process "in-control").

For example, the Shewart Chart shown in Figure 28.1 is for hourly measurements of the Mooney viscosity of an industrial elastomer. The desired target Mooney viscosity value for this grade is 50.000; a standard deviation value of $\sigma = 0.5$ is known to characterize the intrinsic variability associated with the sampling procedure and the rather sensitive Quality Control laboratory equipment used for obtaining these measurements. Table 23.1 contains the original data from which the plot was obtained. The target line is drawn at $y_d = 50$, and the control limits lines are drawn at 50 ± 1.5 since $\sigma = 0.5$. Observe that no data point falls outside the $\pm 3\sigma$ control limits; the process is therefore said to be "in control."

Constructing and Using The Shewart Chart

The following is a typical procedure for constructing and using the Shewart Chart for Statistical Process Control.

1.   Collect historical data indicative of "usual" process operation.

2.   From this, extract a "rational subgroup" — the data set representative of periods of "rational" performance. This data set is used to establish the operating "base": the mean value, as well as a good estimate of σ, the measure of intrinsic process variation needed for setting the control limits.

3.   Draw the centerline at the desired target value (in the absence of a specified target value, the mean value estimated from the rational subgroup may be substituted); set the upper and lower control limits at $\pm 3\sigma$ from this centerline.

4.   Plot process data $y(k)$ sequentially in time on this chart.

5.   A "signal" occurs whenever an observation falls outside the control limits; in such a case, *find* an assignable cause, and take "corrective action."

TABLE 23.1
Hourly Mooney Viscosity Data For Figure 28.1

Time Sequence (in hours)	Mooney Viscosity (Standard Industry Units)
1.0	49.763
2.0	50.050
3.0	51.079
4.0	49.310
5.0	49.855
6.0	51.056
7.0	49.887
8.0	49.806
9.0	49.735
10.0	50.799
11.0	50.699
12.0	50.469
13.0	50.007
14.0	50.341
15.0	49.798
16.0	50.767
17.0	48.678
18.0	50.372
19.0	50.782
20.0	49.648
21.0	49.889
22.0	49.669
23.0	49.544
24.0	50.472
25.0	50.837

Western Electric Rules

While the α risk (the risk of raising a "false alarm" when the process is truly "on-target") is very low with the classic Shewart Chart, the "3-sigma" limits are sometimes considered too wide, because of the increased β risk — the risk of not raising an alarm when the process is "off-target." (A specific example considered in the next subsection illustrates this point.)

The most compelling justification for the low α risk employed by the classic Shewart Chart is that in its original applications, the costs associated with making adjustments were quite significant. The overriding factor was therefore to keep as low as possible the probability of making expensive but unnecessary process adjustments.

As a result of practical experience with the Shewart Chart at the Western Electric Company, four additional rules have been included for use with the

Shewart Chart. These are now commonly known as the "Western Electric" rules. The rules state that a "signal" is declared when:

1. A single data point falls outside the 3-sigma limits.
2. 2 out of 3 consecutive points fall outside the 2-sigma limits.
3. 4 out of 5 consecutive points fall outside the 1-sigma limits.
4. 8 consecutive points fall on one side of the median line.

The first rule is of course recognized as the standard Shewart "one point outside the 3-sigma limits" criterion. Together these four rules reduce the β risk considerably, at the expense of slightly increasing the α risk.

Other Shewart Charts

Although the y (or y-bar) chart discussed above is the most popular of the Shewart Charts, there are many other Shewart Charts: the R-charts (for ranges), used to monitor the intrinsic process variability, σ, itself; m-charts for monitoring medians, p-charts for monitoring variations in data involving percentages (e.g., percent defective parts in a sample), etc. In fact, any statistic that can be computed from process data can be presented in Shewart Chart form, with a line for the nominal value, and one or two other lines indicating limits of acceptable "normal" variability.

28.2.3 The CUSUM Chart

The Shewart Chart is particularly good for detecting large shifts of the order of 3σ or higher; it is less sensitive to shifts of smaller magnitude. The concept of the *average run length* (ARL) is often used to illustrate this point better.

The ARL is defined as the number of sequential observations expected to pass before a "signal" is declared. For example, for the classical Shewart Chart with its 3-sigma limits, when the process is truly "on-target," the probability that an observation from this process will fall outside the control limits is 0.0027 (see Eqs. (28.6) and (28.7)). Thus, *on the average*, we would expect $1/0.0027 \approx 371$ observations to pass before a first "signal" is declared. For this Shewart Chart therefore, the ARL = 371 (370.37 rounded up to the nearest integer).

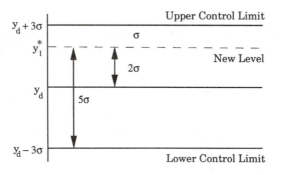

Figure 28.2. Implication of a positive shift of 2σ in the mean level of a process measurement.

Let us now consider a situation in which there is a *positive* shift of 2σ in the mean level of the process measurement (i.e., the true mean has shifted from y^* to $y^*_1 = y^* + 2\sigma$). Under these circumstances, for an observation $y(k)$ to trigger a signal by falling outside the upper limit, normal variation around the new mean will have to exceed one standard deviation, i.e., $(y - y^*_1) > \sigma$; for the observation to fall outside the lower limit will require that $(y^*_1 - y) > 5\sigma$ (see Figure 28.2).

Now from the properties of the normal distribution with mean y^*_1, and variance σ^2, we know that:

$$\Pr\left[(y - y^*_1) > \sigma\right] = 0.1587 \qquad (28.11)$$

while the $\Pr\left[(y^*_1 - y) > 5\sigma\right]$ is negligibly small. Thus for such a shift in mean, the ARL = 1/0.1587 = 6.301, implies that on the average, seven observations will pass before we can expect a signal to be triggered on the Shewart Chart. Since it is not unusual for observations to be made every hour, this implies that we may not see the real upset in y^* of 2 standard deviations until perhaps 7 hours after the event occurred. This may be an unacceptably long time for such a significant deviation in the process output to go undetected.

The CUSUM (for **CU**mulative **SUM**) chart, first introduced by Page [18, 19], is a procedure that is designed to detect smaller shifts in mean value more quickly. The operating model for the CUSUM technique is the same as in Eq. (28.4):

$$y(k) = y^* + e(k) \qquad (28.4)$$

By subtracting the target value from either side, we obtain:

$$[y(k) - y_d] = (y^* - y_d) + e(k)$$

or

$$[y(k) - y_d] = \delta + e(k) \qquad (28.12)$$

where δ represents the discrepancy between the true mean value of the process variable, and the desired target value. Now define $S(k)$ as:

$$S(k) = \sum_{i=1}^{k} [y(i) - y_d] \qquad (28.13)$$

a *cumulative sum* of the deviation of the observation from the desired target value. The CUSUM chart is a plot of $S(k)$ against the time index k. If the true mean value of the observation equals the desired target value, then $\delta = 0$ and from Eq. (28.12):

$$S(k) = \sum_{i=1}^{k} e(i) \qquad (28.14)$$

in which case $S(k)$ will appear as a random walk, wandering randomly around zero. For any nonzero discrepancy δ:

$$S(k) = k\delta + \sum_{i=1}^{k} e(i) \qquad (28.15)$$

and the plot of $S(k)$ will show a sustained, systematic increase or decrease, depending on the whether δ is positive or negative.

To illustrate the difference in sensitivity between the Shewart Chart and the CUSUM chart, let us return to the Mooney viscosity data in Table 23.1 and simulate a deliberate shift in the mean level at $k = 10$ by adding 0.6 to every observation after the ninth one. The resulting Shewart Chart for the "perturbed," y_p as well as the original data set, y, are superimposed and shown in Figure 28.3a. (The empty circles indicate the original data; the solid circles indicate the perturbed data.) Because the shift is relatively small ($0.6 = 1.2\sigma$), the perturbed data still remain within the 3-sigma Shewart bounds, and no signal is triggered.

The corresponding CUSUM chart is shown in Figure 28.3(b), where the empty circles indicate the CUSUM for the original data, S, and the solid circles indicate the CUSUM for the perturbed data, S_p. Note that whereas the simulated shift is quite noticeable by mere inspection of the CUSUM chart, the same is not true for sequential plot on the Shewart Chart. The change in the slope of the CUSUM plot, visible from the tenth observation on, provides a clear indication not only that an "event" has occurred, but also an estimate of when it might have occurred.

For the purpose of objectively ascertaining whether or not a true change in slope has occurred in a CUSUM plot, it is customary to employ the so-called "V-mask" much in the same way that the Shewart Chart employs the parallel UCL and LCL lines. The procedure for constructing and using this "V-mask" is somewhat involved and will not be discussed here. The interested reader may consult, for example, Lucas [9, 10], or Wetherill and Brown [24]. In any event, such a graphical procedure has been largely supplanted by an equivalent, primarily computational, approach discussed, for example, in Lucas and Crosier [12, 13], and also mentioned briefly in MacGregor [15].

Theoretically, the CUSUM method performs what is known as a *Sequential Probability Ratio Test* (SPRT): i.e., it tests the null hypothesis (H_0) that $\delta = 0$ in Eq. (28.12), against the alternative hypothesis (H_a) that $\delta \neq 0$. If we define SPRT as:

$$SPRT = \frac{\Pr[y(1); y(2); \ldots; y(k)] \text{ under } H_a}{\Pr[y(1); y(2); \ldots; y(k)] \text{ under } H_0} \tag{28.16}$$

then H_0 is rejected if the computed value of SPRT exceeds a prespecified α risk value, and H_a is rejected when the computed value of SPRT is smaller than a prespecified β risk value.

Thus, to use the CUSUM method one must specify δ (a measure of the smallest shift in the mean one would like to detect quickly), and values for α and β risks, in addition to specifying y_d, and σ. For this and other reasons, the CUSUM method is considered by many as more complicated to use than the Shewart method.

There are many other variations of the CUSUM method (including a method combining Shewart and CUSUM) that we will not discuss here. The interested reader is referred, for example, to the publications by Lucas and co-workers [11–13]. The important point to note is that in its most basic form, the CUSUM chart is primarily a variation on the Shewart Chart in which instead of evaluating the actual deviation of the observation y from its target value against control limit lines, the cumulative sum of these deviations is plotted.

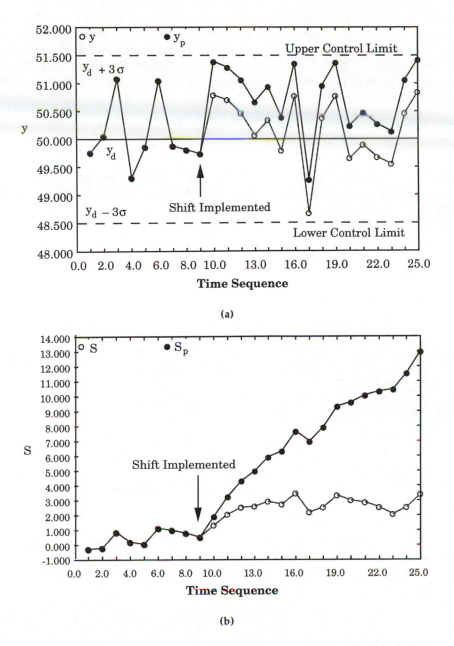

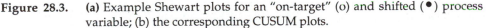

Figure 28.3.   (a) Example Shewart plots for an "on-target" (o) and shifted (●) process variable; (b) the corresponding CUSUM plots.

28.2.4 EWMA: The Exponentially Weighted Moving Average

Each quality control charting technique discussed thus far uses process data differently: the classical Shewart Chart essentially uses $e(k)$, the difference between $y(k)$ and the desired target value, comparing it at each point in time against $\pm 3\sigma$ limits; the CUSUM chart uses the *sum*, $\sum_{i=1}^{k} e(i)$ instead. Thus while the Shewart method focuses entirely on the most current information, paying no attention whatsoever to past information, the CUSUM method pays equal attention to the entire data history as it pays to the most current information. There is a class of "moving average" techniques that fall somewhere in between the Shewart and the CUSUM methods in the manner in which historical data is utilized; the Exponentially Weighted Moving Average (EWMA) method, first proposed by Roberts [20], is the most widely used in this class.

The EWMA of the observation $y(k)$ is given by the expression:

$$\hat{y}(k) = \lambda y(k) + (1 - \lambda)\hat{y}(k - 1) \qquad (28.17)$$

(with $0 < \lambda \le 1$) that is easily recognized by engineers as the computational equivalent of a digital, first-order filter. In terms of $e(k) = y(k) - y_d$, and defining:

$$\hat{e}(k) = \hat{y}(k) - y_d \qquad (28.18)$$

then it follows immediately that Eq. (28.17) may equally well be written as:

$$\hat{e}(k) = \lambda e(k) + (1 - \lambda)\hat{e}(k - 1) \qquad (28.19)$$

By recursive application of this expression, it is easy to show that Eq. (28.19) becomes:

$$\hat{e}(k) = \sum_{i=0}^{k} w(i)\, e(i) \qquad (28.20a)$$

with the weights $w(i)$ given by:

$$w(i) = \lambda (1 - \lambda)^{k-i} \qquad (28.20b)$$

Note how the expression for $\hat{e}(k)$ naturally falls in between $e(k)$ used by the Shewart method (corresponding to $w(i) = 1$ for $i = k$; and zero otherwise), and $\sum_{i=1}^{k} e(i)$ used by the CUSUM method (corresponding to $w(i) = 1$ for all i).

Now, if the errors $e(i)$, $i = 1, 2, ..., k$, are considered as usual to be independent and identically distributed Gaussian random variables with zero mean and constant variance σ^2, then it can be shown that the variance of $\hat{e}(k)$ is given by:

$$\sigma_{EWMA}^2 = \left[[1 - (1 - \lambda)^{2k}]\left(\frac{\lambda}{2 - \lambda}\right) \right] \sigma^2 \qquad (28.21)$$

that has the asymptotic limit (for reasonable values of k):

$$\sigma_{EWMA}^2 = \left(\frac{\lambda}{2 - \lambda}\right)\sigma^2 \qquad (28.22)$$

The parameter λ that, from Eq. (28.20), determines the memory length, and that from Eq. (28.22) also strongly influences the EWMA variance, is to be chosen by the user. Procedures and guidelines for determining "optimum" values for λ are presented, for example, in Hunter [5].

Plotting and Using the EWMA

For the purposes of plotting the EWMA statistic, Hunter [5] suggests that $\hat{y}(k)$ in Eq. (28.17) be considered as a "one-step-ahead forecast" of $y(k + 1)$, say, $y_f(k + 1)$, and plotted as such. The EWMA plot is initialized by setting $y_f(1)$ equal to the target value, i.e., $y_f(1) = y_d$. The three-sigma control limits for the EWMA are obtained from Eq. (28.22), i.e.:

$$y_d \pm 3\sigma_{\text{EWMA}} = y_d \pm 3\sigma \left(\sqrt{\frac{\lambda}{2 - \lambda}} \right) \qquad (28.23)$$

where σ is the standard deviation of the error, $e(k)$, associated with the raw process measurement.

An example plot of the EWMA for the Mooney viscosity data of Table 23.1 is shown in Figure 28.4. Again, the empty circles are used to indicate the EWMA for the original data, while the solid circles indicate the EWMA for the "perturbed" data; λ is chosen as 0.5 in each case. Recall that for this process the target value is $y_d = 50$, the standard deviation, $\sigma = 0.5$, and that for the perturbed data, the shift of 0.6 was implemented after $k = 9$.

For the choice $\lambda = 0.5$, Eq. (28.23) is used to set the 3-sigma limits at:

$$50.00 \pm 3 \times 0.5 \sqrt{\frac{0.5}{1.5}} = 50.00 \pm 0.87$$

Observe that the EWMA forecast triggers a first signal for the perturbed data (the point falls outside the upper control limit) at $k = 12$, two observations after the disturbance was implemented; and in all, 5 out of the 15 EWMA forecasts made after the introduction of the disturbance trigger "out-of-control" signals.

For the purpose of comparison, it is often recommended to combine a Shewart plot (for raw $y(k)$ data) with the corresponding EWMA forecasts (cf.

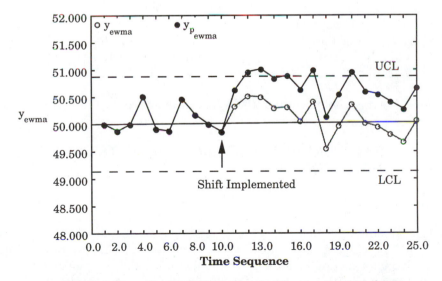

Figure 28.4.   The EWMA plot for the Mooney viscosity data.

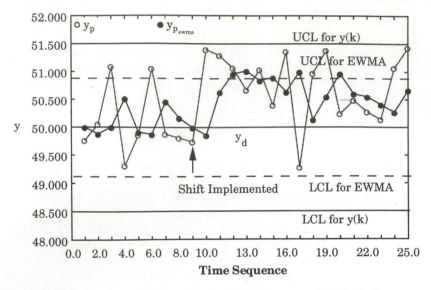

Figure 28.5.   An example combined Shewart and EWMA plot.

Hunter [5]). Such a plot is shown in Figure 28.5 for the perturbed Mooney viscosity data only, plotted as recommended by Hunter [5], and with the respective control limits as indicated. This plot emphasizes once more the point noted earlier: whereas the actual observation in the Shewart plot does not trigger a signal, the EWMA forecast does trigger several signals.

It has been shown by Hunter [6] that the EWMA with the specific choice of $\lambda = 0.4$ is virtually equivalent to the Shewart Chart incorporating the Western Electric rules. Additional information about the EWMA and its various properties are available, for example, in Hunter [5]. The important points to note are:

- It is just as easy to use as the Shewart plot,
- In its use of process data, it falls somewhere in between the Shewart and the CUSUM methods,
- By choosing the "tuning" parameter λ appropriately, it is possible to obtain a procedure that is more sensitive than the classical Shewart Chart in picking up smaller shifts in mean level.

We will have cause to meet the EWMA in a different context in Section 28.5.

28.3   SERIAL CORRELATION EFFECTS AND STANDARD PROCESS CONTROL

28.3.1 The Nature of Industrial Processes and Disturbances

The typical industrial chemical process consists of an interconnected network of individual processing units such as chemical reactors, distillation columns, heat exchangers, mixing tanks, etc. The inherent inertial components present in each of these units, the presence of recycle loops, the transportation lag due to flow through connecting pipes, and the fact that many process variables of importance are sampled and measured frequently, all conspire to guarantee that current observations will be strongly related to past observations. Measurements of controlled process variables will therefore be expected to be strongly serially correlated, but such correlations can be represented reasonably well with (deterministic) dynamic process models: the study of process dynamics in Part II was predicated upon this very fact.

There is the additional factor, however, of the unavoidable variations in raw materials, the build up of impurities, constant fluctuations in ambient conditions, catalyst deactivation, heat exchanger fouling, etc., that combine to "disturb" the process in ways that are virtually impossible to predict, or to represent with any reasonable deterministic model.

These two factors — the inherent process dynamics, and the constant presence of process disturbances — mean that:

> *The typical process variable measurement, $y(k)$, will naturally tend to be very strongly correlated with previous observations; in addition it will also naturally tend to wander and not remain "on-target."*

Thus within the context of the quality control model presented in Eq. (28.4), an observation $y(k)$ obtained from the typical chemical process has:

1. A "deterministic," or explainable part, $y^*(k)$, that is hardly ever constant — or uncorrelated — because of inherent process dynamics.

2. A "stochastic" or unexplainable part, that we will now refer to as $d(k)$, that is hardly the same as the uncorrelated "white noise," $e(k)$; $d(k)$ will usually take the form of a nonstationary drift.

28.3.2 Coping With Serial Correlation

The most important effect of serial correlation in process data is that all the underlying assumptions behind traditional charting methods break down. The Shewart Chart, as well as the EWMA, are based on the so-called "normal theory test" that depends very strongly on the measurement error, $e(k)$, being an independent and identically distributed Gaussian sequence. The sequential probability ratio test carried out in the CUSUM method is also based on $e(k)$ being such a Gaussian sequence.

When correlation is strong enough to undermine the statistical basis of these traditional quality control methods, it is customary to employ one of two approaches. The first approach essentially amounts to applying "band-aids" to the traditional techniques: the limits of the Shewart Chart are widened "appropriately" for positive correlation, or narrowed for negative correlation; the parameters of the CUSUM and EWMA are "adjusted appropriately" to account for the effect of correlation on the ARL. Some of these issues are discussed, for example, in English *et al.* [3], and also in MacGregor, Hunter and Harris [17]. These "band-aid" approaches, however, fail to confront the fundamental fact that all statistical hypothesis tests based on independence of sequential observations are completely invalidated by the presence of correlation.

The alternative approach is more systematic: variations in $y^*(k)$ are modeled appropriately to reflect the dynamics involved in the systematic (explainable) component of the data; disturbance models that go beyond setting $d(k) = e(k)$ are also developed to reflect the nature of the typical disturbances experienced by the process. The use of such techniques specifically for the purpose of developing quality control charts is discussed in detail in Keats and Hubele [7], and also in Harris and Ross [4]. In Section 28.5 we will discuss how this same approach is used for controller design.

28.3.3 Standard Process Control

To be sure, the presence of serial correlation and systematic drifts in the observed value of a process variable severely limit the applicability of traditional QC methods on most chemical processes. But a more serious impediment is the assumption of the traditional QC methods that making adjustments in the manipulated variable to correct for observed deviations from target values *always* has significant associated costs.

In most industrial chemical processes, not only are adjustments in the manipulated variables cost-free, but because of a natural propensity for the process to drift away from target, it is *not* making timely adjustments to correct for observed deviations that has associated costs. This is the motivation for the standard process control paradigm: that the natural dynamics of the process, and the persistent nature of the disturbances, are such that waiting to ascertain the "significance" of an observed deviation before taking corrective control action is often unacceptable. Corrective control action is therefore taken after every observation is made, employing any of the control algorithms discussed in detail in Part IV of the book, the most popular being feedback control, using the PID control algorithm (cf. Chapters 14 and 15).

A well-devised standard control strategy can therefore significantly minimize the effect of persistent disturbances on the process variable of interest, but at the expense of making continuous adjustments in the manipulated variable. However, since the net result of having a key process output variable drift away from desired target value can mean significant financial loss; and since it is relatively easy to make the required changes in the manipulated variables to correct for these process excursions; and since the resulting variation in the manipulated variable value is of no real consequence, it has been said that:

> The primary aim of standard process control is to transfer variations from where it "hurts" (in the process output, y) to where it does not hurt (in the manipulated variable, u) (cf. Downs and Doss [2]).

The most serious criticism of this standard process control philosophy lies in the fact that when the observed deviations from target are truly random, and not statistically significant, the continuous controller adjustments actually make the situation worse. These, and other related issues, have been discussed, for example, in MacGregor [16].

28.4   WHEN IS TRADITIONAL SPC APPROPRIATE?

Under the appropriate conditions, traditional SPC has provided very powerful tools for improving product quality. That these traditional SPC techniques have enjoyed widespread use in many manufacturing operations is a testimony to their simplicity and power. However, it is important to know when these tools are appropriate and when they are not.

28.4.1 Three Basic Process and Control Attributes

In deciding whether or not traditional SPC techniques may be applicable in any particular situation, it is helpful to investigate the status of the following three process and control attributes:

1. *Sampling interval* (*as a ratio of the natural process response time*)
    When the process output variable is measured at a sampling interval that is considered large relative to the natural response time of the process, the natural dynamics of the process are less likely to be manifested in the observations, and measurements are less likely to be serially correlated. The converse is true when the sampling interval is small relative to the natural process response time.

2. *Noise or disturbance structure*
    When the "unexplainable" part of the data is due mostly to measurement noise (no matter how significant this might be), such noise signals are often quite well approximated as random signals from a Gaussian distribution with zero mean. When the process is subject to additional unpredictable disturbances of more complex structure, a Gaussian noise structure will be grossly inadequate in representing such signals realistically.

3. *Cost of implementing control action*
    When the cost of making adjustments on the process is substantial, it is reasonable first to ascertain the significance of any observed deviations from target before taking corrective action; when adjustments are cost-free, such a philosophy is less justifiable, even though control action should still be implemented judiciously.

28.4.2 When Traditional SPC is More Appropriate

Traditional SPC is more appropriate:

1. When the sampling interval is considered *large* relative to the natural process response time (for then the assumption that the process mean is essentially constant and free of dynamics is reasonable).

2. When the process is not subject to significant disturbances, so that any observed variability in the process output is essentially due to measurement noise (for then the assumption that $e(k)$ is Gaussian with zero mean is reasonable), i.e., when the "process disturbances" are dominated by measurement noise.

3. When the cost of making adjustments is substantial.

Condition 1 also permits the operator time to "find" and "eliminate" the assignable cause, as required by the SPC philosophy.

28.4.3 When Standard Process Control is More Appropriate

Standard process control is more appropriate:

1. When the sampling interval is *small* relative to the natural process response time (for then the natural process dynamics are in evidence, and uncorrected deviations from target will tend to persist).

2. When the process is subject to persistent, significant disturbances, so that any observed variability in the process output will likely be due more to the effect of such disturbances than to the effect of zero mean Gaussian measurement noise (i.e., when the "process disturbances" dominate measurement noise).

3. When the cost of making adjustments is negligible.

28.5   STOCHASTIC PROCESS CONTROL

28.5.1 Basic Premise

"Stochastic process control" is a systematic approach to the problem of process — as well as quality — control; it deals simultaneously with the dual issues of process dynamics and stochastic disturbances that are more complex than Gaussian white noise. The basic premise is that, with reasonable models for the explainable (i.e., deterministic), as well as the unexplainable (i.e., stochastic), part of the process data, control decisions that satisfy reasonable performance objectives can be made rationally. There are therefore three essential elements required for the stochastic process control paradigm:

1. A *deterministic* dynamic process model
2. A *stochastic* disturbance model
3. A performance objective

Within this context, it is instructive to observe that the traditional SPC techniques employ the model $y^*(k) = y^*$ (a constant), for the deterministic part of the data, and $e(k)$ = Gaussian white noise for the stochastic part; the performance objective is to minimize the adjustments, subject to the constraints that $y_{LCL} < y(k) < y_{UCL}$.

Classical process control *apparently* does not employ any explicit model either for the process dynamics or for the disturbances (we will see later that it actually does employ some implicit models); and the apparent performance objective is to make $y(k) = y_d(k)$. We will also see later that the classical PID controller's performance objective in fact has a more subtle stochastic interpretation.

28.5.2 Models for Stochastic Process Control

Within the context of the current discussion, linear difference equations models (and the equivalent pulse transfer function models) are the most commonly employed deterministic models. Representing the process output as $y(k)$, as usual, and the process input (manipulated variable) as $u(k)$, the following are some example models:

$$\left(y(k) - y_0\right) = K u(k-1) \tag{28.24}$$

for a pure gain process:

$$\left(y(k) - y_0\right) = \pi_1 \left(y(k-1) - y_0\right) + \zeta_1 u(k-m-1) \tag{28.25}$$

for a first-order process with a time delay of m sample times. (Note that for convenience, the models have been presented in terms of deviation of $y(k)$ from a nominal value y_0.)

In transfer function form, the model in Eq. (28.25) becomes:

$$\left(y(k) - y_0\right) = \left(\frac{\zeta_1 z^{-m}}{1 - \pi_1 z^{-1}}\right) u(k-1) \tag{28.26}$$

where, as we may recall from Part IVC, z^{-1} plays the role of a "backward shift" operator. More generally the transfer function model will be of the form:

$$\left(y(k) - y_0\right) = P(z^{-1}) z^{-m} u(k-1) \tag{28.27a}$$

where

$$P(z^{-1}) = \frac{\zeta(z^{-1})}{\pi(z^{-1})} \tag{28.27b}$$

and $\zeta(z^{-1})$ and $\pi(z^{-1})$ are polynomials in z^{-1}.

Disturbance Models

It is customary to employ time series models (cf. Box and Jenkins [1]) for representing the stochastic disturbances. For example, the following model:

$$d(k) = \phi\, d(k-1) + e(k) \tag{28.28}$$

where $e(k)$ is Gaussian white noise, is known as a first-order autoregressive, or AR(1), model, while:

$$d(k) = e(k) - \theta\, e(k-1) \tag{28.29}$$

is known as a first-order moving average, or MA(1), model. For reasons we will not go into here, these two models are examples of *stationary* stochastic processes. Two *nonstationary* models are: the "random walk" model:

$$d(k) = d(k-1) + e(k) \tag{28.30}$$

and the "integrated first-order moving average," or IMA(1,1) model:

$$d(k) = d(k-1) + e(k) - \theta\, e(k-1) \tag{28.31}$$

The most general linear time series model is the ARIMA(p,D,q) model, the **A**uto**R**egressive, **I**ntegrated, **M**oving **A**verage model, with p as the order of the autoregressive part q as the order of the moving average part, and D as the degree of differencing (or integration). It has the form:

$$\phi(z^{-1})\,\nabla^D d(k) \;=\; \theta(z^{-1})e(k) \tag{28.32}$$

where $\phi(z^{-1})$ and $\theta(z^{-1})$ are, respectively, pth-order and qth-order polynomials in the backshift operator z^{-1}, and ∇^D is the Dth difference operator defined as:

$$\nabla^D d(k) \;=\; d(k) - d(k - D)$$

The interested reader is referred to Box and Jenkins [1] for further details; for our purposes here, we only note an important result due to MacGregor [14] that as the interval between observations increases, the behavior of an ARIMA$(p,1,q)$ model — for any p or q — approaches that of the IMA(1,1) model shown in Eq. (28.31). This is the motivation for restricting our attention to the model in Eq. (28.31).

28.5.3 Minimum Variance Control: General Results

The overall model used for stochastic process control is of the form:

$$\big(y(k) - y_0\big) \;=\; P(z^{-1})z^{-m}\,u(k-1) \;+\; d(k) \tag{28.33a}$$

or

$$\big(y(k) - y_0\big) \;=\; P(z^{-1})\,u(k-m-1) \;+\; d(k) \tag{28.33b}$$

with appropriate form chosen for $P(z^{-1})$, and for $d(k)$; and an appropriate value chosen for m, the process delay. The typical objective is that control action, $u(k)$, should be determined to minimize the *expected* squared deviation of $y(k)$ from its desired target value y_d. Since $E[(y - y_d)^2]$ is the theoretical definition of the variance of y (given that its mean value is y_d), this objective is commonly referred to as the "minimum variance" control objective.

To determine the "minimum variance" control action, without loss of generality, we take the nominal value y_0 around which the model was obtained to be the target value y_d, and note from Eq. (28.33) that:

$$\big(y(k + m + 1) - y_d\big) \;=\; P(z^{-1})u(k) \;+\; d(k + m + 1) \tag{28.34}$$

It can now be shown (cf. Box and Jenkins [1], MacGregor [15]) that the control action, $u(k)$, that minimizes $E\{[y(k + m + 1) - y_d]^2\}$ in Eq. (28.34), is given by:

$$u(k) \;=\; \frac{1}{P(z^{-1})}\,[-\hat{d}(k + m + 1|k)] \tag{28.35}$$

where $\hat{d}(k + m + 1|k)$ is the *minimum variance forecast* of $d(k + m + 1)$ given information up to the present time k. Such a forecast is obtained using procedures discussed in detail in Box and Jenkins [1], and some illustrative examples will be given below. For now, we simply note that the procedure is equivalent to applying a "forecast filter" on the currently available disturbance information, $d(k)$, i.e.:

$$\hat{d}(k + m + 1|k) = F(z^{-1}) d(k) \tag{28.36}$$

and that the specific form taken by $F(z^{-1})$ is determined by the disturbance model.

Whenever an observation $y(k)$ becomes available, we may use Eq. (28.33) to obtain the corresponding value of $d(k)$ as:

$$d(k) = (y(k) - y_d) - \hat{y}_{dm}(k) \tag{28.37a}$$

where

$$\hat{y}_{dm}(k) = P(z^{-1})z^{-m}u(k - 1) \tag{28.37b}$$

is the dynamic model prediction of the deterministic part of the observed data $y(k)$. Thus, Eq. (28.35) may now be rewritten with the aid of Eqs. (28.36) and (28.37) as:

$$u(k) = \frac{1}{P(z^{-1})} \left\{ F(z^{-1}[y_d - (y(k) - \hat{y}_{dm}(k))] \right\} \tag{28.38}$$

In diagramatic form, the strategy represented by Eq. (28.38) is as shown in Figure 28.6 (cf. MacGregor [15]), and is structurally the same as the IMC form developed apparently under the strictly deterministic, standard control paradigm in Chapter 19. However, it is important to observe that a crucial part of the IMC strategy is the requirement of an estimate of the unknown disturbance d, to be determined from plant output data and the process model. Whenever unknown information must be deduced from data, there is always a statistical underpinning, whether explicit or not.

28.5.4 Minimum Variance Control: Specific Results

For certain specific simple forms of dynamic process and disturbance models, the nature of the resulting minimum variance controller can be very instructive. We shall consider four examples.

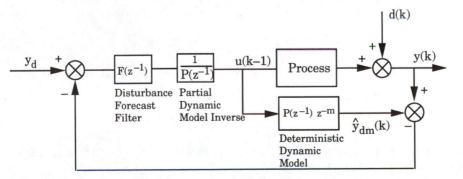

Figure 28.6.   Block diagram of the minimum variance control strategy.

1. Pure gain process with Gaussian disturbance.

The operating model is:

$$\left(y(k) - y_d\right) = Ku(k-1) + e(k) \tag{28.39}$$

and the minimum variance control is given by:

$$u(k) = -\frac{1}{K}\hat{e}(k+1|k) \tag{28.40}$$

$\hat{e}(k+1|k)$, the minimum variance forecast of $e(k+1)$ is obtained as $E[e(k+1)]$, i.e., its expected value. Now, if $e(k)$ is truly a zero mean Gaussian sequence, then:

$$E[e(k+1)] = 0 \tag{28.41}$$

and the minimum variance control is:

$$u(k) = 0 \tag{28.42}$$

i.e., no control action should be taken.

Observe that the conditions implied in this example are precisely those implied by the assumptions upon which the Shewart Chart is based. This justifies the traditional Shewart recommendation to "do nothing" whenever the process is "on-target" (implied by $\hat{e}(k+1|k) = E[e(k+1)] = 0$).

If however, the observed value of the error $e(k)$ takes on a "significant" value, say, δ, the interpretation is usually that there has been a "shift" or disturbance of the same amount; and if this shift is assumed to be constant, then $E[e(k+1)]$ is now equal to the nonzero value, δ. Under these circumstances, the minimum variance control is:

$$u(k) = -\frac{1}{K}\delta \tag{28.43}$$

2. Pure gain process with IMA(1,1) disturbance.

The operating model in this case is:

$$\left(y(k) - y_d\right) = Ku(k-1) + d(k) \tag{28.44}$$

with

$$d(k) = d(k-1) + e(k) - \theta e(k-1) \tag{28.31}$$

and $e(k)$ is zero mean Gaussian noise. The minimum variance control is obtained from:

$$u(k) = -\frac{1}{K}\hat{d}(k+1|k) \tag{28.45}$$

Now, $\hat{d}(k+1|k)$, the minimum variance forecast of $d(k+1)$, is its expected value, $E[d(k+1)]$, which is computed as follows: from Eq. (28.31):

$$d(k+1) = d(k) + e(k+1) - \theta e(k) \tag{28.46}$$

so that:

$$\hat{d}(k+1 \mid k) = E[d(k+1)] = d(k) - \theta e(k) \qquad (28.47)$$

since $E[e(k+1)] = 0$ and $e(k)$, having been observed, is considered a known quantity. By subtracting Eq. (28.47) from Eq. (28.46) we obtain the expression for the "one-step-ahead" forecast error:

$$e(k+1) = d(k+1) - \hat{d}(k+1 \mid k)$$

from which we obtain:

$$e(k) = d(k) - \hat{d}(k \mid k-1) \qquad (28.48)$$

Introducing this now into Eq. (28.47) yields:

$$\hat{d}(k+1 \mid k) = \theta \hat{d}(k \mid k-1) + (1 - \theta)d(k) \qquad (28.49)$$

recognizable either as a first-order filter on $d(k)$, or equivalently, as an EWMA of $d(k)$.

Thus, if $\hat{d}(k+1 \mid k)$ is not significant (in the sense that the EWMA plot does not trigger a signal), then Eq. (28.45) implies that no control action is necessary; if $\hat{d}(k+1 \mid k)$ *is* significant, then the minimum variance control to be implemented is as given in Eq. (28.45).

Example 28.1 MINIMUM VARIANCE CONTROLLER FOR PURE GAIN PROCESS WITH IMA(1,1) DISTURBANCE.

Obtain an expression for the minimum variance controller for the process whose model has been identified as :

$$\left(y(k) - y_d \right) = 5.0u(k-1) + d(k) \qquad (28.50a)$$

with

$$d(k) = d(k-1) + e(k) - 0.6 \, e(k-1) \qquad (28.50b)$$

Solution:

From Eq. (28.45) and the model parameters in Eq. (28.50), we obtain the required minimum variance controller expression as:

$$u(k) = -0.2 \, \hat{d}(k+1 \mid k)$$

The minimum variance forecast required in this specific case is obtained according to:

$$\hat{d}(k+1 \mid k) = 0.6\hat{d}(k \mid k-1) + 0.4d(k)$$

which is an EWMA of the disturbance, $d(k)$, itself obtained from Eq. (28.50a) as:

$$d(k) = [y(k) - y_d] - 5.0u(k-1)$$

3. **Pure gain process with IMA(1,1) disturbance:**
   (Control action to be implemented at every time instant k).

In this case we wish to write an explicit expression for the minimum variance controller we have just derived in Eq. (28.45). First, the forecast formula in Eq. (28.49) is rewritten as:

$$(1 - \theta z^{-1})\, \hat{d}(k + 1 \mid k) = (1 - \theta)d(k)$$

which can be written in the transfer function form:

$$\hat{d}(k + 1 \mid k) = F(z^{-1})d(k) \tag{28.51a}$$

with

$$F(z^{-1}) = \frac{(1 - \theta)}{(1 - \theta z^{-1})} \tag{28.51b}$$

(Note that both the parameter θ and the form of this "forecast filter" $F(z^{-1})$ arise naturally from the IMA(1,1) disturbance model.)

The minimum variance control expression now becomes:

$$u(k) = -\frac{1}{K} \frac{(1 - \theta)}{(1 - \theta z^{-1})} d(k)$$

or

$$(1 - \theta z^{-1})u(k) = -\frac{1}{K}(1 - \theta)[(y(k) - y_d) - Ku(k - 1)] \tag{28.52}$$

where we have made use of the model in Eq. (28.44) to write $d(k)$ in terms of known process information. If we now note that $y_d - y(k)$ is $\varepsilon(k)$, the familiar feedback error, then Eq. (28.52) rearranges to give:

$$u(k) - u(k - 1) = \frac{(1 - \theta)}{K} \varepsilon(k) \tag{28.53a}$$

or

$$u(k) = \frac{(1 - \theta)}{K} \sum_{i=1}^{k} \varepsilon(i) \tag{28.53b}$$

recognizable as a pure integral control algorithm.

> **Example 28.2   MINIMUM VARIANCE CONTROLLER IMPLEMENTED AT EVERY TIME INSTANT k FOR A PURE GAIN PROCESS WITH IMA(1,1) DISTURBANCE.**

Obtain an explicit expression for the minimum variance controller derived for the process given in Example 28.1 if the controller is to be implemented at every time instant k.

Solution:

Recalling from the model in Eq. (28.50) that the process parameters are $K = 5.0$ and $\theta = 0.6$, then from Eq. (28.53b), the required controller is the pure integral controller:

$$u(k) = 0.08 \sum_{i=1}^{k} \varepsilon(i)$$

4. **First-order process with IMA(1,1) disturbance:**
   (Control action to be implemented at every time instant k).

The model in this case is:

$$[y(k) - y_d] = \left(\frac{\zeta_1}{1 - \pi_1 z^{-1}}\right) u(k - 1) + d(k) \tag{28.54}$$

with

$$d(k) = d(k - 1) + e(k) - \theta\, e(k - 1) \tag{28.31}$$

and the minimum variance controller is given by:

$$u(k) = -\left[\frac{1 - \pi_1 z^{-1}}{\zeta_1} \frac{(1 - \theta)}{(1 - \theta z^{-1})}\right] d(k) \tag{28.55}$$

This rearranges to give:

$$(1 - \theta z^{-1}) u(k) = -\frac{(1 - \theta)}{\zeta_1}(1 - \pi_1 z^{-1})\left[y(k) - y_d - \left(\frac{\zeta_1}{1 - \pi_1 z^{-1}}\right) u(k - 1)\right]$$

upon using the model to replace $d(k)$. After further rearrangement, we obtain:

$$u(k) - u(k - 1) = \frac{(1 - \theta)}{\zeta_1}\left[\varepsilon(k) - \pi_1 \varepsilon(k - 1)\right] \tag{28.56}$$

where $\varepsilon(k)$ is the feedback error $y_d - y(k)$. This expression is recognizable as the velocity form of a PI controller; it may be further rearranged to give the familiar position form:

$$u(k) = \frac{(1 - \theta)}{\zeta_1}\left[\varepsilon(k) + (1 - \pi_1) \sum_{i=1}^{k-1} \varepsilon(i)\right] \tag{28.57}$$

The reader should show, as an exercise, that for a second-order model, with IMA(1,1) disturbance, the minimum variance controller is the digital PID controller.

Example 28.3 **MINIMUM VARIANCE CONTROLLER IMPLEMENTED AT EVERY TIME INSTANT k FOR A FIRST-ORDER PROCESS WITH IMA(1,1) DISTURBANCE.**

The deterministic model for a process has been obtained as:

$$\left(y(k) - y_d\right) = 0.5\left(y(k - 1) - y_d\right) + 0.8 u(k - 1) \tag{28.58}$$

with the accompanying stochastic disturbance model given by:

$$d(k) = d(k-1) + e(k) - 0.6\, e(k-1) \qquad\qquad (28.59)$$

Obtain an explicit expression for the minimum variance controller for this process if the controller is to be implemented at every time instant k.

Solution:

Observe that for this process, the model parameters are $\zeta_1 = 0.8$, $\pi_1 = 0.5$, and $\theta = 0.6$. Thus, from Eq. (28.57), the required controller is the PI controller:

$$u(k) = 0.5 \left[\varepsilon(k) + 0.5 \sum_{i=1}^{k-1} \varepsilon(i) \right] \qquad\qquad (28.60)$$

We may now note the following within this stochastic process control framework:

1. Implicit in the classical PID control algorithm is:
    *   A first- or second-order deterministic dynamic model
    *   An IMA(1,1) stochastic disturbance model
    *   A minimum variance performance objective

2. If the EWMA chart is used to implement control action, then implicit in such a strategy is:
    *   A pure gain deterministic model
    *   An IMA(1,1) stochastic disturbance model
    *   A minimum variance performance objective

3. The Shewart Charting paradigm corresponds to:
    *   A pure gain deterministic model
    *   A Gaussian white noise stochastic disturbance model
    *   A minimum variance performance objective

28.6   MORE ADVANCED MULTIVARIATE TECHNIQUES

Realistically, the process engineer is more likely to face a situation in which several process and quality variables are to be monitored and controlled simultaneously. As with multivariable process control, extending the ideas we have presented thus far to the multivariate case is fraught with many complications. Nevertheless, a very efficient multivariate equivalent of the Shewart plot has been developed and presented in Kresta *et al.* [8]. It is based on such tools of multivariate statistical analysis as Principal Component Analysis (PCA) and Projection to Latent Structures (PLS). A full discussion of these multivariate SPC monitoring tools lies outside the intended scope of this chapter; the interested reader is encouraged to consult the cited reference.

The application of these multivariate SPC techniques to the design and implementation of multivariable minimum variance controllers has also been carried out. A specific application to an industrial polymerization process has been reported in Roffel *et al.* [21], to which the interested reader is referred.

28.7   SUMMARY

The industrially important issue of process and quality control in the face of inherent variability has been the main focus of this chapter. We have reviewed the traditional Quality Control (QC) methods — the Shewart, CUSUM, and EWMA charts — highlighting the salient points and the basic assumptions underlying each technique. The role of standard process control within the framework of traditional statistical process control (SPC) was also discussed. Traditional SPC is not applicable in all situations, and neither is standard process control; the circumstances under which each approach is most appropriate have been carefully noted.

Stochastic process control was presented in this chapter as a technique in which appropriate deterministic dynamic models are combined with appropriate stochastic disturbance models and appropriate performance objectives for the purpose of making rational control decisions. It was shown that the apparently diametrically opposed traditional charting methods and the classical PID controllers find ready interpretation within this unifying framework. Higher dimensional, multivariate techniques were briefly mentioned with references provided for the reader interested in pursuing this rapidly developing topic further.

REFERENCES AND SUGGESTED FURTHER READING

1. Box, G. E. P. and G. M. Jenkins, *Time Series Analysis: Forecasting and Control*, Holden-Day, San Francisco (1976)
2. Downs, J. J. and J. E. Doss, "Present Status and Future Needs: The View From Industry," in *Chemical Process Control — CPC IV*, (Ed. Y. Arkun and W. H. Ray), AIChE, New York (1991)
3. English, J. R., M. Krishnamurthi, and T. Sastri, "Quality Monitoring of Continuous Flow Processes," *Computers and Industrial Engineering*, **20**, 251 (1991)
4. Harris, T. J. and W. H. Ross, "Statistical Process Control for Correlated Observations," *Canadian J. Chem. Eng.*, **69**, 48 (1991)
5. Hunter, J. S., "The Exponential Moving Average," *Journal of Quality Technology*, **18**, 203 (1986)
6. Hunter, J. S., "A One-point Plot Equivalent to the Shewart Chart with Western Electric Rules," *Quality Engineering*, **2**, 13 (1989)
7. Keats, J. B. and N. F. Hubele, Eds. *Statistical Process Control in Automated Manufacturing*, Marcel Dekker, New York (1989)
8. Kresta, J. V., J. F. MacGregor, and T. E. Marlin, "Multivariate Statistical Monitoring of Process Operating Performance," *Canadian J. Chem. Eng*, **69**, 35 (1991)
9. Lucas, J. M., "A Modified V-Mask Control Scheme," *Technometrics*, **15**, 833 (1973)
10. Lucas, J. M., "The Design and Use of V-Mask Control Schemes," *Journal of Quality Technology*, **8**, 1 (1976)
11. Lucas, J. M., "Combined Shewart-CUSUM Quality Control Schemes," *Journal of Quality Technology*, **14**, 51 (1982)
12. Lucas, J. M. and R. B. Crosier, "Fast Initial Response for CUSUM Quality Control Schemes: Give Your CUSUM a Head-start," *Technometrics*, **24**, 199 (1982a)
13. Lucas, J. M. and R. B. Crosier, "Robust CUSUM," *Communications in Statistics, Part A — Theory and Methods*, **11**, 2669 (1982b)
14. MacGregor, J. F., "Optimal Choice of Sampling Interval for Process Control," *Technometrics*, **18**, 151 (1976)
15. MacGregor, J. F., "On-line Statistical Process Control," *Chem. Eng. Prog.*, 21 (1988)

16. MacGregor, J. F., "A Different View of the Funnel Experiment," *Journal of Quality Technology*, **22**, 255 (1990)

17. MacGregor, J. F., J. S. Hunter, and T. J. Harris, *SPC Interfaces*, unpublished course notes (1993)

18. Page, E. S., "Continuous Inspection Schemes," *Biometrika*, **41**, 100 (1954)

19. Page, E. S., "Cumulative Sum Charts," *Technometrics*, **3**, 1 (1961)

20. Roberts, S. W., "Control Chart Tests Based on Geometric Moving Averages," *Technometrics*, **1**, 239 (1959)

21. Roffel, J. J., J. F. MacGregor, and T. W. Hoffman, "The Design and Implementation of a Multivariable Internal Model Controller for a Continuous Polybutadiene Polymerization Train," *Proceedings DYCORD+ 89*, IFAC Maastricht, 9–15 (1989)

22. Shewart, W. A., "The Application of Statistics in Maintaining Quality of a Manufactured Product," *JASA*, 546 (1925)

23. Shewart, W. A., *Economic Control of Quality*, Van Nostrand, New York (1931)

24. Wetherill, G. B. and D. W. Brown, *Statistical Process Control: Theory and Practice*, Chapman and Hall, London (1990)

REVIEW QUESTIONS

1. What is Statistical Process Control (SPC) in its broadest sense?

2. Within the context of maintaining a process variable steady in the face of unavoidable, inherent variability, what is the *analysis* problem and what is the *control strategy* problem?

3. What is "common cause" variation, and how is it different from "special cause" variation?

4. Why do the traditional Quality Control (QC) methods pay more attention to the analysis of process variability and not so much to control strategy?

5. Why does standard process control pay more attention to control strategy and not so much to the analysis of process variability?

6. How is the Gaussian probability density function used in the traditional QC methods?

7. In hypothesis testing, what is the difference between a Type I error and a Type II error?

8. What is an α risk and what is a β risk?

9. What is a classical Shewart Chart, and on what premise is it based?

10. When does a "signal" occur in a Shewart Chart?

11. What is the recommended plan of action when a "signal" occur in a Shewart Chart?

12. In employing the standard "3-sigma" limits on a Shewart Chart, what is the risk of raising a false alarm?

13. What is the motivation behind augmenting the Shewart Chart with the Western Electric rules?

14. What is the *average run length*, (ARL), and how does this concept help in assessing the usefulness of a QC method?

15. What process variable is plotted on a CUSUM chart?

16. Other things being equal, why do you think the CUSUM chart is likely to be more sensitive than the Shewart Chart in picking up smaller process shifts?

17. Theoretically, on what test is the CUSUM method based, and how does the test work?

18. What is the EWMA, and in terms of how it uses process data, how does it stand in relation to the Shewart and CUSUM methods?

19. Given the inherent standard deviation of raw process data and the EWMA parameter, how are the "3-sigma" limits set for the EWMA chart?

20. When does the EWMA chart become virtually equivalent to the Shewart Chart incorporating the Western Electric rules?

21. What is serial correlation and how does it affect the applicability of traditional QC methods?

22. What are the typical characteristics of the "deterministic" and the "stochastic" portions of a typical industrial process variable measurement?

23. As summarized by Downs and Doss [2], what is the "primary aim" of standard process control?

24. The status of what three basic process and control attributes should be investigated in deciding whether or not traditional SPC is applicable in a given situation?

25. When is traditional SPC appropriate?

26. When is standard process control appropriate?

27. What is the basic premise of Stochastic Process Control, and what are the three essential elements required for this paradigm?

28. What models are typical used for the deterministic part of stochastic process control models?

29. What models are typical used for the disturbance models in stochastic process control?

30. What is the performance objective of "minimum variance" control?

31. What are the primary elements in the block diagrammatic representation of the minimum variance controller, and how do they compare with the primary elements of the IMC strategy discussed in Chapter 19?

32. Within the context of stochastic process control what is the "performance objective" of the Shewart method?

33. Under what condition does the minimum variance controller become equivalent to the Shewart method?

34. Under what condition does the minimum variance controller become equivalent to the EWMA method?

35. Under what condition does the minimum variance controller become equivalent to the classical PI controller?

CHAPTER

29

SELECTED TOPICS IN
ADVANCED PROCESS CONTROL

Even after devoting individual chapters to *Model Predictive Control* and *Statistical Process Control* here in Part V, there is still a vast array of important advanced process control topics that we have not yet discussed. However, almost all new processes, and many existing processes, are equipped with *Distributed Computer Systems* (DCS) so that digital computer data logging and control are an integral part of the plant environment. As a consequence, *Advanced Process Control* methods are now being implemented broadly throughout the process industries. While the engineers designing these advanced control systems usually have advanced degrees and special training, all engineers involved in plant operations must understand the basic concepts behind these methods. Thus in this chapter, we will introduce the jargon and qualitative ideas associated with some of the advanced process control methods being installed in today's plants.

Essentially all of the methods of advanced process control involve the use of quantitative process knowledge (e.g., a detailed process model) in order to solve a difficult process control problem. For example, this process knowledge can be used to overcome the lack of on-line measurements, to deal with the problem of processes whose dynamic characteristics change with time, to provide control systems for processes with spatial profiles, to allow optimal economic performance, and to maintain safe process operation in case of failure of a key component of the control system (e.g., pump, valve, sensor). In each of the sections of this chapter, we shall describe a different advanced control method and the special problem it is designed to solve. We conclude with a brief discussion of some emerging process control technologies.

29.1 CONTROL IN THE ABSENCE OF GOOD ON-LINE MEASUREMENTS — STATE ESTIMATION

It is very often the case in the process industry that there are key process variables that cannot be measured on-line at frequent time intervals. It is even possible that no measurements at all are available on a routine basis.

Sometimes these measurements can be taken only very infrequently (as in the case of on-line gas chromatography with long elution times) or perhaps samples must be sent to the lab for analysis. In both these cases the measurement result is known only after a significant delay. For these very common situations, the process time constants may be short relative to the time between measurements, and in addition, the measurement result is delayed, indicating the state of the process some time in the past. Hence, direct feedback control based on these measurements is impractical. In such situations, a *State Estimator* may be designed to estimate the values of these process variables between measurements. This state estimator is used to "observe" the values of unmeasured variables and is often called an "observer."

Another type of problem that arises less frequently in process control is a situation where the process variable is measured on-line, but there is a high level of noise in the measurement signal. If the frequency of the noise is widely separated from the frequency of the process measurement itself (as in the case of 60 cycle noise superimposed on slowly varying temperature measurements), then simple filters such as described in Chapter 2 may be used. However, if there is significant overlap of frequencies in the noise signal and in the process measurement, then more sophisticated model-based filtering may be used as part of the state estimator. The most common of these filters is based on the work of Kalman in 1960 [cf. Refs. 1–3] and is termed a *Kalman Filter*.

29.1.1 State Estimator Structure

The general structure of a state estimator is shown in Figure 29.1. The components of the estimator are

1. A dynamic model for the states of the system $\mathbf{x}$ having dimension n:

$$\frac{d\mathbf{x}}{dt} = \mathbf{f}(\mathbf{x}, \mathbf{u}, \alpha) + \xi(t) \tag{29.1}$$

where $\xi(t)$ is some model error. The model depends on a knowledge of the process inputs $\mathbf{u}$ and model parameters α.

2. Initial conditions:

$$\mathbf{x}(0) = \mathbf{x}_0 + \xi_0 \tag{29.2}$$

where ξ_0 represents initial condition error.

3. Measurement devices producing an l vector of signals $\mathbf{y}$:

$$\mathbf{y} = \mathbf{h}(\mathbf{x}, \mathbf{u}, \beta) + \eta \tag{29.3}$$

where η represents measurement error. Usually there are fewer measurements than states ($l < n$). Here the model for the measurement devices $\mathbf{h}(\mathbf{x}, \mathbf{u}, \beta,)$ depends on a set of parameters β. The actual data $\mathbf{y}$ and the model $\mathbf{h}(\mathbf{x}, \mathbf{u}, \beta)$ differ by the error term η.

When on-line measurements are available continuously (as in the case of a temperature, level, or pressure measurement), these components may combine into a state estimator of the form:

$$\frac{d\hat{\mathbf{x}}}{dt} = \mathbf{f}(\hat{\mathbf{x}}, \mathbf{u}, \alpha) + \mathbf{K}(t)[\mathbf{y}(t) - \mathbf{h}(\hat{\mathbf{x}}, \mathbf{u}, \beta)] \tag{29.4}$$

$$\hat{\mathbf{x}}(0) = \mathbf{x}_0 \tag{29.5}$$

where $\mathbf{K}(t)$ is an $n \times l$ dimensional correction gain matrix and $\hat{\mathbf{x}}$ denotes an estimate of the true state $\mathbf{x}$. Note that the last term in the state estimator equation provides a correction to the process model based on the difference between the actual measurement, $\mathbf{y}(t)$, and the value of the measurement signal predicted by the model, $\mathbf{h}(\hat{\mathbf{x}}, \mathbf{u}, \beta)$. Figure 29.2(a) illustrates a typical state estimator trajectory in the case when $y(t)$ is continuous and is the measured value of $x(t)$. When some of the measurements are available only at discrete time intervals $t_1, t_2, t_3, ..., t_k$ and may arrive with some significant delay, then the state estimator consists of two parts. The first part has the form:

$$\frac{d\hat{\mathbf{x}}}{dt} = \mathbf{f}(\hat{\mathbf{x}}, \mathbf{u}, \alpha) + \mathbf{K}_1(t)[\mathbf{y}_1(t) - \mathbf{h}_1(\hat{\mathbf{x}}, \mathbf{u}, \beta)] \tag{29.6}$$

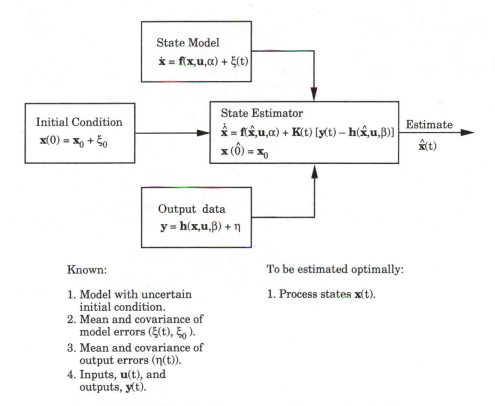

Known:

1. Model with uncertain initial condition.
2. Mean and covariance of model errors ($\xi(t)$, ξ_0).
3. Mean and covariance of output errors ($\eta(t)$).
4. Inputs, $\mathbf{u}(t)$, and outputs, $\mathbf{y}(t)$.

To be estimated optimally:

1. Process states $\mathbf{x}(t)$.

Figure 29.1.   The structure of a state estimator.

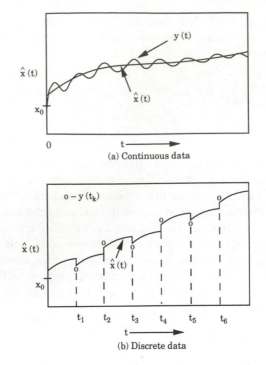

Figure 29.2.   Typical state estimator trajectories for (a) continuous data, and (b) discrete data.

and provides estimates for the time between discrete samples $t_{k-1} < t < t_k$. Here $\mathbf{y}_1(t)$ represents those measurements that are continuous. The second part of the estimator is activated when discrete measurement results becomes known at $t = t_k$ to provide an instantaneous correction according to:

$$\hat{\mathbf{x}}(t_k^{+}) = \hat{\mathbf{x}}(t_k^{-}) + \mathbf{K}_2(t_k)[\mathbf{y}_2(t_k) - \mathbf{h}_2(\hat{\mathbf{x}}, \mathbf{u}, \beta)] \qquad (29.7)$$

Here $\mathbf{y}_2(t_k)$ denotes discrete measurements available at $t = t_k$. The correction gains $\mathbf{K}_1(t)$, $\mathbf{K}_2(t_k)$ may be defined according to the type of estimator one employs. Figure 29.2(b) shows a typical state estimation trajectory for the case of no continuous measurements and only discrete data when $y(t_k)$ is a measured value of $x(t)$.

29.1.2 Applications of the State Estimator

As indicated in the last section, there are two principal functions that can be performed by a state estimator:

1. Observation of unmeasured process states, and
2. Extraction of the process states from noisy measurements.

The first task requires that certain "observability conditions" be satisfied in order for the unmeasured state to be capable of observation. For example, if we have a closed pressure vessel, and we can measure only the temperature history of the gas inside the vessel, can we also estimate the pressure history inside the vessel? This ought to be possible because of an equation of state such as:

$$PV = nRT \qquad (29.8)$$

Thus, by knowing that the volume and quantity of gas in the vessel are fixed, and by measuring the temperature, we may also observe the pressure. In this case the observability condition is satisfied. However, if we also allowed gas flow into the vessel in an unknown way, then the pressure in the vessel could not be estimated based on a temperature measurement because the observability condition is not satisfied. The lesson here is that for successful observation of an unmeasured state, the model must contain enough information to uniquely define that unmeasured state. Many types of observers, as well as Kalman Filters, have been used for observation of unmeasured states. The reader is referred to Refs. [1–3] for further discussion.

To be successful in using a state estimator to extract process states from noisy data requires detailed information about the statistics of the process. For example, one would require the mean and covariance for all relevant quantities such as process modeling error, $\xi(t)$; initial condition error, ξ_0; and measurement error, $\eta(t)$. This information would be provided to statistical evolution equations used in the calculation of $\mathbf{K}(t)$ in Eq. (29.4). Through optimal choices of $\mathbf{K}(t)$, the optimal estimates of $\hat{\mathbf{x}}(t)$ can be extracted from the data $\mathbf{y}(t)$. The particular state estimator used for this task usually involves some form of Kalman Filter (see Refs. [1–3]).

Example 29.1   A CONTINUOUS STIRRED TANK REACTOR (ADAPTED FROM REF. [1])

Let us consider an isothermal continuous stirred tank reactor (CSTR) for the reactions:

$$A \xrightarrow{r_1} B \xrightarrow{r_2} C$$

The rates for each of the reactions are given in terms of the concentrations c_A, c_B in the reactor:

$$r_1 = k_1 c_A$$
$$r_2 = k_2 c_B$$

where k_1, k_2 are rate constants. The modeling equations for the reactor take the form:

$$V \frac{dc_A}{dt'} = F(c_{Af} - c_A) - V k_1 c_A \quad c_A(0) = c_{A0}$$
$$V \frac{dc_B}{dt} = F(c_{Bf} - c_B) + V(k_1 c_A - k_2 c_B) \quad c_B(0) = c_{B0}$$

Here c_{Af}, c_{Bf} are the feed concentrations of A and B respectively and c_{A0}, c_{B0} are the initial concentrations in the reactor at start-up. Now suppose that it is possible to measure the concentration of species B continuously on-line, but we cannot measure the concentration of A even though it is important to know its value. Thus we wish to estimate the concentration of species A using a state estimator.

First let us simplify the equations by noting that $c_{Bf} = 0$ and the quantities V, F, c_{Af}, k_1, k_2 are all constant. Then we can define the dimensionless parameters:

$$Da_1 = \frac{k_1 V}{F} ; \quad Da_2 = \frac{k_2 V}{F} ; \quad t = \frac{t'V}{F}$$
$$x_1 = \frac{c_A}{c_{Aref}} ; \quad x_2 = \frac{c_B}{c_{Aref}} ; \quad u = \frac{c_{Af}}{c_{Aref}}$$

where c_{Aref} is the nominal initial charge of species A to the reactor.
With these definitions the model becomes:

$$\frac{dx_1}{dt} = -(1 + Da_1)x_1 + u \qquad\qquad x_1(0) = x_{10}$$

$$\frac{dx_2}{dt} = Da_1 x_1 - (1 + Da_2)x_2 \qquad\qquad x_2(0) = x_{20}$$

If we have a continuous measurement of species B, then our sensor model is:

$$y(t) = x_2(t) + \eta(t)$$

where $\eta(t)$ is the measurement error.
Then going to Eqs. (29.4) and (29.5) for our state estimator equation, we obtain:

$$\frac{d\hat{x}_1}{dt} = -(1 + Da_1)\hat{x}_1 + u + k_{11}[y(t) - \hat{x}_2(t)] \qquad \hat{x}_1(0) = x_{10}$$

$$\frac{d\hat{x}_2}{dt} = Da_1\hat{x}_1 - (1 + Da_2)\hat{x}_2 + k_{21}[y(t) - \hat{x}_2(t)] \qquad \hat{x}_2(0) = x_{20}$$

To illustrate the performance of this state estimator, numerical computations were carried out for the parameters $Da_1 = 3.0$, $Da_2 = 1.0$, $t_f = 2.0$, true initial conditions $x_{10} = 1.0$, $x_{20} = 0$; and inlet feed concentration $u = 0$. The state estimator was given erroneous initial conditions $\hat{x}_1 = 0.85$, $\hat{x}_2 = 0.15$. The values of k_{11}, k_{21} were determined from Kalman filter equations [1]. Gaussian measurement errors having zero mean and standard deviation $\sigma = 0.05$ were simulated by a random number generator and added to the x_2 values to produce the sensor signals, $y(t)$. The state estimates $\hat{x}_1$, $\hat{x}_2$ are shown in Figure 29.3 along with the "true" values x_1, x_2. Note how the unmeasured and measured states both converge to the "true" values with time.

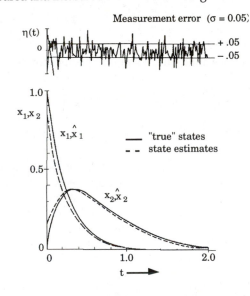

Figure 29.3.   Comparison of "true" states and state estimator estimates for continuous data. $\eta(t)$ is the actual measurement error used ($\sigma = 0.05$).

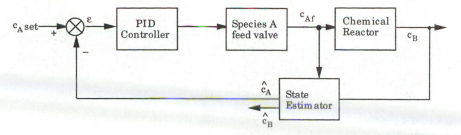

29.1.3 Use of State Estimation in Feedback Control

Now that we understand how to generate state estimates, we may use these estimates in a feedback control loop. For example, suppose that we have the chemical reactor of the last section and we wish to configure a control loop to keep the concentration of species A at its set-point. However, since we cannot measure species A directly, we must use a state estimator to provide these "measurements" to the controller. The control loop is shown in Figure 29.4. Obviously one must have great confidence in the state estimator in order to implement such a control scheme. However, plant experience has shown that with good models, careful instrument calibration and regular control system maintenance, state estimators can be successfully used to improve product quality and control system performance.

29.2   CONTROL IN THE FACE OF PROCESS VARIABILITY AND PLANT/MODEL MISMATCH — ROBUST CONTROLLER DESIGN

One of the issues that has always arisen in control system design is how sensitive the controlled system is to expected variations in the process parameters. For example, if we have tuned a PI controller for a particular loop, how will this control loop behave if the process gain changes by 20% or the time delay changes by 30%? This is a question of process sensitivity and the robustness of the control system design. New design methods have been developed for addressing this issue, and we shall discuss some of them in this section.

29.2.1   SISO Sensitivity and Robust Controller Design

Let us begin our discussion with SISO systems because we can readily illustrate the important concepts. To have a specific example in mind, suppose that we consider the conventional control loop shown in Figure 29.5. If we were to apply classical tuning of a proportional controller to the loop, we would construct a Bode plot of $g(s)$ and determine the process gain at the crossover frequency ω_c (i.e., when the phase lag is 180°), as shown in Figure 29.6. The critical process gain takes the value A, and a proportional controller gain of $K_{cu} = 1/A$, would bring the control loop to the verge of instability. Thus, depending on the

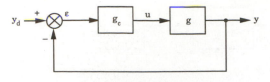

Figure 29.5.   A conventional SISO control loop.

particular controller tuning strategy selected, we use various phase margin and gain margin rules to select the proportional gain. For example, Ziegler-Nichols tuning would call for a gain margin of 2, so that a proportional gain of $K_c = 1/2A$ is recommended. Note that many of our tuning rules involve determining the controller gain which causes the process to go unstable and then reducing the gain by some factor to provide a margin of safety. These time-tested design rules recognize the simple fact that there is always a trade-off between control system *performance*, and robustness to instability (usually termed *robust stability*). Higher controller gains to favor control system performance always reduce the margin of safety for stability so that robustness decreases. As shown in Figure 29.7, one must always seek the proper balance between these two factors for the problem at hand.

In order to determine the best margin of safety for robustness, one would like to know the uncertainty and expected variability in $g(s)$. Suppose the *true process* can be represented exactly by:

$$g(s) = \frac{Ke^{-\alpha s}}{\tau s + 1} \tag{29.9}$$

then we have a process gain K, a time constant τ, and a time delay α, which can vary. Variations in process gain K will cause the entire curve $|g(s)|$ in Figure 29.6 to move up or down by the same factor; variations in the time constant τ will cause $|g(s)|$ to vary at high frequencies and $\angle g(s)$ to vary at medium frequencies; finally, variations in the time delay α will have no effect on $|g(s)|$ and will cause the curve $\angle g(s)$ to vary at higher frequencies. This illustrates the sensitivity of the Bode plot to variations in the process parameters. Note that loop stability is threatened only by changes at higher frequency around the crossover frequency. Thus robust stability tends to be determined by process variability at higher frequencies. This means that for the transfer function in Eq. (29.9), variations in the time delay and process gain are most critical, while variations in the time constant are not so important.

There are two practical situations that lead to problems with robustness. Both could be said to lead to plant/model mismatch and arise as a consequence of the following situations:

1. **Process Variability:** in which an otherwise adequate model no longer matches the observed plant behavior because the *process itself* has changed. Catalyst decay and heat exchanger fouling are two examples of mechanisms underlying such process variability.

2. **Inadequate Modeling:** in which the *model* is an inadequate representation of true plant behavior. This may be because the model structure is a fundamentally poor approximation (as is the case, for example, when a linear model is used to represent a severely nonlinear process), or it may be because of poor estimates of the model parameters in an otherwise adequate model structure.

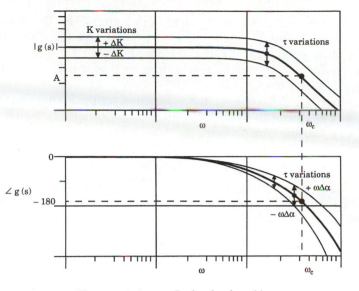

Figure 29.6.   Bode plot for $g(s)$.

Process Variability

Suppose that through trial and error experiment, or the use of a good model, a loop is tuned well with a certain margin of robust stability and put into service. Then, over time, some change occurs in the process to threaten instability. In this case, the changes could be in the process gain (e.g., catalyst decay, heat exchanger surface fouling, etc.), in the time constant, or in the time delay (change in feed pipe diameter, feed flowrate, etc.) In this case let us assume that variations in τ are insignificant but that the variations ΔK, $\Delta \alpha$ are significant so that the true process $g(s)$ lies in a fuzzy region given by the bounds on the transfer function

$$g(s) = \frac{(K + \Delta K)e^{-(\alpha + \Delta \alpha)s}}{\tau s + 1} \qquad (29.10)$$

We will assume that the parameter variations ΔK, $\Delta \alpha$ can be of either sign, and vary between:

$$- \left| \Delta K_{max} \right| \le \Delta K \le \left| \Delta K_{max} \right|$$

$$- \left| \Delta \alpha_{max} \right| \le \Delta \alpha \le \left| \Delta \alpha_{max} \right|$$

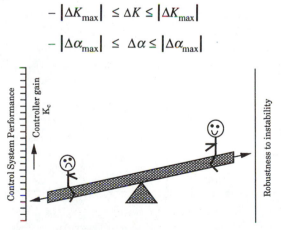

Figure 29.7.   The ever-present balance between control system performance and robust stability in control system design.

The bounds of this fuzzy region can be seen in Figure 29.6. The curve for $|g(s)|$ has possible variability $\pm |\Delta K_{max}|$, while the curve for $\angle g(s)$ has possible variability $\pm \omega |\Delta \alpha_{max}|$. Note the large potential variability in the crossover frequency ω_c and corresponding critical gain, $1/A$, caused by the variability in K and α.

It would be useful to have a transfer function for the bounds on this region of variability. Let us call this transfer function the variability factor, $\psi(s)$. Then we could say that the true process, $g(s)$, varies from the nominal process, $g_m(s)$, by:

$$g(s) = g_m(s)\psi(s) \tag{29.11}$$

Hence for our example here:

$$\psi(s) = \frac{K + \Delta K}{K} e^{-\Delta \alpha s} \tag{29.12}$$

Thus, the nominal process when $\Delta K = \Delta \alpha = 0$ has a variability factor of unity. We may view the Bode plot of this variability factor for $\Delta K = \pm |\Delta K_{max}|$, $\Delta \alpha = \pm |\Delta \alpha_{max}|$ to see how it influences the frequency response of the nominal process, $g_m(s)$ (cf. Figure 29.8).

If one were to convert the Bode plot in Figure 29.6 into a Nyquist diagram (see Figure 29.9), then we can see the uncertainty in gain and phase angle in the same graph. Notice that the variability factor $\psi(s)$ causes the possible domain of $g(s)$ to appear as wedge-shaped regions of uncertainty around $g_m(j\omega)$ when there is variability in only K and α. For example, see the wedge formed around the point $g(j\omega_1)$ as a result of the uncertainty in the actual values of $\angle g(j\omega_1)$

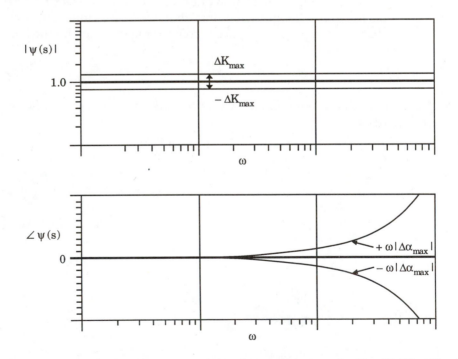

Figure 29.8.   Bode plot for the variability factor $\psi(s)$ given by Eq. (29.12); $\Delta K = \pm 1$, $\Delta \alpha = \pm 2$, $K = \tau = \alpha = 5$.

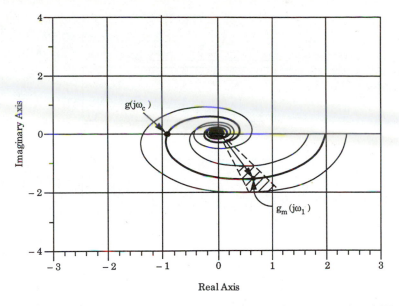

Figure 29.9. Nyquist diagram for $g(s)$, $g_m(s)$ given by Eqs. (29.11) and (29.12).

and $|g(j\omega_1)|$. The envelope of all these regions is the band given by $K \pm |\Delta K_{max}|$ shown in the figure. Notice that tuning rules involving phase margin and gain margin are designed to make sure that no part of the possible domain of $g(s)$ encompasses the point –1 in the Nyquist diagram. The gain margin rules ensure that the gain variations $K \pm |\Delta K_{max}|$ do not cause $g(s)$ to include –1 when $\omega = \omega_c$; the phase margin rules on the other hand ensure that variations in phase lag due to time-delay variability $(\alpha \pm |\Delta\alpha_{max}|)\omega$ do not cause $g(s)$ to encompass –1 for some $\omega < \omega_c$ in the figure. Thus one can see how PID controller tuning rules address the question of robust stability in the face of process variations.

Inadequate Modeling

Suppose that a linear process model, $g_m(s)$, was developed from a set of experimental data such as step test or pulse test data. However, there are always inadequacies in such experimental data sets. This could be due to unmodeled nonlinearities, poor experimental design, or measurement noise problems that prevent the model from representing all features of the process. In normal process identification experiments it is the high-frequency part of the model which is usually in greatest error. This is the process/model mismatch problem. In this situation, the robust stability of a control system depends on how sensitive the control system design is to model inadequacy.

In practice, one finds that the variability factor $\psi(s)$ resulting from process identification experiments has the general shape shown in Figure 29.10. Note that the variability factor is close to unity until the failure frequency ω_f is reached. Above this frequency, the experimental data are inadequate to identify reliable model parameters. If this failure frequency is well beyond the crossover frequency $\omega_f >> \omega_c$, then obviously the model will be adequate for conventional control system design; however, if the failure frequency is close to ω_c, then control systems designed using the model may lead to serious stability problems. We shall discuss some possible remedies for this problem below.

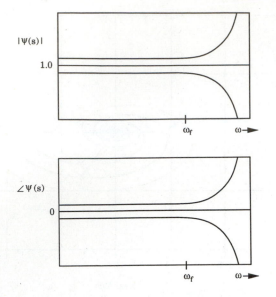

Figure 29.10. Typical frequency dependence of the variability factor, $\Psi(s)$, resulting from process identification experiments. Here ω_f represents the failure frequency beyond which the identification experiments are inadequate.

For the analysis of more complex systems, it is sometimes useful to neglect the specific shape of the variability region $g(j\omega)$ in Figure 29.9 and to approximate them with a circular disk with radius $\overline{l}_a(\omega)$ defined by:

$$\left|g(j\omega) - g_m(j\omega)\right| \le \overline{l}_a(\omega) \tag{29.13}$$

Thus we choose the radius $\overline{l}_a(\omega)$ large enough to encompass the exact shape $g(j\omega)$. Note that for the expected variations in $g(s)$ shown in Figures 29.9 and 29.10, the radius $\overline{l}_a(\omega)$ as a fraction of $\left|g_m\right|$ should be relatively small at low frequencies and increase dramatically at larger frequencies. Hence:

$$\overline{l}_m(\omega) \equiv \frac{\overline{l}_a(\omega)}{\left|g_m(j\omega)\right|} \tag{29.14}$$

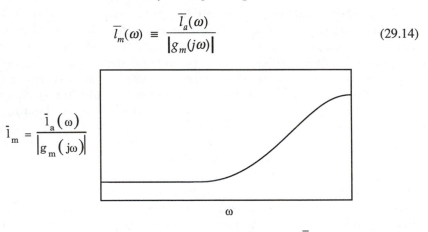

$$\overline{l}_m = \frac{\overline{l}_a(\omega)}{\left|g_m(j\omega)\right|}$$

Figure 29.11. Frequency dependence of the encompassing disk size $\overline{l}_a(\omega)$ as a fraction of the nominal process $\left|g_m(j\omega)\right|$.

should have the frequency-dependent behavior shown in Figure 29.11. Note from Eqs. (29.13) and (29.14) that in terms of the variability factor $\psi(j\omega)$, $\bar{l}_m(\omega)$ must satisfy:

$$\bar{l}_m(\omega) \geq |\psi(j\omega) - 1| \tag{29.15}$$

Robust Controller Design

With this background on the effects of process variability and of model uncertainty, we are now in a position to consider how one designs a controller THAT provides good performance and robust stability. From Figure 29.7, we recall that increasing the proportional controller gain to improve the performance reduces the robustness to instability. However, there is one factor that allows us to improve this trade-off somewhat. Generally, good control system performance requires high controller gains at low frequencies only. Conversely, the question of stability is determined by the controller gain near the crossover frequency. Thus, if we could shape the frequency response of our controller so as to provide high controller gains at low frequencies and a sharp reduction in controller gain as the crossover frequency is approached, we could then guarantee *both* high performance and robust stability from our controller. This would correspond to the curved balance beam in Figure 29.12.

To illustrate this point, let us design a robust PI controller for the system given by Eq. (29.9) with the variability factor $\psi(s)$ given by Eq. (29.12). Let us introduce a frequency-dependent weighting function $W(s)$ in the controller so that:

$$g_c(s) = K_c\left(1 + \frac{1}{\tau_I s}\right)W(s) \tag{29.16}$$

We would like $W(s)$ to reduce the controller gain at high frequency in order to guarantee stability under the worst parameter variations $\Delta K, \Delta \alpha$. This means that ideally $W(s)$ should have a Bode plot of the type shown in Figure 29.13, so that the gain drops off and the phase angle increases at high frequencies.

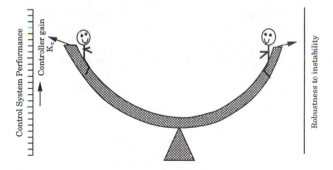

Figure 29.12. Altering the balance between control system performance and robust stability by designing the controller to have the proper frequency dependence.

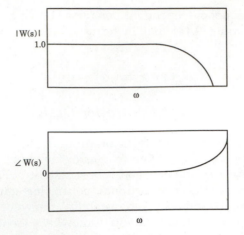

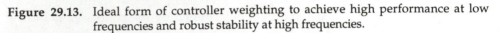

Figure 29.13. Ideal form of controller weighting to achieve high performance at low frequencies and robust stability at high frequencies.

There are difficulties in implementing this ideal controller weighting because simple forms such as:

$$W(s) = \frac{e^{\beta_w s}}{(\tau_w s + 1)^n} \tag{29.17}$$

whose Bode plots have the desired shape, are not causal. However, there are convenient approximations such as:

$$W(s) = \frac{\beta_w s}{(\tau_w s + 1)^n (1 - e^{-\beta_w s})} \tag{29.18a}$$

or

$$W(s) = \frac{(\tau_w s + 1)^n}{(\tau_w s + 1)^n + K_w (1 - e^{-\beta_w s})} \tag{29.18b}$$

which may be used. Let us illustrate with a particular numerical example.

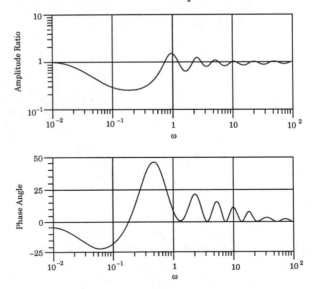

Figure 29.14. Bode plots for the control weightings $W(s)$ given by Eq. (29.19).

**Example 29.2   ROBUST CONTROLLER DESIGN FOR A PROCESS WITH
TIME DELAYS.**

Let us suppose that $K = 5$, $\tau = 5$, and $\alpha = 5$ in the process given by Eq. (29.9). If we
allow $\Delta K = \pm 1$ and $\Delta\alpha = \pm 2$, then the process variability has already been plotted in
Figures 29.6, 29.8, and 29.9. It can be shown that if we choose $W(s)$ to be of the form:

$$W(s) = \frac{5s + 1}{5s + 1 + 2(1 - e^{-5s})} \tag{29.19}$$

then $W(s)$ will have the frequency response shown in Figure 29.14. This is similar to
the ideal form in Figure 29.13 for frequencies $w < 0.5$.

The overall control scheme for the robust controller is represented in Figure 29.15.
The actual process is given by the model $g_m(s)$ combined with the variability factor as
in Eq. (29.11). In a similar way, the controller is a combination of a PI controller and
the input weighting function $W(s)$ as in Eq. (29.16).

For the $W(s)$ chosen (Eq. (29.19)), the Bode plot shown in Figure 29.14 is not as
effective in decreasing amplitude and increasing phase lag at high frequencies as the
idealized form in Figure 29.13. Nevertheless, as we shall see, it does decrease the
process gain and phase lag in the region close to the critical frequency that aids both
performance and stability.

To see the loop shaping effect, the nominal ($\psi(s) = 1$) "effective" process $g_m(s)\,W(s)$
seen by the PI controller has the Bode plot shown in Figure 29.16 for $W(s) = 1$ and for
$W(s)$ given by Eq. (29.17). Note how much the amplitude ratio has decreased at
ω_C. In fact, the gain margin has gone from 0.45 for the uncompensated plant to 1.18 for
the compensated plant. This should allow the PI controller gain to be much lower for
the same performance or the gain to be increased to improve performance with no
deterioration in robust stability.

To illustrate the control system performance the closed-loop step response of the
nominal plant, Eq. (29.9), with uncompensated PI controller ($W(s) = 1$) tuned optimally
to minimize Integral Time Squared Error (ITSE) for $g_m(s)$ is shown in Figure 29.17(a).
The corresponding step responses are shown in Figure 29.17(b) for the PI controller
tuned "manually" when the plant is $g_m(s)\,W(s)$ with $W(s)$ given by Eq. (29.19). Note
that for this "nominal" plant with no parameter variations, compensation with $W(s)$
given by Eq. (29.19) gives much better performance as expected. However, one must test
the effect of parameter variations on the control system.

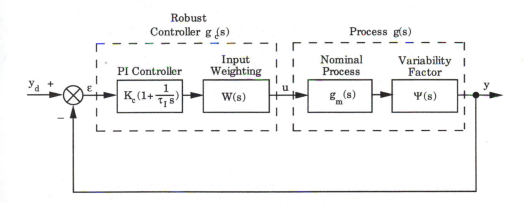

Figure 29.15.   Block diagram for a robust controller $g_c(s)$ consisting of a PI controller and
input weighting, $W(s)$. The process consists of the nominal process, $g_m(s)$,
and the variability factor $\Psi(s)$.

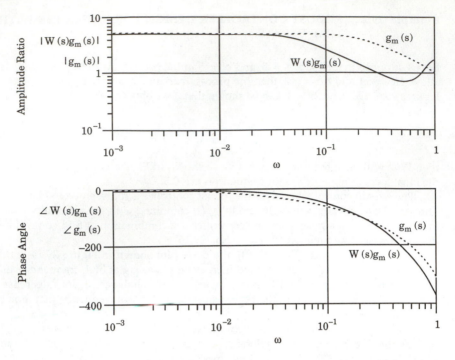

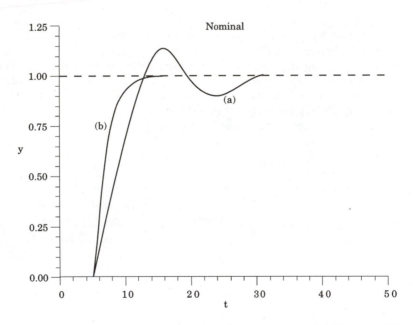

Figure 29.16. Bode plots of the compensated and uncompensated open-loop frequency response of the nominal plot, $W(s)\, g_m(s)$.

Figure 29.17. Closed-loop step response for the process in Eq. (29.9) under PI control with no parameter variations: (a) uncompensated controller tuned to be ITSE optimal for $W(s) = 1 (K_c = 0.184,\ 1/\tau_I = 0.126)$; (b) compensated controller tuned "manually" for good response when $W(s)$ is given by Eq. (29.19)$(K_c = 0.45,\ 1/\tau_I = 0.06)$.

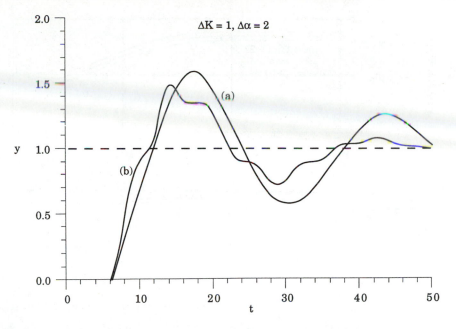

$\Delta K = 1, \Delta\alpha = 2$

Figure 29.18. Closed-loop step response for the process in Eq. (29.9) under PI control with parameter variations $\Delta K = 1$, $\Delta\alpha = 2$; tuning as in Figures 29.17(a) and (b).

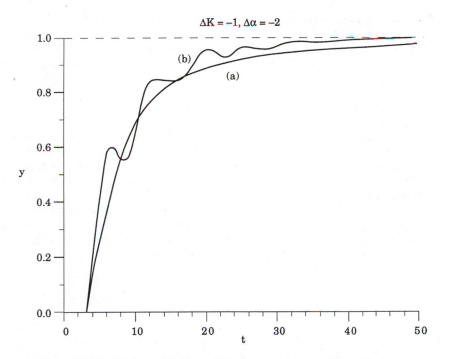

$\Delta K = -1, \Delta\alpha = -2$

Figure 29.19. Closed-loop step response of the process in Eq. (29.9) under PI control with parameter variations $\Delta K = -1$, $\Delta\alpha = -2$; tuning as in Figures 29.17(a) and (b).

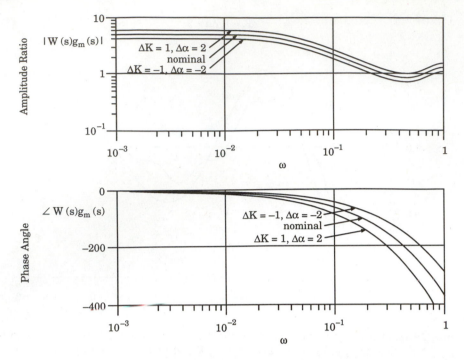

Figure 29.20. Open-loop frequency response of the compensated plant for various process parameter variations.

For parameter variations that cause the plant to be less stable ($\Delta K = 1$, $\Delta \alpha = 2$), the two closed-loop step responses are shown in Figure 29.18. By contrast, parameter variations that cause the plant to be more sluggish ($\Delta K = -1$, $\Delta \alpha = -2$) are applied and the two closed-loop responses shown in Figure 29.19. Observe that the performance of the compensated PI controller is noticeably better for all parameter variations. Thus the compensator chosen has indeed improved both performance and robust stability when the process gain varies by 20% and the process time delay changes by 40%. The reason can be seen from the Bode plots of the compensated process shown in Figure 29.20. In addition to the fact that the amplitude ratio is reduced by the compensator close to the critical frequency ω_c there is another valuable feature. Close to the critical frequency ω_c there is some significant variation in phase lag with parameter variations, but the amplitude ratio is relatively flat. Thus there is robust stability for reasonably large process parameter variations.

Although there are many possible choices for weighting functions $W(s)$ and optimization methods can be used to find good ones, the choice we made in Eq. (29.19) was found by using the equations for a Smith Predictor linked to a proportional controller. This turns out to have the very nice properties noted above.

There is a broad literature describing various approaches for synthesis of robust controllers for specific types of set-point changes or disturbances and for specific forms of model uncertainty, $\bar{l}_m(\omega)$. Those wishing more details can begin with Refs. [4–7].

29.2.2 MIMO Robust Controller Design

When we consider the design of multiple input/multiple output (MIMO) systems, it is much more difficult to characterize the uncertainty in the process and to design robust controllers. To understand the reasons for this, consider the 2 x 2 MIMO system with process model:

$$\mathbf{y} = \mathbf{G}(s)\mathbf{u}$$

where

$$\mathbf{y} = \begin{bmatrix} y_1 \\ y_2 \end{bmatrix}, \qquad \mathbf{u} = \begin{bmatrix} u_1 \\ u_2 \end{bmatrix}$$

and

$$\mathbf{G}(s) = \begin{bmatrix} \dfrac{K_{11}e^{-\alpha_{11}s}}{\tau_{11}s + 1} & \dfrac{K_{12}e^{-\alpha_{12}s}}{\tau_{12}s + 1} \\[3mm] \dfrac{K_{21}e^{-\alpha_{21}s}}{\tau_{21}s + 1} & \dfrac{K_{22}e^{-\alpha_{22}s}}{\tau_{22}s + 1} \end{bmatrix} \qquad (29.20)$$

As compared with the SISO system in Eq. (29.9), where everything could be represented by a Bode or Nyquist plot, this MIMO system has four elements $g_{ij}(s)$, each one with a different frequency response. Thus, in order to design a MIMO controller $\mathbf{G}_c$ (as shown in Figure 29.21) that will achieve good performance and robust stability for this system, more complex methods must be applied.

In keeping with the concept of shaping the frequency response of the control $\mathbf{G}_c$, Doyle and Stein [5] proposed the use of the *singular values* to construct a Bode-like plot for a MIMO system. Recall from Chapter 22 and Appendix D that the singular values of a matrix may be defined as the positive square root of the eigenvalues of the product $\mathbf{G}^H\mathbf{G}$ where $\mathbf{G}^H$ is the complex conjugate transpose of $\mathbf{G}$ — termed the Hermitian of $\mathbf{G}$. That is:

$$\sigma_i(s) = \sqrt{\lambda_i(\mathbf{G}^H(s)\mathbf{G}(s))}; \quad i = 1, 2, ..., N \qquad (29.21)$$

Thus there are N singular values for an $N \times N$ matrix, and the largest and smallest of the singular values provide norms for the matrix $\mathbf{G}(s)$:

$$\sigma_{max}(s) = \| \mathbf{G}(s) \|_2 \qquad (29.22)$$

$$\sigma_{min}(s) = \frac{1}{\|\mathbf{G}(s)^{-1}\|_2} \qquad (29.23)$$

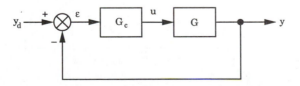

Figure 29.21.   A conventional block diagram for a MIMO control system.

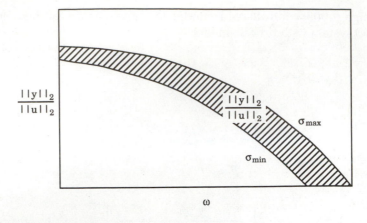

Figure 29.22.   Use of σ_{min}, σ_{max} to construct a fuzzy "Bode plot" for $\|\mathbf{y}\|_2/\|\mathbf{u}\|_2$.

It can therefore be shown that the "Amplitude Ratios" for this multivariable system may be bounded by:

$$\sigma_{min}(s) \leq \frac{\|\mathbf{y}\|_2}{\|\mathbf{u}\|_2} \leq \sigma_{max}(s) \qquad (29.24)$$

Here $\|\mathbf{y}\|_2$, $\|\mathbf{u}\|_2$ denote the L_2 norms of the vectors $\mathbf{y}$ and $\mathbf{u}$.

If we plot Eq. (29.24) as if it were the top half of a Bode plot (see Figure 29.22), we see that the "Amplitude Ratio" of the MIMO system lies between σ_{max} and σ_{min} so that we can view the frequency response of $\|\mathbf{y}\|_2/\|\mathbf{u}\|_2$ as the shaded region between σ_{max} and σ_{min}. Doyle and Stein [5] have worked out a design procedure for MIMO systems that shapes the frequency response of the controller $\mathbf{G}_c$ to guarantee that the product $\mathbf{GG}_c$ has sufficiently high gains at low frequency for good performance and sufficiently low gains at high frequency to guarantee robust stability for the range of model uncertainty expected. To do this, Doyle and Stein construct a singular value plot for $\|(\mathbf{GG}_c))\|_2$ (see Figure 29.23). To achieve the proper controller design, σ_{min} at low frequencies must be sufficiently large, and σ_{max} at high frequencies must be sufficiently small. Although Doyle and Stein prove that this design is sufficient for robust

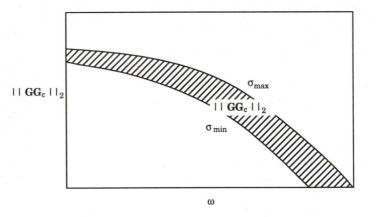

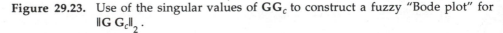

Figure 29.23. Use of the singular values of $\mathbf{GG}_c$ to construct a fuzzy "Bode plot" for $\|\mathbf{G}\,\mathbf{G}_c\|_2$.

stability, it is often too conservative because of all the approximations that are made — this results in controller designs that often give poorer performance than one would want. Obviously, this is even more of a problem the larger the fuzzy region between σ_{max} and σ_{min} for $\mathbf{G}(s)$. Because of this, the ratio of σ_{max} and σ_{min} for $\mathbf{G}(s)$, i.e.:

$$\kappa = \frac{\sigma_{max}}{\sigma_{min}} \tag{29.25}$$

known as the *condition number*, is often used as a measure of the sensitivity of the process to uncertainty. In fact, when evaluating possible control structures for MIMO systems, κ can be used as a screening variable to eliminate highly sensitive structures.

To overcome some of the conservatism of the singular value analysis that averages the model uncertainty over all elements of $\mathbf{G}(s)$ (even, for example, when $g_{11}(s)$ might be known very precisely and $g_{12}(s)$ perhaps very poorly), Doyle [6, 7] proposed the use of the "structured uncertainty" and the concept of "structured singular values." Designs using these as well as other ideas are discussed in detail by Morari and Zafiriou [4]. Unfortunately, the current approaches for robust controller design for MIMO systems still produce rather conservative results in many cases so that controller performance is not the best. Thus, important research remains to be done in order to bring sharper analysis to the problem so that phase lag information and more details of structure can be used in the controller design.

29.3 CONTROL OF SPATIAL PROFILES — DISTRIBUTED PARAMETER CONTROLLERS

Throughout our discussion of control problems to this point, we have assumed that all states, inputs, and measurements were only functions of time. However, there are process control problems for which the process variables are functions of position in space as well as time. For example, in metallurgical processing operations for steel, aluminum, or other types of metal slabs and sheets there are furnaces for creating the proper temperature profile in the metal. Figure 29.24 shows a typical 3-zone furnace in which a desired longitudinal temperature profile is to be imposed on the slab by manipulation of the firing rates in the individual zones of the furnace. In some operations the slab is motionless, and in some cases it is a continuous strip moving through the furnace with velocity v. Note that there are continuous optical pyrometer measurements of temperature at five fixed positions in the furnace. Figure 29.25 illustrates the desired set-point temperature profile and an actual temperature profile history that could arise through application of a good controller.

For this problem, the appropriate model for the process would be in the form of partial differential equations of the form:

$$\frac{\partial x(z, t)}{\partial t} = \alpha \frac{\partial^2 x(z, t)}{\partial z^2} + v \frac{\partial x(z, t)}{\partial z} + \beta x(z, t) + u(z, t) \; ; \; 0 \leq z \leq L \tag{29.26}$$

with boundary conditions:

$$\frac{\partial x(0, t)}{\partial z} = \frac{\partial x(L, t)}{\partial z} = 0 \tag{29.27}$$

3 Furnace Heating Zones

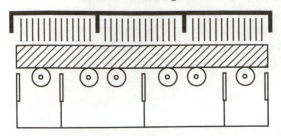

5 Optical Temperature Sensors

Figure 29.24. Heating of a metal slab in a 3-zone furnace having 5 temperature sensors.

Here α is a thermal conduction parameter, v is the velocity of the slab through the furnace, β represents the convective cooling heat transfer coefficient, and $u(z,t)$ is the radiant heat flux from the three zones of the furnace. The boundary conditions Eq. (29.27) indicate that there is negligible heat transfer from the ends of the slab. The measurements are taken at five points z_i where the optical sensors are located:

$$y_i(t) = x(z_i^*, t) \quad i = 1, 2, 3, 4, 5 \tag{29.28}$$

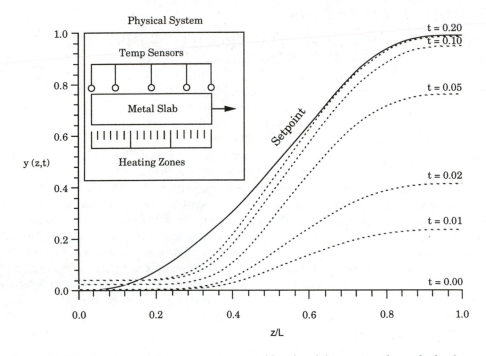

Figure 29.25. Control of the temperature profile of a slab moving through the 3-zone furnace. Start-up from a cold slab.

Note that $u(z,t)$ has the special form:

$$u(z,t) = \begin{cases} u_1(t); & 0 \leq z \leq \dfrac{L}{3} \\[2ex] u_2(t); & \dfrac{L}{3} \leq z \leq \dfrac{2L}{3} \\[2ex] u_3(t); & \dfrac{2L}{3} \leq z \leq L \end{cases} \qquad (29.29)$$

where we can manipulate $u_1(t)$, $u_2(t)$, $u_3(t)$.

The task of the distributed parameter controller could be posed in two ways:

1. **The simple-minded approach:** Forget that there are spatial profiles and simply design a MIMO controller that adjusts u_1, u_2, u_3 in order to bring the five measurements, $y_i(t)$, to their set-point. This approach will work for some problems, but will not in general give the best results because so much information about the spatial structure is neglected.

2. **The more enlightened approach:** Plan to use all the information at hand to design a *distributed parameter controller* that will control spatial profiles. There are several ways in which this may be done (see Ref. [1] for a discussion). If the model equation and parameters (such as Eqs. (29.26 – 29.29)) are known, then the first principles PDE model may be used to design the controller (see Ref. [1] for some examples). Alternatively, if the model equations are unknown, then a spatially distributed transfer function model of the form:

$$x(z,s) = \sum_{i=1}^{N} w_i(z,\, s)\, \sigma_i(s) \int_0^L \overline{v}_i(\zeta,\, s)\, u(\zeta,\, s)d\zeta$$

(where $\sigma_i(s)$ is the singular value and w_i, v_i are right and left singular functions) may be obtained from experiment. Recent work (Refs. [8, 9]) has shown how this approach to modeling captures all of the spatial structure of the distributed system and allows high-quality control systems to be designed after a few process identification experiments. To illustrate the performance of such an approach, the start-up of a slab furnace was carried out with a controller of this type, and the results are shown in Figure 29.25.

Another case where the control of spatial profiles is important is the control of concentration and temperature profiles in a tubular reactor. This singular function approach has been shown to perform well in this case also (cf. Ref. [9])

29.4 CONTROL IN A CHANGING ECONOMIC ENVIRONMENT — ON-LINE OPTIMIZATION

So far in our discussions of control, we have focused on the individual control loops of a multivariable process unit. In practice, of course, the objective of any process operation is to make a profit for the organization. This requires many levels of process scheduling, optimization, and control. To illustrate this in the context of an oil refining company, let us consider Figure 29.26.

The activities at the different levels are as follows:

Level I: At the highest level, the parameters and goals of a multirefinery operation are defined. This could include the expected total production of each product (gasoline, jet fuel, home heating oil, etc.) based on market forecasts, the availability of a variety of crude oils, the capacity of each refinery, operating cost parameters, etc.

Level II: At the second level, mid- and long-term (e.g., 1 month, 3 month, 6 month, 1 year) economic scheduling is carried out using optimization and optimal scheduling algorithms. This scheduling uses geographic market

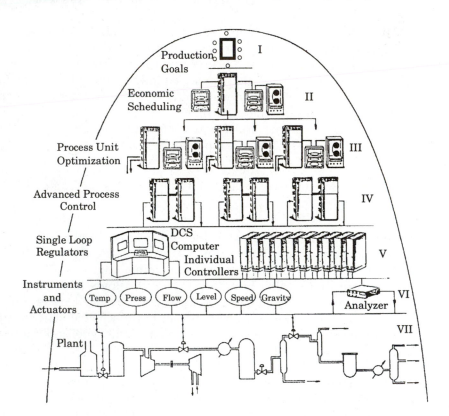

Figure 29.26. The multilevel nature of process operations for an oil refining company (adapted from Ref. [10]).

forecasts, weather predictions, refinery models, cost parameters, crude oil shipments, etc., to set production schedules for each of the refineries for varying time periods into the future. For example, these schedules would call for higher gasoline production rates in the summer and higher home heating oil production in the winter while optimally allocating the manufacture of these products among the various refineries. In this way, an optimal production schedule for each refinery is generated. Because parameters such as crude oil prices and availability, market forecasts, and refinery parameters change frequently, these schedules have to be revised on a weekly to monthly basis.

Level III: At this level, one begins to use detailed steady-state and dynamic process models to optimize the operation of individual process units based on the parameters and production requirements coming from Level II.

Thus, multiproduct pipestills, catalytic reforming units subject to catalyst deactivation, and catalytic cracking operations are all optimized so as to meet the production schedules in the most efficient way. This produces the set-points for each process unit during the scheduling cycle.

Level IV: With the optimization completed in Levels I–III, Level IV is used to implement the more sophisticated advanced process control schemes including multivariable control. These provide set-points to the lower level loops.

Level V: At this level, individual control loops are tuned and maintained. These loops may be implemented via a distributed control system or through individual analog or digital controller hardware. It is at this level that an understanding of SISO control system design is essential.

Level VI: The final interface with the plant (Level VII) consists of the instruments for measuring process variables and actuators such as valves, motors, etc., for carrying out control action.

From this overall picture of process operation, we have only discussed Levels V and VI in any depth in this book; however, we have presented a few examples of advanced process control strategies that would normally be implemented at Level IV.

The process optimization carried out at Level III can be done off-line and must only provide updates to Level IV from time to time, or alternatively can be implemented in real time with a timescale of hours to days for significant changes in set-points to Level IV. A wide variety of optimization methods are used, including linear programming, various forms of nonlinear programming, peak seeking algorithms, and dynamic optimization methods. A survey of such methods may be found in Refs. [11–13].

The economic scheduling tasks of Level II often involve mixed Linear, Integer, and Nonlinear Programs. These are implemented off-line and then provide updated parameters to the lower levels. These appear to the control system as "economic" disturbances that the control system design must accommodate. An overview of these ideas may be found in Refs. [13–14].

29.5 CONTROL IN THE FACE OF COMPONENT FAILURE — ABNORMALITY DETECTION, ALARM INTERPRETATION, AND CRISIS MANAGEMENT

In all of our discussions of control system design, we have assumed that all sensors worked, actuators did not fail, and electric power outages did not cause pumps and compressors to stop, i.e., we designed our control systems for normal operation. However, an increasingly important aspect of control system design is to design control strategies for contingencies such as sensor failure, sticking valves, burnt-out motors, electric power outages, and other possible but rare occurrences. The principal goals of such crisis control schemes are to maintain safe process operation, to minimize the risk of environmental pollution, to take short-term measures to mitigate the effects of the failure, and to identify those steps necessary to restore normal operation.

From the accounts of badly managed crises in the past, we are all aware of the catastrophic consequences of a major explosion and fire, and the impact of the release of toxic or polluting materials into the environment. However, even a minor event not felt outside the process unit can damage a process so severely as to put it out of operation for many months. Thus proper response to only one such event can save lives, jobs, and the economic health of the organization. Clearly, crisis control schemes are an important part of process control system design.

Some aspects of a crisis control scheme would be:

1. **Detection of Abnormal Operation:** Some types of abnormal operation do not produce alarms and are difficult to detect. For example, a serious reactor explosion could be traced to the fact that the heat exchange surface had become fouled by solid material over the course of a few hours. The feedback controller on the coolant flowrate had compensated for this for several hours before the coolant flowrate reached its maximum and the reactor temperature began to rise rapidly. By this time it was too late to take any corrective action, and the reactor exploded. Had the control scheme been designed to detect abnormally high-coolant flowrates and raise an alarm, the problem could have been recognized in time and the explosion prevented.

2. **Interpretation of Alarm Signals:** When there is a crisis, there are usually many alarms that go off cascading one upon the other. For example, a reactor coolant pump failure causes high reactor temperature alarms, high reactor pressure alarms, low reactor feedrate alarms, reactor feedpump failure alarms, cooling jacket pressure alarms, etc. If the plant operators have seen this alarm structure before, then they can identify the primary source of the problem and take the proper corrective action. However, if the situation is new or too complex, the operators might attack a symptom of the problem rather than the source. For example they might try to restart the reactor feedrate pump instead of dealing with the coolant pump failure. Thus, an advisory system to help operators interpret the alarms signals would be a key part of a crisis control system.

3. **Recommendation and Implementation of Control Measures:** After the source of the problem has been identified, there is the need to offer the operators recommendations of possible countermeasures and even take some actions automatically. For example, in our runaway reactor, an automatic countermeasure might be to inject a large dose of catalyst poison or reaction inhibitor into the reactor to stop the chemical reaction. Often this must be done quickly to prevent an explosion. Thus the control scheme could implement this countermeasure automatically above a specified reactor temperature.

Obviously, the design and implementation of such crisis control schemes require a good knowledge of the process so as to be able to determine all possible combinations of failures, their detection and alarm structure, and the appropriate countermeasures in each case. The development of the methodology and software for these control systems is a subject of current research and development in the field of Artificial Intelligence. See Refs. [15, 16] for an overview.

29.6 SOME EMERGING TECHNOLOGIES FOR ADVANCED PROCESS CONTROL

Although there have been powerful tools at hand for applying advanced process control for more than 30 years, the pace of technology is advancing at an accelerating rate, especially in terms of bringing new developments quickly from the research laboratory into industrial practice. In this section, we will briefly introduce some recent and emerging technological developments and indicate how they are being applied in process control.

29.6.1 Dynamic Flowsheet Simulation

For more than 20 years, students in process design courses and engineers involved in process design have had as a major tool *steady-state* simulation programs such as FLOWTRAN™, ASPEN™, PROCESS™, HYSIM™, etc. This has removed the drudgery of routine simulation calculations and allowed process designers to test potential designs through flowsheet simulation. Unfortunately, until recently, there was nothing comparable available for *dynamic* flowsheet simulation. Individual companies have built *ad hoc* dynamic flowsheet simulators that have been applied to specific processes to great benefit, but nothing very general was commercially available. More recently, several new packages for dynamic flowsheet simulation have become available. These are listed in Table 29.1 together with some of their key attributes. Undoubtedly in the years ahead, others will join this list.

Such dynamic simulation packages are essential tools for configuring control schemes that span several units in a flowsheet (e.g., measuring a variable in Unit 3 and manipulating a variable in Unit 1), for studying the propagation of upsets and disturbances through a process, for testing control strategies for transition between product grades in a multiproduct plant, and for designing crisis control schemes for handling alarm situations. Clearly dynamic flowsheet simulation will play an important role in the design of future control systems. A survey of current simulators and some important issues in dynamic simulation may be found in Ref. [17].

TABLE 29.1
SOME DYNAMIC FLOWSHEET
SIMULATION PACKAGES

Package Name	Principal Features	Contact
DIVA	• Preconfigured modules include distillation columns, packed bed reactors, CSTR, etc. • Successful applications in real time for operator training	Prof. E. D. Gilles Institut fur Systemdynamik und Regelungstechnik Universität Stuttgart Pfaffenwaldring 9 70550 Stuttgart Germany
HYSYS	• Preconfigured modules include most unit operations • In use by a number of companies	Hyptrotech Ltd. 300 Hyprotech Centre 1110 Centre Street North Calgary, Alberta T2E 2R2 Canada
PROTISS	• Preconfigured modules include many unit operations • In use by a number of companies	Simulation Sciences Inc. 601 S. Valencia Avenue Brea, CA 92261 U. S. A.
POLYRED	• Preconfigured modules include all polymerization kinetics and most types of reactors, distillation, flash separation, etc. • Handles solids and particle size distribution for many types of polymerization processes • Successfully compared to experimental data from many types of polymerization processes • In use by a number of companies	Prof. W. H. Ray Chemical Engineering Dept University of Wisconsin 1415 Johnson Drive Madison, WI 53706 U. S. A.
SPEEDUP	• Equations entered interactively • Results compiled into executable code • In use by a number of companies	AspenTech Inc. Ten Canal Park Cambridge, MA 02141 U. S. A.

29.6.2 Artificial Intelligence

There has been an explosion of media attention, hype, and optimistic discussion of Artificial Intelligence (AI) regarding its role in technologies of the future. However, at the same time there has been serious research and development in AI directed towards many fields of application. Fortunately, process control is one area where there has been a large amount of recent R&D work so that there are many new applications of AI methods to process control problems. We shall discuss some of these below.

Expert Systems

Perhaps the most mature area of AI applications is that of expert systems. An expert system is a computer program with a user-friendly interface, a knowledge base, and a logic system. These combine to allow the program to provide advice and information, and to answer questions based on the knowledge database and logical rules preprogrammed into the system. Today expert systems can be rapidly configured using powerful development tools which aid the creation of the knowledge base and the rules. In process control, expert systems are being used to provide assistance in control system design (cf. Ref. [20]), as well as in guiding data interpretation, alarm handling, failure detection, and controller action in real time (cf. Refs. [15, 21, 22]).

Neural Networks

A quite different area of AI is that of neural networks. Based on the idea of emulating the human brain, the neural network is a network of input nodes, hidden layers, and output nodes with weighted inputs to each node in the hidden layers (see Figure 29.29(a)). The equation representing the output from a hidden layer is:

$$y_i = f\left(\sum_{j=1}^{n} w_{ij} u_j\right) \quad i = 1, 2, ..., n \quad (29.30)$$

Here the function $f(.)$ is called the "squashing" function and usually has a very simple form such as:

$$f(x) = \frac{1}{1 + e^{-\alpha x}} \quad (29.31)$$

sketched in Figure 29.27(b). The w_{ij} are the input weights. Through the use of an appropriate number of nodes in each hidden layer, one may model anything with these neural networks. In contrast to normal modeling equations that may involve a specific algebraic form with parameters, neural network models have no fixed functional form. One simply specifies the structure of the network and then fits the input weights as parameters. This has the advantage that if one chooses a sufficiently general neural network structure, the model may be better in some cases than if a conventional algebraic functional form were to be used as a model. See Refs. [23–24] for further details. Neural networks have been applied to process control in many ways. One of the most successful has been in interpreting spectra and other data from on-line instrumentation (cf. Ref. [24]).

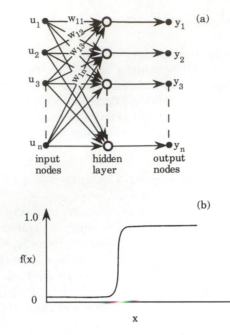

Figure 29.27.   (a) A neural network; (b) a typical squashing function.

Other applications include dynamic modeling, the use of neural networks as controllers, and in-process fault diagnosis (cf. Refs. [25–29]). Fortunately, there is a wide variety of software available to configure neural nets, so that applications are easy to develop.

Learning Algorithms

Perhaps the most exciting aspect of AI methods is the ability of these computer programs to learn. Learning (or "training") is a fundamental part of a neural net. The net is presented with data sets and "trained" by adjusting the weights w_{ij} so as to fit the training data set. Then the net is prepared to recognize the characteristics of new input data and provide appropriate outputs. For example, it is possible to train a neural net to recognize a set of IR spectra for specific compounds and their mixtures. Then the net could be asked to identify the composition of an unknown mixture of these compounds from a new IR spectra (see Ref. [24]). There are many different approaches to training neural nets, but the central feature is that the net is trained to interpret information in a certain way.

Expert systems can also be endowed with learning capabilities. For example, an expert system for control system design could be configured with some basic knowledge and a fundamental set of rules for designing control systems. However, by "training" the expert system on known successful control system designs, it is possible to add information to the knowledge base and add new rules for decision making. Furthermore, in cases where more than one possible design rule might apply, it is possible to adjust the weights for the most successful rules in order to improve the success of the expert system. Again there are many different algorithms for allowing the expert system to learn. See Refs. [22, 30] for a discussion of these.

29.6.3 Parallel Computing

Although process control systems have frequently involved distributed computer architecture with many different CPU's throughout the plant assigned different process control tasks, the communication links have generally required relatively infrequent information transfer. In addition, the tasks allocated to the different CPU's have been well defined and self-contained. More recently, there have been rapid new developments in the field of parallel computing for simple tasks. There are several commercial machines available that allow a network of very fast CPU chips in a single box to be under the control of a single program. The computer architecture can take various forms (see Figure 29.28), but the ideas are the same: the computations for a particular task are divided among the many CPUs operating in parallel and then combined for the final answer. The goal is to increase the speed of the computations by efficient allocation of the subtasks among the various CPU's. In the process control context, this has the possibility of bringing much higher performance computers into the control room so that all the imaginable real-time tasks can be carried out even for processes with very fast dynamics. With the advent of sophisticated expert systems, real-time dynamic process simulation, state estimation, parameter estimation, etc., as part of the control system, large increases in real-time computing power are essential. The natural division of these real-time tasks makes a parallel architecture particularly attractive. For a projection into the future of real-time parallel computing, see Ref. [31].

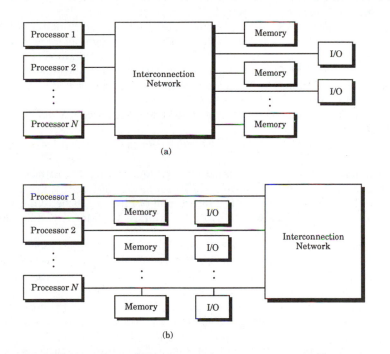

Figure 29.28. Parallel computing architecture [29]. Two multiprocessor structures: (a) all memory and I/O are remote and shared; (b) all memory and I/O are local and private.

29.7   CONCLUDING REMARKS

In this chapter we have only scratched the surface in providing brief introductions to some areas of advanced process control. It is hoped that the reader will be intrigued and excited by the potential of these methods and will probe deeper into the subject either through individual study or through pursuing advanced study in process control. There is still much to be done in applications, and there are many unsolved research problems as fare for the creative mind.

REFERENCES AND SUGGESTED FURTHER READING

1. Ray, W. H., *Advanced Process Control*, Butterworth, Boston (1989)
2. Bryson, A. E. and Y. C. Ho, *Applied Optimal Control*, Halsted Press, (1976)
3. Jazwinski, A. H., *Stochastic Processes and Filtering Theory*, Academic Press, New York (1970)
4. Morari, M. and E. Zafiriou, *Robust Process Control*, Prentice-Hall, Englewood Cliffs, NJ (1989)
5. Doyle, J. C. and G. Stein, "Multivariable Feedback Design: Concepts for a Classical/Modern Synthesis," *IEEE Trans. Auto. Control* , **AC-26**, 4 (1981)
6. Doyle, J. C., "Analysis of Control Systems with Structured Uncertainty," *IEE Proc., Part D*, **129**, 242 (1982)
7. Doyle, J. C., "Structured Uncertainty in Control System Design," *Process IEEE Conf. on Decision & Control*, Ft. Lauderdale (1985)
8. Gay, D. H. and W. H. Ray, "Identification of Control Linear Distributed Parameter Systems through the Use of Experimentally Determined Singular Functions," in *Control of Distributed Parameter Systems* (H.E. Rauch, Ed.), Pergamon (1987)
9. Gay, D. H. and W. H. Ray, "Application of Singular Function Methods for Identification and Model-based Control of Distributed Parameter Systems," in *Model Based Control Workshop*, (Y. Arkun and T.J. McAvoy, Ed.), Pergamon (1989)
10. Garcia, C. E. and D.M. Prett, "Design Methodology Based on the Fundamental Control Problem Formulation," in *The Shell Process Control Workshop*, (D. M. Prett and M. Morari, Ed.), Butterworths, Boston (1987)
11. Biegler, L. T., "Optimization Strategies for Complex Process Models," *Advances in ChE,* **18**, (1992)
12. Hillier, F. T. and G.J. Lieberman, *Operations Research*, Holden-Day, San Francisco (1974)
13. Grossmann, I., "MINLP Optimization Strategies and Algorithms for Process Synthesis," *Foundations of Computer - aided Process Design, FOCAPD – 89*, 105 (1989)
14. Lasdon, L. S. and T. E. Baker, "The Integration of Planning, Scheduling, and Process Control," in *Chemical Process Control III*, (M. Morari, T. J. McAvoy, Ed.), Elsevier, 579 (1986)
15. Moore, R. L. and M. A. Kramer, "Expert System in On-line Process Control," in *Chemical Process Control III*, (M.Morari, T. J. McAvoy, Ed.), Elsevier, 839 (1986)
16. *Proceedings IFAC Symposium on On-Line Fault Detection and Supervision in Chemical Process Industries*, Newark, April 1992
17. Marquardt, W., "Dynamic Process Stimulation — Recent Progress and Future Challenges," in *Chemical Process Control IV*, (Y. Arkun and W. H. Ray, Ed.), AIChE (1991)
18. Taylor, J. H., "Expert Systems for Computer-aided Control Engineering," in *Chemical Process Control III*, (M. Morari and T.J. McAvoy, Ed.), Elsevier, 807 (1986)
19. Niida, K. and T. Umeda, "Process Control System Synthesis by an Expert System," in *Chemical Process Control III*, (M. Morari and T.J. McAvoy, Ed.), Elsevier, 869 (1986)

20. Smith, R. E. and W. H. Ray, "CONSYDEX — An Expert System for Control System Design," (in press)

21. Morris, A. J., G. A. Montague, and M.T.Tham, "Towards Improved Process Supervision — Algorithms and Knowledge-based Systems," in *Chemical Process Control IV*, (Y. Arkun and W.H. Ray, Ed.), AIChE (1991)

22. Stephanopoulos, G., "Towards the Intelligent Controller — Formal Integration of Pattern Recognition with Control Theory, in *Chemical Process Control IV*, (Y. Arkun and W. H. Ray, Ed.), AIChE (1991)

23. Hopfield, J., "Neural Networks — Overview, Programming, and Adoption," in *Chemical Process Control IV*, (Y. Arkun and W. H. Ray, Ed.), AIChE (1991)

24. Piovoso, M. and A. Owen, "Application of Neural Nets to Sensors and Data Analysis," in *Chemical Process Control IV*, (Y. Arkun and W. H. Ray, Ed.), AIChE (1991)

25. Bhat, N. and T. J. McAvoy, "Use of Neural Nets for Dynamic Modelling and Control of Chemical Process Systems," *Comp. Ch.E.*, **14**, 573 (1990)

26. Qin, S. J. and T. J. McAvoy, "A Data-based Process Modeling Approach and its Applications," *Procedings 3rd IFAC Symposium Dynamics and Control of Chemical Reactors, Distillation Columns and Batch Processes (DYCORD 92)*, 321 (1992)

27. Scott, G. M. and W. H. Ray, "Creating Efficient Nonlinear Neural Network Process Models that Allow Model Interpretation," *J. Process Control*, **3** (1993)

28. Scott, G. M. and W. H. Ray, "Experiences with Model-based Controllers Based on Neural Network Process Models," *J. Process Control*, **3** (1993)

29. Venkatsubramanian, V., "A Neural Network Methodology for Process Fault Diagnosis," in *Chemical Process Control IV*, (Y. Arkun and W. H. Ray, Ed.), AIChE (1991)

30. Michalski, R. S., J. G. Carbonell, and T. M. Mitchell, *Machine Learning*, Margen-Kaufman Publishers, New York (1983)

31. Stone, H. S., *High Performance Computer Architecture*, Addison-Wesley (1989)

30

PROCESS CONTROL SYSTEM SYNTHESIS — SOME CASE STUDIES

Thus far in this book, we have focused our presentations on specific tools useful for understanding process dynamics, creating process models, and designing process control systems. In each chapter, we have presented relatively simple examples illustrating how the specific tool under discussion could be applied. However, up to this point there has been no treatment of the *control synthesis* problem for real processes, in which all of the steps in actual control system design and implementation are carried out, and the appropriate analysis and design tools are brought to bear *simultaneously* in solving a real problem. It is the goal of this final chapter to provide such a discussion of control system synthesis, where all the distracting practical details must be dealt with, and real practical difficulties with measurements, controller power, and process understanding must be surmounted in order to design and implement a control system which meets the goals of the project.

Since every control system synthesis problem in real life has its own unique features, we present four case studies which have widely different control system goals and synthesis difficulties. We then conclude with a discussion of how to characterize industrial control problems in order to have guidelines on selecting appropriate tools for a particular problem. Through this mosaic we hope that the reader will appreciate the value of the great variety of tools learned earlier and the versatility and creativity required of a process control engineer in applying these tools successfully.

30.1 THE CONTROL OF DISTILLATION COLUMNS

Because of the importance of chemical separations, distillation columns are widely used in the chemical process industry. For these units, the control of column energy consumption and product concentration variations has a large influence on the economics, ultimate product quality, and environmental impact of the entire process. Thus, having high-performance control systems for distillation columns is very important.

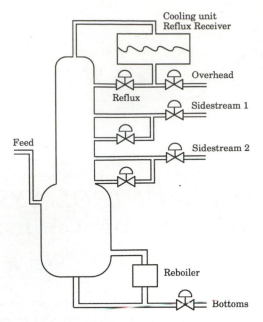

Figure 30.1.   A generic multicomponent, multiproduct distillation column.

The sheer number and variety of distillation columns employed in industry means that there is almost an infinite variety of control system design problems. In the next few sections we shall try to describe the types of problems that arise and discuss some of the controller strategies that can be applied. Necessarily the discussion must be brief; however, the breadth and depth of this topic can be explored further by consulting the enormous literature in this field. Some good starting points are Refs. [2, 6, 11, 12, 16, 22].

30.1.1 Goals of Distillation Control System Design

Because there are so many different types of distillation columns and economic situations, one must first clearly define the goal of the control system to be designed. To illustrate some of the possible goals one might have, consider the generic distillation column shown in Figure 30.1. The column may be multi-component with various product draw-off points and may have auxiliary heaters or coolers in order to achieve the desired temperature profile. Depending on the purpose of the column, the types of disturbances it experiences, the requirements for product purity, and the economics of the operation, one would have different goals for the control system indicated in what follows:

1. Single-Product Stream Columns

If the column has only one product stream — for example, the overhead — which must meet tight specifications, then the other product streams could be feed material to downstream units or products with very loose specifications. In this case the control problem is relatively simple because only the overhead composition control loop needs to be tuned tightly to minimize variability. All other loops are able to tolerate the disturbances arising from column interactions. Simple SISO control methods usually work well here.

2. Multiproduct Stream Columns

When there are multiple product streams that must meet tight specifications, then the control problem is truly multivariable. With strong interactions between the different product streams, as is usually the case, then more sophisticated multivariable controllers are required to meet all of the product stream specifications simultaneously. The ideas of Chapters 20–22, 27, and 29 become important in this case.

3. High-Purity Columns

When product purities required in any stream are *not* extremely high (e.g., below 95%), then linear models relating the product concentrations and the manipulated variables often work well. However, when one or more of the product purities are very high, then the steady-state gain and dynamic parameters in the process model become highly nonlinear as the composition varies. This greatly complicates the control system design because nonlinear models must be used.

4. Energy Minimization Designs

For some very high or very low temperature columns having large throughputs, energy costs dominate the economics. Examples would include crude oil fractionators (heating costs) or cryogenic separation (refrigeration costs). In this case the goal of the control system would be to meet product specifications while minimizing energy costs. This is accomplished by minimizing the use of steam and coolant while taking the necessary control actions to achieve product purity.

5. Disturbance Rejection versus Frequent Grade Changes

The expected tasks of the control system have a strong influence on the type of design selected. For example, if the column is to make only one set of products from one type of feed material and would rarely change its desired steady-state operating point, then the control system would be designed only to reject whatever disturbances might afflict the process. On the other hand, suppose that the column is designed to make five different sets of products every two weeks or to handle eight different types of feed materials every month. In this case the control scheme must be designed to operate over a wide range of operating conditions and to move quickly and economically from one operating point to another in order to avoid significant amounts of "off-spec" product. These two extremes of operation can require very different controller design strategies.

6. Distillation Column Networks

Often the purification train of a process requires multiple distillation columns in a network such as shown in Figure 30.2. In other cases the columns may be energy integrated which greatly increases interactions. In such cases the control system designer may wish to have multivariable cascade controllers which control the entire network as a hierarchical multivariable system.

There are, of course, other situations which have not been mentioned. However, the reader has been given a taste of the issues involved in setting the goals of a distillation column control system.

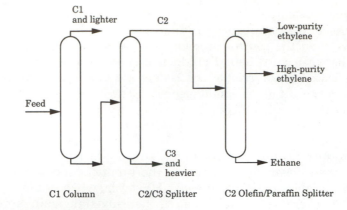

Figure 30.2.   An example of a purification train for light hydrocarbons.

30.1.2 Modeling Issues

As we saw from the distillation column models discussed in Sections 12.3 and 13.5, one may choose to represent the column by a tray-to-tray dynamic model or, alternatively, an approximate input/output model. The rigorous tray-to-tray model requires the solution of $(N + 2)$ $(M - 1)$ differential equations for a column having N trays and M chemical species. Thus for $N = 80$ and $M = 4$ the model would have 246 differential equations. For the second approach, the input/output models make use of low-order linear models (sometimes with time delays) to represent the dynamic relationships between process inputs (feed conditions, recycle ratio, bottoms boilup, sidestream draws, etc.) and process outputs (product stream composition, temperatures on specific trays, etc.). These input/output models are usually the most useful for control system design.

If the input/output data show significant nonlinearities, then nonlinear input/output models may be used; alternatively, a set of linear models (one for each product grade) may be appropriate.

30.1.3 Control System Design Strategies

The choice of control system design strategies depends strongly on the goals of the control system and the dynamic problems being addressed. The influence of control system goals was discussed in Section 30.1.1. Let us indicate the most common dynamic difficulties challenging the control system designer:

1. The process has significant time delays. This usually results from the high-order tray-to-tray dynamics, which appear experimentally as low-order dynamics with time delays. In addition there may be some analysis delay. The most common solution is to implement some type of time-delay compensator in the feedback control scheme.

2. The process has significant loop interactions. This arises from the fact that withdrawing more or less product, or adding or subtracting energy from the column, usually affects *all* product stream compositions simultaneously. The solution in this case is to implement some type of interaction compensator or other multivariable controller design.

3. The process does not have good sensors. It is a common problem that not all stream composition measurements are available on-line; in some cases samples must be sent to the laboratory so that there are long delays in receiving measurement results. The usual solution is to implement some type of inferential control scheme or one based on a composition estimator. The inferential control scheme involves cascade control in which a secondary on-line measurement (such as the temperature on a particular tray) is used as the output being controlled in the inner cascade loop. The outer cascade loop then resets the tray temperature set-point based on the slower stream composition measurements. An alternative scheme involves estimating the composition from other secondary measurements and the process model. Both strategies have been successful in practice.

4. The process is nonlinear. As indicated above, distillation columns may exhibit nonlinear behavior if high purity is required or if there are many widely differing operating steady states corresponding to different product grades or different feed materials. In this case, the solution often can be as simple as incorporating a gain schedule into the controller or designing separate linear controllers for each region of operation. When the nonlinearities involve saturation of some of the manipulated variables, then override or other more complex control strategies may be required.

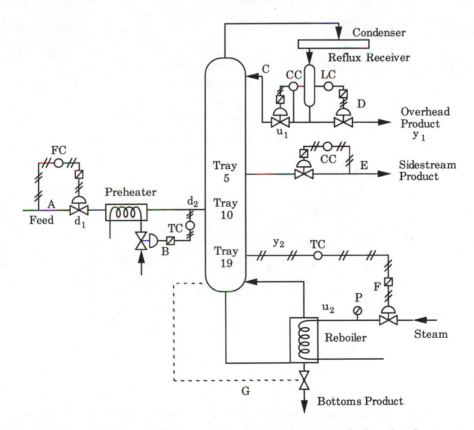

Figure 30.3. A 19-tray distillation column for the separation of ethanol and water.

30.1.4 An Example

As an illustration of a relatively straightforward distillation control system design problem, consider the 19-tray pilot scale distillation column used for the separation of ethanol/water discussed in Section 13.5. The column, shown in Figure 30.3, has overhead, sidestream, and bottoms products; however, for simplicity here, we are only interested in controlling the overhead and bottoms products. Close control of the overhead product is important because it is the feed to other higher purity columns, while control of the ethanol content in the bottoms is important because it is feed material to a final ethanol clean-up column. In order to avoid analysis delays, the overhead composition is measured with an on-line densitometer. Unfortunately, the bottoms composition is so low in ethanol that the densitometer has large relative errors. In this case the sensor is useless, and an inferential control scheme must be used. It is known that the ethanol content of the bottoms is accurately correlated to the temperature of the bottom tray (Tray #19) of the column as shown in Figure 30.4. Thus Tray 19 temperature is a surrogate for the bottoms composition variable.

As discussed in Section 13.5, pulse testing data from the distillation column leads to a linear transfer function model of the form

$$\mathbf{y}(s) \;=\; \mathbf{G}(s)\,\mathbf{u}(s) \tag{30.1}$$

where

$$\mathbf{y} \;=\; \begin{bmatrix} y_1(s) \\ y_2(s) \end{bmatrix} \;=\; \begin{bmatrix} \text{mole fraction ethanol in overhead} \\ \text{Tray 19 temperature} - {}^{\circ}\text{C} \end{bmatrix}$$

$$\mathbf{u} \;=\; \begin{bmatrix} u_1(s) \\ u_2(s) \end{bmatrix} \;=\; \begin{bmatrix} \text{reflux flowrate} - \text{gpm} \\ \text{reboiler steam pressure} - \text{psig} \end{bmatrix}$$

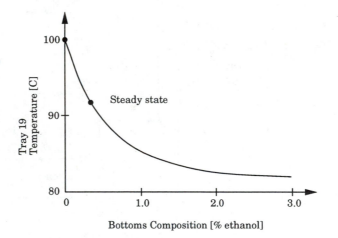

Figure 30.4.   The correlation between column bottoms composition and Tray #19 temperature.

$$\mathbf{G}(s) = \begin{bmatrix} \dfrac{0.66e^{-2.6s}}{6.7s + 1} & \dfrac{-0.0049e^{-s}}{9.06s + 1} \\[3mm] \dfrac{-34.7e^{-9.2s}}{8.15s + 1} & \dfrac{0.87\,(11.6s + 1)\,e^{-s}}{(3.89s + 1)\,(18.8s + 1)} \end{bmatrix} \qquad (30.2)$$

with the time delays and time constants in minutes. More details of the process, its modeling, and control system design are given in Refs. [8, 14].

The control system design for this column must consider both multivariable interactions as well as multiple time delays; thus an advanced model-based control scheme will be applied to the column and compared with simple multiloop PI control. The basic control strategy, shown in Figure 30.5, is the multidelay compensator of Ogunnaike and Jerome [8, 14]. Here $\mathbf{G}(s)$ represents the process, $\tilde{\mathbf{G}}(s)$ the process model, and $\mathbf{G}_c(s)$ a diagonal PI controller of the form

$$\mathbf{G}_c(s) = \begin{bmatrix} K_{c_{11}}\left(1 + \dfrac{1}{\tau_{I_{11}}s}\right) & 0 \\[3mm] 0 & K_{c_{22}}\left(1 + \dfrac{1}{\tau_{I_{22}}s}\right) \end{bmatrix} \qquad (30.3)$$

The matrices $\mathbf{G}_F$ and $\mathbf{D}$ are design parameters for the process. The matrix $\mathbf{D}$ is the same as discussed in Chapter 22 where we must add delays to some of the inputs to ensure causality in our control system. Following the procedure outlined in Chapter 22, we see that

$$\mathbf{D} = \begin{bmatrix} 1 & 0 \\ 0 & e^{-1.6s} \end{bmatrix} \qquad (30.4)$$

will ensure that the control system is causal. The design of the matrix $\mathbf{G}_F$ is determined by the type of multidelay compensator desired. For example, if we wish to ensure the best possible performance, then we would choose $\mathbf{G}_F$ to be the same as $\mathbf{GD}$ with the smallest delay in each row factored out (see Ref. [8] for design details). In this case, the result is

$$\mathbf{G}_F = \begin{bmatrix} \dfrac{0.66}{6.7s + 1} & \dfrac{-0.0049}{9.06s + 1} \\[3mm] \dfrac{-34.7e^{-6.6s}}{8.15s + 1} & \dfrac{0.87\,(11.6s + 1)}{(3.89s + 1)\,(18.8s + 1)} \end{bmatrix} \qquad (30.5)$$

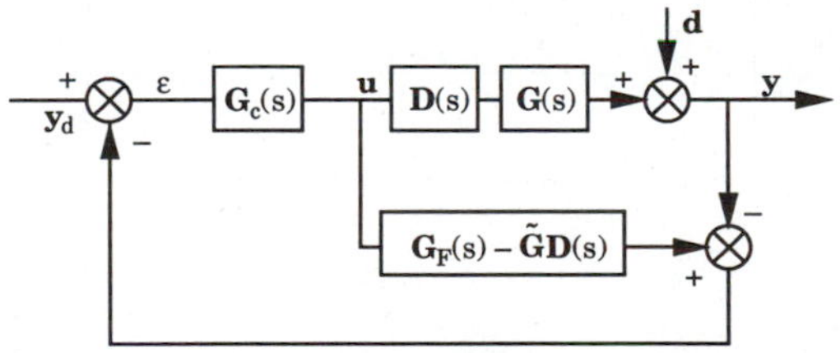

Figure 30.5.   The multidelay compensator structure for the distillation column.

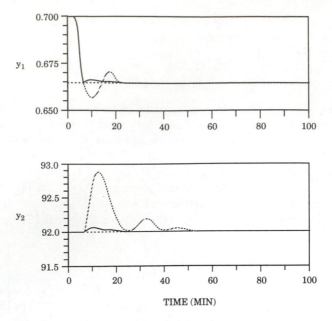

Figure 30.6. Response of distillation column to a set-point change of −0.035 mole fraction ethanol in overhead composition. ⋯⋯ set-points; - - - PI control only; —— PI with multidelay compensator of Jerome [8].

Implementation of this design can lead to extremely good results as shown by the simulation tests shown in Figure 30.6.

Earlier versions of the multidelay compensator were tested experimentally on the actual distillation column of Figure 30.3 and showed excellent results compared to PI control. For example, if we let $\mathbf{D} = \mathbf{I}$ (no delays added) and let $\mathbf{G}_F$ be the same as $\mathbf{G}$ with all delays removed, then the delays are completely removed from the closed-loop characteristic equation of the system. An

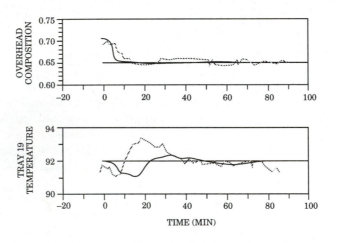

Figure 30.7. Experimental response to a set-point change in overhead composition. —— with the delay compensator; ⋯⋯ with multiloop PI only.

example of the experimental results when a set-point change is made in the overhead composition is shown in Figure 30.7. Further experimental results may be found in Ref. [14].

In this example we were required to deal with process time delays, loop interactions, and the lack of an on-line sensor for bottoms composition. However, these difficulties were overcome, and the final control system design gave excellent results [8, 14].

We have only scratched the surface here in discussing distillation control problems. There are promising new strategies (e.g., [9, 26]) and a host of special situations which we have not been able to mention. However, the references cited here provide a starting point for further study.

30.2 THE CONTROL OF CATALYTIC PACKED BED REACTORS

An important class of chemical processes is the catalytic packed bed reactor shown schematically in Figure 30.8. The catalyst is packed in a tube which may be heated or cooled with a fluid-filled jacket. With a strongly exothermic or endothermic reaction, there may be significant radial as well as axial temperature and concentration gradients. Alternatively, if the tube is adiabatic, then there may be only axial gradients. Sometimes the tube can be made sufficiently thin that radial gradients can be neglected. Since most catalytic processes in industry have multistep kinetics, close control of temperature and composition in the packed bed is essential to achieve the desired reactant conversion and selectivity.

Note that packed bed processes are an example of *distributed-parameter systems* discussed in Chapter 29 and have the special feature that the inputs, states, and outputs can depend on spatial position as well as time. The full control theory for this class of processes is beyond the scope of this book; thus the designs we will discuss here are limited to some of the simpler strategies which are used in practice. For more advanced control system designs, the reader is referred to the references cited in Chapter 29.

30.2.1 Adiabatic Reactors

There are many catalytic packed bed reactors which are operated adiabatically as shown in Figure 30.9. Examples include multistage adiabatic reactors for exothermic reactions (e.g., partial oxidation of SO_2 to SO_3 or ammonia synthesis) with interstage cooling. Alternatively, for endothermic reactions, multistage adiabatic reactors having interstage heating are used (e.g., for naptha reforming). In each case the process loses most of its distributed character because the manipulated variables are the feed temperatures (T_{f_1}, T_{f_2}, T_{f_3} in Figure 30.9) while the outputs are the exit temperatures (T_1, T_2, T_3 in Figure 30.9) and compositions from each adiabatic bed. None of these variables depend on spatial position. Thus adiabatic packed bed reactors, while important, are relatively straightforward to control by conventional control system synthesis and will not be discussed here.

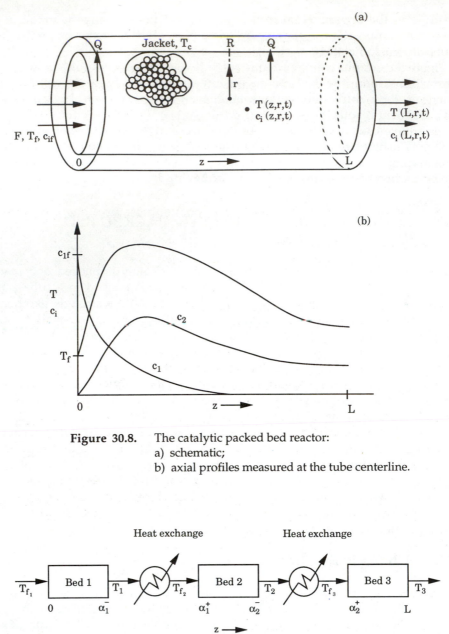

Figure 30.8.   The catalytic packed bed reactor:
a) schematic;
b) axial profiles measured at the tube centerline.

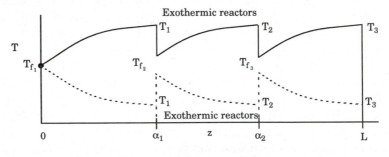

Figure 30.9.   Multistage adiabatic reactors.

30.2.2 Wall-Cooled Reactors

More difficult to control are the jacketed, nonadiabatic reactors such as shown in Figure 30.8. If one considers c_2 to be the concentration of the desired product in Figure 30.8(b), one sees that c_2 passes through a maximum because of side reactions that are accelerated by higher temperatures. Thus by adjusting the feed and jacket conditions to control the packed bed temperature profile one might maximize the yield of c_2 from the reactor. The input variables are T_f, c_{if}, F, T_c (all functions of time only) while the output measurements must come from sensors placed at the exit or within the bed. Some important issues for wall-cooled packed bed reactors are

- The control of spatial concentration, temperature, and pressure profiles
- The maximization of the yield of a desired product
- The dearth of sensors and the best placement of the few available sensors

Some possible design methodologies for the choice of manipulated variables, the choice of sensors and their location, the selection of a control strategy, etc., will become clear through the case study of the next section.

30.2.3 Control of a Jacketed Packed Bed Reactor for the Conversion of Methanol to Formaldehyde

The conversion of methanol to formaldehyde in catalytic packed bed reactors is an important step in the production of high-performance engineering polymers based on formaldehyde. A widely used process operates near atmospheric pressure over a molybdenum oxide/iron oxide catalyst in a jacketed tubular reactor. In a plant reactor, the tubes are of relatively small (~ 2 cm) diameter and are usually in a bundle surrounded by the cooling jacket. On the surface of the catalyst, methanol reacts with excess air to form formaldehyde, the desired product. A complicating factor is that the formaldehyde may be further oxidized to carbon monoxide via an undesirable secondary reaction. The reactions are as follows:

1. $CH_3OH + 1/2\,O_2 \rightarrow HCHO + H_2O$

2. $HCHO + 1/2\,O_2 \rightarrow CO + H_2O$

These oxidation reactions are highly exothermic, thus a liquid jacket for cooling surrounds the tube wall. It is desired to maximize the yield of formaldehyde which is strongly dependent on the temperature profile in the reactor. Thus temperature profile control is a dominant issue.

A pilot scale reactor consisting of a single tube (70 cm long) with a cooling jacket is depicted in Figure 30.10 (taken from the operator's console schematic). Air and methanol are mixed to the desired feed composition (0.05 mole fraction methanol) and heated to the feed temperature of approximately 250°C. The feedrate is 1.4 g/min, and the jacket temperature is about 250°C. Under these conditions the bed temperature and composition profiles at the centerline of the reactor are similar to those shown in Figure 30.11.

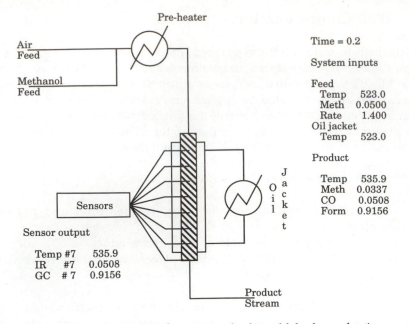

Figure 30.10.   A pilot plant reactor for formaldehyde production.

As indicated above, the objective of the control system is to maximize the yield of the intermediate (formaldehyde) from a reaction mechanism of the form A → B → C. It is desired to complete the first reaction while keeping the extent of the second reaction low, thus maximizing the conversion of methanol to formaldehyde. This may be accomplished by maintaining a temperature profile in the reactor such that the reaction rate is fast in the first half of the reactor where reaction 1 is predominant, and the rate of reaction in the second half of the reactor is slow to minimize the effects of the second reaction which consumes formaldehyde (see Figure 30.11). The control of this packed bed reactor to achieve maximum formaldehyde yield cannot be accomplished by simply monitoring the formaldehyde composition of the product stream and using the maximum yield as a set-point in a control loop. This strategy is not useful for two reasons. First, the formaldehyde outlet concentration is at or near its maximum value at steady state, thus any process disturbances will cause negative deviations in the measured variable. Second, the formaldehyde and methanol concentration measurements are from gas chromatography with a 12.5 minute time delay between measurements. Since most system disturbances are damped out in considerably less time, the GC measurements are obviously unsatisfactory for simple feedback control.

The approach to controlling the packed bed reactor is to maintain optimized steady-state temperature and carbon monoxide concentration profiles in the reactor (Figure 30.11), thus indirectly controlling the formaldehyde yield. To control these profiles, temperature and carbon monoxide concentration must be measured at select locations in the reactor. Temperature measurements may be made using thermocouples; carbon monoxide concentrations are determined using infrared spectroscopy. These measurements provide useful outputs since values are obtained without delays. The need to maintain entire profiles in the reactor using only a few measurements indicates the importance of carefully selecting sensor location.

In configuring the control system, more than just the maintenance of the set-points of exit variables must be considered. Rather, the effects of the control action on entire profiles must be considered, and the influence of these on the yield of formaldehyde must be understood.

The first issue involves the choice of manipulated variables. From the available alternatives:

T_f – Feed temperature
T_c – Jacket temperature
F – Feedrate
y_{mf} – Methanol feed mole fraction

the best choice for inputs seem to be T_c and T_f. Both feedrate F and feed composition y_{mf} affect production rate but are often constrained by upstream or downstream units. Thus we will consider them disturbance inputs.

The choice of outputs must be made from the available sensors and sensor locations. For example, one might have the possibility of placing sensors at one or more of the seven positions denoted by 10 cm increments along the bed (cf. Figure 30.11). Thus with the analytical instrumentation noted above, the possible outputs could be

$T(z_i^*, t)$ – bed temperature at one of seven sensor locations
y_{CO} – CO mole fraction (from the IR) at one of seven sensor locations
y_F – formaldehyde mole fraction (from the GC) at one of seven sensor locations, but with 12.5 min analysis delay

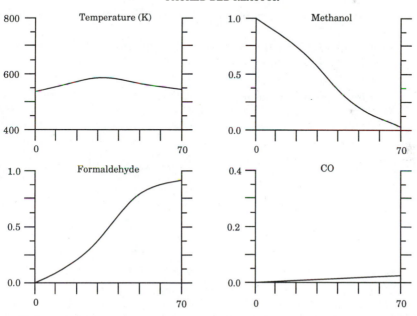

PACKED BED REACTOR

Figure 30.11. Typical axial profiles of temperature and composition; sensors may be located at 10, 20, 30, 40, 50, 60, or 70 cm along the bed.

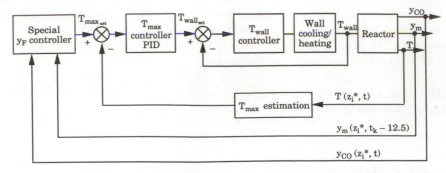

Figure 30.12.   Cascade control schemes for packed bed reactor control.

Obviously this set consists of 21 possible output variables which must be related to two manipulated inputs (T_c, T_f) and two disturbances (F, y_{mf}). A full analysis would require developing 84 dynamic models between inputs and possible outputs. However, the use of engineering judgment can pare this task down to something more manageable as we see below.

Control System Synthesis

A very sophisticated control scheme for this process has been reported by Windes *et al.* [24, 25, 27–29]; however, here we shall discuss only the most basic control system design. Following industrial practice, we shall employ multiple cascade control strategies as shown in Figure 30.12.

Note that we choose to use the cooling jacket wall temperature as the primary manipulated variable. This is due to the fact that adjustments in the feed temperature T_f have the following undesirable effects:

- The steady-state position of the "hot-spot" T_{max} moves depending on the value of T_f.
- Sensors placed downstream from the position of T_{max} exhibit an inverse response to changes in T_f.

Thus T_f is often a secondary manipulated variable in packed bed reactor control.

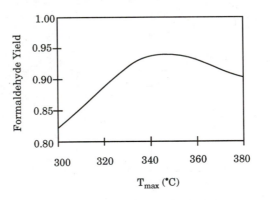

Figure 30.13.   Steady-state formaldehyde yield as a function of hot spot temperature.

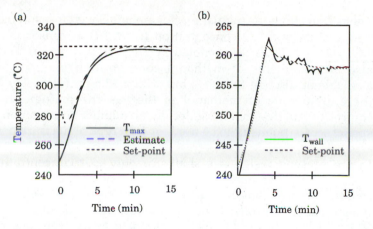

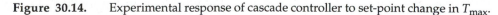

Figure 30.14.   Experimental response of cascade controller to set-point change in T_{max}.

The control scheme, shown in Figure 30.12, works as follows:

1. A special controller, designed to maximize formaldehyde yield, is used to choose the desired hot-spot temperature T_{max} (cf. Ref. [29] for one such design). The design of this controller is complicated by the fact that y_F vs. T_{max} has a maximum (as shown in Figure 30.13) so that a simple feedback controller cannot be used.

2. The desired T_{max} value, $T_{max_{set}}$, is sent to a cascade controller that has as its output, $T_{wall_{set}}$, the desired reactor jacket wall temperature.

3. The value of $T_{wall_{set}}$ is sent to a second controller whose output is the heating/cooling power to the reactor jacket. A jacket wall temperature measurement is used to close the inner loop.

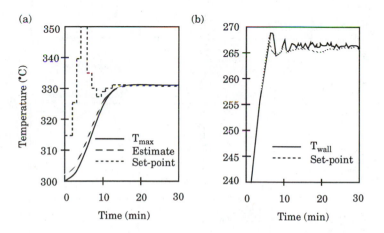

Figure 30.15.   Optimizing controller performance-response of T_{max} and T_{wall}.

4. The outputs from the tubular reactor are undelayed measurements of temperature and CO concentration (from IR) at sensor positions z_i^*; $i = 1, 2, ...$ These are denoted by $T(z_i^*, t)$, $y_{CO}(z_i^*, t)$. There are also measurements of methanol concentration $y_m(z_i^*, t_k - 12.5)$ available at discrete times t_k, but with a 12.5 minute delay due to the elution time of methanol in the gas chromatograph. The control system designer must decide the number and location of the spatial sensor points z_i^* to be used for each type of measurement.

5. In the present design, we shall use only exit concentrations of y_m, y_{CO} every 12.5 minutes, but shall use all available temperature sensors sampled rapidly. The concentration samples will be fed to the special y_F controller in order to adjust the value of $T_{\text{wall}_{set}}$ every 12.5 minutes. Since the time required to move from one reactor steady state to another is 6–8 min, the outer loop in Figure 30.12 is basically for steady-state control.

6. The first inner loop controlling $T_{\max}$ makes use of more frequent temperature profile measurements, and these are used to estimate $T_{\max}$. This can be done by auctioneering control (cf. Chapter 16) or by simple polynomial interpolation between the measurements. Care must be taken not to tune this loop too tightly or it will run faster than the next inner control loop — poor practice for cascade control.

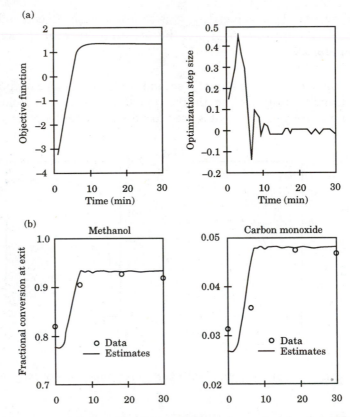

Figure 30.16.   Optimizing controller performance in maximizing formaldehyde yield.

7. The last inner loop regulates the jacket wall temperature and is the fastest of the three loops. It is not instantaneous because of the thermal capacitance of the reactor externals and limited heating/cooling power.

This control scheme has been applied to an experimental reactor (cf. Refs. [27–29]) and has performed quite well. To illustrate, Figure 30.14 shows how T_{max} and T_{wall} respond to a set-point change in T_{max}. When the special optimizing controller was added (cf. Ref. [29]), Figures 30.15 and 30.16 show how the maximized yield of formaldehyde was achieved from cold startup in about 10 minutes.

30.2.4 Concluding Remarks

Although we have been able to mention only some of the key issues in catalytic packed bed reactor control and provide one experimental example, it is hoped that the reader will have a glimpse of the challenges involved in designing control schemes for this class of processes. Much more can be found in the literature cited in Refs. [24, 25, 27–29] for those interested in reading further.

30.3 CONTROL OF A SOLUTION POLYMERIZATION PROCESS

The production of synthetic polymers has grown to the extent that more than 100 million metric tons per year of polymers are now produced worldwide. These products range from low-volume, high-value specialty polymers at several thousand dollars per kg, to extremely high-volume materials which

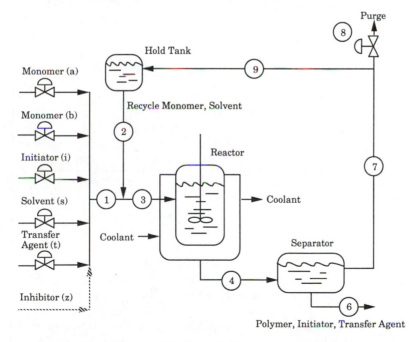

Figure 30.17.   Copolymerization reactor with recycle.

sell for around one dollar per kg. The macromolecular structure and the material properties of these products are largely determined by the kinetic mechanism and the operation of the polymerization reactor. Thus a detailed understanding and close control of these processes is the key to the efficient production of high-quality, tailored polymers with minimal operating problems. One of the most important tools in gaining this understanding and achieving good control of the process is the polymerization process model.

The second essential element required for good polymerization process control is an adequate array of process sensors and good process monitoring algorithms. It is extremely difficult, even with good models, to have long-term, high-quality control system performance without adequate measurements and good process monitoring.

The final element in achieving the safe and efficient operation of a process which yields uniform, high-quality product, is good control system design.

A detailed discussion of the issues involved in creating these three essential elements may be found in Refs. [4, 18–21]. Here we shall only be able to discuss a particular case study of designing a feedforward/feedback control system for a solution copolymerization process.

The process of interest, shown in Figure 30.17, consists of a continuous stirred tank reactor (CSTR) for the copolymerization of methylmethacrylate (MMA) and vinyl acetate (VA) coupled to a separator with recycle of unreacted monomer and solvent. Let us denote MMA as monomer A and VA as monomer B. Many of the details of this control system design are given in Ref. [4].

The basic control problem is that a control system is required that performs well in rejecting feed or recycle disturbances and can quickly achieve set-point changes when there is a transition from one product grade to another. The process is nonlinear (see Ref. [4] for the details of the nonlinear model), but it is thought that approximate linear models may be adequate for control system design. The control strategy is to use feedforward control to adjust the fresh feedrates to ensure that disturbances in the recycle composition can be corrected

---

Table 1.   Control System Design Methodology

Copolymerization Process Model
- Develop a nonlinear process model
- Select a steady-state operating point

Feedforward Control of Recycle
- Design feedforward controllers
- Evaluate disturbance rejection

Feedback Control of Polymer Properties and Rate
- Obtain process and disturbance transfer functions
- Analyze control structure/pairing using singular value decomposition and relative gain array
- Obtain minimum realization linear state-space model
- Tune PI controllers for set-point changes by integration of state-space model
- Check set-point and disturbance rejection on nonlinear model

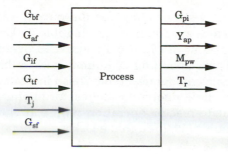

Figure 30.18.   Inputs and outputs of solution polymerization process.

before they enter the reactor. Other disturbances will be handled by feedback control. The overall control system design strategy is given in Table 1. In what follows, we will go through the steps of the design.

The steady-state operating point was selected by the following procedure. The important reactor output variables for product quality control are the polymer production rate (G_{pi}), mole fraction of monomer A in the polymer (Y_{ap}), weight average molecular weight (M_{pw}), and reactor temperature (T_r). The inputs are the flowrate into the reactor of monomer A (G_{af}), monomer B (G_{bf}), initiator (G_{if}), chain transfer agent (G_{tf}), solvent (G_{sf}), inhibitor (G_{zf}), the temperature of the reactor jacket (T_j), and the temperature of the reactor feed (T_{rf}). These are shown in the open-loop block diagram of Figure 30.18. At startup the reactor, separator, and hold tank contained pure solvent preheated to 353.15 K. The inputs to the nonlinear model were then varied in order to

---

Table 2.   Steady-State Operating Conditions

Inputs

Monomer A (MMA) feedrate	$G_{af} = 18$ kg/hr
Monomer B (VAc) feedrate	$G_{bf} = 90$ kg/hr
Initiator (AIBN) feedrate	$G_{if} = 0.18$ kg/hr
Solvent (Benzene) feedrate	$G_{sf} = 36$ kg/hr
Chain transfer (Acetaldehyde) feedrate	$G_{tf} = 2.7$ kg/hr
Inhibitor (m-DNB) feedrate	$G_{zf} = 0$
Reactor jacket temperature	$T_j = 336.15$ K
Reactor feed temperature	$T_{rf} = 353.15$ K
Purge ratio	$\xi = 0.05$

Reactor Parameters

Reactor volume	$V_r = 1$ m^3
Reactor heat transfer area	$S_r = 4.6$ m^2

Outputs

Polymer production rate	$G_{pi} = 23.3$ kg/hr
Mole fraction of A in polymer	$Y_{ap} = 0.56$
Weight average molecular weight	$M_{pw} = 35,000$
Reactor temperature	$T_r = 353.01$ K

obtain acceptable steady-state values for the output variables. The steady-state operating conditions are summarized in Table 2. Under these conditions, the reactor residence time is $\theta_r = 6$ hr and the overall reactor monomer conversion is 20%. These operating conditions ensure that the viscosity of the reaction medium remains moderate. Table 2 also indicates that the reactor feed temperature T_{rf} is practically equal to the reactor temperature T_r, because we have chosen to simulate reactor operation with a preheated feed where the sole source of heat removal is through the jacket.

30.3.1 Feedforward Control of Recycle

In order to compensate for recycle disturbances, a feedforward controller is designed based on measurements of recycle monomer and solvent compositions (pt. ② in Figure 30.17) and adjustment of the fresh feed (pt. ① in Figure 30.17) of these materials so that the compositions at pt. ③ in Figure 30.17 remain constant. In the absence of such a feedforward scheme, the open-loop process response to a 5 min venting of the purge line is shown in Figure 30.19. For the simulations here, the time variable has been scaled and is represented by τ. With the addition of a static feedforward controller, the effects of this disturbance are largely eliminated. This demonstrates the fact that feedforward control can be used to compensate for recycle disturbances before feedback controller action is considered. The feedforward control scheme is shown in Figure 30.20, where we see that the mass flowrate of A, B and solvent in the recycle is subtracted from the desired flowrate in order to determine the fresh feedrate.

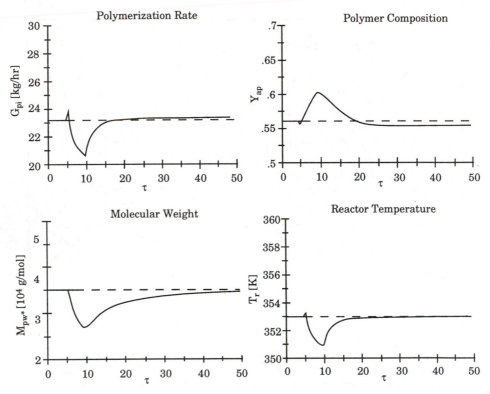

Figure 30.19.   Open-loop output variable response to a purge disturbance.

30.3.2 Feedback Controller Design

Recall from Figure 30.18 that there are six possible inputs. If we consider that one of these, the solvent flowrate, is determined by the total material balance, then we have a five input/four output system. By creating an approximate linear multivariable system model from step test data, one obtains the transfer function matrix (shown in Table 3) relating the four outputs to the five inputs. In order to determine the best loop pairing, we can use two analysis methods presented in the earlier chapters: the relative gain array (RGA) and singular value analysis. Here we will use the RGA and the condition number, $\sigma_{max}/\sigma_{min}$, at steady state to suggest the best pairing for multiloop feedback control. Because scaling is important in evaluating the condition number, both the original scaling and "optimal" scaling was used. Table 5 summarizes the results for the best pairings. Note that the RGA and condition number together favor the CBDE configuration. This would pair the production rate with initiator flowrate, composition with the feedrate of monomer A, molecular weight with the flow of chain transfer agent, and reactor temperature with jacket coolant temperature. Obviously, this pairing makes good physical sense also.

Through the use of engineering insight, it is possible to improve the feedback controller performance by adding ratio control. This has the effect of reducing loop interactions. For example, if one redefines the manipulated variables as shown in Figure 30.21, then the process transfer function would be as shown in Table 4. Note that column A is replaced by column F corresponding to a change in feedrate of B *while all other ratios are kept constant*. Applying RGA and singular value analysis to the transfer function in Table 4 produces the results in Table 5. Note that the RGA and condition number associated with the pairing FBDE are now better than those for the CBDE pairing. Thus we shall choose the FBDE pairing for our final control system design. This means that we choose the total flowrate of monomer B to control production rate, the ratio of flowrates of monomers A and B to control polymer composition, the ratio of chain transfer agent to monomer B flowrates to control molecular weight, and reactant coolant temperature to control reactor temperature. The other variables G_{if}/G_{bf} and G_{sf} are kept constant. Note that while the RGA and singular value analysis led us to the loop pairing employed, a physical understanding of the process would also suggest that this is a good pairing.

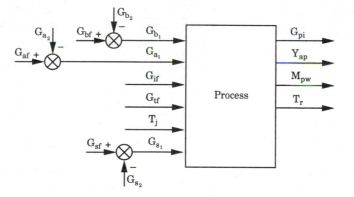

Figure 30.20. Solution polymerization process with feedforward control on recycle stream.

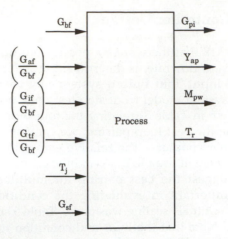

Figure 30.21. Inputs and outputs of the solution polymerization process using ratio control.

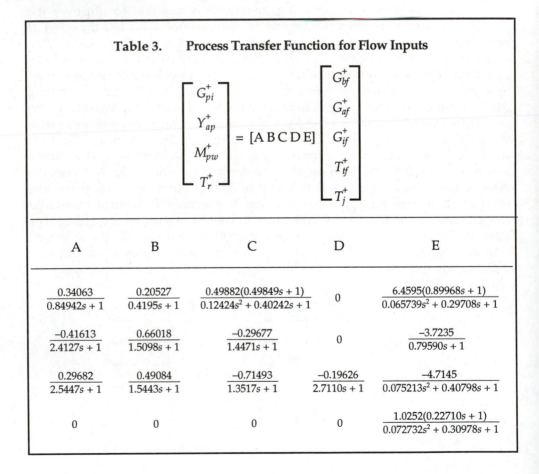

Table 3. **Process Transfer Function for Flow Inputs**

$$
\begin{bmatrix} G_{pi}^{+} \\ Y_{ap}^{+} \\ M_{pw}^{+} \\ T_{r}^{+} \end{bmatrix} = [A\,B\,C\,D\,E] \begin{bmatrix} G_{bf}^{+} \\ G_{af}^{+} \\ G_{if}^{+} \\ T_{tf}^{+} \\ T_{j}^{+} \end{bmatrix}
$$

A	B	C	D	E
$\dfrac{0.34063}{0.84942s + 1}$	$\dfrac{0.20527}{0.4195s + 1}$	$\dfrac{0.49882(0.49849s + 1)}{0.12424s^2 + 0.40242s + 1}$	0	$\dfrac{6.4595(0.89968s + 1)}{0.065739s^2 + 0.29708s + 1}$
$\dfrac{-0.41613}{2.4127s + 1}$	$\dfrac{0.66018}{1.5098s + 1}$	$\dfrac{-0.29677}{1.4471s + 1}$	0	$\dfrac{-3.7235}{0.79590s + 1}$
$\dfrac{0.29682}{2.5447s + 1}$	$\dfrac{0.49084}{1.5443s + 1}$	$\dfrac{-0.71493}{1.3517s + 1}$	$\dfrac{-0.19626}{2.7110s + 1}$	$\dfrac{-4.7145}{0.075213s^2 + 0.40798s + 1}$
0	0	0	0	$\dfrac{1.0252(0.22710s + 1)}{0.072732s^2 + 0.30978s + 1}$

Table 4.   Process Transfer Function for Ratio Inputs

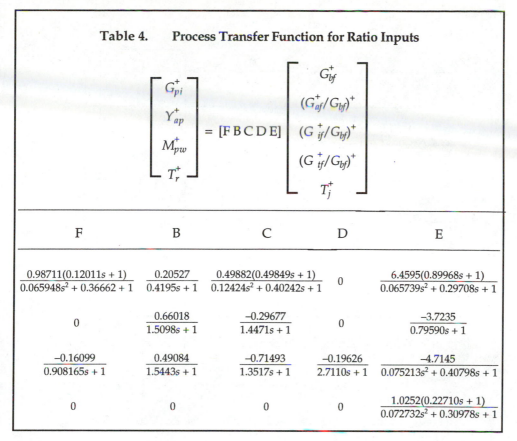

F	B	C	D	E
$\dfrac{0.98711(0.12011s + 1)}{0.065948s^2 + 0.36662 + 1}$	$\dfrac{0.20527}{0.4195s + 1}$	$\dfrac{0.49882(0.49849s + 1)}{0.12424s^2 + 0.40242s + 1}$	0	$\dfrac{6.4595(0.89968s + 1)}{0.065739s^2 + 0.29708s + 1}$
0	$\dfrac{0.66018}{1.5098s + 1}$	$\dfrac{-0.29677}{1.4471s + 1}$	0	$\dfrac{-3.7235}{0.79590s + 1}$
$\dfrac{-0.16099}{0.908165s + 1}$	$\dfrac{0.49084}{1.5443s + 1}$	$\dfrac{-0.71493}{1.3517s + 1}$	$\dfrac{-0.19626}{2.7110s + 1}$	$\dfrac{-4.7145}{0.075213s^2 + 0.40798s + 1}$
0	0	0	0	$\dfrac{1.0252(0.22710s + 1)}{0.072732s^2 + 0.30978s + 1}$

Table 5.   Analysis of Control Structure and Input/Output Pairings

Structure	Best diag (RGA)	Original Scaling		Optimal Scaling	
		σ_{min}	$\sigma_{max}/\sigma_{min}$	σ_{min}	$\sigma_{max}/\sigma_{min}$
ABCD	–	0.0	∞	0.0	∞
ABCE	(0.28, 0.66, 0.57, 1.0)	0.09673	92.37	0.442	5.57
ABDE	(0.72, 0.72, 1.0, 1.0)	0.01629	546.0	0.231	9.837
ACDE	(1.95, 1.95, 1.0, 1.0)	0.02575	346.0	0.1263	18.86
CBDE	(0.84, 0.84, 1.0, 1.0)	0.03765	237.3	0.3142	7.951
FBCD	–	0.0	∞	0.0	∞
FBCE	(1.24, 1.59, 1.79, 1.0)	0.1137	78.87	0.2983	8.321
FBDE	(1.0, 1.0, 1.0, 1.0)	0.09755	91.50	0.4796	4.777
FCDE	(1.0, 1.0, 1.0, 1.0)	0.03649	245.5	0.2890	8.455

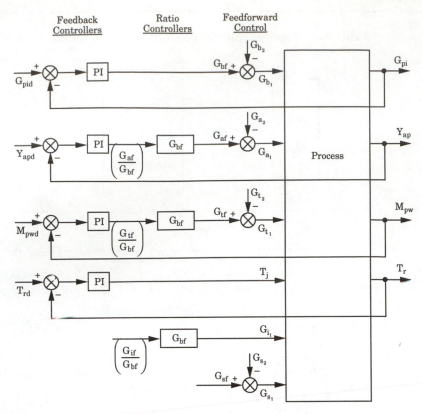

Figure 30.22.   Overall feedforward/ratio/feedback control scheme.

A block diagram of the overall control scheme is shown in Figure 30.22 where the feedback, ratio, and feedforward elements are clearly defined. The tuning parameters for the four PI controllers are given in Table 6. The performance of the control system when applied to a first principles nonlinear model of the process (cf. Ref. [4]) is shown in Figures 30.23 and 30.24. Figure 30.23 demonstrates good performance for a grade change in composition while Figure 30.24 shows the outstanding disturbance rejection capabilities for a significant step increase in inhibitor level in the feed. Thus the control scheme performs very well in simulation.

Table 6.   Tuning Constants for Selected Control Structure*

Manipulated	Output	K	$1/\tau_i$
G_{bf}^+	G_{pi}^+	2	4
$(G_{af}/G_{bf})^+$	Y_{ap}^+	2	1
$(G_{tf}/G_{bf})^+$	M_{pw}^+	−6	1
T_j^+	T_r^+	0.2	6

*Controller reset time τ_i is reported as a fraction of the reactor residence time.

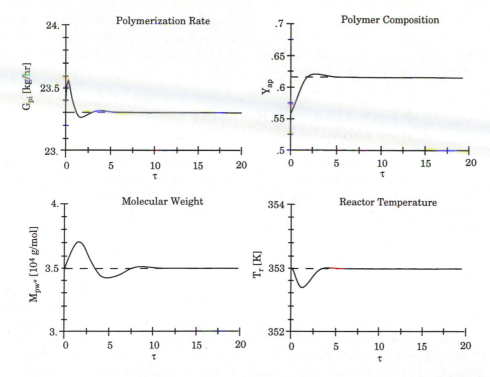

Figure 30.23. Response of output variables to composition set-point change. Feedforward and feedback control are "on:" ——, output; - - -, set-point.

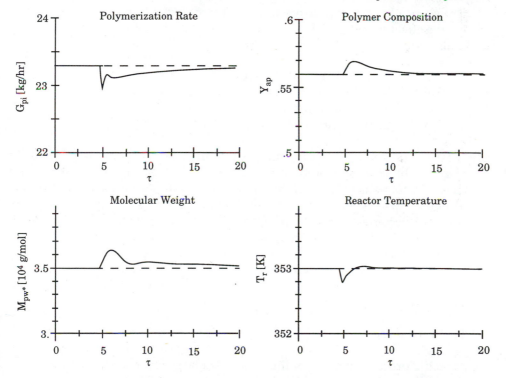

Figure 30.24. Response of output variables to inhibitor disturbance $\tau = 5$. Feedforward and feedback control "on:" ——, output; - - -, set-point.

It is important to realize that even though the control system evaluation carried out here involved model simulation studies, this provided the vehicle for testing a basic control strategy that later was applied successfully to several polymerization plants.

30.4   CONTROL OF AN INDUSTRIAL TERPOLYMERIZATION REACTOR

Large-scale industrial polymer reactors, like any other industrial processing units, are typically multivariable, exhibit nonlinear characteristics, and often suffer from time delays.   However, there are additional unique problems associated with the control of polymer reactors, primarily because polymers are made of *macromolecules* of nonuniform length — hence nonuniform molecular weight — and possibly nonuniform composition (when more than one monomer is involved).   Polymer properties are therefore defined in terms of *distributions* of chain length, molecular weight, and composition.   By altering any of these *fundamental properties*, it is possible to alter significantly the quality of end-use performance or processing characteristics of the polymer product. Unfortunately, these fundamental polymer properties are difficult to measure on-line; and their relationship to the end-use performance characteristics (typically available after long laboratory analysis) is not always easy to determine.   Furthermore, the typical industrial polymer reactor is used to manufacture a variety of *grades* of the same basic product, necessitating frequent startups, shutdowns, or at least, frequent grade transitions.

Thus, in addition to addressing the typical traditional process control issues, the effective industrial polymer reactor control scheme must, at the very least, also address the following issues:

1. The unavailability of on-line measurements (either the fundamental polymer properties, or the end-use performance characteristics, or both).
2. The frequent transitions, startups, and shutdowns.

The first issue is clearly critical, since it is difficult to control what cannot be measured.   Thus the development of reliable on-line sensors and analyzers for polymer reactors has been receiving much attention (cf., for example, Ref. [4], for a review).   However, continuous on-line monitoring of polymer properties has not yet become a widely applicable, commercially viable option; to be effective, many real-time polymer reactor control strategies must still rely on fundamental models capable of predicting the required polymer properties on-line (cf. Refs. [1, 5, 9]).   The remainder of this section is devoted to discussing the design, implementation, and performance of a control system developed for an *industrial* polymerization reactor.

30.4.1 The Process

The specific process involved in this study is a large, evaporatively cooled, continuous stirred-tank reactor used to manufacture 10,000 to 25,000 lb/hr of the terpolymer P from three monomers A, B, and C in the presence of a catalyst. The same reactor is used to manufacture several different grades of the same polymer P. In addition to the production rate, it is desired to maintain at specified values such product quality indicators as "melt" viscosity (a measure of the molecular weight distribution), and polymer composition (in terms of weight %A and %B in terpolymer). These product quality measurements are only available after approximately 30–45 min of laboratory analysis of samples taken every two hours; but reactant composition measurements are available by chromatographic analysis every 5 min. A time delay of approximately 15 minutes is known to be associated with the chromatographic analysis. The process variables that can be manipulated to achieve the desired product quality are the three monomer feedrates, as well as solvent, chain transfer, and catalyst feedrates. The standard operating conditions for the reactor are such that the reactor temperature is maintained constant and cannot be used as a manipulated variable.

The fundamental control problems associated with this reactor are due to the following factors:

- The usual complex, nonlinear, multivariable nature of such processes.
- The 15 min of analysis delay associated with measuring reactant composition.
- The lack of on-line measurements of relevant product quality indicators.

Thus an effective control system will involve a synthesis of the different techniques available for dealing with each of these problems, e.g., a *model-based control strategy* with *time-delay compensation* capabilities to take care of problems 1 and 2, in conjunction with a *modeling* and *estimation* strategy to take care of problem 3. The additional considerations necessary for this specific reactor are considered in the next section.

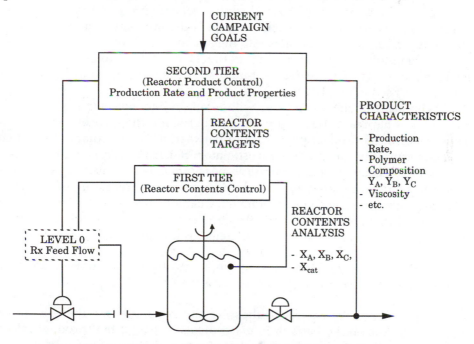

Figure 30.25.   The terpolymerization reactor strategy.

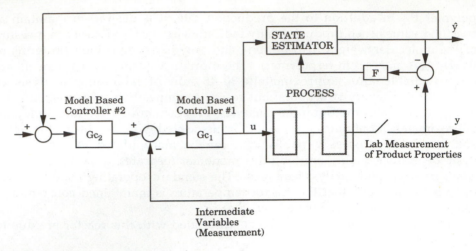

Figure 30.26.   Block diagrammatic representation of the terpolymerization reactor
strategy.

30.4.2 The Overall Control Strategy

Clearly the most significant barrier to good control is due to the fact that
measurements of the very product release properties we wish to control are not
available on-line. In this particular case, the reactor dynamics and the product
quality demands are such that taking control action once every two hours — on
the basis of information delayed by 30–45 minutes — will always be
inadequate. However, "auxiliary" information in the form of reactant
composition measurements is available on-line, and it is possible to exploit this
fact for improving control system performance.

   The broad concept of the overall control scheme proposed for the process
therefore involves a two-tier system in which the monomer, catalyst, solvent,
and chain transfer agent flowrates into the reactor are used to regulate reactant
composition in the reactor at the first tier level, every 5 minutes; at the second
tier level, at a less frequent rate, set-points for the composition of the reactor
contents are used to regulate final product properties.

   At the heart of the second tier will be an on-line, *dynamic first principles
model* (cf. Chapter 12) running in parallel with the process, supplying
estimates of these product properties in between the unacceptably infrequent
two-hour samples. An on-line state estimator (cf. Chapter 29) is used to update
the model predictions when laboratory measurements become available. This
multirate (and multivariable) cascade-type scheme is represented in Figure
30.25; the corresponding block diagram (cf. Chapter 14) is shown in Figure 30.26.
Thus, the scheme proposes to tackle the chronic problem of controlling polymer
product release properties whose measurements are unavailable as follows:

1.   *Indirectly* by focusing first on controlling the more frequently
     measurable auxiliary variables, and then

2.   At a less frequent rate, adjusting the set-points for these auxiliary
     variables on the basis of intersample estimates of the product release
     properties provided by an on-line dynamic first principles model.

3. Because laboratory analysis results are 30–45 minutes behind by the time they become available, they are used primarily to update the model predictions and never *directly* to determine control action.

30.4.3 Model Development and On-Line Implementation

The salient features of the first principles model and its development are now summarized. Full details are available in Ref. [13].

1. Basic Dynamic Model: On the basis of a detailed kinetic scheme, the basic dynamic model was obtained by carrying out material balances on the reactor contents (three monomers, catalyst, cocatalyst, chain transfer agent, and inhibitors). Population balances on growing and dead polymer chains are manipulated in standard fashion to produce a set of *moment equations* (cf. Ref. [17]). Finally, the production rate and product properties — polymer composition, the number average molecular weight M_n, and the weight average molecular weight M_w — are obtained directly in terms of the *bulk moments* and the molecular weights of the three monomers. The complete model consists of 29 ordinary differential equations (seven material balance equations for the reactor contents and 22 moment equations) and five algebraic equations for the polymer properties (cf. Ref. [13]).

2. Empirical Correlations: The most important product release property is a certain "melt" viscosity indicator variable (MVIV) (which we shall represent as η) obtained by laboratory analysis according to the polymer P industry standards; it is somehow related to the molecular weight and molecular weight distribution (MWD) but in no discernibly fundamental manner. Thus, to predict this MVIV from the kinetic model requires correlating MWD data obtained by gel permeation chromatographic analysis with corresponding MVIV data. A power-law relation of the type:

$$\eta = a \left(M_w \right)^b \tag{30.6}$$

was found to be adequate. Thus, estimates $\hat{\eta}$ of the MVIV are obtained using only the kinetic model prediction of M_w.

3. Parameter Estimation: The set of unknown parameters contained in the model fall into the following categories:

- 14 kinetic rate constants
- two power-law viscosity-molecular weight correlation coefficients

A small subset of the rate constants and some reactivity ratios were available in the open literature; the remaining rate constants, along with the other parameters, were estimated from plant startup data using standard principles of parameter estimation in differential equation models (cf. Chapter 12).

Some of the actual off-line results indicating the match between the plant data (represented by the triangles) and the predictions obtained from the complete model (represented by the solid lines) are shown in Figure 30.27.

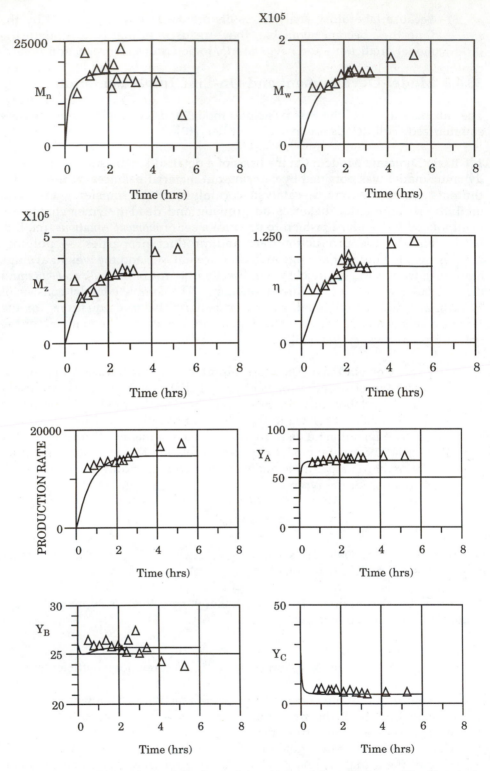

Figure 30.27. Comparison of the dynamic reactor model predictions with plant startup data for production rate and polymer composition.

4. On-line Implementation: By running the reactor model on a real-time process control computer in parallel with the plant, on-line plant input information is used to provide on-line estimates of production rate, polymer composition, molecular weight distribution (particularly M_n and M_w) as well as η, the MVIV, *every minute*. Recognizing that perfect prediction of plant behavior is an unattainable ideal, provision is made for updating model estimates on the basis of actual process measurements whenever these become available.

However, once again, an otherwise standard problem is complicated by the peculiar characteristics of industrial polymer reactors: as a result of a *varying* laboratory analysis delay of between 30 and 45 minutes, t_s (the time at which the product sample was taken), and t_k (the time at which the process data actually become available) are not only different, but their difference, $\tau = t_k - t_s$, is variable. Thus the innovation process in our state estimation scheme must be modified appropriately in order to accommodate this situation.

As a result of these and other such considerations (cf. also, for example, Ref. [23]), the model equations implemented on-line take the form:

$$\frac{d\hat{x}}{dt} = \mathbf{f}(\hat{\mathbf{x}},\mathbf{u},\mathbf{d}) + \mathbf{K}^*(t)\big[\mathbf{y}(t_s) - \hat{\mathbf{y}}(t_k - \hat{\tau})\big] \tag{30.7a}$$

$$\hat{\mathbf{y}} = \mathbf{h}(\hat{\mathbf{x}}) \tag{30.7b}$$

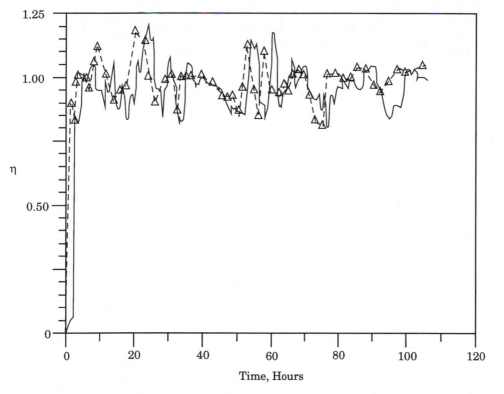

Figure 30.28. Off-line comparison of on-line estimates and actual laboratory measurements: complete 5-day campaign data on MVIV.

where $\mathbf{K}^*(t)$ is a matrix of correction gains defined such that the innovation portion of Eq. (30.7a) is active only when $t = t_k$, i.e., when process data become available; the vector $\hat{\mathbf{y}}$ represents estimates of the product properties (and production rate); $\hat{\tau}$ is an estimate of $t_k - t_s$.

The $\hat{\eta}$ estimates are particularly susceptible to modeling errors since they are estimates of nonfundamental properties based on empirical correlations; polymer composition estimates, on the other hand, are more accurate, being estimates of fundamental properties obtained solely from the kinetic model.

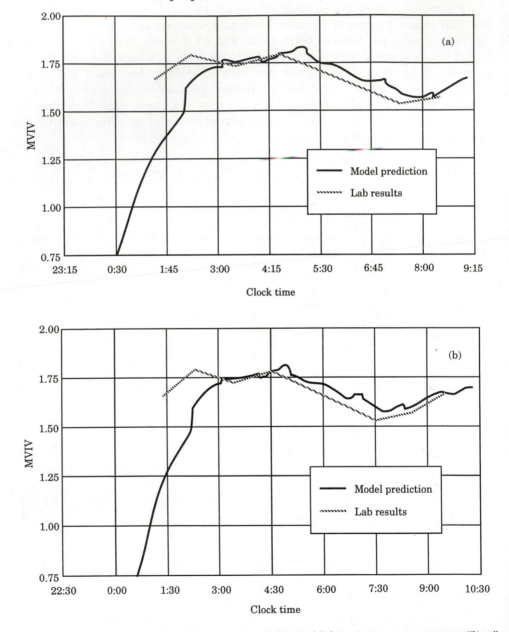

Figure 30.29. (a) Comparison of estimates and actual laboratory measurements: "live" on-line snapshot of MVIV record at about 09:00 hrs.; (b) comparison of estimates and actual laboratory measurements: "live" on-line snapshot of MVIV record just before 10:30 hrs.

Figures 30.28 and 30.29 are indicative of typical on-line performance of the state estimator. It should be noted that with process trends captured on-line (as in Figures 30.29(a, b)) the estimates always lead the laboratory analysis results. Figure 30.28 demonstrates the close agreement between MVIV estimates (solid line) and actual laboratory data (represented by triangles connected by the dotted line).

To illustrate another aspect of the model, Figure 30.29(a) shows an on-line process trend "snapshot" of the MVIV estimates and actual laboratory data, taken sometime after 09:00 hrs as seen on a typical operator console; Figure 30.29(b) shows the same trend 1 hr 15 min later. Apart from the general agreement between the actual laboratory results (the hatched line) and the estimates (the solid line), observe that what was estimated as the current MVIV value (in Figure 30.29(a) *before* the laboratory measurement was available) was confirmed as a very accurate estimate 1 hr 15 min later in Figure 30.29(b).

30.4.4 Control Systems Design

In implementing the reactor control system shown in Figure 30.25, Tier 2 depends on Tier 1 which in turn depends on a lower level of field controllers (Level 0); thus control system design must proceed logically from Level 0 to Tier 1 and finally to Tier 2.

Level 0: Activity at this level is primarily concerned with the tuning of local field controllers responsible for maintaining the various reactor feed flowrates. PI controllers were employed at this level, and the direct synthesis tuning technique discussed in Chapter 15 was employed in selecting the controller parameters (cf. Table 15.5). The Tier 1 control scheme actually sets the set-point for these local controllers. Tuning the reactor level and temperature controllers is also done at this level.

Tier 1: The main characteristics of this first tier are:

1. Multiple inputs (monomers, chain transfer, solvent, and catalyst flowrates) and multiple outputs (mole fraction of reactants in the reactor: X_A, X_B, X_C for the monomers, X_T for the chain transfer agent, and X_{cat} for the catalyst).
2. The time delays introduced by chromatographic analysis.

A model predictive control scheme (cf. Chapter 27) with the following features was therefore designed for this tier:

1. A discrete control-action-implementation/output-sampling interval of 5 minutes (dictated both by the natural "process response time" and by the frequency of the chromatographic measurements).

2. Explicit prediction of reactor content response over a 1-hr time horizon, as well as corrective control action sequence calculation based on a discrete, linear, *parametric* model of the form:

$$\hat{\mathbf{y}}(k + 1) = \hat{\mathbf{A}}\hat{\mathbf{y}}(k) + \hat{\mathbf{B}}\,\mathbf{u}(k - m) \tag{30.8}$$

if we assume, for simplicity, that the time delays associated with each output are identical (a reasonable assumption since the time delay is primarily due to chromatographic analysis).

3. Control structure based on standard MPC (model predictive control) principles with some slight modifications: because reactor dynamic characteristics differ — often significantly — at each of the different operating conditions for manufacturing the various grades, different sets of MPC tuning parameters were used for each grade.

4. Provision for on-line reestimation of approximate linear model parameters according to the strategy outlined in Ref. [13] to accommodate changes in process characteristics during each campaign. (However, these parameters are never updated automatically; the process engineer still retains the responsibility of using his/her discretion in deciding at which point during the campaign to swap models.)

The linear multivariable model in Eq. (30.8) was identified from several pulse test experiments. Pulses were implemented in the multiple inputs (monomers, chain transfer, solvent, and catalyst flowrates) and the responses of the mole fraction of the reactor contents were recorded. The techniques in Section 23.6 of Chapter 23 were used to identify the model parameters. A catalog of these approximate linear models was obtained by repeating this procedure for each different product grade.

Tier 2: This tier has four outputs — production rate, %A, %B in polymer (%C, of course, is determined once these two are fixed), and η, the MVIV — to be regulated by the Tier 1 set-points. The proposed model predictive control scheme at this level is somewhat unusual in that two distinct — and fundamentally different — models are involved: the full-scale, nonlinear first

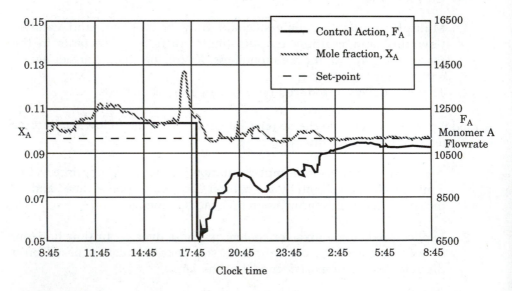

Figure 30.30.   Performance of Tier 1 control system: 24-hr trend of X_A response.

principles model (Model 1), used solely as a surrogate for the reactor to provide estimates of plant output in between the 2-hr samples; the linear parametric model (Model 2) of the type in Eq. (30.8), used in MPC fashion to calculate control action using the Model 1 output in lieu of actual measurement. The proposed control action implementation frequency is once every 15 min, so that the Tier 1 control system operates at three times the rate of the Tier 2 control system.

30.4.5 Some Further Results and Discussion

The typical performance obtained at the Tier 1 level is indicated in Figure 30.30. The hatched line represents the mole fraction X_A; the corresponding control action is represented by the solid line. This plot shows an initial 9-hr period of manual control followed by 15 hr of computer control, using the control scheme in Figure 30.25. The side-by-side comparison provides a clear indication of the effectiveness of the computer control scheme.

The typical performance obtained at the Tier 2 level is illustrated by the transient response shown in Figure 30.31. This 24-hr trend of a product campaign startup demonstrates two things simultaneously: (1) how well the kinetic model represents actual plant behavior, and (2) the effectiveness of the overall control strategy.

The trend in Figure 30.31 shows the MVIV measurements, normalized so that the desired value is 1.0; the solid line represents the 1-minute estimates of η, the MVIV, generated by the state estimator; the other line is the actual measured value (connected by straight lines). First, recalling that the estimates lead the actual measurements, we note that this plot indicates very good agreement between the state estimator and the plant. Furthermore the product MVIV attains the desired value in very reasonable time and is maintained within specifications for the rest of the 24-hr period. Without the

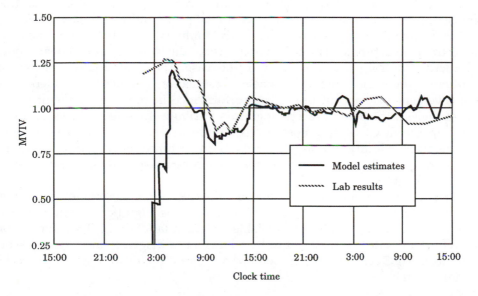

Figure 30.31. Performance of Tier 2 control system: 24-hr campaign startup trend of MVIV transient response.

application of the control system, such startups take almost three times as long to complete, and the MVIV is usually still difficult to maintain at the desired target value thereafter. Similar performance is obtained for the polymer composition as well as for production rate.

30.4.6 Concluding Remarks

The strategy we have presented in this section for achieving good control of a multivariable, nonlinear terpolymerization reactor whose critical output variables are not available for measurement on-line hinges on two points:

1. Being able to separate the reactor control problem into two parts, thereby taking advantage of the availability of some intermediate variable measurements on-line, and controlling these at a faster rate than is possible with the main process output variables which are unavailable for on-line measurements.

2. Utilizing a mathematical model (in a state estimator) as a surrogate for the physical process, to provide on-line estimates of the infrequently available process measurements, and using these estimates to carry out control action at a rate which is slower than that of the first tier control system, but still much faster than if control were to be based only on the laboratory analysis.

By holding the reactant composition in the reactor reasonably constant with the Tier 1 control system, the immediate consequence is a significant reduction in the polymer product property variability; the task of driving these product properties to within specification with the second tier control system is thereby simplified considerably. The result was a significant increase in the amount of "first-quality pounds" of polymer product made at the facility, and this translated into a commensurate increase in the profitability of the business.

Even though the process is decidedly nonlinear, and the same reactor is used to manufacture several different grades of the same product, linear model predictive control proved effective in this case. The additional provision for cautious on-line linear model parameter reestimation, of course, provided some enhancement to the control system performance in the most difficult cases. Nevertheless, the plant process engineers attribute the effectiveness of linear MPC to the fact that the most severe nonlinear effects are associated with temperature variations; these were all but eliminated by the standard operating procedure of fixing the reactor temperature and not using it as a control variable.

It should be noted, finally, that a full-scale nonlinear model predictive control scheme could have been employed; the most significant limitation was the computational facilities available at the plant site. This limitation therefore ultimately turned out to be a blessing in disguise since the full-scale nonlinear MPC scheme may be somewhat more difficult to maintain than the linear scheme.

30.5 GUIDELINES FOR CHARACTERIZING PROCESS CONTROL PROBLEMS

The four case studies we have just presented in this chapter illustrate the fact that the various control problems encountered in various chemical processes can indeed be solved by the many different controller design techniques now available, most of which were discussed in this book. These case studies also attest to the fact that some control strategies are more appropriate than others for certain problems. As a result, it is important for the practicing engineer, charged with the responsibility of designing and implementing industrial control systems, to have a means of matching controller design strategies to the specific problem at hand in a rational manner. In this section we wish to present some concepts to assist in this task.

30.5.1 The Process Characterization Cube

As diverse as chemical processes and processing units are, it is well known that they possess certain characteristics by which they may be grouped into a number of recognizable classes exhibiting similar behavior. (For example, the level of a reboiler in a distillation column, and the exit temperature of certain exothermic tubular reactors, can both exhibit inverse response.) This, in fact, was the main premise of the study of process dynamics in Part II, using ideal process models. Such classifications are particularly useful for control system design since:

> *The characteristics that distinguish one class of processes from another also provide information which, if properly interpreted, indicate how the control problems common to each class ought to be tackled appropriately.*

The appropriate controller design technique may then be selected in a rational fashion.

Following the discussion in Ref. [15], as a possible starting point in the selection of a controller design technique, the process in question may be classified as shown in Figure 30.32, using the following three characteristics defined in the following manner:

1. **Degree of Nonlinearity:** *the extent to which the process behavior is **not** linear* — from essentially linear systems on one extreme, to those that exhibit very strong and acute nonlinear behavior on the other extreme.
2. **Dynamic Character:** *the extent of complexity associated with the dynamic response* — from simple systems exhibiting first-order, or other such low-order dynamic behavior, on the one extreme, to systems exhibiting such difficult dynamics as inverse response and time delays, on the other extreme.
3. **Degree of Interaction:** *the extent to which all the process variables interact with one another* — from those systems whose variables are totally uncoupled, or only weakly coupled, on the one extreme, to those systems whose variables are strongly coupled on the other extreme.

By assessing chemical processes in terms of these three attributes, it is possible to "locate" each process in one of the eight "pigeon holes" in Figure 30.32: four on the "front face," and four on the "back face." For example, essentially linear processes (with low to mild degree of nonlinearity), having simple to moderate dynamic character, and whose variables do not interact significantly with one another, will be located the *bottom left half* "pigeon hole" (or subcube) on the "front face." With a knowledge of the characteristic problems typically associated with each "pigeon hole," and the controller design technique capable of handling each problem, it is possible to use this "characterization cube" as a guide in selecting candidate controller design techniques.

We recognize of course that these three attributes may not capture *all* the possible features of importance in *all* the possible classes of chemical processes; nevertheless, they represent the most significant aspects of most processes, especially as regards the design of effective controllers.

30.5.2 Characteristics of Process Control Problems and Control Scheme Selection

It is well established that effective control of the single-input, single-output (SISO) linear, first-order system requires no more than classical PID feedback control (even though more complicated controller design strategies may also be employed). Thus the need for more sophisticated controllers may be considered as arising primarily to compensate for (or at least to take into consideration) deviations from this "base case" SISO, linear, first-order characteristics; also, the "direction" — as well as the "magnitude" — of such deviation may now be considered as suggestive of the specific type, and extent, of sophistication which might be appropriate.

The following is therefore a catalog of the eight "generic" categories of chemical processes (based on the characterization cube of Figure 30.32), their main problems, and what control schemes are suggested by these characteristics as potential candidates.

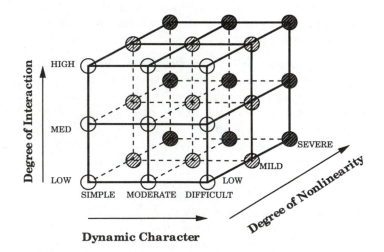

Figure 30.32.   The process characterization cube.

"Essentially Linear" Processes

(These processes are all located on the "front face" of the main cube.)

1. **CATEGORY I PROCESSES (Bottom, Left half, Front face cube)**

Primary Attributes:	Essentially linear processes, with no difficult dynamics, and with little or no interactions among the process variables.
Main Problems:	None.
Suggested Techniques:	Single-loop PID control, with appropriate loop pairing if process is multivariable (cf. Chapters 15 and 21).
Typical Examples:	• Most base case flow, level, and temperature loops.
	• Single-point temperature (or composition) control of medium-purity columns with no significant process time delays.
	• Single-temperature (or composition) control of mildly nonlinear chemical reactors.

2. **CATEGORY II PROCESSES (Bottom, Right half, Front face cube)**

Primary Attributes:	Essentially linear processes, *with difficult dynamics*, but with little or no interactions among the process variables.
Main Problems:	Difficult dynamics (e.g., the presence of inverse response, or time delay, etc.).
Suggested Techniques:	• Single-loop PID control with compensation for the difficult dynamics, e.g., the Smith Predictor for time-delay systems (cf. Chapter 17).
	• Other model-based control techniques such as IMC, Direct Synthesis Control (cf. Chapter 19), or even MPC (cf. Chapter 27).
Typical Examples:	• Loops with significantly delayed measurements or other delays.
	• Exit temperature control of an exothermic tubular reactor having inverse response, or any other such loops with inverse response.

3. **CATEGORY III PROCESSES (Top, Left half, Front face cube)**

Primary Attributes:	Essentially linear processes, with no difficult dynamics, *but with significant interactions among the process variables*.
Main Problems:	Significant Process Interactions.
Suggested Techniques:	• Single-loop PID control with compensation for the loop interactions, e.g., Decoupling, or SVD control (cf. Chapter 22).

- Full-scale multivariable model-based control techniques such as MPC (cf. Chapter 27).

Typical Examples:
- Two-point temperature (or composition) control of single, or coupled medium-purity columns with no significant time delays.
- Simultaneous conversion and composition control of mildly nonlinear chemical reactors.

4. **CATEGORY IV PROCESSES (Top, Right half, Front face cube)**

Primary Attributes: Essentially linear processes, *with difficult dynamics*, *in addition to* significant interactions among the process variables.

Main Problems: Combination of:
- Difficult Dynamics *and*
- Significant Process Interactions.

Suggested Techniques:
- Single-loop PID control with compensation for the loop interactions (e.g., Decoupling, or SVD control (cf. Chapter 22)) *as well as* for difficult dynamics.
- Full-scale multivariable model-based control techniques such as MPC, with capabilities for handling difficult dynamics (cf. Chapter 27).

Typical Examples:
- Two-point composition control of single, or coupled, medium-purity columns with delayed measurements or other time delays.
- Simultaneous conversion and composition control of mildly nonlinear chemical reactors with delayed composition measurements.
- Most mildly nonlinear polymer reactors with multiple control objectives.

"Significantly Nonlinear "Processes

(These processes are all located on the "back face" of the main cube.)

5. **CATEGORY V PROCESSES (Bottom, Left half, Back face cube)**

Primary Attributes: *Significantly nonlinear processes*, with no difficult dynamics, and with little or no interactions among the process variables.

Main Problems: Significant Nonlinearity.

Suggested Techniques:
- Single-loop *nonlinear* model-based techniques, such as Generic Model Control (cf. Chapter 19).
- Custom-designed single-loop PID controllers with process knowledge incorporated, such as gain scheduling (cf. Chapter 18).

Typical Examples:
- Single-point temperature (or composition) control of very high-purity columns without significant time delays.
- Single-temperature (or composition) control of most exothermic chemical reactors.

6. **CATEGORY VI PROCESSES (Bottom, Right half, Back face cube)**

Primary Attributes:	*Significantly nonlinear processes, with difficult dynamics,* but with little or no interactions among the process variables.
Main Problems:	Combination of: • Significant Nonlinearity, *and* • Difficult dynamics.
Suggested Techniques:	• Single-loop *nonlinear* model-based techniques, such as Generic Model Control (cf. Chapter 17) or *nonlinear* MPC (cf. Chapter 27). • Custom-designed single-loop PID controllers with process knowledge incorporated (cf. Chapter 18), and compensation for the difficult dynamics (e.g., time-delay compensation (cf. Chapter 17)).
Typical Examples:	• Any Category V process with significantly delayed measurements or other delays. • Exit temperature control of a highly exothermic reactors having inverse response.

7. **CATEGORY VII PROCESSES (Top, Left half, Back face cube)**

Primary Attributes:	*Significantly nonlinear processes,* with no difficult dynamics, *but with significant interactions among the process variables.*
Main Problems:	Combination of: • Significant Nonlinearity, *and* • Significant Process Interactions.
Suggested Techniques:	• Full-scale multivariable nonlinear model-based control techniques such as Generic Model Control (cf. Chapter 19); or *nonlinear* MPC (cf. Chapter 27).
Typical Examples:	• Two-point temperature (or composition) control of single, or coupled, very high-purity columns without significant time delays. • Simultaneous conversion and composition control of very nonlinear chemical reactors.

8. **CATEGORY VIII PROCESSES (Top, Right half, Back face cube)**

Primary Attributes:	*Significantly nonlinear processes, with difficult dynamics,* **in addition to** *significant interactions among the process variables.*
Main Problems:	Combination of: • Significant Nonlinearity, *and* • Difficult Dynamics, *and* • Significant Process Interactions.
Suggested Techniques:	• Full-scale multivariable, nonlinear, model-based control techniques with capabilities for handling difficult dynamics, such as Generic Model Control, and nonlinear MPC (cf. Chapter 27).

Typical Examples: • Two-point composition control of single, or coupled, very high-purity columns with delayed measurements or other delays.
• Simultaneous conversion and composition control of very nonlinear chemical reactors with delayed composition measurements.
• Some nonlinear polymer reactors with multiple control objectives.

Additional Considerations

There are many situations of practical importance for which it might be necessary to take other factors into consideration in addition to the three attributes employed above. Some of these factors include:

1. **Constraints:** requiring controller design techniques such as MPC with explicit and intelligent constraints handling capabilities (cf. Chapter 27).
2. **Frequent Disturbances:** requiring special controller design techniques specifically for disturbance rejection, such as cascade or feedforward control (cf. Chapter 16).
3. **Ill-Conditioning:** requiring specialized robust multivariable controller design techniques (cf. Chapter 29), since ill-conditioning makes standard multivariable model-based techniques particularly sensitive to the effect of plant/model mismatch, thereby placing limitations on the achievable performance of such schemes.
4. **Significant Inherent Variability in Process Outputs:** requiring controller design techniques specifically for dealing explicitly with the stochastic nature of the measurements, such as the statistical process control schemes discussed in Chapter 28.

30.5.3 Illustrative Example

To illustrate the issues raised in this section consider the following example. The process is a CSTR used to manufacture Cyclopentanol (which we shall represent by B) from Cyclopentadiene (represented by A). The kinetic scheme for this reaction is as follows:

$$A \xrightarrow{k_1} B \xrightarrow{k_2} C$$

$$2A \xrightarrow{k_3} D$$

where C represents Cyclopentanediol, and D represents Dicyclopentadiene. The objective is to make the desired product, Cyclopentanol (B), from pure feed of A, maintaining constant reactor temperature, and a constant concentration c_B of the desired product in the reactor. The feedrate and the reactor jacket temperature are available as manipulated variables.

The modeling equations for this process obtained by material and energy balances are given as follows:

$$\frac{dx_1}{dt} = u_1(x_{10} - x_1) - k_1(x_3)x_1 - k_3(x_3)x_1^2$$

$$\frac{dx_2}{dt} = -x_2u_1 + k_1(x_3)x_1 - k_2(x_3)x_1$$

$$\frac{dx_3}{dt} = \frac{1}{\rho C_p}\left[k_1(x_3)x_1(-\Delta H_1) + k_2(x_3)x_2(-\Delta H_2) + k_3(x_3)x_1^2(-\Delta H_2)\right]$$

$$+ u_1(u_2 - x_3) + \frac{Q}{\rho C_p V_R}$$

along with the outputs given by:

$$y_1 = x_2$$
$$y_2 = x_3$$

Here, x_{10} is the *feed* concentration of A, x_1 is the concentration of A in the reactor, x_2 is the concentration of B, x_3 is the reactor temperature, V_R is the reactor volume, u_1 is the dimensionless feedrate, and u_2 is the reactor jacket temperature.

Using the process parameters reported in Ref. [7] the *steady-state* behavior of the reactor may be investigated as a function of the feedrate u_1. The reactor jacket temperature u_2 is fixed at 130°C, and the *feed* concentration of A, x_{10}, is fixed at 5.1 mole/liter.

Figure 30.33 shows the steady-state behavior of x_2, the concentration of B, as a function of the feedrate. The RGA parameter for this 2 × 2 system, computed as a function of the feedrate, is shown in Figure 30.34. The corresponding steady behavior of x_1, the reactor concentration of A, and x_3, the reactor temperature, are shown in Figures 30.35 and 30.36.

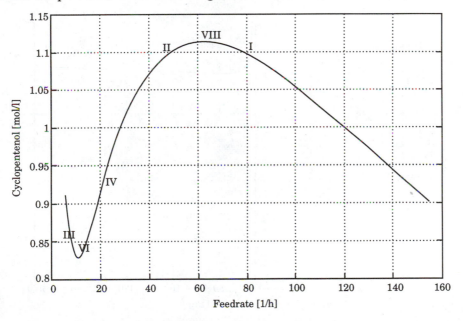

Figure 30.33. Steady-state concentration of Cyclopentanol as a function of reactor feedrate.

Characterizing Various Operating Regimes

By examining Figures 30.33 and 30.34 (and analyzing the linear transfer functions obtained by linearizing the dynamic model around the indicated steady-state operating regime) we see that this simple process shows *six* distinct operating regimes, listed below, each exhibiting radically different dynamic behavior in terms of the three attributes of *Nonlinearity, Dynamic Character*, and *Degree of Interaction*. Thus depending on the operating conditions, this reactor may be placed in six of the possible eight categories discussed above.

Operating Regime A: ($u_1 > 80$ hr^{-1}; y_1 (or x_2) ~ 1.1 moles/liter)

Primary Attributes:	Fairly constant gain; Figure 30.33 indicates a fairly straight line in this region (~ Linear Behavior); Low-order, minimum phase (Simple Dynamics); RGA constant at ~ 1.1 (Virtually noninteracting).
Characterization:	**Category I Process.**
Implication:	2 single-loop PI controllers should perform reasonably well in this operating regime.

Operating Regime B: (u_1 ~ 45 hr^{-1}; y_1 (or x_2) ~ 1.1 moles/liter)

Primary Attributes:	Mild gain variations; Figure 30.33 shows only a slight curve in this region (~ Mildly nonlinear); Nonminimum phase (*Difficult* Dynamics); RGA ~ 0.9 (Virtually noninteracting).

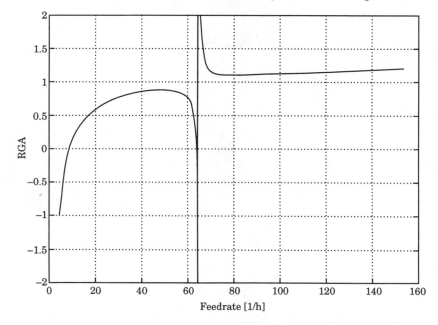

Figure 30.34.   Relative gain values for the process as a function of reactor feedrate.

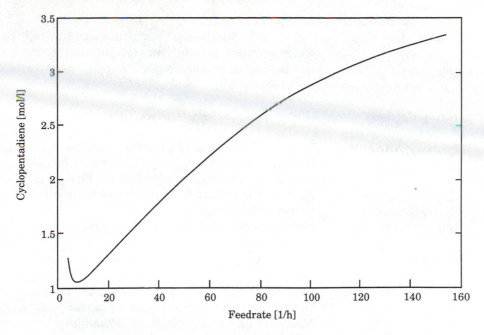

Figure 30.35. Steady-state concentration of Cyclopentadiene as a function of reactor feedrate.

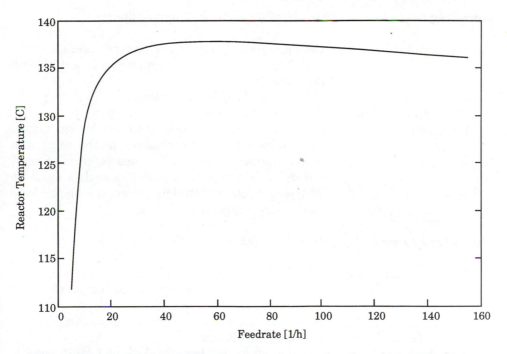

Figure 30.36. Steady-state reactor temperature as a function of reactor feedrate.

Characterization: **Category II Process.**
Implication: Achievable controller performance limited by
                         Nonminimum phase behavior; inverse response
                         compensation required; nonlinearities may further
                         complicate things.

Operating Regime C: $(u_1 \sim 5 \, hr^{-1}; \; y_1 \, (or \; x_2) \sim 0.9 \, moles/liter)$

Primary Attributes: Fairly constant gain; Figure 30.33 indicates a fairly
                         straight line in this region (~ Linear Behavior);
                         Low-order, minimum phase (Simple Dynamics);
                         RGA ~ –1 (Medium-to-High degree of interaction).
Characterization: **Category III Process.**
Implication: Multivariable Control will be helpful.

Operating Regime D: $(u_1 \sim 20 \, hr^{-1}; \; y_1 \, (or \; x_2) \sim 0.9 \, moles/liter)$

Primary Attributes: Very mild gain variations; Figure 30.33 shows a
                         very slight curve in this region (~ Linear);
                         Nonminimum phase (*Difficult* Dynamics);
                         RGA ~ 0.5 (*Severe* interactions).
Characterization: **Category IV Process.**
Implication: Multivariable controller with capabilities for
                         handling nonminimum phase behavior essential.

Operating Regime E: $(u_1 \sim 9 \, hr^{-1}; \; y_1 \, (or \; x_2) \sim 0.82 \, moles/liter)$

Primary Attributes: Gain undergoes sign change (*Severely* nonlinear);
                         Both Minimum and Nonminimum phase (*Difficult*
                         Dynamics);
                         RGA ~ 0 (or 1) (Noninteracting).
Characterization: **Category VI Process.**
Implication: Nonlinear controller required to handle sign change
                         in gain as well as the changes in the minimum
                         phase characteristics. (No linear controller *with
                         integral action* is able to stabilize such a system);
                         Two single-loop controllers are acceptable, as long
                         as they are nonlinear.

Operating Regime F: $(u_1 \sim 60 \, hr^{-1}; \; y_1 \, (or \; x_2) \sim 1.12 \, moles/liter)$

Primary Attributes: Gain undergoes sign change (*Severely* nonlinear);
                         Both Minimum and Nonminimum phase (*Difficult*
                         Dynamics);
                         RGA ~ $\pm \infty$ (*Severe* Interactions + Ill conditioning).
Characterization: **Category VIII Process.**
Implication: Nothing short of carefully designed multivariable
                         nonlinear controllers will work well.

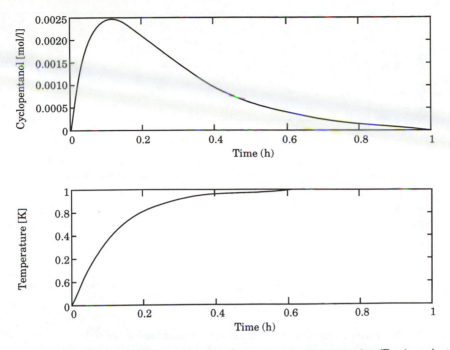

Figure 30.37. Process response to 1° step change in reactor set-point (Regime A, two single-loop PI controllers).

Controller Performance Assessment

To confirm these analyses, the following is an assessment of several controllers designed for the process for operating regimes A, B, and F.

Figure 30.37 shows the response of the control system to a step change of 1° in reactor temperature set-point while operating in the regime A ($u_1 > 80$ hr^{-1}; y_1 (or x_2) ~ 1.1 moles/liter) with two single-loop PI controllers. Note that the behavior is essentially linear, and that the interaction effects are not significant. (The dynamic variations in the reactor concentration of Cyclopentanol are ~0.3% of the steady-state value.) This confirms that in this regime of operation, the reactor behaves essentially as a Category I process for which single-loop PI controllers are adequate.

Figure 30.38 shows the system response to a step change of 1° in reactor temperature set-point while operating in the regime B (u_1 ~ 45 hr^{-1}; y_1 (or x_2) ~ 1.1 moles/liter). The dotted line shows the response under two single-loop PI controllers, while the dashed line shows the response using two single-loop IMC controllers. Note the inverse response exhibited by the Cyclopentanol concentration, while the reactor temperature exhibits no such behavior. Because the reactor temperature has no inverse response, the performance of the PI temperature controller is indistinguishable from its IMC counterpart. However, the performance of the IMC concentration controller is somewhat better.

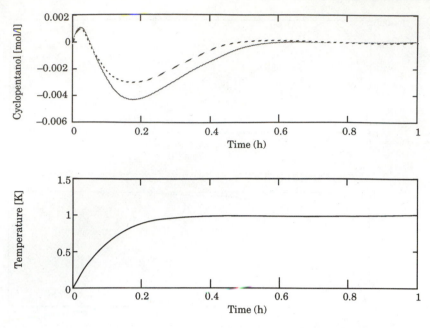

Figure 30.38. Process response to 1° step change in reactor set-point (Regime B); —— two single-loop PI controllers; - - - two single-loop IMC controllers.

The very problematic behavior of the process when operating in the regime F is indicated in Figures 30.39 and 30.40. Figure 30.39 shows the response of the process to a step change of –0.01 mol/l in the Cyclopentanol concentration set-point under two single-loop PI controllers. Observe that the process is unable to attain this new steady-state value; in fact, the reactor proceeds to "washout,"

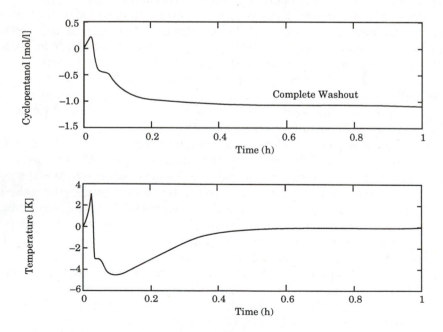

Figure 30.39. Process response to a step change of –0.01 mole/liter in Cyclopentanol concentration (Regime F, two single-loop PI controllers).

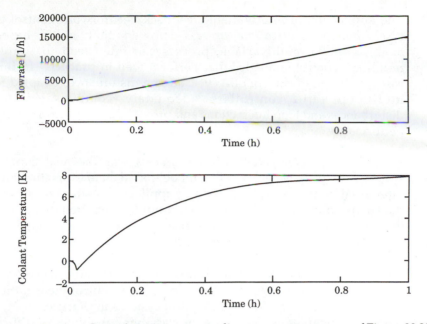

Figure 30.40.   Control action corresponding to process response of Figure 30.39.

and the actual (absolute) Cyclopentanol concentration goes to zero (the initial concentration was 1.1 mol/l). The corresponding control actions taken by each of the two single-loop controllers are shown in Figure 30.40. Note that the behavior of the first controller is essentially unstable.

The reason for this behavior is that the process gain is close to zero at this particular steady state, becoming slightly *positive* for negative deviations from the steady-state flowrate, and changing signs to become slightly *negative* for positive flowrate deviations. Thus if the linear PI controller is designed for a positive process gain, the closed-loop system will become unstable when the actual process gain changes signs.

30.6   SUMMARY

In this final chapter we have provided four different case studies of control system design applied to processes of industrial importance:

1. For *distillation column control*, we discussed how the control system goals and the inherent dynamics of the column strongly influence the choice of the control system to be employed. An experimental example provided one illustration of advanced control system performance.

2. The control of *catalytic packed bed reactors* introduced us to the task of controlling temperature and composition profiles in a process — one of the most common examples of *distributed parameter systems* control. The control system design tasks were illustrated by the application of state estimation, cascade control, and optimizing control to an experimental reactor.

3. The application of multivariable feedback/feedforward control to a *solution polymerization process* showed that model-based analysis of the multivariable process can often lead to simple multiloop control strategies that perform well in practice. In this case, feedforward control of recycle stream disturbances coupled to multiloop ratio controllers adequately compensated for disturbances and loop interactions and provided good control system performance.

4. A more *complex terpolymerization process* was the next control system design challenge. On-line model-based state estimation and inferential control were successfully applied to a situation where the key product quality variables could not be measured except off-line with a significant delay. The process control scheme showed excellent results when implemented in the plant.

Taken together these four case studies illustrate the "art and science," that is, the combination of engineering creativity and process control theory, that can be applied to provide high-performance control systems in practice.

This chapter concluded with a guide for classifying the types of difficulties and specific challenges posed by a specific control problem and some recommendations for remedies to be employed. This guide can be used to provide a starting point for control system design.

REFERENCES AND SUGGESTED FURTHER READING

1. Ardell, G. G. and B. Gumowski, "Model Prediction for Reactor Control," *Chem. Eng. Prog.*, June (1983)
2. Buckley, P. S., W. L. Luyben, and J. P. Shunta, *Design of Distillation Column Control Systems*, Instr. Soc. Amer., Research Triangle Park (1985)
3. Chien, D. C. H. and A. Penlidis, "On-line Sensors for Polymerization Reactors," *Rev. Macromol. Chem. Phys.*, **C30**, (1), 1–42 (1990)
4. Congalidis, J. P., J. R. Richards, and W. H. Ray, "Feedforward and Feedback Control of a Solution Copolymerization Reactor," *AIChEJ*, **35**, 891 (1989)
5. Cozewith, C., "Transient Response of Continuous-flow Stirred-tank Polymerization Reactors," *AIChEJ*, **34** , (2), 272 (1988)
6. Deshpande, P. B., *Distillation Dynamics and Control*, Instr. Soc. Amer., Research Triangle Park (1985)
7. Engell, S. and K.-U. Klatt, "Nonlinear Control of a Nonminimum Phase CSTR," *Proceedings of the American Control Conference*, 2941, San Francisco, June 1993
8. Jerome, N. F. and W. H. Ray, "High-performance Multivariable Control Strategies for Systems Having Time Delays," *AIChEJ*, **32**, 914 (1986)
9. Kaspar, M. H. and W. H. Ray, "Chemometric Methods for Process Monitoring and High Performance Controller Design," *AIChEJ*, **38**, 1593 (1992)
10. McAuley, K. B. and J. F. MacGregor, "Optimal Grade Transitions in a Gas Phase Polyethylene Reactor," *AIChEJ*, **38** (10), 1564 (1992)
11. McAvoy, T. J., *Interaction Analysis*, Instr. Soc. Amer., Research Triangle Park (1983)
12. Nisenfeld, A. E. and R. C. Seemann, *Distillation Columns*, Instr. Soc. Amer., Research Triangle Park (1981)
13. Ogunnaike, B.A., "On-line Modeling and Control of an Industrial Terpolymerization Reactor," *Int. Journal of Control* (in press) (1994)

14. Ogunnaike, B. A., J. P. Lemaire, M. Morari, and W. H. Ray, "Advanced Multivariable Control of a Pilot-plant Distillation Column," *AIChEJ*, **32**, 914 (1986)

15. Ogunnaike, B. A., R. K. Pearson, and F. J. Doyle, "Chemical Process Characterization: With Applications in the Rational Selection of Control Strategies," Proceedings of the 2nd European Control Conference, pp. 1067–1071, Groningen, Netherlands, June 1993

16. Rademacher, O., J. E. Rijnsdorp, and A. Maarleveld, *Dynamics and Control of Continuous Distillation Units*, Elsevier, New York (1975)

17. Ray, W. H., "On the Mathematical Modeling of Polymerization Reactors," *J. Macromol. Sci. — Rev. Macromol. Chem.*, **C8**, (1), 1–56 (1972)

18. Ray, W. H., "Modeling of Polymerization Phenomena," *Ber. Bunsengesellshaft Phys. Chem.*, **90**, 947 (1986)

19. Ray, W. H., "Polymerization Reactor Control," *IEEE Control Systems*, **6** (4), 3 (1986)

20. Ray, W. H., "Computer-aided Design, Monitoring, and Control of Polymerization Processes," in *Polymer Reaction Eng.* (H. Geisler and K. H. Reichert, Ed.), VCH Publ., 105 (1989)

21. Ray, W. H., "Modeling and Control of Polymerization Reactors," in *DYCORD '92* (T. J. McAvoy, Ed.), Pergammon Press, Oxford (1992)

22. Shinskey, F. G., *Distillation Control* (2nd ed.), McGraw-Hill, New York (1984)

23. Schuler, H. and S. Zhang, "Real-time Estimation of the Chain Length Distribution in a Polymerization Reactor," *Chem. Eng. Sci.*, **40** (10), 1891–1904 (1985)

24. Schwedock, M. J., L. C. Windes, and W. H. Ray, "Steady-state and Dynamic Modeling of a Packed Bed Reactor for the Partial Oxidation of Methanol to Formaldehyde. I. Model Development," *Chem. Eng. Comm.*, **78**, 1 (1989)

25. Schwedock, M. J., L. C. Windes, and W. H. Ray, "Steady-state and Dynamic Modeling of a Packed Bed Reactor for the Partial Oxidation of Methanol to Formaldehyde. II. Experimental Results Compared with Model Predictions," *Chem. Eng. Comm.*, **78**, 45 (1989)

26. Skogestad, S., "Dynamics and Control of Distillation Columns — A Critical Survey," *Proceedings DYCORD 92+*, College Park, Maryland (1992)

27. Windes, L. C., A. Cinar, and W. H. Ray, "Dynamic Estimation of Temperature and Concentration Profiles in a Packed Bed Reactor," *Chem. Eng. Sci.*, **44**, 2087 (1989)

28. Windes, L. C. and W. H. Ray, "A Control Scheme for Packed Bed Reactors Having a Changing Catalyst Activity Profile. I. On-line Parameter Estimation and Feedback Control," *J. Proc. Cont.*, **2**, 23 (1992)

29. Windes, L. C. and W. H. Ray, "A Control Scheme for Packed Bed Reactors Having a Changing Catalyst Activity Profile. II. On-line Optimizing Control," *J. Proc. Cont.*, **2**, 43 (1992)

part VI

APPENDICES

"A man should keep his little brain attic st
with all the furniture that he is likely t
and the rest he can put
in the lumber room of his li
where he can get it if he w

Sherlock Holmes, *Five Orang*
(Sir Arthur Conan I

A

CONTROL SYSTEM SYMBOLS USED IN PROCESS AND INSTRUMENTATION DIAGRAMS

In reading Process and Instrumentation (P&I) Diagrams, the process sensors, controllers, and actuators are indicated by special symbols. It is important to be able to recognize these symbols in order to work with design engineers, operators, *et al*. Some of the most commonly employed symbols are indicated in the tables below. In many P&I diagrams, only abbreviated treatment of the control loop is given so that only the measurement, the controller, and the actuator are shown. The standards of the Instrument Society of America (ISA) are followed in this section. Ref. [1] presents a much more comprehensive description of the ISA standards.

[1] **Instrumentation Symbols and Identification**, ISA-S5-1 (ANSI/ISA - 1975, R1981) Instrumentation Soc. of America, in <u>Standards and Practices for Instrumentation</u>, 7th Edition (1983).

Table A.1. Some Line Symbols

Line	Meaning
―――――――――――――――――	Connection to process or mechanical link or instrument supply
--------------------------------------	Electrical signal
—//——————//——————//—	Pneumatic signal or undefined signal

Table A.2. Some Identification Letters

Symbol	First Letter	Succeeding Letter
A	Analysis	Alarm
C	Conductivity	Control
D	Density	
E	Voltage	
F	Flowrate	
I	Current	Indicate
L	Level	
M	Moisture (humidity)	
P	Pressure or Vacuum	
T	Temperature	Transmit
V	Viscosity	Valve

Table A.3. Some Actuator Symbols

Symbol	Actuator
	Manual Control Valve
	Automatic Control Valve
	Motor
	Solenoid
	Pressure control valve, self-contained
	Pressure control valve, external pressure measurement
	Pressure relief or safety valve, angular pattern
	Pressure relief or safety valve, straight through
	Level control valve with mechanical linkage

Table A.4. Some Sensor Symbols

Symbol	Sensor
FT	Flow sensor/transmitter using orifice meter
FT	Flow sensor/transmitter using turbine meter
FT	Flow sensor/transmitter, undefined type
TANK — LT	Level sensor/transmitter, undefined type
PT	Pressure sensor/transmitter, undefined type
TT	Temperature sensor/transmitter, undefined type
VT	Viscosity sensor/transmitter, undefined type
PI	Pressure indicator, undefined type
TI	Temperature indicator, undefined type

Table A.5. Some Example Control Loops

Symbol	Meaning
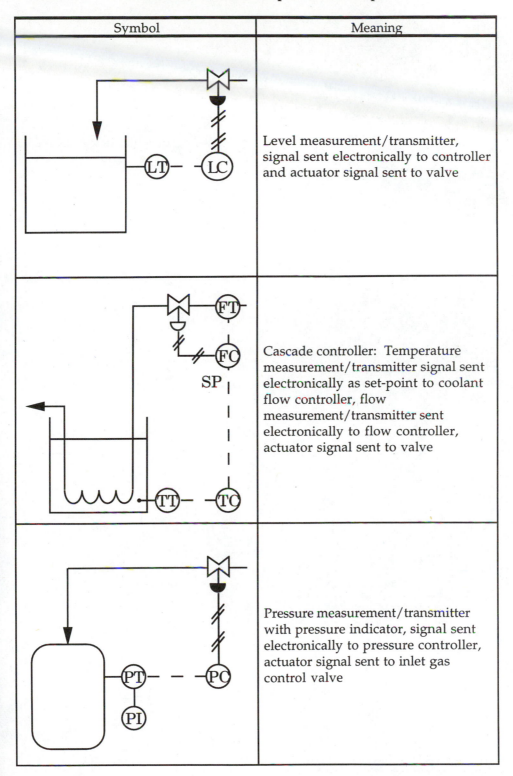	Level measurement/transmitter, signal sent electronically to controller and actuator signal sent to valve
	Cascade controller: Temperature measurement/transmitter signal sent electronically as set-point to coolant flow controller, flow measurement/transmitter sent electronically to flow controller, actuator signal sent to valve
	Pressure measurement/transmitter with pressure indicator, signal sent electronically to pressure controller, actuator signal sent to inlet gas control valve

COMPLEX VARIABLES, DIFFERENTIAL EQUATIONS, AND DIFFERENCE EQUATIONS

Classical process control is known to rest on such traditional fundamental elements as linear ordinary differential equations, the Laplace transform, and the closely related topic of complex variables. The corresponding fundamental elements for computer control are linear *difference* equations, z-transforms, and, again, complex variables. Since Laplace and z-transforms are discussed elsewhere in this book (see Chapters 3 and 24 for the respective introductions and Appendix C for a more complete review) this appendix is therefore devoted to reviewing first the *algebra of complex variables*, then some basic elements of *linear ordinary differential* and *difference equations* that are most relevant to process dynamics and control studies. Brief summaries of pertinent issues regarding nonlinear ordinary differential equations are also included.

B.1   COMPLEX VARIABLES

B.1.1   Definition of a Complex Number

A complex number z is defined in terms of an ordered pair of real numbers (x,y) and the *imaginary number* $j = \sqrt{-1}$ as follows:

$$z = (x,y) \;=\; x + jy \tag{B.1}$$

The real numbers x and y are called, respectively, the real and the imaginary parts of z; and it is customary to express this as:

$$\mathrm{Re}\, z = x \,;\, \mathrm{Im}\, z \;=\; y \tag{B.2}$$

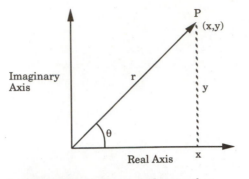

Figure B.1. The complex number.

B.1.2 Representations of Complex Numbers

Since a complex number is defined in terms of an ordered pair of real numbers (x, y), it is natural to associate each complex number with a point P in a plane whose Cartesian coordinates are x and y (see Figure B.1). With such representation, each complex number corresponds to just one point in the plane, and conversely, each point in the plane is associated with just one complex number.

In this context, the Cartesian x-y plane is referred to as the *complex plane*; the Cartesian x-axis becomes known as the "real axis" and the y-axis as the "imaginary axis." It is customary to then refer to diagrams such as in Figure B.1 as *Argand diagrams*. Observe now that the same point P shown in Figure B.1 can be located in the complex plane as the end point of a vector of length r units displaced at an angle θ (radians) from the positive real axis. Thus we may equally well identify the complex number z by the ordered pair (r, θ).

In general it is possible to represent a complex number in any of these two entirely equivalent forms (known respectively as the *Cartesian* and the *polar* forms) and, as will be seen below, one representation is usually more convenient than the other for performing certain algebraic operations.

Cartesian (or Rectangular) Representation

The Cartesian representation of a complex number is as given in Eq. (B.1), i.e.:

$$z = x + jy \tag{B.1}$$

Absolute Value (or Modulus)

The *absolute value* (or *modulus*) of the complex number in Eq. (B.1) is the nonnegative real number denoted $|z|$ and defined by:

$$|z| = \sqrt{x^2 + y^2} \tag{B.3}$$

From Figure B.1 it is obvious that this represents the distance from the point P to the origin. From Eq. (B.2) we now note that:

$$|z|^2 = (\text{Re } z)^2 + (\text{Im } z)^2 \tag{B.4}$$

Complex Conjugate

Associated with every complex number z with Cartesian representation given in Eq. (B.1) is the number $\bar{z}$, its *complex conjugate* defined as:

$$\bar{z} = x - jy \qquad (B.5)$$

That is, $\bar{z}$ is a number represented by the point $(x, -y)$, a mirror image of the number z reflected in the real axis. The following are important relations to note about complex numbers and their conjugates.

1. Their moduli:

$$|\bar{z}| = |z| \qquad (B.6)$$

2. Their sum and their difference:

$$\text{Re } z = \frac{z + \bar{z}}{2} \; ; \; \text{Im } z = \frac{z - \bar{z}}{2j} \qquad (B.7)$$

3. Their product:

$$z\bar{z} = |z|^2 \qquad (B.8)$$

Relation Eq. (B.8) is particularly useful; it indicates that multiplying a complex number by its conjugate produces a (nonnegative) real number.

Polar Representation

The point (x, y) may be represented instead by its polar coordinates (r, θ); in this case, the corresponding complex number may be represented as:

$$z = r \angle \theta \qquad (B.9)$$

where the symbol $\angle$ is used to indicate *implicitly* that θ is an angle. The explicit polar representation corresponding to Eq. (B.1) is obtained by noting that

$$x = r \cos \theta \; ; \qquad \text{and} \quad y = r \sin \theta$$

so that z may now be written as:

$$z = r(\cos \theta + j \sin \theta) \qquad (B.10)$$

The Modulus and the Argument

Note that the modulus of the complex number in this form is given by:

$$|z| = r\sqrt{\cos^2 \theta + \sin^2 \theta} = r \qquad (B.11)$$

The angle θ is called the *argument* of z, and we may write: $\theta = \arg z$. If the complex number z is interpreted as a directed line segment in the complex plane connecting the point (x, y) and the origin, then, as stated earlier, r is the length (or magnitude) of this line and θ the angle (in radians) that this line makes with the positive real axis. Observe, however, that θ is not unique because for any positive integer k, the angles given by $(\theta + 2k\pi)$ equally qualify as arguments of the same complex number. Thus we define the *principal value* of arg z (sometimes denoted by Arg z) as that unique value of θ for which:

$$-\pi < \theta \leq \pi \tag{B.12}$$

The Exponential Polar Form

It is possible to employ Euler's formula:

$$e^{j\theta} = \cos\theta + j\sin\theta \tag{B.13}$$

in Eq. (B.10) to obtain straightforwardly:

$$z = re^{j\theta} \tag{B.14}$$

This is sometimes referred to as the *exponential* polar form for obvious reasons.

Transformations

The following is a summary of the relationships used for transforming one complex number representation to the other.

1. From Cartesian to Polar

Given: $z = x + jy$, the required r and θ are obtained from:

$$r = \sqrt{x^2 + y^2} \tag{B.15a}$$

and

$$\theta = \tan^{-1}\left(\frac{y}{x}\right) \tag{B.15b}$$

In using Eq. (B.15b) care must be taken to ensure that the quadrant in which the point (x, y) falls is taken into consideration. The potential pitfall lies in the fact that the individual signs associated with x and with y — the most critical information needed to determine θ accurately — are lost in the process of evaluating the ratio y/x. The information about the quadrant in which z is located is not "automatically" contained in Eq. (B.15b); it must be "manually" supplied. Observe therefore that as it stands, Eq. (B.15b) will always give θ values in the range $[-\pi/2, \pi/2]$; this will be valid as long as x is positive. When x is negative, the appropriate value of θ is obtained by adding π.

2. From Polar to Cartesian

Given: $z = r \angle \theta$ either in the trigonometric form in Eq. (B.10), or in the exponential form in Eq. (B.14), the required Cartesian x and y numbers are obtained from:

$$x = r \cos \theta \tag{B.16a}$$

and

$$y = r \sin \theta \tag{B.16b}$$

B.1.3 Algebraic Operations

The same basic algebraic operations of addition, subtraction, division, and multiplication possible with real numbers can also be carried out with complex numbers. While it is true, in principle, that *all* the algebraic operations can be performed using *both* the Cartesian *and* the polar forms, it is also true that for each algebraic operation, there is a preferred complex number representation which considerably simplifies the required manipulation.

In what follows, it is without loss of generality that we illustrate algebraic operations using *two* complex numbers z_1 and z_2 given in the Cartesian form by:

$$z_1 = x_1 + jy_1 \tag{B.17a}$$

$$z_2 = x_2 + jy_2 \tag{B.17b}$$

and by:

$$z_1 = r_1 e^{j\theta_1} \tag{B.18a}$$

$$z_2 = r_2 e^{j\theta_2} \tag{B.18b}$$

in (exponential) polar form; the extension to three or more complex numbers is straightforward.

1. **Addition and Subtraction:** (*Cartesian Form Preferred*)

$$z_1 \pm z_2 = \left(x_1 + jy_1 \right) \pm \left(x_2 + jy_2 \right)$$

$$= \left(x_1 \pm x_2 \right) + j\left(y_1 \pm y_2 \right) \tag{B.19}$$

2. **Multiplication:** (*Exponential Polar Form Preferred*)

$$z_1 \cdot z_2 = \left(r_1 e^{j\theta_1} \right) \left(r_2 e^{j\theta_2} \right)$$

$$= r_1 r_2 e^{j(\theta_1 + \theta_2)} \tag{B.20}$$

Using the alternative Cartesian form, we obtain:

$$z_1 \cdot z_2 = \left(x_1 + jy_1 \right) \left(x_2 + jy_2 \right)$$

which, upon expanding and rearranging becomes:

$$z_1 \cdot z_2 = \left(x_1 x_2 - y_1 y_2\right) + j\left(x_1 y_2 + y_1 x_2\right) \tag{B.21}$$

3. **Division:** (*Exponential Polar Form Preferred*)

$$\frac{z_1}{z_2} = \frac{\left(r_1 \, e^{j\theta_1}\right)}{\left(r_2 \, e^{j\theta_2}\right)}$$

$$= \frac{r_1}{r_2} \, e^{\, j(\theta_1 - \theta_2)} \tag{B.22}$$

It is instructive to examine how division is performed using the Cartesian form:

$$\frac{z_1}{z_2} = \frac{\left(x_1 + jy_1\right)}{\left(x_2 + jy_2\right)} \tag{B.23}$$

We must now eliminate the complex number in the denominator and replace it with a real number; this is achieved by multiplying both numerator and denominator in Eq. (B.23) by the complex conjugate of z_2, a procedure known as *rationalization*, i.e.:

$$\frac{z_1}{z_2} = \frac{\left(x_1 + jy_1\right)}{\left(x_2 + jy_2\right)} \frac{\left(x_2 - jy_2\right)}{\left(x_2 - jy_2\right)}$$

with the result that:

$$\frac{z_1}{z_2} = \frac{\left(x_1 x_2 + y_1 y_2\right) + j\left(-x_1 y_2 + y_1 x_2\right)}{x_2^2 + y_2^2}$$

The following important relations regarding the moduli and arguments of the product and quotient of two complex numbers follow directly from the results given above.

1. **Products:**   If $z = z_1 z_2$ then:

$$|z| = |z_1| \, |z_2| \tag{B.24}$$

and

$$\arg(z) = \arg(z_1) + \arg(z_2) \tag{B.25}$$

2. **Quotients:** If $z = \dfrac{z_1}{z_2}$ then:

$$|z| = \frac{|z_1|}{|z_2|} \tag{B.26}$$

and

$$\arg(z) = \arg(z_1) - \arg(z_2) \tag{B.27}$$

B.1.4   Powers and Roots of Complex Numbers

The exponential polar form is useful for representing integral powers of a complex number. If the complex number represented by Eq. (B.14):

$$z = re^{j\theta} \tag{B.14}$$

is multiplied by itself n times, we obtain the result that:

$$z^n = r^n e^{jn\theta} \tag{B.28}$$

Similarly, we have:

$$z^{-n} = r^{-n}e^{-jn\theta} \tag{B.29}$$

In particular, if $r = 1$ in Eq. (B.28), we obtain:

$$(e^{j\theta})^n = e^{jn\theta} \tag{B.30}$$

and upon introducing Euler's formula Eq. (B.13) for both sides of Eq. (B.30), we have:

$$(\cos\,\theta + j\sin\,\theta)^n = \cos\,n\theta + j\sin\,n\theta \tag{B.31}$$

This result is known as *de Moivre's theorem*; it is particularly useful in finding the roots of equations involving the complex number z. To illustrate, we consider the classic problem of finding the n complex roots of the equation:

$$z^n - 1 = 0 \tag{B.32}$$

where n is a positive integer. This is first rearranged as:

$$z^n = 1 = 1e^{j0} \tag{B.33}$$

where we have written 1 in the exponential polar form. Observe now that in this form, the modulus and argument are given by:

$$r = 1 \quad \text{and} \quad \theta = 0 + 2\pi k$$

having recalled the fact that θ, along with any other angle equal to θ plus integral multiples of 2π, are equally valid arguments of the same complex number. Thus, in the trigonometric form Eq. (B.33) becomes:

$$z^n = 1 = \cos\,2\pi k + j\sin\,2\pi k\ ; \ (k = 0, 1, 2, \ldots, n-1) \tag{B.34}$$

so that:

$$z = 1^{1/n} = (\cos\,2\pi k + j\sin\,2\pi k)^{1/n}\ ; \ (k = 0, 1, 2, \ldots, n-1)$$

We may now apply de Moivre's theorem to the RHS and obtain finally:

$$z = \cos\left(\frac{2\pi k}{n}\right) + j\sin\left(\frac{2\pi k}{n}\right)\ (k = 0, 1, 2, \ldots, n-1) \tag{B.35}$$

as the expression for the n roots of the Eq. (B.32).

The references provided at the end of the Appendix (particularly Ref. [6]) are recommended to the reader interested in further information on complex variables.

B.2   LINEAR DIFFERENTIAL EQUATIONS

In general, any equation of the following type:

$$a_0(t)\frac{d^n y}{dt^n} + a_1(t)\frac{d^{n-1}y}{dt^{n-1}} + \dots + a_{n-1}(t)\frac{dy}{dt} + a_n(t)y = q(t) \tag{B.36}$$

involving functions of t (the independent variable), along with y and its derivatives (with respect to t) up to the nth derivative, is known as a *linear, differential equation of order* n. It is said to be linear because so long as the coefficients $a_i(t)$, $i = 0, 1, 2, \dots, n$, and $q(t)$ are functions of t alone (or are constant), the equation is in fact *linear* in the *dependent variable* y and its n derivatives; it is of order n because the nth derivative $d^n y/dt^n$ is the highest differential involved. If in particular $q(t) = 0$, the equation is further classified as *homogeneous*; otherwise it is nonhomogeneous.

The following is a summary of some key methods for solving the most important classes of these linear differential equations.

B.2.1   First-Order Equations

Consider the *linear* first-order homogeneous differential equation:

$$\frac{dy}{dt} + \alpha y = 0 \tag{B.37}$$

where α is a constant; it is easily solved by rearranging and separating the variables to give:

$$y(t) = e^{-\alpha t} y(0) \tag{B.38}$$

Of course not all first-order equations are this simple; but the concepts here illustrated are useful in solving the more general equations.

The general *nonhomogeneous* linear first-order equation is given by:

$$\frac{dy}{dt} + \alpha(t)y = q(t) \tag{B.39}$$

It is usually solved by the method of *integrating factors* and has the more general solution:

$$y(t) = \exp\left[-\int \alpha(t)dt\right]\left\{y(0) + \int \exp\left[\int \alpha(t)dt\right]q(t)dt\right\} \tag{B.40}$$

In the special case when the coefficient α is a constant, the first-order differential Eq. (B. 39) becomes:

$$\frac{dy(t)}{dt} + \alpha y(t) = q(t) \tag{B.41}$$

and the solution will then be obtained immediately from Eq. (B.40) as:

$$y(t) = e^{-\alpha t} y(0) + e^{-\alpha t} \int_0^t e^{\alpha \sigma} q(\sigma) d\sigma$$

or

$$y(t) = e^{-\alpha t} y(0) + \int_0^t e^{-\alpha(t-\sigma)} q(\sigma) d\sigma \tag{B.42}$$

Note that to avoid the potential confusion arising from using the same variable t in the limit of the integral and as the integration variable, it is customary to introduce a dummy time argument (say, σ) under the integral sign, and retain t for current time (and hence in the upper limit).

It is straightforward to show (and the reader is encouraged to do so as an exercise) that if we wish to solve the standard state-space model (Eq. (4.28) from Chapter 4) which includes both control action $u(t)$ and disturbance $d(t)$, i.e.:

$$\frac{dx(t)}{dt} = ax(t) + bu(t) + \gamma d(t) \tag{B.43}$$

$$y = cx(t)$$

then the solution is:

$$y(t) = e^{at} y(0) + c \int_0^t e^{a(t-\sigma)} bu(\sigma) d\sigma + c \int_0^t e^{a(t-\sigma)} \gamma d(\sigma) d\sigma \tag{B.44}$$

B.2.2 Higher Order Equations

The nth-order linear equation with *constant* coefficients:

$$a_0 \frac{d^n y}{dt^n} + a_1 \frac{d^{n-1} y}{dt^{n-1}} + \dots + a_{n-1} \frac{dy}{dt} + a_n y = q(t) \tag{B.45}$$

will have a solution of the form:

$$y(t) = A_0 + \sum_{i=1}^{n} A_i e^{m_i t} \tag{B.46}$$

where the m_i are the n roots of the characteristic equation:

$$a_0 m^n + a_1 m^{n-1} + \dots + a_{n-1} m + a_n = 0 \tag{B.47}$$

The specific form of the solution is most easily found from the method of Laplace transforms (see Appendix C); it is also possible to solve these equations by matrix methods as shown below.

Matrix Methods

Consider the situation in which a new set of variables, $z_1(t), z_2(t), ..., z_n(t)$ are introduced to represent the variable y and its first $n-1$ derivatives as follows:

$$
\left.
\begin{aligned}
z_1(t) &= y \\[4pt]
z_2(t) &= \frac{dy}{dt} \\[4pt]
z_3(t) &= \frac{d^2 y}{dt^2} \\[4pt]
\cdots \quad & \quad \cdots \quad \cdots \\[4pt]
z_{n-1}(t) &= \frac{d^{n-2} y}{dt^{n-2}} \\[4pt]
z_n(t) &= \frac{d^{n-1} y}{dt^{n-1}}
\end{aligned}
\right\}
\tag{B.48}
$$

From here we easily obtain:

$$
\frac{dz_1(t)}{dt} = z_2(t)
$$

$$
\frac{dz_2(t)}{dt} = z_3(t)
$$

$$
\cdots \quad \cdots
\tag{B.49}
$$

$$
\frac{dz_{n-1}(t)}{dt} = z_n(t)
$$

and

$$
\frac{dz_n(t)}{dt} = \frac{d^n y}{dt^n} = \frac{1}{a_0}\left[q(t) - a_1 z_n(t) - ... - a_{n-1} z_2(t) - a_n z_1(t) \right]
$$

where the RHS expression has been obtained by solving Eq. (B.45) for $d^n y / dt^n$ and introducing Eq. (B.48) appropriately. This set of n linear first-order equations now constitute an alternative representation of the linear nth-order equation. Written in the more compact vector-matrix form (see Appendix D) Eq. (B.49) becomes:

$$
\frac{d\mathbf{z}(t)}{dt} = \mathbf{A}\mathbf{z}(t) + \mathbf{f}(t)
\tag{B.50}
$$

where the n-dimensional vectors $\mathbf{z}(t)$, $\mathbf{f}(t)$, and the $n \times n$ matrix $\mathbf{A}$ are given by:

$$\mathbf{z}(t) = \begin{bmatrix} z_1(t) \\ z_2(t) \\ \cdots \\ z_n(t) \end{bmatrix} ; \qquad \mathbf{f}(t) = \begin{bmatrix} 0 \\ 0 \\ \cdots \\ \dfrac{q(t)}{a_0} \end{bmatrix} ;$$

(B.51)

$$\mathbf{A} = \begin{bmatrix} 0 & 1 & 0 & \cdots & 0 \\ 0 & 0 & 1 & \cdots & 0 \\ \cdots & \cdots & \cdots & \cdots & \cdots \\ 0 & 0 & 0 & \cdots & 1 \\ \dfrac{-a_n}{a_0} & \dfrac{-a_{n-1}}{a_0} & \dfrac{-a_{n-2}}{a_0} & \cdots & \dfrac{-a_1}{a_0} \end{bmatrix}$$

It is important to note the similarity between Eqs. (B.50) and (B.41); instead of the scalar functions and coefficients we now have vector functions and matrix coefficients.

The more compact matrix Eq. (B.50) may now be solved by the methods discussed in Appendix D; the solution is:

$$\mathbf{z}(t) = e^{\mathbf{A}(t-t_0)}\mathbf{z}_0 + \int_{t_0}^{t} e^{\mathbf{A}(t-\sigma)} \mathbf{f}(\sigma)\, d\sigma \tag{B.52}$$

where $\mathbf{z}_0$ is the value of the vector $\mathbf{z}$ at the point $t = t_0$. The matrix exponentials, $e^{\mathbf{A}(t-t_0)}, e^{\mathbf{A}(t-\sigma)}$, may be evaluated using the relation Eq. (D.87) in Appendix D.

Note that the general state-space representation, first presented in Chapter 4 (Eq. (4.30)), is a special form of Eq. (B.50):

$$\frac{d\mathbf{x}(t)}{dt} = \mathbf{A}\mathbf{x}(t) + \mathbf{B}\mathbf{u}(t) + \mathbf{\Gamma}\mathbf{d}(t) \tag{B.53}$$

$$\mathbf{y} = \mathbf{C}\mathbf{x}(t)$$

This has the general solution (from Eq. (B.52)):

$$\mathbf{x}(t) = e^{\mathbf{A}(t-t_0)}\mathbf{x}_0 + \int_{t_0}^{t} e^{\mathbf{A}(t-\sigma)} [\mathbf{B}\mathbf{u}(\sigma) + \mathbf{\Gamma}\mathbf{d}(\sigma)]d\sigma \tag{B.54}$$

$$\mathbf{y}(t) = \mathbf{C}\mathbf{x}(t)$$

where the matrix exponentials may be computed using Eq. (D.87) of Appendix D.

In the special case where the coefficient matrices $\mathbf{A}, \mathbf{B}, \mathbf{\Gamma}, \mathbf{C}$ in Eq. (B.53) are functions of t, the solution takes the form shown in Eqs. (D.91) and (D.92) in Appendix D.

B.2.3   Delay-Differential Equations

In modeling the dynamic behavior of certain processes, the resulting equations indicate relations between the dependent variable (say, y) and its derivatives *evaluated at different time arguments*, for example:

$$\frac{dy(t)}{dt} = a_0 y(t) + a_1 y(t - \alpha) \tag{B.55}$$

which involves not only $y(t)$ and its first derivative, but also the function with the delay-time argument. Equations of this type are known as delay-differential (or differential-difference) equations. Even though they are closely related to differential equations, solving them requires specialized techniques. A particularly readable discussion of such equations is contained in Driver (1977) [11].

Of all equations involving delay-time arguments, the types most commonly encountered in process control are usually of the form:

$$\frac{dy(t)}{dt} = ay(t) + bu(t - \alpha) \tag{B.56}$$

(and minor variations thereof) where the delay-time argument is not associated with the unknown function $y(t)$ but with the "forcing function." The solutions to such equations are most straightforwardly obtained using the method of Laplace transforms (see Appendix C).

B.3   LINEAR DIFFERENCE EQUATIONS

Consider the situation in which, instead of dealing with a continuous function, $y(t)$, as in the previous section, we now have a *sequence* $y(0)$, $y(1)$, $y(2)$, $y(3)$, ..., $y(k)$, ..., $y(n)$, where k, the index of the sequence, takes on *only* discrete, integer values. Such sequences generated by difference equations are used to represent discrete-time processes introduced in Chapter 4 and developed in great detail in Chapters 23–26.

B.3.1   First-Order Difference Equations

By analogy with the continuous first-order differential Eq. (B.41), the nonhomogeneous first-order difference equation with constant coefficients is given by:

$$y(k) + \alpha y(k - 1) = q(k) \tag{B.57}$$

where $q(k)$ is a known sequence. To build up a solution from the initial value $y(0)$, we obtain:

$$y(1) = -y(0)\,\alpha + q(1)$$

and subsequent substitution into Eq. (B.57) for consecutive values of k gives:

$$y(2) = y(0)(-\alpha)^2 - \alpha q(1) + q(2)$$

$$y(3) = y(0)(-\alpha)^3 + (-\alpha)^2 q(1) - \alpha q(2) + q(3)$$

from which we see that, in general:

$$y(k) = y(0)(-\alpha)^k + \sum_{i=1}^{k}(-\alpha)^{k-i} q(i) \qquad \text{(B.58)}$$

is the solution to Eq. (B.57). Note the analogy between Eqs. (B.58) and (B.42), the solution to the continuous differential Eq. (B.41).

From Chapter 4 we recall that the standard discrete-time model with control $u(k)$, disturbance $d(k)$, and output $y(k)$ has the form:

$$x(k+1) = ax(k) + bu(k) + \gamma d(k)$$
$$y(k) = cx(k) \qquad \text{(B.59)}$$

Applying the result in Eq. (B.58) we obtain the solution:

$$y(k) = y(0)a^k + c\sum_{i=1}^{k} a^{k-i}[bu(i-1) + \gamma d(i-1)] \qquad \text{(B.60)}$$

If we now compare Eq. (B.59) with Eq. (B.43) and then compare their respective solutions, Eqs. (B.60) and (B.44), we observe that the two solutions are completely analogous, with e^{at} in Eq. (B.44) replaced by a^k, and the integral replaced by the finite sum.

B.3.2   General nth-Order Difference Equations

As was the case with ordinary differential equations, nonhomogeneous, constant coefficient, linear difference equations arise from process models (cf. Chapters 23–26):

$$a_0 y(k) + a_1 y(k-1) + a_2 y(k-2) + \ldots + a_n y(k-n) = q(k) \qquad \text{(B.61)}$$

This nth-order difference equation will have a solution of the form:

$$y(k) = A_0 + \sum_{i=1}^{n} A_i(m_i)^k \qquad \text{(B.62)}$$

where the m_i, $i = 1, 2 \ldots n$, are the n roots of the characteristic equation:

$$a_0 m^n + a_1 m^{n-1} + a_2 m^{n-2} + \ldots + a_n = 0 \qquad \text{(B.63)}$$

Just as we use Laplace transforms with ordinary differential equations, it is possible to use z-transforms (the "discrete counterpart" of Laplace transforms) to solve linear difference equations having constant coefficients. The technique involved is discussed in Appendix C. It is also possible to convert the nth-order difference equation into a system of n first-order difference equations and employ matrix methods.

Matrix Methods

By introducing n new variables $z_1(k), z_2(k), ..., z_n(k)$ defined as follows:

$$\left.\begin{aligned}
z_1(k) &= y(k-n+1) \\
z_2(k) &= y(k-n+2) \\
z_3(k) &= y(k-n+3) \\
\cdots &\quad \cdots \quad \cdots \\
z_{n-2}(k) &= y(k-2) \\
z_{n-1}(k) &= y(k-1) \\
z_n(k) &= y(k)
\end{aligned}\right\} \tag{B.64}$$

it is possible to convert Eq. (B.61) into a system of n linear first-order difference equations. From Eq. (B.64) we easily obtain:

$$\begin{aligned}
z_1(k) &= z_2(k-1) \\
z_2(k) &= z_3(k-1) \\
\cdots &= \cdots \\
z_{n-2}(k) &= z_{n-1}(k-1) \\
z_{n-1}(k) &= z_n(k-1)
\end{aligned} \tag{B.65}$$

and finally, from Eq. (B.61), we obtain, upon solving for $y(k)$, and using the appropriate new $z_i(k-1)$ variables for each $y(k-i)$ term:

$$z_n(k) = \frac{1}{a_0}\left[q(k) - a_1 z_n(k-1) - ... - a_{n-1} z_2(k-1) - a_n z_1(k-1)\right]$$

This system of n equations may now be written in the more compact vector-matrix form as:

$$\mathbf{z}(k) = \mathbf{\Phi}\mathbf{z}(k-1) + \mathbf{f}(k) \tag{B.66}$$

where we see that the n-dimensional vectors $\mathbf{z}(k)$, $\mathbf{f}(k)$, and the $n \times n$ matrix $\mathbf{\Phi}$ are given by:

$$\mathbf{z}(k) = \begin{bmatrix} z_1(k) \\ z_2(k) \\ \cdots \\ z_n(k) \end{bmatrix} ; \qquad \mathbf{f}(k) = \begin{bmatrix} 0 \\ 0 \\ \cdots \\ \dfrac{q(k)}{a_0} \end{bmatrix} ;$$

$$\Phi = \begin{bmatrix} 0 & 1 & 0 & \cdots & 0 \\ 0 & 0 & 1 & \cdots & 0 \\ \cdots & \cdots & \cdots & \cdots & \cdots \\ 0 & 0 & 0 & \cdots & 1 \\ \dfrac{-a_n}{a_0} & \dfrac{-a_{n-1}}{a_0} & \dfrac{-a_{n-2}}{a_0} & \cdots & \dfrac{-a_1}{a_0} \end{bmatrix}$$

This is now the matrix representation of the nth-order linear difference equation. Again, it is important to note the similarity between Eqs. (B.66) and (B.57); instead of the scalar functions and coefficients we now have vector functions and matrix coefficients. By following precisely the same arguments that led to the derivation of the solution in Eq. (B.58) — paying due respect to the peculiar laws of matrix algebra (see Appendix D) — we may obtain the solution to Eq. (B.66) just as straightforwardly; the result is:

$$\mathbf{z}(k) = \Phi^k \mathbf{z}(0) + \sum_{i=1}^{k} \Phi^{k-i} \, \mathbf{f}(i) \tag{B.67}$$

where $\mathbf{z}(0)$ represents the vector of initial conditions on $\mathbf{z}(k)$.

The linear discrete-time model for multivariable systems was presented in Chapter 4 (and discussed in detail in Chapters 24–26). This is a special case of Eq. (B.66) and takes the form:

$$\mathbf{x}(k+1) = \Phi \mathbf{x}(k) + \mathbf{B}u(k) + \Gamma \mathbf{d}(k) \tag{B.68}$$

$$\mathbf{y}(k) = \mathbf{C}\mathbf{x}(k)$$

By extension of the results above, the solution to Eq. (B.68) may be written:

$$\mathbf{x}(k) = \Phi^k \mathbf{x}(0) + \sum_{i=1}^{k} \Phi^{k-i} \left[\mathbf{B}u(i-1) + \Gamma \mathbf{d}(i-1) \right] \tag{B.69}$$

$$\mathbf{y}(k) = \mathbf{C}\mathbf{x}(k)$$

B.4   NONLINEAR DIFFERENTIAL EQUATIONS

As stated earlier, any equation that expresses a relation between a function $y(t)$ and its derivatives is a differential equation; in particular, if the indicated relationship is nonlinear, we have a nonlinear differential equation. For example, the equation:

$$\frac{dy}{dt} = f(t, y) \tag{B.70}$$

is a first-order nonlinear differential equation if $f(t,y)$ is a nonlinear function of y. The general nonlinear nth-order differential equation may be represented as:

$$f\left(\frac{d^n y}{dt^n} ; \frac{d^{n-1} y}{dt^{n-1}} ; \ldots ; \frac{dy}{dt} ; y ; t \right) = 0 \tag{B.71}$$

where $f(\cdot)$ is some nonlinear function of the indicated arguments.

For linear equations, there are well-defined, systematic procedures for finding solutions in all cases, some of which were reviewed in Section B.2. Unfortunately there are no corresponding general procedures for generating solutions to all nonlinear equations. In fact, the determination of analytical solutions for nonlinear equations of any order is not always possible in general; only in a few very special cases are there some specialized methods for obtaining analytical solutions. In the vast majority of cases, one must resort to *numerical* techniques for constructing the approximate solutions. Thus, this section will be devoted to describing the basic approaches to numerical solution of nonlinear differential equations.

Numerical Methods

The basic premise of numerical methods may be summarized as follows:

1. The differential equation in question has an unknown solution which may be represented as:

$$y = \eta(t) \tag{B.72}$$

2. This *continuous* function $\eta(t)$ cannot always be obtained in its "true" functional form (explicitly) by analytical techniques so an attempt is made at generating instead the *discrete* sequence $\eta(t_0)$, $\eta(t_1)$, $\eta(t_2)$, ..., $\eta(t_k)$, ... , representing the numerical values it takes at specific *discrete* points $t_0 < t_1 < t_2 < ... < t_k < ...$ separated by an interval of size Δt.

3. The *continuous* variables in the differential equation are therefore replaced by *discrete* variables and the derivatives by appropriate finite difference quotients. The differential equation is thereby converted to a difference equation (in the form of a recurrence relation) which is easily solved especially when programmed on a high-speed digital computer.

4. Since this process involves some approximation, the discrete sequence generated from the "surrogate" difference equation, which we shall refer to as $y(0)$, $y(1)$, $y(2)$, ..., $y(k)$, ..., will not match the true but unknown values $\eta(t_0)$, $\eta(t_1)$, $\eta(t_2)$, ..., $\eta(t_k)$, ..., precisely; however, approximation errors associated with a good numerical method are usually small.

5. What differentiates one numerical method from another is the technique adopted for deriving the "surrogate" difference equation from the original differential equation.

Before reviewing specific numerical techniques, we note the following important issues and questions raised by numerical methods:

1. The numerical solution $y(0)$, $y(1)$, $y(2)$, ..., $y(k)$, ... constitutes only an *approximation* to the exact but unknown solution because of unavoidable errors introduced in the course of its computation. These errors arise from two separate sources: (a) from the numerical methods formula itself (which must be obtained by introducing some approximations), and (b) from the fact that in any computation, only a finite number of significant

digits can be retained. The former is referred to as *discretization* (or *formula*) error; the latter as *round-off* error.

2. Is it always possible to estimate the magnitude of these errors? This question is important from the point of view of evaluating the integrity of the solutions generated by various numerical methods.

3. As Δt the interval between the discrete points $t_0, t_1, t_2, ..., t_k, ...$ tends to zero, will the values taken by the numerical approximation also tend to the values of the exact, but unknown, solution? This question addresses the issue of *convergence*.

These and other related issues not mentioned here are critical to the successful generation of numerical solutions and are therefore discussed in great detail in texts devoted exclusively to numerical methods. The supplied references [15,16] are recommended to the reader interested in pursuing these aspects further. For now we will review a few of the key techniques used to obtain numerical solutions to differential equations.

The Euler Method

In solving the nonlinear first-order equation:

$$\frac{dy}{dt} = f(t, y) \tag{B.73}$$

given the initial condition $y(t_0) = y(0)$:

1. If we let $y(k)$ represent $y(t)$ at the point $t = t_k$ and therefore replace the continuous $f(t,y)$ with $f(t_k, y(k))$, and

2. If we replace the derivative with the finite difference approximation:

$$\frac{dy(t)}{dt} \approx \frac{y(t_{k+1}) - y(t_k)}{\Delta t} \tag{B.74}$$

then the differential equation becomes:

$$\frac{y(t_{k+1}) - y(t_k)}{\Delta t} = f(t_k, y(k)) \tag{B.75}$$

which is easily rearranged to give:

$$y(t_{k+1}) = y(t_k) + \Delta t f(t_k, y(k)) \tag{B.76}$$

or, simply:

$$y(k+1) = y(k) + \Delta t f(t_k, y(k)) \tag{B.77}$$

Observe now that starting with the initial value $y(0)$, Eq. (B.77) can be used recursively to generate the sequence $y(1)$, $y(2)$, $y(3)$, ... as the required approximate solution to Eq. (B.73).

The formula in Eq. (B.75) is known as Euler's method; it represents one of the simplest schemes available for solving initial-value problems. It has the advantage that it is very straightforward to use, and very easy to program on the computer; its main disadvantage is that it is not very accurate. It can be shown that the *local formula error* for Euler's method is proportional to $(\Delta t)^2$ and to the second derivative of $\eta(t)$, the true, but unknown solution.

While it is true, in general, that the accuracy of Euler's method may be improved by reducing the interval size Δt, unfortunately there is a limit beyond which any further reduction in interval size actually worsens the accuracy. Observe that as Δt is reduced, many more steps are required in going from t_0 to the final point t_N ; but round-off error is introduced at *each* step of the calculation so that the accumulated round-off error actually increases as the interval size Δt is reduced. Clearly it is possible to cross a point where the increased accuracy achieved by the interval size reduction is offset by the increase in the accumulated round-off error.

The Improved Euler Method

The simple Euler method may be improved upon as follows: Let us once again consider the equation:

$$y'(t) = \frac{dy}{dt} = f(t,y) \tag{B.73}$$

with its initial condition $y(t_0) = y(0)$ whose unknown solution is given by $y = \eta(t)$. By integrating from the point t_k to t_{k+1} we obtain:

$$\eta(t_{k+1}) = \eta(t_k) + \int_{t_k}^{t_{k+1}} f(t, \eta(t))\, dt \tag{B.78}$$

Observe therefore that in relation to Eq. (B.78) the Euler formula:

$$y(k+1) = y(k) + \Delta t\, f(t_k, y(k)) \tag{B.77}$$

is equivalent to replacing $f[t, \eta(t)]$ over the entire interval $[t_k, t_{k+1}]$ by $f(t_k, y(k))$, the approximate value taken at the left-hand boundary of the interval. A better approximation may be obtained if $f(t, \eta(t))$ in Eq. (B.78) is replaced by an *average* of its value at the two boundaries of the interval, i.e:

$$f(t, \eta(t)) \approx \frac{f(t_k, \eta(t_k)) + f(t_{k+1}, \eta(t_{k+1}))}{2}$$

and upon replacing $\eta(t_k)$ and $\eta(t_{k+1})$ with their respective approximate values $y(k)$ and $y(k+1)$ we obtain the improved formula:

$$y(k+1) = y(k) + \Delta t \left\{ \frac{f(t_k, y(k)) + f(t_{k+1}, y(k+1))}{2} \right\} \tag{B.79}$$

The only complication now is the appearance of the unknown $y(k+1)$ in the RHS of Eq. (B.79); this is resolved by introducing the simpler Euler formula Eq. (B.77) in its place. Simplifying the notation further by using $y'(k)$ to represent $f(t_k, y(k))$ we obtain the final result:

$$y(k+1) = y(k) + \Delta t \left\{ \frac{y'(k) + f(t_{k+1}, y(k) + \Delta t y'(k))}{2} \right\} \tag{B.80}$$

It can be shown that the *local formula error* for the improved Euler method is proportional to $(\Delta t)^3$, and since the only reasonable values for Δt are usually such that $|\Delta t| \ll 1$, it follows that the improved formula is indeed more accurate. It should be noted that the improved accuracy has been achieved at the expense of increased computation.

The Runge-Kutta Method

The formula:

$$y(k+1) = y(k) + \frac{\Delta t}{6}\left(c_1 + 2c_2 + 2c_3 + c_4 \right) \tag{B.81}$$

where

$$c_1 = f(t_k, y(k)) \tag{B.82a}$$

$$c_2 = f\left(t_k + \frac{1}{2}\Delta t \,;\, y(k) + \frac{1}{2}\Delta t\, c_1 \right) \tag{B.82b}$$

$$c_3 = f\left(t_k + \frac{1}{2}\Delta t \,;\, y(k) + \frac{1}{2}\Delta t\, c_2 \right) \tag{B.82c}$$

$$c_4 = f\left(t_k + \Delta t \,;\, y(k) + \Delta t\, c_3 \right) \tag{B.82d}$$

is similar to the improved Euler formula except that $f(t, \eta(t))$ is now replaced by a weighted average of the values it takes at various points in the interval $[t_k, t_{k+1}]$.

This is the formula for the (fourth-order) Runge-Kutta method; it is a very accurate and very popular method for obtaining numerical solutions to initial-value problems (despite the fact that it is more complicated than any of the Euler methods). It can be shown that the *local formula error* for this method is proportional to $(\Delta t)^5$.

It should be mentioned that there are many other categories of numerical methods (e.g., variable step-size, multistep, etc.) which are discussed at length in standard texts on numerical methods. However, the reader should also note that computer software packages abound which offer a wide variety of these numerical techniques so that it is no longer necessary for one to write one's own program for carrying out the required computations. It is, however, important for one to understand the implications of using one method as opposed to another.

Higher Order Equations

It can be easily shown that any higher order differential equation can be reduced to a system of several first-order equations of the type in Eq. (B.73). Numerical techniques applicable to first-order equations extend quite straightforwardly to systems of first-order equations. Thus problems involving higher order equations, or those involving systems of first-order equations, do not raise any additional issues of consequence. See, for example, Refs. [14, 16].

REFERENCES AND SUGGESTED FURTHER READING

General references that cover several aspects of applied mathematical methods (particularly complex variables and ordinary differential equations) abound; the following, arranged roughly in order of increasing depth, is a small, somewhat representative selection.

1. Stephenson, G., *Mathematical Methods for Science Students,* Longmans, London (1971)
2. Greenberg, M. D., *Foundations of Applied Mathematics,* Prentice–Hall, Englewood Cliffs, NJ (1978)
3. Kreyszig, E., *Advanced Engineering Mathematics*, (3rd ed.), J. Wiley, New York (1972)
4. Wylie, C. R., *Advanced Engineering Mathematics,* (4th ed.), McGraw-Hill, New York (1975)
5. Sokolnikoff, I. and R. Redheffer, *Mathematics of Physics and Modern Engineering,* McGraw-Hill, New York (1966)

The following are two specific references on the subject of Complex Variables: the first is more modern and discusses applications; the second is considered a classic treatise on the *theory* with very little by way of applications.

6. Churchill, R. V., J. W. Brown, and R. F. Verhey, *Complex Variables and Applications,* (3rd ed.), McGraw–Hill, New York (1976)
7. Copson, E. T., *Theory of Functions of a Complex Variable*, Clarendon Press, Oxford (1935)

The following are references on the general topic of Ordinary Differential Equations: Ref. [8] is very pleasant to read, Ref. [9] is a venerable classic, Ref. [10] gets very theoretical at times, and Ref. [11] contains a systematic treatment of Delay-differential equations.

8. Boyce, W. E. and R. C. DiPrima, *Elementary Differential Equations and Boundary Value Problems*, (3rd ed.), J. Wiley, New York (1977)
9. Ince, E. L., *Ordinary Differential Equations*, Dover, New York (1956)
10. Coddington, E. and N. Levinson, *Theory of Ordinary Differential Equations*, McGraw–Hill, New York (1955)
11. Driver, R. D., *Ordinary and Delay-differential Equations*, Springer-Verlag, New York (1977)

Two important references providing detailed and thorough treatments of the topic of Difference Equations are:

12. Goldberg, S., *Introduction to Difference Equations*, J. Wiley, New York (1958)
13. Milne-Thompson, L. M., *The Calculus of Finite Differences*, Macmillan, London (1933)

There are numerous useful references on the *general* subject of Numerical Methods; the following is somewhat of a representative sample in terms of breadth and depth of coverage.

14. Hornbeck, R. W., *Numerical Methods*, Quantum Publishers, New York (1975)
15. Conte, S. D. and C. de Boor, *Elementary Numerical Methods*, (2nd ed.), McGraw-Hill, New York (1972)
16. Carnahan, B., H. Luther, and J. Wilkes, *Applied Numerical Methods*, J. Wiley, New York (1969)
17. Henrici, P., *Elements of Numerical Analysis*, J. Wiley, New York (1964)

The following are references that focus attention on the more computational aspects of Numerical Methods: Ref. [18] is a good introductory text, while the more recent Ref. [19] is a comprehensive and immensely useful (almost indispensable) reference book.

18. Kuo, S. S., *Computer Applications of Numerical Methods*, Addison-Wesley, Reading, MA (1972)
19. Flannery, B. P., S. A. Teukolsky, and W. T. Vettering, *Numerical Recipes (in C or in Fortran): The Art of Scientific Computing*, W. H. Press, Cambridge University Press (1988)

Two of the more specialized references concerned with the numerical solution of Ordinary Differential equations are: Ref. [20], authored by a well-respected practitioner, and Ref. [21], a more advanced treatment of the same subject.

20. Gear, C. W., *Numerical Initial-value Problems in Ordinary Differential Equations*, Prentice-Hall, Englewood Cliffs, NJ (1971)
21. Henrici, P., *Discrete Variable Methods in Ordinary Differential Equations"*, J. Wiley, New York (1962)

APPENDIX

C

LAPLACE AND z-TRANSFORMS

Integral transforms have had a most enduring impact on the analysis and design of linear control systems and are key components of the engineer's toolbox. The most popular of these have been the Laplace transform for continuous systems, and the z-transform for discrete systems. Even though certain aspects of the Laplace and z-transforms have been introduced in Chapters 3 and 24, the material presented here is intended to be a more complete summary. A brief introduction, first to the general family of integral transforms, and then to the Fourier transform in particular, precedes the main subject matter; it is designed to emphasize the close ties between the Laplace transform and the Fourier transform. The unifying theme is extended finally to the z-transform by presenting it as a natural consequence of applying the Laplace transform to sampled functions.

C.1 INTRODUCTION: THE GENERAL INTEGRAL TRANSFORM

The process of multiplying a function of a single variable t, say, $f(t)$, by $K(s,t)$, a function of two variables s and t, and integrating the resulting product *with respect to t* between limits a and b, gives rise to a function of the single surviving variable s, say, $\hat{f}(s)$; the original function, $f(t)$, is thus "transformed" to $\hat{f}(s)$ by the indicated process that may be represented mathematically as:

$$T\{f(t)\} = \int_a^b K(s, t) f(t) dt = \hat{f}(s) \tag{C.1}$$

This is an expression for the *general* integral transform; $K(s, t)$, the function of two variables that serves as the multiplier in the transform formula is called the *kernel* of the transform.

By specifying various kernels and limits of integration, several useful integral transforms can be obtained from Eq. (C.1). For example, when $a = 0$, $b = \infty$, and $K(s, t) = e^{-st}$, we have the Laplace transform defined in Eq. (3.1). Some other examples are:

1. FOURIER TRANSFORM: $a = -\infty$, $b = \infty$, $K(\omega, t) = e^{-j\omega t}$

2. HANKEL TRANSFORM: $a = 0$, $b = \infty$, $K(\lambda, t) = t\, J_\nu(\lambda t)$, where $J_\nu(\lambda t)$ is Bessel's function (of the first kind) of order $\nu > -1/2$

3. MELLIN TRANSFORM: $a = 0$, $b = \infty$, $K(s, t) = t^{s-1}$

4. LAGUERRE TRANSFORM: $a = 0$, $b = \infty$, $K(n, t) = e^{-t}L_n(t)$, where $L_n(t)$ is the Laguere polynomial of order n

Such integral transform formulas are usually accompanied by the "inverse" expression of the kind:

$$T^{-1}\{\hat{f}(s)\} = \int_{a^*}^{b^*} K^*(s,t)\,\hat{f}(s)ds = f(t) \tag{C.2}$$

by which the original function, $f(t)$, may be recovered from its transformed version, $\hat{f}(s)$, using $K^*(s, t)$, the kernel of the inverse transform formula, and integrating out the s variable between the limits a^* and b^*. For example, the corresponding inverse Laplace transform expression has:

$$a^* = \alpha + j\omega,\ b^* = \alpha - j\omega \text{ and } K^*(s, t) = e^{st}2\pi j$$

and, for the inverse *Hankel transform*:

$$a^* = 0,\ b^* = \infty, \text{ and } K^*(\lambda, t) = \lambda\, J_\nu(\lambda t)$$

The integral transform of Eq. (C.1) and its inverse relation in Eq. (C.2) constitute a "transform pair."

Integral transform pairs in general have found significant application in science and engineering; and some transforms are more natural than others for certain applications. For example, in certain aspects of the theory of elasticity, the Mellin transform is the more natural choice. For process control applications, however, the Laplace and Fourier transforms are the most natural and most useful; and it will be shown later that for the analysis and design of discrete-time control systems, it is in fact just a clever adaptation of the Laplace transform that leads to the more natural z-transform.

C.2   THE FOURIER TRANSFORM

The Fourier series expansion for a function $f(t)$ over the interval $[-L, L]$ is given by:

$$f(t) = \frac{a_0}{2} + \sum_{n=1}^{\infty} a_n\cos\left(\frac{n\pi t}{L}\right) + b_n\sin\left(\frac{n\pi t}{L}\right) \tag{C.3}$$

where the coefficients are given by:

$$a_n = \frac{1}{L} \int_{-L}^{L} f(t)\cos\left(\frac{n\pi t}{L}\right) dt \qquad (C.4a)$$

and

$$b_n = \frac{1}{L} \int_{-L}^{L} f(t)\sin\left(\frac{n\pi t}{L}\right) dt \qquad (C.4b)$$

By substituting Eq. (C.4) into Eq. (C.3) and taking limits as $L \to \infty$, after carefully carrying out the required manipulations, and using the exponential representation for the sin and cos functions, we obtain:

$$f(t) = \frac{1}{2\pi} \int_{-\infty}^{\infty} e^{j\omega t} \int_{-\infty}^{\infty} f(\tau)e^{-j\omega\tau} \, d\tau \, d\omega \qquad (C.5)$$

This is the Fourier integral formula; it provides the origin of the Fourier transform and its inverse. For, if we were to define:

$$\hat{f}(\omega) \equiv \int_{-\infty}^{\infty} f(t)e^{-j\omega t} \, dt \qquad (C.6)$$

then Eq. (C.5) immediately becomes:

$$f(t) = \frac{1}{2\pi} \int_{-\infty}^{\infty} \hat{f}(\omega)e^{j\omega t} \, d\omega \qquad (C.7)$$

We may now observe that, in the spirit of the general integral transform pair given in Eqs. (C.1) and (C.2), Eq. (C.6) expresses the transform of the function $f(t)$, and Eq. (C.7) the corresponding inverse. Thus, $\hat{f}(\omega)$ defined as in Eq. (C.6) is called the "Fourier transform" of $f(t)$, and Eq. (C.7) is the corresponding inverse "Fourier transform" formula by which $f(t)$ is recovered from $\hat{f}(\omega)$. Note that the domain of $f(t)$ is infinite.

C.3   THE LAPLACE TRANSFORM

Not all functions are Fourier transformable. In particular, some important functions employed in process dynamics and control — for example, the step function, and the ramp function — are not Fourier transformable because they are functions that do not decay to zero fast enough as $t \to \infty$. Also, the domain of the functions encountered in process dynamics and control is usually the *semi-infinite* region $0 \le t < \infty$ and not the infinite region of the Fourier transform. Thus the Fourier transform may not be the most natural for process dynamics and control problems in general. The Laplace transform is the result of modifying the Fourier transform to better suit such problems.

C.3.1 Definition and Inverse Formula

Consider a function $f(t)$ that satisfies the following conditions:

1.   It is "piecewise regular"; (i.e., it is bounded in every interval of the form $0 \leq t_1 \leq t \leq t_2$, and has at most a finite number of maxima and minima, and a finite number of *finite discontinuities*).

2.   It is of "exponential order" (i.e., there exist constants a, M, and T such that $e^{-at}|f(t)| < M$ for all $t > T$).

Note that if $e^{-at}|f(t)| < M$ then it is also true that $e^{-a_1 t}|f(t)| < M$ for all $a_1 > a$; therefore, the a required by Condition 2 is not unique. Now, let α be the *greatest lower bound* (infimum) of the set of a values for which Condition 2 is satisfied: then α is called the "abscissa of convergence" of the function $f(t)$.

Now, by virtue of Condition 2 and this definition of the abscissa of convergence, for any function satisfying these two conditions, the following composite function:

$$\phi(t) = \begin{cases} 0; & t < 0 \\ e^{-\alpha t} f(t); & t > 0 \end{cases} \tag{C.8}$$

is guaranteed to be Fourier transformable regardless of whether $f(t)$ is or not. Introducing Eq. (C.8) into Eq. (C.5) gives:

$$e^{-\alpha t} f(t) = \frac{1}{2\pi} \int_{-\infty}^{\infty} e^{j\omega t} \int_{0}^{\infty} (e^{-\alpha \tau} f(\tau)) e^{-j\omega \tau} \, d\tau \, d\omega$$

so that:

$$f(t) = \frac{1}{2\pi} \int_{-\infty}^{\infty} e^{(\alpha + j\omega)t} \int_{0}^{\infty} f(\tau) e^{-(\alpha + j\omega)\tau} \, d\tau \, d\omega \tag{C.9}$$

and by introducing the new variable:

$$s = \alpha + j\omega \tag{C.10}$$

Eq. (C.9) becomes:

$$f(t) = \lim_{\omega \to \infty} \frac{1}{2\pi j} \int_{\alpha - j\omega}^{\alpha + j\omega} e^{st} \int_{0}^{\infty} f(\tau) e^{-s\tau} \, d\tau \, ds \tag{C.11}$$

and now, by defining:

$$\hat{f}(s) = \int_{0}^{\infty} f(t) e^{-st} \, dt \tag{C.12}$$

Eq. (C.11) immediately becomes:

$$f(t) = \lim_{\omega \to \infty} \frac{1}{2\pi j} \int_{\alpha - j\omega}^{\alpha + j\omega} \hat{f}(s)\, e^{st}\, ds \qquad (C.13)$$

giving the Laplace transform and its inversion formula.

From Eq. (C.10) we may observe that the Laplace transform variable s is complex. Also, the path of integration indicated in the inversion formula is the straight line in the complex plane from $\alpha - j\omega$ to $\alpha + j\omega$ as $\omega \to \infty$; it is known as the *Bromwich path*, and simply denoted by "C" in the expression in Chapter 3. Thus, as initially stated in Chapter 3:

> The Laplace transform of a function $f(t)$ that is **piecewise regular** and of **exponential order** is given by Eq. (C.12); the corresponding inverse transform is given by Eq. (C.13).

In principle, therefore, according to Eq. (C.13), contour integration in the complex plane must be carried out in order to invert a Laplace transform; however, in practice, it is customary to use tables of Laplace transforms, an example of which is presented at the end of this section.

C.3.2 Properties and Useful Theorems of Laplace Transforms

Basic Properties

1. Linearity

$$L\{c_1 f_1(t) + c_2 f_2(t)\} = c_1 L\{f_1(t)\} + c_2 L\{f_2(t)\} \qquad (C.14)$$

Extension to linear combination of n functions is straightforward.

2. Transform of Derivatives

$$L\left\{\frac{df}{dt}\right\} = s\hat{f}(s) - f(t)\big|_{t=0} \qquad (C.15)$$

$$L\left\{\frac{d^2 f}{dt^2}\right\} = s^2\hat{f}(s) - sf(t)\big|_{t=0} - \frac{df(t)}{dt}\big|_{t=0} \qquad (C.16)$$

or, in general, if $f^{(n)}(t)$ represents the nth derivative of $f(t)$, and by convention, $f^{(0)}(t) = f(t)$:

$$L\{f^{(n)}(t)\} = s^n\hat{f}(s) - \sum_{k=0}^{n-1} s^k f^{(n-1-k)}(t)\big|_{t=0} \qquad (C.17)$$

When all the initial conditions are zero, then:

$$L\{f^{(n)}(t)\} = s^n\,\hat{f}(s) \qquad (C.18)$$

and an nth derivative is turned to an nth power of s. This result is very useful in obtaining transfer functions from differential equation models; it provides the motivation for working with deviation variables, since, by definition, these have zero initial conditions.

3. **Transform of Integrals**

$$L\left\{\int_0^t f(\tau)\, d\tau\right\} = \frac{1}{s}\hat{f}(s) \tag{C.19}$$

Additional Properties Useful In Engineering Applications

If $L\{f(t)\} = \hat{f}(s)$, then:

4. **Time Change of Scale**

$$L\{f(at)\} = \frac{1}{a}\hat{f}\left(\frac{s}{a}\right) \tag{C.20a}$$

Conversely, if $L^{-1}\{\hat{f}(s)\} = f(t)$, then:

$$L^{-1}\{\hat{f}(bs)\} = \frac{1}{b}f\left(\frac{t}{b}\right) \tag{C.20b}$$

5. **Complex Shift** (i.e., shift in s)

$$L^{-1}\{\hat{f}(s-a)\} = e^{at}f(t) \tag{C.21}$$

Typical application: Since $L^{-1}(1/s) = 1$, then by Eq. (C.21):

$$L^{-1}\left\{\frac{1}{s+a}\right\} = e^{-at} \tag{C.22}$$

6. **Time Shift**

$$L\{f(t-a)\} = e^{-as}\hat{f}(s) \tag{C.23a}$$

Conversely:

$$L^{-1}\{e^{-as}\hat{f}(s)\} = f(t-a) \tag{C.23b}$$

This is very useful in obtaining time-domain behavior of time-delay systems.

7. **Complex Differentiation** (or transform of time product)

$$L\{t f(t)\} = -\frac{d}{ds}\hat{f}(s) \tag{C.24a}$$

Conversely:

$$f(t) = \frac{-1}{t} L^{-1} \left\{ \frac{d}{ds} \hat{f}(s) \right\} \tag{C.24b}$$

Typical application: since $L\{e^{-at}\} = 1/(s + a)$, then by Eq. (C.24a):

$$L\{te^{-at}\} = \frac{1}{(s + a)^2} \tag{C.25}$$

Eq. (C.24b) is helpful when the inverse of a transform is not easy to find but the inverse of the derivative is known. For example, if:

$$\hat{f}(s) = \ln\left[\frac{(s + 1)}{(s - 1)}\right]$$

because

$$\frac{d}{ds}\hat{f}(s) = \frac{1}{s + 1} - \frac{1}{s - 1}$$

then by Eq. (C.24b):

$$f(t) = -\frac{1}{t}(e^{-t} - e^{t}) = \frac{1}{t} 2 \sinh t$$

The general form of Eq. (C.24a) is:

$$L\{t^n f(t)\} = (-1)^n \frac{d^n}{ds^n}\hat{f}(s) \tag{C.26}$$

which may be used to deduce, for example, that since $L[1] = 1/s$, then:

$$L\{t^n\} = \frac{n!}{s^{n + 1}}$$

8. **Complex Integration** (or transform of time quotient)

$$L\left\{\frac{f(t)}{t}\right\} = \int_s^\infty \hat{f}(s)\, ds \tag{C.27}$$

9. **Transform of Periodic Functions**

For an arbitrary function $f(t)$ that is periodic with period T (i.e., the behavior of the function is *identical* over the time intervals $[0, T]$, $[T, 2T]$, ..., $[(n - 1)T, nT]$), let:

$$\Phi(s) = \int_0^T f(t)e^{-st}\, dt \tag{C.28}$$

then

$$L\{f(t)\} = \frac{\Phi(s)}{1 - e^{-Ts}} \tag{C.29}$$

For example, if $f(t)$ is a train of spikes of unit height, precisely Δt time units apart, then because $\Phi(s) = 1$, we have in this case that:

$$\hat{f}(s) = \frac{1}{1 - e^{-s\Delta t}} \tag{C.30}$$

(Compare with the z-transform of the sampled unit step function given in Chapter 24.)

10. Final-Value Theorem

$$\lim_{t \to \infty} f(t) = \lim_{s \to 0} [s\,\hat{f}(s)] \tag{C.31}$$

provided $s\,\hat{f}(s)$ does not become infinite for any value of s with a positive real part; i.e., $s\,\hat{f}(s)$ has no positive poles. (The theorem is applicable only to functions for which the limit as $t \to \infty$ exists.) This result is very useful for obtaining the steady-state gain of a (stable) process from its transfer function.

11. Initial-Value Theorem

$$\lim_{t \to 0} f(t) = \lim_{s \to \infty} [s\,\hat{f}(s)] \tag{C.32}$$

applicable only to functions for which the limit as $t \to \infty$ exists. This result was used in Chapter 7 to analyze the behavior of inverse-response systems.

12. The Convolution Theorem

If $L\{f(t)\} = \hat{f}(s)$, and $L\{g(t)\} = \hat{g}(s)$, then:

$$L\left\{\int_0^t f(t - \lambda)g(\lambda)\,d\lambda\right\} = L\left\{\int_0^t f(\lambda)g(t - \lambda)\,d\lambda\right\} = \hat{f}(s)\,\hat{g}(s) \tag{C.33}$$

applicable in obtaining impulse-response models from transform-domain transfer function models. To take full advantage of this theorem, it is helpful also to know the following about the convolution operation.

The convolution of two functions $f(t)$ and $g(t)$ of a real variable t, written symbolically as:

$$\xi(t) = f(t) * g(t) \tag{C.34a}$$

is defined as:

$$\xi(t) = \int_0^t f(t - \lambda)g(\lambda)\,d\lambda \tag{C.34b}$$

It has the following properties:

- Commutativity: $\qquad\qquad\qquad\qquad f(t) * g(t) = g(t) * f(t)$

- Associativity: $\qquad\qquad\qquad [f(t) * g(t)] * h(t) = f(t) * [g(t) * h(t)]$

- Distributivity: $\qquad\qquad [f(t) + g(t)] * h(t) = f(t) * h(t) + g(t) * h(t)$
  (with respect to addition)

The commutativity property is employed, for example, in the first equality in Eq. (C.33).

C.3.3 Solution of Linear Differential Equations

One of the most important applications of Laplace transforms is in the solution of linear differential equations. Consider the general nth-order linear differential equation:

$$a_n \frac{d^n y}{dt^n} + a_{n-1} \frac{d^{n-1} y}{dt^{n-1}} + \dots + a_1 \frac{dy}{dt} + a_0 y = f(t) \tag{C.35}$$

and for simplicity, let us assume that all the initial conditions are zero. Taking Laplace transforms of both sides (and for convenience, simplifying the notation by dropping the ^ used to distinguish the Laplace transformed function), we have, from Eq. (C.18), that:

$$\left(a_n s^n + a_{n-1} s^{n-1} + \dots + a_1 s + a_0 \right) y(s) = f(s) \tag{C.36}$$

which is easily solved for $y(s)$ to give:

$$y(s) = \frac{f(s)}{\left(a_n s^n + a_{n-1} s^{n-1} + \dots + a_1 s + a_0 \right)} \tag{C.37}$$

For any specific forcing function, $f(t)$, with a Laplace transform, $f(s)$, in principle, the solution of the differential equation, $y(t)$, is now that particular function whose Laplace transform appears on the right-hand side of Eq. (C.37).

It is impractical to construct a Laplace transform table general enough to contain *all* possible functions, $y(t)$, and their corresponding transforms, $y(s)$; thus in its present general form, Eq. (C.37) cannot be inverted to obtain the required $y(t)$. However, it is possible to express the right-hand side of Eq. (C.37) as a sum of simpler functions whose inverse transforms are easily found. For example, if for a given $f(s)$, the n roots of the denominator polynomial in Eq. (C.37) are real and distinct, and given as $r_1, r_2, \dots, r_n$, then it is possible to express Eq. (C.37) as:

$$y(s) = \frac{A_1}{s - r_1} + \frac{A_2}{s - r_2} + \dots + \frac{A_n}{s - r_n} \tag{C.38}$$

By the linearity property of Laplace transforms, and by the fact that:

$$L^{-1}\{1/(s-a)\} = e^{at}$$

we now obtain the required $y(t)$ as:

$$y(t) = A_1 e^{r_1 t} + A_2 e^{r_2 t} + \dots + A_n e^{r_n t} \tag{C.39}$$

Observe that the expression in Eq. (C.38) is an expansion of the original complicated function in terms of simpler "partial fractions"; the procedure for converting Eq. (C.37) to the form in Eq. (C.38) is therefore known as *partial fraction expansion*. This is the key step in using Laplace transforms to solve linear differential equations.

C.3.4 Partial Fraction Expansion

This series of procedures involves expressing the rational function, $R(s)$, represented by:

$$R(s) = \frac{P(s)}{Q(s)} \tag{C.40}$$

as a sum of partial fractions. It is assumed that $P(s)$, a polynomial of order p, has no common factors with $Q(s)$, a polynomial of order q; that all the coefficients are real; and that $p < q$; i.e., that $R(s)$ is *strictly proper*. (We will later present a procedure for the situation in which $p = q$, i.e., when $R(s)$ is merely proper.)

The procedures assume that the q roots of $Q(s)$, $r_1, r_2, \dots, r_q$, are known. The possible characteristics of these roots give rise to *three* distinct cases:

1. All q roots are real and distinct
2. A subset of the roots are repeated
3. Some of the roots occur as complex conjugates

CASE 1: Real and distinct roots.

In this case:

$$\frac{P(s)}{Q(s)} = \frac{P(s)}{(s-r_1)(s-r_2)\dots(s-r_q)} = \frac{A_1}{s-r_1} + \frac{A_2}{s-r_2} + \dots + \frac{A_q}{s-r_q} \tag{C.41}$$

and we want to find A_i, $i = 1, 2, \dots, q$. Multiplying by $(s-r_1)$ gives:

$$\frac{(s-r_1)P(s)}{Q(s)} = \frac{P(s)}{(s-r_2)\dots(s-r_q)} = A_1 + \frac{A_2(s-r_1)}{s-r_2} + \dots + \frac{A_q(s-r_1)}{s-r_q} \tag{C.42}$$

and by setting $s = r_1$, we obtain the result:

$$A_1 = \lim_{s \to r_1} \left[(s-r_1) \frac{P(s)}{Q(s)} \right] \tag{C.43}$$

In general:

$$A_i = \lim_{s \to r_i} \left[(s - r_i) \frac{P(s)}{Q(s)} \right] \qquad (C.44)$$

Thus, if we let $\Phi_i(s)$ represent what remains of $R(s)$ after the factor $(s - r_i)$ has been removed from the denominator polynomial $Q(s)$, i.e.:

$$\Phi_i(s) = (s - r_i) R_i(s) \qquad (C.45)$$

then Eq. (C.44) translates to:

$$A_i = \Phi_i(r_i) \qquad (C.46)$$

Example C.1   PARTIAL FRACTION EXPANSION: DISTINCT ROOTS.

Obtain a partial fraction expansion for the function:

$$R(s) = \frac{(s + 5)}{(s + 2)(s + 3)}$$

Solution:

$$\frac{(s + 5)}{(s + 2)(s + 3)} = \frac{A_1}{s + 2} + \frac{A_2}{s + 3}$$

Multiplying by $(s + 2)$ and setting $s = -2$ we obtain immediately $A_1 = 3$. Similarly, multiplying by $(s + 3)$ and setting $s = -3$ gives $A_2 = -2$.

Example C.2   PARTIAL FRACTION EXPANSION: DISTINCT ROOTS.

Obtain a partial fraction expansion for the function:

$$R(s) = \frac{(s^2 + 5s + 1)}{s(s^2 + 5s + 6)}$$

Solution:

Factoring the denominator quadratic gives:

$$R(s) = \frac{(s^2 + 5s + 1)}{s(s + 2)(s + 3)}$$

so that:

$$\frac{(s^2 + 5s + 1)}{s(s + 2)(s + 3)} = \frac{A_1}{s} + \frac{A_2}{s + 2} + \frac{A_3}{s + 3}$$

Clearly:

$$\Phi_1(s) = (s^2 + 5s + 1)/(s + 2)(s + 3)$$
$$\Phi_2(s) = (s^2 + 5s + 1)/s(s + 3)$$
$$\Phi_3(s) = (s^2 + 5s + 1)/s(s + 2)$$

Thus:

$$A_1 = \Phi_1(0) = \frac{1}{6}$$

$$A_2 = \Phi_2(-2) = \frac{5}{2}$$

$$A_3 = \Phi_3(-3) = -\frac{5}{3}$$

are the required coefficients. This is the partial fraction expansion result used in Example 3.3 of Chapter 3.

CASE 2: Repeated roots.

In this case, we consider that $Q(s)$ has $q - k$ distinct roots and one root, r_m, is repeated k times, thus:

$$R(s) = \frac{P(s)}{Q(s)} = \left(\frac{A_1}{s - r_1} + \dots + \frac{A_{q-k}}{s - r_{q-k}} \right) + \frac{P^*(s)}{(s - r_m)^k} \qquad \text{(C.47a)}$$

or

$$R(s) = R^0(s) + R^*(s) \qquad \text{(C.47b)}$$

where $R^0(s)$ corresponds to the sum of unrepeated factors in winged brackets.

Observe that the coefficients A_i, $i = 1, 2, \dots, q - k$ can be found as in Case 1. Now expand the remaining term as follows:

$$R^*(s) = \frac{P^*(s)}{(s - r_m)^k} = \frac{B_k}{(s - r_m)^k} + \frac{B_{k-1}}{(s - r_m)^{k-1}} + \dots + \frac{B_2}{(s - r_m)^2} + \frac{B_1}{(s - r_m)} \qquad \text{(C.48)}$$

Multiplying by $(s - r_m)^k$ in Eq. (C.47a), using Eq. (C.48), and setting $s = r_m$ gives:

$$B_k = \lim_{s \to r_m} \left[(s - r_m)^k \frac{P(s)}{Q(s)} \right] \qquad \text{(C.49)}$$

To simplify the notation, if we define:

$$\Phi^*(s) = (s - r_m)^k \frac{P(s)}{Q(s)} = (s - r_m)^k R(s) \qquad \text{(C.50)}$$

i.e., what is left of $R(s)$ after $(s - r_m)^k$ has been removed, then we have:

$$B_k = \Phi^*(r_m) \qquad \text{(C.51)}$$

The other B_i coefficients are obtained by differentiation. Observe from Eqs. (C.47) and (C.48) that:

$$(s - r_m)^k R(s) = (s - r_m)^k R^0(s) + B_k + (s - r_m)B_{k-1} + (s - r_m)^2 B_{k-2} + \dots +$$

$$(s - r_m)^{k-2} B_2 + (s - r_m)^{k-1} B_1 \qquad \text{(C.52)}$$

And now, by differentiating with respect to s and setting $s = r_m$, we obtain:

$$\lim_{s \to r_m} \frac{d}{ds} \left[(s - r_m)^k R(s) \right] = B_{k-1} \tag{C.53}$$

or

$$B_{k-1} = \left[\frac{d}{ds} \Phi^*(s) \right]_{s = r_m} \tag{C.54}$$

The others are determined similarly by repeated differentiation in Eq. (C.52). The result is:

$$B_j = \frac{1}{(k-j)!} \left[\frac{d^{(k-j)} \Phi^*(s)}{ds^{(k-j)}} \right]_{s = r_m} \tag{C.55}$$

It is straightforward to generalize these results to the case with several sets of repeated roots.

Example C.3 PARTIAL FRACTION EXPANSION: REPEATED ROOTS.

Obtain a partial fraction expansion for the function:

$$R(s) = \frac{1}{s(s^2 + 6s + 9)}$$

Solution:

Factoring the denominator quadratic gives:

$$R(s) = \frac{1}{s(s + 3)^2}$$

which we wish to expand as:

$$\frac{1}{s(s + 3)^2} = \frac{A_1}{s} + \frac{B_2}{(s + 3)^2} + \frac{B_1}{(s + 3)}$$

In this case, the repeated root is at $s = 3$, and:

$$P(s) = 1$$

$$Q(s) = s(s + 3)^2$$

Also:

$$\Phi^*(s) = \frac{1}{s}$$

Thus, we obtain, in the usual manner (multiplying by s and setting $s = 0$):

$$A_1 = \frac{1}{9}$$

Next, we obtain:

$$B_2 = \Phi^*(-3) = -\frac{1}{3}$$

and finally, since:

$$\frac{d}{ds}\Phi^*(s) = -\frac{1}{s^2}$$

then we have:

$$B_1 = -\frac{1}{9}$$

The complete partial fraction expansion is therefore given by:

$$\frac{1}{s(s+3)^2} = \frac{1/9}{s} + \frac{-1/3}{(s+3)^2} + \frac{-1/9}{(s+3)}$$

CASE 3: Complex conjugate roots.

In this case:

$$R(s) = \frac{P(s)}{Q(s)} = \frac{C_1}{(s-r)} + \frac{C_2}{(s-r^*)} + R^+(s) \tag{C.56}$$

where r and r^* are complex conjugate roots:

$$r, r^* = a \pm bj \tag{C.57}$$

and $R^+(s)$ is the remaining portion of $R(s)$. If we let $\Phi^+(s)$ be what is left of $R(s)$ after $(s-r)(s-r^*)$ has been removed, i.e.:

$$\Phi^+(s) = (s-r)(s-r^*)R(s) \tag{C.58a}$$
$$= [(s-a)^2 + b^2]R(s) \tag{C.58b}$$

then C_1 and C_2 are obtained in the usual manner, by multiplying in turn first by $(s-r)$, and setting $s = r$, and then by $(s-r^*)$ and setting $s = r^*$; the result is:

$$C_1 = \left[\frac{\Phi^+(r)}{(r-r^*)}\right] \tag{C.59}$$

$$C_2 = \left[\frac{\Phi^+(r^*)}{(r^*-r)}\right] \tag{C.60}$$

It is easy to show that C_2 will always be the complex conjugate of C_1, so that in actual fact, only one of them need be evaluated.

Example C.4   PARTIAL FRACTION EXPANSION: COMPLEX ROOTS.

Obtain a partial fraction expansion for the function:

$$R(s) = \frac{4}{s(s^2+2s+2)}$$

Solution:

Factoring the denominator quadratic gives:

$$R(s) = \frac{4}{s(s-r)(s-r^*)}$$

with

$$r = -(1+j); \quad r^* = -(1-j)$$

and we wish to expand this as:

$$\frac{4}{s(s-r)(s-r^*)} = \frac{A_1}{s} + \frac{C_1}{(s+1+j)} + \frac{C_2}{(s+1-j)}$$

In this case:

$$\Phi^+(s) = \frac{4}{s}$$

and since $r - r^* = -2j$, we obtain from Eq. (C.59) that:

$$C_1 = -1 - j$$

and from here that:

$$C_2 = -1 + j$$

We obtain A_1 in the usual manner (multiply by s, set $s = 0$) to be 2. The complete partial fraction expansion is therefore given by:

$$\frac{4}{s(s-r)(s-r^*)} = \frac{2}{s} + \frac{(-1-j)}{(s+1+j)} + \frac{(-1+j)}{(s+1-j)}$$

Note from Eqs. (C.59) and (C.60), and also from the example, that C_1 and C_2 will naturally occur as complex numbers. This is awkward if the end result is not merely partial fraction expansion, but the Laplace inversion to obtain $R(t)$. For this reason, Eq. (C.56) is often consolidated to:

$$R(s) = \frac{P(s)}{Q(s)} = \frac{C_1(s-r^*) + C_1^*(s-r)}{(s-a)^2 + b^2} + R^+(s) \tag{C.61}$$

where we have made use of the fact that $C_2 = C_1^*$, the complex conjugate of C_1. It is now easy to show that Eq. (C.61) simplifies to:

$$R(s) = 2\alpha\frac{(s-a)}{(s-a)^2 + b^2} - 2\beta\frac{b}{(s-a)^2 + b^2} + R^+(s) \tag{C.62}$$

where

$$\alpha = \text{Re}(C_1) = \text{Re}\left[\frac{\Phi^+(r)}{(r-r^*)}\right] \tag{C.63a}$$

$$\beta = \text{Im}(C_1) = \text{Im}\left[\frac{\Phi^+(r)}{(r - r^*)}\right] \qquad (C.63b)$$

The Laplace inversion of Eq. (C.62) will therefore be:

$$R(t) = 2e^{at}(\alpha \cos bt - \beta \sin bt) + R^+(t) \qquad (C.64)$$

Thus, for the function in Example C.4, $a = -1$, $b = -1$, $\alpha = -1$, $\beta = -1$, and the inverse transform is:

$$R(t) = 2 + 2 e^{-t} [\sin(-t) - \cos(-t)]$$

or

$$R(t) = 2 - 2 e^{-t} (\sin t + \cos t)$$

CASE 4: R(s) *not* strictly proper.

When $p = q$ (i.e., the numerator and denominator polynomials of $R(s)$ are of equal order), $R(s)$ is now merely proper. The procedure for handling this case calls for expressing $R(s)$ as:

$$R(s) = A_0 + R_1(s) \qquad (C.64)$$

where $R_1(s)$ is now strictly proper and can therefore be expanded as before. It is easy to show that A_0 is given by:

$$A_0 = \lim_{s \to \infty} R(s) \qquad (C.65)$$

Example C.5 PARTIAL FRACTION EXPANSION: R(s) NOT STRICTLY PROPER.

Obtain a partial fraction expansion for the function:

$$R(s) = \frac{(s^2 + 4s + 3)}{(s^2 + 9s + 20)}$$

Solution:

Factoring both the numerator and the denominator quadratics we obtain:

$$R(s) = \frac{(s + 3)(s + 1)}{(s + 4)(s + 5)}$$

which is *not* strictly proper. First we obtain A_0 by dividing numerator and denominator by s^2, before taking the limit as $s \to \infty$; giving the result $A_0 = 1$. Thus, we may expand $R(s)$ as follows:

$$\frac{(s + 3)(s + 1)}{(s + 4)(s + 5)} = 1 + \frac{A_1}{(s + 4)} + \frac{A_2}{(s + 5)}$$

obtaining in the usual manner that $A_1 = 3$, and $A_2 = -8$. The complete partial fraction expansion is now given by:

$$R(s) = 1 + \frac{3}{(s + 4)} - \frac{8}{(s + 5)}$$

The transfer function of the lead/lag system is an example of a rational function which is *not* strictly proper.

Reference [1] should be consulted for more information about the Laplace transform and its application in engineering. A table of Laplace transforms of some common functions is provided below in Table C.1. References [2, 3] contain more extensive tables of Laplace transforms.

Table C.1. Table of Laplace Transforms of Some Common Functions

	$\hat{f}(s)$	$f(t)$
1.	$\dfrac{1}{s}$	1
2.	$\dfrac{1}{s^2}$	t
3.	$\dfrac{1}{s^n}$	$\dfrac{t^{n-1}}{(n-1)!}$
4.	$s^{-3/2}$	$2\sqrt{\dfrac{t}{\pi}}$
5.	$\dfrac{\Gamma(k)}{s^k}\ (k \geq 0)$	t^{k-1}
6.	$\dfrac{\Gamma(k)}{(s+a)^k}\ (k \geq 0)$	$t^{k-1}e^{-at}$
7a.	$\dfrac{1}{s+a}$	e^{-at}
7b.	$\dfrac{1}{\tau s + 1}$	$\dfrac{1}{\tau}e^{-t/\tau}$
8.	$\dfrac{1}{s(\tau s + 1)}$	$1 - e^{-t/\tau}$
9a.	$\dfrac{1}{(s+a)^2}$	te^{-at}
9b.	$\dfrac{1}{(\tau s + 1)^2}$	$\dfrac{t}{\tau^2}e^{-t/\tau}$
10.	$\dfrac{1}{s(\tau s + 1)^2}$	$1 - (1 + t/\tau)e^{-t/\tau}$
11a.	$\dfrac{1}{(s+a)^n}\ (n = 1, 2, \ldots)$	$\dfrac{1}{(n-1)!}t^{n-1}e^{-at}$
11b.	$\dfrac{1}{(\tau s + 1)^n}\ (n = 1, 2, \ldots)$	$\dfrac{1}{\tau^n(n-1)!}t^{n-1}e^{-t/\tau}$
12.	$\dfrac{1}{(s+a)(s+b)}$	$\dfrac{1}{(b-a)}\left(e^{-at} - e^{-bt}\right)$

Table C.1. Continued

	$\hat{f}(s)$	$f(t)$
13.	$\dfrac{s}{(s + a)(s + b)}$	$\dfrac{1}{(b - a)}\left(be^{-bt} - ae^{-at}\right)$
14.	$\dfrac{1}{s(s + a)(s + b)}$	$\dfrac{1}{ab}\left[1 + \dfrac{1}{(a - b)}\left(be^{-at} - ae^{-bt}\right)\right]$
15.	$\dfrac{b}{s^2 + b^2}$	$\sin bt$
16.	$\dfrac{b}{s^2 - b^2}$	$\sinh bt$
17.	$\dfrac{s}{s^2 + b^2}$	$\cos bt$
18.	$\dfrac{s}{s^2 - b^2}$	$\cosh bt$
19.	$\dfrac{b^2}{s\,(s^2 + b^2)}$	$(1 - \cos bt)$
20.	$\dfrac{b}{(s + a)^2 + b^2}$	$e^{-at}\sin bt$
21.	$\dfrac{s + a}{(s + a)^2 + b^2}$	$e^{-at}\cos bt$
22.	$\hat{f}(s + a)$	$e^{-at}f(t)$
23.	$e^{-bs}\hat{f}(s)$	$f(t - b)$; with $f(t) = 0$ for $t < 0$

C.4   THE z-TRANSFORM

Just as the Laplace transform arises from specializing the Fourier transform for a one-sided (semi-infinite) independent variable, we will now show that the z-transform derives from specializing the Laplace transform for sampled functions that take nonzero values only at discrete points in time. Such a preamble is necessary if the *apparent nonsymmetry* between the z-transform and its inversion formula is to be understood properly.

C.4.1 Definition and Inverse Formula

Consider the situation in which the continuous function $f(t)$ in the Laplace transform expression Eq. (C.12) is sampled so that it takes on finite values $f(t_k)$ at discrete points, $t_0, t_1, ..., t_i, ...$ in time, at regular intervals of size Δt. Let this sampled version of $f(t)$ be $f^*(t)$. Then $f^*(t) = f(t_k)$ only at the sample point $t_k = k\Delta t$, for $k = 0, 1, 2, ...$; it is zero elsewhere. It therefore appears as a "train" of spikes at the discrete sample points t_0, t_1, ..., t_i, ..., and may therefore be represented mathematically as:

$$f^*(t) = \sum_{k=0}^{\infty} f(t_k)\, \delta(t - t_k) \tag{C.66}$$

where $\delta(t - t_k)$ is the delta function.

We now wish to find the Laplace transform of this function by introducing Eq. (C.66) into Eq. (C.12):

$$L\{f^*(t)\} = \int_0^{\infty} e^{-st}\left[\sum_{k=0}^{\infty} f(t_k)\, \delta(t - t_k)\right] dt \tag{C.67}$$

Now, by the linearity property of Laplace transforms, we know that the transform of a sum is the sum of the individual transforms; thus Eq. (C.67) becomes:

$$L\{f^*(t)\} = \sum_{k=0}^{\infty}\left[\int_0^{\infty} e^{-st}\, f(t_k)\, \delta(t - t_k)\, dt\right] \tag{C.68}$$

The integral is easily evaluated by invoking the property of the delta function, and Eq. (C.68) becomes:

$$L\{f^*(t)\} = \sum_{k=0}^{\infty} f(t_k)\, e^{-st_k}$$

or, by replacing t_k with $k\,\Delta t$, we obtain finally:

$$\hat{f}^*(s) = L\{f^*(t)\} = \sum_{k=0}^{\infty} f(t_k) e^{-sk\Delta t} \tag{C.69}$$

This should now be compared with the Laplace transform of the continuous function, $f(t)$:

$$\hat{f}(s) = \int_0^{\infty} f(t)\, e^{-st}\, dt \tag{C.12}$$

where we observe that Eq. (C.69) is a discretized version of Eq. (C.12).

If now *for convenience* (to simplify the expression in Eq. (C.69)), we introduce the notation:

$$z = e^{-s\Delta t} \tag{C.70}$$

and refer to the resulting $L[f^*(t)]$ as the "z-transform" of $f^*(t)$, then Eq. (C.69) becomes:

$$\hat{f}(z) = Z\{f^*(t)\} = \sum_{k=0}^{\infty} f(t_k) z^{-k} \tag{C.71}$$

which is the "formal" definition of the z-transform of the sampled signal $f^*(t)$.

Because Eq. (C.69) is a valid Laplace transform, its inverse transform is given by Eq. (C.13); and upon introducing the transformation in Eq. (C.70), we obtain:

$$f(t_k) = f^*(t) = \frac{1}{2\pi j} \int_{\zeta_1}^{\zeta_2} \hat{f}(z) \, z^k \, \frac{dz}{z} \tag{C.72}$$

where

$$\zeta_1 = e^{-(\alpha - j\omega)\Delta t}$$
$$\zeta_2 = e^{-(\alpha + j\omega)\Delta t}$$

Thus unlike the other *integral* transforms for which the transform and its inverse appear symmetric, the z-transform and its inverse appear to be nonsymmetric. However, this is merely an artifact of the notational convenience employed: in its "fundamental" (if not entirely practical) form, the transform is actually as shown in Eq. (C.67), and it is as symmetrical with the inverse in Eq. (C.72), as Eqs. (C.12) and (C.13) are.

As with Laplace transforms, it should be noted that the inverse transform formula is hardly ever used for inverting a z-transform. It is customary to use Tables of z-transforms, an example of which is presented at the end of this section.

C.4.2   Properties and Useful Theorems of z -Transforms

The following are some properties and theorems of the z-transform that are particularly useful in engineering applications. Some of them have precise parallels with the Laplace transform, and some do not. Because of the peculiar nature of discrete signals, some of these properties have no equivalent for the Laplace transform; by the same token some of the properties of the Laplace transform also have no *precise* z-transform equivalent.

1.   Linearity

$$Z\{c_1 f_1(k) + c_2 f_2(k)\} = c_1 Z\{f_1(k)\} + c_2 Z\{f_2(k)\} \tag{C.73}$$

Extension to linear combination of n functions is straightforward.

2.   Time Shift to the Left

If $Z\{f(k)\} = \hat{f}(z)$, then:

$$Z\{f(k+1)\} = z\hat{f}(z) - z f(0) \tag{C.74}$$

and if $f(0) = 0$, then:

$$Z\{f(k+1)\} = z\hat{f}(z) \tag{C.75}$$

Also:

$$Z\{f(k+2)\} = z^2\hat{f}(z) - z^2f(0) - z f(1) \tag{C.76}$$

and in general:

$$Z\{f(k+m)\} = z^m\hat{f}(z) - z^mf(0) - z^{m-1}f(1) - \ldots - zf(m-1) \tag{C.77}$$

or

$$Z\{f(k+m)\} = z^m\left[\hat{f}(z) - \sum_{k=0}^{m-1}f(k)z^{-k}\right] \tag{C.78}$$

3. Time Shift to the Right

$$Z\{f(k-m)\} = z^{-m}\hat{f}(z) \tag{C.79}$$

4. Transform of (Forward and Backward) Differences

Define the *forward difference* operator, Δ, as:

$$\Delta f(k) = f(k+1) - f(k) \tag{C.80}$$

then we may make use of Eq. (C.74) to obtain:

$$Z\{\Delta f(k)\} = (z-1)\hat{f}(z) - zf(0) \tag{C.81}$$

Similarly, since from Eq. (C.80):

$$\Delta^2 f(k) = \Delta(\Delta f(k)) = f(k+2) - 2f(k+1) + f(k) \tag{C.82}$$

then

$$Z\{\Delta^2 f(k)\} = (z-1)^2\hat{f}(z) - z(z-1)f(0) - z \Delta f(0) \tag{C.83}$$

In general:

$$Z\{\Delta^m f(k)\} = (z-1)^m\hat{f}(z) - z \sum_{i=0}^{m-1}(z-1)^{m-i-1}\Delta^i f(0) \tag{C.84}$$

Define the *backward difference* operator, ∇, as:

$$\nabla f(k) = f(k) - f(k-1) \tag{C.85}$$

Then we may make use of Eq. (C.79) to obtain:

$$Z\{\nabla f(k)\} = (1 - z^{-1})\hat{f}(z) \tag{C.86}$$

Similarly, since from Eq. (C.85):

$$\nabla^2 f(k) = \nabla(\nabla f(k)) = f(k) - 2f(k-1) + f(k-2) \tag{C.86}$$

then

$$Z\{\nabla^2 f(k)\} = (1 - z^{-1})^2 \hat{f}(z) \qquad \text{(C.87)}$$

and in general:

$$Z\{\nabla^m f(k)\} = (1 - z^{-1})^m \hat{f}(z) \qquad \text{(C.88)}$$

5. Complex Shift

If the function $f(t)$ is sampled with sample time Δt to obtain $f(k)$, then the sampled version of the function $e^{-at}f(t)$ is $r^{-k}f(k)$ where:

$$r = e^{a\Delta t}$$

We now have the result:

$$Z\{r^{-k}f(k)\} = \hat{f}(rz) \qquad \text{(C.89)}$$

and the converse:

$$Z^{-1}\{\hat{f}(rz)\} = r^{-k}f(k) \qquad \text{(C.90)}$$

Typical application: since $Z^{-1}\{1/(1 - z^{-1})\} = 1$, then by Eq. (C.90):

$$Z^{-1}\left\{\frac{1}{1 - pz^{-1}}\right\} = (p^{-1})^{-k} = p^k \qquad \text{(C.91)}$$

(cf. Example 24.3)

6. Transform of Sums

$$Z\left\{\sum_{i=0}^{k} f(i)\right\} = \frac{1}{1 - z^{-1}}\hat{f}(z) \qquad \text{(C.92)}$$

7. Complex Differentiation

$$Z\{k\,f(k)\} = -z\frac{d}{dz}\hat{f}(z) \qquad \text{(C.93)}$$

and in general:

$$Z\{k^n f(k)\} = (-1)^n\left[z\,\frac{d^n}{dz^n}\,\hat{f}(z)\right] \qquad \text{(C.94)}$$

Typical application: since the z-transform of the unit step function $f(k) = 1(k)$ is known to be $1/(1 - z^{-1})$, then by Eq. (C.93), the z-transform of the unit ramp function $f(k) = k$ is obtained as:

$$Z\{k\} = Z\{k\,1(k)\} = -z\frac{d}{dz}\left(\frac{1}{1-z^{-1}}\right)$$

$$= \frac{z^{-1}}{(1-z^{-1})^{-2}}$$

8. Complex Integration

$$Z\left\{\frac{f(k)}{k}\right\} = \int_z^\infty \frac{\hat{f}(\zeta)}{\zeta}\,d\zeta + \lim_{k\to 0}\frac{f(k)}{k} \tag{C.95}$$

9. Final-Value Theorem

$$\lim_{t\to\infty} f^*(t) = \lim_{k\to\infty} f(k) = \lim_{z\to 1}\left[(1-z^{-1})\hat{f}(z)\right] \tag{C.96}$$

provided $(1-z^{-1})\hat{f}(z)$ has no poles outside of the unit disk. (The theorem is applicable only to functions for which the limit as $k\to\infty$ exists.) As with the Laplace transform equivalent, this result is very useful for obtaining the steady-state gain of a (stable) process from its pulse transfer function (incorporating the ZOH element).

10. Initial-Value Theorem

$$\lim_{t\to 0} f^*(t) = \lim_{k\to 0} f(k) = \lim_{z\to\infty}[\hat{f}(z)] \tag{C.97}$$

applicable only to functions for which the limit as $z\to\infty$ exists.

11. The Discrete Convolution Theorem

If $Z\{f(k)\} = \hat{f}(z)$, and $Z\{g(k)\} = \hat{g}(z)$, then:

$$Z\left\{\sum_{i=0}^\infty f(k-i)g(i)\right\} = Z\left\{\sum_{i=0}^\infty f(i)g(k-i)\right\} = \hat{f}(z)\,\hat{g}(z) \tag{C.98}$$

This is a precise discrete equivalent of the convolution theorem of Laplace transforms given as Property 12 in the previous section. The result is applicable in obtaining impulse-response models from pulse transfer functions.

C.4.3   Solution of Linear Difference Equations

As with the Laplace transform, the z-transform finds application in the solution of linear difference equations. Consider the linear difference equation:

$$y(k) + a_1 y(k-1) + \ldots + a_n y(k-n) = b_1 u(k-1) + \ldots + b_m u(k-m) \tag{C.99}$$

By taking z-transforms of both sides (and dropping the $\wedge$ for convenience), we obtain:

$$\left(1 + a_1 z^{-1} + \ldots + a_n z^{-n}\right) y(z) = \left(b_1 z^{-1} + \ldots + b_m z^{-m}\right) u(z)$$

which is solved for $y(z)$ to yield:

$$y(z) = \frac{\left(b_1 z^{-1} + \ldots + b_m z^{-m}\right)}{\left(1 + a_1 z^{-1} + \ldots + a_n z^{-n}\right)} u(z) \qquad (C.100)$$

Thus the solution to Eq. (C.99) for any given (z-transformable) sequence $u(k)$ is given by that function $y(k)$ whose z-transform is given in Eq. (C.100). Thus, as with the Laplace transform, the main challenge in finding such solutions lies in inverting the indicated z-transform.

The technique of z-transform inversion by long division (or, equivalently expansion as an infinite series) was discussed in Chapter 24. It is not as popular as the technique of partial fraction expansion.

C.4.4   Partial Fraction Expansion

The procedure for obtaining partial fraction expansions is valid for all rational polynomials $R(\bullet)$ regardless of the argument. Thus, what was presented in Section C.3.4 applies directly once we replace s by z (or by z^{-1}). We will therefore merely illustrate with some specific examples.

Example C.6   **SOLUTION OF A FIRST-ORDER DIFFERENCE EQUATION BY z-TRANSFORMS, AND PARTIAL FRACTION EXPANSION.**

Use z-transforms to solve the first-order difference equation:

$$y(k) = 0.7y(k-1) + 1.5u(k-1) \qquad (C.101)$$

given that $u(k) = 1$ for all k. (That is, find the unit step response of the first-order system whose model is as in Eq. (C.101).)

Solution:

By taking z-transforms of both sides and rearranging, we obtain:

$$y(z) = \frac{1.5z^{-1}}{1 - 0.7z^{-1}} u(z)$$

or, introducing the z-transform of the unit step function for $u(z)$:

$$y(z) = \frac{1.5z^{-1}}{(1 - z^{-1})(1 - 0.7z^{-1})} \qquad (C.102)$$

We now wish to expand the rational function of z^{-1} as:

$$\frac{1.5z^{-1}}{(1 - z^{-1})(1 - 0.7z^{-1})} = \frac{A_1}{(1 - z^{-1})} + \frac{A_2}{(1 - 0.7z^{-1})} \qquad (C.103)$$

Since the roots of the denominator polynomial are real and distinct (they are located at $z = 1$ and $z = 0.7$), we proceed as in Section C.3.4: multiplying in Eq. (C.103) by $(1 - z^{-1})$

and setting $z = 1$ we obtain immediately $A_1 = 5$. Similarly, multiplying by $(1 - 0.7z^{-1})$ and setting $s = 0.7$ gives $A_2 = -5$.

Thus, by partial fraction expansion Eq. (C.102) becomes:

$$y(z) = \frac{5}{(1 - z^{-1})} - \frac{5}{(1 - 0.7z^{-1})} \tag{C.104}$$

from where the required solution is obtained from the inverse z-transform of Eq. (C.104) as:

$$y(k) = 5\left[1 - (0.7)^k\right] \tag{C.105}$$

The next example illustrates how to handle repeated roots.

Example C.7   PARTIAL FRACTION EXPANSION: REPEATED ROOTS IN z-TRANSFORMS.

Obtain a partial fraction expansion of the function:

$$y(z) = \frac{0.8 - 0.3z^{-1}}{(1 - z^{-1})^2(1 - 0.5z^{-1})} \tag{C.106}$$

Solution:

Because of the repeated pole at $z = 1$, we wish to expand the RHS of Eq. (C.106) as:

$$\frac{0.8 - 0.3z^{-1}}{(1 - z^{-1})^2(1 - 0.5z^{-1})} = \frac{A_1}{(1 - 0.5z^{-1})} + \frac{B_2}{(1 - z^{-1})^2} + \frac{B_1}{(1 - z^{-1})}$$

We obtain $A_1 = 0.2$ in the usual manner: multiply by $(1 - 0.5z^{-1})$ and set $z = 0.5$. And now:

$$\Phi^*(z) = \frac{0.8 - 0.3z^{-1}}{(1 - 0.5z^{-1})}$$

and from it we obtain B_2 immediately as 1.0 by setting $z = 1$. Next, by differentiation, we obtain:

$$\frac{d}{dz}\Phi^*(z) = \frac{-0.1}{(z - 0.5)^2}$$

from where we now obtain B_1 as -0.4 by setting $z = 1$. The required complete partial fraction expansion is therefore:

$$\frac{0.8 - 0.3z^{-1}}{(1 - z^{-1})^2(1 - 0.5z^{-1})} = \frac{0.2}{(1 - 0.5z^{-1})} + \frac{1}{(1 - z^{-1})^2} + \frac{-0.4}{(1 - z^{-1})} \tag{C.107}$$

which the reader may verify by longhand calculation.

The references given below should be consulted for more information about the theory of the z-transform, and its various applications.

A table of z-transforms of some common functions is provided below in Table C.2; Table C.3 contains some common s transfer functions and the corresponding pulse transfer functions *incorporating the zero-order hold element*.

Table C.2 Table of z-Transforms of Some Common Functions
(Sample Time = Δt)

	$\hat{f}(s)$	$f(t)$	$f(k)$	$\hat{f}(z) = Z\{f(k)\}$
1.	$\dfrac{1}{s}$	$1(t)$	$1(k)$	$\dfrac{1}{1 - z^{-1}}$
2.	$\dfrac{1}{s^2}$	t	$k\Delta t$	$\dfrac{\Delta t z^{-1}}{(1 - z^{-1})^2}$
3.	$\dfrac{(n-1)!}{s^n}$	t^{n-1}	$(k\Delta t)^{n-1}$	$\displaystyle\lim_{a \to 0} (-1)^{n-1} \dfrac{\partial^{n-1}}{\partial a^{n-1}} \left(\dfrac{z}{z - e^{-a\Delta t}} \right)$
4a.	$\dfrac{1}{s+a}$	e^{-at}	$e^{-ak\Delta t}$	$\dfrac{1}{1 - e^{-a\Delta t}z^{-1}}$
4b.	—	—	p^k	$\dfrac{1}{1 - pz^{-1}}$
5.	$\dfrac{1}{(s+a)^2}$	te^{-at}	$k\Delta t e^{-ak\Delta t}$	$\dfrac{\Delta t e^{-a\Delta t}z^{-1}}{(1 - e^{-a\Delta t}z^{-1})^2}$
6.	$\dfrac{s}{(s+a)^2}$	$(1 - at)e^{-at}$	$(1 - ak\Delta t)e^{-ak\Delta t}$	$\dfrac{1 - (1 + a\Delta t)e^{-a\Delta t}z^{-1}}{(1 - e^{-a\Delta t}z^{-1})^2}$
7a.	$\dfrac{a}{s(s+a)}$	$1 - e^{-at}$	$1 - e^{-ak\Delta t}$	$\dfrac{(1 - e^{-a\Delta t})z^{-1}}{(1 - z^{-1})(1 - e^{-a\Delta t}z^{-1})}$
7b.	—	—	$1 - p^k$	$\dfrac{(1 - p)z^{-1}}{(1 - z^{-1})(1 - pz^{-1})}$
8a.	$\dfrac{b-a}{(s+a)(s+b)}$	$e^{-at} - e^{-bt}$	$e^{-ak\Delta t} - e^{-bk\Delta t}$	$\dfrac{(e^{-a\Delta t} - e^{-b\Delta t})z^{-1}}{(1 - e^{-a\Delta t}z^{-1})(1 - e^{-b\Delta t}z^{-1})}$
8b.	—	—	$(p_1)^k - (p_2)^k$	$\dfrac{(p_1 - p_2)z^{-1}}{(1 - p_1 z^{-1})(1 - p_2 z^{-1})}$
9.	$\dfrac{b}{(s^2 + b^2)}$	$\sin bt$	$\sin bk\Delta t$	$\dfrac{z^{-1}\sin b\Delta t}{(1 - 2z^{-1}\cos b\Delta t + z^{-2})}$
10.	$\dfrac{s}{(s^2 + b^2)}$	$\cos bt$	$\cos bk\Delta t$	$\dfrac{1 - z^{-1}\cos b\Delta t}{(1 - 2z^{-1}\cos b\Delta t + z^{-2})}$

Table C.3. Some Common Continuous Transfer Functions and Equivalent Pulse Transfer Functions with Zero-Order Hold Sampling. (Sample Time = Δt)

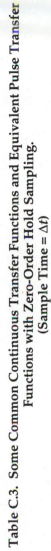

$$g(z) = \frac{b_1 z^{-1} + b_2 z^{-2}}{1 + a_1 z^{-1} + a_2 z^{-2}}$$

	$g(s)$	$g(z)$	$g(k)$
1.	$\dfrac{1}{s}$	$b_1 = \Delta t;\ b_2 = 0$ $a_1 = -1;\ a_2 = 0$	$\Delta t\,1^{k-1}$
2.	$\dfrac{1}{s^2}$	$b_1 = \dfrac{\Delta t^2}{2};\ b_2 = \dfrac{\Delta t^2}{2}$ $a_1 = -2;\ a_2 = 1$	$\dfrac{k\Delta t^2}{2} + \dfrac{(k-1)\,\Delta t^2}{2}$
3.	$\dfrac{1}{s^n}$	$(1 - z^{-1})\,\lim\limits_{a\to 0}\dfrac{(-1)^n}{n!}\,\dfrac{\partial^n}{\partial a^n}\left(\dfrac{z}{z - e^{-a\Delta t}}\right)$	$\dfrac{(k\Delta t)^n}{n!} - \dfrac{[(k-1)\,\Delta t]^n}{n!}$
4.	$\dfrac{1}{\tau s + 1}$	$b_1 = (1 - e^{-\Delta t/\tau});\ b_2 = 0$ $a_1 = -e^{-\Delta t/\tau};\ a_2 = 0$	$b_1 p^{k-1}$ $b_1 = (1 - p);\ p = e^{-\Delta t/\tau}$ **or:** $(p^{k-1} - p^k)$

Table C.3. Continued

	$g(s)$	$g(z) = \dfrac{b_1 z^{-1} + b_2 z^{-2}}{1 + a_1 z^{-1} + a_2 z^{-2}}$	$g(k)$
5.	$\dfrac{1}{s(\tau s + 1)}$	$b_1 = \tau\left(\dfrac{\Delta t}{\tau} - 1 + e^{-\Delta t/\tau}\right);$ $b_2 = \tau\left(1 - e^{-\Delta t/\tau} - \dfrac{\Delta t}{\tau}\, e^{-\Delta t/\tau}\right)$ $a_1 = -(1 + e^{-\Delta t/\tau})$ $a_2 = e^{-\Delta t/\tau}$	$\dfrac{b_1}{1-p}(1 - p^k) + \dfrac{b_2}{1-p}(1 - p^{k-1})$ $b_1 = \tau\left(\dfrac{\Delta t}{\tau} - 1 + p\right)$ $b_2 = \tau\left(1 - p - \dfrac{\Delta t}{\tau}\, p\right)$ $p = e^{-\Delta t/\tau}$ **or:** $\Delta t - \tau(p^{k-1} - p^k)$
6.	$\dfrac{ab}{(s + a)(s + b)}$ $(a \neq b)$	$b_1 = \dfrac{[b(1 - e^{-a\Delta t}) - a(1 - e^{-b\Delta t})]}{(b - a)}$ $b_2 = \dfrac{[ae^{-a\Delta t}(1 - e^{-b\Delta t}) - be^{-b\Delta t}(1 - e^{-a\Delta t})]}{(b - a)}$ $a_1 = -(e^{-a\Delta t} + e^{-b\Delta t})$ $a_2 = (e^{-a\Delta t})(e^{-b\Delta t})$	$\dfrac{b_1}{p_1 - p_2}(p_1^k - p_2^k) + \dfrac{b_2}{p_1 - p_2}(p_1^{k-1} - p_2^{k-1})$ $b_1 = \dfrac{[b(1 - p_1) - a(1 - p_2)]}{(b - a)}$ $b_2 = \dfrac{[ap_1(1 - p_2) - bp_2(1 - p_1)]}{(b - a)}$ $p_1 = e^{-a\Delta t}$ $p_2 = e^{-b\Delta t}$

Table C.3. Continued

	$g(s)$	$g(z) = \dfrac{b_1 z^{-1} + b_2 z^{-2}}{1 + a_1 z^{-1} + a_2 z^{-2}}$	$g(k)$
7.	$\dfrac{a_2}{(s + a)^2}$	$b_1 = 1 - (1 + a\Delta t)\,e^{-a\Delta t}$ $b_2 = e^{-a\Delta t}\,(e^{-a\Delta t} + a\Delta t - 1)$ $a_1 = -2e^{-a\Delta t}$ $a_2 = (e^{-a\Delta t})^2$	$b_1\,k p^{k-1} + b_2\,(k - 1)\,p^{k-2}$ $b_1 = 1 - (1 + a\Delta t)\,p$ $b_2 = p\,(p + a\Delta t - 1)$ $p = e^{-a\Delta t}$ **or:** $[(1 - a\Delta t)\,p^{k-1} - p^k]$
8.	$\dfrac{1}{\tau^2 s^2 + 2\zeta\,\tau s + 1}$	$\omega = \dfrac{\sqrt{1 - \zeta^2}}{\tau} \; ; \; \zeta < 1$ $b_1 = 1 - e^{-\zeta\Delta t/\tau}\left[\cos(\omega\Delta t) + \dfrac{\zeta}{\omega\tau}\sin(\omega\Delta t)\right]$ $b_2 = (e^{-\zeta\Delta t/\tau})^2 + e^{-\zeta\Delta t/\tau}\left[\dfrac{\zeta}{\omega\tau}\sin(\omega\Delta t) - \cos(\omega\Delta t)\right]$ $a_1 = -2e^{-\zeta\Delta t/\tau}\cos(\omega\Delta t)$ $a_2 = (e^{-\zeta\Delta t/\tau})^2$	$p^{k-1}\,[\cos\omega(k - 1)\,\Delta t + \gamma\sin\omega(k - 1)\,\Delta t]$ $- p^k\,[\cos\omega k\Delta t + \gamma\sin\omega k\Delta t]$ $\omega = \dfrac{\sqrt{1 - \zeta^2}}{\tau} \; ; \; \zeta < 1$ $\gamma = \dfrac{\zeta}{\omega\tau}$ $p = e^{-\zeta\Delta t/\tau}$

Table C.3. Continued

$g(s)$	$g(z) = \dfrac{b_1 z^{-1} + b_2 z^{-2}}{1 + a_1 z^{-1} + a_2 z^{-2}}$	$g(k)$
9. $\dfrac{s}{(s+a)^2}$	$b_1 = e^{-a\Delta t}$ $b_2 = -e^{-a\Delta t}$ $a_1 = -2e^{-a\Delta t}$ $a_2 = (e^{-a\Delta t})^2$	$g(k) = kp^k - (k-1)p^{k-1}$ $p = e^{-a\Delta t}$
10. $\dfrac{(s+c)}{(s+a)(s+b)}$ $(a \neq b)$	$b_1 = \dfrac{\left[ab(e^{-a\Delta t} - e^{-b\Delta t}) + bc(1 - e^{-a\Delta t}) - ac(1 - e^{-b\Delta t}) \right]}{(b-a)}$ $b_2 = \dfrac{\left[c(b-a)\, e^{-a\Delta t}\, e^{-b\Delta t} + a(c-b)\, e^{-a\Delta t} + b(a-c)\, e^{-b\Delta t} \right]}{ab(b-a)}$ $a_1 = -(e^{-a\Delta t} - e^{-b\Delta t})$ $a_2 = (e^{-a\Delta t})(e^{-b\Delta t})$	$g(k) = \dfrac{b_1}{p_1 - p_2}(p_1^{\,k} - p_2^{\,k}) + \dfrac{b_2}{p_1 - p_2}(p_1^{\,k-1} - p_2^{\,k-2})$ $b_1 = \dfrac{\left[ab(p_1 - p_2) + bc(1 - p_1) - ac(1 - p_2) \right]}{(b-a)}$ $b_2 = \dfrac{\left[c(b-a)p_1 p_2 + a(c-b)p_1 + b(a-c)p_2 \right]}{ab(b-a)}$ $p_1 = e^{-a\Delta t}$ $p_2 = e^{-b\Delta t}$

REFERENCES AND SUGGESTED FURTHER READING

A general reference that covers several aspects of Laplace transforms theory and applications is:

1. Churchill, R. V., *Modern Operational Mathematics in Engineering*, McGraw-Hill, New York (1958)

More extensive tables of Laplace transforms may be found, for example, in:

2. *CRC Standard Mathematical Tables*, Chemical Rubber Co., Cleveland (1978)
3. Abramowitz, M. and I. A. Stegun, *Handbook of Mathematical Functions*, National Bureau of Standards, Washington, D. C. (1964)

A detailed treatment of the theory and application of z-transforms may be found in Ref. [4]; a more abbreviated but useful discussion may be found in Ref. [5]. Each of these references contains more extensive tables of z-transforms Ref. [6] contains many worked examples of partial fraction expansions for rational polynomials in the z variable.

4. Jury, E. I., *Theory and Application of the z-transform Method*, J. Wiley, New York (1964)
5. Kuo, B. C., *Analysis and Synthesis of Sampled-Data Control Systems*, Prentice-Hall, Englewood Cliffs, NJ (1963)
6. Ogata, K., *Discrete-Time Control Systems*, Prentice-Hall, Englewood Cliffs, NJ (1987)

D

REVIEW OF MATRIX ALGEBRA

The convenience of representing systems of several equations in compact, vector-matrix form — and how such representations facilitate further analysis — combine to make a working knowledge of matrix methods virtually indispensable to a proper study of multivariable control systems. Although most readers will have encountered matrix methods before, it is perhaps useful to briefly review the essential concepts here. Those readers who feel confident of their skills in this area may prefer to skip over this material and move on to another topic.

D.1   DEFINITION, ADDITION AND SUBTRACTION

A *matrix* is an array of numbers arranged in rows and columns, written with the following notation:

$$\mathbf{A} = \begin{bmatrix} a_{11} & a_{12} & a_{13} \\ a_{21} & a_{22} & a_{23} \end{bmatrix} \tag{D.1}$$

Here because $\mathbf{A}$ contains 2 rows and 3 columns, it is called a 2×3 matrix. The elements a_{11}, a_{12}, ..., etc., have the notation that a_{ij} is the element of the ith row and jth column of $\mathbf{A}$.

In general, a matrix with m rows and n columns is said to have dimension $m \times n$. In the special case when $n = m$ (i.e., the number of rows and columns are identical) the matrix, for obvious reasons, is said to be *square*; otherwise the matrix is said to be *rectangular* or *nonsquare*.

As in the case of scalar quantities, matrices can also be added, subtracted, multiplied, etc., but special rules of algebra must be followed.

Only matrices having the same dimensions (i.e., same number of rows and columns) can be added or subtracted. For example, given the matrices $\mathbf{A}$, $\mathbf{B}$, $\mathbf{C}$:

$$A = \begin{bmatrix} a_{11} & a_{12} & a_{13} \\ a_{21} & a_{22} & a_{23} \end{bmatrix}; \quad B = \begin{bmatrix} b_{11} & b_{12} & b_{13} \\ b_{21} & b_{22} & b_{23} \\ b_{31} & b_{32} & b_{33} \end{bmatrix}; \quad C = \begin{bmatrix} c_{11} & c_{12} & c_{13} \\ c_{21} & c_{22} & c_{23} \end{bmatrix} \quad (D.2)$$

only A and C can be added or subtracted; because its dimensions are different, B cannot be added to, or subtracted from, either A or C. The sum of $A + C = D$ can be written:

$$D = A + C = \begin{bmatrix} a_{11} & a_{12} & a_{13} \\ a_{21} & a_{22} & a_{23} \end{bmatrix} + \begin{bmatrix} c_{11} & c_{12} & c_{13} \\ c_{21} & c_{22} & c_{23} \end{bmatrix}$$

$$= \begin{bmatrix} a_{11} + c_{11} & a_{12} + c_{12} & a_{13} + c_{13} \\ a_{21} + c_{21} & a_{22} + c_{22} & a_{23} + c_{23} \end{bmatrix} \quad (D.3)$$

Thus addition (or subtraction) of two matrices is accomplished by adding (or subtracting) the corresponding elements of the two matrices. These operations may, of course, be extended to any number of matrices of the same order. In particular observe that adding k identical matrices A (equivalent to scalar multiplication of A by k) results in a matrix in which each element is merely the corresponding element of A multiplied by the scalar k. Thus, when a matrix is to be multiplied by a scalar, the required operation is done by multiplying each element of the matrix by the scalar; for example, for A given in Eq. (D.2):

$$kA = k \begin{bmatrix} a_{11} & a_{12} & a_{13} \\ a_{21} & a_{22} & a_{23} \end{bmatrix} = \begin{bmatrix} ka_{11} & ka_{12} & ka_{13} \\ ka_{21} & ka_{22} & ka_{23} \end{bmatrix} \quad (D.4)$$

Example D.1   **MATRIX ADDITION, SUBTRACTION, AND SCALAR MULTIPLICATION.**

The following are some simple numerical examples of matrix addition, subtraction, and scalar multiplication:

$$\begin{bmatrix} 1 & 4 \\ 6 & 9 \end{bmatrix} + \begin{bmatrix} 2 & 3 \\ 9 & 5 \end{bmatrix} = \begin{bmatrix} 3 & 7 \\ 15 & 14 \end{bmatrix}$$

$$4 \begin{bmatrix} 1 & 4 \\ 6 & 9 \end{bmatrix} - 3 \begin{bmatrix} 2 & 3 \\ 9 & 5 \end{bmatrix} = \begin{bmatrix} -2 & 7 \\ -3 & 21 \end{bmatrix}$$

D.2   MULTIPLICATION

Matrix multiplication can be performed on two matrices $\mathbf{A}$ and $\mathbf{B}$ to form the product $\mathbf{AB}$ only if the number of *columns* of $\mathbf{A}$ is equal to the number of *rows* in $\mathbf{B}$. If this condition is satisfied, the matrices $\mathbf{A}$ and $\mathbf{B}$ are said to be *conformable* in the order $\mathbf{AB}$.

Let us consider that $\mathbf{A}$ is an $m \times n$ matrix (m rows and n columns); then to form the product $\mathbf{P} = \mathbf{AB}$, $\mathbf{B}$ must be an $n \times r$ matrix (n rows and r columns). The resulting product $\mathbf{P}$ is then an $m \times r$ matrix (m rows and r columns) whose elements p_{ij} are determined from the elements of $\mathbf{A}$ and $\mathbf{B}$ according to the following rule:

$$p_{ij} = \sum_{k=1}^{n} a_{ik}b_{kj} \quad i = 1, 2, \ldots, m; \; j = 1, 2, \ldots, r \tag{D.5}$$

For example, if $\mathbf{A}$ and $\mathbf{B}$ are defined by Eq. (D.2), then $\mathbf{A}$ is a 2×3 matrix, $\mathbf{B}$ a 3×3 matrix, and the product $\mathbf{P} = \mathbf{AB}$ a 2×3 matrix is obtained as:

$$\mathbf{P} = \begin{bmatrix} a_{11} & a_{12} & a_{13} \\ a_{21} & a_{22} & a_{23} \end{bmatrix} \begin{bmatrix} b_{11} & b_{12} & b_{13} \\ b_{21} & b_{22} & b_{23} \\ b_{31} & b_{32} & b_{33} \end{bmatrix}$$

$$= \begin{bmatrix} \left(a_{11}b_{11} + a_{12}b_{21} + a_{13}b_{31}\right) & \left(a_{11}b_{12} + a_{12}b_{22} + a_{13}b_{32}\right) & \left(a_{11}b_{13} + a_{12}b_{23} + a_{13}b_{33}\right) \\ \left(a_{21}b_{11} + a_{22}b_{21} + a_{23}b_{31}\right) & \left(a_{21}b_{12} + a_{22}b_{22} + a_{23}b_{32}\right) & \left(a_{21}b_{13} + a_{22}b_{23} + a_{23}b_{33}\right) \end{bmatrix} \tag{D.6}$$

Example D.2   MATRIX MULTIPLICATION.

The following are some simple examples to illustrate how matrix multiplication is carried out.

1.

$$\begin{bmatrix} 3 & 4 \\ -2 & -1 \end{bmatrix} \begin{bmatrix} 1 & 2 \\ 2 & 5 \end{bmatrix}$$

$$= \begin{bmatrix} (3 \times 1 + 4 \times 2) & (3 \times 2 + 4 \times 5) \\ (-2 \times 1 - 1 \times 2) & (-2 \times 2 - 1 \times 5) \end{bmatrix} = \begin{bmatrix} 11 & 26 \\ -4 & -9 \end{bmatrix}$$

2.

$$[x \ y \ z] \begin{bmatrix} a \\ b \\ c \end{bmatrix} = ax + by + cz; \text{ a scalar quantity}$$

3.

$$\begin{bmatrix} x \\ y \\ z \end{bmatrix} [a \ b \ c] = \begin{bmatrix} ax & bx & cx \\ ay & by & cy \\ az & bz & cz \end{bmatrix}$$

Unlike in scalar algebra, the fact that **AB** exists does not imply that **BA** exists. For example, if **A** is $m \times n$ and **B** is $n \times r$, then **AB** exists and is of order $m \times r$, but **BA** does not exist since the dimension $n \times r$ is not conformable with $m \times n$.

But if instead **B** had been an $n \times m$ matrix, then both **AB** and **BA** exist. Thus, only in cases where the number of columns of **A** (n in this case) equals the number of rows of **B**, *and* the number of rows of **A** (m in this case) equals the number of columns of **B**, can we form both the products **AB** and **BA**. Note, however, that in this case, the order of **AB** will be $m \times m$ while **BA** will be of order $n \times n$, so that, in general:

$$\mathbf{AB} \neq \mathbf{BA} \tag{D.7}$$

In matrix algebra, therefore, unlike in scalar algebra, the order of multiplication is very important.

Those special matrices for which **AB** = **BA** are said to *commute* or to be *permutable*.

Example D.3   NONCOMMUTATIVITY OF MATRIX MULTIPLICATION.

Let us illustrate Eq. (D.7) with the following two matrices:

$$\mathbf{A} = \begin{bmatrix} 1 & 2 \\ 3 & 4 \end{bmatrix}; \quad \mathbf{B} = \begin{bmatrix} 5 & 6 \\ 7 & 8 \end{bmatrix}$$

Since **A** and **B** have the same number of rows and columns one may compute both **AB**:

$$\mathbf{AB} = \begin{bmatrix} 1 & 2 \\ 3 & 4 \end{bmatrix} \begin{bmatrix} 5 & 6 \\ 7 & 8 \end{bmatrix} = \begin{bmatrix} (5+14) & (6+16) \\ (15+28) & (18+32) \end{bmatrix}$$

$$= \begin{bmatrix} 19 & 22 \\ 43 & 50 \end{bmatrix}$$

and **BA**:

$$\mathbf{BA} = \begin{bmatrix} 5 & 6 \\ 7 & 8 \end{bmatrix} \begin{bmatrix} 1 & 2 \\ 3 & 4 \end{bmatrix} = \begin{bmatrix} (5+18) & (10+24) \\ (7+24) & (14+32) \end{bmatrix}$$

$$= \begin{bmatrix} 23 & 34 \\ 31 & 46 \end{bmatrix}$$

from where we see that, indeed, $\mathbf{AB} \neq \mathbf{BA}$.

D.3   TRANSPOSE OF A MATRIX

The transpose of the matrix $\mathbf{A}$, denoted $\mathbf{A}^T$, is the matrix formed by interchanging the rows with the columns. For example, if:

$$\mathbf{A} = \begin{bmatrix} a_{11} & a_{12} & a_{13} \\ a_{21} & a_{22} & a_{23} \end{bmatrix}$$

then

$$\mathbf{A}^T = \begin{bmatrix} a_{11} & a_{21} \\ a_{12} & a_{22} \\ a_{13} & a_{23} \end{bmatrix} \tag{D.8}$$

Thus if $\mathbf{A}$ is an $m \times n$ matrix, then $\mathbf{A}^T$ is an $n \times m$ matrix. Observe therefore that both products $\mathbf{A}^T\mathbf{A}$ and $\mathbf{A}\mathbf{A}^T$ can be formed, and that each will be a square matrix. However, we note again, that in general:

$$\mathbf{A}^T\mathbf{A} \neq \mathbf{A}\mathbf{A}^T \tag{D.9}$$

The transpose of a matrix made up of the product of other matrices may be expressed by the relation:

$$(\mathbf{ABCD})^T = \mathbf{D}^T\mathbf{C}^T\mathbf{B}^T\mathbf{A}^T \tag{D.10}$$

which is valid for any number of matrices $\mathbf{A}, \mathbf{B}, \mathbf{C}, \mathbf{D}, \ldots$

Example D.4   **TRANSPOSE OF A MATRIX PRODUCT.**

To illustrate relation Eq. (D.10), let us define:

$$\mathbf{A} = \begin{bmatrix} 1 & 2 \\ 3 & 4 \end{bmatrix} \qquad \mathbf{B} = \begin{bmatrix} 5 & 6 \\ 7 & 8 \end{bmatrix}$$

and from Example D.3 we recall that:

$$\mathbf{AB} = \begin{bmatrix} 19 & 22 \\ 43 & 50 \end{bmatrix}$$

so that:

$$(\mathbf{AB})^T = \begin{bmatrix} 19 & 43 \\ 22 & 50 \end{bmatrix}$$

Now:

$$\mathbf{A}^T = \begin{bmatrix} 1 & 3 \\ 2 & 4 \end{bmatrix}; \; \mathbf{B}^T = \begin{bmatrix} 5 & 7 \\ 6 & 8 \end{bmatrix}$$

so that:

$$\mathbf{B}^T\mathbf{A}^T = \begin{bmatrix} 5 & 7 \\ 6 & 8 \end{bmatrix} \begin{bmatrix} 1 & 3 \\ 2 & 4 \end{bmatrix} = \begin{bmatrix} 19 & 43 \\ 22 & 50 \end{bmatrix}$$

and Eq. (D.10) may be seen to hold.

D.4   SOME SCALAR PROPERTIES OF MATRICES

Associated with each matrix are several *scalar* quantities that are quite useful for characterizing matrices in general; the following are some of the most important.

D.4.1 Trace of a Matrix

If $\mathbf{A}$ is a square, $n \times n$ matrix, the trace of $\mathbf{A}$, sometimes denoted $\text{Tr}(\mathbf{A})$, is defined as the sum of the elements on the main diagonal, i.e.:

$$\text{Tr}(\mathbf{A}) = \sum_{i=1}^{n} a_{ii} \tag{D.11}$$

For example, given:

$$\mathbf{A} = \begin{bmatrix} 1 & 2 \\ 3 & 4 \end{bmatrix}; \quad \mathbf{C} = \begin{bmatrix} 0 & 2 & 1 \\ -1 & 2 & -1 \\ 3 & 0 & 1 \end{bmatrix}$$

we have that:

$$\text{Tr}(\mathbf{A}) = (1 + 4) = 5$$

and

$$\text{Tr}(\mathbf{C}) = (0 + 2 + 1) = 3$$

D.4.2 Determinants and Cofactors

Of all the scalar properties of a matrix, none is perhaps more important than the *determinant*. By way of definition, the "absolute value" of a *square* matrix $\mathbf{A}$, denoted $|\mathbf{A}|$, is called its determinant. For a simple 2×2 matrix:

$$\mathbf{A} = \begin{bmatrix} a_{11} & a_{12} \\ a_{21} & a_{22} \end{bmatrix} \tag{D.12}$$

the determinant is defined by:

$$|\mathbf{A}| = a_{11}a_{22} - a_{21}a_{12} \tag{D.13}$$

For higher order matrices, the determinant is more conveniently defined in terms of *cofactors*.

Cofactors

The *cofactor* of the element a_{ij} is defined as the determinant of the matrix left when the ith row and the jth column are removed, multiplied by the factor $(-1)^{i+j}$. For example, let us consider the 3×3 matrix:

$$\mathbf{A} = \begin{bmatrix} a_{11} & a_{12} & a_{13} \\ a_{21} & a_{22} & a_{23} \\ a_{31} & a_{32} & a_{33} \end{bmatrix} \tag{D.14}$$

The cofactor, C_{12}, of element a_{12} is:

$$C_{12} = (-1)^3 \begin{vmatrix} a_{21} & a_{23} \\ a_{31} & a_{33} \end{vmatrix} = -(a_{21}a_{33} - a_{31}a_{23}) \tag{D.15}$$

Similarly, the cofactor C_{23} is:

$$C_{23} = (-1)^5 \begin{vmatrix} a_{11} & a_{12} \\ a_{31} & a_{32} \end{vmatrix} = -(a_{11}a_{32} - a_{31}a_{12}) \tag{D.16}$$

The General $n \times n$ Determinant

For the general $n \times n$ matrix $\mathbf{A}$ whose elements a_{ij} have corresponding cofactors C_{ij}, the determinant $|\mathbf{A}|$ is defined as:

$$|\mathbf{A}| = \sum_{i=1}^{n} a_{ij}C_{ij} \quad \text{(for any column } j\text{)} \tag{D.17}$$

or

$$|\mathbf{A}| = \sum_{j=1}^{n} a_{ij}C_{ij} \quad \text{(for any row } i\text{)} \tag{D.18}$$

Eqs. (D.17) and (D.18) are known as the Laplace expansion formulas; they imply that to obtain $|\mathbf{A}|$:

1. Choose any row or column of $\mathbf{A}$.
2. Evaluate the n cofactors corresponding to the elements in the chosen row or column.
3. The sum of the n products of elements and cofactors will give the required determinant. (It is important to note that the same answer is obtained regardless of the row or column chosen.)

To illustrate, let us evaluate the determinant of the 3×3 matrix given by Eq. (D.14) by expanding in the first row. In this case, we have that:

$$|\mathbf{A}| = a_{11}C_{11} + a_{12}C_{12} + a_{13}C_{13}$$

That is:

$$\begin{vmatrix} a_{11} & a_{12} & a_{13} \\ a_{21} & a_{22} & a_{23} \\ a_{31} & a_{32} & a_{33} \end{vmatrix} = a_{11} \begin{vmatrix} a_{22} & a_{23} \\ a_{32} & a_{33} \end{vmatrix} - a_{12} \begin{vmatrix} a_{21} & a_{23} \\ a_{31} & a_{33} \end{vmatrix} + a_{13} \begin{vmatrix} a_{21} & a_{22} \\ a_{31} & a_{32} \end{vmatrix}$$

$$= a_{11}a_{22}a_{33} - a_{11}a_{32}a_{23} + a_{12}a_{31}a_{23} - a_{12}a_{21}a_{33} + a_{13}a_{21}a_{32} - a_{13}a_{31}a_{22} \tag{D.19}$$

It is easily shown that the expansion by any other row or column yields the same result.

Example D.5   DETERMINANT OF A 3×3 MATRIX.

The determinant of the 3×3 matrix:

$$\mathbf{A} = \begin{bmatrix} 1 & 2 & 3 \\ 4 & 5 & 6 \\ 7 & 8 & 9 \end{bmatrix}$$

may be evaluated as indicated above: by expanding in row 1, for example, we have:

$$|\mathbf{A}| = 1 \begin{vmatrix} 5 & 6 \\ 8 & 9 \end{vmatrix} - 2 \begin{vmatrix} 4 & 6 \\ 7 & 9 \end{vmatrix} + 3 \begin{vmatrix} 4 & 5 \\ 7 & 8 \end{vmatrix}$$

$$= 1(45 - 48) - 2(36 - 42) + (32 - 35)$$

$$= -3 + 12 - 9 = 0$$

Thus for this matrix, $|\mathbf{A}| = 0$. We will have cause to refer to this matrix later on.

Properties of Determinants

The following are some important properties of determinants frequently used to simplify determinant evaluation.

1. **Interchanging the rows with the columns leaves the value of a determinant unchanged.**
    The main implications of this property are twofold:
    (a) $|A| = |A^T|$;
    (b) Since rows and columns of a determinant are interchangeable, *any statement that holds true for rows is also true for columns, and vice versa.*

2. **If any two rows (columns) are interchanged, the sign of the determinant is reversed.**

3. **If any two rows (columns) are identical, the determinant is zero.**

4. **Multiplying the elements of any one row (column) by the same constant results in the value of the determinant being multiplied by this constant factor;** i.e., if, for example:

$$D_1 = \begin{vmatrix} a_{11} & a_{12} & a_{13} \\ a_{21} & a_{22} & a_{23} \\ a_{31} & a_{32} & a_{33} \end{vmatrix} \quad \text{and} \quad D_2 = \begin{vmatrix} ka_{11} & a_{12} & a_{13} \\ ka_{21} & a_{22} & a_{23} \\ ka_{31} & a_{32} & a_{33} \end{vmatrix}$$

then $D_2 = kD_1$

5. **If, for any i and j $(i \neq j)$ the elements of row (column) i are related to the elements of row (column) j by a constant multiplicative factor, the determinant is zero;** i.e., if, for example:

$$D_1 = \begin{vmatrix} ka_{12} & a_{12} & a_{13} \\ ka_{22} & a_{22} & a_{23} \\ ka_{32} & a_{32} & a_{33} \end{vmatrix}$$

where we observe that column 1 is k times column 2, then $D_1 = 0$.

6. **Adding a constant multiple of the elements of one row (column) to corresponding elements of another row (column) leaves the value of the determinant unchanged;** i.e., if, for example:

$$D_1 = \begin{vmatrix} a_{11} & a_{12} & a_{13} \\ a_{21} & a_{22} & a_{23} \\ a_{31} & a_{32} & a_{33} \end{vmatrix} \quad \text{and} \quad D_2 = \begin{vmatrix} (a_{11} + ka_{12}) & a_{12} & a_{13} \\ (a_{21} + ka_{22}) & a_{22} & a_{23} \\ (a_{31} + ka_{32}) & a_{32} & a_{33} \end{vmatrix}$$

then $D_2 = D_1$.

7. **The value of a determinant that contains a whole row (column) of zeros is zero.**

8. **The determinant of a *diagonal* matrix** (a matrix with nonzero elements only on the main diagonal; see Section D.5 below) **is the product of the elements on the main diagonal. The determinant of a *triangular* matrix** (a matrix for which all elements below the main diagonal, *or* above the main diagonal are zero; see Section D.5 below) **is also equal to the product of the elements on the main diagonal.**

Alien Cofactors

As indicated in Eqs. (D.17) and (D.18), any expansion in which the elements of any row (column) in a matrix are combined with their corresponding cofactors (i.e., a *Laplace expansion*) gives rise to the determinant of the matrix in question. Any other expansion in which the elements of one row (column) are combined with cofactors of elements of *another* row (column) is called an expansion by *alien cofactors*, and its value is always zero.

Thus, for example, while the Laplace expansion using row 1 of Eq. (D.14) gives:

$$a_{11}C_{11} + a_{12}C_{12} + a_{13}C_{13} = |A|$$

the expansion combining the elements of row 1 with the cofactors of row 2 gives:

$$a_{11}C_{21} + a_{12}C_{22} + a_{13}C_{23} = 0$$

and similarly:

$$a_{11}C_{31} + a_{12}C_{32} + a_{13}C_{33} = 0$$

In general, therefore, for any column j:

$$\sum_{i=1}^{n} a_{ij}C_{ik} = \begin{cases} |A|; & j=k \\ 0; & j \neq k \end{cases} \tag{D.20}$$

and for any row i:

$$\sum_{j=1}^{n} a_{ij}C_{kj} = \begin{cases} |A|; & i=k \\ 0; & i \neq k \end{cases} \tag{D.21}$$

Thus Laplace expansion by corresponding cofactors gives $|A|$; expansion by *alien cofactors* always gives zero.

A very important implication of these facts is that if we form a matrix C from the cofactors of A, then:

$$AC^T = C^T A = \begin{bmatrix} |A| & 0 & 0 & \cdots & 0 \\ 0 & |A| & 0 & \cdots & 0 \\ 0 & 0 & |A| & \cdots & 0 \\ \cdots & \cdots & \cdots & \cdots & \cdots \\ 0 & 0 & 0 & \cdots & |A| \end{bmatrix} \tag{D.22}$$

a diagonal matrix with the determinant of **A** on the main diagonal. We shall refer to this result later.

D.4.3 Minors and Rank of a Matrix

Let **A** be a general $m \times n$ matrix with $m < n$. Many $m \times m$ determinants may be formed from the m rows and any m of the n columns of **A**. These $m \times m$ determinants are called *minors* (of order m) of the matrix **A**. We may similarly obtain several $(m - 1) \times (m - 1)$ determinants from the various combinations of $(m - 1)$ rows and $(m - 1)$ columns chosen from the m rows and n columns of **A**. These determinants are also minors, but of order $(m - 1)$.

Minors may also be categorized as 1st minor, 2nd minor, etc. The minor of the *highest order* is referred to the 1st minor, the next highest minor is the 2nd minor, etc., until we get to the lowest order minor, the one containing a single element. In the special case of a square $n \times n$ matrix, there is, of course, one and only one minor of order n, and there is no minor of higher order than this. Observe, therefore, that the 1st minor of a square matrix is what we have referred to as *the determinant* of the matrix. The second minor will be of order $(n - 1)$. Note, therefore, that the cofactors of the elements of a square matrix are intimately related to the 2nd minors of the matrix.

Rank of a Matrix

The *rank* of a matrix is defined as the order of the highest nonvanishing minor of that matrix (or, equivalently, the order of the highest square submatrix having a nonzero determinant). In Example D.5, for instance, the 1st minor (of order 3) vanished, but none of the 2nd minors (of order 2) vanished; the single elements of the matrix — nine in all — are themselves minors (of order 1), and all are nonzero. Of the nonvanishing minors, therefore, the highest order is 2. Thus, **A** in Example D.5 is of rank 2. A square $n \times n$ matrix is said to be of *full rank* if its rank $= n$; the matrix is said to be *rank deficient* if its rank is less than n.

In several practical applications of matrix methods, the rank of the matrix involved provides valuable information about the nature of the problem at hand. For example, in the solution of a system of linear algebraic equations by matrix methods (see Section D.7 below), the number of *independent* solutions that can be found is directly related to the rank of the matrix involved.

D.5   SOME SPECIAL MATRICES

1.   The Diagonal Matrix

If all the elements of a square matrix are zero except those on the main diagonal from the top left-hand corner to the bottom right-hand corner, the matrix is said to be *diagonal*. Some examples are:

$$\mathbf{A} = \begin{bmatrix} a_{11} & 0 \\ 0 & a_{22} \end{bmatrix}; \qquad \mathbf{B} = \begin{bmatrix} b_{11} & 0 & 0 \\ 0 & b_{22} & 0 \\ 0 & 0 & b_{33} \end{bmatrix}$$

2.   The Triangular Matrix

A square matrix in which all the elements *below* the main diagonal are zero is called an *upper triangular matrix*. By the same token, a *lower triangular matrix* is one for which all the elements *above* the main diagonal are zero. Thus:

$$
\begin{bmatrix} b_{11} & b_{12} & b_{13} \\ 0 & b_{22} & b_{23} \\ 0 & 0 & b_{33} \end{bmatrix} \quad \text{and} \quad \begin{bmatrix} b_{11} & 0 & 0 \\ b_{21} & b_{22} & 0 \\ b_{31} & b_{32} & b_{33} \end{bmatrix}
$$

are, respectively, examples of upper and lower triangular matrices.

3.   The Identity Matrix

A diagonal matrix in which the nonzero elements on the main diagonal are all unity is called the *identity* matrix; it is frequently denoted by $\mathbf{I}$, i.e.:

$$
\mathbf{I} = \begin{bmatrix} 1 & 0 & 0 & \dots & 0 \\ 0 & 1 & 0 & \dots & 0 \\ 0 & 0 & 1 & \dots & 0 \\ \dots & \dots & \dots & \dots & \dots \\ 0 & 0 & 0 & \dots & 1 \end{bmatrix}
$$

The significance of the identity matrix will be obvious later, but for now, we note that it has the unique property that:

$$
\mathbf{IQ} = \mathbf{Q} = \mathbf{QI} \tag{D.23}
$$

for any general matrix $\mathbf{Q}$ of conformable order with the multiplying identity matrix.

4.   Orthogonal Matrices

If $\mathbf{A}^T\mathbf{A} = \mathbf{I}$, the identity matrix, then $\mathbf{A}$ is an *orthogonal* (or *orthonormal*, or *unitary*) matrix. For example, as the reader may easily verify:

$$
\mathbf{A} = \begin{bmatrix} \dfrac{3}{5} & \dfrac{4}{5} \\ -\left(\dfrac{4}{5}\right) & \dfrac{3}{5} \end{bmatrix}
$$

is an orthogonal matrix.

5.   Symmetric and Skew-Symmetric Matrices

When every pair of (real) elements that are symmetrically placed in a matrix with respect to the main diagonal are equal, the matrix is said to be a (real) *symmetric* matrix. That is, for a (real) symmetric matrix:

$$
a_{ij} = a_{ji} \tag{D.24a}
$$

Some examples of symmetric matrices are:

$$A = \begin{bmatrix} 1 & 3 \\ 3 & 4 \end{bmatrix}; \qquad B = \begin{bmatrix} 2 & 1 & 5 \\ 1 & 0 & -2 \\ 5 & -2 & 4 \end{bmatrix}$$

Thus a symmetric matrix and its transpose are identical, i.e.:

$$A = A^T \qquad\qquad \text{(D.24b)}$$

A *skew-symmetric* matrix is one for which:

$$a_{ij} = -a_{ji} \qquad\qquad \text{(D.25a)}$$

The following are some examples of such matrices:

$$A = \begin{bmatrix} 0 & -3 \\ 3 & 0 \end{bmatrix}; \; B = \begin{bmatrix} 0 & 1 & 5 \\ -1 & 0 & -2 \\ -5 & 2 & 0 \end{bmatrix}$$

For skew-symmetric matrices:

$$A = -A^T \qquad\qquad \text{(D.25b)}$$

By noting that a_{ij} can be expressed as:

$$a_{ij} = \frac{1}{2}\left(a_{ij} + a_{ij}\right) + \frac{1}{2}\left(a_{ij} - a_{ij}\right) \qquad\qquad \text{(D.26a)}$$

it is easily established that *every* square matrix A can be resolved into the sum of a symmetric matrix A^O and a skew-symmetric matrix A', i.e.:

$$A = A^O + A' \qquad\qquad \text{(D.26b)}$$

6. Hermitian Matrices

When symmetrically situated elements in a matrix of complex numbers are complex conjugates, i.e.:

$$a_{ji} = \overline{a_{ij}} \qquad\qquad \text{(D.27)}$$

the matrix is said to be *Hermitian*. For example:

$$\begin{bmatrix} 3 & (2-3j) \\ (2+3j) & 5 \end{bmatrix}$$

is a Hermitian matrix. Note that the (real) symmetric matrix is a special form of a Hermitian matrix in which the coefficients of j are absent.

7. Singular Matrices

A matrix whose determinant is zero is said to be *singular*. Thus, for example, the matrix encountered in Example D.5 is a singular matrix. A matrix whose determinant is *not* zero is said to be *nonsingular*.

8. Idempotent Matrices

A matrix $\mathbf{A}$ is said to be *idempotent* if:

$$\mathbf{A}\mathbf{A} = \mathbf{A} \tag{D.28}$$

It is easily verified, for example, that the matrix:

$$\mathbf{A} = \begin{bmatrix} 3 & 2 \\ -3 & -2 \end{bmatrix}$$

is idempotent. Note that the identity matrix is a special type of an idempotent matrix. It is easy to establish that the only nonsingular idempotent matrix is the identity matrix; all other idempotent matrices are singular.

D.6   INVERSE OF A MATRIX

The matrix counterpart of division is the inverse operation. Thus, corresponding to the reciprocal of a number in scalar algebra is the inverse of the matrix $\mathbf{A}$, denoted by $\mathbf{A}^{-1}$, (read as "$\mathbf{A}$ inverse"): a matrix for which both $\mathbf{A}\mathbf{A}^{-1}$ and $\mathbf{A}^{-1}\mathbf{A}$ give $\mathbf{I}$, the identity matrix, i.e.:

$$\mathbf{A}\mathbf{A}^{-1} = \mathbf{A}^{-1}\mathbf{A} = \mathbf{I} \tag{D.29}$$

The inverse matrix $\mathbf{A}^{-1}$ does not exist for all matrices; it exists only if:

1. $\mathbf{A}$ is square,[†] and
2. Its determinant is not zero (i.e., the matrix is *non–singular*).

The inverse of $\mathbf{A}$ is defined by:

$$\mathbf{A}^{-1} = \frac{\mathbf{C}^T}{|\mathbf{A}|} = \frac{\text{adj } (\mathbf{A})}{|\mathbf{A}|} \tag{D.30}$$

where $\mathbf{C}$ is the matrix of cofactors, i.e.:

$$\mathbf{C} = \begin{bmatrix} C_{11} & C_{12} & C_{13} & \cdots & C_{1n} \\ C_{21} & C_{22} & C_{23} & \cdots & C_{2n} \\ \cdots & \cdots & \cdots & \cdots & \cdots \\ C_{n1} & C_{n2} & C_{n3} & \cdots & C_{nn} \end{bmatrix}$$

and, as defined previously, C_{ij} is the cofactor of the element a_{ij} of the matrix $\mathbf{A}$. $\mathbf{C}^T$, the transpose of this matrix of cofactors, is termed the *adjoint* of $\mathbf{A}$ and often abbreviated adj $(\mathbf{A})$. This expression for the matrix inverse may be seen to follow directly from Eq. (D.22) and the definition of the inverse in Eq. (D.29).

[†]Generalizing the concept of an inverse to include nonsquare matrices will be discussed shortly.

Clearly because $|A| = 0$ for singular matrices, the matrix inverse as expressed in Eq. (D.30) does not exist for such matrices.

It should be noted that:

1. The inverse of a matrix composed of a product of other matrices may be written in the form:

$$(\mathbf{A}\,\mathbf{B}\,\mathbf{C}\,\mathbf{D})^{-1} = \mathbf{D}^{-1}\mathbf{C}^{-1}\mathbf{B}^{-1}\mathbf{A}^{-1}$$

2. For the simple 2×2 matrix, the general definition in Eq. (D.30) simplifies to a form easy to remember. Thus given:

$$\mathbf{A} = \begin{bmatrix} a_{11} & a_{12} \\ a_{21} & a_{22} \end{bmatrix}$$

we have that $|A| = a_{11}a_{22} - a_{12}a_{21}$, and it is easily shown that:

$$\mathbf{A}^{-1} = \frac{1}{|A|}\begin{bmatrix} a_{22} & -a_{12} \\ -a_{21} & a_{11} \end{bmatrix} \qquad (D.31)$$

Example D.6   OBTAINING MATRIX INVERSES.

Let us illustrate the foregoing discussion by obtaining the inverses of the following three matrices:

$$\mathbf{A}_1 = \begin{bmatrix} 1 & 2 \\ 1 & 3 \end{bmatrix}; \quad \mathbf{A}_2 = \begin{bmatrix} 1 & 2 & 3 \\ 4 & 5 & 6 \\ 7 & 8 & 9 \end{bmatrix}; \quad \mathbf{A}_3 = \begin{bmatrix} 2 & 2 & 3 \\ 4 & 5 & 6 \\ 7 & 8 & 9 \end{bmatrix}$$

The first is a 2×2 matrix whose determinant is easily obtained as $\left| \mathbf{A}_1 \right| = 3 - 2 = 1$; according to Eq. (D.31) therefore, we easily have that:

$$\left(\mathbf{A}_1\right)^{-1} = \begin{bmatrix} 3 & -2 \\ -1 & 1 \end{bmatrix}$$

from where the reader may want to verify by direct multiplication that $\mathbf{A}_1(\mathbf{A}_1)^{-1} = (\mathbf{A}_1)^{-1}\mathbf{A}_1 = \mathbf{I}$.

Next we observe that determinant of the 3×3 matrix $\mathbf{A}_2$ was obtained in Example D.5 as zero; thus it is a *singular* matrix, and therefore $(\mathbf{A}_2)^{-1}$ does not exist.

The determinant of $\mathbf{A}_3$ may be obtained by expanding in the cofactors of the first row, i.e.:

$$\left|\mathbf{A}_3\right| = 2\begin{vmatrix} 5 & 6 \\ 8 & 9 \end{vmatrix} - 2\begin{vmatrix} 4 & 6 \\ 7 & 9 \end{vmatrix} + 3\begin{vmatrix} 4 & 5 \\ 7 & 8 \end{vmatrix}$$

$$= 2\,(45 - 48) - 2\,(36 - 42) + 3\,(32 - 35) = -3$$

Thus the matrix is nonsingular, and we may proceed to compute the inverse. We recall from the definition of cofactors that the nine cofactors of $\mathbf{A}$ are obtained as:

$$C_{11} = \begin{vmatrix} 5 & 6 \\ 8 & 9 \end{vmatrix} = -3; \quad C_{12} = (-1) \begin{vmatrix} 4 & 6 \\ 7 & 9 \end{vmatrix} = 6; \quad C_{13} = \begin{vmatrix} 4 & 5 \\ 7 & 8 \end{vmatrix} = -3$$

$$C_{21} = (-1) \begin{vmatrix} 2 & 3 \\ 8 & 9 \end{vmatrix} = 6; \quad C_{22} = \begin{vmatrix} 2 & 3 \\ 7 & 9 \end{vmatrix} = -3; \quad C_{23} = (-1) \begin{vmatrix} 2 & 2 \\ 7 & 8 \end{vmatrix} = -2$$

$$C_{31} = \begin{vmatrix} 2 & 3 \\ 5 & 6 \end{vmatrix} = -3; \quad C_{32} = (-1) \begin{vmatrix} 2 & 3 \\ 4 & 6 \end{vmatrix} = 0; \quad C_{33} = \begin{vmatrix} 2 & 2 \\ 4 & 5 \end{vmatrix} = 2$$

Thus the matrix of cofactors is:

$$\mathbf{C} = \begin{bmatrix} -3 & 6 & -3 \\ 6 & -3 & -2 \\ -3 & 0 & 2 \end{bmatrix}$$

so that the adjoint of $\mathbf{A}_3$ is:

$$\text{adj} (\mathbf{A}_3) = \mathbf{C}^T = \begin{bmatrix} -3 & 6 & -3 \\ 6 & -3 & 0 \\ -3 & -2 & 2 \end{bmatrix}$$

The inverse $(\mathbf{A}_3)^{-1}$ is then:

$$(\mathbf{A}_3)^{-1} = \frac{\text{adj} (\mathbf{A}_3)}{|\mathbf{A}_3|} = -\left(\frac{1}{3}\right) \begin{bmatrix} -3 & 6 & -3 \\ 6 & -3 & 0 \\ -3 & -2 & 2 \end{bmatrix} = \begin{bmatrix} 1 & -2 & 1 \\ -2 & 1 & 0 \\ 1 & \frac{2}{3} & -\left(\frac{2}{3}\right) \end{bmatrix}$$

To check, we see by matrix multiplication that:

$$\mathbf{A}_3(\mathbf{A}_3)^{-1} = \begin{bmatrix} 2 & 2 & 3 \\ 4 & 5 & 6 \\ 7 & 8 & 9 \end{bmatrix} \begin{bmatrix} 1 & -2 & 1 \\ -2 & 1 & 0 \\ 1 & \frac{2}{3} & -\left(\frac{2}{3}\right) \end{bmatrix} = \begin{bmatrix} 1 & 0 & 0 \\ 0 & 1 & 0 \\ 0 & 0 & 1 \end{bmatrix} = \mathbf{I}$$

and the reader may in similar fashion verify that $(\mathbf{A}_3)^{-1}\mathbf{A}_3$ is also equal to $\mathbf{I}$.

D.7   LINEAR ALGEBRAIC EQUATIONS

Much of the motivation behind developing matrix operations is due to their particular usefulness in handling sets of linear algebraic equations. For example, the three linear equations:

$$a_{11}x_1 + a_{12}x_2 + a_{13}x_3 = b_1$$

$$a_{21}x_1 + a_{22}x_2 + a_{23}x_3 = b_2$$

$$a_{31}x_1 + a_{32}x_2 + a_{33}x_3 = b_3 \tag{D.32}$$

containing the unknowns x_1, x_2, x_3 can be represented by the single matrix equation:

$$\mathbf{Ax} = \mathbf{b} \tag{D.33}$$

where

$$\mathbf{A} = \begin{bmatrix} a_{11} & a_{12} & a_{13} \\ a_{21} & a_{22} & a_{23} \\ a_{31} & a_{32} & a_{33} \end{bmatrix} ; \quad \mathbf{x} = \begin{bmatrix} x_1 \\ x_2 \\ x_3 \end{bmatrix} ; \quad \mathbf{b} = \begin{bmatrix} b_1 \\ b_2 \\ b_3 \end{bmatrix} \tag{D.34}$$

The solution to these equations can be obtained very simply by premultiplying Eq. (D.33) by $\mathbf{A}^{-1}$ to yield:

$$\mathbf{A}^{-1}\mathbf{Ax} = \mathbf{A}^{-1}\mathbf{b}$$

which becomes:

$$\mathbf{x} = \mathbf{A}^{-1}\mathbf{b} \tag{D.35}$$

upon recalling the property of a matrix inverse, that $\mathbf{A}^{-1}\mathbf{A} = \mathbf{I}$, and recalling Eq. (D.23) for the property of the identity matrix.

There are several standard, computer-oriented numerical procedures for performing the operation $\mathbf{A}^{-1}$ to produce the solution given in Eq. (D.35). Indeed, the availability of these computer subroutines makes the use of matrix techniques very attractive from a practical viewpoint.

It should be noted that the solution Eq. (D.35) exists *only* if $\mathbf{A}$ is nonsingular, as only then will $\mathbf{A}^{-1}$ exist. When $\mathbf{A}$ is singular, $\mathbf{A}^{-1}$ ceases to exist and Eq. (D.35) will be meaningless. The singularity of $\mathbf{A}$ of course implies that $|\mathbf{A}| = 0$, and from property 5 given for determinants in Section D.4, this indicates that some of the rows and columns of $\mathbf{A}$ are *linearly dependent*. Thus the singularity of $\mathbf{A}$ means that only a subset of the equations we are trying to solve are truly linearly independent; as such, linearly independent solutions can be found only for this subset. It can be shown that the actual number of such linearly independent solutions is equal to the rank of $\mathbf{A}$.

Example D.7   SOLUTION OF LINEAR ALGEBRAIC EQUATIONS.

To illustrate the use of matrices in the solution of linear algebraic equations, let us consider the set of equations:

$$2x_1 + 2x_2 + 3x_3 = 1$$

$$4x_1 + 5x_2 + 6x_3 = 3$$

$$7x_1 + 8x_2 + 9x_3 = 5$$

which in matrix notation is given by:

$$\mathbf{Ax} = \mathbf{b}$$

where **A**, **b**, **x** are:

$$
A = \begin{bmatrix} 2 & 2 & 3 \\ 4 & 5 & 6 \\ 7 & 8 & 9 \end{bmatrix}, \quad
b = \begin{bmatrix} 1 \\ 3 \\ 5 \end{bmatrix}, \quad
x = \begin{bmatrix} x_1 \\ x_2 \\ x_3 \end{bmatrix}
$$

Notice that **A** is simply the third matrix whose inverse was computed in Example D.6, i.e.:

$$
A^{-1} = \begin{bmatrix} 1 & -2 & 1 \\ -2 & 1 & 0 \\ 1 & \dfrac{2}{3} & -\left(\dfrac{2}{3}\right) \end{bmatrix}
$$

Thus, from Eq. (D.35), the solution to this set of algebraic equations is obtained as:

$$
x = A^{-1}b = \begin{bmatrix} 1 & -2 & 1 \\ 2 & 1 & 0 \\ 1 & \dfrac{2}{3} & -\left(\dfrac{2}{3}\right) \end{bmatrix} \begin{bmatrix} 1 \\ 3 \\ 5 \end{bmatrix} = \begin{bmatrix} (1 - 6 + 5) \\ (-2 + 3 + 0) \\ \left(1 + \dfrac{6}{3} - \dfrac{10}{3}\right) \end{bmatrix} = \begin{bmatrix} 0 \\ 1 \\ -\dfrac{1}{3} \end{bmatrix}
$$

Thus $x_1 = 0$, $x_2 = 1$, $x_3 = -1/3$ is the required solution.

Let us now alter the coefficient of x_1 in the first equation from 2 to 1 so that we now have the following set of equations to solve for the three unknowns:

$$
\begin{aligned}
x_1 + 2x_2 + 3x_3 &= 1 \\
4x_1 + 5x_2 + 6x_3 &= 3 \\
7x_1 + 8x_2 + 9x_3 &= 5
\end{aligned}
$$

Observe in this case that:

$$
A = \begin{bmatrix} 1 & 2 & 3 \\ 4 & 5 & 6 \\ 7 & 8 & 9 \end{bmatrix}
$$

a matrix whose determinant had previously been obtained as 0 and whose rank was later identified as 2. The immediate implication is that since **A** is singular, we cannot solve this new set of equations for all the three unknowns.

Observe now that by adding the first and third equations we obtain an equation that is *exactly* twice the second equation; thus, the three equations are in fact not completely independent. It is easily shown that we can solve these equations independently only for two of the unknowns; the third unknown can only be obtained as a linear combination of these two. Thus there are only *two* linearly independent solutions to this new set of equations, exactly the same as the rank of **A**.

D.8   GENERALIZED INVERSES

Consider the general problem of solving m linear algebraic equations:

$$a_{11}x_1 + a_{12}x_2 + \ldots + a_{1n}x_n = b_1$$

$$a_{21}x_1 + a_{22}x_2 + \ldots + a_{2n}x_n = b_2$$

$$\ldots \quad \ldots \quad \ldots \quad \ldots \quad \ldots$$

$$a_{m1}x_1 + a_{m2}x_2 + \ldots + a_{mn}x_n = b_m \tag{D.36}$$

containing the n unknowns $x_1, x_2, x_3, \ldots, x_n$. Observe that this system of equations, as in Eq. (D.32) can also be represented by the single matrix equation:

$$\mathbf{Ax} = \mathbf{b} \tag{D.33}$$

except that in this case, $\mathbf{A}$ is an $m \times n$ matrix, $\mathbf{x}$ is an n-dimensional vector, and $\mathbf{b}$ is an m-dimensional vector:

$$\mathbf{A} = \begin{bmatrix} a_{11} & a_{12} & \cdots & a_{1n} \\ a_{21} & a_{22} & \cdots & a_{2n} \\ a_{31} & a_{32} & \cdots & a_{3n} \\ \cdots & \cdots & \cdots & \cdots \\ a_{m1} & a_{m2} & \cdots & a_{mn} \end{bmatrix} ; \quad \mathbf{x} = \begin{bmatrix} x_1 \\ x_2 \\ \cdots \\ x_n \end{bmatrix} ; \quad \mathbf{b} = \begin{bmatrix} b_1 \\ b_2 \\ \cdots \\ b_m \end{bmatrix} \tag{D.37}$$

Three distinct cases of this general problem can be identified:

CASE 1: $m = n$

- Equal number of equations and unknowns (*well-defined system*).
- if $\mathbf{A}$ is nonsingular then n independent solutions exist.

This is the case discussed in Section D.7.

CASE 2: $m > n$

- More equations than unknowns (*overdefined system*).
- No "regular" solutions exist (in the sense that no set of values for the unknowns will *exactly* satisfy *all* the equations simultaneously).

CASE 3: $m < n$

- Fewer equations than unknowns (*underdefined system*).
- An infinite set of solutions exist (in the sense that the excess $n - m$ unknowns can be assigned *any* arbitrary set of values and the remaining m unknowns solved for in terms of these arbitrary values).

For the Case 1 problem, as discussed in Section D.7, whenever $|\mathbf{A}| \neq 0$, the required solution is obtained as given in Eq. (D.35):

$$\mathbf{x} = \mathbf{A}^{-1}\mathbf{b} \tag{D.35}$$

(It is easy to establish that whenever the $n \times n$ matrix $\mathbf{A}$ is singular, the problem reduces to a Case 3 problem with m — now less than n — being equal to the rank of the rank-deficient $\mathbf{A}$.)

In each of the other two cases, solutions of the form given in Eq. (D.35) can be obtained but only after imposing additional conditions as we now show.

Overdefined Systems

The problem here is that there is no vector $\mathbf{x}$ for which $\mathbf{Ax}$ exactly equals $\mathbf{b}$. However, it is possible to obtain a solution in which $\mathbf{Ax}$ is "closest" to $\mathbf{b}$ in the "least square" sense by finding a vector $\hat{\mathbf{x}}$ that minimizes:

$$\Phi = (\mathbf{Ax} - \mathbf{b})^T(\mathbf{Ax} - \mathbf{b}) \tag{D.38}$$

By applying the usual techniques of calculus, it is easy to show that *provided* $(\mathbf{A}^T\mathbf{A})$ *is nonsingular*, this "least squares solution" vector is given by:

$$\hat{\mathbf{x}} = (\mathbf{A}^T\mathbf{A})^{-1}\mathbf{A}^T\mathbf{b} \tag{D.39a}$$

or

$$\hat{\mathbf{x}} = \mathbf{A}^L\mathbf{b} \tag{D.39b}$$

which is of the form in Eq. (D.35) with $\mathbf{A}^L$ defined by:

$$\mathbf{A}^L = (\mathbf{A}^T\mathbf{A})^{-1}\mathbf{A}^T \tag{D.40}$$

now playing the role of $\mathbf{A}^{-1}$.

Note the following about $\mathbf{A}^L$ defined by Eq. (D.40):

1. $\mathbf{A}^L$ is an $n \times m$ matrix while $\mathbf{A}$ is $m \times n$

2. $\mathbf{A}^L\mathbf{A} = \mathbf{I}_n$, the $n \times n$ identity matrix, but $\mathbf{A}\mathbf{A}^L \neq \mathbf{I}$

3. $\mathbf{A}\mathbf{A}^L = \mathbf{L}_1$ an $m \times m$ *idempotent* matrix for which $\mathbf{L}_1\mathbf{L}_1 = \mathbf{L}_1$

4. Premultiplying in (2) above by $\mathbf{A}$ gives:

$$\mathbf{A}\mathbf{A}^L\mathbf{A} = \mathbf{A} \tag{D.41}$$

while postmultiplying by $\mathbf{A}^L$ gives:

$$\mathbf{A}^L\mathbf{A}\mathbf{A}^L = \mathbf{A}^L \tag{D.42}$$

Because only when $\mathbf{A}$ is multiplied on the *left* by $\mathbf{A}^L$ do we get an identity matrix, $\mathbf{A}^L$ is called a *left inverse*.

Underdefined Systems

The problem here is that there is an infinite set of vectors $\mathbf{x}$ for which $\mathbf{Ax}$ exactly equals $\mathbf{b}$. However, it is possible to obtain a solution vector $\mathbf{x}$ of minimum norm by posing the following constrained optimization problem:

$$\text{Min} \quad \Phi = \frac{1}{2}(\mathbf{x}^T\mathbf{x})$$
$$\text{subject to:} \quad \mathbf{Ax} = \mathbf{b}$$

By the usual methods of solving such problems (for example, using Lagrange multipliers) we find that *provided* $(\mathbf{AA}^T)$ *is nonsingular* the required "minimum norm solution" is:

$$\mathbf{x} = \mathbf{A}^T(\mathbf{AA}^T)^{-1}\mathbf{b} \tag{D.43a}$$

or

$$\mathbf{x} = \mathbf{A}^R\mathbf{b} \tag{D.43b}$$

which, once again, is of the form in Eq. (D.35) with $\mathbf{A}^R$ defined by:

$$\mathbf{A}^R = \mathbf{A}^T(\mathbf{AA}^T)^{-1} \tag{D.44}$$

now playing the role of $\mathbf{A}^{-1}$.

Again, we note the following about $\mathbf{A}^R$ defined by Eq. (D.44):

1. $\mathbf{A}^R$ is an $n \times m$ matrix while $\mathbf{A}$ is $m \times n$

2. $\mathbf{AA}^R = \mathbf{I}_n$, the $m \times m$ identity matrix, but $\mathbf{A}^R\mathbf{A} \neq \mathbf{I}$

3. $\mathbf{A}^R\mathbf{A} = \mathbf{L}_2$ another $n \times n$ *idempotent* matrix for which, as usual, $\mathbf{L}_2\mathbf{L}_2 = \mathbf{L}_2$

4. *Post*multiplying the expression in item 2 above by $\mathbf{A}$ gives:

$$\mathbf{AA}^R\mathbf{A} = \mathbf{A} \tag{D.45}$$

while *pre*multiplying by $\mathbf{A}^R$ gives:

$$\mathbf{A}^R\mathbf{AA}^R = \mathbf{A}^R \tag{D.46}$$

In this case, we obtain an identity matrix only when $\mathbf{A}$ is multiplied on the *right* by $\mathbf{A}^R$; therefore $\mathbf{A}^R$ is called a *right inverse*.

Noting the similarities between Eqs. (D.41) and (D.45), and also the similarities between Eqs. (D.42) and (D.46), we can now state the following results:

The general system of linear algebraic equations in Eq. (D.33):

$$\mathbf{Ax} = \mathbf{b} \tag{D.33}$$

with $\mathbf{A}$ as a general $m \times n$ matrix (for which m is not necessarily equal to n) has the solution:

$$\mathbf{x} = \mathbf{A}^+\mathbf{b} \tag{D.47}$$

Here $\mathbf{A}^+$ is referred to as a *generalized* (or *pseudo*) *inverse* of $\mathbf{A}$, and it satisfies the following conditions:

1. $\mathbf{A}\mathbf{A}^+\mathbf{A} = \mathbf{A}$

2. $\mathbf{A}^+\mathbf{A}\,\mathbf{A}^+ = \mathbf{A}^+$

3. *Either* : $\mathbf{A}^+\mathbf{A} = \mathbf{I}$ (for the *left inverse*)
   or: $\mathbf{A}\mathbf{A}^+ = \mathbf{I}$ (for the *right inverse*)

Observe that $\mathbf{A}^{-1}$, the regular inverse of a square, nonsingular matrix satisfies all these conditions. Thus $\mathbf{A}^{-1}$ is actually a special case of $\mathbf{A}^+$, and the above is therefore the most general description of the inverse of *any* matrix. Note, finally, that only $\mathbf{A}^{-1}$ is *both* a left *and* a right inverse; if Eq. (D.33) is an overdefined system of equations, then a least squares solution can be found only via a *left* inverse; for an underdefined system of equations, a minimum norm solution can be found only via a *right inverse*.

Example D.8   SOLVING AN OVERDEFINED SYSTEM OF EQUATIONS BY GENERALIZED INVERSES.

To illustrate the use of generalized inverses in the solution of a system of linear algebraic equations, let us consider the set of equations:

$$
\begin{aligned}
x_1 + 2x_2 &= 25 \\
2x_1 - x_2 &= -25 \\
-2x_1 - x_2 &= -25
\end{aligned}
$$

a set of *three* equations in *two* unknowns. Observe right away that the system is *overdefined* by one equation, and there is no single set of values for x_1 and x_2 that will simultaneously satisfy all three equations. These equations may be written in the form Eq. (D.33):

$$\mathbf{A}\mathbf{x} = \mathbf{b} \tag{D.33}$$

In this case:

$$
\mathbf{A} = \begin{bmatrix} 1 & 2 \\ 2 & -1 \\ -2 & -1 \end{bmatrix}; \quad
\mathbf{x} = \begin{bmatrix} x_1 \\ x_2 \end{bmatrix}; \quad
\mathbf{b} = \begin{bmatrix} 25 \\ -25 \\ -25 \end{bmatrix} \tag{D.48}
$$

and from Eq. (D.47), the "least squares" solution is obtained from:

$$\mathbf{x} = \mathbf{A}^+\mathbf{b} \tag{D.47}$$

where, recalling Eq. (D.40):

$$\mathbf{A}^+ = (\mathbf{A}^T\mathbf{A})^{-1}\mathbf{A}^T$$

Now, $\mathbf{A}^T\mathbf{A}$ is obtained by direct multiplication as:

$$\mathbf{A}^T\mathbf{A} = \begin{bmatrix} 9 & 2 \\ 2 & 6 \end{bmatrix}$$

a nonsingular matrix whose determinant is 50; thus:

$$(\mathbf{A}^T\mathbf{A})^{-1} = \frac{1}{50}\begin{bmatrix} 6 & -2 \\ -2 & 9 \end{bmatrix}$$

so that $\mathbf{A}^+$ is given by:

$$\mathbf{A}^+ = \frac{1}{50}\begin{bmatrix} 2 & 14 & -10 \\ 16 & -13 & -5 \end{bmatrix} \tag{D.49}$$

and

$$\mathbf{x} = \mathbf{A}^+\mathbf{b} = \frac{1}{50}\begin{bmatrix} 2 & 14 & -10 \\ 16 & -13 & -5 \end{bmatrix}\begin{bmatrix} 25 \\ -25 \\ -25 \end{bmatrix}$$

which, upon evaluation, gives:

$$\mathbf{x} = \begin{bmatrix} x_1 \\ x_2 \end{bmatrix} = \begin{bmatrix} -1 \\ 17 \end{bmatrix} \tag{D.50}$$

We now note that the solution in Eq. (D.50) "satisfies" the system of equations only in the least squares sense; i.e., of all possible $\mathbf{x}$ choices, this particular $\mathbf{x}$ given in Eq. (D.50) provides an $\mathbf{Ax}$ that comes closest to $\mathbf{b}$ in the least squares sense.

We may also observe that, as asserted earlier:

$$\mathbf{A}^+\mathbf{A} = \frac{1}{50}\begin{bmatrix} 2 & 14 & -10 \\ 16 & -13 & -5 \end{bmatrix}\begin{bmatrix} 1 & 2 \\ 2 & -1 \\ -2 & -1 \end{bmatrix}$$

$$= \begin{bmatrix} 1 & 0 \\ 0 & 1 \end{bmatrix}$$

a 2×2 identity matrix; also:

$$\mathbf{A}\mathbf{A}^+ = \frac{1}{50}\begin{bmatrix} 1 & 2 \\ 2 & -1 \\ -2 & -1 \end{bmatrix}\begin{bmatrix} 2 & 14 & -10 \\ 16 & -13 & -5 \end{bmatrix}$$

$$= \begin{bmatrix} 0.68 & -0.24 & -0.4 \\ -0.24 & 0.82 & -0.3 \\ -0.4 & -0.3 & 0.5 \end{bmatrix} \qquad \text{(D.51)}$$

It will be a particularly worthwhile exercise for the reader to carefully multiply this matrix given in Eq. (D.51) by itself and therefore confirm that it is indeed an idempotent matrix.

Example D.9   SOLVING AN UNDERDEFINED SYSTEM OF EQUATIONS BY GENERALIZED INVERSES.

Let us now consider the problem of solving the following set of equations:

$$x_1 + 2x_2 + 3x_3 + 4x_4 = 10$$

$$x_1 + x_2 + x_3 + x_4 = 2$$

a set of *two* equations in *four* unknowns, clearly underdefined. Observe right away that we can arbitrarily assign any values to, say, x_1, and x_2, and solve the resulting equations for x_3 and x_4. There is thus an infinite number of solutions.

If these equations are written in the form Eq. (D.33):

$$\mathbf{Ax} = \mathbf{b} \qquad \text{(D.33)}$$

we would have:

$$\mathbf{A} = \begin{bmatrix} 1 & 2 & 3 & 4 \\ 1 & 1 & 1 & 1 \end{bmatrix}; \quad \mathbf{x} = \begin{bmatrix} x_1 \\ x_2 \\ x_3 \\ x_4 \end{bmatrix}; \quad \mathbf{b} = \begin{bmatrix} 10 \\ 2 \end{bmatrix} \qquad \text{(D.52)}$$

and from Eq. (D.47), the "minimum norm" solution is obtained from:

$$\mathbf{x} = \mathbf{A}^+\mathbf{b} \qquad \text{(D.47)}$$

where we now recall from Eq. (D.44) that:

$$\mathbf{A}^+ = \mathbf{A}^T(\mathbf{A}\mathbf{A}^T)^{-1}$$

By direct multiplication, we obtain $\mathbf{A}\mathbf{A}^T$ as:

$$\mathbf{A}\mathbf{A}^T = \begin{bmatrix} 30 & 10 \\ 10 & 4 \end{bmatrix}$$

and hence:

$$(\mathbf{A}\mathbf{A}^T)^{-1} = \frac{1}{20}\begin{bmatrix} 4 & -10 \\ -10 & 30 \end{bmatrix}$$

Thus, we have that $\mathbf{A}^+$ is given by:

$$\mathbf{A}^+ = \frac{1}{20} \begin{bmatrix} 1 & 1 \\ 2 & 1 \\ 3 & 1 \\ 4 & 1 \end{bmatrix} \begin{bmatrix} 4 & -10 \\ -10 & 30 \end{bmatrix}$$

or

$$\mathbf{A}^+ = \begin{bmatrix} -0.3 & 1 \\ -0.1 & 0.5 \\ 0.1 & 0 \\ 0.3 & -0.5 \end{bmatrix} \tag{D.53}$$

and therefore:

$$\mathbf{x} = \mathbf{A}^+\mathbf{b} = \begin{bmatrix} -0.3 & 1 \\ -0.1 & 0.5 \\ 0.1 & 0 \\ 0.3 & -0.5 \end{bmatrix} \begin{bmatrix} 10 \\ 2 \end{bmatrix}$$

which, upon evaluation, gives:

$$\mathbf{x} = \begin{bmatrix} x_1 \\ x_2 \\ x_3 \\ x_4 \end{bmatrix} = \begin{bmatrix} -1 \\ 0 \\ 1 \\ 2 \end{bmatrix} \tag{D.54}$$

The reader may now wish to verify that Eq. (D.54) does in fact satisfy the system of equations exactly, and that for the $\mathbf{A}$ given in Eq. (D.52) and $\mathbf{A}^+$ given in Eq. (D.53), $\mathbf{A}\mathbf{A}^+$ results in the 2×2 identity matrix, while $\mathbf{A}^+\mathbf{A}$ results in the 4×4 idempotent matrix:

$$\mathbf{A}^+\mathbf{A} = \begin{bmatrix} 0.7 & 0.4 & 0.1 & -0.2 \\ 0.4 & 0.3 & 0.2 & 0.1 \\ 0.1 & 0.2 & 0.3 & 0.4 \\ -0.2 & 0.1 & 0.4 & 0.7 \end{bmatrix}$$

D.9 EIGENVALUES, EIGENVECTORS, AND MATRIX DIAGONALIZATION

The eigenvalue/eigenvector problem arises in the determination of the values of a constant λ for which the following set of n linear algebraic equations has nontrivial solutions:

$$a_{11}x_1 + a_{12}x_2 + \dots + a_{1n}x_n = \lambda x_1$$

$$a_{21}x_1 + a_{22}x_2 + \dots + a_{2n}x_n = \lambda x_2$$

$$\dots \quad \dots \quad \dots \quad \dots \quad \dots$$

$$a_{n1}x_1 + a_{n2}x_2 + \dots + a_{nn}x_n = \lambda x_n \qquad (D.55)$$

This can be reexpressed as:

$$\mathbf{Ax} = \lambda\mathbf{x}$$

or

$$(\mathbf{A} - \lambda\mathbf{I})\mathbf{x} = 0 \qquad (D.56)$$

Being a system of linear homogeneous equations, the solution to Eq. (D.55) or equivalently Eq. (D.56) will be nontrivial (i.e., $\mathbf{x} \neq 0$) if and only if:

$$|\mathbf{A} - \lambda\mathbf{I}| = 0 \qquad (D.57)$$

Expanding this determinant results in a polynomial of order n in λ:

$$a_0\lambda^n + a_1\lambda^{n-1} + a_2\lambda^{n-2} \dots + a_{n-1}\lambda + a_n = 0 \qquad (D.58)$$

This equation (or its determinant form in Eq. (D.57)) is called the *characteristic equation* of the $n \times n$ matrix $\mathbf{A}$; its n roots, $\lambda_1, \lambda_2, \lambda_3, \dots, \lambda_n$ — which may be real or imaginary, and may or may not be distinct — are the *eigenvalues* of the matrix.

Example D.10   EIGENVALUES OF A 2 × 2 MATRIX.

Let us illustrate the computation of eigenvalues by considering the matrix:

$$\mathbf{A} = \begin{bmatrix} 1 & 2 \\ 3 & -4 \end{bmatrix}$$

In this case, the matrix $(\mathbf{A} - \lambda\mathbf{I})$ is obtained as:

$$(\mathbf{A} - \lambda\mathbf{I}) = \begin{bmatrix} 1 & 2 \\ 3 & -4 \end{bmatrix} - \begin{bmatrix} \lambda & 0 \\ 0 & \lambda \end{bmatrix} = \begin{bmatrix} (1-\lambda) & 2 \\ 3 & (-4-\lambda) \end{bmatrix}$$

from which the characteristic equation is obtained as:

$$|\mathbf{A} - \lambda\mathbf{I}| = -(1-\lambda)(4+\lambda) - 6 = \lambda^2 + 3\lambda - 10 = 0$$

a second-order polynomial (quadratic) with the solution:

$$\lambda_1, \lambda_2 = -5, 2$$

Thus in this case the eigenvalues are both *real*. Note that $\lambda_1 + \lambda_2 = -3$, and that $\lambda_1\lambda_2 = -10$; also note that $\text{Tr}(A) = 1 - 4 = -3$ and $|A| = -10$. Thus, we see that $\lambda_1 + \lambda_2 = \text{Tr}(A)$ and $\lambda_1\lambda_2 = |A|$.

If we change **A** slightly to:

$$A = \begin{bmatrix} 1 & 2 \\ -3 & 4 \end{bmatrix}$$

then the characteristic equation becomes:

$$|A - \lambda I| = \begin{vmatrix} 1 - \lambda & 2 \\ -3 & 4 - \lambda \end{vmatrix} = \lambda^2 - 5\lambda + 10 = 0$$

whose roots are:

$$\lambda_1, \lambda_2 = \frac{1}{2}\left[5 \pm \left(\sqrt{15}\right)j\right]$$

which are *complex conjugates*. Note again that $\text{Tr}(A) = 5 = \lambda_1 + \lambda_2$ and $|A| = 10 = \lambda_1\lambda_2$.

Finally if we make another small change in **A**, i.e.:

$$A = \begin{bmatrix} 1 & 2 \\ -3 & -1 \end{bmatrix}$$

then the characteristic equation is:

$$|A - \lambda I| = \lambda^2 - 1 + 6 = 0$$

which has solutions:

$$\lambda_1, \lambda_2 = \pm\left(\sqrt{5}\right)j$$

so that the eigenvalues are *purely imaginary* in this case. Once again $\text{Tr}(A) = 0 = \lambda_1 + \lambda_2$ and $|A| = 5 = \lambda_1\lambda_2$.

Returning now to the original problem in Eqs. (D.55) and (D.56), we note that every distinct eigenvalue λ when substituted into Eq. (D.56) gives rise to a corresponding nontrivial solution x_j for which:

$$(A - \lambda_j I) x_j = 0 \tag{D.59}$$

These nontrivial solution vectors are called the *eigenvectors* of **A**. Thus every distinct eigenvalue λ_j of **A** has associated with it a corresponding eigenvector x_j, $j = 1, 2, \ldots, n$. When some eigenvalues are repeated, some specialized considerations — which we will not discuss here — are required. (See the bibliography at the end of the appendix for more details.) We shall restrict ourselves here to the case of distinct eigenvalues.

The following are some useful properties of eigenvectors:

1. Eigenvectors associated with separate, distinct eigenvalues are *linearly independent*.

2. The eigenvectors x_j are solutions of a set of homogeneous linear equations Eq. (D.59) and are thus determined only up to a scalar multiplier; i.e., x_j and kx_j will both be a solution to Eq. (D.59), if k is a scalar constant. We obtain a *normalized* eigenvector $\bar{x}_j$ when the eigenvector x_j is scaled by the vector norm $\| x_j \|$, i.e.:

$$\bar{x}_j = \frac{x_j}{\| x_j \|}$$

3. The normalized eigenvector $\bar{x}_j$ is a valid eigenvector since it is a scalar multiple of x_j; it enjoys the sometimes desirable property that it has unity norm, i.e., $\| \bar{x}_j \| = 1$.

4. The eigenvectors x_j may be calculated by resorting to methods of solution of homogeneous linear equations.

It can be shown that the eigenvector x_j associated with the eigenvalue λ_j is given by *any* nonzero column of $adj(A - \lambda_j I)$; it can also be shown that there will always be one and only one independent, nonzero column of adj $(A - \lambda_j I)$.

Example D.11   EIGENVECTORS OF A 2 × 2 MATRIX.

Let us illustrate the computation of eigenvectors by considering the 2 × 2 matrix:

$$A = \begin{bmatrix} 1 & 2 \\ 3 & -4 \end{bmatrix}$$

for which eigenvalues, $\lambda_1 = -5$, $\lambda_2 = 2$ were computed in Example D.10. For λ_1:

$$(A - \lambda_1 I) = \begin{bmatrix} 6 & 2 \\ 3 & 1 \end{bmatrix}$$

and the adjoint is:

$$adj\,(A - \lambda_1 I) = \begin{bmatrix} 1 & -2 \\ -3 & 6 \end{bmatrix}$$

We may now choose either column of the adjoint as the eigenvector x, since one is a scalar multiple of the other; let us choose:

$$x_1 = \begin{bmatrix} 1 \\ -3 \end{bmatrix}$$

as the eigenvector for $\lambda_1 = -5$. Since the norm of this vector is $\sqrt{1^2 + (-3)^2} = \sqrt{10}$, the *normalized eigenvector* corresponding to $\lambda_1 = -5$ will be:

$$\overline{\mathbf{x}}_1 = \begin{bmatrix} \dfrac{1}{\sqrt{10}} \\[2ex] \dfrac{-3}{\sqrt{10}} \end{bmatrix}$$

Similarly for $\lambda_2 = 2$:

$$(\mathbf{A} - \lambda_2 \mathbf{I}) = \begin{bmatrix} -1 & 2 \\ 3 & -6 \end{bmatrix}$$

and

$$\text{adj}\,(\mathbf{A} - \lambda_2 \mathbf{I}) = \begin{bmatrix} -6 & -2 \\ -3 & -1 \end{bmatrix}$$

Thus:

$$\mathbf{x}_2 = \begin{bmatrix} 2 \\ 1 \end{bmatrix}$$

can be chosen as the eigenvector for eigenvalue $\lambda_2 = 2$. The *normalized eigenvector* in this case is:

$$\overline{\mathbf{x}}_2 = \begin{bmatrix} \dfrac{2}{\sqrt{5}} \\[2ex] \dfrac{1}{\sqrt{5}} \end{bmatrix}$$

Let us return again to the original eigenvalue/eigenvector problem, and choose the calculated n linearly independent eigenvectors $\mathbf{x}_j$, $j = 1, 2, \ldots, n$ to be the columns of a matrix $\mathbf{M}$, i.e.:

$$\mathbf{M} = \begin{bmatrix} \mathbf{x}_1 \vdots \mathbf{x}_2 \vdots \ldots \vdots \mathbf{x}_n \end{bmatrix} \tag{D.60}$$

$\mathbf{M}$ is called a *modal* matrix. Obviously, if the eigenvalues λ_j are distinct, then the eigenvectors $\mathbf{x}_j$ are all independent, and the modal matrix $\mathbf{M}$ will be *nonsingular*, and $\mathbf{M}^{-1}$ will always exist.

This modal matrix is very useful in converting a matrix $\mathbf{A}$ to *diagonal form*. Recall the eigenvalue/eigenvector problem: to determine the values of λ_j for which nontrivial solutions vectors $\mathbf{x}_j$ can be found for:

$$\mathbf{A}\mathbf{x}_j = \lambda_j \mathbf{x}_j; \qquad j = 1, 2, \ldots, n \tag{D.61}$$

For each j, the complete set of equations may be written more compactly as:

$$\mathbf{A}\begin{bmatrix} \mathbf{x}_1 \vdots \mathbf{x}_2 \vdots \ldots \vdots \mathbf{x}_n \end{bmatrix} = \begin{bmatrix} \lambda_1 \mathbf{x}_1 \vdots \lambda_2 \mathbf{x}_2 \vdots \ldots \vdots \lambda_n \mathbf{x}_n \end{bmatrix} \tag{D.62}$$

If we define the diagonal matrix Λ as:

$$\Lambda = \begin{bmatrix} \lambda_1 & 0 & 0 & \dots & 0 \\ 0 & \lambda_2 & 0 & \dots & 0 \\ \dots & \dots & \dots & \dots & \dots \\ 0 & 0 & 0 & \dots & \lambda_n \end{bmatrix} \qquad (D.63)$$

a matrix whose diagonal elements are the n eigenvalues of $\mathbf{A}$, and observe that the square bracket on the left-hand side in Eq. (D.62) is just the modal matrix $\mathbf{M}$, then we see immediately that the square bracket on the right-hand side is just $\mathbf{M}\Lambda$. Thus Eq. (D.62) may be written as:

$$\mathbf{AM} = \mathbf{M}\Lambda \qquad (D.64)$$

and premultiplying both sides by $\mathbf{M}^{-1}$ thus yields:

$$\mathbf{M}^{-1}\mathbf{AM} = \Lambda \qquad (D.65)$$

The $n \times n$ matrix $\mathbf{A}$ is thus reduced to what is known as the *diagonal canonical form* by using the transformation Eq. (D.65). By pre– and postmultiplying both sides of Eq. (D.65) by $\mathbf{M}$ and $\mathbf{M}^{-1}$ respectively, one obtains the *companion relation*:

$$\mathbf{A} = \mathbf{M}\Lambda\mathbf{M}^{-1} \qquad (D.66)$$

These two relations are very useful in the solution of linear algebraic and differential equations as shall be seen later. Note that the latter relation in Eq. (D.66) indicates that a square matrix $\mathbf{A}$ can be decomposed into three matrices involving only its eigenvalues and eigenvectors.

Example D.12   MATRIX DIAGONALIZATION.

To illustrate matrix diagonalization, let us consider the 2×2 matrix:

$$\mathbf{A} = \begin{bmatrix} 1 & 2 \\ 3 & -4 \end{bmatrix}$$

discussed in Examples D.10 and D.11. Recall that the eigenvalues and eigenvectors were determined to be:

$$\lambda_1 = -5, \ \mathbf{x}_1 = \begin{bmatrix} 1 \\ -3 \end{bmatrix}$$

$$\lambda_2 = 2, \ \mathbf{x}_2 = \begin{bmatrix} 2 \\ 1 \end{bmatrix}$$

Thus the modal matrix $\mathbf{M}$ is:

$$\mathbf{M} = \begin{bmatrix} \mathbf{x}_1 \vdots \mathbf{x}_2 \end{bmatrix} = \begin{bmatrix} 1 & 2 \\ -3 & 1 \end{bmatrix}$$

while

$$\mathbf{M}^{-1} = \begin{bmatrix} \dfrac{1}{7} & -\dfrac{2}{7} \\[2mm] \dfrac{3}{7} & \dfrac{1}{7} \end{bmatrix}$$

and the diagonal eigenvalue matrix is:

$$\Lambda = \begin{bmatrix} -5 & 0 \\ 0 & 2 \end{bmatrix}$$

Evaluation of the LHS of Eq. (D.65) yields:

$$\mathbf{M}^{-1}\mathbf{AM} = \begin{bmatrix} \dfrac{1}{7} & -\dfrac{2}{7} \\[2mm] \dfrac{3}{7} & \dfrac{1}{7} \end{bmatrix} \begin{bmatrix} 1 & 2 \\ 3 & -4 \end{bmatrix} \begin{bmatrix} 1 & 2 \\ -3 & 1 \end{bmatrix}$$

$$= \begin{bmatrix} \dfrac{1}{7} & -\dfrac{2}{7} \\[2mm] \dfrac{3}{7} & \dfrac{1}{7} \end{bmatrix} \begin{bmatrix} -5 & 4 \\ 15 & 2 \end{bmatrix} = \begin{bmatrix} -5 & 0 \\ 0 & 2 \end{bmatrix}$$

which verifies Eq. (D.65). Similarly, evaluation of the RHS of Eq. (D.66) gives:

$$\mathbf{M}\Lambda\mathbf{M}^{-1} = \begin{bmatrix} 1 & 2 \\ -3 & 1 \end{bmatrix} \begin{bmatrix} -5 & 0 \\ 0 & 2 \end{bmatrix} \begin{bmatrix} \dfrac{1}{7} & -\dfrac{2}{7} \\[2mm] \dfrac{3}{7} & \dfrac{1}{7} \end{bmatrix}$$

$$= \begin{bmatrix} -5 & 4 \\ 15 & 2 \end{bmatrix} \begin{bmatrix} \dfrac{1}{7} & -\dfrac{2}{7} \\[2mm] \dfrac{3}{7} & \dfrac{1}{7} \end{bmatrix} = \begin{bmatrix} 1 & 2 \\ 3 & -4 \end{bmatrix}$$

verifying the equation.

The following are some general properties of eigenvalues worth noting:

1. The *sum* of the eigenvalues of a matrix is equal to the *trace* of that matrix, i.e.:

$$\mathrm{Tr}(\mathbf{A}) = \sum_{j=1}^{n} \lambda_j$$

2. The *product* of the eigenvalues of a matrix is equal to the determinant of that matrix, i.e.:

$$|\mathbf{A}| = \prod_{j=1}^{n} \lambda_j$$

3.   A singular matrix has at least one zero eigenvalue.

4.   If the eigenvalues of A are $\lambda_1, \lambda_2, \lambda_3, ..., \lambda_n$, then the eigenvalues of A^{-1} are $1/\lambda_1, 1/\lambda_2, 1/\lambda_3, ..., 1/\lambda_n$.

5.   The eigenvalues of a diagonal or a triangular matrix are identical to the elements on the main diagonal.

6.   Given any nonsingular matrix T, the matrices A, and $\tilde{A} = TAT^{-1}$ have identical eigenvalues. In this case, such matrices A and $\tilde{A}$ are called similar matrices.

D.10   SINGULAR VALUE DECOMPOSITION

The concept of matrix diagonalization indicated in Eq. (D.65) and the companion relation in Eq. (D.66) can be extended to all matrices, square and nonsquare, by singular value decomposition (SVD). The main SVD result may be summarized as follows:

For any real $m \times n$ matrix A, it is always possible to find orthogonal (i.e., unitary) matrices W and V such that:

$$W^T A V = \Sigma \tag{D.67}$$

Here, Σ is the $m \times n$ matrix:

$$\Sigma = \begin{bmatrix} S & 0 \\ 0 & 0 \end{bmatrix}; \text{ with } S = \begin{bmatrix} \sigma_1 & 0 & 0 & ... & 0 \\ 0 & \sigma_2 & 0 & ... & 0 \\ ... & ... & ... & ... & ... \\ 0 & 0 & 0 & ... & \sigma_r \end{bmatrix} \tag{D.68}$$

where, for $p = \min(m,n)$, the diagonal elements of S, $\sigma_1 \geq \sigma_2 \geq ... \geq \sigma_r \geq 0$, $(r \leq p)$, together with $\sigma_{r+1} = 0, ..., \sigma_p = 0$ are called the *singular values* of A; these are the positive square roots of the eigenvalues of $A^T A$; r is the rank of A.

W is the $m \times m$ matrix:

$$W = \begin{bmatrix} w_1 \vdots w_2 \vdots ... \vdots w_m \end{bmatrix} \tag{D.69}$$

whose m columns, $w_i, i = 1, 2, ..., m$, are called the *left singular vectors* of A; these are the normalized (orthonormal) eigenvectors of AA^T. V is the $n \times n$ matrix:

$$V = \begin{bmatrix} v_1 \vdots v_2 \vdots ... \vdots v_n \end{bmatrix} \tag{D.70}$$

whose n columns $v_i, i = 1, 2,...,n$, are called the *right singular vectors* of A; these are the normalized (orthonormal) eigenvectors of $A^T A$.

By virtue of their being composed of orthonormal vectors, the matrices W and V are orthogonal (or unitary) matrices, i.e.:

$$W^T W = I = WW^T$$

so that

$$\mathbf{W}^{-1} = \mathbf{W}^T$$

Also:

$$\mathbf{V}^T\mathbf{V} = \mathbf{I} = \mathbf{V}\mathbf{V}^T$$

so that:

$$\mathbf{V}^{-1} = \mathbf{V}^T$$

By applying these properties of unitary matrices, we easily obtain from Eq. (D.67) the companion relation:

$$\mathbf{A} = \mathbf{W}\Sigma\mathbf{V}^T \qquad (D.71)$$

It can be shown that analogously to the eigenvalue/eigenvector expression for square matrices in Eq. (D.61), we have the more general pair of expressions:

$$\mathbf{A}\mathbf{v}_i = \sigma_i\mathbf{w}_i \qquad (D.72a)$$
$$\mathbf{A}^T\mathbf{w}_i = \sigma_i\mathbf{v}_i \qquad (D.72b)$$

The ratio of the largest to the smallest singular value is designated the *condition number* of $\mathbf{A}$, i.e.:

$$\kappa(\mathbf{A}) = \frac{\sigma_1}{\sigma_p} \qquad (D.73)$$

and it provides the most reliable indication of how close $\mathbf{A}$ is to being singular. Note that for a singular matrix, $\kappa(\mathbf{A}) = \infty$; thus nearness to singularity is indicated by excessively large — but finite — values for $\kappa(\mathbf{A})$.

There are several important applications of SVD, particularly in numerical linear algebra. Its applications in process control (discussed in Chapters 20 and 22) typically involves square, invertible matrices. It should be noted however, that the full benefit of SVD is realized in dealing with problems requiring the analysis of nonsquare, possibly rank-deficient matrices. The following example illustrates SVD applied to a nonsquare matrix; keep in mind, however, that efficient computer programs are available for SVD calculations.

Example D.13   **SINGULAR VALUE DECOMPOSITION OF A 3×2 MATRIX**

To illustrate SVD, let us consider the 3×2 matrix:

$$\mathbf{A} = \begin{bmatrix} 1 & 2 \\ 2 & -1 \\ -2 & 1 \end{bmatrix} \qquad (D.74)$$

a slightly modified version of which was studied in Example D.8. For this matrix:

$$A^T A = \begin{bmatrix} 9 & -2 \\ -2 & 6 \end{bmatrix}$$

whose eigenvalues are obtained as 10, and 5; thus the *singular values* of **A** are:

$$\sigma_1, \sigma_2 = \sqrt{10}, \sqrt{5}$$

ordered such that $\sigma_1 > \sigma_2$ as required by SVD analysis. Thus, the 3×2 matrix Σ in this case is given by:

$$\Sigma = \begin{bmatrix} \sqrt{10} & 0 \\ 0 & \sqrt{5} \\ 0 & 0 \end{bmatrix} \tag{D.75}$$

Right Singular Vectors

The first eigenvector of $A^T A$ corresponding to λ_1 is obtained as usual from $\mathrm{adj}\,(A^T A - \lambda_1 I)$, and since:

$$\mathrm{adj}\,(A^T A - \lambda_1 I) = \begin{bmatrix} -4 & 2 \\ 2 & -1 \end{bmatrix}$$

a possible choice for the eigenvector is the second column. Normalizing this with $\sqrt{2^2 + 1^2} = \sqrt{5}$, the norm of the vector, we obtain the first right singular vector v_1 corresponding to $\sigma_1 = \sqrt{10}$ as:

$$v_1 = \begin{bmatrix} \dfrac{2}{\sqrt{5}} \\ \dfrac{-1}{\sqrt{5}} \end{bmatrix}$$

In similar fashion, the second *normalized* eigenvector corresponding to $\lambda_2 = 5$ is obtained as:

$$v_2 = \begin{bmatrix} \dfrac{1}{\sqrt{5}} \\ \dfrac{2}{\sqrt{5}} \end{bmatrix}$$

Thus:

$$V = \begin{bmatrix} \dfrac{2}{\sqrt{5}} & \dfrac{1}{\sqrt{5}} \\ \dfrac{-1}{\sqrt{5}} & \dfrac{2}{\sqrt{5}} \end{bmatrix}$$

and therefore:

$$\mathbf{V}^T \;=\; \begin{bmatrix} \dfrac{2}{\sqrt{5}} & \dfrac{-1}{\sqrt{5}} \\[2ex] \dfrac{1}{\sqrt{5}} & \dfrac{2}{\sqrt{5}} \end{bmatrix} \tag{D.76}$$

The reader may want to verify that this $\mathbf{V}$ is truly a unitary matrix by evaluating $\mathbf{V}^T\mathbf{V}$ as well as $\mathbf{V}\mathbf{V}^T$ and confirming that each matrix product gives the 2×2 identity matrix.

Left Singular Vectors

For the given matrix $\mathbf{A}$:

$$\mathbf{A}\mathbf{A}^T \;=\; \begin{bmatrix} 5 & 0 & 0 \\ 0 & 5 & -5 \\ 0 & -5 & 5 \end{bmatrix}$$

The eigenvalues of this 3×3 matrix are obtained from the characteristic equation which in this case is:

$$(5-\lambda)\big[(5-\lambda)^2 - 25\big] \;=\; 0$$

i.e.:

$$\lambda_1,\, \lambda_2,\, \lambda_3 \;=\; 10,\, 5,\, 0$$

Note that the nonzero eigenvalues of $\mathbf{A}\mathbf{A}^T$ are identical to the eigenvalues of $\mathbf{A}^T\mathbf{A}$.

For $\lambda_1 = 10$:

$$\left(\mathbf{A}\mathbf{A}^T - \lambda_1\mathbf{I}\right) \;=\; \begin{bmatrix} -5 & 0 & 0 \\ 0 & -5 & -5 \\ 0 & -5 & -5 \end{bmatrix}$$

To find the adjoint of this matrix, we first find the cofactors and take the transpose of the matrix of cofactors. In this case, this is easily obtained as:

$$\mathrm{adj}\left(\mathbf{A}\mathbf{A}^T - \lambda_1\mathbf{I}\right) \;=\; \begin{bmatrix} 0 & 0 & 0 \\ 0 & 25 & -25 \\ 0 & -25 & 25 \end{bmatrix}$$

and by normalizing any of the nonzero columns, we obtain the first left singular vector of $\mathbf{A}$ as:

$$\mathbf{w}_1 \;=\; \begin{bmatrix} 0 \\[1ex] \dfrac{1}{\sqrt{2}} \\[2ex] \dfrac{-1}{\sqrt{2}} \end{bmatrix}$$

By a similar procedure the second and third left singular vectors respectively corresponding to the eigenvalues $\lambda_2 = 5$, and $\lambda_3 = 0$ are obtained as:

$$\mathbf{w}_2 = \begin{bmatrix} 1 \\ 0 \\ 0 \end{bmatrix} ; \quad \mathbf{w}_3 = \begin{bmatrix} 0 \\ \dfrac{1}{\sqrt{2}} \\ \dfrac{1}{-\sqrt{2}} \end{bmatrix}$$

with the result that:

$$\mathbf{W} = \begin{bmatrix} 0 & 1 & 0 \\ \dfrac{1}{\sqrt{2}} & 0 & \dfrac{1}{\sqrt{2}} \\ \dfrac{-1}{\sqrt{2}} & 0 & \dfrac{1}{\sqrt{2}} \end{bmatrix} \tag{D.77}$$

Once again, the reader may wish to verify that this is also a unitary matrix.

The matrices given in Eqs. (D.75), (D.76), and (D.77) constitute the *singular value decomposition* of the matrix $\mathbf{A}$ in Eq. (D.71). We now observe that:

$$\mathbf{W\Sigma V}^T = \begin{bmatrix} 0 & 1 & 0 \\ \dfrac{1}{\sqrt{2}} & 0 & \dfrac{1}{\sqrt{2}} \\ \dfrac{-1}{\sqrt{2}} & 0 & \dfrac{1}{\sqrt{2}} \end{bmatrix} \begin{bmatrix} \sqrt{10} & 0 \\ 0 & \sqrt{5} \\ 0 & 0 \end{bmatrix} \begin{bmatrix} \dfrac{2}{\sqrt{5}} & \dfrac{-1}{\sqrt{5}} \\ \dfrac{1}{\sqrt{5}} & \dfrac{2}{\sqrt{5}} \end{bmatrix}$$

$$= \begin{bmatrix} 0 & \sqrt{5} \\ \sqrt{5} & 0 \\ -\sqrt{5} & 0 \end{bmatrix} \begin{bmatrix} \dfrac{2}{\sqrt{5}} & \dfrac{-1}{\sqrt{5}} \\ \dfrac{1}{\sqrt{5}} & \dfrac{2}{\sqrt{5}} \end{bmatrix} = \begin{bmatrix} 1 & 2 \\ 2 & -1 \\ -2 & 1 \end{bmatrix} = \mathbf{A}$$

as in Eq. (D.74). As an exercise, the reader may also wish to verify Eq. (D.67) for this example.

D.11 SOLUTION OF LINEAR DIFFERENTIAL EQUATIONS AND THE MATRIX EXPONENTIAL

It is possible to use the results of matrix algebra to provide general solutions to sets of linear ordinary differential equations. Let us recall a dynamic system with control and disturbances neglected:

$$\frac{d\mathbf{x}}{dt} = \mathbf{Ax} ; \quad \mathbf{x}(0) = \mathbf{x}_0 \tag{D.78}$$

Here $\mathbf{A}$ is a constant $n \times n$ matrix with distinct eigenvalues, and $\mathbf{x}(t)$ is a time-varying solution vector. Let us define a new set of n variables, $\mathbf{z}(t)$, as:

$$\mathbf{z}(t) = \mathbf{M}^{-1}\mathbf{x}(t) \tag{D.79}$$

so that:

$$x(t) = Mz(t) \qquad \text{(D.80)}$$

Here **M** is the modal matrix described above. Substituting Eq. (D.80) into Eq. (D.78) yields:

$$M\frac{dz(t)}{dt} = AMz(t) \; ; \qquad z(0) = M^{-1}x_0 \qquad \text{(D.81)}$$

Premultiplying both sides by M^{-1} gives:

$$M^{-1}M\frac{dz(t)}{dt} = M^{-1}AMz(t)$$

or

$$\frac{dz(t)}{dt} = \Lambda z(t) \; ; \qquad z(0) = z_0 = M^{-1}x_0 \qquad \text{(D.82)}$$

Because Λ is diagonal, these equations may now be written as:

$$\frac{dz_1(t)}{dt} = \lambda_1 z_1(t) \; ; \qquad z_1(0) = z_{10}$$

$$\frac{dz_2(t)}{dt} = \lambda_2 z_2(t) \; ; \qquad z_2(0) = z_{20}$$

$$\cdots \cdots \quad \cdots \cdots \qquad \qquad \cdots$$

$$\frac{dz_n(t)}{dt} = \lambda_n z_n(t) \; ; \qquad z_n(0) = z_{n0}$$

Thus the transformation Eq. (D.80) has converted the original Eqs. (D.78) into n completely decoupled equations, each of which has a solution of the form:

$$z_j(t) = e^{\lambda_j t} z_{j0} \; ; \qquad j = 1, 2, \ldots, n$$

This may be expressed in matrix notation as:

$$z(t) = e^{\Lambda t} z_0 \qquad \text{(D.83)}$$

where

$$e^{\Lambda t} = \begin{bmatrix} e^{\lambda_1 t} & 0 & 0 & \cdots & 0 \\ 0 & e^{\lambda_2 t} & 0 & \cdots & 0 \\ \cdots & \cdots & \cdots & \cdots & \cdots \\ 0 & 0 & 0 & \cdots & e^{\lambda_n t} \end{bmatrix} \qquad \text{(D.84)}$$

is the *matrix exponential* for the diagonal matrix, Λ. The solution Eq. (D.83) may be put back in terms of the original variables, $x(t)$, by using Eq. (D.79):

$$M^{-1}x(t) = e^{\Lambda t} M^{-1} x_0$$

or

$$\mathbf{x}(t) = \mathbf{M}e^{\Lambda t}\mathbf{M}^{-1}\mathbf{x}_0 \tag{D.85}$$

This is often written:

$$\mathbf{x}(t) = e^{\Lambda t}\mathbf{x}_0 \tag{D.86}$$

where the matrix $e^{\mathbf{A}t}$, defined as:

$$e^{\mathbf{A}t} = \mathbf{M}e^{\Lambda t}\mathbf{M}^{-1} \tag{D.87}$$

is called the *matrix exponential* of $\mathbf{A}t$. It is very important to note that Eqs. (D.86) and (D.87) provide the general solution to any set of homogeneous equations Eq. (D.78) provided $\mathbf{A}$ has distinct eigenvalues.

Example D.14   SOLUTION OF LINEAR MATRIX DIFFERENTIAL EQUATIONS.

To illustrate the solution of linear, constant coefficient, homogeneous differential equations, let us consider the system:

$$\frac{d\mathbf{x}}{dt} = \mathbf{A}\mathbf{x} \ ; \quad \mathbf{x} = \mathbf{x}_0 \tag{D.88}$$

where

$$\mathbf{A} = \begin{bmatrix} 1 & 2 \\ 3 & -4 \end{bmatrix}, \quad \mathbf{x}_0 = \begin{bmatrix} 1 \\ 2 \end{bmatrix}$$

Observe that $\mathbf{A}$ is the same matrix we have analyzed in Examples D.10, D.11, and D.12 so we already know its eigenvalues, λ_1, λ_2, and modal matrix, $\mathbf{M}$. Thus, we have that:

$$\Lambda = \begin{bmatrix} -5 & 0 \\ 0 & 2 \end{bmatrix}, \ \mathbf{M} = \begin{bmatrix} 1 & 2 \\ -3 & 1 \end{bmatrix}, \ \mathbf{M}^{-1} = \begin{bmatrix} \frac{1}{7} & -\frac{2}{7} \\ \frac{3}{7} & \frac{1}{7} \end{bmatrix}$$

Using Eq. (D.87) we have that:

$$e^{\mathbf{A}t} = \mathbf{M}e^{\Lambda t}\mathbf{M}^{-1} = \begin{bmatrix} 1 & 2 \\ -3 & 1 \end{bmatrix} \begin{bmatrix} e^{-5t} & 0 \\ 0 & e^{2t} \end{bmatrix} \begin{bmatrix} \frac{1}{7} & -\frac{2}{7} \\ \frac{3}{7} & \frac{1}{7} \end{bmatrix}$$

$$e^{\mathbf{A}t} = \begin{bmatrix} \dfrac{e^{-5t} + 6e^{2t}}{7} & \dfrac{-2e^{-5t} + 2e^{2t}}{7} \\ \dfrac{-3e^{-5t} + 3e^{2t}}{7} & \dfrac{6e^{-5t} + e^{2t}}{7} \end{bmatrix}$$

Thus the solution given by Eq. (D.86) is:

$$\mathbf{x}(t) = e^{\mathbf{A}t}\mathbf{x}_0 = \begin{bmatrix} \dfrac{e^{-5t} + 6e^{2t}}{7} & \dfrac{-2e^{-5t} + 2e^{2t}}{7} \\[2ex] \dfrac{-3e^{-5t} + 3e^{2t}}{7} & \dfrac{6e^{-5t} + e^{2t}}{7} \end{bmatrix} \begin{bmatrix} 1 \\ 2 \end{bmatrix}$$

$$\mathbf{x}(t) = \begin{bmatrix} x_1(t) \\ x_2(t) \end{bmatrix} = \begin{bmatrix} \dfrac{-3e^{-5t} + 10e^{2t}}{7} \\[2ex] \dfrac{9e^{-5t} + 5e^{2t}}{7} \end{bmatrix}$$

The extension of these ideas to provide solutions for the case of *nonhomogeneous linear differential equations*:

$$\frac{d\mathbf{x}(t)}{dt} = \mathbf{A}\mathbf{x}(t) + \mathbf{B}\mathbf{u}(t) + \mathbf{\Gamma}\mathbf{d}(t); \qquad \mathbf{x}(t_0) = \mathbf{x}_0 \qquad \text{(D.89)}$$

is straightforward. Here $\mathbf{u}(t)$, $\mathbf{d}(t)$ represent control variables and disturbances respectively. When the matrices $\mathbf{A}$, $\mathbf{B}$, $\mathbf{\Gamma}$ are *constant* matrices, a derivation completely similar to the one just given leads to the solution:

$$\mathbf{x}(t) = e^{\mathbf{A}(t - t_0)}\mathbf{x}_0 + \int_{t_0}^{t} e^{\mathbf{A}(t - \tau)} \left[\mathbf{B}\mathbf{u}(\tau) + \mathbf{\Gamma}\mathbf{d}(\tau) \right] d\tau \qquad \text{(D.90)}$$

where the matrix exponentials, $e^{\mathbf{A}(t - t_0)}$, $e^{\mathbf{A}(t - \tau)}$, may be evaluated using the relation Eq. (D.87).

Example D.15   SOLUTION OF LINEAR NONHOMOGENEOUS MATRIX DIFFERENTIAL EQUATIONS.

To illustrate the solution of nonhomogeneous equations using Eq. (D.90), let us suppose we have a system Eq. (D.89) with $\mathbf{A}$ the same as the last example:

$$\mathbf{A} = \begin{bmatrix} 1 & 2 \\ 3 & -4 \end{bmatrix}$$

and

$$\mathbf{B} = \begin{bmatrix} 1 & 0 \\ 0 & 1 \end{bmatrix}, \ \mathbf{\Gamma} = \begin{bmatrix} 1 \\ 0 \end{bmatrix}, \ \mathbf{x}_0 = \begin{bmatrix} 1 \\ 2 \end{bmatrix}, \ \mathbf{u} = \begin{bmatrix} u_1 \\ u_2 \end{bmatrix}, \ \mathbf{d} = d_1 \ \text{(a scalar)}$$

In addition assume $u_1 = u_2 = 0, d_1 = 1$ (a constant disturbance) and that $t_0 = 0$. In this case Eq. (D.90) takes the form:

$$\mathbf{x}(t) = e^{\mathbf{A}t}\mathbf{x}_0 + \int_{0}^{t} e^{\mathbf{A}(t - \tau)} \mathbf{\Gamma}\mathbf{d}(\tau) \, d\tau$$

Substituting for $e^{\mathbf{A}t}$, and $e^{\mathbf{A}t}\mathbf{x}_0$ calculated in the Example D.14, we obtain the solution:

$$x(t) = \begin{bmatrix} \dfrac{-3e^{-5t} + 10e^{2t}}{7} \\[3mm] \dfrac{9e^{-5t} + 5e^{2t}}{7} \end{bmatrix} + \int_0^t \begin{bmatrix} \dfrac{e^{-5(t-\tau)} + 6e^{2(t-\tau)}}{7} \\[3mm] \dfrac{-3e^{-5(t-\tau)} + 3e^{2(t-\tau)}}{7} \end{bmatrix} d\tau$$

or

$$x(t) = \begin{bmatrix} x_1(t) \\[3mm] x_2(t) \end{bmatrix} = \begin{bmatrix} -\dfrac{2}{5} - \dfrac{16}{35}e^{-5t} + \dfrac{13}{7}e^{2t} \\[3mm] -\dfrac{21}{70} + \dfrac{48}{35}e^{-5t} + \dfrac{13}{14}e^{2t} \end{bmatrix}$$

For the situation where the matrices $\mathbf{A}$, $\mathbf{B}$, $\boldsymbol{\Gamma}$ may themselves be functions of time, the solution requires a bit more computation. In this case:

$$x(t) = \Phi(t, t_0)\, x_0 + \Phi(t, t_0) \int_{t_0}^t \Phi(\tau, t_0)^{-1} \left[\mathbf{B}u(\tau) + \boldsymbol{\Gamma}d(\tau) \right] d\tau \qquad \text{(D.91)}$$

Here $\Phi(t, t_0)$ is an $n \times n$ time-varying matrix known as the *fundamental matrix solution* which may be found from:

$$\frac{d\Phi(t, t_0)}{dt} = \mathbf{A}(t)\Phi(t, t_0); \quad \Phi(t_0, t_0) = \mathbf{I} \qquad \text{(D.92)}$$

Because $\mathbf{A}(t)$, $\mathbf{B}(t)$, $\boldsymbol{\Gamma}(t)$ may have arbitrary time dependence, the fundamental matrix solution is usually obtained by numerical methods. Nevertheless, once $\Phi(t, t_0)$ is determined, the solution $x(t)$, may be readily calculated for a variety of control and disturbance inputs, $u(t)$, $d(t)$, using Eq. (D.91).

D.12   SUMMARY

In this appendix, the techniques of matrix algebra have been reviewed. These methods find particularly useful application in providing very general solutions to sets of linear algebraic and linear ordinary differential equations. The importance of matrices as a tool for analyzing the dynamics of physical processes should not be undervalued by the reader. As an aid to further study, a bibliography is provided below.

REFERENCES AND SUGGESTED FURTHER READING[†]

1. Kreysig, E., *Advanced Engineering Mathematics* (2nd ed.), J. Wiley, New York (1967), Chap. 7
2. Amundson, N. R., *Mathematical Methods in Chemical Engineering,* Prentice–Hall, Englewood Cliffs, NJ (1966)
3. Wylie, C. R., *Advanced Engineering Mathematics*, J. Wiley, New York (1966)
4. Duncan, W. J., A. R. Collar, and R.Q. Frazer, *Elementary Matrices*, Cambridge University Press (1963)
5. Bellman, R., *Introduction to Matrix Analysis,* McGraw–Hill, New York (1960)
6. Gantmacher, F. R., *The Theory of Matrices*, (Vols. 1 and 2), Chelsea Publishing Company, New York (1960)

[†]Listed in increasing order of depth.

E

COMPUTER-AIDED CONTROL SYSTEM DESIGN

The calculations required for the analysis of process dynamics, the fitting of models to experimental data, and the creation of control system designs can be quite tedious and challenging. Fortunately, there are a host of interactive Computer-Aided Design (CAD) packages which will carry out these calculations with no programming effort required by the user. These packages are available for a reasonable fee, and thus should be used by students and others studying the material in this book. Some of the more widely used packages are listed in Table E.1 below. Reviews of the field of computer-aided design and a discussion of these and other CAD programs can be found in Refs. [1–3].

Table E.1. Some Computer-Aided Analysis and Control System Design Packages

Program	Scope	Source
ACSL	Primarily Simulation	Mitchell and Gauthier Assoc. 200 Baker Avenue Concord, MA 01742
CC	Comprehensive	Systems Technology, Inc. 13766 S. Hawthorne Blvd. Hawthorne, CA 90250
CONSYD	Comprehensive	Prof. W. Harmon Ray Dept. of Chemical Eng. University of Wisconsin 1415 Johnson Drive Madison, WI 53706
Ctrl-C Model-C	Comprehensive	Systems Control Technology 1801 Page Mill Road P. O. Box 10180 Palo Alto, CA 94303
EASY 5	Comprehensive	Boeing Computer Services P. O. Box 24346 Seattle, WA 98124
MATLAB with Control System, Robust Control, and System Identification Toolboxes	Comprehensive	Mathworks, Inc. 24 Prime Park Way Natick, MA 01760
MATRIX-X	Comprehensive	Integrated Systems, Inc. 3260 Jay Street Santa Clara, CA 95054
TUTSIM	Primarily Simulation	Tutsim Products 200 Calif. Ave., #212 Palo Alto, CA 94306
UC-SIGNAL UC-ONLINE	Primarily Simulation	University of California Office of Tech. Licensing Berkeley, CA

REFERENCES

1. Chen, Z.-Y. (Ed.)., *Computer-Aided Design in Control Sytems*, Proceedings 4th IFAC Symposium, Pergamon Press, New York (1988)
2. Holt, B. R., et al., "CONSYD - Integrated Software for Computer-Aided Control System Design and Analysis," *Comput. Chem. Eng.*, **11**, 187 (1987)
3. Linkens, D. A. (Ed.), *CAD for Control Systems*, Marcel Dekker, New York (1993)

author
index

subject index